IMI
Information Centre & Library

D1433234

THE WILEY GUIDE TO MANAGING PROJECTS

THE WILEY GUIDE TO MANAGING PROJECTS

Peter W. G. Morris

Jeffrey K. Pinto

JOHN WILEY & SONS, INC.

This book is printed on acid-free paper. ♾

Published by John Wiley & Sons, Inc., Hoboken, New Jersey.
Published simultaneously in Canada.

For general information on our other products and services or for technical support, please contact our Customer Care Department within the United States at (800) 762-2974, outside the United States at (317) 572-3993 or fax (317) 572-4002.

Wiley also publishes its books in a variety of electronic formats. Some content that appears in print may not be available in electronic books. For more information about Wiley products, visit our web site at www.wiley.com.

Library of Congress Cataloging-in-Publication Data:

Morris, Peter W. G.
The Wiley guide to managing projects / Peter W. G. Morris, Jeffrey K. Pinto.
p. cm.
Includes bibliographical references and index.
ISBN 0-471-23302-1 (cloth)
1. Project management. I. Title: Guide to managing projects. II. Morris, Peter W. G. III. Title.
HD69.P75P552 2004
658.4'04—dc22

2003026695

Printed in the United States of America

10 9 8 7 6 5 4 3 2 1

CONTENTS

SECTION II: THE MANAGEMENT OF PROJECTS

SECTION II.1 STRATEGY, PORTFOLIO, AND PROGRAM MANAGEMENT

SECTION II.2: TECHNOLOGY MANAGEMENT

SECTION II.3: SUPPLY CHAIN MANAGEMENT AND PROCUREMENT

PREFACE

The management of projects represents one of the most significant undertakings in which modern organizations can engage. The economic, social, and technological forces that shape our world are creating an environment that seems, every day, to be oriented more and more towards a project-based approach to getting things done. Everywhere there is evidence of an increased interest in managing projects: Thousands of new members are enrolling every year in project management professional organizations; hundreds of private and public sector enterprises are pushing their operating models toward project-based working; scores of universities and technical agencies are offering courses, certification, and degrees in project management. It is clear that we are experiencing a revolution in the way we organize and manage, happily one not threatening disruption and confusion but proffering improvement and opportunity.

When we, as editors, set out to develop this "handbook," our clear motivation was to create a product that was timely, accessible, and relevant. "Timely" in that project-based work has continued to grow at such an enormous pace, attracting large numbers of new adherents, both as individuals and as organizations. "Accessible" in that we sought also to create a work that spoke the appropriate language to the largest possible audience, appealing to both project management practitioners and academic researchers. But above all "relevant": Too much of project management writing addresses only the basics of time, cost, and scope management (or people and organizational issues) and fails to address the day-to-day nuances that become so important in practice. The reality is that there is far more than this to managing projects successfully. For this book to be useful, it needed to reflect not only well-known and widely used basic project management practices but also the new, cutting-edge concepts in the broader theory and practice of managing projects. To this end we have consciously built on our individual (but in many ways parallel) research to capture the insights of many of the world's leading experts, explicitly organized, as we explain in the Introduction, around a "management of projects" framework.

In short, our goal was to provide a resource that demonstrated the widest possible usefulness for readers seeking to develop and deliver successful projects, regardless of their professional background.

Hence, in *The Wiley Guide to Managing Projects*, we have endeavored to provide a clear view of the cutting edge in project management best practice. Wherever possible, we do this within a soundly based conceptual framework, founded in research and practical experience, which we have endeavored to make explicit. In doing so we have been joined by a truly notable range of authorities, all leaders in their field, drawn from many different industry sectors, practice areas, and countries. Together they address the most significant topics and problems currently facing project managers and project-based organizations today.

Whether you view this book as a comprehensive resource that should be read cover-to-cover or choose a selective subset of the topics that appeal to you directly, we hope you will find the experience rewarding. As a collective whole, we believe the book holds together with clarity and structure; as individual essays, each chapter can provide value to the reader.

It is with genuine gratitude that we would like to acknowledge the efforts of several individuals whose work contributed enormously to this finished product. Bob Argentieri, acquisition editor at Wiley, first conceived of the idea from which this book eventually emerged (the successor to the famous Cleland and King *Project Management Handbook*, as we explain at the outset of the Introduction). It was his energy and enthusiasm that led, in large part, to what you now see. To the contributors of the individual chapters we owe a great debt of thanks as well. To have so many busy professionals first agree to participate in this project and then to contribute work of such outstanding quality, and work with us so patiently in crafting the chapters, has been extremely gratifying. We thank them sincerely. Third, we should especially thank Gill Hypher of INDECO, who has patiently and with good humor shepherded a host of queries and a vast quantum of correspondence in getting the details right to allow publication to proceed, and to Naomi Rothwell of Wiley, who has worked with Gill to embed the emerging document in Wiley's production machinery.

And last, though never least, to our families, we acknowledge a bond that can never be broken and a wellspring that continues to lead to greater and better things. Two people were never better blessed than we have been with this support.

Peter Morris and Jeff Pinto

INTRODUCTION

Peter Morris and Jeffrey Pinto

In 1983 Dave Cleland and William King produced for Van Nostrand Reinhold (now John Wiley & Sons) the *Project Management Handbook*, a book that rapidly became a classic. Now over 20 years later John Wiley & Sons is bringing that landmark publication up-to-date with this, *The Wiley Guide to Managing Projects.*

Why the new title—indeed, why the need to update the original work?

That's a big question, one that goes to the heart of much of the debate in project management today and that is central to the architecture and content of this book. First, why "the management of projects"?

Project management has moved a long way since 1983. If we take the founding of project management to be somewhere between about 1955, when the first uses of modern project management terms and techniques began being applied in the management of the U.S. missile programs, and 1969 to 1970, when project management professional associations were established in the United States and Europe (Morris, 1997), then Cleland and King's book was reflecting thinking that had been developed in the field for about the first 20 years of this young discipline's life. Well over another 20 years has since elapsed. During this time there has been an explosive growth in project management. The professional project management associations around the world now have thousands of members—the Project Management Institute (PMI) itself having over 140,000—and membership continues to grow. Every year there are dozens of conferences; books, journals, and electronic publications abound; companies continue to recognize project management as a core business discipline and work to improve company performance through it; and increasingly there is more formal educational work carried out in universities in teaching programs at both the undergraduate but particularly postgraduate levels and in research.

Yet in many ways all this activity has lead to some confusion over concepts and applications. The basic American, European, and Japanese professional models of project management, for example, are different. PMI's is, not least because of its size, the most influential, with both its *Guide to the Project Management Body of Knowledge* (PMI, 2000) and its newer *Organizational Project Management Maturity Model, OPM3* (PMI, 2003). Yet it is also the most limiting, reflecting an essentially execution, or delivery, orientation. This tendency underemphasizes the front-end, definitional stages of the project, the stages that are so crucial to successful accomplishment. (The European and Japanese models, as you will see, give much greater prominence to these stages.) An execution emphasis is obviously essential, but managing the definition of the project, in a way that best fits with the business, technical, and other organizational needs of the sponsors, is critical in determining how well the project will deliver business benefit and in establishing the overall strategy for the project.

It was this insight, developed through research conducted independently by both the current authors shortly after the publication of the Cleland and King *Handbook* (Morris and Hough, 1987; Pinto and Slevin, 1988) that led to Morris coining the term in 1994 "the management of projects" to reflect the need to focus on managing the definition and delivery of *the project itself* to deliver a successful outcome.

These are the themes that we explore in this book (and to which we will revert in a moment). Our aim is to center the discipline better by defining more clearly what is involved in managing projects successfully, and in doing so, to expand the discipline's focus.

So second, why so big? At around 1,400 pages, this is clearly more than a handbook that will neatly slip into one's bag. It was both John Wiley's desire and our own to produce something substantial—something that could be used by both practitioners and scholars, hopefully for the next 10 to 20 years, like the Cleland and King book, as a reference to the best thinking in the discipline. But why over 1,400 pages? A second driver, significantly responsible for the sheer size of the book, is that it is reflective of the growth of knowledge within the field. The "management of projects" philosophy forces us (i.e., members of the discipline) to expand our frame of reference regarding what projects truly *are* beyond of the traditional PMBOK/OPM3 model.

This, then, is not a short how-to management book but very intentionally a resource book. We see our readership not as casual business readers but as people who are genuinely interested in the discipline and who are seeking further insight and information—the thinking managers of projects. More specifically, the book is intended for both the general practitioner and the student (typically working at the postgraduate level). For both, we seek to show where and how practice and leading thinking is shaping the discipline. We are deliberately stretching the envelope, giving practical examples, and providing references to others' work. The book should, in short, be a real resource, allowing the reader to understand how the key "management of projects" practices are being applied in different contexts, and pointing to where further information can be obtained.

To achieve this aim, we have assembled and worked, at times intensively, with a group of authors who collectively provide truly outstanding experience and insight. Some are, by any count, among the leading researchers, writers, and speakers in the field, whether as academics or as consultants. Others write directly from senior positions in industry, offering their practical experience. In all cases, they have worked hard with us to furnish the relevance, the references, and the examples that the book as a whole aims to provide.

What one undoubtedly gets as a result is a range that is far greater than any individual alone would bring. (One simply can't be working in all these different areas so deeply as all these authors, combined, are.) What one doesn't always get though are all the angles that any one mind might think are important. To an extent this is both inevitable—if, at times, a little regrettable. But to a larger extent, we feel it is beneficial, for two reasons. One, this is not a discipline that is now done and finished. Far from it. There are many examples where there is need, and opportunity, for further research and for alternative ways of looking at things. Rodney Turner and Anne Keegan, for example, in their chapter on managing innovation, ended up positioning the discussion very much in terms of learning and maturity. If we had gone to Harvard, to Wheelwright and Clark (1992) or Christensen (1999), for example, we would almost certainly have gotten something that focused more on the structural processes linking technology, innovation, and strategy. This divergence is healthy for the discipline and, in fact, inevitable in a subject that is so context-dependent as management. Second, it is also beneficial because seeing a topic from a different viewpoint can be stimulating and lead the reader to fresh insights. Hence, we have Steve Simister giving an outstandingly lucid and comprehensive treatment early in the book on risk management, but then later we have Stephen Ward and Chris Chapman coming at the same subject from a different perspective and once again offering a penetrating treatment of it. There are many similar instances, particularly where the topic is not an easy one or may vary in application—for example, with regard to strategy, program management, finance, procurement, knowledge management, performance management, scheduling, competence, quality, and maturity.

In short, the breadth and diversity of this collection of work (and authors) is, we believe, one of our book's most fertile qualities. It represents a set of approximately 60 authors from different discipline perspectives (e.g., construction, new product development, information technology, defense/aerospace) whose common bond is their commitment to improving the management of projects, who provide a range of insights from around the globe. Thus, the North American reader can gain insight into processes that while common in Europe have yet to make significant inroads in other locations, and vice versa. IT project managers can likewise gather information from the wealth of knowledge built up through decades of practice in the construction industry, and vice versa. The settings may change; the key principles are remarkably resilient.

But these are big topics, and it is time to return to the question of what we mean by project management and the management of projects, and to the structure of the book.

Project Management

There are several levels at which the subject of project management can be approached. We have already indicated one of them in reference to the PMI model. As we and several others of the *Guide*'s authors indicate later, this is a wholly valid but essentially delivery-, or execution-, oriented perspective of the discipline: what the project manager needs to do in order to deliver the project "on time, in budget, to scope." If project management professionals cannot do this effectively, they are failing at the first fence. Mastering these skills is the *sine qua non*—the "without which nothing"—of the discipline. Section I addresses this basic view of the discipline, though by no means exhaustively (there are dozens of other books on the market that do this excellently—including some outstanding textbooks: Meredith and Mantel, 2003; Gray and Larson, 2003; Pinto, forthcoming).

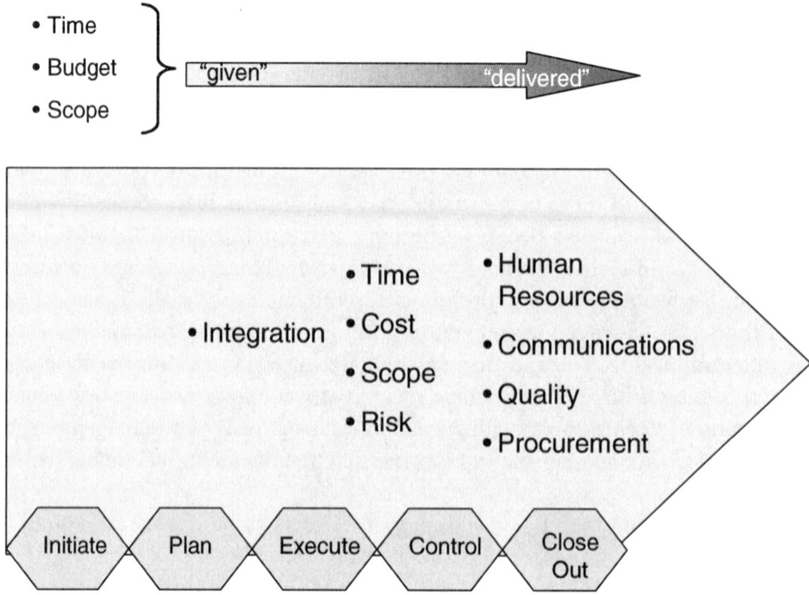

PROJECT MANAGEMENT:

"On time, in budget, to scope" execution/ delivery

The overriding paradigm of project management at this level is a control one (in the cybernetic sense of control involving planning, measuring, comparing, and then adjusting performance to meet planned objectives, or adjusting the plans). Interestingly, even this model—for us, the foundation stone of the discipline—is often more than many in other disciplines think of as project management. Many, for example, see it as predominantly oriented around scheduling (or even as a subset in some management textbooks of operations management). In fact, even in some sectors of industry, this has only recently begun to change, as can be seen toward the end of the book in the chapter on project management in the pharmaceutical industry. It is more than just scheduling, of course: There is a whole range of cost, scope, quality, and other control activities. But there are other important topics, too.

Managing project risks, for example, is an absolutely fundamental skill even at this basic level of project management. Projects by definition are unique: Doing the work necessary to initiate, plan, execute, control, and close out the project will inevitably entail risks. These need to be managed.

Both these areas are mainstream and generally pretty well understood within the traditional project management community, as represented by the PMI PMBOK Guide (PMI, 2000), for example. What is less well covered is the people side of managing projects. Clearly, people are absolutely central to effective project management; without people projects simply could not be managed. There is a huge amount of work that has been done on how organizations behave and perform and how people do too, and much has been written on this within a project management context. (That so little of this finds its way into PMBOK is almost certainly because of its concentration on material that is said in PMBOK to be "unique" to project management.) A lot of this information we have positioned later in the book, around the general area of competencies, but some we have kept here in the earlier "project management" section, deliberately to make the point that people issues are essential in project delivery.

It is thus important to provide the necessary balance to our building blocks of the discipline. For example, among the key contextual elements that set the stage for future activity is the organization's structure, so pivotal in influencing how effectively projects may be run. But organizational structure has to fit within the larger social context of the organization—its culture, values and operating philosophy, stakeholder expectations, socioeconomic and business context, behavioral norms, power and informal influence processes, and so on. This takes us to our larger theme: looking at the project in its environment and managing its definition and delivery for stakeholder success—"the management of projects."

The Management of Projects

The thrust of the book is, as we have said, to expand the field of project management. This is quite deliberate. For as Morris and Hough showed in *The Anatomy of Major Projects* (1987), in a survey of the then existing data on project overruns (drawing on over 3,600 projects as well as eight specially prepared case studies), neither poor scheduling nor even lack of teamwork figured crucially among the factors leading to the large number of unsuccessful projects in this data set. What instead were typically important were items such as client changes; poor technology management; poor change control; changing social, economic, and environmental factors; labor issues, poor contract management; and so on. Basically, the message was that while traditional project management skills are important, they are often not *sufficient* to ensure project success. What is needed is to broaden the focus to cover the management of external and front-end issues, not least technology. Similarly, at about the same time and subsequently, Pinto and his coauthors, in their bespoke studies on project success (Pinto and Slevin, 1988; Kharbanda and Pinto, 1997), showed the importance of client issues and technology as well as the more traditional areas of project control and people.

The result of both pieces of work has been to change the way we look at the discipline. No longer is the focus so much just on the processes and practices needed to deliver projects "to scope, in budget, on schedule" but rather on how we set up and define the project to deliver stakeholder success—on how to manage projects. In one sense this almost makes the subject impossibly large, for now the only thing differentiating this form of management from other sorts is "the project." We need therefore to understand the characteristics of the project development/life cycle but also the nature of projects in organizations. This becomes the kernel of the new discipline, and much in this book is oriented around this idea.

Morris articulated this insight in *The Management of Projects* (1994, 1997), and it significantly influenced the development of the Association for Project Management's Body of Knowledge, as well as the International Project Management Association's Competence Baseline (Morris, 2001). As a generic term, we feel "the management of projects" still works, but it is interesting to note how the rising interest in program management and portfolio management fits comfortably into this schema. Program management is now strongly seen as the management of multiple projects connected to a shared business objective. (See, for example, the chapter by Michel Thiry.) The emphasis on managing for business benefit, and on managing projects, is exactly the same as in "the management of projects." Similarly, the more recently launched Japanese Body of Knowledge, P2M (Program and Project Management), discussed *inter alia* in Lynn Crawford's chapter on project management standards, is explicitly oriented around managing programs and projects to create, and optimize, business value. Systems management, strategy, value management, finance, and relations management, for example, are all major elements on P2M. Few if any appear in PMBOK.

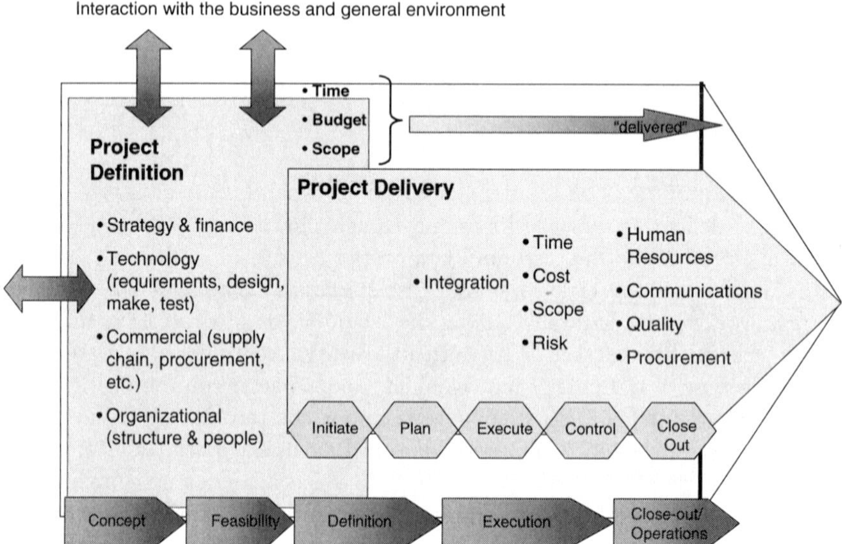

THE MANAGEMENT OF PROJECTS involves managing the definition and delivery of the project for stakeholder success. The focus is on the project in its context. Project and program management – and portfolio management, though this is less managerial – sit within this framework.

("The management of projects" model is also more relevant to the single project situation than PMBOK incidentally, not just because of the emphasis on value but via the inclusion of design, technology, and definition. There are many single project management situations, such as design-and-build contracts, where the project management team has responsibility for elements of the project design and definition.)

Section II addresses at length issues in the management of projects. How?

We've already said that one of the major challenges is how to structure such a broad field. Because so many factors interact, it is difficult to create a placement that, while logical, does not appear to suggest unnecessary dependencies, such as first strategy, then technology, then finance, then organization, and so on. For the truth is that though there may well be a sequence such as this, this need not always be the one to be followed, and in practice there may well be considerable iteration between topic areas. Nevertheless, there must obviously be some structure, and we have adopted the following:

- Strategy, portfolio, and program management
- Technology management
- Supply chain management and procurement
- Control
- Competence development

These are pretty broad headings, however. Let's see what we have in each.

Section II.1: Strategy, portfolio and program management

Strategy represents the fundamental goals and objectives that drive the organization and that, if well understood and delineated throughout it, should affect the manner in which projects are selected, shaped, and executed. The organization's strategy encompasses the way in which it makes sense of its external environment, identifies opportunities, and evaluates its performance. In this manner, projects become, in a term David Cleland coined, "the building blocks of strategy," allowing the organization to operationalize its goals in meaningful, measurable ways (Cleland, 1990). The organization's use of its strategic portfolio of projects and the manner in which it shapes and maintains its programs reflects its commitment to a proactive, rather than reactive, means of achieving its goals.

Acknowledging the links between strategy at the corporate, portfolio, program, and project levels allows organizations to focus on improving their portfolio and program management. These are themes explored in several different chapters in this section, along with a chapter on process modeling, a very useful one on stakeholder management, and two that address how the financing of projects influences their management (the chapter by Graham Ive is particularly important in exploring the influence of "private finance" initiatives on the practice of managing public infrastructure projects.) The result is that our basic "management of projects" model can now be expanded to reflect our increased knowledge of program management, and its concerns—managing for business benefit, managing products (brands, technology), resource allocation, and so on—and portfolio management and its special challenges.

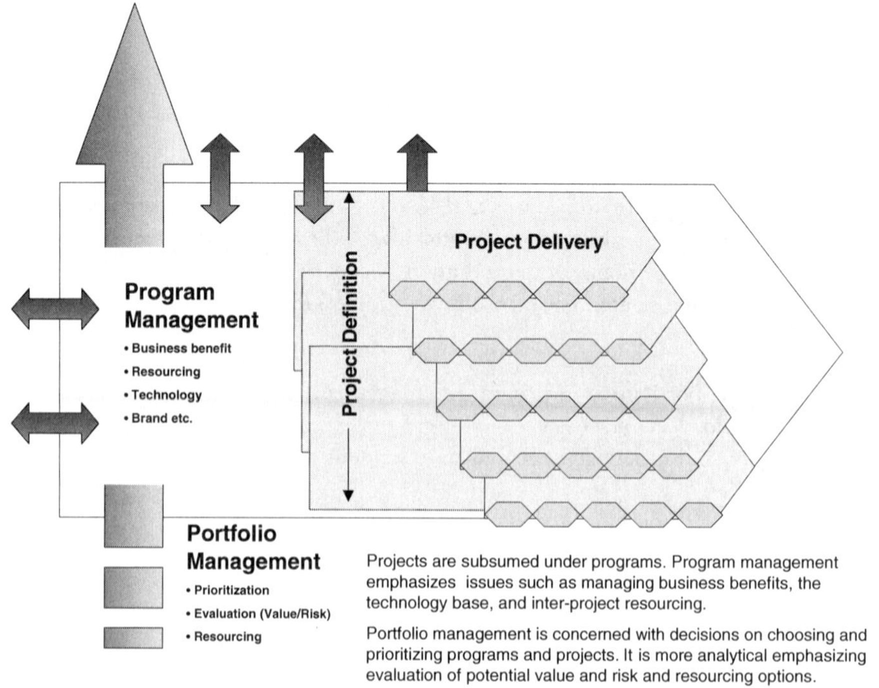
Project Delivery
Program
Management
• Business benefit
• Resourcing
• Technology
• Brand etc.
Project Definition
Portfolio
Management
• Prioritization
• Evaluation (Value/Risk)
• Resourcing
Projects are subsumed under programs. Program management emphasizes issues such as managing business benefits, the technology base, and inter-project resourcing.
Portfolio management is concerned with decisions on choosing and prioritizing programs and projects. It is more analytical emphasizing evaluation of potential value and risk and resourcing options.

Section II.2: Technology management

Unusually perhaps for project management, there is quite a lot in this book on the management of technology. This is partly because our research showed that technology has a major influence on project success, partly precisely because it is so typically under-represented in the literature. There is a huge opportunity, and need, to understand this area better.

Logically the place to begin is requirements. As the seer said, if you don't know what you want, don't be surprised if you don't get it! Requirements management is one of those areas that is often not done well and is generally not well understood. Many people cheerfully talk about business requirements without being very clear about how these should be defined or developed.

Product—and process—design builds off the project's requirements; significantly, its management is also not well understood. Project management's responsibility is not the design itself but to make sure the design is developed as effectively and efficiently in terms of the stakeholders' requirements as possible. Process skills are very important in this. The way development gates are deployed and managed—hard/soft, design reviews, value management, risk management, health and safety/HAZOP management reviews, and so on—are important examples. Modeling, configuration management, and information management are often critical support tools. The proper management of verification and validation is essential.

None of this happens in isolation of the rest of the project, of course; integration with cost/value optimization and with compressed schedule realization is often key. Concurrent engineering practices, for example, try and bundle all this into a project-managed, engineering-organized, business-optimized, integrated approach.

Having thus addressed issues related to strategy, finance, and technology, we then proceed in a logical manner to issues relating to the procurement of resources, beginning first with a piece that links technology, strategy and procurement.

Section II.3: Supply chain management and procurement

Effectively integrating and exploiting the project resource base is a major challenge for project-based organizations. The resource base is partly drawn from within the sponsoring organization—and we shall be discussing people and competencies in Section II.5. However, for most projects an early structural, and continuing dynamic, challenge is to acquire and manage external resources. Among key elements in this process are supply chain management, procurement, and tender and contract management; a long-term integrated logistics support (ILS) approach is also increasingly being recognized as necessary (following new financing pressures, as we see in the Ive chapter).

Rather like concurrent engineering, ILS wraps several "management of projects" topics together—technology, cost, value, time, organization, and so on—but this time under the rubric of supply. Though the emphasis is on looking at long-term, whole-life operations—". . . ensuring that all the elements of design are fully integrated to meet the client's requirements and asset's operational and performance, including availability, reliability, durability, maintainability, and safety at minimum whole-life cost," as David Kirkpatrick and his colleagues quote in their chapter—these need to be planned and managed from the very earliest stages of the project.

Similarly with supply chain management: As we see in Ray Venkataraman's chapter, the manner in which an organization purposefully and proactively manages its supply chain plays a vital role in project delivery, including project features such as functionality, quality, and cost.

A theme that constantly emerges in our research and consulting experiences is the increasingly active role that firms expect their project managers to take in understanding the tender and contracting elements in ongoing projects. The old model, in which all customer disagreements were shunted to lawyers, has been changed to put more emphasis on project managers "working the contract," through better supply chain integration, alignment, team building, and so on. The move toward more partnering-based contracting is just one example of this trend. In general, project managers are these days expected to have much more control over terms and conditions and contract negotiation. There are a variety of reasons for this shift, one obvious one being the need to streamline the development process by leaving as much decision-making authority in the hands of the project manager as possible. The end result is to require project managers to build teams rather than simply rely on contractual law, while at the same time becoming more savvy commercially.

Section II.4: Control

Having developed the project strategy, assembled funding, gotten a project definition, and procured the necessary resources, we now need to ensure we stay on track. "Control" is more than just monitoring, however (as you will see in Section I). In this section we explore a number of issues of a control nature in ways that are not often found in books on project management.

Beyond conventional issues of project planning and control (scheduling, resource management, and critical path development), we consider newer breakthroughs in the manner in which projects are scheduled: Critical chain project management, for example, reflects more than simply a change in convention for how project activities should be sequenced and scheduled: It offers an alternative philosophy on the way in which we fast-track projects while improving team commitment and productivity.

Reporting performance is obviously a key part of project control, but as anyone who has tried to do this soon realizes, it is not easy. The essential challenge is what measures to use and how to report on these in an integrated way. Issues such as what to measure, when to measure and how to interpret these findings play a critical role in establishing metrics for project performance and subsequent control.

Other key elements in the control cycle include risk, value, and quality. We've already mentioned Stephen Ward and Chris Chapman's fresh look at risk (uncertainty) management. Value management (VM), the process of formally optimizing the overall approach to the project (including whether it should be done), could quite legitimately have been placed in the strategy section (II.1). However, because technology and procurement, among many other things, need addressing before VM can really be brought to bear, we decided to delay it till these topics had been discussed. VM is positioned as a strategic process comprising, for Michel Thiry, sensemaking, ideation, elaboration, choice; for Michel, it is "the method of choice to deal with the ambiguity of stakeholders' needs and expectations and the complexity of changing business environment at program level and project initiation."

Quality management is a subject that bears both on strategic planning and operational control. Typically it tends to be described in terms of product quality; there is, however, as Martina Huemann shows, much opportunity for it to be applied to project management processes and practices—and people—not least in quality assurance (project reviews, audits, health checks, etc.).

Section II.5: Competence Development

Ultimately, as we've seen, projects are run, and delivered, by people. How one organizes, motivates, supports, and develops one's people is absolutely critical to people's ability to perform. Foremost among the issues occupying much of the work being done today are those having to do with knowledge, learning, and maturity. But equally important are the broader, more traditional areas like teams, leadership, power and negotiation, and competencies.

For example, the manner in which teams perform on projects represents a number of contractions and conflicts between "ought" and "action." "We have knowledge and wisdom to change, but why do we not do so, or even act in ways contradictory to successful team building and management?" asks Connie Delisle. Among the reasons are the nature of the forces shaping teams and the dynamics of team development and management. Related issues of leadership, power and influence, and negotiation skills development simply reinforce the point that the management of projects cannot proceed without the effective management of people.

Competence has increasingly become recognized as a core descriptor of our ability to manage. The basic idea of competencies is that they are what are required to fulfill a specific role. "Competence is concerned with the capacity to undertake specific types of action and can be considered as a holistic concept involving the integration of attitudes, skills, knowledge, performance and quality of application" as Andrew Gale puts its. Huge efforts have been, and are being, put into developing project management competencies—training, tools, coaching, and the like. But are we giving people the appropriate knowledge and skills; are we supporting their learning and development as effectively as we could? These are the concerns of knowledge management and organizational learning that have become of such interest since the late 1990s.

How do we know, for example, what are the rules, practices, insights, and other knowledge that we may wish to pass on to, or even impose on, others? And how do we get all this to people in the best form, when needed? Clearly one of the things we have to do is define our frame of reference appropriately; this tells us at least what the ballpark looks like. Research on project-based learning suggests that while it ought to be possible to support role-specific learning (at a price), more general standards and accreditation can only go so far and , reinforcing Andrew Gale's point, may have limitations in ensuring real competence. If this is so, what does this mean for the profession and its aspirations for certification?

It's the complexity of projects and the range of issues to be managed that makes the management of projects so difficult to generalize about, to create standards for, and to benchmark. Simple comparisons are generally flawed; strict causality is almost impossible to prove. Hence, even determined efforts to gauge the return on investment of project management, like those of Bill Ibbs and his colleagues at Berkeley on evaluating project management maturity, are frustratingly still some way from delivering really robust conclusions. Terry Cooke-Davies, in a thoughtful chapter, shows why we should be cautious in expecting too much too soon from the application of maturity models to such a complex discipline.

What all these latter chapters emphasize is what so many authors have established in this book: the importance of context in management (Griseri, 1999)—which is why we place so much emphasis on context in this book, particularly in Section III.

Section III: The Management of Projects in Context

The last section of the book is in fact devoted to context. Aaron Shenhar and Dov Dvir begin by reminding us that an organization's shape and character are "contingent" upon its context, and we then have chapters exploring the realities of the management of new product development, pharmaceutical drug development, defense contracting, construction, and auto industry projects.

Across domains such as these—and, indeed, even wider—professionals have been meeting now for over 30 years trying to gather together the principles and practices of good project management. Lynn Crawford concludes the book by surveying these attempts around the globe. Even-handed, she maintains a cautious but fundamentally optimistic tone, as do we.

The Wiley Guide to Managing Projects represents an opportunity to take a step back and evaluate the status of the field, particularly in terms of scholarship and intellectual contributions, some 20 to 25 years after Cleland and King's seminal *Handbook*. Much has changed in the interim. The discipline has broadened considerably—where once projects were the primary focus in just a few industries, today they are literally the dominant way of organizing business in sectors as diverse as insurance and manufacturing, software engineering, or utilities. But as projects have been recognized as primary, critical organizational forms, so has recognition that the range of practices, processes, and issues needed to manage them is substantially broader than was typically seen nearly a quarter of a century ago. The old project management "initiate, plan, execute, control, and close" model once considered the basis for the discipline is now increasingly recognized as insufficient and inadequate, as the many chapters of this book have surely demonstrated.

The shift from "project management" to "the management of projects" is no mere linguistical sleight-of-hand: It represents a profound change in the manner in which we approach projects, organize, perform, and evaluate them.

References

Christensen, C. M. 1999. *Innovation and the general manager.* Boston: Irwin/McGraw-Hill.

Cleland, D. I., and W. R. King. 1983. *Project management handbook.* New York: Van Nostrand Reinhold.

Cleland, D. I. 1990. *Project management: Strategic design and implementation.* Blue Ridge Summit, PA: TAB Books.

Gray, C. F., and E. W. Larson. 2003. *Project management.* Burr Ridge, IL: McGraw-Hill.

Griseri, P. 2002. *Management knowledge: A critical view.* London: Palgrave.

Kharbanda, O. P., and J. K. Pinto. 1997. *What made Gertie gallop?* New York: Van Nostrand Reinhold.

Meredith, J. R., and S. J. Mantel. 2003. *Project management: A managerial approach.* 5th ed. New York: Wiley.

Morris, P. W. G. 1994. *The management of projects.* London: Thomas Telford (distributed in the United States by The American Society of Civil Engineers; paperback edition 1997).

———. 2001. Updating the project management bodies of knowledge. *Project Management Journal* 32(3): 21–30.

Morris, P. W. G., and G. H. Hough. 1987. *The anatomy of major projects.* Chichester, UK: Wiley.

Pinto, J. K., and D. P. Slevin. 1988. Project success: Definitions and measurement techniques. *Project Management Journal* 19(1):67–72.

Pinto, J. K. forthcoming. *Project management.* Upper Saddle River, NJ: Prentice Hall.

Project Management Institute. 2000. *Guide to the project management body of knowledge.* Newtown Square, PA: Project Management Institute.

———. 2003. *Organizational Project Management Maturity Model.* Newtown Square, PA: PMI.

Wheelwright, S. C., and K. B. Clark. 1992. *Revolutionizing new product development.* New York, Free Press.

SECTION I

KEY ASPECTS OF PROJECT MANAGEMENT

INTRODUCTION

As we noted in the introduction, one key element in the management of projects involves the need to grasp the essentials, the execution-based, "on time, in budget, to scope," delivery-focused view of project management. These "walk before we can run" practices tend to revolve strongly around a blend of "control" and people, or organizational, issues that make project management such a challenging activity. The first two chapters in this section take a summary view of project control; the second two of key organizational and people issues.

Peter Harpum in Chapter 1 kicks off with an authoritative look at project control and its way of looking at the discipline. Pete positions project control within a "systems" context, reminding us that, in the cybernetic sense, control involves planning as well as monitoring, and also taking corrective action. All the fundamental levers of project control are touched upon. But as in a proper "systems" way of looking at things, Pete reminds us that projects exist within bigger systems, and hence we need to relate project control to business strategy—or the project's equivalent contextual objectives.

Managing project risks is an absolutely fundamental skill at any level of management—including this "base" level of project management, as PMBOK, of course, recognizes. Projects, by definition, are unique: Doing the work necessary to initiate, plan, execute, control, and close out the project inevitably entails risks. These, as Steve Simister in Chapter 2 succinctly summarizes, will need to be managed, and Steve proposes a "risk strategy-identification-analysis-response-control" process for doing so.

To effectively mobilize the resources needed to manage the project, a great deal needs to be understood about the organizational structures and systems, and roles and responsibilities, that must be harnessed to undertake the project. Erik Larson, in Chapter 3, provides a solid overview of the principal forms of organization structure found in project management. Erik concentrates particularly on the matrix form but shows the types of factors that will affect the choice of organization structure. Each of these choices, deliberate or otherwise, will have a tremendous impact on the resulting likelihood of successful project management.

Then, in Chapter 4, Dennis Slevin and Jeff Pinto provide a broad overview of some key behavioral factors impacting successful project management. Drawing on original research on practicing project managers specially carried out for this book, they summarize these into 12 critical issues that impact the performance of the project manager, ranging from the personal ("micro") to the organizational ("macro").

This first section builds the base for the follow-on chapters. Project management is shown to be best done when we appreciate the required blending of human (behavioral) and control elements, recognizing them not as competing but as complementary challenges, and maintaining our view of the horizon to always incorporate these fundamental issues.

About the Authors

Peter Harpum

Peter Harpum is a project management consultant with INDECO Ltd, with significant experience in the training and development of senior staff. He has consulted to companies in a wide variety of industries, including retail and merchant banking, insurance, pharmaceuticals, precision engineering, rail infrastructure, and construction. Assignments range from wholesale organizational restructuring and change management, through in-depth analysis and subsequent rebuilding of program and project processes, to development of individual peoples' project management capability. Peter has a deep understanding of project management processes, systems, methodologies, and the "soft" people issues that programs and projects depend on for success. Peter has published on design management; project methodologies, control, and success factors; capability development; portfolio and program value management; and internationalization strategies of indigenous consultants. He was a Lecturer, Visiting Lecturer and examiner at UMIST on project management between 1999 and 2003.

Stephen Simister

Dr. Simister is a consultant and lecturer in project management and a director of his own company, Oxford Management & Research Ltd., and an Associate of INDECO Ltd. His specialism is working with clients to define the scope and project requirements to meet their business needs, facilitating group decision support workshops that allow these requirements to be articulated outside suppliers of goods and services, and facilitating both value and risk management workshops. He has experience of most business sectors and has been involved in all stages of project life cycles. As a Fellow of the Association for Project Management (APM), Stephen is currently Chairman of the Contracts & Procurement Specific Interest Group. He is a member of the working party updating APM's Risk Management guide. He is also a Chartered Building Surveyor with the Royal Institution of Chartered Surveyors and sits on the construction procurement panel. Stephen lectures at a number of European universities and has written extensively on the subject of project and risk management. He is coeditor of Gower's *Handbook of Project Management*, 3rd Edition. He received his doctorate in Project Management from The University of Reading.

Erik Larson

Erik Larson is professor and chair of the management, marketing, and international business department at the College of Business, Oregon State University. He teaches executive, graduate, and undergraduate courses on project management and leadership. His research and consulting activities focus on project management. He has published numerous articles on matrix management, product development, and project partnering. He is coauthor of a popular textbook, *Project Management: The Managerial Process*, 2nd Edition, as well as a professional book, *Project Management: The Complete Guide for Every Manager*. He has been a member of the Portland, Oregon chapter of the Project Management Institute since 1984. In 1995 he worked as a Fulbright scholar with faculty at the Krakow Academy of Economics on modernizing Polish business education. He received a BA in psychology from Claremont McKenna College and a PhD in Management from State University of New York at Buffalo.

Dennis P. Slevin

Dennis P. Slevin is Professor of Business Administration at the Katz Graduate School of Business, University of Pittsburgh. He received his education in a variety of university settings, starting with a BA in Mathematics at St. Vincent College and continuing with a BS in Physics at Massachusetts Institute of Technology, an MS in Industrial Administration at Carnegie Mellon University, and a PhD in Business Administration at Stanford University in 1969. Dr. Slevin's research interests focus on entrepreneurship, project management, and corporate governance. He has coauthored the *Total Competitiveness Audit* and the *Project Implementation Profile*; each instrument proposes a conceptual model and a diagnostic tool. He has published widely in a variety of professional journals, including *Administrative Science Quarterly, Academy of Management Journal, Management Science, Sloan Management Review, Project Management Journal,* and numerous other journals and proceedings. His book *The Whole Manager: How to Increase Your Professional and Personal Effectiveness*, New York: AMACOM, 1989 (paperback, 1991), provides concrete tools for use by practicing managers. He was cochair of PMI Research Conference 2000, Paris. This conference gathered project management researchers from around the world and resulted in the book *The Frontiers of Project Management Research,* PMI 2002, of which he is coeditor. He was cochair of PMI Research Conference 2002, Seattle, July 2002. Since 1972, he has also been president of Innodyne, Inc., a management consulting firm specializing in the design and implementation of specially targeted management development programs. He has worked with numerous companies and organizations, such as PPG Industries, General Electric, Alcoa, Westinghouse, GKN plc, IBM, and many other large and small firms.

Jeffrey Pinto

Dr. Jeffrey K. Pinto is the Samuel A. and Elizabeth B. Breene Professor of Management in the Sam and Irene Black School of Business at Penn State Erie. His major research focus has been in the areas of project management, the implementation of new technologies, and the diffusion of innovations in organizations. Professor Pinto is the author or editor of 17 books and over 120 scientific papers that have appeared in a variety of academic and practitioner journals, books, conference proceedings, and technical reports. Dr. Pinto's work has been translated into French, Dutch, German, Finnish, Russian, and Spanish, among other languages. He is also a frequent presenter at national and international conferences and has served as keynote speaker and as a member of organizing committees for a number of international conferences. Dr. Pinto served as Editor of the *Project Management Journal* from 1990 to 1996 and is a two-time recipient of the Project Management Institute's Distinguished Contribution Award. He has consulted widely with a number of firms, both domestic and international, on a variety of topics, including project management, new product development, information system implementation, organization development, leadership, and conflict resolution. A recent book, *Building Customer-Based Project Organizations*, was published in 2001 by Wiley. He is also the codeveloper of SimProject, a project management simulation for classroom instruction.

CHAPTER ONE

PROJECT CONTROL

Peter Harpum

Project control is about ensuring that the project delivers what it is set up to deliver. Fundamentally, the process of project control deals with ensuring that other project processes are operating properly. It is these other processes that will deliver the project's products, which in turn will create the change desired by the project's sponsor. This chapter provides an overview of the project control processes, in order to provide the conceptual framework for the rest of this section of the book.

Introduction

Control is fundamental to all management endeavor. To manage implies that control must be exercised. Peter Checkland connects the two concepts as follows:

> The management process. . .is concerned with deciding to do or not to do something, with planning, with alternatives, with monitoring performance, with collaborating with other people or achieving ends through others; it is the process of taking decisions in social systems in the face of problems which may not be self generated.
>
> Checkland, 1981

In short to

- plan
- monitor
- take action

One may ask what is the difference between project control and any other type of management control? Fundamentally there is little that project managers must do to control their work that a line manager does not do. Managers of lines and projects are both concerned with planning work; ensuring it is carried out effectively (the output from the work "does the job") and efficiently (the work is carried out at minimum effort and cost). Ultimately, managers of lines and projects are concerned with delivering what the customer wants. The line management function is usually focussed on maximizing the efficiency of an existing set of processes—by gradual and incremental change—for as long as the processes are needed. The objective of operations management (or "business–as–usual") is rarely to create change of significant magnitude. Projects, on the other hand, are trying to reach a predefined end state that is different to the state of affairs currently existing; projects exist to create change. Because of this, projects are almost always time-bound. Hence, the significant difference is not in control per se, but in the processes that are being controlled—and in the focus of that control.

Project management is seen by many people as mechanistic (rigidly follow set processes and controlled by specialist tools, apropos a machine) in its approach. This is unsurprising given that the modern origins of the profession lie in the hard-nosed world of defense industry contracting in America. These defense projects (for example, the Atlas and Polaris missiles) were essentially very large systems engineering programs where it was important to schedule work in the most efficient manner possible. Most of the main scheduling tools had been invented by the mid-1960s. In fact, virtually all the mainstream project control techniques were in use by the late 1960s. A host of other project control tools were all available to the project manager by the 1970s, such as resource management, work breakdown structures, risk management, earned value, quality engineering, configuration management, and systems analysis (Morris, 1997).

The reality, of course, is that project management has another, equally important aspect to it. Since the beginning of the 1970s research has shown that project success is not dependent only on the effective use of these mechanistic tools. Those elements of project management to do with managing people and the project's environment (leadership, team building, negotiation, motivation, stakeholder management, and others) have been shown to have a huge impact on the success, or otherwise, of projects (Morris, 1987; Pinto and Slevin, 1987—see also Chapter 34 by Brandon and Chapter 5 by Cooke-Davies). Both these two aspects of project management—"mechanistic" control and "soft," people-orientated skills—are of equal importance, and this chapter does not set out to put project control in a position of dominance in the project management process. Nevertheless, it is clear that effective control of the resources available to the project manager (time, money, people, equipment) is central to delivering change. This chapter explains why effective control is fundamentally a requirement for project success.

The first part of the chapter explains the concept of control, starting with a brief outline of systems theory and how it is applied in practice to project control. The second part of the chapter outlines the project planning process—before project work can be controlled, it is critical that the work to be carried out is defined. Finally, the chapter brings project planning and control together, describing how variance from the plan is identified using performance measurement techniques.

Project Control and Systems Theory

Underlying control theory in the management sciences is the concept of the *system*. The way that a project is controlled is fundamentally based on the concept of system control—in this case the system represents the project. Taking a systems approach leads to an understanding of how projects function in the environment in which they exist. A system describes, in a holistic manner, how groups are related to each other. These may, for instance, be groups of people, groups of technical equipment, groups of procedures, and so on. (In fact, the systems approach grew out of general systems theory, which sought to understand the concept of "wholeness"—see Bertalanffy, 1969.) The systems approach exists within the same conceptual framework as a project; namely, to facilitate change from an initial starting position to an identified final position.

The Basic Open System

A closed system is primarily differentiated from an open system in that the former has impermeable boundaries to the environment (what goes on outside the system does not affect the system), while the latter has permeable boundaries (the environment can penetrate the boundary and therefore affect the system). In a closed system, fixed "laws" enable accurate predictions of future events to be made. A typical example of this is in physical systems (say, for instance, a lever) where a known and unchanging equation can be used to predict exactly what the result will be of applying a force to one part of the system. Open systems do not allow such accurate predictions to be made about the future, because many influences cross the boundary and interact with the system, making the creation of predictive laws impossible.

The key feature of the open system approach that makes it useful in the analysis and control of change is that the theory demands a holistic approach be taken to understanding the processes and the context they are embedded in. It ensures that account is taken of all relevant factors, inputs or influences on the system, and its environmental context. Another key feature of an open system is that the boundaries are to a large extent set arbitrarily, depending on the observer's perspective. Moreover, wherever the boundaries are placed, they are still always permeable to energy and information from the outside. It is this quality that allows the relationships between the system and its environment to be considered in the context of change, from an initial condition to a final one.

A critical part of this theory that is useful when considering management control is that the open system always moves toward the achievement of superordinate goals. This means that although there may be conflicting immediate goals within the creative transformation process(es), the overall system moves toward predefined goals that benefit the system as a whole (see Katz and Kahn, 1969). In a project system these goals are the project objectives.

In simple terms the basic open system model is shown in Figure 1.1.

The Open System Model Applied to Project Control

If the generic system diagram is redrawn to represent a project and its environment, the relationship of control to planning (how the project is going to achieve its objectives) can be made clear (see Figure 1.2).

FIGURE 1.1. THE BASIC SYSTEMS MODEL.

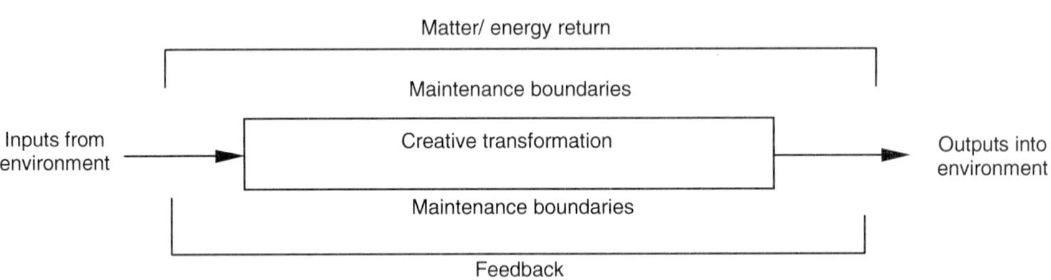

The system: A set of components that are interrelated, acting in a unified way, to achieve a goal.

The creative transformation: The process(es) that act on the energy, resources, and materials from the environment to turn inputs into outputs.

Inputs: Energy, resources, and materials from the environment.

Outputs: Products, knowledge, or services that help the system achieve its goals.

Maintenance boundaries: The mechanism that defines the transformation process(es)—and therefore creates the system's unique identity.

Matter/ energy return: Matter and energy are returned to the environment in order to ensure the system remains in equilibrium.

Feedback: The feedback required to ensure the system stays on course to deliver its goals.

The environment: These are all the influences that act on the system, and the control system attempts to mitigate.

Source: After Jackson (1993).

The value of the open system model of the project shows how the "mechanistic" control meta process attempts to ensure that the project continues toward achievement of its objectives and the overall goal. The "softer" behavioral issues are evident throughout the project system: within the project processes, acting across the permeable boundaries with the project stakeholders and wider environment, and indeed around the "outside" of the project system but nevertheless affecting the project goal—and hence the direction the project needs to move in to reach that goal.

The boundaries of the project are defined by the project plans. These plans define what the project processes need to do to reach the system's goals (defined by the project's environment). The plans also determine *where* in the organizational hierarchy the project exists, because projects have subsystems (work packages) and exist within a supra-system (programs and portfolios of projects). This is shown in Figure 1.3.

FIGURE 1.2. THE PROJECT AS A SYSTEM.

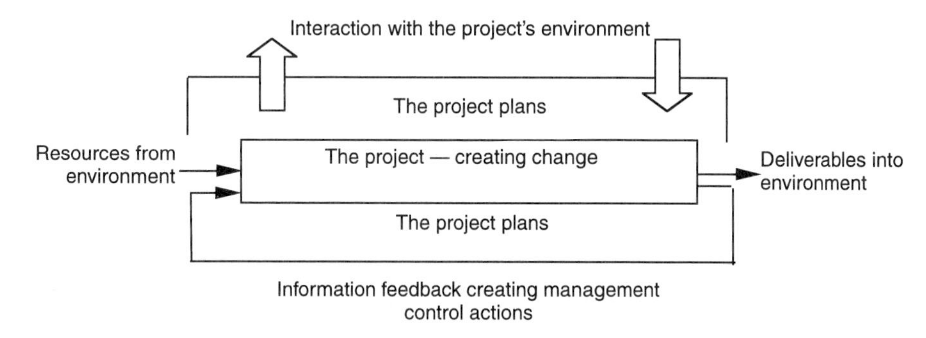

FIGURE 1.3. THE HIERARCHICAL NATURE OF PROJECT CONTROL SYSTEMS.

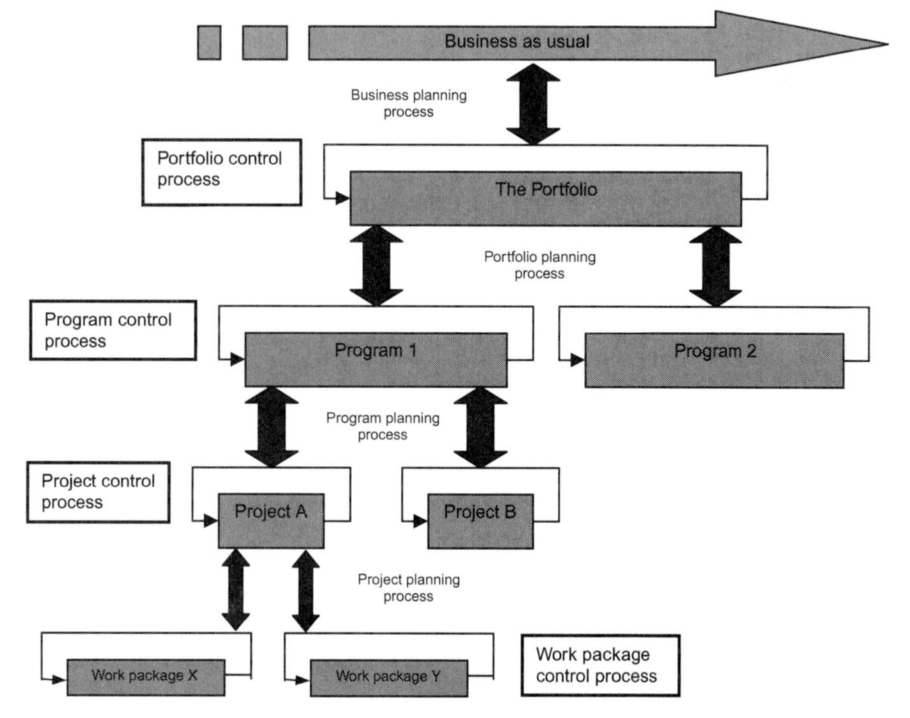

To make the distinctions clear between portfolio and program, their definitions are listed in the following, elaborated for clarity from the Association for Project Management Body of Knowledge, 4th edition (APM, 2000):

Program management Often a series of projects are required to implement strategic change. Controlling a series of projects that are directed toward a related goal is program management. The program seeks to integrate the business strategy, or part of it, and the projects that will implement that strategy in an overarching framework.

Portfolio management In contrast to a program, a portfolio comprises a number of projects, or programs, that are not necessarily linked by common objectives (other than at the highest level), but rather are grouped together to enable better control to be exercised over them.

The feedback loop measures where the project is deviating from its route (the plans) to achieving the project goals and provides inputs to the system to correct the deviation. Control is therefore central to the project system; it tries to ensure that the project stays on course to meet its objectives and to fulfill its goals. The deviation away from the project's goals can be caused by suboptimal project processes (poor plan definition, for instance) or by positive or negative influences from the environment penetrating the permeable boundaries and affecting the processes or goal (poor productivity, failures of technologies to perform as expected, market changes, political influence, project goals being changed, and a host of similar inputs).

At each of these levels of management there is a planning process. This process ensures alignment between objectives of work at different levels, for example, between programs and projects—the program plan. Consequently, the control of these systems is hierarchical in nature.

The abstract models described so far can be used to diagrammatically show the overarching project control process, in which all the system elements are combined. This process is shown in Figure 1.4. In this diagram the way in which the project life cycle stages of initiate (define objectives), plan, and implement (carry out work) are overlaid with the control process is clearly shown. The next part of this chapter describes how the project plan is developed; that is, how the project system is defined.

Defining the Project Objectives

The clear and unambiguous definition of project objectives is fundamental to achieving project success. However, prior to project definition it is necessary to understand the business strategy, or at least that part of the strategy, being delivered (or facilitated) by the project. If the strategic goal is not understood, there is little chance of the project's objectives being accurately defined.

There are various definitions of strategy, viz:

FIGURE 1.4. THE PROJECT CONTROL PROCESS.

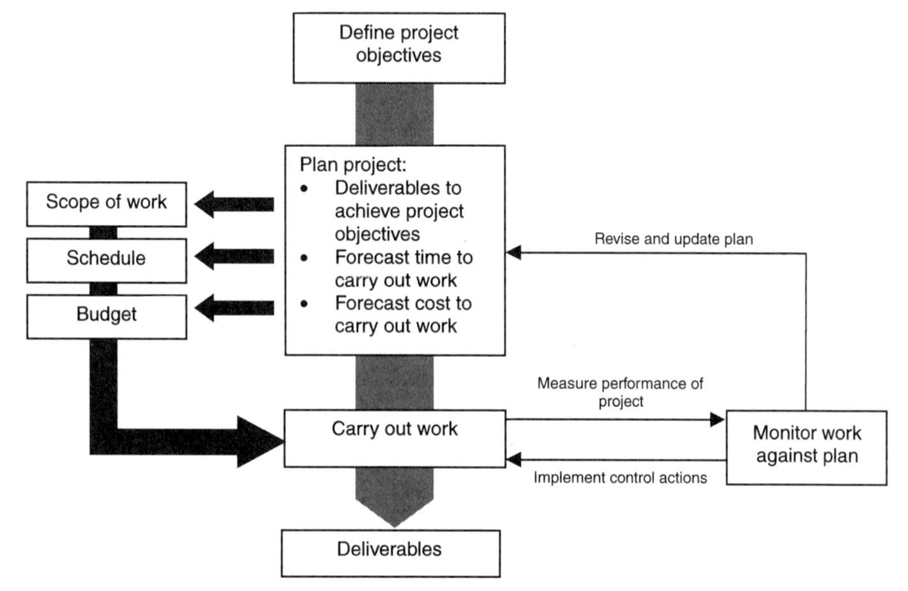

- "Strategic thinking is the art of outdoing an adversary, knowing that the adversary is trying to do the same to you" (Dixit and Nalebuff, 1991).
- ". . .the general direction in which the [company's] objectives are to be pursued" (Cleland and King, 1983).
- " . . . strategies embrace those patterns of high leverage decisions (on major goals, policies, and action sequences) which affect the viability and direction of the entire enterprise or determine its competitive posture for an extended period of time" (Quinn, 1978).

Business strategies have dual functions; firstly to communicate the strategy at a detailed level and identify the method of implementation (in part by programs and projects), and secondly to act as a control device. Both of these functions rely on the strategy having the characteristic of a plan—in other words, strategy is represented in a decomposed and articulated form. The communication aspect of the program informs people in the organization (and those external to it) of the intended strategy and the consequences of the strategy being implemented. They not only communicate the intention of the strategy but also the role that the employees have to take in its implementation (project, nonproject, or business-as-usual work). The control aspect of the strategic program assesses not only performance toward the implementation of strategy but also behavior of the organization as the strategic actions take effect—has the behavior of the firm adjusted as predicted by the strategy? This then forms the feedback loop anticipated in the control system. See Figure 1.5.

FIGURE 1.5. MECHANISM FOR ACHIEVING STRATEGIC CHANGE.

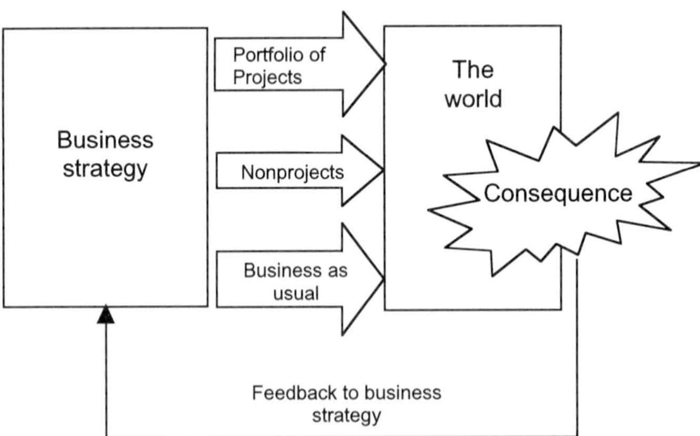

Business strategy provides two or three high-level project objectives, and from these are developed additional project specific objectives and the *project* strategy. Traditionally, project objectives have been defined in terms of the "project triumvirate" of time to complete, cost to complete, and adherence to technical specifications (i.e., quality) (Barnes, 1988). This does not mean that other objectives should not be considered. Objectives for a project to build an oil platform, for example, could be stated as follows:

Primary objectives

- *Safety.* Minimum number of accidents
- *Operability.* Minimum number of days downtime
- *Time.* Maximum acceptable duration before start-up
- *Cost.* Through-life cost for maximum business benefit

Secondary objectives

- *Reliability.* Minimum number of failures per month
- *Ease of installation.* Non-weather-dependent process
- *Maintainability.* Minimum number of maintenance staff required

It is important to reduce uncertainty to the minimum for a project, and setting clear and prioritized objectives is a fundamental part of this process. However, sometimes changes to the objectives become inevitable. Occasionally the environment changes unexpectedly—for example, new legislation may be introduced; economic conditions may change; business conditions may alter. That this may happen is not necessarily in itself a bad thing, or a failure of either the sponsoring organization's management or project management. Such changes may impact on the organization, and its projects, to such an extent that organizational strategy has to be changed and projects either canceled or their objectives changed to meet the needs of the new strategy.

Planning the Project

The essence of project planning is determining what needs to be created to deliver the project objectives (the project deliverables or products), and within what constraints (of time, cost, and quality). Although this may seem like a statement of the obvious, many projects still fail to meet some or all of their objectives because of inadequate definition of the work required to achieve those objectives. Planning must also consider multiple other factors in the project's environment if it is to have any real chance of success—the critical success factors discussed later in the chapter. There are a number of processes that need to be followed to plan projects effectively (see Project Management Institute, 2000):

- Define the deliverables
- Define the work packages
- Estimate the work
- Schedule the work packages
- Manage resource availability
- Create the budget
- Integrate schedule and budget
- Identify key performance indicators
- Identify critical success factors

Each of these processes are briefly described in this section of the chapter (and are described in detail in subsequent chapters of the book).

Defining the Deliverables

Projects are run to create change, and the change is defined by the objectives set for the project. The way in which objectives are achieved is by organizing work (the creative transformation from the basic systems model) to deliver tangible and intangible products into the environment that is to be changed. Therefore, it is central to project success that the specific set of deliverables required is understood and articulated. This set of deliverables forms the project scope.

Accurately defining the project scope entails five subprocesses. Each of these process steps can be highly specialist in nature for projects delivering sophisticated products or services. For this overview chapter they are described briefly in the following paragraphs.

The first of these subprocesses is *requirements definition*—understanding what is required to create the change required from running the project. Requirements are "needs" to be satisfied; they are not the solutions to deliver the change (Eisner, 1997; and see the chapter by Davis et al. later in this book). They are the essential starting point for determining what deliverables need to be made by the project. Poor requirements definition and management has been found to be one of the primary contributory factors leading to project failure. There is little point in managing a project perfectly if the project's deliverables do not solve the right problem, or provide the necessary capability to the organization. Requirements definition consists of the following elements:

Gathering the project requirements This is partly art and partly science (considered by many to be more art usually), particularly when seeking to draw out from the project stakeholders and document as complete a set of desired requirements as possible.

Assessing the requirements The analysis and definition of business and technical requirements to assess the

- project's and organization's technical capability to deliver them
- priorities of the project's requirements, taking into consideration
  —the perceived importance of each requirement to create the change needed
  —the availability of resource (time, people, money, materials) to deliver the requirements
  —the technical capability to deliver the requirements (the requirements may be unachievable technically)
  —the risk profile that the project is able to manage effectively

It is often necessary to iterate the assessment to get a set of requirements that will deliver the entire change desired.

Creating an adequate testing regime In order to be sure that all the conditions to create the change have been met, it must be possible to test that the requirements have been satisfied.

In order for requirements statements to be used efficiently they should be

Structured The project requirements should be clearly linked to the need to create change, and this is done through matching requirements to objectives.

Traceable It should be possible to identify the source of each requirement and trace any changes to the requirements definition, and of the emerging solution to the requirements, as the project evolves.

Testable There should be clear acceptance criteria for each requirement.

After defining the requirements a number of *conceptual designs* are created, the options for delivering the change (see chapter by Harpum on design management provides a more detailed explanation of this and subsequent deliverables definition processes). This process is highly creative and seeks to find efficient solutions to meet the requirements. Whenever solutions are being sought, there are always trade-offs to be considered. Each solution will have with it a set of constraints in terms of what resource is needed to create the solution;

that is to say, each solution will have different needs for money, people, time, and materials. The point of generating a number of solutions is to enable decisions to be taken on what is the most effective trade-off to make for satisfying the project requirements and hence delivering the change required of the project. At this point the *concept design decision gate* is reached—the various concept design options are analyzed in the context of the change that is required to be delivered by the project. There will usually be an economic analysis to determine the following:

- Financial viability of each option
- Schedule to deliver the solution
- Technical capability of the project organization to create the solution
- Availability of suitable materials to create the solution

When the decision is made, it is important that the complete set of deliverables defined by the concept design is clearly documented.

Once the concept design has been selected, the deliverables that form that design must be *specified*—the exact details of the particular set of deliverables must be established. This is obviously important for those carrying out the work to make the deliverables. It is also fundamental to the control process (Reinertsen, 1997). It is against this specification that the project deliverables will be measured; have the deliverables created by the project been made as specified? (This is part of the quality management process). As with the previous subprocesses involved in defining the scope, specifying deliverables *can* be a complex and sophisticated task. There are essentially two ways to specify a deliverable:

Performance specification This type of specification is stated in terms of required results, with criteria for verifying compliance—without stating methods for achieving the required results. (At a work package level the performance specification defines the functional requirements for the deliverable, the environment in which it must operate, and the interface and interchangeability requirements.)

Detailed specification The opposite of a performance specification is a detail specification. A detail specification gives design solutions, such as how a requirement is to be achieved or how an item is to be fabricated or constructed.

After the project scope has been defined as described to this point, a final process to manage change to the scope must be established. *Scope change control* is a critical part of the overall project control meta process. Projects frequently suffer from poor scope change control, leading to the wrong deliverables being produced by the project, which means failing to satisfy the project requirements and ultimately, of course, not delivering the change that the project was set up to create. For this reason, change control is considered one of the "iron rules" of effective project management.

It is also important to realize that scope change is sometimes inevitable within the life cycle of a project. Defining requirements is dependent on having information available on what change the project is set up to achieve. It is rare that all this information is available at the beginning of the project; more usually, as the project scope is developed, additional information becomes available on the true nature of the project requirements, which means that changes to the scope are required.

The management of this scope change needs to be thorough and strictly controlled (Project Management Institute, 2000; Dixon, 2000). The change control process incorporates the following elements:

Identifying changes to scope What new or changed deliverables are required to meet the newly identified, or more clearly understood, requirement. It also includes transmitting the request for scope change to the project's management.

Assessing the need for the scope change This includes deciding whether the change requested is genuinely needed to meet the requirements, any implications on the entire set of project deliverables, and the impact on project constraints (time, money, people, material).

Accepting or rejecting the scope change request This includes documenting the reasons for the decision and communicating the changed set of deliverables (or that part of the deliverables changed) to those making them and to other project stakeholders.

Adjusting the project plan This is done to take account of the changed set of deliverables (meaning changes to budget, schedule, people carrying out the work, etc.).

FIGURE 1.6. EASE OF CHANGE COMPARED TO COST OF CHANGE OVER THE PROJECT LIFE CYCLE.

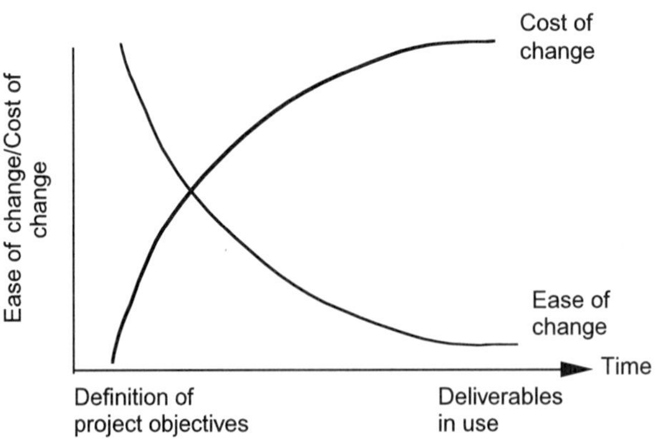

Source: Developed from Allinson (1997).

Figure 1.6 shows how the cost of changes on the project increases dramatically once the project has entered the implementation stages, compared with the much lower cost of change during the concept, feasibility, and design stages. During the early stages, fewer people are involved and the decisions made are more strategic in nature. A simple example is a change fed back from the corporate executive, at the concept stage of an organizational change project, to have separate sales and marketing departments instead of a combined one. This requires reworking the project objectives and reassessing the risk associated with the change on the overall project. It can be carried out by a small number of people relatively quickly. This same change, brought into the project during the implementation stages, will require significant amounts of time and resource to adjust the project plan to meet the new requirement. It may also cause demotivation in the project team, as work already implemented has to be "undone" and the new structure put in place.

It is worth remembering that objectives may need to be changed during the project, reflecting the reality that situations change over time. If this is the case, it may be decided that the best course of action is to complete the project (because some of its objectives are still valid and/or the cost of cancellation would outweigh the benefits of continuing) but accept a lower effectiveness of the deliverables.

A number of specialist project management techniques can be used to help in the scope definition and change control processes and are described in detail in other chapters of this book. They include the following:

Configuration management The definition and control of how all the deliverables are configured; how they all "fit" together

Interface control The exact specification of the interfaces between different deliverables

Systems engineering The way in which a set of deliverables are arranged within a hierarchical "systems architecture"

(The chapter by Cooper and Reichelt addresses the issue of managing changes in more detail.)

Defining the Work Packages

Three tools are used to define the work packages:

- *Product breakdown structure* (*PBS*). What needs to be made by the project.
- *Work breakdown structure* (*WBS*). The work required to make these products
- *Organizational breakdown structure* (*OBS*). Where in the organization the skills reside for doing the work needed

The first part of the work package definition process is to break down the main set of deliverables (identified in the scope process) into their component parts—the deliverables breakdown structure, more commonly known as the *product breakdown structure* (PBS). The disaggregation of the deliverables is developed to the level of detail that is needed by those working on the project, and commensurate with the degree of control that is required to be exercised. The work associated with making the deliverables is also divided into discrete work packages—and documented in the *work breakdown structure* (WBS). The two models must be consistent with each other; the work packages identified in the WBS must be associated with specific deliverables (i.e., products). The PBS and WBS are often combined together, and when this is the case, the diagram is usually called the WBS.

The decomposition of deliverables, and associated work, is fed into the processes for creating the forecasts of time and cost to make them; this is the estimating process. Without a clear understanding of the finite elements that need to be made by the project, it would be very difficult to carry out effective estimation of the duration to complete the tasks required and the cost to make the deliverables.

Fundamental to the planning process is deciding who will be carrying out the project work, documenting this information, and communicating it to the project team. The allocation of people to work packages is recorded in the *organizational breakdown structure* (OBS). The human resources needed to undertake the tasks to make the deliverables are often in short supply. This means that there will rarely be enough suitably skilled people available to create the deliverables as quickly as may be desired. The resources available for the work will ultimately determine the time to make the deliverables.

Estimating the Work

Forecasting how long a work package will take to complete and the cost to carry out that work is essential to effective planning. There are a number of techniques used to estimate time and cost. Essentially, the estimating process is iterative. A number of estimates are produced, reviewed, validated against the availability of resources required for the work packages, and revised accordingly.

In the estimating process, it is important to refer to historical information on the cost and time taken to carry out the same or similar work packages. Since cost is normally directly related to time (because time to complete work packages is mainly dependent on people and materials), time estimates are produced first. The cost estimate is then generated based on the forecast time to complete the work packages and the cost of materials needed:

Time estimating Time estimates are developed by calculating how long the work package will take to complete. The inputs for the estimates of duration typically originate from the person or group on the project team who is most familiar with the nature of the tasks required to complete the work package.

Cost estimating Cost estimating involves calculating the costs of the resources needed to complete project activities. This means the cost of peoples' time must be known, as well as the cost of materials needed to make the deliverables. This includes identifying the project management overhead—the cost of managing the project.

With estimating, there is uncertainty about the exact duration and exact cost of a work package—by definition. The uncertainty can be reflected by estimating the range within which the duration and cost for each work package will fall. The optimistic, most likely, and pessimist values for each can be provided—known as *three-point estimating*. This information can then be fed into the scheduling and budgeting processes to provide a more realistic view of the ranges of outcomes for the project as a whole. (A number of different probability distribution curves can be used.)

A fourth tool used in project planning can now be used—the *cost breakdown structure* (CBS). This documents the cost to carry out each work package, taken from the cost estimate. The information is set out in an integrated way with the PBS, WBS, and OBS to form the fundamental framework for project control. These four fundamental tools describe what has to be made to meet the project requirements, what work is needed to be carried out to make the deliverables, who is allocated to the work packages, and what the cost to create the set of deliverables will be. In large and complex projects, this information can be combined into a three-dimensional matrix (called the *cost cube*) to show the cost per product or deliverable, per resource (Turner and Remenyi, 1995). See Figure 1.7.

One advantage of creating the cost cube is that it provides a framework, or structure, for developing the estimate. For instance, one can more easily identify whether all the appropriate elements of cost associated with resources for a particular product have been included. It also enables the summing along any plane of "cubes" to provide cost information for any of the discrete items on the three axes. For instance, it is a straightforward matter to identify the total cost of resource R4 to the project by adding together the costs in each cube of the plane, as shown on the diagram.

FIGURE 1.7. THE COST CUBE MATRIX.

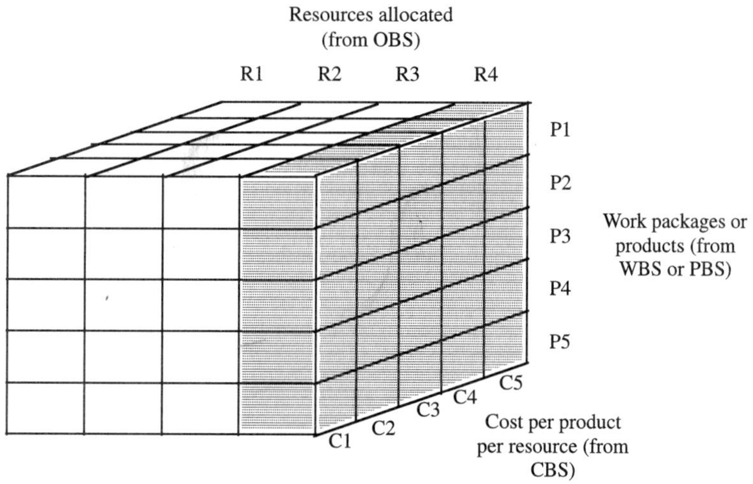

Source: Adapted from Turner and Remenyi (1995).

Scheduling the Work Packages

The essence of scheduling work packages is simple. The following factors must be known:

- Upon what previous work each subsequent work package depends
- The estimated duration of the work package
- How much "float" is available for the work—whether the work package must be carried out within a certain time or whether the time period in which it must be done can float between two known extremes

It is the combination of this information that determines what can be done when.

There are a number of well-known and commonly used techniques for modeling project work (critical path method, program evaluation and review technique, precedence diagramming, amongst others). All these techniques aim to provide flexibility in the manipulation of the model information until the optimum solution is found to suit the particular work packages of the project. Some were invented in the field of operations research, whilst others were developed by organizations for their own use. These modeling techniques are able to create projects schedules, either directly or by feeding into other techniques.

There are two distinct approaches to these scheduling techniques: *activity-on-the-arrow*, depicted in Figure 1.8, and *activity-on-the-node*, depicted in Figure 1.9. They both model the sequence of work packages by using nodes and arrows to build up a diagram that shows dependency and time for all the work packages for the project. It is then a relatively straightforward (and using PC-based software, quick) task to calculate how long it will take to complete all the work packages and, hence, the project's overall schedule.

The route through the network that determines the shortest possible time to complete all the work packages is called the *critical path*. This is important information for the project manager because these work packages will be the focus of management attention, particularly those projects for which completion "on schedule" is critical.

FIGURE 1.8. ACTIVITY-ON-THE-ARROW NETWORK.

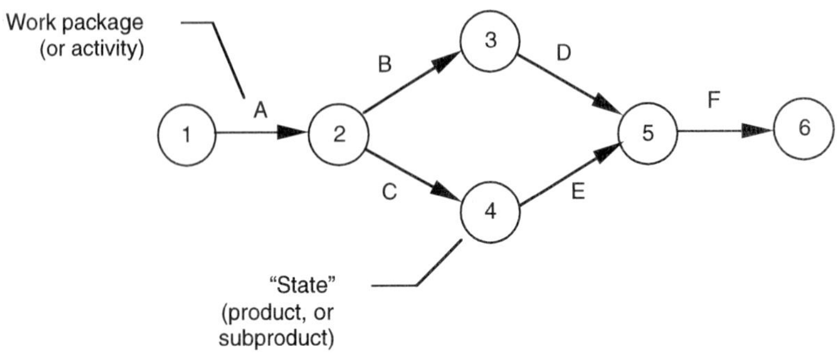

FIGURE 1.9. ACTIVITY-ON-THE-NODE NETWORK.

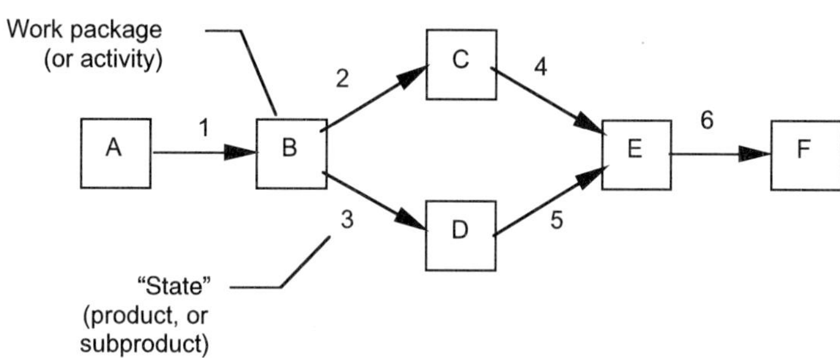

The differences between the two basic network modeling techniques appear trivial at first sight of the diagrams. However, the two approaches have significant differences in the operation of the logic used. Both have advantages and disadvantages.

Once the schedule has been established from the network diagram, it is often presented graphically on a Gantt chart. This format makes it easier to see when the work packages will be carried out in relation to each other. It also allows simple graphical representation of work completed at a given point in time, making reporting of project progress easier to show. Many current software-based scheduling tools allow the user to enter time duration for work packages directly into a Gantt chart, without first going through the process of building a network diagram. This is user-friendly but does not necessarily lead to more effective scheduling!

The schedule must be reassessed in light of the availability of people (and indeed materials—particularly those being supplied by third parties and contractors to the core project team). This part of the scheduling process is called *resource leveling*—making the schedule fit the available resources. There is likely to be an impact on the cost to complete the project, so the budget must also be reassessed (hence, the estimate is progressively elaborated and becomes progressively more accurate). Understanding the causes of variance between the actual time taken to complete the work packages and the schedule is important information for the estimating process in future projects. Similarly, variances between the actual cost to carry out the work packages and the estimated cost will also provide valuable historical estimating information.

The risk management and estimating processes will identify where there is uncertainty in the project. This uncertainty can be modeled in the schedule by allowing extra time for the work packages likely to be affected. It is clear that the entire planning process is iterative, and a number of cycles of scheduling, budgeting, and assessment of resource availability and productivity are required before a final project plan can be established.

Managing Resource Availability

The initial estimates of time and cost to complete the project are ideal estimates; the assumptions are made that sufficient people, materials, facilities, equipment, and services will be available to carry out the project at the maximum efficiency. However, before the schedule and budget can be finalized, the impact of resource availability, and the productivity of those resources, must be taken into account. For example:

- Sufficient people are rarely available (particularly where a few experts must input to many work packages).
- "Ideal" materials are often either not available where and when required, or their cost would make the project untenable (meaning less efficient materials need to be used).
- Equipment is often expensive to use and therefore must be shared across a number of projects.
- The same applies for facilities and services.

In addition to this, the resource "profile" (the types of resources needed at different times in the project) usually changes over the project life cycle. The process for managing resource allocation can be broken down into five stages:

Planning resource allocation Identifying the types of resources required, based on the information defined in the PBS, WBS, OBS, and CBS.

Allocating resources Coordinating the availability of resources with the suppliers of those resources and allocating them to work packages: be they internal to the organization within which the project exists (and this is commonly a major task for people—human resources—where organizations have a matrix structure) or external, such as third-party suppliers of materials, equipment, services, and facilities.

Optimizing the schedule Inputting resource availability in the schedule, which normally means having to use the technique of resource leveling—"smoothing" resource usage to balance schedule and the availability of resources.

Monitoring resource allocation Tracking resource usage and identifying and resolving conflicts associated with resource availability as this, and the project's needs, change over time.

Reviewing and revizing the resource allocation Modeling the impact of changing resource use and availability on the project budget and schedule

Productivity of resources clearly has a significant influence on the schedule, and hence the cost, of the project. Productivity of equipment is often fairly easy to measure and predict; predicting productivity of people is a far more complex thing to do (and predicting productivity of highly creative design resources even more difficult). Productivity information can be gained from historical records of performance on similar work—for people and equipment. Finding and using this information is vital to effective resource allocation and resource "smoothing" of schedules.

Budgeting

The costs to complete all the work packages are identified in the estimating process. Combining the cost information with the schedule allows the cash flow curve to be created. This curve is a key piece of control information. This is particularly the case for organizations that are contracted to deliver projects on behalf of a client. These types of projects have large cash outflows (to pay for material and human resources), which are usually then charged on to the client sometime after the expenditure is incurred. If this is not managed very carefully, the project can become heavily indebted. The difference between what has been spent and what has been recovered by a project, at a given time, can cause the funders of this difference (the owners of the firm running the project) to become insolvent. Hence, effective cash flow management is a highly valued skill in project-based industries, such as construction and engineering, where huge amounts of money flow through the project.

The cash flow curve and the cost forecast are the basis of the project budget. This describes what amounts of money will be spent, on what resources, and when they will be spent. Before the cost forecast becomes a budget (the budget is the *agreed* amount of money that the project manager can spend), the effect of risk on the project needs to be assessed in cost terms and then added to the forecast. This additional amount of money allows the project manager to deal with "certain uncertainty" in the forecasts (the uncertainty can be predicted through the risk management process), and also "uncertain uncertainty" that can affect the project (uncertainty that cannot be assessed—as an example, a key human resource may leave the project without warning). These extra sums of money are the budget contingencies.

The importance of creating accurate budgets, and controlling against the budget, is obvious in a commercial environment—overspending reduces profit; underspending (whilst still making the correct project deliverables) increases profit. In the not-for-profit sector, control of money is clearly still critical. The budget document therefore identifies, line by line (hence the term "line item"), how much money is agreed to be spent per deliverable, or part deliverable. It is this detailed breakdown against which actual costs are measured and reported, and control action initiated.

After the forecasts have been analyzed and the deliverables set has been revisited to seek optimization of all the project constraints, a schedule and budget are agreed upon by the project sponsors. The budget and schedule are absolutely fundamental to the control process. It is against these two documents that the progress of the project in carrying out the work packages, and hence the production of the deliverables, is measured (see Figure 1.4).

However, having two separate documents means there exists a lack of *integrated* information related to deliverables (or products), schedule, and the budget to produce those deliverables—the complete picture of predicted project performance cannot easily be discerned. This can be overcome by combining information on schedule, budget, and the project deliverables. This is called earned value analysis.

Earned Value Analysis

The technique of *earned value analysis* (EVA) allows the actual performance of the project to be compared to the predicted performance. All the information required for this type of analysis should be available from standard project reporting against the schedule and budget (reporting is described later in this chapter). The schedule per work package (or, if preferred, discrete product) is plotted on one axis of a graph, and the budget per work package (or product) is plotted against the other axis. See Figure 1.10.

Common acronyms are used for the information required for the analysis:

- *BCWS.* Budgeted cost of work scheduled (how much money has been allocated to each work package or product in the schedule)
- *BCWP.* Budgeted cost of work performed (how much money was allocated to each work package or product that has been completed)
- *ACWP.* Actual cost of work performed (how much it actually cost for the work package to be performed or the product to be delivered)

The figure for budgeted cost of work performed (BCWP) is the earned value at the point in time that the analysis is being done. The chapter by Brandon looks at this technique in greater detail. At this point it is just worth noting the control information that can be determined from EVA. By manipulation of the data gathered from project performance reporting, project schedule and budget performance can be assessed. The Schedule Performance Index (SPI) and Cost Performance Index (CPI) are calculated as follows:

$$\mathrm{SPI} = \frac{\mathrm{BCWP}}{\mathrm{BCWS}} \qquad \mathrm{CPI} = \frac{\mathrm{BCWP}}{\mathrm{ACWP}}$$

If, at the time of measurement, budgeted cost of work performed and the budgeted cost of work scheduled are the same (i.e., the SPI = 1), the project is exactly on schedule. In the

FIGURE 1.10. EARNED VALUE ANALYSIS PRESENTED GRAPHICALLY.

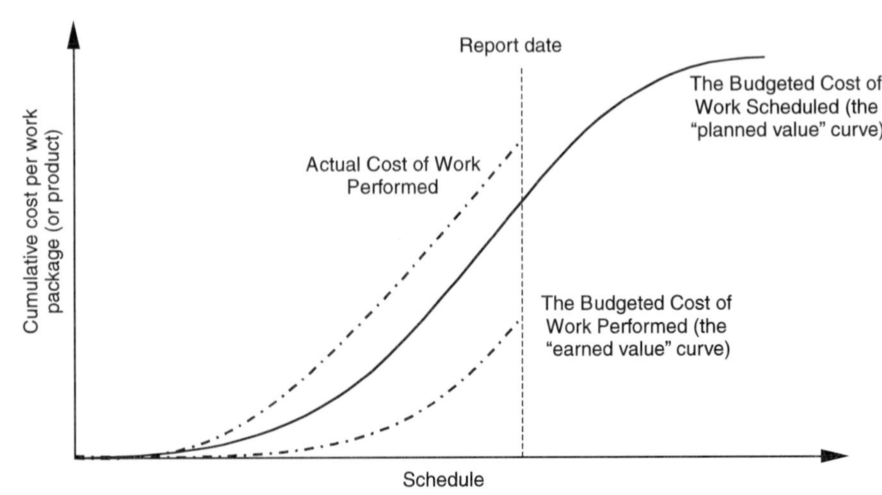

same manner, if the budgeted cost of work performed is the same as the actual cost of work performed (i.e., the CPI = 1) the project is exactly on budget.

If the BCWP is less than the BCWS, the SPI is less than 1; therefore, the project is behind schedule. Equally if the BCWP is less than the ACWP, the CPI is less than 1; there, the project is over budget.

EVA appears relatively straightforward to use, and predictions of future performance can be made using the data (by projecting final time and cost to complete using SPI and CPI values). However, care must be taken to moderate the results from this technique with other project data. There are also inherent dangers in believing that the information provided is a foolproof indicator of current and future progress. EVA does not report the subtleties of project control; it only provides an overview.

Key Performance Indicators

Key performance indicators (KPIs) are used to measure project progress toward achieving objectives, rather than the detail of progress of the work packages. They may be used to

- Measure project performance that is directly related to the change the project is delivering (which could be shareholder value, return on investment, market share, etc.)
- Measure project specific performance—that is, the performance of the project processes (e.g., effectiveness of project control mechanisms, degree of project cost reduction by using designated procurement practices, amount of change occurring in project, etc.).

KPIs must be determined at the beginning of the project and provide direct progress information toward project objectives. The information these measures of performance provide can help the project manager make decisions on trade-offs between the various (usually conflicting) control actions needed.

KPIs also need to be measurable (otherwise how will one know if they have been achieved?). Whilst this sounds obvious, it must be remembered that KPIs can only be useful if the information needed to determine the KPI during and at the end of the project is actually available. This implies that the project management information system must collect relevant data and generate the appropriate information outputs to provide the KPIs to the project's management team—upon which control action will be based.

If the KPIs to be used in a project have been determined by consultation between those needing the change to be delivered by the project (the project sponsor) and the project manager, it is possible to define success as meeting the KPIs at project completion.

Critical Success Factors

Critical success factors (CSFs) are sometimes used synonymously with KPIs. Literally, however, CSFs are the factors that are critical to success. Identification of the CSFs for a project will mean that the project manager and project team know where to concentrate their attention in order to achieve the project objectives. CSFs are therefore the factors that are critical to achieving success, *not* a measure of performance—which is what KPIs are.

A number of studies have been conducted into the factors found to be critical to project success. Many are generic across all projects, but each project will also have its own very specific factors. In their definitive research on success factors in projects Morris and Hough (1987) identified CSFs under the following general headings:

- Project definition
- Politics/social factors
- Schedule urgency
- Legal agreements
- Human factors
- Planning, design, and technology management
- Schedule duration
- Finance
- Project implementation

The development of CSFs for the project (with the involvement of the project manager, project team, project sponsor, and other senior stakeholders) is an important exercise in its own right, since all those associated with the project gain mutual understanding of what is critical to project success.

Chapter 5 by Cooke-Davies address the subject of performance measures in more detail.

Performance Measurement and Control Action

The elements of the project plan have all now been described. On large and complex projects, these elements are often combined into a comprehensive project management control system. These systems also usually aggregate information from many projects, up to their respective program plan, if they are part of a program, and then up to the organization's business management system.

The project plan is constantly adjusted to reflect the reality of what is happening during the project, and so enable the effect of control actions to be predicted on the progress of the current and future work packages. The updated plan provides information to the following:

- *The project team.* Who can then plan their work packages to suit the revised plan
- *The project sponsor.* Who can assess the impact of the new plan on the delivery of the change required
- *Other project stakeholders.* Who may have other areas of work impacted by the changed project plans (including other projects being run—which is particularly important for program managers)

It is critically important, however, that the original plan is not lost—the project plan must be "baselined." This means that while the plan is updated and used to replan future work, it is still possible to compare what *should* have been completed (the baseline plan) with what *has* been completed (the updated plan). Knowledge of the variance between the two plans provides performance information and therefore helps

- Guide the development of the control actions required to bring the project back towards its original plan (if so desired)
- Improve the future control actions to make meeting the new plan more likely
- Gain knowledge to improve future planning (for replanning the same project and for plans for new projects)

The gathering of information to be used for project control is known as *project performance measurement.* The process provides an integrated view of the performance of the project—cost, schedule, technical issues, commercial, and business issues—so that control action can be taken where necessary to correct undesirable variances from the project plan. Equally important is the appropriate reporting of this information—at the right time, to the right people, and in the right format.

The measurement and reporting of progress must fundamentally begin at the work package level. It is here that the information for performance measurement originates. The work package managers must gather information on progress on the specific deliverables that they, or their team, are responsible for creating. The information must be presented in the same manner as the project plan presents it. In this way, variances from the plan are identified at the point where the variance occurs. The work package manager can then instigate control action to bring performance back in line with the project plan—normally a day-to-day management activity. (See Figure 1.3 for a reminder of how all the project control loops nest within a hierarchical control system.)

Performance information is reported to the project manager on a regular basis (weekly and monthly usually). Integrated reporting means that all the work package managers report the same measurements, together with the control actions they have taken to reduce negative variance from the plan. Hence, an aggregated project performance report can be compiled. (Sometimes project performance is reported on an exception basis; i.e., a report is generated only when there is a variance to the project plan requiring the attention of the project manager. Even in those organizations that use such reporting methods, a monthly reporting cycle is common.) From this report the project manager can determine which work packages are underperforming and whether the control action taken is likely to correct the situation. This reporting is the control feedback loop shown in the diagrams earlier in the chapter.

With the overview of project progress afforded by the integrated report, the project manager is in a position to assist the work package managers to improve performance in a manner that does not compromise other work package performance. This is important, since many work packages will be interrelated and may also share resources. The information can be used to replan work packages, and hence the project, and may also mean that the project manager can take action at a level above the work packages. Examples (amongst many possibilities) include the following: work packages could be rescheduled, the specifications of the work package deliverables may be changed, the acceptance criteria of the deliverables against the requirements may be adjusted, or the project scope may be changed.

This entire process is central to the notion of effective project management. The project manager is the single point of integrative responsibility. It is his or her principal function to integrate control action for the greater needs of the project, to ensure that objectives are met and that the desired change is created by the project. After the control action has been taken, the subsequent work package reports will provide evidence of whether variances from the plan have been successfully controlled—and so the process continues.

Organizations often require visibility of project performance at a program or portfolio level. This enables better management of organizational budgets and control of the changes being created by multiple projects and programs. "Rolled-up" performance measurement information, often supported by a program support office (see the chapter by Young and Powell), enables summary reporting to be available to appropriate levels of management in the organization.

Summary

This chapter has outlined the processes that constitute project control, and in so doing introduced many other project processes upon which effective control depends.

Fundamentally, the project must have a

- Set of objectives directly related to the need for the change the project is set up to deliver
- Plan against which the project can be controlled—and so deliver those objectives
- Process to measure performance against the plan—the feedback loop
- Process to control changes to the scope
- Project manager who is truly the single point of integrated control action, with responsibility for delivering the change required of the project

The control process goes on throughout the project life cycle because the internal and external environment of the project is continuously changing. For example:

- Performance of resources is often other than predicted (better or worse).
- New information is generated that may indicate the original plan was not feasible to begin with.
- The objectives for the project may change because the change required to be brought about (by running the project) is itself changed.

In combination, these processes can be seen as forming the "iron rules" for the project. Project control is about

- Good planning of scope, schedule, and budget
- Setting up appropriate metrics to monitor performance

- Reporting the performance against those metrics
- Replanning and instigating corrective action to reduce variance from the baseline plan

References

Association for Project Management. 2000. *Body of Knowledge 4th Edition.* High Wycombe, UK.

Allinson, K. 1997. *Getting there by design.* Oxford, UK: Architectural Press.

Barnes, M. 1988. Construction project management. *International Journal of Project Management* 6 (2, May): 69–79

Bertalanffy, L., von. 1969. *General systems theory: Essays on its foundation and development.* New York: Braziller.

Checkland, P. 1981. *Systems thinking, systems practice.* Chichester, UK: Wiley.

Cleland, D. I., and W. R. King. 1983. *Systems analysis and project management.* International ed. Singapore: McGraw-Hill.

Dixit, A. K., and B. J. Nalebuff. 1991. *Thinking strategically: The competitive edge in business, politics, and everyday life.* New York: W. W. Norton & Co.

Dixon, M. 2000. *Project management body of knowledge.* 4th ed. High Wycombe, UK: The Association for Project Management.

Eisner, H. 1997. *Essentials of project and systems engineering management.* New York: Wiley.

Jackson, T. 1993. *Organisational behaviour in international management.* Oxford, UK: Butterworth-Heinemann.

Katz, D., and R. L. Kahn, 1969. Common characteristics of open systems. In *Systems thinking,* ed. F. E. Emery. 86–104. Harmondsworth, UK: Penguin Books.

Morris, P. W. G. 1994. *The management of projects.* London: Thomas Telford.

Morris, P. W. G., and G. H. Hough. 1987. *The anatomy of major projects.* Chichester, UK: Wiley.

Pinto, J. K., and D. P. Slevin, 1988. Critical success factors across the project life cycle. *Project Management Journal.* 19(3):67–75.

Project Management Institute. 2000. *A guide to the project management body of knowledge.* Newtown Square, PA: Project Management Institute.

Quinn, J. B. 1978. Strategic change: Logical incrementalism. *Sloan Management Review.* (Fall) 7–22.

Reinertsen, D. G. 1987. Managing the design factory. New York: Free Press.

Turner, J. R., and D. Remenyi. 1995. Estimating costs and revenues. In *The commercial project manager* by J. R. Turner. 31–52. London: McGraw-Hill.

Suggested Further Reading

Archibald, R. D. 2003. *Managing high technology programs and projects.* New York: Wiley.

Hartman, F. 2000. *Don't park your brain outside.* Newtown Square, PA: Project Management Institute.

Murray-Webster, R., and M. Thiry. 2000. Managing programmes of projects. In *The Gower handbook of project management.* 3rd ed, *ed.* R. Turner, S. Simister, and D. Lock. Aldershot, UK: Gower.

Smith, N. J. 2002. *Engineering project management.* 2nd ed. Oxford, UK: Blackwell Science.

CHAPTER TWO

QUALITATIVE AND QUANTITATIVE RISK MANAGEMENT

Stephen J. Simister

Risk is present in all projects, and project managers are routinely involved in making decisions that have a major impact on risk. Risk management is concerned with establishing a set of processes and practices by which risk is managed, rather than being dealt with by default. The effective management of risk can only be achieved by the actions of the whole project team, including the client.

Risk management formalizes the intuitive approach to risk that project teams often undertake. By utilizing a formal approach, project teams can manage risk in a more proactive manner. In addition, there needs to be an overall risk management strategy so that this risk management process is implemented in a coordinated fashion. This strategy should include how risk management will be integrated into the project management process on a project.

Definitions

In the context of this chapter, risk management is considered to be a process for identifying, assessing, and responding to risks associated with delivering an objective—for example, completing a construction project—and the focus is on commercial-type risks. Health- and safety-related risks are likely to need separate consideration; Ward and Chapman look at this in their chapter later in the book.

The risk management terms used in this section follow the International Organization for Standardization (ISO) and ISO and International Electrotechnical Commission (IEC)

Guide to Risk Management Terminology (ISO, 2002). The guide covers 29 terms and definitions for risk management, which are categorized into one of four groups: basic terms, terms related to people or organizations affected by risk, terms related to risk assessment, and terms related to risk treatment and control.

The Rationale for Risk Management

Project managers will invariably be called upon to advise as to whether they should undertake risk management on their project. While the benefits of risk management may not be immediately apparent, the direct costs of undertaking the process will be.

The demands of delivering a project are extremely onerous. The desire to deliver projects cheaper and faster while moving closer to the boundaries of innovative design increases the risk profile of a project. In this respect, risk is good: without it there would be no opportunity. It is the very presence of risk that represents the opportunity for a project to go ahead in the first place. Furthermore, the actual impact of risk is unique to a project even though its presence may be commonplace.

Risk management should be flexible, adapting to the circumstances of the client's needs and the project. Some clients require a snapshot of the risks at the outset of the project, with an initial risk assessment, the provision of a one-off risk register, and a quick estimate of the combined effect. Other projects may require a full risk management service, with risk being continually addressed throughout the project.

The Recognition of Uncertainty

By undertaking risk management, the project manager can ensure that clients appreciate just how sensitive their projects are to changing circumstances. This leads to the following:

- Increased confidence in achieving the project objectives and therefore improved chances of success.
- Surprises being reduced, such as cost or time overruns or forced compromises of performance objectives. These surprises usually result in "fire fighting" and the ineffective application of urgent remedial measures.
- Identification of opportunities as risks are diminished, perhaps by relaxing overcautious practices such as duplicating insurances or seeking performance bonds unnecessarily.
- All ranges of parameters that might affect the project being incorporated, rather than a single-point estimate (for example, the cost for a particular component can be between $200 and $300, rather than $225).
- Allowing the team to recognize and understand the composition of contingencies, thus avoiding duplicating any allowance already made.

Justifiable Decision Making

Risk management allows decision making to be based on an assessment of known variables. Judgment can become far more objective and justification for action (or inaction) demonstrable both at the time and at a later date, should it be questioned. As part of risk management, risk is often transferred from one party to another—for example, the main contractor to the subcontractor. The formal application of risk management should ensure that risk transfer is based on the rational assessment of a party's ability to bear and control the risk.

There is a wide range of techniques that can be utilized to facilitate effective decision making. The use of risk management provides an audit trail that can be used should problems occur during later stages of the project.

Team Development

During risk management workshops, a "snapshot" of what might happen to the delivery of a project is taken. At the workshops the project team discuss their concerns and agree on a common way forward. The team must also explore the consequences of risks occurring. Such consequences can be discussed without fear of penalty or contractual restrictions. Discussing the consequence of risk in this way normally reinforces to the team that the project objectives can only be achieved if they work as a team, and not as self-interested parties. As with other types of facilitated workshops, such as value management, such team development-related benefits could on their own justify undertaking risk management.

In summary, to fully realize the benefits of risk management, certain principles should be adhered to:

- Risk management is an awareness of uncertainty.
- Awareness of uncertainty can lead to positive outcomes if recognized early enough.
- Risk management will add both direct and secondary benefits to the delivery of any objective.
- The management of risk must not remove incentive.
- Allocation of risk is to be gauged by the various parties' abilities to bear and control that risk.
- Complete transfer of risk is rarely wholly effective or indeed possible.
- Information about and perception of a risk are fundamental to its assessment and acceptance.
- All risks change with time and any action (or inaction) taken (or not taken) upon them.

The Risk Management Process

A number of risk management processes are publicly available. These can be broken down into three groups: those issued by national standards associations, those issued by professional institutions and those issued by government departments. Information on the various publications is provided in the further reading section at the end of this chapter. In addition, the Web site address for the organizations is also provided.

National standards associations

- *British Standards Institute (2000)*. The UK national standards association. The risk management standard forms part of a wider project management standard (www.bsi-global.com).
- *Canada Standards Association (1997)*. Canada's national standards association (www.csa.ca).
- *Standards Australia (1999)*. Australia's national standards association (www.standards.com.au).

Professional institutions

- *ICE (1998)*. Produced by the UK-based Institution of Civil Engineers (ICE) in partnership with the Institute of Actuaries. The risk management process is designed specifically for infrastructure projects, such as roads and so on (www.ice.org.uk).
- *Japan Project Management Forum (2002)*. This Japan-based forum describes risk management as an input into how project management can best stimulate innovation and generate improved business value to a company (www.enaa.or.jp/JPMF/).
- *PMI (2000)*. The Project Management Institute's *A Guide to the Project Management Body of Knowledge* has a chapter on risk management (www.pmi.org).
- *APM (Dixon, 2000)*. The Association for Project Management's body of knowledge includes a section on risk management. The APM also produces a specific risk management guide (Simon, Hillson, and Newland, 1997; www.apm.org.uk/).

Government departments

- *DoD (2002)*. The U.S. Department of Defense has a process showing specifically how risk management is to be applied to defense projects (www.defenselink.mil/).
- *The UK Office of Government Commerce (2002)*. The OGC has a generic guide to managing risk in a project environment (http://www.ogc.gov.uk)

All these publications propound their own risk management process. While the processes differ from each other to some extent, there is a common process that runs through them all. This common risk management process consists of five basic steps: risk strategy, risk identification, risk analysis, risk response, and risk control. These five steps are iterative in nature, and it is this iteration that constitutes risk management.

The five steps are detailed in the following:

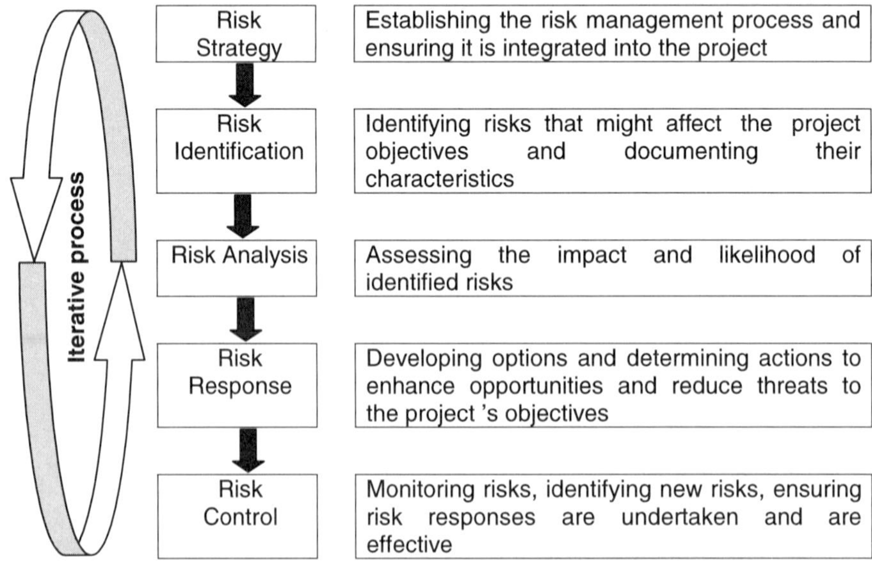

The process proposes a formal application of risk management. The dangers of an informal process are that stages are missed. Projects are typically undertaken within demanding time frames. In such circumstances there is a temptation to deal with risk reactively. In such cases it is only possible to deal with the consequences of the risk occurring; there is no opportunity to mitigate or even avoid the risk.

The risk management process should be commenced as early as possible in a project life cycle. Since any risk management assessment is a snapshot in time, the process has to be undertaken on an iterative basis.

Risk Strategy

The risk strategy needs to set out how risk management will be undertaken on a project. The risk strategy needs to be integrated with the project strategy as well as the wider project management process.

As part of the risk strategy, a risk management plan should be developed. The risk management plan is similar in nature to the project execution plan and may form part of that document. Key areas of a risk management plan might include the following (Simon, Hillson, and Newland, 1997):

- Scope and objectives of risk process
- Roles and responsibilities of participants in the process
- Approach & process to be used
- Deliverables of the process
- Review and reporting cycle
- Tools to be used

The risk strategy should also describe how the risk management process would contribute to any project evaluation exercise that might take place at the completion of the project.

Risk Identification

The identification stage should commence as early as possible, preferably as part of any feasibility study for the project. One of the key principles is that the identification of risks should be undertaken at various stages of the project life cycle. Identification is not just undertaken at the start of the project and then simply monitored during execution. The execution phase itself will generate new risks that need to be identified through the risk identification process.

Risk Identification Methodology

(i) Structured identification of all sources of risk to the project. The risk management process is based on the concept of providing a structured approach to dealing with risk. It may seem obvious that the availability of suitably qualified labor is one source of risk, but how would this impact other risks that might be identified for the project? It is this interrelationship between risks that can only be drawn out through the use of a structured workshop and identification methods, where the whole project team has the ability to contribute.

(ii) Preliminary analysis to establish probable major risks for further investigation. The identification stage is designed to prompt team thinking and to bring out all possible risks that might impact the project. Several hundred risks could be identified, and the team may not have time to analyze and develop responses to them all. It is normal during the identification stage to begin by sifting and removing those risks that the team feels are not worth investigating further.

(iii) The true risk needs to be identified. Risks are caused by background conditions that are essentially "givens," for example, location of site. It is these background conditions that cause risks to occur. If the risk does occur, it will have an effect on the project. This is shown diagrammatically in Figure 2.1.

It is all too easy to mistakenly manage the effect rather than the risk itself. When the team identifies and manages the true risk, their effort is expended managing the risk rather than the effect.

Risk Identification Techniques

(i) Research. While exactly the same project will not have been executed before, something similar will have been. Investigation into projects on neighboring sites or projects previously undertaken by the client should be instigated.

(ii) Structured interviews/questionnaires. Interviews with key members of the project team (including client and suppliers) will elicit the greatest insight into risks to the project and how they are perceived by individuals. Interviews allow a greater depth of understanding to be achieved than group discussions but do take up a lot of time.

FIGURE 2.1. CAUSE, RISK, AND EFFECT.

Question	Example
What background condition is driving the risk?	The project is being undertaken in a particular city.
What is the area of uncertainty?	There may not be enough suitably qualified labor in the city.
What is the consequence of the risk occurring?	Additional labor from outside the city is required, which adds cost to the project.

(*iii*) *Checklists/prompt lists.* A simple and effective way to stimulate the team into thinking about risk is to use a checklist. An easy way to start a checklist is for each team member to write down all the variation orders on their last project, with the reasons for their issue.

(*iv*) *Brainstorming in a workshop environment.* Bringing the team together in a focused workshop is a powerful environment in which to discuss risk. The team can have a better understanding of how each member perceives risk differently. One important aspect of such workshops is that lateral thinking is encouraged. The workshop environment gives the team an opportunity to experiment with different viewpoints that individuals might normally reject out of hand if working alone.

(*v*) *Risk register.* A risk register is probably the most useful tool in the risk management process. It enables risks to be logged and tracked through the life of a project. While a risk register can be maintained by hand, it is much more useful to use a computer spreadsheet package. This allows the information to be sorted into different categories. Specialist risk management software is also available that integrates the risk register with risk analysis tools.

Typical column headings for a risk register are as follows:

- *Risk number.* A unique identifying number for the risk.
- *Risk description.* A written description of the risk.
- *Ownership.* Who is responsible for the management action in responding to the risk?
- *Probability.* How likely is the risk to occur?
- *Impact.* What happens if the risk does occur?
- *Risk factor.* Probability multiplied by impact.
- *Response.* What actions need to be taken to deal with the risk?

- *Status.* The status of the risk can be shown as:
  - *Done.* The risk has arisen and been dealt with.
  - *Active.* The risk is currently being managed.
  - *Monitor.* The risk has been identified, but no analysis or response has yet been developed for it.
- *Comments.* Allows notes to be kept on the risk.

It is also possible to place identified risks into categories, for example, "client-retained risks." The use of categories allows risk to be bundled together and can help responses to be tailored to deal with a category, rather than an individual risk.

An example risk register for a construction project is shown in Figure 2.2.

The risk register is a very useful format for showing a wide range of information in the risk management process. If the register is placed on a computer spreadsheet, it is easy to sort the risks in a variety of ways—for instance, by category or magnitude of probability. The risk register allows risks to be logged but can also display additional information linked to the project management process.

(vi) Database of historic risks. Experience is a great teacher, and the provision of a risk register allows information to be stored in a convenient format ready for use on the next project. The benefit of historical risk registers is that you know if a risk actually occurred and whether the appropriate response was set in place. The historic database should be used as a starting point in the risk management process, but although it may save some time in the identification stage, this stage cannot be completely omitted.

Risk Analysis

The analysis stage is concerned with trying to achieve a better understanding of the nature of the risks identified during the previous stage. After analysis, it will be possible to directly compare risks on a like-for-like basis. This is crucial in establishing the prioritization of risks in order to best apply organizational resources where they are most needed or can provide the biggest positive impact.

Risk analysis is generally divided into two parts: qualitative and quantitative. A qualitative risk analysis should always be undertaken as part of the risk management process. As part of the risk strategy, consideration should be given as if a quantitative analysis is required.

Qualitative Risk Analysis Methodology. Some form of qualitative risk analysis should be undertaken on all projects. It is the most basic form of risk analysis and is the foundation in understanding project risks.

Qualitative risk analysis can be performed at a number of levels. On the simplest level, the project manager can sit down with a few key stakeholders and discuss how risk should be managed on the project. A complex approach might involve having a dedicated subteam who manage the risk management process on behalf of the project manager.

No matter what level is chosen, a number of techniques can be used.

FIGURE 2.2. EXAMPLE RISK REGISTER.

ID	Risk Description	Ownership	Current Status D=Done A=Active M=Monitor	Probability Very high: 0.9 Very low: 0.1	Impact Very high: 0.8 Very low: 0.05	Risk Factor (Prob. x impact)	Response
100	Third-party influence						
110	Revisions to planning permission may be required.	Architect	D	0.7	0.05	0.035	Discussions with planning authority.
120	Contaminated ground hot spots may have to be removed from site.	Civil engineer	A	0.5	0.1	0.05	Discussions with environment agency.
130	Fire officer requirements not yet fixed.	Design team	A	0.7	0.1	0.07	Discussions with fire officer.
200	Client influence						
205	Protracted decision process — Client may not agree designs in line with design program.	Project manager	M				No response yet in place.
300	Design team						
310	Consultant appointments may not provide a clear definition of responsibilities.	Project team	D	0.1	0.05	0.005	Review consultants agreements.
320	Structural unknowns: composition of existing building (walls, roof).	Design team	M				No response yet in place.
325	Existing electrical power feed into building may not be of sufficient capacity.	Design team	M				No response yet in place.

(i) *Analysis of risks to assess the severity of impact and the probability of occurrence.* Risks have two dimensions that need to be assessed and analyzed:

- *Probability.* This is the likelihood of the risk occurring and is generally expressed as a percentage.
- *Impact.* If the risk did occur, what impact would it have on meeting the project's objectives?

For each of the risks identified, its probability and impact need to be assessed, normally in a workshop environment with key stakeholders present. In addition, before the risks can be entered into the risk register, a scale has to be set for the probability and impact dimensions. Setting a common scale will ensure a consistent approach to placing newly identified risks on the matrix at a later stage in the project life cycle. An indicative layout for defining scales is shown in Figure 2.3.

While the probability scale shown in Figure 2.3 may be utilized for any project, the impact scale needs to be set for each project. In Figure 2.3, the cost and time impacts are shown as a percentage of the original cost and time. Hence, a very high impact would mean the cost of the project would increase by more than 50 percent of the original budget.

Although time and cost impacts have been shown separately, it is possible to combine them, or to use other criteria such as performance. Again, the scale has to be set according to the needs of a project. Once set, the scales should not be changed during the life of a project. This allows risks to be compared on a common scale throughout a project and for the risk profile of a project to be accurately monitored.

A higher level of sophistication can be applied to the ranking of the risks on the probability/impact matrix. When generic units are assigned to a matrix, the relative importance of each cell can be seen, as shown in Figure 2.4.

The method of having the probability scale linear and the impact scale nonlinear emphasizes the significance of low-probability/high-impact risks. The matrix allows each risk

FIGURE 2.3. SCORING OF PROBABILITY AND IMPACT SCALES.

	PROBABILITY	*IMPACT*	
		Cost	**Time**
Very high	70%–95%	>50%	>50%
High	50%–70%	20–50%	20–50%
Medium	30%–50%	10–20%	10–20%
Low	10%–30%	5–10%	5–10%
Very low	5%–10%	<5%	<5%

FIGURE 2.4. MATRIX OF PROBABILITY VERSUS IMPACT.

PROBABILITY							
Very high	0.9	0.045	0.09	0.18	0.36	0.72
High	0.7	0.035	0.07	0.14	0.28	0.56
Medium	0.5	0.025	0.05	0.10	0.20	0.40
Low	0.3	0.015	0.03	0.06	0.12	0.24
Very low	0.1	0.005	0.01	0.02	0.04	0.08
		0.05	0.1	0.2	0.4	0.8
		Very low	Low	Medium	High	Very high
		IMPACT				

Key:

High Risk >0.20: Review risk in great detail. Amend project strategy to reduce
Medium Risk 0.08 – 0.20: Develop contingency plans. Monitor risk development
Low risk <0.08: Maintain record of risk. Consider contingency measures in outline

to be ranked and compared on an even basis. Where resources are scarce, it also allows attention to be focused on those risks that sit in the high-risk category of the matrix.

(ii) *Synthesis of all risks to predict the most likely project outcome.* Risk tends to have a "knock-on" effect, and so it is important to consider not only the effect of individual risks but their cumulative effect as well.

(iii) *Investigation of alternative course of actions.* Analysis allows for different scenarios to be developed to provide a high degree of confidence that the appropriate course of action has been chosen. The alternative courses of action may be plotted on the probability/impact matrix to obtain a clearer understanding of their relative merits in relation to the risk profile.

Qualitative Risk Analysis Tools and Techniques. The probability/impact matrix is the easiest technique to use and allows all project team members to participate in the process via a risk workshop. This matrix, shown in Figure 2.5, indicates very quickly those risks that require detailed consideration and those that only need a cursory glance. So, a risk with a high probability of occurring, which does occur, would have a very high impact on the project and would require detailed consideration. This allows the right amount of resource to be used in managing risk.

The probability/impact matrix allows a risk profile for a project to be developed. For instance, if most of the risks are plotted in the top left section of the matrix, it indicates a high-risk project. This matrix is very useful for providing a graphical representation of the risk on a project. The risk management strategy is to manage these risks proactively through the project life cycle. During later risk assessment exercises, the profile should show risks plotted more in the bottom right section of the matrix.

Qualitative risk analysis allows an insight into the risks that a project faces and how those risks can be managed. Many of the tools and techniques allow risks to be scored and

FIGURE 2.5. MATRIX OF PROBABILITY VERSUS IMPACT.

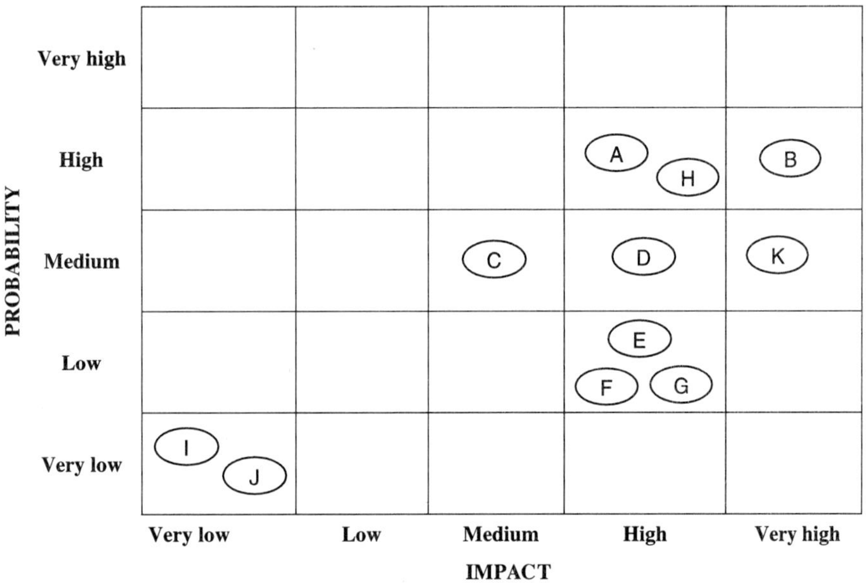

PROBABILITY \ IMPACT	Very low	Low	Medium	High	Very high
Very high					
High				A, H	B
Medium			C	D	K
Low				E, F, G	
Very low	I, J				

ranked against each other. It should be remembered that the process relies on mainly subjective data. Often much of the data will be a perceptual assessment of a risk provided by project stakeholders.

It can be seen that qualitative risk analysis provides a useful understanding of project risk. The subjective nature of this understanding must not be lost sight of, especially when you are making decisions based solely on qualitative analysis.

A more detailed insight can often be provided by undertaking a quantitative risk analysis.

Quantitative Risk Analysis. Quantitative risk analysis allows project risks to be modeled. Quantitative analysis applies statistical theory to the risk management process. Because of the often complex calculations involved, many of the quantitative analysis techniques are supported by various computer software packages. The UK-based Association for Project Management provides a software directory of available risk management software packages that can be accessed via its Web site.

The use of statistics in quantitative risk analysis tends to enhance the credibility of the output even though there is often inadequacy in the underlying data and assumptions that the model is built upon. For this reason, it is always beneficial to use qualitative techniques to obtain a firm understanding of project risks prior to making the resource intensive investment in the quantitative process. Quantitative techniques should only be used when the case for using them is fully justified.

Decision trees. A *decision tree* is a visual representation of a problem situation and the alternative options that are available for its solution. In Figure 2.6, the project manager must make a decision as to whether the company should stick with an existing contractor who is performing badly or replace the contractor with a new contractor.

The probability of each potential outcome together with its associated cost can be placed on the decision tree. This shows the expected outturn cost of each decision, hence allowing a more informed choice to be made.

Monte Carlo simulation. Monte Carlo simulation generates a number of possible scenarios based on input to the simulation model. Monte Carlo effectively accounts for every possible value that each variable could take and weights each possible scenario by the probability of its occurrence.

The objective of a Monte Carlo simulation is to calculate the combined impact of the model's various uncertainties to determine a probability distribution of the possible model outcomes. The technique involves the random sampling of each probability distribution within the model to produce large numbers of scenarios (often called iterations or trials).

FIGURE 2.6. DECISION TREE.

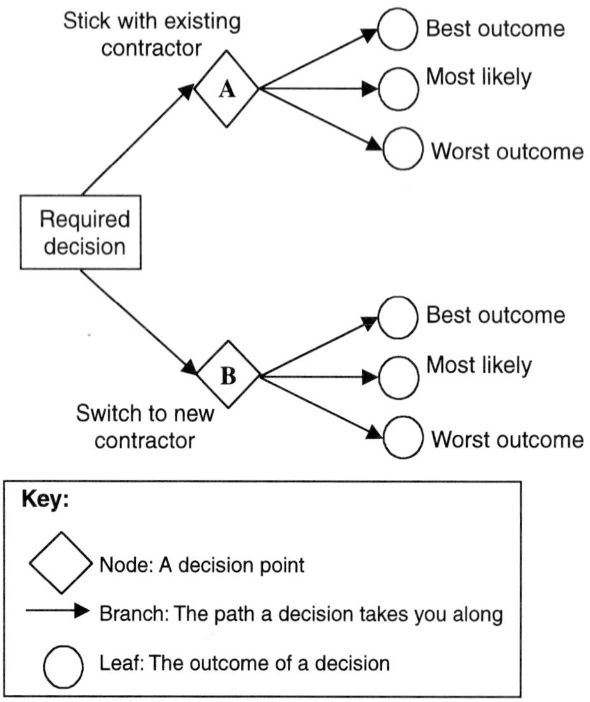

The output from a Monte Carlo simulation is usually displayed as a S curve, as illustrated in Figure 2.7.

Figure 2.7 shows the Monte Carlo output on a cost analysis for a project. This information can be used to set a project budget. The Monte Carlo analysis shows that there is a 50 percent probability the project will cost $229,970. This figure becomes the project budget. A risk allowance can be set at the 60 percent probability level, or $249,180. This would provide the project manager with a risk allowance of $19,210. Of course, the qualitative risk analysis will have identified the risks, so the risk allowance can be spent against known risk events if they occur.

Monte Carlo simulation can also be applied to project schedules. A key output here is the provision of *criticality indexes* for each activity. An activity with a criticality index of 100 percent means that during the simulation, activity was always on the project's critical path. An activity with a criticality index of 75 percent means that activity was on the project's critical path for 75 percent of the simulations. Activities with criticality indexes between 1 and 99 percent are called subcritical. This analysis identifies activities that under certain circumstances may become critical.

Quantitative risk analysis is a complex area that cannot be adequately covered in this introductory chapter. Further guidance is available in Vose (2000), which provides a good overview of quantitative risk analysis and, in particular, its application to project management.

FIGURE 2.7. COST-RISK ANALYSIS S CURVE.

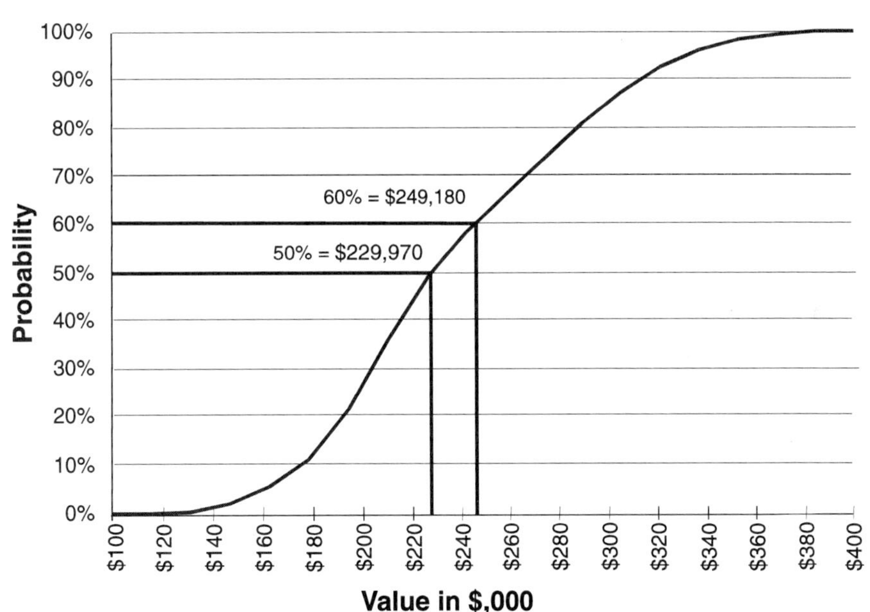

Quantitative analysis provides quite detailed information on the uncertainty that surrounds projects. The modeling process allows the influence of risk to be varied so the impact can be better understood. As with all models, the output is only as good as the input. The byword here is GIGO (garbage in, garbage out).

Risk Response

The analysis stage provides the team with a better understanding of the risks. The next stage is to develop a response to those risks.

Risk Response Methodology. The key concern in this stage of the process is to choose the appropriate course of action. The principle is to choose the right response based on available information. It might be that the response changes with time as more information becomes available. The response will generally fall into one of the following categories:

- *Avoid.* Identifying responses to put in place to sidestep a risk.
- *Transfer.* Transferring a risk from one party to another, for example, from a client to a contractor.
- *Mitigate.* The party who carries a risk should identify responses to lower both the probability of the risk occurring and the impact should it occur.
- *Control.* Responses need to be monitored to ensure they are appropriate in changing circumstances.

The four responses are used in combination with each other, since one will never cover all risk on a project. An important point to consider when developing responses concerns the generation of secondary risk. When a response is proposed, its full implications have to be assessed to ascertain if a secondary risk arises out of implementing the response. If the sum of the secondary risk plus the reduced risk (the original risk with the response in place) is greater than the original risk, an appropriate response has clearly not been identified and an alternative response should be found (see Figure 2.8).

FIGURE 2.8. SECONDARY RISKS.

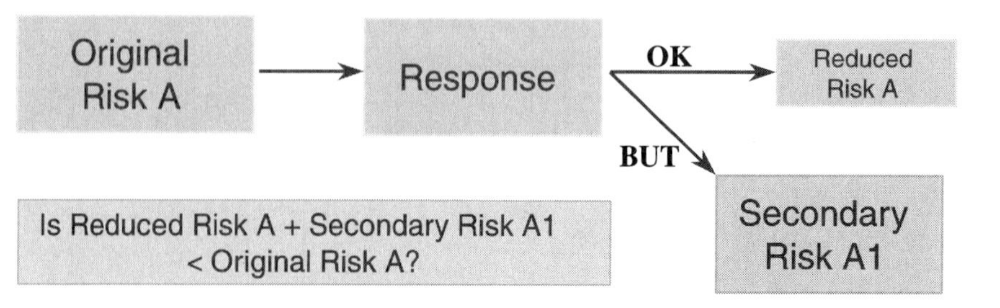

Risk Response Tools and Techniques. The techniques available are primarily integrated with the project management process, for example:

- *Contract acquisition (procurement) plan.* Ensuring that responses to risk are placed into relevant work packages.
- *Contingency management.* Ensuring monies are only released if a predefined event occurs (contingencies are not there to allow for poor budgetary planning).
- *Project controls.* These allow the project to react to changing circumstances, which includes risk.

The tools available are not usually related purely to risk management. For instance, most project managers would recommend to their clients that a contractor has third-party insurance for the project. This is a risk management response designed to reduce clients' exposure to claims for damages if a particular incident were to occur. This is a risk management tool, although most project managers would consider it to be common project procedure. Typical tools available are as follows:

- *Insurances/bonds/warranties.* These are often used to cover the impact of a risk occurring.
- *Contingency plans.* This ensures that should a risk occur, a preprepared plan is put into action;
- *Forms of contracts.* These should be drafted so risks are apportioned as intended by the risk management process;
- *Contingency drawdown models.* Contingency monies should be allocated to specific risks and only released if that risk is within agreed parameters.
- *Special cost allowance.* Not all risk can be foreseen and there is a requirement to allow for unforeseen risks.
- *Training.* This involves ensuring the team understands their role in implementing the risk responses.

All responses to risks that are put in place will need to be managed. This forms part of risk control. Unless responses are effectively managed, the whole risk process fails. Until a response is actioned, the risk may still occur, regardless of how well it has been identified and analyzed.

Risk Control

The whole risk management process relies on there being a control process in place that ensures the risk process is effectively implemented. As part of risk control, a regular review of the risk management process should be undertaken:

- *Risk strategy.* Have any changes to the project been made that would require the risk strategy to be altered?
- *Risk identification.* Have any new risk been identified?

- *Risk assessment.* Assess any new risks that have been identified. Are the assessments for existing risk still valid or should they be revised?
- *Risk response.* Have responses been implemented? Are future responses still valid?

Summary

Risk management should be undertaken as part of a structured, formal process that needs to be aligned to the overall approach to project management. In conclusion, key elements that need to be considered as part of a risk management process are as follows:

- Identify staff and resources assigned to the risk management process.
- Define lines of reporting and responsibility for the risk management process.
- Link the risk management plan to other project tools such as safety, quality and environmental management, and planning and reporting systems.
- Consolidate all risks identified into an appropriate and digestible response strategy in order that cumulative effects can be perceived.
- State risk audit intervals and key milestones.
- Include risk milestones in project plan.
- Identify possible response strategies and programs for each risk category, including contingency plans and how to handle new or unresolved risks.
- Assess cost involved.
- Monitor success of responses strategies, and produce feedback for reporting into future projects.

References and Further Reading

Akintoye, A. 2003. *Public private partnerships: Managing risks and opportunities.* Oxford, UK: Blackwell Science.

British Standards Institute. 2000. *BS 6079—Guide to project management—Part 3 Risk management.* London: British Standards Institute.

Canada Standards Association. 1997. *CAN/CSA-Q850-97 Risk management guidelines for decision makers.* Toronto: Canada Standards Association.

Chapman, C., and S. Ward. 1997. *Project risk management: Processes, techniques, and insights.* Chichester, UK: Wiley.

Chicken, J., and T. Posner. 1998. *The philosophy of risk.* London: Thomas Telford, London.

Godfrey, P. S. 1996. *Control of risk: A guide to the systematic management of risk from construction: SP125.* London: CIRIA.

DoD. 2002. *Risk management guide for DoD acquisition.* 5th ed. Fort Belvoir, VA: Department of Defense, Defense Acquisition University.

ICE. 1998. *RAMP: Risk analysis and management for projects.* London: Institution of Civil Engineers and Institute of Actuaries.

International Organization for Standardization. 2002. *ISO/IEC Guide 73: Risk management—Vocabulary—Guidelines for use in standards.* Geneva: ISO.

Japan Project Management Forum. 2002. *Project management body of knowledge (PMBOK)*. Tokyo: Japan Project Management Forum.

Office of Government Commerce. 2002. *Management of risk: Guidance for practitioners*. London: The Stationery Office.

Project Management Institute. 2000. *Project management body of knowledge (PMBOK)*. Newtown Square, PA: Project Management Institute.

Simon, P., D. Hillson, and K. Newland. 1997. *Project risk analysis and management (PRAM)*. High Wycombe, UK: Association for Project Management.

Smith, N. J. 1998. *Managing risk in construction projects*. Oxford, UK: Blackwell Science.

Standards Australia. 1999. AS/NZS 4360:1999 Risk management. Strathfield, Australia: Standards Association of Australia.

Vose, D. 2000. *Risk analysis: A quantitative guide*. 2nd ed. Chichester, UK: Wiley.

CHAPTER THREE

PROJECT MANAGEMENT STRUCTURES

Erik Larson

A project management structure provides a framework for launching and implementing project activities within a parent organization. A good structure appropriately balances the needs of both the organization and the project by defining the interface between the project and parent organization in terms of authority, allocation of resources, and eventual integration of project outcomes into mainstream operations (Gray and Larson, 2003).

In the past, many business organizations have struggled to create an effective system for implementing projects. One of the major reasons for this struggle is that projects contradict fundamental design principles embedded in traditional organizations. Projects are unique, one-time efforts with a distinct beginning and end. Most organizations are designed to efficiently manage ongoing activities. Efficiency is achieved primarily by breaking down complex tasks into simplified, repetitive activities, as characterized by assembly line production methods. Projects by their very nature are not routine and are therefore an anomaly in these work environments.

A second reason businesses find it difficult to effectively organize projects is that most projects are multidisciplinary in nature. For example, a new-product development project will likely involve the combined efforts of people from design, marketing, manufacturing, and finance. However, most organizations have been departmentalized according to functional expertise, with specialists from design, marketing, manufacturing, and finance residing in different units. Many researchers have noted that these groupings naturally develop unique customs, norms, values, and working styles that inhibit integration across functional boundaries (Lawrence and Lorsch, 1969; Harrison and Beyer, 1993; Majchrzak and Wang, 1996). Not only are there departmental silos, but managing projects poses the additional dilemma of who is in charge of the project. In most organizations, authority is distributed hierarchically across functional lines. Because projects span functional areas, identifying and legitimizing project management authority is often problematic.

In recent years there has been a dramatic shift in the management of projects. Global competition and technological advances have led to the compression of the product life cycle, the emergence of speed as a competitive advantage, and corporate downsizing that has placed a premium on implementing projects. Projects are no longer the exception but are central to the success of firms. As a result, organizations are being reengineered to support project implementation. Businesses are taking advantage of the latest information technology to coordinate and track the efforts of professionals both within and across organizations.

Four different approaches to project management organization are discussed here: functional organization, dedicated project teams, matrix structure, and network organization. Although not exhaustive, these structures and their variant forms represent the major approaches for organizing projects. The advantages and disadvantages of each of these structures are discussed, as well as some of the critical factors that might lead a firm to choose one form over others.

Organizing Projects within the Functional Organization

One approach to organizing projects is to simply manage them within the existing functional hierarchy of the organization. Once management decides to implement a project, the different segments of the project are delegated to the respective functional units, with each unit responsible for completing its segment of the project (see Figure 3.1). Coordination is maintained through normal management channels. For example, an optical scope manufacturing firm decides to differentiate its product line by offering high-end binoculars designed for avid bird watchers. Senior management authorizes the project, and different segments of the project are distributed to appropriate areas. The industrial design department is responsible for designing the binoculars. The production unit is responsible for devising the means for producing binoculars according to these new design specifications. The marketing department is responsible for gauging demand and price, as well as identifying distribution outlets. The overall project will be managed within the normal hierarchy, with the project being part of the working agenda of top management.

The functional organization is also commonly used when, given the nature of the project, one functional area plays a dominant role in completing the project or has a dominant interest in the success of the project. Under these circumstances, a high-ranking manager in that area is given the responsibility for coordinating the project. For example, the transfer of equipment and personnel to a new office would be managed by a top-ranking manager in the firm's facilities department. Likewise, a project involving the upgrading of the management information system would be managed by the information systems department. In both cases, most of the project work would be done within the specified department and coordination with other departments would occur through normal channels.

There are advantages and disadvantages for using the existing functional organization to administer and complete projects (Stuckenbruck, 1981; Youker 1977; Verma, 1995). The major advantages are the following:

FIGURE 3.1. FUNCTIONAL ORGANIZATION.

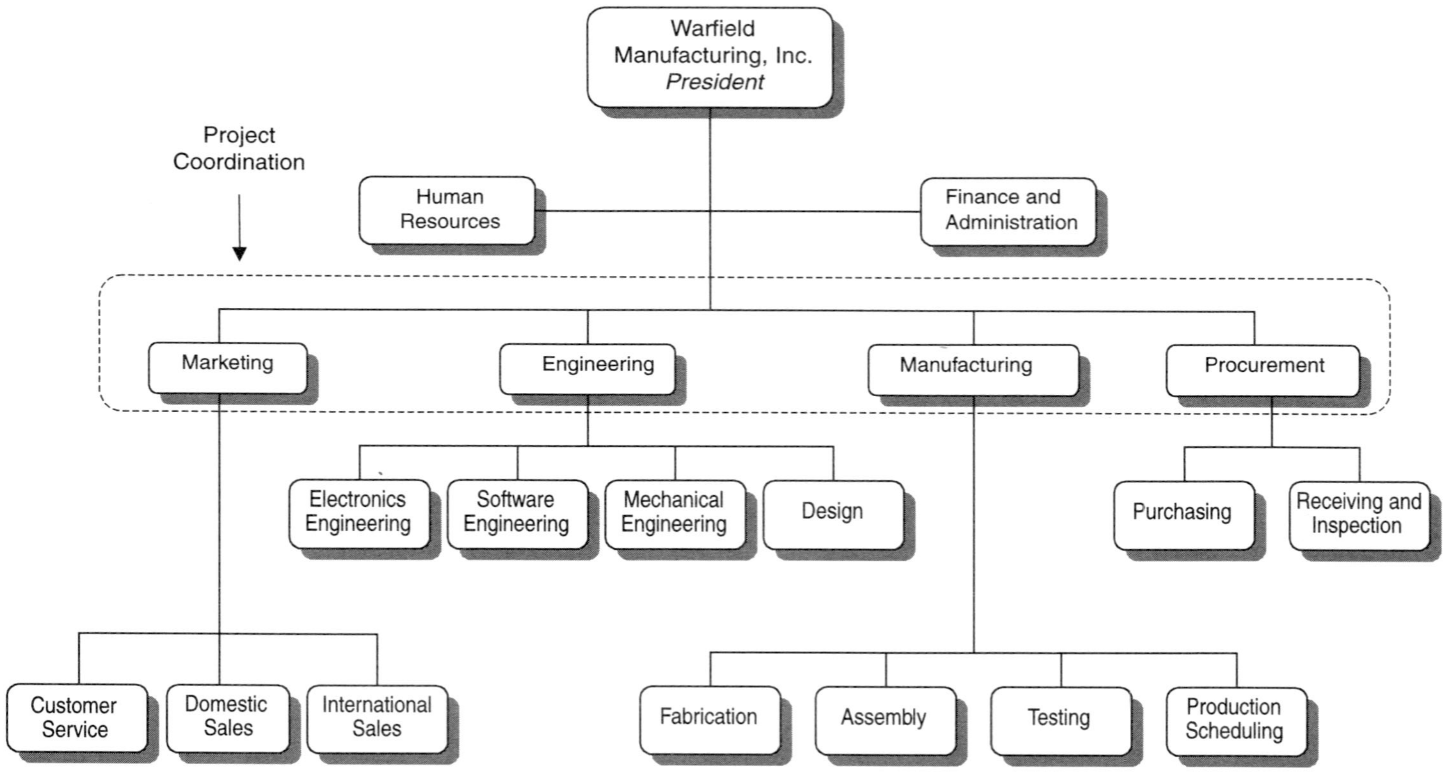

- *No change.* Projects are completed within the basic functional structure of the parent organization. There is no radical alteration in the design and operation of the parent organization.
- *Flexibility.* There is maximum flexibility in the use of staff. Appropriate specialists in different functional units can temporarily be assigned to work on the project and then return to their normal work. With a broad base of technical personnel available within each functional department, people can be switched among different projects with relative ease.
- *In-depth expertise.* If the scope of the project is narrow and the proper functional unit is assigned primary responsibility, in-depth expertise can be brought to bear on the most crucial aspects of the project.
- *Easy post-project transition.* Normal career paths within a functional division are maintained. While specialists can make significant contributions to projects, their functional field is their professional home and the focus of their professional growth and advancement.

Just as there are advantages for organizing projects within the existing functional organization, there are also disadvantages. These disadvantages are particularly pronounced when the scope of the project is broad and one functional department does not take the dominant technological and managerial lead on the project:

- *Low commitment.* Each functional unit has its own core routine work to do; sometimes project responsibilities get pushed aside to meet primary obligations. This difficulty is compounded when the project has different priorities for different units. For example, the marketing department may consider the project urgent, while the operations people consider it only of secondary importance.
- *Poor integration.* There may be poor integration across functional units. Cross-functional communication and coordination are slow, and limited at best, in most hierarchical organizations. Furthermore, there is a tendency to suboptimize the project, with respective functional specialists being concerned only with their segment of the project and not the total project.
- *Slow.* It generally takes longer to complete projects through this functional arrangement. This is in part attributable to slow response time—project information and decisions have to be circulated through normal management channels. Furthermore, the lack of horizontal, direct communication among functional groups contributes to rework as specialists realize the implications of others' actions after the fact.
- *Lack of ownership.* The motivation of people assigned to the project can be weak. The project may be seen as an additional burden that is not directly linked to their professional development or advancement. Furthermore, because they are working on only a segment of the project, professionals do not identify with the project. Lack of ownership discourages strong commitment to project-related activities.

Organizing Projects as Dedicated Teams

At the other end of the spectrum is the creation of dedicated project teams. These teams operate as independent units from the rest of the parent organization. Usually a full-time project manager is designated to recruit a core group of specialists who work full-time on the project. The project manager recruits necessary personnel from both within and outside the parent company. The subsequent team is physically separated from the parent organization and given marching orders to complete the project (see Figure 3.2).

The interface between the parent organization and the project teams will vary. In some cases, the parent organization prescribes administrative and financial control procedures over the project. In other cases, firms allow the project manager maximum freedom to get the project done. Such was the case of the original *Skunk Works* established by Kelly Johnson at Lockheed Martin. Kelly and a small, isolated band of Lockheed mavericks developed the revolutionary U-2 spy plane during the early 1950s. Lockheed went on to use this approach to develop a series of high-speed planes, including the F-117 Stealth Fighter (Johnson, Smith and Geary, 1990; Miller, 1996).

In the case of firms where projects are the dominant form of business, such as a construction firm or a consulting firm, the entire organization is designed to support project teams. Instead of one or two special projects, the organization consists of sets of quasi-independent teams working on specific projects. The main responsibility of traditional functional departments is to assist and support these project teams. For example, the marketing department is directed at generating new business that will lead to more projects, while the human resources department is responsible for managing a variety of personnel issues, as well as recruiting and training new employees. This type of organization is referred to in the literature as a *project* organization and is graphically portrayed in Figure 3.3.

FIGURE 3.2. DEDICATED PROJECT TEAM.

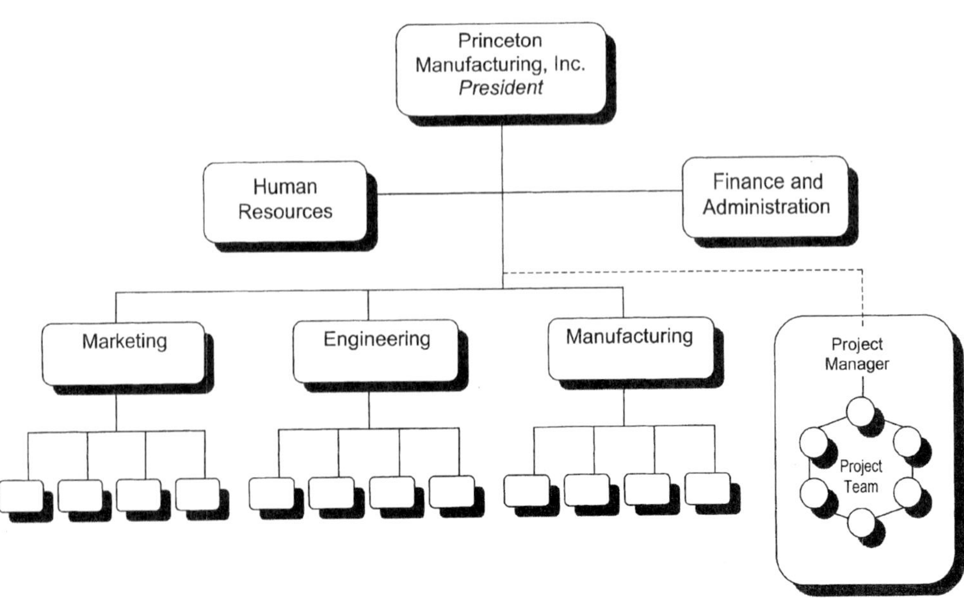

FIGURE 3.3. PROJECT ORGANIZATION STRUCTURE.

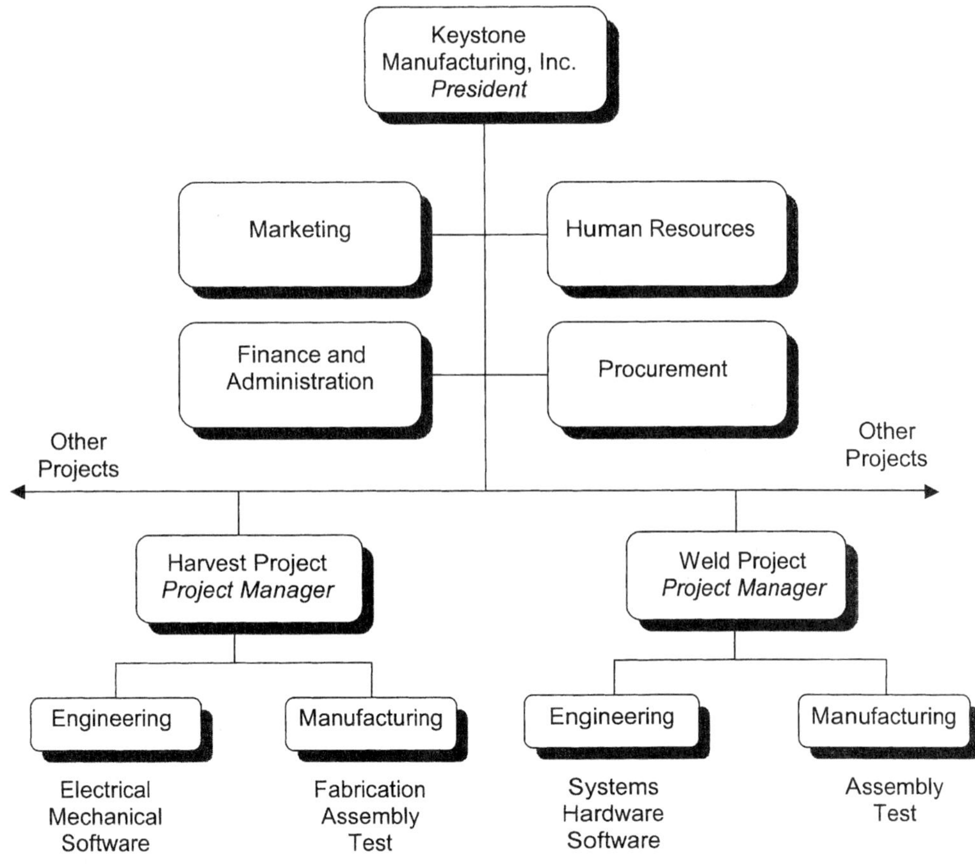

As in the case of functional organization, the dedicated project team approach has strengths and weaknesses (Stuckenbruck, 1981; Youker, 1979, Verma, 1995). The following are recognized as strengths:

- *Simple.* Does not directly disrupt ongoing operations. Other than taking away resources in the form of specialists assigned to the project, the functional organization remains intact with the project team operating independently.
- *Speed.* Perhaps the main reason for this is that participants devote their full attention to the project and are not distracted by other obligations and duties. Furthermore, response time tends to be quicker under this arrangement because most decisions are made within the team and are not deferred up the hierarchy.
- *Cohesion.* Participants share a common goal and personal responsibility toward the project and the team.

- *Cross-functional integration.* Specialists from different areas work closely together and, with proper guidance, become committed to optimizing the project, not their respective areas of expertise.

In many cases, the project team approach is the optimum approach for completing a project, when you view it solely from the standpoint of what is best for completing the project. Its weaknesses become more evident when the needs of the parent organization are taken into account:

- *Expensive.* Not only have you created a new management position (project manager), but resources are also assigned on a full-time basis. This can result in duplication of efforts across projects and a loss of economies of scale.
- *Internal strife.* A strong we/they divisiveness sometimes emerges between the project team and the parent organization. This divisiveness can undermine not only the integration of the eventual outcomes of the project into mainstream operations but also the assimilation of project team members back into their functional units once the project is completed.
- *Limited technological expertise.* Creating self-contained teams inhibits maximum technological expertise being brought to bear on problems. Technical expertise is limited somewhat to the talents and experience of the specialists assigned to the project. While nothing prevents specialists from consulting with others in the functional division, the we/they syndrome and the fact that such help is not formally sanctioned by the organization discourage this from happening.
- *Post-project assimilation.* Assigning full-time personnel to a project creates the dilemma of what to do with personnel after the project is completed. If other project work is not available, the transition back to their original functional departments may be difficult because of their prolonged absence and the need to catch up with recent developments in their functional areas.

A good example of internal strife is the saga of the successful Macintosh project team at Apple Computer. Steve Jobs, who at the time was both the chairman of Apple and the project manager for the Mac team, pampered his team with perks including at-the-desk massages, coolers stocked with freshly squeezed orange juice, a Bosendorfer grand piano, and first-class plane tickets. No other employees at Apple got to travel first-class. Jobs considered his team to be the elite of Apple and had a tendency to refer to everyone else as "bozos" who "didn't get it." Engineers from the Apple II division, which was the bread and butter of Apple's sales, became incensed with the special treatment their colleagues were getting.

One evening at Ely McFly's, a local watering hole, the tensions between Apple II engineers seated at one table and those of a Mac team at another boiled over. Aaron Goldberg, a long-time industry consultant, watched from his barstool as the squabbling escalated. "The Mac guys were screaming, 'We're the future!' The Apple II guys were screaming, 'We're the money!' Then there was a geek brawl. Pocket protectors and pens were flying. I was waiting for a notebook to drop, so they would stop and pick up the papers" (Carlton, 1997).

Although comical from a distance, the discord between the Apple II and Mac groups severely hampered Apple's performance during the 1980s. John Sculley, who replaced Steve Jobs as chairman of Apple, observed that Apple had evolved into two "warring companies" and referred to the street between the Apple II and Macintosh buildings as "the DMZ" (demilitarized zone) (Sculley, 1987).

Organizing Projects within a Matrix Arrangement

Matrix management is a hybrid organizational form in which a horizontal project management structure is overlaid on the normal functional hierarchy. In a matrix system, there are usually two chains of command, one along functional lines and the other along project lines. Instead of delegating segments of a project to different units or creating an autonomous team, project participants report simultaneously to both functional and project managers.

Companies apply this matrix arrangement in a variety of different ways. Some organizations set up temporary matrix systems to deal with specific projects, while matrix may be a permanent fixture in other organizations. Let us first look at its general application and then proceed to a more detailed discussion of finer points. Consider Figure 3.4. There are three projects currently under way: Silver, Gold, and Rust. All three project managers report to a director of project management, who supervises all projects. Each project has an administrative assistant, although the one for the Rust project is only part-time.

The Silver project involves the design and expansion of an existing production line to accommodate new metal alloys. To accomplish this objective, the project has assigned to it 5.5 people from manufacturing and 6 people from engineering. These individuals are assigned to the project on a part-time or full-time basis, depending on the project's needs during various phases of the project. The Gold project involves the development of a new product that requires the heavy representation of engineering, manufacturing, and marketing. The Rust project involves forecasting changing needs of an existing customer base. While these three projects, as well as others, are being completed, the functional divisions continue performing their basic, core activities.

The matrix structure is designed to optimally utilize resources by having individuals work on multiple projects as well as being capable of performing normal functional duties. At the same time, the matrix approach attempts to achieve greater integration by creating and legitimizing the authority of a project manager. In theory, the matrix approach provides a dual focus between the functional/technical expertise and project requirements that is missing in either the project team or functional approach to project management. The project manager is responsible for integrating functional input and overseeing the completion of the project. Functional managers usually "own" the resources in their area and are responsible for overseeing the functional contribution to the project. See Table 3.1 for a further delineation of the two roles.

Different Matrix Forms

In practice there are really different kinds of matrix systems, depending on the relative authority of the project and functional managers (Larson and Gobeli, 1985; Smith and

FIGURE 3.4. MATRIX ORGANIZATION STRUCTURE.

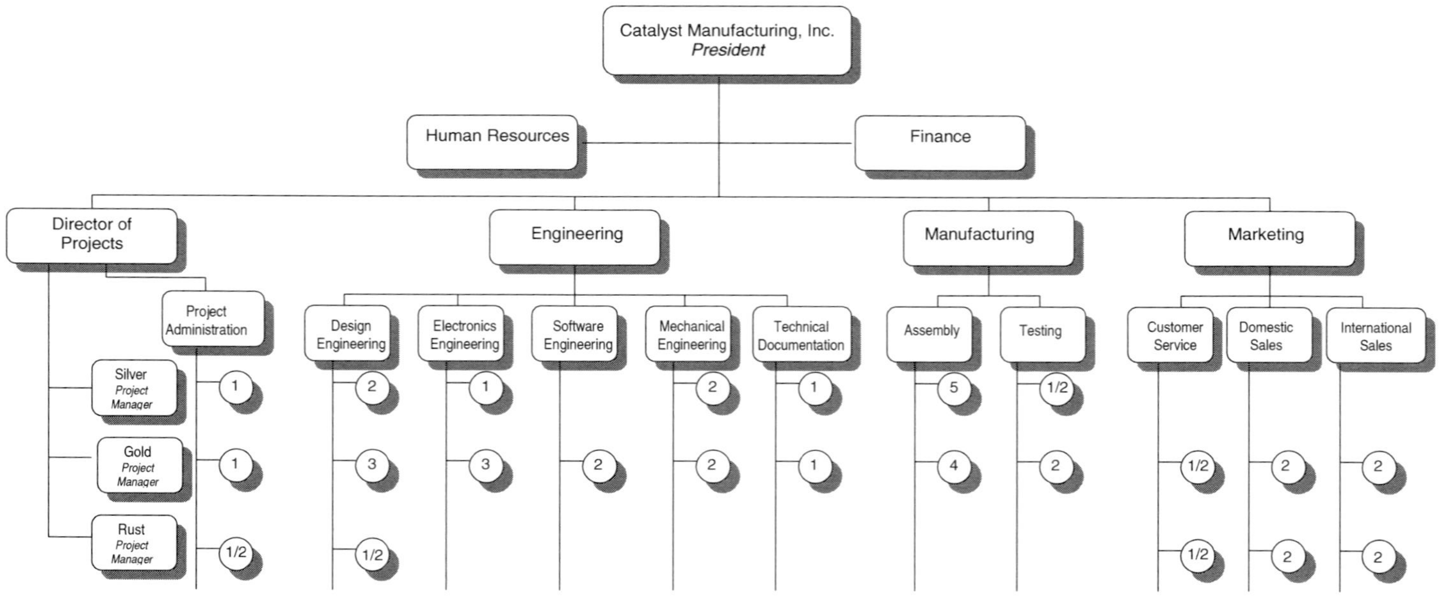

TABLE 3.1. DIVISION OF PROJECT MANAGER AND FUNCTIONAL MANAGER RESPONSIBILITIES IN A MATRIX STRUCTURE.

Project Manager	Negotiated Issues	Functional Manager
What has to be done?	Who will do the task?	How will it be done?
When should the task be done?	Where will the task be done?	How will the project involvement impact normal functional activities?
How much money is available to do the task?	Why will the task be done?	How well has the functional input been integrated?
How well has the total project been done?	Is the task satisfactorily completed.	

Reinertsen, 1995; Bowen, Clark, Holloway, and Wheelwright, 1994). *A weak matrix,* also called a lightweight or functional matrix, is one in which the balance of authority strongly favors the functional managers. A *balanced matrix,* also called a middleweight matrix, is used to describe the traditional matrix arrangement. *A strong matrix,* also called heavyweight or project, is one in which the balance of authority is strongly on the side of the project manager.

The relative difference in power between functional managers and project managers is reflected along a number of related dimensions. One such dimension is level of reporting relationship. A project manager who reports directly to the vice president of product development has more clout than a marketing manager who reports to a regional sales manager. Location of project activities is another subtle but important factor. A project manager wields considerably more influence over project participants if they work in his office than if they perform their project-related activities in their functional offices. Likewise, the percentage of full-time staff assigned to the project contributes to relative influence. Full-time status implies transfer of obligations from functional activities to the project.

One other significant factor is who is responsible for conducting performance appraisals and compensation decisions. In a weak matrix, the project manager is not likely to have any direct input in the evaluation of participants who worked on the project. This would be the sole responsibility of the functional manager. Conversely, in a strong matrix, the project manager's evaluation would carry more weight than the functional manager's. In a balanced matrix, either input from both managers is sought, or the project manager makes recommendations to the functional manager, who is responsible for the formal evaluation of individual employees. Often companies will brag that they use a strong, project-oriented matrix only to find upon closer examination that the project managers have very little say over the evaluation and compensation of personnel.

Ultimately, whether the matrix is weak or strong is determined by the extent to which the project manager has authority over participants. Authority may be determined informally by the persuasive powers of managers involved and the perceived importance of the project, or formally by the prescribed powers of the project manager. Here is a thumbnail sketch of the three kinds of matrices:

- *Weak matrix.* This form is very similar to a functional approach with the exception that there is a formally designated project manager responsible for coordinating project activities. Functional managers are responsible for managing their segment of the project. The project manager basically acts as a staff assistant who draws the schedules and checklists, collects information on status of work, and facilitates project completion. The project manager has indirect authority to expedite and monitor the project. Functional managers call most of the shots and decide who does what and when the work is completed.
- *Balanced matrix.* This is the classic matrix in which the project manager is responsible for defining what needs to be accomplished, while the functional managers are concerned with how it will be accomplished. More specifically, the project manager establishes the overall plan for completing the project, integrates the contribution of the different disciplines, sets schedules, and monitors progress. The functional managers are responsible for assigning personnel and executing their segment of the project according to the standards and schedules set by the project manager. The merger of "what and how" requires both parties to work closely together and jointly approve technical and operational decisions.
- *Strong matrix.* This form attempts to create the "feel" of a project team within a matrix environment. The project manager controls most aspects of the project, including scope trade-offs and assignment of functional personnel. The project manager controls when and what specialists do, and has final say on major project decisions. The functional manager has title over his/her people and is consulted on a need basis. In some situations, a functional manager's department may serve as a "contractor" for the project, in which case they have more control over specialized work. For example, the development of a new series of laptop computers may require a team of experts from different disciplines working on the basic design and performance requirements within a project matrix arrangement. Once the specifications have been determined, final design and production of certain components (i.e., power source) may be assigned to respective functional groups to complete.

Both matrix management in general and in its specific forms have unique strengths and weaknesses (Stuckenbruck, 1981; Youker, 1979; Larson and Gobeli, 1987; Verma, 1995). The advantages and disadvantages of matrix organizations in general are noted in the list that follows, which also briefly highlights specifics concerning different forms.

- *Efficient resource utilization.* Resources can be shared across multiple projects as well as within functional divisions. Individuals can divide their energy across multiple projects on an as-needed basis. This reduces duplication required in a pure project team structure.
- *Dual project/functional focus.* A strong project focus is provided by having a formally designated project manager who is responsible for coordinating and integrating contributions of different units. This helps sustain a holistic approach to problem solving that is often missing in the functional organization. At the same time, functional input reinforces rigor and high quality standards.
- *Post-project assimilation.* Because the project organization is overlaid on the functional divisions, the project has reasonable access to the entire reservoir of technology and ex-

pertise of functional divisions. Furthermore, unlike dedicated project teams, specialists maintain ties with their functional group, so they have a home port to return to once the project is completed.

- *Flexible.* Matrix arrangements provide for flexible utilization of resources and expertise within the firm. In some cases, functional units may provide individuals who are managed by the project manager. In other cases, the contributions are monitored by the functional manager.

The strengths of the matrix structure are considerable. Unfortunately, so are the potential weaknesses. In large part, this is because a matrix structure is more complicated and the creation of multiple bosses represents a radical departure from the traditional hierarchical authority system. Furthermore, one does not simply install a matrix structure over night. Experts argue that it takes three to five years for a matrix system to fully mature (Davies and Lawrence, 1977; Graham and Englund, 1997). So many of the weaknesses described in the following represent growing pains:

- *Dysfunctional conflict.* The matrix approach is predicated on creative tension between functional managers and project managers who bring critical expertise and perspectives to the project. Such tension is viewed as healthy and a necessary mechanism for achieving an appropriate balance between complex technical issues and unique project requirements. Unfortunately, sometimes legitimate conflict can spill over to a more personal level, resulting from conflicting agendas and accountabilities. Worthy discussions can degenerate into heated arguments that engender animosity among the managers involved.
- *Infighting.* Any situation in which equipment, resources, and people are being shared across projects and functional activities lends itself to conflict and competition for scarce resources. Infighting can occur among project managers, who are primarily interested in what is best for their project.
- *Stress.* Matrix management violates the management principle of unity of command. Project participants have at least two bosses—their functional head and one or more project managers. Working in a matrix environment can be stressful if you are being told to do three conflicting things by three different managers.
- *Slow.* In theory, the presence of a project manager to coordinate the project should accelerate the completion of the project. In practice, decision making can get bogged down, as agreements have to be forged across multiple functional groups. This may be especially true for the balanced matrix.

The advantages and disadvantages are not necessarily true for all three forms of matrix. The strong matrix is likely to enhance project integration, diminish internal power struggles, and ultimately improve control of project activities and costs. On the downside, technical quality may suffer because functional areas have less control over their contributions. Finally, internal strife may erupt as the members develop a strong team identity and relations outside the project become strained.

The weak matrix is likely to improve technical quality as well as provide a better system for managing conflict across projects because the functional manager assigns personnel to different projects. The problem is that functional control is often maintained at the expense of poor project integration. The balanced matrix can achieve better balance between technical and project requirements, but it is a very delicate system to create and manage and is more likely to succumb to many of the problems associated with the matrix approach (Larson and Gobeli, 1987).

Organizing Projects within Network Organizations

The turn of the century has seen a radical shift in the organizational architecture of business firms. Corporate downsizing and cost control have combined to produce what some have called "network organizations" (Miles and Snow, 1995; Miles, Snow, Mathews, Miles, and Coleman, 1997). In theory, a network organization is an alliance of several organizations for the purpose of creating products or services for customers. This collaborative structure typically consists of several satellite organizations beehived around a "hub" or "core firm." The core firm coordinates the network process and provides one or two core competences, such as marketing or product development. For example, Cisco Systems mainly designs new products and utilizes a constellation of suppliers, contract manufacturers, assemblers, and other partners to deliver products to their customers. Likewise, Nike, another prime example of this kind of organization, provides marketing expertise for its sports footwear and apparel. The key organizing principle is that, instead of doing everything in-house, a firm outsources key activities to other businesses with the requisite competencies.

The shift toward network organizations is easily apparent in the film industry. During the golden era of Hollywood, huge, vertically integrated corporations made movies. Studios such as MGM, Warner Brothers, and Twentieth Century Fox owned large movie lots and employed thousands of full-time specialists—set designers, camera people, film editors, directors, and even actors. Today, most movies are made by a collection of individuals and small companies who come together to make films project-by-project. This structure allows each project to be staffed with the talent most suited to its demands rather than choosing from only those people the studio employs. This approach also disperses financial risk across many organizations.

The network approach is now being applied to a wide range of projects. For example, see Figure 3.5.

Figure 3.5 depicts a situation in which a new reclining chair is being developed. The genesis for the chair comes from a mechanical engineer who suffers from a bad back. She developed the prototype in her garage. The inventor negotiates a contract with a furniture firm to develop and manufacture the recliner. The furniture company, in turn, creates a project team of manufacturers, suppliers, and marketing firms to create the new chair. Each participant adds requisite expertise to the project. The furniture firm provides its brand name and distribution network to the project. Tool-and-die firms provide customized parts, which are delivered to a manufacturing firm that will assemble the chairs. Marketing firms refine the design and test-market potential names and options. A project manager is assigned by the furniture firm to work with the inventor and the other parties to complete the project.

Collectively, the project is the summation of different structures. For example, the tool-and-die firm may assign a dedicated team to create the process for producing the custom

FIGURE 3.5. NETWORK PROJECT.

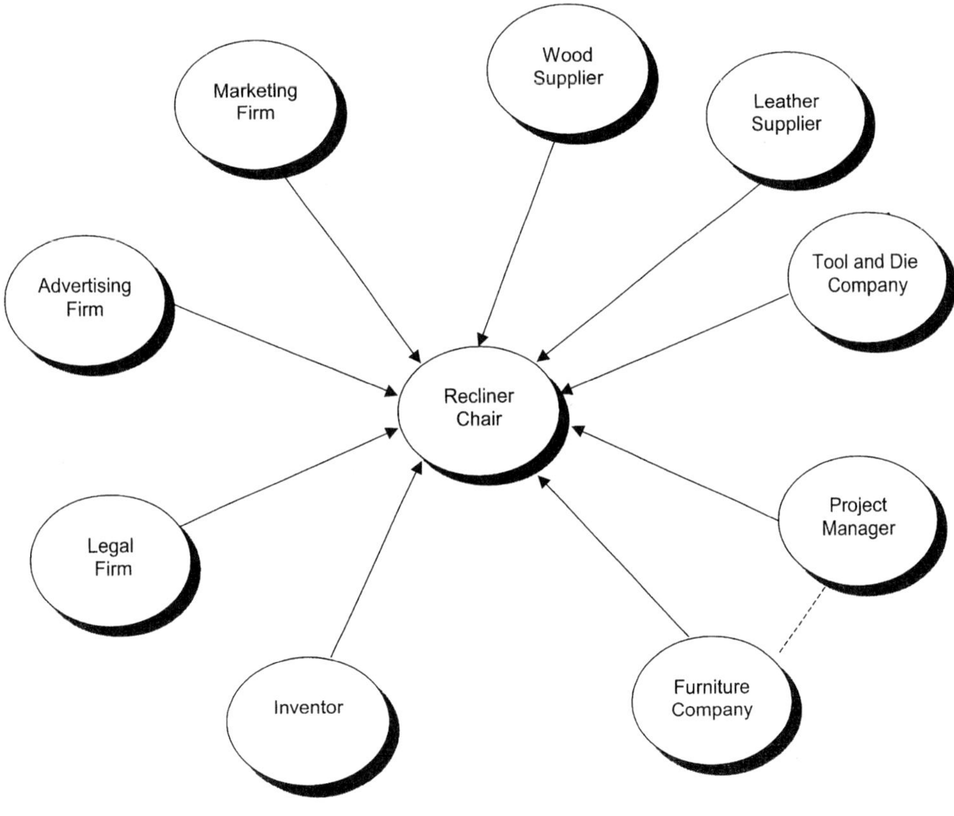

parts. The marketing work may be performed within functional departments, while the project manager works on this project part-time.

The advantages of network projects are many:

- *Cost reduction.* The most noteworthy is cost reduction. Companies can secure free-market prices for contracted services, especially if the work can be outsourced offshore. Furthermore, overhead costs are dramatically cut, since the company no longer has to maintain internally the contracted services.
- *High level of expertise.* A high level of expertise and technology can be brought to bear on the project. A company no longer has to keep up technological advances. Instead, it can focus on developing its core competencies and hire firms with the know-how to work on relevant segments of the project.

- *Increased flexibility*. Organizations are no longer constrained by their own resources but can pursue a wide range of projects by combining their resources with talents of other companies. Small companies can instantly go global by working with foreign partners.

The disadvantages of network projects are less well documented:

- *Breakdowns in coordination.* Synchronizing the work of professionals from different organizations can be challenging, especially if the project work requires close collaboration and mutual adjustment. This form of project management structure tends to work best when each party, as in the case of most construction projects, is responsible for a well-defined, independent deliverable.
- *Loss of control.* The core team depends on other organizations over which they do not have direct authority. While long-term survival of participating organizations depends upon performance, a project may falter when one partner fails to deliver.
- *Conflict.* Finally, networked projects are more prone to interpersonal conflict, since the different participants do not share the same values, priorities, and culture. Trust, which is essential to project success, can be difficult to forge when interactions are limited and people come from different organizations.

Most networked projects operate in a virtual environment in which people are linked by computers, faxes, computer-aided design systems, and video teleconferencing and sometimes rarely, if ever, see one another face-to-face. Many projects are being networked across time zones so that work never stops. For example, members from one organization work on a software project in New York and then pass their work at the end of day to another organization in Hawaii. The Hawaiian team passes their work to a team in India, which in turn passes their work to a Dutch firm. Although it is too early to say how applicable this 24-hour tag team approach to project management will be, it exemplifies the potential that exists given the information technology that is available today.

Within networked projects, people come and go as services are needed, much like a matrix structure. They are not formal members of one organization, just technical experts who form a temporary alliance with an organization, fulfill their contractual obligations, and then move on to the next project.

Choice of Project Management Structure

There is growing empirical evidence that project success is directly linked to the amount of autonomy and authority project managers have over their projects (Gray, Dworatschek, Gobeli, Knoepfel, and Larson, 1990; Gobeli and Larson 1987; Brown and Eisenhardt, 1995). For example, Larson and Gobeli (1989) studied 546 development projects and found that projects relying on functional organization or a functional matrix were less successful than those that used a balanced matrix, project matrix, or project team. Furthermore, the project matrix outperformed the balanced matrix in meeting schedules and outperformed the dedicated project team in controlling cost. However, this and other studies have focused only on what is best for managing specific projects. It is important to remember what was stated in the beginning of the chapter—that the best system balances the needs of the project with those of the parent organization. So what project structure should an organization use?

This is a complicated question with no precise answers. A number of issues need to be considered at both the organization and project level.

Organizational Considerations

At the organization level, the first question that needs to be asked is how important is project management to the success of the firm? What percentage of core work involves projects? If over 75 percent of work involves projects, an organization should consider a fully projectized organization. If an organization has both standard products and projects, a matrix arrangement would appear to be appropriate. If an organization has very few projects, a less formal arrangement is probably all that is required. Temporary task forces could be created on an as-needed basis and the organization could outsource project work.

A second key question is resource availability. Remember, the matrix evolved out of the necessity to share resources across multiple projects and functional domains, while at the same time creating legitimate project leadership. For organizations that cannot afford to tie up critical personnel on individual projects, a matrix or network system would appear to be appropriate. An alternative would be to create a dedicated team but outsource project work when resources are not available internally.

A third consideration is whether the organization has a firm grasp of its priorities and can effectively communicate the relative importance of different projects. This is particularly true for matrix structures in which resources are shared across projects and functions. An effective project priority system needs to be in place to guide resource assignments and avoid the infighting that can unravel a matrix into a chaos. If priorities are not established, dedicated or network project teams are advised to sidestep the resource contention issue. However, this can carry an expensive price tag for top management failing do to their job (Graham and Englund, 1997).

The final consideration is the culture of the firm. The metaphor that is used to describe the relationship between organizational culture and project management is that of a riverboat trip. Culture is the river and the project is the craft. Organizing and completing projects within an organization in which the culture is conducive to project management is like paddling downstream: Much less effort is required, and the natural force of the river generates progress toward the destination. In many cases, the current can be so strong that steering is all that is required. Such is the case for projects that operate in a project-friendly environment where teamwork and cross-functional cooperation are the norms, where there is a strong commitment to excellence, and where healthy conflict is voiced and dealt with quickly and effectively.

Conversely, trying to complete a project in an organization in which the dominant culture inhibits effective project management is like paddling upstream: Much more time, effort, and attention are needed to reach the destination. This would be the situation in cultures where cross-functional competition is high, where risks are to be avoided, and where getting ahead is based less on performance and more on cultivating favorable relationships with superiors. In such cases, the project manager not only has to overcome the natural obstacles of the project but also has to overcome the prevailing negative forces inherent in the culture of the organization.

The implications of this analogy are obvious but important. Greater project authority and resources are necessary to complete projects that encounter a strong, negative cultural current. Conversely, less formal authority and fewer dedicated resources are needed to complete projects in which the cultural currents generate the behavior and cooperation essential to project success. The key issue is the degree of interdependency between the parent organization and the project team and the corresponding need to create a unique project culture conducive to successful project completion (Kerzner, 1997; Brown, Grove, Kelly, and Rana, 1997; Jassawalla and Shashittalk, 2002).

Project Considerations

At the project level, the question is how much autonomy does the project need in order to be successfully completed. Hobbs and Menard (1993) identify seven factors that should influence the choice of project management structure:

- Size of project
- Strategic importance
- Novelty and need for innovation
- Need for integration (number of departments involved)
- Environmental complexity (number of external interfaces)
- Budget and time constraints
- Stability of resource requirements

Hobbs and Ménard advise that the higher the levels of these seven factors, the more autonomy and authority the project manager and project team needs to be successful. This translates into using either a dedicated project team or a project matrix structure. For example, these structures should be used for large projects that are strategically critical and are new to the company, thus requiring much innovation. These structures would also be appropriate for complex, multidisciplinary projects that require input from many departments, as well as for projects that require constant contact with customers to assess their expectations.

Shenhar and his colleagues have taken a slightly different approach through their efforts to develop a meaningful framework for classifying different kinds of projects (Shenhar, Dvir, Lechler, and Poli, 2002; Shenhar, 1998; Shenhar and Dvir, 1996—see also the chapter by Shenhar and Dvir toward the end of this book). They use three dimensions to distinguish different kinds of projects: *U*ncertainty at the moment of project initiation, *C*omplexity as reflected in the number and variety of elements/disciplines required to complete the project, and *P*ace with regard to speed and criticality of time. Together, these dimensions form the *UCP model* for distinguishing different kinds of projects. Shenhar and his colleagues argue that, depending upon how a project is configured on these dimensions, different project management techniques and styles are required for success. For example, "critical/blitz projects," which are most urgent and essential to the success of the firm, require dedicated project teams and strong top management support. Conversely, projects that lack inherent time-to-market pressure tend to take longer than planned without tight management.

Evidence of this contingency approach can be found in firms that have created a flexible management system that organizes projects according to project requirements. For example, Chaparral Steel, a mini-mill that produces steel bars and beams from scrap metal, classifies projects into three categories: advanced development, platform, and incremental (Bowen, Clark, Holloway, and Wheelwright, 1994) Advanced-development projects are high-risk endeavors involving the creation of a breakthrough product or process. Platform projects are medium-risk projects involving system upgrades that yield new products and processes. Incremental projects are low-risk, short-term projects that involve minor adjustments in existing products and processes. At any point in time, Chaparral might have 40 to 50 projects under way, of which only one or two are advanced, three to five are platform projects, and the remainder are small, incremental projects. The incremental projects are almost all done within a weak matrix, with the project manager coordinating the work of functional subgroups. A strong matrix is used to complete the platform projects, while dedicated project teams are typically created to complete the advanced-development projects. It is anticipated that more and more companies will be using this mix-and-match approach to managing projects.

Summary

Four different kinds of project management structures have been described and their relative strengths and weaknesses discussed. No one structure is optimum. Choice depends upon the needs of the organization and the requirements of the project. The future should see more and more organizations adopt a flexible project management system in which different structures will be used for different projects. Sophisticated information systems will be used to provide real-time progress reports and optimally utilize resources across projects. Still, the importance of organizational culture cannot be underestimated. Matrix management can flourish in one organization and be a total disaster in another. The reason for this was not matrix per se, but differences in the culture of the two organizations.

References

Bowen, H. K., K. B. Clark, C. A. Hollaway, and S. C. Wheelwright, eds. 1994. *The perpetual enterprise machine*. New York: Oxford Press.

Brown, P., S. Grove, R. Kelly, and S. Rana. 1997. Is cultural change important in your project? *PM Network*. January: 48–51.

Brown, S., and K. M. Eisenhardt. 1995. Product development: Past research, present findings, and future directions. *Academy of Management Review* 20(2):343–378.

Carleton, J. 1997. *Apple: The inside story of intrigue, egomania, and business blunders.* New York: Random House.

Davies, S. M., and P. R. Lawrence. 1977. *Matrix.* Reading, MA: Addison-Wesley.

Gobeli, D. H., and E. Larson. 1987. The relative effectiveness of different project management structures. *Project Management Journal.* 18(2):81–85.

Graham, R. J., and R. L. Englund. 1997. *Creating an environment for successful projects: The quest to manage project management.* San Francisco: Jossey-Bass.

Gray C., and E. Larson. 2003. *Project management: A managerial approach.* New York: McGraw-Hill/Irwin.

Gray, C., S. Dworatschek, D. H. Gobeli, H. Knoepfel, and E. Larson. 1990. International comparison of project organization structures: Use and effectiveness. *International Journal of Project Management.* 8(1):26–32.

Harrison, M. T., and J. M. Beyer. 1993. *The culture of organizations.* Upper Saddle River, NJ: Prentice Hall.

Hobbs, B., and P. Ménard. 1993. Organizational choices for project management. In *The AMA handbook of project management,* ed. Paul Dinsmore. New York: AMACOM.

Jassawalla, A. R., and H. C. Sashittal. 2002. Cultures that support product-innovation processes. *Academy of Management Executive.* 15(3):42–54.

Johnson, C. L., M. Smith, and L. Geary. 1990. *Kelly: More than my share of it all,* Washington D.C.: Smithsonian Institute Publications.

Kerzner, H. 1997. *In search of excellence in project management.* New York: Van Nostrand Reinhold.

Larson, E., and D. Gobeli. 1985. Project management structures: Is there a common language? *Project Management Journal.* 16(2):40–44.

———. 1989. Significance of project management structure on development success. *IEEE Transactions in Engineering Management.* 36(2):119–125.

Lawrence, P. R., and J. W. Lorsch. 1969. *Organization and environment.* Homewood, IL: Irwin.

Majchrzak, A., and Q. Wang. 1996. Breaking the functional mind-set in process organizations. *Harvard Business Review.* (September–October): 93–99.

Miles, R. E., and C. C. Snow. 1995. The new network firm: A spherical structure built on a human investment philosophy. *Organizational Dynamics.* Spring: 5–18.

Miles, R. E., C. C. Snow, J. A. Mathews, G. Miles, and H. J. Coleman. 1997. Organizing in the knowledge age: Anticipating the cullular form. *Academy of Management Executive.* 11(4):7–24.

Miller, J. 1996. *Lockheed Martin's skunk works.* New York: Speciality Publications.

Sculley, J. 1987. *Odyssey: Pepsi to Apple.* New York: Harper & Row.

Shenhar A. J. 1998. From theory to practice: Toward a typology of project management styles. *IEEE Transactions in Engineering Management.* 41(1):33–48.

Shenhar, A. J., and D. Dvir. 1996. Toward a typological theory of project management. *Research Policy* 25:607–632.

Shenhar, A. J., D. Dvir, T. Lechler, and M. Poli. 2002. One size does not fit all: True for projects, true for frameworks. *Frontiers of Project Management Research and Application. Proceedings of PMI Research Conference,* 99–106. Seattle.

Stuckenbruck, L. C. 1981. *Implementation of project management.* Upper Darby, PA: Project Management Institute.

Smith, P. G., and D. G. Reinersten. 1995. *Developing products in half the time.* New York: Van Nostrand Reinhold.

Verma, V. K. 1995. *Organizing projects for success: The human aspects of project management.* Newtown Square, PA: Project Management Institute.

Youker, R. 1977. Organizational alternatives for project management. *Project Management Quarterly.* 8:24–33

CHAPTER FOUR

AN OVERVIEW OF BEHAVIORAL ISSUES IN PROJECT MANAGEMENT

Dennis P. Slevin, Jeffrey K. Pinto

Glaciers move. If the geological conditions are right, they move inexorably toward the sea. Anyone who has visited and stood on a glacier in areas such as Alaska's Inland Passage is impressed by the dynamism and fluidity of the process. Glacial moraines are amazing flowing rivers of rock. Analogously, project management has been engaged in an inexorable movement toward the human side of the enterprise. The field of project management is one that has always been characterized by its joint emphasis on a blend of technical elements (e.g., PERT charts, beta distributions, earned value analysis, resource leveling) coupled with its vital connection to behavioral and management concepts. While numerous tools, techniques, and quantitative aids were developed in the 1960s and 1970s, people and teamwork have become crucial issues at the turn of the millennium.

Our interest in this topic has been an enduring one, starting almost two decades ago (Slevin and Pinto, 1986, 1988; Slevin and Pinto, 1992). We have long noted that projects are not successful because of the use of the latest project management techniques; they occur as the result of understanding the role that people play in fostering an environment for success. Many of these issues are addressed in a recent collection of new research studies published by the Project Management Institute (Slevin, Cleland, and Pinto, 2002). Research continues to bear out this position as recent studies clearly show. Interest in the behavioral side of successful project management continues to be keen and will continue to grow in this decade (Kloppenborg and Opfer, 2002). The purpose of this chapter is to provide a broad overview of some key behavioral factors impacting successful project management. We sought clarity and guidance in this task by interviewing a number of practicing project managers, many of whom wished to remain anonymous. Their insights and observations were invaluable in helping shape our thinking on these issues and providing the framework for this chapter.

A 12-Factor Model

As one examines the current literature, a number of key behavioral factors emerge as central to successful project management. We have identified 12 factors that we believe are crucial in impacting behavioral issues of project management. We developed this list by doing a quick scan of a variety of recent texts in project management, selected journal articles and papers presented at 2000 and 2002 PMI frontiers of Project Management Research Conferences. Also, a review was conducted of our own recent project management course syllabi. We then sorted the list of factors on the micro–macro continuum in an attempt to provide a useful structure for comments provided to us by practitioners. While we do not argue that this is an exhaustive list, we do feel that it takes a big bite out of the universe of key behavioral issues impacting the project manager. Our analysis has been primarily from the perspective of the individual project manager (the person on the firing line), as opposed to broader perspectives, such as a project management office or general organizational structure. These factors are listed in Figure 4.1.

As one goes down the list, the focus transitions from more micro (individual) issues to more macro (organization wide) issues. Changes in the environment over the past decade have generated increasing challenges in each of these 12 areas for the modern project manager. It is likely, in fact, that the changes we are observing at this point are simply milestones in the overall movement of the field, much as the glacier's movements may be easy to track from point to point, though the eventual destination of the glacier will always remain in question.

Key Comments from Practitioners

In an attempt to make this chapter as pragmatic as possible, we have solicited comments from practitioners concerning these key issues. Via e-mail and telephone interview techniques, a number of practicing project managers were asked to share their opinions con-

FIGURE 4.1. TWELVE KEY BEHAVIORAL FACTORS FOR SUCCESSFUL PROJECTS.

MICRO ↑

1. *Personal Characteristics of the Project Manager*
2. *Motivation of the Project Manager*
3. *Leadership and the Project Manager*
4. *Communications and the Project Manager*
5. *Staffing and the Project Manager*
6. *Cross-Functional Cooperation and the Project Manager*
7. *Project Teams and the Project Manager*
8. *Virtual Teams and the Project Manager*
9. *Human Resource Policies and the Project Manager*
10. *Conflict and Negotiations and the Project Manager*
11. *Power and Politics and the Project Manager*
12. *Project Organization and the Project Manager*

MACRO ↓

cerning trends in project management over the past decade, and where they felt the field was going in the future. Among others, this panel represents an opportunistic selection of individuals who attended PMI Research Conference 2002 (July 14 to 17, 2002, Seattle, Washington; Conference Co-Chairs: Dennis P. Slevin, Jeffrey K. Pinto, and David I. Cleland).

We believe that selected practitioner comments provide some interesting insights into current trends in the field. They also serve as an important perspective to the academic view of these constructs. In conducting these interviews, we deliberately sought a mix of project managers from a variety of organizational or business settings, including traditional production organizations, service industries, and governmental agencies. The range of responses provided additional evidence that more and more organizations are becoming "totally projectized" as they attempt to cope with rapidly changing technology and turbulent business environments (Lundin and Hartman, 2000).

1. Personal Characteristics of the Project Manager

It has been suggested for some time that project management skills are related closely and directly to key general management skills (*PMBOK Guide,* 2000 Edition). While it is clear that general management is a challenging profession, it is also obvious that the project manager often faces special challenges (Meredith and Mantel, 2003). While the general manager often has formal authority and considerable power, the project manager often faces the challenge of working from a low-power, informal position. The unique setting of project management has given rise to a literature that attempts to identify the characteristics most conducive to running successful projects (Posner, 1987; Einsiedel, 1987; Petterson, 1991). Based on these studies, a number of perspectives, traits, and features of project leadership have begun to emerge. Though there is by no means a general consensus of the specific traits of successful project managers, our evaluation of the relevant characteristics from the universe of 32 behavioral dimensions would include the following list (Byham, 1981; Slevin, 1989):

Dimensions

- *Planning and organizing.* Establishing a course of action for self and/or others to accomplish a specific goal; planning proper assignments of personnel and appropriate allocation of resources.
- *Control.* Establishing procedures to monitor one's own job activities and responsibilities or to regulate the tasks and the activities of subordinates. Taking action to monitor the results of delegated assignments or projects.
- *Technical/professional knowledge.* Level of understanding of relevant technical/professional information.
- *Oral communication.* Effective expression in individual or group situations (includes organization, gestures, and nonverbal communication).
- *Listening.* Use of information extracted from oral communication. The ability to pick out the essence of what is being said.
- *Written communication.* Clear expression of ideas in writing in good grammatical form; includes the plan or format of the communication.

- *Sensitivity.* Actions that indicate a consideration for the feelings and needs of others. Awareness of the impact of one's own behavior on others.
- *Group leadership.* Utilization of appropriate interpersonal styles and methods in guiding a group with a common task or goal toward task accomplishment, maintenance of group cohesiveness, and cooperation. Facilitation of group process.
- *Job motivation.* The extent to which activities and responsibilities available in the job overlap with activities and responsibilities that result in personal satisfaction; the degree to which the work itself is personally satisfying.
- *Analysis.* Identifying issues and problems, securing relevant information, relating and comparing data from different sources, and identifying cause-and-effect relationships.
- *Judgment.* Developing alternative courses of action and making decisions that reflect factual information, are based on logical assumptions, and take organizational resources into consideration.
- *Initiative.* Originating action and maintaining active attempts to achieve goals; self-starting rather than passively accepting. Taking action to achieve goals beyond what is necessarily called for.

Two things come to mind when selecting personal characteristics of the project manager:

- *The list tends to be long and diverse.* Project management is an intellectually and physically challenging profession. It requires a wide range of capabilities.
- *Technical skills are important.* While a general manager might surround him- or herself with technical experts, the project manager must be intimately involved with the technology concerning his or her project. As technology advances, this technical proficiency challenge becomes even more significant.

It has been suggested that successful project managers are both born and made (Melymuka, 2000). Some have suggested that project managers have key management styles that account for success, focusing primarily on their ability to function well as facilitators and communicators (Montague, 2000). Others have suggested that full-time leadership skills are essential (Schulz, 2000). In fact, though many of these theories offer some face validity and surface appeal, they also reinforce the problem of trying to isolate the type of person who makes an effective project manager.

Key Comments from Practitioners

> "For the most part, I find that today's project managers tend to be very achievement-oriented with strong characteristics towards working together with others in a cooperative manner, as opposed to the classical superior/subordinate relationship that was characterized ten years ago."
>
> "Organizations that recognize this are beginning to put more investment in team building and teamwork training."

"In our organization, we don't look for the ideal project manager. Too much of this job is learned as you go. We find it better to select likely candidates and work with them, giving them small assignments and testing their abilities. Can they make decisions? Are they intelligent enough to ask the right questions? Do they know what they don't know?"

2. Motivation and the Project Manager

The project manager must be a motivational genius. The project manager must have a high level of self-motivation and also be quite skillful at motivating the project team, often under situations of insufficient resources, low team member commitment and morale, and little formal authority. The self-motivation of the project manager is often an intrinsic thing. NASA has been quoted as saying "we don't work very hard on motivating astronauts, but we certainly are extremely careful in selecting astronauts." In that sense, self-motivated project managers are born, not made. However, the organization can do a variety of things to make sure that the motivational structures for project managers are as well developed and carefully executed as those for functional managers (Dunn, 2001). Job satisfaction can obviously be enhanced through appropriate motivational techniques. Further, having clear career ladders for project managers is essential. In many organizations, project managers form a subclass of manager. Because they do not belong to any department, it is easy for their careers to be overlooked in favor of functional standouts. As one wag put it to us, "In our organization, there are two career ladders, but only one has rungs!" Project managers will be self-motivated to the degree they perceive that their performance is likely to earn them advancement or other positive reinforcers.

Concerning the motivation of the project team, often highly creative and unusual techniques must be exploited. Feedback to the team is often long in coming in terms of project success. A typical salesperson receives information concerning sales progress every month; however, in complex projects, there are often unclear measures of progress. While information on schedule, budget, earned value analysis, and other dimensions of project progress may be available, the typical global feedback concerning project success occurs after it is completed, transferred, and used. This presents a particular challenge to a project manager concerning motivation. Recently, one of the authors attended a Christmas party at which the president of an entrepreneurial IT consulting company presented awards to ten employees, many of whom manage projects off-site. Each person received a very nice bronze plaque indicating that he or she had been awarded the President's Excellence in Performance Award for the year. In addition, each individual was given a very stylish briefcase. The president then said, "Each of you should open your briefcase. Who knows? One of them might have $1,000 in it." To the astonishment of the recipients, each briefcase contained 1,000 one-dollar bills—not an inexpensive approach to motivation, but in a competitive IT world where turnover is often a problem, this had a major impact on these lucky employees.

Another area of motivational import for the project manager concerns the management of risk. In a rapidly changing technological world, risk management becomes increasingly important. The importance of risk has been identified by a number of researchers in the field (Turner, 1993; Chapman and Ward, 1997; Wideman, 1998). The *PMBOK Guide,* 2000 Edition contains a significantly revised chapter on project risk management. A new edition of a major textbook in the field contains a substantially enhanced treatment of risk (Meredith and Mantel, 2003). It is important that the organization develop an open and cooperative attitude toward risk, along with approaches that reduce the motivation for concealing risks (Schmidt and Dart, 1999).

Key Comments from Practitioners

> "While organizations that recognize project management as a valid discipline and process have put into place motivational inducers such as specific job descriptions and career paths, this is not the norm in the corporate environment but rather the exception. Hence, I see more motivation coming from the individual project manager's need to achieve any personal satisfaction than from the classical motivational means. While this serves well in most cases, the lack of tangible motivational rewards in many organizations leads to disappointment and apathy when the position of project manager is not recognized and does not lead to a specific career objective."

> "When we ask practitioners, 'what is your motivation to stay in project management?' the comments often include a passion, a challenge, opportunity to influence, growth, finding better ways, the variety, ability to impact, being a change agent, able to achieve results, et cetera. The people who stay in project management are often self-motivated, at least until a cumbersome management grinds them down."

3. Leadership and the Project Manager

Leadership is crucial for effective project management in two ways:

- Leadership determines the effectiveness of the project planning process.
- Leadership style has a crucial impact on the effectiveness of the project team.

Leadership is important at the onset of the project because it provides key inputs to the project planning process. For example, leadership is crucial in definition of the project scope and the development of the project plan (Globerson and Zwikael, 2002). The implications of this finding are key: The leadership of the project manager immediately sets the stage for not only project team development but the metamorphosis of the project. Hence, effective organizations attempt to develop a positive leadership environment to enhance project success (Jiang, Klein, and Chen, 2001).

As Peg Thoms and John Kerwin explore in their chapter, leadership style issues present a particular problem for the project manager (Slevin, 1989; Slevin and Pinto, 1991). One of the key challenges of the project manager is the need to use consensus leadership approaches in working with the project team. One approach to the clarification of consensus issues is the two-dimensional Bonoma-Slevin Leadership Model. The two dimensions are *information input* and *decision authority*. Information input is represented by the subordinate groups' degree of information inputted into the decision-making process. The decision authority dimension determines whether the leader makes the decision solely by him- or herself or shares the decision making with the group. The grid below helps to define four leadership decisions (see Figure 4.2).

The four extremes of leaders (depicted in the four corners of the grid) are the following:

- *Autocrat* (100, 0). Such managers solicit little or no input from their groups and make the managerial decisions solely by themselves.
- *Consultative autocrat* (100, 100). In this managerial style, intensive input is elicited from the members, but these formal leaders keep all substantive decision-making authority to themselves.

FIGURE 4.2. BONOMA-SLEVIN LEADERSHIP MODEL.

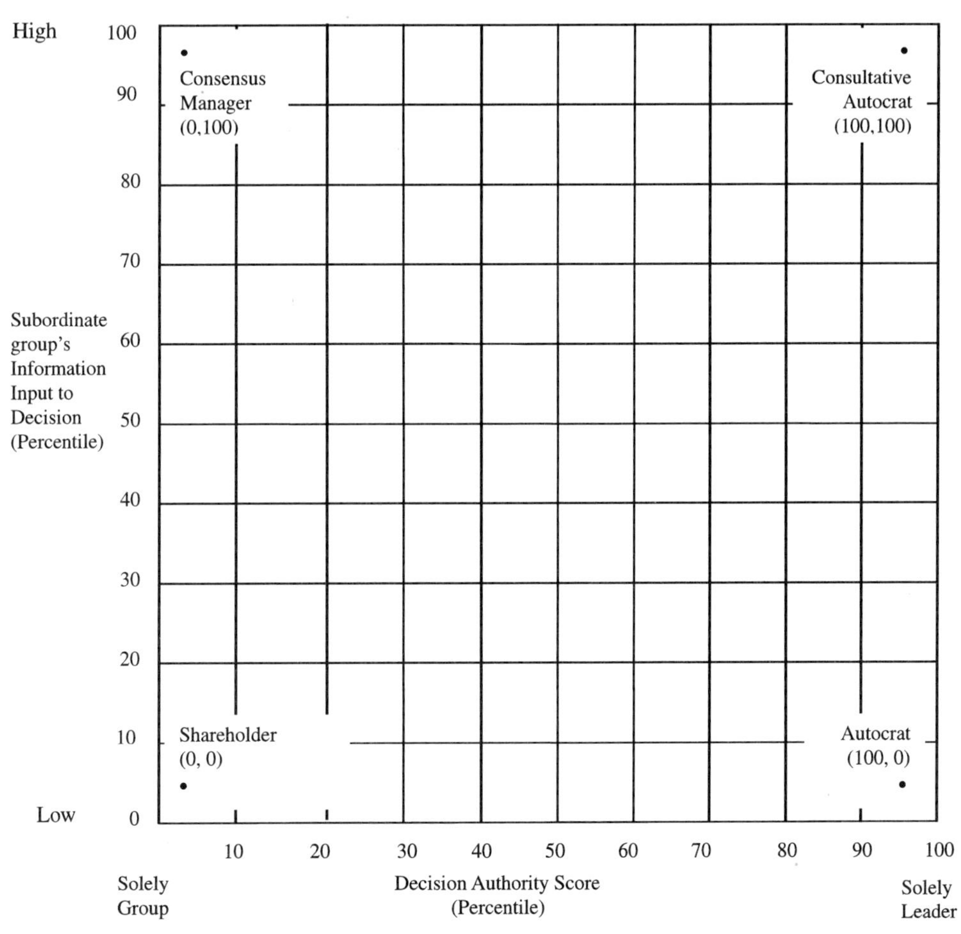

- *Consensus manager* (0, 100). Purely consensual managers throw open the problem to the group for discussion (input) and simultaneously allow or encourage the entire group to make the relevant decision.
- *Shareholder manager* (0, 0). This position is literally poor management. Little or no information input and exchange take place within the group, while the group itself has ultimate authority for the final decision.

The leadership style challenge for the project manager concerns heavy pressures, driving the decision making in the consensus area (northwest corner) of the grid, while often time pressures induce one to behave in an autocratic fashion. The good news from this leadership model is that leadership behaviors are flexible. It would be a mistake to assume that once identified as possessing a certain style, it is impossible to alter that style for different circumstances or situations. In fact, successful project managers have been shown to employ a great deal of flexibility in their use of leadership approaches. They may employ more authoritative practices when dealing with troublesome team members, consensus styles when working with technical people to collectively solve a problem, or a consultative autocrat style when developing project plans, duration estimates, or PERT (program evaluation and review technique) schedules. The typical realities of the project environment suggest that maximum returns derive from a flexible, thoughtful approach to project leadership styles. The reactive, one-dimensional project manager will find his or her leadership style may work well under some situations but is totally unsuited for others (Kangus and Lee-Kelly, 2000).

Key Comments from Practitioners

> "The project manager is becoming more of a leader than the classical view of a manager. As a leader, the project manager is beginning to meld together the leadership and management responsibilities into a well-rounded capability to not only see the trees but also see the forest . . . and what's on the other side of the forest."
>
> "Leadership in our company usually consists of being out in front, being able to talk the technical talk with the engineers, the financial talk with the accountant, and the management talk with the administrators. Our project leaders have to have the respect of the rest of the team, and it is never given easily—it has to be earned!"
>
> "The best leader I ever saw seemed to instinctively understand how to relate to different groups and different situations. I've watched him go from a sweet person to an SOB and back to a sweet person again in about five minutes, depending upon the situation and the person he was dealing with. It wasn't an act—he just knew the tune he had to play with each person."

4. Communications and the Project Manager

As technology advances, communications challenges will increase. Each time information is exchanged, time is expended and project resources are consumed (Back, 2001). Communication of the project vision to all affected stakeholders can be a tremendously important step in the process (Reed, 2002). The Internet can provide a major tool for project management communications but also can be a significant consumer of time and energy (Giffen,

2002). As a result, we are faced with a conundrum: Research demonstrates the importance of maintaining clear lines of communication with all project stakeholders to improve the chances of success, and yet, because of the manner in which many communication mediums operate, it is not always clear how to generate the most effective messages and communicate them for maximum benefit.

A number of factors have emerged that make communications a greater challenge every year for the project manager, including the following:

- Increased project complexity
- Globalization of projects
- The Internet and all of its ramifications
- E-mail
- Virtual teams

Example: A pharmaceutical company assembles a virtual project team of 30 scientists and managers in six different countries. They interact using the most advanced teleconferencing technologies. The e-mail load is enormous. Even though the technology is marvelous, scheduling and executing a meeting is a challenge because of global time zone differences. As a result, the communications network and initially approved modes for communication do not perform nearly as well as hoped, causing project delays and numerous face-to-face meetings to clarify differences: something the project manager had hoped to avoid by investing in state-of-the-art networking!

The *PMBOK Guide,* 2000 Edition suggests four major communication processes:

- Communications planning
- Information distribution
- Performance reporting
- Administrative closure

The manner in which each of these communications steps is approached can have a huge impact on the viability of project team and stakeholder communications. Each of these steps must be carefully considered, its strengths and weaknesses assessed, and fallback positions identified. Done well, project communications processes are a hugely important factor in project success. Done poorly, they may result in conflicting messages, priorities, isolated pockets within the team, and an information vacuum.

Key Comments from Practitioners

> "Mulenburg's 4th Law: All people problems are problems of communication. And all (at least most) project leader problems are people problems—the team, the line manager, upper management, the customer, suppliers, et cetera—or involve people that have to be convinced, persuaded, stroked, or put on the right path."
>
> "The project manager has become more of a communicator and facilitator than experienced years ago. With the increase in technological changes, the higher emphasis on 'better, quicker, cheaper,' and more complex projects, this has occurred more out

> of necessity than from design. However, project managers who are good communicators are becoming more and more difficult to find."
>
> "There are two keys to communication: speaking well and writing well. I see some project managers that simply cannot write coherently. In fact, their writing is embarrassing. On the other hand, some project managers lack the ability to put two sentences together once they are up in front of people. We don't have the luxury of hiding project managers who can't communicate."
>
> "The key to our business is keeping the customer happy. The person who is responsible for that is the project manager. Communication skills are essential!"

5. Staffing and the Project Manager

Careful staffing of organizations has long been known as a secret to success. There are two common problems found in staffing project teams: (1) taking the first available resource regardless of their level of motivation, skill, or background in the project being considered or (2) having functional managers use project teams as a dumping ground for their poorer performers. Unfortunately, while successful project teams should be staffed by the best and brightest available, often the reverse is true. It is not difficult to see the end result from creating a team made up of personnel with low motivation or the suddenly discovered news that their boss considers them expendable. Alternatively, research suggests that when care is taken to staff project teams from available talent pools, the end result is much more promising for creating an environment for success. It has been suggested that the interview process can be made more effective by following these ten steps:

1. Write the job description.
2. Conduct the job analysis.
3. Select the behavioral dimensions.
4. Construct an interview form.
5. Recruit qualified candidates.
6. Study the résumés or applications.
7. Interview the applicants, record the data, defer judgment.
8. Score the interviews.
9. Use multiple interview consensus.
10. Make the hiring decision (Slevin, 1989, p. 287).

The implications are clear: Project teams that are staffed carefully, based on the hunt for the best talent available, are more likely to perform well. At a time when turnover in many critical knowledge-based industries is high (Abdel-Hamid, 1992), it would be a mistake to approach project team staffing in a way that will turn off potential valuable contributors to the project. Research suggests that individuals should be selected not only for their skills but also for the interpersonal capabilities and diversity that they bring to the team (McDowell, 2001; Melymuka, 2002).

Key Comments from Practitioners

> "I have not seen many changes. Most project managers within organizations indicate that they have little, if any, impact on staffing of the project. In most cases, PMs inherit their staff and simply have to live with who they have. While this works in some cases, in some cases it does not work. High success has been shown in organizations who involve the project manager in determining, negotiating, and in some cases actually hiring staff for the project."
>
> "As mentioned, project management is still pretty much an accidental profession in [our organization], which appears consistent with much of industry from the literature."
>
> "Project managers (and team members) are selected based on technical expertise, not on managerial skills, especially not on communication skills, unless they have already shown capability in project management."

6. Cross-Functional Cooperation and the Project Manager

Most projects have long required a team that includes members of different functional groups or members with diverse backgrounds. The cultures of their departments and differentiated manner in viewing the world often combine to make it extremely difficult to achieve cross-functional cooperation. Because cross-functional teams can greatly facilitate the successful implementation of projects, it is critical to better understand the mechanisms and motivations by which members of different functional groups are willing to collaborate on projects. Research suggests that four antecedent constructs can be important in accomplishing cross-functional team effectiveness (Pinto, Pinto, and Prescott, 1993):

- *Superordinate goals.* The need to create goals that are urgent and compelling, but whose accomplishment requires joint commitment and cannot be done by any individual department.
- *Accessibility.* Project team members from different functional departments cooperate when they perceive that other team members are accessible, either in person or over the telephone or e-mail system.
- *Physical proximity.* Project team members are more likely to cooperate when they are placed within physically proximate locations. For example, creating a project office or "war room" can enhance their willingness to cooperate.
- *Formal rules and procedures.* Project team members receive formal mandates or notification that their cooperation is required.

Cross-functional/multifunctional members of the project team can present a challenge for harmonious and enthusiastic teamwork, but able leadership can overcome the challenge (Rao, 2001). Cross-functional teams have been found particularly useful the greater the novelty or technical complexity of the project (Tidd and Bodley, 2002).

Key Comments from Practitioners

> "Project managers tend to be very cooperative with cross-functional relationships. I do not believe this is a problem. The problem I have seen, and continue to see, is just the reverse. The cross-functional individuals—especially functional managers and personnel—tend not to cooperate with the project managers. This creates serious problems, since cooperation is a two-way street."
>
> "In our organization, marketing and engineering do not get along. We have developed this mentality where these two groups actively work to discredit each other. Of course, no one stops to think about who the real loser in this situation is!"

7. Project Teams and the Project Manager

Organizations of the future are relying more and more on project teams for success. This movement implies that the team-building processes themselves may be a subobjective of the project (Bubshait and Farooq, 1999). One important discovery in team research in recent years has been the work of Gersick (1988; 1989), who investigated the manner in which groups evolve and adapt to each other and to the problem for which they were formed. Her research suggests that the old heuristic of "forming, storming, norming, performing, and adjouning" (Tuchman, 1965) that has been used to guide group formation and development for decades does not stand close scrutiny when examined in natural settings. Rather, coining a term from the field of biology, "punctuated equilibrium," she found that groups tend to derive their operating norms very quickly, working at a moderate pace until approximately the midpoint of the project, at which time a sense of urgency, pent-up frustrations, and a desire to re-address unacceptable group norms lead to an internal upheaval. The result is to create a better-performing project team. Gersick's work has been important for helping project managers understand how to better and more proactively manage the process by which their teams develop.

The chapter by Larson refers to the strengths and weaknesses of organizing projects as dedicated teams, an important issue. Similarly, Chapter 5 by Cooke-Davies identifies a capable team of task-oriented individuals, led by a competent leader as a key factor critical to the achievement of project management success.

Another crucial element to team success is knowledge management (Drew, 2003). Assembling knowledge management teams and distributing knowledge across the players can be extraordinarily important. The use of project management offices (PMOs) has been a useful tool in maintaining this center for knowledge management. Though the transition to PMOs is not always a smooth one, the ability to apply a centralized base of project knowledge to ongoing problems makes this process a useful one for promoting project success and helping develop the expertise of project team members. (See the chapter by Powell and Young.)

Key Comments from Practitioners

> "A proliferation of books and articles address the issue of teams without much emphasis on the project manager's role for how to become the leader of these teams.

Everyone is supposed to 'get along' somehow. The only well-oiled teams I have seen are that way because of a project leader with the skills to make it work."

8. Virtual Teams and the Project Manager

The world is going virtual. Two major universities recently received a multimillion-dollar grant to perfect the use of supercomputers in an application that enables people several thousand miles apart to work jointly together as if they were in the same room. With the increasing globalization of project management, teams comprising individuals who may never directly interact with each other are becoming commonplace. Their primary means of communication is through Internet, e-mail, and virtual meetings. This increase in virtual teamwork creates an entirely new level of complexity to the challenge of team building for project success (Adams and Adams, 1997; Townsend, DeMarie, and Henrickson, 1998).

Issues of cost, transportation, globalization, skill distribution, and a variety of other pressures have hastened the movement toward virtual team use in recent years (Elkins, 2000). Likewise, more and more global companies are experimenting with the process of partnering, which implies additional pressures on the virtual team (Bresnen and Marshall, 2002). However, because of a variety of factors beyond the control of project managers, such as organization structure or corporate culture, the degree to which project teams quickly acclimate to the virtual environment is quite variable. Some organizations have been able to develop and employ effective virtual teams, including adopting quite effective virtual team-building processes, while others have continued to find the technology difficult to master (Delisle, 2002). (See the chapter by Delisle.)

Key Comments from Practitioners

"With improved conferencing technology, virtual teams have become a reality (no pun intended). Distance and cost combine to make virtual teams a necessity in many instances. A major change that seems obvious is that to work, they need face time together initially, and periodically throughout the project."

"The big challenge we face is trying to make virtual teams act and work just like real teams. When you lose the sense of proximity to others working on the project, there is a feeling of disconnect. We require our virtual teams to make up in frequency of communications what they lack in proximity of communications."

9. Human Resource Policies and the Project Manager

For decades, human resource policies have been designed primarily to fulfill the needs of line management activities. Recent experiences have shown that the human resource function can become a full business partner with a project management process without losing integrity to line managers (Clark, 1999). In other words, the HR function is being designed more carefully to expedite project team development and staffing. (See the chapter by Heumann, Turner, and Keegan.) The *PMBOK Guide,* 2000 Edition suggests the following major processes concerning project human resource management.

- Organizational planning
- Staff acquisition
- Team development

Another way in which HR is becoming more attuned to project management needs is through legal issues, compliance, and safety and health in the workplace. As projects are occasionally created in less than optimum work conditions, such as harsh environmental conditions or to work on projects with health or safety risks, human resource expertise has been tapped by project managers so they have a clearer understanding of issues of corporate liability and due diligence regarding safety and hiring practices.

Key Comments from Practitioners

> "In general, organizations have not responded to the needs of project management with respect to human resource policies. Most HR policies address the organizational needs from an ongoing, functional aspect with few addressing the particular aspects of project management. Key 'holes' exist in addressing the temporary nature of projects, matrix management, and the classical problem in accounting of accruals (organization) versus committed (project) costs."

10. Conflict and Negotiations and the Project Manager

The project manager is in a constant environment of conflict and negotiations (Kellogg, Orlikowski, and Yates, 2002). As Jeff Pinto and John Magenau explain in their chapter, the need to exercise the influential side of project management occurs for a number of reasons. For example, many organizations run projects within structures where departmental heads retain all control over project resources, requiring project managers to negotiate for their team resources. Other reasons for conflict and negotiation occur within projects where it is vital that the project manager and key team members understand important terms and conditions of contracts. The result is a circumstance in which project managers routinely exercise influence, deal with conflict, and negotiate with parties both inside and outside their own organization. Consequently, negotiation skills are considered to be an important part of the project manager's tool kit (Pravda and Garai, 1995). Project managers face a constant dilemma of determining how they are to acquire the authority to overrule resource and line managers in order to accomplish project objectives (Pinto et al., 1998; Vandersluis, 2001).

Key Comments from Practitioners

> "Project managers continue to be very involved with conflict management and negotiations. However, very little formal or informal training and development in these areas is prevalent. There still tends to be the concept of the 'accidental project manager,' resulting in throwing individuals into situations for which they are really not prepared."

"As with most elements of project management, organizations need to have specific and structured training and development programs that address project management in general and the specific areas of conflict management and negotiations."

"Our most productive project managers are the ones who instinctively understand that their job does not start and stop with the scheduling and administrative duties. They have to handle the hard duties, like negotiation and conflict, every day."

"Our brand-new project managers (the people who have never run projects before) are always shocked at how little power they have in this company. If they want to succeed, they learn that they better sharpen up their negotiation skills real fast!"

11. Power and Politics and the Project Manager

One of the least-talked-about aspects of project management duties involves the necessity of mastering the art of influence and political behavior. Attitudes regarding the use of politics, in this sense, point to an interesting dichotomy among managers in organizations. On the one hand, by a margin of almost 4 to 1, successful mid-level managers acknowledge that politics and influence are vital to performing their jobs effectively. On the other hand, by the same margins, these managers routinely affirm that the use of politics wastes company resources, is unpleasant to engage in, and is personally repugnant to them. The implications are interesting: On the one hand, managers do not like to use politics in their jobs, and yet on the other hand, they recognize that in order to successfully manage their projects, it is a vital skill to master.

Often negotiating the political terrain can be a greater challenge than the technical details of the project itself. All projects have numerous stakeholders (as Winch demonstrates in his chapter on stakeholder management). The political processes that characterize interactions between project managers and top managers are becoming evermore important in the success of new forms of organizations such as the project management office (Vandersluis, 2002). One solution to enhancing the project management process from the power and politics perspective is the institutionalization of an executive champion (Wreden, 2002). Champions can often serve to alleviate some of the political headaches that project managers accrue by serving as the point man for the project with key stakeholder groups. Champions exert their own kind of influence on behalf of the project. The difference is that because of the authority of status of champions, they are in a better position to help the project along.

Key Comments from Practitioners

"Many organizations are beginning to focus on the results of the two types of power: organizational power resulting in either compromise or compliance, and portable power resulting in commitment or loyalty. When faced with the question "would you rather have a project team that is committed and loyal or one that compromises and complies?" most recognize that the former creates a stronger project team and leads to a higher success rate. By focusing on this analysis, power and authority concerns tend to become resolved."

"I get a real kick out of the reaction of people who join our organization right out of college and are confronted with their first real taste of company politics. They can argue until they are blue that their opinion should win out because their way is 'the best,' but until they learn how to get things accomplished around here through the back door, they will never really be successful. All our successful project managers are successful politicians."

12. Project Organization and the Project Manager

The *PMBOK Guide,* 2000 Edition suggests two extremes of organizational form to a project:

- *The functional organization.* People and positions are grouped together according to the work they perform.
- *The projectized organization.* People are grouped together by project commitments, regardless of the functional background or expertise they possess.

As one moves toward the projectized organization, *PMBOK Guide,* 2000 Edition suggests four levels of matrix:

- *Weak matrix organization.* Limited project manager authority; 0 to 25 percent of personnel time dedicated to project management work; part-time project management administrative staff.
- *Balanced matrix organization.* Low to moderate project manager authority; 15 to 60 percent of personnel time dedicated to project management work; part-time project management administrative staff.
- *Strong matrix organization.* Moderate to high project manager authority; 50 to 95 percent of personnel time dedicated to project management work; full-time project management administrative staff.
- *Projectized organization.* High to almost total project manager authority; 85 to 100 percent of personnel time dedicated to project management work; full-time project management administrative staff.

The argument regarding the optimal type of organization suggests that there is a transition period in which organizations move from less-than-optimal structures, such as the functional structure, to those that are better able to support and sustain a project focus. Wheelwright and Clark (1992) refer to this movement as the drive toward "heavyweight" project organizations, in which power and decision-making authority are no longer shared between project and function, but rest solely in the hands of project managers. Research has clearly demonstrated the benefits to project success from crafting an organization form that supports these activities (Gobeli and Larson, 1987). Experience in interacting with senior managers indicates that more and more organizations are moving in the "projectized" direction (Lundin and Soderholm, 1995; Lundin and Midler, 1998).

Key Comments from Practitioners

> "More and more I see project organizations aligned towards managing multiple projects as opposed to the single, large stand-alone project. This is especially true in internal corporate IT organizations, product development, and internal organizational support functions. It is also more prevalent in organizations that do projects for profit for external customers. Unfortunately, the concept of the project or program office in support of this is just now starting to catch on but suffers from a lack of documented experience, research, literature, and project management software tools that focus on multiple project management."

Conclusions

We have spent the better part of the past two decades researching, teaching, and consulting in project management and project organizations. Over that time, we have had the opportunity to witness the advent of a number of important innovations in the project management field in a variety of areas: scheduling, project monitoring and control, structural changes, and so forth. While all of these ideas have doubtlessly had a positive affect on the way projects are being run today, we find ourselves, in some sense, coming full circle as we note that the "true" determinants of successful project management are in many ways as clear today as they were two decades ago. Successful projects are those in which the "people side" has been well managed. All the technology in the world cannot overcome poor leadership, motivation, communications skills, team building, and so forth. On the other hand, project managers who take the time to perfect their skills in these critical areas continue to demonstrate that successful project management depends first and foremost on our ability to effectively manage the human resources for which we have been made responsible.

This chapter has offered a brief overview of some of the important themes in managing projects and the behavioral challenges that this process involves. As the chapter makes clear, the challenges are diverse; they broadly cover the gamut of individual and interpersonal relationships all the way to larger, organization theory issues of organization structure and cultural processes. As a result, it should be apparent that the types of skills needed to master the discipline of project management, whether from a practitioner or academic research perspective, requires both a depth of understanding and a breadth of knowledge that makes project management a truly unique undertaking. Successful project managers must learn first an appreciation of the myriad behavioral challenges they are going to face, as well as develop a commitment to pursuing knowledge in these diverse areas.

Acknowledgment

The authors are indebted to Walter Bowman, Gerald Mulenburg, and a number of other anonymous, practicing project managers for substantial insights and helpful comments in the preparation of this chapter.

References

Abdel-Hamid, T. K. 1992. Investigating the impacts of managerial turnover/succession on software project performance. *Journal of Management Information Systems* 9:127–145.

Adams, J. R., and L. L. Adams. 1997. The virtual projects: managing tomorrow's team today. *PM Network,* 11(1):37–41.

Back, W. E. 2001. Information management strategies for project managers. *Project Management Journal* 32:10–20.

Bresnen, M., and N. Marshall. 2002. The engineering or evolution of co-operation? A tale of two partnering projects. *International Journal of Project Management* 20:497–505.

Bubshait, A. A., and G. Farooq. 1999. Team building and project success. *Cost Engineering* 41:37–42.

Byham, W. C. 1981. *Targeted selection: A behavioral approach to improved hiring decisions.* Pittsburgh: Development Dimensions International.

Chapman, C. B., and S. Ward. 1997. *Project risk management: Processes, techniques and insights.* Chichester, UK: Wiley.

Clark, I. 1999. Corporate human resources and "bottom line" financial performance. *Personnel Review.* 28:290–307.

Delisle, C. 2002. Success and communication in virtual project teams. PhD diss., Department of Civil Engineering, University of Calgary. Calgary, Alberta.

Drew, R. 2003. Assembling knowledge management teams. *Information Strategy: The Executive's Journal* 19:37–42.

Dunn, S. C. 2001. Motivation by project and functional managers in matrix organizations. *Engineering Management Journal* 13:3–10.

Einsiedel, A. A. 1987. Profile of effective project managers. *Project Management Journal* 18(5):51–56.

Elkins, T. 2000. Virtual teams. *IIE Solutions.* 32:26–32.

Gersick, C. 1988. Time and transition in work teams: Towards a new model of group development. *Academy of Management Journal.* 31:9–41.

———. 1989. Making time predictable transitions in task groups. *Academy of Management Journal.* 32: 274–309.

Giffin, S. D. 2002. A taxonomy of internet applications for project management communication. *Project Management Journal.* 33:32–47.

Globerson, S. and O. Zwikael. 2002. The Impact of the *Project* Manager on *Project Management* Planning Processes. *Project Management Journal.* 33:58–65.

Gobeli, D. H., and E. W. Larson. 1987. Relative effectiveness of different project management structures. *Project Management Journal.* 18(2):81–85.

Jiang, J. J., G. Klein, and H. Chen. 2001. The relative influence of IS project implementation. *Project Management Journal.* 32(3):49–55.

Kangis, P., and L. Lee-Kelley. 2000. Project leadership in clinical research organizations. *International Journal of Project Management.* 18:393–342.

Kellogg, K., W. Orlikowski, and J. Yates. 2002. Enacting new ways of organizing: Exploring the activities and consequences of post-industrial work. *Academy of Management Proceedings.*

Kloppenborg, T. J. and W. A. Opfer. 2002. Forty years of project management research: Trends, interpretations, and predictions. In *The frontiers of project management research,* ed. D. P. Slevin, D. I. Cleland, and J. K. Pinto. Newtown Square, PA: Project Management Institute.

Loo, R. 2002. Journaling: A learning tool for project management training and team-building. *Project Management Journal.* 33:61–68.

Lundin, R. A., and F. Hartman. 2000. Pervasiveness of projects in business. In *Projects as business constituents and guiding motives,* ed. R. A. Lundin and F. Hartman. Dordrecht, Germany: Kluwer Academic Publishers.

Lundin, R. A., and C. Midler. 1998. *Projects as arenas for renewal and learning processes.* Norwell, MA: Kluwer Academic Publishers.

Lundin, R. A., and A. Soderholm. 1995. A theory of the temporary organization. *Scandinavian Journal of Management.* 11(4):437–455.

McDowell, S. W. 2001. Just-in-time project management. *IIE Solutions.* 33:30–34.

Melymuka, K. 2000. Born to lead projects. *Computerworld.* 34:62–64.

———. 2002. Who's in the house? *Computerworld.* 36.

Meredith, J. R. and S. J. Mantel. 2003. *Project Management: A Managerial Approach.* New York: Wiley.

Montague, J. 2000. Frequent, face-to-face conversation key to proactive project management. *Control Engineering,* Vol. 47:16–17.

Petterson, N. 1991. What do we know about the effective project manager? *International Journal of Project Management.* 9:99–104.

Pinto, J. K. and D. P. Slevin. 1988. Critical success factors across the project life cycle. *Project Management Journal* 67–75.

———. 1992. *Project implementation profile (PIP),* Tuxedo, NY: XICOM INC.

Pinto, J. K., P. Thoms, P., J. Trailer, T. Palmer, and M. Govekar. 1998. *Project leadership from theory to practice.* Newtown Square, PA: Project Management Institute.

Pinto, M. B., J. K. Pinto, and J. E. Prescott. 1993. Antecedents and consequences of project team cross-functional cooperation. *Management Science.* 39:1281–1298.

PMBOK Guide 2000 *A guide to the project management body of knowledge.* Newtown Square, PA: Project Management Institute.

Posner, B. Z. 1987. What it takes to be a good project manager. *Project Management Journal.* 18(1):51–54.

Pravda, S. and G. Garai. 1995. Using skills to create harmony in the cross-functional team. *Electronic Business Buyer.* 21:17–18.

Rao, U. B., 2001. Managing cross-functional teams for project success. *Chemical Business.* 5:8–10.

Reed, B. 2002. Actually making things happen. *Information Executive.* 6:10–12.

Schmidt, C., and P. Dart. 1999. Disincentives for communicating risk: A risk paradox. *Information and Software Technology.* 41:403–412.

Schulz, Y. 2000. Project teams need a qualified full-time leader to succeed. *Computing Canada.* 26:11.

Slevin, Dennis P. 1989. *The whole manager.* Innodyne, Inc., Pittsburgh, PA.

Slevin, D. P., D. I. Cleland, and J. K. Pinto, eds. 2002. *The frontiers of project management research.* Newtown Square, PA: Project Management Institute.

Slevin, D. P. and J. K. Pinto. 1987. Balancing strategy and tactics project implementation. *Sloan Management Review.* 29(1):33–41.

———. 1991. Project leadership: understanding and consciously choosing your style. *Project Management Journal.* 22(1):39–47.

Tidd, J. and J. Bodley. 2002. The influence of project novelty on the new product development process. *R&D Management.* 32:127–139.

Townsend, A. M., S. DeMarie, and A. R. Hendrickson. 1998. Virtual teams: technology and the workplace of the future. *Academy of Management Executive.* 12(3):17–29.

Tuchman, B. W. 1965. Developmental sequence of small groups. *Psychological Bulletin.* 63:384–399.

Turner, J. R. 1993. *The handbook of project-based management.* New York: McGraw-Hill.

Vandersluis, C. 2001. Projecting your success. *Computing Canada.* 27:14–16.

Wheelwright, S. C., and K. Clarke. 1992. Creating project plans to focus product development. *Harvard Business Review.* 70(2):70–82.

Wideman, R. M. 1998. Project risk management. In *The Project Management Institute project management handbook,* ed. J. K. Pinto. Jossey-Bass Publishers and Project Management Institute.

Wreden, N. 2002. Executive champions: Vital links between strategy and implementation, *Harvard Management Update.* 7:3–6.

SECTION II

THE MANAGEMENT OF PROJECTS

SECTION II.1

STRATEGY, PORTFOLIO AND PROGRAM MANAGEMENT

INTRODUCTION

Morris and Hough's (1987) work signaled the beginning of our recognition that an important sea change was occurring in the manner with which we might begin to make sense of the project management discipline. Their "broadening out" of the context of project-based work, as articulated in the introduction, forced researchers and practitioners alike to reorient the traditional views of project management into a more inclusive and comprehensive body of knowledge regarding the management of projects. Because projects are typically initiated with the goal of achieving corporate strategic objectives, it became necessary to better understand the positioning of projects to align with strategic portfolios and programs. Likewise, until and unless we can come to a firmer recognition of the goals a project is intended to support, it will be difficult and frustrating to ascribe notions of success or failure to any particular project venture. Is it true, to put it into Shakespeare's words, that "ripeness is all," or must we not have a firmer basis for understanding the causes and results of what we attempt?

A good place to start, as we've already seen, is the whole debate about project success. In Chapter 5 Terry Cooke-Davies reviews the studies that have been carried out in this area since the landmark work by Baker, Murphy, and Fisher in 1974. He concludes that there are essentially three levels of success measure: project management success (was the project done right?), project success (was the right project done?), and consistent project success (were the right projects done right time after time?). While warning that there are no silver bullets, Terry nevertheless identifies the half dozen or so key factors that he believes, from his research and that of his colleagues, are critical at each level.

Roland Gareis, in Chapter 6, offers a completely different approach. Refreshing, ambitious, and comprehensive, Roland discusses the characteristics of organizations that are project- (and program-) oriented. Again building on years of original research as well as practical consulting, Roland encapsulates most of the ideas this resource book addresses, although using his own distinctive "management by projects" framework (which is just slightly different, as the wording would suggest, from our "management of projects").

In Chapter 7, Karlos Artto and Perttu Dietrich offer another comprehensive and very thorough overview, but this time on strategic business management through multiple projects. Covering a wealth of academic work in the area, Karlos and Perttu examine the way that companies manage the relationships between portfolios, programs, and projects in different situations. They then generalize this into an overall theoretical model, which they illustrate from a series of research projects they have undertaken with industry.

Ashley Jamieson and Peter Morris, in Chapter 8, similarly survey the literature on moving from corporate strategy to project strategy, again emphasizing the sequence of moving via portfolios and programs into projects (and subprojects/tasks). Like Artto and Dietrich, their chapter introduces fresh research evidence to substantiate their findings, this time from a project funded largely by PMI. The project uses case study data as well as evidence from a questionnaire survey of PMI members and shows that most of those who replied routinely work in programs and projects, and use techniques to value-optimize the strategy (as is later discussed in Chapter 36).

David Cleland builds on this argument in Chapter 9, drawing on a wealth of experience to extend the arguments regarding corporate planning and programs with more detail at the project planning end—for example, explaining how project strategic planning feeds into work packages via the work breakdown structure.

Joe Lampel and Pushkar Jha, in their chapter on project orientation, Chapter 10, contend that many projects fail as a result of poor strategic orientation—of not achieving a proper fit between the enterprise and the project. They hypothesize three types of project/enterprise orientation—project based, project led, and core operations led—and propose that project scoping, programming, and autonomy shape the interaction between the project and its corporate parent. They conclude by presenting research findings that explore these interactions.

Having reviewed aspects of the linkages between strategy at the corporate, portfolio, program, and project levels, the next two chapters focus more purposefully on, first, portfolio management, and second, program management.

Norm Archer and Fereidoun Ghasemzadeh are two of the world's leading authorities on portfolio management (in the project, as opposed to the financial, sense). Their Chapter 11 brings out the importance particularly of managing risk and outsourcing options at the portfolio level, and the need for a framework for classifying project type. They then proceed to look at the different characteristics that affect portfolio choice and develop a generic process model for portfolio selection.

Michel Thiry, in Chapter 12, develops a number of interesting perspectives to better understand the characteristics of program management. Building on the ideas already presented in the previous chapters on strategy and portfolio management, Michel emphasizes the importance of learning in developing a strategic response to evolving conditions; in his process model of program management, learning needs understanding as clearly as performance delivery. (This leads him to define project management in the more specific sense of being primarily about uncertainty reduction.) He then elaborates this into a two-dimensional phase model linking strategy, programs, projects, and operations cyclically around the activities of formulation, organization, deployment, appraisal, and dissolution.

Ali Jaafari, in Chapter 13, focuses on the characteristics of large (engineering) projects, emphasizing in particular how they are subject to environmental uncertainty and may need much more front-end, strategic management than the smaller project. Ali then walks the reader through a high-level process model of the major things that need doing in the management of large projects.

Graham Winch, in Chapter 14, provides an enormously useful practical account of the importance of stakeholder management in achieving successful project outcomes. Taking a systems development project as an example (the computerization of share dealing on the London Stock Exchange), Graham shows how the failure to identify and manage different parties' expectations can not only lead to "academic" discussions of whether and for whom the project was or was not a success, but can in a very real sense lead to loss of control and ultimately project disaster. Graham concludes by drawing out nine key lessons for managing stakeholders effectively.

Finance is an important dimension to the strategic development of projects. The availability of money will affect what can be done, and when. In the public area particularly, changes in the way projects are funded have had a huge effect on the whole way projects are set up and carried out. The next two chapters, by Rodney Turner and Graham Ive, illustrate this.

In Chapter 15 Rodney Turner gives a masterful overview of the characteristics and means of financing projects as well as the process of financial management on projects. One of the newer forms of project finance to have developed over the last 20 or so years (coming out of the oil sector in the 1970s) has been that of basing the project's funding solely on the revenues generated specifically by the project itself, with no other security from other parties. This form, strictly termed "project financing," has become very important in many parts of the world in bringing ways of using private funds to finance public infrastructure projects. It is no exaggeration to say that this has had a revolutionary impact on the management of public sector projects where it has been applied.

Graham Ive, in Chapter 16, discusses this form of project financing in detail, with particular reference to its application in the British public sector, which is widely regarded as leading the field in this area. He outlines the origin of this form of financing, known in the United Kingdom as PFI (for Private Finance Initiative), and shows how it impacts on the management of projects, for example by requiring increased clarity on project objectives, risk management, value management, securing stakeholder consent, capturing users' requirements, and on procurement and bidding practices. (All issues either already addressed at this point or to be covered later in the book.) Interestingly, with regard to the procurement challenges, Graham uses an economic tool (agency theory) to analyze the problems of devising the reward structure, selection of resources, and moral hazard. Throughout, he illustrates his points with reference to a real PFI project, a new hospital.

About the Authors

Terry Cooke-Davies

Terry has been a practitioner of both general and project management continuously since the end of the 1960s. He is the Managing Director of Human Systems Limited, which he founded in 1985 to provide services to organizations in support of their innovation projects and ventures. Through the family of project management knowledge networks created and supported by Human Systems, he is in close touch with the best project management practices of more than 70 leading organizations globally. The methods developed in support of the networks are soundly based in theory, as well as having practical application to members, and this was recognized by the award of a PhD to Terry by Leeds Metropolitan University in 2000 for a thesis entitled, "Towards Improved Project Management Practice: Uncovering the Evidence for Effective Practices through Empirical Research." He is now an Adjunct Professor of Project Management at the University of Technology, Sydney and an Honorary Research Fellow at University College, London. Terry is a regular speaker at international project management conferences in Europe, North America, Australia, and Asia and has published more than 30 book chapters, journal and magazine articles, and research papers. He has a bachelor's degree in Theology, and qualifications in electrical engineering, management accounting, and counseling, in addition to his doctor's degree in Project Management.

Roland Gareis

Dr. Roland Gareis graduated from the University of Economics and Business Administration, Vienna; had habilitation at the University of Technology, Vienna, Department of Construction Industry; was Professor at the Georgia Institute of Technology in Atlanta; and was Visiting Professor at the ETH, Zurich, at the Georgia State University, Atlanta and at the University of Quebec, Montreal. He is currently Professor of Project Management at the University of Economics and Business Administration, Vienna and Director of the postgraduate program "International Project Management." He owns the firm ROLAND GAREIS CONSULTING.

Karlos A. Artto

Dr. Karlos Artto is professor at the Department of Industrial Engineering and Management at the Helsinki University of Technology (HUT), Finland, where he is responsible for developing research and education in the field of project-oriented activities in business. His current research interests cover the management of project-based organizations (including the area of project portfolio management and strategic management of multiple projects in organizations); the management of innovation, technology, R&D, new product development, and operational development projects in different organizational contexts; project networks and project delivery chains; and risk management (with the emphasis on management of business opportunities and considerations of project success and related criteria).

Perttu H. Dietrich

Perttu Dietrich works as a project manager and researcher at the BIT Research Centre at the Helsinki University of Technology (HUT), Finland. He has pursued research and development in several companies in the area of project portfolio management. His research interests include strategic management, multiproject management, and organizational development in different organizational contexts.

Ashley Jamieson

Ashley Jamieson worked for many years as a business manager, senior program manager, and project manager with global aerospace and defense companies on British, European, North American, Middle East, Southeast Asia, and Australasian aircraft programs and projects. He then took up a career in research and academia. He has worked at the Centre for Research in the Management of Projects (CRMP) at UMIST, where he carried out research into design management in major construction projects and was a visiting lecturer in project management. More recently, he has worked at University College, London, where through a PMI-funded research project, he investigated how strategy is moved from corporate planning to projects and is currently conducting research for APM on its Bok revision with Peter Morris. He holds an MSc in Business Management and UMIST and is the coauthor (with Peter Morris) of *Translating Corporate Strategy into Project Strategy* (2004), published by PMI.

Peter W. G. Morris

Peter Morris is Professor of Construction and Project Management at University College, London, Visiting Professor of Engineering Project Management at UMIST, and Director of the UCL/UMIST-based Centre for Research in the Management of Projects. He is also Executive Director of INDECO Ltd, an international projects-oriented management consultancy. He is a past Chairman and Vice President of the UK Association for Project Management and Deputy Chairman of the International Project Management Association. His research has focused significantly around knowledge management and organizational learning in projects, and in design management. Dr. Morris consults with many major companies on developing enterprise-wide project management competency. Prior to joining INDECO, he was a Main Board Director of Bovis Limited, the holding company of the Bovis Construction Group. Between 1984 and 1989 he was a Research Fellow at the University of Oxford and Executive Director of the Major Projects Association. Prior to his work at Oxford, he was with Arthur D. Little in Cambridge, Massachusetts, and previously with Booz Allen Hamilton in New York and with Sir Robert McAlpine in London. He has written approximately 100 papers on project management, as well as the books *The Anatomy of Major Projects* (Wiley, 1988), *The Management of Projects* (Thomas Telford, 1997) and *Translating Corporate Strategy into Project Strategy* (PMI, 2004). He is a Fellow of the Association for Project Management, Institution of Civil Engineers, and Chartered Institute of Building and has a PhD, MSc and BSc, all from UMIST.

David Cleland

David I. Cleland is currently Professor Emeritus in the School of Engineering at the University of Pittsburgh. He is the author/editor of 36 books in the fields of project management, engineering management, and manufacturing management. He has served as a consultant for both national and foreign companies, and has been honored for his original and continuing contributions to his disciplines. Dr. Cleland is a Fellow of the Project Management Institute (PMI) and has received the Distinguished Contribution to Project Management Award from PMI in 1983, 1993, and 2001. In 1997 he was honored with the establishment of the "David I. Cleland Excellence in Project Management Literature Award" sponsored by PMI.

Joseph Lampel

Joseph Lampel is a Professor of Strategy at Cass Business School, City University, London. He obtained his undergraduate degree in Physics from McGill University, Canada, and later pursued his MSc in Technology Policy at the Institut d'Histoire et Sociopolitique des Sciences at Université de Montréal, Canada. After working for the Science Council of Canada and the Ontario government, he returned to McGill University to pursue doctoral studies in Strategic Management. His dissertation "Strategy in Thin Industries" won the Best Dissertation Award from the Administrative Science Association of Canada in 1992. He was Assistant Professor at the Stern School of Business, New York University from 1989 to 1996. Subsequently, he was Reader at the University of St. Andrews from 1996 to 1999, and Professor of Strategic Management at University of Nottingham Business School from 1999 to 2001. He has also taught at McGill University, Concordia University, Montreal, and the University of Illinois at Urbana-Champaign. Joseph Lampel is the author with Henry Mintzberg and Bruch Ahlstrand of the *Strategy Safari* (Free Press/Prentice Hall, 1998). He is also the editor with Henry Mintzberg, James Brian Quinn, and Sumantra Ghoshal of the fourth edition of *The Strategy Process* (Prentice Hall, 2003). He edited a Special Issue of *International Journal of Project Management,* on Strategic Project Management (2001, Vol. 19, No. 8). He has published in *Strategic Management Journal, Sloan Management Review, Organization Science, Journal of Management, Journal of Management Studies, R&D Management, International Journal of Technology Management,* and *Fortune Magazine.*

Pushkar P. Jha

Pushkar Jha holds an MPhil in process engineering from the University of Newcastle-upon-Tyne and degrees in management and commerce. He is currently a research fellow at Cass Business School, London, working in the area of project-based organizational learning. He was a member of the Advanced Process Control Group at the University of Newcastle upon Tyne, and prior to this he worked on development projects in India.

Norman P. Archer

Dr. Norm Archer is Professor Emeritus in the Management Science and Information Systems Area of the Michael G. DeGroote School of Business at McMaster University and is Special Advisor to the McMaster eBusiness Research Centre (MeRC). Dr. Archer consults, teaches, and supervises graduate student research projects. He is active in the study of organizational problems relating to the implementation of e-business approaches in existing organizations, and the resulting impacts on processes, employees, customers, and suppliers. Current research projects include the study of knowledge transfer in network organizations, supply chain management issues, change management in organizations, and management of e-business projects. Together with his students, he has published more than 70 papers in refereed journals and conference proceedings, primarily on project management, business-to-business e-commerce, intelligent agents, and the human-computer interface, in *Decision Support Systems; Internet Research; International Journal of Management Theory and Practice; IEEE Transactions on Systems, Man, and Cybernetics; International Journal of Human-Computer Studies; Journal of the Operational Research Society, Communications of the ACM;* and many others.

Fereidoun Ghasemzadeh

Dr. Fereidoun Ghasemzadeh is an assistant professor in the Management and Economics School at Sharif University of Technology. Dr. Ghasemzadeh has been involved in teaching MIS, DSS, and e-commerce courses to MBA students, and supervises graduate students in their research projects. He is the cofounder and currently the CEO of Afranet, a leading Internet and e-commerce company in Iran, and is heavily involved in advanced e-commerce and e-government research and applied projects. Articles by, or interviews with, Dr. Ghasemzadeh have appeared more than 50 times in the media, and he has organized many conferences, symposia, and workshops. With his colleagues and students, Dr. Ghasemzadeh has published ten academic papers in refereed journals and conference proceedings, primarily on project management, e-commerce, and e-government.

Michel Thiry

Michel Thiry is managing partner of Valeuse Ltd, a European-based organizational consultancy. He has 30 years' experience in project management in North America and Europe and has worked in Canada, the United States, Australia, the United Kingdom, and continental Europe. He is currently adjunct professor for the Lille Graduate School of Management (France) and seminar leader for PMI Seminars World. He is also visiting professor at UIS (Australia) and external lecturer at Reading University (UK). He holds a MSc in Organizational Behaviour from the School of Management and Organizational Behaviour, University of London and is currently reading for a PhD at University College London. He regularly speaks and publishes at the international level. He has also provided value and project or program management expertise to major organizations, in various fields, including construction, pharmaceutical, IT and IS, telecom, water treatment, transportation (air and rail), and others. He has authored the book *Value Management Practice* and coauthored the "Managing Programmes of Projects" chapter in the *Gower Handbook of Project Management,* 3rd Edition. He also authored the program management and value management chapters in *Project Management Pathways,* published by the Association for Project Management (UK) in early 2003. In addition, he regularly writes and reviews for *the International Journal of Project Management.* Mr. Thiry is also past Director of the Project Management Institute's Montreal and UK chapters and past President of the European Governing Board for Value Management Certification and Training, based in Paris.

Ali Jaafari

Professor Jaafari was, until his recent retirement, Professor of Project Management at the University of Sydney. He received his PhD in Business Economics (Quantitative Modelling and Forecasting) from Surrey University in the United Kingdom (Joint SSRC-SRC scholarship holder and Swan Award) in 1977, and his Master of Science (Distinction) in Highway Engineering and Transportation Management in 1974 from the same university. He has a Master of Engineering (Distinction) from Tehran University awarded in 1968. He has more than 15 years of executive experience gained on major projects and programs in some 20 countries in Europe, mid East, Australia, and Asia. He has acted as an expert consultant to industry and governments worldwide for more than 15 years. In 1994 he acted as a special consultant to the European Community on the management of the Productivity Initiative Programme as part of TACIS. He has, to date, authored over 130 publications in project and program management, including strategy-based project management, whole-of-life framework and philosophy, concurrency, management of technology and innovations, information management systems, Total Quality Management, risk, opportunity and uncertainty analysis and management, and education of professionals. Since 1982 he has conducted courses and seminars for over 3,000 executives, managers, and professionals in Australia, Asia, and Europe. He specializes in graduate education and professional development, and has developed innovative online graduate programs that have won three Excellence Awards. He has been a member of the International Project Management Association since 1984, and a regular contributor to the World Congresses on Project Management. Professor Jaafari has chaired many functions both nationally and internationally and is the winner of many prizes and awards. Professor Jaafari has held visiting professorial appointments at a number of universities, in the United States, the United Kingdom, Europe, and Asia.

Graham Winch

Graham Winch is Professor of Construction Project Management at the Manchester Centre for Civil and Construction Engineering, UMIST, where he is head of the Project Management Division. He taught for ten years at the Bartlett School, University College, London after a career in management research in business schools and managing construction projects. He is author of *Managing Construction Projects: An Information Processing Approach* (Blackwell, 2002), as well as three other books and over 30 refereed journal articles, complemented by numerous book chapters, conference papers, and research reports. His research currently focuses on the strategic management of projects and on innovation in the construction industry. In addition to the work on stakeholder management, he is also investigating the processes of risk identification using cognitive mapping.

Rodney Turner

Rodney Turner is Professor of Project Management at The Lille Graduate School of Management and at Erasmus University, Rotterdam, in the Faculty of Economics. He is also an Adjunct Professor at the University of Technology Sydney, and Visiting Professor at Henley Management College. He studied engineering at Auckland University and did his doctorate at Oxford University, where he was also for two years a post-doctoral research fellow. He worked for six years for ICI as a mechanical engineer and project manager on the design, construction, and maintenance of heavy process plant, and for three years with Coopers and Lybrand as a management consultant. He joined Henley in 1989, Erasmus in 1997, and Lille in 2004. Rodney Turner is the author or editor of seven books, including *The Handbook of Project-Based Management,* the best-selling book published by McGraw-Hill, and the *Gower Handbook of Project Management.* He is editor of *The International Journal of Project Management* and has written articles for journals, conferences, and magazines. He lectures on and teaches project management worldwide. From 1999 and 2000 he was President of the International Project Management Association, and Chairman for 2001 to 2002. He has also helped establish the Benelux Region of the European Construction Institute as foundation Operations Director. In addition, he is a Fellow of the Institution of Mechanical Engineers and the Association for Project Management.

Graham Ive

At University College London since 1977, Graham Ive is responsible for the overall direction of the MSc Construction Economics and Management (CEM) course, and for economics teaching on the masters program. His research has focused on aspects of the industrial economics of the construction sector, embracing the structure of the construction industries; the strategies, behaviour, and performance of construction firms; and the complex and specific economic institutions of the construction process, specifically contracting and procurement systems. Current research within this theme focuses on the Private Finance Initiative (build-own-operate-transfer projects). Much of his research has been undertaken with partners from the UK construction sector, coordinated through the United Kingdom's Construction Industry Council, to whom Graham is an economic advisor. He is author/coauthor of two studies of PFI for the Construction Industry Council (*The Constructors' Key Guide to PFI* and *The Role of Cost-Saving and Innovation in PFI Projects*), published by Thomas Telford, and two books, *The Economics of the Modern Construction Sector* and *The Economics of the Modern Construction Firm*, published by Palgrave Macmilllan.

CHAPTER FIVE

PROJECT SUCCESS

Terry Cooke-Davies

Few topics are more central to the art and science of managing projects than project success. It would seem to be self-evident that every person involved in the management of a project will be striving to make it successful. In the world of the twenty-first century, "success," like its close relative "winning," seems to be an unquestioned "good." So surely there can be nothing too difficult about measuring project success?

Unfortunately, behind this rather obvious-sounding question, there lies a seething mass of complex assumptions and interrelated concepts that have led one author almost despairingly to ask, "Measuring success—can it really be done, and if carried out, what purpose does it serve?" (De Wit, 1988). Difficulties abound for many reasons: the different viewpoints, interests, and expectations of groups of stakeholders involved in any given project; the subjective nature of perceptions of "success"; the tendency of perceptions to evolve over extended periods of time; the difficulty of assessing complex phenomena using simple metrics—the list is a lengthy one. On closer examination, project "success" turns out to be a rather slippery subject.

And yet the need remains. Every project is undertaken to accomplish some purpose, and it is both natural and right to seek to assess the extent to which that purpose has been achieved. Equally, if the art and science of project management is to advance, then practices that lead to success are to be encouraged over those that lead to failure. Indeed, these two aspects of the need to understand project success each lead to a different aspect of the topic that will be covered more fully in this chapter. Success criteria are the measures against which the success or failure of a project will be judged, and success factors are those inputs to the management system that lead directly to the success of the project. Each is important, but the two should not be confused.

A Brief Survey of the Literature on Project Success

Many of the practitioner-focused textbooks on project management define project success criteria in terms of the time, cost, and product performance (expressed as quality, or scope, or conformance to requirements) compared to the plan. Indeed, this is so widely accepted that one popular book aimed at practitioners is subtitled "How to plan, manage, and deliver projects on time and within budget" (Wysocki, Beck, and Crane, 1995). As a headline, it commands attention, although in the body of their book, the same authors acknowledge the need to define success criteria more completely during the early stages of project definition. The difference of emphasis, however, serves to highlight a distinction that is well expressed by De Wit (1988), who differentiates between the success of project management (for which measures of time, cost, and quality might be broadly appropriate) and the success of the project, which will depend on a wider range of measures. This distinction is important, although often ignored.

The importance of the distinction is emphasized by Munns and Bjeirmi (1996), who draw attention to the short-term goals of the project manager (in delivering the required product or service to schedule and within budget) as opposed to the long-term goals of the project (to deliver the promised business benefits). Kerzner makes a similar distinction between "successful projects" and "successfully managed projects." "Successful implementation of project management does not guarantee that individual projects will be successful . . . Companies excellent in project management still have their share of project failures. Should a company find that 100 percent of their projects are successful, then that company is simply not taking enough business risks" (Kerzner, 1998; p. 37).

De Wit, as it happens, is following Baker, Murphy, and Fisher's classic analysis of 650 completed aerospace, construction, and other projects (1974), which was subsequently developed further by the same authors (1988). They concluded (p. 903) that "if the project meets the technical performance specifications and/or mission to be performed, and if there is a high level of satisfaction concerning the project outcome among key people in the parent organization, key people in the client organization, key people on the project team and key users or clientele of the project effort, the project is considered an overall success." A definition that includes elements of both project management success (technical performance specifications; satisfaction of key people on the project team) and project success (meets mission to be performed; satisfaction in parent and client organization).

This tendency to blur the distinction is also followed in work subsequent to Baker, Murphy, and Fisher by authors writing both before and after De Wit's article. Morris and Hough (1987), for example, in their seminal work on major projects, make a convincing case for the popular perception that an excessively large number of "major projects" are perceived by the public to fail, and then argue on the basis of both a comprehensive survey of the literature and also eight meticulously conducted case studies for three or possibly four dimensions to project success criteria: project functionality, project management, contractors' commercial performance, and possibly, in the event that a project was canceled, whether the cancellation was made on a reasonable basis and the termination handled efficiently. Project functionality, as defined by Morris and Hough, includes an assessment of both project technical performance, which forms a part of "project management success," and other aspects of performance, which presages the much more recent language of benefits management.

More recently, a survey of 127 Israeli project managers (Shenhar, Levy, and Dvir, 1997) concluded that there are four dimensions to project success: project efficiency (broadly De Wit's "project management success"), impact on customer, business success, and preparing for the future. The latter three fall within De Wit's category of "project success," as well as being remarkably similar to Baker, Murphy, and Fisher's conclusions.

A backdrop to the discussion on success criteria is provided by an understanding of the different parties to the project that have a legitimate interest in its success or failure. Baker, Murphy, and Fisher (1988; pp. 903*ff.*) emphasize the importance of perceptions and name the "client" and the "parent" in addition to the project team. Morris and Hough (1987, pp. 194*ff.*) refer to "sponsors," contractors, owners, regulators, financiers, and governments, as well as citizens and environmentalists. DeWit (1988, pp. 167–168) reviews the breadth of possible project "stakeholders," as does Geddes (1990). Authors generally acknowledge that each stakeholder group can have different criteria for the success of the project, thereby introducing greater complexity to the subject.

The literature on project success factors is more extensive than that on success criteria (Crawford, 2001), although much of it is based on anecdotal evidence or studies with very small sample size: The state of current understanding can perhaps best be illustrated by considering three representative studies: Baker, Murphy, and Fisher's (1988) considered findings from their analysis of 650 aeronautical, construction and other projects; Pinto and Slevin's studies (1988b; 1988a) of answers provided by 418 project managers from various industries, and Lechler's survey (1998) of 448 projects in Germany. These three have been chosen as representative because of their large samples of empirical data, because they include projects from different industries, because they use complementary data analysis methods, and because they cover the past three decades, during which time 99 percent of all the articles published about project management have been written. (Kloppenborg and Opfer, 2000).

Baker, Murphy, and Fisher

Baker, Murphy, and Fisher adopted the definition of success that has already been cited. It includes a number of factors and the perceptions of success of different groups of stakeholders. Their conclusion is that there are 29 factors that strongly affect the perceived failure of projects; 24 factors that are necessary, but not sufficient, for perceived success; and 10 factors that are strongly linearly related to both perceived success and perceived failure (i.e., their presence tends to improve perceived success, while their absence contributes to perceived failure).

The output measure (whether the project was successful or not) was a simple categorization of projects into three success "bands," based on a multiple of the factors contributing to their definition of success, which has already been discussed.

The ten factors are as follows:

1. Goal commitment of project team
2. Accurate initial cost estimates
3. Adequate project team capability
4. Adequate funding to completion
5. Adequate planning and control techniques
6. Minimal start-up difficulties
7. Task (vs. social) orientation
8. Absence of bureaucracy
9. On-site project manager
10. Clearly established success criteria

Pinto and Slevin

Pinto and Slevin derived from Baker, Murphy, and Fisher an understanding of the factors that influence project success, and then developed from it a more explicit definition of the criteria for judging project success (see Figure 5.1).

They then assessed the opinions of 418 PMI members who responded to a questionnaire about which factors were critical to which elements of project success (just over half of them

FIGURE 5.1. PINTO AND SLEVIN'S MODEL OF PROJECT SUCCESS CRITERIA.

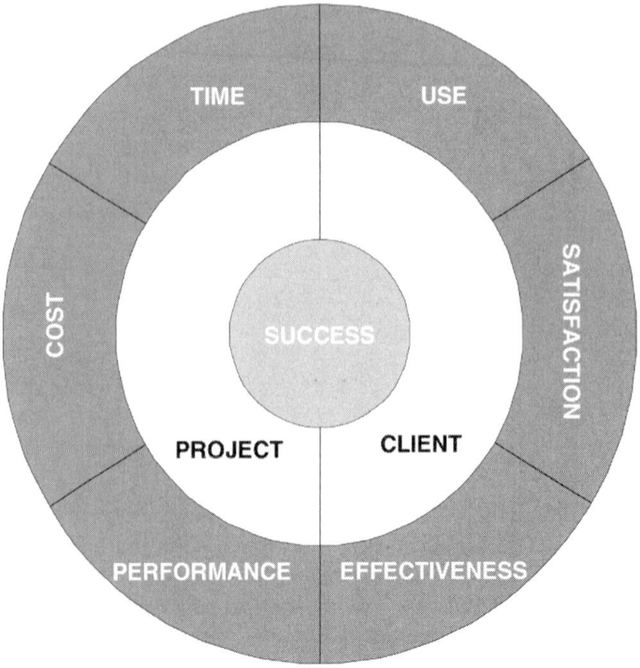

Source: Pinto and Slevin (1988a, p. 69). Project Management Institute, *Project Management Journal.* Project Management Institute, Inc., 1988. Copyright and all rights reserved. Material from this publication has been reproduced with the permission of PMI.

were project managers and nearly a third were members of project teams). They also related the results to the particular phase of the project's life cycle within which each of the factors were significant, using a simple four-phase model: conceptualization, planning, execution, and termination. Participants were instructed to "think of a project in which they were involved that was currently under way or recently completed. This project was to be their frame of reference while completing the questionnaire. The four-phase project life cycle . . . was included in the questionnaire, and was used to identify the current phase of each project" (Pinto and Slevin, 1988a; p. 70).

The results identified ten "critical success factors," which were then developed into an instrument to allow project managers to identify how successful they were being in managing their project. The ten factors are as follows:

1. *Project mission.* Initial clarity of goals and general direction.
2. *Top management support.* Willingness of top management to provide the necessary resources and authority/power for project success.
3. *Project schedule/plans.* A detailed specification of the individual action steps required for project implementation.
4. *Client consultation.* Communication, consultation, and active listening to all impacted parties.
5. *Personnel.* Recruitment, selection, and training of the necessary personnel for the project team.
6. *Technical tasks.* Availability of the required technology and expertise to accomplish the specific technical action steps.
7. *Client acceptance.* The act of "selling" the final product to its ultimate intended users.
8. *Monitoring and feedback.* Timely provision of comprehensive control information at each phase in the implementation process.
9. *Communication.* The provision of an appropriate network and necessary data to all key factors (*sic*) in the project implementation.
10. *Troubleshooting.* Ability to handle unexpected crises and deviations from plan.

Lechler

Lechler, in the most recent of the three empirical studies, also started from an analysis of the literature. His starting point was that "cause and effect" is rarely taken into consideration, but rather that the "critical success factors" are analyzed as separate, independent variables. He reviewed 44 studies, covering a total of more than 5,700 projects, and from them deduced that 11 discrete key success factors could be identified. Out of these, he chose the eight that were most frequently cited for his own empirical analysis.

Working from Pinto and Slevin's questionnaire, Lechler isolated 50 questions that corresponded to his chosen eight critical success factors, and distributed them to members of the German Project Management Society (Gesellschaft für Projektmanagement—GPM). Each respondent was sent two questionnaires and asked to complete one for a project that they considered to be successful and one for a project that they considered to be unsuccessful. They were invited to assess the project as successful if "all people involved" regarded the process (social success), the quality of the solution (effectiveness), and the adherence to time and cost objectives (efficiency) as overall positive. A total of 448 questionnaires were received and analyzed; 257 of them relating to "successful" projects and 191 to "unsuccessful" ones.

The first step in Lechler's analysis was to seek correlations between individual technical factors included in the questionnaire and overall project success. Only four factors were found to have significant correlations:

- The appropriate technology (equipment, training programs, etc.) has been selected for the project.
- Communication channels were defined before the start of the project.
- All proceeding methods and tools were used to support the project well.
- The project leader had the necessary authority (a composite of four different questions).

The second step in the analysis was to carry out a LISREL analysis (Linear Structural Relationships) for the eight critical success factors. This resulted in the path diagram shown in Figure 5.2.

Weightings were calculated for the various different paths of "causality," which, cumulatively, gave a value for r^2 of 0.59: 0.47 from the "people" factors (top management, project leader, and project team) and only an incremental 0.12 from the "activities" (participation, planning and control, and information and communication) and "barriers" (conflicts and changes of goals). Lechler indicates the importance that he attaches to this

FIGURE 5.2. LECHLER'S CAUSAL ANALYSIS.

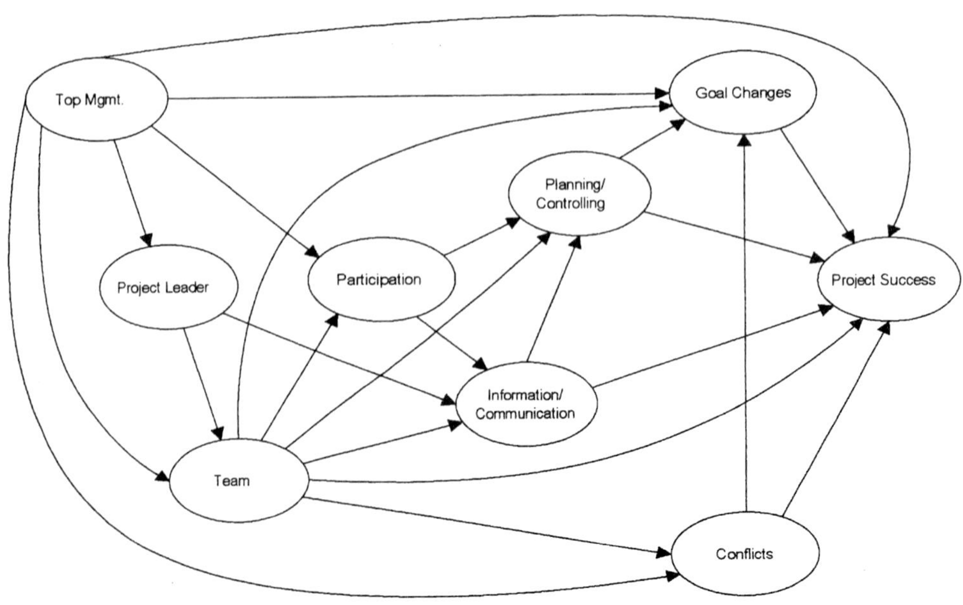

conclusion through his choice of title for the paper: "When it comes to project management, it's the people that matter."

The weighting given to each of the eight factors is shown in Table 5.1.

The Three Studies in Summary

The three studies show a certain commonality. Each of them emphasizes the importance of clear and doable project goals, of careful and accurate project planning, of adequate resources provided through top management support, and of what would today be referred to as stakeholder management. Perhaps this is not too surprising, since each of the latter two builds on the earlier work. What it has meant, however, is that these factors have become "accepted wisdom" within the world of project management practice. There are, moreover, some serious questions to be asked about how generally valid the results are for all types of projects under all circumstances. After all, each study ultimately employed a single "composite" criterion for success and based conclusions about the extent to which it was achieved on the answers provided by the same respondents who identified the role of different factors in contributing to that success. Further discussion is clearly called for.

Distinguishing Three "Levels" of Success

Regardless of what criteria are used to assess project success, and even with the broad agreement within the literature on the kinds of factors that are essential prerequisites to success, the fact must be faced that a disproportionately large number of projects are unsuccessful (Morris and Hough, 1987; O'Connor and Reinsborough, 1992; KPMG, 1997; Cooke-Davies, 2001). This suggests that there is something missing from the debate on project success, and continuous action research with more than 70 multinational or large national organizations in the United States, Europe, and the Asia-Pacific suggests that even the distinction between project success and project management success may be insufficient (Cooke-Davies, 2002a).

TABLE 5.1. WEIGHTINGS OF EIGHT FACTORS.

Factor	Direct	Indirect	Total
Top management	0.19	0.41	0.60
Project leader	—	0.18	0.18
Project team	0.16	0.36	0.52
Participation	—	0.10	0.10
Planning/controlling	0.16	0.01	0.17
Information/communication	0.12	0.06	0.18
Conflicts	−0.21	−0.08	−0.29
Goal changes	−0.20	—	−0.20

It has been argued elsewhere that the question of "which factors are critical to project success depends on answering three separate questions: 'What factors lead to project management success?' 'What factors lead to a successful project?' and 'What factors lead to consistently successful projects?'" (Cooke-Davies, 2002a, p185). The same article describes the relationship between business success and project success.

So what can be gained by regarding these three questions as pertaining to three different "levels"? Are there essential characteristics that can be used to distinguish each level from the other two? Or is this simply another conceptual framework to further bedevil a field of practice that already could be said to suffer from a surfeit of conceptual models along with a paucity of empirical data? The answer to these questions will emerge as each level is considered in turn, first, in the next section, with regard to success criteria and then subsequently with regard to success factors.

Three Levels of Success Criteria

1. Project Management Success—Was the Project Done Right? This is the measure of success that has dominated the practitioner-oriented literature on project management. In the folklore of the project manager, it is about managing time, cost, and quality. In reality, project objectives are rarely this simple. There will often be a business case to be borne in mind or a gross profit to be made; there may be health, safety, and environmental objectives to be accomplished; if the project is a technical one, or a "platform" new product development, there could be scientific or technical goals to reach. Nevertheless, the principle is simple: the endeavor is to deliver the project so that it meets the objectives within the constraints. If anything changes, which is likely given the inherent uncertainty that is involved in any new endeavor, techniques such as project risk management and project change control can be called into play as appropriate. As a guided missile seeks its target, adjusting its trajectory as appropriate along the way, so the project team seeks to achieve the project objectives. Is this then an appropriate level at which to measure the success of a project? There are three different kinds of arguments that suggest that it is.

First, modern project management has developed from a base of managing relatively "discrete" projects, each with its own organization and each established to accomplish specific purposes (Morris, 1994). The kind of success criteria that are broadly used as measures of "project management success" have not only been those most commonly applied in the history of project management (e.g., *A Guide to the Project Management Body of Knowledge*, PMI, 1996), but they also allow the project team as a coherent organizational unit to be accountable for its own performance, and the practice of aligning accountability with authority is one of the well-attested principles of good management practice.

Second, the underlying concept behind measuring success at this level is based on the well-understood principles of first-order cybernetics (Schwaninger, 1997) in much the same way that a thermostat or a guided missile operate. This is clearly appropriate for projects in which both the goals and the methods of achieving them are relatively clear at the outset (Turner and Cochrane, 1993). Third, the capture of data about the extent to which projects within the same enterprise are successful in terms of project management success enables the enterprise to compare and contrast the practices that are generally associated with successful projects with those associated with unsuccessful ones. This in turn provides the enterprise with valuable information about which project management practices are in need of improvement within project teams.

These are convincing arguments that support the case for continuing to measure project management success for many projects in many organizations. It is far from being the whole story, however, and for the second level of success, it is necessary to turn to the second of De Wit's levels—what he calls "project success."

2. Project Success—Was the Right Project Done? This level of project success is perhaps the one that is of most interest to the owner or sponsor of the project. It is, in a sense, a measure of "value for money" in its broadest sense. The assumption is that the project will be successful only if it successfully delivers the benefits that were envisaged by the people and organizations (i.e., the stakeholders) that agreed to undertake the project in the first place. In an attempt to isolate those core elements that are central to the way a project manager thinks about his or her work, a detailed analysis of the topics contained in six recent "bodies of knowledge" (Cooke-Davies, 2001, pp. 51 to 90 and Appendices P1 and P2) has shown that they can be clustered into 11 topic areas and related to each other in narrative fashion through a "systemigram" (see Figure 5.5). Viewed in this way, it becomes clear that "anticipated benefits" become the touchstone not only for formal "stage gate" reviews of projects but also for the continuous "informal assessment" of the likely success

FIGURE 5.3. THE INVOLVEMENT OF BOTH PROJECT MANAGEMENT AND OPERATIONS MANAGEMENT IN THE ACHIEVEMENT OF "PROJECT SUCCESS".

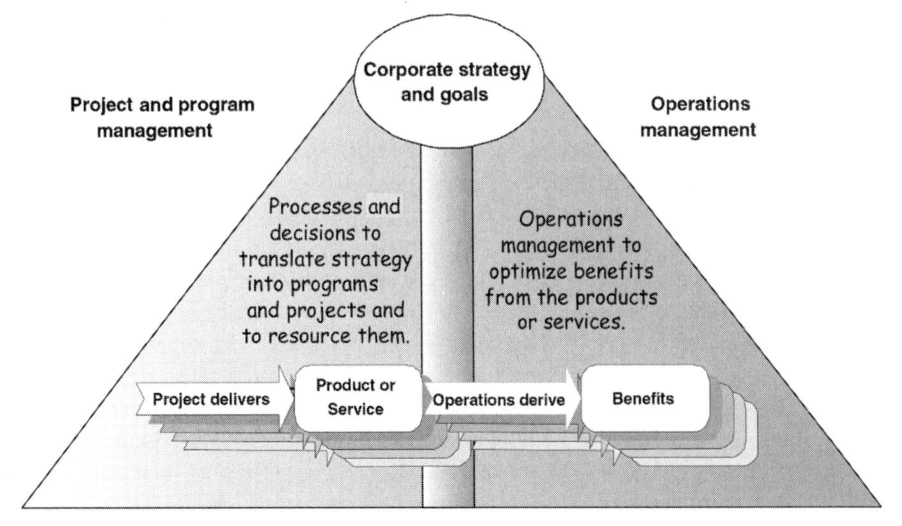

of projects carried out by owners, sponsors, or senior management and for influencing decisions about priorities and resource allocation.

Comparison of the 11 topic areas with previously published research about project success reveals a silence about "benefits" (Cooke-Davies, 2001; p. 90, Figure 7) perhaps because little has been written about benefits management or benefits realization until recently, and perhaps because the subject of benefits has been subsumed in the general discussion about project purpose or project goals. Nevertheless, there are three reasons why this is an appropriate level at which to measure the success of a project separately from the first level that was discussed, project management success.

First, as Figure 5.3 shows, benefits are not delivered or realized by the project manager and project team; they require the actions of operations management. This calls for a close cooperation between the project team on the one hand and the sponsor, customer, and/or user(s) on the other. Thus, the discussion of project success involves dialogue with a wider cross section of the organization than is appropriate or necessary for project management success. Second, delivering project success is necessarily more difficult than delivering project management success, because it inevitably involves "second-order control" (both goals and methods liable to change) and thus brings into play an additional set of corporate processes to those that are involved in delivering project management success. And third, the extent of project success is unlikely to become clear during the life of the project itself, whether success is measured quantitatively in terms of financial benefits or qualitatively in some less tangible form. For these three reasons, project success is itself a viable level at which to establish success criteria.

It is not being suggested here that project success is somehow a "better" level at which to establish success criteria. Both project success and project management success are important to any project. If a project achieves project success without project management success, there is the inevitable conclusion that even greater benefits could have been realized. On the other hand, if project management success is achieved without project success, then the owner or sponsor has failed to obtain the benefits that the project was designed to provide. And that brings us to the third level of success.

3. Consistent Project Success—Were the Right Projects Done Right, Time after Time? As the focus moves from project management success, through project success to consistent project success, a completely new set of criteria come into play, as adjudged by different groups of stakeholders. Projects are the means by which all organizations accomplish business change, as well as the means by which some organizations deliver profits to their shareholders. The consistency with which projects accomplish both project success and project management success is thus a matter of interest to every organization that is competing in markets for scarce resources, such as customers or finance.

At this level, a discussion of the criteria by which consistent project success is achieved is one that embraces the whole organization, and that will inevitably be influenced by its chosen strategy. For operations-driven organizations (such as financial services companies or mass manufacturers), consistent project success in such areas as effective overall IT expenditure and new product development can lead to competitive advantage. For project-based organizations (such as engineering contractors, defense suppliers, or turnkey IT systems providers), consistent project success can lead to profitable expansion. In either case, as the proportion of total work that is carried out in the form of projects increases, so consistent project success assumes an increasing strategic significance.

In recent years there has been a growing interest in project portfolio management for new product development (e.g., Cooper, Edgett, and Kleinschmidt, 2001), specifically for R&D (e.g., Matheson and Matheson, 1998) or generally for project spend in organizations (e.g., Artto, Martinsuo, and Aalto, 2001). But many organizations, particularly in traditional project-based industries, do not adopt a portfolio approach. For such organizations, as for all others, the effective and efficient use of scarce resources (particularly, but not only, people and money) remain of paramount importance. Thomas and Jugdev (2002) in their award-winning article on project management maturity models emphasize that long-range competitive advantage is enjoyed by those organizations that make the best use of their strategic assets (i.e., resources). Further, they conclude that maturity models are not in and of themselves sufficient to enable organizations to capitalize on their intangible assets, such as strength in project management, they do, however, go some way toward establishing the value of "project management maturity" as a further criterion of success at this third level.

A Word about Project Metrics

One practical implication of this discussion of three different levels of success criteria is that an organization would do well to monitor its performance using a "suite" of project metrics that incorporates all three levels of success, if it is serious about understanding and improving its success in the field of projects. As Figure 5.4 shows, each of the different levels of success is of interest to different levels of the corporate hierarchy, and each is visible after different amounts of time have elapsed relative to the project duration. No significant studies have as yet been published about the nature and extent of project metrics, although Atkinson (1999) and Lim and Mohamed (1999) each argue that the need for multilayered project success criteria is intimately linked to the need for more comprehensive metrics. Unpublished research (Egberding and Cooke-Davies, 2002), however, indicates that very few organizations are happy with the metrics that they use.

After considering which factors influence success at which level, a framework for a hierarchical suite of metrics will be suggested at the conclusion of this chapter.

Factors Contributing to Success at Each of the Three Levels

Although much has been written about project success factors, the distinction between different "levels of success" is a recent addition to the conceptual language of the project management research community. The predominant tenor of the discussions is to construct (or, at worst, to imply) some overall measure of "success" and then to establish by primary research, by secondary research, or by personal observation those factors that seem to correlate to success or to failure. The three chosen examples from the literature that were

FIGURE 5.4. MEASUREMENT OF PROJECT SUCCESS—ORGANIZATIONAL COMMITMENT AND TIME ELAPSED

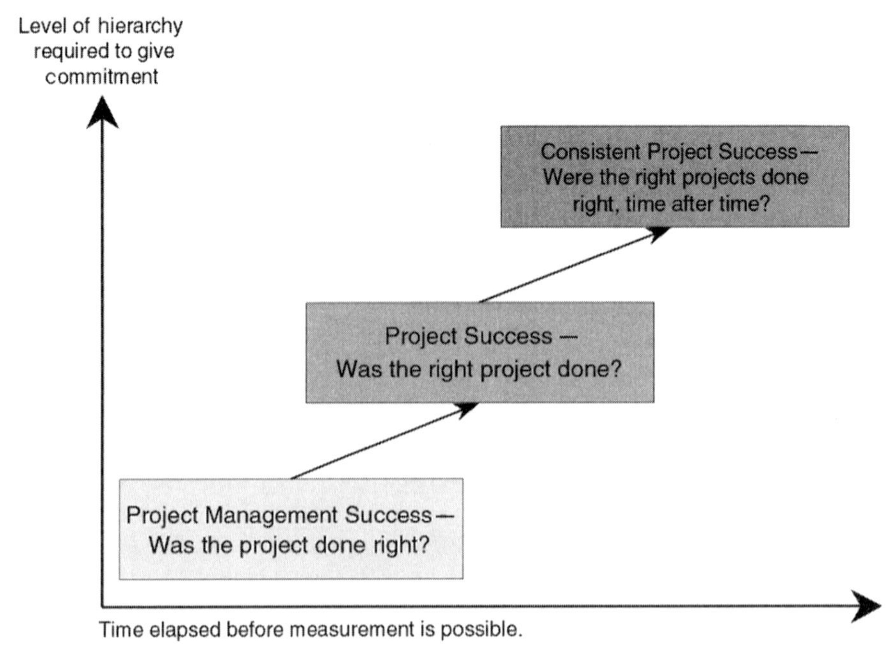

reviewed earlier in this chapter illustrate this point (Baker, Murphy, and Fisher, 1988; Pinto and Slevin, 1988a; Lechler, 1998). This is not the only criticism that can be leveled at the whole body of research into project success. Much of it uses survey techniques to collect answers from respondents both about the success or failure of individual projects and about the factors that contributed to that success or failure. It is thus better presented as research into the opinions of the project management community about success factors than as absolute success factors themselves. That is not to say that it is not useful—it is—but it is less useful than could be wished for.

Lest the pendulum be pulled too far in the opposite direction at this point, it is worth reflecting on the danger of what accountants call "spurious accuracy" in quantitative research into project success. Any assessment of project success will be carried out by specific stakeholders at specific times, and this will inevitably be influenced by many factors that are not directly related to the project itself. Business transformation or new product development projects, for example, may well be at the mercy of unforeseen and even unforeseeable developments that an assessor may or may not to take into account when judging success. And the longer the delay between project initiation and the point of assessment, the more difficult it becomes. It can be very difficult to distinguish between "luck" and "success" for any single project!

This is not a counsel for despair of ever producing any useful quantitative data—other "soft" science disciplines such as economics suffer from the same difficulties. But it does suggest the need to discern patterns or laws within large quantities of data, and thus as a prerequisite to create semantic frameworks that allow data to be compared on as near as possible a like-for-like basis.

Morris and Hough (1987), Belassi and Tukel (1996), and Crawford (2001, Appendices C and D) include excellent tabular listings of published research that between them account for 44 different research-based studies. Each of these three tables shows the breadth of conclusions that different researchers have reached concerning which factors are truly "critical" to success, although Crawford (who includes the Morris and Hough work in her own table, as well as all three studies described earlier) categorizes them into 24 groups of similar factors. Nevertheless, 24 is a very large number of "critical" factors, and if so many things are all equally important, it is also fair to conclude that nothing is especially important. What can the perplexed practitioner conclude from all this?

The first legitimate conclusion is that this is a genuinely difficult field of study that is bedeviled by at least three dimensions of difficulty. The first of these is the absence of generally accepted definitions for all the terms used to describe the subject, and it has already been noted what a slippery topic it is. Variations in language occur in different places: between researchers both as they frame the research questions and as they describe the results; between project managers and teams as they provide the data for analysis; between stakeholder groups with differing interests in the same project, and even between any given stakeholder group as its perceptions change over time; between organizations in their own internal project management guidance literature; and between industries and markets that each have their own distinct vocabulary (try talking to a research chemist in pharmaceutical R&D about "project scope management").

But if that were not enough, it is still only a part of the story. A second dimension of difficulty is the multifactorial variability of projects themselves, which makes comparisons between any two projects fraught with uncertainty. Projects are undertaken by unique temporary organizations, using unique combinations of resources (human and other) to undertake a unique, novel, and temporary endeavor that is faced with unique inherent uncertainty in order to deliver unique beneficial objects of change (Turner and Müller, 2002). On top of this, it may be the case that projects, like the weather or stock markets, are subject to "sensitive dependence on initial conditions" (Richardson, Lissack, and Roos, 2000). The third dimension of difficulty is the problem of developing robust research methods that need to encompass three worlds as varied as the physical, the social, and the personal, each of which plays an important part in the management of projects. Taken together, these three dimensions present a call to action to the project management research community that it needs to raise its game if it is to offer practical assistance to project management practitioners and the organizations that employ them.

The second conclusion is perhaps more helpful in terms of improving project management practice—there is no silver bullet with which instant success can be achieved. As in the majority of challenging human endeavors, achieving project success comes through a combination of factors, and an organization can be sure that it understands them only when it begins to see improvements in its level of consistent project success.

The third conclusion is more helpful still. Although success can be achieved only through a combination of factors, there is a relatively high degree of agreement on the kind of factors that are critical to project success. It is these that the remainder of this section will address.

But first, there is a need to map the existing research onto the three levels of success that have been defined in this article. This can be done with the aid of the project manager's worldview analysis (Cooke-Davies, 2001) that was mentioned earlier, in the discussion of the criteria for project success. Figure 5.5 shows the relationship of the 11 groups of topics expressed as a systemigram.

Of these, an examination of the detailed concepts incorporated into each allows seven of the groups to be associated with criteria for project management success as follows (numbered to correspond to those used in Figure 5.5):

2. *Project goals.* Establishing, specifying, and achieving the projects goals.
3. *Product or service.* Defining, specifying, ensuring, manufacturing, and delivering the product or service.
4. *Project work.* Identifying, structuring, planning, executing, and controlling the work to be carried out.

FIGURE 5.5. ELEMENTS OF A PROJECT MANAGER'S WORLDVIEW.

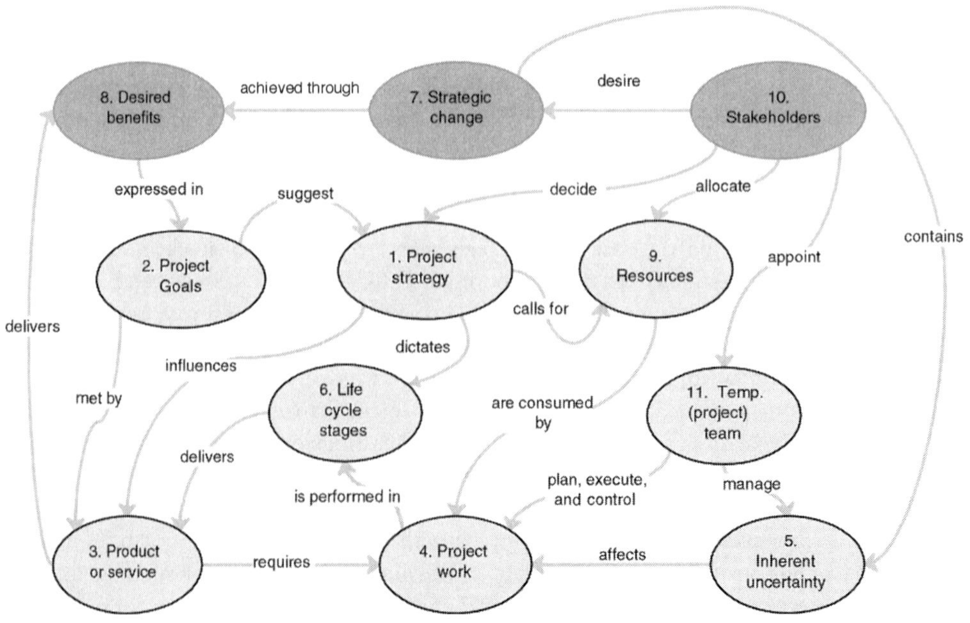

5. *Inherent uncertainty.* Managing the uncertainty that is inherent in the uniqueness of the project.
6. *Life cycle stages.* Practices relating to managing the stages that the project will need to pass through.
9. *Resources.* Allocating organizational resources to the project.
11. *Temporary team.* Creating, leading, and managing the temporary team that will initiate, plan, control, execute, and close the project

The remaining four (along with project goals, which acts as the hinge that links the two levels) can best be associated with project success as follows:

1. *Project strategy.* Establishing a strategic framework for the project.
2. *Project goals.* Establishing, specifying and achieving the projects goals.
7. *Strategic change.* Practices relating the project to the elements of business strategy to which it contributes.
8. *Benefits.* Defining, quantifying, and harvesting organizational benefits as a result of carrying out the project.
10. *Stakeholders.* Identifying and aligning the interests of the project stakeholders.

The detailed analysis underlying the worldview identified no groups of topics that could be associated directly with consistent project success, although individual topics that are contained at a lower level such as quality, culture, and organizational learning clearly contribute to this third level.

Project Management Success—How to Ensure That the Project Is Done Right?

The criteria for project management success, as has been seen, may include cost, time, quality, scope, commercial performance, technical achievements, or safety record. Although these can all be said to be indicators of project management success, the achievement of each of them is likely to depend on different factors, as one piece of recent research has indicated (Cooke-Davies, 2002a). In other words, if cost matters much more than time, then different factors are likely to be critical to the project team. Having said that, taking all the published research into account, six groups of factors can be identified as contributing to success at this level:

1. Achieve and maintain clarity about the goals of the project. Define the project in a way that clarifies both the goals of the project and the needs of stakeholders. Minimize changes to the goals once the project has started.
2. Select and assemble a capable project team of task-oriented individuals, led by a competent leader. Ensure that the team contains the right capabilities, is appropriately structured, communicates well, and has good processes for teamwork, problem solving, and decision making.

3. Ensure that the project is resourced adequately to the project scope and objectives. Mobilize top management support and ensure that there is adequate support from the organization and effective project administration.
4. Establish clarity at the outset about the technical performance required from the product and manage the scope of work tightly, using a mature change management process.
5. Plan meticulously, using well-established estimating procedures, and to a sufficient level of detail to allow effective monitoring and control. Maintain excellent metrics that relate the technical content of work done to the elapsed time and expenditure incurred.
6. Employ established risk management practices that are well understood by all project participants, including effective risk response development and control.

A summary mapping the origin of these factors onto the seven relevant topic groups from the project managers' worldview is shown in Table 5.2.

Project Success—How to Ensure That the Right Project Is Done?

Before the "success" of any individual project can be measured, the benefits that it is intended to deliver must be considered, and these can vary considerably as the following partial list of project types indicates:

- Successful business process reengineering projects (which have a notoriously low rate of achievement of their objectives) can lead directly to improved competitiveness.
- If the organization is essentially project-based (as is the case in many of the traditional project management environments such as engineering, defense, petrochemical exploration, construction, or IT/IS systems integration), then successful project performance translates directly into an improved bottom line.
- If the organization is operations-based, then successful projects to support or to improve operations (such as marketing projects, plant shutdowns, or production engineering projects) lead indirectly to improved bottom-line performance.
- Successful research projects and (in the case of some industries such as pharmaceuticals) development projects lead to a maximized return on R&D spend, leading directly to the creation of new streams of operating revenue.
- Successful development projects improve time-to-market and can enhance competitive position, product sales, or product margins.
- Successful IT/IS projects deliver improved financial benefits (either directly or indirectly) and/or reduced wastage from aborted projects (Standish, 1995).
- Successful projects to design, procure, and construct new capital assets can enhance time-to-market, return on investment, reduced operating costs, or some combination of all three.

In spite of these complexities, recent work (e.g., Cooke-Davies 2002a) on benefits- and stakeholder-management supports the main body of literature in suggesting that there are

TABLE 5.2. CRITICAL FACTORS FOR PROJECT MANAGEMENT SUCCESS.

Worldview	Level	Baker et al.	Pinto and Slevin	Lechler	Crawford	Cooke-Davies
1. Project goals	1	Goal commitment Good cost estimates Clear success criteria	Project mission	Goal changes	Project definition	
2. Product or service	1				Technical performance	Project scope management
3. Project work	1	Planning and control	Schedule/plans Monitoring and feedback	Planning/controlling	Planning Monitoring and control	Performance management
4. Inherent uncertainty	1		Trouble-shooting		Monitoring and controlling (risk)	Project risk management
5. Life cycle stages	1	Few start-up problems				
6. Resources	1	Adequate funding	Top mgmt support Technical tasks	Top management	Organizational support Administration	
7. Temporary team	1	Adequate capability Task orientation On-site project manager	Personnel	Project leader Project team Participation	Team selection Communication Leadership Team development Organization structure Task orientation Decision making and problem solving	

in fact fewer factors that are critical. There are four of them, including the one that is also critical to project management success:

1. Achieve and maintain clarity about the goals of the project. Define the project in a way that clarifies both the goals of the project and the needs of stakeholders. Minimize changes to the goals once the project has started.
2. Establish and maintain active commitment to the success of the project and its mission on the part of all significant stakeholder groups, such as sponsors, clients, owners, operations management, parent company, and so on. Establish effective communication and conflict resolution methods.
3. Develop and sustain effective processes during the project and after completion to deliver the anticipated benefits of the project and ensure that they are realized. Ensure that a close link is developed and maintained between anticipated benefits, the business case for the project, and the explicit project goals.
4. Develop a project strategy, or "trajectory" in the words of Miller and Hobbs (2000), that is appropriate to the unique environment and circumstances of the project. (Trajectory is a term that encompasses both strategy and life cycle model, and is derived from a detailed study of 60 megaprojects.)

A summary mapping the origin of these factors onto the seven relevant topic groups from the project managers' worldview is shown in Table 5.3.

Consistent Project Success—How to Ensure That the Right Projects Are Done Right, Time after Time?

This third level of success has received little attention in the literature to date. The factors that are identified in the list that follows are thus necessarily more speculative than those for either of the other levels. These three have been identified from a variety of elements of my own continuous action research described elsewhere (Cooke-Davies, 2001, 2002b; Egberding and Cooke-Davies, 2002).

1. An effective means of learning from experience on projects, that combines explicit knowledge with tacit knowledge in a way that encourages people to learn and to embed that learning into continuous improvement of project management processes and practices. Indeed, in a number of recent project management maturity models (e.g., Kerzner, 2001; Fahrenkrog et al., 2003), continuous improvement represents the fifth and highest stage of project management maturity in an organization.
2. Portfolio and program management processes that allow the enterprise to resource fully a suite of projects that are thoughtfully and dynamically matched to the corporate strategy and business objectives. These processes include the dynamic allocation of scarce resources to competing projects, in a way that serves the enterprise as a whole.
3. A suite of project, program, and portfolio metrics that provides direct line-of-sight feedback on current project performance, and anticipated future success, so that project,

TABLE 5.3. CRITICAL FACTORS FOR PROJECT MANAGEMENT SUCCESS.

Worldview	Level	Baker et al.	Pinto and Slevin	Lechler	Crawford	Others
1. Project strategy	2					Project trajectory*
2. Project goals	1	Goal commitment; project mission Good cost estimates Clear success criteria		Goal changes	Project definition	
3. Strategic change	2		Project mission		Strategic direction	
4. Benefits	2					Benefits realized† Business Case‡ Benefits delivery & management§
5. Stakeholders	2	No bureaucracy	Client consultation Client acceptance Communication	Conflicts Information / communication	Stakeholder management	

*Miller and Hobbs, 2000
†KPMG; Thorp, 1998
‡Beale, 1991
§Cooke-Davies, 2002

TABLE 5.4. THE ELEMENT OF PROJECT SUCCESS.

Success "Level" Accountable	Typical Criteria for Success at This Level	Possible Factors Critical for Success at This Level	Organizational Level
Level 1: Project management success. "Was the project done right?"	Time Cost Quality Technical performance Scope Safety	1. Clear project goals 2. Well-selected, capable and effective project team 3. Adequate resourcing 4. Clarity about technical performance requirement 5. Effective planning and control 6. Good risk management	Project manager Project team
Level 2: Project success. "Was the right project done?"	Benefits realized Stakeholder satisfaction	1. Clear project goals 2. Stakeholder commitment and attitude 3. Effective benefits management and realization processes 4. Appropriate project strategy	Project sponsor "Client", "owner," or "operator" (recipient of benefits)
Level 3: Consistent project success. "Are the right projects done right, time after time?"	Overall success of all projects undertaken Shareholders (or equivalent) Overall level of project management success Top managers Productivity of key corporate resources. Directors of project management. Effectiveness in implementing business strategy.	1. Continuous improvement of business, project and support processes. 2. Efficient and effective portfolio, programme and resource management processes. 3. Comprehensive and focused suite of metrics covering all three levels. Business unit managers. Portfolio managers.	

portfolio, and corporate decisions can be aligned. Since corporations are increasingly recognizing the need for upstream measures of downstream financial success through the adoption of reporting against such devices as the balanced scorecard (Kaplan and Norton, 1996), it is essential for a similar set of metrics to be developed for project performance in those areas where a proven link exists between project success and corporate success. (See the chapter by Brandon.) For the project management community, it is also important to make the distinction between project success (which cannot be measured until after the project is completed) and project performance (which can be measured during the life of the project). No system of project metrics is complete without both sets of measures (performance and success) and a means of linking them so as to assess the accuracy with which performance predicts success.

Summary

Table 5.4 summarizes the points made in this chapter in tabular form. The table indicates clearly the different organizational levels that are involved in the assessment of project success and shows how each of the three levels is necessary but, on its own, not sufficient for any organization that is serious about achieving project success consistently. The table as a whole represents a framework for thinking and talking about project success—a framework such as is necessary to underpin any attempts to advance the art and science of project management.

References

Abdel-Hamid, T., and S. Madnick. 1991. *Software project dynamics: An integrated approach.* Upper Saddle River, NJ: Prentice Hall.

Artto, K. A., M. Martinsuo, and T. Aalto. 2001. *Project portfolio management: Strategic management through projects.* Helsinki, Finland: Project Management Association Finland.

Baccarini, D. 1999. The logical framework method for defining project success. *Project Management Journal* 30(4):25–32.

Baker, B. N., D. C. Murphy, and D. Fisher. 1974. Determinants of project success. NGR 22-03-028. National Aeronautics and Space Administration.

———. (1988). Factors affecting project success. In *Project Management Handbook.* ed. D. I. Cleland and W. R. King, 2nd ed. 902–919. New York: Wiley.

Belassi, W., and O. I. Tukel. 1996. A new framework for determining critical success/failure factors in projects. *International Journal of Project Management.* 4(3):141–151.

Construction Industry Institute. 1993. Cost-trust relationship. Austin, TX: Construction Industry Institute.

———. 1995. Quantitative effects of project change. Austin, TX: Construction Industry Institute.

Cooke-Davies, T. J. 2000. Discovering the principles of project management. *IRNOP IV*, Sydney: University of Technology, Sydney.

———. 2001. *Towards improved project management practice: Uncovering the evidence for effective practices through empirical research.* Available at www.dissertation.com.

———. 2002a. The "real" success factors on projects. *International Journal of Project Management* 20(3): 185–190.

———. 2002b. Establishing the link between project management practices and project success. *PMI Research Conference.* Newtown Square, PA: Project Management Institute.

Cooper, K. G. 1993. The rework cycle: benchmarks for the project manager. *Project Management Journal* XXIV (1).

Cooper, R. G. 2000. *Product leadership: Creating and launching superior new products.* Cambridge, MA: Perseus Books.

Cooper, R. G., S. J. Edgett, and E. J. Kleinschmidt. 2001. *Portfolio management for new products.* Cambridge, MA: Perseus.

Crawford, L. 2001. Project management competence: The value of standards. Henley-on-Thames: Henley Management College.

———. 2000. Profiling the competent project manager. *Proceedings of PMI Research Conference.* Newtown Square, PA: Project Management Institute.

Crawford, L., and P. Price. 1996. *Project team performance: A continuous improvement methodology.* Paris.

De Wit, A. 1988. Measurement of project success. *International Journal of Project Management* 6(3):164–70.

Duncan, W. R. 1996. *A guide to the Project Management Body of Knowledge.* Newtown Square: Project Management Institute.

Egberding, M., and T. J. Cooke-Davies. 2002. *GTN Metrics Survey: Preliminary report on Findings.* Available at www.humansystems.net.

Fahrenkrog, S., C. M. Baca, L. M. Kruszewski, and P. R. Wesman. 2003. Project Management Institute's Organizational Project Management Maturity Model (OPM3). *PMI Global Congress 2003—Europe.* Newtown Square, PA: Project Management Institute.

Freeman, M., and P. Beale. 1992. Measuring project success. *Project Management Journal* XXIII(1):8–17.

Geddes, M. 1990. Project leadership and the involvement of users in IT projects. *International Journal of Project Management* 8(4):214–216.

Haalien, T. M. 1994. Managing the cultural environment for better results. *Internet '94 12th World Congress*, Oslo.

Hayfield, F. 1979. Basic factors for a successful project. *Proceedings of 6th Internet Congress.* Garmish-Partenkirchen FRG: I.P.M.A. (formerly "Internet").

Jiang, J. J., G. Klein, and J. Balloun. 1996. Ranking of systems implementation success factors. *Project Management Journal* 27(4):50–55.

Jiang, J. J., G. Klein G. and H. Chen. 2001. The relative influence of IS project implementation policies and project leadership on eventual outcomes. *Project Management Journal* 32(3):49–55.

Kaplan, R. S., and D. P. Norton. 1996. *The balanced scorecard: Translating strategy into action.* Cambridge, MA: Harvard Business Press.

Kerzner, H. 2001. *Strategic planning for project management using a project management maturity model.* New York: Wiley.

———. 1998. *In search of excellence in project management. Successful practices in high performance organizations.* New York: Van Nostrand Reinhold.

Kharbanda, O. P., and J. K. Pinto. 1996. *What made Gertie gallop? Lessons from project failures.* New York: Van Nostrand Reinhold.

Kharbanda, O. P., and E. A. Stallworthy. 1983. *How to learn from project disasters.* Aldershot, UK: Gower.

———. 1986. *Successful projects with a moral for management.* Aldershot, UK: Gower.

Kloppenborg, T. J., and W. A. Opfer. 2000. Forty years of project management research: Trends, interpretations and predictions. *Proceedings of PMI Research Conference.* Newtown Square, PA: Project Management Institute.

Kotter, J. P., and J. L. Heskett. 1992. *Corporate culture and performance.* New York: Free Press.

KPMG. 1997. Profit-focused software package implementation. *Profit-Focused Software Package Implementation.* London: KPMG.

Laufer, A., and E. J. Hoffman. 2000. *Project management success stories: Lessons of project leaders.* New York: Wiley.

Lechler, T. 1998. When it comes to project management, it's the people that matter: An empirical analysis of project management in Germany. *IRNOP III. The Nature and Role of Projects in the Next 20 Years: Research Issues and Problems.* Calgary: University of Calgary.

Matheson, D., and J. Matheson. 1998. *The smart organization: Creating value through strategic R&D.* Boston: Harvard Business School Press.

Miller, R., and B. Hobbs. 2000. A framework for managing large complex projects: The results of a study of 60 projects. *Proceedings of PMI Research Conference.* Newtown Square, PA: Project Management Institute.

Morris, P. W. G. 1988. Managing project interfaces: Key points for project success. In *Project Management Handbook,* 2nd ed., ed. D. I. Cleland, and W. R. King, pp. 16–55. New York: Wiley.

———. 1994. *The management of projects.* London: Thomas Telford.

———. 2000. Benchmarking project management bodies of knowledge. *IRNOP IV,* Sydney: University of Technology in Sydney.

Morris, P. W. G., and G. H. Hough. 1987. *The anatomy of major projects. A study of the reality of project management.* Chichester: Wiley.

Munns, A. K. and B. F. Bjeirmi. 1996. The role of project management in achieving project success. *International Journal of Project Management* 14(2).

O'Connor, M. M., and L. Reinsborough. 1992. Quality projects in the 1990s: A review of past projects and future trends. *International Journal of Project Management* 10(2):107–14.

Pettersen, N. 1991. What do we know about the effective project manager? *International Journal of Project Management* 9(2).

Pinto, J. K. 1990. Project implementation profile: A tool to aid project tracking and control. *International Journal of Project Management* 8(3).

Pinto, J. K., and D. P. Slevin. 1988a. Critical success factors across the project life cycle. *Project Management Journal* 19(3):67–75.

———. 1988b. Project success: definitions and measurement techniques. *Project Management Journal* 19(1):67–72.

———. (1998). Critical success factors. 379–395. San Francisco: Jossey-Bass.

Richardson, K. A., M. R. Lissack, and J. Roos 2000. Towards coherent project management. *IRNOP IV,* Sydney: University of Technology in Sydney.

Robins, M. J. 1993. Effective project management in a matrix-management environment. *International Journal of Project Management* 11(1).

Schwaninger, M. 1997. Status and tendencies of management research: A systems oriented perspective. *In Multimethodology: Towards theory and practice and mixing and matching methodologies,* ed. J. Mingers and A. Gill. Chichester, UK: Wiley.

Shenhar, A. J., O. Levy and D. Dvir. 1997. Mapping the dimensions of project success. *Project Management Journal* 28(2):5–13.

Sommerville, J., and V. Langford. 1994. Multivariate influences on the people side of projects: stress and conflict. *International Journal of Project Management* 12(4).

Thamhain, H. J. 1989. Validating technical project plans. *Project Management Journal* 20(4):43–50.

Thomas, J. and Jugdev, K. 2002. Project management maturity models: The silver bullets of competitive advantage? *Project Management Journal* 33(4):4–14.

Thorp, J. 1998. *The information paradox: Realizing the business benefits of information technology.* New York: McGraw-Hill.

Turner, J. R., and R. A. Cochrane. 1993. Goals-and-methods matrix: Coping with projects with ill defined goals and/or methods of achieving them. *International Journal of Project Management* 11(2):93–102.

Turner, J. R., and R. Müller. 2002. On the nature of the project as a temporary organization. *Proceedings of IRNOP V. Fifth International Conference of the International Network of Organizing by Projects.* Rotterdam: Erasmus University.

Wateridge, J. 1998. How can IS/IT projects be measured for success? *International Journal of Project Management* 16(1):59–63.

———. 1995. IT projects: A basis for success. *International Journal of Project Management* 13(3):169–72.

Wysocki R. K., R. Beck, Jr., and D. B. Crane. 1995. *Effective project management.* New York: Wiley.

CHAPTER SIX

MANAGEMENT OF THE PROJECT-ORIENTED COMPANY

Roland Gareis

The general topic of the World Congress of the IPMA (International Project Management Association) in June 1990 in Vienna, Austria, was "Management by Projects." Since then the vision to cope with the complexity and the dynamics of companies by projects has become a reality. Management by projects is today the most appropriate organizational practice in many industries. Research results about the specific strategies, structures, and cultures of the "project-oriented company" (POC) have been published, even the project orientation of regions and nations have been assessed and benchmarked.

This chapter introduces the POC as a social construct and describes models for the organizational differentiation and integration in the POC, such as projects, programs, expert pools, the PM office, and the project portfolio group. Further, a *POC Maturity Model*, based on the specific business processes of POCs, is presented, which can be applied to assess and to benchmark the competences of POCs.

It is not intended to describe the specific business processes of the POC in detail. This is done in the literature on project and program management. The intention is rather to elaborate on the new perceptions of projects and programs as temporary organizations and social systems, and to present an integrative model for the POC.

The POC: A Social Construct

Companies are becoming more project-oriented. Projects and programs are applied in all industries and in the nonprofit sector. To perceive a company as a POC is a social construction. Any company (or parts of a company, such as a division or a profit center) that frequently applies projects and programs to perform relatively unique business processes of large scope can be perceived as being project-oriented. (To simplify the further reading, the term project will be used instead of "project and program." Many of the presented concepts apply to projects as well as programs.)

A POC can be defined as an organization that

- defines "management by projects" as an organizational strategy;
- applies temporary organizations for the performance of business processes of medium and large scope;
- manages a project portfolio of different project types;
- has specific permanent organization units, such as a PM office and a project portfolio group;
- applies a "new management paradigm"; and
- perceives itself as being project-oriented.

Observing the project orientation of a company requires that we put on a special pair of "project orientation" glasses to view the practices of project, program, and project portfolio management and to observe the organizational design and the personnel management practices to support these approaches. These observations are the basis for management interventions needed to optimize the maturity as a POC.

Organizational Fit of the Strategies, the Structures, and the Cultures of the Project-Oriented Company

According to the *organizational fit model*, a company can be described by its strategies, structures, and cultures. These have to fit in order to provide good-quality services and to be cost- and time-efficient.

The specific organizational strategy of the POC is one of "management by projects." Further, it is characterized by permanent and temporary organization structures, and by a culture based on a "new management paradigm." Projects as temporary structures can only be performed successfully if appropriate strategic and cultural provisions exist.

"Management by Projects": The Organizational Strategy of the Project-Oriented Company

Project-oriented companies consider projects not only as tools to perform business processes of medium and large scope but as a strategic option for the organizational design of the company. By applying a management-by-projects approach, the following organizational objectives are pursued:

- Organizational differentiation and decentralization of management responsibility
- Quality assurance by project teamwork and holistic project definitions
- Goal orientation by defining and controlling project objectives

- Personnel development in projects
- Organizational learning by projects

For the implementation of management-by-projects symbolic management measures, showing the importance of projects, are required. Such measures include the following:

- Showing in the organization chart of the company not only the permanent organization structures but also temporary organizations (see Figure 6.1).
- Including project-related functions in job descriptions of all managers and top managers
- Including a statement on the strategic importance of project management in the company mission statement
- Marketing and promoting project management.

Projects and Programs: Temporary Organizations of the POC

For business processes with different characteristics, different organizations are adequate. Line functions such as procurement or production will generally be responsible for routine

FIGURE 6.1. ORGANIZATION CHART OF A PROJECT-ORIENTED COMPANY.

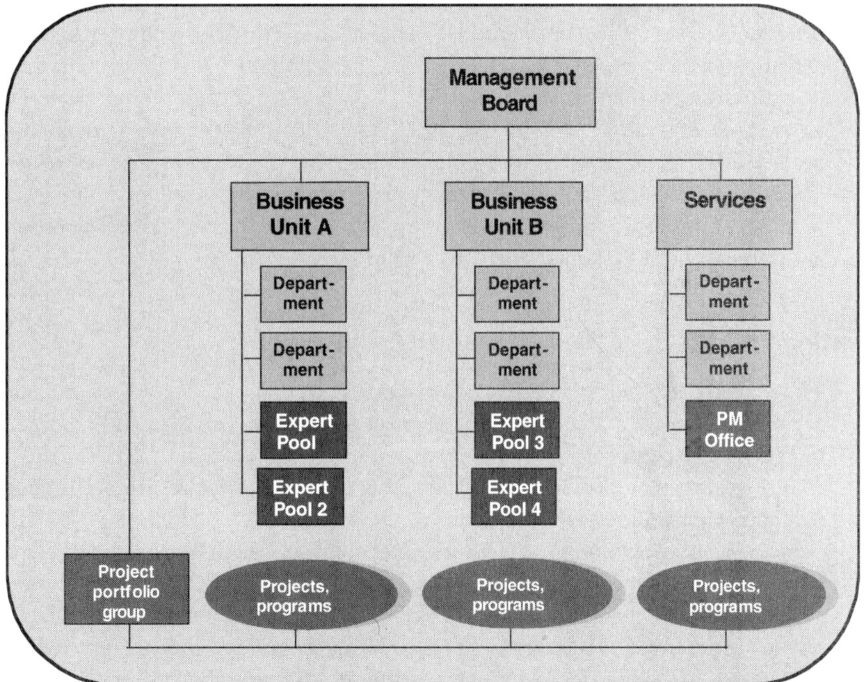

types of business processes. However, for relatively unique business processes of medium or large scope, and of short to medium duration, projects are the appropriate organizations. Projects can be defined for the performance of "contracts" for external clients, as well as for product developments, marketing campaigns, investments in the company infrastructure, or for reengineering activities for internal clients.

A program is a temporary organization for the performance of a business process of large scope. (See the chapters by Thiry, Jamieson and Morris, Archer and Ghamazadeh, Arttos and Deitrich, and Shenhar and Dvir, among others, for a further discussion of programs and program management.) A program consists of several closely coupled projects and activities. It has a time dimension and is medium or long term in duration. Typical programs are the development of a "product family" (and not of a single product), the implementation of a comprehensive IT solution for an international concern, the reorganization of a group of companies in a holding structure, and infrastructure investments considering several investment objects.

Projects and programs allow us to further differentiate companies. In addition to the permanent organizations of companies, such as divisions, profit centers, and departments, temporary organizations can also be added.

Clusters of Projects (and Programs)

For the integration of the different projects performed simultaneously in a project-oriented company, projects have to be clustered. Clustering according to the sequence in which projects are performed results in a "chain of projects." By relating projects to each other according to a defined criterion, such as the technology applied, a common client, or a geographic region, a "network of projects" results. By considering all projects performed by an organization, the "project portfolio" results. A *project portfolio* is defined as a set of all projects a POC holds at a given point in time and the relationships between these projects.

In a project portfolio, different project types, such as internal and external projects; unique and repetitive projects; marketing, contracting, organizational development projects; and infrastructure projects, might be included.

Supporting Projects by a "New Management Paradigm"

The project-oriented company is characterized by the existence of an explicit project management culture—that is, by a set of project management-related values and norms. For the project and the program management processes, specific procedures exist, creating a common understanding for the performance of these processes, the roles involved, and the management methods to be applied.

The application of this "new management paradigm" supports efficiency in the performance of projects. Traditional management approaches, based on a mechanistic management paradigm such as that of Taylorism, emphasize detailed planning methods, focus on the assignment of clearly defined work packages for individuals, rely on contractual agreements with clients and suppliers, and use the hierarchy as a central integration instrument. "New" management concepts, such as lean management, Total Quality Management, the learning organization, and business process reengineering, introduce new approaches. Among the common features of these "new" management approaches are the following:

- The use of the organization to create competitive advantage
- The empowerment of employees
- Process orientation and teamwork in flat organizations
- Continuous and discontinuous organizational change
- Customer orientation, and networking with clients and suppliers

Of course, projects can be performed within a traditional management culture. But this often results in costly, time-consuming, and for the project team members, frustrating experiences. The real benefits, the added values of project management, can only be achieved if some concepts of the new management paradigm are applied in the project-performing companies.

Management of Dynamics and Complexity in the Project-Oriented Company

POCs have dynamic boundaries and contexts. On the one hand, as the number and the sizes of the projects performed are constantly changing, permanent and temporary resources are employed and cooperations with clients, partners, and suppliers are organized in virtual teams. On the other hand, varying strategic alliances are established and relationships with different social environments of different projects have to be managed.

The greater diversity of projects that a company holds in its project portfolio, the more differentiated it becomes organizationally and the greater its management complexity will be. In order to support the successful performance of single projects as well as to ensure the compliance of the objectives of the different projects with the overall company strategies, specific integrative structures, such as a strategic center, expert pools, a PM office, and a project portfolio group are required.

Specific Business Processes of the Project-Oriented Company

The POC is characterized by specific business processes. The processes can be grouped by those that relate to single projects, those that relate to the project portfolio, and those that relate to the permanent organization. The processes of project management, program management, and assurance of the management quality of projects relate to projects. (See the chapter by Heumann on quality assurance.) The assignment of projects, the project portfolio coordination, and the networking between projects can be defined as project portfolio management processes. The personnel management and the organizational design relate to the POC overall.

The specific business processes of the POC can be visualized in a spiderweb graph (see Figure 6.2), which can be used for the analysis of the maturity of a POC (see *Maturity of the*

FIGURE 6.2. POC MATURITY MODEL BASED ON ROLAND GAREIS' MANAGEMENT OF THE PROJECT-ORIENTED COMPANY.®

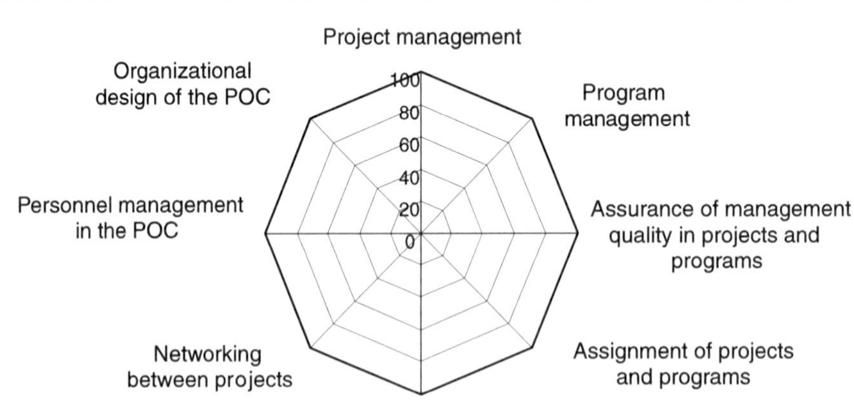

Project-Oriented Company later in the chapter). For each of these processes, which will be described briefly in following paragraphs, the POC requires individual and organizational competences.

Project Management

The perception of projects influences the project management approach. The traditional perception of projects as complex, goal-determined, and risky tasks supports above all the use of project planning methods. Typical examples are the work breakdown structure, the CPM/network scheduling, CPM-supported resource and cost plans, and risk management.

When projects are defined as temporary organizations, the formal establishment of a project, its integration into the overall company organization, and the development of a project specific culture are emphasized. The perception of projects as social systems further promotes the emphasis on, and orientation toward, context in project management. The relationship of a project to company strategies to the other projects performed simultaneously, to the relevant social environments, and to the business case of the investment initialized by the project become a concern (see the chapters by Jamieson and Morris, and by Arttos and Deitrich in this regard). "Social" project controlling—that is, the controlling of the relationships to relevant project environments and the relationships in the project organization—is considered in addition to controlling the hard project facts (progress, schedule, costs).

ROLAND GAREIS Project and Programme Management® represents a such a systemic-constructionistic approach to project management. Projects are perceived as temporary organizations and social systems; the development of project plans is considered as a constructionistic process. The objects of consideration in the project management process are not only the scope of work, the project schedule, and the project costs, but also the project objectives, the project income, the project organization, the project culture, the project context dimensions—including relationships to the relevant environments, to other projects, and to the company strategies—and the business case.

Project management is defined as a business process of the POC, which includes the subprocesses: project start, continuous project coordination, project controlling, resolution of a project discontinuity, and project closedown. (A *project discontinuity* is a discontinuous development of a project. In the case of an existential threat to a project, we talk about a "project crisis." Another type of a project discontinuity, which also requires a change of the project identity—project objectives, strategies, organization, and culture—is the "project chance.") The project management process starts with the formal project assignment and ends with the approval of the project results by the project owner.

The project start is the most important project management subprocess, because in it the basis for the other project management subprocesses is established. The project plans, the project communication structures, the relationships to relevant environments, and so on are developed and defined in the project start process. For each project management subprocess, the objectives, functions, methods, responsibilities, and deliverables can be described; this enables the quality of the project management process to be measured.

The objectives of the project management process are to

- successfully perform the project according to the project objectives,
- contribute to the optimization of the business case of the investment, initialized by the project,
- manage the project complexity and project dynamics,
- continuously adjust the project boundaries, and
- manage the project-context relationships.

The project management objective, to contribute to the optimization of the business case of the investment, initialized by the project, is of great concern in product development and capital investment projects and of less concern in contracting projects. Decisions in projects can influence the business case of the investment—for instance, in a new office building—to a large extent. Therefore, the project manager and the project team have to take on the responsibility for the optimization of the business case of the investment too.

A project needs an appropriate degree of complexity to relate appropriately to its environment. It is a project management function to manage—to build up and to reduce—the project complexity. The differentiation of project roles, the creation of subteams, as well as the consideration of different functional disciplines and hierarchical levels in the project team, are organizational possibilities for building up complexity. The application of different project management methods (i.e., work breakdown structure, the schedule, the cost and resources plan, risk analysis, project environmental analysis, etc.) offers different perspectives of the project. This multimethod approach further contributes to the development of the project complexity.

A reduction of project complexity occurs by the application of project management standards and by agreements. The definition of project-specific rules and norms, the development of project plans, and agreements in project owner meetings and project team meetings provide an orientation for the project work.

The project boundaries define what belongs to the project and what does not. The project boundaries are determined by the project scope and by the project start event and end event. Defining the project start and end event allows the preproject phase and the post-project phase to become, or at least limit, the project context. The social context of a project is determined by its social environments. For a project, these are those environments that we can expect will influence the success of the project.

Projects evolve over time. In the course of this, we can differentiate between continuous and discontinuous developments. Discontinuous developments can take place when project crises or chances occur.

Program Management

A program, as we have seen, consists of several projects and activities that are closely coupled by common program objectives, strategies, and rules. Usually, some of the projects in a program are performed sequentially and some are performed in parallel.

Program management has to be performed in addition to the management of the single projects of a program. The program management process has the same structure as the project management process (though see the chapter by Thiry regarding different emphases in program management). This includes the subprocesses of starting, coordinating, controlling, and closing down a program, and possibly resolving a program discontinuity. Also, program management methods are similar to the project management methods—in other words, there is a program work breakdown structure, a program bar chart, a program environment analysis, and so forth.

To allow for autonomous projects on the one hand but to ensure the benefits of organizational learning, economies of scale, and networking synergies in a program on the other hand, a specific program organization has to be designed. Typical program roles are program owner team, program manager, program office, and program team (see Figure 6.3). The program owner assigns the program to the program manager, who is responsible for the program management. He or she is supported by the program team members and the program office. Typical program communication structures are program owner meetings and program team meetings. The function of the program organization is to integrate the different projects of a program, in order to fulfil overall program objectives and strategies.

The advantages of designing a program organization instead of defining a large "project" with several subprojects are as follows:

- A less hierarchical organization
- A clear terminology: a program manager and several project managers instead of one project manager and additional "project managers" of the subprojects

FIGURE 6.3. PROGRAM ORGANIZATION CHART.

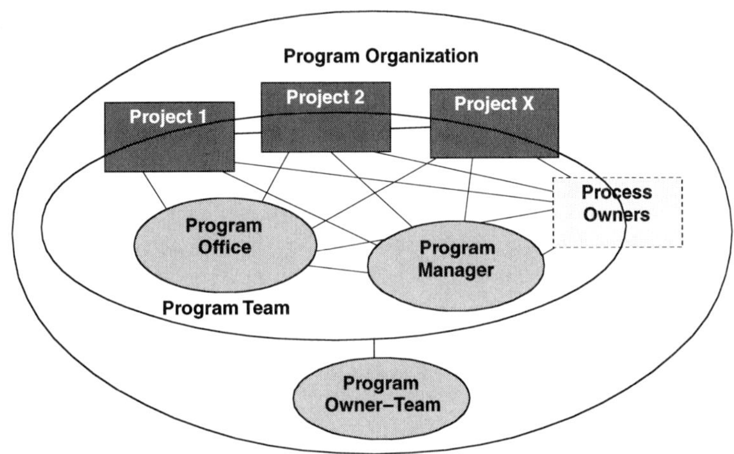

- Empowerment of the projects (of the program) by allowing for specific project cultures, specific relationships to social environments, and specific project organizations
- Differentiation between program ownership and ownerships for the different projects

Assurance of the Management Quality in Projects and Programs

Projects and programs are relatively autonomous organizations of the POC. The management of projects and programs is supposed to be performed according the general project and program management procedures of the POC. Management consulting and management auditing have to be performed in order to ensure the application of these general management procedures and to ensure the management quality in the projects and programs.

Management Consulting of Projects and Programs. The objective of management consulting of a project is to further develop the project management competence of this project. Not only are the competences of the project manager or the project team further developed (in which case this service would be defined as "coaching") but the temporary organization overall is consulted. The project as a temporary organization becomes the object of the management consulting service. This means that not only a permanent organization, such as a company or a profit center, but also a project can be the client in the consulting process.

By definition, projects are complex, risky, and dynamic. Therefore, they need high management attention and management quality. This can be ensured by involving project-external consultants. In a consulting assignment, any of the management subprocesses of a project (start, controlling, resolution of a discontinuity, closedown) or all of them can be considered. Specific management consulting services for programs might include the establishment of the program office, the development of a program marketing plan, or the definition and description of program- specific business processes.

The quality of the management consulting process can be measured. The key quality metric is the management competence of the project. It can be measured if there are improvements in the management competences of the members of the project organization, in the quality of the project meetings, in the project management documentation, in the project image, and in the relationships of the project to the relevant project environments.

The management consulting process starts with the initial contact between the potential consultant and the project. It includes analyzing the situation, moderating meetings and workshops, documenting these meetings and workshops, supporting the development and the updating of project management documentations, and so forth. Among the important intervention methods of consultants are interviews, documentation analyses, observations of meetings, and feedback.

The management consultants are assigned by the project owner. They are project-external people, but not necessarily company-external individuals. In many project-oriented companies, management consulting on projects and programs is considered an opportunity for job enlargement for senior (project) managers. Therefore, internal management consultants for projects and programs are developed in special training and coaching programs.

Management Auditing of Projects and Programs. Often a management audit on a project or a program is performed because of performance problems. Actually, management auditing is an instrument of quality management in projects and programs. It is an instrument of organizational learning of the POC.

The NEN-EN ISO 19011 (2002) defines auditing as "systematic, independent and documented process for obtaining audit evidence and evaluating it objectively, to determine the extent to which the audit criteria are fulfilled." Criteria for a management audit of projects are the project management approaches against which the projects are audited. These approaches are documented either in company-specific or in generic project management procedures.

While in a project audit the business processes for the contents as well as the project management process are considered, management auditing focuses on the project management competences only. The objects of consideration of a management audit are the project management subprocesses.

In the management audit of projects, we can differentiate the roles that the audit owner, auditor(s), and representatives of the audited project or program assume. The auditor(s) are project-external, but might be recruited from within the POC. The role of a management auditor of projects and programs is also sometimes seen to be an option for job enlargement for senior (project) managers of the POC.

Assignment of Projects and Programs

Companies invest in their infrastructure, in new products or services, in new markets, in the organization, or in their personnel. A project or a program might involve initializing such an investment. Therefore, an investment decision is often the basis for the decision to pursue a project or a program. For the investment decision, a business case analysis and initial project (or program) plans have to be developed. These documents are part of the investment proposal.

We must ensure the alignment of an investment with the company strategies. A formal instrument for this alignment is an investment scorecard. In such a scorecard, the investment decision criteria are documented (see Figure 6.4). Among the criteria to be considered for scorecards are financial data, customer relations, processes and resources, and innovations. (See also the chapter by Brandon on the balanced scorecard and project management.)

The investment decision is the basis for the decision about the appropriate organization for its implementation. A decision board has to decide if the investment is to be implemented by a project, by a program, or by organizational units of the permanent organization.

It is also necessary to analyze the fit of the project into the existing project portfolio. Each new project added to the portfolio means the mix has changed and a new portfolio is created. The relationships of the new project to the existing projects have to be considered and optimized. The decision about the project owner has to be made. The assignment of key personnel to the project, and the selection of partners and contractors for the project have to be made in consideration of the relationships to other projects in the project portfolio. (See the chapter by Archer and Ghasemzadeh.)

The authority to make the investment decision and the project assignment decision may either be divided between an investment decision board and a project portfolio group (PPG) or might be with the PPG only. The role of the PPG is described in *Organizational Design of the POC* later in the chapter.

FIGURE 6.4. INVESTMENT SCORECARD.

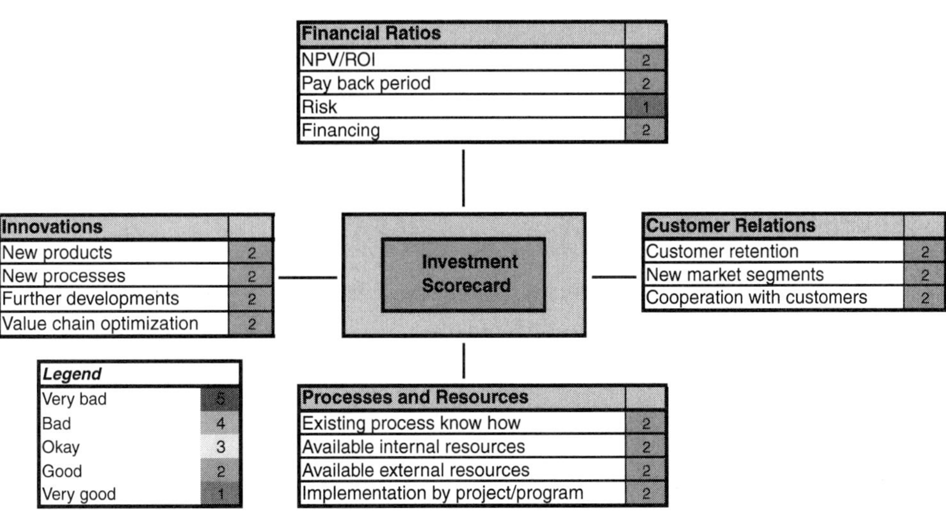

Project Portfolio Coordination

The objectives of the project portfolio coordination process are as follows:

- Optimization of the results of the project portfolio, and not of the single projects
- Definition of project priorities
- Coordination of internal and external resources
- Organization of learning of and between projects

The basis for the coordination of the project portfolio is a project portfolio database, which typically includes data about the project types, relations of projects to other projects, the project organizations, relevant project environments, and project ratios. The project portfolio database is not a project information system but contains aggregated project data only. It might be integrated in a project information system.

The project portfolio database allows the development of project portfolio reports. Typical project portfolio reports are the project portfolio bar chart, the project portfolio profit-

FIGURE 6.5. PROJECT PORTFOLIO SCORECARD.

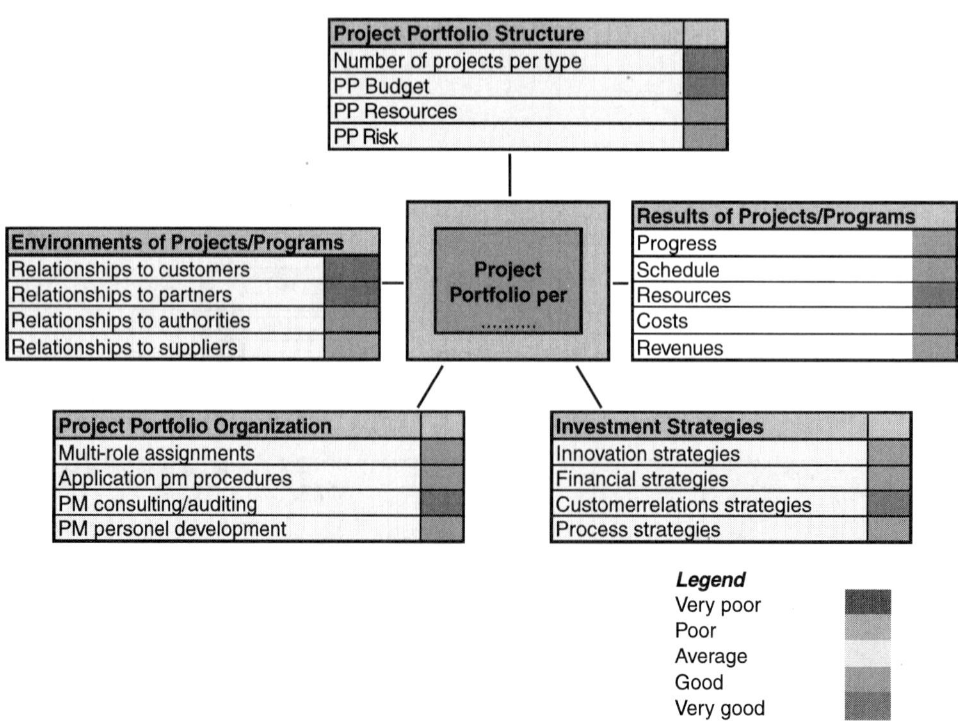

risk graph, and the project portfolio progress chart. An integrative project portfolio reporting tool is the project portfolio scorecard (see Figure 6.5). It shows how the actual project portfolio contributes to the implementation of the company strategies, reporting on the structure of the project portfolio and on the project portfolio status overall. Visualizing the project portfolio reports contributes to their acceptance as communication instruments for management and top management.

Networking between Projects

A set of closely coupled projects can be defined as a network of projects. Examples of the criteria which might relate projects in a network are a common technology applied, a common client, a common partner or supplier, or a common geographic region. The construction of a network of projects occurs at a point in time, in order to resolve a common problem or use a common opportunity. Therefore, a network of projects is not an organization with a common objective and a manager, such as a program, but it is an ad hoc communication structure.

The objective of constructing networks of projects is to identify synergies and potential conflicts between projects and to define strategies and measures to resolve the conflicts and to use the synergies. The networking between projects might result in a redefinition of the objectives of one or more projects of the network, in an assignment of common resources to two or more projects, or in renegotiations of contracts with the clients, partners, or suppliers.

The networking might be promoted by the PM office to generate the added value. The construction of the network of projects, the description of existing relationships between projects, and the establishment of new relationships is organized through ad hoc communications (meetings, workshops) of the project managers of the projects of the network. Also, project team members and representatives of relevant project environments, which might contribute relevant information, might participate. The communication between the projects requires a trust and openness between these partners.

An example of networking between projects is presented in Figure 6.6.

The demand for networking between projects of an Austrian consulting company, visualized in Figure 6.6, occurred because of a crisis in the project "eSupported PM-Seminars." In this project the cooperation with the IT supplier caused major problems. The project portfolio database showed that this IT supplier was also supplier or cooperation partner in other projects. Measures to resolve this problem could not be decided under consideration of the eSupported PM-Seminars project only, but the relationships to the other projects the IT supplier was involved in had to be considered too.

To analyze the consequences for all the projects the IT-supplier was involved in, a networking workshop with the project managers of the projects was organized, with the objective to analyze the relationships between the projects and to develop common strategies and measures to resolve the problem. The strategy agreed on was to further cooperate with the IT company but to reduce the dependency on it as partner and supplier. Measures agreed on were to resolve the crisis of the eSupported PM-Seminars project, to assign the

FIGURE 6.6. NETWORK OF PROJECTS PERFORMED WITH AN IT COMPANY.

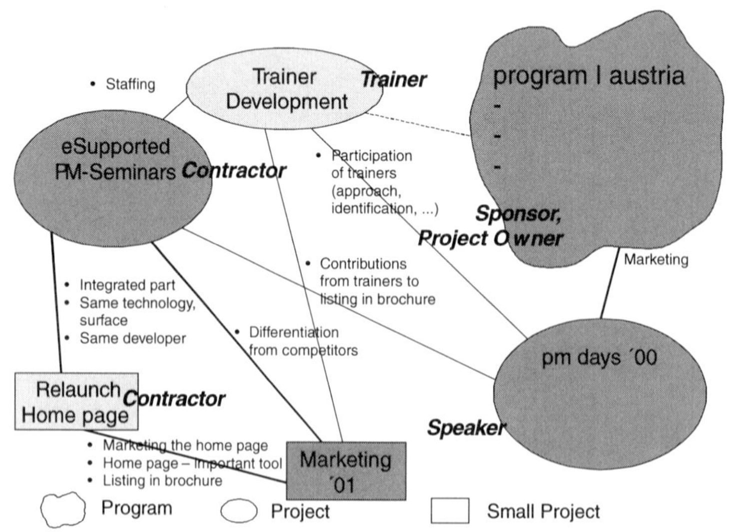

same developer to two projects, and to cancel the invitation of employees of the IT supplier to join the trainer team.

Personnel Management in the Project-Oriented Company

Personnel management processes in the POC include the recruiting, leading, developing, and releasing processes for project personnel. (See the chapter by Heumann, Turner, and Keegan for more on personnel management and project management.)

In POCs a project management career path exists. This is based on definitions of competences for the different roles in the POC. *Competence* can be defined as knowledge, skills, behaviors, and experience required for the performance of a business process. The specific competences that are required in a POC relate to the performance of the specific business processes of the POC. The competences are required by individuals and by teams.

Competences of Project Management Personnel. The project management competences required differ according to the project roles to be fulfilled by individuals. The following project roles can be performed by individuals: project owner, project manager, project management assistant, project team member, and project contributor. The project contributor contributes to the performance of work packages but does not (compared with the project team member) participate in project team meetings. The project management functions to be performed by project personnel can be described in project role descriptions (see, for example, Figure 6.7).

FIGURE 6.7. PART OF THE DESCRIPTION OF THE ROLE "PROJECT MANAGER."

Role Description: Project Manager
Objectives
• Representation of the project interests
• Contribution to the realization of the project objectives and to the optimization of the business case
• Leading the project team and the project contributors
• Representation of the project towards relevant environments
Organizational Position
• Member of the project team
• Reports to the project owner team
Tasks
During the project assignment process
• Formulating the project assignment together with project owner team
• Nominating the project team members
During the project start process
• Know-how transfer from the pre-project phase into the project
• Development of adequate project plans
• Design of an adequate project organization
• Performance of risk management
• Design of project-context relations
• etc.
During the project controlling process
• Determination of the project status
• Redefinition of project objectives
• Development of project progress reports
• etc.

FIGURE 6.7. (*Continued*).

During the resolution of a project discontinuity
• Analysis of the situation and definition of ad hoc measures
• Development of project scenarios
• Definition of strategies and further measures
• Communication of the project discontinuity to relevant project environments
• etc.
During the project closedown process
• Coordination of the final contents work
• Transfer of know-how into the base organization
• Dissolution of project-environment relations
• etc.

The project manager requires knowledge and experience not only to apply project management methods but also to creatively design the project management process. These design functions include the following:

- Selection of the project management methods appropriate for a given project
- Selection of the appropriate communication structures
- Facilitation of the different workshops and meetings
- Decision to involve a project management consultant
- Selection of the appropriate IT and telecom infrastructure
- Definition of the appropriate form for the project management documentations

The project management competence of a project manager is the capability to fulfill all functions specified in the role description. Besides the project management knowledge, skills, behaviors, and experience for a given project type, a project manager needs product, company, and industry knowledge. In international projects, cultural awareness and language knowledge are also prerequisites.

Competences of Project Teams. To perform a project successfully, a project team requires team competence. The competence of a project team can be defined as the competences of the project team members plus the social knowledge and experience of the team to create the "big project picture," to produce synergies, to solve conflicts, and to ensure learning in the team.

A project team cooperates in workshops and meetings. The application of project plans, such as a work breakdown structure, a schedule, a project environment analysis, and so on, are tools, to support the communication in the project team.

Organizational Design of the Project-Oriented Company

To integrate the different projects performed simultaneously, a POC requires specific permanent organization structures, such as a PM office, a project portfolio group, and expert pools.

The PM Office. To ensure that the different ongoing projects apply a common management approach, somebody in the POC has to take on the ownership for the project management process. The project management competences of the POC have to be institutionalized. The PM office is the organizational unit that can take on this responsibility. (See the chapter by Young and Powell for more on the PM office.)

The PM office is a permanent structure and is part of the base organization of the POC. It provides services for all projects of the POC and also for the project portfolio group. The PM office has to be differentiated from project offices and program offices, which are temporary, are part of the project or program organization, and provide services for one project or one program only.

The organization chart of a PM office (see Figure 6.8) includes the roles PM office manager and personnel for project, program, and project portfolio management services. The expert pools "PM-Personnel" and "PM-Trainer, PM-Consultants" might be coordinated by the PM office manager.

Services provided by the PM office might include the following:

- Development and maintenance procedures for project and program management, management auditing of projects and programs, and project portfolio management
- Development and maintenance of standard project plans (standard WBSs, work package specifications, milestone lists, etc.)
- Provision of project management support services
- Project and program marketing
- Organization of project management training, coaching, consulting, and auditing
- Promotion of project management as a profession by establishment of a project management career path, project management certification programs

FIGURE 6.8. ORGANIZATION CHART OF THE PM OFFICE.

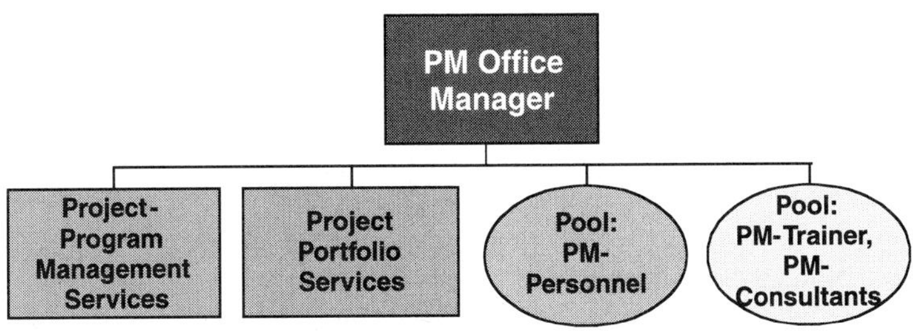

- Assurance of a project management infrastructure (meeting rooms, ICT tools, moderation tools)
- (Internal) benchmarking of the project and program management processes
- Maintenance of the project portfolio database
- Development of project portfolio reports
- Support of the meetings of the project portfolio group

The Project Portfolio Group. In a POC many projects are performed simultaneously. Synergies and possible conflicts between these projects have to be managed. The results of the project portfolio have to be optimized. Because of the high organizational differentiation of the POC, the management of the project portfolio might not be taken care of in the usual meeting structures, such as management board meetings, department head meetings, and so on. It might be preferable to delegate this responsibility to a specific communication structure, the project portfolio group.

The project portfolio group is a permanent organization structure of the POC. It could be considered as a staff or a line position reporting to the management board. Five to eight managers of profit centers and departments being strongly involved in projects and programs should be members of the project portfolio group. In a major Austrian telecommunication with some 1,500 employees and 70 to 80 projects at any given time in the project portfolio, the project portfolio group consists of the department heads for marketing, engineering, call center, IT, and PM office. The head of the group is the director of finance. The project portfolio group meets every week for two to three hours.

Depending on the duration and the dynamics of the projects in the portfolio, not more than 100 projects should be managed by one project portfolio group. In large POCs several project portfolio groups, differentiated by project types or by business units, might be required. The major services of the project portfolio group are the selection of projects, to be started and integrated into the project portfolio, the creation of synergies (common use of resources, economies of scale, organizational learning) and the resolution of conflicts within the project portfolio, and the quality assurance and the provision of early warnings in the project portfolio. Criteria for the assurance of the project portfolio quality are, for example:

- The number of projects for one project manager
- The number of projects for one project owner
- The number of projects per project type
- The project portfolio budget (per project type)

Expert Pools. Experts, to perform the work packages in projects, are required in POCs. Depending on the type of industry a POC is in and the type of projects it performs, different expert categories are required. Typical expert pools are pools of engineers (differentiated by mechanical, electrical, engineering, etc.), procurement experts, and marketing experts. These experts most of all need competence in the discipline they represent, but they also need project management competence in order to cooperate in teams. The roles in an expert pool are the expert pool manager, the experts, and exchange of experience groups and possibly supervision groups.

One important expert pool of a POC is that pertaining to "project management." The experts of this pool might be differentiated in relation to a project management career path into project management assistants, junior project managers, project managers, senior project managers, and program managers.

An expert pool manager has personnel management, knowledge management, and infrastructure management functions. He or she has to recruit, develop, and allocate personnel, has to further develop standards and provide ethics of work, and has to provide the infrastructure for the performance of the work packages. On the other hand, the quality control of the work packages, which are performed in the projects and programs, is not his or her responsibility.

Maturity of the Project-Oriented Company

Not only individuals but also organizations have the capability to acquire knowledge and experience and to store it in a "collective mind" (Senge, 1994; Weik and Roberts, 1993). Organizational principles, which might be stored in the collective mind of a POC, are project management procedures, project management templates, standard project plans, procedures for the management auditing of projects, as well as project portfolio management procedures, structures for a project portfolio database, and standard project portfolio reports.

Assessing and Benchmarking the Maturity of a POC

The organizational competences of the POC can be assessed with the *POC Maturity Model.* This model is based on a POC questionnaire and is visualized in a "POC spiderweb." The axes of the spiderweb represent the business processes of the POC. The maturity of a POC in the performance of each of these processes can be assessed by the application of a set of questions relating to the business process.

Questions relating to the project start process are grouped regarding the planning of project objectives, the project risk, the project context relationships, the project organization, and the project culture. They are not assessed for the application of a given project management method, but for the resulting project management documents.

The overall maturity of a POC can be represented by the area in the spiderweb (see Figure 6.9).

The maturities of POCs can be benchmarked and further developed. Instruments to develop the competences of individuals performing roles in the POC are (self-) assessments, trainings (classroom, on the job), and coachings. Instruments to develop the organizational competences of the POC are assessments, benchmarkings, and organizational development projects—for instance, to implement project portfolio management in order to establish a PM office.

Summary

"Management by projects" is the organizational strategy of the POC. The application of a "new management paradigm" supports the efficient application of projects in the POC.

FIGURE 6.9. MATURITY OF A POC.

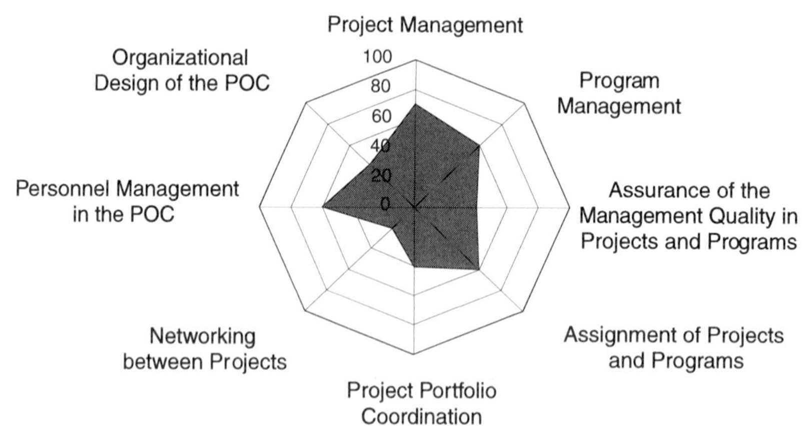

Specific organizational structures, such as expert pools, a project portfolio group, and a PM office, are required to perform integrative functions in the POC. For the performance of the specific business processes of the POC, specific competences are required by individuals, teams, and the POC overall.

It is important to differentiate between the project management process and other specific processes of the POC. Even though project management is the most important business process of the POC, competences for the performance of the other specific business processes of the POC have to be further developed too. The *POC Maturity Model* supports the assessment as well as the further development of these competences.

References

Ashby, W. R. 1970. *The process of model building in the behavioral sciences.* Columbus, Ohio: Ohio State University Press.

CCTA. 1996. *Managing successful projects with PRINCE2.* London: The Stationery Office.

Cleland, D., and R. Gareis. 1994. *Global project management handbook.* New York: McGraw-Hill.

Cleland, D., and L. Ireland. 2002. *Project management: Strategic design and implementation.* 4th ed. New York: McGraw-Hill.

Englund, R. L., R. J. Graham, and P. C. Dinsmore. 2003. *Creating the project office: A manager's guide to leading organizational change.* The Jossey-Bass Business & Management Series. New York: Wiley.

Gareis, R. (Hg.) 1994. *Erfolgsfaktor Krise: Konstruktionen, Methoden, Fallstudien zum Krisenmanagement.* Wien: Signum-Verlag.

———. 2000. Managing the project start. In *The Gower handbook of project management,* ed. J. R. Turner and S. J. Simister. Aldershot, UK: Gower.

Gareis, R., ed. 2002. *PM baseline knowledge elements for project and programme management and for the management of project-oriented organisations,* July, version 1, Prokekt Management Austria, Vienna, Austria.

Gareis, R., and M. Huemann. 2000. *PM-competences in the project-oriented organization.* In *The Gower handbook of project management, eds.* J. R. Turner and S. J. Simister. Aldershot, UK: Gower.

Gareis, R. and M. Huemann. 2003. *Project management competences in the project-oriented company.* In *People in Project Management,* ed. J. R. Turner, Aldershot, UK: Gower.

Hartmann, F. 2000. Don't park your brain outside. Newtown Square, PA: Project Management Institute.

Jones, M. O. 1996. *Studying organizational symbolism.* Thousand Oaks, CA: Sage Publications.

Kaplan, R. S. 1996. *The balanced scorecard: Translating strategy into action.* Boston: Harvard Business School Press.

Knutson, J. 2001. *Succeeding in project-driven organizations: People, processes, and politics.* New York: Wiley.

NEN-EN ISO 19011. 2002. *Guidelines for quality and/or environmental management systems auditing.* Brussels: European Committee for Standardization.

PMI. 2000. *A guide to the project management body of knowledge (PMBOK Guide).* Newtown Square, PA: Project Management Institute.

Schein, E. 1985. *Organizational culture and leadership.* San Francisco: Jossey-Bass.

Senge, P. 1994. *The fifth discipline fieldbook: Strategies and tools for building a learning organization.* New York: Doubleday.

Turner, J. R. 1999. *The handbook of project based management.* 2nd ed. London: McGraw-Hill.

Weik, A. and K. Roberts. 1993. Collective mind in organizations heedful interrelating on flight decks. *Administrative Quarterly* 38.

CHAPTER SEVEN

STRATEGIC BUSINESS MANAGEMENT THROUGH MULTIPLE PROJECTS

Karlos A. Artto, Perttu H. Dietrich

Effective management of single projects does not suffice in today's organizations. Instead, the managerial focus in firms has shifted toward simultaneous management of whole collections of projects as one large entity, and toward effective linking of this set of projects to the ultimate business purpose. This approach is contained in concepts of project-based management, programs, and portfolios. Portfolios of different project types are typically positioned under the governance of organizational units or responsibility areas (see Figure 7.1). Management processes above projects must link projects to business goals and assist in reaching or exceeding the expectations set by company strategy.

One major starting point for the development of business-oriented management of projects in a company context was introduced in the end of 1980s in an expert seminar in Vienna, where the contribution of project management to the general world of management was discussed as contained in the concept of "management by projects" (see the chapter by Gareis). Since that time there have been an increasing number of studies on the broader role available for project-based management, project-based organizations, and project business. Recent examples of such studies include Turner (1999), Turner and Keegan (1999, 2000), Turner et al. (2000), Gareis (2000a, 2000b), Artto (2001), Artto et al. (2002), and Elonen and Artto (2003).

Early theories of organizational strategy saw "strategy as an action of intentionally and rationally combining selected courses of action with the allocation of resources in order to carry out organizational goals and objectives in order to achieve strategic fit and thereby obtain competitive advantage" (Hatch, 1997). This is based on the idea that strategy involves creating a match between organization and environment (Ansoff, 1965). Galbraith (1995) proposed that strategy establishes the criteria for choosing among alternative organizational forms. Each organizational form enables some activities to be performed well while hinder-

FIGURE 7.1. TWO COMPANIES (OR TWO BUSINESS UNITS WITHIN ONE COMPANY) WITH NETWORKED PROJECTS AND PORTFOLIOS. THERE ARE CROSS-ORGANIZATIONAL PROCESSES IN THE SHARED NETWORK AT STRATEGIC, PORTFOLIO, AND PROJECT LEVELS.

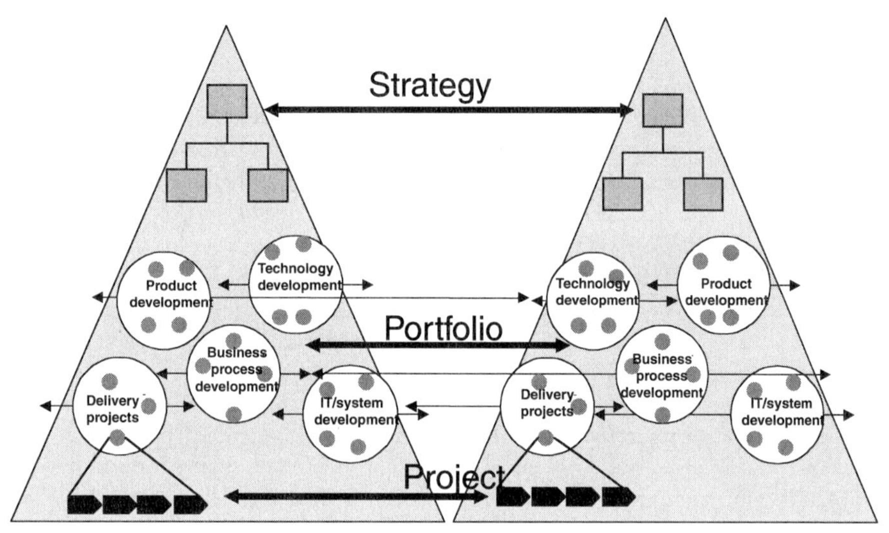

Source: Artto et al. (2002).

ing others. Choosing between organizational alternatives involves trade-offs. Strategy can help with this by pointing to those activities that are most necessary, thereby providing a basis for making the best trade-offs.

The purpose of this chapter is to introduce managerial practices relevant to strategic business management in multiple-project environments. Multiproject environments are introduced in terms of different project types, programs, and portfolios and their management. Based on the knowledge from this, we introduce issues that serve as guidelines to the theme of strategic business management of multiple projects. We conduct an analysis of content and process of strategy and how these relate to the setting of goals and objectives, and to effective decision making with multiple projects. Based on this, the chapter identifies effective managerial practices for the strategic business management in multiple-project environments. We also combine strategy with management from an applications viewpoint by looking at four case organizations.

Different Project Types and Their Different Strategic Importance

Different project types have different strategic importance; each type typically requires different management approaches. Crawford et al. (2002), Shenhar et al. (2002), and Youker

(1999) are studies of project classification that attempt to address this issue. (See Shenhar and Dvir's chapter in this book.) These are valuable in understanding not only different project types and their characteristics but also the different success criteria and respective strategic importance, and accordingly, different successful managerial practices associated with each project type.

Shenhar et al. classify projects into external and internal types, where the position or closeness of the customer (external or internal) provides the basis for the classification. This classification also considers the ultimate customer in the external markets in relation to how direct or indirect the relationship of the ultimate customer is to the project deliverable. Their starting point is innovation management literature that makes a distinction between incremental and radical innovation. Thus, according to Shenhar et al., projects can be either strategic or operational in their nature, depending on the project type.

External projects typically relate to developing products for customers in the market. Shenhar et al. distinguish between derivative, platform, and breakthrough projects, all as external projects. Wheelwright and Clark (1992) call these three project types *commercial development projects*. Based on Shenhar et al.'s considerations, derivative projects relate to extending, improving, or upgrading existing products. They typically aim at short-term benefits, and they are thus more operational than strategic in their nature. Platform and breakthrough projects relate to new product development or production processes where there is a longer-term perspective, and, accordingly, a reaching for a more strategic nature. Another interpretation of an external project is that of a delivery project where the project is in a commercial setting, and where an organization is running projects for other organizations (Turner and Keegan, 1999). Such external delivery projects are often mere production or manufacturing devices that run more or less predetermined work for an organization according to a contract between the customer and project supplier (Artto, 2001). The similarity of project-based operations with both external and internal customers is demonstrated by Turner and Keegan (1999), who defined a project-based organization as a stand-alone entity that makes products for external customers, or a subsidiary of a business unit of a larger firm that makes products for internal or external customers.

Shenhar et al. (2002) divide internal projects into problem solving, utility, maintenance and research projects. Wheelwright and Clark (1992) distinguish between internal projects based on research and development, which are a precursor to commercial development, and alliances and partnerships, which can be commercial or basic research directed. Figure 7.2 describes Wheelwright and Clark's view on different types of development projects (the figure includes four types; the fifth type—alliances and partnerships—can include any of the other four types). Mikkelsen et al. (1991) define internal projects as organizational or operational development projects, such as systems planning and implementation, the introduction of new manufacturing technology, and organizational change. Shenhar et al.'s utility and research projects usually come from a long-term perspective and can be considered as strategic projects. Problem-solving and maintenance projects usually focus on the shorter term, typically aim at performance improvements, and can be seen as operational projects (Shenhar et al., 2002).

We appreciate the consideration of strategic importance now given to project types but consider that the "strategic versus tactical" importance given to these also depends on parameters other than project type as defined by existing project classification literature. Fur-

FIGURE 7.2. MAPPING THE TYPES OF DEVELOPMENT PROJECTS.

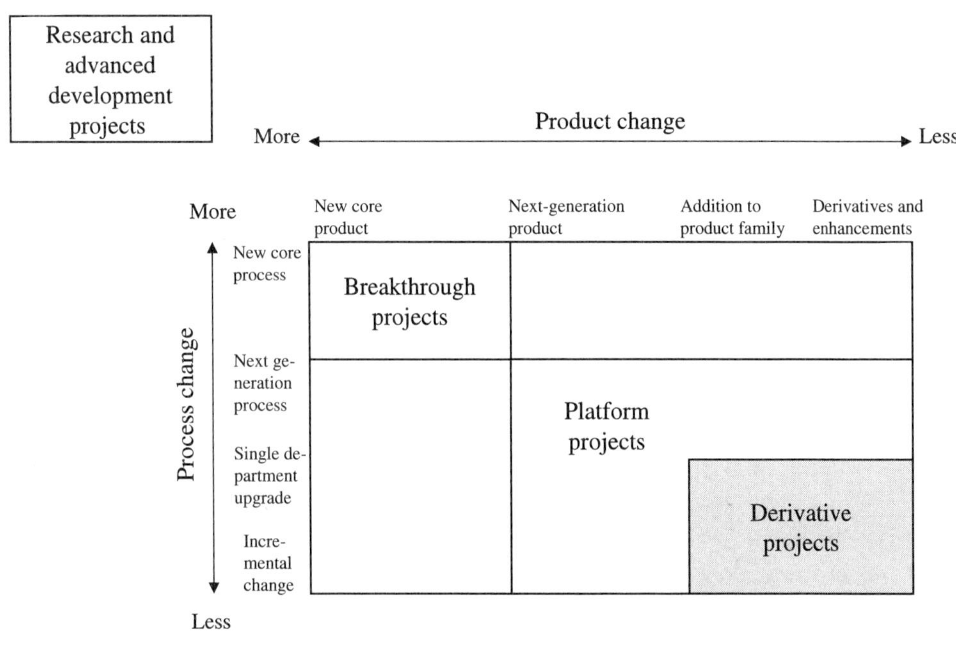

Source: Wheelwright and Clark (1992).

thermore, the strategic importance cannot be evaluated in a straightforward manner, such as presuming that long-term projects are always more strategic, as is widely argued in the literature.

Programs and Portfolios

Guidance for the management of multiple projects in organizations can be derived from several different theoretical and practically oriented discussion arenas. The program management and project portfolio management contents are outlined in the following section.

Archer and Ghasemzadeh (1999) define a project portfolio as a group of projects that are conducted under the sponsorship or management of a particular organization (see their chapter in this book). They point out that these projects compete for scarce resources. The three well-known objectives of portfolio management are as follows (Cooper et al., 1998):

- Maximizing the value of the portfolio
- Linking the portfolio to the strategy
- Balancing the portfolio

Dye and Pennypacker (1999) define project portfolio management as the art and science of applying a set of knowledge, skills, tools, and techniques to a collection of projects to meet or exceed the needs and expectations of an organization's investment strategy. In PMBOK (2000), project portfolio management refers to the selection and support of project investments or program investments that are guided by the organization's strategic plan and available resources. A strategic task of project portfolio management is to maintain corporate identity and ensure linkages between projects and constrain the impact of individually implemented projects with no links to other projects (Lundin and Stablein, 2000). According to Platje et al. (1994), a portfolio is a set of projects that are managed in a coordinated way to deliver benefits that would not be possible if the projects were managed independently. This definition is similar to many definitions introduced for a project program. For example, Turner (1999) and Poskela et al. (2001) emphasize that projects in a program are a coherent group that is managed in a coordinated way for added benefit. Murray-Webster and Thiry (2000) define a program as a collection of change actions (projects and operational activities) purposefully grouped together to realize strategic and/or tactical benefits. (See the chapter by Thiry on program management.)

From the strategic management point of view, the main driver for the management of multiple projects in different forms—for instance, programs—is the change in the business environment of an organization (OGC, 2002). Changes in the environment imply system or organizational changes (Ackoff, 1999). In these changes, program management provides a framework for the management of complexity and risk with the general intent of implementing business strategies and initiatives, or large-scale change (OGC, 2002).

The management of risk and uncertainty can appear in different ways. For example, in the R&D area, the important task of a business manager may be to increase risk to balance the portfolio of projects for business benefit. We can see this from findings illustrating how radical projects with high risk have the highest business potential (Loch, 2000).

Programs usually represent entities that have a determined purpose, predefined expectations related to the benefits scheme, and an organization, or at least a plan for organizing the effort. A program is set up to produce a specific outcome that may be defined at a high abstraction level of a "vision." According to PMBOK (2000), a program consists of several associated projects that will contribute to the achievement of a strategic plan. Many programs also include elements of ongoing operations. Program management helps to organize, manage, accommodate, and control adaptation and changes such that the eventual outcome meets the objectives set by the business strategy (OGC, 2002). Program management includes the management of interfaces between projects, prioritization of resources, and a reduction in overall management effort (Turner, 1999). The objectives of projects under the same project program are interdependent (Platje et al., 1994). Turner (1999) emphasizes the importance of the overall strategic resource sharing scheme related to program management. Such strategic resource sharing is implemented through a well-organized balance of responsibility, where the program directors' responsibility is to link programs with corporate objectives, the overall corporate plan, and corporate resource plan. OGC (2002) defines program management as the coordinated management of a portfolio of projects that change organizations to achieve benefits that are of strategic importance.

Constructing a Theoretical Framework

The previous sections introduced aspects of existing knowledge on multiproject environments and attempted to show the need for new knowledge in the area of strategic business management of multiple-project environments. Based on this analysis, and the needs reflected by it, we can identify the following issues:

1. How can multiple projects be collectively aligned with business strategy in a manner that generates enhanced benefits for the whole business?
2. What is the role of specific projects in implementing, creating, and renewing business strategies?
3. How best can strategic business management be applied in organizations with multiple projects, and what are the relevant managerial practices for accomplishing this?

The preceding three questions are addressed in this remainder of this chapter via current strategy, business administration and project management literature, as well as findings from four case organizations.

Strategy and Strategic Management

In ancient military terminology in Athens, *Strategos* referred initially to attributes of the general commander in the army. The word strategy later was expanded to include the art of managerial skills for employing forces to overcome opposition (Mintzberg et al., 1995). Ancient military terminology and early strategic management literature emphasize the relative position of an organization to its external competitive environment, with emphasis on activities necessary to achieve a desired position (Chaffee, 1985). The concept of strategy has also used contributions from other disciplines, such as industrial organizational economics approach, resource-based approaches, ecologist-evolutionary approaches, and systems thinking approaches (Pavón, 2002). These emphasize, among others, the importance of rational decision making, and learning as an issue that shapes strategies.

The variety of attempts to express the specific nature of the strategy has led different authors to create different strategic schools (see, for example, Chaffee, 1985; Mintzberg et al., 1995). Mintzberg et al. (1995) introduced five Ps of strategy as a means to show the complex nature of strategy, where the Ps include definition of strategy as a Plan, Ploy, Pattern, Position, and Perspective. This last P—perspective—emphasizes sharing visions and mental images inside the organization to form a common understanding and culture, as strategies are abstractions in individuals' minds. This is consistent with Chaffee's (1985) *interpretative view,* which focuses on corporate culture and symbolic management as essential means to motivate participants and potential participants in ways that can favor an organization. This view makes a clear distinction with traditional strategic literature (see, for example, Chandler, 1962; Andrews, 1971; Hofer, 1975; Mintzberg, 1978) by suggesting that organizations' behavior is rather irrational in nature.

Early studies on strategic management focused on the content of strategy. Later literature distinguished between the content of the strategy and the process of strategy formulation and implementation (Chaffee, 1985). A distinction was made between an analytically objective strategy formulation process and a behavioral implementation process (see Andrews, 1971; Fredrickson, 1984; Pettigrew, 1992). Organization theorists tended to emphasize the meaning of *human processes (e.g., decision styles) in strategy making* (Burgeois, 1985), which started with rationality as a principal assumption of strategy process (see, for example, Andrews, 1971; Ansoff, 1965; Porter, 1980). From this, strategy management research introduced ideas of bounded rationality as a means to circumvent the reality of aspects of "organizational anarchy" within an organization (Simon, 1957; Cohen et al., 1972). This emerging recognition of an existing imperfect rationality in organizations has shifted toward emphasizing the extent and type of involvement of individuals in the organization or its environment (stakeholders) in strategy process (Hart, 1992). For example, Chaffee's (1985) emerging school of interpretative perspective on strategic management is an example of seeing the importance of individuals' involvement in strategy making. We can conclude from this that strategy is, and is accepted as, an important concern of the whole organization, not just its top management, and that motivation arises as a more crucial element of strategy realization.

Strategy Formulation and Implementation

Strategic processes comprise both strategy formulation and implementation. The strategic management literature mainly focuses on the strategy formulation aspect, with less attention given to strategy implementation (Aaltonen and Ikävalko, 2001). Andrews (1995) identifies organizational structures as requirements for the efficient implementation of intended actions. These structures include elements such as information systems and relationships enabling execution and management of subdivided activities. Moreover, Andrews (1995) states that one critical requirement for the successful implementation of strategy is to ensure that decisions made by managers and senior managers are consistent with the organization's goals and objectives. Leading the organization to intended goals and objectives requires measurement of the current state or performance of actions, analyzing the gap between the current and intended state, and making corrective actions. Diagnostic control systems are traditionally recognized as an important means of controlling the intended performance of the organization (see, for example, Simons, 1995). These systems are supported by defined performance characteristics called *critical performance variables* or *key success factors* that serve as indicators for achievement of organizational goals in the means of efficiency and effectiveness (Simons, 1995). (See the chapter by Brandon.)

The balanced scorecard method introduced by Kaplan and Norton (1992) is a good example of a way to measure the performance of an organization in enhancing the achievement of organizational goals and strategy implementation. The scorecard can be used to derive objectives and measures related to company vision and strategy that can be derived to further project-specific objectives that are well aligned with business strategy. The strategic objectives to be measured fall into four perspectives:

- Customer
- Financial

- Internal business process
- Learning and growth

Employee capabilities, technology, and corporate climate contribute to the organization's capability for learning and growth (Kaplan and Norton 2001).

The success domains/dimensions in some project success studies are analogous to the four perspectives of balanced scorecard introduced by Kaplan and Norton (1992, 1996). For example, Shenhar et al. (1997) introduce the following four dimensions of project success:

- Project efficiency
- Impact on customer
- Business success
- Preparing for the future

In general, project success studies contribute to definition of requirements for decision-related information used, for instance, in project selection criteria or in performance measures.

Another contribution of project success studies is their indication of the most relevant managerial areas and even managerial practices that can serve as enablers for success (see the chapter by Cooke-Davies). Although many project success studies still limit their views to the success and successful management of one single project only, they can also introduce the important aspect of the overall context where a single project occurs. This extends the evaluation of success toward strategic issues that take a viewpoint of the whole business. According to Saravirta (2001) and Kotsalo-Mustonen (1996), the relevant success domains are related to the following:

- Strategy (e.g., new competitive advantage, reference value)
- Relationship (e.g., client satisfaction)
- Situation (e.g., learning by doing, unlearning)
- Product/service (e.g., commercial success, quality)
- Project implementation (e.g., cost, time, process quality)

Furthermore, evaluation of success depends on the stakeholder and its perspective on the project. From Morris and Hough (1987) and Rouhiainen (1997), we can derive the following synthesis of what the important success domains are:

1. Technical performance, project functionality, client satisfaction, and technical and financial performance of the deliverable for the sponsor/customer
2. Project management: on budget, on schedule, and to technical specification
3. Supplier's commercial performance: commercial benefit for the project service providers
4. The learning that project stakeholders acquire

Emergent Strategies

Mintzberg (1978) examined the relation between an organization's intended strategy and its realized strategy. Mintzberg showed that in addition to intentional strategies, strategies can also include unintentional, emergent components. Strategies emerge from different sources and from different levels of organization. Mintzberg proposes that the concept of realized strategy consists of intentions that lead to deliberate strategy, intentions that lead to unrealized strategy, and emergent strategies that develop in the organization without a priori intentions. Simons (1995) explains that an emergent strategy process consists of actions of individuals at all organizational levels to seize the opportunities and deal with the problems.

The emergent perspective of the strategy process seems to focus now on organizational learning (see the chapter by Lampel and Jha) and works to identify strategy as the cumulative impact of operative decisions taken by management (Christensen, 2000). Lindblom (1959) explained strategic management from the policy formation viewpoint, by seeing policies as consisting of small, politically acceptable, disjointed decisions. Moreover, Quinn's (1995) logical incrementalism proposed that strategies should rely on flexibility and experimental applications to move from broad concepts toward specific commitments, and strategic decisions should be made at the last possible moment in order to utilize the most topical and available information for minimizing risks. Quinn's argument is based on recognition of the biases that are found in reality among the formal "systems planning" and "power-behavioral" approaches of strategy formation in organizations. Good strategies are not formulated in a comprehensive master plan. According to Quinn, the formal systems planning approach relies on quantitative data ignoring vital qualitative, organizational, and power-behavioral factors, which often tend to represent the dynamic, time-related attributes of organizational success. Power-behavioral perspectives focus on psychological issues, trying to understand the influence of human dynamics, power relationships, and organizational processes in strategy formation. However, power-behavioral approaches can introduce drawbacks associated with ignoring the normative component of rationality in strategic decision making. Quinn thus emphasizes the importance of "process limits" in strategic decision making and management.

Process limits deal with issues such as timing and sequencing, building comfort levels, developing consensus, and selecting and training people. These imperatives can become the determinants of the system itself, and they finally determine the outcome of the decisions. This resembles Mintzberg and Waters' (1985) umbrella strategy perspective, where top-managers define boundaries and guidelines for the organization to operate, and where within these boundaries individuals in the organization can take initiatives. Mintzberg and Waters's study illustrates that even if the goals and objectives for the organization are predetermined at the top level of the organization, managers at the middle level can, by their actions and decisions, affect the formation of strategy. Burgelman (1983) supports this while proposing that in addition to induced strategic behavior, there exists also an autonomous strategic behavior within the organizations, and that behavior develops outside of the strategic umbrella defined by top management. This autonomous behavior appears when people at the operational level notify the resources provided by the organization as a means to utilize new opportunities (Floyd and Wooldridge, 2000). In his later study, Burgelman (1991) reported evidence from a longitudinal case study of Intel Corporation. The findings indicate that successful firms are characterized by both top-down strategic intent and bottom-up experimentation and selection process. Hart (1992) further developed the idea of organization-wide involvement in strategy formation and claimed that strategy making is an organizational capability that determines an organization's success or failure.

The preceding can be summarized as confirming that the role of individuals can be extremely important in viable strategy formulation and implementation. Projects and the individuals who work on them are particularly important. This is supported in the literature concerning product development and internal development projects, which emphasizes the project manager's role as a champion, gatekeeper, facilitator, or coach, and the top management representative's involvement and supporting role (Loch, 2000; Terwiesch et al., 1998, Brown and Eisenhardt, 1995; Eisenhardt and Tabrizi, 1995; and Mikkelsen et al. 1991). An important managerial problem is to encourage projects and individuals in their role in emerging strategies to create new ideas and renew existing strategies.

Thus, the challenge of successful strategic management may lie in managing the tension between creative innovation and predictable goal achievement. This tension occurs by

- reconciling unlimited opportunities with managers' limited attention;
- implementing top-down strategies while allowing bottom-up strategies to emerge;
- creating predictable environments while maintaining innovativeness; and
- controlling actions while simultaneously allowing the organization to learn new ones (Simons, 1995).

The ability to learn is raised as one major sources of sustainable competitive advantage in many companies. The study by De Geus (1988) provides a good example of the impact of learning to the success of companies. He examined the survival of Fortune 500 companies and found that one-third of the companies listed in 1973 had vanished by 1983. A key source of the success of the survivors was their ability to learn by continuously exploring opportunities for new business and organizational development. The emphasis should be placed on focusing that organizations are doing the right things, rather than doing things right. This capability of an organization to question its underlying policies and goals is called *double-loop learning*. Senge (1990) proposes creative tension in organizations as a principal building block of learning organizations. Creative tension is created by integrating pictures of desired future and current reality. However, this creative tension differs from solving existing problems in an undesired state of current reality. Rather, it comes from individuals' intrinsic motivation and generative learning with its emphasis on continuous experimentation and feedback. Brown and Eisenhardt (1997) argue that managers learn from possible futures. Small losses through experimental products that fail, or futurists' predictions that do not come true, are probably the most effective learning devices. A variety of probes creates hands-on experiences (experimental products and experimental strategic alliances) and indirect experiences (meetings). Eisenhardt's 1997 study suggested semi-structures that would ensure responsibilities, ownership, prioritization, and communication. Semi-structures relate to quasi-formal structures (committees, teams, task forces, information exchange relationships and arrangements) introduced by Schoonhoven and Jelinek (1996). Hence, in board meetings that represent gates or reviews, practical issues such as agendas, visual aids, and other decision support mechanisms, together with chairing, coaching, facilitating, and communication issues may play an important role as knowledge-sharing meetings and meetings for learning.

Organizational Design and Decision Making from a Strategy Perspective

As we have already indicated, any individual, and especially managers at the middle level (e.g., project managers), can, by their actions and decisions, affect the formation of strategy. The early strategic literature suggested that this approach of strategy formation by individuals at the lower organizational levels may not be effective. Instead, the early strategic literature suggested that strategic issues must be placed as part of a higher-level strategy process at the top level of the organization (e.g., Mintzberg, 1978; Ansoff, 1965). Shendel and Hofer (1979) extended this executive-focused view of strategic management to include other organizational levels. They specified three distinct organizational levels where strategic consideration should happen. First, at the corporate level, the main question is what business the organization should be in. Second, at business unit level, the focus is more on how to compete in that given business. Third, there is the integration of subfunctional activities and the integration of functional areas with the environment. The focus and perspective on strategy thus changes by levels.

Hart (1992) studied different models of strategy-making processes and classified five principal models of the strategy-making process according to the distribution of power in the organization: command, symbolic, rational, transactive, and generative modes. The command mode represents one extreme, where the role of top management is dominant and the participation of other members of the organization is limited to strategy implementation. At the other extreme, in a generative mode, the role of the top manager is to sponsor new ideas—for instance, project proposals emerging from the bottom of the organization—and guide those initiatives to a strategic direction. Moreover, Hart (1992) proposed that the three middle modes of strategy making (symbolic, rational, and transactive), characterized by better use of resources and organizational capabilities, led to higher levels of performance than the two extreme modes. He concluded that the strategy process should be considered as an issue that concerns the whole organization. Moreover, Hart (1992) proposed that strategy making is a capability of an organization that influences its overall performance, and organizational success requires multiple modes in strategy making.

Loch's (2000) study of a European technology manufacturer provides an excellent example of how the organizational setting is arranged in a multiproject environment in terms of distribution of power. It also emphasizes the importance of decision making as an important part of organizational design. Loch identified three different project clusters that defined how the manufacturer initiated and executed product development. An interesting finding was that there was no actual difference in success among the three clusters. Each of the three approaches had its strength. The first cluster, "formal process" projects, used the company's institutionalized product development process and relied on the Stage-Gate process recommended by Cooper (1994), and the formal process supports professional execution of the majority of all new product development projects (Cooper and Kleinschmidt 1987; Cooper, 1994); The second cluster, "under-the-table-projects," represented small teams or "skunk works" (Wolff, 1987) that supported organizational experimentation for new and unstructured ideas and flexibility (Quinn, 1985). The third cluster, comprising "pet projects," or "sacred cows," (projects determined by a powerful senior manager; see, for example, Meredith and Mantel, 1999), can be effective for difficult actions that need management support from a high level, and patience.

Two important weaknesses of undifferentiated process use were what Loch called "rigidity" and "lack of linkage." First, rigidity appears as the formal process where a company follows a relatively rigid Stage-Gate process and is perceived as inflexible in adjusting to specific project needs. Employees resorted to under-the-table projects because the formal process was too rigid and no alternative structure was available. Loch argued that the formal process may be too heavy-handed for incremental projects and too structured for radical projects. Second, lack of linkage occurs where there is a lack of structure for feeding unofficial under-the-table projects into the formal process. Loch argues that many companies suffer from the problem of new-product development not being integrated with strategy. He suggests that the company should develop a customized project portfolio with strategic positioning of projects, and a corresponding mixture of processes to meet its strategic innovation needs. Moreover, Loch considers that the lack of training of business unit managers in general strategy and technology management limit their ability to link strategic context and new-product development approach.

Our analysis of the role of managerial boards, and project and other teams pointed to an emphasis on meetings and reviews that relate to appropriate cross-organizational communication and decision-making processes. From an organizational design viewpoint, Ackoff (1978, 1981) introduces boards and board meetings as major organizational vehicles for participation and communication in what he calls a *circular organization.* McGrath (1996) provides an example of how cross-organizational cooperation is organized through teams, boards, or committees in a managerial model with practical orientation for product management and new product development. A product development project is conducted by a cross-functional core team. The core team is directly responsible for the success of the project, and the team is empowered with full authorization. The core team generally consists of five to eight individuals with different skills and a core team leader. The core team does not have the classical hierarchical approach to organization. Product development decisions are made by the product approval committee designated with the authority and responsibility to make them. The committee members are representatives of senior management representatives. Because the committee is a decision-making group, it should remain small. Four to five executives is an appropriate size. In some cases the committee is the company's executive committee. The decisions are made at phase reviews that are decision-making sessions that occur at specific milestones of the product development. Specifically, the product approval committee initiates new product development projects, cancels and reprioritizes projects, ensures that products being developed fit the company's strategy, and allocates development resources. While the core teams and the product approval committee are for short-term product development, the mid-term technology development is organized in a similar manner through technology development teams and a senior review committee. The senior review committee is a decision-making body of senior scientists and business managers that oversees technology development projects via technology phase reviews. Technology transfer teams with evolving team membership transfer the technology to product development projects (McGrath, 1996).

Important factors—or enablers—for project success often represent issues that are significant from the viewpoint of organizational design. For example, Mikkelsen et al. (1991) studied internal organizational and operational development projects and reported that the characteristics and roles of project managers and top managers were important drivers for project success. Furthermore, according to Brown and Eisenhardt (1995), important success factors of product development include cross-functional teams enabling cross-organizational integration, effective internal and external communication, powerful project leader, and senior management support. Brown and Eisenhardt also discuss the important role of team tenure that reflects the effectiveness of the pattern of working together, the important role of gatekeepers who are individuals that supply external information to the team, and the important role of a team group process that enables effective internal and external communication within the team and with customers, suppliers, and other individuals in the organization. Loch (2000) investigated a larger body of work on new product development and concluded that the following success drivers would represent good management practices: customer orientation and demand pull, cross-functional cooperation, top management support, existence of a champion, good planning and execution with a strong project manager, and the use of a well-defined process with formal measures. The success factors of new product development have slight differences according to the industry, though (e.g., Eisenhardt and Tabrizi, 1995; Terwiesch et al., 1998).

Goal Setting in Time and Aspects of Timing in Relation to Doing the Right Thing

Recent project management and business management literature has raised various aspects of managing time as one important issue in determining how overall efficiency can be achieved (Yeo and Ning, 2002; Steyn, 2002; DeMarco, 2001; Perlow, 1999; Goldratt, 1997). This literature, however, often argues that efficiency of timely performance would contribute to other indirect benefits in terms of efficiency and even effectiveness in overall performance. However, when discussing the management of time, the literature too often emphasizes the aspect of just doing the work efficiently instead of the more strategic dimension of doing the right things. This issue is introduced by Rämö (2002) by focusing on different notions of chronological and nonchronological time in organizational settings. He refers to Drucker's (1974) well-known discussion on efficiency and effectiveness, arguing that efficiency is concerned with doing things right. This is reflected in managerial approaches such as Taylor's scientific management or Deming's Total Quality Management, which both are concerned with doing things right and [just] in time (Rämö, 2002; Drucker, 1999). Such approaches emphasize exact clock time—*chronos*. They require efficiency and doing things right, which requires management and improvement of what is already known. Effectiveness, instead, is doing the right things (Drucker 1974). Rämö (2002) suggests that Drucker's discussion on the difference between efficiency and effectiveness also implies a dualism of time: clock time (*chronos*) emphasizes the chronological sequences of activities and, accordingly, rules efficiency, while the nonchronological aspect of time (*kairos*) relates to right timing and, accordingly, is essential for effectiveness. Seizing windows of opportunities requires a good sense of timing. *Chronos*, or clock time, does not govern such a sense of timing which, instead, it is based on a *kairic* feeling for the right moment (Rämö 2002).

Ackoff (1999) discussed the introduction of different types of systems, with particular attention to organizations as systems. A system may have a memory that can increase its efficiency over time in producing the outcome that is its goal. A purposeful system changes its goals under constant conditions; it selects ends and means and thus displays a will. An ideal-seeking system is a purposeful system that, on attainment of its goals or objectives, then seeks another goal and objective that more closely approximates its ideal. An ideal-seeking system is thus one that has a concept of "perfection" or the "ultimately desirable" and pursues it systematically—that is, in interrelated steps. The time that it would take to reach the ideal could be considered as infinite. Ackoff (1999) introduces the concept of "ends planning" that takes an approach to different perspectives in terms of three types of desired outcomes, each related to different timely perspective. First, "goals" represent ends that are expected within the period of a plan. The goals may be related to entities like, for instance, projects. Second, "objectives" represent ends that provide right directions but are not expected to be obtained until after the period planned for. Our interpretation is that such objectives may be achieved through a collection of projects, such as programs or portfolios, during a longer time span. Third, "ideals" are ends that are believed to be unattainable but toward which continuous progress is thought to be possible and is expected. Ideals provide strategic directions that enable good portfolio decisions for selecting the right projects—and indeed whole collections of well-balanced projects—with the right strategic intent. Goals are means with respect to objectives, and objectives are considered with respect to ideals. Ackoff's ends planning includes four steps: first, selecting a mission; second, specifying the desired properties of the system planned for; third, idealized redesign of that system; and fourth, selecting the gaps between this design and the reference scenario that planning will try to close. One additional related issue that Ackoff (1994) introduces is backward planning, and within backward planning, working backward from the present—that is, from where one wants to be right now to where one is right now.

Aalto et al. (2003) provide an example of what different timely perspectives mean in the R&D context for the management of projects and their portfolios, and the linkage between projects and portfolios. This is illustrated in Figure 7.3, as adapted from Aalto et al. (2003), with modifications to the figures presented by Groenveld (1997, 1998) and Kostoff and Schaller (2001). Aalto et al. use the term R&D to include research, technology development, and product development. Product development is the shortest-term activity. Technology development is more volatile by nature and the projects are typically focused on producing certain technologies or their combinations in the medium term. Such technologies are used in short-term product development. Research is the longest-term activity. It provides technology development with a potential for paradigm shifts and, thus, new points of departure. The interrelatedness of different projects with different time spans and purposes introduces challenges to successful R&D management in terms of how projects and project portfolios are managed.

Summary of the Theoretical Framework Construction

Table 7.1 summarizes the preceding theoretical analysis on strategic business management through multiple projects. The right column of the table presents existing artifacts that are

FIGURE 7.3. INTERRELATEDNESS OF DIFFERENT PROJECTS AND THEIR PORTFOLIOS WITH DIFFERENT TIMELY PERSPECTIVES IN THE R&D FIELD.

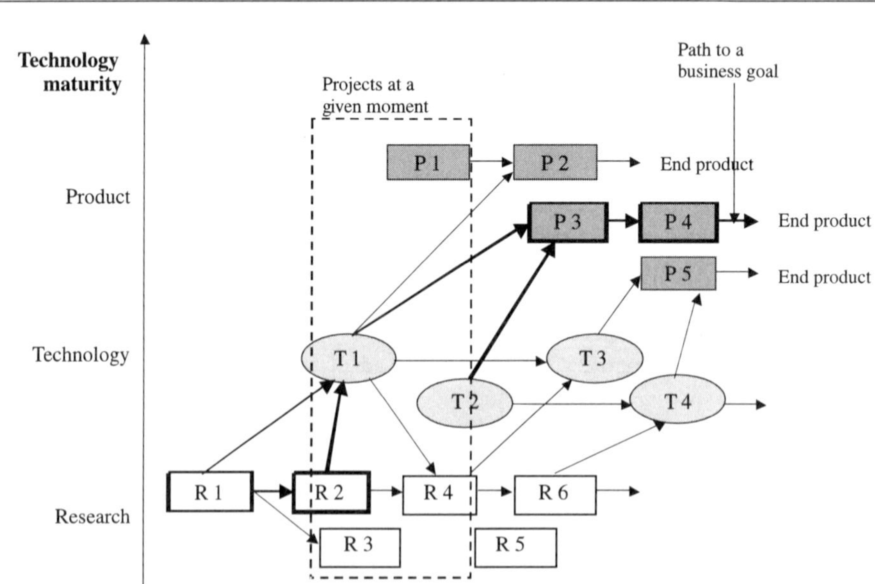

Source: Adapted from Aalto et al. (2003).

relevant to this topic. The left column of the table groups the artifacts by their content in the focal areas. These focal areas can be considered as important prerequisites for successful strategic business management in multiple projects environment. The focal areas are as follows:

1. Categorizing projects by their type
2. Supporting structured and flexible decision making
3. Ensuring effective communication and information transparency
4. Linking projects and strategy process
5. Establishing an organizational design to support strategic management in the multiple-project environment
6. Setting and measuring goals for different time spans in the future
7. Evaluating strategic contents, distinguishing between effectiveness and efficiency

Table 7.1 is self-explanatory. Concerning the table, only two additional explanations are raised here. First, concerning categorizing projects, we argue that the strategic content is partly specific to single projects. This occurs as the project itself is a fundamental managerial entity that interacts with its environment by producing, transferring, and receiving strategic

TABLE 7.1. EXISTING ARTIFACTS FROM THE THEORETICAL FRAMEWORK CONSTRUCTION GROUPED TO SEVEN FOCAL AREAS.

Focal Area	Existing Artifacts
1. Categorizing projects by their type	• Different project types are of different strategic importance. • Different project types require different managerial approaches.
2. Supporting structured and flexible decision making	• Structured decision-making practices (e.g., in board meetings with project-specific decisions) are important for adopting a view on whole portfolios and for linking strategy with projects. • Structured decision-making practices support the realization of strategic intentions of the organization. • Flexible decision-making practices enhance the emergence of innovative ideas and learning. Use of various types of processes in decision making (both formal and informal) increase an organization's ability to succeed. • Decision-making structures such as meetings are important for communication among top management, middle management, and project management. In addition to communicating the intended strategy component top-down, the meetings and their communication serve to foster grounds for the bottom-up emergent strategy component.
3. Ensuring effective communication and information transparency	• Information transparency, both vertical transparency across organizational levels and horizontal transparency across projects and organizational boundaries, and open communication help with building linkages. • Effective communication and information transparency enhances creativity and appearance of new strategic ideas. • Effective communication and information transparency enhances quality and optimality of decisions. • Communication enables learning. • Open sharing of information and information transparency results in better commitment and involvement among individuals and groups in organization.
4. Linking projects and strategy process	• Linking projects and the strategy process enables top management to acquire a holistic picture of ongoing project activities and new innovative ideas emerging from different organizational levels. This holistic picture of project activities and new ideas increase top management's ability to manage organization in a concrete manner toward desired direction. • Linking projects and the strategy process ensures that projects positioned at lower levels become aware of their status in the whole picture of implementing business strategy. This means that project managers are aware of why each of their projects exists and what should be accomplished in the end. A prerequisite for this is that the project manager understands clearly the intended strategy and the ways the project manager is capable of adjusting his or her project's direction. • Linking projects and the strategy process ensures that strategic initiatives are introduced both top-down and bottom-up. • Linking projects and the strategy process ensures that resources are allocated to "strategically right activities." • Linking projects and the strategy process ensures that those activities as a whole contribute in an optimal manner to the whole business.

TABLE 7.1. (*Continued*)

Focal Area	Existing Artifacts
5. Establishing an organizational design to support strategic management in the multiple-project environment	• Because organizational design and related structures partly determine the strategy and strategic capabilities (e.g., controllability and innovativeness) of the organization, it is essential to establish an organizational design that supports successful management schemes. • Hierarchy and boundaries of portfolios in the organization determine: which project activities must be viewed as a whole, how different kinds of portfolios contribute the strategic aims of the organization, and what is the relationship between different portfolios in the organization. • Power structures in the organization determine the organizational decision-making practices. • Management culture and project culture in the organization are important issues. The managerial practices must match the culture, and on the other hand, the culture can be changed by introducing new managerial practices. • An important enabler for the emergent strategic component is that there is a fluent interaction between different organizational levels, that projects are put into strategic perspective by being viewed as whole entities throughout the organization, and that there is communication about how these entities contribute to new strategic dimensions.
6. Setting and measuring goals for different time spans in the future	• Long-term strategic objectives of any organization differ from how short-term objectives are set. Further, different projects may be established simultaneously for different time spans. Accordingly, these objectives and their associated projects must be managed simultaneously by taking both the long term and the short term into account. • Especially in the long perspective, the future is uncertain, and the basic aim is to take advantage for different possible futures. This occurs through managing options toward the uncertain future. • In the case of different planning horizons, concepts such as risk, uncertainty, imperfect knowledge, and ambiguity become important parameters for how projects are managed successfully. • Organizational levels may relate to the length of projects, at least in that top management must have a long-term view of the future. Thus, at higher organizational levels, many projects may be established to pave the way for the long-term mission of the organization.
7. Evaluating strategic contents, distinguishing between effectiveness and efficiency	• For the successful management of multiple projects, it is important to distinguish whether the projects are established for effectiveness or for efficiency. Effectiveness refers to doing the right thing, and efficiency refers to doing the thing right. Effectiveness often means creating something new; efficiency means perfecting something that is already known.

information. Second, concerning support for structured and flexible decision making, flexible practices are needed to allow freedom to adjust the project management approach to fit the project type or its strategic importance. For example, creativity and the emergence of new strategic directions should be allowed in innovative project schemes. This could be achieved by avoiding too centralized and/or too formal management schemes in such projects.

Finally, the following sections of this chapter discuss empirical examples, and the conclusion section at the end of this chapter introduces a framework for strategic business management with suggestions for managerial practices as derived from theoretical and empirical reasoning.

Empirical Examples from Four Case-Study Organizations

The empirical examples discussed in the following pages are based on a study carried out with four case organizations—two private and two public. The investigated project environments included organizational and operational development projects and product development projects. Depending on the case study corporation, our empirical study focused either on organization-unit-/business-unit-specific project portfolios, or on cross-functional project portfolios of a certain project type (e.g., projects with strong IT orientation) across the whole corporation. We call the four organizations C-service, D-engineering, E-maintenance, and M-service. C-service is a large public organization delivering services locally for society and individuals. D-engineering is a large public organization delivering services for society. E-maintenance is a medium-sized private organization with engineering services, systems, and equipment deliveries for industrial customers. M-service is a large private organization with mainly service product deliveries for individual and industrial customers. The empirical examples discussed in the following are partly derived from the extensive empirical analysis and documentation of company-specific data, produced by our colleagues in our research team. Lindblom (2002), Elonen (2002), Hongisto (2002), and Nurminen (2003) are examples of such material. The majority of the documents are proprietary. The object of the research was the broad area of the management of project portfolios in the case study organizations. The methodology was modified from the developmental workshop scheme suggested by Järvinen et al. (2000). The empirical data gathering occurred in 2001 to 2004. We and our colleagues acted in the role of university researchers representing external change agents and facilitators for workshops, piloting efforts, and other developmental schemes. This way, our empirical study represented an organizational and operational change project from the company representatives' viewpoint.

Decision-Related Processes

Figures 7.4 and 7.5 and the following discussion illustrate a multiproject management process as generalized from case organization-specific processes. In the processes with all the case study organizations, the role of making decisions both at the single-project level and at the level of multiple projects was central. In the organization-specific processes, however,

FIGURE 7.4. DECISION-MAKING FROM THE PERSPECTIVE OF ONE SINGLE PROJECT.

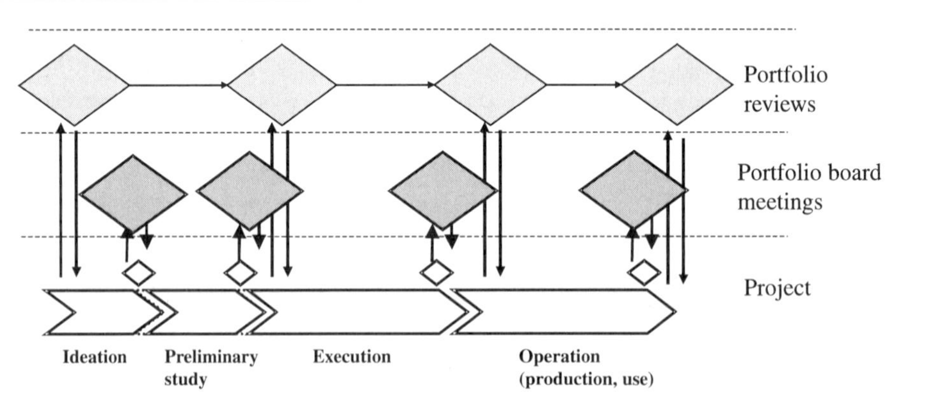

FIGURE 7.5. DECISION-MAKING WITH MULTIPLE PROJECTS.

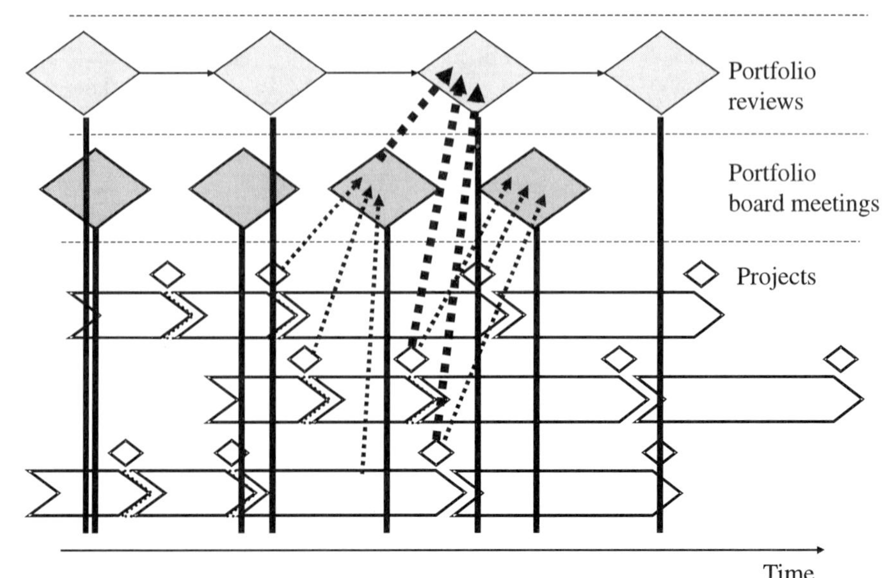

the extent and role of top-down and bottom-up components of the strategy varied depending on the organization. Our process model divides decision-making power at three different levels and emphasizes the communication and information sharing between those levels (see Figure 7.4).

The portfolio board consists of managers from different organizational units or responsibility areas. The cross-organizational composition of the board increases the variety of perspectives on organization within the board and enhances communication and discussion between different organizational areas. This develops a holistic view not only for selecting and prioritizing the right projects, but also for appropriate decisions on resource allocation and timing issues. Thus, the portfolio board is responsible for go/no-go decisions and decisions on major project-specific modifications. The strategic considerations at this decision level concentrate on ensuring that the set of projects under execution and new project ideas provide the best possible basis to achieve organizational objectives and goals. The portfolio board decision meetings strongly support the implementation of intended strategies through projects. The decision flow must be structured in a way that the meeting enhances the possibility of new ideas being introduced effectively, and in that sense also enforces the emergence of new strategies.

While the portfolio board meetings are strongly focused on making project-specific decisions project by project, whole project portfolios are considered in portfolio review meetings (see Figure 7.5). Portfolio reviews adopt a view on whole portfolios rather than single projects as strategic vehicles of action to link current reality and intended positions. Portfolio review meetings are held by boards/bodies consisting of managers representing a higher organizational level than the members of portfolio boards. The objective of the portfolio review meeting is to create a strategic snapshot of the portfolio of projects under execution and new project ideas, and to use this whole entity as a roadmap in planning guidelines for future objectives. The aim of the strategic decisions at this level is to create a feasible match between the organization's capabilities and the resources and opportunities and risks of today and the future.

The vertical arrows up from the level of single projects and down from upper-level decision-making points in Figure 7.4 indicate information and communication flows that are essential for the whole decision-oriented process for the strategic management of multiple projects. The arrows describe information and communication flows that are essential inputs and outputs for respective decision points. Figure 7.5 develops this scheme further by illustrating how decisions made at the level of single projects, and information derived from lower levels, serve as important triggers for decision making at upper levels and simultaneously for strategy implementation and emergence. Figure 7.5 illustrates also how portfolio board meetings and portfolio reviews occur at discrete points in time. Projects and project ideas enter into those upper-level meetings and reviews. The strategic management of multiple projects toward the achievement of business benefits/advantage requires the dynamic comparison of portfolios of project ideas, ongoing projects, and already completed projects. New project ideas, ongoing projects, and completed projects (e.g., internal IT systems, or existing offering of products) potentially affect decision making on any portfolio or any project. Thus, they should all be considered as belonging to the same pool of interlinked activities and opportunities.

Decision-Making Flows in Board Meetings

The following examples of board meetings in M-service and C-service describe the content of decision-making practices at the multiple-projects level. In M-service, the two management boards (i.e., portfolio board for gate decisions, and top management board for reviews) are preexisting boards, but the portfolio focus and related systematic managerial practices introduced new responsibilities and tasks for these managerial bodies. The portfolio review meetings take place two or three times a year and serve as a forum for top management members to discuss and determine the strategic guidelines and objectives for the portfolio. Our discussion here concerns the monthly portfolio board meetings that provide a vital basis for strategy implementation and the emergence of new ideas. The decision flow of the portfolio board meeting in M-service starts by reviewing the information of project reports (including project reports for new project ideas) provided by responsible project managers or owners of the projects/ideas. There must be adequate information that is considered as sufficiently valid and reliable to proceed to a scoring discussion aimed at achieving consensus against a variety of decision criteria. The meeting proceeds by comparing the project's score with the average score of all projects in the portfolio. This prepares the next step of accepting or rejecting projects, and for allocating priorities and resources among projects. The aim of the balancing is to compare ongoing projects to new project ideas by taking into account at least the most important parameters related to strategic importance, benefits, risk, and resources. The actual balancing considerations may use visual graphs as inputs for discussion (for an example of such graph, see Figure 7.6). Decisions on resources are the most important outputs of the discussion. The final step of the meeting includes deriving feedback to be delivered to projects. The feedback includes both written information and information to be explained orally to responsible project individuals. The information comprises the most relevant decision issues related to progress of the project and an explanation on reasons for the decision.

As the focus in C-service's application is on IT project portfolios, the portfolio board consists of managers responsible for IT projects from different functional areas. New project ideas and ongoing projects are evaluated and prioritized in the meeting, and decisions on resource allocations are made. The meeting agenda is not as structured/formal as in the case of M-service, but the discussions are stimulated by using visual Web-based IT tool as an catalyst to capture different views of the current situation of the portfolio of projects. The chair of the portfolio board facilitates the decision-making situation in the meeting. The Web-based application includes important parameters of projects recorded by the responsible project managers or project owners prior to the meetings. The IT tool provides semistructured project-related information as an input for the decision making. This information enforces discussion, and with the help of the facilitator, consensus is achieved and decisions made. An important advantage of structuring decision meetings around the Web-based IT tool is that it provides a shared communication channel to enhance two-way information sharing between projects and the management levels above them. This channel is used to integrate the organizational vision, strategies, and the actual project work.

Information Contents and Decision Criteria

The important information content for the information flows between the project and portfolio board levels was investigated separately for each decision point through the project

FIGURE 7.6. VISUAL GRAPH ON SCORE PROFILES OF MULTIPLE PROJECTS.

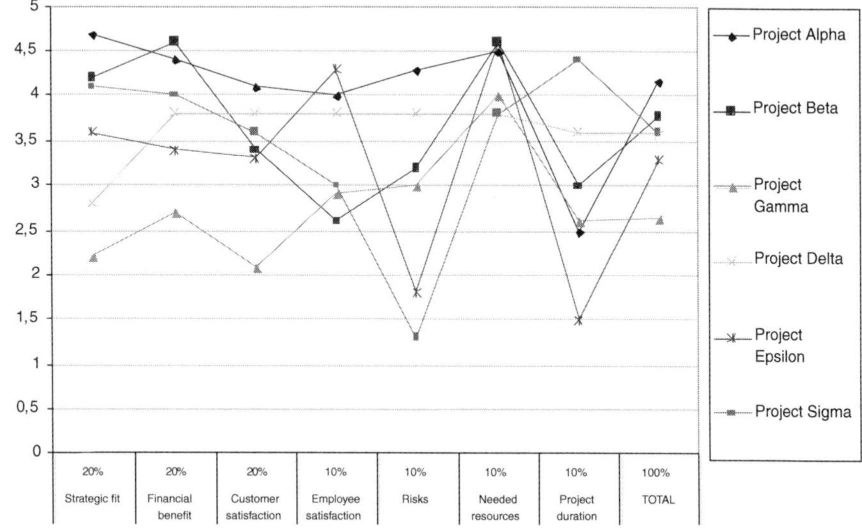

process. The strategic information contents in decision points change during the project process. In the early ideation phase, the information is strongly focused on result-related issues of the project from the business perspective. In the execution phase, the information relates more to monitoring pre-estimated strategic issues and updating estimates. Furthermore, in the post-project phase, the information relates more to the start of the application (going operational or production start) and to gathering feedback information from the operation phase for the purpose of learning for future projects.

The practical applications in organizations with explicit information displays are simplified schemes from contents of strategic information. For example, M-service applies a condensed set of simple criteria for project selection, covering the following categories of criteria: strategic importance, benefits, risk, and resources. As the major current challenge in M-service is to deal with the problems of organizing and resourcing its multiproject efforts vis-à-vis its extensive number of ongoing projects and scarce resources, the criteria set in M-service includes also many issues that relate to the project execution phase. This orientation is reflected by criteria with a major aim of monitoring issues that relate to execution and the successful implementation of the project work.

The portfolio board meetings in M-service are provided with additional structure by allowing the individuals to encode their opinions and beliefs with many criteria in terms of estimating quantitative scores for each criteria prior to the meetings. The scores and weights for criteria allow wide possibilities for preparatory calculations and visual illustrations that can be used as catalysts for communication and decision making that occurs in the actual portfolio board meeting.

Figure 7.6 is a visual look-alike sample graph from M-service's application that illustrates how weighted scores can be used for creating a simultaneous view of many projects at one point of time. This kind of figure can also serve as a tool for evaluating whether intended projects as a whole are effective for fulfilling the strategic objectives. Scoring, quantitative data, and visual illustrations can be used to structure the discussion among meeting attendants on how each project and the project portfolios as a whole contribute to the overall achievement of business benefits, and how well they fulfill strategic business objectives. Such aggregate visual information is helpful especially in portfolio review meetings among top management representatives, where the focus should be more on the information that relates to portfolios as an entity, rather than just to project-specific information. Illustrations with aspects of timing and timely perspectives, for example, roadmap-like presentations, are effective in many decision situations. However, in addition to recording scores for each criteria in M-service, the explicit information recorded for each criterion also includes qualitative and other quantitative information (e.g., monetary figures from economic/financial calculations, resource usage/needs, timely units indicating the schedule). The importance of qualitative information (i.e., documentation, explanations) is emphasized in the M-service's decision process. Finally, we believe that the true content for decision making and related important communication and learning occurs in board meeting events themselves rather than in explicit information contents in practical applications.

The explicit information contents of the project-specific parameters in C-service's application do not represent in a straightforward manner decision criteria as such, but rather represent the relevant information contents to be communicated for decision making and other management purposes. Furthermore, as compared to M-service, C-service emphasizes even more the importance of recording explicit qualitative information for each parameter. Such qualitative information is recorded in the Web-based IT tool. However, for enhanced communication and for increased clarity in comparing projects in a portfolio, many qualitatively expressed parameters (e.g., status, or priority, benefit, and risk) are categorized into classes. Such subdivision into classes, marked with integer figures in C-service, has some analogies with the application of scores in M-service. The qualitative and quantitative information contents in C-service's www tool include, among others, the following important themes: priority, benefit, risk, interconnectedness and linkages to other projects, contact data of responsible project manager and project owner, start date and complete date, cost, resource usage (person-hours), percent complete, and status. Much of the information in C-service's current application reflects the need to organize effectively ongoing projects and manage project progress, resources, cost, and time in a multiproject scheme, where the constraint of scarce resources and the interrelatedness of projects play an important role.

Conclusion: A Framework for Strategic Business Management

Figure 7.7 summarizes the preceding analysis by presenting a framework for strategic business management in multiple-projects environment. For successful multiproject management, it is essential that the managers and decision makers understand the sphere of ultimate

FIGURE 7.7. FRAMEWORK FOR STRATEGIC BUSINESS MANAGEMENT THROUGH PROJECTS.

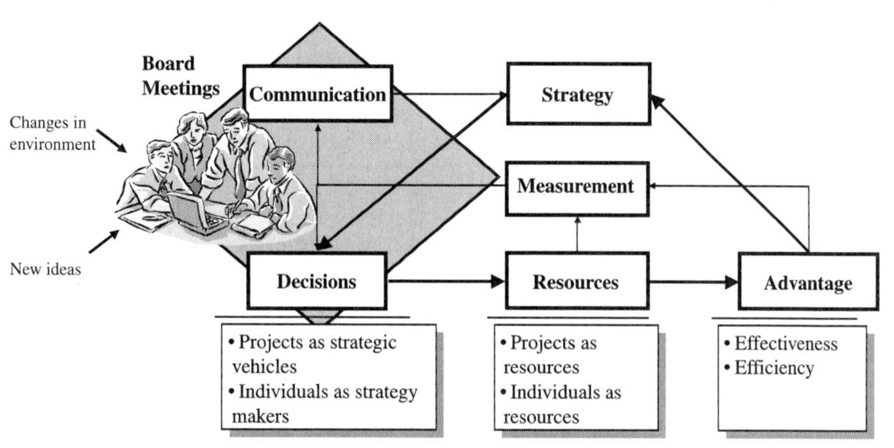

potential purposes of any project or idea. This occurs only if a mature understanding is in place that makes a clear distinction between effectiveness and efficiency.

Meetings, reviews, workshops, or other communication platforms where a group of individuals are collected together are central elements in strategy formation. Such occasions and situations are not only fostering grounds for communication and creative implicit or explicit decisions but also for new ideas. New ideas often arise simultaneously while a well-planned structure is applied for decision making in a group of individuals. However, such a structure should leave enough room for communication and/or expressing new and even radical ideas that fall outside the scope of the actual and concrete decision-making situation at hand.

Figure 7.7 illustrates the central role of a board meeting when managing strategies through multiple projects. When the issue concerning a set of multiple projects is brought to such a meeting through structured information and/or appropriate visual display of such information, and through a well-structured meeting agenda or well-facilitated meeting flow, both explicit and implicit decisions over multiple projects occur in an effective manner. Such information, visualization, agendas, and flows serve as structures that guarantee effective decisions and support for the realization of strategic intentions of the organization, produced in a creative manner, while simultaneously allowing the appearance of creative and innovative new ideas that reformulate strategies and strategic directions.

The corporate strategy (or business strategy) provides the individuals with guidelines, goals, and objectives for decision making. The dynamic nature of strategy implementation is supported by measuring both the achievement of advantages and the resources as the organization's internal capabilities, in relation to the requirements set by the project initia-

TABLE 7.2. MANAGERIAL PRACTICES FOR STRATEGIC BUSINESS MANAGEMENT THROUGH MULTIPLE PROJECTS BY FOCAL AREAS.

Focal Area	Managerial Practice
1. Categorizing projects by their type	• Form specific portfolios or buckets based on strategic guidelines. Consider strategic goals and responsibility areas while doing this. • Establish specific and tailored management models both at the project level and at levels above projects for each portfolio. Ensure that these models enable strategic management in an appropriate manner.
2. Supporting structured and flexible decision making	• Establish meetings, reviews, or workshops where a group of individuals are collected together, to serve as central elements for decision making. • Define specific levels where decision making is expected to occur. Distinguish the different roles of top management, middle management, and project management. • Differentiate operational single project decisions from the strategic ones. Authorize project managers or middle managers to make most operational project decisions. Furthermore, extend strategic considerations to include simultaneous consideration of multiple projects. • Distinguish between two types of portfolio-level decision making. Portfolio board meetings serve as a frequent forum for active monitoring and decision making to ensure that the structure of portfolio aligns with intended strategic guidelines. Portfolio reviews are typically organized few times in a year, and their focus is in strategic future-oriented planning and monitoring the overall situation of the portfolio. • Establish clear and limited roles and responsibilities for decision making. Assign a responsible individual (e.g., portfolio coordinator) for each portfolio, to take the overall responsibility of introducing the situation of the portfolio. • Avoid unnecessary rigidity while following the intended strategy. • Appropriate visual display of structured information, a well-structured meeting agenda, and well-facilitated meeting flows enhance not only effective explicit and implicit decisions but also the appearance of creative and innovative new ideas that reformulate strategies and strategic directions. • Establish criteria that enable comparison, selection, and prioritization of projects. Include strategic issues and project-success-related issues in those criteria. • Organize for measurement of projects, activities, and portfolios. Ensure that measurement is in line with established criteria and strategic guidelines. • Leave room for interactive discussion as principal element of decision making in meetings and boards.

TABLE 7.2. (*Continued*)

Focal Area	Managerial Practice
3. Ensuring effective communication and information transparency	• Support communication by establishing and using systematic structures such as: • Structured meeting agendas for board meetings • Systematic follow-up and measurement of portfolios of projects • Project type specific criteria and decision tools for project prioritization, and for stimulus for discussion • Top-down flow of feedback information down to projects • IT tools that enable the availability of project-related information vertically and horizontally in the organization. • Enhance learning, both directly by allowing experimental schemes or small probes that may fail and indirectly by exchanging experiences in meetings. • Use projects themselves as structured communication platforms, similar to meetings: Both meetings and projects as such bring individuals together to the same structured sphere of communication, decision making, and fostering of new ideas.
4. Linking projects and strategy process	• Make strategic portfolio review meetings timely (e.g., meetings related to the annual strategy process). Use effectively the advantage of knowing the situation of the current portfolio of projects when forming strategic guidelines for the organization. • Ensure a fluent interaction between different organizational levels, and make sure that projects are put into strategic perspective by looking at project entities as a whole (e.g., in portfolio review meetings) and by looking at how these entities contribute to new strategic dimensions. The interaction between organizational levels can be achieved through a cascade of meetings across lower and higher organizational levels. • Introduce top-management-originated intended strategies to lower-level boards and organizational bodies. Furthermore, emphasize the importance of feedback by providing the lower level bodies and boards with top management's decisions concerning projects or portfolios. • Use visualization methods in portfolio review meetings among top management representatives to reflect the project portfolios' role in the adaptation of new strategic directions through the set of projects and their strategic content as a whole. The visualization can, for example, be a time phased, roadmap-like view that paves the potential paths for the future.

TABLE 7.2. (*Continued*)

Focal Area	Managerial Practice
5. Establishing an organizational design to support strategic management in the multiple-project environment	• Define the role of different bodies and individuals at different organizational levels, especially the responsibilities and authorization for decision making. • Use (preferably) existing board structures that are assigned new tasks and responsibilities. • Pay attention to appropriate level of openness, trust, and encouragement of individuals. Top management support is an important factor. • Create clear ownership for each functional activity, cross-organizational process activity, and portfolio of projects. Recognize the overlapping areas of responsibility, and deal with such complexity by organizing for effective communication and information sharing. • Establish project-office-like organizational bodies or responsibilities for such supportive activities that ensure effective support for the overall complex setting of managing multiple projects. • Plan carefully how centralized or how decentralized different decision-making-related activities are. Match the level of centralization/decentralization to fit the organizational culture.
6. Setting and measuring goals for different time spans in the future	• Use roadmap-like presentations that put the projects and their mutual interrelations into timely perspective. • Use supportive illustrations that emphasize the life cycle perspectives (both product life cycle and project life cycle), in order to understand the relationship of new or existing projects to current products or systems and their life cycles. • Analyze stated assumptions carefully, and make different scenarios of the business environment in the uncertain future. This is especially important for understanding the potential outcomes in the long-term future. • Establish effective risk management or uncertainty management procedures for coping with the imperfect knowledge, ambiguity, and uncertainty. Manage options in an effective manner; sometimes one important strategy is to keep options open as long as possible.
7. Evaluating strategic contents, distinguishing between effectiveness and efficiency	• Make a clear distinction between those projects that are driven by improving effectiveness and those that are driven by improving efficiency. • Evaluate the strategic importance of each project for understanding the type of strategic impact produced by the project. Furthermore, estimate the managerial challenge and need for a specific management style that relates typically to newness and risk dimensions in the project.

tives and the external environment. In our framework, projects are used for strategy implementation and emergence. The framework also emphasizes the role of individuals as strategy makers. Individuals' commitment and motivation often guarantee that the intended strategies are in fact realized. Furthermore, the role of projects and their individuals are important strategy makers and remarkable introducers of new ideas. This occurs as projects serve as structured communication platforms with similar impacts to what was discussed previously regarding meetings and group sessions. Both meetings and projects as such bring individuals together to the same structured sphere of communication.

Although we emphasize the role of projects and individuals both at the project level and at levels above projects as *strategy makers*, our framework shows that projects and individuals also can be seen in another role: *resources*. It may be clear that projects in the execution phase can be interpreted as resources, but we emphasize here that even ideas at a very early pre-execution phase are important resources that carry important issues related to the future, often in terms of the potential or strategic business content embedded in the idea. When projects are thought of as resources, measuring resources can be seen as measuring current project-based activities and new projects ideas. Having a clear picture of the current situation of the organizational realities at the project level (capabilities), and comparing that with the desired state of the future, provides a frame for successful decision making in the multiple-projects environment.

Our framework emphasizes the role of face-to-face discussions and communications that takes place in certain specific contexts—for example, meetings or projects. An important managerial challenge is to avoid unnecessary rigidity and sometimes even too much discipline while following the intended strategy. Discussions characterized by various perspectives and stimulated by measures that support flexibility and creativity, are the necessary components for emerging of new strategies. Decisions—small and big—then determine future outcomes in terms of effectiveness and efficiency. The framework suggests that advantages and benefits result primarily from individuals' (company and business unit managers') decisions on projects and project ideas, and individuals' (project managers' and project team members') decisions made in single-project contexts. Finally, Table 7.2 concludes this chapter with what we perceive to be the most important managerial practices in the framework of strategic business management in a multiproject environment.

Acknowledgments

Professor David L. Hawk participated in meetings in the early stages when writing this chapter. Furthermore, he provided valuable comments to the early versions of the manuscript. Our research colleague M.Sc. (Eng.) Merja I. Nurminen helped the writing process with fruitful discussions, while simultaneously conducting her own research scheme on strategic management with project portfolios. In the editor's role, Professor Peter W. G. Morris put in extensive effort by providing us with helpful suggestions and constructive comments for how to make the chapter much better. Professor David L. Hawk, M.Sc. (Eng.) Merja I. Nurminen, and Professor Peter W. G. Morris all deserve our greatest thanks for their most valuable help.

References

Aalto, T., M. Martinsuo, K. A. Artto. 2003. Project portfolio management in telecommunications R&D: Aligning projects with business objectives. 99–147. In *Handbook of product and service development in communication and information technology*, ed. T. O. Korhonen and A. Ainamo. Boston: Kluwer Academic Publishers.

Aaltonen, P., and K. Ikävalko. 2001. Implementing strategies successfully. *XII World Productivity Congress. Track One: Cultivating Innovation.* Hong Kong and Beijing, November 5–10.

Ackoff, R. L. 1978. *The art of problem solving.* New York: Wiley.

———. 1981. *Creating the corporate future.* New York: Wiley.

———. 1994. *The democratic corporation: A radical prescription for recreating corporate America and rediscovering success.* New York: Oxford University Press.

———. 1999. *Ackoff's best: His classic writings on management.* New York: Wiley.

Andrews K. R. 1971. *The concept of corporate strategy.* Homewood, IL: Irwin.

———. 1995. The concept of corporate strategy. In *The strategy process.* H. Mintzberg, J. B. Quinn, and S. Ghoshal. Prentice Hall London Exerpted from K. R. Andrews. 1980. *The concept of corporate strategy.* Copyright Richard D. Irvin, Inc.

Ansoff, H. I. 1965. *Corporate strategy.* New York: McGraw-Hill.

Archer, N. P., and F. Ghasemzadeh. 1999. An integrated framework for project portfolio selection. *International Journal of Project Management* 17(4):207–216.

Artto, K. A. 2001. Management of project-oriented organisation: Conceptual analysis. In *Project portfolio management: Strategic management through projects*, ed. K. A. Artto, M. Martinsuo, and T. Aalto, 5–20. Helsinki, Finland: Project Management Association Finland.

Artto, K. A., P. H. Dietrich, and T. Ikonen. 2002. Industry models of project portfolio management and their development. In *Proceedings of the PMI Research Conference 2002,* ed. D. P. Slevin, J. K. Pinto, and D. I. Cleland. 3–13. Seattle, July 14 to 17. Newtown Square, PA: Project Management Institute.

Bourgeois, L. J. 1985. Strategic goals, perceived uncertainty, and economic performance in volatile environments. *Academy of Management Journal* 28(3):548–573.

Brown, S. L., and K. M. Eisenhardt. 1995. Product development: Past research, present findings, and future direction. *Academy of Management Review* 20(2):343–378.

———. 1997. The art of continuous change: Linking complexity theory and time-paced evolution in relentlessly shifting organizations. *Administrative Science Quarterly* 42(1):1–34.

Burgelman, R. A. 1983. A model of the interaction of strategic behavior, corporate context, and the concept of strategy. *Academy of Management Review* 8(1):61–70.

———. 1991. Intraorganizational ecology of strategy making and organizational adaptation: Theory and field research. *Organizational Science* 2(3):239–262.

Chaffee, E. E. 1985. Three models of strategy. *Academy of Management Review* 10(1):89–98.

Chandler, A. D. 1962. *Strategy and structure: Chapters in the history of the American industrial enterprise.* Cambridge, MA: MIT Press.

Cohen, M. D., J. G. March, and J. P. Olsen. 1972. A garbage can model of organizational choice. *Administrative Science Quarterly* 17:1–25.

Cooper, R. G. 1994. Debunking the myths of new product development. *Research Technology Management* (July–August): 40–50.

Cooper, R. G., and E. J. Kleinschmidt 1987. New products: What separates winners from losers. *Journal of Product Innovation Management* 4.

Cooper, R. G., S. J. Edgett, and E. J. Kleinschmidt. 1998. *Portfolio management for new products.* New York: Perseus Books.

Crawford, L., J. B. Hobbs, and J. R. Turner. 2002. Investigation of potential classification systems for projects. *PMI Research Conference 2002*. pp. 181–190. Seattle, July 14–17. Newtown Square, PA: Project Management Institute.

De Geus, A. P. 1988. Planning as learning. *Harvard Business Review* (March–April): 70–74.

DeMarco, T. 2001. *Slack*. New York: Random House.

Drucker, P. F. 1974. *Management. tasks, responsibilities, practices*. London: Heinemann.

———. 1999. *Management challenges for the 21st century*. New York: Harper Business.

Dye, L. D., and J. S. Pennypacker. 1999. An introduction to project portfolio management. In *Project portfolio management: Selecting and prioritizing projects for competitive advantage*, ed. L. D. Dye and J. S. Pennypacker xi–xvi. West Chester, PA: Center for Business Practices.

Eisenhardt, K. M., and S. L. Brown. 1998. Time pacing: Competing in markets that won't stand still. *Harvard Business Review* (March–April):59–69.

Eisenhardt, K. M., and B. N. Tabrizi. 1995. Accelerating adaptive processes: Product innovation in the global computer industry. *Administrative Science Quarterly* 40:84–110.

Elonen, S. 2002. Project portfolio management: Managerial problems and solutions for business development portfolios. Master's thesis, Helsinki University of Technology.

Elonen, S., K. A. Artto. 2003. Problems in managing internal development projects in multi-project environments. *International Journal of Project Management* 21(6):395–402.

Floyd, S. W., and B. Woodridge. 2000. *Building strategy from the middle: Reconceptualizing strategy process*. Thousands Oaks, CA: Sage Publications.

Fredrickson, J. 1984. The competitiveness of strategic decision processes: Extension, observations, future directions. *Academy of Management Journal* 27:445–466.

Galbraith, J., 1995. *Designing organizations: An executive briefing on strategy, structure and process*. San Francisco: Jossey-Bass.

Gareis, R. 2000a. Programme management and project portfolio management: New competences of project-oriented companies. *Fourth International Conference of the International Research Network on Organising by Projects IRNOP IV*. Sydney, January 10th–12th.

———. 2000b. Competences in the project-oriented organization. In *Project Management Research at the Turn of the Millennium: Proceedings of the PMI Research Conference, ed.* D. P. Slevin, D. L. Cleland, and J. K. Pinto. 17–22. Newtown Square, PA: Project Management Institute.

Goldratt, E. M. 1997. *Critical chain*. Great Barrington, MA: North River Press.

Groenveld, P. 1997. Roadmapping Integrates Business and Technology. *Research-Technology Management* (September–October):48–55.

———. 1998. The roadmapping creation process, Presentation at the Technology Roadmap Workshop, Washington, D.C., October 29.

Hart, S. L. 1992. An integrative framework for strategy making process, *Academy of Management Review* 17(2):327–351.

Hatch, M. 1997. *Organization theory: Modern, symbolic and postmodern perspectives*, New York: Oxford University Press.

Hofer, C. W. 1975. Toward a contingency theory of business strategy. *Academy of Management Journal* 18:784–810.

Hongisto, J. 2002. Project selection and decision making for successful project portfolio management. Master's thesis, Helsinki University of Technology.

Järvinen, P., K. A. Artto, and P. Aalto. 2000. Explorations on the integration of fractured process improvement: The 3A-workshop procedure. *Project Management* 6(1):77–83.

Kaplan, R. S., and D. P. Norton. 1992. The balanced scorecard: Measures that drive performance. *Harvard Business Review* (January–February).

———. 1996. *The balanced scorecard: Translating strategy into action*. Boston: Harvard Business School Press.

———. 2001. Transforming the balanced scorecard from performance measurement to strategic management: Part I. *Accounting Horizons* 15 (1, March): 87–104

Kostoff, R. N., and R. R. Schaller. 2001. Science and technology roadmaps. *IEEE Transactions on Engineering Management* 48(2):132–143.

Kotsalo-Mustonen, A. 1996. Diagnosis of business success: Perceptual assessment of success in industrial buyer-seller business relationship. PhD. diss., Helsinki School of Economics and Business Administration, Publications A-117, Helsinki, Finland.

Lindblom, C. E. 1959. The science of muddling through. *Public Administration Review*, pp. 79–88.

Lindblom, L., 2002. A decision support system for project prioritization and selection: A case study. Seminar study, unpublished company report, Helsinki University of Technology.

Loch, C. 2000. Tailoring product development to strategy: Case of a European technology manufacturer. *European Management Journal* 18 (3, June): 246–258

Lundin, R., and R. Stablein. 2000. Projectisation of global firms: Problems, expectations and meta-project management. Fourth International Conference of the International Research Network on Organising by Projects IRNOP IV, Sydney, Australia, January 10–2th.

McGrath, M. E., ed. 1996. *Setting the PACE in product development: A guide to Product and cycle-time excellence.* Boston: Butterworth-Heinemann.

Meredith, J. R., and S. J. Mantel, Jr. 1999. Project selection. In *Project portfolio management: Selecting and prioritizing projects for competitive advantage*, ed. L. D. Dye and J. S. Pennypacker. 135–167. West Chester, PA: Center for Business Practices.

Mikkelsen, H., W. Olsen, J. O. Riis. 1991. Management of internal projects. *International Journal of Project Management* 9(2):77–81.

Mintzberg, H. 1978. Patterns in strategy formation. *Management Science* 24:934–948.

Mintzberg, H. and Waters, J. A. 1985. Of strategies, deliberate and emergent. *Strategic Management Journal* 6:257–272.

Mintzberg, H., J. B. Quinn, and S. Ghoshal. 1995. *The strategy process.* Prentice Hall London.

Morris, P. W. G., and G. H. Hough. 1987. *The anatomy of major projects: A study of the reality of project management.* Chichester, UK: Wiley.

Murray-Webster, R., and M. Thiry M. 2000. Managing programmes of projects. In *Gower handbook of project management*, ed. J. R. Turner and S. J. Simister. 3rd ed. 47–63. Aldershot, UK: Gower.

Nurminen, M. 2003. Strategic management with project portfolios. Master's thesis, Helsinki University of Technology.

Office of Government Commerce (OGC). 2002. *Managing successful programmes.* 3rd ed. London: The Stationery Office.

Pavón, R. C. C. 2002. Systemic intra-organizational industry demand analysis to support an organization's strategy making. PhD diss., Helsinki University of Technology, Institute of Strategy and International Business.

Perlow, L. A. 1999. The time famine: Toward a sociology of work time. *Administrative Science Quarterly* 44:57–81.

Pettigrew, A. M. 1992. On studying managerial elites. *Strategic Management Journal* 13:163–182.

Platje, A., H. Seidel, and S. Wadman. 1994. Project and portfolio planning cycle: Project based management for multiproject challenge. *International Journal of Project Management* 12(2):100–106.

Project Management Institute. 2000. *A guide to the Project Management Body of Knowledge.* Project Management Institute Standards Committee. Newtown Square, PA: Project Management Institute.

Porter, M. E. 1980. *Competitive strategy: Techniques for analyzing industries and competitors.* New York: Macmillan.

———. 1996. What is strategy. *Harvard Business Review.* (November–December).

Poskela J., M. Korpi-Filppula, V. Mattila, and I. Salkari. 2001. Project portfolio management practices of a global telecommunications operator. In *Project portfolio management: Strategic management through projects*, ed. K. A. Artto, M. Martinsuo, and T. Aalto. 81–102. Helsinki: Project Management Association.

Quinn, J. B. 1985. Managing innovation: Controlled chaos. *Harvard Business Review*, (May–June):73–84.

———. 1995. Strategies for change. In *The strategy process*. H. Mintzberg, J. B. Quinn, and S. Ghoshal. Prentice Hall London, Europe. Excerpted from J. B. Quinn. 1980. *Strategies for change: Logical incrementalism*. Copyright Richard D. Irwin, Inc.

Rouhiainen, P. 1997. Managing new product development: Project implementation in metal industry. PhD diss., Tampere University of Technology, Publications 207, Tampere, Finland.

Rämö, H., 2002. Doing things right and doing the right things: Time and timing in projects. *International Journal of Project Management* 20 (7, October): 569–574.

Saravirta, A. 2001. Project success through effective decisions: Case studies on project goal setting, success evaluation, and managerial decision making. PhD. diss., Acta Universitatis Lappeenrantaensis 121, Lappeenranta University of technology, Lappeenranta, Finland.

Schoonhoven, C. B., and M. Jelinek. 1996. Dynamic tension in innovative, high technology firms: Managing rapid technological change through organizational structure. In *Managing strategic innovation and change: A collection of readings*, ed. M. Tushman and P. Anderson. 233–254. New York: Oxford University Press.

Senge, P. M. 1990. The leader's new work: Building learning organizations. *Sloan Management Review* (Fall): 7–23.

Shenhar A. J., O. Levy, and D. Dvir. 1997. Mapping the dimensions of project success. *Project Management Journal* 28(2):5–13.

Shenhar, A. J., D. Dvir, T. Lechler, and M. Poli. 2002. *One size does not fit all: True for projects, true for frameworks*. 99–106. PMI Research Conference, Seattle, July 14–17. Newtown Square, PA: Project Management Institute.

Simon, H. A. 1957. *The new science of management decision*. New York: Macmillan.

Simons, R. 1995. *Levers of control: How managers use innovative control systems to drive strategic renewal*. Boston: Harvard Business School Press.

Steyn, H. 2002. Project management applications of the theory of constraints beyond critical chain scheduling. *International Journal of Project Management* 20 (1, January): 75–80.

Terwiesch, C., C. H. Loch, and M. Niederkofler. 1998. When product development performance makes a difference: A statistical analysis in the electronics industry. *Journal of Product Innovation Management* 15:3–15.

Turner, J. R., and A. Keegan. 1999. The management of operations in the project-based organization. In *Managing business by projects*, ed. K. A. Artto, K. Kähkönen, and K. Koskinen. Vol. 1, pp. 57–85, and Vol. 2. Helsinki: Project Management Association Finland and Nordnet

Turner, J. R. 1999. *The handbook of project-based management: Improving the processes for achieving strategic objectives*. 2nd ed. London: McGraw-Hill.

Turner, J. R., and A. Keegan. 2000. Processes for operational control in the project-based organization. In *Project management research at the turn of the millennium: Proceedings of the PMI Research Conference*, ed. D. P. Slevin, D. I. Cleland, and J. K. Pinto. 123–134. Newtown Square, PA: Project Management Institute.

Turner, J. R., A. Keegan, and L. Crawford. 2000. Learning by experience in the project-based organization. In *Project management research at the turn of the millennium: Proceedings of the PMI Research Conference*, ed. D. P. Slevin, D. I. Cleland, and J. K. Pinto. 445–456. Newtown Square, PA: Project Management Institute.

Wheelwright, S. C., and K. B. Clark. 1992. Creating project plans to focus product development. *Harvard Business Review* (March–April): 70–82.

Wolff, M. F., 1987. To innovate faster, try the skunk works. *Research Technology Management* (September–October: 7–8.

Yeo, K. T., and J. H. Ning. 2002. Integrating supply chain and critical chain concepts in engineer-procure-construct (EPC) projects. *International Journal of Project Management* 20 (4, May): 253–262.

Youker, R., 1999. The difference between different types of projects. *Proceedings of Project Management Institute Annual Seminars and Symposium.* Newtown Square, PA: Project Management Institute.

CHAPTER EIGHT

MOVING FROM CORPORATE STRATEGY TO PROJECT STRATEGY

Ashley Jamieson, Peter W. G. Morris

Developing and implementing corporate strategy is one of the most actively researched, taught, and talked about subjects in business today. Projects and project management are often quoted as important means of implementing strategy, but the way this implementation happens in practice is rarely the subject of detailed review. This chapter addresses this relationship head-on.

Many organizations move from corporate strategy to project strategy using highly structured and fully integrated sets of business management, strategic management, and project management processes, practices, and methods, deployed by highly skilled professional staff. This chapter looks at these means. The key findings from a series of studies of the way major organizations do this, together with a survey that we recently undertook of project management professionals' practice in this area are also included.

Developing Corporate Strategy

Corporate strategy is created as a means of thinking through and articulating how an organization's corporate goals and objectives will be pursued and achieved. This strategy is then typically cascaded through several strategic business units (SBUs) and then ends up being represented as collections—portfolios—of programs or projects. These become the vehicles for implementing the approved strategic initiatives.

Various processes are employed for doing this. Highly structured business management models are used by some of organizations; these identify the major "value delivery" processes and/or the key business processes that enable the organization to operate effectively. Much of traditional management writing tends only to cover the strategic management processes that formulate and implement strategy at the corporate level; there is a real dearth of writing about how corporate strategy gets translated into comprehensive program or project management strategies. The two sets of activities are, in practice, well interconnected and are the main means by which strategy is moved and aligned through corporate, business unit, and project levels. Where they often work less well, however, is in the feedback and adjustment process as events unfold and as program or project definition needs realigning (or vice versa: as project or program definition creates new situations that, in turn, impact the enterprise's strategic intent, thereby requiring the process to iterate!).

Developing effective strategy for major programs or projects from corporate and business strategies is a complex activity. It involves the use of key project management processes such as project definition and includes strategic elements from a wide range of project management practices, such as risk management, value management, and procurement management, and encompasses the entire project life cycle. (See the chapters by Ward and Chapman, Thiry, Venkataraman, Langford, and others.) The roles, responsibilities, and accountabilities for those operating these processes and practices are usually identified within the process documentation, and their competency levels specified in competency frameworks or job descriptions. Recent research that we conducted shows that some companies have developed and implemented highly effective means of doing this. Many other organizations use less ambitious and effective approaches, however.

Positioning Project Management within a Business Management Context

To understand the way corporate strategy is translated into project strategy, it is important to start by considering the business management context and the position of project management within it, and how the business management functions perceive the project management function. While project management professionals and practitioners may think their function is central to the success of a company, it has little meaning unless it is clearly established and embedded within the organizational structure and business management processes of the enterprise. An indication of the position and the relative prominence given to project management can be gauged from the business management models used by organizations.

Watson (1994) gives McKinsey the credit for developing the generic business enterprise model shown in Figure 8.1. This model shows the structure of an organization in terms of processes rather than the traditional functional or matrix form and serves as a process framework for an organization.

The model also identifies what are considered to be the major processes of the value delivery system of a company and the key business processes that enable them, and implies a high level of connectivity between the two. Recent research findings by Morris and Jamieson (2004) show that many major organizations regard the management of projects as a key business process, and Figure 8.1 has been modified to reflect this fact.

FIGURE 8.1. GENERIC BUSINESS ENTERPRISE MODEL.

Value Delivery System	Business Planning System	Product Generation System	Customer Delivery System	Customer Support System
Key Business Processes	• Corporate Governance • Strategic Planning • Information Architecture • Competitive Analysis • Technology Acquisition • Strategic Alliance • **Management of Projects**	• Market Research • Product Design • Production Process • Supplier Management • Product Assembly • Product Launch • **Management of Projects**	• Channel Management • Account Management • Sales Process • Order Fulfilment • Collection Process • Satisfaction Measurement • **Management of Projects**	• Delivery Installation • Customer Training • Expectation Setting • Preventative Maintenance • Unplanned Service • Reliability Assessment

Source: Watson (1994); adapted by Morris and Jamieson (2004).

Creating Corporate and Business Strategies

The strategy management process is dealt with extensively by numerous authors. Most include the concepts and processes associated with strategy analysis, strategy creation (formulation), strategy evaluation, and strategy implementation. But few explicitly connect corporate and business unit strategy with project strategy or suggest that it should be taken into account at the strategy formulation stage or when determining the capability of an organization at the strategy implementation stage.

Strategy management is a dynamic process; and Mintzberg and Quinn (1996) show that "emergent" strategy is a key factor—that is, strategy that becomes evident as it, and events, emerge with time—in influencing realized strategy. Hill and Jones (2001) demonstrate how emergent strategy influences intended strategy through the components of the strategic management process, as, for example, those shown in Figure 8.2. This model indicates that strategy formulation flows from an organization's mission and goals through functional, business, and corporate levels. But, in common with much of the strategic management literature (Mintzberg, Ahlstrand, and Lampel, 1998, for example, point out that there are hundreds of different strategic planning models), Hill and Jones do not explicitly take account of the influence or impact of projects or project management activities on the creation and implementation of strategy, although it is clearly implied.

The findings presented in this chapter, however, show that most of the components of the strategic planning process, such as internal analysis, organizational structures, and con-

FIGURE 8.2. COMPONENTS OF THE STRATEGIC MANAGEMENT PROCESS.

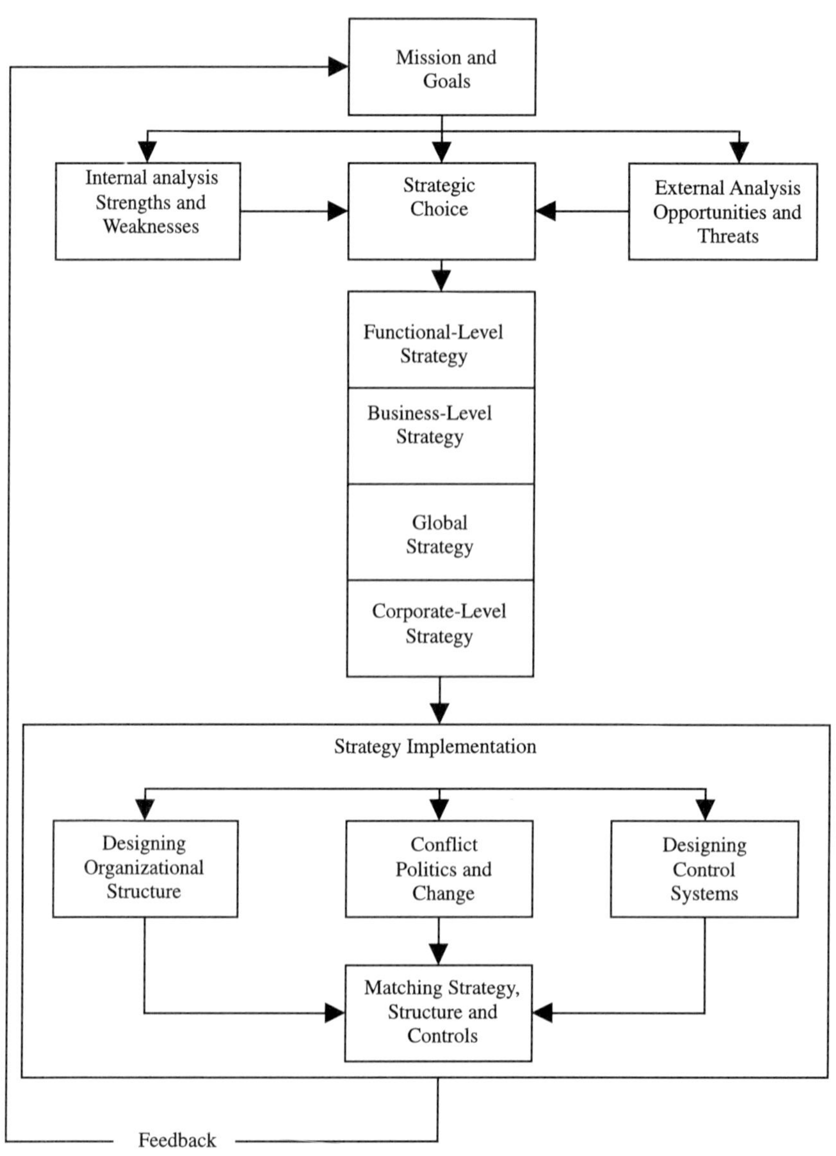

Source: Hill and Jones (2001).

trol systems, have strong links to project management processes and activities, and thereby strongly influence intended corporate and business strategies.

The availability of resources is a major factor in deciding the strategy of an organization. Figure 8.3 shows a schematic used by Mintzberg and Quinn (1996) to analyze strategy and serves to indicate some of the key activities involved in the formulation and implementation of strategy, one of which is determining the managerial resources of the organization. A fundamental responsibility of project management in organizations managing major projects is to manage and/or coordinate effectively whatever company resources are at its disposal. A company's skill at managing projects, and therefore its project management capability, is an essential managerial resource that can and does influence corporate and business strategies strongly—particularly in companies that see themselves in dynamic change situations where agility is important or that are driven by major projects (as in aerospace, construction, new product development, etc.). (See the chapters by Archer and Ghasemzadeh, Shenhar and Dvir, Milosevic, and others.)

Hierarchy of Objectives and Strategies

A hierarchy of objectives and strategies can be formed as a result of using a strategy planning process; this can be a very effective means of structuring and managing strategy and communicating it to the organization. One such model, used by Cleland (1990), is Archibald's hierarchy of objectives, strategies, and projects, shown in Figure 8.4.

This model maps out in detail the structure and relationship of objectives and strategies at the policy, strategic, operational, and project levels. Specific objectives and strategies are developed at each level from higher-level ones and cascaded down, thereby ensuring alignment and continuity of strategy. Projects and their objectives, strategies, and plans are shown at the operational level.

Hierarchy of Strategy Plans

The corporate and business strategies created by organizations using a strategic planning process are incorporated into a hierarchy of strategy plans. Kerzner (2000) provides an example of a typical hierarchy, as shown in Figure 8.5.

This model shows typical strategic plans cascading corporate strategy to SBU level from a single corporate strategic plan and supporting plans cascading business strategy from each SBU. Another example of a hierarchy of strategic plans is the Stanford Research Institute's (SRI's) "system of plans," which is shown in Mintzberg, Ahlstrand, and Lampel (1998).

If this, roughly, is how corporate strategy is created from an organization's mission, goals, and objectives and how this strategy is used to create business unit objectives and strategies, the next stage is to consider how business unit strategy is moved through portfolios and programs into projects.

FIGURE 8.3. STRATEGY AS A PATTERN OF INTERRELATED DECISIONS.

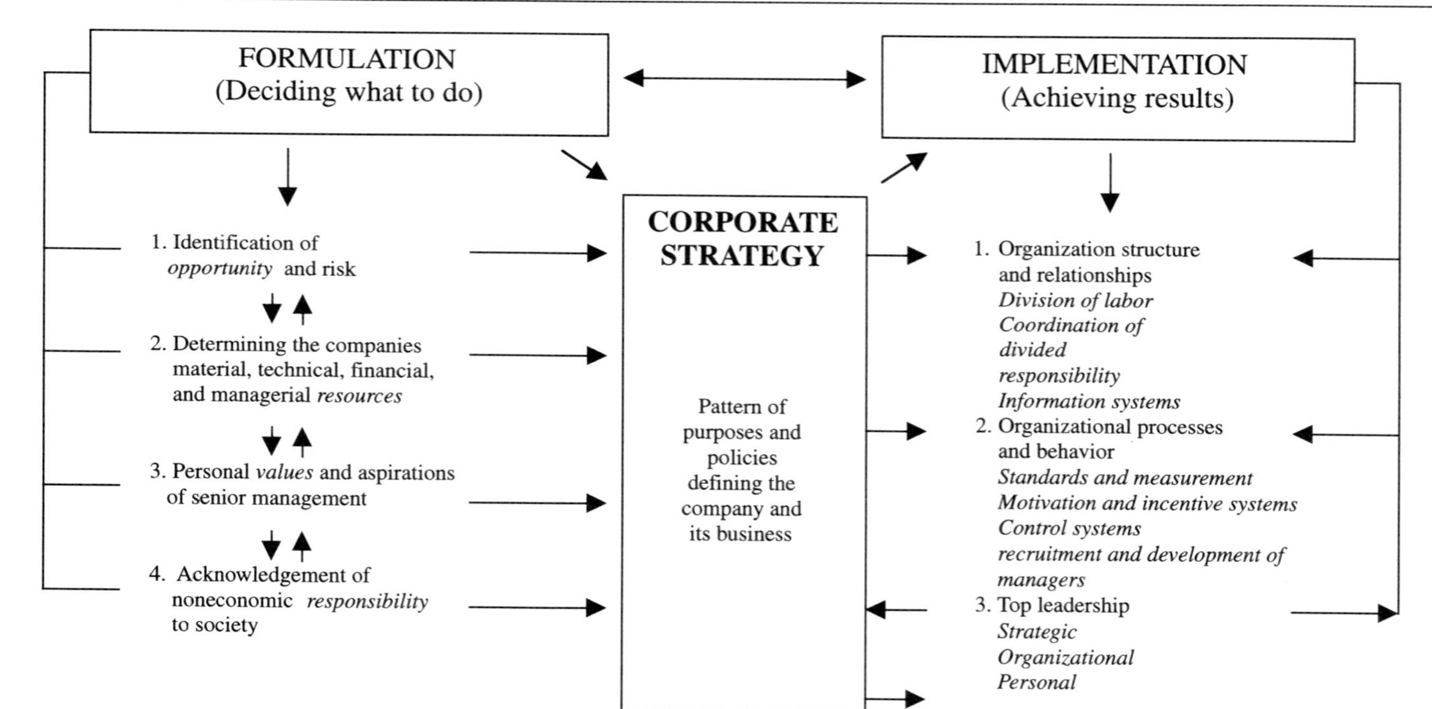

Source: Mintzberg and Quinn (1996).

FIGURE 8.4. HIERARCHY OF OBJECTIVES, STRATEGIES, AND PROJECTS.

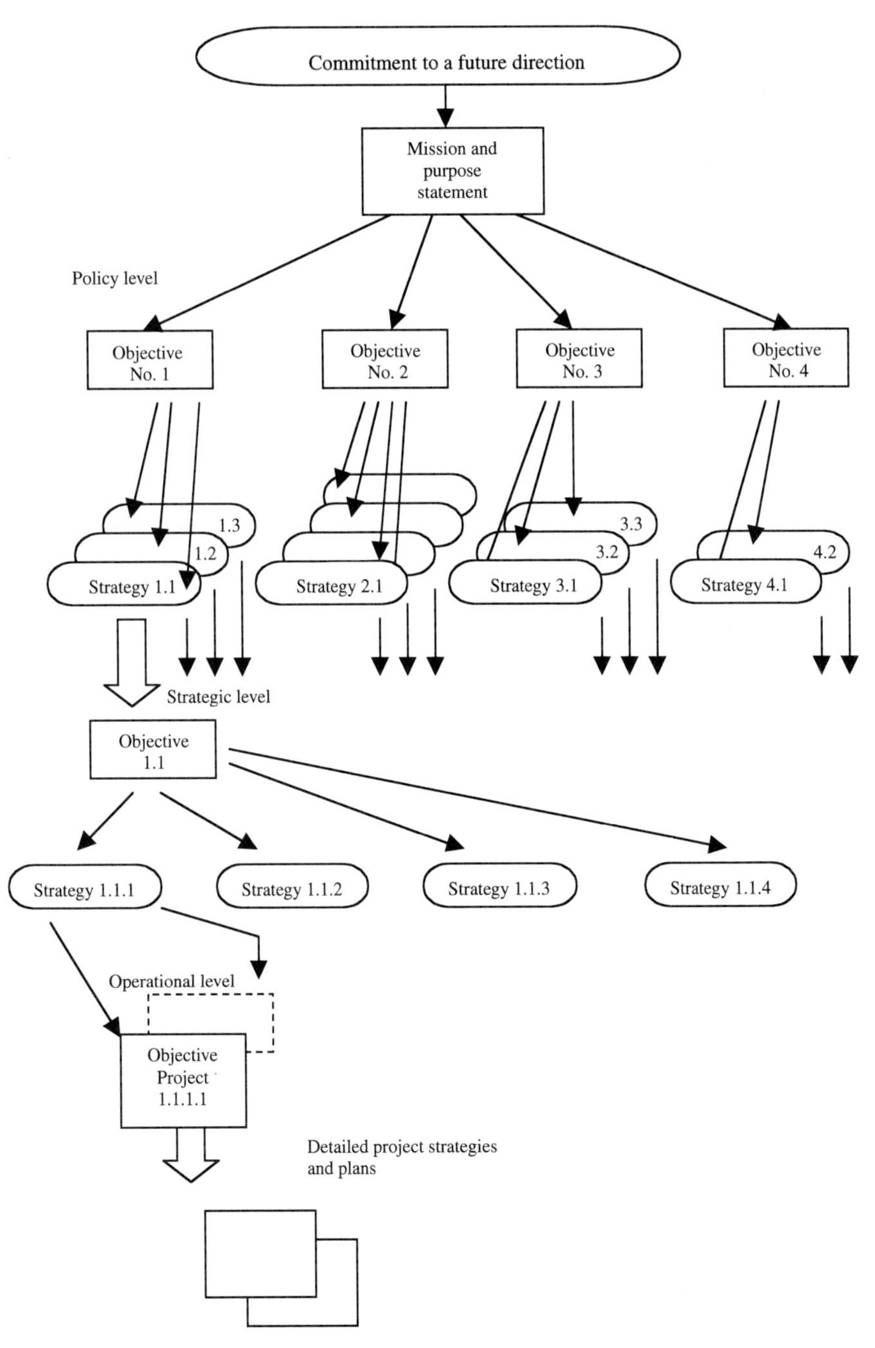

Source: R. D. Archibald (1987).

FIGURE 8.5. HIERARCHY OF STRATEGIC PLANS.

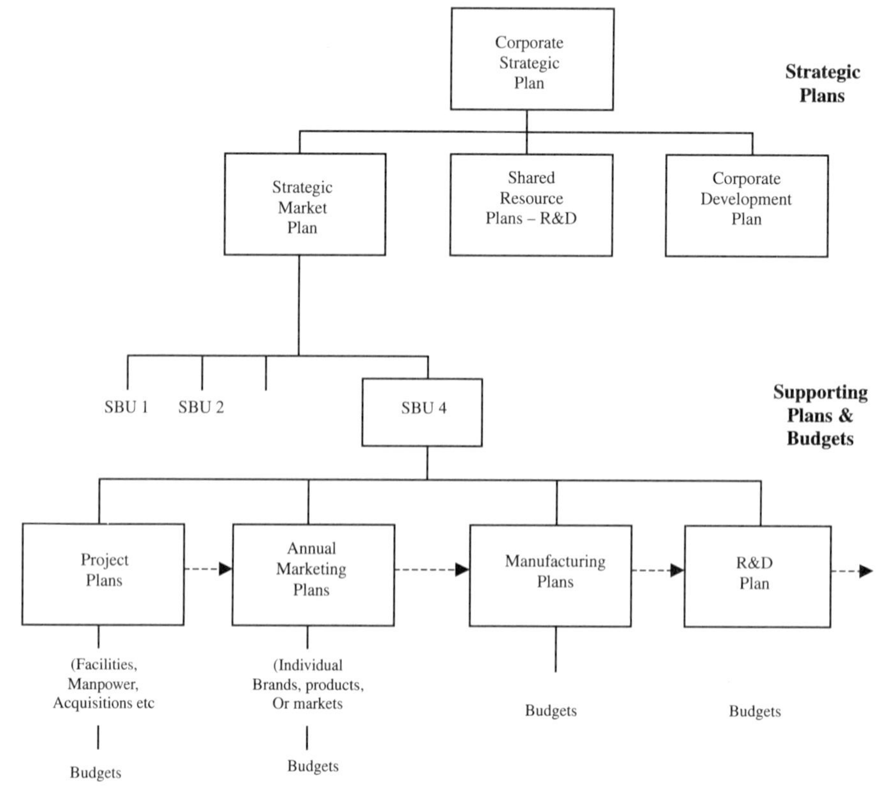

Source: Kerzner (2000).

Moving Business Strategy through Portfolios, Programs, and Projects

Turner (2000) points out that the majority of projects take place as part of a portfolio of several projects. Program management is the way of coordinating projects that have a shared business aim (see the chapter by Thiry). Both portfolio management and program management focus on prioritizing resources and optimizing the business benefit (Kerzner, 2000; Morris and Jamieson, 2004) Portfolio management, as Archer and Ghasemzadeh outline in their chapter, tends to be about the selection and prioritization of projects or programs. Program management is about the day-to-day management of programs (products, platforms, brands, or multiple projects) to deliver business value (OGC, 1999; see also the chapter by Midler and Navarre).

Turner (1999) cites Youker to illustrate how organizations undertake programs and projects to achieve their development objectives. Morris and Jamieson (2004) have adapted this model to include business strategy and portfolios, as shown in Figure 8.6, and to indicate that a portfolio may comprise of groups of programs and/or projects.

Portfolios

Project portfolio management is predominantly about choosing the right project, whereas project management is about doing the project right. It has also been described as the activity of aligning resource demand with resource availability, to achieve a set of strategic goals (see the chapter by Archer and Ghasemzadeh). Crawford (2001) believes that making the conceptual leap from the tools-and-techniques-focused variety of project management to portfolio management (and indeed program management), with its broader focus on business strategy and enterprise-wide integration, is a special challenge and one that many now face with little in the way of standards, best practices, or other generally accepted knowledge to guide them.

Knutson (2001) provides a generic project portfolio management process model, schematized in Figure 8.7, which takes strategy issues into account. The solicitation stage of the process ensures that a potential program or project has a credible strategy that is aligned with the organization's objectives and strategy; and that a business case for the project, containing these details, is developed. The relative value of the project and its overall synergy

FIGURE 8.6. LINKING CORPORATE AND PROJECT STRATEGY.

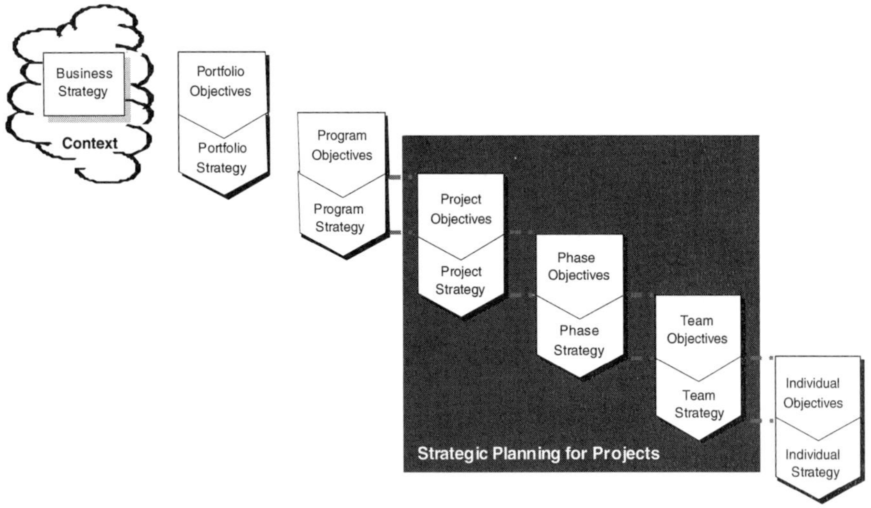

Source: The Handbook of Project-Based Management, 2nd ed. J. R. Turner © 1999 Reproduced by kind permission of the Open University Press, McGraw-Hill Publishing Company.

FIGURE 8.7. A GENERIC PORTFOLIO MANAGEMENT PROCESS.

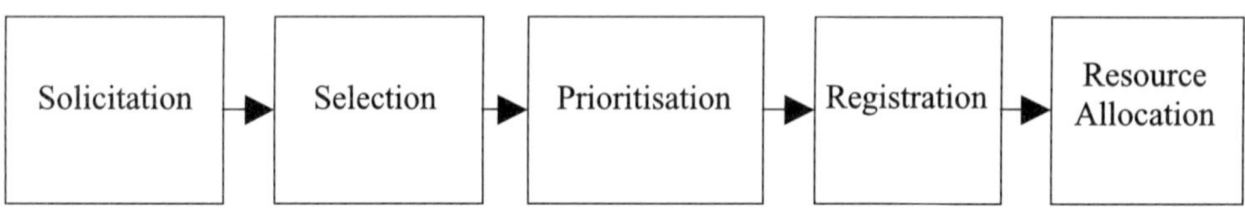

Source: J. Knutson (2001) and schematized by Morris and Jamieson (2004).

with the organization's strategy is evaluated during the selection process. If a project is selected, the next step is its prioritization. A scoring system is used to determine the priority of the project with respect to other projects, during the prioritization stage (see Archer and Ghasemzadeh's chapter); and subsequently, depending upon availability, resources are allocated to the project to allow it to proceed.

Archer and Ghasemzadeh (1999) also provide a general framework for project portfolio selection that demonstrates the need for strategy to be set at corporate level and then filtered down to a project level.

Programs

Murray-Webster and Thiry (2000) suggest that the discipline of program management is emerging as a fundamental method of ensuring that an organization gains the maximum benefit from the integration of project management activities. They also suggest that program management involves more iteration and strategic reflection than the more "single-shot" project management, and they describe programs as "the missing link"—the means of effectively bridging the gap between a strategy, subjected to emergent change, and projects. (See also Thiry's chapter in this book.)

The UK Office of Government Commerce (OGC) (1999) considers the alignment between strategy and projects to be one of the main benefits of program management. The program management process they use comprises the following stages:

- Identifying, defining and establishing a program
- Managing the portfolio
- Delivering benefits
- Closing the program

They perceive the environment of program management to be that shown in Figure 8.8, which indicates programs emanating from business strategies and initiatives, and an iterative hierarchy of programs, projects, and business operations cascading from them.

The objectives and strategies for the programs are created and aligned with the objectives and strategies of the organization; and the objectives and strategies for individual projects are created and aligned with their respective programs. We will consider in detail how project strategy is created within projects.

FIGURE 8.8. THE PROGRAM MANAGEMENT ENVIRONMENT.

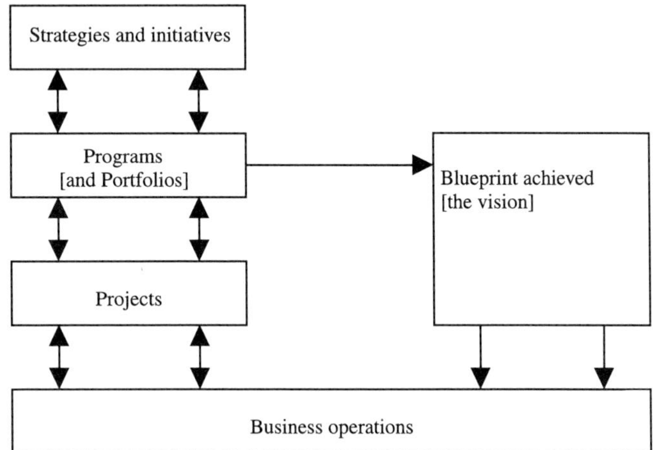

Source: UK Office of Government Commerce (OGC) (1999). Adapted by Morris and Jamieson (2004).

Projects

To move business strategy into project strategy most effectively, whether a project interfaces directly with business units or indirectly through portfolios and programs, there needs to be coherent project management processes that integrate seamlessly with the strategic management processes. We (the authors) developed a schematic to show the way business strategy is translated into project strategy at the front-end of a project, predicated on sections of the PMI PMBOK (2000) and the APM BoK (2000). This model, shown in Figure 8.9, comprises two stages and a number of key project management activities or processes within each stage. The stages are as follows:

- Translating business strategy stage, including:
  - Project definition (PMBOK)
  - Project scope management (PMBOK
  - Requirements management (APM BoK)
  - Strategic framework (APM BoK)
- Creating project strategy stage, including:
  - Project management planning and integration processes (PMBOK)
  - Project plans development process (PMBOK)
  - Generic project management knowledge and competencies (APM BoK)
  - Elements of project strategy

Translating Business Strategy in Projects. The project definition process is an essential part of developing project strategy from business strategy. Turner (1999) builds on a model for the strategic management of projects developed by Morris (1987). This model indicates that

FIGURE 8.9. CREATING PROJECT STRATEGY FROM CORPORATE AND BUSINESS STRATEGIES.

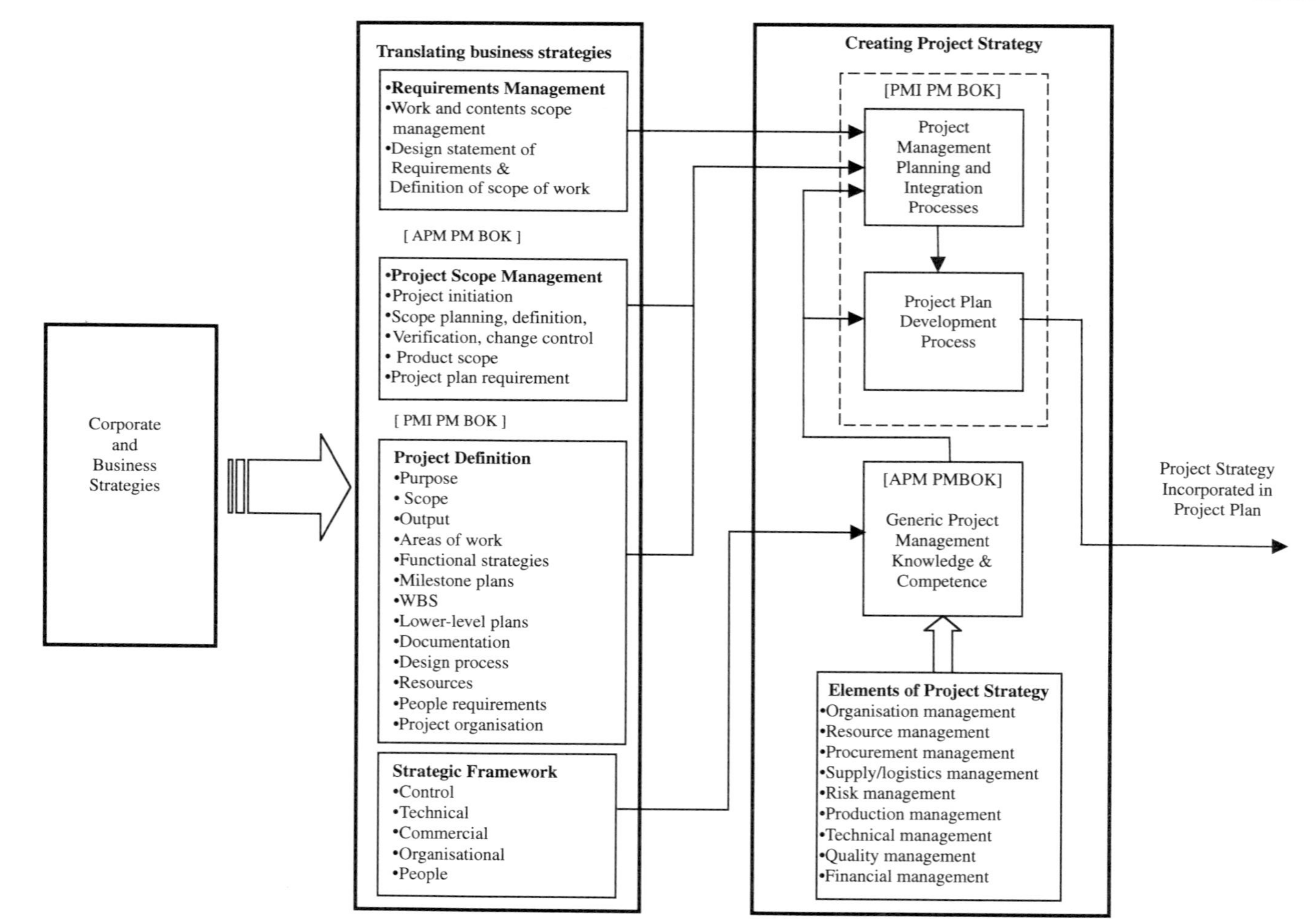

Source: Morris and Jamieson (2004).

projects are subjected to seven forces and the two that relate to strategy are identified as attitudes and definition.

Turner also advocates developing a comprehensive definition of a project at the start of the project, and is achieved by:

- setting the project objectives;
- defining the scope through a strategic, or milestone plan;
- setting the functional strategies for assessing technical risk;
- carefully managing the design process; and
- managing resources and the context.

Through the project definition process:

- the vision of the project is created;
- the purpose of the project is defined;
- the project plans are aligned with the business plans; and
- the basis of cooperation for the project is agreed upon.

The project scope management processes described in the PMI PMBOK (2000) identify strong links with corporate and business strategies. The initiation process (the first process of the PMBOK project scope management) links a project to the ongoing work in the performing organization and recognizes that a project can be authorized as a result of market demand or a business need. The inputs to the initiation process include the strategic plan of the performing organization, which is also used in project selection decisions and is seen as providing the link through which the project strategy is aligned to the organization's strategy. The output of the initiation process reflects the organization's strategy and is fed as an input to the second process of project scope management.

The second process is "scope planning," the output of which is the project scope statement containing the project justification, product, deliverables, and objectives. All of these are seen as key elements of project strategy; and are inputs to "scope definition" process. Thus, the scope of the project is driven by the key elements of project strategy that are directly linked to business strategy. Figure 8.9 lists what is generally accepted in the PMI PMBOK (2000), the APM BoK (2000), and Turner (1999) to be the key elements constituting the definition or scope of a project and what is usually contained in the project definition document for a project.

The APM BoK (2000) addresses work content and scope management as a single topic and places requirements management within the technical context of a project.

Requirements management covers the process of defining the user/customer technical requirements and building the system requirements for a project. The technical requirements of a project frequently affect project strategy significantly, and for this reason, it is identified as a key project management process. (See also the chapter by Davis et al.)

Creating Project Strategy in Projects. Figure 8.9 shows the project requirements, project scope, and project definition from the "translating business strategies" stage being inputs to the project management planning and integration processes in the "creating project strategy" stage, and used in the project plan development process to create project strategy. The project plan development process also takes the outputs from the planning processes from other project management knowledge areas (not shown in Figure 8.9), such as resource plans, risk plans, procurement plans, and so on, and subsidiary plans such as scope management and schedule management plans, and incorporates them in a consistent, coherent project plan document or set of documents. A summary of the contents of these key documents and management plans is incorporated into the project plan, which, in the PMBOK schema, encapsulates the project strategy.

Strategic Framework and Project Management Practices

The APM BoK identifies a broad range of topics essential to the effective management of projects. These topics are grouped into a number of sections, one of which is labeled "Strategic." This section covers the topics that manage the strategic framework of a project. Such a framework is an important factor in the process of creating project strategy, and for this reason it is incorporated in Figure 8.9. The strategic framework provides the overall integrative framework for managing projects efficiently and effectively and potentially includes strategic elements from almost all the topics contained in the Control, Technical, Commercial, Organizational, and People sections of the APM BoK. Figure 8.9 identifies a number these elements, all of which are also commonly referred to as project management practices.

A strategic framework containing these elements can be developed using generic project management knowledge and competence, predicated on the APM BoK, and applied to the project integration management processes identified in the PMBOK, thereby enhancing the quality of the inputs to the processes from which project strategy is created, and ultimately the quality of the project strategy. Some may perceive there to be a risk of overlap between the APM's strategic framework and project management practices, and the planning processes of project management knowledge areas described in the PMBOK, and used in Figure 8.9. This is not the case. The idea is that by using the PMBOK project management planning and integration process model, within the context of a much broader generic strategic (APM) framework, a more effective project strategy can be created.

A Structured Approach to Creating and Moving Strategy within Projects

Figure 8.9 demonstrates the large number of factors involved in creating project strategy at the front end of a project. This highlights the need for organizations to have an effective way to manage the whole process of project strategy creation, which covers not only the front end of a project but the entire project life cycle. Our research (Morris and Jamieson, 2004) shows that companies have developed structured approaches for creating and managing project strategy that cover the entire project life cycle and are integrated with the business strategy development processes. Figure 8.10 is an example of an actual structured approach of a major manufacturing company. It comprises the following:

- A business process model
- The managing major projects process
- The project management process

The project management process comprises five stages covering the entire project life cycle. The key tasks undertaken during each of the process phases are stipulated in the supporting documentation and reflect many of the front-end elements shown in Figure 8.9. Defining project strategy is one of the key tasks undertaken in all of the phases and is done in accordance with a list of key topics by those working on the project. This approach enables project strategy to be managed throughout the life cycle of the project, by all those involved, in a manner that is specific, comprehensive and dynamic, and highly visible.

Competencies, Roles, Responsibilities, and Accountabilities for Moving Strategy

Moving strategy by means of sophisticated processes, such as those shown in Figure 8.9, requires an extensive range of competencies, highly skilled staff, and a clear definition of their roles, responsibilities, and accountabilities. Armstrong (1999) points out that the descriptions of competencies may be called competency frameworks, competency maps, competency profiles, or competency clusters—and that competency frameworks define the competency requirements that cover all the key jobs in the organization or all the jobs in a job family. An example of a generic approach in a project/technical area, is the UK Institution of Civil Engineers' competency framework structure (2000). Marchington and Wilkinson (2002) have observed that a competency framework provides a set of performance criteria at organization and individual levels, and that it identifies the expected outcomes of achieving those criteria. Whiddett and Hollyforde (1999) provide further information on competency frameworks. (And the chapter by Gale later in this book also provides more information.)

Behavioral competencies describe how we behave while performing our work tasks. Armstrong (1999) reports that a survey of 126 organizations shows the most common behaviors sought by the organizations, one of which is strategic capability. Crawford (2000) reveals a number of knowledge, skills, and personal attributes of project managers, including that of strategic direction. Our research (Morris and Jamieson, 2004) shows the core behavioral competences for project directors and project managers in one global organization, as shown in Figure 8.11. It is clear that a high percentage of these competencies, to a greater

FIGURE 8.10. A STRUCTURED APPROACH TO CREATING AND MOVING PROJECT STRATEGY.

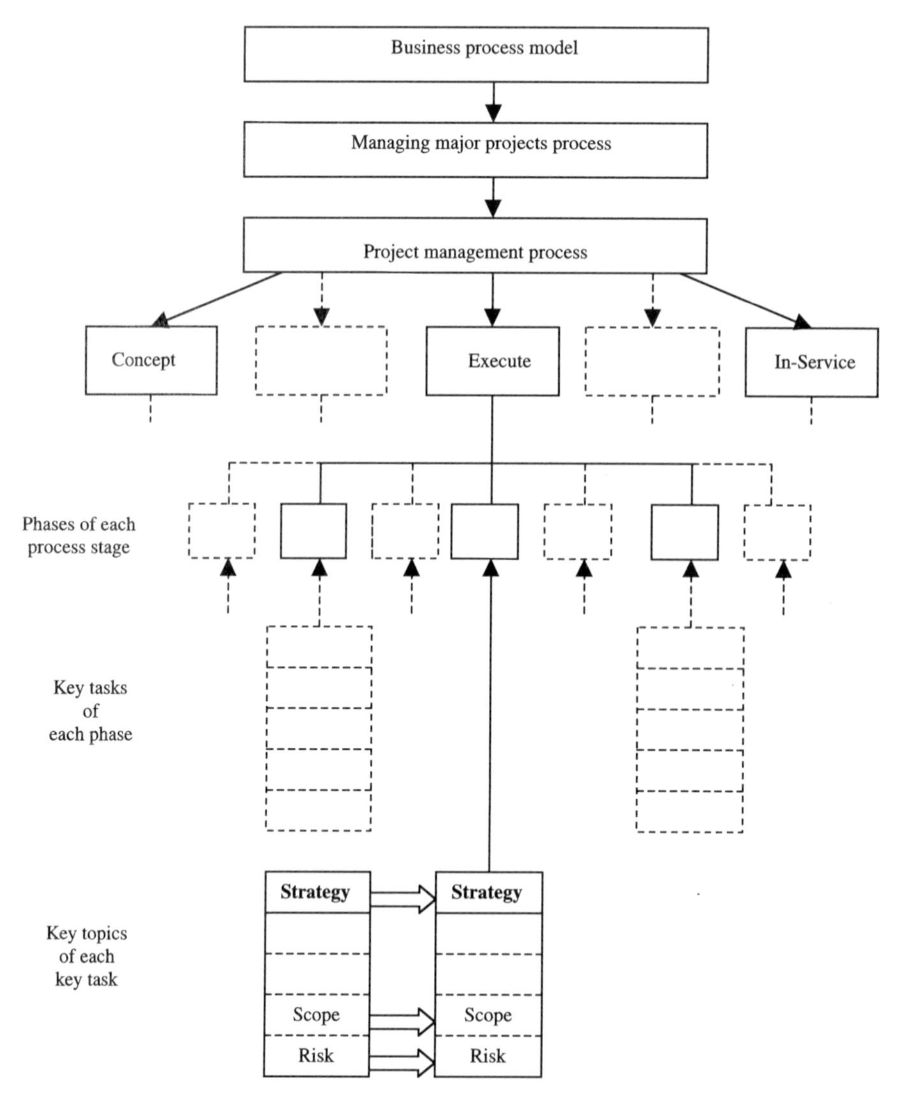

Source: Morris and Jamieson (2004).

FIGURE 8.11. COMPETENCIES FOR PROJECT DIRECTORS AND PROJECT MANAGERS.

Core Competencies	Project Director	Project Manager
Managing vision and purpose	X	
Business acumen	X	
Customer focus	X	X
Priority setting	X	X
Directing others	X	X
Leading from the front	X	X
Drive for results	X	
Dealing with ambiguity	X	
Composure	X	
Comfort around higher management	X	X
Negotiating	X	X
Building effective teams	X	X
Conflict management		X
Timely decision making		X
Motivating others		X
Organizing		X

Source: Morris and Jamieson (2004). This material is used by permission of the Project Management Institute.

or lesser extent, are required to move strategy from corporate to projects, particularly at the project director level. A number of the competencies for a project manager are the same as those for a *project* director—but not program director (see below)—with two notable exceptions: managing vision and purpose and business acumen.

In addition to the behavioral competencies, this company has also defined project management functional competencies—that is, knowledge and skills, based on the APM's BOK (2000). These functional competencies, grouped into several categories, are those required to develop the elements of a strategic framework, shown in Figure 8.9, and subsequently used to create project strategy, using processes similar or equivalent to those identified in

Figure 8.9. The categories of strategic management competencies, and the processes to which they are applicable, are as follows:

- *Strategic.* Including strategy/project management plan and value management
- *Control.* Including work content and scope management
- *Technical.* Including requirements management
- *Commercial.* Including business case, financial management, and procurement
- *Organizational.* Including life cycle design and management, and organizational structure and roles

These competencies are also related to the job requirements for project directors, project managers, and other professional project management staff. The levels of competency for a project director and a project manager, for the elements mentioned previously, in most instances are the same, but in one or two cases, such as requirements management, the project manager is required to have a higher level of competency.

A second example from our research (Morris and Jamieson, 2004) reveals how a family of job descriptions focuses on strategy. The role and purpose of a global program director are expressed in terms of leading and coordinating the resources required to implement a strategic change program for the organization. This involves *inter alia* using project management disciplines, ensuring the initiatives are aligned to organizational strategy and values, and the delivery of the operational vision on behalf of the sponsor. A program normally involves multiple projects and working across lines of business and function. The emphasis on leading and coordinating resources to implement strategic change endorses the point made previously—namely, that it is a fundamental responsibility of project management to effectively manage and/or coordinate whatever company resources are at its disposal, using well-structured project management processes, practices, and methods, and professional project management staff.

A global program director's responsibilities, in the company studied in the research, include the following strategy related areas:

- Global strategic direction of the program and its alignment with business unit vision direction and priorities;
- Business case management;
- Release management of change program operational implementation;
- Definition management; aligning the solution to the business unit vision and ensuring the solution remains aligned;
- Management of the implementation of the solution, and its scope, resource, and schedule requirements; and
- Development of program plans to reflect all progress, change, and problematic issues.

To effectively discharge these responsibilities, a global program director in this company typically possesses the following knowledge, skills, and experience:

- Business and commercial management
- Task management
- Planning and organization
- Project management
- Time management
- People/relationship management, including leadership

Additionally, as with the previous example, the company also specifies behavioral competencies for the role holder, of which the following are relevant to project strategy:

- Strategic thinking
- Conceptual thinking
- Innovativeness
- Analytical thinking

The knowledge, skills, behaviors, and experience required for project managers in the company fall into similar categories as those for a global program director, but the levels of each are within the context of a project and not a program, and accordingly are at a commensurately lower level. As with the previous example, these knowledge, skills, experience, and behaviors are integrated with the project management process and described in the associated documentation.

Companies document the roles, responsibilities, and accountabilities for the staff involved in managing projects using many different formats. Figure 8.12 provides an example of the approach taken by one company (the example could be expanded to show the many staff involved in managing projects). It is essential that the roles and responsibilities are defined and rolled out in a way that is user-friendly and are bought into by those using the processes.

FIGURE 8.12. ROLES, RESPONSIBILITIES, AND ACCOUNTABILITY MATRIX.

Tasks / person accountable or responsible	Project Board	Business Manager	Project Manager	-	-	-	-	-
Understand and assess the brief								
Covers 4 subactivities, e.g., Define project success criteria		R	A/R					
Review team composition								
Covering 1 activity			A/R					
Gather business case information								
Covers 5 activities.			A					
Finalise project definition document								
Covers 5 activities.		R	A/R					

Source: Morris and Jamieson (2004). This material is used by permission of the Project Management Institute.

The matrix covers one subprocess, associated with creating project strategy of a major projects process that comprises numerous subprocesses. The person or bodies involved in the subprocess are identified; in this case there are three, but normally all those involved are identified, which frequently amounts to a much larger number. All the activities of the subprocess are also identified, and either an A or R (there are other indicators) on the matrix indicates who has the accountability and/or the responsibility for them. (the company defines accountability as having full responsibility for the task, and responsibility is limited to completing the task.) The matrix approach is a very quick and effective means of identifying the roles, responsibilities, and accountabilities of all those involved in managing major projects, although it may be perceived by some as being very prescriptive.

Four Studies of Moving from Corporate to Project Strategy

In our research we studied four companies: a global aerospace company, a division of a global pharmaceutical company, a group within a global financial services company, and an international airport owner and operator. The resulting case studies provide valuable evidence and insight into the way corporate strategy is created and moved through SBUs, portfolios, and into programs and projects by means of, and through the activities and support of, processes, practices, and people. A number of the key findings of the case studies are summarized in the following.

Business Models

Two of the companies used business models that are equivalent to the McKinsey model outlined previously. These top-level frameworks were used to manage the activities of all the business units of the companies concerned and included strategy management, portfolio management, and program and project management processes.

Cascading Corporate Strategies into Projects and Strategy Plans

The companies created corporate objectives, goals, and strategies using processes that are typically like those strategic management processes described by Mintzberg, Hill, and Thompson. Like the model shown in Turner (1999), these objectives, goals, and strategies were cascaded to the SBUs or equivalent organizational entities, which in turn, and in conjunction with corporate strategy planners, developed their own objectives, goals, and strategies, in some instances using additional processes that were fully integrated with the business strategy processes. The SBUs subsequently developed objectives, goals, and strategies with and for their respective program and project teams, again, in some instances using fully interconnecting business and project management processes. In all four cases, the program and/or project teams developed strategies that aligned with the SBU and corporate strategy using project strategy or similar processes. The outputs of the processes containing the objectives, goals, and strategies included strategy plans, business plans, deployment plans, and project plans, the hierarchy of which, in most cases, was similar to Archibald's hierarchy of objectives, strategies, and projects, and SRI's system of plans. The most comprehensive case is shown in Figure 8.10.

Portfolio Management

The importance of project portfolio management was recognized by all the companies, and one had developed a dedicated project portfolio management practice. Within the companies, portfolio management was used primarily to select and prioritize programs and projects, and not to manage programs or projects. Corporate and business units assembled a strategic portfolio of programs and projects, or measured the strategic contribution of a program or project, using strategic management and project management processes, tools, and techniques. Company management boards or committees of senior managers adopted or rejected projects based on this information. (This is very similar to the pattern described by Artto and Dietrich in their chapter.)

Program Management

Program management was also practiced by the majority of companies, primarily within the context of managing a large group of high-value projects with a common aim and/or of delivering regular benefits over a protracted period of time. Program management and project management activities were carried out using the same set of common processes, variously called integrated program management, program management, or even project management; and accordingly, the development of program strategy and its alignment with corporate and business strategy was achieved in a similar way to that for projects.

Business Cases and Project Strategy

The creation of business cases was a key element of the business and project management interface within all the companies. An outline project strategy was developed during this activity and was aligned with corporate and business strategies. Subsequently, business strategy, in most of the companies, was translated into a comprehensive project strategy using project management processes, similar to those used in Figure 8.9 (see also Cleland's chapter in this book). Project strategy was generally in the form of a diversity of management plans and project plans rather than a single comprehensive document.

A Structured Approach to Creating and Managing Project Strategy

Two of the companies used a very structured approach to create and manage project strategy. One had institutionalized a project strategy management practice that was equivalent to, for example, risk management or technical management. The other had identified specific project-strategy-related issues for each phase and stage of the project management process and the project life cycle, as shown previously in Figure 8.10. Both companies assigned roles and responsibilities for the execution of the processes. The other companies used a less structured approach and developed management plans for their projects, but they tended not to summarize the plans nor develop a single project strategy statement from them. They also tended not to use the term project strategy in their project management processes. (There is a research issue left open here—namely, whether it would be beneficial to manage project strategy as a more formal, single document and process.)

Two out of the four companies manage project strategy for effectively the entire project life cycle and not just at the front end of a project.

Processes and Procedures

Of the processes identified in the preceding text, the ones that were most consistently used were those in which the structure and content were described at a practical level—for instance, flowcharts with inputs and outputs for key processes—and those who were accountable and responsible for carrying the process activities were identified. Conversely, when the procedures for these processes were described in too much detail, staff tended not to use them. The best examples of the deployment of the business models and associated processes were those that were fully documented and incorporated within a company's quality management system, and that were Web-based and available online throughout the organization. (Again, see Artto and Dietrich's chapter.) The companies that had not implemented such sophisticated systems or extensively integrated business and project management processes nevertheless linked the activities of their business units and projects to ensure alignment of strategy.

Roles, Responsibilities, and Accountabilities

All of the companies specified the roles, responsibilities, and accountabilities of all those involved in the business management and project management processes within the process documentation, some using comprehensive sets of tables and matrices that were linked directly to the processes. These tables and matrices covered in detail all the phases and stages of the project management process and project life cycle, including those for creating and maintaining project strategy, and identified who does what and when at any point along the process. In one company, a family of project management job descriptions was used in conjunction with the tables and matrices. These identified the job holder's roles, responsibilities, and accountabilities for specific project management process activities and outputs; they provided an unusually high degree of integration between the process and the individual or team.

Competencies and Frameworks

The companies also employed a number of other methods to identify and specify the skills, knowledge, behaviors, and experience required to manage projects and project strategy. These included, for example, company-wide competency frameworks that defined the competency requirements for all the key jobs and comprised families of job descriptions, including those for project management staff; core behavioral competencies for senior project management staff, such as managing vision and strategy; and project management functional competencies covering knowledge and experience of strategy-related areas like scope management.

Survey Data on How Companies Move Strategy from Corporate to Projects

The case study data briefly summarized in the preceding text provides a rich qualitative context in which to explore whether companies do in fact move from corporate to program and project strategy as we have hypothesized. But the data sample is obviously small. To provide a bigger data sample, we carried out a survey of the way members in a number of PMI chapters in European countries moved from corporate to program and project strategy (Morris and Jamieson, 2004).

A series of 32 questions were developed and used to examine the processes, practices, and people issues involved in moving strategy from the corporate level to projects, covering business management, strategic management, portfolio management, program management, and project management. The questionnaire was sent out randomly to PMI chapter members to obtain their views, based on their experience and knowledge. Seventy-five responses (about 50 percent from the UK) were received from people, at various levels of seniority, in small, medium, and large enterprises in a diverse range of business sectors, such as aerospace, automotive, IT, telecommunications, pharmaceuticals, retail, transportation, publishing, and academia and consultancy.

The findings of the survey, summarized in Exhibits 8.1 to 8.5, are the result of the analysis of the answers to questionnaire. The exhibits cover the following areas:

1. Business management
2. Program management and portfolio management
3. Project management and project strategy
4. Value management
5. Project management competencies

Most items in each Exhibit have a population % figure, which is a percentage of the total number of organizations represented in the survey. For example, in Exhibit 8.1, Item 1, 67 percent of the organizations in the survey indicated that they used a generic business model. Or in Exhibit 8.2, Item 3, 85 percent of organizations surveyed use programs to implement change.

The first major area the survey explored was the extent to which processes were used within organizations and to what degree was the continuity of strategy achieved through them.

EXHIBIT 8.1. SURVEY FINDINGS—BUSINESS MANAGEMENT.

1. The extent to which a business model was used:

Model used	Generic*	Equivalent
Population %	67	23

*Including project management processes

2. 80% of the organisations indicated they were process-oriented organisations as follows:

Level of process-oriented	Extensively	Adequately	Inadequately
Population %	40	40	17

3. Those extensively or adequately process-oriented indicated they had processes for moving corporate goals and objectives to project strategy as follows:

Level of processes	Extensive	Adequate
Population %	50	33

4. The level of interconnection between the corporate, business, and project management processes for the organizations with extensive processes was as follows:

Level of interconnection	Extensive	Adequate	Inadequate
Population %	40	50	10

5. Organizations having extensive processes/subprocesses for moving corporate goals and objectives to project strategy consider continuity of strategy is achieved as follows:

Level of continuity achieved	Fully	Well	Inadequate or poor
Population %	20	55	25

6. Hierarchy of objectives and strategies developed and deployed for structuring strategy:

Hierarchy span	Corporate through to project	SBU to project
Population %	53	68

How organizations perceive and use program management and portfolio management and the extent to which they moved strategy through programs to projects was the second major area investigated.

EXHIBIT 8.2. SURVEY FINDINGS—PROGRAM MANAGEMENT AND PORTFOLIO MANAGEMENT.

	Population %
1. Program management was defined as the management of a portfolio of projects sharing a business objective of strategic importance, probably utilizing shared resources.	95
2. Programs were considered to comprise:	
Groups of projects	20
Number of separate projects	20
Combination of both	70
3. Programs were used to implement change:	85

EXHIBIT 8.2. (*Continued*)

4. Some form of portfolio management was implemented:	50
5. Portfolio management was considered as:	
Selecting the right project quantitatively	50
Maintaining a balanced portfolio	60
Managing projects grouped around a common theme	66
6. Hierarchy of objectives and strategies were developed and deployed at program and project level	75
Of which, the levels deployed were as follows:	
Program	90
Program and project	60
Program, project, and project team	45
Program, project, project team, and individual	28
7. Program management was implemented	90
Of which, program management included the following:	
A. [i] Managing an integrated set of projects to achieve a common theme, aim, or working off a common platform;	55
[ii] Integrated project teams	
[iii] Managing resources in an integrated manner	
B. [i] and [ii]	10
C. [i] and [iii]	15
8. Program management implied the management of business benefits	75
Of which:	
It was normal practice to formally identify a benefits process within the overall program management process.	70
Those who do not incorporate a benefits process believed they should.	60
Nonfinancial measures were used to track benefits in programs.	70
9. Program management implied the aggregation of risks	75
Of which:	
It is normal practice to formally identify risk aggregation as part of the overall risk management activity.	80
Those who do not identify risk aggregation believed it should be incorporated into program management.	60

The third area investigated, project management, followed on naturally from program management. The survey focused on identifying the key strategy inputs and some of the project management activities, which were employed to create project strategy.

EXHIBIT 8.3. SURVEY FINDINGS—PROJECT MANAGEMENT AND PROJECT STRATEGY.

	Population %
1. Organizations had extensive or partially integrated project management processes to help manage project strategy, which contained:	Almost all
Project strategy management	85
Requirements management, project strategy, project definition and project scope management	75
Requirements management, project definition and project scope management.	85
2. Organizations had specific strategy inputs to integrated project management processes	
Which included:	most
Corporate strategy	75
Corporate strategy and business strategies	65
Corporate, business, and portfolio strategies	50
Corporate, business, portfolio, and program strategies	45
Portfolio and program strategies only	55
Program strategy	75
3. The integrated project management processes delivered the following outputs:	
A project or program plan and strategy plan	50
Other project management plans	75
A project or program plan, strategy plan and other plans	45
4. Organizations with integrated project management processes managed project strategy dynamically	65
5. The roles and responsibilities for developing, implementing and updating project strategy were specified in:	
Project management procedures	60
Project plans	55
6. Project plans were formally reviewed at project "gates"	85
Those who did not and thought they should	85
7. Peer groups formally reviewed project plans	75
Those who did not and thought it would be sensible to do	65
8. It was clear who approved and signed off project strategy	75
9. In broad terms, a project sponsor was the individual or group within the performing organisation who provides the financial resources, in cash or in kind, for the project; and as the owner of the business case, represents the funder's interests.	90
The relationship between the project sponsor and the project management team was normally defined in project plans.	70
10. Strategy was expected to be upgraded and reviewed:	
During the development of the project	65
Systematically as projects develop from concept to execution	55
Of which: it was systematically undertaken at project review gates	85

The survey also explored a two other areas:

- Project value management and its link to project strategy
- To what extent do companies define competencies to manage program and project strategy, and incorporate them in competency frameworks and job descriptions.

EXHIBIT 8.4. SURVEY FINDINGS—VALUE MANAGEMENT.

	Population %
1. A process was used for optimizing the value of proposed project/program strategy	55
Of which:	
Value was expressed as benefit over resources used	80
The process was formalized as value management	55
Of which value management workshops were held at strategic stages in the life of the project.	40
Those not using a process for optimizing the value of project/program strategy believed they should.	55
2. Value engineering was practiced on programs and projects	25
Of which:	
Value engineering (optimizing the value of the technical configuration) was distinguished from value management.	80
Those not practicing value engineering on programs and projects thought they should.	56
3. The value optimization process was integrated with risk management	75
Those that did not thought it should be done.	40

EXHIBIT 8.5. SURVEY FINDINGS—PROJECT MANAGEMENT COMPETENCIES.

	Population %
1. Project management skills and knowledge competencies required to manage programs or projects were formally defined.	80
Of which included:	
Those required to develop program and project strategy	75
Linking the competencies to personal appraisal and development systems	80
Linking personal objectives to project objectives	65
2. Those that did not formally define the project management skills and knowledge competencies incorporated the management of project strategy in job descriptions or job specifications.	50
3. Organization-wide behavioral competency frameworks were used	60
Those that did not use them considered they should.	45
4. Competency support programs for program and project managers were provided	70
Of which covered support for project strategy development.	66

Summary

Project and program management is widely used as a means of implementing corporate strategy. Normatively, we can expect strategies to be aligned and moved from the corporate level through programs and projects in a systematic and hierarchical manner that provides cohesion, visibility, and an effective means of communication. There is a cascade of moving from corporate planning at the enterprise level through portfolio management into programs and projects. Within this framework, project strategy is managed dynamically.

Enterprise-wide business models play an important part in effecting this transformation. Many organizations have project management as a core process.

Programs are important vehicles for implementing corporate strategy and for implementing change. Most companies consider that program management implies the management of business benefits (as well as the ideas of product, brand, or platform management).

Project strategy management is widely recognized as an important project management practice that systematically relates project definition and development to corporate goals and strategies. Structured project management approaches are now being used by organizations, covering the entire project life cycle, with project strategy development, review, and optimization occurring at specific points. Value management is quite widely used in optimizing the strategy, often in combination with risk management.

Project management resources and capabilities are key factors in creating, deploying, and maintaining enterprise, portfolio, program, and project strategies. The project management roles, responsibilities, and accountabilities required for this are generally well defined. A high percentage of those organizations surveyed define the personal project management competencies required to develop project strategy.

Project strategy management is an underexplored and insufficiently described subject in the business and project literature but is in fact a relatively well-trodden area within industry and commercial practice, and deserves more recognition.

References and Further Reading

Andrews, K. R. 1999. *The concept of corporate strategy.* Custom ed. New York: McGraw-Hill Companies, Inc.

Archer, N. P., and F. Ghasemzadeh, F. 1999. An integrated framework for project portfolio selection. *International Journal of Project Management* 17:207–216.

Armstrong, M. 1999. *A handbook of human resource management practice.* 7th ed. London: Kogan Page Ltd.

Association of Project Management. 2000. *Project management body of knowledge.* 4th ed. High Wycombe, UK: Association of Project Management.

Cleland, D. I. 1990. *Strategic design and implementation.* New York: McGraw-Hill Companies, Inc.

Crawford, J. L. 2001. Portfolio management: Overview and best practices. In Project Management for Business Professionals. J. Knutson (ed). New York: Wiley.

Crawford, L. 2000. Profiling the competent project manager. *Proceedings of PMI Research Conference.* Newtown Square, PA: Project Management Institute.

Hill, C. W. L., and G. R. Jones. 2001. *Strategic management: An integrated approach.* 5th ed. Boston: Houghton Mifflin Co.

Kerzner, H. 2000. *Applied project management: best practices on implementation.* New York: Wiley.

Knutson, J. 2001. *Succeeding in project-driven organizations: People, processes and politics.* New York: Wiley.

Marchington, M., and A. Wilkinson. 2002. *People management and development: Human resource management at work.* 2nd ed. London: Chartered Institute of Personnel and Development.

Mintzberg, H., B. Ahlstrand, and J. Lampel. 1998. *Strategy safari.* London: Pearson Education.

Mintzberg, H., and J. B. Quinn. 1996. *The strategy process: Concepts, contexts and cases.* 3rd ed. Upper Saddle River, NJ: Prentice Hall.

Morris, P. W. G. 1997. *The management of projects.* 2nd ed. London: Thomas Telford.

Morris, P. W. G., and G. H. Hough. 1987. *The anatomy of major projects.* Chichester, UK: Wiley.

Morris, P. W. G., and H. A. Jamieson. 2004. *Translating corporate strategy into project strategy.* Newtown Square, PA: Project Management Institute.

Murray-Webster, R., and M. Thiry. 2000. Managing programmes of projects. In *The Gower Handbook of Project Management.* 3rd ed., ed. R. Turner and S. Simister. (Aldershot, UK: Gower.

Project Management Institute. 2000. *A guide to the Project Management Body of Knowledge.* Newtown Square, PA: Project Management Institute.

Thompson, J. L. 2001. Strategic management. 4th ed. London: Thomson Learning.

Turner, R. J. 1999. The handbook of project-based management. Maidenhead, UK: McGraw-Hill. Reproduced by kind permission of the Open University Press/McGraw-Hill Publishing Company.

Turner, R., and S. Simister, eds. 2000. *The Gower handbook of project management.* 3rd ed. Aldershot, UK: Gower.

The Institution of Civil Engineers. 2000. Management development in the construction industry: Guidelines for the construction professional. London: The Institution of Civil Engineers.

The Office of Government Commerce. 1999. *Managing successful programmes.* London: The Stationery Office. Crown copyright material is reproduced with the permission of the Controller of HMSO and Queen's Printer for Scotland.

Watson, G. H. 1994. Business systems engineering: Managing breakthrough changes for productivity and profit. Chichester, UK: Wiley.

CHAPTER NINE

STRATEGIC MANAGEMENT: THE PROJECT LINKAGES[1]

David I. Cleland

"There is nothing permanent except change."

HERACLITUS OF GREECE, 513 B.C.

The purpose of this chapter is to present an overview of what is involved in the development of project strategy. Project strategy does not stand alone; rather, it is an integral part of the overall strategic plan for the management of change in the enterprise. A general philosophy of what project strategy is, as well as how and where project strategy falls in the overall scheme of the strategic management of the enterprise, is presented.

This chapter also provides an overall perspective—a philosophy—to guide the design and development of project strategy. A *philosophy* is defined as a system of thought based on some logical relationship between concepts and principles that explains certain phenomena, and supplies a basis for rational solutions of related problems (Davis, 1951). A philosophy, taken in its most basic sense, is simply a way of thinking about a field of endeavor. A sound philosophy of project strategy is a precondition of starting the project planning process.

The dictionary defines *strategy* as "The essence of art or military command as applied to the overall planning and conduct of large-scale combat operations; a plan of action resulting from the practice of strategy, the art or skill of using stratagems, especially in politics and business (*Webster's II New College Dictionary*, 1999).

The details of project strategy, such as scheduling, networking techniques, scope planning, risk assessment, resource planning, cost estimating, life cycle planning, project specifications, and so forth, will not be presented in this chapter. Rather, what will be presented is an overall description of project strategy, including how it fits into the overall scheme of

[1] In the development of this chapter I have drawn material from David I. Cleland & Lewis R. Ireland, *Project Management: Strategic Design & Implementation*, 4th Edit., McGraw-Hill, New York, NY 2002.

enterprise strategy, and some general guidelines on the "work packages" that should be developed in the preparation of project strategy.

We tend to live, worship, work, socialize, and so forth, based on some philosophy—some way of thinking about our world. Such a philosophy may be nebulous and fleeting, or it may be well defined and provide the performance and behavior standards of our life.

The key question that the reader needs to ask is this: What philosophy do I have regarding the development of project strategy? Can the philosophy that is presented in this chapter help me—and the people with whom I work—in the development of project strategy?

Background

Project management is currently being used worldwide in a wide range of applications, both in developed and developing countries. Theses applications include the following::

- *Industrial.* For new or improved products, services, or organizational processes
- *Social.* To support new or enhanced programs for society
- *Economic.* For stimulation of local, regional, national, and international economies
- *Technological.* To advance the state-of-the-art of organizational products, services, or processes
- *Legal.* For the development of new or modified laws and their application
- *Political.* Governmental initiatives to support local, regional, and national strategies
- *Military.* For campaigns to support military strategies
- *Discovery.* To find new territories and worlds

Pinto has stated: "Many of the products that are being created today in a variety of industries, from children's toys to automobiles, are becoming more technically complex to develop, manufacture, and use. Technologically driven innovation presents a tremendous challenge for organizations in the areas of engineering, design, production, and marketing. As a result, many organizations are relying on project teams composed of cross-functional groups to create and move to market these products in as efficient a time frame as possible" (Pinto, 1998).

Project teams are being applied to a wide variety of uses within the individual organization. These uses are shown in Table 9.1 (Cleland, 1996). It should be noted that four of these teams are ongoing rather than ad hoc, as would be expected of a traditional project team. However, setting up the ongoing teams is often handled through traditional projects. The continuous challenge of coping with the inevitable changes facing contemporary organizations will likely expand the use of these teams in the future.

A broad definition of a *project* is that it is a combination of organizational resources being pulled together to create something that did not previously exist and that will, when completed, provide a performance capability to support strategic management initiatives in the enterprise. Four key considerations are always involved in a project:

TABLE 9.1. CLASSIFICATION OF TEAMS.

Type	Output/Contribution	Time Frame
Reengineering teams	Handle business process changes	Ad hoc
Crisis management teams	Manage organizational crises	Ad hoc
Product and process development teams	Handle concurrent product and process development	Ad hoc
Self-directed production teams	Manage and execute production work	Ongoing
Task forces and problem-solving teams	Evaluate/resolve organizational problems/opportunities	Ad hoc
Benchmarking teams	Evaluate competitors/best-in-industry performance	Ongoing
Facilities construction project teams	Design/develop/construct facilities/equipment	Ad hoc
Quality teams	Develop/implement total quality initiatives	Ongoing
General-purpose project teams	Develop/implement new initiatives in the enterprise	Ad hoc
Audit teams	Evaluate organizational efficiency and effectiveness	Ad hoc
Plural executive teams	Integrate senior-level management decisions	Ongoing
New business development teams	Develop new business ventures	Ad hoc

Source: David I. Cleland, *The Strategic Management of Teams* (New York: Wiley, 1996, p. 10).

1. What will be the expected cost of designing and creating the expected project results?
2. How much time is required to develop, produce, and place the project results in an operational environment?
3. What capability will the project results provide the project owner?
4. How will the project results fit into the operational or strategic capability of the project owner's enterprise?

Project Evaluation and Selection

In this chapter it has been assumed that the decisions concerning which projects to develop to support strategic management initiatives have been made by the responsible executives of the enterprise. Indeed, the selection of projects to support strategic management initiatives is a most important decision—certainly an important decision in the development of project strategy.

The proper choice of how enterprise resources will be used to support the strategic management is crucial to the long-term survival and growth of the enterprise. Many of the existing texts on project management suggest procedures and processes for how projects can best be selected. A few criteria to use in determining which projects to select to support strategic and operational initiatives in the enterprise are suggested. These criteria are presented in the form of questions.

- Will there be a customer for the expected results of the project?
- Will the project results provide the enterprise a competitive edge?
- Will the project results make a distinctive contribution to existing products, services, or organizational processes?
- Will the enterprise be able to handle the risk and uncertainty likely to come forth as the project is undertaken?
- What is the probability of the project being completed on time, within budget, and at the same time satisfy its technical performance objectives?
- Will the project results provide value to the expected customer?
- Will the project provide a satisfactory return on investment to the customer's organization?
- Will the project have a high probability of supporting the enterprise's strategic initiatives?

These are key questions to be considered during the project selection process. In addition, these questions should be examined for each project during its life cycle, particularly during major reviews of the project's status.

Relationship to Strategic Management

Strategic management is the management of the organization as if its future mattered. Strategic management has two interrelated elements: (1) strategic planning and (2) strategic implementation.

In the design and execution of strategic management initiatives, an early assessment of expected real and potential environmental changes is required. This assessment should consider the general environmental conditions that currently exist and what should be expected for the future. In addition, the current and expected strengths and weaknesses of the competition should be evaluated. Once a meaningful database has been established regarding the present and forthcoming environmental conditions, the strengths and weaknesses of the organization should be compared to what is expected in the competitive future in which the enterprise will exist.

The assessment of current and expected environmental change is usually done in the following areas:

- *Political.* What political conditions might have an impact on the development of the project? For example, a Department of Defense (DoD) defense contractor would carefully watch the political developments in Washington, D.C., as such developments might impact an existing or expected DoD market. The current debate under way regarding whether or not the United States military force should buy or lease military transport aircraft is being watched carefully by aircraft manufacturers and lessors.

- *Social.* What social and cultural considerations might have an impact on the development and use of the project results in an emerging country?
- *Economic.* Will there be sufficient resources available to develop and sustain the project results in its operational environment?
- *Technological.* Will the project be supported by the required state-of-the-art resources in its development? For example, a project to build a new manufacturing plant would include a consideration of the current state-of-the-art in design and production processes.
- *Legal.* What legislation might exist, or can be expected to develop, regarding the use of the project results? Enterprises that develop projects in the construction industry need to be mindful of what environmental impact requirements need to be satisfied in the use of the project.
- *Competitive.* What are the strengths, weaknesses, and probable strategies of competitive firms performing in the same marketplace? Benchmarking of competitive firms is particularly important in the defense and construction industries.

As the assessment of the possibilities and probabilities of the expected future facing the project results is carried out, key decisions can be made through the project planning process that will provide a strategy for the project results..

Strategic management is inextricably interwoven into the entire fabric of the management process; it is not something separate and distinct from the process of management of the organization. Strategic management keynotes the shift of organization management from operations to long-range strategy. Stated another way, strategic management should not be distinguished from the rest of the organization management process, in particular the project management process where the strategic management of change is carried out.

In earlier times the managerial emphasis was primarily on operations being sufficiently efficient to maximize the return on investment in operational resources. Then came the need to develop organization strategies to deal with the growing turbulence and rapidly changing environments facing the enterprise. Today, there is a continuing emphasis on effective strategic management and efficient operational management, but an increasing emphasis on the use of project management to better facilitate the management of both operational and strategic change.

Organizational Changes

Strategic management provides a focus for the development and providing of support for the future *product*, *service*, and *organizational processes* of the enterprise. It is in these areas that competition becomes most critical for the enterprise through providing a capability for the following:

- *New or improved products* offered by the organization
- *New or improved services* offered by the organization

- *New or improved organizational processes* used to support the organization's products and services. This would include marketing, manufacturing, financial, R&D, and other organizational functions required to maintain the enterprise functions as entities capable of competitive performance in the marketplace.

As an ongoing entity, the competitive enterprise needs to maintain a balance in its utilization of resources. The key is to maintain a balance between operational competence, strategic effectiveness, and functional excellence. Figure 9.1 portrays this balance (Cleland, 1996).

Operational competence concerns the ability to use resources to provide customers with quality products and/or services that are delivered on time and provide value for customers. Operational competence requires the ability to use resources in a cost-effective manner to produce and deliver products and services that exceed what competitors are offering.

The final measure of operational competence is profit or, in the case of a not-for-profit entity, greater value to the customers than was required to develop, produce, and deliver the products and services to the customer. To maintain operational competence, the enterprise depends on the strategic effectiveness with which the enterprise is managed.

Strategic effectiveness is the ability to assess what may be possible and probable in the enterprise's future products and services, as well as the organizational processes required to

FIGURE 9.1. BALANCE MUST BE ACHIEVED AND MAINTAINED AMONG STRATEGIC MANAGEMENT CHALLENGES.

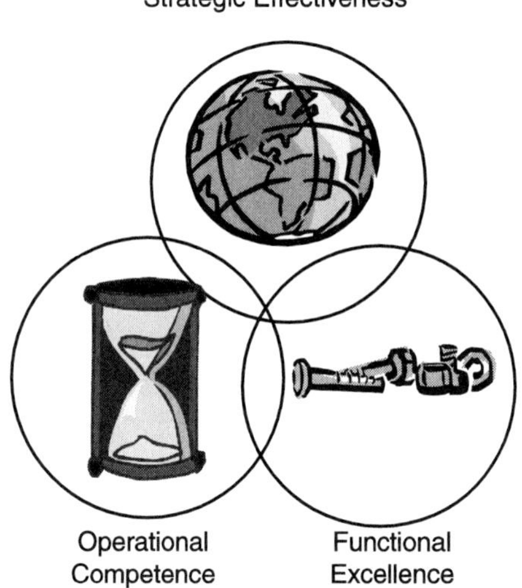

Source: David I. Cleland/Lewis R. Ireland, *Project Management: Strategic Design and Implementation,* 4th ed. (New York: McGraw-Hill, 2002, p. 8). Reproduced with permission of the McGraw-Hill Companies.

support future purposes. Strategic effectiveness is concerned with doing the "right things" to prepare the organization for its future. Project teams are crucial in preparing the enterprise for its future, thereby contributing to the strategic effectiveness of the enterprise.

Functional excellence is the use of state-of-the-art resources in the disciplines that support the enterprise's organizational processes. Functional excellence includes ongoing improvement in employees' capability as well as other organizational resources so that the effective and efficient use of resources is sustained for the organization. Project teams can be used to develop employee training programs, improve the use of resources, and design and develop new initiatives in the application of functional resources to the enterprise.

The bottom line of strategic management is to maintain balance among strategic effectiveness, operational competence, and functional excellence. Project management can provide critical support to the maintenance of this balance.

Context of Project Strategy

The development of project strategy has to be done in the context of the strategic management of the enterprise. Projects do not stand alone in the enterprise. Rather, they are part of the overall organizational initiatives. Thus, project strategy has to be developed to support larger organizational plans. In Figure 9.2, projects are portrayed in this larger context of the enterprise—essentially as building blocks in the "choice elements" of the enterprise. In the material that follows, these choice elements are briefly described. A more detailed explanation can be seen in the source cited in Figure 9.2.

FIGURE 9.2. CHOICE ELEMENTS OF STRATEGIC MANAGEMENT.

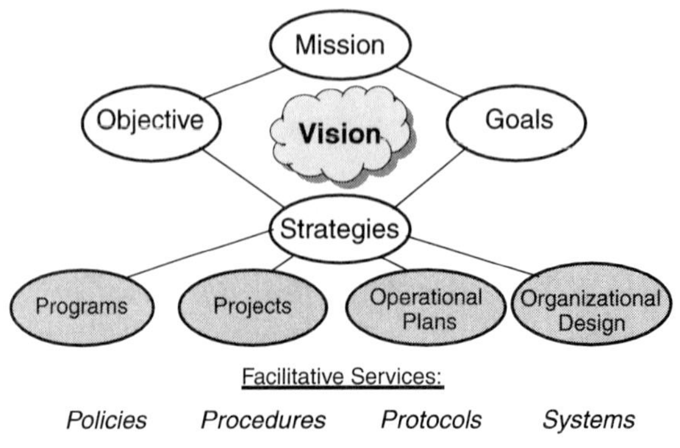

Source: David I. Cleland/Lewis R. Ireland, *Project Management: Strategic Design and Implementation,* 4th ed. (New York: McGraw-Hill, 2002, p. 8).

The choice elements are as follows:

- *Vision.* The development of intelligent and relevant foresight of probable future opportunities. One company sees its vision to be a "world-class competitor and we keep it that way—we have programs and projects in place to do just that." Another company states its vision as "People working together as a global enterprise for aerospace leadership" (Boeing, 2002). Another company included in its vision statement: "We will enhance our competitiveness by being first in the development of advanced technology that supports our world-class products and services." A statement of the vision of an enterprise is a mental image of that organization's current and future reason for existing.
- *Mission.* A broad, enduring intent that an organizational entity pursues—essentially the "business" that the organization pursues. One drug manufacturer stated its mission as "The development of model drug absorption systems for therapeutic companies that provide distinctive benefits for the physician and patient."
- *Objectives.* The desired future destination of the enterprise stated in quantitative and/or qualitative terms. A computer company describes one of its objectives as "Leading the state-of-the-art of technology in our product lines."
- *Goals.* Specific time-sensitive milestones to be accomplished using organizational resources. Attainment of a goal signifies that progress has been made toward attaining an organizational objective. One company stated that "We will, by the end of 2002, complete the transition from a predominantly R&D services company to an industrial manufacturer." Goals are supported by projects within the organization.
- *Strategies.* The design of the means through the use of resources to accomplish organizational purposes. Strategies include designation of the use of resources to design and implement organizational programs, projects, operational plans, and organizational design arrangements. Sometimes a strategy for an enterprise is stated in general terms. For example, one major aerospace company described their overall strategy as "...to excel in all principal aerospace markets to reduce our dependence on the cyclical commercial airplane market."
- *Facilitative services.* Those plans for the development of policies, procedures, protocols, and systems to support organizational resources. For example, many companies have published a project management guidebook to provide guidance on how projects will be managed within the enterprise.

Every failure in the development and implementation of resources directed to these choice elements will impact the other choice elements. A failure in the development of credible project plans can cause problems in the execution of these plans and will ultimately impact other choice elements, in particular the goals of the enterprise. From an organizational perspective, completion of a project means that an organizational goal has been completed. Indeed, the breakdown of the choice elements for the enterprise where the relative position of the vision, mission, objectives, goals, programs, and projects could be considered a conceptual work breakdown philosophy for the strategic management of the organization. For example, if a project (an organization goal) is not being developed and produced, then

- the objective of the organization will be adversely affected;
- the organization mission can be compromised;
- failure to obtain a full realization of the vision can occur; and
- the overall performance of the organization in its marketplace will be impacted.

Truly "everything is related to everything else" in the management of an organization as well as in the management of the appropriate supporting choice elements in the organization.

In the material that follows, the role of project planning is presented.

The Importance of Project Planning

Writers in project management have recognized the importance of project planning:

> During the early 1960s, after hundreds of projects had been completed, it became apparent that many projects successfully achieved their basic project objectives, whereas some failed to achieve budget, schedule, and performance objectives originally established. The history of many of these projects was carefully reviewed to identify conditions and events common to successful projects, vis-à-vis those conditions and events that occurred frequently on less successful projects. A common identifiable element on most successful projects was the quality and depth of early planning by the project management group. Execution of the plan, bolstered by strong project management control over identifiable phases of the project, was another major reason why the project was successful (Duke, 1977).

The author of what is believed to be the first book on project management stated: "No other aspect of project management is so essential to success as planning. Most of the troubles that confront project management on the rocky road to completion of the project are traceable directly to faulty planning—unrealistic, incomplete, too-broad, or just plain lack of plans" (Baumgartner, 1963). Several hundred books on project management have appeared since Baumgartner published his leading contribution to the project management discipline. No other books have stated the importance of project planning any better than this first book in the field.

Basically, project planning is a process for achieving success in the future of the project and of the organization. It is a plan of action for getting the best return from the resources that are going to be used on the project during its life cycle. The project plan is an expected arrangement for dealing with the ever-changing environment facing the project and the enterprise. Project planning starts with the development of work breakdown structure (WBS).

The Work Breakdown Structure (WBS)

The most basic consideration in the development of project strategy is the WBS. The WBS divides the overall project into work units that can be assigned either to the organization or to an outside agency. The underlying philosophy of the WBS is to divide the project into work packages that are assigned for which accountability can be expected. Each work package becomes a performance control unit; it is negotiated and assigned to a specific organizational professional or manager. The person to whom the work package is assigned becomes the "work package manager." The process of developing the WBS is to establish a scheme for dividing the project into major groups and then dividing these groups into tasks, subtasks, and so forth. Projects should be planned, organized, monitored, and controlled around the lowest level of the WBS. If a small work package is not on schedule or is running over its budget or has failed to provide the expected performance objective, the question can be raised concerning the impact on the overall project.

With an aircraft project the WBS might look like the model that is shown in Figure 9.3.

A WBS can also be portrayed as shown in Figure 9.4.

The WBS provides for the following:

- Summarizing all products, services, and processes that make up the project
- Displaying how the work packages are related to each other, to other activities within the organization, including support, as well as to outside agencies such as contractors and other stakeholders
- Providing the basis for establishing the authority-responsibility patterns in the organization, usually reflected in some form of matrix structure

FIGURE 9.3. WORK BREAKDOWN STRUCTURE CODING SCHEME FOR AIRCRAFT (EXAMPLE).

X-33 Aircraft

1. **System**
   - 1.1 Airframe
   - 1.2 Tail Section
   - 1.3 Wings
   - 1.4 Engine
   - 1.5 Avionics
2. **Documentation**
   - 2.1 Operator's manual
   - 2.2 Repair manual
3. **Test and Demonstration**
   - 3.1 Static system test on ground
   - 3.2 Dynamic air test
     - 3.2.1 Initial flight for aerodynamics
     - 3.2.2 Initial flight maneuver test
     - 3.2.3 Endurance flight test
4. **Logistics**
   - 4.1 Maintenance tools
   - 4.2 Repair parts (spares)

Source: David I. Cleland/Lewis R. Ireland, *Project Management: Strategic Design and Implementation*, 4th ed. (New York: McGraw-Hill, 2002, p. 319).

FIGURE 9.4. WBS IN A GRAPHIC DIAGRAM (EXAMPLE).

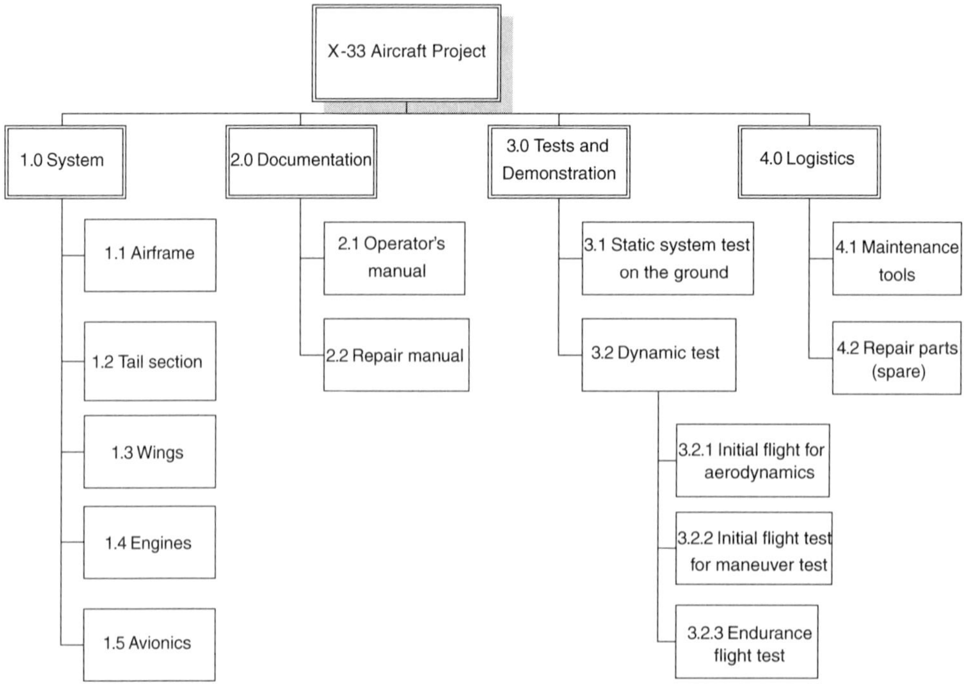

Source: David I. Cleland/Lewis R. Ireland, *Project Management: Strategic Design and Implementation*, 4th ed. (New York: McGraw-Hill, 2002, p. 320).

- Estimating cost
- Performing risk analysis
- Scheduling the work packages
- Building the project management information system
- Monitoring, evaluating, and controlling the application and use of resources on the project
- Providing a reference point for getting people committed and motivated to support the project

Work packages are the goals to be accomplished on the project. There are a few key criteria that should be applied to the WBS work packages:

- Are the work packages clear and understandable?
- Are they specific?

- Are they time-based and capable of being scheduled in the assignment of work on the project?
- Are the work packages measurable?

If an adequate and accurate WBS has not been established for the project, the project cannot be effectively and efficiently managed.

Project planning can be carried out through a representative model of the work packages. Figure 9.5 depicts these work packages. The subsequent discussion provides basic descriptions of these packages.

Project Planning Work Packages

A project plan can be broken down into work packages. The following is a general guide to these work packages:

- *Establish the strategic and/or operational fit of the project.* Ensure that the project is truly a building block in the design and execution of organizational strategies and that it provides

FIGURE 9.5. PROJECT PLANNING WORK PACKAGES.

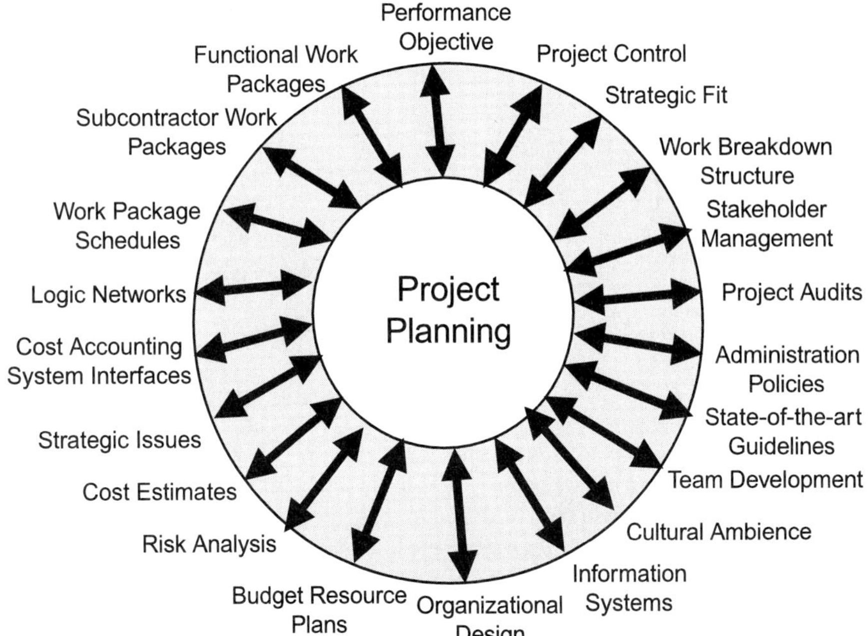

the project owner with an operational capability not currently existing or improves an existing capability.

- *Create a WBS.* Develop a product-oriented family tree division of hardware, software, services, and other tasks to organize, define, and graphically display the product to be produced, as well as the work to be accomplished to achieve the specified results.
- *Determine who the project stakeholders are and plan for the management of these stakeholders.* Determine how these stakeholders might change through the life cycle of the project.
- *Plan for the nature and timing of the project audits.* Determine the type of audit best suited to get an independent evaluation of where the project stands at critical junctures.
- *Design project administration policies, procedures, and methodologies.* Administrative considerations are often overlooked. Take care of them during early project planning, and do not leave them to chance.
- *Integrate contemporaneous state-of-the-art project management philosophies, concepts, and techniques.* The art and science of project management continue to evolve. Take care to keep project management approaches up-to-date.
- *Develop the project team.* Establish a strategy for creating and maintaining effective project team operations.
- *Assess the organizational cultural ambience.* Project management works best where a supportive culture exists. Project documentation, management style, training, and attitudes all work together to make up the culture in which project management is found. Determine what project management training would be required. What cultural fine-tuning is required?
- *Provide for the project management information system.* An information system is essential to monitor, evaluate, and control the use of resources on the project.
- *Select the organizational design.* Provide the basis for getting the project team organized, including delineation of authority, responsibility, and accountability. The linear responsibility chart (LRC) is a useful tool to determine individual and collective roles in a project-oriented matrix organization.
- *Develop the project budgets, funding plans, and other resource plans.* Establish how the project funds should be utilized, and develop the necessary information to monitor and control the use of funds on the project.
- *Perform risk analysis.* Establish the degree or probability of suffering a setback in the project's schedule, cost, or technical performance parameters.
- *Estimate the project costs.* Determine what it will cost to design, develop, and manufacture (construct) the project, including an assessment of the probability of staying within the estimated costs.
- *Identify the strategic issues that the project is likely to face.* Develop a strategy for how to deal with these issues.
- *Ensure the development of organizational cost accounting system interfaces.* Since the project management information system is tied in closely with cost accounting, establish the appropriate interfaces with the function.
- *Develop the logic networks and relationships of the project work packages.* Determine how the project parts can fit together in a logical relationship.
- *Develop the master and work package schedules.* Use the appropriate scheduling techniques to determine the time dimension of the project through a collaborative effort of the project team.

- *Identify project work packages that will be subcontracted.* Develop procurement specifications and other desired contractual terms for the delivery of the goods and services to be provided by outside vendors.
- *Make provisions for the assignment of the functional work packages.* Decide which work packages will be done in-house, obtain the commitment of the responsible functional managers, and plan for the allocation of appropriate funds through the organizational work authorization system.
- *Develop the project technical performance objective.* Describe the project end product(s), services, and/or processes that satisfies a customer's needs in terms of capability, capacity, quality, quantity, reliability, efficiency, and such performance standards.
- *Develop project control concepts, processes, and techniques.* How will the project's status be judged? On what basis? How often? By whom? How? Ask and answer these questions prospectively during the planning phase.

The Project Plan

Of course, the size and organization of the project plan depends on the project. The *key essentials* of a project plan, regardless of the size of the project should include the following:

- A summary of the project that states briefly how the project is to be developed and the processes and techniques that will be used. A clear statement of the expected results of the project should be provided so that these results can be identified and compared to the standards used on the project during its life cycles.
- A list of tangible goals of the project, usually expressed in terms of a work element or work package without ambiguity so that it can be easily determined that a goal has been achieved.
- A brief outline of how organizational resources will be used to accomplish the project ends.
- A Gantt chart and activity network that shows the sequence and relationship of work packages to provide a roadmap of how project work is to be done.
- Budgets and schedules for all the work packages of the project.
- A description of how authority and responsibility is delegated to the project team members, as well as other to project stakeholders.
- Policy guidelines on how the project work will be monitored, evaluated, and controlled.
- A list of the key project participants and their assignments relative to the project WBS.

The results of the development of project plans should be reflected in project and organizational documentation.

Project Strategy Documentation

Every project plan that is prepared requires documentation that provides a written record of the content of the resources committed and the action initiatives required to start the project and bring it to a successful conclusion through its life cycle. During the planning and implementation of a project, the required documentation should include the following elements:

- Citation of resources required
- A schema for the allocation of resources
- A project plan
- Individual and collective roles of project team
- Assigned project responsibility and authority
- Work breakdown structure
- Cost estimates
- Schedules
- Anticipated contents of project guidebook
- Summary of project planning work packages
- Quotes or bids
- Financial plan
- User requirements
- Marketing plan
- Preliminary construction/manufacturing plan
- Project evaluation, monitoring, and control strategies
- Protocol for progress payments
- Project team performance review strategies
- Customer reporting strategies
- Configuration baselines
- Communication plan
- A project risk assessment plan
- Logistic support plans
- Stakeholder support strategies

The initial documentation should include summary material about the project during its life cycle. For example, the documentation required during the closeout of the project could include such things as:

- Final report to project owner
- Technical drawings
- As-built drawings
- User guidebooks
- Project history to include "lessons learned"
- Closeout of resources used on project

Understanding Linkages

An effective and clear link between the strategic management and project management of the enterprise depends on several considerations:

- Recognition that strategic management involves a continued examination of how enterprise resources can be used in the future.
- An understanding of the mutuality between strategic management and project management.
- Acceptance of a reference focus for the use of choice elements in allocating organizational resources in strategic management.
- Recognition that project results are building blocks in the design of new or improved products, services, or organizational processes.
- Acceptance that projects provide the most reasonable means for dealing with environmental changes facing the enterprise.
- The existence of policy and procedures documentation that clearly links strategic management and project management—and provides a protocol for how such linkages can be maintained and strengthened.

Project Planning Principles

This chapter concludes with the citation of a few principles of project planning. A principle is a basic truth, law, or assumption; a fixed or predetermined policy or mode of action; the essence or vital philosophy to deal with phenomena.

Henry Fayol, truly one of the founding fathers of modern management, put forth his General Principles of Management in his book (Fayol, 1949). He noted that

> . . . I shall adopt the term principles whilst dissociating it from any suggestion of rigidity, for there is nothing rigid or absolute in management affairs . . . Therefore principles are flexible and capable of adaptation to every need; it is a matter of knowing how to make use of them, which is a difficult art requiring intelligence, experience decision and proportion.

It is in the spirit inherent in Fayol's description of principles that the following principles of project planning are presented:

- Planning for projects is a key responsibility of project managers.
- The project planning process should be carried out with those stakeholders that have an interest in the project.
- A project plan will typically require some change during the life cycle of the project as new information becomes available or basic project requirements change.
- Project planning is a process that pulls information from the organization as well as from other key stakeholders.
- The quality of project planning relates directly to the probability of success on the project.
- All plans will require revision to meet emerging changes and new information.

Summary

This chapter presented an overview of a philosophy of project planning. Such a philosophy is a way of thinking about how project planning should be carried out to support project and enterprise purposes. Project planning must support strategic management protocol and processes of the enterprise. Projects are building blocks in the design and execution of enterprise strategies that provide for the principal means for dealing with the change facing the enterprise.

References

Barkley, B. T., and J. H. Saylor. 1993. *Customer-driven project management: A new paradigm in total quality implementation.* Boston: McGraw-Hill.

Baumgartner, J. S. 1963. *Project management.* Homewood, IL: Richard D. Irwin.

Cleland, D. I. 1996. *The strategic management of teams.* New York: Wiley.

Cleland, D. I., and L. R. Ireland. 2000. *Project manager's portable handbook.* New York: McGraw-Hill.

———. 2002. *Project management: Strategic design and implementation.* 4th ed. New York: McGraw-Hill.

Davis, R. C. 1951. *The fundamentals of top management.* New York: Harper & Brothers.

Duke, R. K. et al. 1977. Project management at Fluor Utah, Inc. *Project Management Quarterly* 3:33.

Fayol, H. 1949. *General and industrial management.* London: Pitman.

Hamilton, A. 1997. *Management by projects: Achieving success in a changing world.* Dublin: Oak Tree Press.

Knutson, J., ed. 2001. *Project management for business professionals: A comprehensive guide.* New York: Wiley.

Lewis, J. P. 2000. *Project planning, scheduling and control: A hands-on guide to bringing projects in on time and on budget. 3rd ed.* New York: McGraw-Hill.

Meredith, J. R. and S. J. Mantel. 1985. *Project management: A managerial approach.* New York: Wiley.

Pinto, J. K., ed. 1998. *Project management handbook.* San Francisco: Jossey-Bass Publishers.

Randolph, W. A., and B. Z. Posner. 1988. *Effective project planning and management: Getting the job done.* Englewood Cliffs, NJ: Prentice Hall.

Steiner, G. A. 1979. *Strategic planning: What every manager must know: A step-by-step guide.* New York: Free Press.

Turner, J. R., ed. 1999. *Handbook of project-based management.* 2nd ed. London: McGraw-Hill.

CHAPTER TEN

MODELS OF PROJECT ORIENTATION IN MULTIPROJECT ORGANIZATIONS

Joseph Lampel, Pushkar P. Jha

More than 30 years of research on project management have produced a relatively coherent body of principles for effective project planning and execution. Over the same period, there have been consistent reports of the failure of managers and organizations to follow these prescriptions. The obvious response to this failure has been to look for "best practices" that can bridge the gap between prescriptions and outcomes. The rationale behind this effort is to comparatively analyze the conduct and performance of different projects, and then translate effective managerial practices into a set of robust practices that are widely accepted.

While projects may fail because of factors that are internal to the project itself, such as poor leadership, conflict, poor coordination, and so forth, organizational factors that are external to the projects also play a role in the failure of prescriptions to produce expected outcomes. More specifically, researchers hypothesize that problems in the relationship between organizations and projects will hamper the implementation of project management prescriptions. Thus, to understand why project management prescriptions often fail, it is necessary to identify these problems. Research that deals with new product innovation (Wheelwright and Clark, 1992; Lewis and Welsh, 2002) suggests that projects usually fail because they are insufficiently resourced, insufficiently empowered, and/or, generally lack wider organizational support.

When these problems are examined in some depth, the normative conclusion that follows is that organizations that take the trouble to do these things properly are more likely to successfully plan and execute projects than organizations that do not. The ideal organization, from this point of view, is one in which structures, empowerment, and resources are aligned as far as possible with the needs of projects. In practice, most projects take place in organizations that depart from this ideal. And by the same token, the further they depart from this ideal, the more likely are the project management specialists to encounter problems when translating prescriptions into practice.

While this point is often raised in the literature (Raz et al, 2002; Shenhar et al., 2001), what is clearly lacking is a construct that allows going beyond simple observations of the shortcomings of organizations that do not properly resource, support, and empower projects. This is a task made difficult by the very nature of projects, as organizations often have projects that serve different clients, meet different needs, require different resources, and develop along different trajectories (Lampel, 2001).

In this chapter, we define the degree to which organizations resource, empower, and support projects as "project orientation". The term project orientation has been used in the literature to examined a multitude of interconnected concepts, from project leadership to team and project management competencies (Gareis and Hueman, 2000; Gareis, 2002) and management design for project coordination (Gareis, 1992). The term is used here to define an integrated construct that is likely to be of considerable utility to the project management specialist. The concept of project orientation allows project management specialists to discuss and measure the constraints in which they operate with greater clarity and precision. This may not only be useful for communication among managers that are directly involved in planning and implementing projects but is also useful in discussions with support functions and top management.

The construct of project orientation, however, has wider utility and can be of importance when it comes to managing the overall strategy of the firm. The starting point for considering the use of the construct is the interrelationship between project selection and management on the one hand and organizational strategy on the other (Cleland, 1994, p. 81). Attaining a "fit" between strategy and project management activities is an important aspect of the search for overall sustainable advantage (Porter, 1996). According to Porter (1996), the search for fit operates on three levels. First-order fit is the consistency of different activities with overall strategy. Second-order fit occurs when activities reinforce each other. And third-order fit is the optimization of efforts across activities.

Attaining fit on any of these levels, let alone on all three, requires organizations to choose between different combinations of structures, systems, and processes (Mintzberg, Ahlstrand, and Lampel, 1998, pp. 302–347). The choice is often a function of organization type (Miller and Mintzberg, 1983). For example, small law firms are likely to approach the choice differently from large, diversified consumer goods firms. In the case of projects, we would argue that the search for fit between strategy and project management activities is likewise dependent on the type of organization. Research suggests (Wheelwright and Clark, 1992) that when it comes to projects, organizations generally fall into one of the following types:

- *Project-based.* Organizations with centralized operations tuned to support projects. Here, a formal body (for example, a project office) may lead project planning. This is typical of organizations with large-scale projects and with a project portfolio characterized by a large number of projects for clients' external to the organization. Organizations whose projects are characterized by extensive post project support in client handover also tend to fall in this category.

- *Project-led.* Organizations where operations provide support for projects. Here core operations in functions or areas such as manufacturing, information systems, human resources and product development, to name several, tend to have a significant say in project initiation and design. The term project-led typically describes an organization that may or may not have a project portfolio aligned to service an external client base but is very likely to be characterized by a reasonable density of internal change/process improvement projects. Complex alliancing requirements in project initiation and design stages may also be a feature of projects in such an organization.
- *Core-operations-led.* Organizations where projects are support mechanisms to augment/improve the efficiency of core operations. Projects are likely to be internal and could be low down in top leadership involvement beyond the initiation and conceptualization stage. This is in comparison to organizations that can be categorized as project-based and project-led.

In general, project orientation tends to be high in project-based organizations, moderate in project-led organisations, and relatively low in core-operations-led organizations. Strategically, however, organizations tend to move within this typology. Project-based organizations may change their portfolio of projects and clients. Project-led organizations may become focused on one type of projects. And finally, core-operations-led organizations may decide that their future depends on a specific type of transformational projects.

In all these cases, organizations may experience a mismatch between their strategy and their project orientation. Put differently, their strategy may demand a different project orientation than the one that has evolved over time. To become more effective strategically, they would need an audit of their project orientation; first, in order to determine their project orientation, and second, in order to decide to what extent it should change.

Models of Project Orientation

In the sections that follow, two models of project orientation are presented. The first is the project perspective of project orientation and the second the organizational perspective. The interfacing environment between the projects and the organization is viewed from both sides of this interface: the project side and the organization side, respectively. The primary basis for these models are insights drawn from the first phase of exploratory interviews with seven partner organizations from different sectors in the EPSRC research project on "Organisational Learning and Business Performance in Project Based Organisation" (PROBOL; 2001–2003). The exploratory interviews attempted to understand the project portfolio of these organizations and how they made sense of project experiences to configure organizational strategy for projects.

The aim of the study is to explore and validate project orientation as a construct. We have followed the standard approach of using principal component analysis to evaluate construct validity (Carmines and Zeller, 1979). An outline of the data and analysis is provided after the discussion on the two models. The interpretations drawn from the analysis are at present exploratory. They are outlined in the form of some key relationships that may dominate the interfacing environment that are defined as project orientation.

Model 1: Project Perspective of Project Orientation

Our reading of the literature (Bugetz, 1992; Lawrence and Johnson, 1997; DeFillippi and Arthur, 1998; Bugetz, 1992; Rose et al, 2000; Nobelius, 2001; Sydow and Staber, 2002) suggests that seen from the perspective of projects, three project dimensions shape the interaction between projects and organizations. These are, project scoping, project programming, and project autonomy. Figure 10.1 uses these dimensions to visualize a space inhabited by the project portfolio of the organization. This is the "project to organisation" perspective of project orientation as it has evolved over time.

Project scoping consists of the basic definition of the project plus the extent to which the concept is fully mapped during the early phase of the project, as opposed to allowing these details to emerge subsequently. The contrast is between fixing the scope in some detail and curtailing exploration very early in the life -cycle of the project, as opposed to simple sketching of the scope and allowing considerable room for exploration as further information and learning is gained.

Project programming consists in the extent to which the project is tightly constrained by budgets, schedules, and targets, as opposed to being loosely constrained by these factors. A tightly programmed project is often a project in which every effort is made to foresee

FIGURE 10.1. PROJECT PERSPECTIVE ON PROJECT ORIENTATION.

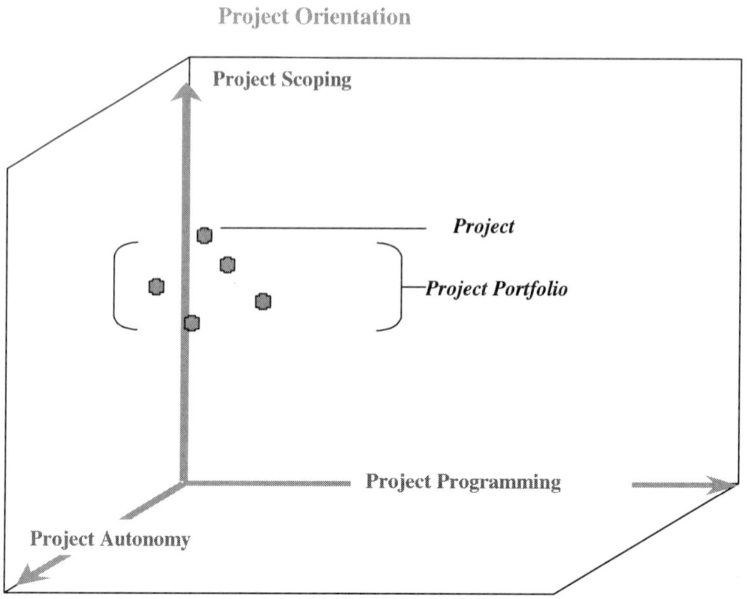

contingencies, break down the process into discrete steps (i.e., gating the process), and insert strong incentives for compliance with schedules and targets. A loosely programmed project is one in which schedules are simple and approximate, and incentives are tied to achievement of generally formulated rather than highly specified targets.

Project autonomy refers to the degree to which the project is allowed to evolve without constant report and input from the organization. Here we have two extremes. At one extreme we have projects that (a) constantly report progress to other parts of the organization (in particular, to higher organizational levels) and (b) accept input and request for changes. At the other extreme we have projects that are (a) given high latitude to move forward with little reporting and (b) do not expect or easily accept input and suggestions for changes.

These three dimensions constrain and define project evolution within the confines of the organization. They are in effect evolutionary trajectories that are shaped by the interface with the rest of the organization. With this in mind, project orientation from the perspective of projects is determined by the extent to which these dimensions are set by the needs of the projects, as opposed to the needs of the organization as a whole. Such needs do not privilege specific polarity along these dimensions.

In some industries it is better if projects are fully scoped in advance, while in others basic scoping with plenty of room for exploration tends to produce better results. In some industries tight programming is regarded as best practice, whereas in other industries loose programming is often seen as preferable. The same can be said of project autonomy. Here not only will industry practice vary, but company practice will likewise vary, since autonomy, as you will see, is as much a political as a strategic issue.

A note of caution should be sounded at this point. From a statistical point of view, the dimensions are probably not independent. In general, fully scoped projects often go together with tightly programmed projects; and tightly programmed projects often go together with projects that have only low autonomy. Having said this, it is also worth emphasizing that this statistical observation is not necessarily strategic.

Organizations do have a choice when it comes to how they design, program, and control their projects. In some industries and organizations, one finds fully scoped projects with loose programming and projects with high autonomy that are tightly programmed. As Table 10.1 shows, the key issue is the evolutionary trajectories that are defined by these dimensions. Some may be rare, but they are all in principle possible.

Model 2: Organizational Perspective of Project Orientation

Organizations are systems of power and resources. From the point of view of projects, most decisions regarding power and resources should devolve to the level of the project. From the point of view of organizations, this is not always possible or indeed desirable. There is frequently a tug-of-war here between what is best for the project and what is best for the organization. The tension may be at odds with prescriptive models of project management that ordain that projects should be aligned with the strategy of the organization as a whole (Cleland, 1994), but it is a tension that persists for a number of reasons.

First, organizations often have multiple projects with different clients and different needs. Second, projects overlap in terms of planning and execution. Today's strategy may give rise to new projects, but the organization may still be structured to serve projects that

TABLE 10.1. TYPES OF PROJECTS.

			Autonomy Low or High (Interpreted by a combination of:)	
Illustrative Projects Defined by Some Lead Characteristics	**Scoping**	**Programming**	**Resource Allocation**	**Monitoring Light (low frequency, loosely structured) or heavy (high frequency and very structured)**
1. Client-specified solution where similar projects for the client have been carried out successfully in the past	Rigid	Detailed	Up front	It could be either light or heavy.
2. Problem-solving R&D project with high sanction from the top	Rigid	It could be either detailed or sketchy.	Up front	It could be either light or heavy.
3. New-product development targeted at specific market needs	Rigid	It could be either sketchy or detailed.	Either up front or gradual	It could be either light or heavy.
4. Exploratory R&D project for Product technology/core technology improvements anticipated to provide competitive edge and championed from the top	Flexible	Sketchy	Up front	It could be either light or heavy.
5. Internal change project where solution specifications are provided by external R&D	Rigid	Detailed	Gradual	It could be either light or heavy.
6. Internal project where solution specifications are to be developed in-house for a given need; also, utility projects that involve critical resources and are intended "to keep the lights on" (Shenhar, et al., 2002)	Flexible	Detailed	Gradual	Heavy
7. Client-specified "problem only" projects where detailed specifications for output are missing, e.g., software development projects and consultancy projects	Rigid	Sketchy	Gradual	Heavy
8. New-product development characterized by a fast-changing market	It could be either rigid or flexible.	Sketchy	Gradual	Heavy

were conceived some time back. Third, firm strategy is often interpreted and enacted differently in different part of the organisation (Gann and Salter, 2003). Finally, projects are not always aligned with the strategy because strategy itself is frequently an umbrella that permits a range of options rather than a clearly and tightly defined set of goals (Mintzberg and Waters, 1985). This is also highlighted in numerous industry specific case studies (Drake, 1999; Westney, 2003) in recent times.

All of the preceding leads to a perspective of project orientation that is centered on the organization, as distinct from model 1, which is centered on the project. From an organizational point of view, the following issues matter most when it comes to making key project decisions: First, what resources does the project need to perform effectively and how should these resources be allocated? And second, how much political backing does the project need from top management (Green, 1995) and support systems to perform effectively?

These two issues are clearly related. For example, resource allocation often requires top management support. And accessing support systems is often a matter of resource allocation. But beyond this, the ability of projects to gain resource and political support at the front end and subsequently their ability to gain backing and more resources from both top management and support systems is a measure of the organization's project orientation.

In this respect, organizations are likely to be different. In some organizations, projects receive proportionately limited attention and support. They are regarded as one area of action among others. In other organizations, projects are central to the strategy of the company and thus are more likely to top the agenda and have first claim to resources.

As Figure 10.2 suggests, the distinction between these two types can be perceived by an alignment typifying high project orientation in the latter case against low project orientation in the former case. This alignment evolves over time and may need to be redefined by deliberate action. This may be through key transformational projects when the organization feels the need to develop this emerged pattern to fit its interpretation of business needs from projects. Some organizations may also engage in experimental projects where they test the performance of deliberate action used to change this alignment, before setting in motion organization-wide change in the way projects are done.

Project Orientation Variables

Exploratory research during the first phase of the PROBOL research and the theorization for the models of project orientation have been used to generate a list of variables that manifest themselves in the interfacing environment between the projects and the organization. Based on these variables, an attempt to investigate the key relationships in the project organisation interface has been undertaken. A list of these variables, which forms the basis of the data collection instrument, is as follows:

- Treatment of past project performance
- Top leadership's support for change
- Basic characteristics of project systems, including volatility of the technology involved
- Baseline for project management performance (time, cost, and scope)

FIGURE 10.2. DEGREES OF PROJECT ORIENTATION.

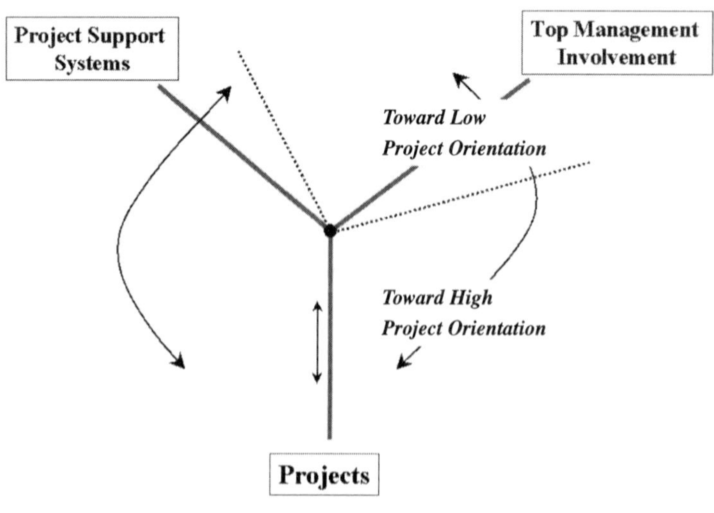

- Top leadership involvement in project planning including role in giving visibility to projects
- Experience and role clarity of multiproject leadership
- Role span of the project manager
- Capacity building for project roles
- Project alliancing decisions (e.g., contractor alliancing and inter-organizational alliancing.)
- Support department's involvement in project control
- Senior management's involvement in project control
- Approach toward project teams in terms of physical location and performance assessment
- Project resourcing modes
- Incentive for gaining project experience
- Nature of communities of practice (and networking of knowledge management structures) for projects

Project Orientation Data: A Profile of the Sample

Using a structured instrument, quantitative data has been collected from 54 respondents over six of the seven organizations in the second phase of the PROBOL research. All seven organisations had contributed to the qualitative data generated from exploratory interviews in the first phase of the research.

The six partner organizations that compose the data in the second phase of the research are drawn from the sectors of information technology, pharmaceuticals and healthcare, poverty alleviation, banking and finance, construction, aerospace and energy.

Nearly three-fourths of the respondent sample was composed of project practitioners and one-fourth was composed of personnel who are working with project support departments. Some project leadership and top leadership personnel are also included in this stratification. The work experience range for the entire sample was from 1 to 36 years of experience with the organization. This provides a comprehensive "perception map" of how the aforesaid variables influence project orientation. It has also allowed for an identification of the key driving relationships for each organization. These key drivers may significantly aid organizational efforts to mold the evolved project orientation profile to match the strategic needs of the organization from projects.

Instrument Testing

The structuring of the variables and the design of indicators to measure them was based on inputs from the literature survey and insights about partner organizations from the first-phase exploratory data collection under the research. Inter-rater reliability, based on rank order correlation across respondents for each organization, was at acceptable levels. This implies that different respondents with their respective profiles understood the indicators as measuring the same phenomenon in their respective organisations.

Multiple indicators were designed to measure each variable for reasons of validity. A total of 57 indicators make up the questionnaire that measures these 15 variables. The responses were generated using a Likert scale [A scale where the respondent specifies a level of agreement or disagreement with statements that express a favorable or unfavourable attitude towards the concept under study/being measured]. For content validity purposes, questions were worded both negatively and positively (based on what constituted high or low project orientation) to make sure that the right phenomenon were being measured and each question represented an aspect of the variables under investigation.

Content validity was supported by interim data analysis that investigated relationships between the indicators. These relationships were presented to the respective organizations and were recognized by them as phenomena that characterize the interfacing environment between the project and the organization, thus establishing that the expected values for project orientation were being measured.

Data and Analysis

The 57 questionnaire items were aggregated over the 15 variables under consideration. The main part of the analysis has been carried out using principal component analysis to reduce dimensionality because of a large number of variables as enumerated earlier.

Simulation was carried out for reasons of data limitation. Standard deviation in the original or base sample was used as a factor to increase the data set for statistically significant analysis. The simulation was carried out for the sample as a whole and also for individual organization data sets to achieve a significant ratio between variables and number of samples.

Open-ended responses and post-instrument-administration group discussions were also used to further augment the understanding of project orientation. The interpretations were discussed in partner organization group sessions and received confirmation as a reliable interpretation of the phenomenon of project orientation.

Limitations of sample size and the extensive range of factors involved in configuring an organization's project orientation make the results and implications of the study less definitive than we would like them to be. Nevertheless, the findings provide useful insights to illustrate the concept of project orientation.

Findings and Implications

Principal component analysis was used on the sample as a whole and also to individually process patterns for each organization. These findings support the introduction of the third element, top management involvement at the project-organization interface level as also enumerated in the discussion leading to the second model (Figure10.2). This was because variables related to this aspect (top leadership and multi-project leadership) load strongly on the principal components.

The two models of project orientation have been related to six of the seven organizations that contributed data toward the second phase of the research. These organisations are profiled in the following:

Organization 1: Characterized by external-client-focused projects, fast-changing technology, and projects with small- to medium-length[1] life cycles.

Organization 2: Characterized by internal projects, product development, infrastructure, and/or change focus.

Organization 3: Characterized by long life cycle and fast-changing technology; projects move in clusters using centralized resources, product development, and external client focus.

Organization 4: Characterized by large-scale external-client-focused projects, moderate to high project life cycle, widespread in physical location of project sites, and an ongoing restructuring to improve the interface between project roles and top management roles.

Organization 5: Characterized by internal change and process efficiency improvement projects, relatively smaller project life cycle.

Organization 6: Characterized by clusters of highly interconnected projects within divisions, which almost always exist as part of larger programs.

[1] Length of lifecycle is discussed as relative over the organisations in the sample.

Table 10.2 links the preceding organization types and the project orientation models. The implications for each of the organizations in the sample lie in the key relationships enumerated here. As an illustration, these may be understood as levers that can be effectively used to affect project orientation and thus, by extension, also affect organizational objectives essential to the projects.

The table is essentially a matrix that has several possible manifestations of organizational perspective of project orientation in columns and manifestations of the project perspective in rows. For categorization purposes here, high, moderate, and low qualifiers are used. The six partner organizations in the PROBOL research that contributed to the data in the second phase are placed in the respective intersection of the two models as in this matrix. The categorization of organizations is based on exploratory interviews and background research on the organizations to understand the organizations in the sample and the projects they do. Principal component analysis outputs are used to express relationships that dominate the interfacing environment between the organization and the projects.

Conclusions

Morris (2003) has argued that we should make a distinction between project management and the management of projects. The first deals primarily with practices and principles that are internal to the discipline of project management as it was traditionally conceived. The second looks to a wider perspective of the relationship between projects and a more holistic view of their context—whether their context is the organization or the larger economic and social environment. The construct of project orientation belongs to the management of projects, rather than to project management. It is put forward as a core construct that defines the relationship of the organization and its projects and perspectives from both sides of this interface (i.e., the project-to-organization perspective and the organization-to-project perspective, respectively) are employed.

The results presented from ongoing research (PROBOL, 2001–2003) that partners from seven global firms in different business areas provide an illustration of the utility of the construct of project orientation. There are statistical limitations because of sample size in the second-phase quantitative data used in the research (that was drawn from six of the seven organizations). The range of variables that impact the project organization interface also contributed to limitations in analysis. However, use of simulation techniques has allowed us to partly overcome the limitations of a small sample. This has been further aided by use of associated qualitative data from discussions with respondents from the partner organizations and by sharing findings with the partner organisations to get feedback on the accuracy of the diagnosis and constructs, which has been affirmative.

Projects have an ambiguous relationship to the corporate environment in which they evolve (Garbher, 2002). On the one hand, projects often underpin the organization's main business and are a direct expression of its strategy. On the other hand, however, they often stretch and change both operations and strategy. This tug-of-war between projects and organizations is inevitable and, when properly managed, can be constructive. Often, however, the tension translates into friction and failure—in part because such tensions trigger

TABLE 10.2. PROJECT ORIENTATION PERSPECTIVES.

Organizational Perspective / Project Perspective	Low Project Orientation	Moderate Project Orientation	High Project Orientation
Low spread in the project portfolio cluster	**Organization 5.** A strengthening of project manager role is instrumental in relaxation in project standards of cost, time, and scope.		**Organization 4.** Top leadership support for projects critically influences variation in core technology in projects and independence of individual project planning in relation to the larger portfolio. Clearer career progression paths from project to project leadership levels are vital for coherent project strategy for the organization
Moderate spread in the project portfolio cluster	**Organization 6.** Strengthening of the project manager's role encourages seeking project experience. Top leadership support is critical to knowledge sharing across projects and developing communities of practice.	**Organization 1.** Incentive for gaining project experience depends how project performance is treated (buried or magnified). Clear role definition for multiproject leadership may help reduce support department's control on project execution.	**Organization 3.** Senior–middle management involvement may be critical in influencing top management support and sponsorship for projects. Clear role definition for multiproject leadership may help reduce support department's control on project execution.
High spread in the project portfolio cluster	**Organization 2.** Incentive for gaining project experience depends upon a greater boundary spanning in the project manager's role. Top leadership support is critical to knowledge sharing across projects and developing communities of practice.		

power politics but also because there is an insufficient understanding of the causes and dynamics of the tension between projects and organizations in the first place. Project orientation is a useful way of looking at the relationship between projects and the corporate environment. It opens the way to deeper understanding and better management of the factors that shape the interface between project management at the project level and the management of the organization at the higher, strategic level.

References

Burgetz, B. A. 1992. Project design: The critical step to successful systems. *CMA Magazine* 66(4):10.

Carmines, E. G., and R. A. Zeller. 1979. *Reliability and validity assessment.* Beverly Hills: Sage Publications.

Cleland, D. I. 1994. *Project management: Strategic design and implementation.* 2nd ed. p. 81. New York: McGraw-Hill.

Drake, D. L. 1999. Projects must fit into company strategies. *Business News New Jersey* 12(1):22–23.

DeFillipi, R., and M. Arthur. 1998. Paradox in project based enterprise: The case of filmmaking. *California Management Review* 40(2):125–139.

Gann, D. M., and A. J. Salter. 2003. Project baronies: Growth and governance in project-based firm. 19th EGOS Colloquium, Copenhagen.

Gareis, R. 1992. Management of network of projects. *AACE transactions.* www.wu-wien.ac.at/pmg/pos/docs/pub_portfolio_management.pdf.

Gareis, R., and M. Hueman. 2000. Project management competencies in the project-oriented organisation. In *The Gower Handbook of Project Management.* ed. J. R. Turner and S. J. Simister. 709–721. Aldershot, UK: Gower Publishing.

Grabher, G. 2002. Cool projects, boring institutions: Temporary collaboration in social context. *Regional Studies* 36(3):205–214.

Green, S. G. 1995. Top management support of R& D projects: A strategic leadership perspective. *IEEE Transactions on Engineering Management* 42(3):223–232.

Lampel, J. 2001. The core competencies of effective project execution: The challenge of diversity. *International Journal of Project Management* 19(8):471–483.

Lawrence, B., and B. Johnson. 1997. The project scoping gamble. *IEEE Transactions* 14(3):107–109.

Miller, D., and H. Mintzberg. 1983. The case for configuration. In *Beyond Method,* ed. G. Morgan. Beverly-Hills: Sage.

Mintzberg, H., and J. Waters. 1985. Of strategies, deliberate and emergent. *Strategic Management Journal* 6:257–272.

Mintzberg, H., B., Ahlstrand, and J. Lampel. 1998. *Strategy safari.* London: Prentice Hall.

Morris, P. W. G. 2003. Irrelevance of project management as a professional discipline. IPMA 17th World Congress on Project Management, Moscow.

Nobelius, D. 2001. Empowering project scope decisions: Introducing. R&D content graphs. *R&D Management* 31(3):265–274.

Porter, M. E. 1996. What is strategy? *Harvard Business Review.* (November–December): 61–78.

PROBOL 2000–2003. UK government-funded (EPSRC) research project on Organisational learning and business performance in project based organisations. www24.brinkster.com/probol.

Raz, T., A. J. Shenhar, and D. Dvir. 2002. Risk management, project success, and technological uncertainty. *R&D Management* 32(2):101–109.

Rose, K. H., J. K. Pinto, J. W. Trailer, and K. Rose. 2000. Essentials of Project Control. *Project Management Journal* 31(2):60–61.

Shenhar, A. J., D. Dvir, T. Lechler, and M. Poli. 2002. One size does not fit All: True for projects, true for frameworks. Project Management Institute Research Conference. Seattle.

Shenhar, A. J., D. Dvir, D., O. Levy, and A. C. Maltz. 2001. Project success: A multidimensional strategic concept. *Long Range Planning* 34:699–725.

Sydow, J., and U. Staber. 2002. The institutional embeddedness of project networks: The case of content production in German television. *Regional Studies* 36(3):215–227.

Turner, R., R. Peyami. 1996. Organizing for change: A Versatile Approach. *In The Project Manager as a Change Agent*, Composed and edited by J. R. Turner, K. V. Grude, and L. Thurloway, 62–75. New York: McGraw-Hill.

Westney, R. 2003. Setting the strategy for successful projects. *Offshore* 63(5):133–134.

Wheelwright, S. C., and K. B. Clark. 1992. *Revolutionizing Product Development: Quantum Leaps in Speed, Efficiency, and Quality*. 175–196. New York: Free Press.

CHAPTER ELEVEN

PROJECT PORTFOLIO SELECTION AND MANAGEMENT

Norm Archer, Fereidoun Ghasemzadeh

A project portfolio is a group of projects to be carried out under the sponsorship of a particular organization. These projects must compete for scarce resources (labor, finances, time, etc.), since there are usually not enough resources to carry out every proposed project. Project portfolio selection is the periodic activity involved in selecting a portfolio from the set of available project proposals and from projects currently under way. Compared to the managerial and operational decisions that are usually involved in managing individual projects, portfolio selection is a strategic decision. To ensure a maximum return on selected projects, the selection process must be linked to the business strategy of the organization.

The purpose of this chapter is to outline portfolio selection and management as a strategic process, using known techniques and tools in a logical and organized manner. We begin with a literature review, and then discuss the characteristics of projects affecting their selection and outline some portfolio selection methodologies. Choosing among these methodologies and organizing a logical portfolio selection process requires a framework that can be followed easily, and we describe one such framework in detail. Finally, we review some current issues in portfolio management.

Literature Review

Project portfolio management is an essential concept in many industries, and applications include new product development (NPD), construction, pharmaceuticals, process development, product maintenance, fundamental research, and so on. These tend to differ along certain dimensions, and we will demonstrate some of their differences by examples.

The majority of innovation-based manufacturing industries generate products obtained predominantly via innovative R&D processes. NPD portfolios are more likely to be successful if they include a limited number of carefully selected, positioned, and balanced projects (Wheelwright and Clark, 1992; Cooper, Edgett et al., 2000). A portfolio is balanced if there is a suitable distribution of projects on dimensions such as technology and market risk, completion time, and return on investment. If there are too many projects in the pipeline, there is a high degree of conflict over existing resources, slowing project progress and reducing successful completion rates. This may result in missed market opportunities and limited investment returns.

On the other hand, discovery-based research activities dominate in the development of new pharmaceuticals. This involves a search and a trial-and-error test or screening process, whereas NPD involves a concept, proof of feasibility, a product design, and an evaluation. In the pharmaceutical industry, the risk is so high in the initial (discovery) phase that literally hundreds of selected compounds or molecules may be investigated in order to improve the probability of finding a few that are worthy of the next phase of development (Prabhu, 1999).

In the construction industry, technology risk is relatively low, but there are other risks resulting from uncertainties in weather, political decisions, and labor availability. Another feature of portfolios in the construction industry is that most projects are outsourced, and this industry is characterized by a lack of long-term relationships between supplier and client. Welling (Welling and Kamann, 2001) suggests that cooperation can be improved when the same individuals, representing the general contractor and subcontractors, respectively, deal with each other in a series of projects, rather than dealing with different individuals for each project. This suggests an increasing emphasis on long-term cooperation, networking, and strategic alliances.

Most recent advances in portfolio selection and management have occurred in the new product development area. A comprehensive review of this field has been completed by Cooper, Edgett, and Kleinschmidt (Cooper, Edgett et al., 2001). In addition, this team did an extensive survey of NPD industry practice of portfolio management (Cooper, Edgett et al., 1999) in 205 U.S. companies. From their study, they were able to cluster companies into four groups, with "benchmark" businesses as the top performers. This group's new product portfolios consistently scored the best in terms of performance. These companies tended to choose high-value projects and their portfolios were aligned with the company's business's strategy, with the right balance and the right number of projects. Benchmark businesses employed a much more formal, explicit method in managing their portfolios. They relied on clear, well-defined portfolio procedures, were consistent in applying their methods to all projects, and there was strong management support for portfolio selection and management.

Risk and outsourcing are currently having the most active impact on changes and advances in portfolio selection and management. For these reasons, the following two subsections are devoted to these particular topics.

Risk

There are multiple sources of risk, including technology, market, schedule, cost, legal, political and so on, depending on the field of application. Sophisticated methodologies to assess risk amount and impact prior to project commitment, through risk identification, quantification, response development, and control, have been identified as one of the areas in which major advances have recently occurred (Pinto, 2002). A key criterion for successfully applying risk evaluation in portfolio selection is that risk assessment and quantification be uniformly applied across all projects and teams in order to distinguish among projects that have acceptable and unacceptable levels of risk. For construction portfolios, since there is less uncertainty in the expected costs, schedule, and performance, financial approaches are the accepted norm for project selection. However, there is often a significant uncertainty in such factors as weather, labor availability, and political environment. Disagreement frequently occurs among parties to lump-sum construction projects, because of a lack of advance agreement in apportioning risk between client and contractor (Hartman, Snelgrove et al., 1998).

Assessing and managing risk in NPD programs is discussed in detail by Githens (Githens 2002). Although risk is one of the most important characteristics of projects, it is often difficult to evaluate. The frequency with which project risk analysis has been applied in industry (Raz, Shenhar, et al., 2002) has been disappointing, although when it is used, the chances of success are much improved.

The risks of failure of individual research and development projects can be very high when truly new technology is involved in new product development. A recent survey indicates that it takes an average of 6.6 ideas to produce one successful new product, while on average 50 to 60 percent of NPD projects fail (Griffin, 1997). Companies that rely heavily on internal R&D for new products need to be able to take risks without compromising the profitability of the company. It is critical to link company strategy to portfolio development when company strategy involves both a high degree of innovation and a high rate of growth (Wadlow, 1999).

Risk-value project selection procedures that seek to minimize risk for a certain specified value tend to bias against higher risks, both for technical risk and commercial risk uncertainties. Another option is to maximize value for a certain level of risk. If innovation and growth are major strategic goals, project selection must first select on the basis of growth without unduly compromising other important objectives. However, formal product portfolio selection may bias against the inclusion of more highly innovative projects in a portfolio. For example, Roseneau (Roseneau, 1990) notes a reduction in the degree of innovativeness of new products through internal competition, with the attractive short lead times associated with low-risk, less innovative projects. Speed to market and high degrees of innovation in new products tend to be conflicting objectives.

Typical models used in analyzing risk include Monte Carlo simulation, decision theory and Bayesian statistical theory (Hess, 1993; Riggs, Brown et al., 1994; Martino, 1995), and decision theory combined with influence diagram approaches (Rzasa, Faulkner et al.,1990). One new approach to portfolio selection in NPD is risk strategy analysis (Wadlow, 1999). This focuses on risk as a key factor in the development of new technological products. A risk strategy, in this context, describes a specific extent and type of involvement in research in terms of the different levels of technical and commercial risk associated with the mix of projects making up the portfolio. Accordingly, a risk strategy in terms of technology and commercial risk is analyzed in terms of statistical risk of failure as opposed to probable outcome. Given a number of basic project descriptors, this approach can be used to design as well as analyze risk strategies while screening projects for portfolio selection, based on the firm's strategic business objectives. While there is uncertainty in almost every project parameter, an interesting application of cost uncertainty in portfolio analysis is provided by Stamelos (Stamelos and Angelis, 2001).

The unique nature of projects, and their range in objectives, size, complexity, and variety of technological content, is almost limitless and not confined by industry boundaries. Shenhar et al. (Shenhar, 2001) have derived a general classification system for project management commonalities, which can be helpful in establishing portfolio risk profiles, since an optimal choice of projects for a portfolio may need to span these dimensions. Although market risk also reflects the potential acceptance of a new product, two dimensions they suggest are technological uncertainty and contribution to business. Table 11.1 reflects these two dimensions, although some of the examples given in the table might in fact extend across more than one of the given categories on one or both dimensions. But there are other uncertainties in addition to technological uncertainty, particularly in the construction industry.

The applications given in Table 11.1 reflect classifications of (a) construction projects that tend to have uncertainties in areas other than technology, (b) derivatives or enhancements of existing products or processes, (c) platform projects or pharma (pharmaceutical) line extensions that entail new combinations and extensions of existing technologies, (d) positioning or breakthrough projects that incorporate revolutionary new technologies or manufacturing processes, possibly enhancing the company's ability to develop new markets, (e) scouting or probing offerings to early adopters in new markets, and (f) stepping-stone R&D and pharma R&D discovery, which are long-term, high-risk projects in areas where both markets and technologies are relatively undeveloped (Shenhar and Wideman, 1997; Wheelwright and Clark, 1992; MacMillan and McGrath, 2002; Bunch and Schacht, 2002).

Outsourcing

The construction industry almost invariably outsources projects, so the selection and management of portfolios of construction projects by general contractors is an established dis-

TABLE 11.1. PROJECT CHARACTERISTICS.

	Technological Uncertainty			
Contribution to Business	**Established**	**Mostly Established**	**Advanced**	**Highly Advanced**
Immediate	Construction			
Short Term		Derivatives; Enhancement		
Medium Term			Pharma line extensions; Platform development	Scouting
Long Term			Positioning; Breakthrough	Pharma R&D discovery; Stepping-stone R&D

cipline. On the other hand, the practice of outsourcing technology development in other industries (Kimzey and Kurokawa, 2002) is playing an increasingly important role in achieving competitive advantage, providing access to larger technology resource pools, increasing ability to develop products that could not be developed using internal resources, shortening product cycle times, and reducing development costs. However, technology outsourcing tends to be on a tactical case-by-case basis rather than forming a part of corporate strategy.

A typical application of outsourcing is the pharmaceutical industry, where firms face high R&D risks and costs as well as intense new product competition. Here, larger companies need new blockbuster products that smaller biotechnology companies have to offer, but lack the pharmaceutical company research funding and marketing muscle. This often leads to alliances that combine the core strengths of both types of firm (Bunch and Schacht, 2002). Smaller pharma firms often try to mitigate their risks and leverage limited R&D resources by contracting out upstream (laboratory scale) research for a portfolio of R&D projects to smaller technology institutions. These firms can then concentrate limited R&D resources on downstream (commercial scale) R&D that utilizes the limited stream of successful upstream research outputs received from their collaborators (Prabhu, 1999).

Although portfolio selection may have increased flexibility through outsourcing, it must be evaluated carefully for the potential gains. There are many differences in managing external projects or subcontracts versus managing internal projects. For example, the Royal Bank of Canada has won awards over the years in the quality of its IT portfolio development and operations. It consistently handles development operations internally, with hundreds of projects across some 12 product groups under development in any given year (Melymuka 1999). Outsourcing their IT portfolio to possibly inferior developers would unnecessarily increase costs and risks.

Characteristics of Projects Affecting Portfolio Choice

The most important characteristics of projects that affect their selection depend upon the industry sector. Common across all fields are project scope, the total investment, labor, required expertise, materials, facilities, estimated return on investment, and timing. Alignment with the firm's business strategy is also important in each case. In the NPD and pharmaceutical sectors, technological and market risk becomes a critical issue, and whether the proposal involves an incremental or platform project (Tatikonda, 1999) is a major consideration in the selection decision.

The availability of outsourcing resources plays a crucial role in construction, and it is becoming of more importance in NPD and pharmaceuticals. The division of risk between client and contractor is an increasingly important topic in construction. If this is not decided in advance, it can lead to costly and lengthy legal disputes over responsibility when time, cost, or quality shortfalls are in question.

At the portfolio level, constraints exist that affect project selection. Reaching a proper balance of risk, timing, diversity, total investment, and return is critical to the overall portfolio selection. In addition, sharing resources (labor, facilities, materials) among projects, project priorities, and interdependencies all play a role in the selection process.

The ground needs to be well prepared in advance of the project selection process, to improve the likelihood that the best kinds of projects are being proposed. As project descriptions are generated, there must be a continuous dialogue between management and project proponents, covering issues such as (a) management's strategic intent, (b) implementation issues, (c) related projects that are competing for resources, and (d) corporate technology practices and rules (Bordley, 1998).

Clearly, in order that an informed judgment be made in selecting the portfolio, reliable project data are required to apply models such as those discussed in the following section. In the construction industry, for example, estimating costs and timing is based on experience with many similar projects. This may also be the case in NPD for maintenance projects, but for new and even incremental product development, estimates are needed for market, revenue, pricing, manufacturing or operations, risk, and resource requirements (Cooper, Edgett et al., 2001, Chapter 8). These data tend to be highly uncertain, particularly at the time the selection decision must be made.

Portfolio Selection Models and Methodologies

The use of specific project characteristics for portfolio selection is situation-dependent. For example, a product development organization may use market research, economic return, and risk analysis to develop project characteristics that can be useful in selection exercises. Or a government agency may use economic and cost benefit measures. Measures used may be qualitative or quantitative, but regardless of the techniques used to derive them, common measures must be used during portfolio selection so projects can be compared equitably.

An important consideration is that while there are many possible methodologies that can be used in selecting a portfolio, there is no consensus on which are the most effective. As a consequence, each organization tends to choose, for the project classes being considered, the methodologies that suit its culture and that allow it to consider the project attributes it believes are the most important. Also, the methodologies most useful in developing a portfolio for one class of projects may not be the best for another (e.g., good estimates of quantitative values such as costs and time may be readily available for certain construction projects, but qualitative judgment is more likely to be used for development of advanced new products). A major difficulty with portfolio selection is the number of possible combinations of projects that can be selected, with resources and schedules for each project affecting those available for the remainder of the portfolio. This number of possible solutions grows geometrically as the number of projects increases. Many companies use more than one approach in the selection process. For example, the NPD portfolio selection process utilizes an average of 2.4 techniques in the typical firm (Cooper, Edgett et al., 1999). The following is an outline of major classifications of methodologies used in portfolio selection.

Economic Return

These techniques typically require financial estimates of investment and income flows over the time frame of the project, often based on experience with similar projects. They are the best choice in construction portfolios, where most of the costs and schedules can be estimated with reasonable accuracy. Results from the calculations can be used in ranking or displaying information for decision making. Techniques include net present value (NPV), discounted cash flow (DCF), internal rate of return (IRR), return on original investment (ROI), return on average investment (RAI), payback period (PBP), and expected value (EV) (Martino, 1995; Souder, 1984). The latter allows a consideration of expected risk at various project stages, usually based on either IRR or NPV. A variation is the productivity index approach, popularized by the Strategic Decisions Group (SDG) (Matheson, Matheson et al., 1994). It tries to maximize the economic value of a portfolio for given resource constraints. NPV and DCF approaches tend to penalize risky projects because they do not include a provision for early project termination. Options pricing theory has recently become more popular for financial calculations because it takes this possibility into account (Rzasa, Faulkner et al., 1990). Economic techniques are the most widely used in establishing NPD portfolios but at the same time tend to yield poor outcomes (Cooper, Edgett et al., 1999).

Market Research

Market research can be used to collect data for forecasting the demand for new products or services, based on concepts or prototypes presented to potential customers, to gauge the potential market. Techniques used include consumer panels, focus groups, perceptual maps, and preference mapping, among many others (Wind, Mahajan et al., 1981).

Portfolio Matrices

Portfolio matrices ("bubble diagrams") are popular for displaying parameter values on three or four project dimensions. For example, probability of success can be plotted against net present value for projects in the portfolio, with the size of the icons used on the chart representing expected return on investment. A wide variety of bubble diagrams used to plan new product portfolios in a number of companies is demonstrated by Cooper et al. (Cooper, Edgett et al., 2001, Chapter 4). Although bubble diagrams are popular for graphical representations and comparisons, they have little theoretical or empirical support, and they may lead decision makers to overlook profit maximization (Armstrong and Brodie, 1994). They should therefore be used in conjunction with other tools, and primarily to illustrate relative characteristics of projects and the outcomes of balancing processes.

Comparative Approaches

Included in this classification are Q-sort (Souder, 1984), pair wise comparison (Martino, 1995), the Analytic Hierarchy Process (AHP) (Saaty, 1990), and Data Envelopment Analysis (DEA) (Linton, Walsh et al., 2002). Q-sort is the most adaptable of these in achieving group consensus. In these methods, first the weights of different objectives are determined, then alternatives are compared on the basis of their contributions to these objectives, and finally a set of project benefit measures is computed. The *Expert Choice®* commercial software package can be used to support the use of AHP in portfolio selection. Once the projects have been arranged on a comparative scale, decision makers can use these measures for guidance in selecting projects. An advantage of these techniques is that quantitative, qualitative, and judgment criteria can be considered. A major disadvantage is the large number of comparisons involved, making them difficult to use for analyzing large portfolios. Also, anytime a project is added or deleted from the list, the process must be repeated.

Scoring Models

Scoring models use a relatively small number of decision criteria, such as cost, workforce availability, probability of technical success, and so on, to specify project desirability (Martino, 1995). The merit of each project is determined with respect to each criterion. Scores are then combined (when different weights are used for each criterion, the technique is called "weighted factor scoring") to yield an overall benefit measure for each project. A major advantage is that projects can be added or deleted without recalculating the merit of other projects. Scoring models are probably the easiest to use of all the models, and a major advantage is that they can combine scores for both quantitative and qualitative measures. They are the third most preferred in NPD, after economic and strategic approaches (Cooper, Edgett et al., 1999).

Optimization Models

Optimization models select from the list of candidate projects a set that provides maximum benefit (e.g., maximum net present value). These models are generally based on some form of mathematical programming, to support the optimization process and to include project interactions such as resource dependencies and constraints, technical and market interactions, or program considerations (Martino, 1995). Some of these models also support sensitivity analysis, but most do not seem to be used extensively in practice. Probable reasons for disuse include the need to collect large amounts of input data, the inability of most such models to include risk considerations, and model complexity. Optimization models may also be used with other approaches that calculate project benefit values. For example, 0-1 integer linear programming can be used in conjunction with AHP to handle qualitative measures and multiple objectives, while applying resource utilization, project interaction, and other constraints (Ghasemzadeh, Archer et al., 1999).

Portfolio Decision Support Systems

Decision support systems in portfolio selection are typically based on a mathematical optimization approach that begins by selecting a portfolio from a set of candidate projects, providing a maximum overall benefit (e.g., net present value based on financial measures, benefit based on AHP calculations, or results from scoring models) (Ghasemzadeh, Archer et al., 1999; Ghasemzadeh and Archer, 2000; Dickinson, Thornton et al., 2001). If a mathematical programming approach is used to support the optimization process, other considerations may also be included, such as resource dependencies and constraints, precedence dependencies among the projects, risk, timing, technical and market interactions, or program considerations. Unless there are only a few projects to be considered, it is virtually impossible to take such constraints into account without the assistance of such models. For example, almost all other portfolio selection models assume that all projects begin at the same time, whereas it is simple to use mathematical programming to schedule individual projects in a way that minimizes competition for scarce resources. Solving the model optimally gives a beginning point. This solution, when combined with an interactive display of results, gives decision makers the ability to make adjustments to the portfolio, based on nonquantifiable judgments such as balancing risk. This approach can be effective if decision makers receive feedback on the resulting consequences, in terms of optimality changes and effects on resources. However, decision support approaches such as this are rarely used, partly because of the perception that a large amount of detailed financial data is required.

Cooper and his co-workers (Cooper, Edgett, et al., 1999) recently investigated the relative popularity of various portfolio selection methods in the NPD field. They found that financial methodologies were the most popular and dominated in portfolio decisions. However, the businesses that achieved the best results from their portfolios placed less emphasis on financial approaches and more on alignment of their portfolios with corporate strategy, and they tended to use more multiple approaches than other firms. Strategic methods that aligned the portfolio with business strategy, along with scoring approaches, yielded the best portfolios; financial methods yielded poorer results.

Process Model for Portfolio Selection

Portfolio selection is usually a committee process, where objective criteria such as predicted rate of return and expected project cost are mingled with subjective criteria relating to the needs of the different organizations represented on the project selection committee. All committee members should have access to information with which project intercomparisons are made, as well as information on the project portfolio as a whole. Portfolio selection can be managed most effectively in the context of an integrated framework that decomposes the process into a flexible and logical series of activities, involving full participation by the portfolio selection committee. Such an approach can take advantage of the best characteristics of a combination of existing methods, well grounded in theory. Because of the variance in interests and experience of responsible decision makers, it is important to include a flexible choice of techniques and interactive system support for those involved. The following discussion outlines such an integrated framework, suited to either distinct manual use of methodologies or to potential decision support system (DSS) application for portfolio selection.

Stages in the Project Portfolio Selection Framework

The process of portfolio selection can be highly complex unless approached logically and carefully. With process simplification in mind, the portfolio selection process should be organized into a series of stages, allowing decision makers to move logically toward an integrated consideration of projects most likely to be selected, based on sound theoretical models. However, each step should have a sound theoretical basis in modeling and should generate suitable data to feed the following step. Users need access to data underlying the models, with "drill-down" capability to develop confidence in the data being used and the decisions being made. At the same time, users should not be overloaded with unneeded data; it should be available only when needed and requested. Users also need training in the use of techniques that specify project parameters to be used in making decisions. An overall balance must be achieved between the need to simplify and the need to generate well-founded and logical solutions. A series of discrete stages for portfolio selection is depicted in Figure 11.1, where *process* stages are represented by the five outlined boxes linked horizontally in the center of the diagram. The ovals in the diagram represent *preprocess* activities, including methodology selection and strategy development. *Post-process* stages (below and to the right of the portfolio selection process) are also shown for completeness, since these may result in data generation and project evaluation during project development, affecting future portfolio selection activities and decisions.

Table 11.2 summarizes the stages in the framework, the associated activities, and some of the potential methodologies previously mentioned, for each stage. Each of the three phases and the included stages are considered in detail in the following discussion.

FIGURE 11.1. PROJECT PORTFOLIO SELECTION PROCESS.

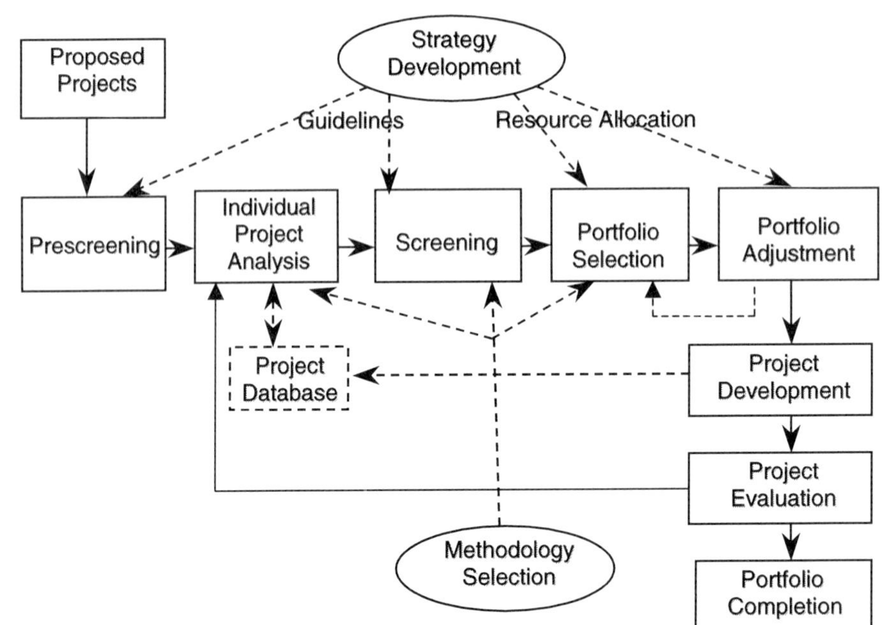

Source: Adapted from Archer and Ghasemzadeh (1999).

TABLE 11.2. ACTIVITIES AND METHODOLOGIES IN THE PORTFOLIO SELECTION FRAMEWORK.

Process Stage	Selection Stage	Activity	Potential Methodologies
Pre-process	Methodology selection, strategy development	Choice of modeling techniques, development of strategic focus, budgeting, resource constraints	Business strategy correlation and allocation, cluster analysis, etc.
Portfolio selection process	Prescreening	Rejection of projects that do not meet portfolio criteria	Manually applied criteria; strategic focus, champion, feasibility study availability.
	Individual project analysis	Calculation of common parameters for each project	Decision trees, risk est., NPV, ROI, resource req'ts., etc.
	Screening	Rejecting non-viable projects	Ad hoc techniques
	Portfolio selection	Integrated consideration of project attributes, resource constraints, interactions	AHP, constrained option, scoring models, sensitivity analysis
	Portfolio adjustment	User-directed adjustments	Matrix displays Sensitivity analysis
Post-process	Final portfolio	Project development	Project management techniques, data collection

Source: Adapted from Archer and Ghasemzadeh 1999.

Preprocess Phase

Methodology Selection. Methodology selection is a strategic process that clearly should be done in advance of any other activities in portfolio selection. It need only be done once for all time, with minor adjustments from time to time if other methodology choices appear to be better matches for the task at hand before each cycle of portfolio selection. Methodology selection should be flexible, based on decision maker understanding of the candidate methodologies or willingness to learn new approaches. Each stage involves choices, which are at the discretion of users, in order to gain maximum acceptance and cooperation of decision makers with the portfolio selection process. Choosing and implementing techniques suitable to the project class at hand, the organization's culture, problem-solving style, and project environment may also depend upon previous experience. It is also critical that common measures (e.g., NPV, scoring attributes, valuation of risk, etc.) are chosen so they can be calculated separately for each project under consideration, allowing an equitable comparison of the projects.

Strategy Development. Strategic decisions concerning portfolio focus and overall budget considerations should be made in a broader context that takes into account both external and internal business factors, *before* the project portfolio is selected. The strategic implications of portfolio selection are complex and varied (Cooper, Edgett, et al., 2001, Chapter 5) and involve considerations of many factors, including the marketplace and the company's strengths and weaknesses. These considerations can be used to build a broad perspective of strategic direction and focus, and specific initiatives for competitive advantage. This strategy can be used to develop a focused objective for a project portfolio and the level of resources needed for its support. Firms engaged in contract competition often use consensus-driven approaches to decide budget allocations for research and development. Consensus is derived at meetings among key personnel, based on informal arguments and assumptions regarding customer priorities, selection criteria, competitive position, and political priorities (Vepsalainen and Lauro, 1988). All may bear on the probability of successfully completing the projects being considered.

The front-end planning process is often done poorly (Khurana and Rosenthal, 1997). However, project portfolio matrices (Hax and Majluf, 1996) and graphs (Cooper, Edgett, et al., 2001) are useful tools for evaluating the strategic positioning of the firm, where various criteria for a firm's position are shown on one or more displays on two or three descriptive dimensions. These displays can be used by decision makers to evaluate their current position, and where they would like their firm to be in the future.

Overall resource allocation to different project categories also involves high-level decisions that must be made before the portfolio selection process, since this in turn determines the resource levels available for projects to be undertaken. Rules regarding the admissibility of projects into the portfolio need to be decided in advance of the selection process.

Process Phase

Prescreening. Prescreening precedes portfolio calculations. It is based on guidelines developed in the strategy development stage and ensures that any project being considered for the portfolio fits the strategic focus of the portfolio. Projects should be classified in advance of portfolio selection, according to criteria that can override other considerations. Frequently there are criteria that override all other considerations and mandate selection of specific projects, including sacred cow (mandated by influential stakeholders), operating necessity, and competitive necessity. Chien (Chien, 2002) also suggests that projects should be classified as independent, interrelated, or synergistic in advance of the selection process.

The upfront work to prepare for a project's evaluation is critical to its acceptability (Bordley, 1998). Essential requirements before the project passes this stage should include a feasibility analysis and estimates of parameters needed to evaluate each project, as well as a project champion who will be a source of further information. Elimination of projects not ready for serious consideration at this time helps to simplify following activities by reducing the number of projects to be considered. Mandatory projects are also identified, since they will be included in the remainder of the portfolio selection process. Mandatory projects are projects agreed upon for inclusion, including improvements to existing products no longer competitive, projects without which the organization could not function adequately, and so on.

Individual Project Analysis. Projects are analyzed individually after prescreening. Here, a common set of parameters required for equitable comparisons in following stages is calculated separately for each project, based on estimates available from feasibility studies and/

or from a database of previously completed projects. For example, project risk, net present worth, return on investment, and so on can be calculated at this point, including estimated uncertainty in each of the parameter estimates. Scoring, benefit contribution, risk analysis, market research, or checklists may also be used. Note that current projects that have reached certain milestones may also be reevaluated at this time, but estimates related to such projects will tend to have less uncertainty than those projects that are proposed but not yet under way. The output from this stage is a common set of parameter estimates for each project. For example, if the method to be used were a combination of net present value combined with risk analysis, data required would include estimates of costs and returns at each development stage of a product or service, including the risks. Uncertainty could be in the form of likely ranges for the uncertain parameters. Other data needed could include qualitative variables such as policy or political measures. Quantitative output could be each project's expected net value, risk, and resource requirements over the project's time frame, including their calculated uncertainties.

Screening. The screening stage is shown in Figure 11.1 following the individual project analysis stage. Here, project attributes from the previous stage are examined in advance of the actual selection process, to eliminate any projects or interrelated families of projects that do not meet preset criteria such as estimated rate of return, except for those projects that are mandatory or required to support other projects still being considered. The number of projects that may be proposed for the portfolio may be quite large, and the complexity of the decision process and the amount of time required to choose the portfolio increases geometrically with the number of projects to be considered. In addition, the likelihood of making sound business choices may be compromised if large numbers of projects must be considered unnecessarily. Screening may be used to eliminate projects that do not match the strategic focus of the firm, do not yet have sufficient information upon which to base a logical decision, do not meet a marginal requirement such as minimum internal rate of return, and so on. Care should be taken to avoid setting thresholds that are too arbitrary, to prevent the elimination of projects that may otherwise be very promising.

Another consideration at this point is the optimal number of research projects to be developed for NPD portfolios (Lieb, 1998). Lieb developed a model that views research and development as a two-stage process, where the task of research is to reduce the uncertainty for eventual development. Research projects require both technical and business evaluation, including marketing research. The number of research projects undertaken that maximizes the likelihood of development success depends on the cost-effectiveness of the research effort and the ability of the organization to support the development effort. Lieb concludes that the optimal fraction of research projects to be undertaken is critically dependent on the relative average research project cost and effectiveness compared to development.

Optimal Portfolio Selection. Optimization is performed in the second-to-last stage, as a beginning point for final portfolio selection. Here, interactions among the various projects are considered, including interdependencies, competition for resources, and timing, with the value of each project determined from a common set of parameters that were estimated for each project in the previous stage. AHP, scoring models, and portfolio matrices are popular among decision makers for portfolio selection, because they allow users to consider a broad range of quantitative and qualitative characteristics as well as multiple objectives. However, none of these techniques consider multiple resource constraints and project interdependence. AHP, pairwise comparison, and Q-sort also become cumbersome and unwieldy for larger numbers of projects. We suggest a two-step process for the portfolio selection stage.

In the first step, the relative total benefit is determined for each project. A comparative approach such as Q-sort, pairwise comparison, or AHP may be used in this step for smaller sets of projects, allowing qualitative as well as quantitative measures to be considered. This may involve extensive work by committee members in comparing potential project pairs. For large sets of projects, scoring models are more suitable, as these do not involve comparison of large numbers of project pairs. The result of either of these approaches would be to establish the relative worth of the projects on some scale that could combine both quantitative and qualitative attributes.

In the second step of this stage, all project interactions, resource limitations, and other constraints should be included in an initial optimization of the overall portfolio, based on the relative worth established for each proposed project. At this point, relationships among projects, and other constraints must be considered (Verma and Sinha, 2002). For example, (a) projects compete for scarce resources, including funding, labor, and facilities; (b) one project may be dependent upon the completion of another (e.g., construction of a new subdivision must await the construction of access roads); (c) projects may be mutually exclusive (e.g., a choice must be made among alternative solutions to the same requirements); and so on. Other constraints on project inclusion may include (a) a project must be completed prior to a particular date, (b) a project is mandatory (e.g., maintenance work), and so on.

If all the project measures could be expressed quantitatively, the first step in this stage could be omitted, since optimization could be performed directly in a mathematical program in the second step. In the unusual case where interdependence and timing constraints were not important, and there is only one resource that is binding, it might be tempting in the second step to simply select the highest-valued projects until available resources are used up. However, this does not necessarily select an optimal portfolio (combinations of certain projects may produce a higher total benefit than individual projects with higher individual benefits). The relative worth of each project should be input to a computerized process, which can be a 0-1 integer linear programming model that applies resource, scheduling, interdependence, and other constraints to maximize total benefit (Ghasemzadeh, Archer, et al., 1999).

Portfolio Adjustment. This is the final stage, using as input the initial optimal portfolio from the preceding stage. The end result is to be a portfolio that meets the objectives of the organization optimally or near-optimally, but the approach must have provisions for final judgmental adjustments, which may be difficult to model. This is a strategic decision, and the relevant information must be presented so it allows decision makers to evaluate the portfolio without being overloaded with unnecessary information. The portfolio adjustment stage needs to provide an overall view, where the characteristics of projects of critical importance in an optimized portfolio (e.g., risk, net present value, time-to-complete, etc.) can be represented, using matrix-type displays, along with the impact of any suggested changes on resources or selected projects (Cooper, Edgett, et al., 2001, Chapter 4). It is important to use only a limited number of such displays to avoid confusion (cognitive overload) while the final decisions are being made. Decision makers should be able to make changes at this stage, and if these changes are substantially different from the optimal portfolio developed in the previous stage, it may be necessary to recycle back to recalculate portfolio parameters such as project schedules and time-dependent resource requirements. In addition, sensitivity analysis should be available to predict and provide feedback to decision makers on the impact of their suggested changes (addition or deletion of projects) on resources and portfolio optimality.

There are a variety of ways in which portfolio balance can be attained; this requires the right mix of project size and/or duration, and a risk profile that is suited to the company environment. Balance on project size is important, because the commitment of a high proportion of resources to a few large projects can be catastrophic if more than one fails. And too many long-term projects, no matter how promising they are, may cause cash flow problems. In achieving portfolio balance on the risk dimension, the greater the risks taken, the greater the potential rewards should be. This should not preclude the greatest care in choosing the portfolio so that risky projects are balanced with less risky projects to avoid jeopardizing the overall company strategy of maximizing profitability and ensuring long-term survival. The average risk of projects should be such that decision makers view the portfolio as manageable, without dominance of high-risk projects. At the same time, there is danger in reducing risk that there will be a bias against selecting breakthrough projects that may provide a large long-term profit to the company. In NPD portfolio selection, an overall objective may be to maximize value while minimizing risk. In NPD portfolios, balancing the portfolio is second in importance only to having the right number of projects in the pipeline at one time (Cooper, Edgett et al., 2001).

Post-Process Phase

Project Development and Project Evaluation. Development and evaluation are post-process activities that can generate data from experience that are highly useful to learning and project evaluation, for future portfolio selection exercises. This could involve both the evaluation of existing projects that may be candidates for termination or generation of data for future use in estimating parameters for contemplated projects that are similar to already completed projects. Current projects that have reached major milestones or gates can be reevaluated at the same time as new projects being considered for selection. This allows a combined portfolio to be generated within available resource constraints at regular intervals because of (a) project completion or abandonment, (b) new project proposals, (c) changes in strategic focus, (d) revisions to available resources, and (e) changes in the environment.

The framework we have outlined can be used for project portfolio selection in an environment that is partially supported by computerized modeling and databases, since users are given the flexibility of choosing their own techniques or models at each stage. However, many of the stages in the framework (see Figure 11.1) can be integrated into a decision support system, including a model management module to handle models of the many different types that may be chosen. To be effective as a decision support system, the system requires a common interface, a database through which data may be interchanged among the models, and a user interface that allows user control and overrides of model calculations (Archer and Ghasemzadeh, 1999).

Portfolio Management

A program manager is the organizational leader charged with executing a portfolio of projects. In managing a portfolio, the problems are not centered on just managing individual projects, but they extend to a variety of complexities, including multiple, cross-functional, global, overlapping, interdependent projects; resource allocation; politics; sponsorship; and culture. Addressing these issues requires centralized support, increasingly provided in larger organizations through a project office. Global competency standards and best practices for project organizations such as this have been provided by Toney (Toney, 2002; see also the chapter by Lynn Crawford). Project offices can fulfill portfolio needs by giving support and gathering data that can be used for monitoring project progress and providing estimates for future portfolio selection activities. They provide standard tools and methodologies, as well as coaching. They collect information by means of project management tools that give a current view of the company-wide IT project portfolio, including staff deployment (Melymuka, 1999).

After some time in operation, a project office can provide a historical view of project estimates compared with actual performance, enabling management to see where projections part with reality. Some staffs see their project offices as centers of excellence for project management; see, for example, the chapter by Powell & Young. Theoretically, a project office forces the business to face the reality of limits on staff, time, and budget, and requires it to prioritize projects. It manages the portfolio selection process and assists in evaluating ongoing projects. It gives project managers the documentation they need to demand adequate staffing, funding, and reasonable deadlines. This results in better planning, portfolio selection, coordination and execution, and more projects completed on time and within budget. Virtual project management also becomes an issue when there is more than one site involved in projects within the portfolio, and communications arrangements can be supported through the project office (face-to-face visits, e-mail, video, telephone conferencing). Virtual projects are almost certainly an issue when projects are outsourced. When portfolios are developed internally, management can direct its attention to understanding and managing the relationships among the resources and project managers involved. When projects are outsourced, the firm has to cooperate with one or more external organizations in managing the portfolio, and it is confronted with the task of managing not just a portfolio of projects, but a portfolio of new and evolving relationships.

Summary

In this chapter we provided a review of project portfolio selection and management, along with a suggested logical approach to the selection process. In recent years, there has been a great deal of interest in this topic as project approaches have become more widely used but project failure rate experience in poorly managed operations has continued to be unacceptable. Improved portfolio selection and management practice can greatly enhance the likelihood of success, contributing to the rate of innovation in new products or controlling other problems in industries such as construction where disputes over performance or quality have long been a problem.

There are some areas of portfolio selection and management that require further research. For example, because of the increased emphasis on outsourcing, the management of outsourced and high-technology risk portfolios needs more study. Collaborative commerce (c-commerce), a major area of e-commerce application in which a Web-based system is used for communication, design, planning, information sharing, and information discovery (Turban, McLean et al., 2002), has significant potential for further research. This includes extensions of online project management and partner relationship management (PRM) tools and techniques for portfolios that include outsourced projects. In spite of new technological support for these types of cooperation, a problem arising from this trend is that knowledge sharing across organizational boundaries becomes more difficult due to legal, cultural, and virtual management issues

References

Archer, N. P., and F. Ghasemzadeh. 1999. An integrated framework for project portfolio selection. *International Journal of Project Management* 17(4):207–216.

Armstrong, J. S., and R. J. Brodie. 1994. Effects of portfolio planning methods on decision making: Experimental results. *International Journal of Research in Marketing* 11:73–84.

Bordley, R. F. 1998. R&D project selection versus R&D project generation. *IEEE Transactions on Engineering Management* 45(4):407–413.

Bunch, P. R., and A. L. Schacht. 2002. Modeling resource requirements for pharmaceutical R&D. *Research Technology Management* 45(1):48–56.

Chien, C.-F. 2002. A portfolio evaluation framework for selecting R&D projects. *R & D Management.*

Cooper, R. G., S. J. Edgett, and E. J. Kleinschmidt. 1999. New product portfolio management: Practices and performance. *Journal of Product Innovation Management* 16:333–351.

———. 2000. "New problems, new solutions: Making portfolio management more effective." *Research Technology Management* 43(2):18–33.

———. 2001. *Portfolio Management for New Products.* Cambridge, MA: Perseus Publishing.

Dickinson, M. W., A. C. Thornton, and S. Graves. 2001. Technology portfolio management: Optimizing interdependent projects over multiple time periods. *IEEE Transactions on Engineering Management* 48(4):518–527.

Ghasemzadeh, F., and N. Archer. 2000. Project portfolio selection through decision support. *Decision Support Systems* 29:73–88.

Ghasemzadeh, F., N. Archer, and P. Iyogun. 1999. A zero-one model for project portfolio selection and scheduling. *Journal of the Operational Research Society* 50:745–755.

Githens, G. D. 2002. How to assess and manage risk in NPD programs: A team-based risk approach. In *The PDMA Book for New Product Development,* ed. P. Belliveau, A. Griffin, and S. Somermeyer, 187–214. New York, Wiley.

Griffin, A. 1997. PDMA research on new product development practices: Updating trends and benchmarking best practices. *Journal of Product Innovation Management* 14:429–458.

Hartman, F. T., P. Snelgrove, and R. Ashrafi. 1998. Appropriate risk allocation in lump-sum contracts—Who should take the risk? *Cost Engineering* 40(7):21–26.

Hax, A., and N. S. Majluf. 1996. *The Strategy Concept and Process: A Pragmatic Approach.* Upper Saddle River, NJ: Prentice Hall.

Hess, S. W. 1993. Swinging on the branch of a tree: Project selection applications. *Interfaces* 23(6):5–12.

Khurana, A., and S. R. Rosenthal. 1997. Integrating the fuzzy front end of new product development. *Sloan Management Review* 38:103–120.

Kimzey, C. H., and S. Kurokawa. 2002. Technology outsourcing in the U.S. and Japan. *Research Technology Management* 45(4):36–42.

Lieb, E. B. 1998. How many R&D projects to develop? *IEEE Transactions on Engineering Management* 45(1):73–77.

Linton, J. D., S. T. Walsh, and J. Morabito. 2002. Analysis, ranking, and selection of R&D projects in a portfolio. *R & D Management* 32(2):139–149.

MacMillan, I. C., and R. G. McGrath. 2002. Crafting R&D project portfolios. *Research Technology Management* 45(5):48–59.

Martino, J. P. 1995. *Research and Development Project Selection.* New York: Wiley.

Matheson, D., J. E. Matheson, and M. M. Menke. 1994. Making excellent R&D decisions. *Research Technology Management* 37(6):21–24.

Melymuka, K. 1999. The project office: A path to better performance. *Computerworld.*

Pinto, J. K. 2002. Project management 2002. *Research Technology Management* 45(2):22–37.

Prabhu, G. N. 1999. Managing research collaborations as a portfolio of contracts: A risk reduction strategy by pharmaceutical firms. *International Journal of Technology Management* 18(3,4):207–231.

Raz, T., A. J. Shenhar, and D. Dvir. 2002. Risk management, project success, and technological uncertainty. *R & D Management* 32(2):101–109.

Riggs, J. L., S. B. Brown, and R. P. Trueblood. 1994. Integration of technical, cost, and schedule risks in project management. *Computers and Operations Research* 21(5):521–533.

Roseneau, M. D. J. 1990. *Faster New Product Development: Getting the Right Product to Market Quickly.* New York: AMACOM.

Rzasa, P. V., T. W. Faulkner, and N. L. Sousa. 1990. Analyzing R&D portfolios at Eastman Kodak. *Research Technology Management* 33(1):27–32.

Saaty, T. L. 1990. *The Analytic Hierarchy Process.* Pittsburgh: RWS Publications.

Shenhar, A. J. 2001. Contingent management in temporary, dynamic organizations: The comparative analysis of projects. *Journal of High Technology Management Research* 12:239–271.

Shenhar, A. J., and R. M. Wideman. 1997. Toward a fundamental differentiation between project types. *Innovation in Technology Management: The Key to Global Leadership—PICMET '97 conference.* Portland, Oregon, July.

Souder, W. E. 1984. *Project Selection and Economic Appraisal.* New York: Van Nostrand Reinhold.

Stamelos, I. and L. Angelis. 2001. Managing uncertainty in project portfolio cost estimation. *Information and Software Technology* 43:759–768.

Tatikonda, M. V. 1999. An empirical study of platform and derivative product development projects. *Journal of Product Innovation Management* 16:3–26.

Toney, F. 2002. *The Superior Project Organization.* New York: Marcel Dekker.

Turban, E., E. McLean, and J. Wetherbe. 2002. *Information Technology for Management: Transforming Business in the Digital Economy.* New York, Wiley.

Vepsalainen, A. P. J., and G. L. Lauro. 1988. Analysis of R&D portfolio strategies for contract competition. *IEEE Transactions on Engineering Management* 35(3):181–186.

Verma, D., and K. K. Sinha. 2002. Toward a theory of project interdependencies in high tech R&D environments. *Journal of Operations Management* 20(5):451–468.

Wadlow, D. 1999. The role of risk in the design, evaluation and management of corporate R&D project portfolios for new products. *Sensors Research Consulting* 32.

Welling, D. T., and D.-J. F. Kamann. 2001. Vertical cooperation in the construction industry: Size does matter. *Journal of Supply Chain Management* 37(4):28–34.

Wheelwright, S. C., and K. B. Clark. 1992. Creating project plans to focus product development. *Harvard Business Review* 70(2):70–82.

Wind, Y., V. Mahajan, and R. N. Cardozo. 1981. *New Product Forecasting.* Lexington, MA: Lexington Books.

CHAPTER TWELVE

PROGRAM MANAGEMENT: A STRATEGIC DECISION MANAGEMENT PROCESS

Michel Thiry

Recent large-scale studies have demonstrated that 30 percent of projects are canceled before the end (Standish, 1996; KPMG, 1997); or that large, long-term projects—more than three years—are "significantly less predictable" in terms of time and scope (Cooke-Davies, 2002). These studies and others have sorely exposed the failure of project management to respond to emergent inputs, as well as the lack of integration between strategic intents and the results generated by projects. Recently, a number of authors (Frame, 2002; Thomas et al., 2000; Kendall, 2001) have advocated the evolution of the role of project manager toward a more business-focused "function," or a stronger focus on business benefits for projects (Cooke-Davies, 2002; Morris, 1997). This is probably part of the answer, but it will not be sufficient if not supported by a robust framework. The business world is moving toward a systemic process-based framework for the management of strategic decisions (Kaplan and Norton, 2000; EFQM, 1999).

This chapter develops a view in which program management (PgM) may effectively link strategic decision making with its successful implementation through projects. It first discusses the need to combine a performance (project-based) and a learning (value-based) process in the management of programs. The chapter then explains the need for a complete strategic decision management process, which includes decision making and decision implementation, and suggests an integrated learning-performance model for program management. The chapter then outlines in detail a program management life cycle, based on that model and grounded in a long experience of managing stakeholders' needs and expectations through value, project, and program management in a variety of industries.

Program Management Paradigms

Program management is an emerging discipline that has its roots in project management; however, as the need for better strategic focus is recognized, a project paradigm based purely on performance efficiency becomes insufficient. In their research on selling project management to executives, Thomas et al. (2000) have argued that "successful messages reflect the buyer's needs as the buyer understands them." Program management needs to reflect the concepts and rhetoric of strategic long-term management, rather than the tactical short-term view of project management, in order to gain executive management support and truly be able to support strategic decision management. Projects, as currently defined (PMI, 2000; APM, 2000), are time-framed and require clear objectives; they are based on a performance paradigm (Thiry, 2002a) embedded in an "uncertainty-reduction" process (Winch et al., 1998). This does not sell well to executives.

A number of textbooks and papers (CCTA, 1999; Reiss, 1996; Gareis, 2000; DSMC, 2000; NASA, 2001) have suggested program processes that, albeit their different designation, are in most instances just transpositions of the project paradigm into program management. Although it is now generally agreed that programs need to produce business-level benefits and are a link between strategy and projects, few management concepts and little rhetoric have made their way into the program management literature and practice. In the report of an ongoing survey of organizations practicing program management, Reiss and Rayner (2001; 2002) point out that most respondents still consider "organization," "issues and risk," "planning," and "accounts and finance" as key to the success of programs, whereas "achievement of benefits," "stakeholder management," "communications," and "configuration management," which were identified by the developers of the model as key program components, seem less important. Generally speaking, organization, risk, planning, and cost management are key elements of project management. Benefits, stakeholders, communications, and configuration management are about the management of stakeholders' needs and expectations and emergent change. All are linked to value management, which is often not part of the current PMI-based project management paradigm (PMI, 2000), although it has been included in the Association for Project Management's Body of Knowledge (APM, 2000).

Johnson and Scholes (1997) have written that strategic management is "ambiguous and complex; fundamental and organization-wide; and has long-term implications." Wijnen and Kor (2000) write that a program strives for the achievement of a number of, sometimes conflicting, aims and has a broader corporate goal than projects, which aim to achieve single predetermined results. Görög and Smith (1999) argue that strategic management is based on continuous reformulation and is a form of continuous adjustment, whereas projects concentrate on achieving one single particular result within set time and cost constraints. Partington (2000) argues that programs require integration across strategic levels, controlled flexibility, team-based structures, and especially, an organizational learning perspective, which is able to accept paradox and uncertainty. Murray-Webster and Thiry (2000) advocate a vision that includes mechanisms to identify and manage emergent change; they use an idea developed by Hurst (1995) to promote the concept of a "learning loop," which completes the project "performance loop."

Programs are often ongoing or long-term and are subjected to both uncertainty and ambiguity. They require a *strategic decision management* paradigm, taking into account a strategic perspective, organizational effectiveness, a systems view, and a learning (ambiguity reduction) approach. Whereas projects are essentially "deliberate" (planned) strategies, programs combine both deliberate strategies and "emergent" (unplanned) strategies.

As outlined in the preceding text, strategic decision making and change situations often mean that multiple stakeholders, with conflicting needs and expectations, are competing with each other, which creates ambiguity. Effective strategic decision making requires an "ambiguity reduction" process, based on a learning paradigm, to take place before any attempt is made at uncertainty reduction (see Figure 12.1); otherwise, it will lead to results that are not necessarily in line with stakeholders' needs. This process can effectively be supported by value management (VM), which uses a range of "soft" methodologies and techniques like sensemaking, stakeholder analysis, functional analysis, ideation, soft systems analysis, and others (as more fully described in the chapter on value management).

In a complex, continually changing environment, a program management paradigm must integrate both a learning-based value loop and a performance-based project loop to address both ambiguity reduction and uncertainty reduction to form a full strategic decision management framework. You will also see later that the learning loop must be part of both

FIGURE 12.1. THE UNCERTAINTY-AMBIGUITY RELATIONSHIP IN CHANGE SITUATIONS.

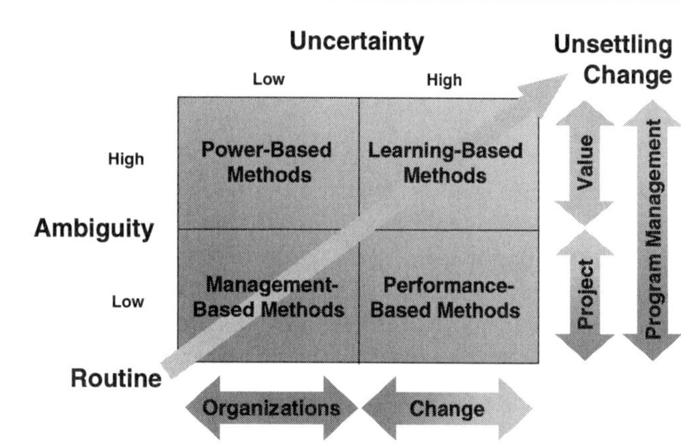

Source: © Michel Thiry (2000–2002).

project reviews and program appraisal in order to achieve strategic benefits and stakeholders' satisfaction at delivery.

Definitions

Following the description of program paradigms, it is important to define some basic terms: strategy, program management (PgM); portfolio management, value management (VM), and project management (PM). Although these definitions might not gain general consensus, they are sufficiently documented to gain acceptance by most scholars and practitioners.

Strategy

In a program management context, *strategy* is essentially the organization's response to external or internal pressures to change. For the purpose of this chapter, strategies are grouped in two major types—*deliberate* strategies, which are planned, and *emergent* strategies, which are responses to unplanned inputs (Mintzberg and Waters, 1994)—and include also the concept of *configuration* strategy (Mintzberg et al., 1998), a combined use of the two.

A deliberate strategy is submitted to a formal process of analysis, design, and planning; its implementation is subjected to a formal baseline control process. An emergent strategy is an unplanned response to an emergent input originating from the environment or from within the organization. Such strategies are usually based on a leader's intuitive decision (based on vision and/or experience), negotiation, a collective process, or, more simply, a knee-jerk reaction.

Project Management

There are currently two paradigms in the project management community; one, based on the PMI's PMBoK Guide (PMI, 2000), views project management essentially as a delivery process based on a performance paradigm; the other, based on the view developed by members of IPMA (International Project Management Association), views project management as a broader concept that includes preinitiation phases and project definition and, in that sense, is closely linked to program management. In any case, there is one point, once expected outcomes have been clearly defined, when project management becomes a performance process intended to deliver with the highest possible efficiency—best scope-quality versus lowest cost-time—though the *management of projects* concept is obviously broader (Morris 1997).

Value Management

In the context of this chapter, value management is based on a learning paradigm aimed at *ambiguity reduction* (see the chapter on value management in this book). Value management is a group decision-making process expected to develop a shared understanding of a complex situation and an agreement on options to resolve it. It uses creative thinking concepts.

Portfolio and Program Management

Program and portfolio management are both emerging disciplines and are often confused. The two are different, and it is therefore important to describe how they can be clearly distinguished. In Europe, researchers and practitioners have looked into this distinction for the last few years. In the United States, there seems to be less use of the word portfolio management, and program management is often associated with very large projects (DSMC, 2000; NASA, 1998) or the management of large, multifunctional projects as a portfolio of independent projects through a project office (CBP, 2002). A number of authors still associate program management with multiproject coordination or portfolio management, which is often associated with resource management (Patrick, 1999) or account/client management; this view is supported by so-called program management computer software, which is essentially designed to support resource management, planning, and cost/time control across many projects.

A sound, widely acceptable definition for both portfolio and portfolio management is this: "A project portfolio is a collection of projects to be managed concurrently under a single management umbrella where each project may be related or independent of the others" (Martinsuo and Dietrich, 2002). Portfolio management consists of "the management of a multi-project organization and its projects in a manner that enables the linking of the projects to business objectives" (Artto et al., 2002). Writing in the context of decision making, Spradlin (1997) links portfolio management to the need to choose among a number of options where resources are insufficient to fund all options. Portfolio management is a management approach where projects could be related or not to a common objective.

There are a number of definitions of *program*; all these definitions have elements in common as well as differences. The main common points are that programs usually cover a group of projects, that their management must be coordinated, and that they create a synergy, which will generate greater benefits than projects could do individually. The main differences are that the elements must have a common objective (APM, 2000), that they also cover ongoing operations (PMI, 2000), and that their impact is at the organizational level and concerns change (CCTA, 1999). A widely acceptable definition of program should integrate those three elements. The definition that follows does that with the concept of purposefulness that is related to setting common objectives; the word "*actions*," which refers to ongoing operations as well as to projects; and the concept of strategic or tactical benefits, which are always measured at the organizational level.

Programs are "a collection of change actions (projects and operational activities) purposefully grouped together to realize strategic and/or tactical benefits" (Murray-Webster and Thiry, 2000). Program management consists of the "purposeful and integrated direction and coordination of a group of actions, their interface and consequences for strategic effectiveness and/or tactical efficiency" (Thiry, 2002a); it is a management process that addresses both decision-making and decision implementation.

In summary, portfolio management is mostly concerned with the ongoing prioritization and management of a group of existing projects or programs, to efficiently support a project-oriented organization. Program management, on the other hand, is mainly a purposeful strategic decision management process, grounded in change and aimed at the effectiveness of solutions. It can include both projects and non-project actions. Figure 12.2 shows how program management draws on three main areas of management to combine processes belonging to recognized fields of knowledge and practice.

The Link between Programs and Strategy

Authors and researchers in various fields have argued that one of the major problems concerning the failure of decision implementation lies within inaccurate understanding of

FIGURE 12.2. PROGRAM MANAGEMENT AS AN INTEGRATING PROCESS.

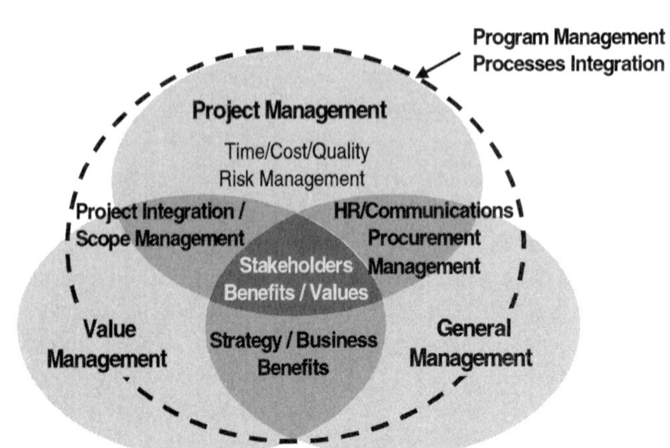

Source: © Michel Thiry (2001).

stakeholders' needs and expectations or the inability to respond to their changing needs (Carver and Scheier, 1990; Kubr, 1996; Standish, 1996; Kirk, 2000; Hartman and Ashrafi, 2002; see also the chapter by Winch in this book). The need to change, which triggers a strategic decision, is usually caused by an unsatisfactory condition. Quinn (1978) talks about the "change deficit" as a measure of the need for an organization to put in place change measures that go beyond incremental change or simple evolution. The change deficit represents the difference between the rate of change of an organization and its environment: competitors, market, and so on. Generally, internal pressures to change are addressed in an incremental way (Quinn, 1978), but it may sometimes be necessary for an organization to implement more fundamental changes to increase or maintain its effectiveness. An organization that falls back in terms of change deficit will quickly lose its competitive edge and market share. The problem with large, heavily structured organizations is that they cannot adjust quickly and are more adapted to incremental change than fundamental change. In a fast-changing environment, organizations have to find a way to be more flexible in making and implementing strategic decisions.

In this context, program and project management are effective means of modifying an unsatisfactory situation or pursuing an already ongoing change process. These strategic change processes generally aim to improve organizational effectiveness or competitiveness or to respond to a change deficit.

Authors have suggested an abundance of criteria to measure an organization's *effectiveness*, but it is generally accepted that effectiveness can be defined as the match between results and stakeholders' satisfaction, and that satisfaction is dependant on clearly agreed and understood objectives (EFQM, 2000; Kaplan and Norton, 1996; Porter, 1998). Goals and objectives must be clearly outlined as part of a strategic decision-making process, as satisfaction will be measured by the rate of progress from the actual situation toward the desired outcome (Carver and Scheier, 1990).

A number of authors (Guba and Lincoln, 1989; Kubr, 1996; Stake, 1975) also argue that when situations are complex and fluid, control needs to be more qualitative, based on negotiation between the stakeholders and with criteria evolving in time. This is the case for strategic decision implementation.

These concepts require that a strategic decision management framework be systemic and flexible, based at the same time on deliberate and emergent strategies.

Program Management Process

Because of their long-term characteristic, programs will be subjected to contextual change and therefore must include not only deliberate strategies (typically projects) but also elements of emergent strategies. In emergent strategies the focus must be on expected benefits and the process must "display consistent patterns of actions over time" (Mintzberg and Waters,

1994). This requires a learning approach where results are regularly appraised against benefits during implementation and changes are managed against these stated benefits.

Recent research has shown that there is a need for a decision management process in which anticipated results are directly linked to the justification for the decision (expected benefits) and the means to support their delivery (resources) (Thiry, 2002b; Hartman and Ashrafi, 2002). It has also been observed that there is a lack of communication between the strategic and tactical levels of management (Hatch, 1997; Neal, 1995; Pellegrinelli and Bowman; 1994, Thomas et al., 2000). This points to the need for a "strategic decision management process," to link strategic analysis and choice with strategy implementation. Some authors have suggested program management could rise to this challenge (Murray-Webster and Thiry, 2000; Partington, 2000; Wijnen and Kor, 2000).

A Program Management Framework

The program environment is complex: There are multiple stakeholders with differing and often conflicting needs, emergent inputs are always affecting the process, and both uncertainty and ambiguity can be high at the same time. Processes that are applicable to project management cannot always be readily applied to program management, as programs often have a finality that cannot be defined clearly and require processes that are both cyclic and aimed at reducing ambiguity.

Hurst (1995) described organizational change as a cyclic "Change Eco-Cycle" composed of a renewal loop (emergent strategy), which follows a crisis (emergent input), and a conventional loop (deliberate strategy), which follows a choice (deliberate input). Mintzberg et al. (1998) have argued that organizations have to adapt to both periods of stability and periods of transformation, and they identified this pattern as the *configuration* school of strategy. Since program management is linked to organizational change and can be seen as a strategic decision management process, a consistent program management framework will include both a *performance loop*, associated with a "deliberate" strategy process, based on a precise plan consistent with project management, and a *learning loop*, which can be likened to a largely responsive "emergent" strategy process and related to value management. This process is shown in Figure 12.3.

The learning loop is in fact a decision-making process; its first step is to make sense of the emergent input that justifies a change. The second step, ideation, is based on a lateral thinking concept (de Bono, 1990) and consists of identifying as many alternatives as possible to increase the quality of the decision. The third step, elaboration, consists of evaluating and combining alternatives to develop them into viable options, which can then be assessed and prioritized.

The performance loop is the decision implementation process; it can only be effective if a clear decision has been made and objectives clearly defined. It is only through the iteration of a number of learning-performance cycles that program management can be the framework to successful implementation of strategic decisions. The model in Figure 12.4

FIGURE 12.3. THE INTEGRATED PROGRAM MANAGEMENT CYCLE MODEL.

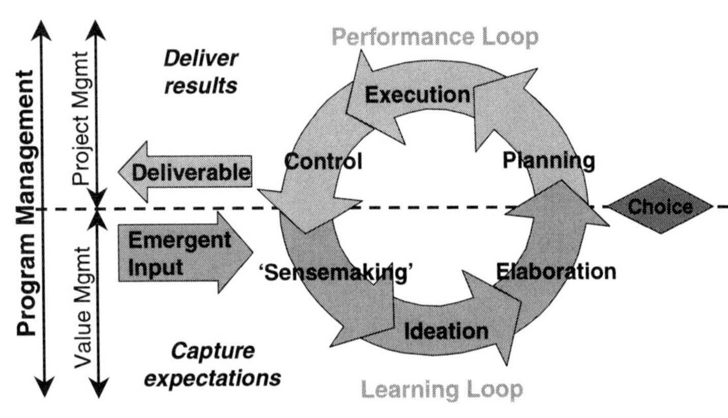

Source: © Michel Thiry (1999–2002).

FIGURE 12.4. STRATEGIC DECISION MANAGEMENT MODEL.

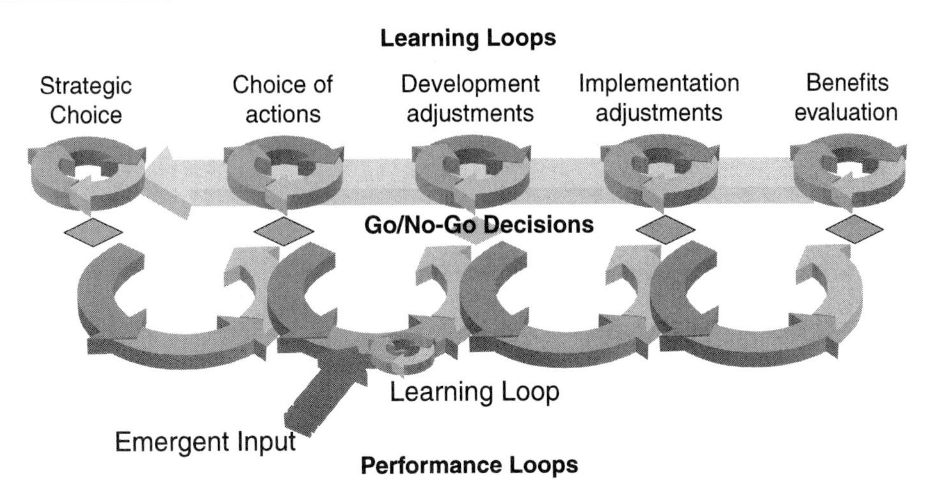

shows how iteration of the program management cycle allows regular evaluation of project outcomes, in regards of organizational benefits, and offers the flexibility to readjust expectations, as required by emergent inputs, or to readjust the expected outcomes to fit the circumstances.

The final element of the strategic decision management framework concerns evaluation paradigms. According to Guba and Lincoln (1989), evaluations can be "summative" (to assess) or "formative" (to improve). Quinn (1996) argues that, in complex situations, each stakeholder will construct different evaluation criteria and evaluation measures will therefore need to be negotiated. Guba and Lincoln (1989) and Stake (1975) talk about "responsive" evaluation, where parameters and boundaries are determined through an interactive negotiated process.

Most proponents of a combined learning/performance program management paradigm (Wijnen and Kor, 2000; Görög and Smith, 1999; Murray-Webster and Thiry, 2000) argue for a continuous reevaluation or "reformulation" of the program, in regards of the realization (or nonrealization) of benefits. On the other hand, proponents of the performance paradigm (Bartlett, 1998; CCTA, 1999; Reiss, 1996) argue for summative control, based on performance parameters such as time, cost, resources, risks, and so on.

In the context of complex strategic situations, a valid decision model must offer *both* summative evaluation at the project level and formative evaluation at the value-strategy level, allowing for negotiated evaluation as program objectives are redefined.

A Program Management Life Cycle

There are two characteristics that will make program management the most suitable methodology to ensure successful implementation of strategies:

- The fact that it is a cyclic process, which enables regular assessment of benefits, evaluation of emergent opportunities, and pacing of the process
- The emphasis on the "interdependencies" of projects, which ensures strategic alignment and delivery of strategic benefits

To support strategic decisions, reflect a management discourse, and acknowledge the two preceding characteristics, the program life cycle must be iterative, rather than linear; include periods of stability, where benefits can impact the organization and therefore be measured; and have a learning and systems perspective. The program management life cycle in Figure 12.5 is iterative in nature and reflects the extended and evolving nature of strategic decisions.

It is composed of five phases (Thiry 2002c):

FIGURE 12.5. THE PROGRAM MANAGEMENT LIFE CYCLE.

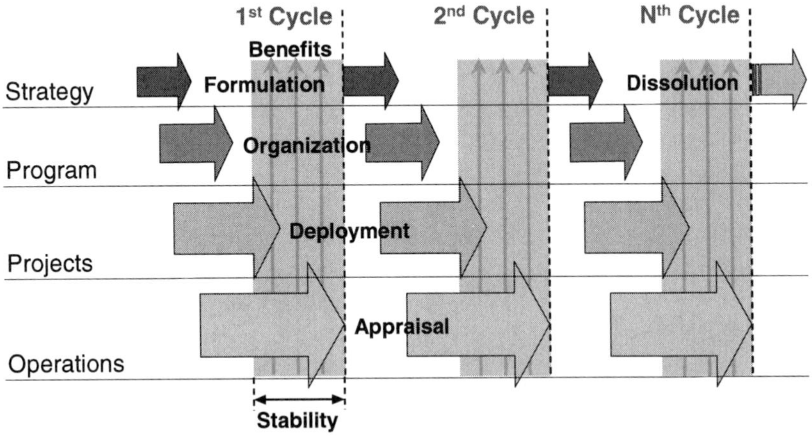

Source: © Michel Thiry (2002).

- *Formulation* is the stage where purpose is defined and stakeholders, along with their needs and expectations, are identified. It is also where program benefits are determined and critical success factors (CSFs) and key performance indicators (KPIs) defined. Contrary to project initiation, it is a complex process, where ambiguity is high. It is the initial learning cycle of programs where sensemaking, ideation, and evaluation of alternatives take place and that ends with the decision to undertake the program. It is iterated regularly during the program and confirms or redefines the direction of the program and the actions required to support it. The main objective of this phase is to identify opportunities and select the best course of actions.
- *Organization* is the process of selecting and prioritizing projects and other actions required to deliver benefits and set up the program team and structures. It includes the installation of operational procedures and structures that will enable project interdependencies and interrelationships to be managed, as well as pacing the program and ensuring ongoing benefits delivery.
- *Deployment* involves the actual initiation of projects and other actions; management of interdependencies and resources; project sponsor type control, including scope verification and closeout; and benefits delivery. It is made up of review, pacing, and approval of project outputs and change/configuration control, including realignment and reprioritizing of resources and projects.
- *Appraisal* essentially concerns the program-level assessment of benefits. It is a process that requires constant reevaluation of the program's circumstances and CSFs and typically

corresponds to a period of stability, enabling benefits to be measured in regards of their impact on the organization.

- *Dissolution* happens when the rationale for the program no longer exists. Uncompleted work, projects, and resources are reallocated to other programs, which are reformulated as needed; a post-program feedback is carried out; and knowledge is recycled. It is a phasing-down process, much more extensive than project closing.

The following five sections detail each of the program life cycle phases.

The Formulation Phase

The formulation phase aims to identify internal or external pressures to change and determine the best way to address them to add value for the stakeholders. For programs to successfully support strategic decisions, it is essential to take a value approach to formulation.

Stakeholders Management

To reduce ambiguity, the formulation phase must first address stakeholder issues. It requires identifying their needs and expectations, estimating achievability of these, gaining agreement on objectives, and in the appraisal phase, assessing results against these and reiterating all the steps on a regular basis. Figure 12.6 illustrates this process.

Benefits Management and Value

Benefits management (Ward and Murray, 1997; CCTA, 1999) is the process of identification, prioritization, quantification, and delivery of expected benefits throughout the program. Benefits are the translation of stakeholders' needs and/or expectations into measurable outputs; the sum of these outputs constitutes the "value" of the program to the organization. *Value management,* and in particular the function breakdown structure (FBS), can be used to manage benefits very successfully (see the chapter on value management in this book). The FBS is to the program what the WBS is to the projects.

Benefits identification is a key process of formulation; both tangible and intangible benefits should be identified, and although it may be more difficult for some, measures of success should be determined for all so that the success of the program can be appraised objectively. Benefits must refer to an organizational-level result. It is a mistake to consider project deliverables as measures of benefits, as it is the impact of the deliverables on the organization, not the deliverable itself, that constitutes the benefit. For example, the delivery

FIGURE 12.6. STAKEHOLDERS MANAGEMENT.

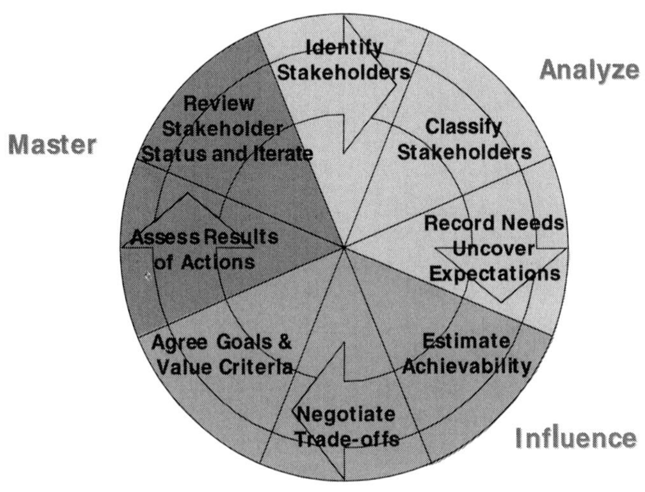

Source: © Michel Thiry (2002).

of training to a number of participants is not a benefit, although it is a deliverable. The benefit occurs when participants apply the knowledge they have gained to their work, and it is the improvement of performance or of work results that is the measure of the program's benefit.

Benefits and their measures of success are best expressed as CSFs and KPIs. *Sensemaking,* as described in the value management chapter of this book, is an effective method to identify, prioritize, and quantify benefits. Once CSFs and KPIs have been agreed and prioritized, the ideation process is used to actively and creatively seek as many as possible alternative courses of actions.

The elaboration process consists of evaluating alternatives qualitatively and quantitatively, combining them and/or further developing them to form options. These options are potential projects or actions that will form the program. When elaborating an option, one must identify both direct and indirect values, and both hard (tangible) and soft (intangible) benefits. *Direct values* are financial or nonfinancial impacts directly related to the choice of an option; they are usually easy to measure. *Indirect values* are consequences of the option, valued by stakeholders and, especially, decision-makers; they are usually more difficult to measure because generally they are not clearly expressed. *Hard benefits* are economic, technical, and operational; *soft benefits* are linked to power, politics, and communications and are more difficult to identify and measure.

Selection/Prioritization of Actions

All the options that are clearly acceptable are ranked in order of their contribution to expected benefits (CSFs) of the program. In addition, actions must be evaluated in terms of their achievability, which is based on factors like financial resource availability (funding, capital cost, cash flow, life cycle costs, etc.), parameters and constraints (size, budget, time-scale, type of work), human resources (expertise, spread, external vs. internal), people factors (availability, competence, customer perception), and complexity (innovativeness, interdependencies, stakeholders). Weighted matrices are typically used to rank actions against CSFs (see the value management chapter for detailed use) and achievability factors. Selection/prioritization matrices are then employed to classify the portfolio of potential actions (see Figure 12.7).

If economic factors are not part of the CSFs or of the achievability factors, options are then assessed in terms of their economic contribution to the business. Typical methods include economic value added (EVA), net present value (NPV), internal rate of return (IRR), and return on investment (ROI).

Choice consists of deciding the best course of action for the program; it is the last step of this phase. The decision is based on the results of options prioritization combined with

FIGURE 12.7. EXAMPLE OF SELECTION/PRIORITIZATION MATRIX.

ACHIEVABILITY					
VHI 0.90 *801-1000*	0.05	009	0.18	0.36	0.72
HI 0.70 *601-800*	0.04	0.07	0.14	0.28	0.56
MED 0.50 *401-600*	0.03	0.05	0.10	0.20	0.40
LO 0.30 *201-400*	0.02	0.03	0.06	0.12	0.24
VLO 0.10 *001-200*	0.01	0.01	0.02	0.04	0.08
	0.05 *001-100* VLO	0.10 *101-250* LO	0.20 *251-450* MED	0.40 *451-700* HI	0.80 *701-1000* VHI
	BENEFITS				

Weighted Matrix Scores

Set thresholds

> 0.15
HIGH PRIORITY
Implement

0.05 - 0.15
MODERATE PRIORITY
Improve and reassess

< 0.05
LOW PRIORITY
Reject

Source: © Michel Thiry (2002).

pacing and synergy factors. Priorities are set between different programs within the organization, and "business cases" are compiled. The choice process also seeks to secure approval and support for the program and allocation of the funding and authority to undertake it.

Especially in programs, it must be acknowledged that decisions are based on available information and expressed value *at the time of the decision.* As these may evolve during the course of the whole program, assumptions and choice factors must be clearly documented and reassessed regularly; this is the main purpose of the appraisal stage.

The Organization Phase

Once funds are allocated to a program, or to the first "cycle" of a program, and the authority of the program manager is secured, the organization phase can begin. It essentially consists of the strategic planning of the program and the definition of actions that constitute its current phase.

Strategic Plan

The objective of strategic planning is to develop a resource management and communications plan, pace the program actions, and develop a change management system. The hierarchy of the FBS shows how each action contributes to a CSF (benefit) and how these support the strategy, thus allowing the program manager to clearly define responsibilities for each CSF and identifying interactions and interdependencies.

The FBS also allows identification of the communication channels for the program; how information is filtered and sorted to suit different organizational levels and how access to systems is allowed or restricted. Using the stakeholder analysis, the program manager develops a marketing plan that clearly identifies expected benefits and goals in measurable terms and, as the program progresses, reports on them. This is a key element of maintaining support throughout the duration of the program.

Resource Planning. The efficient management of resources across the organization is a key component of program management, but this process must also foster the effective use of resources in each project. Resource planning essentially consists of matching required resources with available resources (O'Neill, 1999); it is a value-based concept. "Required resources"—or demand—consists of the prioritized workload necessary to implement the program (see preceding Formulation Phase Section). "Available resources"—or supply—consists of the assessment of the capacity, capability, and availability of existing resources. The program manager's role is to make sure that supply is always equal to or greater than demand, or that demand is adjusted to match supply. Regular reprioritization and reevaluation provides enhanced flexibility and proactive, rather that reactive, resource allocation so that the best resources are used on the most significant assignments. The concepts developed in the book *The Critical Chain* (Goldratt, 1997)—having a holistic view, sequential tasking rather than multitasking, focus on a "drum" resource, and the use of buffers—can also be effectively applied to program resource planning (see the chapter by Leach).

Pacing. The distribution and pace of projects, duration of cycles, and the interval between periods of stability are based on a minimum acceptable level of performance, which cannot be crossed (see Figure 12.8). Periods of stability, or lesser pace, during which benefits are allowed to impact the organization, are spread significantly throughout the program to pace it. Their distribution depends on benefits sought and the significance of those benefits for the business. Financial issues, such as cash flow and/or funding, and human resource issues, like the organization's culture (e.g., risk seeking or risk averse) and the resistance to or acceptance of change, determine this minimum level. Pacing, which includes prioritization of actions, must focus on early benefits, positive cash flow, and maintenance of the motivation of stakeholders. The program appraisal process is also built around these program "cycles" (see Figure 12.5).

FIGURE 12.8. THE PACING OF PROGRAMS.

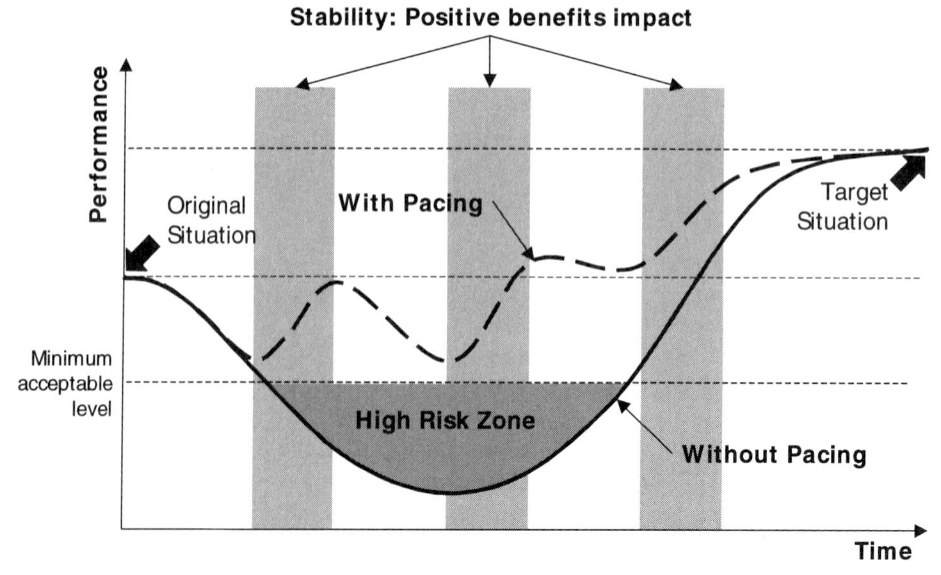

Gateways and Change. Major project changes and benefits appraisal, both of which involve program managers in their his or her project sponsor role, are usually planned around the gateways of projects; those must be significantly spread to correspond both to major project deliverables and to expected benefits delivery for the program. They are not related to the program cycles but rather to project milestones.

The last element of the program's strategic plan is the change management system. In programs, the change process must be value-based. This signifies that, because the program's specific objectives and actions are very likely to be modified as it progresses, emergent opportunities and threats are to be evaluated on a regular basis, as well as the capacity and capability of the program organization to respond to them. This system has to be clearly established from the beginning.

Definition of Actions

The definition of actions is the extension of the formulation phase's preselection process, leading to the initiation of actions. It consists of identifying and prioritizing projects and operational actions for the current phase and developing project initiation documents, if possible in collaboration with the project managers. During this phase and deployment, the program manager acts as the sponsor of projects.

Interdependencies. The selection of actions will take into account the interdependencies between projects, in terms of effects on each other (usually input-output dependencies but also synergy) and in terms of their combined contribution to the benefits sought by the organization. The decision to implement must not be made solely on the individual value of each project or action but also on their capability to contribute to the program as a whole and to align with the strategy. In practical terms, a milestone chart with interdependency activities-on-arrow or a natural network (Marion and Remine, 1997) will effectively represent these interactions.

Prioritization. The prioritization of actions in the overall program must ensure that synergies between projects are taken into account, that there are no overlapping or conflicting actions, and that benefit delivery is optimized. Actions are first prioritized against the critical success factors of the program, but prioritization also takes into account achievability and elements discussed in pacing. Aside from cash flow and funding, financial factors can include payback period, cost-benefit and risk elements. The objective is to make early gains to "fund" the rest of the program and motivate stakeholders, especially investors, to continue supporting the program. In terms of human resources, the program team must take into account not only culture and responsiveness to change but also the perception of benefits by the users, the assessment of expectations of those affected by the change, and the measures put in place to support smooth transition from the original situation toward the target situation. Finally, the prioritization plan will be flexible enough to allow regular reprioritization as the program progresses and priorities or objectives change; if the initial plan has been well documented, this poses no problem.

Constraints and Assumptions. The last point is the identification of constraints and assumptions and their documentation. It is crucial that the constraints and other factors on which the assumptions have been based are well identified, because they will evolve and change as the program progresses and will need to be reassessed regularly. This is an integral part of the risk management process of the program.

The Deployment Phase

This phase includes the actual initiation of actions as well as the management of interdependencies between projects. Once actions are initiated, the program manager will have the role of project sponsor, although in a number of organizations, this role is shared with customers, resource managers, or senior managers. For effectiveness and efficiency purposes, it is advisable that the sponsor's authority is not shared and others play an advisory role. As such, the program managers need to manage project reviews (called *gate keeping*), considering project deliverables against CSFs and KPIs. Other responsibilities include continuous assessment and management of stakeholders' needs and expectations, resource prioritization, and change management, using a formative approach.

Planning and Execution of Actions

During this phase of the program, the main task of the program manager is to allocate or reallocate funds and other resources according to priorities. In developing and approving project management plans, the program manager authorizes or confirms the allocation of resources to undertake the detailed planning of projects.

The program manager also manages the communications to and from project managers and especially acts as a buffer to contain senior management's direct influence of the projects. It is not reasonable to expect project managers and project teams to, at the same time, focus on delivering specific project objectives and develop effective responses to emergent change; the latter is clearly the role of the program team.

The identification of emergent (unplanned) inputs on the program, which could trigger the need for changes, also needs to be monitored and managed in an orderly way. This is where the use of risk and value management techniques becomes essential.

Value Management. During the deployment phase, VM is the process through which project gateways will be handled. Project gateways correspond to go/no-go decisions that are typically situated at significant milestones and concern major project deliverables or phases. At each of these stages, the program manager will apply VM methodology to assess results and decide if changes are required to the project management process.

VM is the preferred way to handle stakeholders' needs and expectations and change management as it is a formative approach based on the definition of CSFs and KPIs.

Risk Management. At the program level, risks generally can be categorized into three levels:

- *Program risks.* Risks that directly affect the program and are handled directly at the program level.
- *Aggregated risks.* Project risks that are common to more than one project and therefore considered from a program-level point of view: project-level response with program support or program-level response with project-level follow-up.
- *Project risks.* Risks that affect only one project.

Risk impact is assessed against all the weighted CSFs to identify those risks that must be dealt with at the program level. The risk response management includes coordination of responses throughout the program in order to optimize results.

Interdependencies and Resource Management

The program manager role requires focus on project interdependencies, rather than on project activities. Even in terms of the project product, the program-level intervention consists of assessing benefits of major project deliverables and output-input relationships between projects—not solving the technical or operational-level problems.

Because program managers are required to manage interdependencies, it is advisable to revert to a variation of the ADM (arrow diagramming method) to represent the program-level interactions between projects. Combined with milestone scheduling and critical chain concepts, it can offer a high level and effective view of a number of projects to program managers and enable them to monitor only major tasks and deliverables, while keeping an eye on both resources and interdependencies (see Figure 12.9).

Resource planning must take into account resources offered (supply) versus resources required (demand); demand should always be less than or equal to supply. This is one of the challenges of good program management, as it is often not the case. Demand consists of the prioritized requirements; supply is an evaluation of the prioritized workload and includes information on resource capacity, availability, capabilities, competence, and signif-

FIGURE 12.9. PROGRAM-LEVEL NETWORKS.

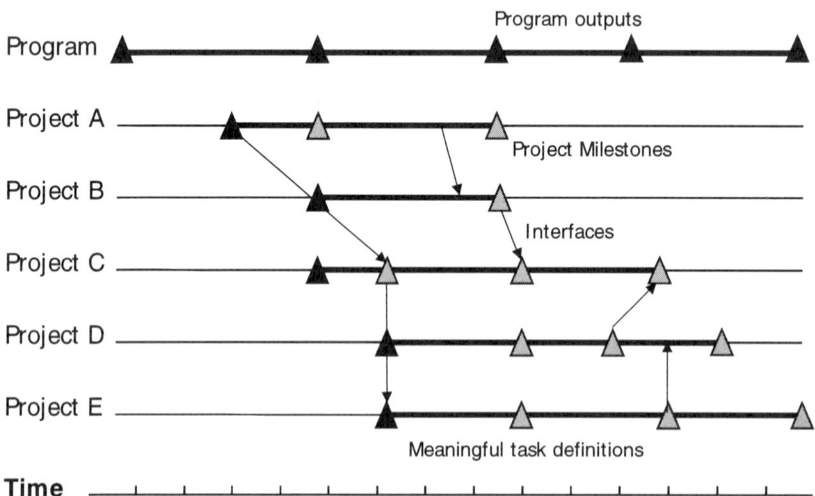

icance. The objective is to create a flexible system that enables quick, objective reaction to emergent inputs or changing priorities.

Project Review and Control

At the program level, project review/control goes beyond baseline control to include continuous reevaluation and reprioritization of projects in regards of the achievement of organizational benefits. Reviews generally correspond to project gateways, or major milestones. Control of projects concentrates on the impact of deliverables on the overall program benefits and strategic alignment, rather than a simple assessment of quality, cost, and schedule—or even earned value. All these measures become valid only as a means to make the necessary adjustments to achieve overall objectives. The program manager controls project progress with a "system's" view of all the actions included in the program, taking into account resources (see the chapter by Leach) and major project deliverables. The key is to control outputs and milestones, not activities.

Because program management should be formative rather than summative, one of the roles of the program manager is to assess the need for project plan review or readjustment, propose or implement changes in projects, and assess their impact on the critical success factors. Project performance is analyzed with a view to the reallocation or reprioritization of resources and contingencies amongst projects or scope changes required to respond to results. The output of this process is a decision to continue, realign, or stop individual projects. Because of this, a sound aggregated information management system needs to be in place, both at project level and at program level. Program-level information management addresses the information between project- and program-level stakeholders, especially the reporting system, as well as the information circulating among projects that needs to be managed at the program level.

The Appraisal Phase

The periods of stability that mark the end of each cycle are an ideal opportunity to appraise the program. This is the time when the organization must evaluate the need to carry on with the program, review its purpose, or stop it. In this phase, results are compared with expected business benefits; emergent threats and opportunities are managed and overall program performance assessed. One of the key objectives is to evaluate the opportunity to continue and/or repace or reprioritize work and actions if required.

In essence, this learning/formative process outlines the program team's success in achieving expected benefits and produces useful feedback for the next phases. The first step is to ask a number of questions:

- Have the expected benefits for this cycle been achieved?
- Have the program or business circumstances changed?
- Have stakeholders' needs or expectations changed?
- Does the rationale for the program still exist?
- If not, should the program be reviewed; should it be stopped?

The appraisal process requires the program team to loop back to the formulation phase in order to reassess the validity of the original needs in regards of external or internal developments since the program was started; this includes positive or negative impact on the business of outcomes from previous cycles and knowledge management. Any change in the CSFs must be identified and examined to understand how it affects the expected benefits.

Following that exercise, the actual benefits of the cycle are evaluated against the expected benefits and a gap analysis is performed. If gaps are identified, the team defines how the program plan is to be modified to take them into account. Using value management, the team examines alternatives and evaluates options. Finally, a decision is made on how to modify the plan for the next cycle. If everything has gone as planned and nothing has changed in the expected benefits, the decision is made to carry on according to the original plan.

Benefits Assessment

When assessing the delivery of benefits from a program management point of view, the program team must have a broad perspective and examine two different levels:

1. At the organizational level, the team's main feedback is the effective delivery of expected benefits and satisfaction of CSFs, specifically:
   - Management of changing corporate or client objectives
   - Prioritization of shared and/or limited resources
   - Interface between functional and project managers
   - Clear definition of roles and responsibilities and mutual support to achieve corporate goals
   - Effectiveness of project review and approval process
   - Project managers' focus on key business issues
2. At the project level, the team reviews the relevance of projects that are spanning over a number of cycles and the aggregated benefits generated by all the projects that are part of that cycle. In particular:
   - Assess overall performance of projects against business benefits, including emergent factors.
     - Quality and timeliness of deliverables
     - Resource usage and budget
     - Use of contingencies
     - Interfaces and interdependencies
   - Identify new threats and opportunities and implement changes, if required.
   - Replan work and relative priorities, at the business level, for the next projects or phases of projects.
   - Loop back to project definitions and readjust, if required (learning loop).
   - Ensure information is recycled into a feedback loop, for the next phases or future programs.

Program Validation

The elements on which the justification for continuation will be established are as follows: progress toward achievement of expected benefits; response to emergent change in business environment, including changing needs and bottom-up initiatives; overall performance measured against CSFs and KPIs, including identification of threats and opportunities; and efficient and effective management of resources in general and of line and project manager's complementary roles in particular.

Following appraisal, the team reviews work and relative priorities for the next cycle, specifically, resource allocation, compare pace of projects, and the need to replace project managers. Table 12.1 outlines major decisions in regards of evaluation outcomes.

TABLE 12.1. MAJOR PACING DECISIONS.

Evaluation	Decision
Reassess significance of benefits for business (C5Fs). —Internal pressures change focus. —External pressures require adjustment.	Reprioritize as needed.
Test integration of change on ongoing basis. —Changes integrated slower/faster. —Resistance to change identified.	Adjust pace as necessary; hold information workshops, training sessions, etc.
Measure delivery of benefits. —Benefits not delivered as planned. —Rate of benefits impact too low.	Reassess distribution of islands of stability as required.
Monitor achievement of minimum level of performance. —General performance drops. —Specific areas are in difficulty.	Review priorities and reallocate funds to urgent matters accordingly.

Programs take place over a period of many months (IT-supported change programs, business process reengineering, etc.) or many years (drug development, transportation infrastructure refurbishment, new product development, etc.). Some programs are even ongoing (performance appraisal schemes, account management, continuous improvement, etc.). The appraisal process is the first step toward dissolution; even in ongoing programs it must be carried out on a regular basis, and, every time, the program team must ask itself: Should the program be stopped or is it worth continuing?

The Dissolution Phase

If the team realizes that the rationale for the program no longer exists, it implements the dissolution process. This decision can be based on a performance paradigm (initial expected benefits have been achieved; cost of the program is greater than the benefits it is bringing to the organization; cost-benefit ratio of program is greater than that of independently run projects) and/or on a learning paradigm (the environment or context have changed and the benefits that the program was seeking to achieve are no longer required, or the implementation of the first cycle or cycles has demonstrated that the program's ultimate purpose cannot be achieved).

As for projects, the closing, or dissolution, of the program is not an easy task; when a decision to stop a program is made, a number of people and funds are reallocated to other ventures, and therefore there is likely to be resistance from the team to "let go." For this reason, some organizations even choose to involve a team external to the program in the closing phase to make it more efficient.

There will always be some uncompleted work that needs to be either completed within a reasonable period of time or transferred to other programs. The team needs to identify and agree on uncompleted work and clarify what can be completed within a reasonable period; it must agree on and secure resources to carry out a post-program feedback, transfer residuals to other programs, allocate outstanding work, reformulate programs, and reallocate resources as required. The program team estimates the resources required to deliver outstanding benefits and the value to do so (benefits/cost analysis). All the documentation must be updated and filed and a post-program review conducted. The data is then fed back into the organization through a learning/innovation loop.

Once all this has been accomplished, the program dissolution team is disbanded and reassigned. This is also the time when overall knowledge, gained from the process, must be collated, although knowledge management takes place at each cycle, especially appraisal.

The Program-Based Organization: Framework and Support

Program Culture

A number of factors currently influence the management of organizations that run projects: Change is accelerating and becoming more complex, influences affecting projects are often of high level, project objectives conflict with one another, projects compete for the same resources, project management is too product-centric, the link between projects and strategies is not apparent, and there is often no coordination of deliverables to realize benefits. The results are ineffectiveness of overall solutions, difficulty to achieve corporate benefits, lack of assessment of the organization's capability, inefficient use of resources, ill response to unplanned (emergent) change, and overruns caused by a lack of coordination. This requires the development of a different view of organizations that includes a strategic/systemic view and the capability to quickly adjust and/or respond to emergent change and to change the relative priority of projects or realign project objectives.

Program and portfolio management help address these challenges, but that often means changing the organizational culture. To succeed in the implementation of a program culture, an organization will require the support of senior management and a clear distinction between project and program paradigms to establish its foundation; structures supporting knowledge management as well as a systemic view of resource management (contingencies, personnel, funding, priorities); and, finally, an attitude aligned on vision and strategy, aimed at stakeholders benefits, and an openness to change.

Merritt and Helmreich (1996) have defined a few keys to successful culture change:

- *Role modeling.* Significant models, senior management
- *Mentoring.* To explain implicit cultural norms
- *Language/discourse.* Manuals, guidelines, logos, mission statements
- *Attractive membership.* Success stories, belonging to a club, fostering pride, and other incentives
- *Proactive approach.* Show concern and willingness to support culture
- *Timing.* Use of emergent inputs and successes to reinforce culture
- *Training and development.* Build culture into training and development; use training and development to reinforce culture

These can be part of the introduction a program culture.

Program-Based Organization

For large programs and organizations that run a number of programs, the organizational structure shown in Figure 12.10 is probably appropriate. For smaller organizations, some roles can be combined.

In this organizational context, the roles of each stakeholder are as follows:

- Top Management and Executive Leadership establish strategic direction for organization.
- The Corporate Program Office is responsible for the whole enterprise portfolio and ensures the strategy is met.

FIGURE 12.10. PROGRAM-BASED ORGANIZATION TYPICAL STRUCTURE.

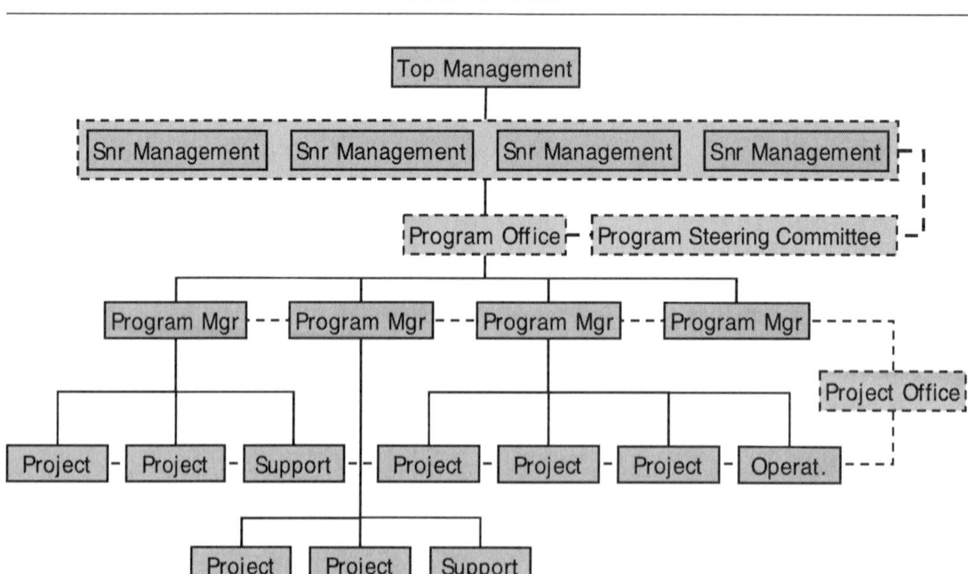

- Individual Senior Managers are usually the program sponsors. Collectively, Senior Management is accountable for the organization's portfolio.
- The Program Steering Committee is the executive group responsible for individual program oversight, guidance, and barrier removal. It links with key senior stakeholders, typically strategy, finance, commercial, and production. It can be combined with the program management office.
- The Program Management Office—or Program-Level PMO—supports the program manager in large, complex programs, prioritizes program resource allocation, and takes a strategic perspective with a focus on pacing and program appraisal. (See the text that follows for the detailed role.)
- Program Managers have accountability for the program, act as project sponsors, and ensure that benefits are delivered. More specifically, their specific responsibilities are listed in the text that follows.
- The Project Management Office provides assistance to project managers on large projects, provides tracking and oversight of projects, coordinates information and reports for the program level, and sets organizational level project management processes and procedures.

- The Project Manager is the single point of accountability for projects, managing project resources within set parameters, as well as day-to-day project activities, and delivering agreed outcomes at milestones.

Concerning the program management office, a series of recent papers (EDS Web site, 2002, Richards, 2001; Kendall, 2001; Moore, 2000; and the chapter by Powell and Young in this book) have redefined its role to give it a more active role. It is now generally acknowledged that this role can cover two major areas: support and delivery. Key areas of support cover the improvement of program and project management efficiency and effectiveness: becoming a corporate program knowledge management center, offering consultancy and advisory services, and supporting program and project managers in their roles and responsibilities. Key areas of delivery include ensuring that corporate strategy is owned and delivered and that it is on track, ensuring delivery of specific measurable outcomes and identifying gaps in portfolio of projects, and analyzing past situations and future trends for knowledge management.

Specific responsibilities of the program manager include, but are not limited to:

- Prioritizing projects and actions in regards of CSFs
- Initiating projects and other actions
- Allocating or reallocating resources within program
- Managing contingencies and coordination of resources
- Balancing project scope and quality versus cost and time (a value ratio)
- Ensuring that benefits are delivered
- Managing projects' interfaces and interdependencies
- Assessing deliverables against expected benefits at gateways

Summary and Conclusions

"Unaligned organization is a waste of energy, whereas commonality of direction develops resonance and synergy."

TSUCHIYA, 1997

Project and program management depend on different paradigms. Whereas project management is subjected to a performance paradigm and has proven effective in delivering short-term tactical-level deliverables, it has not proven its ability to deliver strategic change or improvement programs. Experience with strategic programs or soft organizational change programs has demonstrated that programs need to take into account a learning paradigm that comes from strategic management and value.

A robust program management approach will increase the efficiency of organizational processes (financial, resources, knowledge), support deliberate change (strategies, reengineering), capture bottom-up inputs (innovation, continuous improvement), and enable the proto-typing of emergent strategies by shortening feedback from experience and limiting risks.

Some key issues for program management success are as follows:

- If programs are using a project (performance only) management mind-set and methodology, benefits are lost.
- Programs should exist only if they generate benefits over and above those that projects generate on their own.
- Organizations must understand why program management ought to be implemented and how it can be sustained.

Program management will fall short as a strategic decision management process if the organization implements programs without taking the organizational culture into account, if it does not adequately quantify expected benefits or link them to project deliverables, or if there is a loss of focus of stakeholder expectations over time.

References

Artto, K. A., M. Martinsuo, and T. Aalto. 2001. *Project portfolio management.* Project Management Association, Helsinki, Finland. ISBN: 951-22-5594-4.

Association for Project Management. 2000. *The Association for Project Management Body of Knowledge.* High Wycombe, UK: APM.

Association for Project Management. 2003. Programme management. Chap. 11 in *Project Management Pathways.* High Wycombe, UK: APM.

Bartlett, J. 1998. *Managing programmes of business change.* Wokingham, UK: Project Manager Today Publications.

Carver, C. S., and M. F. Scheier. 1990. Origins and functions of positive and negative affect: A control process view. *Psychological Review* 97:19–35.

Central Computer and Telecommunications Agency. (1999) *Guide to programme management.* London: The Stationery Office.

Center for Business Practices. 2002. Top 500 project management benchmarking forums. *PM Network* (November): 6–8.

Cooke-Davies, T. 2002. Establishing the link between project management practices and project success. *Proceedings of the 2nd PMI Research Conference.* Seattle, July. Newtown Square, PA: Project Management Institute.

de Bono, E. 1990. *Lateral Thinking: A Textbook of Creativity*. 3rd ed. Harmondsworth, UK: Penguin.

Defense Systems Management College. 2000. *Program management 2000, Know the way: How knowledge management can improve DoD acquisition*. Fort Belvoir, VA: Defense Systems Management College Press.

EDS 2002. Program Management Office according to EDS. Retrieved June 2003 from: http://www.eds.com/services_offerings/so_project_mgmt.shtml.

European Foundation for Quality Management. 1999. *The EFQM Excellence Model*. Brussels, Belgium. www.efqm.org/.

Frame, J. D. 2002. How PMI is keeping up with rapid change in the profession. *PMI Today*. Newtown Square, PA: Project Management Institute.

Gareis, R. 2000. Program management and project portfolio management: New competences of project-oriented organizations. *Proceedings of the Project Management Institute 31st Annual Seminars and Symposium*. Drexel Hill, PA: PMI Communications.

Goldratt, E. 1997. *The Critical Chain*. Great Barrington, MA: North River Press.

Görög, M., and N. Smith. 1999. *Project management for managers*. Sylva, NC: Project Management Institute.

Guba, E. G., and Y. S. Lincoln. 1989. *Fourth generation evaluation*. Newbury Park, CA: Sage Publications.

Hatch, M. 1997 Strategy and goals. Chap. 4 in *Organization theory: Modern symbolic and postmodern perspectives*, 101–119.Oxford: Oxford University Press.

Hartman, F., and R. A. Ashrafi. 2002. Project management in the information systems and information technologies industries. *Project Management Journal* 33(3):5–15.

Hurst, D. K. 1995. *Crisis and renewal: Meeting the challenge of organizational change*. Boston: Harvard Business School Press.

Johnson, G., and K. Scholes. 1997. *Exploring corporate strategy*.4th ed. Hemel Hempstead, UK: Prentice Hall Europe.

Kaplan, R. S., and D. P. Norton. 2000. Having trouble with your strategy? Then map it. *Harvard Business Review*. (September–October): 167–176.

Kendall, G. I. 2001. New executive demands of projects and the PMO. *Proceedings of the 2001 PMI: Seminars and Symposium*. Newtown Square, PA: Project Management Institute.

KPMG. 1997. *What went wrong? Unsuccessful information technology projects*. http://audit.kpmg.ca/vl/surveys/it_wrong.htm.

Kubr, M. 1996. *Management consulting: A guide for the profession*. 3rd ed. Geneva: International Labour Office.

Marion, E. D., and E. W. Remine. 1997. Natural networks: A different approach. *Proceedings of the Project Management Institute 28th Annual Seminars and Symposium*. Newtown Square, PA: Project Management Institute.

Martinsuo, M., and P. Dietrich. 2002. Public sector requirements towards project portfolio management. *Proceedings of PMI Research Conference 2002*. Seattle, July. Newtown Square, PA: Project Management Institute.

Merritt, A. C., and R. L. Helmreich. 1996. Creating and sustaining a safety culture. *CRM Advocate* 1:8–12

Mintzberg, H., and J. A. Waters. 1994. Of strategies, deliberate and emergent. Chap. 10 in *New thinking in organizational behavior, ed.* T. Hardimos, 188–208. Oxford, UK: Butterworth-Heinemann. Previously published in *Strategic Management Journal* 6:257–272.

Mintzberg, H., B. Ahlstrand, and J. Lampel. 1998. *Strategy safari*. London: Prentice Hall.

Moore, T. J. 1999). An evolving program management maturity model: Integrating program and project management. *Proceedings of the PMI Seminars and Conference 1999.* Newtown Square, PA: PMI.

Morris, P. W. G. 1997. *The management of projects.* London: Thomas Telford.

Murray-Webster, R., and M. Thiry. 2000. Managing programmes of projects. Chap. 3 in *Gower handbook of project management, 3rd ed.* Ed. R. Turner and S. Simister. 47–64. Aldershot, UK: Gower.

NASA. 1998. NASA program and project management processes and requirements. NPG: 7120.5A.

Neal. R. A. 1995 Project definition: The soft systems approach. *International Journal of Project Management* 13(1):5–9.

Partington, D. 2000. Implementing strategy through programmes of projects. Chap. 2 in *Gower handbook of project management, 3rd ed.* Ed. R. Turner and S. Simister. Aldershot, UK: Gower.

Patrick, F. S. 1999. 'Program management: Turning many projects into few priorities with TOC. *Proceedings of the PMI-99 30th Annual Seminars and Symposiums.* Newtown Square, PA: Project Management Institute.

Pellegrinelli, S., and C. Bowman. 1994. Implementing strategy through projects. *Long Range Planning* 27(4):125–132.

Porter, M. 1985. *Competitive advantage.* New York: Free Press.

Project Management Institute. 1996. *A guide to the Project Management Body of Knowledge.* Newtown Square, PA: Project Management Institute.

Quinn, J. B. 1978. Strategic change: Logical incrementalism. *Sloane Management Review* 1(20):7–21.

Quinn, J. J. 1996. The role of 'good conversation' in strategic control. *Journal of Management Studies.* 33(3):381–394.

Reiss, G. 1996. *Programme management demystified.* London: Spon.

Reiss, G., and P. Rayner. 2001. The programme management maturity model. *Proceedings of the 4th PMI-Europe Conference,* London.

Reiss, G., and P. Rayner. 2002. The programme management maturity model: An update on findings. *Proceedings of the 5th PMI-Europe Conference.* Cannes.

Richards, D. 2001. Implementing a corporate programme office. *Proceedings of the 4th PMI-Europe Conference.* London.

Spradlin, T. 1997. A lexicon of decision making. *Decision Analysis Society.* http://faculty.fuqua.duke.edu/daweb/lexicon.htm.

Stake, R. E. 1975. *Evaluating the arts in education.* Columbus, OH: Merryl.

Standish Group International. 1996 *A Standish group research on failure of IT projects.* Yarmouth, MA: The Standish Group.

Thiry, M. 2001. Sensemaking in value management practice. *International Journal of Project Management.* Oxford, UK: Elsevier Science.

———. 2002a. Combining value and project management into an effective programme management model. In *International Journal of Project Management.* Oxford, UK: Elsevier Science.

———. 2002b. How can the benefits of PM training programs be improved? *Proceedings of the 5th PMI Europe Conference.* Cannes, June. *International Journal of Project Management.* Oxford: Elsevier Science. 22, 13–18 (January 2004).

———. 2002c. FOrDAD: A program management life-cycle process. *Proceedings of the 5th PMI Europe Conference.* Cannes, June.

Thomas, J., C. Delisle, C., K. Jugdev, and P. Buckle. 2000. Selling project management to senior executives: What's the hook? *Proceedings of the Project Management Institute 1st Research Conference.* Newtown Square, PA: Project Management Institute.

Tsuchiya, S. 1997. Simulation/gaming: An effective tool for project management. *Project Management Institute 28th Annual Seminars and Symposium Proceedings*. Drexel Hill, PA: PMI Communications.

Weick. K. E. 1995. *Sensemaking in organizations*. London: Sage Publications.

Winch G., A. Usmani, and A. Edkins. 1998. Towards total project quality: a gap analysis approach. *Construction Management and Economics* 16:193–207.

Wijnen, G., and R. Kor. 2000. *Managing unique assignments*. Aldershot, UK: Gower.

CHAPTER THIRTEEN

MODELING OF LARGE PROJECTS

Ali Jaafari

This chapter presents an integrated approach to the conceptualization, planning, and implementation of large, complex projects. The perspective is on the whole project life cycle, which includes creation, definition, initiation, planning and documentation, execution, commissioning and start-up, operation, and recycling. (The operation phase is considered only in terms of managerial decisions that need to be taken during the preceding phases.) On most large projects it is not possible to separate the project's end product from that of project delivery and management activities (PMCC, 2001; Brook, 2000; Forsberg and Mooz, 1996); thus, any reference to the project life cycle is taken to imply both product and project life cycles.

Generally speaking, the project life cycle can be divided into three distinct phases: the project strategic (promotion) phase (all activities up to and including project approval and funding), the project implementation phase (comprising initiation, planning, detailed design, documentation, execution, and commissioning activities), and the project operation phase (including operation and eventual recycling). Some authors divide the life cycle into two phases only: development and operation. The former includes all activities prior to the start-up and operation phase; the latter includes the utilization phase, including project recycling. It is worth emphasizing that the project as whole is the focus, not the functions of individual players within the project life cycle.

Examples of large projects are aerospace, defense, mining, infrastructure, large telecommunication systems, large software, power, and transportation schemes—all must be recognized in terms of their complexity and managed accordingly. Thus, one would expect to see a similar approach to the management of this class of projects regardless of their industry, yet this is not necessarily the case. For example, in aerospace and defense projects, typically, the emphasis has been on systems engineering and procurement functions; in the construction industry, the emphasis has been on contract and resource management; software and information systems projects have tended to be approached from a technical perspective.

Objectives

This chapter portrays the complex and uncertain internal and external environments within which large projects are typically developed and implemented. A broad classification of project types (in terms of both the characteristics of these projects and their environmental complexity) is presented and the position of capital projects highlighted. The chapter will show that project strategies must relate to project types and environmental complexity (uncertainty). While an integrative framework is needed to manage the evolution of the project concept, management of risks and uncertainty will have to guide the entire process (Jaafari, 2001). This discussion leads to the presentation of a framework appropriate for modeling large projects. The critical criteria for successful management of these projects are highlighted and their realization through the adoption of appropriate strategies is demonstrated. The chapter presents a brief overview of techniques that aid the quantitative and qualitative management of large projects. It emphasizes the need for a holistic approach as far as possible.

Characteristics of Large Projects

There is no universal definition for large projects. Complexity is a common feature in these projects. Complexity stems from two sources: the project's external environment and the complex make-up of the project itself. Miller and Lessard (2001) state: "Large engineering projects are high-stakes games characterized by substantial irreversible commitments, skewed reward structures in case of success, and high probabilities of failure." The environmental complexity is normally created because of the changing market and regulatory regimes impacting both implementation and operation of these projects. Project complexity can be understood in terms of relevant interlocking subsystems of hardware, software, of project-specific and temporary human and social systems, of related technical and technological systems, of financial and managerial systems, of specialized expertise and information sets, and so on that are typically created and managed to realize the project objectives (Jaafari, 2001; Yeo, 1995; Yeo and Tiong, 2000). The cost to promote these projects up to the implementation point is high, of the order of 5 to 10 percent of the total capital expenditure (Merna et al., 1993; McCarthy, 1991). A recent study by Hobbs and Miller (1998) puts the front-end costs up to 30 percent of total costs. Risks are high and the project delivery method is normally shaped to achieve a reasonable outcome in respect of the promoters' and community objectives (Hobbs and Miller, 1998; Wang and Tiong, 2000). Many infrastructure projects are nowadays delivered under build-own-operate (BOO) arrangements (see the chapters by Turner and by Ive). In the resources and industrial sectors, projects are normally fully owned and operated by the private sector.

The risk profile on large projects is complex. Some risks arise from the clash of social, political, and commercial interests and values of project promoters and those of the wider stakeholders that surface during the project development phase. Others relate to project functionality and fitness for purpose. A third set relate to project delivery dynamics. Often there is a window of opportunity in which a project can be favorably launched, as delays may see either the project concept becoming less relevant or even obsolete, or competitors moving in to fill the market need. Major projects are often dependent on novel technologies and innovative solutions; this in itself is a major source of risk.

Miller and Lessard (2001) classified risks on large engineering projects as market-related, technical risks and institutional/sovereign risks. See also Yeo and Tiong (2000).

Exposure to risks can change with time; new risks can be encountered and seemingly unimportant risks pose new threats. On the positive side, there can be opportunities, too, that may provide conditions for improving the project's base value (Miller and Lessard, 2001). This narrative suggests that risk management must inform all decisions and guide all strategies adopted for the creation and delivery of these projects. Hobbs and Miller (1998) undertook a study of a sample of 60 projects in 4 continents (31 power, 5 petroleum, 20 urban infrastructure, and 4 technology projects). They found that the front-end part of the project life cycle was particularly risky. This phase was often marred by serious setbacks that put projects as a whole at risk. Some key findings by these authors have been summarized in the following list to shed light on the dynamics of capital projects worldwide:

- No distinct phases (e.g., feasibility studies, design, and construction) could be identified on these projects; instead, a series of milestones were found to be common. In addition, a front-end part (referred to as the *strategic phase* in this chapter) could be distinguished from the engineering-procurement-construction (EPC) part (refered to as the *implementation phase*);
- There were wide variations in terms of the length of time taken to develop the projects in the sample to the point of implementation: 55 percent took more than five years. Projects that had shorter front ends were required to fulfill urgent needs.
- The promotion phase was found to be a dynamic play, which, in addition to those involved in more technical aspects of the project, saw participation of communities affected, environmental organizations, pressure groups, financial institutions, politicians, regulators, and government agencies.
- The decisions made during the front-end and the institutional, organizational, and financial framework that were put in place by and large controlled the success or failure of these projects and profoundly influenced the implementation stage.
- Some projects in the sample had a defined technical solution right from the outset, while for others the technical solution either evolved along the path or was deliberately held off until late to accommodate changes until the implementation stage
- The influence of the environmental, social, political, and community aspects on the sample projects were found to be increasing. This is largely because large projects epitomize the current profound restructuring of the institutions of society and government machinery. The principles of social equity, privatization, user pay, sustainable development, legal

legitimacy, and community ownership all exert varying degrees of influence on the creation and execution of large projects.

According to Hobbs and Miller (1988):

> "In recent years, the process has become much more complex. This increase in complexity is due to several factors including: the globalisation of competition, the trend toward deregulation, the changing role of governments under dual influences of free market doctrine and debt loads that prohibit further borrowing, and the actions of the pressure groups locally and internationally. In their search for solutions in this highly complex context, organisations have developed highly complex solutions which often include some form of coalition building. Often the initiators of projects do not have all the political, social, technical, financial and organisational resources and skills that are needed to deal with the multiple risks that the highly complex context presents. Therefore, they search for partners who can bring needed resources or skills, or that can control or support various risks. In the process of searching for a feasible solution, the skills of managing political and social interfaces, of organisational and financial engineering, and of deal-making were often critical.
>
> Planning in this context is very difficult. A deductive and linear plan to get to a solution is not workable because the solution is not identifiable at the outset, in fact, the problem is not usually well defined at this stage. At the outset, it is not obvious who the important players will be, and there is some trial and error in the search for partners and solutions to the many problems posed by the project. The activities of risk identification, analysis, mitigation and partitioning among players dominate the process. Negotiations are constant throughout this phase. Further, if the process was not complex and unpredictable enough, it is highly likely that during the search for a solution, a new problem will materialise and send the project off track at least once."

Evidence from other sources is just as revealing; Jaafari and Schub (1990) carried out a field study of large and complex projects in Germany and concluded that the development of such projects was substantially impacted by the resolution of risks and uncertainties. These authors showed that risks were not only increased by a poor choice of concept at the outset but also by community demands, changing regulations, political and social forces, and dynamics of the project environment itself. Morris and Hough (1987) have also shown the complexity and dynamics of these influences in their case studies of a number of large and complex projects executed in the United Kingdom.

This brief review is intended to show that project conceptualization and implementation is a complex, dynamic, and evolving process; that it should be managed on the basis of a set of objectives, which themselves would be subject to change, on a fully fluid and flexible basis (Jaafari, 2001; Chaaya and Jaafari, 2001; Jaafari, 2000, Miller and Lessard, 2001; Yeo and Tiong, 2000; Morris and Hough, 1987). Further, that a holistic and integrative framework is needed in which not only planning and proactive management of technical and financial factors receive attention but equally the social, environmental, political, and community aspects are placed at the center of attention (Jaafari, 2001; Jaafari and Manivong, 2000; Miller and Lessard, 2001). The objectives chosen should embrace the project's viability in its broadest sense, over its entire life and should facilitate management of the process using continuous risk and uncertainty resolution within a fluid and flexible management framework. This is very much an open-systems approach to the management of large projects of this nature (Yeo, 1995). Scott (1992) states that "systems are interdependent activities linking shifting coalitions of participants; the systems are embedded in—dependent on continuing exchanges with and constituted by—the environments in which they operate." For a more detailed understanding of complex systems' theory refer to Scott (1992). See also www.brint.com/Systems.htm.

Environmental Complexity and Influence on Strategic Direction of Large Projects

Large projects occupy the complex side of the project population space (Figure 13.1). Numerous forces impact project environments; viz (a) increased demand from owners and

FIGURE 13.1. CLASSIFICATION OF PROJECTS IN TERMS OF PROJECT AND ENVIRONMENTAL COMPLEXITY.

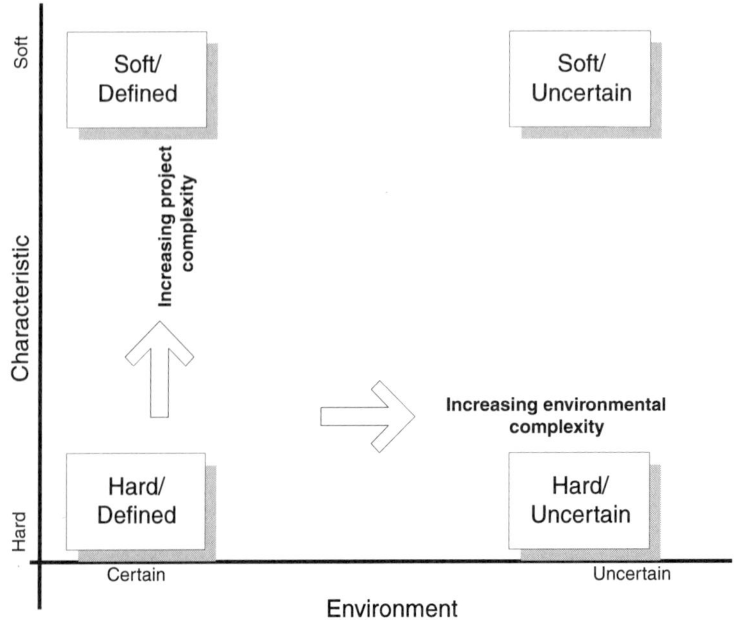

customers for solutions that deliver definitive advantage to them along their business objectives; (b) operation of markets, which nowadays shift a lot in a chaotic manner; (c) rapid rate of change in the underlying technology and scale of operation, which in many instances require novel solutions; (d) the influence of the regulatory bodies, who tend to aim for zero-risk solutions; (e) the information technology revolution enabling global collaboration and streamlined managerial processes; and (f) rising influence of community and pressure groups (Miller and Lessard, 2001; Jaafari, 2001; PMCC, 2001; Dixon, 2000).

As an example of environmental complexity, the following is an excerpt from Byers and Williams (2000). This excerpt illustrates the complex commercial and regulatory environment for electrical utility industries in the United States.

> "As the world-wide economy evolves, electric utility companies in the United States and most industrialized nations of the world are under increasing pressure. In the U.S., deregulation of the electric utility industry has led to significant business and management changes. Corporate reorganizations, staff downsizing, outsourcing of services, and reengineering of business processes have had a profound impact on the industry.
>
> The traditionally conservative U.S. electric utility industry, which had previously considered itself almost impervious to outside influences, was feeling the effects of a global economy and was under pressure to become more efficient and cost effective. Clearly, there was a continuing world-wide transformation going on, moving faster than most people had anticipated; now, the world was our market place with new opportunities and new competition.
>
> Nation-wide, a vigorous move toward deregulation had, in just three years, changed the industry's view of its customers, its competition, and itself. Customers formerly bound to the company by geographic monopoly now had to be courted and costs had to be reined in to help meet low competitive pricing."

Other sectors of the economy, too, experience rapid changes in a similar fashion. Commercial and sociopolitical environments will set the scene for projects, as these are often the foundation for reshaping the competitiveness of firms or whole industry. As Struples (2000) writes:

> Today's large engineering projects can involve significant organisational and operational complexity covering joint ventures, consortium working, international involvement, conflicting stakeholders, shareholders, special purpose companies, project financing, prime contracting, etc. Often such projects face diverse political pressures with difficult-to-define socio-economic impacts, and can be subject to risk-sharing contractual arrangements with pain/gain sharelines. As the project moves through its life cycle, organisational and operational issues and engineering change together can have catastrophic effects on the project schedules and costs, often with dire consequences. Many recent projects have been reported as experiencing cost overruns in excess of 300% and 150% to 200% is becoming the norm.

The Defence Evaluation and Research Agency (DERA) in the United Kingdom has portrayed a hierarchy of environments impacting projects and systems, shown in Figure 13.2. As seen, the project environment takes place within specific social and organizational environments (DERA, 1997). The social environment defines legal requirements, social norms, fiscal rules, environmental requirements, and business competitiveness. The impacts of the social environment on project formulation and management can be profound over its life.

The enterprise environment comprises both the suppliers and the customers and end users of large projects. This environment has a profound influence on project environment, particularly when projects are sponsored and implemented by a coalition of firms with complementary resources and expertise. There are two main enterprise environments: that of suppliers and those of the sponsors or client/end users. Often the actions and responses of these organizations create a dynamic setting that influences many project decisions and outcomes.

DERA (1997) cites examples of these as

- the business scope of the enterprise; this determines the markets, application areas, and opportunities that systems' ventures pursue
- business policy that defines how, on behalf of the stakeholders in the enterprise, resources will be invested. It influences decisions to bid, invest, or proceed with a product development; it determines the nature and allocation of corporate resources capacity; and so on.
- market strategy that influences product system families, intended product lifetimes and support policy
- investment strategy; this impacts product novelty, introduction or use of technology, supporting infrastructure for system design, the capabilities and training of personnel, the manufacturing locations and capacities, and so on.
- business practices that lead to mandated or recommended process standards, business and technical process improvement actions, common methods, and tools

FIGURE 13.2. THE HIERARCHY OF ENVIRONMENTS.

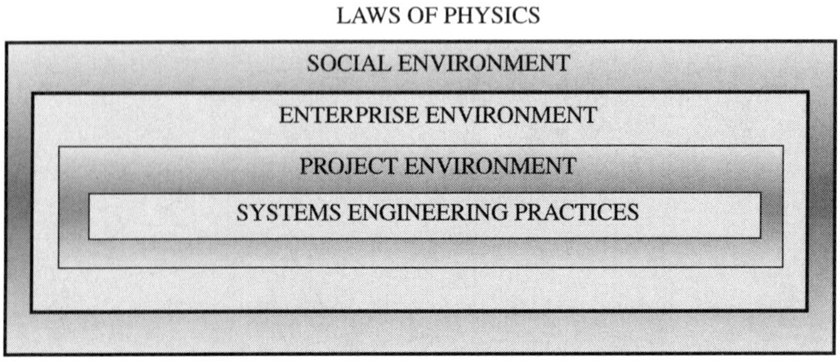

Source: DERA (1997).

FIGURE 13.3. INFLUENCES ON PROJECT DECISIONS.

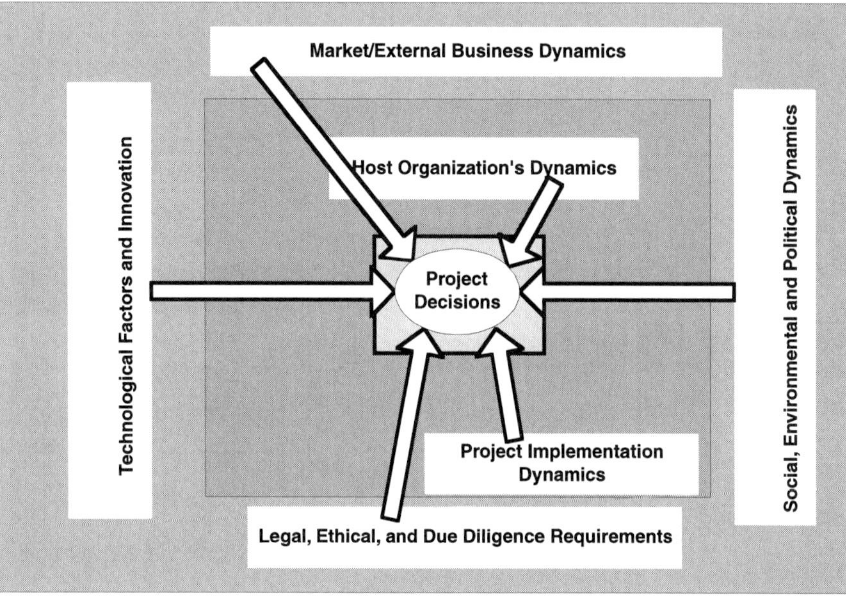

The project environment is in reality a subset of the enterprise environment, though in the case of large projects influenced by different enterprise cultures, the project environment may be more complex. Figure 13.3 indicates the influences on project decisions in a complex setting.

The project environment has to respond to a number of challenges, e.g. the aspirations of the sponsor organisations, the needs of wider stakeholders, the legal and due diligence requirements, the social and environmental requirements and so on (Manivong et al., 2002).

DERA (1997) states that

> In the project environment:
>
> - teams are built to provide capacity and breadth of understanding and experience
> - plans are devised to guide the technical endeavours
> - achievement is monitored to ensure that resources are effectively applied
> - decisions are made, selecting alternatives to most successfully achieve objectives
> - uncertainty is contained, limiting the commercial exposure of the host enterprise.

Project Life Cycle

Figure 13.4 shows a typical life cycle of large projects. As seen, one or a number of business and/or social needs or changes must be satisfied. A project idea is then born to respond to

FIGURE 13.4. PROJECT LIFE CYCLE PHASES, LARGE PROJECTS.

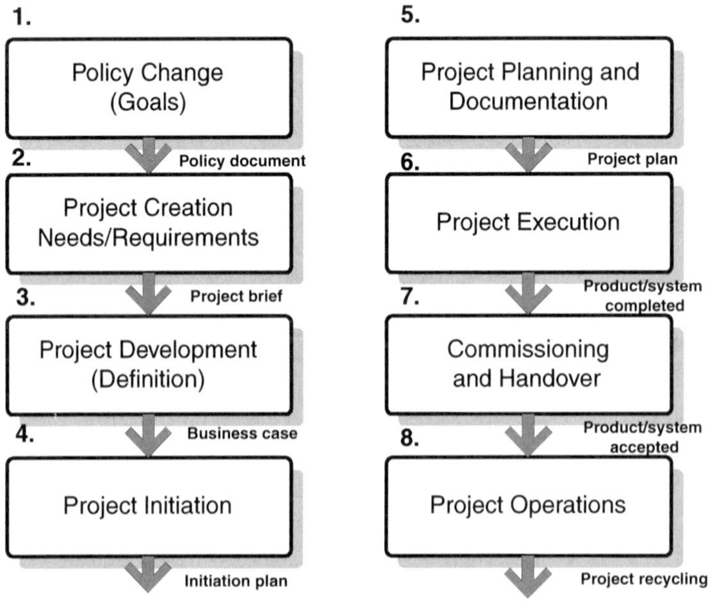

these needs. The responsibility for the project (or more accurately the needs to be fulfilled) is assigned to an operational unit (client) who typically sets up a project directorate to handle the project formally and properly. The next phase is the project development phase; this is the basis for detailed investigation of alternative ways to respond to the project goals or satisfy the stated needs. The outcome of this phase is very critical in the sense that it identifies the optimum way to respond to the relevant business needs and requirements and to formulate a clear business case for the selected approach (which may or may not lead to the scope originally foreshadowed). The outcome is normally captured in a project definition report and is used for approval and funding of the project.

The implementation phase is really about finding an optimum way to deliver the business case (Adams and Brown, 2002). It starts with the initiation phase that formulates a set of strategies that will guide the planning and execution phases. If all decisions taken during the development phase are optimal and provided that there are no major changes in the business or operational environment, one would expect the project to progress to a successful conclusion, leading to the operational phase.

The project life cycle may be portrayed using different terms—for example, that of the Asian Development Bank (see Figure 13.5). This representation is based on project identification, preparation, appraisal, funding (loan negotiation and board approval), implementation, and evaluation.

Another way to look at a project is from the underlying system point of view, where the following can be noted (DERA 1997):

FIGURE 13.5. THE ADB PROJECT LIFE CYCLE REPRESENTATION (www.adb.org/projects).

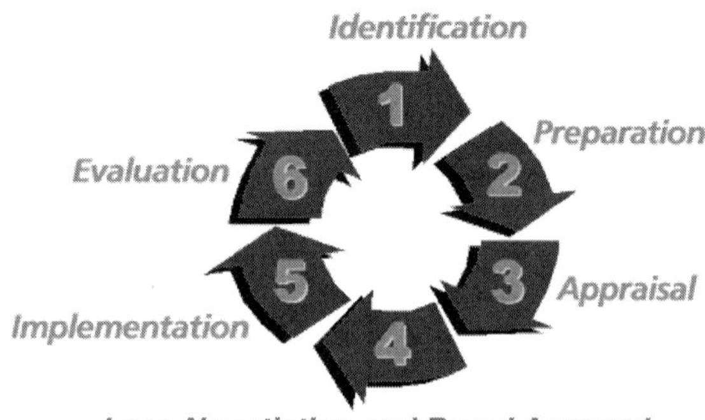

- Conception
- Creation
- Utilization
- Disposal
- Whole of life approach

Decisions made during the front-ends of projects have a profound impact upon the success or failure of the project during both the implementation and operation phases, not only in terms of scope, cost, and time but more importantly the underlying operational capability and business viability. So it is important to consider all implementation and operational aspects, and combine upstream and downstream information before formulating the project concept. It is also important that a consistent and integrated framework/model of the project is set up and all decisions evaluated against a set of criteria representing the whole project life cycle. This needs a modeling approach that enables the project team to develop an optimal project solution initially, coupled with a capability for real-time adjustment of the same throughout project life cycle.

Management of Risks and Uncertainties

The challenge of planning, and successfully delivering, large projects principally lies in the effective recognition and management of project risks and uncertainty, given relevant environmental complexities.

Many business and strategic considerations motivate project promoters, including securing a presence in a particular market, entering global competition, and maintaining technological supremacy. However, more often than not, the overriding objective is to gain financial rewards for a relatively small financial outlay. This requires a core competency in risk and uncertainty management. The promoters' first challenge is to obtain permits to construct and operate their schemes. Promoters are not always investors, and the investors' interests may be different in the sense that many institutional and individual investors are not active participants in the management of the process but invest in, or lend funds to, the project in the expectation of future returns. Put differently, the promoters' objective is to create a long-term financially viable and balanced business entity. The eventual facility is a compromise between the promoters' interest and the interests of the community at large. If the objective is to create a viable business entity, then the processes of development and decision making must also be shaped primarily by the same consideration (Jaafari, 2000, 2001).

It is interesting to note that managers on the sample of 60 large engineering projects studied by Miller and Lessard (2001) ranked market-related risks (and uncertainty) at 41.7 percent, followed by technical risks at 37.8 percent and institutional/sovereign risks at 20.5 percent. The latter includes social acceptability risks.

Public sector projects are somewhat different in the sense that they are primarily dependent on budgetary restrictions and stakeholders' consensus. Changing stakeholders' expectations imposes substantial challenges on the project success in terms of scope changes and shifting priorities over the project life cycle. While these projects are subject to different risks and challenges, the need for creation of value and reduction of risks and liabilities over the project life remains unchanged. The value proposition on these projects is often expressed in terms of minimization of the total life cycle costs for each function fulfilled over the project life. Alternatively, another measure representing the service value generated by the project over its life can be defined as the basis for project value creation and optimization.

If the general project environment is subject to change, the project objectives must also change in line with relevant dynamics. This means that the entire project management philosophy must be opportunistic and driven by risk/rewards prospects throughout the project life. The necessity for adopting such a flexible and fluid framework for projects lies partly in the turbulent environments (particularly shifting markets) and partly in the rapid rate at which technology changes. A third factor is the increasing influence of the host communities and stakeholders, as well as the complex requirements often imposed by legal, environmental, social, safety, and fiscal regulations (Hobbs and Miller, 1998). Thus, it is not possible to close a project's options too quickly by freezing everything in the form of fixed design/specifications and/or lump-sum fixed-scope/-price contracts (Miller and Lessard, 2001; Laufer, et al., 1996). Decisions have to be analyzed and optimized continuously using the life cycle objective functions (LCOFs) as the basis of evaluation (Jaafari, 2000, 2001). Miller and Lessard (2001) argue that "Sponsors strategize to influence outcomes by using four main risk-management techniques: (1) shape and mitigate; (2) shift and allocate; (3) influence and transform institutions; and (4) diversify through portfolios." Note that these strategies are not mutually exclusive. Of these, the last strategy, namely diversification through the acquisition of a portfolio of similar projects, is not often open to sponsors. The exception is large multinationals who operate globally (typically in mining, resources, and industrial sectors). While risk management processes will enable project sponsors to approach large, complex projects systematically, criteria for evaluation must always be life cycle objectives, as these projects take a relatively long period to eventuate and then a longer period to operate, to retire the investment and return a profit to sponsors.

Life cycle objective functions are those that

- determine the project's financial status and its profitability;
- represent the operability, quality, or performance of the facility or the utility of the product to customers; and
- will influence the owner's short- and long-term liabilities, including occupational health and safety (OH&S) risks during both construction and operation, environmental impact, and third-party liabilities.

The financial LCOFs vary from one project to another. These may include cost/performance ratio (total life cycle cost/unit output), which typically suits production or extraction facilities; cost/worth ratio, which suits public projects; internal rate of return; and profitability index (ratio of the net present value over capital expenditure). There may be other (secondary) objectives, viz early cash flow generation, debt reduction, and so on (Woodward, 1997). On privately sponsored infrastructure projects, the concession award, the environmental impacts statement (EIS), and the finance deals may well contain appropriate target values that can be used as project life cycle objectives. Targets set for operability, quality, or facility performance will directly or indirectly affect profitability, while OH&S and environmental objectives influence profitability and long-term liabilities.

Traditionally, all LCOFs, objectives are assessed at the time of project planning and definition (or feasibility studies and conceptualization). Also, on major projects, an environmental impact statement is normally required for the issuance of a permit by the relevant authorities. Such documents typically contain recommendations for environmental management and safeguards against adverse environmental impacts. Under a whole of life approach, the project status and its decisions are evaluated continuously in terms of the relevant LCOFs and in comparison with the targets set for the LCOFs at the start of the project.

Under this approach, project time and cost are not to be treated as the main objective functions even for the management of the implementation phase, as these do not directly represent the LCOFs. As an example, a modest rise in the capital expenditure on a project may well be justified if it leads to a shorter delivery timescale and an increase in the project's internal rate of return. In general, the status of the life cycle objectives, including compliance with the statutory requirements and exposure to risks and/or future liabilities, will determine project success and should therefore be the basis for ongoing evaluation. Put differently, project decisions must at all times be aimed at improving the base value of the project, its fitness for purpose, and due compliance, while at the same time minimizing the impact of potential risks and uncertainties.

Critical Success Criteria and Framework for Adoption

Successful synthesis, development, and delivery of complex projects typically involve a large number of professional people working within a range of organizations who are party to the project via contracts or other means. The owner or sponsor has, however, the highest stakes in the project and must adopt a set of criteria that will enable him or her to solicit and integrate the diverse inputs from all participants and steer the project to a successful completion in an optimum manner. The following criteria are recommended:

- *Focus on the whole business of which project hardware and software are only parts.* This means that all decisions throughout the project life should be judged in terms of their impacts, whether positive or negative, on the business and strategic objectives that the project is required to deliver. Thus, intermediary objectives such as scope, time, and cost should be used for communication and expediting of the project implementation, not as the ultimate criteria for decision making.
- *Maximize opportunities for value creation.* As the project moves from the concept to development and implementation phases, the opportunity to add value and reduce exposure to risks and liabilities will decrease and the cost to implement change will increase exponentially (see Figure 13.6). It is important that the project team is encouraged and given due assistance to develop breakthrough solutions early in the life of the project in a manner that maximizes the project's chance for success and minimizes its exposure to

FIGURE 13.6. VALUE CREATION AND RISK REDUCTION POTENTIAL VERSUS PROJECT PHASES.

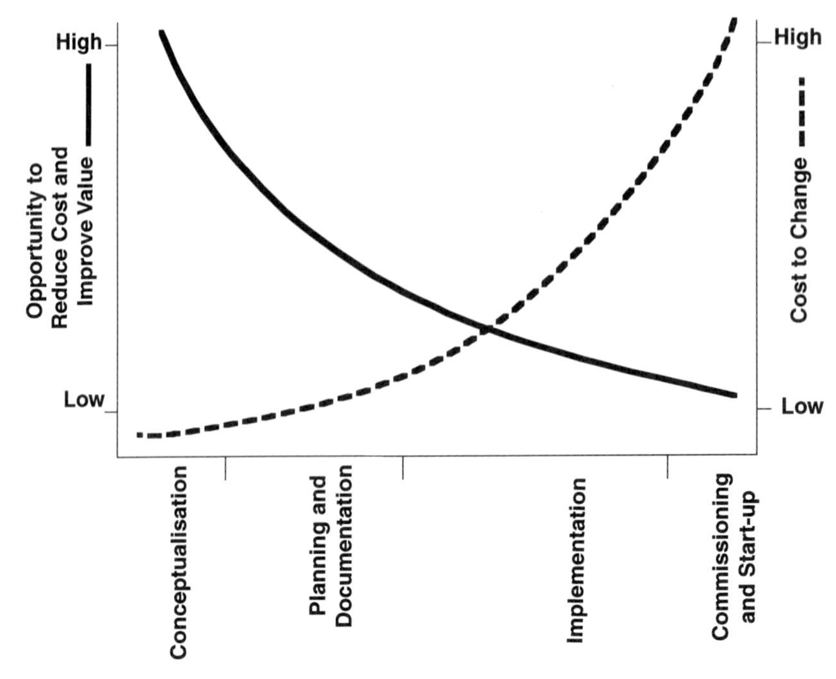

risks and liabilities. Value engineering is most effective at the concept and design phase; though it should be the basic focus of project team in all phases and in a creative manner (Merna, 2002). (See the chapter by Thiry.)

- *Institute a proactive approach to handling risks and uncertainties.* Large, complex projects are subject to multiple risks and uncertainties, particularly in their formation phase (Miller and Lessard, 2001). Most projects suffer significant cost and time overrun, or experience performance setbacks because of the unresolved risks surfacing or new risks arising unexpectedly. Prudent risk and uncertainty management is the key to successful project management. (See the chapters by Chapman and Ward, and by Simister.)
- *Incorporate and manage community and stakeholders' interests.* While this sounds like a self-evident statement, its implementation is quite complex. Many projects touch the life and economic well-being of many people in their host communities and have multiple stakeholders who are not necessarily financial party to the project. The interest and influence of these parties have to be recognised and factored in at the time of project conceptualization and planning. These have to be managed systematically throughout project life. Appropriate resources must be allocated to ensure success on this front. (See the chapter by Winch.)
- *Create synergy among participants.* A successful project needs to capture and effectively utilize the energy and intellect of all its participants in an effective manner. This is not easy, as project participants often come from different organizations, each with its own unique culture, norms, and standards. Synergy must be created in terms of congruence of project objectives and contractual terms that commit participating contractors and consultants to project objectives. Formulation of the actual terms and conditions of contracts is strategically very critical to the project success. In recent times, contractual terms have even included obligations to attend team facilitation workshops in order to create a teamwork spirit. (See the chapters by Venkataraman, by Langford, and by Lowe.)

Modeling and Integrated Life Cycle Planning

Figure 13.7 is a simplified representation of modeling for a large project. As can be seen, the model should ideally

- provide a consistent and efficient framework for development of the project from concept to completion and through to facility operation;
- integrate project information related to all project life cycle phases;
- integrate all project management functions, including both hard and soft functions;
- support scenario analysis and offer an integrated environment to effectively and interactively apply what-if planning;
- have the potential to accommodate modeling and simulation of the operation of the end facility (Jaafari & Doloi, 2002);
- provide graphical simulation of the proposed implementation plan in a manner that helps the optimization of relevant work plans;
- provide a means for interdisciplinary communication and teamwork;

FIGURE 13.7. SIMPLIFIED REPRESENTATION OF PROJECT MODEL.

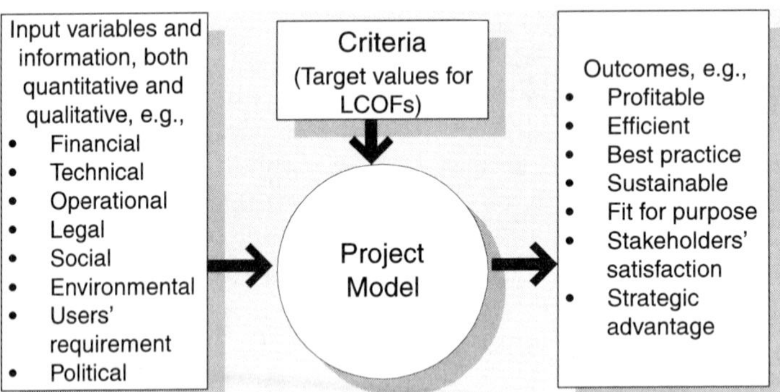

- provide an effective means for conveying the planning results to the field;
- allow early problem detection, including removal of clashes and interferences; and
- integrate the processes of planning, engineering, documentation, procurement, and execution throughout the project life cycle (dynamic process).

Realization of such sophisticated modeling systems is still some way off. The project model has to expand from the concept to the implementation phase in terms of information sets and linking of these within the model so as to enable the holistic evaluation of all project decisions. What follows is a study of large projects in terms of the different phases—that is, the strategic phase and the implementation phase (see Figure 13.8).

Project Strategic Phase

Phases 1 to 3 inclusive in Figures 13.4 and 13.8 are part of the strategic planning phase (project promotion phase). The strategic planning happens at two levels: organizational and project.

Project Creation. Figure 13.9 shows the organizational planning framework. Very often policy changes (particularly at government and large corporations) will give rise to the project concept (Beder, 1991; Kelley, 1982). Discussion regarding grounds for such policy changes and the processes that are followed to introduce such changes are outside the scope of this chapter. These are typically the domain of strategic planning for the whole organization, government department/agency, and/or relevant communities.

Generally speaking, the need must be expressed in terms of business or strategic needs of the organization, not acquisition of new assets or increased capacity (Artto et al., 2001). The criteria for fulfilment of this need should also be spelled out in terms that can easily be cast into target values for LCOFs. This approach fulfills two specific purposes: It provides

FIGURE 13.8. PROJECT LIFE CYCLE.

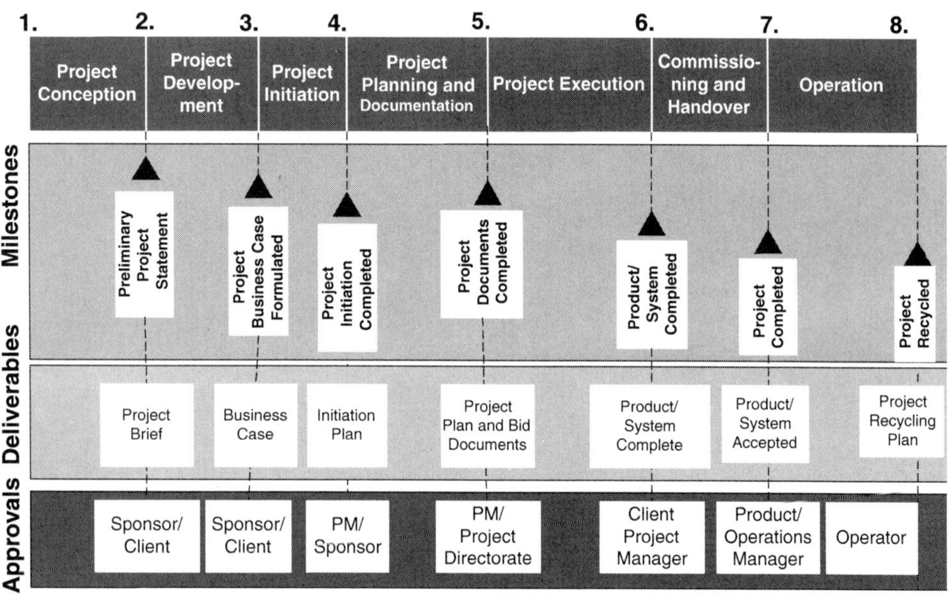

a basis for search and/or development of alternative business solutions that may or may not involve a new project, and it enables risk and uncertainty evaluation be carried out from the outset in terms of LCOFs. (See the chapters by Artto and Dietrich, and by Jamieson and Morris.)

Some of the activities in this phase are (1) determination of organizational and decision making processes and structures at owner level, (2) conduct of high-level business/strategic consultations and deliberations to ensure that business/strategic needs are correctly determined and documented, (3) setting of initial targets for LCOFs, and (4) notional budget and designated time frame to fulfill the stated needs. Thus, the main purpose is to define both the needs and the required business hurdles (target LCOFs) that must be met in subsequent project development and implementation phases. The study and decisions outcomes are captured in a document typically referred to as a *project brief*.

A project brief may or may not contain several suggested options for responding to the stated needs, but it should really avoid giving a prescriptive solution or preempting the creative process that needs to be followed subsequently to locate and develop an optimum solution to relevant business/strategic needs. Generally speaking, a brief is the basis for conducting a systematic project definition studies and delineation/optimization of the project business case for the preferred solution. Many alternatives (including non-project options) should be considered, developed, and appraised to locate the optimum business solution.

FIGURE 13.9. BROAD REPRESENTATION OF PROJECT CREATION PHASE.

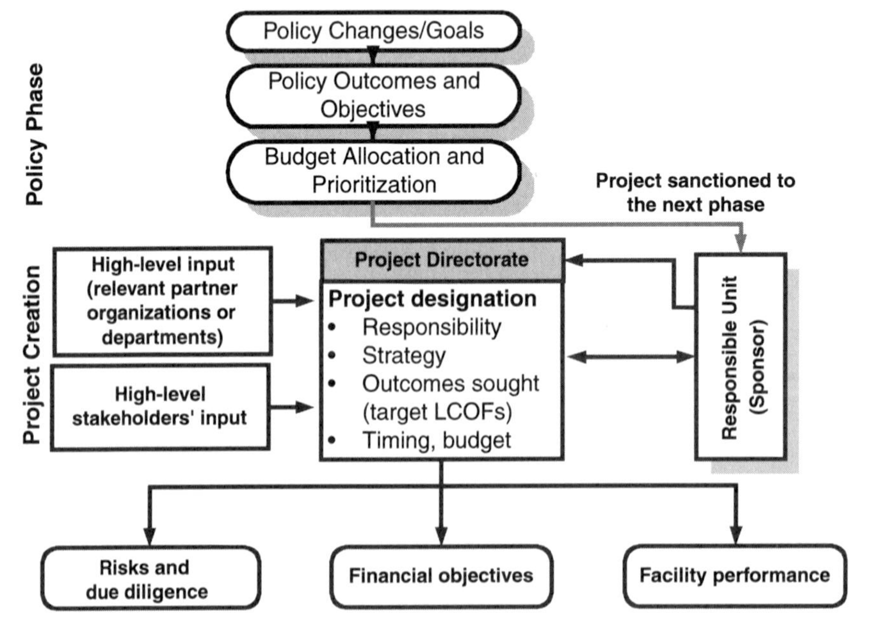

Some owners assume that the project brief is a sufficient basis for proceeding to the project implementation stage. This has to be resisted, as it may lead to suboptimal solutions or increased exposure to hidden risks and other traps (Miller and Lessard, 2001; Jaafari, 2001). The purpose of a project brief is to have enough information on the problem (business needs and requirements) so that a multi-disciplinary team of experts can be engaged to conduct project definition studies; locate, define, and refine an optimum solution; and evaluate its value proposition vis-à-vis target LCOFs. As a minimum, a project brief should contain the following key information:

- Strategic needs and business requirements
- Commercial opportunities
- Objectives (target LCOFs)
- Perceived constraints
- Funding options
- Value proposition in terms of client's business needs
- Nominated project organization and governance
- Anticipated project budget and duration
- Major resources required to realize the objectives and perceived risks
- Proposed plan to conduct project definition studies and establish the project business case

Project Development (Definition). Figure 13.10 shows a broad representation of this phase, whose aim is to locate, define, and refine an optimum solution with a firm business case vis-à-vis targets set for LCOFs. The project business case is the basis for subsequent project implementation (initiation, planning, execution, and commissioning) (Gray et al., 1985; Morris, 1983).

The project development phase is sometimes referred to as the *project definition phase.* This phase results in a project definition report, which is often the basis for project funding and approval. The project development phase is the most creative phase of the project life cycle. The project study team needs to explore all plausible options that can be thought of in order to satisfy target LCOFs and stated business needs.

Typically a project business case must provide the following key information:

- Executive summary
- Needs (of end users and customers) assessment and facility/product definition, priorities
- Target LCOFs to be achieved and means of assessment
- Background
- Stakeholders' requirements
- Options generated and evaluated, and selection of the preferred option
- Product description
- Financials
- Project organization and governance
- Commercial risks
- Delivery strategies

FIGURE 13.10. BROAD REPRESENTATION OF PROJECT DEVELOPMENT PHASE.

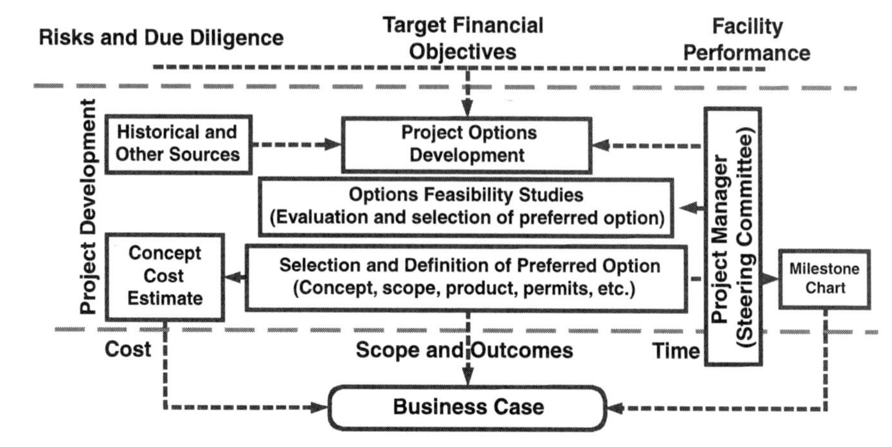

- Project Quality Management model
- Intellectual property and licensing issues
- Resources
- Statutory requirements and due diligence
- Operational issues
- Handover issues
- Time-to-market
- Knowledge management
- Community and stakeholders' interests
- Financing
- Teamwork
- Value creation opportunities

It must be noted that while Figure 13.10 shows a linear process for obtaining an optimum solution as the basis for project scope, in reality the whole process is recursive—that is, evaluation of potential options may necessitate reverting to the original assumptions, target LCOFs, and business needs or strategic goals. In the light of the knowledge gained from the first cycle of evaluation, it may be necessary to clarify or modify the original assumptions, hurdles, and business objectives, including targets set for LCOFs, and then start the process again to see if the solution will work. The test as to whether or not the solution is optimum is the extent to which target LCOFs can be met assuming a successful project completion. However, there may be secondary objectives or constraints that must also be met, and these are typically addressed in the business case statement.

As found by Miller and Lessard (2001), the front-end of large projects takes a considerable amount of time (typically ten years on the sample of 60 large engineering projects they studied). During this period, the project may be subject to several evaluations at different junctures with regard to new political developments, market shifts, stakeholders' shifting priorities, and so on. This period often provides an opportunity to develop creative solutions that may also make the project ultimately feasible and attractive to invest in. An example is the Northwest Shelf Liquefied Petroleum Gas (LPG) project, constructed in the 1980s in Australia. The original design proposed construction of a cooling tower requiring a large steady supply of fresh water that was not available locally. The project was delayed considerably, and in the meantime, the dry cooling technology developed further. The eventual solution was a dry cooling process that made the LPG project attractive to invest in. A major design change of this type will normally trigger a fresh round of project evaluation and replanning.

Project Implementation Phase

The project implementation phase can be thought of as the part of the project life cycle that starts after the project funding and approval and concludes with the successful handover of the end product to the client organization, including the contractual closeout of the project, lessons-learned documentation, and archiving of the project documents. Ideally an integrated project team oversees the entire implementation phase (Chaaya and Jaafari, 2001; Jaafari, 2000; Jolivet and Navarre, 1996).

One of the theses advanced by this chapter is that all project implementation decisions must ultimately satisfy the project's business needs and requirements. Implementation strategies and scenarios must be evaluated continuously using target LCOFs as the criteria. This is so even though, traditionally, scope, time, and cost management have been the focus of the project implementation phase. When the project environment is dynamic, it is necessary to regularly review the criteria against which decisions must be evaluated. In the case of a life cycle project management framework, the appropriate criteria are the targets set for LCOFs; these should be continuously reviewed and revised downward or upward as deemed appropriate to ensure alignment with the market, realism in terms of what is achievable, optimality in terms of balancing needs and requirements, and consistency across the project life cycle.

Why should decision criteria be reviewed and adjusted continuously? The reason is that unlike small to medium-size projects, which take a relatively short period of time to conceptualize and deliver, large, complex projects span many years and are subject to change (Jaafari and Schub, 1990; Morris and Hough, 1987). Thus, a chief function of the project management team (PMT) is to continuously monitor the project's underlying business appeal from an owner's and operator's overall perspective and implement changes to the target LCOFs as well as the project business case in response to shifts in the business, social, political, and regulatory environments.

It is ironic that few large projects are managed in such a dynamic and systematic fashion. There is still a belief that during the implementation phase, project managers should focus on the management of the delivery phase as generally characterized by the management of cost, time, scope, and quality. So it is not uncommon to read reports in the press on cost and time overruns experienced on public projects. Very little discussion centers on whether or not the project's business case has been enhanced because of positive changes introduced, notwithstanding cost and time overrun. The emphasis on value creation extends to contractors who can also search for a better outcome from their perspective purely for self-interest, such as adoption of an accelerated completion strategy that may increase the direct costs somewhat but lead to a significant reduction in the total indirect costs (Jaafari, 1996a, 1996b). Indirect costs are generally a function of project duration (Jaafari, 1996a, 1996b). In addition to self-interest, contractors will need to compete increasingly on the basis of their capabilities to deliver a strategic advantage to their client organizations through value creation opportunities that often come from the reengineering of projects early in their implementation phase. In this way, they can enter into partnership deals with major clients and share any potential gains.

The competency of the PMT is of paramount importance to the project implementation success. While assessment, acquisition, and enhancement of the team's competencies, and the delineation of competency gaps, are outside the scope of this chapter, it is important to note that key team competencies fundamentally determine the fate of the project. Note that the emphasis should be on *team* competencies not just individual competencies. The array of competencies required normally includes technical and commercial acumen, people skills, and project and organizational management, to mention a few. (See the chapters by Delisle and by Gale.) If the PMT lacks the necessary competencies to respond to the project challenges optimally, then it must set about to remedy its deficiencies in an appropriate and timely manner, such as acquisition of new staff, intensive training, hiring of competent consultants, and so on. One way to acquire this is to bring the range of expertise needed on board through an alliance mode of project delivery (Scott, B., 2001; Black et al., 2000; Halman and Braks, 1999).

Life Cycle Project Management

The life cycle project management (LCPM) approach for the entire implementation phase of large, complex projects is recommended. Traditionally, projects are packaged into multiple contracts, and each package is given out to a contractor or consultant to deliver. Each contractor sees his or her role as that of delivering the scope of the contract with minimal concerns about the impacts his or her work may have on the rest of the project. A contractor's main focus typically centers on achieving the target profit margin while capping or eliminating the corresponding risks and liabilities. Integration of works delivered by a multitude of contractors and consultants is a major challenge to the PMT/owner and is often prone to serious errors and omissions, delays, and cost overruns. Under the traditional mode of project delivery, the energy and intellect of the owner or his or her PMT are typically absorbed on the management of contracts and interfaces, and not necessarily spent on the attainment of the best overall project results (Halman and Braks, 1999).

LCPM addresses all of these shortcomings, as it shifts the focus of decision making and optimization from traditional scope, time and cost management in each contractual package to reaching or exceeding targets set for life cycle objective functions for the whole project. It also provides a firm basis for a more efficient management and integration of the entire implementation process. Life cycle project management is based on the following components:

- A culture of collaboration based on strategic partnership and unity of purpose (partnership for achieving or exceeding target LCOFs and sharing the resultant rewards)
- A life cycle philosophy and framework and an integrated single-phase approach for the entire implementation activities, covering initiation, planning and documentation, execution, commissioning, and finalization
- A concurrent teamwork approach, facilitated by a real-time communication system to cut the project delivery timescale
- A fully integrated project organization structure, run by a *project board* constituted from executives of the relevant project participating organisations
- An integrated project management information system that facilitates real-time evaluation of LCOFs as the basis for decision making

Fundamental to the success of the life cycle project management approach is the selection of the right partners (consultants and contractors) who can augment the owner and his or her PMT in terms of the missing competencies. Normally the project board has the ultimate decision-making oversight (strategic role) over the entire project implementation activities, including negotiation with the client body and relevant competent authorities. It appoints its own PMT with delegated authority to run the project on a day-to-day basis.

It must be noted that in recent times there has been a prevalent shift to partnering and alliance mode of project delivery. However, most alliance deals are still based on delivery objectives of time and cost. Also, integrated product development teams are generally missing in these arrangements. A full life cycle project management approach requires true open collaboration and working in terms of a true integrated product development teams.

Implementation of the life cycle project management approach also requires a project management information system (PMIS) that has real-time capability and can facilitate concurrent teamwork that underpins LCPM methodology. Such a system will assist the PMT to stage, run, and effectively integrate the contributions by the relevant consultants and designers organized in specific integrated (product development) teams. The PMT has a central and pivotal role in the evolution of the project, accumulation of parts, and integration of these into a whole viable project outcome. In this respect, the PMIS does not remove this responsibility from the project management team but facilitates this process by real-time analysis of the status of the LCOFs versus submissions of the suggested solutions received from teams. However, the PMIS can automate or expedite a number of tasks—for instance, sharing of input data, estimation of costs and LCOFs values, and production of numerous reports. It can also reduce redundant data entries, replace multiple pieces of software, and economize on human resources.

In summary, the ideal LCPM approach is based on the following key strategies:

- An integrated organization structure of the key partners on the project (owner, project manager, operator, contractors, designers, and major suppliers) whose executives make up the project board, responsible for major and strategic decisions and approvals, and appointing an integrated project management team headed by a competent project manager for the day-to-day management of the project.
- A continuous and integrated approach to the management of the implementation phase of the project as a whole (working back from the project business case).
- Simultaneous inclusion of the relevant information for decision making, including engineering, design, approval, manufacturing, construction, commissioning, operation, recycling, and so on (Jaafari, 1997; Laufer et al., 1996).
- Formation of integrated product development teams, each having a representative from the pertinent parties to the project, including, where appropriate, the appointed architect, engineers, manufacturers, constructor, operator. and facility manager. In some cases, representatives of governments and statutory authorities may also be considered.
- Division of the work into defined parts (product) and allocation of each to a single integrated product development team.
- Proactive management of the project and its parts, specifically, planning, staging, and managing all project activities continuously to maximize attainment of the target LCOFs.
- Integration of the work of the teams into a single project.
- Establishment of direct and real-time intra- and inter-team communication and document integration systems to facilitate the whole process.

As can be seen from Figure 13.11, the project board and the PMT need to drive the whole implementation phase in an integrated and systematic manner. They have the highest influence on the outcome of the implementation phase and, as such, must set up to discharge their responsibility objectively and optimally. If the LCPM approach is applied prudently and provided project participants possess appropriate competencies and do not behave opportunistically during the course of project implementation, it can be expected that optimal results will be achieved through the LCPM approach. The most important gains will be the enhanced project value and its fitness for purpose (in terms of meeting the underlying business objectives), as the entire project organisation will work in harmony to achieve breakthrough solutions that can meet or exceed targets set for LCOFs. This is because many non-essential (duplicate) activities, typical in multiple phase delivery, are eliminated, information and decisions are integrated and optimized against LCOFs, and commercial objectives of the partners are aligned with those of project objectives. This state of tightened

FIGURE 13.11. BROAD REPRESENTATION OF PROJECT IMPLEMENTATION.

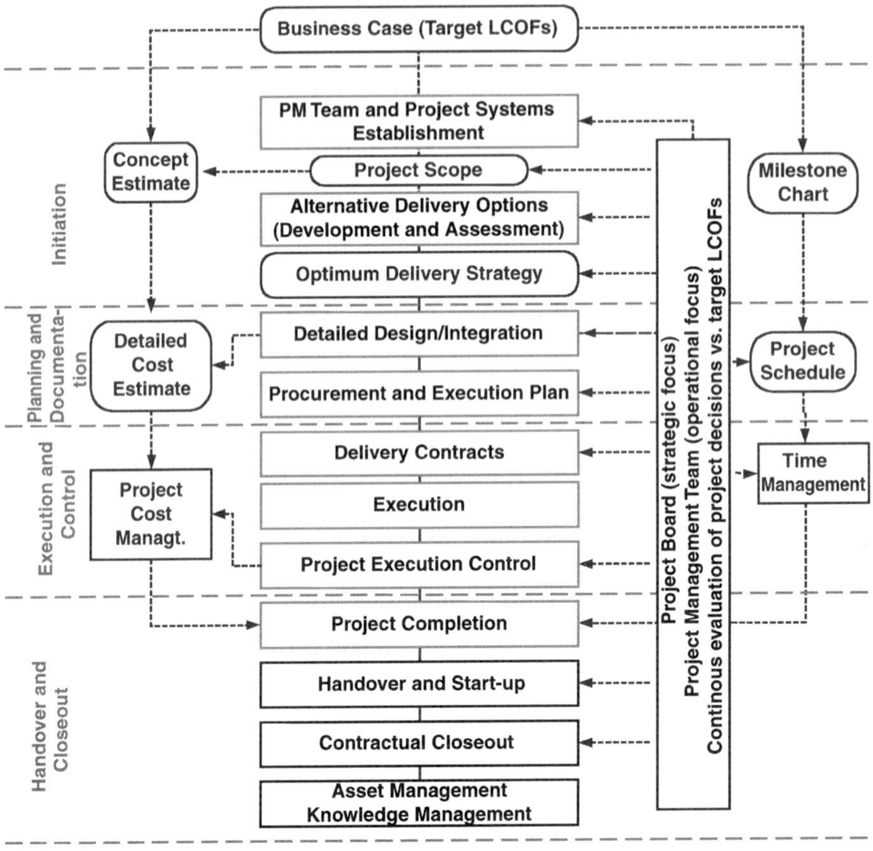

collaboration is often referred to as *cocreation* because of its emphasis on value creation and waste elimination.

Once the owner selects the relevant contractors and consultants, the owner should take them on board as project partners and tie their fortunes on the project to the attainment of target LCOFs through an appropriate gain-share/pain-share scheme. Note that each party comes to the project with a different mind-set and from a different corporate culture; so it is necessary to forge the parties into a unified project organization and develop the project culture as the dominant culture. The parties are then formed into multidiscipline teams with tough targets to achieve in terms of LCOFs (known as *stretched targets*). An example is asking a team to deliver a solution for a power station cooling system that not only meets relevant environmental and permit requirements but is also more efficient in terms of total life cycle cost per unit load.

It should be noted that despite the welcome trend from hard-dollar contracting to relationship-based contracting in the construction industry, there is still some distance to go to set up true alliances based on life cycle objectives and true integrated product development teams, rewarded on the basis of the life performance of the project. Most alliances are based on cost and schedule performance of the project and following a functional project organization structure. For example, on Wandoo B-Platform Oil Project, completed in 1997 in the Western Australia's Northwest Shelf, an alliance scheme based on total project cost was devised and applied, as seen in Figure 13.12. However, in order to encourage solutions that were cost-effective in terms of life performance of the project, this scheme contained a

FIGURE 13.12. THE WANDOO ALLIANCE PAIN-SHARE/GAIN-SHARE SCHEME.

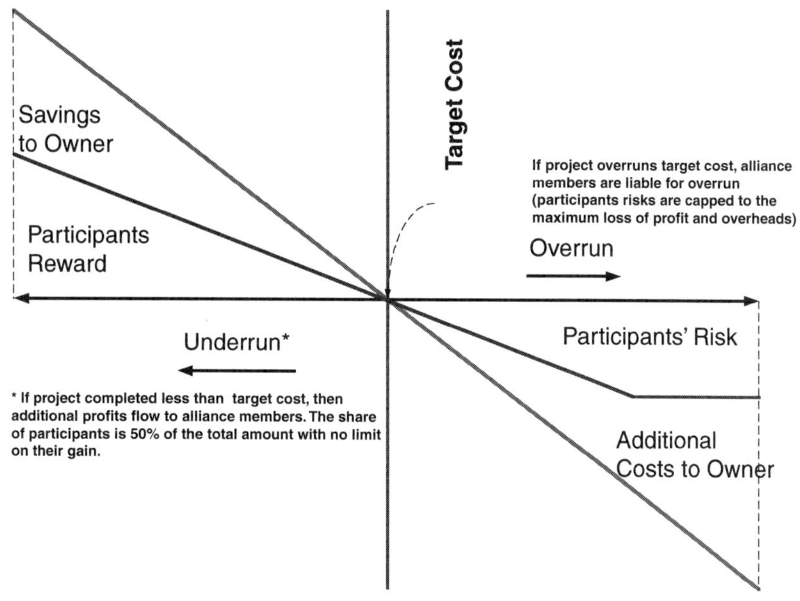

bonus for the alliance parties payable during the operation phase, provided that the financial performance of the project exceeded certain thresholds.

Figure 13.13 shows typical alliance organizational arrangements used on most large projects, which were also applied to the Wandoo project. As seen, this is far from a true integrated product development approach needed to foster creativity and development of break-through solutions in terms of true life performance of the project. For more information on alliance and relationship contracting, see Scott, B., 2001; Walker et al., 2000; Black et al. 2000; and Halman and Braks,1999.

Integrated product development teams can then go through a creative process to come up with optimal solutions that can be evaluated holistically at project level, using LCOFs as the basis. To achieve breakthrough solutions, one needs to consider all stages—in other words, not only the operation phase but also the implementation phase. Together, these comprise initiation, planning and documentation, execution and control, and commissioning and handover. The recycling of the end facilities at the end of the project's useful life should also be considered.

Project Initiation

Traditionally, projects go through distinct phases; for example, the PMT typically conducts the initiation phase; goes out to procure the services of consultants for project design, planning, and documentation; and then goes out to tender to select and appoint contractors to deliver the project in the manner foreshadowed in the contract documents (Jaafari, 1997).

The LCPM model is somewhat different. Project initiation is conducted centrally by the project board and coordinated by the PMT. It is the project board that makes strategic decisions on the implementation phase as a whole, with the input coming from all partners. For planning, documentation, and execution (even pre-installation commissioning), the preference is to engage dedicated teams made up of the participating organizations (see Figure 13.14). Each will be required to come up with its own breakthrough solution in a creative manner to meet simultaneously the global criteria of LCOFs and the specific criteria set for the part under consideration. The project board together with the PMT have the responsibility to preside over the evolution of the project as a whole, including integration of all solutions forwarded by teams and oversight of project commissioning and finalization activities.

Decisions made by the project board and the PMT at the initiation phase have a profound impact on the subsequent success or failure of the whole project implementation process. The main purpose of project initiation is to determine an optimum strategy that will achieve the intent of the business case of the project. This stage also considers the policy and regulatory issues, risks, and due diligence. Successful project initiation will require the generation of multiple options for the realization of project business case, evaluation of these, and selection of an optimum project implementation strategy. Note that while its chief function is to locate or develop an optimum solution to realize the intent of the business case, project initiation activities may lead to further adjustment of the business case as part

FIGURE 13.13. TYPICAL ALLIANCE ORGANIZATION STRUCTURE.

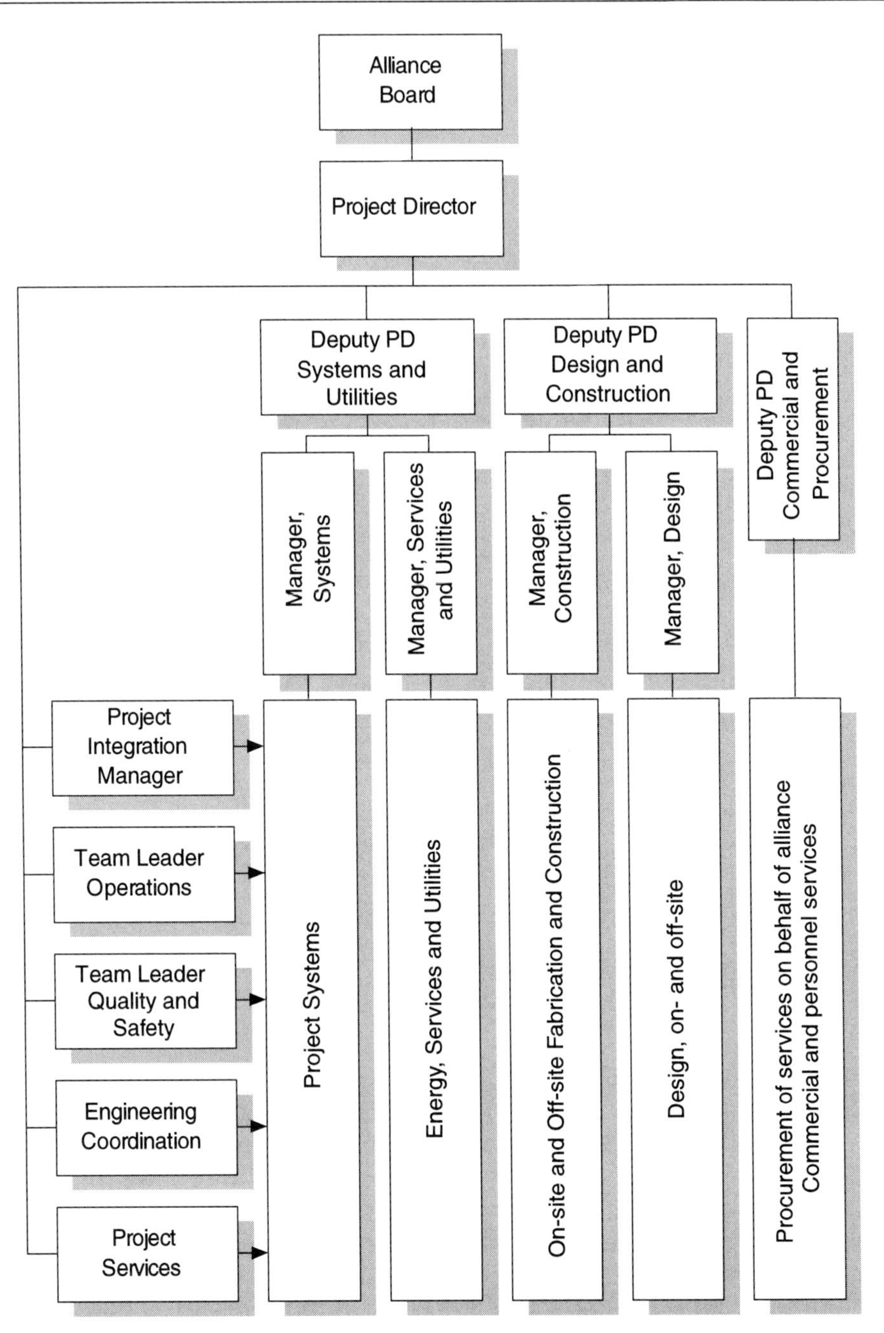

FIGURE 13.14. LCPM PROCESS FOR DESIGN AND DOCUMENTATION PHASE.

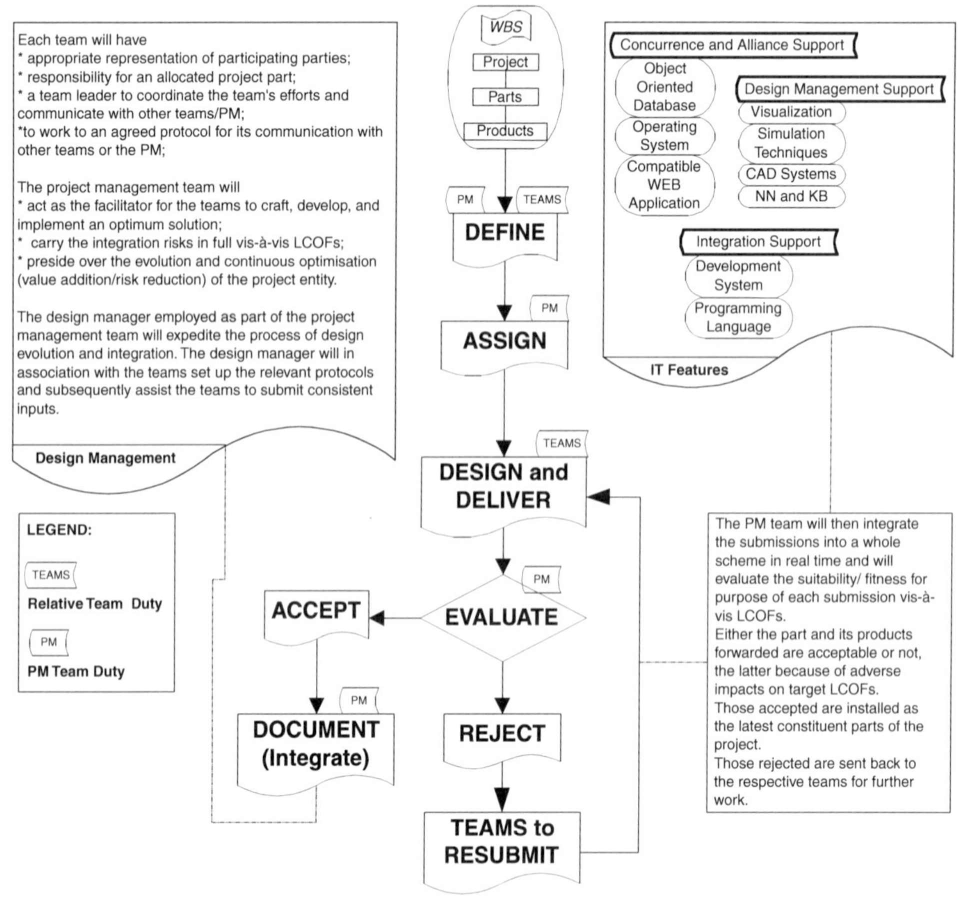

of a dynamic process and because of learning that comes with the first cycle of planning. This will lead to the identification of critical determinants and their influence on the project concept. However, the temptation to totally modify the project business case should be avoided unless changed circumstances demand a total revisit of the business case.

For example, if there has been a significant delay between the conclusion of project definition studies and the start of the implementation phase, it may be necessary to reconfirm the project business case first before proceeding to project initiation. In particular, target values set for LCOFs need careful review to ensure that these are still achievable and the business climate has not experienced significant shifts since the conclusion of project definition studies. Other situations that can give rise to a changed business case are as follows:

- Product changes or changed business priorities
- Changed financial model and/or funding model
- New commercial risks
- Changed legislation and permit conditions
- Constraints because of the intellectual property and technology licensing matters
- Shortage of resources because of other new developments under way or expected
- Stricter due diligence
- Uncovered new handover and operational issues
- New community and stakeholders' demands

Project initiation starts with the establishment of project office, setting up of the relevant project development and management systems, hiring, and formation of the project organization structure, project board, and project management team. The board and the PMT need to thoroughly review, confirm, and internalize the business case requirements and then through a creative process generate options for delivering these. A multicriteria approach may be necessary to evaluate the available options. However, the three main determinants are financial attractiveness (not necessarily least capital expenditure but highest total gain or lowest total life cycle cost over the entire project life cycle; Woodward, 1997), quality and/or fitness for purpose, performance, and likely impacts because of pertinent risks and uncertainties. The outcome is a preferred (optimum) implementation strategy that shows maximum potential and is least risky. As has been emphasised, risk and uncertainty management will drive the whole process.

Project initiation phase typically results in the following outcomes:

1. Adoption of a set of policies and standards that will meet the project needs and requirements cost-effectively (e.g., standards for design, quality, risk, and procurement management)
2. Specification of key performance indicators (from the target LCOFs), framework, and criteria for evaluation of project implementation decisions
3. Details of the preferred project delivery method, including division into parts, criteria or targets for delivery and acceptance, integration strategy, and associated managerial structures
4. Assignment of each part to a team with an appropriate set of protocols for teamwork
5. Design of a system for life cycle integration of all decisions, information, and functions on the project, including systems for project quality management, communication management, and procurement management
6. Methods/strategies for the evaluation and mitigation of risks and uncertainties at both global and local levels
7. Articulation of these in a project initiation report that can be a main source document for subsequent planning, documentation, and management of the project

Project Planning and Documentation

As stated, project planning and documentation is generally conducted by teams and facilitated by the PMT/project board in response to the stretched targets set during the project initiation stage in a concurrent manner. Figure 13.14 shows the process used under the LCPM approach, viz:

1. All teams will convene separately; each team will generate and develop its own alternative solutions, carry out some preliminary investigation, and come up with a preferred solution.
2. These preferred solutions are assembled by the relevant design discipline and analyzed to ensure the system's integrity (at discipline level) and conformity with the relevant statutory requirements.
3. The results from discipline investigation/detailing are fed back to teams to integrate the same into their solutions in the form of products and submit their products to the PMT, including the relevant information on life cycle aspects.
4. The project manager evaluates the LCOFs, installing those products that meet all the LCOFs and returning the remainder to the relevant team for further consideration.
5. The preceding steps are repeated in a dynamic process until a baseline design is evolved. The baseline design is the basis for the documentation of the whole project and procurement of the products/parts and monitoring of the execution process.
6. During the life of a project, the preceding process is maintained and managed continuously, as the responsibility for the eventual realization of the LCOFs will remain collectively with the project board, respective teams, and the PMT right up to the time of facility operation and beyond (when operation and maintenance are also part of the scope).

Project Execution and Control

The LCPM model applies the principles of integrated teamwork to the execution phase and on to commissioning and handover. The following general strategies expedite the application of LCPM methodology to the execution and control phase:

1. Apply the decision-making and risk management methods that the project board/PMT has decided upon for the execution phase of the project; this means that the intention of the execution phase is to apply the plans and strategies that the partners have collectively developed in the preceding phases in a united manner with the reporting and responsibility allocation unchanged throughout the implementation phase.
2. Develop a proactive management philosophy. The execution and control framework must be designed to measure the effectiveness of plans and decisions throughout the project life cycle by evaluating their impacts on the LCOFs. As an illustration, the potential benefit associated with shortening of the execution timescale must be evaluated through the impact it has on the target LCOFs.
3. Use a continuous objective-focused approach for major parts and the project as a whole. Such an approach embodies maximum flexibility in terms of innovation in the underlying concept, timing, resources, and other factors.

4. Focus should be on problem anticipation and resolution. Tap opportunities and monitor the status of risks continuously. In the event of a risk materializing, put in place recovery plans and minimize the impacts on the project outcome (LCOFs).
5. Employ an integrated information management system to assess progress and to provide feedback on the performance of the execution phase for each part and/or the project as a whole using appropriate key indicators but generally using the target LCOFs for project performance and project control functions, such as time and cost and/or earned value for project progress monitoring and reporting.
6. Use a dynamic scheduling and real-time reporting system so as to facilitate the management of anticipated changes. Maintaining maximum flexibility and attempting to add value to the project base value as a whole throughout its life cycle may mean many significant changes during the execution phase. However, most changes in this phase are due to imperfect design information, changed product specification, and so on.
7. Source and apply knowledge relevant to the optimal execution of the parts and the project as a whole.

Project Handover and Closeout

In LCPM the handover and closeout phase is generally planned both strategically and in detail during the project initiation and planning phases. The activities associated with project handover and contractual closeout should be planned as an essential part of an integrated approach to the entire project implementation phase. This means determining early in the initiation phase what strategies need to be put in place for project handover and contractual closeout. In the planning stage, relevant activities to apply the preceding strategies are planned, and these are applied in parallel to the execution phase. The major emphasis in this phase is the project end result or facility performance verification, OH&S and environmental compliance, and due diligence.

Typical activities in this phase include operational strategies, operator training, handover, start-up, commissioning and testing, defects identification and rectification, performance demonstration and validation, operational manuals, parts catalogue, as-built drawings, maintenance of project documents and records including materials and manufacturing records, contractual closeout (i.e., between the project alliance board and the client and between the alliance and the parties forming the same), financial settlement, asset management, and knowledge feedback.

Under the LCPM approach, project handover and closeout can be conceived as comprising two distinct phases: hand over of parts and hand over of the entire project. As noted, each major part will have been entrusted to a team who will deliver the same to the project board and the PMT and through them to the client. Where relevant, each part can be pre-commissioned and tested, and after meeting all the required performance hurdles, it can be certified as meeting the relevant performance criteria. The project board and the PMT will need to develop a clearly articulated and systematic acceptance scheme that spells out how parts are to be tested and accepted. There is no doubt that eventual facility performance criteria must underline the approach. Pretesting and acceptance of parts saves valuable time and minimizes the incidence of errors and last-minute discovery of major defects; it is also easier to implement. Once relevant parts are delivered, the entire facility can be commissioned and tested for performance verification and certification. (See the chapter by Mooz.) Normally, a team of commissioning experts will take over this task, which may take anything from one to three months.

Configuration and asset management practices need to be addressed from the outset and pursued to completion in this phase. (See the chapter by Kidd and Burgess.) The project solution will have embodied significant technical and managerial innovations; these must be captured systematically for future reference. Knowledge thus created should be recognized as a valuable asset in its own right and managed accordingly.

Summary

Creation, structuring, optimization, and implementation of large, complex projects require a systematic approach within a whole of life perspective. Large projects touch the lives of communities, require considerable investment and execution resources, are subject to regulatory and political pressures, and generally involve multiple stakeholders, with different needs and aspirations.

The author has focused on developing an integrated approach to whole of life planning, evaluation, and implementation of these projects. The result is presented as life cycle project management (LCPM) philosophy and framework in contrast with many contemporary project management approaches.

References

Adams, J. D., and A. W. Brown. 2002. Does project management add value to public sector construction projects: A critical perspective. *Proceedings of the International Conference on Project Management—ProMAC 2002.* 117–124. Singapore, July 31–August 2.

Artto, K. A., J. M. Lehtonen, and J. Saranen, 2001. Managing projects front-end: incorporating a strategic early view to project management with simulation. *International Journal of Project Management* 19:255–264.

Beder, S. 1999. Controversy and closure: Sydney's beaches in crisis. *Social Studies of Science* 21 (May 1991): 223–256.

Black, C., A. Akintoye, and E. Fitzgerald. 2000. An analysis of success factors and benefits in construction. *International Journal of Project Management* 18:423–434.

Brook, P. 2000. Project management and systems engineering: An evolving partnership. *15th World Congress on Project Management Organised by the Association for Project Management and the International Project Management Association.* London, May 22–25.

Byers, M. P. and F. L. Williams. 2000. Transforming electric utility project management for the new millennium business environment. *15th World Congress on Project Management organised by the Association for Project Management and the International Project Management Association.* London, May 22–25.

Chaaya, M., and A. Jaafari. 2001. Cognisance of visual design management in life cycle project management. *Journal of Management in Engineering* 127(1):66–75.

Defence Evaluation and Research Agency. 1997. *DERA systems engineering practices model.* Farnborough, UK: Defence Evaluation and Research Agency, GU14 6TD.

Dixon, M., ed. 2000. *APM Body of Knowledge.* 4th ed. High Wycombe, UK: Association for Project Management (www.apm.org.uk).

Gray, K. G., A. Jaafari, and R. J. Wheen, eds. 1985. *Macroprojects: Strategy, planning, implementation.* p. 146. Sydney: The Warren Centre for Advanced Engineering, The University of Sydney.

Halman, J. I. M., and B. F. M. Braks, 1999. Project alliancing in the offshore industry. *International Journal of Project Management* 17 (2, April): 71–76.

Hobbs, B., and R. Miller. 1998. The international research programme on the management of engineering and construction projects. Vol. 1. 302–310. *14th World Congress on Project Management,* Ljubljana, Slovenia, June 10–13.

Jaafari, A., and H. K. Doloi. 2002. A simulation model for life cycle project management. *Journal of Computer-Aided Civil and Infrastructure Engineering* 17:162–174.

———. 2001. Management of risks, uncertainties and opportunities on projects: Time for a fundamental shift. *International Journal of Project Management* 19:89–101.

———. 2000. Life cycle project management: A new paradigm for development and implementation of capital projects. *Project Management Journal* 31(1):44–53.

———. 1998. Perspectives on risks specific to large projects. Keynote paper p. 20 presented to National Infrastructure Strategy 98. August 20–21, 1998, The Institution of Engineers, Australia.

———. 1997. Concurrent construction and life cycle project management. *Journal of Construction Engineering and Management* 123 (4, December): 427–436.

———. 1996a. Twinning time and cost in incentive-based contracts. *Journal of Management in Engineering* 12 (4, July/August): 62–72.

———. 1996b. Time and priority allocation scheduling technique for projects. *International Journal of Project Management* 14 (5, October): 289–299.

Jaafari, A., and K. K. Manivong. 2000. Synthesis of a model for life cycle project management. *Journal of Computer-Aided Civil and Infrastructure Engineering* 15(6):26–38.

Jaafari, A., and K. K. Manivong. 1998. Towards smart project management information systems. *International Journal of Project Management* 16:249–265.

Jaafari, A., and A. Schub. 1990. Surviving failures: The lessons from a field study. *Journal of Construction Engineering and Management* 116 (1, March): 68–86.

Jolivent, F., and C. Navarre. 1996. Large-scale projects, self organising and meta-rules: Towards new forms of management. *International Journal of Project Management* 14(5):265–271.

Kelley. 1982.

Laufer, A., G. R. Denker, and A. J. Shenhar. 1996. Simultaneous management: The key to excellence in capital projects. *International Journal of Project Management* 14 (4, August): 189–199.

Manivong, K. K., A. Jaafari, and D. Gunaratnam. 2002. Games people play with proactive management of soft issues on capital projects. *Proceedings of the International Conference on Project Management, ProMAC,* 141–146. Singapore, July 31–August 2.

McCarthy, S. C. 1991. BOT and OMT contracts for infrastructure projects in developing countries. PhD diss., University of Birmingham.

Merna, T. 2002. Value management. Chap. 2 in *Engineering Project Management, ed.* N. J. Smith. 2nd ed. 16–28. Oxford, UK: Blackwell Science.

Merna, A., and N. J. Smith. 1993. Guide to the preparation and evaluation of build-own-operate-transfer (BOOT) project tenders. Manchester, UK: University of Manchester Institute of Science and Technology.

Morris, P. W. G. 1983. Managing project interfaces: Key points for project success. In *Project management handbook.* D. I. Cleland and W. R. King. New York: Van Nostrand Reinhold.

Morris, P. W. G, and G. H. Hough. 1987. *The anatomy of major projects.* New York: Wiley.

PMCC. 2001. *A guidebook of project and program management for enterprise innovation.* www.enaa.or.jp/PMCC/.

Scott, B. 2001. *Partnering in Europe.* London: Thomas Telford/European Construction Institute.

Scott, W. R. 1992. *Organizations: Rational, natural, and open systems.* 3rd ed. Upper Saddle River, NJ: Prentice Hall.

Stupples, D. W. 2000. Using system dynamics modelling to understand and address the systemic issues on complex engineering projects. *15th World Congress on Project Management Organised by the Association for Project Management and the International Project Management Association,* London, May 22–25.

Walker, D. H. T., and K. D. Hampson, eds. 2002. *Procurement strategies: A relationship based approach.* Oxford, UK: Blackwell Science.

Walker, D. H. T., K. D. Hampson, and R. Peters, R. 2000. *Relationship-based procurement strategies for 21st century.* Canberra, Australia: AusInfo.

Wang, S. Q., and L. K. Tiong. 2000. Case study of government initiatives for PRC's BOT power plant project. *International Journal of Project Management* 18 (1, February): 69–78.

Woodward, D. G. 1997. Life cycle costing:- Theory, information acquisition and application. *International Journal of Project Management* 15 (6, December): 335–344.

Yeo, K. T. 1995. Planning and learning in major infrastructure development: systems perspectives. *International Journal of Project Management.* 13(5):287–293.

Yeo, K. T., and R. L. K. Tiong. 2000. Positive management of differences for risk reduction in BOT projects. *International Journal of Project Management* 18:257–265.

CHAPTER FOURTEEN

MANAGING PROJECT STAKEHOLDERS

Graham M. Winch

*"*Taurus *meant an awful lot of different things to different people, it was the absolute lack of clarity as to its definition at the front that I think was its Achilles' heel."*

PETER RAWLINS, FORMER CHIEF EXECUTIVE OF THE LONDON STOCK EXCHANGE, INTERVIEWED IN THE MARCH 7, 1995, EDITION OF THE *FINANCIAL TIMES*, ON THE DEMISE OF THE MASSIVELY OVERRUN AND UNSUCCESSFUL PROJECT TO COMPUTERIZE SHARE DEALING ON THE EXCHANGE.

The challenges for the project management team are growing more complex. This point is illustrated in many different ways throughout this book and is the fundamental insight of the "management of projects" perspective. The aim of this chapter is to address one of the more important elements in that complexity: the increasing diversity and power of *project stakeholders*. The chapter starts by briefly identifying some of the sources of this growing complexity, before formally defining the concept of project stakeholder. It then goes on to propose a framework for mapping the stakeholders on the project as a prerequisite for analyzing their ability to influence the definition of the project mission and to disrupt its execution. The framework presented here was used to analyze publicly sponsored construction projects in Winch and Bonke (2002). Here the analysis of power is developed further and the framework used to analyze the case of a private-sector-sponsored IT project for back-office settlement on the London Stock Exchange (LSE). It will be argued that one of the major differences between the failed project (TAURUS) and the successful project (CREST) was the effectiveness of stakeholder management. Some implications of the analysis are then drawn out for the effective management of project stakeholders by the project management team (PMT).

The Growing Complexity of Stakeholder Management

Projects have always had stakeholders, but they have usually been either the funders of the project as client or suppliers of the project as members of the project coalition. Inherently, these stakeholders have had an interest in the effective delivery of the project with the minimum capital investment for the functionality required by the business case, and the project management team could focus on this objective. However, long-run changes in the social, political, and economic environment of projects have meant that this is no longer necessarily the case, for a number of reasons:

- Since 1945, most projects were financed from the general revenue streams of the client organization—whether streams derived from profits on turnover or the raising of taxes. Increasingly, for both private and public sector clients, projects are financed by loans or equity raised by a special project vehicle (SPV) with the returns on that investment generated directly by the revenue stream from the asset created by the project (see the chapters by Ive and Turner). This immediately introduces financiers as a new class of project stakeholder, as well as creating a wholly new type of project actor in the SPV itself.
- Traditionally, the client—that is, the party with which contracts are made by the principal supply-side members of the project coalition—and the project sponsor were the same entity. This is no longer necessarily the case. In urban regeneration projects, for example, the sponsors may be local political elites, who then choose a public body to be the client, as in Manchester's sports-led schemes to host the Olympic Games and (successfully) the Commonwealth Games (Cochrane et al., 2002), or on the Central Artery/Tunnel (CA/T; Hughes, 1998) in Boston, Massachusetts.
- Regulators are growing ever more insistent on the project definition taking into account wider social objectives than the effective exploitation of the asset being created by the project. At least one authoritative study has concluded that the interventions of regulators are a principal source of budget overruns on projects (Merrow et al., 1981). The most obvious example here is environmental protection, institutionalized through environmental impact assessments, but operational safety, local purchasing and labor requirements, and land-use policy issues can also figure large in regulators' concerns.
- Direct action by environmentalists or local loser groups can also be highly disruptive for the project during the execution phase, and the only way to address this issue aside from confrontation is to address the concerns of such groups—to the extent that they can be considered as legitimate—during project definition.

These points indicate the diversity of parties that can be considered to be project stakeholders. Cleland (1998, p. 55) has defined "project stakeholder" thus:

> Stakeholders are people or groups that have, or believe they have, legitimate claims against the substantive aspects of the project. A stake is an interest or share or claim in a project; it can range from informal interest in the undertaking, at one extreme, to a legal claim of ownership at the other extreme.

There are two important aspects of this definition worthy of note. First, stakeholders only have to *believe* that they have a claim on the project to cause problems—that claim might be perceived as illegitimate by the client and PMT. As the sociologist William I. Thomas put it, "if men define situations as real, they are real in their consequences" (cited Coser,

1978, p. 315). Second, those claims are usually met by adjusting the project mission in some way, unless the claimant is too weak to press their claim. Even direct-action opponents whose claims are not considered legitimate by most parties can force changes in project definition by enforcing additional expenditure on site security and the like. As Fred Salvucci, a member of the Boston political elite sponsoring the CA/T, put it, these mitigations of stakeholder claims on the project are the modern equivalent of "delivering some chunk of mastodon meat back to the tribe" (Hughes, 1998, p. 221).

Project stakeholders are a diverse group—some are formally members of the project coalition, others not. The first group is usually defined as the primary or *internal* stakeholder group (Calvert, 1995; Cleland, 1998). They are defined by having a contractual relationship with the client or a subcontract from another internal stakeholder. They usually enter willingly into the project coalition, and are, by definition, positive about the project even if they negotiate toughly for their share of the value added by the project. Their claims are usually enforceable directly as breach of contract. The second group is usually defined as the secondary or *external* stakeholder group. They may have little choice about whether the project goes ahead and may be either positive or negative about the project. They rarely have a directly enforceable claim on the project and are therefore reliant upon regulators to act on their behalf, the mobilization of political influence either covertly or through public campaigns, or, occasionally, direct action.

Internal stakeholders can be broken down to those clustered around the client on the demand side and those on the supply side. External stakeholders can be broken down into private and public actors. This categorization, with some examples, is shown in Table 14.1.

On the demand side, a complex array of interests is indicated, and there can be no assumption that they are all aligned. One of the largest differences of interest within the demand side is often between the client and its employees. Particularly where the project is associated with reengineering business processes, employees may face significant changes in their work, or even lose their jobs, as a result of the project. Even where this is not the case, failure to adequately capture the needs of users as they perceive them can cause difficulties or even failure during the commissioning phase of the project. Similarly, the interests of financiers, clients, and sponsors may be divergent. Where project sponsors or

TABLE 14.1. SOME PROJECT STAKEHOLDERS.

Internal Stakeholders		*External Stakeholders*	
Demand Side	**Supply Side**	**Private**	**Public**
Client	Consulting engineers	Local residents	Regulatory agencies
Sponsor	Principal contractors	Local landowners	Local government
Financiers	Trade contractors	Environmentalists	National government
Client's employees	Materials suppliers	Conservationists	
Client's customers	Employees of the above	Archaeologists	
Client's tenants			
Client's suppliers			

Source: Adapted from Winch (2002) Figure 4.1. Used with permission of the Project Management Institute.

clients do not finance the project from their own resources, there is a great temptation to underestimate the costs and overestimate the benefits of the project (Flyvbjerg et al., 2003). For example, the finance of the Boston CA/T using the "10-cent dollar," where the U.S. federal government matches every ten cents put in by local taxpayers with 90 cents from federal taxpayers (Hughes, 1998), created a major misalignment of incentives. Much the same would appear to have happened on the West Coast Main Line project in the United Kingdom, which has suffered program and budget overruns similar to Boston CA/T. Where the project sponsors are on the supply side—as on the Channel Fixed Link project—this can cause great suspicion among financiers regarding the integrity of the decision making of the client (Winch, 1996).

On the supply side, a whole coalition of interests is arrayed. The supply side satisfies its claim on the project through the income stream generated by working on the project and the learning acquired through solving project problems (Winch, 2002). It is immediately clear that there is an inherent conflict of interest between the stakeholders on the demand side and those on the supply side as they compete to appropriate the income stream from the project—what Porter (1985) calls margin—in the project value system. Managing this conflict is the central task of project governance (Winch, 2001).

Among the external stakeholders, there is even more diversity. By and large, the internal stakeholders will be in support of the project, although there may be factions within the client that are backing alternative investments. External stakeholders may be in favor, against, or indifferent. In the private sector, those in favor may be local landowners who expect a rise in the value of their holding and local residents supporting a rise in the general level of amenity. Those against may also be local residents and landowners who fear a fall in amenity and hence the value of holdings. Such objectors are known as NIMBYs (not in my back yard) and can delay infrastructure projects such as airports for decades, if not stop them all together. Environmentalists and conservationists may take a more principled view than local losers, while archaeologists are concerned about the loss of important historical artifacts.

The public external stakeholders—in those situations where the public sector is not also the client—will tend to be indifferent. Their interest arises from the general level of economic activity, rather than from any particular project, and their claim is met through the taxes generated by this economic activity. Regulatory agencies that enforce regulatory arrangements such as those for urban planning, quality of specification, and heritage assets will tend to be indifferent to any particular project definition, so long as it complies with the codes. National and local government may, however, wish to encourage development, particularly in regeneration areas. At times, there may be conflicts of interest within the public sector between its role as a project sponsor and its responsibilities as a regulator, as happened with the Sheffield Arena project (Winch, 2002).

Managing Stakeholders

The successful management of stakeholders by the project management team requires the following processes (cf. Cleland, 1998):

- Identify those stakeholders with a claim on the project
- Specify the nature of each stakeholder's claim
- Assess each stakeholder's ability to press that claim
- Manage the response to that claim so that the overall impact on the definition and execution of the project are minimized.

Stakeholder mapping is proposed as a valuable aid to completing these management processes successfully. The first step in managing the stakeholders is to map their interest in the project (Winch and Bonke, 2002). This can be done using the framework illustrated in Figure 14.1. The focus of the approach is the project mission as represented by the asset to be created. Stakeholders can be considered as having a problem or issue with the project mission and as having a solution (tacit or explicit) that will resolve that problem. Where such solution proposals are inconsistent with the client's proposals, they can be defined as being opponents to the project. An important part of stakeholder management is to find ways of changing opponents to proponents by offering appropriate changes to the project mission and preventing possible proponents defecting to the opponent camp by offering to accommodate more explicitly their proposed problem solutions.

Once the stakeholder map has been drawn up, the power/interest matrix can be used to develop a strategy toward managing the different stakeholders (Johnson and Scholes, 2002). It consists of two dimensions: the power of the stakeholder to influence the definition of the project and the level of interest that the stakeholder has in that definition. The level

FIGURE 14.1. MAPPING STAKEHOLDERS.

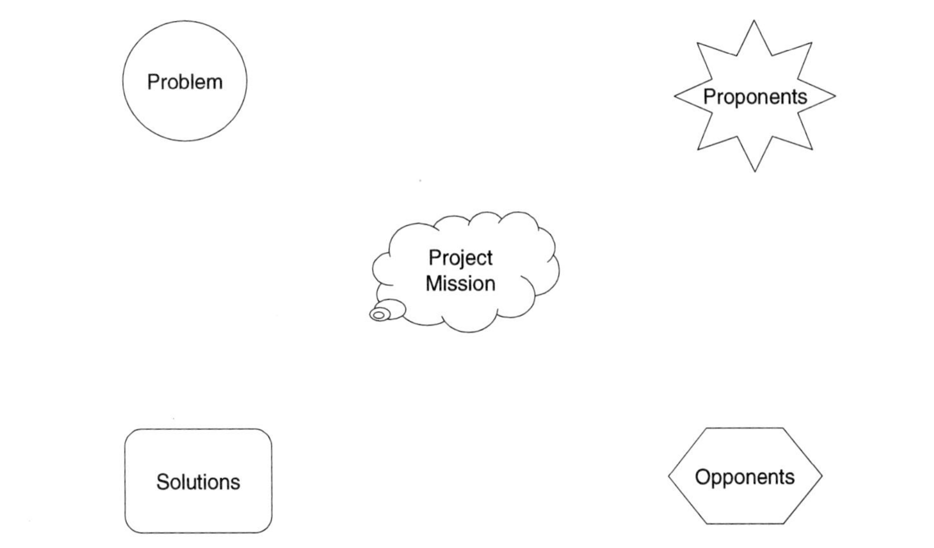

Source: Bonke (1996).

of interest is conceptually simple—it is a function of the expected benefit or loss from the project. Power is a more slippery concept and is discussed more fully in the text that follows.

This matrix categorizes the stakeholders into one of four types, but the discussion here can only be indicative—where a particular stakeholder sits in relation to the project depends entirely on the specific context of that project. The first group is those who require *minimal effort,* such as the client's customers, or local and national government. A public relations approach to this group will often suffice, aimed at ensuring that those who might be opposed to the project stay in the low-interest category while those who are likely supporters are tempted to move to the high-interest category.

The second group are those who need to be *kept informed.* Groups who may be opposed to the project, such as local residents, conservationists, or environmentalists, need to be carefully managed. If such groups coalesce into well-organized movements and are able to mobilize the press behind them, they may well be able to move into the key player category, causing the project severe disruption, or even cancellation. To a certain extent, such groups can be bought off to prevent this happening, with inevitable consequences for the business plan. For instance, it is now standard practice for clients building in the City of London—which contains important Roman and mediaeval remains—to finance an archaeological dig prior to works commencing on site. Similarly, the concept of planning gain within the UK regulatory system is common, where a project promoter provides additional utility for the benefit of the local community to defuse potential opposition. Some groups—typically environmentalists—cannot, however, be bought off and can go on to disrupt physically the project during execution on-site. The impact of these groups is twofold:

- They upset the calculations underlying the business plan because of delays in the schedule and additional costs of security;
- They dissuade future clients from coming forward with similar projects.

Those who need to be *kept satisfied* usually fall into two main groups—regulatory bodies and the supply-side stakeholders—which require very different management approaches. Regulatory bodies are, in essence, the institutionalized interests of the low-power stakeholders. They provide forums in which local residents, landowners, and government can have their voice (planning enquiries), the safety of the client's or operator's employees, tenants, and customers is ensured (safety codes), and environmental and conservationist interests are heard (environmental impact assessments). The latter even allows the purported claims of stakeholders that do not yet exist—future generations—to be heard. The first task of the PMT is to ensure compliance with the regulatory requirements, supported by lobbying tactics where the requirements are open to interpretation. Supply-side stakeholders are placed in this category, rather than the key players category, for two reasons, First, most of them are mobilized after the project is defined; second, they will typically have a portfolio of projects at any one time—while their power to influence the outcome of any one project is very high, their interest in the definition of any one project is typically limited.

The final category is that of the *key player.* Here the client and sponsor are central; the analytic questions revolve around which of the other demand-side stakeholders are also in this category. Where finance is raised from the traditional sources—equity and debt secured through a floating charge for the private sector, and the taxpayers for the public sector—financiers are typically in the keep-satisfied category. However, where project finance techniques are used as the source of capital—as is increasingly common in both the public and private sectors—such financiers move into the key player category. In commercial property development where the asset is pre-let, the client's tenant can become a key player in definition, while in the provision of social housing this is rarely the case. The client's customers are also usually in the key player category, but through the proxy voice of the corporate marketing department; if they misunderstand the market for the facility, it is unlikely to be successful because those customers will simply use other competing facilities. Whether the client's employees are in the low- or high-power categories depends upon the internal organization of the client and its understanding of its business processes.

Understanding the Power of Project Stakeholders

Awareness of the importance of power relationships in project management has been growing (Pinto, 1996; and see the chapter by Magenau and Pinto later). A definitive analysis of power as a relationship is provided by Lukes (1974). He identified three facets of power, as illustrated in Figure 14.2.

- *Facet 1, overt power.* The ability of A to persuade B to choose the option A prefers
- *Facet 2, agenda-setting power.* The ability of A to set the agenda so that B's preferred option is "off the agenda"

FIGURE 14.2. THE THREE FACETS OF POWER.

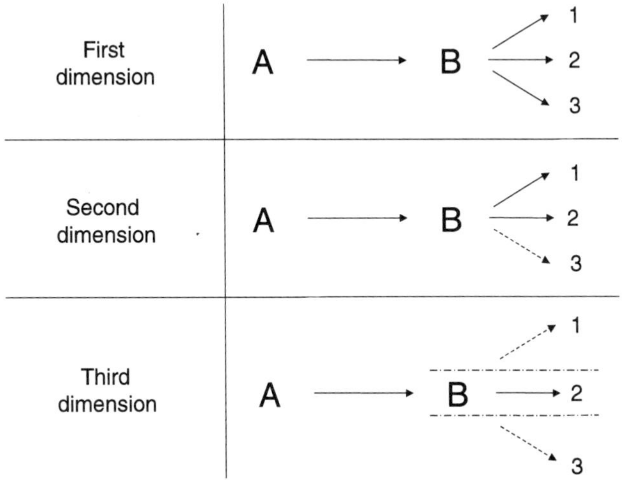

Source: Winch (2002). Used with permission of the Project Management Institute.

- *Facet 3, hegemonic power.* The ability of A to define the issues in such a way that B sees no alternative but to make choices favorable to A

There are five main sources of power in organizations (Handy, 1994), which can be illustrated with examples from the overt facet of power:

- *Physical* power. This is rarely used and of little relevance to project stakeholder management.
- *Positional power.* This is derived from the powerful actor's ability to deploy power derived from position in an organizational hierarchy or legal authority. For instance, the power of regulators usually derives from their backing in national legislation to protect the interests of relatively powerless external stakeholders. Similarly, the power of clients to have the last word on key decisions on the project is enshrined in the contracts used between the parties.
- *Resource power.* Those that are providing the resources for the project—especially capital—typically have considerable say in how those resources are used. Indeed, this can often amount to a veto on some options.
- *Expert power.* This is the power wielded by specialist advisors hired to provide expertise not otherwise available; their power is strongly linked to their reputation for competence.
- *Personal power.* This is the ability of charismatic individuals to sway opinions and win arguments simply by force of personality. This type of power should not be underestimated.

A telling example of the agenda-setting facet of power comes from the appraisal of privately financed public projects in the United Kingdom. The business case for all such projects must be subject to a public sector comparison to see whether private or public finance is the more cost-effective. However, should the comparison show that public finance is more cost-effective, it does not follow that public finance will be available for the project. Project sponsors are, therefore, tempted to manipulate the comparison to show that private finance is more cost-effective in order to obtain the support of financiers. An example of the hegemonic facet of power is the sponsorship of IT projects to tackle the so-called Y2K or "millennium bug" problem. This was the fear that many mission-critical IT systems would cease to function properly after 1999 because of shortcuts taken in the coding of many software programs. With the benefit of hindsight, the problem can now be seen to have been greatly exaggerated, and countries such as Italy that spent relatively little tackling the problem did not suffer significant adverse consequences (Finkelstein, 2000). The sponsors of Y2K projects successfully promoted them as TINA (there is no alternative) projects, and thereby distorted corporate capital budgeting.

The Case of Settlement on the London Stock Exchange

In March 1993, around £80m of investment in a computerized trading settlement system was canceled by the Board of the London Stock Exchange (LSE), followed by the resignation of the chief executive responsible. Many different analyses of this project failure are possible, for it is rich in lessons for project management. Here the focus is on the way in which poor stakeholder management led to project scope escalation to a level of unmanageable complexity, and hence project failure. The sources for this case are Drummond (1996) and Head (2001). Neither author bears any responsibility for the interpretation of their research presented here.

In October 1986, the so-called Big Bang had swept away the cozy, clubbable world of the LSE, constituted as a mutual organization for the benefit of its members. In essence, this was a government deregulation initiative designed to remove price fixing in trading commissions so as to ensure that the City of London retained its premier position in global trading markets. In preparation, the LSE had implemented nine separate IT projects for front-office trading, which were successfully launched on Big Bang day with a synchronized switch-on. IT was also seen as central both to minimizing back-office settlement costs and to speeding up the settlement process. The stock market crash of October 1987 had revealed the fragility of the existing paper-based system, and a systemic risk was identified of default during the lengthy settlement process. In November 1989, the Council of the LSE appointed a new chief executive, committed to reforming the LSE and ensuring that the "dinosaurs" retired to their clubs. However, like all stock exchanges, the LSE was simply a forum for its members:

- The "jobbers," who traded on behalf of brokers using an "open cry" trading floor and who were replaced by "market-makers" using computer screens
- The brokers, who dealt with investors and were rapidly being taken over by the major banks

The TAURUS (Transfer and Automated Registration of Uncertified Stock) system was aimed at computerizing the settlement system. This is the post-trade system by which stocks are exchanged for cash between seller and buyer. Settlement is far from the front-office buzz of trading, but an indispensable element in the process. Full computerization of settlement implies *dematerialization*, or the replacement of paper stock certificates, which can be placed in a bank vault or under the mattress, with asserted "rights" of ownership contained in an electronic register. This was a cultural change of immense significance, particularly to the private equity investors who accounted for around 75 percent of trades but only around 25 percent of value traded. In 1979, "Talisman" had been implemented and was making considerable progress in computerizing settlement. Instead of certificates being physically carried around The City from broker to broker, Talisman acted as an electronic clearing-house, and perhaps most importantly, centralized the clearing process through the LSE itself. It thereby came to provide 50 percent of LSE income. TAURUS represented the next stage of back-office computerization.

The agenda for TAURUS was set by the self-appointed Group of 30 (G30) senior City figures who had become increasingly concerned with back-office capabilities and were strong advocates of TAURUS. "The G30, with its exclusive membership of top-level policymakers and business leaders, is the most eminent of financial think-tanks" (January 24, 2003 edition of the *Financial Times*). The TAURUS project was launched in May 1986 with a schedule of 36 months and a budget of £6m (all budget figures are in then current prices). The crash of 1987 prompted the program to be speeded up, and a powerful committee—SISCOT (Securities Industry Steering Committee on TAURUS)—was established to oversee development. SISCOT decided to abandon TAURUS 1 in 1989. The main problem was the proposal to create a single central register of share ownership. This would have made the registrars, which had traditionally held such registers, redundant. While historically small firms, many of them had been recently been taken over by the major banks, which were exercising their muscle in the post-Big Bang environment and did not wish to see their investment wasted. The registrars thereby moved from a low-power to a high-power position.

However, the perceived urgency remained, and Coopers and Lybrand were appointed as project managers, led by a highly experienced and respected IT project manager who demanded complete autonomy for the PMT in August 1989. At the same time, major changes in the overall management of the LSE were taking place as it evolved from a mutual society to a corporate entity. The new, reforming chief executive took up his post in November 1989.

Although he tried to kill the project at this point because he could not see the rationale for it, he was advised that all was in order and that the newly appointed PMT was in full control. He therefore addressed himself to what he perceived as more fundamental problems with the LSE than back-office systems. In March 1990, the prospectus for TAURUS 8 was announced—a 19-month schedule costing around £47.5m. The decision to adapt an existing computer package, rather than develop a bespoke one, was then taken, and a contract was signed with Vista Corporation of the United States for their package, implemented on IBM mainframe hardware. Adaptations to the package were to be made on a cost-plus basis.

Learning from the experience of TAURUS 1 suggested sensitivity to stakeholder interests. The system was, therefore, designed from the outside in, so as to enable consultation with key stakeholders. This attempt to engage with stakeholders spawned over 30 committees around The City, and while "everybody was giving input to Taurus" and "everybody was being promised the earth" (all citations are from Peter Rawlins, LSE chief executive, 1989 to 93), it was not at all clear who was actually responsible for it on the client side. Formally it was overseen by a TAURUS board that met once a month for under two hours, while the external monitors that were intended to act on behalf of the LSE Council as on the Talisman project were sacked as part of a cost-cutting exercise. Building the system on the basis of the Vista package started in earnest in early 1990 with delivery due in October 1991. To move with the perceived urgency, information was sent to stakeholders so that they could start building their own systems, but as this information was received, stakeholders started to demand changes in the specification to suit their own interests, which then had knock-on implications for other stakeholders. SISCOT was the forum in which these debates were held and became known to those involved in the project as the "Mad Hatters' Tea Party."

As 1990 wore on, it became increasingly clear that dematerialization would require primary legislation. The regulatory body concerned—the UK Department of Trade and Industry (DTI)—produced a bill amounting to over 100 pages of legislation. It also required that the LSE indemnify investors for any errors in attribution of ownership caused by the system. It also refused to allow the LSE to make membership of the TAURUS system compulsory, which had been a major assumption of system design. The DTI, therefore, played the classic regulatory role of champion for low-power/high-interest stakeholders such as private investors.

The first public announcements of problems came in January 2001 as the delivery date was delayed, the blame for which was put on the problems with the regulator. However, the diversion of the project director, who now had to go round the country to sell participation in TAURUS now that it was voluntary, was causing problems of internal project management, despite the long hours and dedication of the project team. By the autumn of that year, budget sanctions had been increased and delivery was promised for April 1993. By mid-1992, 50 percent of the Vista package code had been rewritten, and the IBM servers for the state-of-the-art security interfaces had failed to work. The LSE refused to fund the secondment of more Coopers and Lybrand personnel to the PMT as the project slipped into crisis. Andersen Consulting had been commissioned in early 1992 to provide a general review of the LSE's trading and settlement systems. Formally, TAURUS was outside their brief, but in the October they were asked to include it in their scope. It was on the basis of their report that the chief executive took his decision to cancel the project in March 1993 and then resigned. Three-hundred and fifty other people also lost their jobs, and stakeholders lost an estimated £400m of associated investment.

The failure stimulated those stakeholders with a primary concern for the viability of The City to action. The Bank of England—the UK's central bank—set up a tight project team to develop what became known as CREST, to be run by a not-for-profit organization called CRESTCo. The central mission of the CREST project was the same as that of the TAURUS project—dematerialized back-office settlement. However, the mission was more restrictive in one crucial way: Membership of CREST was to be voluntary. It deliberately used outsiders to The City—career civil servants—to engineer the settlement processes that *had* to happen, rather than would be nice to have. When stakeholders lobbied against the proposed systems design, they were told to "take it or leave it." The team did not use a formal systems methodology but stuck rigorously to a design-then-build approach, supported by strong configuration management. Even here, not all stakeholders could be kept at bay. The UK Treasury had offered to give up £3bn of stamp duty on the purchase of stocks in order to encourage dematerialization with TAURUS, but reneged on this halfway through the CREST project. Nevertheless, the system was delivered on schedule and budget in July 1996. Two redesigns were required to meet system limitations identified during commissioning, thanks to higher-than-expected use of the system.

Apparently, the TAURUS PMT project did many things in accordance with project management best practice. Taking the ten critical success factors (Pinto and Slevin, 1998):

1. The original definition of the mission was clear—a dematerialized settlement system with a central register to enhance the international competitiveness of the LSE.
2. The project was publicly supported by everybody from the G30 down, and those who privately expressed reservations such as the incoming LSE chief executive were warned against interference.

3. Project scheduling and planning techniques were used.
4. Extensive user consultation was undertaken.
5. The LSE appointed a strong project manager and team with an exemplary track record in managing similar projects, and the team worked very hard to achieve project success.
6. There was no shortage of technical competence, and the PMT opted to use an existing package rather than designing a bespoke system from scratch.
7. Users were kept continually informed of progress through SISCOT.
8. Monitoring and feedback arrangements did exist, but the project manager had demanded complete autonomy so as to reduce the possibility of interference by the client.
9. Extensive communication with the user community was undertaken, and great efforts were made to accommodate their views.
10. Troubleshooting arrangements appeared satisfactory.

However, the implementation of these PM disciplines was vitiated by the inability of the client to manage the stakeholders in the project. Figure 14.3 shows the stakeholder map for the TAURUS and CREST projects, with the mission of achieving a dematerialized back-off settlement system at its heart. The essence of the problem was that the PMT was prepared to listen to too many stakeholders, each with its own agenda. This was in the context of TAURUS being defined as a TINA project by the G30, with vociferous backing from the financial press. Those who had doubts were reluctant to express them in this febrile context.

The TAURUS power/interest matrix shown in Figure 14.4 illustrates that the key players were a disparate group, and, as the case shows, had different agendas. In other words, the Council of the London Stock Exchange was not powerful enough in relation to its stakeholders (Drummond, 1996, p. 82).

The importance of this power relationship is shown by the success of the CREST project, as shown in Figure 14.5. Here the team was backed by a supremely powerful actor—the Bank of England—as opposed to a relatively powerless actor, the Council of the London Stock Exchange. The weakness of the Council was a larger problem for the LSE as it strove to meet the challenges of the post Big Bang era, and many of the broader organizational changes implemented by the CEO during the life of the project were intended to strengthen the executive against the members of the exchange. As a result, the strength of the LSE itself relative to its stakeholders had also grown because of reform efforts of the ousted chief executive, and the Council had been replaced by a Board by the time the CREST project was implemented.

Approaches to Managing Stakeholders

Drawing on the lessons of the preceding analysis, together with those cases presented in Winch and Bonke (2002), the following suggestions can be made for the improvement of stakeholder management.

The Alignment of Incentives. The most effective way of managing internal stakeholders is to make sure that their incentives are aligned—the problem of project governance (Winch

FIGURE 14.3. TAURUS/CREST STAKEHOLDER MAP.

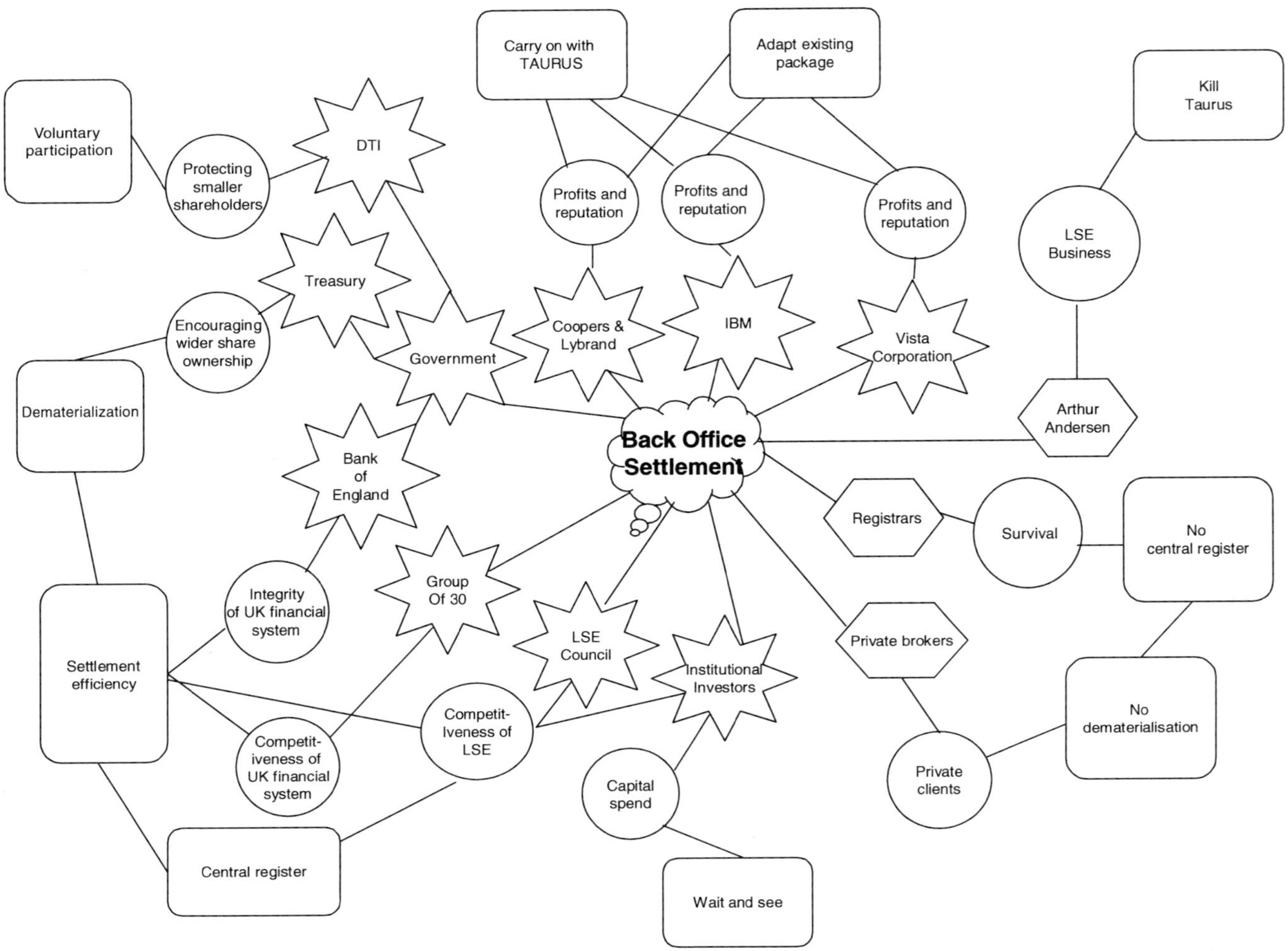

FIGURE 14.4. TAURUS POWER INTEREST MATRIX.

Level of Interest

Power to Influence	Low	High
Low	***A: Minimal effort***	***B: Keep informed*** Supplier staff
High	***C: Keep satisfied*** UK Government Group of 30 IBM Vista Corporation Private brokers Registrars	***D: Key players*** LSE Members LSE Council Coopers & Lybrand Institutional Investors

FIGURE 14.5. CREST POWER INTEREST MATRIX.

Level of Interest

Power to Influence	Low	High
Low	***A: Minimal effort*** supplier staff	***B: Keep informed*** Registrars Private brokers Institutional investors
High	***C: Keep satisfied*** Department of Trade & Industry HM Treasury Group of 30	***D: Key players*** LSE Board Bank of England

2001; see also Miller and Lessard, 2000). On the demand side, this can be done by making sure that all key players have stakes in the project that move in step as project budget and schedule vary. A good model of this is where financiers, sponsors, clients, and facility operators all have an equity stake in an SPV—a good example of this approach is the successful Second Severn Crossing in the UK (Campagnac, 1996). Between the demand side and the supply side, this can be done through using incentive contracts as opposed to lump-sum or fee-based contracts..

Early Development of a Mitigation Strategy. Stakeholder mapping by the PMT should identify the claims that are likely to be made by external stakeholders and their power to press them. This should form the basis for a consistent strategy regarding which claims can be accepted without undermining the business plan and which are too costly. If likely claims by key players cannot be met without threatening the integrity of the business case, then not proceeding with the project is the wise option. The TAURUS project clearly lacked such a strategy and was overwhelmed by claims, while the CREST project adopted a take-it-or-leave-it strategy. The failure to identify the airlines as key players on the Denver International Airport project (Applegate, 2001; Montealegre et al., 1996) was, arguably, a major factor in the major schedule and budget overruns experienced.

Friends in the Right Places. This is often the responsibility of the project sponsors—for instance, the ability of members of the Boston political elites sponsoring the Boston CA/T to lobby effectively in Washington for funds was critical to its successful financing (Hughes, 1998). The project director of the Apollo mission was appointed for his contacts in Washington, not his technical expertise (Sayles and Chandler, 1993). Project management teams can be twin-headed, with one project director responsible for managing external stakeholders and another responsible for internal delivery, as on the Tate Modern project in London and Boston CA/T (Winch, 2002). One of the problems on TAURUS was that the project director was obliged to shift his attention to external stakeholder management, leaving a vacuum with internal delivery.

An Ethical Approach. Corporate social responsibility can be defined as the extent to which "an organization exceeds its minimum required obligations to stakeholders" (Johnson and Scholes, 2002, p. 216). One of the strongest arguments for an ethical approach to business is the damage to the brand that can occur when corporations fail to act responsibly, so much of the debate about business ethics has concerned the consumer goods sector. However, the issues are, arguably, equally relevant to the capital goods sector. In this spirit, perhaps *project social responsibility* can be defined as

> the extent to which the definition of the project mission exceeds the minima established in the NPV calculation and those required to obtain regulatory consents (Winch, 2002, p.78).

Explicit adoption of such an approach may well help to reduce the level of claims from external stakeholders who can see that their claims are being fairly and dispassionately considered. An ethical approach should avoid the bribe-and-ignore approach to consent management

Effective Consent Management. The management of consent (Stringer, 1995) within the regulatory framework is a strategic matter within project definition—for projects such as the fifth terminal at London's Heathrow airport, the management of consent amounts to a major project is its own right. Three basic approaches can be identified (Winch, 2002):

- Define-and-enquire
- Consult-and-refine
- Bribe-and-ignore

Where the regulations are unambiguous and prescriptive, then the define-and-enquire approach is appropriate—the codes are published and simply require to be interpreted. The consult-and-refine approach is more appropriate where codes are not prescriptive, or where there are significant uncertainties in their interpretation. The bribe-and-ignore strategy is, unfortunately, widespread—zoning codes are routinely ignored in many countries as recent tragedies where shantytowns have been engulfed by mudslides show. Recent earthquakes in a number of countries have indicated widespread ignorance of—or at least failure to implement—the codes relating to structural integrity. Its use is likely to undermine the other approaches to stakeholder management discussed here.

A Strong Client. A client that knows what it wants to achieve and how much it is prepared to invest to achieve it is more likely to be able to effectively manage stakeholders. A major difference between the TAURUS and the CREST projects was that the client was powerful in relationship to the project stakeholders on the latter and knew exactly what it was trying to achieve. A weakness of the Channel Fixed Link project was that it was "assembled round a hole like a Polo Mint . . . [there was] no client driving it forward with a vision of what the operator needed to have" (September 9, 1995 edition of the *Financial Times*). Where the client is selected by the sponsor on the grounds of political expediency rather than project management capability—as on Boston CA/T (Hughes, 1998)—then problems are likely to follow.

Getting the Concrete on the Table. The words of the project manager of the Storebælt Link in Denmark (cited Bonke, 1998, p. 10) are fundamentally wise: "we had to have the concrete on the table in a hurry." Stakeholders can change their minds about supporting the project—especially where political support is required—but the more capital that has been sunk in the project, the more likely it is to be pushed through to completion, even in the face of mounting doubts about the business case. This is the phenomenon of project escalation, and the tactic of making large spends early in the project life cycle on nonfungible assets can be used by project managers to keep their projects rolling. This was used on the Channel Fixed Link project to manage the risk of cancellation following a Labour win in the 1987 general election (Winch, 1996).

Public Relations. Keeping the external stakeholders informed using PR techniques can pay significant dividends, if only to avoid misleading rumors circulating. The press tend to focus on projects that go wrong, as the PMTs on the West Coast Main Line and Boston CA/T know well. Releasing good news in a systematic manner can help keep stakeholders on-side, both in press releases and on the project Web site. For example, the Web site for the C/AT (www.bigdig.com/) was exemplary, even if its stakeholder management overall was not.

Visualization. In earlier generations, the only way to visualize the asset being created was through artist's impressions or scale models. Digital techniques are making the visualization

TABLE 14.2. TAURUS AND CREST COMPARED.

Strategy	TAURUS	CREST
Client power	The LSE Council was weak in relation to the stakeholders.	The Bank of England was powerful in relation to the stakeholders. The LSE Council had also been replaced by the LSE Board.
Mitigation	A forum was set up—SISCOT—which encouraged claims to be made by stakeholders.	A take-it-or-leave-it approach made possible by voluntary membership of CREST.
Consent management	The project was hit by regulatory surprises, particularly from DTI, which undermined the business case.	The business case was not dependent on regulatory consent, and the need for such consent was reduced because of the voluntary nature of CREST membership.
Alignment of incentives	The principal contractors were awarded a cost-plus contract, and stakeholders were able to act as free riders.	Tightly managed process with strong in-house team. Free riding stakeholders (notably registrars and institutional investors) were moved to the low-power category.

task much easier. Regulators and other external stakeholders can see a digital image of the proposed facility in situ and can come to a more informed assessment of its real impact on their interests. User groups among the internal stakeholders can interact with virtual reality (VR) images of the proposed facility to provide their input to the design of a building, or rapid prototyping techniques can be used in the development of user interfaces for IT systems.

A number of strategies are, therefore, available to the PMT to help them in the management of project stakeholders. Table 14.2 summarizes the differences between the failed TAURUS project and successful CREST project in terms of their deployment of these suggested strategies.

Summary

"The only good stakeholder that I can recall is Professor Van Helsing in the Hammer horror films, who would drive a stake into the vampire's heart".

Alastair Ross-Goobey, an experienced fund manager and project financier cited in the *Financial Times* (February 21, 2003, edition), takes a rather jaundiced view of the claims of stakeholders compared to shareholders. However, at least so far the PMT is concerned, it is hoped that this chapter has convinced them that the Hammer Films approach to stakeholder management is less viable today than it ever was.

Projects are not very likely to keep all stakeholders happy. Projects are, fundamentally, vehicles of planned change in business and society (Winch, 2002), and changes nearly always benefit some stakeholders more than others. In many cases, this may simply means a number of employees losing their jobs; in a few, the perception may be that the irreparable destruction of the natural environment or heritage assets is at stake. In all cases, project budgets have to be won and defended against the claims of sponsors of different projects competing for the same resources. Capital budgeting techniques are designed to allow these decisions to be taken rationally, but the inherent uncertainty associated with outturn capital costs and future returns from the assets created by the project leave plenty room for opinion and argument.

The growing complexity of project stakeholder management makes the balancing act required by the PMT between potential gainers and losers from the project ever more difficult. Mitigating losers too much may undermine the business case, while mitigating inadequately may lead to the mobilization of opposition through political pressure or direct action and oblige the cancellation of the project. The skills of project managers in internal delivery are well honed and can offer significant returns on investment for clients. However, projects still overrun program and budget, and fail to deliver a properly working asset. There is growing consensus that the problems lie in the setting of schedule and budget objectives and specifying the facility (Merrow, et al., 1981; Morris, 1994)—in other words, in the definition of the project mission. This chapter has argued that stakeholder management plays a vital part in defining the project mission, and has offered one way of aiding such management processes.

References

Applegate, L. M. 2001. *BAE automated systems (A) and (B)*. Harvard Business School Teaching Note 5-399-099.

Bonke, S. 1996. Technology management on large construction projects. London, Le Groupe Bagnolet Working Paper 4.

———. 1998. The storebælt fixed link: The fixing of multiplicity. London, Le Groupe Bagnolet Working Paper 14.

Calvert, S. 1995. Managing stakeholders. In *The commercial project manager*, ed. J. R. Turner. London: McGraw-Hill.

Campagnac, E. 1996. La maîtrise du risque entre différences et coopération: Le cas du severn bridge. London, Le Groupe Bagnolet Working Paper 12.

Cleland, D. I. 1998. Stakeholder management. In *Project management handbook*, ed. J. Pinto. San Francisco: Jossey-Bass.

Cochrane, A., J. Peck, and A. Tickell. 2002. Olympic dreams: Visions of partnership. In *City of revolution: Restructuring Manchester,* ed. J. Peck and K. Ward. Manchester: Manchester University Press.

Coser, L. A. 1978. American trends. In: *A history of sociological analysis*, ed. T. Bottomore and R. Nisbet. London: Heinemann.

Drummond, H. 1996. *Escalation in decision-making: The tragedy of TAURUS.* Oxford: OUP.

Finkelstein, A. 2000. Y2K: A retrospective view. *Computing and Control Engineering Journal* 11:156–159.

Flyvbjerg, B., N. Bruzelius, and W. Rothengatter. 2003. *Megaprojects and risk: An anatomy of ambition.* Cambridge, UK: Cambridge University Press.

Handy, C. 1993. *Understanding organizations.* 4th ed. Harmondsworth, UK: Penguin.

Head, C. H. 2001. *TAURUS and CREST: Failure and success in technology project management.* Henley, UK: Henley Management College.

Hughes, T. P. 1998. *Rescuing Prometheus: Four monumental projects that changed the modern world.* New York: Vintage Books.

Johnson, G., and K. Scholes. 2002. *Exploring corporate strategy.* 6th ed. London: Prentice Hall.

Lukes, S. 1974. *Power: A radical view.* London: Macmillan.

Merrow, E. W., K. E. Phillips, and C. W. Myers. 1981. *Understanding cost growth and performance shortfalls in pioneer process plants.* Santa Monica, CA: RAND Corporation.

Miller, R., and D. R. Lessard. 2000. *The strategic management of large engineering projects: Shaping institutions, risks, and governance.* Cambridge, MA: MIT Press.

Montealegre, R., H. J. Nelson, C. I. Knoop, and L. M. Applegate. 1996. BAE automated systems (A): Denver international airport baggage-handling system. Harvard Business School Case 9-396-311.

Morris, P. W. G. 1994. *The management of projects.* London: Thomas Telford.

Pinto, J. 1996. *Power and politics in project management.* Newtown Square, PA: Project Management Institute.

Pinto, J., and D. P. Slevin. 1998. Critical success factors. In *Project management handbook*, ed. J. Pinto. San Franciso, Jossey-Bass.

Porter, M. E. 1985. *Competitive advantage.* New York: Free Press.

Sayles, L. R., and M. K. Chandler. 1993. *Managing large systems.* New Brunswick, NJ: Transaction Publishers.

Stringer, J. 1995. The planning enquiry process. In *The commercial project manager*, ed. J. R. Turner. London, McGraw-Hill.

Winch, G. M. 1996. The channel tunnel: Le projet du siècle. Le Groupe Bagnolet Working Paper 11.

———. 2001. Governing the project process: A conceptual framework. *Construction Management and Economics* 19:799–808.

———. 2002. *Managing construction projects: An information processing approach.* Oxford, UK: Blackwell Science.

Winch, G. M. and S. Bonke. 2002. Project stakeholder mapping: Analyzing the interests of project stakeholders. In *The frontiers of project management research*, ed. D. P. Slevin, D. I. Clelend, and J. K. Pinto. Newtown Square, PA: Project Management Institute.

CHAPTER FIFTEEN

THE FINANCING OF PROJECTS

Rodney Turner

This chapter describes the financing of projects. No project can take place without funding, and yet the raising of finance is something most project managers do not get involved in. However, the financing can have a huge influence on their project, and so project managers do need to contribute to the development of the financial strategy for their project, which in itself is a vital element of the overall project strategy.

The chapter is titled "The Financing of Projects," rather than "Project Finance." The vast majority of projects are paid for by the parent organization, out of revenue expenditure, as in the case of maintenance or research projects, or out of capital expenditure, as in the case of new investments. The capital is raised from the sources of finance described in the *Types and Sources of Finance* section, based on the reputation of the parent organization, and secured against its assets. The parent organization needs to repay the finance independent of the success or failure of the project. A few, usually large projects are financed as entities in their own right. This is called unsecured, nonrecourse, or off-balance-sheet financing. It is unsecured because it is only secured against the project's assets and its revenue stream. If the project fails, the lenders have no recourse against which to recover their money. It is off-balance-sheet, because if a parent organization invests in a project in this way, the capital invested does not appear in the company's balance sheet. Limited-recourse financing is a mixture of nonrecourse and recourse financing, where the parent organization invests some equity in the project, which will appear on its balance sheet and will be lost in the event of project failure, but the majority of finance will be loans secured only against the project's assets. Lenders prefer this because with some of their ownmoney invested in the project, the sponsor will have a greater interest in the success of the project. The term project finance is usually reserved for nonrecourse or limited-recourse financing. The other (usual) case is the financing of projects. This chapter considers both cases, and the sources of finance described in *Types and Sources of Finance* section can be used for financing the parent organization or an individual project. However, the chapter mainly focuses on the second case. In the majority of projects, the money will be made available by the parent organization. In this case, money may be borrowed to finance a specific project, but it will be guaranteed against other assets of the parent organization, and so will be available at lower interest rates and will appear on the company's balance sheet. The role of the project manager, champion, or sponsor is to justify the expenditure using investment appraisal techniques, (Aston and Turner, 1995; Lock, 2000; Akalu, 2001). It is only in the case of non- or limited-recourse financing that the project team will be involved in developing a financial package specifically for the project and approaching the financial markets to finance the project.

The financing of projects, including project finance, is a vast subject; whole books are devoted to the subject. Thus, in the limited space of a single chapter, we can only focus on a number of key issues. The next section begins by explaining characteristics of project finance. Then sources of finance are described. These sources are broadly the same for both the parent organization and individual projects. Types of finance are considered, and conventional and unconventional sources of those different types described. The next section describes project finance, the financing of individual projects on an unsecured, nonrecourse basis. Finally, the process of creating a financial package for a project to meet the investment needs while responding to the risk is considered.

Characteristics of the Financing of Projects

There are a number of features of the financing of projects that affect the design of a particular financial strategy. Some of these features relate to the financing of all projects; the latter ones refer more specifically to project finance. The features include the following:

- Finance is the largest single cost on a project.
- There is no project without finance.
- Financial planning begins at feasibility.
- The projects involved are often complex.
- Financial planning adds to project complexity.

Largest Single Cost

Finance is often the largest single cost on a project. On a large construction project, the material costs may be 30 percent of the total capital cost, construction costs 30 percent, and design, project management, commissioning, working capital, and contingency about 10 percent each. On the other hand, on a project that takes two years to build, on the day it is commissioned, the total cost of the financial package (interest paid to providers of debt and returns to equity holders) may amount to 20 percent, and 60 percent by the time the debt has been paid off. Finance is therefore twice as much as the next largest cost. On infrastructure costs taking longer to build, they will be even larger. Thus, the financial costs may be almost as great as the project costs, and so good financial management is at least as important as good project management.

This may not be obvious to many project managers, especially on projects financed by the parent organization, because the cost of project finance is not included in the estimate. It is allowed for in the investment appraisal process through the discount factor applied (see Aston and Turner, 1995; Lock, 2000; Akalu, 2001). Unfortunately, this may create distortions. Decisions will be made to minimize the capital cost, not minimize the financial cost or total cost. Usually, lower capital cost leads to lower financial cost. But sequencing the cash flow can make a difference. A lower capital cost solution with the expenditure front-loaded in the project may have higher finance charges than a higher capital cost solution with expenditure later in the project. Whole-life costing techniques are being developed to address this problem. For nonrecourse financing, the financiers and financial agents will be more concerned with optimizing the cash flows to minimize financial costs.

No Project without Finance

Consider three projects: the first one has the design package complete, but only 90 percent of the finance has been sourced; on the second both the design package and the financial package are 95 percent complete; and on the third, the design package is 90 percent finished, but the financial package is totally in place. Only the third project can receive the go-ahead to begin. No project can begin without the financial package in place. Yet many project managers will focus on design completion and almost ignore the need for the financial package. On projects financed by the parent organization, preparing the design package may be part of the investment appraisal process, and so there is no finance until the design package is complete. However, for nonrecourse financing, it may be necessary to obtain finance before design can start, and the financiers may want to influence the design solution adopted, preferring less risky solutions.

Financial Planning in the Feasibility Study

For this reason, financial planning for a project should begin at the same time as the technical solution or earlier. Regrettably, it is often begun at the last minute, when most other features are already determined. The most successful projects are ones in which the financial planning is a key part of the project strategy from the start and the aim of the project is to minimize the whole-life cost, including the financing charges and technical solutions adopted considering their impact on the financial solution. The lenders, whether the parent organization or external financiers, may have a view about which options should be selected. They may have a view about which options minimize the risk and therefore provide the safest haven for their money. They may not want the project's promoter to choose what appears to be the best value for the promoter or champion, but instead the one that provides the best value for them (the lenders), taking account of the risks. It is therefore useful to engage the financiers as early as possible, so options are not selected that have to be rejected.

Complexity

Problems with complexity will be particularly the case with projects using project finance, because they have several features that will make them especially complex:

- They will usually be very large.
- They may cross national boundaries.
- They often exceed the capacity of a single organization to plan, supply, and construct.
- They are technically complex, demanding skills not widely available.
- They are dedicated to a single purpose, a major project rather than a program.
- They are located at remote sites.
- They take place over long timescales, with the return on investment often taking decades.

Financial Planning Adds Complexity

Projects requiring project finance tend to be complex, but the finance planning adds to the complexity. Three issues that need to be emphasized, because they exert significant influence on the financial planning:

1. The sources of finance need to be identified before the technical specification of the equipment. The financial package often imposes a sense of compromise. But further, the lenders will impose constraint as to the level of risk they are willing to bear and may indeed look for higher rates of interest with higher-risk technical solutions.
2. The structuring and acceptability of the financial package is different from the perspective of lender and borrower. Each will take a different perspective of the risk, each giving a different emphasis to the risks and the compromises involved. The lenders may want higher returns for certain risks, and thus it is sensible to take account of their concerns in the project's feasibility and planning stages.
3. However, if you do involve the lenders early, be aware that the availability of finance and the terms on which it is available can be subject to significant and rapid change, for reasons beyond the control of the project sponsor or project manager. The impact of these changes needs to be tracked throughout the feasibility and planning of the project.

Types and Sources of Finance

This section considers types of finance and their sources. These types and sources may be used by the parent organization to finance itself, or they may be used in non- or limited-recourse financing to finance the individual project

Types of Finance

There are two main types of finance: equity and debt.

Equity. *Equity* is money subscribed by investors or shareholders. The shareholders get returns from dividends and capital growth in the value of their equity. With equity, there is no guarantee that a dividend will ever be paid, nor that the money itself will ever be repaid. A dividend can only be paid after the interest and scheduled repayments of loans have been paid. And the equity can only be drawn out of the company or venture after all the obligations to the providers of debt have been met. If the project performs badly, the providers of equity may receive nothing, but if it performs well, the returns may be huge. Because of the higher risks involved, the providers of equity expect higher returns on average than the providers of debt.

Mezzanine Debt. *Mezzanine*, or *subordinate, debt* is loans made by the holders of equity. It is sometimes treated as equity, especially in the calculation of debt/equity ratios and so is also sometimes called *quasi-equity*. It differs from equity in that it is repayable against a schedule of payments and the returns to investors are in the form of interest payments at a predetermined rate. However, those interest payments and scheduled repayments can only be made after all the obligations to the providers of senior debt have been met. Thus, mezzanine debt is higher risk that senior debt and so commands higher interest rates. Mezzanine debt can take the form of debentures, preferred stocks, and other instruments.

Senior Debt. *Senior debt* is money borrowed from a number of possible sources but particularly banks. It is repayable against agreed schedules of payment, including the predetermined interest payments. Senior debt must be repaid before all other forms of finance (hence, the name), and the providers of senior debt have the first claim on all assets of the venture if the borrower goes into liquidation. Debt can take the form of loans, bonds, and nonconvertible debentures. It can also take the form of equipment provided under supply contracts with repayment to be made over the life of the plant. There are two types of senior debt:

1. *Secured debt.* This is debt secured against assets or collateral easily convertible to cash. It must almost by definition be money lent to the parent organization, secured against its assets, and is repayable regardless of the performance of any projects for which it is intended. It commands a lower interest rate than unsecured debt—the interest rate only being dependent on the reputation of the parent organization and the security of its assets.
2. *Unsecured debt.* This is debt lent for a specific project, secured only against the assets of the project and its predicted revenue streams. It is unsecured, nonrecourse, off-balance-sheet project financing. Given that it is dependent on the project's success, the interest rate will be higher than for secured debt and will be linked to the riskiness of the project.

Eurotunnel, the construction of the channel tunnel between England and France, was an example of a limited-recourse financed project, involving a mixture of debt and equity. Equity was provided by the project promoters, mainly contractors involved in the construction of the tunnel, and by private investors. But the vast majority of the finance (in excess of 80 percent) was loans provided by banks. During construction of the tunnel, interest on the loans was added to the debt. But in the early stages, equity holders received some returns on their investment in that the share price steadily rose as the commissioning date, and hence the expected returns, became closer. In fact, the early revenue streams were not sufficient to cover the schedule of debt and interest payments to the banks. The shares now had no value, so the equity investors lost their investment, and the project effectively became nonrecourse finance. But there was no point the banks foreclosing on their loans, because the hole in the ground (the project's main asset) had no value if it was not being used. So the banks converted much of their loans to equity, and are now receiving their returns over the life of the project in the form of dividend payments.

Cost of Capital

Each form of finance has a cost associated with it: the cost of borrowing that type of capital. The *cost of capital* is the average cost of all forms of finance used by a project or company. The discount factor used when performing investment appraisal (Aston and Turner, 1995; Lock, 2000; Akalu, 2001) is the cost of capital inflated by any allowance for risk.

Cost of Equity. The *cost of equity* (returns gained by shareholders) is the dividends they receive plus capital growth of the equity. (During construction of the channel tunnel, the equity investors did not receive dividends, but there was growth in value of the equity, providing initial early returns.) The capital growth is not predetermined, and so in calculating the cost of equity, a guess has to be made as to its likely size. The most common model for calculating the cost of equity is now the Capital Asset Pricing Model (CAPM). According to this model, the cost of equity is

$$\text{Cost of equity} = \text{Risk free rate of return} + \text{Beta} \times \text{Equity risk premium}$$

where:

Risk free rate of return is the most secure rate of return anyone could obtain from investing their money, for example, by investing in government bonds.

Equity risk premium represents the excess return expected from investing in equity given that it is repayable after all other debt. It represents the average risk of the stock market. A value of 3 percent to 4 percent is often used, which in the past has underestimated returns.

Beta represents the risk premium of a particular stock, measured (obviously) as the ratio of that stocks riskiness compared to the average of the stock market. For a particular company, it will represent the riskiness of the industry that company operates in and its own features that make it more or less risky than the average for the industry. For a project it will represent the riskiness of the project.

Cost of Debt. The cost of equity (dividends and capital growth) is payable out of untaxed income. The cost of debt (interest) is payable out of taxed income. Thus, the *cost of debt* is the interest rate minus an allowance for the tax not paid:

$$\text{Cost of debt} = \text{Interest} \times (1 - \text{Tax rate})$$

Cost of Capital. The total *cost of capital* is the weighted average of the different types of capital. (If there are several forms of debt, this sum obviously must be done over all the forms of debt.)

$$\text{Cost of capital} = \text{Ratio of equity} \times \text{cost of equity} + \text{Ratio of debt} \times \text{Cost of debt}$$

Conventional Sources of Finance

Shareholders. These are public or private investors, institutions, or individuals who provider the equity or quasi-equity in a company. Sources of equity include the following:

- Retained profit of a company
- Funds raised through the stock market
- Venture capital companies
- Joint venture partners
- International investment institutions such as the World Bank

Banks. Banks and other financial institutions are the main providers of debt. Commercial banks are the most readily available to most project investors. They split into retail banks, which provide finance in the local main-street and merchant banks. There is a large choice available for companies raising finance, and this has led to intense competition. In choosing a bank, the decision will not be so much based on the interest rate charged as on the following factors:

- The size of the bank
- The experience in financing that type of project
- Any support they may offer with the financial engineering

International Investment Institutions. Another form of bank is the international investment institutions, including the World Bank and other development banks (see Table 15.1). They will provide both debt and equity.

International Financial Markets. International financial markets offer an alternative to domestic markets, giving easy access to foreign sources of funds. There are many, but the two most important are the Eurocurrency and Eurobond Markets. The Eurocurrency markets are the most efficient in the world and provide for smooth movement of funds. They provide short-term finance at competitive rates but are primarily for large organizations. The Eu-

TABLE 15.1. INTERNATIONAL INVESTMENT INSTITUTIONS.

African Development Bank, AfDB
Asian Development Bank, ADB
Commonwealth Development Corporation, CDC
European Development Fund, EDF
European Investment Bank, EIB
European Bank for Reconstruction and Development, EBRD
Inter-American Development Bank, IDB
International Bank for Reconstruction and Development, IBRD, or World Bank
International Development Association, IDA
International Finance Corporation, IFC

robond markets provide promissory notes or bonds issued outside the United States. The bonds can be issued in small denominations, making them attractive to the small investor. They provide anonymity, so interest rates are usually lower than on domestic markets, making them the most competitive in the world.

Suppliers

Suppliers have been placed under conventional forms of finance because it is becoming increasingly common for suppliers to be paid out of the revenue stream of a large project rather than up front on supply of their piece of equipment. They may be paid under buyer or supplier credit, or by becoming part of the joint venture investing in a project.

Export Credit, Buyer and Seller credit

Most developed countries have saturated home markets and so encourage companies to export. On the other hand, developing countries need access to advanced technologies. Many OECD (Organisation for Economic Co-operation and Development) countries have established an export financial agency national interest lender to facilitate trade with developing countries. This agency often also provides insurance to cover export risks. The finance it provides may also be linked to the country's aid budget. Examples include the following:

- *Canada.* The Export Development Corporation (EDC)
- *United Kingdom.* The Export Credit Guarantee Department (ECGD)
- *United States.* Tthe Export-Import Bank of the United States (EIBUS), the Private Export Funding Corporation (PEFCO), and the Overseas Private Investment Corporation (OPIC).

The United Kingdom's ECGD does not directly provide finance but provides a number of related services:

- Access to cheap finance, allowing companies to offer competitive terms and win contracts they might otherwise lose

- Insurance against political, legal, economic, and social risk
- Export credit insurance to insure a national supplier against nonpayment
- Support for a performance bond required by a national supplier to win a contract overseas

In providing or arranging finance for equipment for export, the export credit agency may lend money to the national supplier, known as *supplier credit* (see Figure 15.1), or to the overseas buyer, known as *buyer credit* (see Figure 15.2).

Unconventional Sources of Finance

There are several less conventional sources of finance.

Leasing. Rather than buying an asset, the project promoter leases it. The financier pays for the asset and receives their return in the form of rental for the asset. This makes the asset available to the project promoter through off-balance-sheet financing. Under the terms of the deregulation of the electricity industry in Ireland, this is now effectively the form of financing used for investment in the national grid. The national grid is operated by a privatized company, called EirGrid, which acts as promoter for new investment in the grid. However, investments are paid for and constructed by the generating company, the Electricity Supply Board, and EirGrid pays a lease for the assets.

Counter Trade. The seller accepts goods or services in lieu of cash. This can be expensive and cumbersome, especially as the seller has to sell the goods or services to receive their returns, and so is not well liked by project contractors and project financiers. However, it is quite common under buyer or seller arrangements (see preceding text) or project financing arrangement, as discussed in the next section, for all of the revenue from operation of the project's facility post commissioning to be paid into an escrow account under the control of the project's financiers. All debt and interest payments are the drawn from this account, before any surplus is made available to the project's promoter.

FIGURE 15.1. SUPPLIER CREDIT.

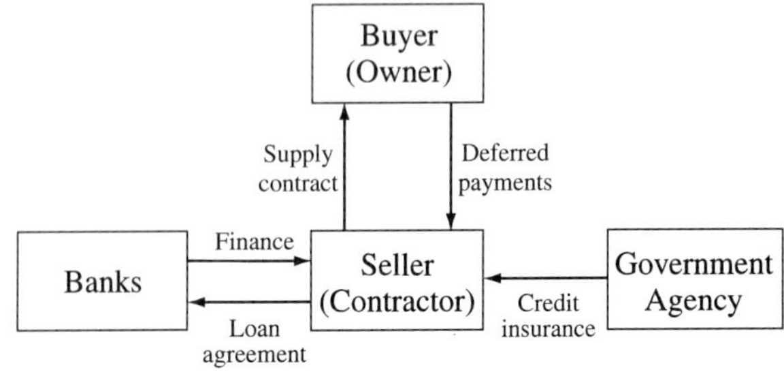

FIGURE 15.2. BUYER CREDIT.

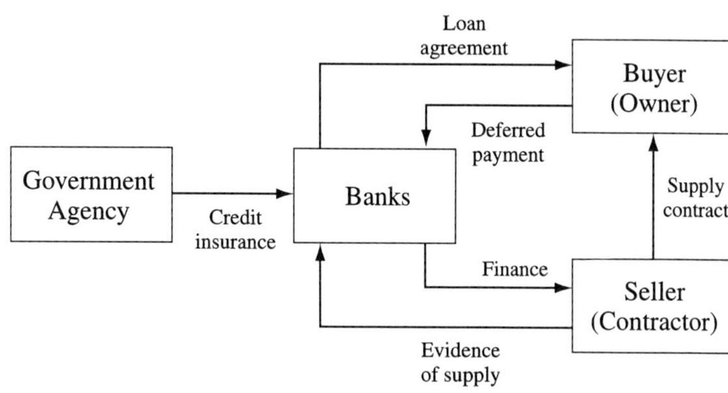

Forfaiting. Finance is made available through the sale of financial instruments due to mature at some time in the future. Finance is provided by trading in assets in the financial futures market. This can be a very expensive form of finance.

Switch Trade. Switch trade makes use of an uncleared credit surplus arising from bilateral trade arrangements. For example, if Country A has a credit surplus with Country B, exports from C to A can be financed with payments from B to C.

Debt/Equity Swapping. Debt/Equity swapping is designed to encourage investment in developing countries by owners of technology. A multinational company buys host country debt at a discount. This is redeemed in local currency at favorable rates of exchange and is used to set up local companies. These are used to transfer technology, generate foreign exchange, replace imports, and create employment.

Islamic Banks. These banks provide finance according to Sharia law, under which interest is banned. They provide finance and share in the profits of the investment. They appear to be like equity holders, except the payments are made against agreed-upon schedules and are repayable ahead of equity.

Project Finance

This section describes project finance—that is, unsecured, nonrecourse, off-balance-sheet financing of a project as a stand-alone entity. This has become popular recently under what is known as either the Private Finance Initiative (PFI) or Public-Private Partnership (PPP). Under this approach, a national government either works with the public sector or delegates responsibility to the private sector to design, construct, finance, operate, and maintain infrastructure projects that would normally be the responsibility of the national government working on its own (see Ive's chapter). The section begins by describing different forms of project undertaken in this way and then gives an overview of the structures commonly adopted, describing the roles of various parties and the contracts adopted.

Forms

Project finance is usually adopted for large infrastructure projects undertaken by the private sector on behalf of a national government, or working with the national government, under what is no known as the Private Finance Initiative (PFI) or Public Private Partnership (PPP). During the nineteenth century, it was common for infrastructure projects to be built by the private sector and paid for out of the revenues generated. In the early twentieth century, many national governments, under a sense of public spiritedness, took infrastructure development on themselves. However, during the 1960s and 1970s, companies in the mining and oil and gas industries began using limited-recourse financing for the following reasons:

- They had few assets other than the minerals in the ground with which to secure the loan.
- Off-balance-sheet financing offered tax advantages.

In the late 1970s and 1980s, many governments, especially developing economies, found they could not afford to undertake infrastructure development without involving the private sector. The Turkish government of the 1980s took a lead, but the decision to build the Channel Tunnel between England and France gave prominence to the approach. It became increasingly popular during the 1990s because governments

- saw it as a way of reducing public debt
- believed inherent efficiencies in the private sector would result in a cheaper product even though the private sector pays higher interest rates

PFI should be distinguished from privatization, which either

- transfers assets to the private sector that were previously owned by the public sector
- provides for services to be undertaken by a private company that were previously undertaken by the public sector

Project finance is used where a private sector company needs money for the construction of public infrastructure on the basis of a contract or license:

- An off-take contract where the public sector contracts to by the outputfrom the plant, as in the case of a power station
- A concession agreement in which the facility is constructed to provide a public service, as in the case of a hospital

- A concession agreement in which a facility is constructed to provide a service to the general public, as in the case of a toll road
- A license in which a facility is constructed to provide a new service to the public, such as a mobile phone network

Not all projects undertaken by the private sector on behalf of the government necessarily involve project finance. Some are paid for by the government. Others may be paid for out of the capital assets of the sponsoring company. These are likely to be smaller projects, which is why it was stated previously that it is larger projects for which project finance will be used.

Is there a difference between PPP and PFI? Some people treat them as synonymous. The generally agreed distinction is as follows (ECI, 2003):

- PFI is a subset of PPP and are projects financed by the private sector.
- PPP is all projects involving a collaboration between the public and private sector. The United Kingdom's treasury has issued a directive that all projects undertaken by government departments must be done either as PFI, prime contracting, or design and build, unless special dispensation is given. PFI and prime contracting are both PPP, but prime contracting is paid for by government funds. In the United Kingdom, design and build is treated as neither PFI nor PPP.

There are a number of different forms of such arrangements.

***Build-Own-Operate-Transfer* (BOOT).** This method was the approach used on the Channel Tunnel. The private sector company constructs the project and owns and operates it for the concession period. It earns the revenues during that period, to repay the finance and make a profit, and at the end of the concession period, the company hands the asset over to the government.

***Build-Own-Operate* (BOO).** Sometimes the asset is not handed over to the government at the end of the concession period. This situation may be in the case of a power station where at the end of the concession period, the plant has no residual value, or for a mobile phone network where the project company gets the benefit of the residual value.

The UK government prefers that all PFI projects use one of these two forms. For reasons of motivation and risk sharing, the promoter should own the asset through the concession period.

***Build-Operate-Transfer* (BOT), *Design-Build-Finance-Operate* (DBFO), *and Build-Lease-Transfer* (BLT).** This form of financing is used when for some reason it is not appropriate for the private sector company to own the asset. The private sector company designs and builds the asset and obtains the finance to do so, and then operates it for the concession period, obtaining revenues to repay the finance and to cover its costs. Under BLT, the government raises the finance and then leases the asset to the project company. This circumstance was described previously under unconventional sources of finance and is used for construction of extensions to the national electricity grid in Ireland, as described previously.

***Build- Transfer-Operate* (*BTO*).** This is identical to the previous form, except the project company owns the asset during construction. At commissioning, ownership transfers to the government, but the project company operates it.

Structures

Figure 15.3 illustrates a typical finance structure used for these types of project. At the heart is the project company, sometimes called the special purpose or project vehicle (SPV). One or more of the project stakeholders will be partners in this joint venture company. Almost certainly the operator will be a partner. All the others may or may not be partners.

Finance consists of a mixture of debt and equity. Project finance usually involves what would normally be considered high debt-to-equity ratios. A ratio of debt to equity of 4:1 is common. For a company, the reverse, a debt equity ratio of 1:4, is considered a large amount of debt. The higher the debt, the higher the return to the project sponsors. However, the banks insist that there should be some equity, because if the project fails, the equity

FIGURE 15.3. PROJECT FINANCE STRUCTURE.

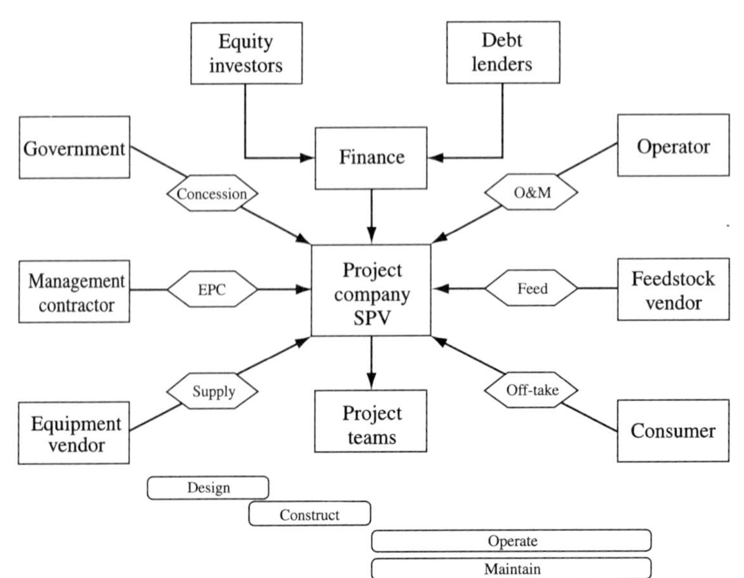

holders lose their money first, and so they will impose good management on the project company to maximize their own returns. The equity will primarily be provided by the partners to the SPV, but some equity may be raised from private investors.

Roles and Contracts

Figure 15.3 also illustrates some of the contracts involved. There are two main contracts that that enable the project to happen:

- The concession agreement with the government
- The off-take contract with the consumer of the project's outputs

In some cases the consumer will be the government, and so the off-take contract will be with the government. In other cases it will be with other bodies. In some cases the revenue from the off-take will be paid into an escrow account controlled by the lenders of debt. Interest and debt service in accordance with the agreed schedule of payments will be drawn from the escrow account before any surplus cash is paid to the SPV for payment of its costs and distribution to the equity partners. There are also a number of ancillary contracts:

- Design and build or EPC (engineering procurement and construction) contract with the managing contractor
- Equipment supply contracts with material suppliers
- Operation and maintenance agreement with the operator
- Fuel and other supply contracts with the feedstock suppliers

All of these people may be paid directly, or they may be given delayed payments, by either being given equity in the SPV or by having their payments converted into loans. If they are also equity holders, the loans will be subordinate debt to the bank loans.

There are also several support contracts:

- Permits and other rights
- Insurance
- A support agreement with the government

The support agreement with the government will impose a number of duties and obligations on both the government and concessionaire to facilitate the project completion.

The Process of Financial Management

You saw previously that the process of financial management begins at the feasibility study and continues right through the project to commissioning and operation. The key steps are as follows:

1. Conducting the feasibility study
2. Planning the project finance
3. Arranging the financial package
4. Controlling the financial package
5. Managing the risk

Sometimes the project champion, the project manager, and the organization they work for have little experience in the financing of projects. They will then be looking for considerable support from the lenders and may employ a financial consultant to support them.

Conducting the Feasibility Study

The financial assessment of the project begins at the feasibility study. The project will be assessed using financial appraisal techniques (Aston and Turner, 1995; Lock, 2000; Akalu, 2001). Options will be considered to maximize the value of the project. But the financial implications of those choices also need to be considered. Many decisions have financial implications, and those need to be taken account of from the start. During the feasibility, the financial objectives of the project will be set, and these will be different for the borrowers and the lenders, who will have different concerns. The sponsors concerns are as follows:

- Raising the necessary funds at times and in the currencies required by the project
- Minimizing costs and maximize revenues from the project
- Sharing risk between the project stakeholders, including the financiers
- Maintaining flexibility and control, including rescheduling repayments if necessary
- Being able to pay dividends to equity holders

On all projects, the lenders concerns are as follows:

- Repayments will be made in accordance with the schedule of interest and repayments.
- There is adequate security in the event of default by the borrower.
- Satisfactory dividends will be paid to equity holders.
- The risks are understood and will be managed.
- The basis of the cash flows, the nature of the business and market for products.
- There are adequate arrangements for insurance and maintenance to protect the value of the assets and maintain the cash flow.

In the case of project finance, the lenders will have a wider list of concerns:

- How robust are the project's cash flows?
- Do any other parties have claim on the cash flows?
- Are the project's assets dedicated only to the project?
- Can they be pledged as security?
- What is the market value of the assets in the case of project failure?
- Who are the project stakeholders?

Planning the Project Finance

Having undertaken the feasibility study, the project sponsor and their advisers will plan the financial package. This will involve the following issues:

1. From the feasibility study, the sponsor and their agents will have identified the total cost of the facility and the total amount of money that needs to be borrowed. Critically, this will include the scheduling of the cash flows, the time at which the money needs to be borrowed. It is also important to remember to include working capital and inflation.
2. Next, they will need to develop a financial strategy. They will need to plan the debt-to-equity ratio and develop a financial strategy involving a mixture of debt and equity to optimize profits while maintaining support of lenders.
3. They need to identify potential sources of debt and equity. They need to identify the sources of senior debt and determine the mixture of domestic and international finance. They also need to define the insurance required and the need for export credits. In the case of project finance, they need to consider who will be partners in the project company (SPV), how much equity will each provide, and will they provide any subordinate debt. They need to consider whether shares will be sold to private investors.
4. Having identified sources of finance, it can be worthwhile revisiting the cash flows to see if interest rates can be rescheduled to optimize interest rates.

Arranging the Financial Package

The financial package then needs to be assembled. The main supplier of senior debt will almost certainly be involved in this step as an adviser to the project promoter and will do most of the work. They will help the sponsor by:

- raising equity; a merchant bank will also be involved with this step.
- identifying additional sponsors and suppliers of subordinate debt.
- raising money through the Eurocurrency and Eurobond markets.
- liaising with government export agencies and arranging buyer or supplier credit.
- arranging insurance.
- arranging more sophisticated and less conventional financial packages.

Controlling the Financial Package

As the project progresses, the financial arrangements need to be managed, and again this will probably be done by the financial agent. This will involve monitoring progress against the plan and taking action to eliminate deviations from the plan. This may include the following:

1. *During project execution.* Monitoring expenditure to ensure it follows the predicted schedule and that debt and equity are drawn in accordance with the plan. It will be necessary to work with the cost controllers to forecast cost to complete, and make arrangements for additional finance if necessary.

2. *During commissioning.* Monitoring initial operating and maintenance costs and initial sales revenues to ensure they are as predicted and that the financial plan will be achieved. Variances need to be identified immediately and eliminated to ensure the project performs in accordance with its financial objectives.
3. *During operation.* Helping the sponsor with their management accounting to ensure the asset is operated at its peak profitability, and to provide reports to the lenders to avoid nasty surprises.

All of this is good project control, but it takes place at a level above the project. It draws on the project control and performance data, as discussed elsewhere in this book, to provide financial and business control for the project sponsor.

Controlling Risks

Risks on a project break into three broad categories:

1. Financial risks
2. Design and construction risks
3. Health and safety risks

The first of these is primarily the responsibility of the project sponsors and their agents. They will be manages using a risk management process, one of which will be described elsewhere in this book. However, a generic process follows six steps:

1. *Focus on risk management.* The planning of the financial packages should be done is such a way as to facilitate risk analysis and management. This has to begin during the project feasibility study.
2. *Identify risks.* Possible financial risks are listed in Table 15.2; most of these are self-explanatory.
3. *Qualitative assessment.* The risks will be assess in terms of likelihood and impact and prioritized for further assessment.
4. *Quantitative analysis.* The larger risks will be analyzed using techniques such as Monte Carlo analysis to determine the overall impact on the project and to help determine mitigation strategies.
5. *Develop a mitigation strategy.* A strategy for reducing the impact of the risks will be developed. For each significant risk, there is one of four possible actions:
   - Avoid the risk.
   - Reduce the likelihood or impact of the risk.
   - Pass the risk on to a contractor, insurance company, lender, or other stakeholder.
   - Develop a contingency plan.
6. *Monitor the risks.* The risks are monitored as the project progresses to ensure success of the mitigation strategy.

Finance is complex and the largest single cost on a project, so financial risk management is one of the most important success factors for a project.

TABLE 15.2. POSSIBLE FINANCIAL RISKS.

Type of Risk	Examples
Macroeconomic risks	Inflation
	Interest rates
	Currency and exchange
Political risks	Country risk
	Legislation and regulation
	Change of government
Commercial risk	Viability and feasibility
	Cost and schedule completion
	Performance and operation
	Revenue
	Availability, reliability, and maintainability
Contractual risks	Management
	Equipment supply
	Feedstock supply
	Concession and licenses
	Sales agreements

Summary

The cost of providing finance on a project is one of the largest single costs, being equal to or even greater than other major costs such as materials, design, or implementation over the life of the project. Further, as the arrangement of the financial package can have a direct impact on many of the design and technological choices on the project, it is necessary for the financial strategy to be considered as an integral part of the overall project strategy, beginning at the feasibility study. In this chapter, we considered the financing of projects, looking at the financing of projects (recourse finance) and project finance (non- or limited-recourse finance). We considered the features of project finance, and the sources of finance to companies, for both recourse and nonrecourse financing of projects. Both conventional and unconventional sources were described. The specific case of project finance was then described—that is nonrecourse, off-balance-sheet finance for projects, particularly as it applies to public-private partnership, with the private sector undertaking infrastructure development on behalf of governments. Finally, we considered the process of financial management on projects and how the financial strategy can be made an integral part of the project strategy, to minimize the cost of finance on the project.

References and Further Reading

Akalu, M. M. 2001. Re-examining project appraisal and control: developing a focus on wealth creation. *International Journal of Project Management* 19(7):375–384.

Aston, J., and J. R. Turner. 1995. Investment appraisal. In *The commercial project manager, ed.* J. R. Turner. London: McGraw-Hill.

Dingle, J., and A. Jashapara. 1995. Raising project finance. In *The Commercial Project Manager,* ed. J. R. Turner. London: McGraw-Hill.

ECI. 2003. *Public private partnerships: A review of the key issues.* Loughborough, UK: European Construction Institute.

Lock, D. 2000. Project apppriasal. In *The Gower handbook of project management,* ed. J. R. Turner and S. J. Simister. Aldershot, UK: Gower.

Merna, A. 2000. Managing finance. In *The Gower handbook of project management,* ed. J. R. Turner and S. J. Simister. Aldershot, UK: Gower.

Nevitt, P., and F. J. Fabozzi. 2000. *Project financing.* London: Euromoney Institutional Investor.

Turner, J. R., ed. 1995. *The commercial project manager.* London: McGraw-Hill.

Walker, C., and A. Smith. 1995. *Privatised infrastructure: The BOT approach.* London: Thomas Telford.

Yescombe, E. R. *Principles of project finance.* San Diego: Academic Press.

CHAPTER SIXTEEN

PRIVATE FINANCE INITIATIVE AND THE MANAGEMENT OF PROJECTS

Graham Ive

The first section of the chapter defines what is meant by the Private Finance Initiative (PFI). We then proceed to explore the special or key challenges that PFI as a structure of organization of the project process poses for the management of projects. This is done using the framework of generic project processes proposed by Winch (2000; 2002), namely: (1) defining the project mission, (2) mobilizing the resource base, (3) riding the project life cycle and leading the project coalition, and (4) maintaining the resource base.

In PFI there is a project management challenge both for the public sector client and its advisors and for the private sector provider. Accordingly, this chapter looks at the issues from both perspectives. However, where shortage of space makes it impossible to be comprehensive, priority has been given to the client's problems.

Throughout, issues will be illustrated by reference to a "stylized" and simplified account of an example of a PFI project—a 40-year contract to design, build, finance, and operate a major new teaching hospital, called the Gower Street Redevelopment Project, for University College London Hospitals (UCLH) National Health Service (NHS) Trust. The outline business case for this project was developed in 1994 and involved consolidating four hospitals (UCH, Middlesex, Tropical Diseases, EGA for Women), occupying many buildings, into a single new complex.

While the chapter tries where possible to give normative advice, its main aim is to describe the tensions and trade-offs that PFI project owners and managers must resolve, given their own priorities and their project's attributes.

What Do We Mean by PFI?

The name PFI (Private Finance Initiative) is the term used in the United Kingdom, but also elsewhere, for what, for project management purposes, would more clearly be labelled as DBFO (Design-Build-Finance-Operate) procurement and contracting.

DBFO is potentially applicable to any kind of project for delivery of a service underpinned by provision of a specific capital asset or set of assets (facility or facilities). In the United Kingdom, for instance, it has been used for public sector projects requiring new or refurbished buildings or infrastructure assets, military vehicles and equipment, and IT systems. It can even be used for projects with private sector clients, though the contract period will then tend to be much shorter than the 25-to-40-year operating periods found in pubic sector DBFOs.

PFI and Concession Contracts Compared

PFI is a form of a broader phenomenon, BOOT (build-own-operate-transfer) contracting. The acronyms DBFO and BOOT are sometimes used as interchangeable synonyms. I propose that it is more useful to restrict the use of DBFO (and thus PFI) to mean the following version or variant of BOOT and to distinguish it clearly from concession contracts.

Concessions. In non-DBFO BOOT contracts, the essence of the matter is that the role of the public sector becomes restricted to that of granting and regulation of a *concession* (see Campagnac and Winch, 1997), which gives the concession company monopoly rights to use and appropriate income from a specified asset or set of assets for a defined period. There may in addition be either a public subsidy paid to the concession company or the reverse, participation by a public body in the distribution of profits arising, but neither of these is essential. What is essential is that the concession company obtains most of its revenues by charges to users of the services it provides. What distinguishes this from the ordinary commercial production and sale of private goods and services is that in this case the producer requires either the use of public property or the exercise on its behalf of rights and powers belonging solely to the state; and that, in return, it accepts more or less close regulation by an agency of that state, the form of which regulation is articulated in the concession agreement between the public and private sector parties.

DBFO. In PFI or DBFO, on the other hand, the essence of the matter is that the public sector (and not the users) purchases the services and pays the PFI company (the private provider) for providing to a set of users (which can be the general public or a defined subset thereof); they remain therefore, in a political-economic sense *public services.* The financing of the assets required in order to provide those services becomes private (hence, private *finance* initiative). But the *funding* of the service remains public—proximately, government pays the PFI company and it is taxpayers, not users, who ultimately pay for provision (HM Treasury Task Force, 1997).

In comparison with conventional public services, two things have changed: (1) instead of both purchasing and providing those services to the public, the public sector now only purchases some or all of them, while the PFI company does some or all of the providing; and (2) the public sector no longer directly purchases (and therefore owns) the assets required to provide those services—instead, ownership rights in those assets are split between the PFI company (which can, as a slight simplification, be thought of as owning those assets for the length of its PFI service-provision contract with the public sector) and the public sector body (to which ownership of the assets will normally pass, without further payment, at the end of the PFI contract).

In the great majority of UK PFI projects, service provision is split between public and private sector bodies. A public sector body continues to provide *final* services to users (for example, education services, clinical services) and therefore to employ the producers of such services and control the main interface with users, while the PFI company provides *intermediate* or *support* services (some or all of the range of what can be described as facility management or FM services). The exceptions to date are highway and custodial projects. In these cases the PFI company provides the final service and there is no public sector involvement in provision.

UK-type DBFO projects are found around the world, but most especially to date in countries with "Anglo-Saxon" political, legal, and business cultures, such as Ireland, the United States, Canada, Australia, New Zealand, and South Africa, but also in such diverse countries as Chile and the Netherlands. Many international banks, service providers, and management, legal, and technical advisers with experience gathered mostly to date in the United Kingdom are actively selling the "UK model" around the world, and the UK government and its agents are actively advising many other governments on its use. Thus, a study such as contained in this chapter, based on UK experience and practice, is of far more than parochial relevance.

In the United Kingdom and elsewhere, PFI has been used by government where it does not wish to transform public services and public assets into private services and private assets, as would be achieved by a full marketization of services and privatization of the service provider, but where it does nonetheless wish to introduce an element of private sector provision of services (on efficiency or comparative value-for-money grounds) and private sector finance (on macroeconomic, reduction of public sector borrowing requirement grounds).

Thus, PFI has been applied mostly in the field of core public services such as defence, custodial, justice, police, education, and health services, rather than in the standard BOOT fields of utilities.

If we stick to this clear and useful distinction, then we must say that UK projects such as the Severn, Skye and Thames Crossings (toll bridges), and city mass transit systems with passenger fares (Docklands Light Railway, various tram and light-rail systems), though undoubtedly UK examples of BOOT or concession, are not instances of what is new and unique about PFI or DBFO. They therefore lie outside the main focus of this chapter, though many of its arguments apply to them (and all concession projects) as well as to PFI projects.

There is a third type of arrangement also sometimes classed together with PFI. In this case, there is a public sector provider of marketed goods or services (a "nationalized industry") that acts as client by placing a BOOT contract. This would include, for example,

BOOT projects placed by London Underground Ltd. or by Scottish (publicly owned) utility providers. In project management terms, this has much in common with "narrow" PFI (payment by a public client for asset-based services delivered by a private provider under an output specification). Consequently, almost all of the following analysis also applies to such projects.

For project management purposes, the difference is clear between a private concession company as the client for the project in one case (with government and regulator as non-client stakeholders) and a public sector body as the client for the project in the other, with this client defining and specifying the services to be provided and for which it will pay.

Aims and Objectives of Parties to PFI Contracts

The basic idea of PFI, as defined by the National Audit Office (National Audit Office, 2002; see also HM Treasury Task Force, 1997) is that, while the public sector still pays for the project, they do so

- by entering into a long-term contract to purchase services, not facilities
- by leaving to the private sector the procurement, ownership, and operation of the underlying facilities required to deliver the preceding service
- by specifying the services to be purchased in terms of outputs, not inputs
- by linking service payment to delivery
- by defining delivery in terms of performance measures set out as part of the output specification (performance specification).

UCLH

In the case of UCLH, for example, the NHS Trust and the Department of Health used PFI because they hoped thereby

- to get expenditure on a capital project approved by the Treasury (ministry of finance), in a context of tight limits on approvals for capital spending requiring public borrowing
- to get better value-for-money-spent (VFM) than they would from conventional procurement by getting a price for provision of services that reflected lower private than public sector costs of provision
- to get greater certainty at the precontract stage about construction time and price
- to get greater certainty of long-run service delivery

Meanwhile, the successful PFI bidder, Health Management Group (HMG), at bid stage a joint venture between AMEC plc (a major construction and engineering services group) and Building and Property Group Ltd (a major UK facility management service provider), hoped, as putative owners of the PFI project company, the special purpose vehicle (SPV), would get

- a high return on equity capital invested in the project
- a long-run stable flow of profit income
- an increased presence in and experience of a fast-growing market

As construction and FM companies, respectively, they hoped to get

- large and profitable contracts from the SPV
- experience in a fast-growing market
- opportunities to find innovative, lower-cost ways of providing construction and FM solutions that would raise their profit margins

Consequences of Using Project Finance

PFI almost invariably involves *project* rather than corporate finance. That is, a special purpose vehicle or company (SPV) is set up whose only asset and only revenue is the PFI contract with the public sector body and the revenues due under that contract. These revenues are "ring fenced" and kept strictly separate from the other revenues and expenditures of the companies owning the equity of the SPV. Thus, they cannot be used to cover losses or liabilities elsewhere in the parent firms. Private finance is provided through this SPV, in the form of equity stakes (normally held by subsidiaries of the same parent companies that own the firms that will take the principal construction and operating contracts from the SPV) and debt (either bonds issued by the SPV or loans to the SPV by banks). Project finance debt is in principle distinguished from corporate finance debt by its "nonrecourse" character. That is, the lender has no recourse to the other corporate assets or revenues of the equity investors in the SPV in the event of a financial default by the SPV. However, in practice this distinction is blurred by guarantees given by those parent companies (these become "contingent liabilities" for the companies giving the guarantees; unlike simple liabilities to service debt, these need not appear on the balance sheets of the companies in question). The SPV is usually a "non-stick" entity, retaining neither risks nor profits. All profits are passed to the companies owning its equity, and all risks transferred to it from the public client by the PFI contract are promptly passed on to the companies with which it enters into supply contracts, namely, the principal constructor and principal operator, or to other third parties. One exception is that sometimes the SPV retains responsibility for replacement of components needed during the life of the contract, though sometimes this is passed on to the operator.

One further defining characteristic of UK-style PFI is the debt-to-equity ratios of the project finance structures. It is unusual for equity to amount to as much as 15 percent of the capital structure. Moreover, the main equity-providers normally hedge their exposure to financial risk arising from fluctuations in interest rates. It is widely asserted by practitioners that these high proportions of debt and hedging (a) have the effect of reducing the weighted average cost of capital (in part, but not only, because interest payments attract corporation tax relief) and thus the amount of gross profit that must be covered by PFI contract prices and (b) reflect the unwillingness of PFI company sponsors to risk more equity capital, or to expose that capital to greater financing risk.

However, such high levels of financial gearing leave lenders exposed if there are substantial risks affecting projected revenue and cost streams (small proportional reductions in revenues or increases in costs could wipe out the equity, put the SPV in financial default, and leave lenders to bear any further losses). Thus, the lenders seek to assure themselves that the probabilities of such risks are very low. Increasingly, therefore, PFI has come to be seen as suitable for low-risk projects and unsuitable for projects with a high intrinsic total risk, unless the majority of that risk is borne by either the public sector client or the parent company owners behind the SPV, by means of additional guarantees to the lenders. This somewhat militates against one of the perceived rationales for the public sector adopting PFI, which is the transfer to the private sector of "significant" project risks.

PFI and Project Risks

PFI project failures, for all stakeholders, to date have been greatest in projects with high intrinsic risk:

- High "demand risk" ('if we build it, will they come?'); for example, the Royal Armouries Museum
- High "technical risk" (a technical specification at frontier of technical capabilities of the provider); for example, the National Physical Laboratory

PFI has been more successfully applied to projects with only "normal" levels of construction risk and no demand-risk transfer to the private sector. The PFI company, under direction from its lender, manages these normal risks by aiming for certainty of construction price and time. It achieves this certainty by

- specifying "tried and tested" solutions
- exerting strong "change control"
- transferring "all construction risks" in its contract to the design-and-build (D&B) supplier

"Time" risks are then controlled by direct monitoring of the contractor's performance and by incentives (penalties for late completion), and "price" risks are controlled by contract and by guarantees to lenders from parent of the D&B company.

UCLH

The lender, Abbey National Treasury Services (ANTS), and the SPV, Health Management Group (HMG) expressed concern that the innovative heating and cooling system proposed by HMG's designers had not been used previously in a UK hospital and perceived that unproven technology could increase the likelihood of availability deductions. To overcome this concern, HV plant for the entire building were "pooled," so that all plants in effect served all areas of the hospital. This meant that if one plant failed, it would merely reduce overall capacity of the system but not cause unavailability in any one part of the hospital. Once convinced that several plants would have to fail simultaneously for there to be non-availability, ANTS and HMG approved the solution. This raised costs but increased revenue certainty.

During project development, the SPV is carefully monitored by advisers appointed by, and reporting to, the principal lender. The project monitors' task is to help identify key risks in advance and then monitor the PFI company's performance in managing each risk, advise the PFI company on what it has to do to keep the bank happy, and report (or not) to the bank that performance is good enough for the banks to release their next tranche of finance. By controlling the flow of project finance in this way, the banks seek to manage design and construction risks. The banks, and their monitors, are not particularly concerned about risks of cost overrun, because the SPV will have passed these risks to their D&B contractor, whose balance sheet should be strong enough to bear them. Rather, their main concerns are with anything that could delay the start or halt the flow of service payments by the public sector client. This could be either a construction delay or a risk that the project would fail to pass quality (safety; fit-for-purpose) tests at commissioning stage, or that it subsequently becomes "unavailable" to any significant extent.

It is characteristic of PFI (and concessions) that the SPV's cash flow position is at its worst during construction. Once service delivery payments (or, in concessions, user payments) begin, monthly operating cash flows are strongly positive, and sufficient to cover negative financing cash flows (interest payments). But the SPV initially has only just enough finance (and then only if the bank releases tranches of finance as scheduled) to cover construction expenditure (principally, payments under contract to its D&B contractor). Interest nominally due on the construction loan normally has to be "capitalized" (i.e., added to the amount of loan principal that has eventually to be repaid, and on which interest is charged). Thus, a delay in the start of service delivery payments (because the project is late reaching operational stage) is likely to be fatal to the SPV's ability to service its debt and force it to seek a rescheduling or debt-to-equity conversion from the bank. Later, the SPV may be able to sue the D&B contractor (probably its principal initial shareholder) for substantial damages, but by that time the ownership of the SPV will probably have passed to the bank, and the other equity owners (besides the constructor that failed to perform) will also have lost their investment.

BOOT thus creates one of the strongest incentives to the PFI or concession contract holder to deliver on time of any procurement mechanism yet devised. The SPV and the bank will not simply rely on the contract passing responsibility for risk of delay to the D&B contractor but will actively manage this risk.

Note the asymmetry in the treatment of construction budget and schedule risks. If construction costs overrun, the contractor may in theory attempt to sue to reclaim these later from the SPV (in effect, from the other shareholders in the SPV), on the grounds that these overruns were caused by changes in the SPV's requirements. However, in the meantime the completed facility is in the hands of the SPV, and it is collecting operating revenue. Thus, the SPV will probably survive.

Construction-based PFI projects have achieved an excellent track record of becoming operational at or before the contracted date (National Audit Office, 2003). We do not have public information on cost-control performance. There will have been a number of cases where construction costs have exceeded contract price agreed between an SPV and its D&B contractor. In such cases, the normal consequence will be for the parent company of the D&B contractor to absorb the loss.

The PFI contract transfers project delivery risks to the PFI company. However, it has proved better (for value-for-money) not to transfer to a risk-averse private firm risks that the latter is in no position to manage or control (HM Treasury Task Force, 1997). Thus, for example, general inflation risk is retained by the public sector (payments due are indexed to changes in the Retail Price Index or some similar inflation index). Also, the public sector bears the risk of the impact of decisions under its control, such as new "specific" (though not general) legislation or regulations affecting PFI, or the price of changes it requests in the specification of the service after the contract is signed. However, other construction risks, and many operating risks, are transferred under the terms of the PFI contract, the exact allocation of risks varying from the PFI sector norm in each project and being negotiated normally after appointment of the preferred bidder.

PFI usually involves transfer of ownership of the asset, at a nominal price, at expiry of the contract, though it needs to be noted that much earlier official guidance suggested it might not be the case (Private Finance Panel, 1996D and 1997). More exactly, the public sector normally has an option to take over the assets at the end of the contract, which it need not exercise if it judges the facilities to have become a net liability (to have negative net future value). Since the assets are usually highly specialized, the client's *ex ante* estimation of their residual value-in-use at contract expiry normally exceeds the bidder's *ex ante* assessment of possible residual value to them, and it is thus optimal to contract for ownership to revert.

In what follows I shall use the specialized terminology of the PFI procurement process. For fuller definitions see Construction Industry Council (1998) and HM Treasury Task Force (1999a).

Defining the Project Mission

Here the key and distinctive challenges of PFI are all consequences of the use of output and not input specification (Private Finance Panel, 1996c; Public Private Partnerships Programme, undated; Department of Health, 2002).

Assessing Value-for-Money

In UK PFI it is an HM Treasury policy requirement that the final proposal of the winning bidder for each PFI project be appraised for value-for-money against a public sector comparator (PSC). Only if it is deemed to offer the public sector better VFM than the PSC (after adjusting for the value of risks transferred under the PFI but retained by the public sector under the PSC) should the proposal be accepted and a PFI contract signed.

The PSC is an attempt to estimate the price of a "conventionally procured" solution to meet the same business case need and output specification. It therefore should comprise both a construction price estimate for the design and construction solutions that would be produced by following current public sector procurement best practice and a public sector operating cost estimate for the same period as the PFI contract, converted to present value terms.

The PSC should ideally be based on a realistic "reference project", and this should be a "real" alternative option that could be developed, publicly financed, and chosen instead of the PFI bid. However, as the National Audit Office reports into PFI have pointed out, often one or both of these conditions are not met.

On the other hand, it is equally important that the PSC should be for a truly comparable level and quality-of-service delivery over the project life. Often, the costing of the "operating" element of the PSC is subject to query on these grounds.

In the absence of a price for a PSC that is a real alternative, *ex post* assessment of VFM (for example, by the National Audit Office) becomes a matter of assessing the effectiveness of competition achieved in the PFI project procurement process and of identifying where (and how securely) the projected benefits or DBO cost savings are coming from, in a particular project case, to offset the extra financing costs and procurement costs of PFI (CIC, 2000).

In practice, to date few PFI proposals have been rejected at final business case stage on grounds of lack of VFM compared to the PSC, after putting a value on risk transfer. Rather, more projects have fallen on "affordability" grounds, or from inability to agree an allocation of risks acceptable to all parties.

UCLH

The outline business case was based on the financial (net present value of savings in operating expenditure and capital receipts from land, less capital costs) and nonfinancial (improved healthcare, education, and research provision) benefits. Eight alternative project solutions were appraised. Highest ratio of benefit-to-cost was found to be for centralization of all accommodation in the locality of the existing UCH site. This version of the project, known as the 4box solution, from the four new buildings envisaged, was put to NHS Executive for project approval.

The Trust then initiated a PFI procurement, using estimated costs and benefits of its designers' proposals for the 4box as the public sector comparator—the basis for judging whether the best PFI bid was better value-for-money than conventional procurement—and including it in bid documentation as a point of reference for bidders, who were allowed, however, to submit variant proposed solutions.

Writing the Output Specification

The first challenge for the public sector's project managers is to turn the project mission statement and outline business case into an output specification. The difference between output and input specification is part of the basic idea of PFI (see *Aims and Objectives of Parties to PFI Contracts*). This goes beyond "performance" or "functional" specifications, as those might be found in the brief for a turnkey design-and-build project, because they are not primarily specifications of performance or functionality to be achieved at the point of handover or commissioning of the project, but standards to be achieved throughout the n-year life of the PFI contract. Thus, in addition to an output specification of the facility required, there are specifications of operating services to be provided. Moreover, in the fields in which UK PFI has been chiefly applied (for example, schools, hospitals, prisons), there is no recent UK history of the type of "turnkey" or "full" design-and-build procurement that is based on performance specification of facilities. Recently, in UK public sector building, if any form of integrated procurement has been used, it has been "develop and construct" (where the client's consultant prepares concept drawings, site layout, building dispositions, and plan forms) rather than full design-and-build.

Capturing users' requirements in a brief is always one of the hardest parts of a project to accomplish. In conventional building procurement there is at least a feedback process when architects' interpretation of and proposed solution to that brief can be reviewed by users. It is often pointed out that users may find it difficult to "read" and fully understand the implications of architectural drawings. Might a written/numeric specification be easier for lay persons to understand? On the other hand, what if the problem lies not in the medium of communication (drawings or written specifications of outputs) but in the articulation of implicit knowledge or in inappropriately "rational" models of thinking that actually proceeds by use of image? We all know how we like our coffee, but many of us would find it hard to describe what we are looking for in terms of attributes (units of bitterness, and the like) and would find it easier to say "like a real Italian espresso" and rely on an "architect" sharing our understanding of what we mean by that. Likewise, army generals know what they mean by "consistent with the military ethos" but find it hard to put it into words in an output specification. Elephants are easy to recognize but hard to define.

Note also that in the competitive stages of PFI procurement there are only very limited opportunities for bidders' designers to consult with users. Partly this arises from constraints on the time available to develop designs at tender stage, known as invitation to negotiate (ITN) stage, and partly from the procedural requirement for even-handed treatment of all bidders.

UCLH

UCLH proposed a radical shift from grouping in-patients by clinical speciality or department to grouping provision according to level of clinical need and dependency. In its output specification, it provided a draft set of operational policies for healthcare provision. It also included reference to a set of design guidance documents and principles. It indicated the number of acute (537) and low-dependency (60) beds that it thought it required bidders to provide. However, the final project company's proposals, accepted by the Trust, were for 669 acute and no low-dependency beds.

Assessing Affordability

The next challenge is to produce accurate budget estimates and thus assess affordability of the project at an early enough stage to avoid either: (a) late aborting of projects, with consequent write-offs of costs incurred to that point; or (b) late-stage downward rescoping of projects. Here the problem to be avoided is as follows. If, when bids are received, the lowest bid is higher than the public sector client can afford, either the project must be aborted or a preferred bidder appointed with whom a price for a reduced-scope specification will then be negotiated. The latter involves the client, in effect, in negotiating a service-content and a price with an incumbent monopolist, not under competitive pressure from rival bidders. Official advice is that the client should only proceed to negotiate scope and price with a preferred bidder if the bid price is "within negotiating distance" of affordability. The preferred situation for the public sector is that only relatively small details of risk allocation should be negotiated after receipt of competitive bids.

It is open to the client to publicize, at ITN stage, the amount of annual service payment they can afford and thus discourage non-affordable bids. However, if, in the view of bidders, the client has "specified a Rolls-Royce" but "provided a budget for a Ford," mere announcement of the budget limit will hardly solve the problem. In this situation, the choices open to a bidder are these: withdraw from bidding so as to avoid further unrecoverable bidding costs on a project that may be aborted; take the output specification at face value and ignore the budget limit; or take the budget limit seriously and "reinterpret" the output specification.

UCLH

The Trust made explicit to bidders both the design solution of its public sector comparator and its annual affordability constraint.

Whereas project managers can draw on a formidable body of information and experience in order to estimate market prices for an input specification for a building (with all its corresponding design drawings and technical descriptions), the price estimation, by clients' technical advisers, of output specifications for construction-based PFI services has involved a higher degree of uncertainty and error and yielded ranges rather than single-value estimates.

Ideally, projects should be checked for affordability before going out to the market, that is, at outline business case (OBC) stage. The OBC should involve "a clear definition in terms of service delivery of what is sought: the output specification" and "should incorporate a Reference Project (RP), i.e. a particular possible solution to the output requirement. . .which is worked-up in sufficient detail to provide full and adequate costing. . .including quantification of key risks. . .and hence is *prima facie* affordable" (HM Treasury Task Force, 1999). However, at this stage the specifications are often in practice extremely sketchy and therefore impossible to price accurately. It seems that a common practice is instead to estimate the price of a reference project without a detailed specification, simply using (updated by indexation) historic averages for simple descriptors (such as "construction cost per bed place" plus annual facility management cost per place in existing facilities). Cost estimates derived in this way are, of course, subject to a high degree of error. (As yet unpublished research I've directed has found evidence of wide discrepancies, for UK schools' PFI projects, between technical advisers' price estimates for the DBO costs of the output specification and the actual prices contained in lowest bidders' financial models.) Thus by deferring design until after calling for bids, PFI projects lose the advantage of relatively accurate prebid estimates (permitting early checks for affordability) offered by the traditional construction process of design-estimate-then-tender.

The process of formation of an output specification for a PFI project is too often in practice one in which initially, as more users are more fully consulted, more and more desiderata are added in to the specification, only, once bids have been received, for more and more desiderata to be cut out as unaffordable.

Securing Stakeholder Consent

The final distinctive challenge posed by PFI to the process of defining the project mission concerns the management of stakeholder consent. PFI is, of course, politically controversial. In locally devolved projects, stakeholders will include councillors in local authorities and members of boards of trustees or governors who may be either opposed to the whole idea of PFI (as "creeping privatization") or at least suspicious of it. Government has sought to overcome opposition of this kind by offering local authorities Hobson's choice—a PFI project or nothing. There have been a few cases (the pathfinder Pimlico School project is perhaps the best known) where local representatives have chosen "nothing." This way of forcing PFI on reluctant stakeholders has not been good for its reputation.

> "He who's convinced against his will
> is of the same opinion still"

A relatively recent development in the PFI market has been the "bundling" of smaller projects, especially refurbishments and partial rebuilds of groups of schools. This increases the number of stakeholders in the project enormously, and there are several important instances where the project client coalition fell apart (for example, the Brent secondary schools project) before a contract could be agreed.

Perhaps the strongest opposition to PFI has come from the trade unions representing public sector employees. Opposed to PFI on principle, these unions are most particularly concerned when existing union members will be transferred to become employees of the private sector. From this perspective, PFI projects fall into two kinds: where the business case is essentially to expand total capacity; and where the case is essentially to replace existing capacity. In the former, the PFI companies essentially recruit new employees. In the latter, they take over the existing public sector employees. UK legislation says that any transferred employees must work on wage levels, terms, and conditions that are "no worse" than they previously enjoyed. For private sector project managers, this often creates a problem of managing a two-tier workforce, comprising a tier of transferees and a tier of new recruits, potentially doing similar jobs on different pay and conditions, and may reduce the firms' ability to introduce operating regimes based around "flexible working."

Although client requirements are meant to be captured by the output specification, both the client and other public regulatory stakeholders will have technical requirements embodied in technical guidance notes and codes, to which they must be convinced the project conforms before they will give consent. Often in the United Kingdom these standards, requirements, and codes are prescriptive in form and therefore simply require to be interpreted, rather than be consulted or negotiated upon. However, their prescriptiveness may greatly reduce the opportunity for the PFI consortium to develop unusual or innovative solutions.

Finally, in the United Kingdom, PFI projects, like any building project, require first outline and then detailed planning (zoning) consent. The normal practice is for the client to obtain outline consent on the basis of a maximal "footprint," mass and floor area, before proceeding to ITN. This is done to reduce the risk to bidders that consent may not be forthcoming. After appointment of the preferred bidder (PB), it becomes their responsibility to obtain detailed consent for their proposed solution. Nevertheless, it is not unknown for PFI bidders to propose solutions on sites other than those owned and proposed by the client.

There are sometimes, therefore, effectively two interdependent projects for the PFI company to manage: the PFI project itself, as defined by the client, and a property development or dealing project, involving some acquisition and some disposal of sites. If proceeds from sales of sites are to be used to fund part of the PFI project, this may mean that separate "bridging finance" has to be arranged to cover the period until sites are vacated and can be sold. Otherwise, in the simple case where the boundary of the site required is identical to the boundary of the site owned and proposed for use by the public sector, PFI requires that the site be leased by the PFI company.

UCLH

The Trust went out to ITN with a four-site planning consent (the so-called 4box proposal, which had also been used for the OBC), only for the winning bidder to propose a single-site two-towers-and-podium solution, not requiring some of these sites but requiring some adjacent land not included in the 4box, on which the bidder had obtained an option to buy from its private owner. In such cases the bidder shares the risk that planning consent will not be forthcoming, but, of course, reduces competitors' ability to match their proposal.

Mobilizing the Resource Base

Under this heading we will consider the forming and the motivating of the project coalition.

The client's procurement problem is synonymous with the principal/agent problem identified in economic theory, and it is that theory that I will use here. The principal/agent problem arises when an agent (in this case, a PFI company) knows more about its real competence or about the effort it will expend than does the principal (in this case, the public sector client). The theory of principal and agent represents the main attempt by economists to generate advice to clients about how to design the *reward structure* of their contractors, such as a PFI company, so as to obtain optimal outcome. The rest of the problem can be decomposed into the *adverse selection* problem and the *moral hazard* problem—the former to do with selecting and the latter to do with opportunistic behavior by a supplier (Milgrom and Roberts, 1992; Douma and Schreuder, 1998).

Devising the PFI Co.'s reward structure

One of the most important tasks, therefore, for the client and their project manager, is to think carefully about just how to link payment to the PFI company to project outcomes, using the output specification to define those outcomes.

In PFI, payment to the agent is a mixture of a fixed element (agreed in advance and unlikely to vary according to outcomes or outputs achieved) and a variable element that is explicitly tied, via specific performance measures, to the quantity and quality of service-delivery outputs achieved.

PFI payment mechanisms divide the client's unified payment for services into two, or sometimes three, parts:

- A part for availability of the facility or of major parts thereof, and therefore with deductions for nonavailability for use
- A part for level of performance of the facility management (FM) or other service functions, with deductions therefore for performance below a set standard
- Sometimes, a part linked to a variable volume of use (as with early "shadow toll" highways, but also, say, laundry services payments made pro rata to volume of bed use in hospitals).

The availability payment invariably dominates the mix, not only in PFIs for "accommodation service" (where the costs of constructing and then maintaining the facility are the majority of the PFI company's costs, so that the structure of payments is broadly aligned to the structure of costs), but even in PFIs for "final service" (like prisons) where the cost structure is very different. Until commissioning is achieved, the timing and amount of the expected availability payment must be regarded as subject to risk. Once commissioning is complete, however, the client has agreed that the project is, at that point, meeting all its availability requirements. Thereafter, this part of the payment can only vary (downwards) if all or part of the facility subsequently seriously deteriorates in condition. Thus, the risk, to the PFI company, becomes a matter of preventing complete and minimizing partial nonavailability through deterioration leading to failure to remain available for use. The initial questions facing the client are as follows:

- By how much, and for how long, should a measured condition have to deteriorate in order to constitute "nonavailability" in terms of the payment mechanism?

- Shall payment be linked mainly to "aggregate" availability or separately to the availability of each part of the facility?
- How big or small a part of the availability payment will be deducted in such a case?

The PFI Co and its lenders want to be able to regard the availability element of the payment as "virtually fixed", once commissioning has been achieved. Then, on the basis of the security offered by this "fixed" stream of revenue, low-risk/low-interest debt finance can be arranged. Given that the client, too, should expect to share in the savings from lower interest costs, they are likely to answer the questions posed above in ways that allow this treatment of availability payments as being "mostly" fixed.

The economics of optimizing incentives via payment mechanisms is, in principle, straightforward (Private Finance Panel with HMTreasury, 1997; p. 8), but in practice fiendishly difficult. Penalties for nonperformance or nonavailability should be set so that the monetary value of the penalty is just less than the monetary imputed difference of value of the benefits to users between the two states (of availability and nonavailability, or of performance to standard X and performance to level Y). If this penalty is greater than the cost to the PFI company of remedying the poor performance or nonavailability, then the company will be induced to undertake the remedy; and if not, then nonremedy is preferable to remedy, because marginal cost of remedy exceeds marginal value of the resulting benefit (I have assumed, for simplicity, that the cost of remedy is the same as the cost of prevention; if not, the prescription becomes somewhat more complex).

This requires, first and foremost, that the client knows the value of each marginal benefit. Lack of such knowledge is one of the biggest impediments to effective design of incentive-payment systems. If this problem is overcome, there remains the problem for the PFI company of obtaining knowledge of the whole-life cost and revenue implications of each of its design and operating choices, without which it will be unable to respond rationally to any structure of incentives embodied in a payment mechanism.

However, the client may feel that it is simply unacceptable for some events of nonavailability to occur at all (such as failure of power supplies to operating theaters) and thus penalize such events so heavily that the contractor has very strong incentives to avoid it happening (by paying for the cost of backup systems, for example). Whereas, for other nonavailability events, the incentives will be set weaker (failure of power supply to office areas, for example) and therefore lower relative to the costs of remedy. Such "strong" incentives need to be used sparingly, partly because they will significantly add to overall cost and thus price (and by adding to the SPV's risk may make the project nonbankable, i.e., unattractive to lenders) and partly because they run the risk of distorting the PFI company's allocation of resource and managerial effort.

How many and what kind of separate outputs should be measured and linked to the payment? The "equal compensation principle" of agency theory requires that

> if the agent is required to allocate time between two or more activities, then [for an incentive contract to work] the marginal return to the agent from time spent on each

> activity must be equal, or else the activity with the lower marginal return will receive no time (Millgrom and Roberts, 1992; p 228).

Simplifying more than a little, the moral of the equal-compensation principle story is that the client should ensure that each and every desired output or outcome is given its own place in the payment mechanism. However, as you shall see, there are countervailing arguments in favor of simpler (or different) payment mechanisms.

The first of these arguments derives from the "informativeness principle." This states that the client should exclude from the PFI company's compensation formula any performance measures that partially reflect factors outside the agent's control and include in the formula any measure that increases the likelihood that the size of payment will relate to the actual effort of the PFI company.

One corollary might be that it would be desirable to benchmark each PFI company's performance against other comparable PFI projects and link payment mainly to benchmark (i.e., relative) scores. This would be so *if* "factors outside the PFI Co.'s control" that impact project outcomes (such as the common procedures and practices of the client) are likely to impact also on all those other PFIs, while differences in performance between SPVs are more likely to relate to differences in quality and quantity of their effort. This has hardly as yet been attempted in PFI.

How large should the deductions for nondelivery of each output be?

To answer this, project managers and clients need to think about the costs of monitoring. The issue can be thought of in terms of the "monitoring-intensity principle." This states that the level of monitoring of performance or output must be proportionate to the intensity of the incentive. If contracts set up large deductions in relation to an output, then the client must be prepared to spend more in total on the costs of monitoring of that output. Other things being equal, the more *costly* it is to monitor accurately a particular output, the weaker the strength of the incentive that should be created by linking payment to that output. Ideally, the amount of measurement and the intensity of incentives should be chosen together, as a package. Where outputs are heterogeneous and intrinsically hard to measure (because they are services, not commodities), as in PFI projects, and *a fortiori* if they are intermediate rather than final services, again as in most PFIs, so that accurate measurement is very costly, it may therefore be best to set up only relatively weak incentives to produce those outputs.

In practice, most PFI clients use "weak" incentives for service quality and rely more on there being a shared understanding of what constitutes "good provision" with, and an organizational culture of quality management in, their providers. It is normal for the providers themselves to be asked to do most of the monitoring of quality of service, because it is much less costly for them to undertake.

There are some further concerns for clients. What if methods of delivering outcomes are to a large extent technologically fixed, so that the same methods will be used despite creation of incentives? What if the things the client can measure are only more or less crude approximations of or proxies for the things they really wish to be provided? What if the PFI company (or, in a PFI context, their provider of capital, i.e., the bank) is *highly* risk-averse? Or, finally, what if the PFI company, or one of its principal suppliers, is, in terms of organizational behavior, conservative and habitual rather than progressive and innovative, again despite creation of incentives?

All the preceding issues are key matters a PFI client has to judge before deciding on an incentive structure. To guide such judgments, we need research to clarify what payment mechanisms work best in practice in what PFI project contexts. A highly provisional summing-up might be this: Beware overly strong incentives unless outcomes are few and simple to measure.

In practice, it may be better in PFI projects to decouple the issues of incentive and replacement. Incentives alone (because "weak" and incomplete) may be inadequate to prevent serious nonperformance. The real remedy, in extreme cases, will lie in replacement. It should not be the case that performance measures' importance is judged solely by the size of impact on payment. Any performance measures, however, need to be objective and accurate enough to be used to begin moves to replace the agent, if necessary. Replacement of FM providers is often possible, within a PFI, because the latter may have only relatively short-term contracts from the SPV. However, it can be very difficult indeed to make sure that the performance measurement of the FM provider only measures their own effort and not that of the D&B provider.

Adverse Selection

In a construction context, adverse selection means that suppliers who intend to offer inferior (but hidden) quality or to behave opportunistically with respect to claims and variations have the greatest incentive to charge the lowest prices. Thus, simple reliance on price competition does not work and, if insisted upon by the client, poor-quality and opportunistic suppliers will drive out good quality and trustworthy ones. The classic solution in construction was to divide procurement of the design from that of construction, on the basis that, while quality in design could not be assured in advance and defects could be hidden, and variations in client requirements during design development could not be prevented, the same would be less true of construction, if the construction contract consisted of an agreement to execute, under inspection, a completely predeveloped design. Designers therefore had to be selected in ways that assured the client that they could be trusted, whereas contractors did not and could be selected by lowest-price tender competition, within only weak parameters of prequalification to eliminate the egregiously insolvent or incompetent. Designers might well work in an in-house department of the client or, if not, would be appointed on the basis of reputation, track record, and an ongoing relationship with the client over past and projected future projects (Winch, 2002).

Now, PFI defies the logic of this approach, by making the client choose the designer, constructor, and operator all together, by the same method, and as a combined (take-it-or-leave-it) set. At the same time, it rules out use of the in-house provision solution for the client. This means, in effect, that the client needs to have the same level of confidence in their PFI consortium (confidence that they will not try to deliver and conceal inferior quality, and confidence that they will not exploit opportunities to extract monopoly rents) that they classically had to have in their designer. Alternatively, if they do not have such confidence but are still determined to use PFI, they must develop much more effective *contractual* means of inspecting and enforcing quality, and of ensuring that they do not require variations or changes not explicitly provided for in the contract.

However, as will be shown, contracts alone are unlikely to be efficacious in securing client objectives. Trust, and also a shared conception of project mission and a collaborative rather than confrontational climate between client and PFI company, is the *sine qua non* for effective PFI projects.

PFI uses a form of selective competitive tendering, followed by negotiation with a preferred bidder. Normally three (never more than four) consortia or firms are short-listed to bid. These are the firms that score highest in prequalification evaluations of firms that responded to the invitation to express interest (EOI). Track record in PFI projects is one key criterion for short-listing. Ideally, but in practice by no means always, clients should only short-list firms that are not and will not become too busy with other PFI projects, to avoid uncompetitive bids or bidders withdrawing. A review of the literature of PFI project case studies reveals at once that in a large proportion of cases, firms that were part of a short-list of two or three bidders subsequently withdrew, or submitted uncompetitive, non-conforming bids. Thus, it has been common for the client to have only two, or even only one, bid(s) to consider (see NAO PFI project case study reports, various dates).

UCLH

> The Trust chose a short-list of three consortia. One then dropped out because it became overcommitted on other PFI projects. Another collapsed after its FM partner withdrew following a conflict with its contractor. Thus, the Trust received just one bid.
>
> The unitary payment proposed in this bid exceeded the Trust's announced affordability constraint. However, the variant nature of its proposed solution was judged to yield further savings in operating costs to the Trust of healthcare provision, thus increasing the amount the Trust would have available for the unitary payment. Thus, the Trust judged the bid to be both affordable and better VFM than the PSC, because it provided more benefits in terms of healthcare provision, though at higher annual cost.

Bid evaluation should either proceed in two stages (with stage 1 being a technical appraisal of bids for ability to meet requirements, with only bids passing the standards required proceeding to stage 2, selection of lowest-price bid, where lowest-price is judged across all scenarios, allowing for differences in risks accepted or contract period proposed) or comprise a "weighted" evaluation in which price, risk acceptance, robustness, and scores on other quality criteria are all combined into a weighted composite score (HM Treasury Task Force, 1999b). Where the weighting method is used, invariably in practice by far the greatest weight is given to price (at most, to this author's knowledge, bids with a high score for "quality" but a higher price have been selected over bids with a merely satisfactory score for quality if the price difference is of the order of 5 percent or less).

Moral Hazard: "Holdup"

In principle, negotiations with the preferred bidder (PB), appointed following clarification of initial bids and evaluation of variant bids departing from proposed "standard" commercial terms and risk allocations, should be

> limited to fixing the final detail of the documentation and satisfying the reasonable requirements of the supplier's financiers . . . by insisting the PB's financiers have indicated their comfort with the risk allocation embodied in their bid at a stage where there is still the lever of competition (HM Treasury Task Force, 1999a).

In practice, however, it is common for one or other party to seek materially to amend the terms of the proposed contract. On the client's side, this is most likely to occur because of problems of nonaffordability (see preceding text). On the PB's side, it may reflect either changes proposed by their financiers, once the latter come to examine risks in detail, or an attempt to take advantage of the easing of competitive pressure. Public sector guidance recommends that the client should, where possible, keep a "second choice" bidder ready in reserve, as a counter to such opportunism, and reserve the right to require the PB to conduct a funding competition at this stage (Office of Government Commerce with Partnerships UK, 2002) if the source of the problem lies with their proposed financier.

The costs of preparing contracts designed to guard against later holdup of either party by the other in PFI have been much higher overall than in any other mode of construction procurement (Construction Industry Council, 1998 and 2000). The publication recently of standard forms of contract for use in some of the main PFI market sectors, as well as standard specifications and payment mechanisms, may help to reduce these transaction costs (Gruneberg and Ive, 2000, Chapter 5). However, this will depend on the extent to which clients and bidders feel obliged to vary from these standard terms.

It has proved very expensive, both in terms of advisers' fees and of the impact on contract price, for PFI clients to attempt to foresee as a contingency every relevant "state-of-the-world" and every possible change in their requirements over a period of several decades ahead. Thus, in practice, PFI clients are likely to find it better to commit to a loss of flexibility in their PFI-contracted requirements, leaving their non-PFI projects to buffer and absorb any such changes.

A large part of the fee costs—fees paid by the client to technical, legal, and financial advisers—arise from the perceived need to prepare complete and unambiguous tender documents. If the client is trying to achieve a complete contract and depend upon their ability to enforce this contract in the event of disputes, they must incorporate ways of measuring performance and specify methods of resolving disagreements over actual levels of performance.

In some cases, clients appear to have lost sight of the big picture in a mass of detail, paying for the development of very elaborate performance measures and payment mechanisms that actually are unlikely to achieve any key client goals, because of the small value (and perhaps also contestability) of the penalties resulting, relative to the unified service charge or to the likely costs of providing remedies.

In the case of one hospital PFI (not UCLH), a very complex payment mechanism was developed. It comprised 14 components in the payment, with some components based on actual or "notional" bed occupation units (BOU) and non-bed attendances (NBA) for each area of the hospital, including marginal rates of £3.67p per NBA, along with a complex system of deductions, retentions, and deficiency points with seven performance bands. All of the bands led to a potential difference between most optimistic and most pessimistic feasible scenarios, of an amount equal to 1.44 percent of the total payment stream to the SPV (Parker, 2000).

Riding the Project Life Cycle and Leading the Project Coalition

The PFI company is in a position potentially to take a holistic view of project costs and revenues across, if not the whole life of the facility, at least the life of the PFI contract. However, various barriers stand in the way and need to be overcome by its management of the project before the opportunities offered can be seized.

First, the project must be run by an integrated coalition and led by a project director keen to optimize the project's returns over its contract life cycle, and not prepared to accept subgoal optimizing behavior by members of that coalition (where the contractor tries to minimize construction cost regardless of operating costs or impact on revenues, for example). Each organization involved in the coalition needs to be responsive and adaptable to the fact that it has become a shareholder in its own client (the SPV) and thus, in addition to its "usual" interest (in maximizing profits from a construction or FM contract) has another interest, in maximizing profits for the SPV. This means, first, contributing to improving the chances of its consortium of winning the bid competition and, second, contributing to the *whole-life* PV (discounted present value) of the net revenues (profits) expected from that bid if successful. This latter is, of course, the PV of the difference between service payments receivable from the client and the whole-life costs of providing that service.

Devising solutions so as to maximize profit over the whole life of the contract, however, requires that the bidder's project management team have good information about the WLC (whole-life cost) and revenue implications of their choices of design solution and operating regime. If the managers lack confidence in the quality of that information, they may regard options that will certainly increase initial construction cost but that are "expected" to reduce subsequent operating costs as, in effect, "riskier" than the alternative, and thus reject them.

By improving the feedback flow of information about costs-in-use, actual replacement lives of components, and the like, well-managed project organizations can in principle reduce the uncertainty with which WLC forecasts are regarded. Stable consortia, operating in the same market and comprising the same partners working together over many successive projects, ought to begin to receive detailed feedback from their operator partners or departments on actual operating and replacement costs, in forms sufficiently robust to convince their bankers. However, this data will only provide a clearer picture of the "path most traveled." The other paths will continue to disappear into dark forests of ignorance, unless there are some pioneer PFI projects that are allowed to explore those paths.

Maintaining the Resource Base

It is under this heading that you can best consider the intersection of issues regarding the management of a single project with issues of strategic resource planning. In the context of the management of *programs* of projects, especially, resource-maintenance activities, including maintaining a sufficient number of resource-holding firms, become a key part of the successful management of projects.

Maintaining the Number of Firms

All government departments funding PFI programs have had to concern themselves with first the creation and then the maintenance of a population of provider firms that own or employ the relevant specialized resources and possess the relevant organizational experience and knowledge. All PFI markets in the United Kingdom have to date shown strong tendencies toward supply-side concentration. Economies of experience appear to be high. Barriers to "late" entry appear to be substantial. In a market comprising only three or four firms (consortia), exit by even one firm would pose major problems for client procurement strategy, which is to rely upon competition at tender stage to obtain prices that give value-for-money. A similar, though temporary, problem arises if one or more of such a small number of suppliers are too loaded with other projects to be able to bid keenly for a particular project.

The program client must deliberately develop the market in ways that minimize the risk of a firm exiting. This may require it to disadvantage temporarily any firm that is threatening to achieve market dominance.

Since there will be PFI markets in which competitive pressure on bidders is sometimes weak, clients cannot afford to rely solely on competition, but have to be able to negotiate effectively, from a basis of knowledge of the costs of an efficient provider and from the bargaining power of a real alternative option. For this, public clients need to continue to put some projects of each type out to "conventional" procurement.

Encouraging Investment in the Resource Base

Opportunistic behavior all round is understood to be kept in check mainly by the adverse impact such behavior would have on a player's profits in future plays of the game as other players retaliated (dangers of exclusion of a consortium from future short-lists, dangers of exclusion of a partner from future consortia). The greater the threat or possibility of opportunistic behavior, undoubtedly the more the damage that will be done to the efficient maintenance of the resource base, as all become reluctant to make contract-specific investments in resource development. It is therefore important from this perspective that PFI is perceived to comprise a program that will continue indefinitely.

Acknowledgments

I am grateful to Kai Rintala for permission to draw on his case study doctoral research into University College London Hospital PFI project. This research will be published in 2004 by VTT (Valtion teknillinen tutkimuskeskus) of Finland as "The Economic Efficiency of Accommodation Service PFI Projects." I am also grateful for comments by the editors, and by Chris Field, Kai Rintala, and Hedley Smyth.

References

Audit Commission. 2003. *PFI in schools: The quality and cost of buildings and services provided by early PFI schemes.* London: Audit Commission.

Campagnac, E., and G. M. Winch. 1997. The social regulation of technical expertise: the corps and the profession in France and Great Britain. In *Governance and work: The social regulation of economic relations in Europe*, by R. Whitley and P. H. Kristensen, Oxford, OUP.

Construction Industry Council. 2000. *Role of cost saving and innovation in PFI projects.* London: Thomas Telford.

Construction Industry Council. 1998. *Constructors' key guide to PFI.* London: Thomas Telford.

Department of Health. 2002. *Standard output specification.* London: Department of Health.

Douma, S., and H. Schreuder. 1998. *Economic approaches to organisations.* London: Prentice Hall.

Gruneberg, S., and G. Ive. 2000. *Economics of the modern construction firm.* Basingstoke, UK: Macmillan.

HM Treasury Task Force. 2000A. *Technical note no. 7: How to achieve design quality in PFI projects.* London: HM Treasury.

HM Treasury Task Force. 2000B. *Technical note no. 6: How to manage the delivery of long term PFI contracts.* London: H M Treasury.

———. 2000c. *Technical note no. 5: How to construct a public sector comparator.* Series 3—technical notes. London: HM Treasury.

———. 1999a. *A step-by-step guide to the PFI procurement process.* Revised version. Series 1: Generic guidance. London: HM Treasury.

———. 1999b. *Technical note no. 4: How to appoint and work with a preferred bidder.* Series 3—technical notes. London: HM Treasury.

———. 1999c. *Technical note no. 3: How to appoint and manage advisers to PFI projects.* London: HM Treasury.

———. 1999d. *Technical note no. 1: How to account for PFI transactions.* Series 3—technical notes. London: HM Treasury.

———. 1997. *Partnerships for prosperity: The private finance initiative.* Series 1—generic guidance. London: HM Treasury.

Milgrom, P., and J. Roberts. 1992. *Economics, organization and management.* Englewood Cliffs, NJ: Prentice Hall.

Mumford, M. 1998. *Public projects, private finance.* Welwyn GC, UK: Griffin.

National Audit Office. 2003. *PFI: Construction performance.* Report by Comptroller and Auditor General, HC 371, Parliamentary Session 2002-3.

National Audit Office. 2002. *PFI and value for money.* Conference presentation by Mr Jeremy Colman, Assistant Auditor General.

Office of Government Commerce and Partnerships UK. 2002.

OGC guidance on certain financing issues in PFI contracts. London: Private Finance Unit, OGC.

Parker, M. 2000. *The importance of the payment mechanism in the allocation of risk in PFI projects.* MBA diss., University of Warwick.

Private Finance Panel with Highways Agency. Undated. *DBFO: Value in roads.* London: Highways Agency.

Private Finance Panel with HM Prison Service. 1996a. *Report on the procurement of custodial services for the DCMF prisons at Bridgend and Fazakerley.* London: HM Prison Service.

Private Finance Panel with HM Treasury. 1997. *Further contractual issues.* London: Private Finance Unit, HM Treasury.

———. 1996b. *PFI in government accommodation.* London: Private Finance Unit, HM Treasury.

———. 1996c. *Writing an output specification.* London: Private Finance Unit, HM Treasury.

———. 1996d. *Risk and reward in PFI contracts.* London: Private Finance Panel.

Private Finance Panel with HM Treasury. 1995. *Private opportunity, public benefit.* London: Private Finance Unit, HM Treasury.

Public Private Partnerships Programme (4Ps). Undated. *Output specification for PFI projects: A 4Ps guide for schools.* London: 4Ps. www.4ps.co.uk/publications.

Winch, G. M. 2002. *Managing construction projects: An information processing approach.* Oxford, UK: Blackwell Science.

———. 2000. The management of projects as a generic business process. In *Projects as business constituents and guiding motives*, ed. R. A. Lundin and F. Hartman. Dordrecht, Netherlands: Kluwer.

SECTION II.2

TECHNOLOGY MANAGEMENT

INTRODUCTION

At first glance, including an entire section on technology management might seem strange; in fact, for many project management books and texts, the topic is still given short shrift. Yet it is also a well-documented fact that technology represents a major issue in the effective management of projects. Technology can be broadly (and oftentimes confusingly) defined to evoke a wide range of meanings—some helpful and others not. For our purposes, a working definition of technology within the context of the management of projects involves not so much actually *doing* the "technical" elements of the project as managing the processes and practices needed to ensure the technical issues by which projects are transformed from concepts into actual entities—and doing this effectively within the time, cost, strategic, and other constraints on the project. In this regard, this section includes a number of chapters that guide you through the key life cycle issues that define the project, ensure its viability, manage requirements, and track changes; in short, that highlight the key steps in transforming and realizing the technical definition of the project.

In Chapter 17 Al Davis, Ann Hickey, and Ann Zweig take us carefully and systematically through the different types of requirements, basically from a strong IT systems perspective, showing how requirements have to be elicited and selected (triage), and how this then leads to specification. The sequencing of the requirements management process is then examined for different types of (systems) projects, and different types of requirements management tools are discussed. The chapter concludes by looking at trends in requirements management research and practice. (Being so oriented toward systems projects, there is no discussion of equivalent practices in other industries—construction "briefing," for example. Instead, this topic is discussed by Peter Morris in Chapter 54.)

From the elicited and triaged requirements and the resulting specifications, the project design can be elaborated. Peter Harpum, in Chapter 18, discusses design management in another comprehensive overview covering the nature of design and how designers design, systems engineering, and whole-life design (life cycle management), along with design management "techniques" such as design for manufacturing, concurrent engineering, CAD/CAM, risk management, as well as several for scheduling, cost, and quality design management.

Hans Thamhain extends this discussion with his treatment, in Chapter 19, of concurrent engineering. Concurrent engineering (CE) is one of those terms whose meaning varies from industry to industry and firm to firm. It is in reality a combination of several things; as Hans says, CE "is a systematic approach to integrated project execution that emphasizes parallel, integrated execution of project phases, replacing the traditional linear process of serial engineering." Hans lists the following characteristics as typifying good CE: a uniform process model, integrated product development, gate functions, standard project management process, quality function deployment, early testing, and organizational involvement and transparency. He concludes with an extensive list of recommended practices for different phases of the project development cycle.

In Chapter 20 Rachel Cooper, Ghassan Aouad, Angela Lee, and Song Wu broaden the discussion into process and product modeling and the management of projects. They begin with a formal review of process modeling techniques, showing how process models can be used to represent the development of new product development and construction projects. Product modeling is then introduced with particular emphasis on object modeling. It is shown how product models might ultimately be integrated with activity models so that product and project information could be drawn off the same integrated model.

The emphasis on information modeling is taken further in Chapter 21 by Callum Kidd and Tom Burgess on configuration management (CM). They begin by rehearsing what CM is, showing that it has applications in a wide spectrum of industries ranging from power plants to new product development, as well as systems/IT projects. The heart of their chapter, however, centers around the way contemporary information management practices are shaping CM. The chapter concludes on a somewhat ambivalent note, with research data from the authors suggesting that often Tier 2 and 3 suppliers (in aerospace) use CM more to be compliance with Tier 1 requirements than for real business benefit.

The next two chapters may both seem rather industry-specific, though in fact both have wide-ranging and important implications. Both are extremely authoritative. In Chapter 22 Alistair Gibb discusses the management of safety, health and environmental (SHE) issues. As in so many of the chapters, Alistair argues the need to consider these issues from the very early stages of a project. He identifies the requirements for early definition work on SHE policy and objectives, risk assessment, designer actions, sustainability assessment, and SHE plan preparation and goes on to discuss the implications of method statements, procurement strategy, and resources and competence development. He then discusses key issues in implementation throughout the project life cycle. Though the chapter is written from a predominantly construction perspective, nearly all the points raised apply more broadly, for example, to most manufacturing situations and even, albeit to a lesser extent, in many IT projects.

Hal Mooz, in Chapter 23, discusses verification—what in old language would probably have been termed testing. (Today, verification and validation have quite specific meanings;

Hal begins by defining these along with several other key terms.) The chapter is set in a predominantly systems development context, but systems being defined here quite broadly. (One of Hal's memorable examples is from Harley-Davidson.) Verification and validation have to be understood in terms of their position in the systems development/integration process. Here Hal introduces the Vee++ systems development cycle that he and his colleagues Kevin Forsberg and Howard Cotterman introduced in *Visualizing Project Management* (1996). Validation and verification techniques are described together with some key insights developed over years' of experience in managing verification. Again, while the terminology may seem a little strange to some people, particularly those not working in manufacturing or systems, the applicability of Hal's chapter is instructive and relevant.

Rodney Turner and Anne Keegan conclude the section on managing technology with some insights, in Chapter 24, on managing innovation. They do so, as we said in the Introduction, very much from an organizational learning perspective, emphasizing the importance of creating an environment supportive of, and the management conditions most likely to be conducive to, innovation in a project context. They then introduce four practices—systems and procedures, project reviews, benchmarking, project management communities—that have been adopted for the selection, retention, and distribution of technological developments and conclude by showing how these relate to different stages of organizational learning and of project management maturity. (Both topics are discussed later, in Chapter 44 by Bredillet, Chapter 45 by Morris, and Chapter 49 by Cooke-Davies.)

About the Authors

Alan M. Davis

Al Davis is Professor of Information Systemsat University of Colorado at Colorado Springs and is president of The Davis Company, a consulting company. He has consulted for many corporations over the past 27 years. Previously, he was a member of the board of directors of Requisite, Inc.; Chairman and CEO of Omni-Vista, Inc., a software company in Colorado Springs; Vice President of Engineering Services at BTG, Inc., a Virginia-based company that went public in 1995 and was acquired by Titan in 2001; a Director of R&D at GTE Communication Systems in Phoenix, Arizona; and Director of the Software Technology Center at GTE Laboratories in Waltham, Massachusetts. He has held academic positions at George Mason University, University of Tennessee, and University of Illinois at Champaign-Urbana. He was Editor-in-Chief of IEEE Software from 1994 to 1998. He is an editor for the *Journal of Systems and Software* (1987 to present). He is the author of *Software Requirements: Objects, Functions and States* (Prentice Hall, 1990 and 1993) and the best-selling *201 Principles of Software Development* (McGraw-Hill, 1995). He is founder of the IEEE International Conferences of Requirements Engineering. He has been a fellow of IEEE since 1994, and earned his PhD in Computer Science from University of Illinois in 1975.

Ann Hickey

Ann Hickey is an Assistant Professor of Information Systems at the University of Colorado at Colorado Springs. She worked for 17 years as a program manager and senior systems analyst for the Department of Defense before beginning her academic career. She teaches graduate and undergraduate courses in systems analysis and design, enterprise systems, and information systems project management. Her research interests include requirements elicitation, elicitation technique selection, collaboration, and scenario and process modeling. Her work has been published in the *Journal of Management Information Systems*, the *Database for Advances in Information Systems, the Journal of Information Systems Education*, the *Requirements Engineering Journal*, and national and international conference proceedings. She received her BA in mathematics from Dartmouth College and her MS and PhD in Management Information Systems from the University of Arizona.

Ann Zweig

In 1997, Ms. Zweig cofounded the start-up software company, Omni-Vista, in Colorado Springs, Colorado. Omni-Vista provided products and services that assist software development companies in making informed business decisions about existing or planned products and projects by performing intelligent trade-off analyses that incorporate critical business factors as well as technology. Most recently, Ms. Zweig served as president and COO. Before becoming a computer scientist, Ms. Zweig was a biologist with The Nature Conservancy and also at the Rocky Mountain Biological Laboratory in Gothic, Colorado. Ms. Zweig served two years as a Peace Corps volunteer in the Kingdom of Tonga, where she taught biology, chemistry, and physics. Ms. Zweig has an MS in Computer Science from the University of Colorado and a BS in Biology from the University of Kansas.

Peter Harpum

Peter Harpum is a project management consultant with INDECO Ltd, with significant experience in the training and development of senior staff. He has consulted for companies in a wide variety of industries, including retail and merchant banking, insurance, pharmaceuticals, precision engineering, rail infrastructure, and construction. Assignments range from wholesale organizational restructuring and change management, through in-depth analysis and subsequent rebuilding of program and project processes, to development of individual peoples' project management capability. Peterhas a deep understanding of project management processes, systems, methodologies, and the "soft" people issues that programs and projects depend on for success. Peter has published on design management; project methodologies, control, and success factors; capability development; portfolio and program value management; and internationalization strategies of indigenous consultants. He was a Lecturer, Visiting Lecturer and examiner at UMIST on project management between 1999 and 2003.

Hans J. Thamhain

Dr. Hans J. Thamhain specializes in technology-based project management. Currently a Professor of Management and Director of Project Management Programs at Bentley College, Waltham/Boston, his industrial experience includes 20 years of high-technology management positions with GTE/Verizon, General Electric, and ITT. Dr. Thamhain has PhD, MBA, MSEE, and BSEE degrees. He is well known for his research on technology-based project control and team leadership. He has written over 70 research papers and 5 professional reference books on project and technology management. Dr. Thamhain is the recipient of the Distinguished Contribution Award from the Project Management Institute in 1998 and the IEEE Engineering Manager of the Year 2000 Award. He is certified as New Product Development Professional, NPDP, and Project Management Professional, PMP.

Rachel Cooper

Rachel Cooper is Professor of Design Management at the University of Salford, where she is Director of the Adelphi Research Institute for Creative Arts and Sciences, and also Co-director of the EPSRC Funded Salford Centre for Research and Innovation in the Built and Human Environment. She has been undertaking design research for the past 20 years. Her work covers design management, new product development, design in the built environment, design against crime, and socially responsible design. Projects include Engineering & Physical Sciences Research Council study of Requirements Capture, Cost and Benefits of Partnering, Generic Design & Construction Process Protocol, Future Scenarios for Distributed Design Teams, three projects for the Design Council/Home Office on Design Against Crime, a study of the use of design in government departments for the Design Council and government, and an 18-country study of New Product Development in High Technology Industries. Professor Cooper was Founding Chair of the European Academy of Design and is also Founding Editor of *The Design Journal.* Professor Cooper has written over 100 papers and six books; her latest, coauthored with Professor Mike Press, *The Design Experience,* was published June 2003.

Ghassan Aouad

Professor Ghassan Aouad is Head of the School of Construction & Property Management and Director of the Centre for Research and Innovation in the Built and Human Environment at the University of Salford. He also leads the prestigious research (from 3D to nD modeling). Professor Aouad's research interests are in modeling and visualization, development of information standards, process mapping and improvement, and virtual organizations. He has published extensively in these areas.

Angela Lee

Dr. Angela Lee is a research fellow at the University of Salford and has worked on numerous projects, including 3D to nD Modelling, PeBBu (EU Thematic Performance Based Building Network), and Process Protocol II. Her research interests include performance measurement, process modeling, process management, and requirements capture, and she has published extensively in these fields. She completed a BA (Hons) in Architecture at the University of Sheffield and her PhD at the University of Salford.

Song Wu

Song Wu is a research fellow on the 3D to nD Modelling project, and he previously worked on the Process Protocol II project at the University of Salford. He was trained as civil engineer and worked as a Quantity Surveyor in Singapore and China for three years. Song was awarded an MSc in Information Technology in Construction in 2000 and is currently completing his PhD in IT support for construction process management at the University of Salford. His research interests include data modeling, database management, information management, and IFC (industry foundation classes) implementation.

Alistair G. F. Gibb

Alistair Gibb is Program Director at Loughborough University of the Construction Engineering Management Programme, sponsored by 13 major construction organizations. His research work falls primarily into two main areas: off-site fabrication and health and safety. In off-site fabrication, Alistair has managed a string of major research projects and has recently secured the primary academic role in *prOSPa*, the prestigious UK government-funded Pii Programme on the subject. In the health and safety area, he is Director of the APaCHe partnership for construction health and safety, working closely with the HSE and industrial collaborators. He also has a leading role in both European and international networks in health and safety. Since 1995 he has been Project Director of the European Construction Institute's Safety, Health and Environment task force, and in 2001 he joined the main board of ECI-ACTIVE as a nonexecutive director. He has been a member of numerous committees and task groups, including the Association of Planning Supervisors (coopted board member). He is a member of the Conseil International de Batiment (CIB) working commission on health and safety, ICBEST (the international council for building envelope), ISSA (International Social Security Association), and WSIB (Canadian Workplace Safety & Insurance Board) Research Advisory Council.

Callum Kidd

Having worked in managing configurations for over ten years in the process and then aerospace industries, Callum Kidd moved to UMIST to set up and run the Industrial Management Centre. He later moved to Leeds University, where in 1994 he established the Configuration Management Research Group and was awarded research grants from BT, Royal Mail, European Commission, Ericsson, and Vickers Defence to research the future of CM in a variety of contexts. He moved back to UMIST in 2000 to manage the Project Management Professional Development Programme, where he is currently carrying out research into the synergy between CM and PM.

Thomas F. Burgess

Thomas Burgess is Senior Lecturer in Operations and Technology Management at Leeds University Business School. After qualifying as an engineer, Tom worked in roles connected with operations management, information systems, and project management in a number of consulting and engineering companies prior to entering academia. His MBA (Bradford) is in Production Management and his PhD (Leeds) is in Computer Studies. His research has focused on major management-related process innovations and their impact on organizations, and the use of modeling and simulation methods to assist in understanding and supporting these new innovations. Lately his research has centered on improving the processes for development projects in the chemicals and pharmaceuticals industry.

Hal Mooz

Hal Mooz is founder and CEO of the Center for Systems Management—a company dedicated to training, mentoring, consulting, and culture building in project management and systems engineering and related disciplines. His experience covers being a Chief Systems Engineer, Program Manager, and Deputy Director of Programs for intelligence satellites at Lockheed Missiles and Space Company. He is coauthor of *Visualizing Project Management*, published 1996 by Wiley, and a contributing author of *The Handbook of Managing Projects* to be published by Wiley in 2004. He is a member of PMI and presenter of papers and tutorials at PMI, ProjectWorld, and international project management conferences; he is a Certified Project Management Professional (PMP) by Project Management Institute (PMI). Recertified in 2001, he is a member of the International Council on Systems Engineering (INCOSE) and a presenter of papers and tutorials at international INCOSE conferences. Several papers were judged best of conference. He is also a recipient of the CIA Seal Medallion for contributions in project management, as well as a recipient of the INCOSE 2001 Pioneer Award for furthering the cause of Systems Engineering.

Rodney Turner

Rodney Turner is Professor of Project Management at the Lille Graduate School of Management and at Erasmus University, Rotterdam, in the Faculty of Economics. He is also an Adjunct Professor at the University of Technology, Sydney, and Visiting Professor at Henley Management College. He studied engineering at Auckland University and did his doctorate at Oxford University, where he was also for two years a post-doctoral research fellow. He worked for six years for ICI as a mechanical engineer and project manager, on the design, construction, and maintenance of heavy process plant, and for three years with Coopers and Lybrand as a management consultant. He joined Henley in 1989 and Erasmus in 1997. Rodney Turner is the author or editor of seven books, including *The Handbook of Project-based Management*, the best-selling book published by McGraw-Hill, and the *Gower Handbook of Project Management.* He is editor of *The International Journal of Project Management* and has written articles for journals, conferences, and magazines. He lectures on and teaches project management worldwide. From 1999 and 2000 he was President of the International Project Management Association, and Chairman for 2001 to 2002. He has also helped to establish the Benelux Region of the European Construction Institute as foundation Operations Director. In addition, he is also a Fellow of the Institution of Mechanical Engineers and the Association for Project Management.

Anne Keegan

Anne Keegan is a University Lecturer in the Department of Marketing and Organisation, Rotterdam School of Economics, Erasmus University, Rotterdam. She delivers courses in Human Resource Management, Organisation Theory, and Behavioural Science in undergraduate, postgraduate, and executive-level courses. She has been a member of ERIM (Erasmus Research Institute for Management) since 2002. In addition, she undertakes research into the project-based organization and is a partner in a European-wide study into the versatile project-based organization. Her other research interests include human resource management in knowledge-intensive firms, new forms of organizing, and critical management theory. Dr. Keegan has published in *Long Range Planning* and *Management Learning* and is a reviewer for journals such as the *Journal of Management Studies* and the International Journal of Project Management. She is a member of the American Academy of Management, the European Group for Organisation Studies (EGOS), and the Dutch HRM Network. Dr. Keegan studied management and business at the Department of Business Studies, Trinity College, Dublin, and did her doctorate there on the topic of management practices in knowledge-intensive firms. Following three years postdoctoral research, she now works as a university lecturer and researcher. Dr. Keegan has also worked as a consultant in the areas of Human Resource Management and Organizational Change to firms in the computer, food, export, and voluntary sectors in Ireland and the Netherlands.

CHAPTER SEVENTEEN

REQUIREMENTS MANAGEMENT IN A PROJECT MANAGEMENT CONTEXT

Alan M. Davis, Ann M. Hickey, and Ann S. Zweig

Project success is the result of proper planning *and* proper execution. Fundamental to proper planning is making sure that the work to be performed by the project is well understood and that the amount of work is compatible with available resources. Requirements management is all about learning and documenting the work to be performed by the project, and ensuring compatibility with resources. A well-executed on-time project that does not meet customer needs is of no use to anybody.

Requirements

Requirements define the desired behavior of a system[1] to be built by a development project. More formally, a *requirement* is an externally observable characteristic of a desired system. The two most important terms of this definition are *externally observable* and *desired.* Externally observable implies that a customer, user, or other stakeholder is able to determine if the eventual system meets the requirement by observing the system. Observation here could encompass using any of the five senses, as well as any kind of device or instrument. Next, a requirement must state something that is desired by some stakeholder of the system. Stakeholders include all classes of users, all classes of customers, development personnel, managers, marketing, product support personnel, and so on (see the chapter by Winch). It

[1]A *system* is any group of interacting elements that together perform one or more functions. The elements could be electronic hardware, mechanical devices, software, people, and/or any physical materials.

is not so easy to determine if a candidate requirement is a valid requirement from this perspective. In fact, the only way to make the determination is to ask the stakeholders. The word *desired* was chosen purposefully and is meant to encompass both wants and needs (see *Wants vs. Needs* later in the chapter).

Requirements Management

This chapter is all about how project managers and analysts manage requirements. *Requirements management* is the discipline of

- learning what the candidate requirements are—the learning aspects of requirements management are generally called *elicitation*;
- selecting a subset of those candidate requirements that are compatible with the project's goals, budget, and schedule—the selecting aspects of requirements management are generally called *triage*;
- documenting the requirements in a fashion that optimizes communication and reduces risk—the documenting aspects of requirements management are generally called *requirements specification*; and
- managing the ongoing evolution of those requirements during the project's execution.

On large projects, the individuals who perform requirements management are generally called analysts, requirements analysts, requirements managers, requirements engineers, systems analysts, business analysts, problem analysts, or market analysts. In companies that mass-market the products of their development projects, these individuals are generally within the marketing organization of the company. In companies that build custom products for their customers, these individuals are generally within either the marketing or the development organizations of the company. In IT organizations where the products of development projects are used within the company, these individuals are within the IT organization itself and interface with the internal customers, or are within the internal customer organization and interface with the IT organization.

On smaller projects, the project manager often performs a majority of the requirements management activities because these strategic activities are so critical to project success.

Requirements Management and Project Management

Much of requirements management can be thought of as part of (or preceding) project planning, because one goal of requirements management is the decision concerning *what* system is to be built. However, because needs of customers are often in constant flux, requirements must be addressed throughout the project. At project inception, the project manager is often intimately involved in defining requirements. Because any subsequent change to requirements affects project scope, the project manager tends to stay involved in the requirements management process throughout development.

Project management of requirements activities is unique among most project responsibilities because of two factors: (1) the strong customer focus and (2) the "softness" of the discipline. In most aspects of project management, the constraints upon the task are predefined, known, and finite. The project manager's job is to control the project in such a way that the short-term and long-term project goals are achieved. In the case of requirements, none of that is true. The stakeholders who are the source of the requirements may not be available when needed. Even worse, their needs are constantly in flux. The very act of asking the stakeholders for their needs induces the stakeholders to conceive of new requirements hitherto not thought of. Every time a requirement is stated, the stakeholders will think of many more. Every time a prototype is constructed and demonstrated to the stakeholders, they will think of dozens of additional requirements. The phenomenon is likened to a continuous application of Maslow's hierarchy of needs. Every time any need is satisfied, more needs appear. Thus, the actual performance of requirements management causes the project to expand in scope.

Most activities being planned, controlled, and monitored by project management tend to appeal to the left side of the brain. Everything is (or should be) well defined, concrete, measurable, and to a large degree controllable. Requirements management requires a large dose of both left-side and right-side brain function. For example, the skills required to perform requirements elicitation primarily reside in the right side of the brain. Such skills deal with communication, feeling, and listening. On the other hand, the skills needed to record and manage the changes to requirements (including the use of so-called requirements management tools) reside primarily in the left side of the brain. These skills deal with specification, attention to detail, and precision. For this reason, requirements management is more like project management than like the other tasks performed by the individuals reporting to the project manager. Requirements management, like project management, require a very diverse set of skills.

Types of Requirements

We defined a requirement as an externally observable characteristic of a desired system. Although this sounds fairly specific, in practice requirements come in a wide variety of flavors and serve a wide variety of purposes. The following sections describe some of this richness.

User/Customer vs. System (Problem vs. Solution)

Some authors demand that requirements describe a problem purely from the perspective of the customer and must omit any reference to any solution system. Other authors demand that requirements specifically describe the external behavior of the solution system itself (IEEE, 1993). We have found that most practitioners divorce themselves from either extreme and recognize that as the requirements process proceeds, requirements naturally evolve from descriptions of the problem to descriptions of the solution. When requirements are stated in terms of the problem without reference to a solution, they look like this:

> We need to reduce billing errors by 50 percent.

When requirements are stated in terms of the external behavior of the solution, they look like this:

The system shall provide an "audit" command, which verifies the accuracy of bills.

There is only a fine line separating the problem and the solution. In the preceding examples, one *could* argue that the former is actually within the solution domain. After all, reducing billing errors is just one way of trying to accomplish some real goal, such as increasing collections, increasing revenue, or maximizing cash flow.

Lauesen (2002) differentiates between user requirements and system or software requirements. He states that user requirements are supposed to address just the needs of the user, and system or software requirements are supposed to address the expected behavior of the solution system. However, he also correctly points out that in practice, most requirements describe external behavior of the solution system anyway, and that the term user requirements is generally applied to any requirements that are written in a language that users can understand.

Systems of Systems vs. Single Systems

By their very nature, systems are composed of other systems, as shown in Figure 17.1. For such systems, requirements are written for every system, usually starting with the top one. When requirements are written for the topmost system, they are written from a perspective outside that system, thus ensuring that all its requirements are externally observable. After these requirements are documented in a *system requirements specification*, system design (generally not considered part of requirements) is performed to decompose the system into its constituent subsystems and then to document those subsystems. Then requirements are written for

FIGURE 17.1. SYSTEMS ARE COMPOSED OF SYSTEMS.

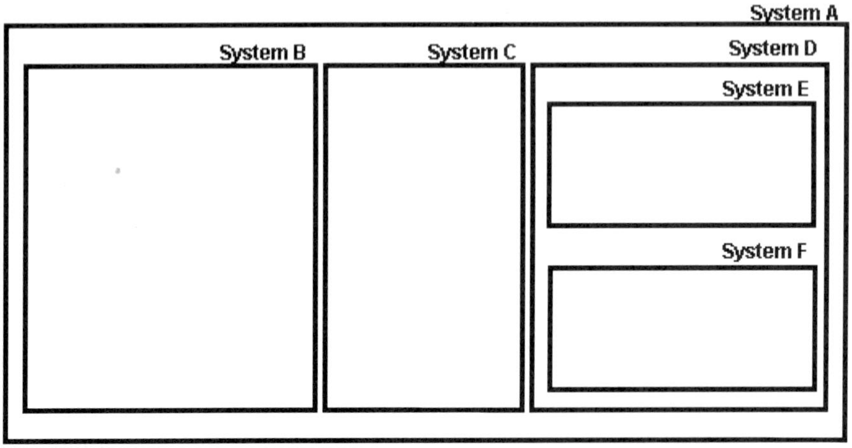

each subsystem, from a perspective outside each of those subsystems, and the process repeats itself. As we get toward the lower-level systems, the system requirements are often replaced with two documents, a *software requirements specification* and a *hardware requirements specification,* each of which defines the requirements for its part of the system.

When a system is simple enough to not require decomposition into subsystems, the most common approach is to write a *system requirements specification* for the overall system, allocate each of the requirements to either software or hardware or both, and then proceed to write a software requirements specification and a hardware requirements specification.

When a system is composed entirely of either software or hardware, just one document is usually written—either a software requirements specification or a hardware requirements specification.

Primary vs. Derived

Thayer and Dorfman (1994) differentiate between requirements that are defined initially and requirements that are derived from those original requirements because of design decisions. For example, once the decision is made to include this requirement:

> The system shall provide service *x* to the customers.

it becomes evident that we must also include this requirement:

> The system shall bill the customers for using service *x*.

Project vs. Product

IEEE Standard 830 (1993) and Volere (Robertson and Robertson, 2000) make a clear distinction between requirements that constrain the solution system itself, for instance:

> When the button is pressed, the system shall ignite the light.

and requirements that constrain the project responsible for creating the product, for instance:

> The product must be available for commercial sale no later than April 2004.

IEEE Standard 830 calls the former *product requirements* and the latter *project requirements*. Volere differentiates between two types of *product* requirements: functional and nonfunctional; and three types of *project* requirements: project constraints, project drivers, and project issues.

Much agreement exists in the industry that product requirements are requirements, but little agreement exists concerning whether project requirements are really requirements. We happen to believe they are not requirements, but it is only a semantic issue. The fact is that during requirements activities, the team *will* need to perform trade-off analyses between both types of "requirements."

Behavioral vs. Nonbehavioral

Some requirements describe the inputs into and the outputs from a system, and the relationships among the inputs and outputs. Others describe general characteristics of the system without defining inputs, outputs, and their interrelationships—that is, the functions that the system is intended to support. The former requirements are called *behavioral requirements,* although they have also been called *functional requirements* by the Robertsons (2000) and Davis (1993). The latter requirements are called *nonbehavioral requirements*, although they have also been called *developmental quality requirements* by Faulk (1997) and by the quite ambiguous and almost deceptive term *nonfunctional requirements*, by the Robertsons (2000) and Davis (1993).

Following are examples of behavioral requirements:

> When the button is pressed, the system shall ignite the light. If the power is on and the on-off button is pressed, the system shall turn power off. When the user enters the command *xyz*, the system shall generate the report shown in Appendix H.

Examples of nonbehavioral requirements include all aspects of performance, reliability, adaptability, throughput, response time, safety, security, and usability, and they include such requirements as the following:

> The system shall handle up to 25 simultaneous users. All reports shall be completely printed by the system within five minutes of the request by the user. The user interface shall conform to Microsoft standard *xxx*.

Wants vs. Needs

Many requirements writings seem to imply that one of the responsibilities of the analyst is to remove from consideration any requirements that are deemed to be "wants" rather than "needs" of the customers/users (IEEE, 1983; Swartout and Balzer, 1982; Siddiqi and Shekaran, 1996). Common wisdom and experience contra indicates this. Marketing studies have shown that people decide to buy or use a system because it satisfies their wants as well as their needs.

Requirements vs. Children of Those Requirements

When requirements are documented, they often are recorded more abstractly than is desirable, for example,

> The system shall be easy for current system users to use.

This may be sufficient for early discussions, but it must be refined before the parties should agree to the effort. The most common way to do this is to document the refined requirements as subrequirements of the parent requirement, as in the following:

The system shall be easy for current system users to use.

(a) The system shall include conventional keyboard and mouse.
(b) The system shall exhibit the same "look and feel" of the existing legacy system.

Requirements should be refined whenever a discussion arises concerning the meaning or implications of a requirement.

Original Requirements vs. Modified Requirements

According to Standish Group Reports (1995), 58 percent of all requirements defined for software-based systems will change during the development process. According to Reinertsen (1997), a similar rate of change occurs for all products in general. This constant flux requires us to recognize that requirements evolve not only toward increasing detail but also toward altered functionality. We must clearly differentiate between requirements that were originally documented and requirements that become apparent only after development began.

Requirements in One Release vs. Requirements in Another

Almost all products evolve. Many requirements stated for, and implemented in, release *n* will undergo change in subsequent releases. This observation makes it clear that we must record the relationship between specific requirements and specific product releases.

Requirements Activities

Three distinct types of activities are performed under the auspices of requirements: elicitation, triage, and specification. The following subsections elaborate on these.

Elicitation

The first major set of activities within requirements management is called *elicitation*. Elicitation is the process of determining who the stakeholders are and what that they need—in other words, what their requirements are. Some of these needs can be "gathered"—that is, they are known and understood by the stakeholders, and all the analyst needs to do is "pick them up" from the stakeholder. Others may surface only as the result of stimulating the stakeholders; this type of activity most closely corresponds to the dictionary definition of "elicitation." Other requirements need to be learned through study, experimentation, reading, or consultation with subject matter experts. Still others are discovered via observation. Regardless of the process used, and regardless of what the activity is called, the analysts must find out what the stakeholders needs are. Elicitation includes not just obtaining the needs but also analyzing and refining those needs to improve the team's understanding of them. Once elicited, analyzed, and refined, these needs should be recorded as a list of candidate requirements, as shown in Figure 17.2

FIGURE 17.2. ELICITATION CREATES A LIST OF CANDIDATE REQUIREMENTS.

The user starts the RLM by placing it within the border of a defined lawn and pressing BEGIN MOWING from the Main Menu.

The RLM shall determine if it is in a defined lawn. If not, the RLM shall sound the error tones and display the message MOWER NOT IN RECOGNIZED LAWN on the first line and RETURN on the second line.

If correctly placed, the RLM shall beep once and wait for the user to step back beyond the safe distance range. After the user has moved beyond this range, the RLM shall move to a starting location within the lawn and begin mowing.

While mowing, the RLM's panel shall display nothing except in the event of an error condition, dump or refueling required, or an obstacle comes within the minimum safe distance.

The RLM shall check the grass height, grass type, grass density, and moisture of the lawn to determine the settings proper for cutting. Adjustments to the blade position and speed shall be made as required. When a swath is properly cut, the RLM shall move to an uncut area.

The cutting pattern shall begin with the perimeter of the lawn and work inward to the lawn's center. Each pass shall overlap the previous pass by a width less than or equal to 33% of the RLM's swath but greater than or equal to 25% of the RLM's swath.

This normal cutting pattern may be altered by obstacle avoidance maneuvers but shall resume when avoidance maneuvers are complete.

During avoidance maneuvers, the RLM may, for the sake of fuel efficiency, temporarily shut off its blades if over an area that has been properly cut. Obstacle avoidance is discussed in Requirement 510.

The RLM shall shut off the blades if fouling occurs to the degree that the RLM may damage itself. Should blade fouling occur, the RLM shall sound the error tones and display the message BLADES FOULED on the first line of the display. Should there be more than one blade . . .

The individual who conducts elicitation is generally called an *analyst*. An experienced analyst is adept at using a wide variety of elicitation techniques and possesses the sensitivities and skills necessary to assess the political, technical, and psychological characteristics of a situation to determine which elicitation technique to apply (Hickey and Davis, 2003; and Hickey and Davis, 2003a). Some of the classic techniques used during elicitation are as follows:

- *Interviewing* is the process of repeatedly prompting one or more stakeholders to verbalize their thoughts, opinions, concerns, and needs. The most effective prompts are open-ended questions, which force the stakeholder to think and respond in nontrivial ways. For example, prompts such as these are open-ended: "Would you please elaborate upon the problems you are experiencing now?" and "Why do you consider this a problem?" Other important aspects of effective interviewing include listening, taking notes, and playing back what you heard to verify that it was what was intended. Because over half of communication among individuals is nonverbal (Knapp and Hall, 1997), face-to-face interviewing is best. However, interviewing can also be performed over a telephone, though less efficiently. Gause and Weinberg (1989) provide a wealth of ideas on how to perform interviewing.
- *Brainstorming* is the process of gathering multiple stakeholders in a room, posing an issue or question, encouraging the stakeholders to express their ideas aloud, and having those ideas recorded somehow. The reason for demanding that ideas be expressed aloud is to encourage people to piggyback their own ideas on top of others' ideas. Criticism is generally discouraged. A wide range of variations of such meetings exists. Some variations

enforce anonymity via a tool; some have stakeholders record their own ideas, while others utilize a single scribe to record all ideas; and some discourage voicing the ideas aloud.

- *Conducting collaborative workshops* involves gathering multiple stakeholders together in structured, facilitated workshops to define the requirements for a system. Workshops may run from several hours to several days. During the workshops, facilitators lead stakeholders through a series of preplanned activities designed to produce the requirements deliverables needed. For example, participants may brainstorm on a variety of issues; create or review models, prototypes, or specifications; or negotiate and prioritize requirements. JAD (Wood and Silver, 1995) is probably the most widely known type of collaborative workshop, but there are many other variations, some of which use collaborative tools to increase efficiency (Dean et al., 1997). Gottesdeiner (2002) provides the best compendium of ideas on how to use collaborative group workshops for requirements elicitation.
- *Prototyping* is the process of creating a partial implementation of a system, demonstrating it to stakeholders, and perhaps allowing them to play with it. The bases for prototyping are that customers (a) can often think of new requirements only when they can visualize more basic requirements and (b) often can identify what they don't want more easily than what they do want. Davis (1995) provides the best overall summary of prototyping techniques and effects.
- *Questionnaires* are composed of series of questions that are then distributed to many stakeholders. Their responses are then collected, compiled, and analyzed to arrive at an understanding of general trends among the stakeholders' opinions. Unlike interviews and brainstorming, questionnaires assume that the relevant questions can be articulated in advance. For this reason, they are most effective at confirming well-formulated hypotheses concerning requirements, rather than assisting with the requirements synthesis process itself.
- *Observation* is an ethno-methodological technique where the analyst observes the users and customers performing their regular activities. In such cases, the analyst is passive and aims to not affect the activities in any way. It is the ideal technique for uncovering tacit knowledge possessed by the stakeholders. The best survey of techniques involving observation can be found in Goguen and Linde (1993).
- *Independent study* includes reading about problems and solutions, performing empirical studies, conducting archeological digs (Booch, 2002), or consulting with subject matter experts. Independent study is effective when others have addressed a similar problem before but the problem is relatively new to you.
- *Modeling* involves the creation of representations of the problem or its solutions in a notation that increases communication and provides fresh insights into the problem or solution. A wide range of modeling approaches exist, including object diagrams, data flow diagrams (DFD), the Unified Modeling Language (UML), Z, finite-state machines (FSMs), Petri nets, the System Description Language (SDL), statecharts, flowcharts, use cases, decision tables and trees, and so on. See (Davis, 1993; Kowal, 1992; Wieringa, 1996) for descriptions of most of these modeling notations. Although each provides the analyst with unique insights into the problem or its solution, the largest benefit often comes from using more than one. This is because each induces the analyst to ask (or answer) a certain class of questions, and the combination of multiple models induces more questions than the sum of using each one separately.

Triage

It is a rare project that has sufficient resources to address all the candidate requirements. To overcome this problem, project managers or teams need to conduct a scoping exercise typically called *triage*. Triage is the process of determining the appropriate subset of candidate requirements to attempt to satisfy, given a desired schedule and budget (Davis and Zweig, 1990; Davis, 2003). It is an activity conducted for an individual project that is quite similar to the performance of portfolio management, in which a set of projects are competing for the same finite set of resources and the project manager must choose from among them. See Chapter 2 in Meredith and Mantel (2003).

Triage is conducted in a formal meeting, usually led by the project manager, product manager, or independent facilitator. The participants must include representatives of at least three groups:

- *Primary stakeholders* need to determine the relative priority of candidate requirements and ensure that the voices of all classes of users and customers are expressed. Ideally, these representatives should be customers and users themselves, but often they are composed of marketing personnel, analysts, or subject matter experts.
- *Development* needs to be present to ensure that the requirements selected for inclusion in any release are reasonable relative to the realities of schedule and budget demands.
- *Financial support* must also be present. Otherwise, it is too easy for the other two parties to solve the triage problem by simply increasing available budgets.

Triage can be conducted by viewing the problem as one of balancing a multiarmed seesaw (see Figure 17.3). The three arms are the selected candidate requirements, the available budget, and the desired schedule. These three variables must be repeatedly manipulated until they are in balance. In this case, balance implies that there is a reasonably acceptable probability that the selected requirements can be satisfied by the project within the budget and schedule. Although the traditional development project manager's goal is to ensure

FIGURE 17.3. TRIAGE BALANCES A SEESAW.

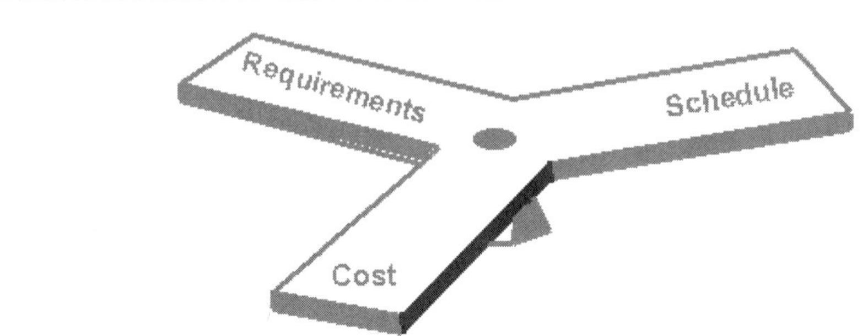

completion on schedule and within budget, an even more responsible project manager takes a larger view. Just because the selected requirements *can* be built within the budget and schedule constraints does not mean that the project *should* be undertaken. A responsible project manager thus considers additional arms of the seesaw, which capture the risks associated with and the effect on achievement of business goals of the selected requirements. Thus, if the product is to be sold externally, additional arms include aspects of marketing, finance, personnel, and other factors as shown in Figure 17.4, adapted from Chapter 2 of Meredith and Mantel (2003). If the product is to be used internally, fewer factors must be considered, as shown in Figure 17.5.

The result of triage is a pruned version of the list shown in Figure 17.4. Although most practitioners think of this as a pruned list, a more reasonable way to visualize it is as the full original list, with each requirement annotated by whether or not it is included in the next release, as shown in Figure 17.6.

FIGURE 17.4. ADDITIONAL SEESAW ARMS.

- Marketing Factors
  - Size of Potential Market
  - Likely Market Share
  - Time Entering Market Window
  - Impact on Existing Products
  - Consumer Acceptance
  - Estimated Life of Product
  - Spin-Off Potential
  - Degree to Which We Understand Market
- Financial Factors
  - Revenue Expectation
  - Profitability (Net Present Value)
  - Cash Flow Impact
  - Payout Period
  - Cash Requirements
  - Time to Breakeven
- Personnel Factors
  - Training Needs
  - Labor Skill Needs
  - Level of Resistance
- Other Factors
  - Impact from Government Standards
  - Impact on Other IT Systems
  - Reaction from Stockholders (if a Corporation)
  - Reaction from Securities Markets (if Publicly Held Company)
  - Patent and Trade Secret Protection
  - Potential for New Patent Creation
  - Impact on Brand
  - Impact on Image with Customers and Competitors
  - Degree to Which We Understand New Technology
  - Ability to Direct and Control New Process
  - Experience We Gain from this Project to Be Applied to Future Projects
  - Average Order Size

Source: Adapted from Meredith and Mantel, 2003.

FIGURE 17.5. ADDITIONAL SEESAW ARMS FOR INTERNAL DEVELOPMENT.

- Demand Factors
  - Size of Potential Use
  - Customer Acceptance
  - Estimated Life of Product
- Financial Factors
  - Increased Revenue Expectation
  - Decreased Cost Expectation
  - Cash Flow Impact
  - Payout Period
  - Cash Requirements
- Personnel Factors
  - Training Needs
  - Labor Skill Needs
  - Level of Resistance
- Other Factors
  - Impact on Other IT Systems
  - Degree to Which We Understand New Technology
  - Ability to Direct and Control New Process
  - Experience We Gain from this Project to Be Applied to Future Projects

Specification

Once a subset of requirements is selected and agreed to by all parties, those requirements need to be refined and documented. This process is often called *requirements specification.*

Forms of Specification. A variety of common practices exist in the industry for documenting requirements, including the following:

- *A polished word-processed document.* Such a document typically follows one of the many standards available in the industry (e.g., IEEE, 1993 and Robertson and Robertson 2000) and is typically called a *software requirements specification* (SRS). Like all technical documents, it is composed of chapters and paragraphs. The biggest advantage of this approach is that all parties can read the document with a minimum of training. On the other hand, the biggest disadvantages are that (a) often many resources are expended polishing the noncritical parts of the document, (b) triage is almost impossible, (c) natural language can prove to be ambiguous, and (d) it is awkward to annotate each requirement *in situ.* This is a popular approach for constructing large embedded real-time critical applications, where "critical" usually means *life*-critical, *financial*-critical, or *security*-critical.
- *A hierarchical list of requirements.* Whether the list is packaged within the constraints of a formal SRS or not, it appears as a two-dimensional table, with each row corresponding to a single requirement and each column corresponding to an attribute of that requirements, including a unique identifier, the text, the priority, estimated development cost, and so on. The biggest advantages of this approach are that (a) all parties can read the list with a minimum of training, (b) fewer words means less time spent polishing, (c) triage

FIGURE 17.6. TRIAGE CREATES A LIST OF SELECTED CANDIDATE REQUIREMENTS.

Requirement	
The user starts the RLM by placing it within the border of a defined lawn and pressing BEGIN MOWING from the Main Menu.	NO / YES
The RLM shall determine if it is in a defined lawn. If not, the RLM shall sound the error tones and display the message MOWER NOT IN RECOGNIZED LAWN on the first line and RETURN on the second line.	NO / YES
If correctly placed, the RLM shall beep once and wait for the user to step back beyond the safe distance range. After the user has moved beyond this range, the RLM shall move to a starting location within the lawn and begin mowing.	NO / YES
While mowing, the RLM's panel shall display nothing except in the event of an error condition, dump or refueling required, or an obstacle comes within the minimum safe distance.	NO / YES
The RLM shall check the grass height, grass type, grass density, and moisture of the lawn to determine the settings proper for cutting. Adjustments to the blade position and speed shall be made as required. When a swath is properly cut, the RLM shall move to an uncut area.	NO / YES
The cutting pattern shall begin with the perimeter of the lawn and work inward to the lawn's center. Each pass shall overlap the previous pass by a width less than or equal to 33% of the RLM's swath but greater than or equal to 25% of the RLM's swath.	NO / YES
This normal cutting pattern may be altered by obstacle avoidance maneuvers but shall resume when avoidance maneuvers are complete.	NO / YES
During avoidance maneuvers, the RLM may, for the sake of fuel efficiency, temporarily shut off its blades if over an area that has been properly cut. Obstacle avoidance is discussed in Requirement 510.	NO / YES
The RLM shall shut off the blades if fouling occurs to the degree that the RLM may damage itself. Should blade fouling occur, the RLM shall sound the error tones and display the message BLADES FOULED on the first line of the display. Should there be more than one blade . . .	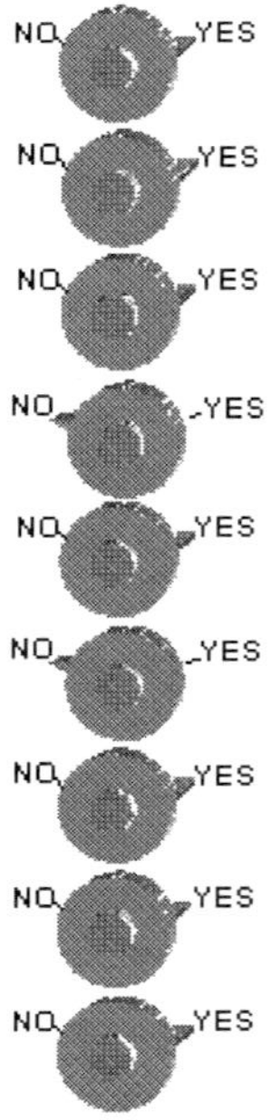 NO / YES

can be performed easily, and (d) it is trivial to annotate the requirements. On the other hand, the biggest disadvantage is that natural language can prove to be ambiguous.

- *Few or no documented requirements.* In this scenario, documentation of requirements is seen as a detractor from getting the product out. In effect, the code is the requirements, or more correctly, the code implies the requirements. The biggest advantage of this approach is that (in theory) no time is required to write or review the requirements, and thus total development time can be reduced by, say, 15 percent. However, this advantage does not come without the considerable risk of building the wrong product altogether. The proponents of this approach possess a variety of motivations. For example, some of those in the entrepreneurial world feel that getting to market fast with an innovative product is so critical to its market success they cannot afford to spend the time "investigating" the requirements—and they may be right! Meanwhile, those in the agile development community (Cockburn, 2002; Highsmith and Cockburn, 2001) claim that they build such small increments of the product, and if they make a mistake in such an iteration, it is easy to back it out and try again. Justification for recording requirements can be found in Hoffman and Lehner (2001).
- *The model is the requirements.* In some industries, requirements are not documented in natural language but are instead captured adequately in a model (see previous discussion of models). For example, in some business applications, a majority of the requirements can be captured using use cases, data flow diagrams, and entity relation diagrams. In some user-interface-intensive applications, a majority of requirements can be captured using use cases. And in some real-time systems, a majority of the requirements can be captured using Petri nets, finite-state machines, or statecharts. The unified modeling language (UML; Booch, 1999) is an attempt to capture all these models in one notation. The biggest advantage to this approach is that systems people (on the IT side and the customer side) can read the notations easily. The biggest disadvantages are that (a) nonsystems people on the customer side have difficulty understanding the notations; (b) no model is sufficient to represent *all* requirements, so they must be augmented in some way (for example, few of the aforementioned notations provide the ability to capture nonbehavioral requirements as described previously); (c) triage is likely to be difficult; and (d) it is almost impossible to annotate individual requirements.
- *The prototype is the requirements.* In this case, a prototype system is constructed and the customer likes it. Then the real system is constructed to mimic the behavior of the prototype. The biggest advantage to this approach is that customers can witness the intended system's behavior first hand. The biggest disadvantages are that, (a) by definition, a prototype does *not* exhibit all the behaviors of the real system, so it must be augmented in some way, (b) triage is likely to be difficult, and (c) it is almost impossible to annotate individual requirements.

All of the approaches can be followed in an incremental manner (i.e., document a little, build a little, validate a little, then repeat) or a full-scale manner (i.e., document a lot, build a lot, validate a lot). Table 17.1 summarizes the advantages and disadvantages of the five approaches. In this table, notice that just because a technique has more check marks in its

TABLE 17.1. DISADVANTAGES OF VARIOUS REQUIREMENTS DOCUMENTATION APPROACHES.

	Documentation Approach				
Disadvantages	Document	List	Few/None	Model	Prototype
Natural language is inherently ambiguous	√	√			
Challenging for multinational efforts	√	√			
Notation not already known by customer				√	
Difficult to annotate individual requirements	√		√	√	√
Difficult to select subset of requirements for inclusion	√		√	√	√
Insufficient to represent *all* requirements				√	√
Could imply unintentional requirements					√
High risk of building the wrong product			√		
Risk of incurring unnecessary up-front (perhaps nonrecoverable) costs	√	√		√	√
Could be challenging to maintain	√	√		√	√
Difficult to trace to origins and be traced from downstream entities	√			√	√
Difficult to diagnose reasons for misunderstandings			√		

column does not necessarily make it a worse approach; each comes with its own inherent risks. Only the project manager can decide which risks are worth taking.

As requirements are documented using any of the precedinig approaches, disagreements will naturally arise concerning what individual requirements mean. In such cases, three solutions exist: (a) document the requirement in less ambiguous terms but using the same general approach, (b) supplement the requirement with another approach that has less ambiguity, and (c) refine the requirement into its constituent subrequirements, as described previously.

Attributes of a Specification. As work proceeds on requirements, they should evolve toward increased value to the project team. That means they should become less ambiguous, more correct, more consistent, and more achievable. For a more complete list of attributes that requirements should exhibit see Davis (1995). The activities involved in determining if the requirements are evolving toward increased quality are generally called *validation and verification*, or V&V for short (Wallace, 1994). There appears to be some confusion within the industry concerning the differences between the two terms as applied to requirements, for example, see Christensen and Thayer (2001), Leffingwell and Widrig (2000), Wiegers (1999); and Young (2001). The confusion arises from the use of the terms in latter phases of system development. In later phases, *verification* of that phase's output is the process of ensuring that the output is correct relative to the outputs of the previous phase, and *validation* of that phase's output is the process of ensuring that the output is correct relative to the requirements (IEEE, 1986). Since requirements are usually considered the first phase of a system development life cycle, those definitions do not apply. However, if you consider that these words imply that verification ensures that the product is being built right and validation ensures that the right product is being built (Boehm, 1982), then we can extrapolate their meanings to requirements, as follows:

- *Requirements verification* ensures that the requirements themselves are written in a quality manner.
- *Requirements validation* ensures that the requirements as documented reflect the actual needs of the users/customers.

Then, to *verify* the quality of requirements, the following attributes must be addressed:

- *Ambiguity* is the condition in which multiple interpretations are possible given the identical requirement. Ambiguity is inherent to some degree in every natural-language statement. Thus, the parties can easily spend their entire project budget attempting to remove every bit of ambiguity. A more successful project will reword or refine a requirement only when the potential for adverse consequences is evident if the requirement stays as is. Another way to decide on whether a requirement statement is "good enough" is to determine if *reasonable, knowledgeable, and prudent* individuals would make different interpretations of the requirement.
- An SRS is *inconsistent* if it contains a subset of requirements that are mutually incompatible. For example, if two requirements are incompatible, or are in conflict with each other, then the SRS is inconsistent. Furthermore, an SRS should also be consistent with all other documents that have been previously agreed to by the parties.
- Requirements should also be *achievable*, which means it is possible to build a system with available technology, and within existing political, cultural, and financial constraints.

To *validate* requirements, the following attribute must be addressed:

- A requirement is *correct* if it helps to satisfy some stakeholder's need. Obviously, if a candidate requirement fails this test, it should be triaged out of the product.

Variations of Requirements Management Practices

Requirements management practices vary based on many aspects of the project. Let's look at some of these aspects and see how they effect requirements management.

Size of iterations

All product development efforts are iterative because as soon as customers start using any product, new requirements appear, thus driving another iteration. The differences lie in how big each iteration is and whether or not the team tries to satisfy "all the known requirements" in each iteration. As iterations increase in size (either in terms of elapsed time or sheer number of requirements), risks increase. In particular, the risks that increase include the likelihood of exceeding the budget, of completing after the desired delivery date, and of failing to meet customer needs. On the other hand, as iterations decrease in size, the effort for overhead tasks become a larger proportion of the total effort. With larger iterations, more effort must be expended during the requirements phases of each iteration.

Relationship of Iterations to Planning

In some cases, an entire product's requirements are explored and documented at project inception, and a product rollout strategy is developed that incorporates successively larger subsets of requirements in each iteration. In other cases, limited requirements activity occurs up front. The initial product is released primarily to acquire requirements feedback. Each successive iteration's requirements are defined based on the feedback acquired from the previous iteration.

Use of Throwaway Prototypes

Any iteration can be prefaced with the construction of a prototype. The purpose of the prototype is to remove the risk of building the wrong iteration. By seeing a prototype, stakeholders can provide valuable feedback concerning whether or not the development team is on the right track. Such an approach reduces the risk of the next iteration. When a prototype is used, minimal requirements effort is expended at project inception. Most requirements are defined after the initial prototype but before the development for the first real iteration begins.

Manufacturing Needed

Some systems require a manufacturing phase after development. This is primarily a function of the media involved. Pure software systems require no manufacturing (other than the trivial creation of CD-ROMs), whereas systems that include physical components do. When manufacturing is required, care must be taken during requirements elicitation and specification to ensure manufacturability and testability.

Research Needed

Some systems require research, invention, or innovation prior to starting the development activities. Usually, requirements are difficult to express when innovative research is needed. In such cases, a set of goals is stated (which are rarely termed requirements). Then the research is performed. Requirements efforts do not commence in earnest until after the research effort is complete.

Management Demand for Sequentiality.

If management enforces the idea that no task may be started until the previous task is completed, then elicitation must be completed before triage begins, and triage must be completed before specification can begin. Only the most conservative of management organizations still adhere to this ancient custom.

Iterative Nature of Requirements Process Itself

Hickey and Davis (2002) describe requirements as an iterative process where each iteration uncovers additional requirements, and changes the current situation. These changes to the situation, and the new requirements uncovered, drive the analysts to modify their approach for the next iteration. This is a more realistic view of the requirements process than attempting to do all elicitation on one phase.

Software-Intensive Applications

Traditionally, software had been developed using large iterations, with all the planning up front, with the assumption of high sequentiality. This approach was termed the *waterfall* model. It is typically represented by a linear PERT chart, as shown in Figure 17.7. Figure 17.8 shows where the requirements activities are performed during the development.

More modern software development projects use the so-called iterative model of software development (also called incremental). There are two general ways to plan the requirements for each iteration: by fixed time and by logical functionality sets. In the former, the length of time for each iteration is set in advance, and then the requirements are managed to ensure that only those requirements that can be satisfied in that time frame are included. Iteration length varies typically from a few weeks to a few months. In the latter way, logical subsets of requirements are grouped together and each iteration is scheduled to be reasonable with respect to the functions it is satisfying. In either case, the iterative method is typically represented as shown in Figure 17.9. Figure 17.10 shows where the requirements activities are performed during the development.

A more recent approach to software development is generally called *agile*. The agile movement (Cockburn, 2002; Highsmith and Cockburn, 2001) proposes a significant decrease in the power of project management and general management, and instead pushes many responsibilities down to the individual contributors. Readers wishing to learn the details of agile development should refer to the sources cited in the previous sentence. Here we discuss the implications of agile methods on requirements management itself. Instead of attempting to elicit requirements at the beginning of the development process, agile devel-

FIGURE 17.7. A WATERFALL MODEL.

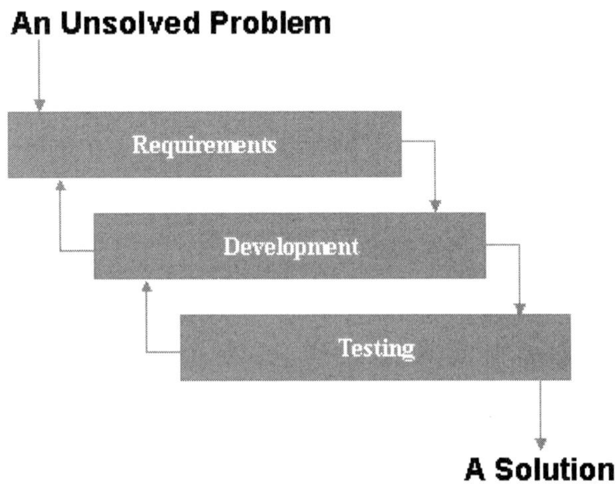

FIGURE 17.8. REQUIREMENTS ACTIVITIES WITHIN A WATERFALL MODEL.

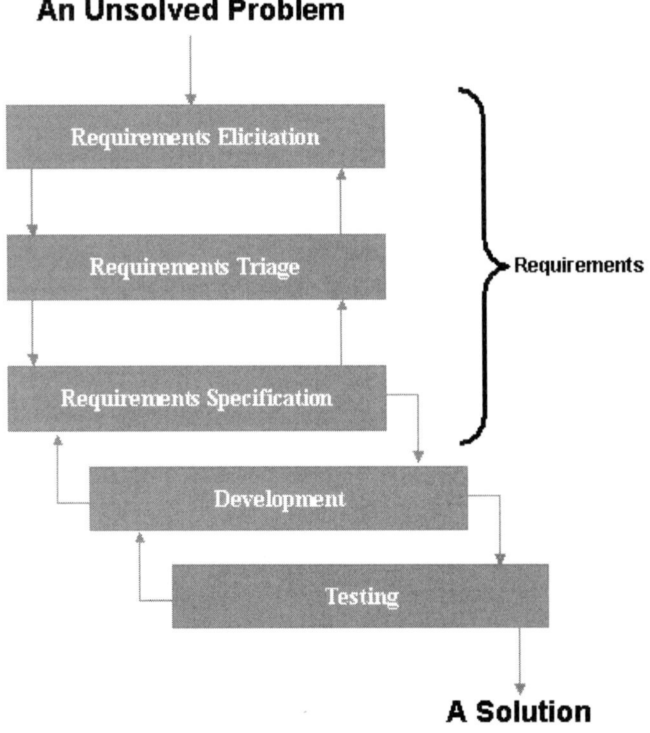

FIGURE 17.9. AN ITERATIVE DEVELOPMENT MODEL.

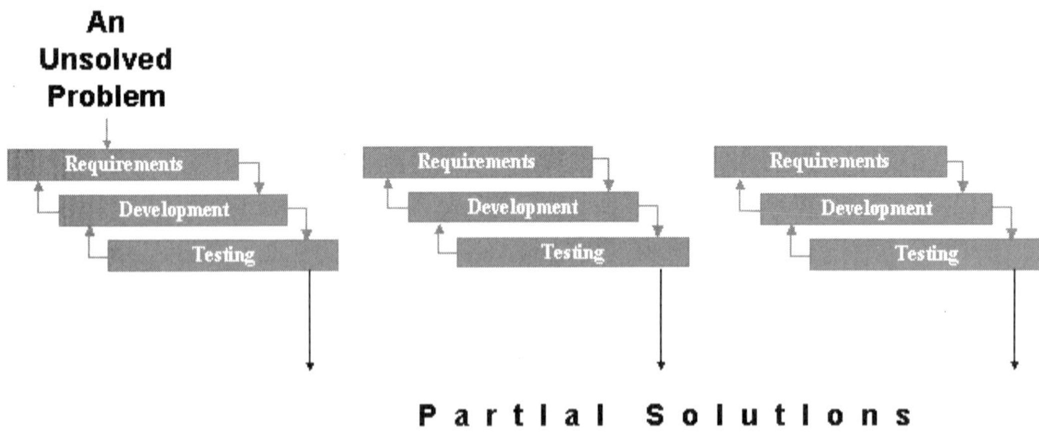

opment recommends that systems be built immediately. Agile developers construct iterations of the system in rapid succession, even as short as every day. A customer is required to be on-site with the development team at all times. Thus, requirements elicitation is performed constantly and is based primarily on the stimulation resulting from seeing system iterations. The omnipresent customer also has exclusive authority to select which requirements to include in each iteration. Thus, elicitation and triage are performed constantly, and specification is not performed *per se.*

Agile development is a reaction by software developers to what they perceive as too much control. The fact is that software development *is* difficult, and it requires a great deal of coordination. Agile development is likely to work well in situations where (a) the requirements are not changing, (b) there is only one customer (or there are more than one customer, but no conflicts exist among the stakeholders), (c) the problem is relatively simple, so that few misunderstandings concerning requirements are likely to arise.

Maintenance Projects

Once a system is deployed, the life of the system, in the eyes of the user, has just begun. Now that the user has had an opportunity to put the product through its paces, there will likely be plenty of feedback regarding the software. This feedback falls into two general categories: (a) failures of the product to meet the intended requirements and (b) requests for new features. The demand for new features will accelerate in any system that is being used (Belady and Lehman, 1976). Rather than allowing the system to be under constant flux, system evolution should be managed as a series of well-planned releases. The length of time between subsequent releases is a function of (a) the rate of arrival of new requirements, (b)

FIGURE 17.10. REQUIREMENTS ACTIVITIES WITHIN AN ITERATIVE DEVELOPMENT MODEL.

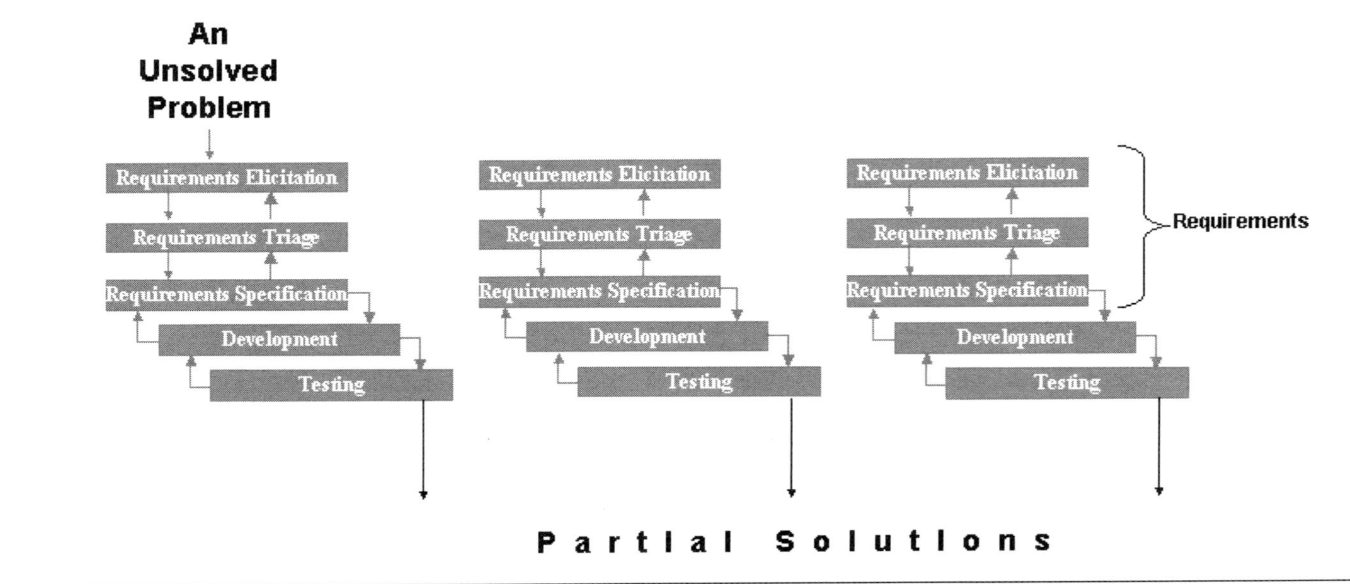

the overhead involved in producing and maintaining a release, and (c) the demand for early satisfaction. As each new requirement is discovered, it should be annotated just like the original requirements and documented in the same way that all previously approved requirements were. When the time arrives to initiate development of a new release, a triage meeting should be held. In principle, the management of post-deployment maintenance releases is no different than the management of predeployment iterations.

After a requirement is approved for a new release, multidirectional traces should be maintained between the change request, the new requirement, and all changes to the product and its documentation made in response to the change request. This enables the development team to (a) undo the changes if they prove erroneous and (b) reconstruct the history of changes made to the product.

Even with the best of processes in place, a product's entropy increases as it evolves (Lehman, 1978). The length of time that a system can survive is a function of the resiliency of the original architecture and the number of changes made over time. shows how the same system could last 7, 12, or 18 years before its entropy renders it no longer maintainable, based solely on the quality of the original architecture.

System Procurement

Many projects are commissioned to solve a problem by procuring, or acquiring, an available system from a third party. In such cases, requirements should still be elicited as described earlier. However, rather than performing an explicit triage step, the team generally prioritizes the elicited requirements and performs a "best fit" analysis with the available solutions.

Tool Issues

A requirements tool is a software application designed to assist the team in performing some combination of requirements elicitation, triage, and specification. Here is a list of the kinds of things such tools could do:

Elicitation

- Collect candidate requirements.
- Allow analysts to record lists of requirements as they are ascertained.
- Allow stakeholders to record their recommended requirements.
- Enforce discipline and/or protocol during elicitation sessions.
- Provide for anonymity during elicitation.
- Prompt for key missing information.

Triage

- Collect priorities and effort estimations.
- Allow analyst to record inclusion/exclusion of each requirement.

FIGURE 17.11. LONGEVITY OF A PRODUCT IS A FUNCTION OF ORIGINAL ARCHITECTURE'S RESILIENCY.

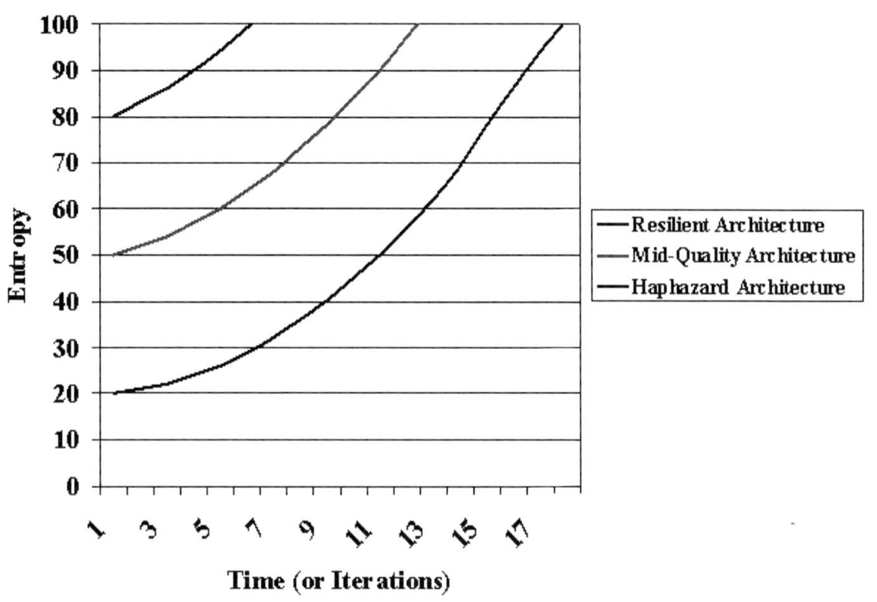

- Determine probability of completing a set of requirements within a given budget.
- Determine probability of completing a set of requirements within a given schedule.
- Allow analyst to refine requirements.

Specification

- Store requirements in a database.
- Determine ambiguities.
- Determine inconsistencies.
- Allow analyst to sort requirements based on multiple criteria.
- Allow analyst to cross-reference[2] requirements among themselves.
- Allow analyst to cross-reference[2] requirements to other products of the development effort (e.g., tests, designs).
- Provide the stakeholders with a simulation of the requirements (i.e., a prototype of the system).

[2] Also termed "traceability."

Requirements tools range from such basic tools as spreadsheets and word processors to extremely sophisticated tools such as special-purpose requirements-based simulation tools. In general, they fall into the following categories:

- *General-purpose tools that happen to be useful during requirements activities.* Word processors allow you to record requirements in natural language either in paragraph form or tabular form. Spreadsheets and databases provide the same capability but also give you the ability to easily define and record attributes such as effort, priority, and inclusion easily. Examples of these tools are Microsoft Word or any other word processor, Microsoft Excel or any other spreadsheet, and Microsoft Access or any other database.

  A majority of projects use these low-cost tools because they are already readily available on desktops with no additional cost. They also present no learning curve for the analysts, stakeholders, or project managers.
- *Meeting facilitation tools.* These tools are particularly helpful during elicitation. They enable stakeholders to record their suggested requirements easily, and even anonymously. They help to keep the discussion on-topic, can sort and filter the candidate requirements easily, and in some cases, can populate a requirements database tool. Two examples of such tools are Ventana's GroupSystems and Meetingworks' Connect.

  Facilitation tools have had surprisingly little impact on most companies. Analysts performing elicitation tend to either interview stakeholders or hold group sessions without tools.
- *Requirements database and traceability tools.* These tools include a database view that is already populated with common requirements attributes. They provide special sorting and filtering capabilities unique to requirements management. Many also provide a word-processed view, so you can update requirements in either the word-processed view or the database view and the other updates automatically. Furthermore, all of these tools make cross-referencing and establishing relationships among requirements easy. Some of these tools are integrated into a full development environment, thus facilitating referencing to and from requirements, designs, and tests. Examples of these tools include RequisitePro from IBM Rational Software, Caliber RM from Borland Software Corporation, and DOORS from Telelogic.

  Approximately 25 percent of all software development projects use requirements database and traceability tools. They significantly reduce the effort expended by analysts in recording and maintaining requirements, but have little impact directly on the stakeholders. One of their biggest advantages is to the project manager who can make intelligent and useful queries such as "Which requirements are high priority, included in the next release, and which are related to software components that Sally is working on."
- *Requirements risk analysis tools.* These tools help the project manager assess the likelihood that the selected requirements will be completed on schedule and within budget. Examples include OnYourMark Pro from Davis and the EstimatePro from Software Productivity Solutions, and part of Caliber RM from Borland Software Corporation.

  These tools have been in existence only since the late 1990s. Early adopters have started experimenting with them, but their adoption has been slow. The primary benefactor is the project manager and, indirectly, the company.

- *Requirements simulation tools.* These tools allow the requirements analyst to simulate the requirements after they have been written. In all cases, the requirements must first be written in a relatively formal notation. One example is Statemate Magnum from I-Logix.
  These tools have been in existence since the early 1970s. All of the vendors have had a hard time finding their niche. The primary benefactor of such tools appears to be the engineering analyst.

In summary, requirements tools can assist analysts in all aspects of requirements management. But no tool makes any aspect of requirements management easy. Elicitation still requires great listening skills. Triage still requires great diplomacy, and specification still requires incredible precision. The tools simply offload the more mundane aspects of the discipline.

Trends in Requirements Management

Research

The field of requirements research is one of the most active in universities. Recent research surveys (Finkelstein, 1994; Hsia et al., 1993; van Lamsweerde et al., 2000; Nuseibeh et al., 2000; and Potts, 1991) have defined the following trends:

- *Data and process modeling* is viewed as a critical activity in requirements. Much of the research since the 1970s has focused on the creation and analysis of modeling notations and techniques. Two somewhat contradictory trends occurring in this area include (1) the increasing emphasis on object-oriented modeling notations (e.g., UML) that focus on the system and (2) the recognition that modeling cannot focus on the system in isolation but must occur in an organizational context (Nuseibeh et al., 2000; Goguen and Jirotka, 1994; and Zave and Jackson, 1997). More recent emphasis has been on techniques to detect errors in models. See the special issue of the *Requirements Engineering Journal* guest edited by Easterbrook and Chechik (2002).
- *Increasing formality* to improve the quality and testability of requirements specifications has been a goal of requirements research (Hsia et al., 1993), especially for process control and life- and safety-critical systems (van Lamsweerde et al., 2000). For example, in the area of reactive systems for process control, specification notations and languages such as SCR Heninger, 1980), CORE (Faulk, 1992), and RSML (Leveson et al., 1994) have been developed to support automated consistency and completeness checking. Formal specification languages such as Z (Spivey, 1990) and others are designed to support requirements verification, visualization, and simulation.
- *Viewpoints* explicitly capture different perspectives or views of multiple stakeholders. Viewpoint integration can be used to check for consistency and aid in the resolution of conflicts among stakeholders (Easterbrook, 1994; Nuseibeh and Easterbrook, 1994). The earliest references to using viewpoints date back to 1981 (Orr, 1981).
- Since the beginning of requirements research, attempts have been made to *reduce ambiguity* in requirements. Obviously, the aforementioned activities of modeling and increasing

formality are aimed at this goal. Additional research has been done to either reduce or detect ambiguity in natural-language specifications. This includes work as early as 1981 (Casey and Taylor, 1981) and extends to the current day (Duran et al., 2002).

- *Goal-oriented requirements elicitation* takes an organizational approach to completeness and consistency checking of requirements by explicitly identifying and representing organizational goals for the system, and then checking the requirements against those goals (van Lamsweerde et al., 2000). Research in this area has resulted in a variety of methods and notations for representing, analyzing, and resolving conflicts between goals including KAOS (Dardenne et al., 1993; van Lamsweerde et al., 1998) and NFR (Mylopoulos, 1992).
- Behavioral requirements have always been the primary emphasis in requirements research. However, *nonbehavioral requirements* have also been addressed for many years and continues to be the focus of many research efforts. Some efforts have spanned the wide range of nonbehavioral requirements, for instance Chung et al. (1993), Chung (2000), Cysneiros and Leite (2002), Kirner and Davis (1995) Mostert and van Solms (1995), and Mylopoulos (1992), and others emphasize specific kinds of nonbehavioral requirements such as security (Shim and Shim, 1992), safety (Berry, 1998; Hansen et al., 1998), and performance (Nixon, 1993).
- *Scenarios* are concrete descriptions of the sequence of activities that users engage in when performing a specific task (Carroll, 1995). Studies have shown than scenarios are extremely useful for requirements elicitation when users are having difficulty specifying goals or using more abstract modeling techniques (Weidenhaupt, 1998; Jarke, 1999; van Lamsweerde, 2000). Scenarios have also proven useful in systems design and testing, for example, in user interface design (Carroll, 1995), and for generating test cases (Hsia, 1994). Other scenario uses are described in an *IEEE Transactions on Software Engineering* special issue on scenarios in (Jarke and Kurki-Suonio, 1998). Finally, scenarios are closely related to the Jacobson's use cases (Jacobson et al., 1992) in object-oriented analysis and the user stories, which are a key component of XP (Beck, 2000).
- With the wide variety of requirements techniques now in existence, some researchers are focusing on the *criteria for technique selection.* For example, Hickey and Davis (2003, 2003a) describe the best way to select the right elicitation techniques. Similar research still needs to be conducted for model selection.
- The field of software (design and code) reuse has settled into a status quo now; modern programming languages include large libraries of reusable entities whose use has become standard. However, *requirements reuse* has not yet reached this level of maturity. Perhaps this is because reusing requirements has little direct benefit to increasing quality or productivity. Instead, the real potential benefit of requirements reuse comes from the second-order effect of reusing design and code components associated with the reused requirements. See Castano and Antenellis (1993), Homod and Rine (1999), van Lamsweerde (1997), and Maiden and Sutcliffe (1996) for some of the latest ideas on requirements reuse.

Practice

It is surprising how little of the current research in the requirements field is making its way to practice (Davis and Hickey, 2002). From the inception of software engineering as a discipline in the 1970s until the current day, (a) the standard for documenting requirements has been the word-processed SRS, (b) analysts in specialized applications have advocated the use of models, and (c) a counterculture has existed that is firmly convinced that writing requirements is primarily a waste of time.

In spite of the enormity of these invariants, a few changes have occurred. Two of these changes are in the evolution of the modeling notations themselves. The first is the introduction of new notations that provide unique perspectives of the system under specification. Classic among these are the introductions of statecharts by Harel (Harel, 1988; and Harel and Politi, 1998). Second is the tendency for the industry to move from sets of specialized notations (which in theory force analysts to become skilled in multiple languages) to all-encompassing notations (which in theory force analysts to become skilled in just one language, albeit enormous), and back to the specialized languages in a cycle. We expect this cycle to continue indefinitely into the future.

Another trend is in the isolation of optimal "starting points" for requirements activities. For many years, analysts have struggled with the question of where to start because of the sheer enormity of requirements. We have thus seen structured analysis (DeMarco, 1979) augmented by events as starting points (McMenamin and Palmer, 1984), and object-oriented analysis (Booch, 1994) augmented with use cases as starting points (Jacobson et al., 1992). This trend will continue. Unfortunately, every situation demands starting points that are a unique function of situational characteristics.

Summary

Project management cannot succeed without careful attention to requirements management. Requirements management is responsible for determining the real needs of the customers, as well as clearly documenting the desired external behavior of the system being constructed by the project. If either of these goals is ignored, the project is guaranteed to result in failure.

References

Beck, K. 2000. *Extreme programming explained.* Boston: Addison-Wesley.

Belady, L., and M. Lehman.1976. A model of large program development. *IBM Systems Journal* 15 (3, March): 225–252.

Berry, D. 1998. The safety requirements engineering dilemma. *Ninth International Workshop on Software Specification and Design.* 147–149. Los Alamitos, CA: IEEE Computer Society Press.

Boehm, B. 1982. *Software engineering economics.* Upper Saddle River, NJ: Prentice Hall.

Booch, G., 1994. *Object-oriented analysis and design.* Redwood City, CA: Benjamin/Cummings.

———. Personal conversation with two of the authors; September 17, 2002, Colorado Springs, Colorado.

Booch, G., et al. 1999. *The Unified Modeling Language user guide.* Reading, MA: Addison-Wesley.

Borland Software Corporation, Inc. 2003. See www.borland.com/products or www.starbase.com/products.

Carroll, J., ed. 1995. *Scenario-based design: Envisioning work and technology in system development.* New York: Wiley.

Casey, B., and B. Taylor. 1981. Writing requirements in English: A natural alternative. 95–101.*IEEE Software Engineering Standards Workshop*. Los Alamitos, CA: IEEE Computer Society Press.

Castano, S., and V. De Antonellis. 1993. Reuse of conceptual requirements specification. 121–124. *International Symposium on Requirements Engineering,* January. Los Alamitos, CA: IEEE Computer Society Press,

Christensen, M., and R. Thayer. 2001. *The project manager's guide to software engineering's best practices*. Los Alamitos, CA: IEEE Computer Society Press.

Chung, L. 1993. *Representing and using non-functional requirements: A process-oriented approach*. Department of Computer Science. PhD. thesis, University of Toronto.

Chung, L., et al. 2000. *Non-functional requirements in software engineering*. Norwell, MA: Kluwer.

Cleland, D., and L. Ireland. 2000. *Project manager's portable handbook*. New York: McGraw-Hill.

Cockburn, A. 2002. *Agile software development*. Boston: Addison-Wesley.

Cysneiros, M., and J. Leite, 2002. Non-functional requirements: From elicitation to modeling languages. 699–700. *Twenty-fourth International Conference on Software Engineering*. Los Alamitos, CA: IEEE Computer Society Press.

Dardenne, A., et al. 1993. Goal-directed requirements acquisition. *Science of Computer Programming* 20:3–50.

Davis, A., 1993. *Software requirements: Objects, functions, and states*. Upper Saddle River, NJ: Prentice Hall.

———. 1995. Software prototyping. *Advances in Computers 40*. 39–63. New York: Academic Press.

———. 2002. Requirements management. In *Encyclopedia of software engineering*. 2nd ed., ed. J. Marciniak. New York: Wiley-Interscience.

———. 2003. Secrets of requirements triage. *Computer* 36 (3, March): 42–49.

Davis, A., and A. Zweig. 2000. The missing piece of software development. *Journal of Systems and Software* 53 (3, September): 205–206.

Davis, A., et al. 1993. Identifying and measuring quality in software requirements specifications. 141–152. *IEEE-CS International Software Metrics Symposium*. Los Alamitos, CA: IEEE Computer Society Press.

Davis, A., and A. Hickey. 2002. Requirements researchers: Do we practice what we preach? *Requirements Engineering Journal* 7(2):107–111.

Dean, D., et al. (1997–1998. Enabling the effective involvement of multiple users: Methods and tools for collaborative software engineering. *Journal of Management Information Systems* 14 (3, Winter): 179–222.

DeMarco, T. 1979. *Structured analysis and system specification*. Upper Saddle River, NJ: Prentice Hall.

Duran, A., et al. 2002. Verifying software requirements with XSLT. *ACM Software Engineering Notes* 27: 39 ff.

Easterbrook, S. 1994. Resolving requirements conflicts with computer-supported negotiation. In *Requirements engineering: Social and technical Issues*, ed. M. Jirotka and J. Goguen. 41–65. London: Academic Press.

Easterbrook, S., and M. Chechik 2002. Guest editorial: Special issue on model checking in requirements engineering. *Requirements Engineering* 7(4):221–224.

Faulk, S. 1997. Software requirements: A tutorial. In *Software Requirements Engineering*, ed. R. Thayer and M. Dorfman. 128–149. Los Alamitos, CA: IEEE Computer Society.

Faulk, S., et al.1992. The CORE method for real-time requirements *IEEE Software* (September): 22–33.

Finkelstein, A.1994. Requirements engineering: A review and research agenda. 10–14. *First Asia-Pacific Software Engineering Conference*. December. Los Alamitos, CA: IEEE Computer Society.,

Gause, D., and J. Weinberg 1989. *Exploring requirements: Quality before design*. New York: Dorset House.

Goguen, J., and C. Linde 1993. Software requirements analysis and specification in Europe: An overview. 152–164. *First International Symposium on Requirements Engineering*. Los Alamitos, CA: IEEE Computer Society Press.

Goguen, J., and M. Jirotka, eds.1994. *Requirements engineering: Social and technical issues.* Boston: Academic Press.

Gottesdeiner, E. 2002. *Requirements by collaboration.* Reading, MA: Addison-Wesley.

Hansen, K., et al. 1998. From safety analysis to software requirements. *IEEE Transactions on Software Engineering* 24 (7, July): 573–584.

Harel, D.1988. On visual formalisms. *Communications of the ACM* 31 (5, May): 514–530.

Harel, D., and M. Politi 1998. *Modeling reactive systems with statecharts.* New York: McGraw Hill.

Heninger, K.1980. Specifying software requirements for complex systems: New techniques and their application. *IEEE Transactions on Software Engineering* 6(1):2–13.

Hickey, A., and A. Davis. 2002. The role of requirements elicitation techniques in achieving software quality. *International Workshop on Requirements Engineering: Foundations for Software Quality (REFSQ).* Los Alamitos, CA: IEEE Computer Society Press.

———. 2003a. Requirements elicitation and requirements elicitation technique selection: A model of two knowledge-intensive software development processes. *Proceedings of the Thirty-Sixth Hawaii International Conference on System Sciences.* Los Alamitos, CA: IEEE Computer Society Press.

———. 2003b. Elicitation technique selection: How do the experts do it?" International Joint Conference on Requirements Engineering (RE03). September. Los Alamitos, CA: IEEE Computer Society Press.

Highsmith, J., and A. Cockburn. 2001. Agile software development: The business of innovation. *Computer* (September): 120–122.

Hofmann, H., and F. Lehner 2001. Requirements engineering as a success factor in software projects. *IEEE Software* 18 (4, July/August): 58–66.

Homod, S., and D. Rine. 1999. Building requirements repository using requirements transformation techniques to support requirements reuse. *World Multi-Conference on Systemics, Cybernetics and Informatics,* Volume 2.

Hsia, P., et al. 1993. Status report: Requirements engineering. *IEEE Software* 10 (6, November): 75–79.

Hsia, P., et al. 1994. Formal approach to scenario analysis. *IEEE Software* 11(2):33–41.

IEEE. 1983. *IEEE standard glossary of software engineering terminology.* IEEE Standard 729. New York: IEEE Press.

———. 1986. *IEEE standard for software verification and validation plans.* IEEE Standard 1012. New York: IEEE Press.

———.1993. *A guide to software requirements specifications.* Standard 830-1993. New York: IEEE Press.

I-Logix Corporation. www.ilogix.com/products/magnum/index.cfm.

Jacobson, I., et al. 1992. *Object-oriented software engineering.* Reading, MA: Addison-Wesley.

Jarke, M., and R. Kurki-Suonio. 1998. Guest editorial: Introduction to the special issue. *IEEE Transactions on Software Engineering* 24(12):1033–1035.

Jarke, M. 1999. Scenarios for modeling. *Communications of the ACM* 42(1): 47–48.

Kirner, T., and A. Davis. 1996. Nonfunctional requirements for real-time systems. *Advances in Computers.*

Knapp, M., and J. Hall. 1997. *Nonverbal communication in human interaction.* Austin, TX: Holt, Rinehart and Winston.

Kotonya, G., and I. Sommerville. 1997. Integrating safety analysis and requirements engineering.259–271. *Fourth Asia-Pacific Software Engineering Conference.* Los Alamitos, CA: IEEE Computer Society.

Kowal, J. 1992. *Behavior models.* Upper Saddle River, NJ: Prentice Hall.

Lam, W., et al. 1997. Ten steps towards systematic requirements reuse. 6–15. *IEEE International Symposium on Requirements Engineering.* January Los Alamitos, CA: IEEE Computer Society Press. Also appears in *Requirements Engineering Journal* 2(2):102–113.

van Lamsweerde, A. 2000. Requirements engineering in the year 00: A research perspective. *Proceedings of the 22nd International Conference on Software Engineering.* 5–19. New York: ACM Press, van Lamsweerde, A., et al.1998. Managing conflicts in goal-driven requirements engineering. *IEEE Transactions on Software Engineering* 24 (11, November): 908–926.

Lauesen, S. 2002. *Software requirements: Styles and techniques.* London: Addison-Wesley.

Leffingwell, D., and D. Widrig. 2000. *Managing software requirements.* Reading, MA: Addison-Wesley.

Lehman, M. 1978. *InfoTech State of the Art Conference on Why Software Projects Fail.* Paper #11, April.

Leveson, N., et al. 1994. Requirements specification for process-control systems. *IEEE Transactions on Software Engineering* 20 (9, September): 684–706.

McMenamin, J., and J. Palmer. 1984. *Essential systems analysis.* Upper Saddle River, NJ: Prentice Hall.

Maiden, N., and A. Sutcliffe. 1996. Analogical retrieval in reuse-oriented requirement engineering. *Software Engineering Journal* 11(5):281–292.

Meetingworks, Inc. 2003. www.meetingworks.com.

Meredith, J., and S. Mantel. 2003. *Project management: A managerial approach.* 5th ed. New York: Wiley.

Microsoft, Inc. 2003. www.microsoft.com..

Mostert, D., and S. von Solms. 1995. A technique to include computer security, safety, and resilience requirements as part of the requirements specification. *Journal of Systems and Software* 31 (1, October): 45–53.

Mylopoulos, J., et al. 1992. Representing and using nonfunctional requirements: A process-oriented approach. *IEEE Transactions on Software Engineering* 18(6, June): 483–497.

Nixon, B. 1993. Dealing with performance requirements during the development of information systems. 42–49. *IEEE International Symposium on Requirements Engineering.* Los Alamitos, CA: IEEE Computer Society Press.

Nuseibeh, B., et al. 1994. A framework for expressing the relationships between multiple views in requirements specifications. *IEEE Transactions on Software Engineering* 20 (10, October): 760–773.

Nuseibeh, B., and S. Easterbrook. 2000. Requirements engineering: A roadmap. *Proceedings of the 22nd International Conference on Software Engineering.* 35–46. New York: ACM Press.

Opdahl, A. 1994. Requirements engineering for software performance, *International Workshop on Requirements Engineering: Foundations of Software Quality.* June.

Orr, K. 1981. *Structured requirements definition.* Topeka, Kansas: Ken Orr and Associates.

Project Management Institute. 2000. *A guide to the project management body of knowledge.* Newtown Square, PA: Project Management Institute.

Potts, C. 1991. Seven (plus or minus two) challenges for requirements research. *Sixth International Workshop on Software Specification and Design.* Los Alamitos, CA: IEEE Computer Society.

Rational Software Corporation, Inc. 2003. www.rational.com/products.

Robertson, J., and S. Robertson. 2000. *Mastering the requirements process.* Reading, MA: Addison-Wesley.

Reinertsen, D. 1997. *Managing the design factory.* New York: Free Press.

Shim, Y., H. Shim, et al. 1997. Specification and analysis of security requirements for distributed applications. 374–381. *Ninth IEEE International Conference on Software Engineering and Knowledge Engineering.* June. Skokie, IL: Knowledge Systems Institute.

Siddiqi, J., and C. Shekaran. 1996. Requirements engineering: The emerging wisdom. *IEEE Software* 13(2):15–19.

Software Productivity Centre, Inc. 2003. http://www.spc.ca/products/estimate.

Spivey, J. 1990. An introduction to Z and formal specifications. *Software Engineering Journal* 4:40–50.

The Standish Group. Undated. *The CHAOS Chronicles* www.standishgroup.com.

Swartout, W., and R. Balzer 1982. On the inevitable intertwining of specifications and design. *Communications of the ACM* 25 (7, July): 438–440.

Telelogic, Inc. 2003. www.telelogic.com/products.

Thayer, R., and M. Dorfman 1994. *Standards, guidelines, and examples on system and software requirements engineering.* Los Alamitos, CA: IEEE Computer Society Press.

Ventana, Inc. 2003. www.ventana.com.

Wallace, D. 1994. Verification and validation. In *Encyclopedia of Software Engineering*, ed., J. Marciniak. 1410–1433. New York: Wiley.

Weidenhaupt, K., et al. 1998. Scenarios in system development: Current practice. *IEEE Software* 15(2): 34–45.

Wieringa, R. 1996. *Requirements engineering*. Chichester, UK: Wiley.

Wiegers, K. 1999. *Software requirements*. Redmond, WA: Microsoft Press.

Wood, J., and D. Silver 1995. *Joint application development*. 2nd ed. New York: Wiley.

Young, R. 2001. *Effective requirements practices*. Boston: Addison-Wesley.

Zave, P., and M. Jackson. 1997. Four dark corners of requirements engineering. *ACM Transactions on Software Engineering and Methodology* 6 (1, January): 1–30.

CHAPTER EIGHTEEN

DESIGN MANAGEMENT

Peter Harpum

Design is of primary importance in the project and is carried out throughout the project life cycle. Design begins with the business case formulation for a project—how the project can most effectively and efficiently deliver the benefits to the organization that are required of it. At the other end of the life cycle, when the project's deliverables are being decommissioned (whether it is a nuclear power station, a financial service product, or computer software), design work is required to ensure that the products that the project made are effectively removed from the environment.

Central to the notion of design is creativity—creation of the business case, outline design, in-service improvements, and work in all other stages of the project that have some element of design. Creativity, however, is notoriously hard to define, and in many people's opinion even harder to manage. Much of the difficulty in managing design is found at the psychological interface between what is seen as an "instrumentalist" project management paradigm—that is, a tool used to predict the future—and the creative flair, and at times genius, needed for great design work (Allinson, 1997).

This chapter describes the following:

- Design in the context of projects, discussing its importance and specific characteristics
- The strategic design management considerations, including the philosophical approaches that can be taken
- Control of the design process, in terms of scope, schedule, budget, and quality.

Design in the Context of Projects

The Importance of Design in the Project

There are many different views on the role and process of design management. At one end of the spectrum is the approach that design is *the* dominant business process. This is common in industries that depend on a continuous supply of new products for sustaining profitability, such as consumer goods and computer software. In such industries the design function is often represented at Board level. Project managers in these companies are relatively junior compared to those managing design, often reporting to the design manager of the new product (Topalian, 1980; Cooper, 1995). At the other end of the spectrum, project management is the dominant process, and design managers have only a coordination role between different design groups. This situation is more likely to be found in industries that have a history of being implementation-oriented—that is, they are focused on making the project's products (the project "deliverables"). Construction projects are more likely to follow this arrangement (though signature architects may dominate project managers).

These differences in perception of how design is managed within the corporate context reflect on the various models of design management. Where design dominates the organizational culture, the strategies and tactics of design management center on the relationships between business-as-usual and individual projects to ensure corporate value is created through design. When the *delivery* of projects to external buyers is the dominant paradigm design management models more commonly address information flow, interface control, and the logistics of producing error-free, high-quality design work, on time and on budget (Gray et al., 1994). These types of projects often "buy-in" design services from third-party design consultants, since the core capabilities of firms that deliver such projects are in the management of the implementation phases.

Characteristics of Design in Projects

Project management has its roots very much in the management science and systems engineering fields. The earliest modern tools for managing projects evolved from the scientific school of management thinking during the early part of the twentieth century: the Gantt chart and the little known Harmonygraph. These were later to form the foundation of scheduling techniques such as program evaluation and review technique (PERT) and the critical path method (and more recently Gantt–chart-based software), developed primarily to help plan large systems engineering programs in the U.S. defense sector (Morris, 1994).

This meant that from the beginnings of modern project management, designers were being asked to work to explicit schedules—schedules they often perceived to be inflexible and that did not reflect the reality of their work, and designers believed design could not be scheduled. This view persists to this day. Many designers resent having time constraints imposed on them, based on what they consider to be a mechanistic, and hence unrealistic, management paradigm: project management.

The separation of design from "making" is still only recent. Up to the beginning of the twentieth century it was normal for craftspeople to design and make whatever it was they were producing. In this way of working, the delivery of the "product" was the responsibility of one person. This approach, where design and making are inseparable, also applied to most of the large engineering projects of the time (one thinks of Brunel, Telford, Stephenson, and other great engineers of the seventeenth and eighteenth centuries). Many of these engineers were better at design than at the management of those responsible for the building of their creations. In short, many of these people are viewed in retrospect as more artist than scientist. Around the end of the nineteenth century, when the work required to turn the design solution into a physical reality became increasingly sophisticated, these two fundamental aspects of projects began to be separated (Lawson, 1997). Since then, the project manager has been striving to *reintegrate* designing and making, made challenging by the intrinsically different mind-set required for design in comparison to implementation (Harpum and Gale, 1999).

This difference in mind-sets between the two groups can be shown as a steady change across the life cycle of the project (see Figure 18.1). The designer works on the left of the diagram, where there is a greater freedom (a larger "action space"). This means multiple perspectives can be generated in order to solve a problem—therefore creating many possible solutions. As the project life cycle moves inexorably through its early stages, the degree of freedom gets smaller as the final design solution becomes clear. The action space continues to become smaller as the stages in the life cycle move into implementation and completion; there is less and less ability for the solution to be changed (and it becomes increasingly expensive in time and money to make any changes).

FIGURE 18.1. THE RELATIONSHIP BETWEEN THE PROJECT LIFE CYCLE AND THE DESIGNER'S ACTION SPACE.

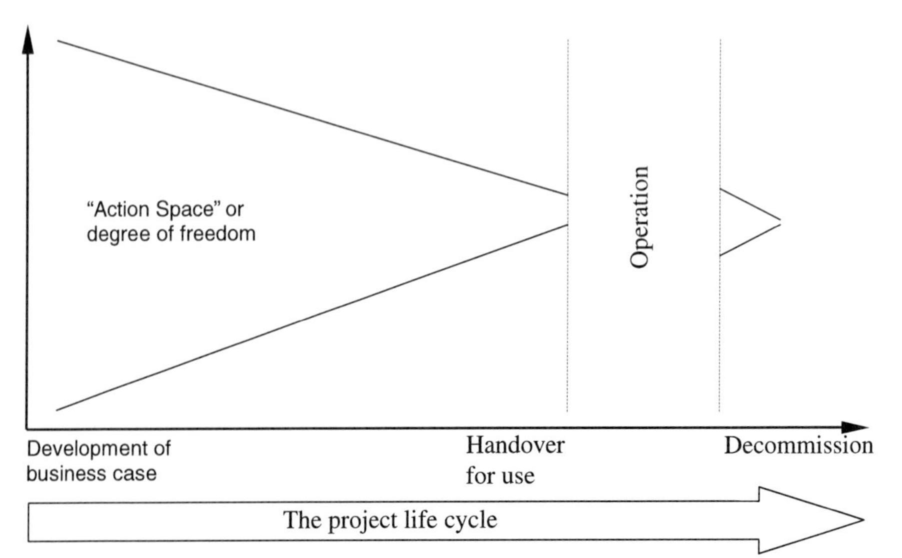

Source: Developed from an earlier version in Smith (2002).

Recent research shows that there can be significant advantages for projects where the power relationship between the project manager, the design manager, and the manager of the implementation phases of the project is more balanced than has been the case historically. This is being driven in part by increasing emphasis on the project front-end by project sponsors and other significant players in the project community, seeking to shorten delivery times and maximize value by reducing rework (and indeed aborted implementation work). It is also driven by the increasing projectization of many sectors of industry, including those where design management has traditionally played a strong role in product development—meaning in these industries that the project manager's authority has increased in relation to the design manager. An increasing awareness of the fundamental importance of the design phases of projects is also no doubt reshaping the relationship between creative designers and action-oriented implementation people.

Fundamentally, design is about creativity—creating solutions to problems. Therefore, managing design means managing creative people. This is inherently difficult. The large degree of freedom that design needs to be most effective, the requirement for room to create multiple solutions before any final one is chosen, tends to preclude artificial restriction. Creativity is about making connections between ideas that are often not obviously connected, and we are only just beginning to understand how this happens in our minds. Yet design within the project context (which actually means almost all design work carried out) must have some element of control placed on it. Nearly all projects have *some* time and cost constraints placed on them. And this is the conundrum at the heart of managing design. Design must be managed to ensure project success, but by its very nature, design rejects the concept of management (Allinson, 1997).

How Designers Design

Effective management requires a comprehension of the activity that is to be managed. Managing design is no different. However, the traditional mechanistic, and predominantly implementation-oriented, paradigm of project management has tended not to acknowledge the creative (hence artistic) aspect of design. Therefore, it is important to have an understanding of the unpredictable nature of the creativity required to design to enable more effective management of this work. This means acknowledging that the design process is unpredictable.

A number of theories exist to describe how the creative process works. None has yet been able to make the process more predictable in terms of the time frame required to find a particular solution to a particular problem. However, the theories do help us understand the process through which creative thought moves. Perhaps the most common model of the creative process is the Assess-Synthesize-Evaluate model (Lawson, 1997), shown in Figure 18.2.

Each step is described as follows:

- *Assess.* A number of information inputs are considered by the creator in relation to the problem to be solved. Some of these inputs will be well known—for instance, known solutions to similar problems—others will be more fragmented in the thinker's mind, without any great clarity about their relevance to the problem.

FIGURE 18.2. THE ASSESS-SYNTHESIS-EVALUATE MODEL OF CREATIVITY.

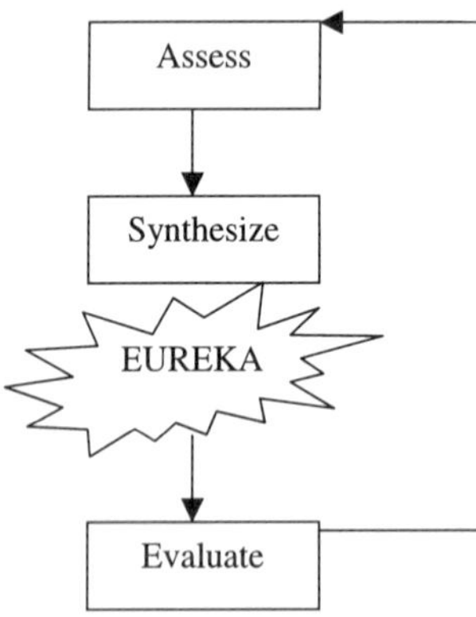

Source: After Lawson (1997). Copyright (1997), with permission from Elsevier.

- *Synthesize.* Conscious and subconscious mental activity in the thinker tries to make the linkages between the various information inputs that will provide solutions to the problem. This period can become intensely frustrating, as often the more the solutions are forced, the more difficult their crystallization becomes. The thinker may move away from actively trying to solve the problem, while the subconscious continues to ponder the solutions.
- *Eureka.* This is an intense, usually short period of time when what often seems to be a fully formulated solution comes into the thinker's mind.
- *Evaluate.* The thinker now evaluates the solution, usually finding that it is not quite as complete as at first seemed to be the case. The flaws in the solution are explored and the information is fed back into the assess stage of the process.

This creative process may cycle many times until a final solution is found; at other times a complete solution is found at the first attempt. There are many examples that demonstrate how difficult it is to know when, or how, a solution will be found (Csikszentmihalyi, 1996). Murray Gell-Mann, the Nobel-prize-winning physicist, articulates the reality of this unpredictability when recollecting a meeting of a group of physicists, biologists, painters, and poets and their discussion on creativity:

> First we had worked, for days or weeks or months, filling our minds with the difficulties of the problem in question and trying to overcome them. Second, there had come a time when further conscious thought was useless, even though we continued to carry the problem around with us. Third, suddenly, while we were cycling or shaving or cooking . . . the crucial idea had come. We had shaken loose of the rut we were in.'
>
> Gell-Mann, 1994

The unpredictability comes from the fact that it can be very difficult to forecast how long the synthesis phase of the process will take. Clearly some problems are harder to solve than others, and not all problems necessarily need great creative thought to produce acceptable solutions. The more difficult problems when designing in the project context are usually found early in the life cycle. It is normally the case that greater creativity is needed when developing concept designs than when working on the detailed solutions to the chosen concept. This accords with Figure 18.1, since in detail design, there is less freedom for the designer to work within; the solution has become constrained. (This is not always the case, of course. During detail design, a solution that was expected to be relatively easy may turn out to be very difficult indeed. This is where many projects encounter their first significant delays, as much more time is used up on the particularly difficult design task than was forecast.)

Strategic Design Management Considerations

The first part of the chapter discussed the inherent difficulty of managing the unpredictable nature of creative design work. Project management is viewed as mechanistic and unsympathetic to design work. However, working within explicit processes can *facilitate* creativity, as less attention needs to be paid by the designer to ensure ad hoc processes are in place to meet the needs and constraints of the project (Luckman, 1984; Pugh, 1990). The next part of this chapter examines the strategic approaches and techniques associated with design management processes.

Aside from the creative aspect of design, there are also various ways of looking at the overall design process. The way design is approached affects the way that the design processes are organized. This in turn determines the effectiveness of the management control that can be brought to bear.

Philosophical Approaches to Organizing Design Work

Organizing the way that the design work is carried out by teams, and the people in them, is not a trivial activity. From a high-level and generic point of view, the design process options are as follows (Simon,1981; Kappel and Rubenstein, 1999):

Design Team Approach	*Designer's Personal Approach*	
	Depth First	**Intuitive**
Bottom-up		
Top-down		
Meet-in-the-middle		

- *Bottom-up.* Basic elements of the solution are created and then put together, changing them until an overall fit is achieved.
- *Top-down.* The desired end solution is conceptualized, and the designer then works backwards until all the basic elements have been completed.
- *Meet-in-the-middle.* As the name implies, top-down and bottom-up approaches are combined until the design is fully integrated.
- *Depth first.* The designer takes whichever possible solution is conceptualized first and attempts immediately to make it work.
- *Intuitive.* The designer considers several possible solutions but takes the one that intuitively seems to offer the best hope of working.

Hence, there are six possible approaches to the way in which the design work can be organized. The option chosen will have implications for the selection of team members. It is almost inevitable that the design process chosen will not perfectly match all the members of the design team. This does not spell disaster for the team, but it does mean that the manager of design ought to be aware of the possibility for mismatch, and motivate and lead individuals accordingly. It is worth pointing out here that even in groups that are inherently highly motivated, such as is found in "skunk works" environments, great care and effort is needed by the manager to get the best from each team member.

The models described previously are by necessity fairly abstract. Developing a design process for a particular project context requires "mapping" design activity in a more detailed way, showing generally what work must be done at consecutive points in the process. Such a map will include processes for (at the least) the following:

- Determining functions of the design solution (and often their physical structures)
- Elaborating specifications
- Searching for solution principles
- Developing layouts
- Optimizing design forms;
- Dividing design work into realizable modules.

Systems Engineering

Some sectors of industry use systems engineering as a fundamental and core part of their design process. Indeed, in computer software and hardware design it is synonymous with the design process. Other industries, such as aerospace and electronics, are similar. However, formal systems engineering is little known in other industries. In some cases, this seems surprising, since the design solutions are quite similar in nature to computers, aircraft, and electronic circuits. For example, building design is clearly about creating a system with multiple subsystems (heating, ventilation, water, waste disposal, lighting, power), yet the discipline has made little contribution so far to building design (Groák, 1992).

There are two reasons for including a review of the subject in this chapter:

1. Significant parts of industry use formal systems engineering as a design approach.
2. Most design solutions (some would say all) are of the nature of a system, and knowledge of the formal approach may be beneficial to those not currently using it.

Many of the design solutions that are needed to satisfy project objectives can be classified as systems (indeed, in a purist sense, every solution is a system, or at the very least becomes part of a system). It is not easy to define a system in a readily understandable way, while at the same time being totally clear and unambiguous about what is meant. The term system (in the context of a design solution to a problem) implies that

- a number of elements must work together to deliver a consistent output;
- those elements are dependent on each other for their proper functioning.

Systems can be "open" or "closed." That is, they may be impacted by their external environment (open) or may be independent of their environment (closed). Open systems are usually part of a larger supra-system and also contain subsystems. Almost all systems that form the output from a project are open in some way or another, even if only because they are subject to climatic changes (an electrical circuit is affected by temperature, for instance). Many "softer" systems, such as financial products, customer service products, and the like, are by nature open to multiple environmental inputs.

The essence of the systems approach is well captured by Howard Eisner (1997) in his description of the key features and results of taking a systems approach:

1. Follow a systematic and repeatable process.
2. Emphasize interoperability and harmonious system operations.
3. Provide a cost-effective solution to the customer's problem.
4. Ensure the consideration of alternatives.
5. Use iterations as a means of refinement and convergence.
6. Satisfy all user and customer requirements.
7. Create a robust system.

Points 3, 4, 5, and 6 are well within the remit of much current design management. The other points, however, are not so obviously in the domain of much design work that is carried out. Taking a systems approach to design management is done in many technical industries, particularly defense contracting, aerospace, and computer hardware and software. The design solution to most project objectives in these industries is an engineering system. The approach has benefits, though, in many other less technically oriented sectors. As an example, a systems view of the design for a piece of clothing is not necessarily obvious. Careful consideration, however, shows that a clothes designer already works in a systems way—following much of the advice in the preceding list, although perhaps without being consciously aware of doing so. Specifically:

- Designing clothes follows a well-determined process.
- The sleeves, collar, cuffs, and panels of a blouse or shirt must obviously work together and be harmonious.
- The system needs to be robust; it must be easy to put on, cleanable, repairable, and work correctly with other clothes that will be worn with it.

There are two distinct aspects of a system. At a high level there is a system architecture, and below this there are the subsystems that together form the functioning system. Broadly speaking, the relationship between the development of the system and the design stages is shown in Table 18.1.

The architecture of the system defines the best combination of subsystems to meet the business and technical requirements, as well as the definition of the functions to be carried out by each subsystem. However, system architecting is more than providing the framework for subsystems to work within. It includes defining the approach that should be taken toward the creation of the functional subsystems, as well as identifying the most cost-effective arrangement of these subsystems. This means that a specification for each subsystem must be written, and the interfaces between the subsystems clearly delineated and documented (Eisner, 1997).

Developing the subsystems is very much the domain of designers expert in their particular field, and this applies whatever system is being delivered to meet project objectives (engineering, financial services, organizational change, etc.). See Figure 18.3.

There are two main advantages to creating a system architecture:

1. Thinking carefully about the system, as distinct from a collection of individual deliverables to be put together at the end of the project, can help enormously to improve the

TABLE 18.1. THE RELATIONSHIP BETWEEN SYSTEM DEVELOPMENT AND LIFE CYCLE STAGES.

Life Cycle Stages	System Development
Concept design	Decide what type of system is most likely to meet the business and technical needs, expressed by the statement of requirements (or design brief).
Feasibility studies	Assessing whether the type of system decided on can be created successfully, by measuring against carefully set criteria (see the description of generic design stages in the *Design Management Techniques* section later in the chapter).
Outline design	The system architecture is developed with reference to the requirements, by understanding the various functions required of the system and how these can be achieved (including deciding on any trade-offs needed between the various requirements).
Detail design	The way in which each subsystem of the final delivered system will provide the function required of it.

FIGURE 18.3. ILLUSTRATIVE SYSTEM ARCHITECTURE FOR AN INSURANCE PRODUCT.

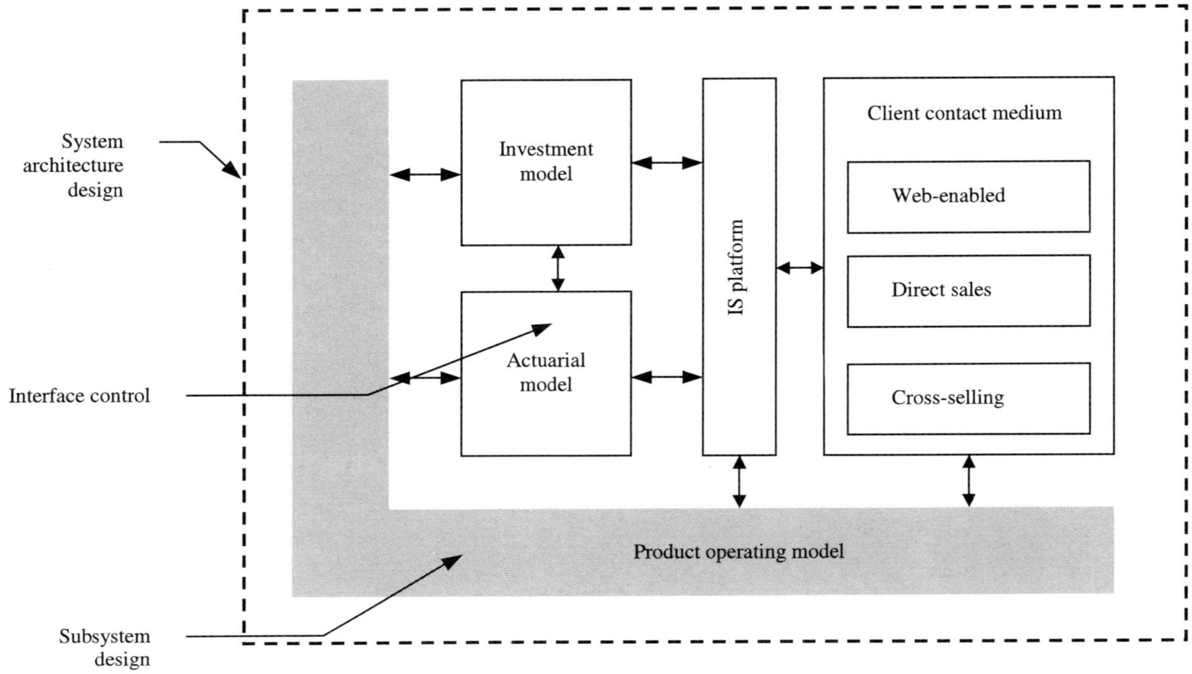

System architecture design

Interface control

Subsystem design

Investment model

Actuarial model

IS platform

Client contact medium

Web-enabled

Direct sales

Cross-selling

Product operating model

way that the overall deliverable works—ensuring it provides a better solution to meet the project's objectives;

2. Designing the system architecture requires different skills than designing each functional subsystem, in whatever sector of industry the project exists; thinking systemically at an early stage can bring significant improvement in the overall solution that is decided on.

The need for effective interface definition and control becomes apparent as the subsystem design begins. Setting, and subsequently "freezing," the interface requirements between subsystems means that the designers of the subsystems can then work on designing their part of the overall solution without further reference to those working on adjoining systems. Each subsystem design only needs to satisfy the interface constraints. If these are met by the subsystem, the operation of the internal components in the subsystem is not relevant to other interfacing subsystems—hence, the term "black box." The need for information to constantly flow between the designers working on the separate systems is removed.

The work of defining interfaces is not trivial. The degree to which the overall design is broken down, and the size of the subsystems, is fundamental to effective system (and hence design) management. The crucial interface control issues are as follows:

- Level of disaggregation of the system to subsystems—which determines the number of interfaces.
- Amount of compromise that can be tolerated for each interface constraint (since subsystems frequently have conflicting interface constraint needs).
- Tolerance that the constraints should have: If the constraints are too tightly specified, optimization of subsystem design can be reduced dramatically; if too loosely specified, the overall design solution is likely to perform poorly.
- Need to freeze interfaces, and their constraints, at an appropriate time in the design project's life cycle. Freezing too soon will lead to suboptimization of the overall system, since not enough is known about the system's properties, whereas freezing too late will prevent the designers from making the technical (and quite likely commercial) decisions needed to deliver the subsystem on time.

A schedule of the interfaces showing freeze dates and required delivery dates for subsystem designs is a valuable design management tool.

Life Cycle Management

Different approaches to the management of the project life cycle lead to different emphasis being put on design. There are two fundamentally different types of project life cycle. They are differentiated by what is considered to be the work of project management. The task-oriented life cycle includes the major activities that require to be managed, vis business case, feasibility studies, concept design, detail design, implementation, commissioning, handover, operations, and decommissioning. (And there are often others included such as procurement and testing.) These life cycles are usually drawn in a circular, or spiral, way. An example is shown in Figure 18.4.

FIGURE 18.4. TASK-ORIENTED PROJECT LIFE CYCLE.

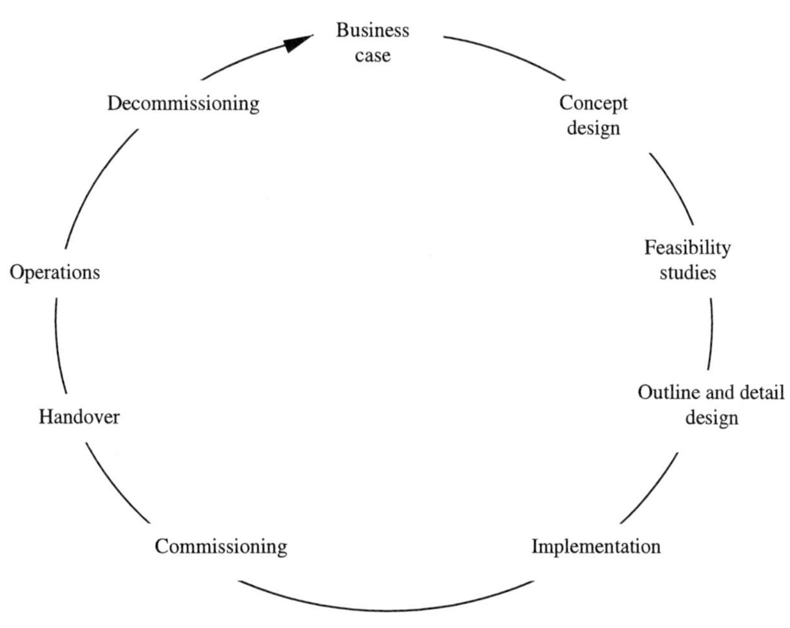

Source: After Wearne (1989).

In contrast to the task-oriented life cycle, the product-oriented life cycle de-emphasizes design (along with other processes). The life cycle only describes the management of a strictly limited set of "pure" project management tasks: start-up, plan, implement, closedown. All other tasks associated with the project's work packages, including design, are considered to be part of the *product* life cycle. The diagram typical of this type of life cycle is shown at Figure 18.5.

The danger is that the disassociation of design from project management implied by the product-oriented life cycle leads to insufficient attention being paid to the management of design (and indeed the management of other processes such as testing, handover, procurement, etc.). Design requires much attention. Decisions made at business case through to detail design fundamentally define the project's outputs. This means the cost of the project, the time likely to be needed to carry out the project, the type of resources needed, and the quality requirements of the products. If the conceptual design of the project's deliverable does not reflect the context of the project as a whole, and therefore the wrong design solution is chosen, the project has little chance of success.

The Inputs to Design

The primary input at the highest level into the design process is the project objective. What change has the project been set up to create? This applies whether the project is internal

FIGURE 18.5. THE PRODUCT-ORIENTED PROJECT LIFE CYCLE.

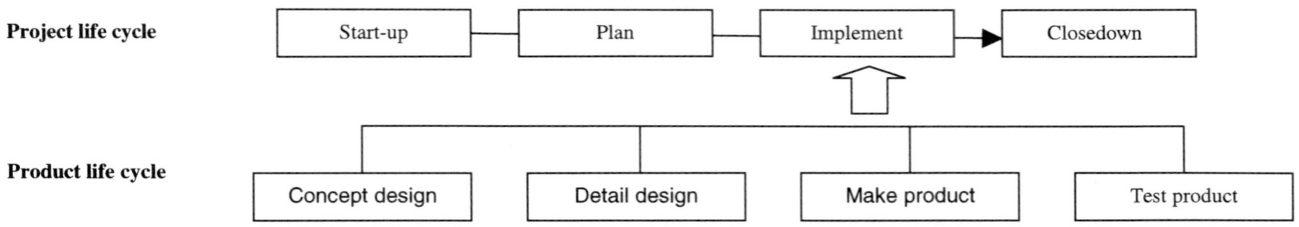

to the organization or is an external project, delivering change to a client's organization. From an understanding of the project objectives, the primary deliverables can be deduced. This sounds easy but in fact can be quite difficult. The process that links objectives to primary deliverables is requirements capture (requirements capture is discussed in detail in the chapter by Davis, Hickey, and Zweig). This means understanding what *both* the business and technical needs of the organization are to enable the project objectives to be satisfied. The requirements are a clear and concise statement of the problem that the design is to overcome, completely devoid of any suggestion of the solution.

It is clear that involving experienced designers in the capture of business requirements can significantly improve the understanding of the needs of the project. The reason for this is the designer brings knowledge of the ways in which similar business needs have been satisfied in the past.

It is more obvious that designers need to be involved in capturing technical requirements of the project, since they

- know when sufficient technical requirements have been captured to be able to proceed to concept design;
- bring knowledge of how similar technical needs have been met in the past; and
- will have a first-hand understanding of the requirements, enabling them to match them with solutions more quickly and easily in later design stages.

The documented output from requirements capture is usually called the "statement of requirements." In a number of industries where the idea of explicitly capturing business and technical requirements without an implied solution is relatively new, the input to the design process has usually been called the "design brief" (Barratt, 1999). This document is in many ways similar to the statement of requirements but is more often used when briefing professional design consultants with whom a contract will be placed to deliver design to a project. The brief is often more directive than a statement of requirements in that it specifies the expected design solution (for instance, that an office building is to be designed, normally with quite a lot of detail as to the expected final design[1]). In this sense it is often a contract document, and so has a different purpose to the statement of requirements.

Design Management Techniques

Stage Gate Control

The life cycle shown previously identified a number of design stages at the front end of the project. A more detailed explanation of these stages will help us to understand how the design solution is managed as it evolves through the life cycle. What matters is that the

[1] A statement of requirements might say that the business needs to increase the number of workers it employs—for which the solution *could* be more office space, or could be more home working, or hot desking using the existing office space.

evolving design is best controlled if the work is managed in discrete stages (British Standards Institute, 1996). The generic design stages can be described as follows:

Concept A number of high-level design proposals are developed that will all lead to the project objectives being accomplished; each concept design must satisfy the business case developed in first stage of the project. The designs are at a low level of detail but are sufficiently well developed that the overall cost of the project can be estimated.

Feasibility The feasibility of the various options are considered against a number of criteria, typically:

- Cost to make the project deliverables
- Amount of time that would be needed to complete the project
- Capability of the organization to make the deliverables
- Congruence with the technology strategies of the project participants
- Environmental impact the deliverables will have

(There are, of course, many more criteria that may be used to assess the design solutions proposed.)

Outline The concept design, which may or may not have been further extended during the feasibility stages, is now developed to the outline level of detail. The major parts of the deliverables are defined in terms of form and function (and "delight" in most consumer-oriented industries). Outline design includes the following:

- Process design
- Space planning
- General arrangement drawings
- System architecture
- Design specification for major components/subsystems

Detail The individual elements of the overall project deliverable are now broken down to a great level of detail. Each element of the design at this level will probably form a discrete work package in the implementation stage, as well as being a design work package in its own right. Individual components are designed, then integrated to form the work package.

The progression of the design work is controlled by "stage gates." These are shown in Figure 18.6.

A gated design process means that at certain points in the life cycle, the evolving design must pass through stage gates. Part of setting up the design management framework must include deciding which types of gates will be employed, and between which stages they are needed. The basic rules for passing through the gates are noted in the box in Figure 18.6. However, the specific rules that will be applied to the gates will differ according to industrial sector, and usually the criticality of the project to the organization.

Commonly there are three types of gates: "hard," "soft," or "fuzzy." A hard gate is where the design cannot be progressed to the next stage if the gate is not passed. The design process may not move into the following stage until sufficient rework has been done to allow the design to pass through the gate. Soft gates are ones in which the design is allowed to

FIGURE 18.6. THE GATED DESIGN PROCESS AND STAGE GATE RULES.

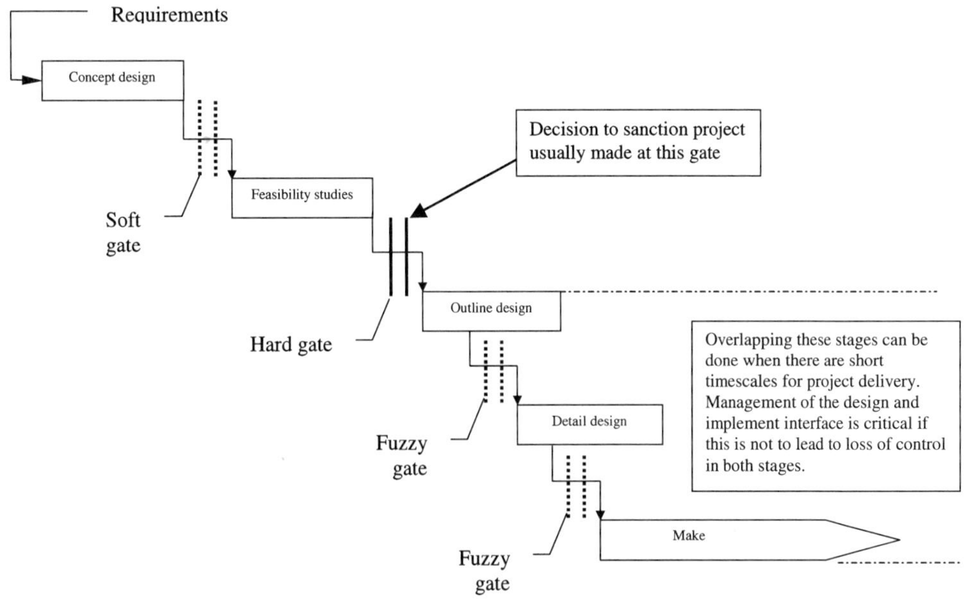

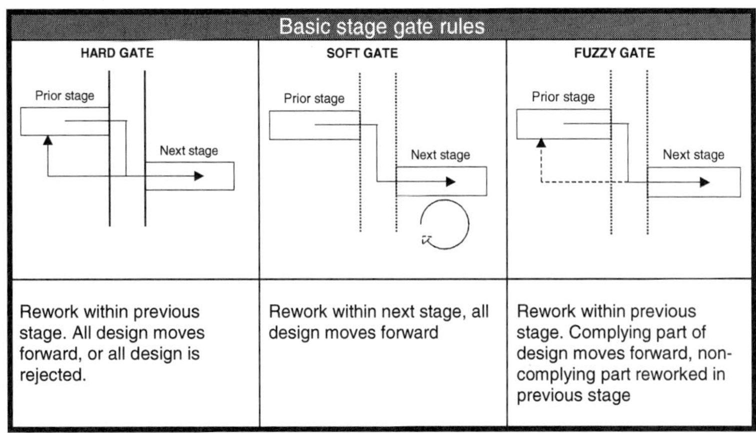

progress to the next stage, even if not being accepted as "compliant" (dependent on the gate's rules). However, a commitment must be made by the person responsible for the design to make changes to the design to ensure that it becomes compliant before the next gate. It is also possible to have "fuzzy" gates, which are essentially a combination of hard and soft gates. In a typical fuzzy gate process, parts of the design may be progressed to the next stage (those that comply with the rules), while the noncomplying parts must be reworked in the previous stage until they do comply. Fuzzy gates are typically used where stages are being overlapped to shorten the overall time to delivery for the project. This type of gate ensures proper attention is given to the rework needed, while not stopping the work in the next stage from progressing.

Concurrent Engineering

Concurrent engineering is used to keep control of the design and implementation stages when they are overlapped to reduce overall project duration. It means that the early stages of the making of some or all of the deliverables begins before the final design of those deliverables has been decided. This has become common practice in industries supplying consumer goods where time-to-market is one of the dominant project success factors (Shtub et al., 1994). When you are overlapping project stages, it becomes crucial to take a holistic view of the overall project, including:

- Understanding and reviewing the strategic issues that drive the solution to the problem the project has been set up to solve
- Assessing the level of sophistication required in the project deliverable (which in most projects, in most industries, includes deciding on the level of technological innovation to be incorporated in the design solution);
- Assessing process capability to make the various possible solutions
- Deciding the appropriate level of compromise between core project control issues of schedule, budget, and quality and performance
- Determining the through-life costs of the deliverable to the owner—essentially, initial capital, operating, maintenance, and disposal costs (whether internal or external to the organization)
- Understanding the strategy for extending the value the deliverable could bring to the owner during its life

The processes that could help ensure that all these aspects are appropriately managed include value management, project strategy development, quality management, technology management, design for manufacturing/design for assembly, project control, testing, maintainability of the deliverable, product liability, uncertainty management, and others. Moreover, these different aspects of the management of the project design stages need to be managed simultaneously. Concurrent engineering is discussed in detail in the chapter by Thamhain.

Design for Manufacturing

Generally speaking, the majority of a project's cost is incurred in the implementation stages of the life cycle. This applies whether the project deliverable is a physical object or artifact (typically in construction, mechanical and electrical engineering, electronics, computer hardware and software, and new product development) or nonphysical (such as a changed organization, a financial services product, or other service industry product). The "making" stages of projects typically account for between 75 and 90 percent of the total project expenditure, depending on industrial and technological factors affecting the project. Therefore, anything that reduces the cost of creating the project deliverables should be pursued. One of the biggest cost drivers in projects making physical deliverables is design work that does not take account of the most cost-effective processes for making the deliverables. In industries where there are long production runs for the product created during the project, or where the cost of production is very high (due, for instance, to stringent quality requirements), effective design can significantly improve production costs. Hence, it is clear that manufacturing specialists need to have significant input at the design stages.

The process of bringing in this expertise to design is called design for manufacturing (DFM). DFM aims to optimize the design at the earliest stages to take account of the processes that will be used to make the deliverables. This is not an easy or comfortable approach to design for many designers and manufacturing specialists alike. Figure 18.1 reminds us of the fundamental difference in mental models between designer and implementer. Getting these groups of people to work together effectively is a key task for the person managing the design work. It is important to recognize that for optimal effectiveness, DFM needs to be started at the earliest stages of design, when concepts are being generated for the various solutions to the design problem. There is little point in choosing a concept design to progress into detailed design work if the concept chosen cannot be supported by the existing capability of the organization to make the deliverable (or at least the high-value components of the deliverable). At the least, DFM allows a logical debate to take place about trading off the costs of new manufacturing capability against the attributes of the design that can create extra value in the final product.

The success of DFM can be ensured by recognizing, and acting on, the realization that differing cultures within design and manufacturing exist. The primary obstacle that this difference creates is that of effective communication. There are two key ways to improve communication between these two groups:

- *Plan for communication.* This means identifying where in the design project life cycle DFM will have most effect (invariably early on) and then ensuring appropriate DFM processes are created in time to be used most effectively. It also implies that DFM workshops and review meetings are built into the schedule.
- *Ensure common understanding.* It is far from obvious to designers that the manufacturing process capability required to actually make a design solution may not exist—particularly when an external client is doing the making. However, this lack of knowledge of manufacturing capability is also frequently found when the design will be made in-house. Equally, manufacturing specialists are rarely aware of the specific reasons why a particular feature of the design is necessary to create added value to the client.

The differences in awareness between designers and manufacturing specialists are to be expected. It is up to managers of design to manage the DFM process effectively for the greater good of the organization itself and, where applicable, the external client.

A related design management process is design for assembly (DFA). A major part of the "making" cost for a design solution is the time needed for the assembly of the various components forming the overall product. In such industries as aerospace, power engineering, electronics, and the manufacture of consumer goods, assembly time is heavily influenced by the ease of assembly of the product that will be sold. Consequently, the specialists in assembly processes must be brought into the design process in the same way as the manufacturing experts are involved in DFM. Unsurprisingly, the differences in culture between the designers and assembly specialists are just as evident in the DFA process as for DFM. Communication between the two groups is facilitated in the same way as for DFM: Plan to communicate, and create a situation where common understanding can be gained.

The management processes of DFA and DFM clearly interact with, and affect, the design solution chosen. It is quite possible that the design of a component that has been optimized for manufacturing is very difficult to assemble, adding time (and therefore expense) to the processes that will deliver the final product. Conversely, a design optimized for assembly may be expensive, or even impossible, to make using existing manufacturing process capability. It is incumbent on the design manager to ensure that the correct trade-offs are made between designing for maximum client value, low-cost manufacturing, and ease of assembly.

Computer-Aided Design and Computer-Aided Manufacturing

Both computer-aided design (CAD) and computer-aided manufacturing (CAM) are part of the wider area of technology of computer-aided engineering (CAE). CAD is part of the fabric of much design work that is carried out, particularly for technically oriented projects (as opposed to business change and other "softer" projects). The initial manifestation of CAD in the mid-1970s was to replace the designer's drawing board, making the production, updating, storage, and transmission of technical design drawings more efficient. The rapid increases in computing power and associated increase in the sophistication of software means that the nature of design work in architecture, new product development, and all sectors of engineering has changed. Current CAD software packages are very powerful tools to help designers generate and test design ideas, working in three dimensions and allowing virtual models to be created. As such, the creative process in design has been affected by the ability to move through the assess-synthesize-evaluate cycle more quickly, and in more detail (particularly the evaluate stage). This means more options for the solution to the design problem can be generated before a concept design is chosen. The ease with which CAD systems share information is another factor that has impacted the way that design is managed. Specifications, drawings, and other design information is transmitted electronically between all groups involved in the design process, from the project owner or sponsor, via the project and design teams, to suppliers of equipment and end users.

Computer-aided manufacturing takes advantage of many aspects of CAD and integrates them with aspects of manufacturing that are computerized. CAM allows design information to be fed directly into such processes as material ordering, manufacturing scheduling, resource management, and testing and quality management. It is commonplace for design information to be fed directly into the manufacturing process and products made, tested, and quality checked without hardcopy information being generated, or indeed any solid "real" prototype being produced.

Essentially one must be aware of how CAD/CAM changes the way people work. The critical issue is in creating design organizations that can make maximum use of the technology available. Frequently this means dispersed "virtual" teams work together on the design processes. Document management is completely redefined with few hard copies of drawings made. Techniques for control and tracking of the design itself are different from those used previously. The fundamentals of design management are not changed by the technology, but the detailed way in which designers and design is managed must suit the tools used.

Uncertainty (Risk) Management in the Design Process

Uncertainty in the design stages of a project should be actively managed. This is normally done by carrying out risk identification and assessment, and then implementing action plans to reduce the risks or minimize the effects of risks if they actually occur. It is becoming increasingly common to manage opportunities as well as risks, and there are often many opportunities to be found in the design stages. Some of the common risks and opportunities are shown in Table 18.2.

At the project level it can bring significant benefit if those involved in the design stages participate in the risk (and opportunity) management process. Often risks to work in the implementation stages of the project are not identified as having a possible effect on the design process. Design involvement in the overall risk management process can help to ensure these risks are picked up and mitigating actions incorporated into the design schedule and budget.

Controlling the Design Process

The design work can be controlled as though it was a project in its own right—a project within a project. This approach is fiercely resisted by some designers, for the reasons given earlier.

There is merit, however, in using mechanistic control techniques, so long as

- the design manager (and the project manager as well) do not expect design work to be as predictably controlled as implementation work;
- the techniques are used sensitively—that is, there is explicit recognition of the inherent difficulties posed by controlling design in this way.

Planning for the project and planning for the design stages are inextricably linked. Many inputs to project planning will flow from the earlier stages of the design work. Likewise, these earlier stages will also define the plans for the remaining design stages.

TABLE 18.2. COMMON RISKS AND OPPORTUNITIES IN THE DESIGN PROCESS.

Risks	
Technology—How well understood is the technology that the design solution is based on?	If the technology is mature, and there is great experience and knowledge in the design firm of working with the technology, there is probably little risk in this area. Conversely, if the technology is new, or the designers have little experience of working with it, then the risk of overrunning the time to produce the design deliverable is high.
Change—To requirements or brief	In some sectors, change to the requirements (and design brief) are quite likely as the market is very volatile: New product development is typical. Fast response and flexibility are needed to cope with this situation.
Process capability—For both design and making processes	Often the process capability to make the product that has been designed is unknown or untested (typically in precision engineering and similar sectors). Design-for-manufacturing and 'Designfor-assembly are therefore important techniques in this environment.
Opportunities	
Step change in capability of product	If the right environment and context can be created for designers to work in, there is the possibility of designing a product with a step change improvement over existing products. (Skunk works design environments can help with this, isolating the design team and ensuring they are greatly motivated).
Early development of new products that use knowledge gained from the project	The insights and knowledge gained during the design stages of projects about new technology, the application of new manufacturing process capability, and also improved design processes themselves can all contribute to the fast development of additional products, whether they are for the use of internal or external clients.
Reduced overall project schedule and budget	There are often opportunities in the design stages to identify ways of delivering the design solution quicker, or of changing the design solution to make it faster to make or implement.
Increase value delivered by the project	The value management process begins at the earliest stages of the design phase of the project, and most of the outputs from the process will impact the design solution. Hence, it is vital that designers make a full contribution to all stages of the value process.

Chapter 1 describes the project control process. The control diagram can be redrawn in the context of design, as shown in Figure 18.7.

The diagram shows that three fundamental control documents need to be generated: the scope of the design work, the schedule to carry out the design work, and the budget for this work. The specific nature of creating these documents for design is briefly described in the following. (However, the *processes* for each one are identical to those described in some detail in my earlier chapter on project control.).

Scope of Design Work

The work required to carry out the outline and detailed design is defined by the solution chosen to deliver the project objective—the concept design. The exact nature of the work required to produce the design is dependent on the nature of the product's deliverables. Software projects involve writing code, creating system architectures, creating and documenting module interface requirements, and so forth. This is very different in *nature* to the work needed to create a new financial product (market analysis, actuarial calculations, investment risk strategies, etc.). What is fundamentally important is to work out what discrete deliverables are needed to make the final project product, then assign design work packages to each of these deliverables, and document this information in the outline design. This process can be quite complicated, although different industries have developed techniques to help this process.

FIGURE 18.7. THE DESIGN PROCESS CONTROL.

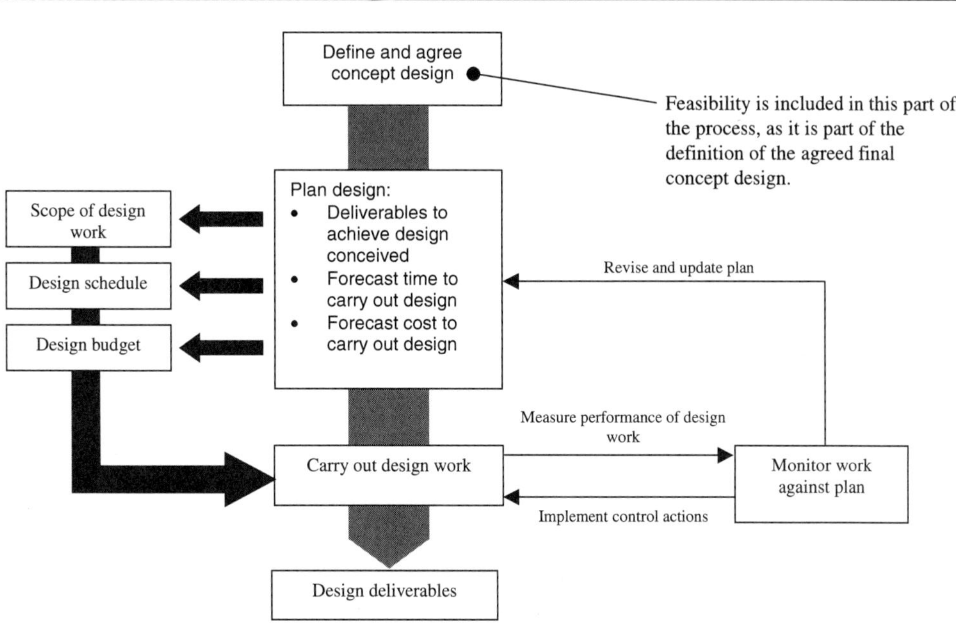

A work breakdown structure (WBS) for design can be created, showing how the individual elements of the design are related to each other. In many technical industries, the design WBS is in fact a description of the system architecture. As such, it will contain information on how each design element is configured—broadly speaking, a description of the interface between each part of the design and the other elements it is directly connected to. However the design work required to be carried out is captured, it forms the central part of the plan for how that work will be carried out. Changes to the scope should be managed by the project change control mechanism (see earlier chapter on project control).

Design Work Schedule

Making sure that the design stages are completed in a time frame that is to some extent or other predictable is the key challenge stemming from the creative nature of the work. Forecasting the time durations to complete each work package identified in the design scope is difficult. The durations for some work packages are more difficult to forecast than others—for instance, those that are innovative or in some other way new, or those where a great deal of iterative work is known to be required.

The process for scheduling design work must be based on knowledge of the individual deliverables (drawings, calculations, reports, specifications, and other documents) from the WBS. Reliance on previous experience and considered thought by experts in the field about the work required to be done leads to a set of forecast durations being established.

Scheduling the creation of the design deliverables means putting the work packages in a logical sequence and then calculating the total time required to complete all the work. The sequencing of the work is done by creating a dependency network (see the earlier chapter on project control). The most common tool then used to establish the time to carry out the work on the network's critical path is a Gantt chart—that is, linear scheduling. The difficulty with linear scheduling is that the iteration that is a fundamental aspect of the design process cannot be modeled effectively. Hence, it is not a very satisfactory way of scheduling design, leading to continually adjusting the schedule as the iterative cycles in the design work unfold.

The amount of iteration required between design work packages must be "built in" to the schedule in some way. Traditionally, this is done by adding time to the schedule where there is significant doubt about the likely duration of difficult design work. However, this rather defeats the purpose of creating a logically consistent schedule and can help lead to loss of control of the design stages of the project.

There is a method of scheduling that can overcome this difficulty by using a dependency structure matrix (DSM). The development of DSM originated with systems modeling (which is also where linear scheduling techniques were developed in the 1950s and 1960s). The essence of the technique is the creation of a matrix showing the activities within a system and their dependence on information from each other. The matrix can then be manipulated to show the most effective route through the activities based on information dependency (and that also identifies critical decisions in terms of their impact on other decisions). This means that iterative processes can be more clearly understood and management attention focused on critical information flows. When the project team is managing design schedules, critical design information flows can be spotted and where necessary educated guesses can be made at certain points to keep the overall information flow moving. The guesses are then validated when the true information becomes available and limited and more predictable amounts of rework can be carried out than would otherwise have been likely to be the case. Much work continues on making user-friendly stand-alone and Web-based software available that will carry out DSM scheduling (Austin et al., 2000).

One of the critical considerations when planning design work is to decide the extent of "front-end loading" that will be carried out during the project. *Front-end loading* is the practice of employing a significantly higher number of designers earlier in the design phases of the project than has normally been the case. This concentrates project resource in the project life cycle where there is the greatest opportunity to reduce the overall project time scale. The iterative cycles can be moved through rapidly and the final outline design solution articulated in a much shorter time. In essence, this means concentrating effort at the stage of the project where uncertainty can be removed most effectively—specifically, outline design and early detail design.

Ensuring that many experienced people work on these various options simultaneously helps to reduce the overall time taken at this stage and, with careful management, should lead to a more robust final design being arrived at. Detail design can then be started with less risk that the outline design will have to be revisited (which often means that the design process has to be stopped while the implications of technical risk in the outline design are reassessed). Loading of extra resource is also done at concept, feasibility, and detail stages, as shown in Figure 18.8. However, finding a large number of experienced designers is quite difficult, so there is likely to be a natural limit on how much front-end loading can be carried out.

Design Work Budget

Forecasting the cost for carrying out design work is a straightforward process, since it is almost entirely the cost of designers' time, usually defined in terms of cost per hour. There is a very small cost element for fixed material costs. There are also overheads to consider (for equipment, offices, management, etc.). This means that the cost to produce the design is directly linked to the time taken to create the design, and, of course, the number of designers employed on the work. When the forecast has been developed, a cost breakdown structure can be built up to allocate budget to specific parts of the work breakdown structure.

Quality

It is during the design phases of the project that much of the quality of the ultimate project deliverable is established or enabled. The quality process used in the design stages must also ensure high-quality design work per se. A number of aspects need to be covered:

- Accurately capturing the requirements of the client
- Putting in place a design process capable of developing an appropriate solution
- Ensuring that the solution developed satisfies the client's requirements.

FIGURE 18.8. THE DESIGN RESOURCE PROFILE OR A PROJECT WITH FRONT-END LOADING.

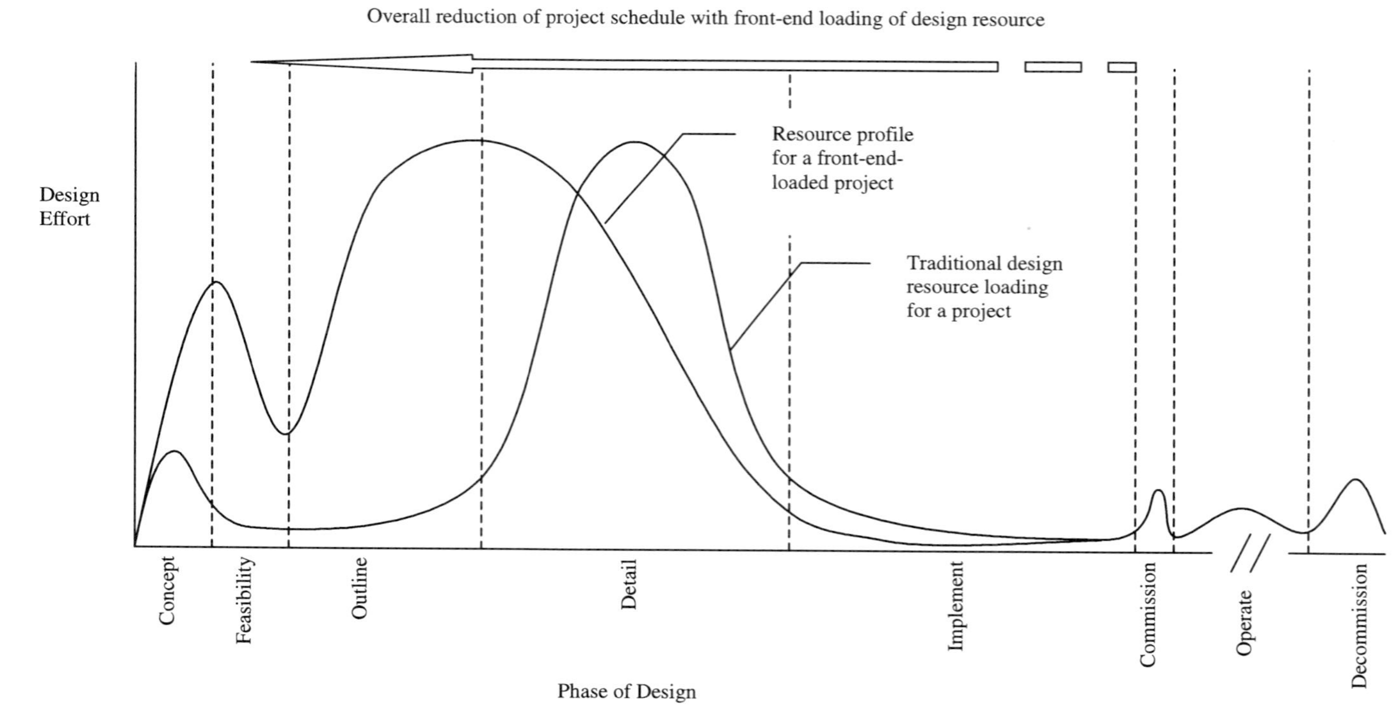

Source: Harpum (2002).

Carrying out these activities effectively is dependent on an integrated process for achieving high-quality design. The most well established and comprehensive quality system for the design phases of a project is known as quality function deployment (QFD). QFD monitors the transformation of the client's requirements into the design solution, to ensure that quality is inherent in the solution (Hauser and Clausing, 1988). To do this, QFD integrates the work of people in the project's participant organizations in the following areas:

- Requirements capture (to understand client's business and technical requirements)
- Technology development (to understand what technology is available to be used)
- Implementation (typically DFM and DFA)
- Marketing (to understand the client's perceptions of the solution that satisfies the requirements)
- Management (to understand how the processes to ensure quality can be operationalized)

The primary set of considerations for the QFD team are as follows:

1. *Who* are the clients. In the broadest terms (i.e., the users of the project deliverable, the owner, other stakeholders).
2. *What* are the customer's business requirements. Which may or may not be explicitly stated in a design brief.
3. *How* will these requirements be satisfied. Including an evaluation at the highest level of abstraction, such as should the project actually build a road or a railway to meet the requirement to transport people from A to B?

Figure 18.9 shows how client requirements are matched to the individual elements (subsystcms) of the design solution. The importance of the client (or user of the project's

FIGURE 18.9. THE QUALITY FUNCTION DEPLOYMENT MATRIX.

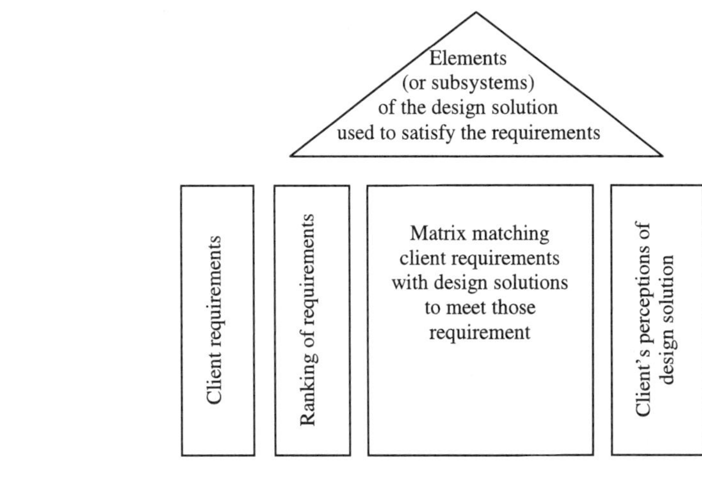

deliverables) to the design process is clear from the diagram. There is a clear and auditable trail from the collection of client requirements through to the client's perception of whether those requirements have been met in the solution proposed.

The client requirements are scored in order of their relative importance and ranked, after a weighting criteria is used. The "roof" of the matrix (QFD is also called the "house of quality") contains the elements of the design solution that will satisfy the requirements of the client. This part of the matrix represents the system that has been designed to meet the project objectives, whether that system is formally recognized as one (a system architecture with sub-systems) or not. The ability of the elements of the design solution to satisfy the requirements is then estimated using experience and judgment in the central matrix. The final aspect of the QFD matrix is to evaluate the client's acceptance of the design solution both at an overall and an elemental level of the matrix. It is important to understand that QFD does not generate the design solution; it enables the quality of the chosen solution to meet the client's requirements to be monitored with rigor and accuracy.

Summary

This chapter set out to provide the context for design in projects, the strategic management considerations that arise, and how design can be controlled effectively. The way in which design is managed depends on the focus of value creation for the business sector. The relationship between design and the project, and hence between design management and project management, varies tremendously. In some sectors, project management is subservient to design management; in others, project management dominates design management.

The major challenge in managing design work in projects is ensuring the necessary level of integration is achieved with the "making" phases of the work. Because of the differing mental models of the people that work in these two fundamentally different stages of the project, this is not an easy task. Creativity *is* difficult to manage—not the least for the person who is doing the creating! However, there are a number of approaches to organizing design work at the personal and team level, both strategic and tactical, that can help to bring control to the process without threatening the freedom required to be creative.

References

Allinson, K. 1997. *Getting there by design: An architects guide to design and project management.* Oxford, UK: Architectural Press.

Austin, S., A. Baldwin, B. Li, and P. Waskett. 2000. Application of the analytical design planning technique to construction project management. *Project Management Journal* 31(2):48–59.

Barratt, P. 1999. *Better construction briefing.* Oxford, UK: Blackwell Science.

British Standards Institute. 1996. *BS 7000: Design management systems: Part 4: Guide to managing design in construction.*

Cooper, R. 1995. *The design agenda: A guide to successful design management.* Chichester, UK: Wiley.

Csikszentmihalyi, M. 1996. *Creativity: Flow and the psychology of discovery and invention.* New York: HarperCollins.

Eisner, H. 1997. *Essentials of project and systems engineering management.* New York: Wiley.

Gell-Mann, M. 1994. *The quark and the jaguar*. London: Abacus.

Gray, C., W. Hughes, and J. Bennett. 1994. *The successful management of design.*, Reading, UK: Center for Strategic Studies in Construction, University of Reading.

Groák, S. 1992. *The idea of building: Thought and action in the design and production of buildings*. London: Spon.

Harpum, P. 2002. In *Engineering Project Management*, 2nd ed., ed. N. J. Smith. Oxford, UK: Blackwell Science, 238–263.

Harpum, P., and A. W. Gale. 1999. Achieving success by early project planning and start-up techniques. In *Managing business projects*, ed. K. A. Artto, K. Kähkönen, and K. Koskinen. Espoo, Finland: Project Management Association of Finland.

Hauser, R., and D. Clausing. 1988. The house of quality. *Harvard Business Review* 66 (May–June): 63–73.

Kappel, T. A., and A. H. Rubenstein. 1999. Creativity in design: The contribution of information technology. *IEEE Transactions on Engineering Management* 46 (2, May).

Lawson, B. 1997. *How designers think: The design process demystified.* 3rd ed. Oxford, UK: Elsevier Science.

Luckman, J. 1984. An approach to the management of design. In *Developments in Design Methodology*, ed. N. Cross. Chichester, UK: Wiley.

Morris, P. W. G. 1994. *The management of projects*. London: Thomas Telford.

Pugh, S. 1990. *Total design: Integrated methods for successful product engineering*. Wokingham, UK: Addison-Wesley.

Shtub, A., J. B. Bard, and S. Globerson. 1994. *Project management: Engineering, technology, and implementation.* Upper Saddle River: Prentice Hall.

Simon, H. A. 1981. *The science of the artificial.* Cambridge, MA: MIT Press.

Smith, N. J., ed. 2002, *Engineering project management.* Oxford, UK: Blackwell Science.

Topalian, A. 1980. *The management of design projects*. London: Associated Business Press.

Wearne, S. H., ed. 1989. *Control of engineering projects*, 2nd ed. London: Thomas Telford.

CHAPTER NINETEEN

CONCURRENT ENGINEERING FOR INTEGRATED PRODUCT DEVELOPMENT

Hans J. Thamhain

When Benjamin Franklin said "time is money," he must have anticipated our fiercely competitive business environment where virtually every organization is under pressure to do more things faster, better, and cheaper. Indeed, for many companies, speed has become one of the great equalizers to competitiveness and a key performance measure. New technologies, especially computers and communications, have removed many of the protective barriers to business, created enormous opportunities, and transformed our global economy into a hypercompetitive enterprise system. To survive and prosper, the new breed of business leaders must deal effectively with time-to-market pressures, innovation, cost, and risks in an increasingly fast-changing global business environment. Concurrent engineering has gradually become the norm for developing and introducing new products, systems, and services (Haque et al., 2003; Yam, 2003).

The Need for Effective Management Processes

Whether we look at the implementation of a new product, process, or service or we want to build a new bridge or win a campaign, project management has traditionally provided the tools and techniques for executing specific missions, on time and in a resource-efficient manner. These tools and techniques have been around since the dawn of civilization, leading to impressive results from Noah's ark, ancient pyramids, and military campaigns to the Brooklyn Bridge and Ford's Model T automobile. While the first formal project management processes emerged during the Industrial Revolution of the eighteenth century, with focus on mass production, agriculture, construction, and military operations, the recognition of project management as a business discipline and profession did not occur until the 1950s with the emergence of formal organizational concepts such as the matrix, projectized organizations, life cycles, and phased approaches (Morris, 1997).

These concepts established the organizational framework for many of the project-oriented management systems in use today, providing a platform for delivering mission-specific results. Yet the dramatic changes in today's business environment often required the process of project management to be reengineered to deal effectively with the challenges (Denker et al., 2001; Nee and Ong, 2001; Rigby, 1995; Thamhain, 2001) and to balance efficiency, speed, and quality (Atuahene-Gima, 2003). As a result, many new project management tools and delivery systems evolved in recent years under the umbrella of integrated product development (IPD). These systems are, however, not just limited to product developments but can be found in a wide spectrum of modern projects, ranging from construction to research, foreign assistance programs, election campaigns, and IT systems installation (Koufteros et al., 2000; Nellore and Balachandra, 2001). The focus that all of these IPD applications have in common is the effective, integrated, and often concurrent multidisciplinary project team effort toward specific deliverables, the very essence of *concurrent engineering processes*.

A Spectrum of Contemporary Project Management Systems

Driven by the need for effective multidisciplinary integration and the associated economic benefits, many contemporary project management systems evolved with a focus on cross-functional integration. Many of these contemporary systems evolved from the traditional, well-established *multiphased approaches to project management.* They often focus on specific project environments such as manufacturing, marketing, software development, or field services (Gerwin and Barrowman, 2002). Many mission-specific project management platforms emerged under the umbrella of today's integrated product development (IPD), including design for manufacture (DMF), just-in-time (JIT), continuous process improvement (CPI), integrated product and process development (IPPD), structured systems design (SSD), rolling wave (RW) concept, phased developments (PD), Stage-Gate processes, integrated phase reviews (IPR), and voice-of-the-customer (VOC), just to name a some of the more popular concepts. What all of these systems have in common is the emphasis on effective cross-functional integration and incremental, iterative implementation of project plans. This is precisely the focus of concurrent engineering (CE), perhaps one of the most widely used IPD concepts, today.

Concurrent Engineering—A Unique Project Management Concept

Concurrent engineering, CE, is an extension of the multiphased approach to project management. At the heart of its concept is the concurrent execution of tasks segments, which creates overlap and interaction among the various project teams. It also increases the need for strong cross-functional integration and team involvement, which creates both managerial benefits and challenges (Wu, Fuh, and Nee, 2002). While concurrent engineering was originally seen as a method for primarily reducing project cycle time and accelerating product developments (Prased et al., 2003; Prased, 1998), today, the concept refers quite generally to the most resource- and time-efficient execution of multidisciplinary undertakings.

Moreover, the CE concept has been expanded from its original engineering focus to a wide range of projects, ranging from construction and field installations to medical procedures, theater productions, and financial services (Dimov and Setchi, 1999; Pilkinton and Dyerson, 2002; Skelton and Thamhain, 1993). The operational and strategic values of concurrent engineering are much broader than just a gain in lead time and resource effectiveness, but include a wide range of benefits to the enterprise, as summarized in Table 19.1. These benefits are primarily derived from effective cross-functional collaboration and full integration of the project management process with the total enterprise and its supply chain (Prasad et al., 1998, 2003). In this context, concurrent engineering provides a process template for effectively managing projects. Virtually any project can benefit from this approach as pointed out by the Society for Concurrent Product Development (www.soce.org).

As a working definition, the following statement brings the management philosophy of concurrent engineering into perspective:

> Concurrent engineering provides the managerial framework for effective, systematic, and concurrent integration of all functional disciplines necessary for producing the desirable

TABLE 19.1. POTENTIAL BENEFITS OF CONCURRENT ENGINEERING.

- Better cross-functional *communication* and *integration*
- Decreased *time-to-market*
- Early detection of *design problems*, fewer *design errors*
- Emphasizes human side of *multidisciplinary teamwork*
- Encourages *power sharing, cooperation, trust, respect,* and *consensus building*
- Engages all stakeholders in *information sharing* and *decision making*
- Enhances ability to support *multisite manufacturing*
- Enhances ability for coping with *changing requirements, technology,* and *markets*
- Enhances ability for executing *complex projects* and *long-range* undertakings
- Enhances *supplier communication*
- Fewer *engineering changes*
- High-level of *organizational transparency*, R&D-to-marketing
- Higher *resource efficiency* and *personnel productivity*; more resource-effective project implementation
- Higher *project quality*, measured by customer satisfaction
- Minimizes "downstream" *uncertainty, risk, and complications*; makes the project *outcome more predictable*
- Minimizes design-build-rollout *reworks*
- Ongoing *recognition and visibility of team accomplishments*
- Promotes *total project life cycle thinking*
- Provides a *template or roadmap* for guiding multiphased projects from concept to final delivery
- Provides *systematic approach* to multiphased project execution
- Shorter *project life cycle* and execution time
- *Validation of work in progress* and deliverables

project deliverables, in the least amount of time and resource requirements, considering all elements of the product life cycle.

In essence, concurrent engineering is a systematic approach to integrated project execution that emphasizes parallel, integrated execution of project phases, replacing the traditional linear process of serial engineering and expensive design-build-rollout rework. The process also requires strong attention to the human side, focusing on multidisciplinary teamwork, power sharing, and team values of cooperation, trust, respect, and consensus building, engaging all stakeholders in the sharing of information and decision making. In addition, the process must start during the early project formation stages and continue over the project life cycle.

The concurrent engineering process is graphically shown in Figure 19.1, depicting a typical product development. In its basic form, the process provides a template or roadmap for guiding multiphased projects from concept to final delivery. One of the prime objectives for using concurrent engineering is to minimize "downstream" uncertainty, risk, and complications, and hence make the project outcome more predictable (Iansiti and MacCormack, 1997; Liker and Ward, 1998; Moffat, 1998; Noori, Munro, and Deszca, 1997; O'Connor, 1994; Sobek, 1998). However, concurrent execution and integration of activities does not just happen by drawing timelines in parallel but is the result of carefully defined cross-functional linkages and skillfully orchestrated teamwork. Moreover, concurrent phase exe-

FIGURE 19.1. GRAPHICAL PRESENTATION OF CONCURRENT PROJECT PHASE EXECUTION.

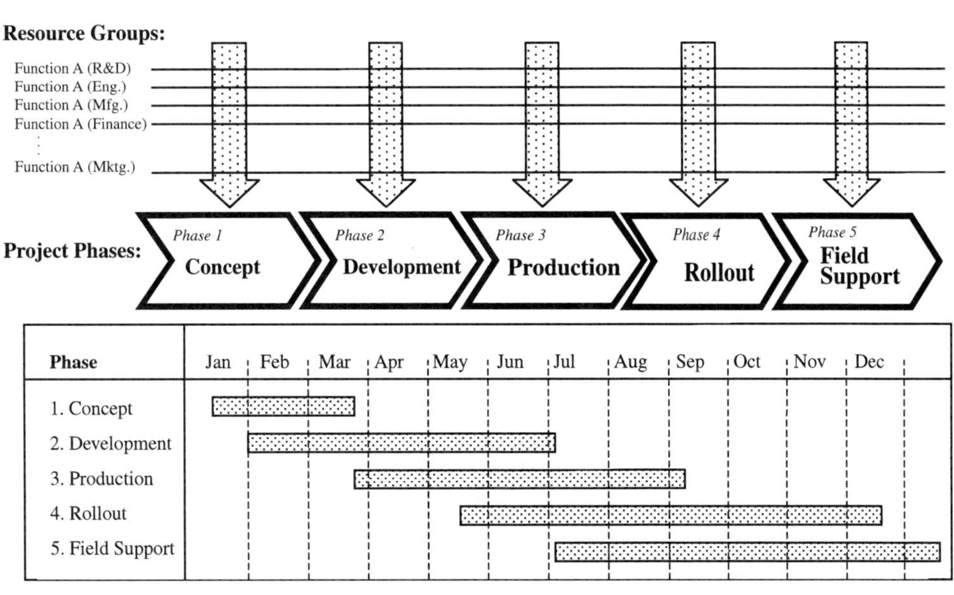

cution makes several assumptions regarding the organizational system and its people, as summarized in Table 19.2 and discussed in the next section.

For many managers and researchers the concept of concurrent engineering is synonymous with integrated product development because using concurrent engineering is to minimize "downstream" uncertainty, risk and complications, and hence make the project outcome more predictable (Iansiti and MacCormack, 1997; Liker, and Ward 1998; Moffat, 1998; Noori, Munro, and Deszca). For simplicity, concurrent engineering is often shown as a linear process, with overlapping activity phases, scheduled for concurrent execution, such as shown in Figure 19.1.

TABLE 19.2. CRITERIA FOR SUCCESSFULLY MANAGING CONCURRENT ENGINEERING PROJECTS.

Organizations and concurrent engineering teams must be able to (listed in approximate chronological order) do the following:

- Allocate sufficient time and resources for up-front planning.
- Identify major task teams, their mission, and interfaces at the beginning of the project cycle.
- Work out the logistics and protocol for concurrent phase implementation.
- Lay out the master project plan (top level) covering the project life cycle.
- Establish consensus on project plan among project team members.
- Be willing to work with partial, incremental inputs, and evolving requirements throughout the team organization, throughout the project life cycle.
- Identify all project-internal and -external "customers" of its work, and establish effective communication linkages and ongoing working relations with these customers.
- Work flexibly with team members and customers, adjusting to evolving needs and requirements.
- Share information and partial results regularly during the project implementation.
- Identify the specific deliverables needed by other teams (and individuals) as inputs for their part of the project, including the timing for such deliverables.
- Establish effective cross-functional communication channels and specific methods for work transfer.
- Establish techniques and protocols for validating the work and its appropriateness to its "customers" on an ongoing basis.
- Work with partial results (deliverables) and incremental updates from upstream developments.
- Reiterate or modify tasks and deliverables to accommodate emerging needs of downstream task teams and to optimize the evolving project outcome.
- Prepare for its mission prior to receiving mission details (e.g., manufacturing is expected to work on pilot production setup prior to receiving full product specs or prototypes).
- Work as an integrated part of a unified and agreed-on project plan.
- Have tolerance for ambiguity and uncertainty.
- Establish reward systems that promote cross-functional cooperation, collaboration, and joint ownership of results.
- Have top management buy-in and support to concurrent engineering process.
- Have established a uniform project management system throughout the concurrent engineering team/organization.

Criteria for Success

To make such concurrent project phasing possible, the organizational process must be designed to meet specific criteria that establish the conditions conducive to concurrent, incremental implementation of phased activities, such as summarized in Table 19.2. By its very definition, concurrent engineering is synonymous with cross-disciplinary cooperation, involving all project teams and support groups of the enterprise, internally and externally, throughout the project life cycle. The CE process relies on organizational linkages and integrators that help in identifying problems early, networking information, transferring technology, satisfying the needs of all stakeholders, and unifying the project team. It is important to include all project stakeholders in the project team and its management, not only enterprise-internal components, such as R&D, engineering, manufacturing, marketing, and administrative support functions but also external stakeholders, such as customers, suppliers, regulators, and other business partners.

Taken together, the core ingredient of successful concurrent engineering is the development and effective management of organizational interfaces. For most organizations, these challenges include strong human components that are more difficult to harness and to control than the operational processes of project implementation (Prased, 1998). They involve many complex and constantly changing variables that are hard to measure and even more difficult to manage, especially within self-directed team environments that are often required for realizing the concurrent engineering process (Banly and Nee 2000; Hall et al., 1996). While procedures provide (1) the baseline and infrastructure necessary to connect and integrate the various pieces of the multidisciplinary work process and (2) an important starting point for defining the communication channels, and are necessary for effectively linking the core team with all of its support functions, the resulting process is only as good as the team that implements it.

Defining the Process

After reaching a principle agreement with major stakeholders, the concurrent engineering process should be defined, showing the major activity phases or stages of the project to be executed. Even more advantageous for future projects would be the ability to define phases that may be common to a *class of projects* that is being executed by the enterprise over time. To illustrate, let us use the example of a new product development, shown in Figure 19.1, which proceeds through five project phases: (1) concept development; (2) detailed development; (3) pilot production; (4) product rollout, launch, and marketing; and (5) field support. Each phase or stage is defined in terms of principle scope, objectives, activities, and deliverables, as well as functional responsibilities. Each project phase must also include cross-functional interface protocols, defining the specific collaborations and organizational linkages needed for the concurrent development. While the principle cross-functional interfaces can be summarized graphically, as shown in the upper part of Figure 19.1, more sophisticated group technology tools, such as the quality function deployment (QFD) matrix, shown in Figure 19.2, are usually needed for defining (1) the specific cross-functional requirements,

FIGURE 19.2. QUALITY FUNCTION DEPLOYMENT (QFD) MATRIX, A TOOL FOR DEFINING INTERFACES.

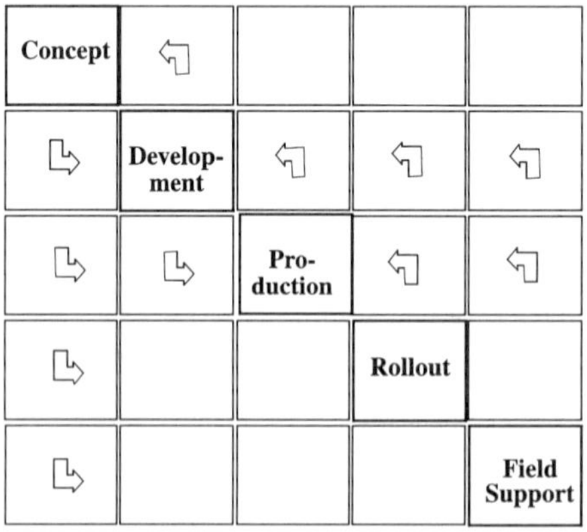

Concept	↰			
↳	**Develop-ment**	↰	↰	↰
↳	↳	**Pro-duction**	↰	↰
↳			**Rollout**	
↳				**Field Support**

(2) the methods of work transfer (often referred to as technology transfer), and (3) the stakeholder interactions necessary for capturing and effectively dealing with the changes that ripple through the product design process.

The best time for setting up these interface protocols is during the definition phase of a specific project when the team organization is most flexible regarding lines of responsibility and authority. To illustrate, Figure 19.2 shows the specific inputs and outputs required during the various phases of a product development process. Each arrow indicates that a specific input/output requirement exists for that particular interface. Most likely, some interface requirements exist for each project phase to each of the others. In our example, the total number of potential interfaces is defined by the 5 × 5 matrix, which equals 25 interfaces (this explains why the QFD Matrix is also referred to as the N-Squared Chart). The QFD Matrix is a useful tool for identifying specific interface personnel and input/output requirements. That is, for each interface, key personnel from both teams have to establish personal contacts and negotiate the specific type and timing of deliverables needed. In many cases, multiple interfaces exist simultaneously, necessitating complex multiteam agreements over project integration issues. An additional challenge is the incremental nature of deliverables resulting from the concurrent project execution. For downstream phases, such as production, to start their work concurrently with earlier project phases, such as product development, it is necessary for all interfaces to define and negotiate (1) what part of the phase deliverables can be transferred "early," (2) the exact schedule for the partial deliverable, and (3) the validation, iteration, and integration process for these partial deliverables.

Yet, another important condition for concurrent engineering to work is the ability of "downstream task leaders" to guide the "upstream" design process toward desired results, and to define the upstream gate criteria on which they depend as "customers." This interdisciplinary integration is often accomplished by participating in project and design reviews, by soliciting and providing feedback on work-in-progress, and by cross-functional involvement with interface definitions and technology transfer processes. Interface diagrams, such as the QFD Matrix shown in Figure 19.2, can help to define the cross-functional roadmap for establishing and sustaining the required linkages for each task group.

Hidden Challenges and Benefits

In spite of all its potential benefits for more effective project implementation, including higher quality, speed, and resource effectiveness, project implementation, concurrent engineering holds many organizational challenges regarding its management. Some of the toughest challenges relate to the compatibility of concurrent engineering with the organizational culture and its values. Concurrent engineering requires a collaborative culture and a great deal of organizational power sharing, which is often not present in an enterprise to the degree required for concurrent engineering to succeed. Designing, customizing, and implementing a new project management system usually affects many organizational subsystems and processes, from innovation to decision making, and from cross-functional communications to the ability of dealing effectively with risk and organizational conflict. Hence, integrating concurrent engineering into a business process and its physical, informational, managerial, and psychological subsystems without compromising business performance is an important issue that must be dealt with during the implementation phase. Strong involvement of people from all levels throughout the organization is required for concurrent engineering to become institutionalized and to be used effectively by the people in the organization.

Why are companies doing it? Few companies go into a major reorganization of their business processes lightly. At best, introducing a new process is painful, disruptive, and costly. At worst, it can destroy established operational effectiveness and the ability to compete successfully in the marketplace. Obviously, companies that adopt concurrent engineering have powerful reasons for using this contemporary concept of project management. These companies are able to use concurrent engineering as an organizational platform to *increase project effectiveness, quality, and ultimately reduce recourse needs and cycle time.*

Understanding the Organizational Components

The preceding benefits are not always obvious, looking at the basic concept of concurrent engineering, because they are often derivatives of more subtle organizational characteristics that unfold within a well-executed concurrent project management system. These characteristics need to be understood and skillfully exploited for project leaders and managers to gain the full benefits *of concurrent engineering.*

1. *Uniform Process Model.* The concurrent engineering concept provides a uniform process model or template for organizing and executing a predefined class of projects, such as specific new product developments.

   Primary benefits: Time and resource savings during the project/product planning and start-up phase.

Secondary benefits: Standardized process model breaks the project cycle into smaller, predefined modules or phases, resolving some of the project complexities, predefining potential risks and areas requiring managerial interactions and support. Standardized platform for project execution provides basis for continuous process improvement and organizational learning.

2. *Integrated Product Development (IPD).* Because of its focus on cross-functional cooperation, concurrent engineering promotes an integrated approach to product development and other project work.

   Primary benefits: Promotes unified, collective understanding of project challenges and search for innovative solutions. Helps in team integration: identifying organizational interfaces, lowering risks and reducing cycle time.

   Secondary benefits: Responsibilities for team and functional support personnel are more visible.

3. *Gate Functions.* The concurrent engineering platform is similar to other multiphased project management concepts, such as Stage-Gate, structured systems design, or rolling wave concepts, hence encouraging the integration of predefined gates, providing for performance reviews, sign-off criteria, checkpoints, and early warning systems.

   Primary benefits: Ensures incremental guidance of the product/project execution and early problem detection, provides cross-functional accountability, helps in identifying risk and problem areas, minimizes rework, highlights organizational interfaces and responsibilities.

   Secondary benefits: Stimulates cross-functional involvement and visibility; identifies internal customers, promotes full life cycle planning, focuses on win strategy.

4. *Standard Project Management Process.* The concurrent engineering concept is compatible with the standard project management process, its tools, techniques and standards. Predefined gates provide performance and sign-off criteria, checkpoints, and early warning systems, ensuring incremental guidance of the product development process and early problem detection.

   Primary benefits: Provides cross-functional accountability, helps in identifying risk and problem areas, minimizes rework, highlights organizational interfaces and responsibilities.

   Secondary benefits: Stimulates cross-functional involvement and visibility, identifies internal customers, promotes full life cycle planning, focuses on win-strategy.

5. *QFD Approach.* Using the quality function deployment (QFD) concept, built into the concurrent engineering process, helps to define cross-functional interfaces and provides pressures on both the performing and receiving organization toward closer cooperation and "upstream" guidance of the product development.

   Primary benefits: Provides an input/output model for identifying work flow throughout the project/product development process, identifies organizational interfaces and responsibilities.

   Secondary benefits. Stimulates cross-functional involvement and visibility; identifies internal customers, promotes full life cycle planning.

6. *Early testing.* Concurrent engineering encourages early testing of overall project or product functionality, features, and performance. These tests are driven by team members of both downstream and upstream project phases. Downstream members seek assurances for problem-free transfer of the work into their units, and upstream members seek smooth transfer and sign-off for their work completed.

   Primary benefits: Early problem detection and risk identification, opportunity to "fail early and cheap," less rework.

   Secondary benefits: Stimulates cross-functional involvement and cooperation, assists system integration.
7. *Total organizational involvement and transparency.* Because of its emphasis on mutual dependencies among the various phase teams, strong cross-functional involvement and teamwork is encouraged, enhancing the level of visibility and organizational transparency.

   Primary benefits: Total development cycle/system thinking, enhanced cross-functional innovation, effective teamwork, enhanced cross-functional communications and product integration, early warning system, improved problem detection and risk identification, enhanced flexibility toward changing requirements.

   Secondary benefits: Total team recognition; enhanced team spirit and motivation, conducive to self-direction and self-control.

Taken together, the top benefits of concurrent engineering refer to time, resource, and risk issues that ultimately translate into increased project performance: (1) reducing project start-up time, (2) reducing project cycle time, (3) detecting and resolving problems early, (4) promoting system integration, (5) promoting early concept testing, (6) minimizing rework, (7) handling more complex projects with higher levels of implementation uncertainty, (8) working more resource effective, and (9) gaining higher levels of customer satisfaction.

Recommendations for Effective Management

A number of specific suggestions may help managers understand the complex interaction of organizational and behavioral variables involved in establishing a concurrent engineering process and managing projects effectively in such a system. The sequence of recommendations follows to some degree the chronology of concurrent engineering system design-implementation-management. Although each organization is unique with regard to its business, operation, culture, and management style, field studies show a general agreement on the type of factors that are critical to effectively organizing and managing projects in concurrent multiphase environments (Denker, 2001; Harkins, 1998; Nellore, 2001; Pilllai, 2002; Prasad, 1977; Thamhain and Wilemon, 1998).

Phase I: Organizational System Design

Take a Systems Approach. The concurrent engineering system must eventually function as a fully interconnected subsystem of the organization and should be designed as an integrated part of the total enterprise (Harque, Pawar, and Barson, 2003). Field studies emphasize consistently that management systems function suboptimal, at best, or fail because of a poor understanding of the interfaces that connect the new system with the total business process (Kerzner, 2001; Moffat, 1998). System thinking, as described by Senge (2001), Checkland (1999), and Emery and Trist (1965), provides a useful approach for front-end analysis and organization design.

Build on Existing Management Systems. Radically new methods are usually greeted with anxiety and suspicion. If possible, the introduction of a new organizational system, such as concurrent engineering, should be consistent with already established project management processes and practices within the organization. The more congruent the new operation is with the already existing practices, procedures, and distributed knowledge within the organization, the more cooperation management will find from their people toward implementing the new system. The highest level of acceptance and success is found in areas where new procedures and tools are added incrementally to already existing management systems. These situations should be identified and addressed first. Building upon an existing project management system also facilitates incremental enhancement, testing, and fine-tuning of the new concurrent engineering process. Particular attention should be paid to the cross-functional workability of the new process.

Custom-Design. Even for apparently simple situations, a new concurrent engineering process should be customized to fit the host organization, its culture, needs, norms, and processes (Hull, Collins, and Liker, 1996). For reasons discussed in the previous paragraph, the new system has a better chance for smooth implementation and for gaining organizational acceptance if the new process appears consistent with already established values, principles, and practices, rather than a new order to be imposed without reference to the existing organizational history, values or culture (cf. Swink, Sandvig, and Mabert, 1996; Kerzner, 2001).

Phase II: System Implementation

Define Implementation Plan. Implementation of the new concurrent engineering system is by itself a complex, multidisciplinary project that requires a clear plan with specific milestones, resource allocations, responsibilities, and performance metrics. Further, implementation plans should be designed for measurability, early problem detection and resolution, and visibility of accomplishments, providing the basis for recognition, and rewards.

Pretest the New Technique. Preferably, any new management system should be pilot tested on small projects with an experienced project team. Asking a team to test, evaluate, and fine-tune a new concurrent engineering process is often seen as an honor and professional challenge. It also starts the implementation with a positive attitude, creating an environment of open communications, candor, and focus on actions toward success.

Ensure Good Management Direction and Leadership. Organizational change, such as the implementation of a concurrent engineering system, requires top-down leadership and support to succeed. Team members will be more likely to help implement the concurrent engineering system, and cooperate with the necessary organizational requirements, if management clearly articulates its criticality to business performance and the benefits to the organization and its members. People in the organization must perceive the objectives of the intervention to be attainable and have a clear sense of direction and purpose for reaching these goals. Senior management involvement and encouragement are often seen as an endorsement of the team's competence and recognition of their efforts and accomplishments (Thamhain and Wilemon, 1998). Throughout the implementation phase, senior management can influence the attitude and commitment of their people toward the new concept of concurrent engineering. Concern for project team members, assistance with the use of the tool, enthusiasm for the project and its administrative support systems, proper funding, help in attracting and retaining the right personnel, support from other resource groups—all will foster a climate of high involvement, motivation, open communications, and desire to make the new concurrent engineering system successful.

Involve people affected by the new system. The implementation of a new management system involves considerable organizational change with all the expected anxieties and challenges. Proper involvement of relevant organizational members is often critical to success (Barlett and Ghoshal, 1995; Nellore and Balachandra, 2001). Key project personnel and managers from all functions and levels of the organization should be involved in assessing the situation, evaluating the new tool, and customizing its application. While direct participation in decision making is the most effective way to obtain buy-in toward a new system (Pham, Dimov, and Setchi, 1999), it is not always possible, especially in large organizations. Critical factor analysis, focus groups, and process action teams are good vehicles for team involvement and collective decision making, leading to ownership, greater acceptance, and willingness to work toward successful implementation of the new management process (Thamhain, 2001).

Anticipate Anxieties and Conflicts. A new management system, such as concurrent engineering, is often perceived as imposing new management controls, seen as disruptive to the work process and creating new rules and administrative requirements. People responses to such new systems range from personal discomfort with skill requirements to dysfunctional anxieties over the impact of tools on work processes and performance evaluations (Sundaramurthy and Lewis, 2003). Effective managers seem to know these challenges intuitively, anticipating the problems and attacking them aggressively as early as possible. Managers can help in developing guidelines for dealing with problems and establishing conflict resolution processes, such as information meetings, management briefings, and workshops, featuring the experiences of early adopters. They can also work with the system implementers to foster an environment of mutual trust and cooperation. Buy-in to the new process and its tools can be expected only if its use is relatively risk-free (Stum, 2001). Unnecessary references to performance appraisals, tight supervision, reduced personal freedom and autonomy, and overhead requirements should be avoided, and specific concerns dealt with promptly on a personal level.

Detect Problems Early and Resolve. Cross-functional processes, such as concurrent engineering, are often highly disruptive to the core functions and business process of a company (Denker, Steward, and Browning, 2001; Haque, 2003). Problems, conflict, and anxieties over technical, personal, or organizational issues are very natural and can be even healthy in fine-tuning and validating the new system. In their early stages, these problems are easy to solve but usually hard to detect. Management must keep an eye on the organizational process and its people to detect and facilitate resolution of dysfunctional problems. Round-table discussions, open-door policies, focus groups, process action teams, and management by wandering around are good vehicles for team involvement leading to organizational transparency and a favorable ambience for collective problem identification, analysis, and resolution.

Encourage Project Teams to Fine-Tune the Process. Successful implementation of a concurrent engineering system often requires modifications of organizational processes, policies, and practices. In many of the most effective organizations, project teams have the power and are encouraged to make changes to existing organizational procedures, reporting relations, and decision and work processes. It is crucial, however, that these team initiatives are integrated with the overall business process and supported by management. True integration, acceptance by the people, and sustaining of the new organizational process will only occur through the collective understanding of all the issues and a positive feeling that the process is helpful to the work to be performed. To optimize the benefits of concurrent engineering, it must be perceived by all the parties as a win-win proposition. Providing people with an active role in the implementation and utilization process helps to build such a favorable image for participant buy-in and commitment. Focus teams, review panels, open discussion meetings, suggestion systems, pilot test groups, and management reviews are examples for providing such stakeholder involvement.

Invest Time and Resources. Management must invest time and resources for developing a new organizational system. An intricate system, such as concurrent engineering, cannot be effectively implemented just via management directives or procedures, but instead requires the broad involvement of all user groups, helping to define metrics and project controls. System designers and project leaders must work together with upper management toward implementation. This demonstrates management confidence, ownership, and commitment to the new management process. This will also help to integrate the new system with the overall business process. As part of the implementation plan, management must allow time for the people to familiarize themselves with the new vision and process. Training programs, pilot runs, internal consulting support, fully leveraged communication tools such as groupware, and best-practice reviews are examples of action tools that can help in both institutionalizing and fine-tuning the new management system. These tools also help in building the necessary user competencies, management skills, organization culture, and personal attitudes required for concurrent engineering to succeed.

Phase III: Managing in Concurrent Engineering

Plan the Project Effectively. As for any other project management system, effective project planning and team involvement is crucial to success. This is especially important in the concurrent engineering environment where parallel task execution depends on continuous cross-functional cooperation for dealing with the incremental work flow and partial result transfers. Team involvement, early in the project life cycle, will also have a favorable impact on the team environment, building enthusiasm toward the assignment, team morale, and ultimately team effectiveness. Because project leaders have to integrate various tasks across many functional lines, proper planning requires the participation of all stakeholders, including support departments, subcontractors, and management. Modern project management techniques, such as phased project planning and Stage-Gate concepts, plus established standards such as PMBOK, provide the conceptional framework and tools for effective cross-functional planning and organizing the work toward effective execution.

Define Work Process and Team Structure. Successful project management in concurrent engineering requires an infrastructure conducive to cross-functional teamwork and technology transfer. This includes properly defined interfaces, task responsibilities, reporting relations, communication channels, and work transfer protocols. The tools for systematically describing the work process and team structure come from the conventional project management system; they include (1) a *project charter,* defining the mission and overall responsibilities of the project organization, including performance measures and key interfaces; (2) a *project organization chart,* defining the major reporting and authority relationships; (3) *responsibility matrix* or *task roster;* (4) *project interface chart,* such as the N-Squared Chart discussed earlier; and (5) *job descriptions.*

Develop Organizational Interfaces. Overall success of a concurrent engineering depends on effective cross-functional integration. Each task team should clearly understand its task inputs and outputs, interface personnel, and work transfer mechanism. Team-based reward systems can help to facilitate cooperation with cross-functional partners. Team members should be encouraged to check out early feasibility and system integration. QFD concepts, N-Square charting, and well-defined phase-gate criteria can be useful tools for developing cross-functional linkages and promoting interdisciplinary cooperation and alliances. It is critically important to include into these interfaces all of the support organizations, such as purchasing, product assurance, and legal services, as well as outside contractors and suppliers.

Staff and Organize the Project Team. Project staffing is a major activity, usually conducted during the project formation phase. Because of time pressures, staffing is often done hastily and prior to defining the basic work to be performed. The result is often team personnel that is suboptimally matched to the job requirements, resulting in conflict, low morale, suboptimum decision making and ultimately poor project performance. While this deficiency will cause problems for any project organization, it is especially unfavorable in a concurrent engineering project environment that relies on strong cross-functional teamwork and shared decision making, built on mutual trust, respect, and credibility. Team personnel with poorly matched skill sets to job requirements is seen as incompetent, affecting their trust, respect, and credibility and ultimately their "concurrent team performance." For best results, project leaders should *negotiate the work assignment* with their team members one-to-one, at the outset of the project. These negotiations should include the overall task, its scope, objectives, and performance measures. A thorough understanding of the task requirements develops often as the result of personal involvement in the front-end activities, such as requirements analysis, bid proposals, project planning, interface definition, or the concurrent engineering system development. This early involvement also has positive effects on the buy-in toward project objectives, plan acceptance, and the unification of the task team.

Communicate Organizational Goals and Objectives. Management must communicate and update the organizational goals and project objectives. The relationship and contribution of individual work to overall business plans and their goals, as well as of individual project objectives and their importance to the organizational mission, must be clear to all team personnel. Senior management can help in unifying the team behind the project objectives by developing a "priority image" through their personal involvement, visible support, and emphasis of project goals and mission objectives.

Build a High-Performance Image. Building a favorable image for an ongoing project, in terms of high-priority, interesting work; importance to the organization; high visibility; and potential for professional rewards are all crucial for attracting and holding high-quality people. Senior management can help develop a "priority image" and communicate the key parameters and management guidelines for specific projects (Pham, Dimov, and Setchi, 1999). Moreover, establishing and communicating clear and stable top-down objectives helps in building an image of high visibility, importance, priority, and interesting work. Such a pervasive process fosters a climate of active participation at all levels and helps attract and hold quality people, unifies the team, and minimizes dysfunctional conflict.

Build Enthusiasm and Excitement. Whenever possible, managers should try to accommodate the professional interests and desires of their personnel. Interesting and challenging work is a perception that can be enhanced by the visibility of the work, management attention and support, priority image, and the overlap of personnel values and perceived benefits with organizational objectives (Thamhain, 2003). Making work more interesting leads to increased involvement, better communication, lower conflict, higher commitment, stronger work effort, and higher levels of creativity.

Define Effective Communication Channels. Poor communication is a major barrier to teamwork and effective project performance, especially in concurrent engineering environments, which depend to a large degree on information sharing for their concurrent execution and decision making. Management can facilitate the free flow of information, both horizontally and vertically, by workspace design, regular meetings, reviews, and information sessions (Hauptman and Hirji, 1999). In addition, modern technology, such as voice mail, e-mail, electronic bulletin boards, and conferencing, can greatly enhance communications, especially in complex organizational settings.

Create Proper Reward Systems. Personnel evaluation and reward systems should be designed to reflect the desired power equilibrium and authority/responsibility sharing needed for the concurrent engineering organization to function effectively. Creating a system and its metrics for reliably assessing performance in a concurrent engineering environment is a great challenge. However, several models, such as the Integrated Performance Index (Pillai, Joshi, and Rao, 2002), have been proposed and provide a potential starting point for customization. A QFD philosophy, where everyone recognizes the immediate "customer" for whom a task is performed, helps to focus efforts toward desired results and customer satisfaction. This customer orientation should exist, both downstream and upstream, for both company-internal and -external customers. These "customers" should score the performance of the deliverables they received and therefore have a major influence on the individual and team rewards.

Ensure Senior Management Support. It is critically important that senior management provides the proper environment for a project team to function effectively (Prasad, 1998). At the onset of a new project, the responsible manager needs to negotiate the needed resources with the sponsor organization and obtain commitment from management that these resources will be available. An effective working relationship among resource managers, project leaders, and senior management critically affects the credibility, visibility, and priority of the engineering team and their work.

Build Commitment. Managers should ensure team member commitment to their project plans, specific objectives, and results. If such commitments appear weak, managers should determine the reason for such lack of commitment of a team member and attempt to modify possible negative views. Anxieties and fear of the unknown are often a major reason for low commitment (Stum, 2001). Managers should investigate the potential for insecurities, determine the cause, and then work with the team members to reduce thesc negative perceptions. Conflict with other team members and lack of interest in the project may be other reasons for such lack of commitment.

Manage Conflict and Problems. Conflict is inevitable in the concurrent engineering environment with its complex dynamics of power and resource sharing, and incremental decision making. Project managers should focus their efforts on problem avoidance. That is, managers and team leaders, through experience, should recognize potential problems and conflicts at their onset, and deal with them before they become big and their resolutions consume a large amount of time and effort (Haque, 2003).

Conduct Team Building Sessions. A mixture of focus team sessions, brainstorming, experience exchanges, and social gatherings can be powerful tools for developing the concurrent work group into an effective, fully integrated, and unified project team (Thamhain and Wilemon, 1999). Such organized team-building efforts should be conducted throughout the project life cycle. Intensive team-building efforts may be especially needed during the formation stage of the concurrent engineering team. Although formally organized and managed, these team-building sessions are often conducted in a very informal and relaxed atmosphere to discuss critical questions such as (1) how are we working as a team? (2) what is our strength? (3) how can we improve? (4) what support do you need? (5) what challenges and problems are we likely to face? (6) what actions should we take? and (7) what process or procedural changes would be beneficial?

Ensure Personal Drive and Involvement. Project managers and team leaders can influence the concurrent engineering environment by their own actions. Concern for their team members, the ability to integrate personal needs of their staff with the goals of the organization, and the ability to create personal enthusiasm for a particular project all can foster a climate of high motivation, work involvement, open communication, and ultimately high engineering performance.

Provide Proper Direction and Leadership. Managers can influence the attitude and commitment of their people toward concurrent engineering as a project management tool by their own actions. Concern for the project team members, assistance with the use of the tool, and enthusiasm for the project and its administrative support systems can foster a climate of high motivation, involvement with the project and its management, open communications, and willingness to cooperate with the new requirements and to use them effectively.

Foster a Culture of Continuous Support and Improvement. Successful project management focuses on people behavior and their roles within the project itself. Companies that effectively manage projects, and reap the benefits from concurrent engineering, have cultures and support systems that demand broad participation in their organization developments. Ensuring organizational members to be proactive and aggressive toward change is not an easy task, yet it must be facilitated systematically by management. Our continuously changing business environment requires that provisions are being made for updating and fine-tuning the established concurrent engineering process. Such updating must be done on an ongoing basis to ensure relevancy to today's project management challenges. It is important to establish support systems—such as discussion groups, action teams and suggestion systems—to capture and leverage the lessons learned and to identify problems as part of a continuous improvement process.

Summary

In today's dynamic and hypercompetitive environment, proper implementation and use of concurrent engineering is critical for expedient and resource-effective project execution. The full range of benefits of concurrent engineering is in fact much broader than just a gain in lead time and resource effectiveness, but includes a wide spectrum of competitive advantages to the enterprise, ranging from increased quality of project deliverables to the ability of executing more complex projects and to higher levels of customer satisfaction. These benefits are primarily derived from effective cross-functional collaboration and full integration of the project management process with the total enterprise and its supply chain. However, these benefits do not occur automatically!

Designing, implementing, and managing in concurrent engineering requires more than just writing a new procedure, delivering a best-practice-workshop, or installing new information technology. It requires the ability to engage the organization in a systematic evaluation of specific competencies, such as for concurrent engineering, assessing opportunities for improvement, and designing a project management system that is fully integrated with the overall enterprise system and its strategy. Too many managers end up disappointed that the latest management technique did not produce the desired result. Regardless of its conceptual sophistication, concurrent engineering is just a framework for processing project data, aligning organizational strategy, structure, and people. To produce benefits for the firm, these tools must be fully customized to fit the business process and be congruent with the organizational system and its culture.

One of the most striking finding, from both the practice and research of concurrent engineering, is the strong influence of human factors on project performance. The organizational system and its underlying process of concurrent engineering is equally critical, but must be effectively integrated with the human side of the enterprise. Effective managers understand the complex interaction of organizational and behavioral variables. During the *design and implementation* of the concurrent engineering system, they can work with the various resource organization and senior management, creating a win-win situation between the people affected by the intervention and senior management. They can shake up conventional thinking and create a vision without upsetting established cultures and values. To be successful, both *implementing* concurrent engineering and *managing* projects through the system requires proactive participation and commitment of all stakeholders. It also requires congruency of the system with the overall business process and its management system.

Taken together, leaders must pay attention to the human side. To enhance cooperation among the stakeholders, managers must foster a work environment where people see the significance of the intervention for the enterprise and personal threats and work interferences are minimized.

One of the strongest catalysts, to both the *implementation of concurrent engineering* and the *management of projects*, is professional pride and excitement of the people, fueled by visibility and recognition of work accomplishments. Such a professionally stimulating environment seems to lower anxieties over organizational change, reduce communications barriers and conflict, and enhance the desire of personnel to cooperate and to succeed, a condition critically important for developing the necessary linkages for effective cross-functional project integration. Effective project leaders are social architects who understand the interaction of organizational and behavioral variables and can foster a climate of active participation and minimal dysfunctional conflict. They also build alliances with support organizations and upper management to ensure organizational visibility, priority, resource availability, and overall support for the project undertaking.

While no single set of broad guidelines exist that guarantees success for managing in concurrent engineering, project management is not a random process! A solid understanding of modern project management concepts, their tools, support systems, and organizational dynamics, is one of the threshold competencies for leveraging the concurrent engineering process. It can help managers in both, developing better project management systems and in leading projects most effectively through these systems.

References

Atnahene-Gimo, K. 2003. The effects of centrifugal and centripetal forces on product development speed and quality. *Academy of Management Journal* 43(3, June):359–373.

Bauly, J. and A. Nee. 2000. New product development: implementing best practices, dissemination and human factors. *International Journal of Manufacturing Technology and Management* 2:(1/7):961–982.

Bishop, S. 1999. Cross-functional project teams in functionally alligned organizations. *Project Management Journal* 30(3, September):6–12.

Chambers, C. 1996. Transforming new product development. *Research Technology Management* 39(6, November/December):32–38.

Checkland, P. 1999. *Systems thinking, systems practice.* Hoboken, NJ: Wiley, 1999.

Cooper, R., and Kleinschmidt, E. 1993. Stage-Gate systems for new product success. *Marketing Management* 1(4):20–29.

Denker, S., D. Steward, and T. Browning. 2001. Planning concurrency and managing iteration in projects. *Project Management Journal* 32(3, September):31–38.

Emery, F. 1969. *Systems thinking.* Harmondsworth, UK: Penguin.

Emery, F., and E. Trist. 1965. The causal texture of organizational environments. *Human Relations* 18(1):21–32.

Gerwin, D., and N. Barrowman. 2002. An evaluation of research on integrated product development. *Management Science* 48(7, July):938–954.

———. 2002. An evaluation of research on integrated product development. *Management Science* 48(7, July):938–954.

Githens, G. 1998. Rolling wave project planning. *Proceedings of the 29th Annual Symposium of the Project Management Institute.* Long Beach, CA, October 9–15.

Goldenberg, J., R. Horowitz, and A. Levav. 2003. Finding your innovation sweet spot. *Harvard Business Review* 81(3, March):120–128.

Haddad, C. J. 1996. Operationalizing the concept of concurrent engineering. *IEEE Transactions on Engineering Management* 43(2, May):124–132.

Haque B., K. Pawar, and R. Barson. 2003. The application of business process modeling to organizational analysis of concurrent engineering environments. *Technovation* 23(2, February):147–162.

Harkins, J. 1998. Making management tools work. *Machine Design* 70:(12, July):210–211.

Hauptman, O. and K. Hirji. 1999. Managing integration and coordination in cross-functional teams. *R&D Management* 29(2, April):179–191.

Hull, F., P. Collins, and J. Liker. 1996. Composite form of organization as a strategy for concurrent engineering effectiveness. *IEEE Transactions on Engineering Management* 43:(2, May):133–143.

Kerzner, H. 2001. *The project management maturity model.* Hoboken, NJ: Wiley.

Koufteros, X., M. Vonderembse, and M. Doll. 2002. Integrated product development practices and competitive capabilities: The effects of uncertainty, equivocality, and platform strategy. *Journal of Operations Management* 20(4, August):331–355.

LaPlante, A., and A. Alter. 1994. Corning Inc.: The stage-gate innovation process. *Computerworld* 28(44, October):81–84.

Litsikas, M. 1997. Break old boundaries with concurrent engineering. *Quality* 36(4, April):54–56.

Moffat, L. 1998. Tools and teams: Competing models of integrating product development projects. *Journal of Engineering and Technology Management* 1(1, March):55–85.

Morris, P. W. G. 1997. *The management of projects.* London: Thomas Telford.

Nee, A. and S. Ong. 2001. Philosophies for integrated product development. *International Journal of Technology Management* 21(3):221–239.

Nellore, R., and R. Balachandra. 2001. Factors influencing success in integrated product development (IPD) projects. *IEEE Transactions on Engineering Management* 48(2, May):164–174.

Neves, T., G. L. Summe, and B. Uttal. 1990. Commercializing technology: what the best companies do. *Harvard Business Review* (May/June):154–163.

Noori, H., M. Hugh, and G. Deszca. 1997. Managing the P/SDI process: best-in-class principles and leading practices. *Journal of Technology Management* 13(3, 1997):245–268.

O'Connor, P. 1994. Implementing a stage-gate process: A multi-company perspective. *The Journal of Product Innovation Management* 11: (3, June): 183–200.

Paashuis, V., and D. Pham. 1998. *The organisation of integrated product development.* Berlin: Springer-Verlag.

Pham, D., S. Dimov, and R. Setchi. 1999. Concurrent engineering: A tool for collaborative working. *Human Systems Management* 18(3/4):213–224.

Pilkinton, A., and R. Dyerson. 2002. Extending simultaneous engineering: electric vehicle supply chain and new product development. *International Journal of Technology Management* 23(1,2,3,):74–88.

Pillai, A., A. Joshi, and K. Raoi. 2002. Performance measurement of R&D projects in a multi-project, concurrent engineering environment. *International Journal of Project Management* 20(2, February):165–172.

Prasad, B. 1976. *Concurrent engineering fundamentals: Integrated product and process organization.* Vol. 1. Englewood Cliffs, NJ: Prentice Hall.

———. 1977. *Concurrent engineering fundamentals: Integrated product development.* Vol. 2. Englewood Cliffs: Prentice Hall.

———. 1998. Decentralized cooperation: a distributed approach to team design in a concurrent engineering organization. *Team Performance Management* 4(4):138–146.

———. 2002. Toward life-cycle measures and metrics for concurrent product development. *International Journal of Computer Applications in Technology* 15(1/3):1–8.

———. 2003. Development of innovative products in a small and medium size enterprise. *International Journal of Computer Applications in Technology* 17(4):187–201.

Prasad, B., F. Wang, and J. Deng. 1998. A concurrent workflow management process for integrated product development. *Journal of Engineering Design* 9(2, June):121–136.

Rasiel, E. 1999. *The McKinsey way.* New York: McGraw-Hill.

Rigby, Darrel K. 1995. Managing the management tools. *Engineering Management Review* (*IEEE*) 23(1, Spring):88–92.

Senge, P. M. 1990. *The fifth discipline: The art and practice of the learning organization.* New York: Doubleday/Currency.

Senge, P. and G. Carstedt. 2001. Innovating our way to the next industrial revolution. *Sloan Management Review* 42(2):24–38.

Shabayek, A. 1999. New trends in technology management for the 21 century. *International Journal of Management* 16(1, March):71–76.

Skelton, T., H. Thamhain, J. Hans. 1993. Concurrent project management: A tool for technology transfer, R&D-to-market. *Project Management Journal* 24(4, December).

Sobek, D. K., K. Jeffrey, K. Liker, C. Allen, and A. Ward. 1998. Another look how Toyota integrates product development. *Harvard Business Review* (July–August):36–49.

Stum, D. 2001. Maslow revisited: Building the employee commitment pyramid. *Strategy and Leadership* 29(4, July/August):4–9.

Sundaramurthy, C. and M. Lewis. 2003. Control and collaboration: Paradoxes of governance. *Academy of Management Review* 28(3, July):397–415.

Swink, M., J. Sandvig, and V. Marbert. 1996. Customizing concurrent engineering processes: Five case studies. *Journal of Product Innovation Management* 13(3, May):229–245.

Thamhain, H. 2003. Managing innovative R&D teamsteams. *R&D Management* 33(3, June):297–311.

Thamhain, H. 1994. A manager's guide to effective concurrent project management. *Project Management Network* 8(11, November):6–10.

———. 1996. Best practices for controlling technology-based projects. *Project Management Journal* 27(4, December):37–48.

———. 2001a. Leading R&D projects without formal authority. In *Management of Technology: Selected Topics*, ed. T. Khalil. Oxford, UK: Elsevier Science.

———. 2001b. The changing role of project management. Chap. 5 in *Research in management consulting*, ed. Anthony Buono. Greenwich, CT: Information Age Publishing.

———. 2002. "Criteria for effective leadership in technology-oriented project teams," Chapter 16 in *The Frontiers of Project Management Research* (Slevin, Cleland and Pinto, eds.), Newtown Square, PA: Project Management Institute, pp. 259-270.

Thamhain, H. and D. Wilemon. 1998. Building effective teams for complex project environments. *Technology Management* 4: 203–212.

Wu, S., J. Fuh, and A. Nee. 2002. Concurrent process planning and scheduling in distributed virtual manufacturing. *IIE Transactions* 34(1, January):77–89.

Yam, R, W. Lo, H. Sun, and P. Tang. 2003. Enhancement of global competitiveness for Hong Kong/ China manufacturing industries. *International Journal of Technology Management* 26(1):88–102.

CHAPTER TWENTY

PROCESS AND PRODUCT MODELING

Rachel Cooper, Ghassan Aouad, Angela Lee, and Song Wu

It is widely recognized that modeling processes and information is a complex task. This chapter looks at the various techniques that can be used for both process and product modeling. Beginning with a discussion of the importance of process management in managing projects, the chapter then defines a process, describes the various approaches to modeling processes, and illustrates the development of different process map generations. It also provides examples of processes in the management of new products in the manufacturing and construction industry sectors.

The second part of the chapter covers product modeling and object modeling paradigms. Included is the use of 3D and virtual reality to support visual modeling through work developed for the construction industry. The last part of the chapter discusses current research and how trends in the use of information communication technologies will influence development and management of both process and product modeling.

The Importance of Managing Projects

Today, companies introduce new products in a variety of ways, ranging from chaotic to systematic. However, it is unwise to constantly rely on luck to salvage the organizational procedure of the work at hand (Peppard and Rowland, 1995). There are still companies that mistakenly believe that an idea or impetus will easily become a successful new product. Furthermore, once a superficially attractive idea has been articulated, many companies push ahead into development but may forget or overlook some important steps, and so will consequently slip from the desired schedule and will incur increased costs. Unstructured development, a chaotic or random approach, usually leads to problems, especially when change in a new product's specification occurs. It has been shown that without a formal structure in which to freeze a specification or evaluate changes, "creeping elegance often runs amok" (Elzinga et al., 1995). (See also the chapters by Archer and Ghasemzadeh, Cooper and Reichelt, and Milosevic.)

In addition, those companies that have a process outlook—large and small, public and private, domestic and global—are now finding themselves in an era of inherent competition. Firms operate in dynamic environments and not stable ones, as both external competition and internal environments evolve over time. White (1996) proposes that in the near future around 75 percent of many organizations currently in business will no longer exist, either due to takeover or decline, whilst the others will emerge as international giants. Within such an environment, processes must also continuously change to enable firms to remain effective and profitable throughout changing conditions (Moran and Brightman, 1998). Those organizations undertaking improvements in productivity, quality, and operations need to reconsider their working practices (Elzinga et al., 1995). Katzenbach (1996) reports that organizational change is becoming everyone's problem and that customers require it, shareholder performance demands it, and continued growth depends upon it. Modeling and managing organizational processes are critical factors contributing to successful organizational change.

Definition of a Process

Research has found that every successful organization needs a ". . .formal blueprint, roadmap, template or thought process" for driving a new project (Cooper, 1994). Table 20.1 illustrates the various approaches to defining a 'process'.

More simply stated, a *process* has an input and an output, with the process receiving and subsequently transforming the input into the desired output (see Figure 20.1). A process can be visible, and at the same time, it can be invisible. We all tend to do familiar things in the same way, in a manner we are used to, and do not reflect upon the fact that "now I am performing an activity" or "now I have completed this task." However, in order to model a task or a process, we need to describe the "what happens," thus providing a simplified description of real-world phenomena. Often, nouns, verbs, and adjectives are used to depict a process (Lundgren, 2002). The noun usually refers to a person, place, or object; a verb is a word or a phase that describes a course of events, conditions, or experiences; and the adjective specifies an attribute of the noun (see Figure 20.2). There is a flow relation between the noun, the verb, and the adjective—a car is painted and the result is a painted car.

Approaches to Process Modeling

An understanding of processes can be reached in different ways. The project process is often depicted/modeled to enhance team coordination and communication through simple mech-

TABLE 20.1. DEFINITION OF A PROCESS.

Author	Definition
Davenport (1993)	A process is simply a "structured, measured set of activities designed to produce a specified output for a particular customer or market" and that they are "the structure by which an organisation follows that is necessary to produce value for its customers."
Cooper (1994)	Provides the thinking and action framework for transforming an idea into a product, and the processcan either be tangible or intangible, functionally based or organizationally based.
Oakland (1995)	"The transformation of a set of inputs, which can include actions, methods, and operations, into outputs that satisfy customer needs and expectations, in the form of products, information, services or—generally—results"
Zairi (1997)	"A process is an approach for converting inputs into outputs. It is the way in which all the resources of an organisation are used in a reliable, repeatable and consistent way to achieve its goals."
Bulletpoint (1996)	Suggests that regardless of the definition of the term "process," there are certain characteristics that this process should have: • Predictable and definable inputs • A linear, logical sequence of flow • A set of clearly definable tasks or activities • A predictable and desired outcome or result

anisms such as flow and Gantt charts (a flowchart that encompasses time). To model more complex scenarios of real-world phenomena, techniques such as IDEF0 (Integrated Definition Language) and analytical reductionism/process decomposition are commonly used (Koskela, 1992).

IDEF0

During the 1970s, the U.S. Air Force Program for Integrated Computer-Aided Manufacturing (ICAM) sought to increase manufacturing productivity through systematic application of computer technology. The ICAM program identified the need for better analysis and communication techniques for people involved in improving manufacturing productivity and thus developed a series of techniques known as the IDEF family (IDEF, 2002):

FIGURE 20.1. A PROCESS.

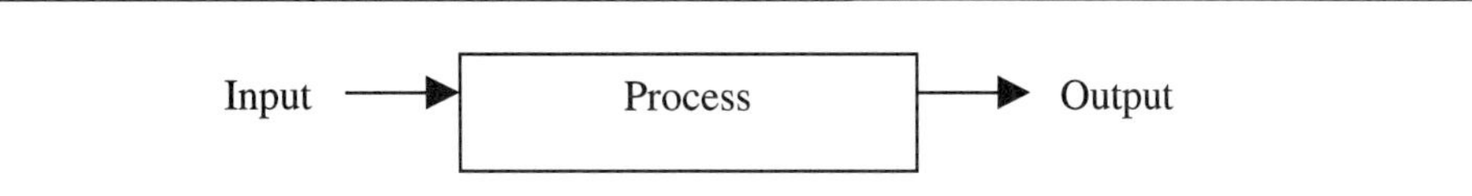

Source: Vonderembse and White (1996).

FIGURE 20.2. DESCRIPTION OF A PROCESS.

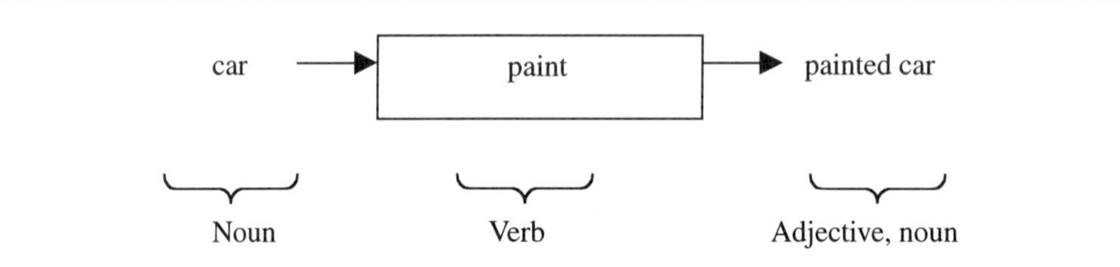

Source: Lundgren (2002).

1. IDEF0, used to produce a "function model"—a structured representation of the functions, activities, or processes within the modeled system or subject area.
2. IDEF1, used to produce an "information model," represents the structure and semantics of information within the modeled system or subject area.
3. IDEF2, used to produce a "dynamics model," represents the time-varying behavioral characteristics of the modeled system or subject area.

In 1983, the U.S. Air Force Integrated Information Support System program enhanced the IDEF1 information modeling technique to form IDEF1X (IDEF1 Extended), a semantic data modeling technique. Currently, IDEF0 and IDEF1X techniques are widely used in the government, industrial, and commercial sectors, supporting modeling efforts for a wide range of enterprises and application domains. For the purpose of this chapter, IDEF0 will be described as it most closely relates to the "functional" new product development process.

The Integrated Definition Language 0 for function modeling is an engineering technique for performing and managing needs analysis, benefits analysis, requirements definition, functional analysis, systems design, maintenance, and the baseline for continuous improvement (IDEF, 2002). IDEF0 models provide a "blueprint" of functions and their interfaces that must be captured and understood in order to make systems engineering decisions that are logical, integratable, and achievable, to provide an approach to:

- performing systems analysis and design at all levels, for systems composed of people, machines, materials, computers, and information of all varieties—the entire enterprise, a system, or a subject area;
- producing reference documentation concurrent with development to serve as a basis for integrating new systems or improving existing systems;
- communicating among analysts, designers, users, and managers;
- allowing team consensus to be achieved by shared understanding;
- managing large and complex projects using qualitative measures of progress; and
- providing a reference architecture for enterprise analysis, information engineering, and resource management.

The modeling language itself makes explicit the purpose of a particular activity and is composed of a series of boxes and arrows (see Figure 20.3). The boxes of the IDEF0 tech-

FIGURE 20.3. THE BASIC CONCEPT OF THE IDEF0 SYNTAX.

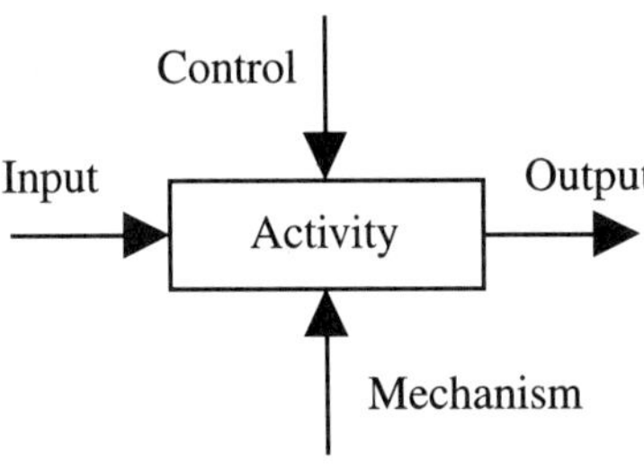

Source: IDEF (2002).

nique represent functions, defined as activities, processes, or transformations. Each box should consist of a name and number inside the box boundaries; the name is of an active verb or verb phrase that describes the function, and the number inside the lower right corner is to identify the subject box in the associated supporting text.

The arrows in the diagram represent data or objects related to the functions and do not represent flow or sequence as in the traditional process flowchart model. They convey data or objects related to functions to be performed. The functions receiving data or objects are constrained by the data or objects made available. Each side of the function box has a standard meaning in terms of box/arrow relationships. The side of the box with which an arrow interfaces reflects the arrow's role. Arrows entering the left side of the box are inputs; inputs are transformed or consumed by the function to produce outputs. Arrows entering the box on the top are controls; controls specify the conditions required for the function to produce correct outputs. Arrows leaving a box on the right side are outputs; the outputs are the data or objects produced by the function. Arrows connected to the bottom side of the box represent mechanisms; upward-pointing arrows identify some of the means that support the execution of the function.

The functions in an IDEF0 diagram can be broken down or decomposed into more detailed diagrams, until the subject is described at a level necessary to support the goals of a particular project (see Figure 20.4). The top-level diagram in the model provides the most general or abstract description of the subject represented by the model. This diagram is followed by a series of child diagrams providing more detail about the subject. Each subfunction is modeled; on a given diagram, some of the functions, none of the functions, or all of the functions may be decomposed individually by a box, with parent boxes detailed by child diagrams at the next lower level. All child diagrams must be within the scope of the top-level context diagram/parent box. In turn, each of these subfunctions may be decomposed, each creating another, lower-level child diagram.

Analytical Reductionism / Process Decomposition

Analytical reductionism/process decomposition involves breaking the process down into levels of granularity, as demonstrated in Figure 20.5, with the lower-level subprocesses further

FIGURE 20.4. IDEF DECOMPOSITION STRUCTURE.

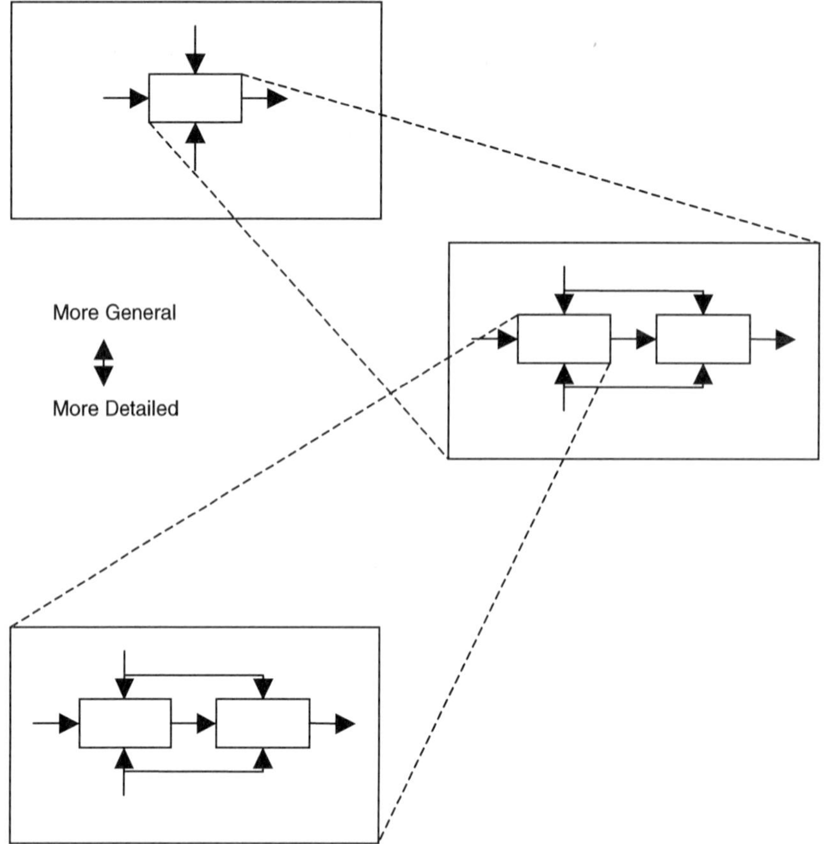

defining its corresponding upper-level process. It shares similarities with IDEF0, in that it breaks the parent process into subsequent more detailed child/subprocesses and then onto procedures and activities. The level at a process that differentiates from a procedure is, however, still a topic of discussion in the process management field.

A process (Koskela, 1992; Cooper, 1994; Vonderembse and White, 1996):

- converts inputs into outputs;
- creates a change of state by taking the inputs (e.g., material, information, people) and passing it through a sequence of stages during which the inputs are transformed or their status changed to emerge as an output with different characteristics. Hence, processes act

FIGURE 20.5. PROCESS LEVELS.

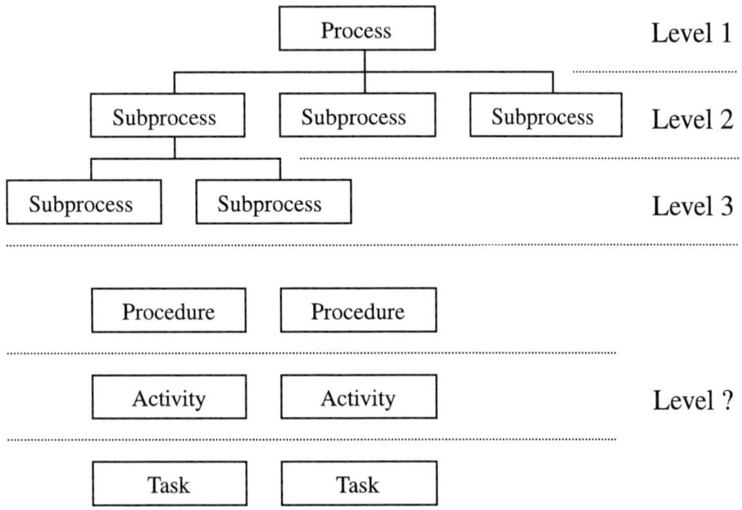

upon input and are dormant until the input is received. At each stage the transformation tasks may be procedural but may also be mechanical, chemical, and so on;
- clarifies the interfaces of fragmented management hierarchies;
- helps to increase visibility and understanding of the work to be done;
- defines the business/project activities across functional boundaries.

A procedure (Lee et al., 2001):

- is a sequence of steps. It includes the preparation, conduct, and completion of a task. Each step can be a sequence of activities, and each activity a sequence of actions. The sequence of steps is critical to whether a statement or document is a procedure of something else;
- is required when the task we have to perform is complex or is routine and is required to be performed consistently;
- defines the rules that should be followed by an individual or group to carry out a specific task; their definition is usually rigid, leaving no opportunity for individual initiative;
- supports the process.

Process Management and New Product Development

According to Davenport (1994), the design of the project process should start with a high-level model that engages the management team. This is to avoid a too-detailed description of processes in the initial creative stage; a detailed model will only lower the motivation of the management team. The core process model should be used as a tool to communicate a shared project view on a high level of abstraction. The following section describes how this is applied to the NPD (new product development) process that is used in manufacturing.

The development of new products and services that can successfully compete in local, national, and global markets has become a key concern for a large majority of organizations (Cooper, 1992). The NPD process is fundamental for organizations to support this growth (HM Treasury, 1998). The process has received, and continues to receive, much attention by academia and practitioners to improve its effectiveness and efficiency, and its development has been examined with growing interest across various industrial sectors in lieu of the changing nature of the economic climate.

A new product is one that has not been previously manufactured by a company and is a necessary risk that companies must undertake (Crawford, 1992). Technological developments, shorter product life cycles, complexity of products, increasingly changing market demands, and globalized competition means that companies face a limited space in which they can succeed. NPD is a critical means by which the whole organization as a business as well as its employees can adapt, diversify, and in some cases, reinvent their firms to match evolving market and technical conditions. The fundamental aim of the NPD program is to get the right product to the market or customer as quickly as possible. The NPD process is composed of a number of activities (Crawford, 1944), initiated by the identification of the need or the adoption of an idea. A number of technical, financial, and business preliminary evaluations are then performed, followed by further detailed technical development follows. Finally, after a series of company and market tests, the finished product is launched onto the market (Crawford, 1994). Generically, these activities can be separated into three main broad categories (Cooper and Kleinschmidt, 1995):

- *Predevelopment activities.* Idea generating/establishing the need, followed by a number of preliminary market, technical, financial, and production assessments
- *Development activities.* The physical development of the product
- *Post-development activities.* The final launch of the product into the marketplace

From a historical point of view, NPD models can be classified into three main groups: sequential, overlapping, and Stage-Gate (Cooper and Kleinschmidt, 1995).

NPD Process Models

In the 1960s, the NPD process was still in its first generation, following a simple linear sequential structural model whereby the development moved through different, almost mutually exclusive, phases in a logical step-by-step fashion (McGarth, 1996). These phases are shown in Figure 20.6.

The development proceeded to the next phase only after all the requirements of the preceding phase were satisfied, and in each succeeding phase, different intermediate results were created, with the outputs of one phase forming the major inputs to the next (Coughlan, 1991). In this sense, the major activities of the process were isolated from each other, creating an over-the-brick-wall effect, whereby each discrete activity played little or no regard to the

FIGURE 20.6. TYPICAL SEQUENTIAL NPD PROCESS.

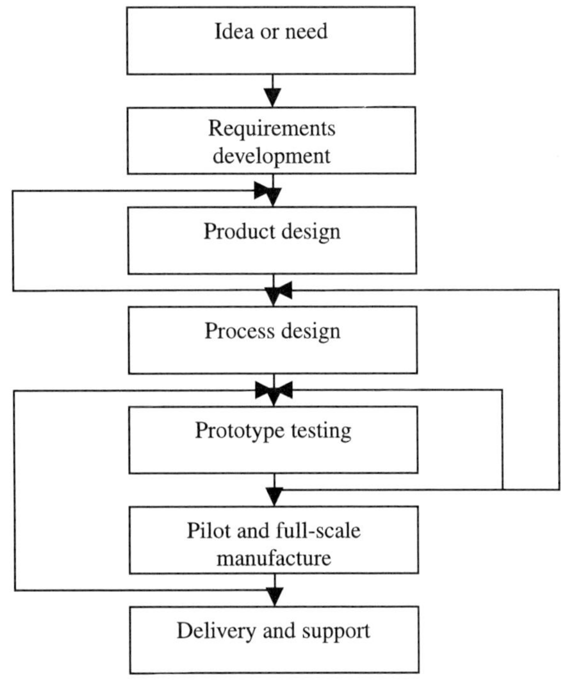

Source: Kagioglou (1999).

next activity (see Figure 20.7). This led to long lead times, late product launch, increased development costs, lack of effective information flow, and lack of flexibility for change in the process (Turino, 1990). However, this approach does offer high staff utilization in departments; it is favorable for breakthrough projects that require a revolutionary innovation or for very big projects where the shear size of personnel involved limits extensive communications between the members, and when product development is masterminded by a

FIGURE 20.7. SEQUENTIAL OVER THE "BRICK WALL" APPROACH.

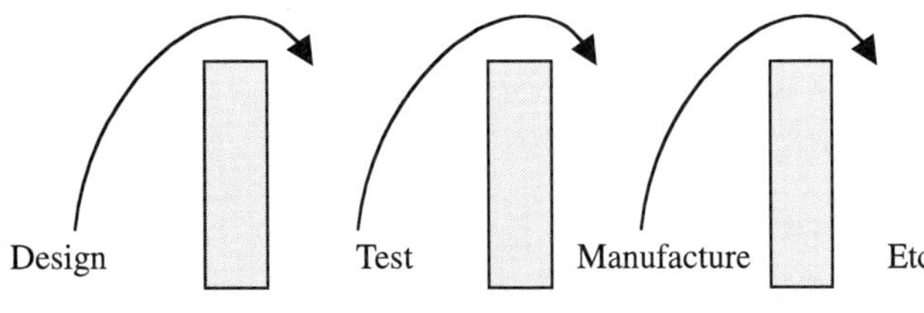

genius who makes the invention and hands down a well-defined set of product specifications (Takeuchi and Nonaka, 1986).

The need for change from the sequential to a more concurrent approach to NPD became increasingly apparent in the last two decades, where the manufacturing function had to be integrated into the design function so as to improve coordination and communication in the project. Thus, the NPD process steadily evolved into a more increasingly complicated second generation "coupling" model (Tidd, et al., 1997). Robert Cooper's NPD Stage-Gate model gained wide acceptance, as illustrated in Figure 20.8 (Cooper, 1990). It is presented as a series of stages and gates, which can vary from typically four to seven (Cooper, 1993). Each stage represented a number of activities that needed to be performed before progressing to a "decision" gate before the next stage; the stages represented multifunctional activities, involving a number of people from various departments relevant to the activities. These gates were clearly defined as "yes" or "no" decision points that provided organizations with the capability to measure and control the process and match subsequent funding to meeting the requirements at each gate (McGarth, 1996).

The Stage-Gate process was found to reduce development time and produce marketable results and optimized internal resources (Anderson, 1993). However, while enabling a higher degree of control and understanding of the progression of a project process, such gates required variable tasks to be checked off against predetermined lists. This often made the process both cumbersome and slow (Cooper and Kleinschmidt, 1992). Projects were forced to wait at each gate until all tasks were completed and not to stray from a process through which all projects had to progress. Any overlapping of activities was impossible (Devinney, 1995). Therefore, in order to overcome unnecessary delay and to enable smoother progression, the more recently developed third-generation "parallel" processes have sought to accommodate the need for certain tasks to overlap during a NPD program (Cooper, 1994). The main characteristic of the new process was the overlapping of the stages. Go/kill decisions were delayed to allow for flexibility and speed—the previous "hard" gates were replaced with "fuzzy" gates. (See also the chapter by Thamhain below on Concurrent Engineering.) In essence, these fuzzy gates allowed conditional-go decisions, enabling a degree that permitted the overlap of certain stages (see Figure 20.9). In addition, by being more outcome-focused, these processes have permitted organizations to build prioritization models that enabled projects to move through the process with more flexibility.

FIGURE 20.8: SECOND-GENERATION NPD STAGE GATE PROCESS.

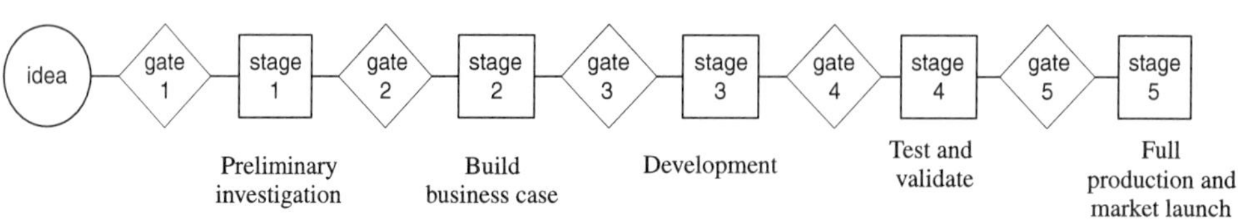

Source: Cooper (1990).

FIGURE 20.9. THIRD-GENERATION FUZZY STAGE GATE OVERLAPPING NPD PROCESS.

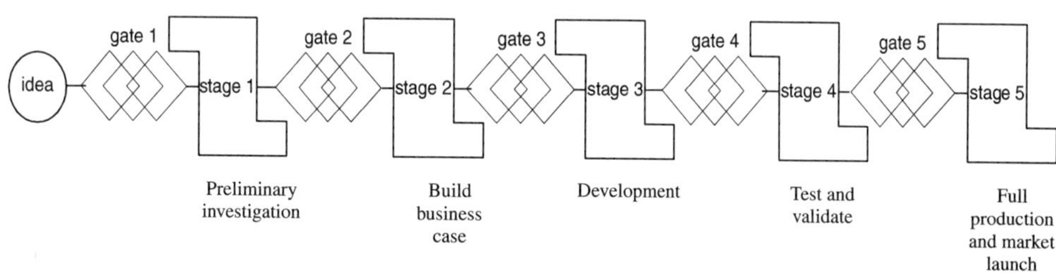

Source: Cooper (1994).

Process Management in the Construction Industry

The evolution of the NPD process has often been cited as a learning point for other industries to improve their practice, in particular the construction industry (Howell, 1999). The UK construction industry is under increasing pressure to improve its practices (Hill, 1992; Howell, 1999). Indeed the construction industry has been criticized for its poor performance by several government and institutional reports, such as Emmerson (1962), Banwell (1964), Gyles (1992), Latham (1994), and more recently, Egan (1998). Most of these reports conclude, time and time again, that the fragmented nature of the industry, lack of coordination and communication between parties, the informal and unstructured learning process, adversarial contractual relationships, and lack of customer focus are what inhibits the industry's performance. (See also the chapter by Morris where the argument is broadened into the United States and process engineering context.)

The Traditional Design and Construction Process/RIBA Plan of Work

In 1959, the United Nations defined the building (project) process as ". . . the design, organisation and execution of building project' that has come to be recognized as . . . normal practice in any country or region . . . it is characterized by the fact that all operations follow a set pattern known to all participants in the building operation" (United Nations, 1959). However, this description is essentially untrue today. The nature of the design and construction process has grown in complexity since the 1950s, thus leading to an increased number of actors in the project.

The term largely associated with the "traditional building process" today usually refers to the practice where, upon perceiving a need for a new facility, a building client approaches an architect/engineer to initiate a process to design, procure, and construct a building to meet his or her specific needs. The process, in turn, almost invariably consists of the project being designed and built by two separate groups of disciplines who collectively form a temporary multiorganization for the duration of the project: the design group and the construction group (Mohsini and Davidson, 1992). The design group, typically, is coordinated by an architect/engineer. Depending upon the circumstances of the project at hand, it may also include other design professionals and specialists such as engineers, estimators, quantity surveyors, and so on. The principal function of this group is to prepare the design specifications of the work and other technical and contractual documents. The construction group, on the other hand, is usually coordinated by the main contractor and consists of a host of subcontractors and suppliers/manufacturers of building materials, components, hardware, and subsystems. This group is primarily responsible for the construction of the building project.

The two groups typically do not work coherently together (Kagioglou et al., 1998). The design activities in construction are usually isolated from the realities of the real issues facing production, as each function is expected to play a specific and limited role in any phase, thus contributing to the industry's problems, as highlighted by the many governmental and industrial reports (Emmerson, 1962; Banwell, 1964; Gyles, 1992; Latham, 1994; Egan, 1998). This factor has contributed to the problems of construction of poor supply chain coordination and fragmented project teams with adversarial relationships (Mohsini and Davidson, 1992). The Royal Institute of British Architects' (RIBA) Plan of Work (RIBA, 1997) fundamentally represents this practice. The model (see Figure 20.10) was originally published in 1963 as a standard method of operation for the construction of buildings, and it has become widely accepted as the operational model throughout the building industry (Kagioglou et al., 1998). However, it was designed from an architectural perspective, which has in some way restricted its applications to specific forms of UK construction contracts and is increasingly inappropriate for the newer types of contracts being used both in the United Kingdom and elsewhere, such as "partnering" frameworks.

Generic Design and Construction Process Protocol (GDCPP)

The development and use of more generic and comprehensive process models for the new product development in construction has been a concern for researchers and the construction industry itself since the early 1990s. There is now wider use of such models in the industry. The Process Protocol is a generic model developed by the authors in conjunction with leaders in the construction industry and is an attempt to drive construction toward the third/new generation. This approach is one that, in light of increasing outsourcing and supply partnering in manufacturing, can be used to address process management in any extended enterprise.

The protocol is ". . . a common set of definitions, documentation and procedures that provides the basis to allow a wide range of organisations involved in a construction project to work together seamlessly' (Kagioglou et al., 1998b). It maps ". . . the entire project process from the client's recognition of a new or emerging need, through to operations and maintenance" (Cooper et al., 1998; Kagioglou et al., 1998c) by breaking down the design and construction process into four broad stages—Preproject, Preconstruction, Construction, and

FIGURE 20.10. RIBA PLAN OF WORK.

Predesign	A	B											
Design			C	D	E								
Preparing to build						F	G	H					
Construction									J	K	L		
Post-construction												M	

Stage A:	Inception	Stage B:	Feasibility
Stage C:	Outline proposals	Stage D:	Scheme design
Stage E:	Detail design	Stage F:	Production info
Stage G:	Bills of quantities	Stage H:	Tender action
Stage J:	Project planning	Stage K:	Operations on site
Stage L:	Completion	Stage M:	Feedback

FIGURE 20.11. DETAIL OF THE PROCESS MAP.

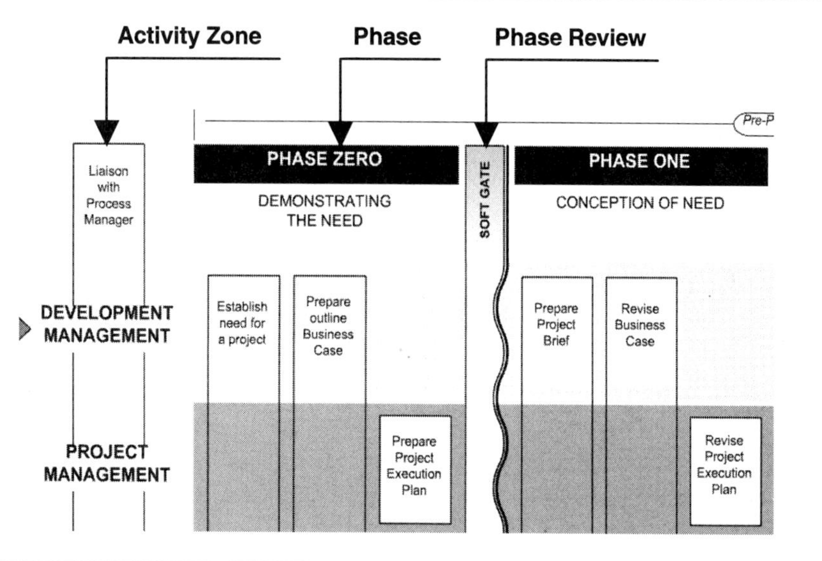

Post-construction stages—and ten phases (demonstrating the need; conception of need; outline feasibility; substantive feasibility study and outline financial authority; outline conceptual design; full conceptual design; coordinated design procurement and full financial authority; production information; construction; operation; and maintenance). These are represented by vertical columns (see Figure 20.11) separated by gates, soft and hard. As in Cooper's (1994) third-generation fuzzy Stage-Gate model, the soft gates allow flexibility of control, while the hard gates ensure that all work is progressing to program and are usually related to finance, and production signoff. The horizontal bands (see Figures 20.11 and 20.12) represent coordinated activities—namely, Development, Project, Resource, Design, Production, Facilities, Health and Safety and Legal, and Process Management—because in construction these are undertaken by numerous professional consultants and subcontractors as well as disciplines. In defining the activities rather than the disciplines, the Process Protocol emphasizes the need for team collaboration and coordination representing the fact that most construction projects are completed by a virtual enterprise of organization works.

The Process is based on six key principles (Sheath et al., 1996; Aouad et al., 1998; Kagioglou et al., 1998a; Cooper et al., 1998):

- *Whole project view.* The process has to cover the whole life of the project, from recognition of a need to the operation of the finished facility. This approach ensures that all the issues are considered from both a business and a technical point of view. The separation between the design and production functions, as described previously, has been pronounced as a key contributor to the inadequacies of construction (Harvey, 1971). The NPD process brought about the integration of multifunctions, thereby introducing those who do the building earlier into the design phase. Gunaskaran and Love (1998) argue that this will be invaluable. These specialist organizations have specific knowledge concerning the capability of the life cycle of materials, the overall performance of a product, and the programming of site operations.
- *Progressive design fixity.* Drawing from the "Stage-Gate" approach in manufacturing processes, the protocol adapts a phase review process that applies a consistent planning and review procedure throughout the project. The benefit of this approach is fundamentally the progressive fixing and approval of design information throughout the process. This is particularly useful in bringing the risk and cost of late changes to the attention of clients who are not familiar with the impact of their design changes on a project cost.
- *A consistent process.* The generic properties of the Process Protocol will allow a consistent application of the phase review process. This, together with the adoption of standard approach to performance measurement, evaluation, and control, will help facilitate the process of continual improvement in design and construction. Using the same basic generic process uniformly yields the most productive results (Kuczmarski, 1992). Everyone involved in the process develops a comfortable and consistent level of working; they see why category analysis works, and they understand the purpose of strategic roles better. The most important underlying factor to any development process is making it understandable and actionable by all people concerned.

FIGURE 20.12. THE PROCESS PROTOCOL HIGH-LEVEL MAP.

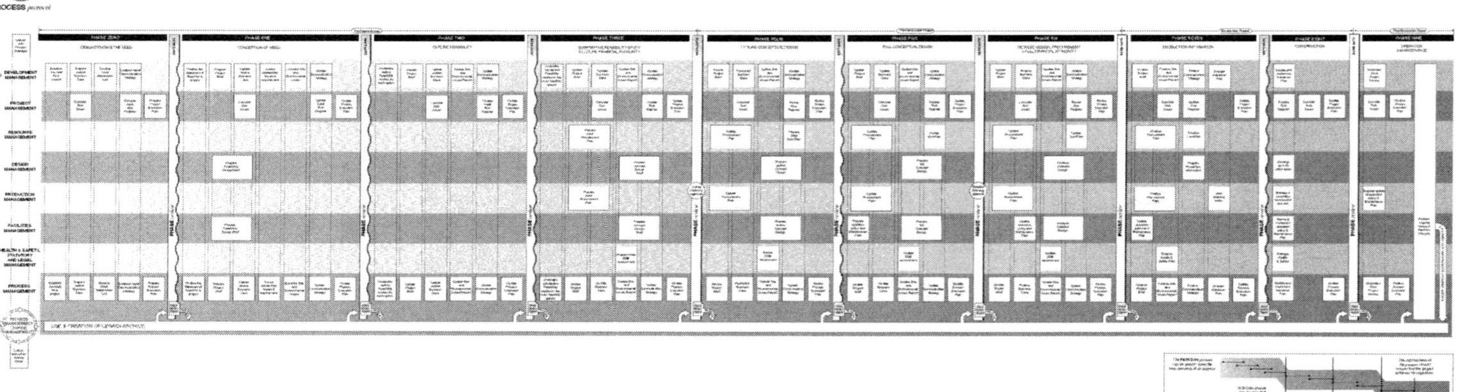

Source: www.processprotocol.com.

- *Stakeholder involvement/teamwork.* Project success relies upon the right people having the right information at the right time. The proactive resourcing of phases through the adoption of a "stakeholder" view will help to ensure that appropriate participants from each of the functions are consulted earlier in the process than is traditionally the case. Again, research suggests that full team participation improves the process by bearing all project input simultaneously, hence avoiding or reducing further revisions as to reduce time and money (Jassawalla and Sashittal, 1998).
- *Coordination.* Researchers have long argued that the employees are the critical building block of an organization (Crawford, 1977b). Successful teams bring together diverse information on every aspect that impacts customer satisfaction, and they overcome the shortcomings of hierarchical structures and generate quality decisions (Hoffman, 1979). Therefore the process map emphasizes the need to coordinate across activities the key actors in the process
- *Feedback.* Because of the nature of the supply chain in construction, rarely is knowledge or lessons learned in construction systematically incorporated back into the process. According to Li and Love (1998), construction problem-solving reliant on tacit knowledge has traditionally moved with individuals from project to project; cumulative project knowledge is not collected. Therefore, real benefits in cost, schedule, quality, and safety for future projects can only emerge if construction knowledge can be effectively harnessed in planning and executing future work is to be incorporated into the process (Kartam, 1996; Kumaraswamy and Chan, 1998). The Process Protocol recommends the use of a legacy archive; a central repository or information spine (Hinks et al., 1997) that can take the form of an electronic information management system. There have traditionally been such systems available to manufacturing industry but only recently have they been introduced to aid the collection and coordination of project knowledge and information (usually Web-enabled) to connect disparate suppliers and subcontractors.

These principles based on 20 years of research in manufacturing and construction.

Product Modeling

As discussed in the previous section, process modeling involves the modeling of processes in a project and can often include the data and material that flows between them. Conversely, product modeling is used to model the elements specific to a product and the related process relationships; visual models are commonly produced through the mapping of conceptual data and process models and describe the information infrastructure of the product under development. The rapid prototyping of buildings/products using 3D/virtual-reality (VR) technologies enable developers and clients to quickly assess and evaluate their requirements before committing fully to the project. This section of the chapter considers aspects of the way information is used in product modeling and, by example, the use of IT specifically to model the construction product is detailed.

The UK construction industry has currently not fully adopted and envisaged the benefits of product modeling, unlike other industries such as aerospace and manufacturing. This is largely attributed to its deployment on an ad hoc basis without context or framework, leading to the development of unreliable information models that become unusable over time. Thus, efforts and resources of product modeling are wasted. In addition, the construction industry is divided for historical rather than logical reasons (see Morris's chapter at the end of this book). These divisions tend to reflect the traditional roles performed by the disciplines (as discussed previously), and not the information required to complete the project. This leads to problems associated with information and project team integration.

It has recently been cited by a number of leading researchers (Lee et al., 2003; Dawood et al., 2003; Fischer, 2000; Graphisoft, 2003; Rischmoller et al., 2000) that object technology, coupled with client server applications and the Web environment, will provide the best way forward to enable project collaboration and information sharing, thus evoking a central project-based information database (building information/product model) and exchange between professionals. Graphical schema languages such as Entity Relationship Diagrams, NIAM, and IDEF1X were commonly used to undertake information modeling within the construction industry (Bjork and Wix, 1991; Rasdorf and Abudayyeh, 1992) until the early 1990s. Now UML (Unified Modeling Language) has become more popular because of its wide use in the software industry. However, the use of such modeling techniques is not advocated as appropriate for the industry, as they imply a separation between the data and the processes performed on the data. To overcome this problem, object-oriented models can be developed to describe the static information as well as the behavior of objects. This has proven to be more advantageous, as the resulting information model is richer and more natural, thus more usable for construction and other industries. This, it is anticipated, will enable effective coordination and communication of information among all project team members.

Object Modeling

Unlike traditional data modeling techniques, the object-oriented paradigm models can be viewed as a collection of objects "talking" to each other via messages. The behavior of one object may result in changes in another object. This is done through message sending. For example, if the object "column" has been moved, it should send a message to the object "beam" (to which it is related) to tell it to resize itself reflecting the "object" change. This way of modeling is very powerful and is peculiar to the object-oriented world. In such a world, objects can be composed of other objects. These objects can be images, speech, music, or possibly a video. The object-oriented paradigm also supports the notions of encapsulation, abstraction, inheritance, and polymorphism (Martin and Odell, 1992) that were considered as critical in handling the complex task of information modeling. Encapsulation permits objects to have properties (data) and actions (operation). For instance, an object "beam" can have properties such as "length," "width." and so on and behaviors or actions such as "move beam," "calculate load on beam," and so on. Abstraction allows the analyst to abstract information according to requirements. For instance, the information about a beam can be abstracted in terms of properties, shape, materials, and so forth. Inheritance allows information in the parent object (beam) to be inherited by the child object (cantilever beam). Polymorphism allows objects to have one operation that can have different implementations. For instance, an operation such as "calculate area" can be attached to an object called "beam." However, the implementations of this operation differ according to whether the beam is a "rectangular beam," a "T beam," and so on.

Another major benefit of object orientation is the support of the notion of reusability. With such a notion, integrated databases can be developed from reusable object-oriented components that can be assembled as required. This is very similar to the way a building designer uses reusable plans that can be configured to his or her requirements. The object-oriented paradigm also supports the notion of "perspectives." This notion allows the construction professional to view the information from their own perspective. For instance, the architect is interested in features such as color, aesthetics, and texture, whereas the construction planner is interested in features such as time and resources. To illustrate this point, take the concept of a wall. This can be viewed from different perspectives. An architectural wall has attributes such as dimensions, color, and texture. A construction planner wall has attributes such as dimensions time and cost. It is therefore logical to store common information such as length and width in "wall" that can be inherited by the architectural wall, and so on through inheritance.

Object modeling is aimed at the identification of concepts/data—relationships between the concepts, attributes, and operations that are to be supported by the database. This task should be done independently of any implementation platform. Figure 20.13 shows an illustration of an object model incorporating objects, relationships, attributes, and operations.

Activity/Process Modeling

As described earlier in the chapter, the activities performed within a construction project can be modeled using techniques such as activity hierarchy, data flows, IDEF0 techniques, and flowcharts. These techniques describe the information flows between processes. This is useful in understanding how information is communicated between processes. Figure 20.14 shows different representations of different process, data, and matrix models.

Product Data Technologies

In the context of this chapter, product data technology (PDT) refers to techniques of data modeling, data exchange, and data management, which are aimed at the integration of product information through standard data models. Historically, the initial requirement for a standardized data model came from the need for different versions of CAD application to share their graphic files. IGES (the Initial Graphics Exchange Specification) was devel-

FIGURE 20.13. AN ILLUSTRATION OF AN OBJECT MODEL.

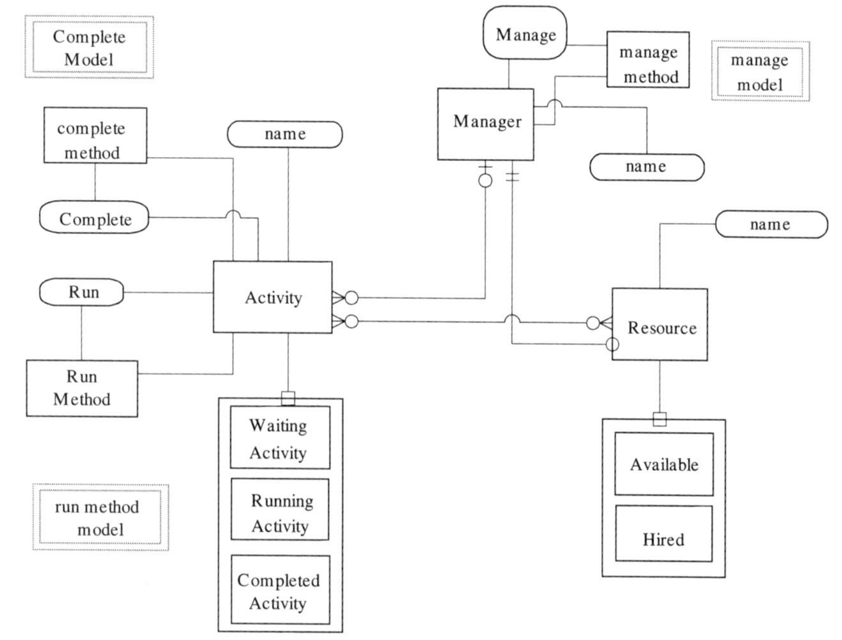

oped as a protocol for this purpose. However, graphical and geometrical data is only part of the information required in a building project. IGES is not able to support the exchange of other type of data such as construction, thermal, light, and so on. Therefore, a new project, PDES (Product Data Exchange Specification), was proposed in the United States in the early 1980s to overcome these limitations. In the same period, similar efforts were made in other countries, for example, the SET (Standard d'Echange et de Transfert) in France and the VDAFS (Verband der Deutschen Autombilindustrie Flaechen Scnittstelle) in Germany. In 1983, all these initiatives were coordinated into a major international program under the umbrella of the International Standard Organisation, Standard for Exchange of Product data (STEP). Thus, this became a comprehensive ISO standard (ISO 10303) that describes how to represent and exchange digital product information. In the construction industry, IFC (Industry Foundation Classes) was developed as a standard for exchange building product data, which is compliant with STEP. IFCs are an interoperable data standard that are linked to any proprietary software application.

A Methodology for Modeling Information

Figure 20.15 illustrates how product models can be produced starting at a strategic/contextual level. This type of approach to developing product models helps in deriving a framework that ensures that all models are developed within a context. Activity hierarchy techniques

FIGURE 20.14. MODELING TECHNIQUES.

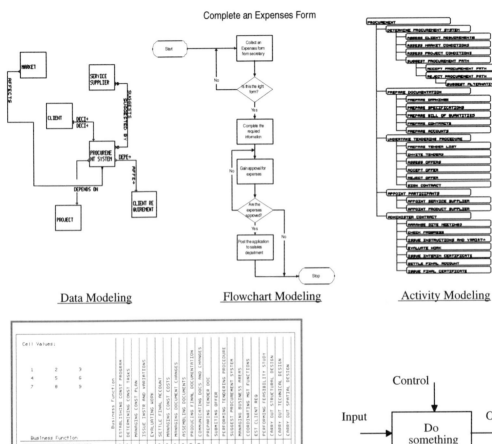

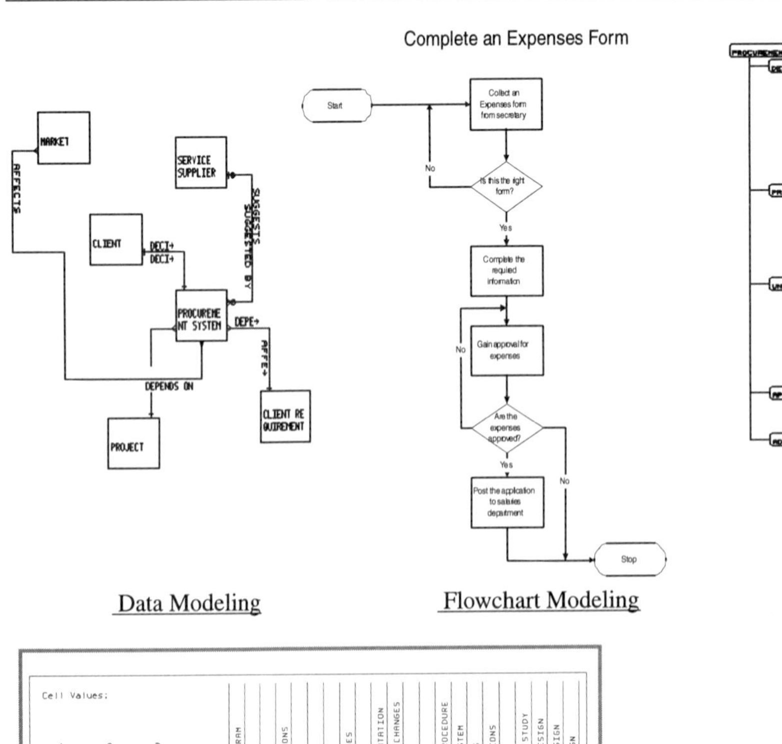

Data Modeling

Flowchart Modeling

- PROCUREMENT
  - DETERMINE PROCUREMENT SYSTEM
    - ASSESS CLIENT REQUIREMENTS
    - ASSESS MARKET CONDITIONS
    - ASSESS PROJECT CONDITIONS
    - SUGGEST PROCUREMENT PATH
      - ACCEPT PROCUREMENT PATH
      - REJECT PROCUREMENT PATH
        - SUGGEST ALTERNATIVE
  - PREPARE DOCUMENTATION
    - PREPARE DRAWINGS
    - PREPARE SPECIFICATIONS
    - PREPARE BILL OF QUANTITIES
    - PREPARE CONTRACTS
    - PREPARE ACCOUNTS
  - UNDERTAKE TENDERING PROCEDURE
    - PREPARE TENDER LIST
    - INVITE TENDERS
    - ASSESS OFFERS
    - ACCEPT OFFER
    - REJECT OFFER
    - SIGN CONTRACT
  - APPOINT PARTICIPANTS
    - APPOINT SERVICE SUPPLIER
    - APPOINT PRODUCT SUPPLIER
  - ADMINISTER CONTRACT
    - ARRANGE SITE MEETINGS
    - CHECK PROGRESS
    - ISSUE INSTRUCTIONS AND VARIAT+
    - EVALUATE WORK
    - ISSUE INTERIM CERTIFICATE
    - SETTLE FINAL ACCOUNT
    - ISSUE FINAL CERTIFICATE

Activity Modeling

Cell Values:

1 2 3
4 5 6
7 8 9

Business Function	ESTABLISHING CONST PROGRAM	DETERMINING CONST TASKS	MANAGING CONST PLAN	ISSUE INSTR AND VARIATIONS	EVALUATING WORK	SETTLE FINAL ACCOUNT	MANAGING CONST COSTS	MANAGING DOCUMENT CHANGES	ASSEMBLING DOCUMENTS	PRODUCING FINAL DOCUMENTATION	COMMUNICATING DOCS AND CHANGES	PREPARING TENDER DOC	SUBMITING OFFER	PERFORMING TENDERING PROCEDURE	SUGGEST PROCUREMENT SYSTEM	MANAGING BUSINESS AREAS	COORDINATING MGT FUNCTIONS	EST CLIENT REQ	PERFORMING FEASIBILITY STUDY	CARRY OUT STRUCTURAL DESIGN	CARRY OUT TECHNICAL DESIGN	CARRY OUT SPATIAL DESIGN
MANAGING DOCUMENT CHANGES	6	9				6	8	9	6	5	5											
ASSEMBLING DOCUMENTS		5						6	9	9	9	6	5									
PRODUCING FINAL DOCUMENTATION						6		5	9	9	9	6										
COMMUNICATING DOCS AND CHANGES		5						5	9	9	9	7	5			5	5					
PREPARING TENDER DOC									6	6	7	9	8	7								
SUBMITING OFFER									5		5	8	9	9	5							
PERFORMING TENDERING PROCEDURE												7	9	9								
SUGGEST PROCUREMENT SYSTEM													5		9	5	5					
MANAGING BUSINESS AREAS							6				5				5	9	9					
COORDINATING MGT FUNCTIONS											5				5	9	9					
EST CLIENT REQ																		9	7			
PERFORMING FEASIBILITY STUDY																		7	9	7	7	7
CARRY OUT STRUCTURAL DESIGN																			7	9	9	9
CARRY OUT TECHNICAL DESIGN																			7	9	9	9
CARRY OUT SPATIAL DESIGN																			7	9	9	9
FINALISING REQUIREMENTS																		6	7	9	9	9
ESTABLISH MAINTENANCE POLICY																			5	8	9	9
EST AND OBTAIN PLANNING REQ																						

Model name: CRUD MATRIX	21 Jul. 1993	MTX
Subset name: ALL	14:47	Page: 1

Matrix Modeling

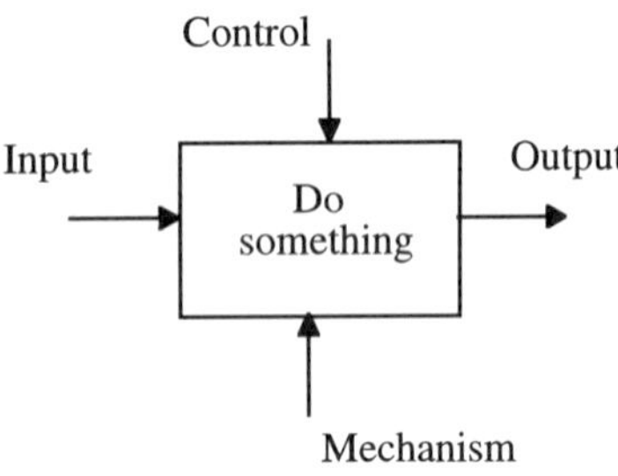

IDEF0/Process Modeling

FIGURE 20.15. METHODOLOGY FOR MODELING INFORMATION.

1. Strategic/Contextual Modeling: To provide a framework for modeling

2. Domain Specific Model:

Activity Decomposition: To understand structure

Domain Object Modeling: To model static and dynamic information

Integration Modeling: To define models that need to share information

3. Object Model implementation

Define Implementation Environment: Technologies required

Produce Implementation Perspectives: Define specific implementation Requirements (object databases)

4. Produce Applications: interface development

and matrix modeling could be used to define a contextual model (Graphisoft, 2003). The next step is to model information for specific domains using object-oriented techniques. Object as well as process models can be developed at this stage. The models should comply with standards such as the IFCs (Industry Foundation Classes). Following this, an environment for implementation must be defined; this includes technologies such as databases, interfaces, and so on. The last step is to implement the models in databases and define the interface requirement for developing and linking into applications. The interfaces to VR/3D applications can be developed at this stage.

The work on information modeling and integration in construction was initiated more than 30 years ago. However, fruitful practical results have failed to emerge because of the complex structure of the construction industry and because information technology has not been exploited properly. This chapter emphasized the importance of establishing a framework into which models from different domains and of varying abstraction levels can fit. The inclusion of object types viewed from different perspectives but shared across different domains and abstraction levels is seen as a major step forward in integrating information throughout the construction industry. Such structuring is considered essential for the development of accurate, understandable models. "nD modeling" has recently emerged in construction as the next step forward—the integration of process and product modeling (Lee et al., 2003).

nD modeling

The building design process is complex, encapsulating a number and variety of factors in order to satisfy the client's requirements. In fact, it is rarely the case that there is one homogeneous "customer," but a number and variety of stakeholders who will be the end users of the building. These stakeholders are increasingly demanding the inclusion of design features such as maintainability, environmental friendliness, accessibility, crime deterrence, acoustical soundness, and energy performance. Each of these design parameters that the stakeholders seek to consider will have a host of social, economic, and legislative constraints that may be in conflict with one another. Furthermore, as each of these factors vary—in the amount and type of demands they make—they will have a direct impact on the time and cost of the construction project. Perspectives of design are usually balanced between aesthetics, ecology, and economics—a three-dimensional view of design that acknowledges its social, environmental, and economic roles is now necessary. The criteria for successful design therefore will include a measure of the extent to which all these factors can be coordinated and mutually satisfied to meet the expectations of all the parties involved (Lee et al., 2002a).

The volume of information required to interplay these scenarios to enable the client to visualize design changes and to assist with decision making—changing the design, planning schedules, and cost estimates—can be laborious, time-consuming, and costly (Lee et al., 2002b). There is now a need to allow users to create, share, contemplate and apply knowledge from multiple perspectives of user requirements. Conceptually, this will involve taking three-dimensional modeling in the built environment to an *n* number of dimensions, and thus integrates the process and information flow within a construction environment. Indeed market, regulatory, social, economic, and environmental factors are becoming so complex in the development of product in both construction and manufacturing that nD modeling is becoming a necessity.

This chapter has only been able to touch on the main approaches to process and product modeling in two industries; however, it does illustrate how critical both are to effective and efficient futures.

The Future

Imagine a system that:

> given an idea, illustrates alternatives, illustrates constraints, and enables the understanding of both quantitative (time, cost, legislation) and qualitative (aesthetics, usability) dimensions. It enables all the stakeholders to participate and allows users to virtually experience the product concept. The system will determine the build specification, the manufacturing resources, and the production processes, and it will provide the drawings and the tool sets. It is a system where we can use our knowledge, in conjunction with the other stakeholders, to achieve the best solution, at the right cost, in a faster time, and in a sustainable manner.

This is the Holy Grail for process and product modeling. The need will not go away. Indeed, as companies, systems products, and markets get even more complex, we need the models to guide and help us make decisions. However, organizational behavior illustrates that we are not automatons; we will never work to a detailed and prescribed process and procedure, when situations demand innovation, creativity, and constant change to enable us to compete. Yet we do need systems to help us work through a complex world for the benefit of its inhabitants. The challenge is to understand what systems are the most appropriate, how we can best introduce them into organizations, and the impact that they will have on our work behavior and the future of the organizations who use them.

References

Alshawi, M. 1996. *SPACE: Integrated environment.* Internal paper. University of Salford, July 1996.

Ammermann, E., R. Junge, P. Katranuschkov, and R. J. Scherer. 1994. *Concept of an object-oriented product model for building design*; Technische Universität, Dresden.

Anderson, E. J. 1994. *Management of manufacturing, models and analysis.* Wokingham, UK: Addison-Wesley.

Aouad G., M. Betts, P. Brandon, F. Brown, T. Child, G. Cooper, S. Ford, J. Kirkham, R. Oxman, M. Sarshar, and B. Young B. 1994. *Integrated databases for design and construction: Final report.* Internal report. University of Salford, July 1994.

Aouad, G. 1999. *Trends in information visualisation in construction.* IV 99, London, 1999.

Aouad, G., P. Brandon, F. Brown, T. Child, T., G. Cooper, S. Ford, J. Kirkham, R. Oxman, and B. Young. 1995. The Conceptual modelling of construction management information. *Automation in Construction* 3:267–282.

Aouad, G., J. Hinks, R. Cooper, D. Sheath, M. Kagioglou, and M. Sexton. 1998a. An information technology IT map for a generic design and construction process protocol. *Journal of Construction Procurement,* 4 (1, November): 132–151.

Aouad, G., M. Kagioglou, and R. Cooper. 1999. IT in construction: A driver or and enabler? *Journal of Logistics and Information Management* 12:130–137.

Aouad, G., J. Kirkham, P. Brandon, F. Brown, G. Cooper, S. Ford, R. Oxman, M. Sarshar, and B. Young. 1993. Information modelling in the construction industry: The information engineering approach. *Construction Management and Economics.* 11(5):384–397.

Arditi, D., and H. M. Gunaydin. 1998. Factors that affect process quality in the life cycle of building projects. *Journal of Construction Engineering and Management* 124(3):194–203.

ATLAS 1992. *Architecture, methodology and tools for computer integrated large scale engineering: ESPRIT Project 7280.* Technical Annex Part 1: General Project Overview.

Augenbroe, G 1993. *COMBINE.* Final report. Delft University.

Banwell, H. 1964. *Report of the Committee on the Placing and Management of Contracts for Building and Civil Engineering Work.* HMSO, London.

Bjork, B. C., 1989. Basic structure of a proposed building product model. *Computer Aided Design,* 21 (2, March): 71–78.

———. 1991. A unified approach for modelling construction information. *Building and Environment* 27(2): 173–194.

Bjork, B. C. and J. Wix. 1991. *An Introduction to STEP.* VTT and Wix McLelland Ltd., Bracknell, England.

Brandon, P., and M. Betts. 1995. *Integrated construction information.* London: Spon.

Bulletpoint. 1996. *Creating a change culture: Not about structures, but winning hearts and minds.* Wesley, New York.

Burbidge, J. L. 1996. *Period batch control.* Oxford, UK: Oxford University Press.

Cooper, R. G. and E. J. Kleinschmidt. 1987a. New products: What separates winners from losers? *Product Innovation Management Journal* 4:169–184.

———. 1987b. Success factors in product innovation. *Industrial Marketing Management Journal* 7:9–21.

———. 1995. Benchmarking the firm's critical success factors in new product development. *Journal of Product Innovation Management* 12:374–391.

Cooper, R. G. 1984. The performance impact of product innovation strategies. *European Journal of Marketing* 18(5):223–229.

———. 1990. Stage-Gate system: A new tool for managing new products. *Business Horizons* (May–June): 44–54.

———. 1993. *Winning at new products: Accelerating the process from idea to launch.* Reading, MA: Addison-Wesley.

———. 1994. Third-generation new product processes. *Journal of Product Innovation Management* 10(6–14).

———. 1999. From experience: The invisible success factors in product innovation. *Journal of Production Innovation Management* 16:115–33.

Cooper, R., M. Kagioglou, G. Aouad, J. Hinks, M. Sexton, and D. Sheath. 1998. Development of a generic design and construction process. *European Conference on Product Data Technology*, BRE, 205–214.

Coughlan, P. D. 1991. Differentiation and integration: The challenge of new product development. *Proceedings of the 5th Annual Conference of the British Academy of Management.* June 28.

Crawford, C. M. 1977a. *New products management.* Homewood, IL: Irwin.

———. 1977b. Product development: Today's most common mistakes. *University of Michigan Business Review* 6:7–8.

———. 1992. The hidden costs of accelerated product development. *Journal of Product Innovation Management* 9(3):161–176.

Crawford, K. M. and J. F. Fox. 1990. Designing performance measurement systems for just-in-time operations. *International Journal of Production Research* 28(11):2,025–2,036.

Davenport, T. H. 1993. *Process innovation: Reengineering work through information technology.* Boston: Harvard Business School Press.

Dawood, N., E. Sriprasert, and Z. Mallasi. 2003. Product and process integration for 4D visualisation at construction site level: A uniclass-driven approach. In *Developing a vision of nD-enabled construction.* A. Lee. et al.. Construct IT report, Salford, 64–68.

Devinny, T. M. 1995. Significant issues for the future of product innovation. *Journal of Product Innovation Management* 12:70–75.

Egan, J. 1998. *Rethinking construction.* Report from the Construction Task Force, Department of the Environment, Transport and Regions, UK.

Elzinga, D. J., T. Horak, L. Chung-Yee, and C. Bruner. 1995. Business process management: survey and methodology. *IEEE Transactions on Engineering Management* 24(2):119–128.

Emmerson, H. 1962. *Studies of problems before the construction industries.* HMSO, London.

Fenves, S. J. 1990. Integrated software environment for building design and construction. *Computer-aided design* 22 (1, January/February).

Fischer, M. 1997. 4D Modelling. *Proceedings of Global Construction IT.*

———. 2000. Benefits of 4D models for facility owners and AEC service providers. *Construction Congress VI.* ASCE. Orlando, FL. February, 990–995.

Froese, T. and B. Paulson. 1994. OPIS: An object model-based project information system. *Microcomputers in Civil Engineering* 9:13–28.

Graphisoft 2003. *The Graphisoft virtual building: Bringing the information model from concept into reality.* Graphisoft white paper.

Griffin, A. 1997. PDMA research on new product development practices: updating trends and benchmarking best practices. *Journal of Product Innovation Management* 14:429–458.

Gunasekaran, A. and P. E. D. Love. 1998. Concurrent engineering: A multidisciplinary approach for construction. *Logistics Information Management* 11(5):295–300.

Gyles, R. 1992. *Royal commission into productivity in the New South Wales building industry.* R. Gyles QC, Government Printer, London.

Harvey, J. P. 1971. *The master builders: Architecture in the Middles Ages.* London: Thames and Hudson.

Hill, T. J. 1992. Incorporating manufacturing perspectives in corporate strategy. In *Manufacturing Strategy.* C. A. Voss. Oxford, UK: Chapman & Hall.

Hinks, J., G. Aouad, R. Cooper, D. Sheath, M. Kagioglou, and M. Sexton. 1997. IT and the design and construction process: A conceptual model of co-maturation. *The International Journal of Construction* (July): 56–62.

HM Treasury 1998. *Innovating for the future.* Department of Trade and Industry. HMSO, London.

Hoffman, L. R. 1979. *The group problem solving process: Studies of a valance model.* New York: Praeger.

Howard, H. C. 1991. Linking design data with knowledge-based construction.

Howell, D. 1999. Builders get the manufacturers in. *Professional Engineer* (May): 24–25.

IDEF 2002. www.idef.com.

Jassawalla, A. R. and H. C. Sashittal. 1998. An examination of collaboration in high-technology new product development processes. *Journal of Product Innovation Management* 15:237–254.

Kagioglou, M. 1999. *Adapting manufacturing project processes into construction: A methodology.* Unpublished PhD thesis. Salford, UK: University of Salford.

Kagioglou, M., R. Cooper, G. Aouad, J. Hinks, M. Sexton, and D. Sheath. 1998a. *Final report: Generic design and construction process protocol.* Salford, UK: The University of Salford.

———. 1998b. *A generic guide to the design and construction process protocol.* Salford, UK: The University of Salford.

———. 1998c. Cross-industry learning: The development of a generic design and construction process based on the Stage-Gate new product development process found in the manufacturing Industry. *Proceedings of the Engineering Design Conference.* Brunel, UK.

Kartam, N. 1994. ISICAD: Interactive system for integrating CAD and computer-based construction systems. *Microcomputers in Civil Engineering* 9:41–51.

———. 1996. Making effective use of construction lessons learned in project life cycle. *Journal of Construction Engineering and Management* (March): 14–21.

Katzenbach, J. 1996. *Real change leaders.* London: Nicholas Brealey.

Khurana, A. and S. R. Rosenthal. 1998. Towards holistic "front ends" in new product development. *Journal of Product Innovation and Management* 15:57–74.

Koskela, L. 1992. *Application of the new production philosophy to construction.* Technical report no. 72. Center for Integrated Facility Engineering, Stanford University.

Kuczmarski, T. D. 1992. *Managing new products: The power of innovation.* Upper Saddle River, NJ: Prentice Hall.

Kumaraswamy, M. M. and D. W. M. Chan. 1998. Contributors to construction delays. *Construction Management and Economics Journal* 16(1):17–29.

Latham, M. 1994. *Constructing the team: Final report of the government/industry review of procurement and contractual arrangements in the UK construction industry.* London: The Stationery Office.

Lee, A., M. Betts, G. Aouad, R. Cooper, S. Wu, and J. Underwood. 2002b. Developing a vision for an nD modelling tool. Key note speech. *Proceedings of CIB W78 Conference—Distributing Knowledge in Building (CIB w78)*, 141–148. Denmark.

Lee, A., A. J. Marshall-Ponting, G. Aouad, S. Wu, I. Koh, C. Fu, R. Cooper, M. Betts, M. Kagioglou, and M. Fischer. 2003. *Developing a vision of nD-Enabled construction.* Construct IT report. Salford, UK.

Lee, A., S. Wu, G. Aouad, and C. Fu. 2002a. Towards nD Modelling. Submitted to the *European Conference on Information and Communication Technology Advances and Innovation in the Knowledge Society.* E-sm@art, Salford, UK.

Li, H. and P. E. D. Love. 1998. Developing a theory of construction problem solving. *Construction Management and Economics* 16:721–727.

Lundgren 2002. Process. Unpublished proposal.

Martin, J. and J. Odell. 1992. *Object oriented analysis and design.* Upper Saddle River, NJ: Prentice Hall.

McGarth, M. E. 1996. *Setting the pace in product development.* Boston: Butterworth-Heinemann.

MOB 1994. *Rapport final. Modeles objet batiment, appel d'offres du plan construction et architecture.* Programme Communication/Construction.

Mohsini, R. A. and C. H. Davidson. 1992. Detriments of performance in the traditional building process. *Journal of Construction Management and Economics* 10:343–359.

Moran, J. W. and B. K. Brightman. 1998. Effective management of healthcare change. *The TQM Magazine* 10(1):27–29.

Oakland, J. S. 1995. *Total quality management: The route to improving performance.* 2nd ed. Boston: Butterworth-Heinemann.

Peppard, J. and P. Rowland. 1995. *The essence of business process re-engineering.* Upper Saddle River, NJ: Prentice Hall.

Plossl, K. R. 1987. *Engineering for the control of manufacturing.* Upper Saddle River, NJ: Prentice-Hall.

Powell, J. 1995. Virtual reality and rapid prototyping for Engineering. *Proceedings of the Information Technology Awareness Workshop.* University of Salford, Salford, UK.

Rasdorf, N. J. and O. Abudayyeh. 1992. NIAM conceptual database design in construction management. *Journal of Computing in Civil Engineering,* 6(1):41–62

Rezgui, Y. A, G. Brown, R. Cooper, A. Aouad, J. Kirkham, and P, Brandon. 1996. An integrated framework for evolving construction models. *The International Journal of Construction IT* 4(1):47–60.

RIBA. 1997. *RIBA plan of work for the design team operation.* 4th ed. London: Royal Institute of British Architects Publications.

Riedel, J. C. K. H., and K. S. Pawar. 1997. The consideration of production aspects during product design stages. *Integrated Manufacturing Systems* 8(4):208–214.

Rischmoller, L., and R. Matamala. 2003. Reflections about nD Modelling and Computer Advanced Visualisation Tools (CAVT). In *Developing a vision of nD-enabled construction.* A. Lee, et al. Construct IT report. Salford, UK, 92–94.

Rischmoller, L., M. Fisher, R. Fox, and L. Alarcon, L. 2000. 4D planning and scheduling (4D-PS): Grounding construction IT research in industry practice. Proceedings of CIB W78 Conference on Construction Information Technology: Taking the construction industry into the 21st century. Iceland, June.

Schonberger, R. J. 1982. *Japanese manufacturing techniques: Nine hidden lessons in simplicity.* New York: Free Press.

Sheath, D. M., H. Woolley, R. Cooper, J. Hinks, and G. Aouad. 1996. A process for change: The development of a generic design and construction process protocol for the UK construction industry. *Proceedings of the CIT Conference.* Institute of Civil Engineers. Sydney, Australia, April.

Sower, V. E., J. Motwani, and M. J. Savoie. 1997. Classics in production and operations management. *International Journal of Operations and Production Management* 17(1):15–28.

Takeuchi, H. and I. Nonaka. 1986. *The new product development game.* Cambridge, MA: Harvard Business Press.

Tidd, J., J. Bessant, and K. Pavitt. 1997. *Managing innovation.* Chichester, UK: Wiley.

United Nations 1959. *Government policies and the cost of building*. Geneva: ECE.

Vonderembse, M. A. and G. P. White. 1996. *Operations management: Concepts, methods and strategies*. New York: West Publishing.

Watson, A. and A. Crowley. 1995. CIMSteel integration standard. In .., *Product and process modelling in the building industry*, ed. R. J. Scherer. 491–493. Rotterdam: A. A. Balkema.

White, A. 1996. *Continuous quality improvement: A hands-on guide to setting up and sustaining a cost effective quality programme*. Gloucester: Judy Piakus.

Zairi, M. 1997. Business process management: A boundary-less approach to modern competitiveness. *Business Process Management* 3(1):64–80.

CHAPTER TWENTY-ONE

MANAGING CONFIGURATIONS AND DATA FOR EFFECTIVE PROJECT MANAGEMENT

Callum Kidd, Thomas F. Burgess

Configuration management (CM) has a severe image problem in many modern organizations: It is too often viewed as nothing more than glorified change control or version management—a costly exercise in form filling, with little or no technical content. As a value-added business activity, configuration management is, almost invariably, rated as less significant than, for example, quality management or project management (Kidd, 2001). The irony is that neither of these activities is possible without an effective configuration management process. Quality management, for example, requires us to know when configurations meet stated requirements. But how can we be sure that we are measuring against the most current list of requirements? Can we be sure that the reasons for making any changes were identified and impacts assessed prior to a decision being made? What effect will those changes have on the project schedule, and on total cost? Answering these questions is the business of configuration management. Effective configuration management is an essential part of an overall project management activity. To treat it as anything less is a recipe for disaster.

What Is Configuration Management?

Configuration management is a technique used by many companies to support the control of the design, manufacture, and support of a product. ISO10007 (ISO 1997) defines configuration management as:

> a management discipline that applies technical and administrative direction to the development, production and support life cycle of a configuration item. This discipline is

> applicable to hardware, software, processed materials, services, and related technical documentation.

It is important to understand at this point that the term configuration is a generic name for anything that has a defined structure or is composed of some predetermined pattern. Software, hardware, buildings, process plant, assets, and even the human body comes under the broad definition of a configuration. From a management perspective, it is often better to use the generic name of configuration, as it often avoids the software/hardware bias that causes confusion within the organization. Managing the definition of that pattern or structure from concept through to disposal is commonly termed configuration management.

According to Daniels (1985):

> Very simply Configuration Management is a management tool that defines the product, then controls the changes to that definition.

In essence, the key to configuration management is founded on good business sense and straightforward practice in handling documentation. However, regardless of the routines practiced by some adept companies, in general, many companies have been comparatively poor in their control of the depth and uniformity of the relevant documentation. These deficiencies came in to focus in the United States during the late 1950s in the arms race to produce reliable, working defense materiel. As with any substantial program that is faced with tight deadlines and severe competition, the magnitude of change that was generated by the various collaborators to ensure compatibility among elements was enormous. The emphasis on hitting deadlines meant that when the various parties took stock after a successful missile flight was finally made, the realization dawned that adequate technical documentation to complete an identical missile was not available. Records of part identification, build statements, changes applied, changes implemented, and technical publications reflecting the build standard were missing. Such situations generated the impetus to systematically deal with product specifications and their modifications throughout the development and build life cycle.

The impetus to improve management in this key area was pushed forward by the customers, who generally were governments or their armed forces, and the developers, who often were large companies involved in defense work. To establish better control, the involved parties drew up configuration management standards that decreed how the projects were to be managed. In standards such as the EIA649 standard (EIA 1998), configuration management covers the full product life cycle from "concept through to de-commission."

The majority of case studies and written examples of configuration management come from highly technical and complex environments. Perhaps the following household example will demonstrate the application of CM in an environment that will be familiar to the majority of us. Consider a washing machine in your home. Bought in 1998, it has provided some years of trouble-free, reliable service. The 12-month warranty passed some years ago, but to date, there have been no problems with the appliance. Over the past few days, you notice that a patch of water has appeared in front of the machine. It appears during the wash cycle and looks to be getting worse. You phone the service engineer, and he tells you that it is most likely the seal on the pump. He asks the make and model, then arranges a visit. He arrives with the new seal and detaches the pump assembly. He checks the product identification number (PIN) on the side of the unit, then checks his catalogue. The seal he has brought does not look like the one on this pump. But how can that be? Surely each model will have common components? Not necessarily so. In Figure 21.1, you can see a simplified product breakdown structure (PBS) for the washer. The model, XYZ, consists of a number of assemblies common across the full model range. One of those assemblies, though, had a seal problem that was not identified until late 1998. Depending on when your machine was manufactured, it may contain the old seal on the pump assembly. But how do you know? Simple. The PIN gives detailed information on date of manufacture and batch number. The PIN, not the model number, will tell you which seal is on your pump unit. A further check in the catalogue will determine if the old seal and the new one brought by the service engineer are interchangeable. If not, it will provide another alternate part. This is simple configuration management in action—the same principle keeps aircraft in the air, cars on the road, and software working.

The Configuration Management Process

Configuration management is probably best seen as a process for managing the following:

- The composition of a product
- The documentation and other data and products defining the product that supports it

The process may be related to a single product or to an associated collection of products, often referred to as systems and subsystems.

FIGURE 21.1. PRODUCT BREAKDOWN STRUCTURE OF A WASHING MACHINE.

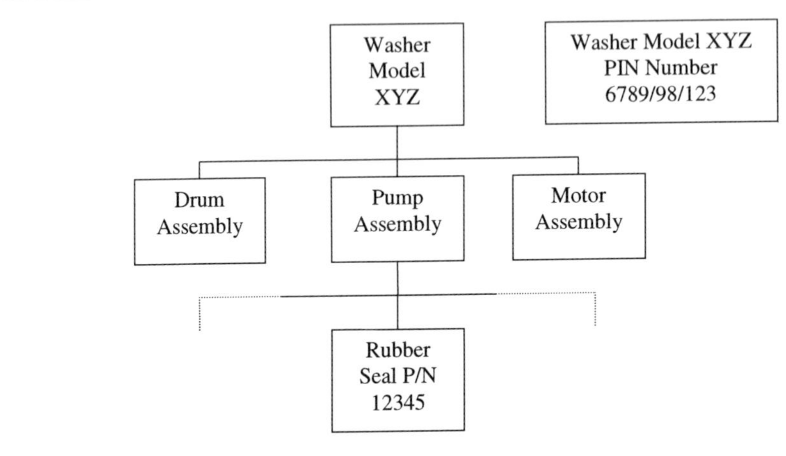

Efforts to develop a global consensus standard of best CM practice have resulted in the publication of ANSI/EIA 649, the most widely used CM practice model to date (Kidd, 2001). Configuration Management is traditionally defined in terms of the four interrelated activities:

- Configuration identification
- Configuration change management
- Configuration status accounting
- Configuration verification and audit

This structure (see Figure 21.2) is followed in ANSI/EIA 649, where the four activities sit beneath the overall planning activity. Each of these four areas is dealt with next.

Configuration Identification

Configuration identification is the key element of the CM process. According to ANSI/EIA 649, configuration identification is the basis from which the configuration of products are defined and verified, products and documents are labeled, changes are managed, and accountability is maintained. Typical activities include the following:

- Define product structure and select elements to be managed
- Assign unique identifiers
- Define product attributes, interfaces, and details in product information

Configuration identification can be problematic because of the way in which we dissociate the development of the product and system structuring from change management. The difficulty of identifying configurations is further exacerbated by the fact that there may be several structures, or views, of each configuration, depending on which phase of the life cycle is under consideration.

FIGURE 21.2. GENERIC CONFIGURATION MANAGEMENT ACTIVITIES.

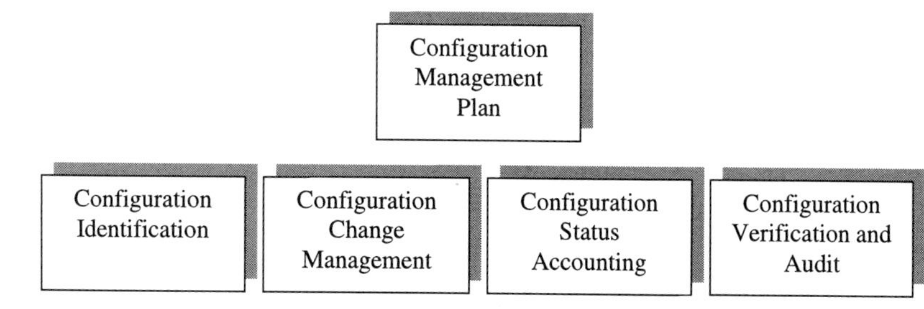

Configuration Change Management

Configuration change management is a process for managing product changes and variances. According to ANSI/EIA 649, the purpose and benefits of the change management process include the following:

- Enable decisions to be based on knowledge of complete change impact
- Limit changes to those that are necessary or offer significant benefit
- Facilitate evaluation of cost savings and trade-offs
- Ensure customer interests are considered
- Provide orderly communication of change information
- Preserve control at product interfaces
- Maintain and control a current baseline
- Maintain consistency between product and information
- Document and limit variances
- Facilitate continued supportability of the product after change.

Change management is the most commonly recognized aspect of configuration management. It is also, unfortunately, the principal source of the reputation of CM being cumbersome and overly bureaucratic. There needs to be a documented process for change, through which all changes must progress. The processing of all changes through a single change board activity is where most organizations see unnecessary bureaucracy in the configuration management process. For this reason, it is important that clear rules exist whereby change classifications can help streamline the approval/implementation process, and changes that are considered minor, or low impact changes, can be directed to those empowered to do so.

Configuration Status Accounting

Configuration status accounting allows the organization to view the current configuration at any stage of the life cycle. It is the means by which a company ensures that its product data and documentation are consistent. At certain points in the life cycle of a project, the configuration status information may need to be reported directly to the customer. Arguably, a more important reason for performing configuration status accounting is to report on the effectiveness of the configuration management process. This should start early in the life cycle of the project by defining target goals that are measurable. The reports from configuration status accounting should then be used throughout the project to identify areas for process improvement.

ANSI/EIA 649 states the typical configuration status accounting activities as the following:

- Identify and customize information requirements
- Provide availability and retrievability of data consistent with needs of various users
- Capture and reporting of information concerning
  - Product status
  - Configuration documentation

- Current baselines
- Historic baselines
- Change requests
- Change proposals
- Variances

Configuration Verification and Audits

Configuration verification and audits are performed on two levels. First, configuration management is responsible for the functional and physical audits of the product. This determines if the product meets the requirements defined by the customer in terms of form, fit, and function. Second, the process itself is subject to audit. Few organizations have applied effective metrics to assess the CM process. Cost of change, cycle time of change, and defect analysis are all ways of assessing the effectiveness of the CM process.

ANSI/EIA 649 defines the purpose and benefits of the verification activity as including the following:

- Ensure that the product design provides the agreed to performance capabilities
- Validate the integrity of the configuration information and data
- Verify the consistency between the product and its configuration information
- Provide confidence in the establishment of baselines
- Ensure a known configuration is the basis for operation and maintenance and life cycle supportability documentation

Configuration and Data Management

The relationship between physical documentation and digital data has been one of great debate in configuration management circles for many years. Historically, many found it hard to manage both with the same process, and as such there has been a rise in the development of both the hardware configuration management and software configuration management practices, with the latter being the life cycle management of both digital data and software. Essentially, the process was the same; what differed in the majority of cases were terminology, perception, and practice. The technological advances in digital product modeling and a growing interest in the management of product data meant that the divide between managing physical and digital representations was fast becoming an issue that needed a resolution.

To establish a common platform for the practice of managing configurations, it may be helpful at this point to understand the nature of managing data, information, and knowledge, as distinct from a physical, tangible product. The terms "data," "information," and "knowledge" are frequently used or referred to in much of the literature relevant to data management or information management (Checkland and Howell, 1998). Data, according to Tricker (1990) is an entity, and is used to refer to things that are known. Taggart and Silbey (1986) consider data to be groups or strings of characters recognized and understood by people. Data can be either "hard"—that is precise, verifiable, and often quantitative—

or "soft"—that is judgmental and often qualitative. Data has a cost; it can be sold, lost, or stolen and is considered to be an entity that is precise and verifiable and forms the foundation (building blocks) for information.

If data are the building blocks for information, it can be considered that information is formed from individual pieces of data knitted together in a cohesive manner. Taggart and Silbey (1986) provide the view that information is data that has usefulness, value, or meaning. Tricker (1990) states that information is a function resulting from the availability of data, the user of that data, and the situation in which it is used. Information, therefore, can be considered data that has meaning and usefulness and occurs as a result of a process and is understood to be the legacy of human endeavor.

Knowledge on the other hand can be made up of a number of factors, including experiences, education, and acquired information. Davenport et al. (1998) consider that knowledge is information combined with experience, context, interpretation, and reflection. Tricker puts it another way and suggests that it is the aggregate of data held together with understanding. In other words, it is the sum of what is known. Earl (1996) suggests that data is gathered from events and that this data, through manipulation, interpretation, and presentation, produces information. By testing and validation, the information leads to the acquisition of knowledge. From these interpretations of data, information, and knowledge, it can be considered that data is a verifiable and precise entity recognized by people. As a class, data should be easier to manage and control than the other, fuzzier categories.

A level of confusion stems from the different use of the term "data responsibility," particularly with regard to "data owner" and "data custodian." In a conventional representation of an information chain, the author, as originator or creator of the material, and therefore the owner of the information, sits at the top of the list (Basch, 1995). In other words, someone or some organization has to create the data initially and therefore has the authority over its attributes and use. As Van Alstyne et al. (1995) points out, ownership is a critical factor in the successful operations of information systems.

Within organizations, it is generally accepted that, legally, employees do not own the data they create on behalf of that organization. However, the creators of the data or the business function they work within would normally have the responsibility on behalf of the organization for ensuring that "their" data is not abused or misused; in effect, such employees are data owners in the nonlegal sense. Employees, other than the data owners, will also probably use the data; these can be designated data custodians. In some writing on data management, the distinction between custodianship and ownership is not drawn and the term data owner is used loosely to cover both categories of data responsibility.

One of the problems faced by large organizations is that potentially there are many data owners (and custodians) scattered throughout the organization (Brathwaite, 1983) who may adopt piecemeal approaches to data responsibility. Levitin and Redman (1998) point out that data is rarely managed well in organizations. Goldstein (1985) outlines the traditional solution to the problem of ensuring a consistent approach to data responsibility within organizations; he suggests that organizations should introduce a staff function, which he terms "information resource management" (IRM). Goldstein's reasoning behind this is that information is a basic organizational resource in the same way that people and money are. As such, information like these other resources should have a professional, high-level management group responsible for its effective use throughout the organization. The implication of this suggestion is that as data is the foundation for information, then if the organizational information has an owner, so has the data making up the information.

However, data responsibility issues are not simply contained within the organization's boundaries; ownership of data can be, and often is, protected via such as patents and copyright. In his paper "Ownership of Data," Cameron (1995) looked at some of the legal issues surrounding data where the oft-asked question is, "Is data a property or not?" Cameron puts forward the view that

> to be treated as proprietary data, the data must have been created by the owner, been created for the owner, or been purchased from its creator.
>
> Cameron, 1995; p. 47

In many commercial situations, the item being purchased is a product and not necessarily the underlying or ancillary data. In such circumstances, the ownership of the data is not usually transferred—as is the case with many software licenses where the purchaser has a license to use the software but does not own the underlying code (data). However, in circumstances where a customer purchases a "project" rather than a mass-produced product, as is more like the situation in the aerospace industry, for example, then the issue of transferring data ownership does becomes more of an issue.

Life Cycle Management and Configuration Management

A lot of configuration management's work comes, as in the previous washing machine example, from component parts of a product entering into the project, or program, at different stages of its life cycle (see Figure 21.3 for a typical life cycle with stages). As the life cycle matures from concept through to disposal, the amount of information that comes

FIGURE 21.3. LIFE CYCLE MANAGEMENT PHASES.

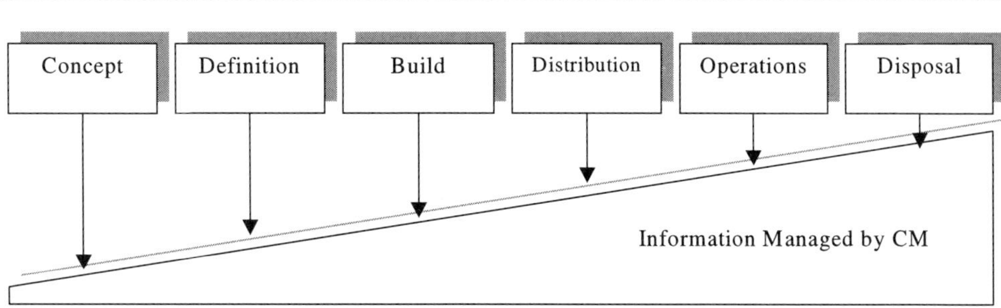

under configuration control becomes greater and greater. Indeed, it is true to say that the life cycle can be defined in terms of information, in that it begins with the first release of information and ends with retiring the last definitive information sets. The product itself may not appear for some time into that cycle and may be withdrawn long before the information is retired. In practice, "life cycle phases" are developed to suit the product type, the development method, the company doing the developing, and the industry; and they operate quite successfully despite such modifications. In the aerospace industry, life cycles can exceed 50 years, and such longevity does pose problems (Osborne, 2001). It is not uncommon for other industries with long product/program life cycles, such as process and nuclear power, similarly to emphasize configuration management. Indeed, the regulatory bodies of FDA (Food and Drug Administration), NIRMA (Nuclear Information and Records Management Agency) as well as CAA and FAA (aerospace regulatory bodies) make strict demands of the application of configuration management activities, although not always referring to them as "configuration management." Configuration management is increasingly accepted as a major factor in the design, development, and production of products and is becoming a major requirement in the in-service life of such essential products.

Within the field of configuration management, it can be argued that there are no new concepts as such; the elements of configuration management are the same as 20 years ago. The important issue concerns how the configuration management processes are implemented. One could argue on two fronts: (a) in the past, the implementation of the configuration management processes has not been as effective as it might be, and (b) the organizational context in which configuration management processes are implemented has changed and therefore the nature of the implementations need to change to reflect this fact.

The first level of argument includes the assertion that configuration management "paper" processes have been simply automated in a piecemeal fashion in nonintegrated software tools. Organizational philosophies in the past often consisted of managing the product within a "functional" environment. Thus, "islands of information" (Morton, 1994) were created, as piecemeal implementations were made on functionally based computer systems that rarely interacted sufficiently well with each other. However, configuration management activities cut across functional domains within the organization and thereby often create major problems in terms of contiguous data management across both the organization and throughout the product life cycle. Integrating the systems to ensure that data is available to the users at the required time and in the correct format can pose major challenges for the organization. It is also worthwhile recognizing that users are not necessarily passive recipients—quite often they are active in creating and modifying the data. This then raises issues about responsibility for where configuration management activities lie.

The second level of argument has a number of major strands that link to the prior point about functional organization and the lack of integration, and to the whole discussion about data, information, and knowledge.

- First, the availability of improved IT has enabled the implementation of a more integrated configuration management process (Osborne, 2001).
- Second, the adoption of teamworking by organizations with the consequent breaking down of functional barriers means that integrated approaches are required more than

ever before. As an illustration, the advent of Total Quality Management (TQM) (Oakland, 1992) has surfaced issues about the integration of quality responsibility with teams rather than with traditional functions.

- Third, the move toward increased levels of participation within industry—for instance, the "extended enterprise" (Schonsleben and Buchel, 1998) also unleashes an increasing pressure for integration of the configuration management processes, but this time the focus is *between* rather than *within* the organizations.
- A fourth strand relates to the relaxation by customers of a demand for adherence to their own specific configuration management standards, permitting organizations to put in place generic solutions to the CM "problem." This relaxation has originated, in part, from the dwindling of the public sector and the dominance of a business (private sector) ethos in many developed economies.

All these combine to create an organizational environment that could be characterized by an increased awareness of the role that configuration management could play and have heightened pressures to substantially alter configuration management practices. In particular, they accentuate the need for more integrative and effective CM, facilitated particularly through the use of IT. However, in practice, reports on configuration management practices do not feature prominently in the general CM literature, nor in the more specialized domain of software configuration management (Davies and Nielsen, 1992). In short, a knowledge gap exists.

The *Oxford English Dictionary* defines a configuration as "an arrangement of things." It follows, therefore, that managing configurations is concerned with managing arrangements or patterns. Questions arise as to where in the life cycle we manage those patterns and where exactly we stop managing the arrangement. To answer those questions, we need to consider the life cycle of the configuration itself.

Today, organizations in most business sectors place great emphasis on managing the product life cycle. However, this is only a small part of a much bigger picture. The information life cycle is much broader in scope and operation, and the product life cycle, system life cycle, project life cycle, and asset life cycle all lie within its phases. In many manufacturing industries, such as aerospace and automotive, there is a considerable period where the product exists in a purely digital form. Digital mock-ups and 3D models are representations of a product in an information form. The need to manage these representations with a common set of configuration management processes is still regarded as a major challenge.

At the other end of the configuration life cycle, when do we stop managing the pattern or arrangement? Prior to BOT/PFI (see the chapters by Ive and by Turner for a discussion of this type of project), the majority of projects closed out at the delivery phase, when we handed over to the client. The supportability of products has now become of strategic importance to many companies. In the aerospace community, there is considerable financial return for the supportability of in-service aircraft. Benefits are only realized when we ensure that we are certain of the current configuration status of each of those products at the time of service. The concept of the information life cycle has taken on a new level of importance. The flow from concept through to end of life must be well managed and maintained. The failure to do so will result in a catastrophic impact on the bottom line.

From a different viewpoint, in the majority of cases the product life cycle finishes with the disposal phase. In many cases it may be important to maintain the information post-disposal. There may be a legislative requirement, as in the medical device and pharmaceutical industries. Alternatively, it may be to assist in future development projects through support for modularization. Although many products are considered unique, the need to limit development costs means considerable benefits can be gained from cataloguing modules for future design projects.

In short, therefore, the configuration, or information, life cycle is the dominant life cycle when we are assessing the strategic impact of the full life of products and systems. It begins with the release of the first definitive set of information and ends with the retirement of the last. For many, the beginning will be the opening of the bid phase to the end of the contract; for the owner, however, it will be early in the development cycle and go through to the disposal phase. The salient point is that it is information, not the physical product itself, that has to be managed.

Organization of Configuration Management

The existence of different organizational structures leads to a variety of views as to who owns the configuration management process; for example Sage (1995) describes CM as being owned by system engineering. However, other views exist where PM, quality management, engineering management, and logistics management all have a stake in the ownership of the enterprise configuration management process.

Integrated product development (IPD) practices are a recent, and significant, advent in organizational structures (see the chapters by Archer and Ghasemzadeh, and by Milosevic). The IPD philosophy is implemented through integrated product teams (IPTs) and encompasses concurrent engineering where the effective configuration of the system's life cycle takes on special significance, since simultaneous development activities need to be carefully coordinated and managed early on (see the chapter by Thamhain). Integrated product teams cause problems in identifying responsibilities for the different elements of the CM process. A typical response is that the development of the CM process and techniques within a company become the responsibility of a core CM discipline, whereas the day-to-day operational tasks become the responsibility of the IPT leaders. To discharge the CM responsibilities in such circumstances over the project life cycle requires a flexible, responsive structure.

While the above holds true for major manufacturing organizations, the software community has developed a very similar pattern for configuration management application. The majority of the changes to code and structure are carried out by the developers themselves. Organizationally, the planning of the CM activity is managed by a centralized CM function. It is fair to say at this point, however, that in the IT community, configuration management is a far more automated activity, with tools facilitating the change and versioning process.

Changing Nature of Configuration Management in the Aerospace Industry

Aerospace is perceived as an industry that has a well-established and documented use of configuration management practice. (See the chapter by Roulston.) Further analysis, however, shows that the depth of such practice and innovation in application of the CM process is varied across the sector (see the research study that follows). Some of the world's highest-profile collaborative development projects such as Airbus and Eurofighter have suffered major cost overruns and schedule problems, due in part to CM not being coordinated at the outset. A European Commission Framework 4 project (AdCoMS; Project No. 22167) sought to establish a common CM platform for all partners in collaborative development programs, based on commercial best practice, rather than existing standards. Sadly, after a 2000 completion, little has been utilized by the consortium partners who developed it.

Interestingly, the Perry Initiative in the U.S. defense industry encouraged the use of commercial standards to replace those of the U.S. Department of Defense (DoD), where appropriate (Ciufo, 2002). Many years after this, it is still not uncommon to see traditional defense industry practices in configuration management being adopted. For many, the comfort of using an overly defined and regimented process became a barrier to change. Not surprisingly, therefore, the perception of value and organizational recognition of configuration management was, in many cases, poor.

Recent research has looked at the changing nature of CM application in the aerospace industry in Europe. Historically, aerospace has been a key player in the development and innovation of the configuration management process (Kidd, 2001), and the research identified the changing nature of its application within the rapidly changing environment of the aerospace industry. In the study, organizations were categorized as Tier 1, developing and manufacturing at a high level, and Tier 2, suppliers to Tier 1 organizations.

A population of 210 organizations were surveyed, with the nominated configuration managers being asked a total of 50 questions. A follow-up interview was undertaken with a cross section of the organizations in the initial investigation. A summary of the results was as follows:

1. *How do aerospace industry players define configuration management?* The use of international CM standards was evident across the whole of the product life cycle and drew on a good mix of standards. Customer needs and standards are perceived as important factors in defining the design of the CM process, while IT is identified as a key mechanism to support this. Overall, the responses indicated that the use of the CM plan was a major activity within the companies, with 78 percent of companies having a CM plan and 50 percent referring to it frequently. However, given the key role of the plan in CM thinking, even higher levels of reliance on a CM plan were to be expected, and therefore there is some evidence that companies are treating CM as a compliance issue rather than wholeheartedly believing in it. Sixty percent of respondents indicated that CM activities were generally not the responsibility of the quality function but were the province of a separate CM function, which had reporting links to other areas of the orga-

nization such as project management. Leaving aside the responsibility for CM activities, individual functional departments typically carry out the activities needed for CM within their own domain. The interviews highlighted strong views for a more active role for CM personnel. Views were expressed that companies should move to an organization form where a CM discipline managed the CM activities and requirements across the product life cycle rather than the functional fragmentation indicated previously. At present, little evidence is apparent of career progression, education, and training; the latter was particularly lacking in second-tier companies.

2. *How do the companies value configuration management?* The responding companies demonstrated their low reliance (25 percent) on metrics to measure the performance of the CM activities and the low incidence of risk assessment. Again, this suggests CM is seen as a passive compliance activity rather than an area that, if managed properly, could deliver benefit. The CM processes within these companies were claimed to be flexible and supportive of customer and project requirements. External auditing of the CM process was undertaken against required standards by both tiers; however, first-tier companies were open to more external scrutiny.
3. *How do the companies carry out the configuration management process?* Only just over half of the first-tier companies who responded to the questionnaire claimed to have an "end-to-end" process; however, there was a clear spread of activities across the differing company functions. (An end-to-end process is taken to cover the whole of the product life cycle rather than supporting limited parts such as the design process.) The second-tier companies were in a different position, because their activities often represented only part of the CM life cycle. The 80 percent view from both tiers indicated that there was a conceptual process for CM that was documented in line with the appropriate standards. The companies indicated that their CM procedures were developed in-house and supported the individual function's requirements. Companies indicated the uniqueness of their processes (76 percent) despite the common principles that underpin CM. The interviews probed further the view that an end-to-end process was employed and that this comprehensive scheme interacted with many other processes externally to the company. With the extra depth of information, it soon became clear that the end-to-end process is an intention rather than what is actually happening in companies. A life cycle process for configuration management is mainly a vision that most of the companies wish to attain but at present do not have.
4. *Is configuration management recognized within the organization?* The key message here is the lack of recognition of the CM function, which stood at only 31 percent of companies overall but was particularly poor in second-tier companies (21 percent). Responsibility for CM is not vested in a specific senior manager; indeed, even at lower levels in the organization (20 percent), there is a clear absence of a designated CM manager. Fragmentation of CM activities is evident across individual functions. There is a lack of clear career progression and a lack of education and training provision. In the interviews, organizational structure and career recognition linked to education and training strategies was seen as a major requirement for the developing CM world and were viewed as much more important than apparent in the results of the questionnaire.

5. *Is configuration management covered by IT means?* Eighty percent of the questionnaire responses indicated that IT within the organizations was not fully covering the requirements for CM. The majority of the respondents indicated that the use of both IT systems and paper were the means to record and report the CM requirements. In the interviews, the development of the CM process was seen to be standards-driven, with best practice and experience adding to this development. Hence, the general feeling was that IT had not been a driving influence for CM. However, this appeared to be changing, and respondents saw IT as a driving force for process development, given the advances in the technology employed within the companies. This increased influence of technology was changing the manner in which processes were developed and deployed within the organization, and therefore this was having a major impact on configuration management in terms of process, data management, and status accounting.
6. *Is configuration management a stand-alone process, or is it covered by other, separate processes?* Questionnaire respondents were near unanimous in indicating that the CM process was not stand-alone and instead connected to many other wider spread processes, including processes external to the company. It was evident from the interview responses that the CM process cuts across all the different company functions and links to many required activities, particularly in first-tier companies. Thus, the CM process is not viewed as a single process but the interaction of many processes. Within all the companies who responded to the questionnaire, there appeared to be a good understanding of the requirements for CM and a good knowledge of the standards that were used. But the variation in the manner that the process was employed suggested that there was no single process that could have been developed that would fit the needs for all of them. There are many aspects that influence the requirements for the process for CM; one of these is the way that product development is organized. This could be a single integrated product development team, a separate function, an individual company, or a mixture of them all. Therefore, the process that needed to be employed was seen to differ significantly according to the requirement of the organization.
7. *Does configuration management add any value to the business?* This question was mainly addressed by looking to see views on the level of knowledge that CM personnel had and how CM data contributed to business activities. The positive responses indicated CM did add value to the business, with the CM personnel being seen as making a valuable contribution. CM activities were not seen as restricted to those individual functions with clear CM responsibilities within the organization, and CM personnel fulfilled a valuable role in advising on the requirements for projects that need to be undertaken. In total, 87 percent of those surveyed felt that CM added value to both program and the business as a whole. It can be surmised that that the remaining 13 percent felt that their efforts were either unrewarded or the CM process they were working with was inadequate for the purpose.

In summary, the preceding study of CM in the aerospace industry provided some interesting perceptions of the value of configuration management in the development, build, and maintenance of highly complex products. Seen as an industry that relies heavily on such practices to ensure integrity and reliability, it would appear that CM still carries a perception of being cumbersome and administrative to many. Part of the reason for this may be to do with the regulatory nature of CM in the defense sector. Many of the standards used in this sector were indeed user-unfriendly and relied on the use of prescribed documentation and process. However, this is changing, with the encouragement of companies to use commercial standards where appropriate and to innovate their own processes to include best practice. The focus on managing life cycles in the aerospace industry has breathed new life into configuration management. Many now see it is a part of their everyday work and not just the job of the configuration manager (Kidd, 2001). It could well be argued that CM is at a point today where quality management was in the late 1970s: transitioning from a control process to an enterprise-wide activity. Clearly, whether CM ultimately follows the same trajectory taken by QM over the last 30 years to reach such prominence depends upon the actions of all in organizations and not just CM professionals. Fostering this change argues for configuration management to be treated in a similar way to quality—that is, where everybody in the organization is exhorted to think about quality and be responsible for quality. Of course, this comparison with quality also points to the potential downside that people see CM as a bureaucratic impediment to be dealt with simply on a compliance basis.

Summary

Configuration management is not just about managing products. It is about managing everything that defines the product or system across the full life cycle. When do we start doing CM? When we issue the first definitive information, not when we have a "configuration" to manage—by then it is too late. When do we stop doing CM? When we no longer have a need for the information, and we retire the last definitive information set. We live in a world characterized by rapid innovation. This also means rapid change, and we must develop better methods of incorporating change into products, systems, and services. As more and more organizations seek to exploit the benefits of life cycle support and service agreements, the role of configuration management becomes pivotal in maximizing value. If we do it badly, then the costs of maintaining poorly defined products will heavily impact the bottom line. For those who do it well, the benefits will set them apart from the competition.

For those of us working in project management, the role of CM should now be clear. How beneficial would it be to have the right information, in the right format, in the right place and at the right time? Would this assist in the decision-making process of managing projects? The clear answer is yes.

Acknowledgment

We would like to acknowledge the research work carried out by Dave McKee and Colin Hillman from BAE SYSTEMS, and Kevin Byrne from CSC, while working on their master's theses with the CM Research Group at Leeds University. The results of this research contributed to this chapter.

References

Anon. 1998. *EIA-649 National Consensus Standard for Configuration Management*, Electronics Industries Alliance.

Anon. 1997. *ISO 10007 Guidelines for Configuration Management*, ISO Geneva.

Basch, R. 1995. *Electronic information delivery*. Aldershot, UK: Gower.

Brathwaite, K. S. 1983. Resolution of conflicts in data ownership and sharing in a corporate environment. *Database* 15(1):37–42.

Cameron, D. M. 1995. Ownership of data: The evolution of "virtual" property, data as property. Presented at Toronto, Ontario, Canada, January.

Checkland, P., and S. Howell. 1998. *Information, systems and information systems: Making sense of the field*. London: Wiley.

Ciufo, C. A., 2002. Editorial. *COTS Journal* (Spring): 78.

Daniels, M. A. 1985. *Principles of configuration management*. Advanced Applications Consultants.

Davenport, T. H., D. W. De Long, and M. C. Beers. 1998. Successful knowledge management projects. *Sloan Management Review* (Winter).

Davies, L., and S. Nielsen. 1992. An ethnographic study of configuration management and documentation practices. *IFIP Transactions A—Computer Science and Technology* 8:179–192.

Earl, M. J. 1996. *Information management: The organizational dimension*, London: Oxford University Press.

Goldstein, R. C. 1985. *Database: Technology and management*. London: Wiley.

Kidd, C. R. 2001. The case for configuration management. *IEE Review* (September).

Levitin, A.V., and T. C. Redman. 1998. Data as a resource: properties, implications and prescriptions. *Sloan Management Review* 40(1):89–98.

Oakland, J. S. 1992. *Total Quality Management*. Oxford, UK: Butterworth-Heinemann.

Osborne, J. 2001. Avoiding potholes on the data highway. *Professional Engineering* (July): 39–40.

Sage, A. P. 1995. *System management for IT and software engineering*. London: Wiley.

Schonsleben, P., and A. Buchel, eds. 1998. *Organizing the extended enterprise*. London: Chapman and Hall.

Scott, M. A. 1994. *Information technology and the corporation of the 1990s*. New York: Oxford University Press.

Taggart, W. M., and V. Silbey. 1986. *Information systems: People and computers in organisations*, New York: Allyn and Bacon Inc.

Tricker, R. I. 1990. The management of organizational knowledge, Paper presented at the 1990 Conference on Systems Management, Hong Kong.

Van Alstyne, M., E. Brynjolfsson, and S. Madnick. 1995. Why not one big database? Principles of data ownership. *Decision Support Systems* 15:267–284.

CHAPTER TWENTY-TWO

SAFETY, HEALTH, AND ENVIRONMENT

Alistair Gibb

Many readers may be wondering why safety, health, and environment (SHE) are included in a book about project management. Sadly, this view is not unusual. Even in "developed" countries, there is still a paucity of consideration of SHE issues for projects in all industrial sectors. This chapter introduces the reader to some of the key issues as they affect the overall management of a project. All tasks in all industrial and commercial sectors involve SHE risks; however, the intrinsic nature of most projects is such that steady state has not been achieved and the project conditions and environs are continually changing. This is particularly true for construction projects. Therefore, to provide a focus for this chapter, SHE issues have been considered mainly from a construction project perspective, although reference is made to other project scenarios where appropriate. The key principles apply to both large and small projects, although the implementation of them may vary (CII, 2001).

In the European Union, "construction" has been defined as all works associated with the project, including demolition and decommissioning. "Health" covers occupational health issues of construction workers, which are often overlooked in efforts to address the more immediate challenges of "safety." Safety and health implications of the completed buildings or facilities are also important but are outside the scope of this chapter except for maintenance aspects. "Environment" has become a much-used term, covering a broad spectrum of issues of the sustainability on the built environment. The sustainability of a project covers issues from construction and throughout the life cycle of the completed facility. Sustainability itself is a broad subject typically considered as relating to three main areas: environmental (planet), social (people), and economic (prosperity).

Once again, to maintain focus, this chapter concentrates on construction site aspects of the environment. Health and safety are typically covered together in much legislation and many publications. While environmental issues are different, there is often an overlap with health and safety in terms of management strategies and techniques. SHE is considered an integrated management task in many large, global organizations, although those responsible for it are often biased by background and training at least toward one particular aspect, often safety. It therefore cannot be taken for granted that all three aspects will be given the appropriate emphasis.

The causes of accidents, ill health, and environmental disasters are multifactorial and should not be considered simplistically (Hide et al., 2002; Reason, 1990; and others); however, it is accepted that effective project management will have a positive affect on SHE risks. The saying "if you can't manage health and safety, you can't manage" is supported by most writers on the subject. Griffith and Howard (2001) stress that the "management of health and safety is without doubt the most important function of construction management." Notwithstanding, SHE is still absent from many general management texts.

The chapter explains the importance of SHE, introduces SHE objectives and strategy, and highlights design and procurement activities and an action plan for construction. It briefly introduces life cycle issues and the measurement of success. This structure has been taken from the European Construction Institute's SHE manual (ECI, 1995). The ECI also has guidance documents dealing specifically with health and the construction environment (Gibb et al., 1999 and 2000), and readers may consult these publications for a more complete coverage of the subject.

Why Are Safety, Health, and Environment (SHE) Essential Project Management Considerations?

Moral Responsibility for SHE Management

International comparison of SHE performance is impossible, and it is decidedly unwise to even attempt it. Griffith and Howarth (2001) argue that there will be "considerable differences in, for example, economic climate, market forces, political environment, construction methods and availability of resources." Nevertheless, through my involvement with the international research network, Conseil Internationale de Batiment, it is obvious that the statistics throughout the world are unacceptably high. It really is not acceptable in the twenty-first century that someone working in construction cannot expect to complete a career in the industry without sustaining some form of injury or occupational disease. Furthermore, the issue of the environment has passed from a pressure group topic into the mind-set of the average person in the street—although they might not understand all the complexities, they believe that companies should take a responsible attitude toward caring for the environment.

Legal Responsibility for SHE Management

In an international publication like this it is inappropriate to describe the legal arrangements of one particular country. Nevertheless, throughout the world, enshrined in the law of most countries is a duty of care to others, and in particular an employer's duty of care to those employed to work on their behalf.

Since the early 1990s European states have had the further legal requirement to ensure that designers overtly consider the health and safety of construction workers (EC, 1992). This same directive requires effective health and safety management systems to be used. A similar requirement for environmental management is enshrined in the ISO 14001 standards (ISO, 1996) and is expanded by Griffith (1994). A good summary of environmental law in the United Kingdom is provided by Stubbs (1998). In general, many other countries, including the United States, have not brought together the legal requirement for safety, occupational health, and environmental protection.

Financial Necessity for SHE Management

"Humanitarian factors alone are more than enough to justify the effort required to eliminate worker injury. . . . . however, the significant cost of worker injury cries out for exposure to those who worry about the cost of safety programs. . . . Eliminating injury makes good business sense." (Nelson, Shell Oil Company, 1993). In the United Kingdom, the Egan report, "Rethinking Construction" (DTI, 1998), stated that "accidents can account for 3 to 6 per cent of total project cost." Nelson (1993) estimates that "the total cost of injury for the $450 billion U.S. construction industry ranges from $7 billion to $17 billion annually." The hidden costs can be many times more than the visible costs.

"In the Piper Alpha explosion (North Sea oil rig disaster), 167 lives were lost and £746 million (US$ 1243 million) was paid out by insurers, but estimates put the total loss, including business interruption, investigation costs, hiring and training replacement personnel and the like at over £2 billion (US$ 3.3 billion)" (Clarke, 1999).

The cost of accidents or environmental incidents include the following:

- *Management and organization.* Resources, administration, and accident investigation
- *Damage to reputation.* Adverse publicity and impact on industrial relations; impact on future tenders; liability; and compensation;
- *Loss of productivity.* On the day of the incident and for some time thereafter
- *Litigation and legal fees.*
- *Fines from statutory authorities and similar bodies.*
- *Delays to the project.* While the situation is normalized
- *Sick pay to injured personnel.*
- *Damage to property and materials.*
- *Increased insurance premiums.* Some countries make a direct correlation between SHE performance and insurance rates. In the United States this is called the EMR (Experience Modification Rating) and can have a significant financial effect
- *Medical costs.* Liability for these will vary between countries, but the costs can be substantial irrespective of who has to pay them.

Hinze has studied cost aspects of construction health and safety in the United States for many years (e.g., Hinze and Appelgate, 1991; Hinze, 1991 and 1996; CII, 1993a), and he argues strongly for serious consideration of the real costs of accidents and incidents, including the very substantial hidden costs. Those looking for a fuller discussion of financial issues can review the proceedings of the Conseil Internationale de Batiment W99 conference dedicated to the subject (Casals, 2001).

Any cost exercise should include the costs associated with setting up an effective SHE management system and procedures, where all parties recognize the implicit and explicit costs. If possible, SHE activities will be included in the contract agreements between all parties. Many of the explicit costs can then be linked to specific project activities such as scaffolding, or asbestos removal, but they can also be in the form of a SHE specification, priced as part of the contract.

The link between health and safety performance and project profitability has been debated for many years. In 2001, the UK construction and development organization Taylor Woodrow compared health and safety audit scores with the profitability of each project. Figure 22.1 shows the results with the main vertical scale being the audit score and the shades of the columns representing varying degrees of profitability (more than 2 percent above the expected return; within ±2 percent of expected return; more than 2 percent below expected return). The graph demonstrates that many of the poorer-performing projects from a health and safety viewpoint were also performing badly financially (shown as

FIGURE 22.1. COMPARISON OF HEALTH AND SAFETY AUDIT SCORES WITH FINANCIAL PERFORMANCE.

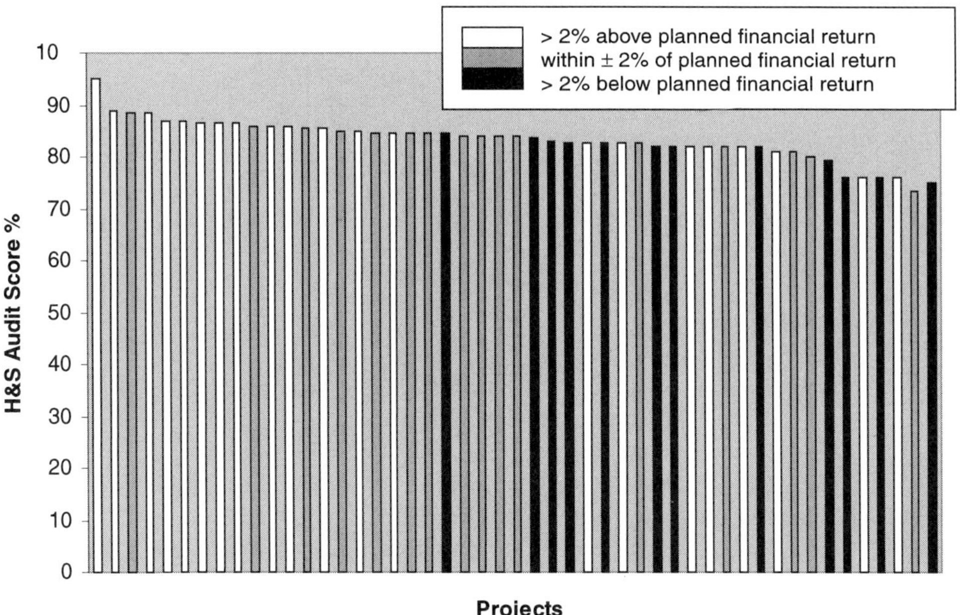

black columns that are grouped toward the lower end of the safety performance spectrum). Clearly the sample would need to be enlarged to argue this point more strongly; however, that there is a clear indication of a link did not go unnoticed within Taylor Woodrow's senior management team. This aspect is discussed further by Kunju-Ahmad and Gibb (2003).

Cultural Challenge for SHE Management

There is a real cultural challenge for SHE management, with an acceptance among many involved in construction that the industry is inevitably dirty, unhealthy, and unsafe. Construction is often seen as a "macho" industry where brute force and a bravado attitude pervade. Those complying with the need to wear personal protective equipment, working with care, and putting safety first are too frequently taunted by their coworkers or branded as "difficult" by management. Furthermore, there are some national cultures that place less worth on the well-being of certain individuals, for example, those who work with their hands. We are part of a global human culture that tends to ignore the waste it produces and care for the environment only if it does not affect our everyday lives. These cultural misconceptions should be challenged and a positive safety culture cultivated at a project as well as a community level. Much has been written about achieving such a culture and the precise methods promoted vary with the application, but the principal remains where everyone looks out for the safety and health of themselves and of others along with having consideration for the environment.

SHE Policy, Objectives, and Strategy

SHE Policy and Objectives

The project's SHE objectives and strategy will be based upon a sound SHE policy for the stakeholders' organization. The policy will be a public-domain document, emanating from the executive board, which states the organization's corporate SHE philosophy in the context of its overall business activities. HSE (1997) stresses that "effective health and safety policies contribute to business performance by

- supporting human resource development;
- minimizing the financial losses which can arise from avoidable unplanned events;
- recognizing that accidents, ill-health and incidents result from failings in management control and are not necessarily the fault of individual employees;
- recognizing that the development of a culture supportive of health and safety is necessary to achieve adequate control over risks;
- ensuring a systematic approach to the identification of risks and the allocation of resources to control them; and
- supporting quality initiatives aimed at continuous improvement."

ECI (1995) advise that a "SHE policy should be clear, concise and motivating. The content should clearly express

- what the company intends to PREVENT (using words such as prevent, limit, protect, eliminate);
- what the company intends to IMPROVE (using words such as create, develop, carry out, replace); and
- what the company intends to COMPLY with (using words such as comply, demand, require)."

For instance, for the environment, the policy may aim to pursue progressive reduction of emissions, effluents, and discharges of waste materials that are known to have a negative impact on the environment with the ultimate aim of eliminating the negative impacts.

Typical strategic SHE objectives may include the early identification of major hazards, the examination of the impact on construction of SHE considerations during design, the development of a SHE framework for construction and the project life cycle, the development of a SHE plan by the principal contractor before site work begins, and compliance with this plan thereafter. SHE objectives must be achievable and therefore be in-line with other project management objectives such as time, cost, and quality. It is important to note, however, that it may be necessary to amend the time and cost parameters so that the SHE objectives can be achieved. This is another reason why SHE should be considered along with other project-wide issues as part of an overall project strategy rather than as a stand-alone issue. Many large organizations now incorporate SHE management holistically, within an overall quality management system. However, there is still some debate on this approach (e.g., Smallwood, 2001; Griffith and Howarth, 2001; CIRIA, 2000, Gibb and Ayode, 1996; Rwelamila and Smallwood, 1996).

Project SHE Concept, Initial Risk Assessment and SHE Plan

The overall policy and objectives will be worked through at project level. Griffith and Howarth (2001) state that "project health and safety planning and management should be considered in two parts. The first part focuses on the client's project evaluation and design processes with the objective of producing a 'pre-tender' health and safety plan. The second part focuses on the site production processes with the objective for the appointed principal contractor to produce a construction phase health and safety plan." They go on to say that "it is the essential part of planning within each part which forms the basis for a systematic management approach, within which risk assessment is an important theme."

At this early stage, the emphasis will be on major hazards, with the output being an initial risk assessment and a preliminary SHE plan. The risk assessment process is described in more detail in the next section, although at this phase the exercise will be done at a fairly high level. Typical risks to be considered at the concept stage include those shown in Table 22.1 (ECI, 1995).

TABLE 22.1. TYPICAL SHE RISKS TO BE CONSIDERED AT THE PROJECT CONCEPT STAGE.

Safety Risks	Health Risks	Environment Risks
Climate	Infections	Emission
Natural hazards	Hygiene	Effluents
Transport	Worker accommodation	Wastes
Security factors	Medical facilities	Noise and vibration
Unskilled labor	Potable water	Light
Major risk factors, e.g., heavy lifts; excavations; demolition	Chemicals	Damage to surroundings
		Contaminated ground
		Heat
Concurrent operations		Electricity
		Pressurized systems

Source: After ECI (1995).

Many companies in the engineering construction sector (petrochemical/power generation construction) use the HAZCON procedure. This is a two-part, formal procedure for early identification and assessment of SHE hazards in construction to enable all reasonably practicable steps to be taken to reduce or eliminate the risk. HAZCON 1 identifies major hazards to owner personnel, contractors, visitors, or the general public, along with actions and recommendations for hazard elimination or reduction. Risks may occur within the site or beyond its boundaries. HAZCON 1 uses checklists to aid the evaluation, and it is done as early as possible in the project, at least before the project scope and site details are finalized. HAZCON 2 is done later in the process, to provide a detailed assessment of construction hazards based on the completion of a significant level of engineering definition, at least including plans and elevations together with a draft overall construction method statement, contract plan, project schedule, and site layout drawings. It should also include a review of HAZCON 1 results to see whether the development of the scope has added or removed any major construction hazards. The HAZCON procedure and checklists are explained further in ECI's SHE manual (ECI, 1995). The follow-on procedure, HAZOP, relates to operating aspects of the constructed facility.

The SHE plan will include strategies for design, procurement, construction, commissioning, maintenance, decommissioning, and demolition. The SHE plan will also cover the following issues at a strategic level (ECI, 1995):

- "SHE management and leadership
  - including organization; communications and meeting schedule.
- SHE organization and rules
  - including policy statement; legislation; standards; procedures; basic rules; health; medical and welfare program; auditing; environmental; and sub-contractor strategy.
- SHE risk assessment and management
  - including, hazard identification; risk assessment; SHE performance and measurement; and emergency response procedure.

- SHE training
  - including employee orientation program; promotion and awareness; training program; and involvement of professionals.
- Personal protective equipment (PPE)
  - including risk assessment; PPE requirements and use.
- Incident/accident/injuries records and data
  - including reporting procedures
- Equipment control and maintenance
  - including SHE equipment and inspection; hygiene and housekeeping'.

Design and Preconstruction activities

Risk Assessment and Risk Avoidance

Risk assessment is an essential part of all business processes and again also necessary for SHE issues. In Europe, risk assessment and management is mandatory during both the design and construction phases. Designers are required to identify hazards and their associated risks and then to eliminate, reduce, or control the risks they have created. The designer's role in generating risk and identifying solutions has not yet been fully acknowledged outside Europe, and in many states risk assessment and control is left to the construction team. The design team will review the hazards identified at concept stage (through HAZCON 1 or similar) and develop the risk assessment in more detail, checking that no new hazards have become apparent.

It is important to understand two key terms: hazard and risk. According to the United Kingdom's Management of Health and Safety at Work regulations (1999), a hazard is "something with the potential to cause harm" and risk expresses "the likelihood that the harm from the hazard is realized."

Beilby and Dean (2001) identify "five steps to risk assessments:

- Step 1: Look for the hazards.
- Step 2: Decide who might be harmed and how.
- Step 3: Evaluate the risks and decide whether the existing precautions are adequate or whether more should be done.
- Step 4: Record your findings.
- Step 5: Review your assessment and revise it if necessary."

Most risk assessment methods follow a similar format. I favor a simple three-point scale where both hazard severity and likelihood are given a score of 1, 2, or 3. The risk is then the product of the hazard severity and the likelihood of occurrence. Often a risk matrix such as that shown in Figure 22.2 is used. More complicated systems are available, but they do not necessarily produce more accurate results.

As an example of this process, for health and safety (HSE, 1997) the following levels would apply:

FIGURE 22.2. RISK ASSESSMENT MATRIX.

Severity of hazard			
3	1 x 3 = **3**	2 x 3 = **6**	3 x 3 = **9**
2	1 x 2 = **2**	2 x 2 = **4**	3 x 2 = **6**
1	1 x 1 = **1**	2 x 1 = **2**	3 x 1 = **3**
	1	2	3
	Likelihood of occurrence		

Severity of hazard

- *Level 3.* Major—death or major injury or illness causing long-term disability
- *Level 2.* Serious—injuries of illness causing short-term disability
- *Level 1.* Slight—all other injuries or illness

Likelihood of occurrence

- *Level 3.* High/probable—where it is certain that harm will occur
- *Level 2.* Medium/possible—where harm will often occur
- *Level 3.* Low/improbable—where harm will seldom occur

The following hierarchy of risk actions are taken from the European Directive (89/391/EEC) by Griffith and Howarth (2001) but have international applicability:

- "avoiding risks;
- evaluating the risks which cannot be avoided;
- combating the risks at source;
- adapting the work to the individual, especially as regards the design of workplaces, the choice of work equipment and the choice of working and production methods, with a view, in particular, to alleviating monotonous work and work at a pre-determined work rate and to reducing their affect on health;
- adapting to technical progress;

- replacing the dangerous by the non-dangerous or the less dangerous;
- developing a coherent overall prevention policy which covers technology, organization of work, working conditions, social relationships and the influence of factors relating to the working environments;
- giving collective protective measures priority over individual protective measures; and
- giving appropriate instructions to employees."

Recent work at Loughborough University (ConCA, 2002) studying 100 construction accidents has found that many risk assessments are virtually useless, in that they have little or no effect on the actual task operation itself. Too often the risk assessment is done as a "tick-box" exercise rather than a thoughtful assessment of the risk. Frequently the style, language, and length of the documents is such that they are not accessed at the workface by the operatives and supervisors but are retained in the site office "gathering dust." There is a real need for task-based risk assessments. A few organizations have started to address this shortfall. For example, the Channel Tunnel Rail Link (CTRL) project in England included a task risk evaluation as part of the supervisors' briefing and discussion with operative gangs each morning. Other cultures, such as the Japanese, include daily orientation as part of a start of the day routine for all workers. This can provide the opportunity for specific health and safety aspects to be raised and dealt with.

Designer's Role

"Construction worker safety is impacted by the designer's decisions" (Hinze, 1998). The European Directive, leading to the Construction (Design and Management) Regulations (CDM) in the United Kingdom, have formalized the requirements for designers to consider health and safety in their designs. While not mandatory outside of Europe, this strategy has realized support from researchers and industry leaders worldwide (Gibb, 2000; Hinze and Gambatese, 1996; Tenah, 1996; Oluwoye and MacLennan, 1996). However, despite market leaders emphasizing the importance of designing-in safety and health over many years (e.g., CII, 1996), the take-up of the strategies where not driven by legislation has been very limited.

In the United Kingdom, the CDM regulations require designers to

- inform clients/owners of the CDM regulations;
- apply the hierarchy of risk control to their designs;
- cooperate with other designers;
- cooperate with the "planning supervisor" (who is charged with coordinating H&S effort, particularly during design—the similar EC role is called design phase coordinator); and
- provide information about their design for inclusion in the health and safety file (a document that should form the central core of the health and safety management of a project).

While environmental issues are not covered in the CDM legislation, many projects take the opportunity to deal with them in the same manner as health and safety. In fact, many designers are more comfortable addressing environmental challenges than those associated with health and safety, which are often seen as the responsibility of the construction team alone.

Recommendations from the early concept stage risk assessments (HAZCON 1 or similar) will be made available to the design team and should influence site layouts, detailed design drawings, schematics, and specification. ECI (1995) stresses that these design assessments "must include identification of design errors, ambiguities and/or omissions. Questions of ambiguity and omission are especially important since the definition of design work and its separation from the construction phase is not always clear. In some disciplines, for example structural engineering, parts of the design are not fully detailed by the designer but are subsequently completed or amplified during fabrication and construction." Designers will often need to obtain advice from other domain experts in order to adequately assess the risks, and the contractual arrangements must facilitate this dialogue and knowledge exchange. It is at this stage that SHE benefits from integrated teams can be realized. Effective design risk assessment is still in its infancy, but various guidance documents have been published to assist (e.g., Cooks et al., 1995; CIRIA, 1999; Ove Arup, 1997). One of the challenges for these documents is how to guide a process such as design in a way that is both effective and does not stymie the design creativity.

Most designers will first of all consider the SHE issues of the permanent works, and this is not inappropriate; however, the risks present during construction must also be specifically addressed. Such risks should be systematically identified and removed or reduced during the design phase. A flowchart strategy is described in Figure 22.3

Designers can affect SHE on-site in a number of areas, for example site access. Here they will consider access to site for delivery, offloading, collection, and disposal of materials; access across site to facilitate safe movement of materials and personnel to and from the workplace; and plant/people separation during all construction activities. Another example would be hazardous materials, where designers should ensure that these are used only where necessary and that all materials are classified, with data sheets produced showing all associated risks, including delivery, use, and disposal. The ConCA project (2002) has identified that increased use of preassembly is one of the ways that designers can best improve SHE performance on-site.

Sustainability during Design

Sustainability is a broad subject dealing with the impact of the built environment on the environment as a whole. It is also often a politically motivated concept, with organizations and even countries playing games with statistics to defend their particular viewpoint. A full discussion of this important subject is clearly outside the scope of this chapter, and this section concentrates only on the issues relating to the design phase of the project. Key considerations for designers are embodied energy of the building elements (covering the energy used to extract, form. and fashion the elements; deliver them to site; install them; and ultimately dispose of them), energy consumption in use, emissions, hazardous materials, and ultimate demolition and disposal of the elements that make up the building. Designing

FIGURE 22.3 FLOWCHART FOR SYSTEMATIC IDENTIFICATION AND REDUCTION OF RISKS.

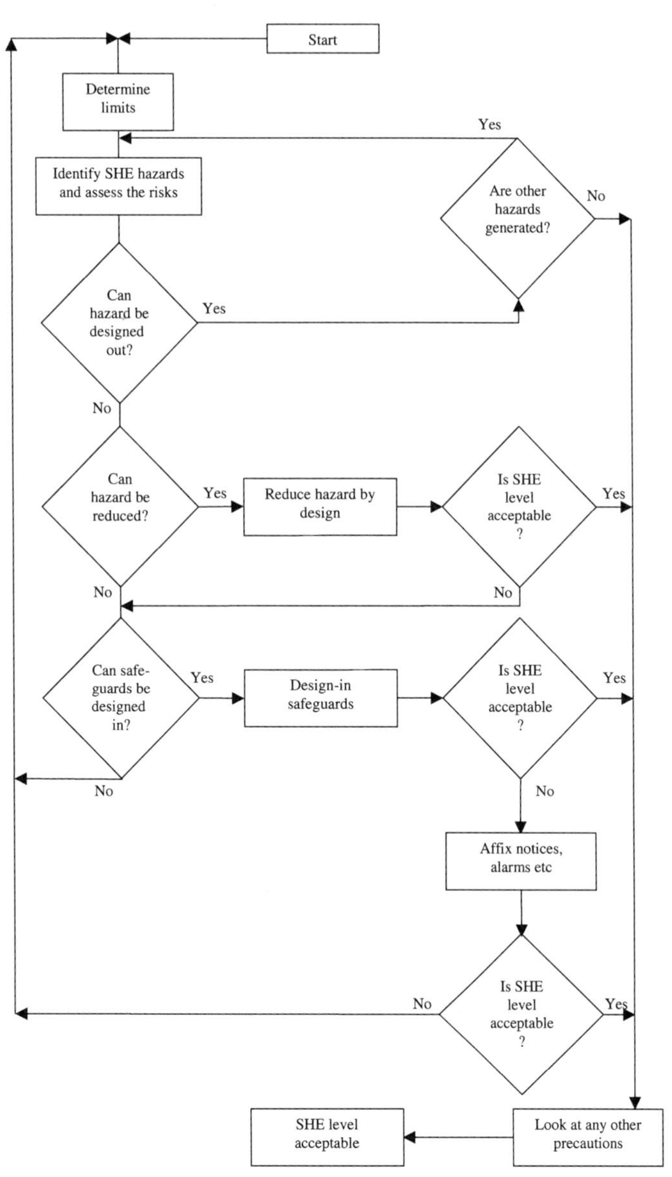

Source: Adapted from Pilz GmbH (1993).

for sustainability also covers broader issues such as site location (near public transport to reduce car use, on a previously used "brownfield" site) and stimulating the sustainable community (use of local labor, etc.).

According to Halliday (1998) "materials and products within buildings should not

- endanger the health of building occupants, or other parties, through exposure to pollutants, the use of toxic materials or providing host environments to harmful organisms;
- cause damage to the natural environment or consume disproportionate amounts of resources during manufacture, use or disposal;
- cause unnecessary waste of energy, water or materials due to short life, poor design, inefficiency or less than ideal manufacturing, installation and operating procedures;
- create dependence on high impact transport systems with their associated pollution; and
- further endanger threatened species or environments."

She claims that these are "issues of pollution and toxicity and a strategic approach starting at inception of a project is required to create a truly healthy environment".

Nath et al. (1998) have produced a useful book covering the methods and "tools" of environmental management, with chapters by individual specialist contributors. The first volume commences with the global aspects of environmental management and goes on to cover environmental planning, standards, exposure, and ecological risk assessment, and topics such as environmental risk assessment, life cycle assessment environmental auditing, and environmental accounting. Later sections cover economic and financial instruments for environmental management. The book contains summaries of international, European, and American environmental law, and finishes with chapters on environmental communication and education. Gibb et al. (2000) have produced a glossary of publications on the subject, particularly covering construction implications. The Construction Industry Research and Information Association have also published much on the subject (www.ciria.org.uk). Interested parties are advised to consult these other texts.

SHE Plan and SHE File

This section is drawn from the European practice where a specific plan for health and safety is central to the effective health and safety management (with larger organizations often including environmental issues as part of SHE) as shown in Figure 22.4. This plan is formally presented as the SHE file, which, as a document, evolves throughout the project process until it is handed over to the end user as a record that tells those who might be responsible for the structure (or facility/building) in the future about the risks that have to be managed during maintenance, repair, or renovation (HSE, 1994).

ECI (1995) explain that a SHE plan is required

- "to fulfill the statutory duty;
- to ensure tenderers take SHE into account and explain their proposals for managing SHE and that clients/owners provide their objectives and background information for the project;

FIGURE 22.4. CENTRALITY OF THE HEALTH AND SAFETY PLAN IN EUROPEAN PRACTICE.

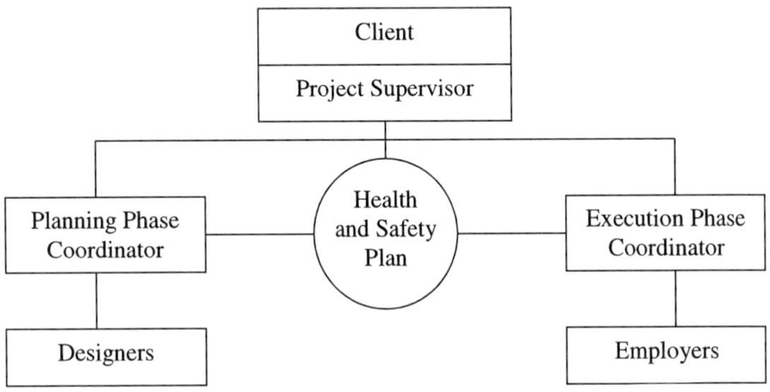

Source: Adapted from the EC Directive 92/57/EEC (1992).

- to ensure that all persons involved with the project (client, designers, planning supervisors, principal contractors and subcontractors) provide information to the plan and agree to the SHE management controls;
- to reduce the risk of accidents/incidents both during construction and for the lifetime of the facility;
- to reduce the losses associated with accidents/incidents;
- to protect the health of all project personnel and subsequent employees; and
- to reduce pollution and protect the environment."

The plan covers all construction work that, in Europe at least, is deemed to include maintenance and demolition. The plan is initially developed from information from the client/owner and designers and then developed in detail by the construction team, resulting in one plan rather than two separate documents. The level of detail will depend on the size and nature of the project and the procurement route adopted. Inputs to the plan from the various parties must be carefully coordinated. In Europe, this is a formal role, performed by the planning supervisor/design phase coordinator. ECI (1995) outline the main components of the SHE plan, as shown in Table 22.2.

Method Statements

Typically, method statements are required by the contract rather than legislation. They are, however, often confused with risk assessments and used interchangeably. Furthermore, as with risk assessments, it is essential that the target readership is acknowledged in the style and delivery of the material—too often the method statements just stay "on the shelf." Clarke (1999) explains the benefits of an effective method statement:

TABLE 22.2. MAIN COMPONENTS OF A SHE PLAN.

Section	Subsection
Project summary	Objectives
	Management organization and responsibilities
	Schedule of activities
	Existing environment
	Contract strategy
Design plan	SHE information
	Organization and responsibilities
	Hazard identification
	Designers' risk assessment
Procurement plan	Material hazards
	Construction risks
	Selection of principal contractor and key suppliers
Construction plan	Management organization and responsibilities
	Selection of contractors, subcontractors, and other suppliers
	Site rules and procedures
	Welfare arrangements
	Training
	Hazard identification/risk assessments/method statements
	Environmental control
	Handover of documents and SHE file
	Monitoring, auditing, and review

Source: Adapted from ECI (1995).

- "in getting people to write things down, it encourages them to think about the task in hand;
- it encourages them to commit to what they are writing;
- it helps communicate the planner's thoughts and intentions to operatives;
- it serves as a basis for coordination with other activities and for planning; and,it establishes an audit trail."

Procurement Strategy

Clarke (1999) cites the following procurement issues as impacting on health and safety management:

- "lowest price mentality of clients;
- competitive tendering;
- dutch auctioning (adversarial leverage to knock down tender prices);
- adversarial contracts;
- subcontracting; and
- design separation".

Citing "experienced commentators," Clarke (1999) claims that the "extensive and increasing use of self-employment" (especially labor-only subcontracting) in construction is an important factor in its poor safety record in construction. Other commentators add that the adversarial nature of many construction contracts also makes cooperation on SHE issues more problematic. Integrated teams are better placed to address the challenges together from a project-wide or even business-wide perspective.

Whatever the procurement strategy, the contract documents must adequately and unambiguously address SHE issues. Risks, rights, and obligations should be clearly spelled out. Efforts to hide important requirements within pages of text goes against the cooperative culture supported by this book and will ultimately lead to SHE problems either during construction or through the facility's life cycle. In most countries there will be specific legislation relating to SHE issues and construction contracts—for instance, in Europe, legislation is explicit about the roles of clients/owners, designers, planning supervisors (coordinators), principal contractors, and other contractors. Any contract strategy must be consistent with the relevant legislation. A number of other countries are considering strategies to draw the owner and designer into this process; however, there is considerable resistance to this move, with some being keen to retain the full responsibility for SHE issues with the contractor, who they argue is the organization best placed to solve the problems. Whereas this may be valid regarding the ability to control risk, the opportunity to remove or reduce the risk is best taken before work starts on site, and the preconstruction team should play a major role in this.

Assessment of Competence and Resources

"Competence" is an important concept in the recent legislation emanating from the European Union. According to this legislation, for European projects, key staff must have a knowledge and understanding of the work involved in the management and prevention of risk and of relevant SHE standards. They must also have the capacity to apply this knowledge and experience to their role on the project. The client/owner has a duty to ensure that all parties employed on a construction project are competent to perform their duties under the legislation. The client also establishes the extent and adequacy of the resources that have been, or will be, allocated. To assess competence, the key personnel need to be identified at an early stage. This may be hard for some organizations and may require a change in culture, away from the "day-to-day" approach often adopted in construction staff allocation. Where deficiencies exist, they may be addressed by further training. The specific requirements listed here are obviously only legally required on European projects; however, project managers are advised to take this model seriously in their considerations regarding project personnel and resources.

In addition to individuals, each company should be assessed for competence and any deficiencies in their organization and administration arrangements identified. Screening arrangements may include questionnaires, evaluation of previous experience, general reputation within the industry sector, SHE policy review, and specific service provision. Sample competence questionnaires are provided by ECI (1995).

SHE Training And Education

Training is an essential part of effective project management, both preconstruction and for site-based personnel of all types. A detailed discussion on training is outside of the scope of this chapter; nevertheless, following the assessment of competence, training is often needed to address the identified shortfalls and inadequacies. Designer training rarely moves beyond a cursory coverage of the necessary legislation, and this situation must be changed if improvements are to be made. Construction training will include, but not be limited to, inductions for all personnel, toolbox talk addressing topical issues, and strategic training based on a personal development plan to increase the base level of knowledge and expertise.

There is increasing pressure to include SHE issues in the education of all construction-related professionals. However, in the United Kingdom, progress is slow, confounded by a lack of knowledge of most educators and difficulties with knowing how to include extra information into an already crowded curriculum. A recent survey concentrating on health and safety (Carpenter, 2001) showed that, although there are some exceptions, many higher education establishments have still not begun to address the issues.

Construction Action Plan

Planning

Market leaders take a "planned and systematic approach to implementing a SHE policy through an effective SHE management system" (adapted from HSE, 1997), where planning is a continuum throughout the project life cycle. SHE planning starts with a general, high-level plan at concept stage and develops in detail as the availability of detailed information increases. At the start of the construction phase, the initial SHE plan is reviewed and updated, as it is important to build on the foundation already laid and benefit from the knowledge gained by the design team. Once again, integrated teams will achieve this more easily, and the earlier that the construction team becomes involved in the planning process, the better the plan will be. In UK practice, the role of the planning supervisor, who has been coordinating health and safety matters during design, will overlap with the principal contractor, who is responsible for the construction phase. Typically, on most large projects, they will develop and expand the SHE plan jointly as the design is finalized and the construction methods are decided.

In an ideal world the design would have been completed prior to the start of construction. However, in reality, there is always a degree of overlap, and effective management of continued design development during the construction phase is essential for the success of the overall project. This is equally the case in SHE matters.

The SHE plan will be sufficiently complete and detailed to cover the part of the construction work that is to be executed and should be completed as soon as possible. However, planning does not stop with the production of the overall project SHE plan. Individual contractors and subcontractors work is also planned, with special consideration given to the interfaces between the packages (Pavitt and Gibb, 2003).

Management, Leadership, and Organization

Changes in European legislation in the 1990s have brought the client/owner into the safety and health management process, and ultimately management and leadership starts with the owner. This view is supported by observation, where high-profile clients have achieved much improved SHE performance on their projects. This has also proved to be the case with client/owners such as DuPont, bringing strategies and culture from the hi-tech manufacturing sector to apply pressure on construction. The "zero-accidents" drive that was very prevalent in the 1990s was initiated by informed and influential client/owners (CII, 1993b) and is still prevalent today (CII 2003a, CII 2003b). While client/owners do not do the design or construction work, they clearly produce the brief and requirements and set the overall project culture, and these have a major affect on SHE performance.

Obviously, the "sharp–end" of SHE management is met by site-based managers and supervisors. CIRIA has produced an excellent site safety handbook targeted at site managers (Bielby and Read, 2001). It has also produced many publications on environmental issues for site managers and are now planning an occupational health manual (for more information see www.ciria.org.uk). Beilby and Read (2001) have produced a useful diagram providing a framework for individuals charged with the management of site safety (see Figure 22.5). This shows the effect on an individual's actions of the overlapping requirements of legislation, company policy, specific site rules for safe systems of work (that may be influenced by client/owner requirements), and professional codes of conduct and ethics. The same framework can be applied to health and environmental issues.

Figure 22.6 has been adapted from Griffith and Howarth (2001) to show the outline organization of the project health and safety management for a principal contractor (the main organization responsible for the on-site construction work). This figure shows the roles

FIGURE 22.5. INFLUENCE FRAMEWORK FOR MANAGING SITE SAFETY.

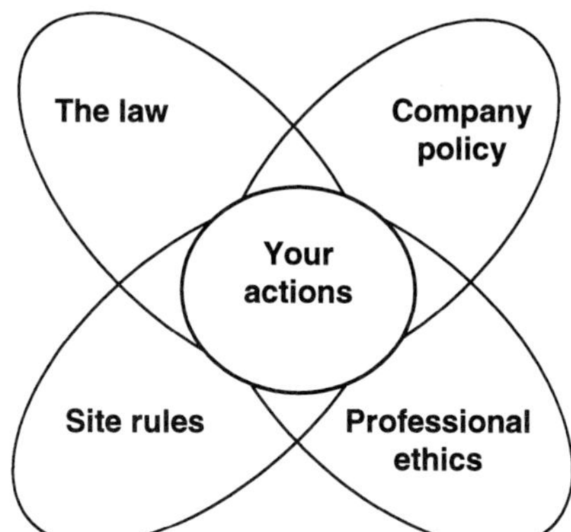

Source: Adapted from Beilby and Read (2001).

FIGURE 22.6. PRINCIPAL CONTRACTOR'S OUTLINE ORGANIZATION FOR PROJECT HEALTH AND SAFETY MANAGEMENT.

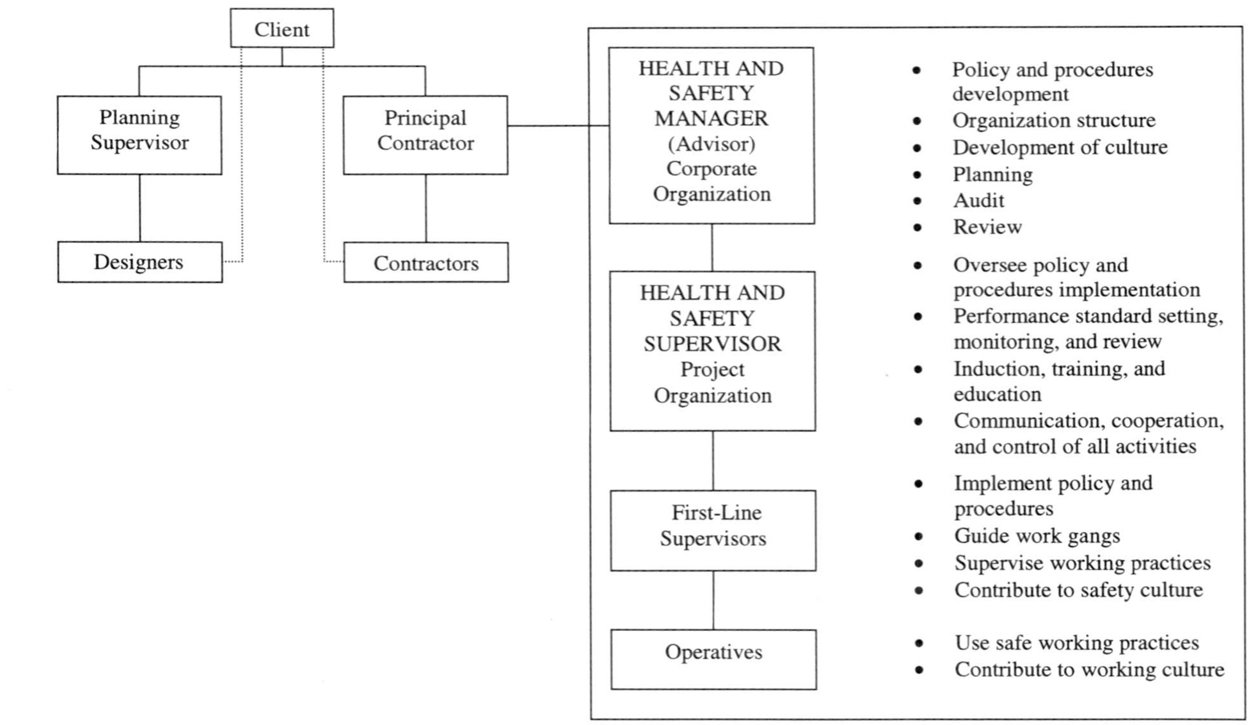

Source: After Griffith and Howarth (2001).

of the main players in the principal contractor's team. In most cases in the United Kingdom, the operatives and possibly the supervisors would actually be employed by the contractors or subcontractors rather than the principal contractor.

Sustainability during Construction

This section concentrates on sustainability issues relating to the construction phase of the project. One of the most significant areas of environmental management for construction organizations is waste management. In many countries, governments have sought to put pressure on industry to reduce waste by taxing its removal and disposal. When the "landfill" tax was applied in the United Kingdom, the hire of a typical rubbish skip used on construction sites increased from around £25 (US$42) to more than £70 (US$117). Thus, waste management becomes an important factor in the overall financial success of the project.

There are many useful publications on environmental management during construction, and Gibb et al. (2000) have produced a glossary of publications on the subject. Of particular note is a manual by Coventry et al. (1999) covering general site rules; managing materials; water; waste; noise and vibration; dust, emissions, and odors; ground contamination; the natural environment; and archaeology. There is also a companion book and training video for site staff.

Working Procedures

The method statements developed earlier in the process must be brought down to working procedures such that they can be implemented. Unfortunately, this will be done down to a certain level but rarely taken, in an integrated manner, to the level of the workplace and operative instructions. As a result, the actual impact at the "sharp end" is significantly reduced.

Procedures should be in place to ensure that all contractors and subcontractors comply with the SHE plan and allocate necessary resources. A site layout plan should be developed showing temporary accommodation, storage space, access routes for vehicles and pedestrians, preassembly areas, and emergency access/egress routes. Specific SHE hazards must be identified, following a review of the initial risk assessment and procedures developed for addressing the construction risks including, but not limited to, those shown in Table 22.3.

Audits

Audits should be part of any management process, and this is equally true for SHE issues. Through audits and reviews, the "organization learns from all relevant experience and applies the lessons" (HSE, 1999). Clarke (1999) states that "the performance of all systems, and of people, changes over time. It usually deteriorates, unless something is done to maintain it." He adds that the purpose of auditing is to "maintain performance and ensure relevance and effectiveness." Watkins (1997) stresses that regular auditing of management systems is vital to sustaining those systems, together with the policies and performance. The

TABLE 22.3. PRELIMINARY LIST OF TOPICS FOR DEVELOPING PROCEDURES FOR CONSTRUCTION WORKS.

Area	Primary Impact		
	Safe	Health	Environment
Abrasive wheels	X	X	
Asbestos		X	X
Cartridge-operated tools	X	X	
Cladding and the building envelope	X	X	
Confined spaces	X		
Contaminated ground		X	X
Crane operation	X		X
Demolition	X	X	X
Diving	X	X	
Drainage	X	X	X
Electricity	X		
Ergonomics and human factors	X	X	
Excavations and groundworks	X	X	X
Explosives	X		X
Falsework	X		
Fit out and finishes	X	X	
Flammable materials	X		X
Hazardous materials		X	X
Heavy lifts	X	X	
Hoists	X		
Lead burning		X	X
Lifting gear	X		
Noise		X	X
Pressure testing	X	X	
Radiography		X	X
Roof work/work at height	X	X	
Structural frame work	X	X	
Transport	X		X
Woodworking machinery	X	X	
Work over water	X	X	
Work within/near live facilities	X	X	X

Taylor Woodrow approach mentioned earlier is based on periodic audits of key issues, carried out by visiting auditors.

Another reason for audits is to ensure that the systems devised keep up with the needs and challenges of a changing society. Watkins (1997) explains that "if it were possible to establish the perfect system today, by tomorrow it would begin its long descent into obsolescence. Slowly at first, almost imperceptivity, but steadily. The world moves on. New work practices emerge, legislation is superseded, people change. Unless your systems move along with the rest of the world they will inevitably fall out of step with the demands of the law. Auditing is one of the ways to guard against this".

Life Cycle Issues

Operation, Maintenance, and Facilities Management

"Attention to SHE issues during design does not only provide safer construction but will result in more efficient operation, safer maintenance and facility management" (ECI, 1995). This aspect of "construction" varies dramatically depending on the nature of the built facility. Process plants will, by their nature, require more consideration for their operation than, say, speculative office blocks, in that the severity of unplanned events from process plants will be much more serious for health and safety of those in the vicinity as well as for the environment as a whole. *Human Factors in Industrial Safety* (HSE, 1999) stresses the important role that design should play. Reason (1990) describes some of the well-known disasters that have involved human error during operation and/or maintenance—for instance, Bhopal in 1984 or Chernobyl in 1986. In all cases, operational systems should be "fail-safe" and must take into account human error. ECI (1995) provide a list of key considerations for maintenance, particularly for process plants:

- "analysis of the operator-critical tasks and risks of failure;
- evaluation of decisions to be made between automatic and physical controls;
- consideration of emergency actions required and the display of process information;
- arrangement for maintenance access; and
- provision of working environment for lighting, noise and thermal considerations."

ECI (1995) goes on to explain that "the maintenance criteria may be on a routine preventative basis or left to a breakdown/replacement regime. If frequent access to plant controls is required then access can be permanently designed for the facility. If breakdown maintenance is accepted then equipment installed to assist safe and fast turnaround is the designer's consideration".

The SHE issues for other construction projects, such as offices or schools, may appear less crucial when compared to the process sector; however, they are still important. A particular safety issue is maintenance and cleaning access, especially for the building envelope. On the environmental side, emissions from buildings and use of energy are requiring more serious consideration, as are the ultimate demolition and disposal of the elements that make up the building. As already noted, the designer's role in achieving a good SHE performance throughout the life cycle of the project is critical.

One factor that has changed the typical approach toward maintenance issues, at least in the United Kingdom, is the increased use of private/public partnerships (see the chapters by Turner and by Ive). In these projects, the constructing consortium is also responsible for maintenance and operation of the road or hospital or prison for a considerable period after the completion of construction. This does not alter the legal situation, nor should it affect the moral obligation to care for maintenance workers, but it does provide a clearer feedback loop on maintenance issues to designers and constructors.

As explained earlier, the SHE file, prepared by the design and construction team, should be available, identifying SHE implications for maintenance. It is essential that the format and usability of this document is carefully considered to ensure that it can be effectively used throughout the life cycle.

Demolition and Decommissioning

Demolition and decommissioning are explicitly included as "construction" activities by the European Directives on health and safety issued since the early 1990s. Nevertheless, it has taken some time for designers to address this aspect of design risk assessment. Environmental life cycle strategies, as the sector responds to the sustainability lobby, now commonly have to include demolition and final disposal or, ideally, reuse of the materials from the completed building or facility.

An additional challenge for the construction sector is that most of the built environment has been designed before these considerations were even suggested. This has resulted in a major legacy issue for construction SHE. For instance, the ubiquitous and uncontrolled use of asbestos in all forms of construction now presents one of the biggest challenges for all societies. The health issues for its removal and the long-term environmental risk are leading many building owners to just cover up and leave it in place, perhaps hoping for some miracle solution to be developed. However, all that is happening is that the problem is just being stored up for a future generation. The industry must ensure that an equivalent catastrophe cannot occur in the future.

Driving Change and Measuring Success

Driving Change

This chapter has argued that there is a real need to drive change in the SHE performance of construction sector. No one party can deliver this change alone: it requires buy-in of all the stakeholders. If the client/owner is not committed to it, then there will not be enough resources allowed in the brief to adequately manage the risks. The designers have a major influence, and all this previous effort will come to naught unless the construction team, including suppliers and subcontractors, have ownership of the SHE solutions.

Measuring Success

Measurement is essential to maintain and improve performance. There are two ways to generate information on performance (adapted from HSE 1997):

- Reactive systems that monitor accidents, ill health, and incidents
- Active systems that monitor the achievement of plans and the extent of compliance with standards.

Reactive Measurement: Quantitative Lagging Indicators. The most common form of health and safety performance measures are quantitative, lagging indicators. These are reactive and form the basis of most governmental measurement systems. Laufer (1986) suggested that "safety measuring methods are characterized primarily by the manner in which they relate to the criteria of safety effectiveness, the events measured and the method of data collection." Kunju-Ahmad and Gibb (2003) explain that the "frequency element of the undesirable event usually splits up into four categories:

1. Lost day cases—cases which bring absence from work;
2. Doctor's cases—non-lost workday cases that are attended by a doctor;
3. First aid cases—non-lost workday cases requiring only first aid treatment; and
4. No-injury cases—accidents not resulting in personal injury but including property damage or productivity disruption."

There are a number of additional problems with this approach—for example, the practice of citing only directly employed (and usually office-based) staff in statistics returned, rather than including all the people involved in the project. As most of the people who are injured or suffer ill health are "workers," and many of the owner organizations do not directly employ the workers, this can produce very misleading project statistics. The practice should be to include *all* personnel involved in the project and generally exclude home-office staff from project figures to avoid skewing the statistics. Another dilemma is that where safety culture is poor, there is a tendency to heavily underreport. This leads to the issue of dealing with a perceived increase in incidents once the safety culture starts to improve. These are often caused simply by an increase in the number of incidents being recorded, which may then mask an actual decrease in the incidents themselves.

Environmental performance for specific projects is sometimes also measured, often when a client/owner wants to use the score as a business marketing advantage. In the United Kingdom, the BREEAM technique, developed by the Building Research Establishment (BRE) is typically used. BREEAM assesses the performance of buildings in the following areas:

- *Management.* Overall management policy, commissioning site management, and procedural issues
- *Energy use.* Operational energy and carbon dioxide (CO_2) issues
- *Health and well-being.* Indoor and external issues affecting health and well-being
- *Pollution.* Air and water pollution issues
- *Transport.* Transport-related CO_2 and location-related factors
- *Land use.* Greenfield and brownfield sites
- *Ecology.* Ecological value conservation and enhancement of the site
- *Materials.* Environmental implication of building materials, including life cycle impacts
- *Water.* Consumption and water efficiency

Developers and designers are encouraged to consider these issues at the earliest opportunity to maximize their chances of achieving a high BREEAM rating. Credits are awarded in each area according to performance. A set of environmental weightings then enables the credits to be added together to produce a single overall score. The building is then rated on a scale of PASS, GOOD, VERY GOOD, or EXCELLENT, and a certificate is awarded that can be used for promotional purposes. More information on this technique can be found at http://products.bre.co.uk/breeam.

Active Measurement: Behavior, Culture, and Process Management. Kunju-Ahmad and Gibb (2003) argue that "proactive measures should be used to evaluate SHE performance rather than backward-looking techniques. These techniques concentrate on evaluating behavior, culture and process management. An industry-wide technique is a potential vision for the future, however, difficulties in applying a single tool to construction remain and, for the foreseeable future, individual organizations are likely to continue to develop their own systems. These individual organizations can derive considerable benefits internally despite being unable to accurately compare their performance with others."

Lingard and Rowlinson (1994) describe behavioral safety management as a "range of techniques which seek to improve safety performance by setting goals, measuring performance and providing feedback." This concentration on behavior is also supported by other research such as Duff et al. (1994). Cameron (1998) describes an audit system as a means to develop goals, implement checks, and provide ongoing feedback.

The United Kingdom's Health and Safety Executive (HSE 1999) describe three aspects of human factors that influence human behavior:

- *Individual.* Competence, skills, personality, attitudes, risk perception
- *Organization.* Culture, leadership, resources, work patterns, communications
- *Job.* Task, workload, environment, display and controls, procedures.

In its work, described earlier in the chapter, the Loughborough ConCA has developed the causality model shown in Figure 22.7. This clearly shows the various levels of influences on a particular accident, and these can be adapted to suit an ill-health or environmental event. The basic point here is that the effective project management approach will address issues much further back up the process chain, rather than leaving all the responsibility for SHE management to those who inherit the problems on-site.

Summary

This chapter has argued the need to consider SHE from an early stage as part of project management from a moral, legal, financial, and cultural standpoint. It has identified requirements for SHE policy and objectives, SHE concept, initial risk assessment, and the SHE plan. It has described SHE actions required during the design and preconstruction phases, namely risk assessment and risk avoidance, designers actions, sustainability assessment, SHE plan and SHE file development, method statements, procurement strategy, assessment of competence and resources, and SHE training and education. The chapter then introduced the key aspects of the construction action plan, including planning; management, leadership, and organization; sustainability; working procedures; and audits. It has raised issues for the life cycle of the project such as operation, maintenance, and facilities man-

FIGURE 22.7. FACTORS IN ACCIDENT CAUSALITY.

Originating Influences

client requirements, economic climate, construction education

permanent works design, project management, construction processes
safety culture, risk management

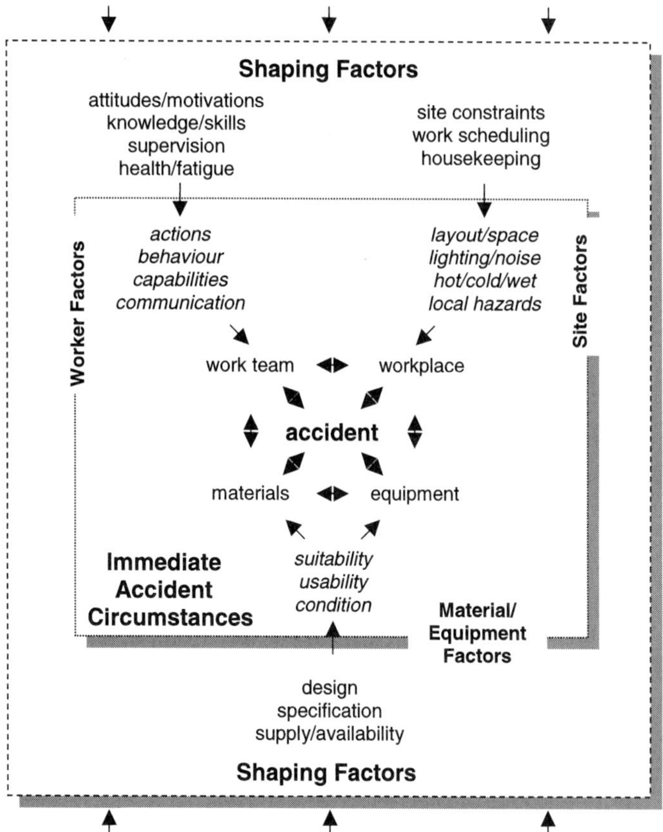

permanent works design, project management, construction processes
safety culture, risk management

client requirements, economic climate, construction education

Originating Influences

Source: ConCA (2002).

agement, as well as demolition and decommissioning. Finally, the chapter looked at driving change and measuring success, using both reactive and proactive techniques.

References

Beilby, S. C., and J. A. Read. 2001. *Site safety handbook.* 3rd ed. London: Construction Industry Research and Information Association (CIRIA). ISBN: 0-86017-800-5.

Cameron, I. 1998 Pilot study proves value of safety audits. *Construction Manager*, Chartered Institute of Building. London: Thomas Telford.

Carpenter, J. 2001. Identification and management of risk in undergraduate courses. Contract Research Report 392/2001. London: HSE Books.

Casals, M. 2001. Costs and benefits related to quality and safety and health in construction. *Proceedings of the CIB W99 International Conference*, Barcelona, October.

CII. 1993a. *Zero injury economics.* Document No. SP32-2. Austin, TX: Construction Industry Institute.

———. 1993b. Zero injury techniques. Document No. RS32-1. Austin, TX: Construction Industry Institute.

———. 1996. Design for safety. Document No. 101-2. Construction Industry Institute.

———. 2001. Small projects toolkit. Document No. 161-2. Austin, TX: Construction Industry Institute.

———. 2003. The owners' role in construction safety. Document No. RS190-1. Austin, TX: Construction Industry Institute.

———. 2003b. Safety plus: Making zero accidents a reality. Document No. RS160-1. Austin, TX: Construction Industry Institute.

CIRIA. 1998. CDM training pack for designers. Pub. C501. London: Construction Industry Research and Information Association.

———. 2000. Integrating safety, quality, and environmental management. Pub. C509. London: Construction Industry Research and Information Association.

Clarke, T.. 1999. *Managing health and safety in building and construction.* Oxford, UK: Butterworth-Heinemann.

Cooks, J. et al.. 1995. CDM regulations: Case study guidance for designers. Pub. R145. London: Construction Industry Research and Information Association (CIRIA).

ConCA and Loughborough University. Forthcoming. Study of 100 construction accidents to identify causal relationships. Funded by the Health and Safety Executive. Final report awaiting publication. For more information contact a.g.gibb@lboro.ac.uk.

Coventry, Woolveridge, and Kingsley. 1999. Environmental good practice: Working on site. C502 (hardback manual), C503 (pocket book), C525V (training video). London: Construction Industry Research and Information Association (CIRIA),

Croner. 1994. Croner's *management of construction safety.* Surrey, UK: Croner Publications.

DTI. 1998. *Rethinking construction,* Department of Trade and Industry (formerly DETR), Construction Task Force. London: The Stationery Office.

Duff, A. R., I. T. Robertson, R. A. Phillips, and M. D. Cooper. 1994. Improving safety by the modification of behavior. *Construction Management and Economics* 12(6):67–78.

EC. 1989. *Directive concerning the introduction of measures to encourage improvements in the health and safety of workers at work.* Directive 89/391/EEC, European Commission. London: The Stationery Office.

———. 1992. *Directive concerning temporary and mobile construction sites.* Directive 92/57/EEC, European Commission; the basis of the United Kingdom's Construction (Design and Management) Regulations (CDM). London: The Stationery Office.

ECI. 1995. *Total project management of construction safety, health and environment.* 2nd ed., ed. D. Tubb and A. G. F. Gibb. European Construction Institute. London: Thomas Telford.

Gibb, A. G. F., ed. 2000. Designing for safety and health. *Proceedings of the CIB W99/ECI International Conference.* Various papers on designing for safety and health. European Construction Institute. London, June.

Gibb, A. G. F., and A. I. Ayode. 1996. Integration of quality, safety, and environmental systems. In *Proceedings of the First International Conference of CIB Working Commission W99, Portugal, September: Implementation of Safety and Health on Construction Sites,* ed. L. M. Alves Dias, and R. J. Coble. pp. 11–20. Rotterdam: A. A. Balkema.

Gibb, A. G. F.. D. E. Gyi, and T. Thompson., eds. 1999. *The ECI guide to managing health in construction.* London: Thomas Telford. ISBN: 0-7277-2762-1.

Griffith, A., and T. Howarth, T. 2001. *Construction health and safety management.* Pearson Education.

Gibb, A. G. F., J. Slaughter, and G. Cox. 2000. The ECI guide to environmental management in construction. Interactive CD-ROM. Loughborough, UK: European Construction Institute.

Griffith, A.. 1994. *Environmental management in construction.* Basingstoke, UK: Macmillan.

Griffith, A., and T. Howarth. 2001. *Construction health and safety management.* London: Pearson Education. ISBN: 0-582-41442-3.

Halliday, S. 1998. Construction health and safety: Materials impact. In *Proceedings of the International Conference of CIB Working Commission W99: Environment, Quality and Safety in Construction,* ed. Alves, Dias, and Coble. pp. 9–20. Lisbon, Portugal.,

Hide, S., A. G. F. Gibb, R. A. Haslam, D. E. Gyi, S. Hastings, and R. Duff. 2002. ConCA: Preliminary results from a study of accident causality. *Proceedings of the Triennial International Conference of CIB Working Commission W99,* ed. Rowlinson. pp. 61–68, CIB Pub. 274. Hong Kong, May.,

Hinze, J. 1998. Addressing construction worker safety in the design phase. *Proceedings of the International Conference of CIB Working Commission W99: Environment, Quality and Safety in Construction,* ed. Alves, Dias, and Coble. pp. 46–54. Lisbon, Portugal.

———. 1996. Quantification of the indirect costs of injuries. In *Safety and health on construction sites.* CIB Working Commission W99. Pub. 187, ed. Coble, Issa, Elliott. pp. 307–321. ISBN: 1-886431-04-03.

Hinze, J. 1991. *Indirect costs of construction accidents.* Source Document No. 67 Austin, TX: The Construction Industry Institute.

Hinze, J., and L. Appelgate, L. 1991. Costs of construction injuries *Journal of Construction Engineering and Management* 117 (3, September).

Hinze, J., and J. Gambatese. 1996. Design decisions that impact construction worker safety. In *Safety and health on construction sites,* ed. Coble, Issa, and Elliott. pp. 219–231. CIB Working Commission W99. Pub. 187. ISBN: 1-886431-04-03.

HSE. Human factors in industrial safety. HSG 48. *Health and Safety Executive.* London: The Stationery Office.

———. 1994. CDM regulations: How the regulations affect you. *Health and Safety Executive.* London: The Stationery Office.

———. 1997. Successful health and safety management. HSG 65. *Health and Safety Executive.* London: The Stationery Office.

———. 1999. Reducing error and influencing behavior. HSG 48. *Health and Safety Executive.* London: The Stationery Office.

ISO. 1996. ISO 14001: Environmental management systems: Specification with guidance for use. Geneva: International Standards Organization.

Kunju-Ahmad, R., and A. G. F. Gibb. 2003. Towards effective safety performance measurement: Evaluation of existing techniques and proposals for the future. In, ed. S. Rowlinson. London: Spon.

Laufer, A. 1986. Assessment of safety performance measurement at construction sites. *Journal Construction Management and Economics* 112(4):530–542.

Lingard, H., and S. Rowlinson. 1994. Construction site safety in Hong Kong. *Construction Management and Economics* 12(6):501–510.

Nath et al., eds. 1998.*Instruments for environmental management.* Vol.1 of *Environmental management in practice.* London: Routledge.

Nelson, E. J. 1993. *Zero injury economics.* Special Pub. 32-2. Austin, TX: Construction Industry Institute.

Oluwoye, J., and H. MacLennan, H. 1996. Pre-planning safety in project buildability. In *Safety and health on construction sites,* ed. Coble, Issa, and Elliott, pp. 239–248. CIB Working Commission W99. Pub. 187.

Ove Arup and Partners. 1997. *CDM regulations: Work sector guidance for designers.* Pub. R166. London: Construction Industry Research and Information Association (CIRIA).

Pavitt, T. C., and A. G. F. Gibb. 2003. Interface management within construction: in particular the building façade. *Journal of Construction Engineering and Management.* American Society of Civil Engineers. Vol. 129, No. 1: 8–15. ISSN: 0733-9364.

Pilz GmbH and Co. 1997. *Guide to machinery standards.* p. 57.

Reason, J. 1990. *Human error.* Cambridge University Press.

Rwelamila, P., and J. Smallwood. 1996. Total Quality Management (TQM) without safety management? In *Safety and health on construction sites,* ed. Coble, Issa, and Elliott. pp. 83–100. CIB Working Commission W99. Pub. 187.

Smallwood, J. 2001. Total Quality Management (TQM)—the impact?, Costs and benefits related to quality and safety and health in construction. *Proceedings of the CIB W99 International Conference* pp. 289–297. Barcelona, October.

Stubbs, A. 1998. *Environmental law for the construction industry.* London: Thomas Telford.

Tenah, K. 1996. Incorporating safety mechanisms into engineering design. In *Safety and health on construction sites,* ed. Coble, Issa, and Elliott, pp. 249–259. CIB Working Commission W99, Pub. 187.

Watkins, G. 1997. *The health and safety handbook.* London: Street and Maxwell.

CHAPTER TWENTY-THREE

VERIFICATION

Hal Mooz

"Proof of compliance with specifications. Verification may be determined by test, inspection, demonstration, or analysis."

MOOZ, FORSBERG, COTTERMAN (2002)

When you are managing projects, it is usually necessary to prove that the solution satisfies both the specifications and the users. The process called *verification* develops this proof. Verification encompasses a family of techniques and can be applied irrespective of whether the project is completely hardware, software, a combination of both, or an operations-only solution. While verification methods may differ according to project disciplines, some method of verification is usually required, ranging from full measured compliance of every aspect to the random sampling of production units. It is critical that the verification approach is developed early in the project cycle in conjunction with requirements determination and represents consensus between the solution provider and the customer.

The Context

Verification is closely associated with other project management terms that address the proof that solutions satisfy one or more requirements. The family of terms includes verification; validation; qualification; certification; integration; independent verification and validation (IV&V); integration, verification, and validation (IV&V); and independent integration, verification, and validation (IIV&V).

As with all of project management and system engineering communication, it is imperative that these terms are well understood and are properly communicated among the team members to avoid unintended consequences. The following definitions are the baseline for the remaining discussions of this chapter (Mooz, Forsberg, and Cotterman, 2002).

Validation. Proof that the user is satisfied.

Qualification. Proof that the design will survive in its intended environment with margin. The process includes testing and analyzing hardware and software configuration items to prove that the design will survive the anticipated accumulation of acceptance test environments, plus its expected handling, storage, and operational environments, plus a specified qualification margin. Qualification testing often includes temperature, vibration, shock, humidity, software stress testing, and other selected environments.

Certification. To attest by a signed certificate or other proof to meeting a standard.

Integration. The successive combining and testing of system hardware assemblies, software components, and operator tasks to progressively prove the performance and compatibility of all components of the system.

Independent verification and validation (IV&V). The process of proving compliance to specifications and user satisfaction by using personnel that are technically objective and managerially separate from the development group. The degree of independence of the IV&V team is driven by product risk. In cases of highest risk, IV&V is performed by a team that is totally independent from the developing organization.

Integration, verification, and validation (IV&V). The combining of system entities, the proving the system works as specified, and the confirming that the right system has been built and that the customers/users are satisfied.

Independent integration, verification, and validation (IIV&V). The integration, verification, and validation sequence conducted by objective personnel separate from the development organization.

To clarify the interrelated contexts of integration, verification, and validation, system development Vee model illustrations will be used (Forsberg, Mooz, and Cotterman, 2001; Forsberg and Mooz, 2001).

The Vee Model

Any phased project cycle is composed of three aspects.

- The business aspect represents the pursuit of the business case.
- The budget aspect represents the pursuit and management of the funding.
- The technical aspect represents the technical development strategy.

Projects start with high-level conversations with users/sponsors about the problem to be solved and the tangible proof needed at acceptance to prove that the problem has been solved. The technical process proceeds from those high-level discussions down through solution decomposition with progressively lower-level concepts and designs, and then ascends up to operations and final high-level discussions with the users/sponsor relative to their satisfaction with the solution. The image of this technical aspect of the cycle is best depicted as a Vee where the elaboration of the evolving solution baseline forms on the core of the Vee (see Figure 23.1).

This Vee format accurately illustrates levels of decomposition, from solution requirements and concepts down to the lowest replaceable unit in the left Vee leg and then upward consistent with fabrication and integration of the solution elements into the completed system

FIGURE 23.1 PROJECT CYCLE VEE+ MODEL.

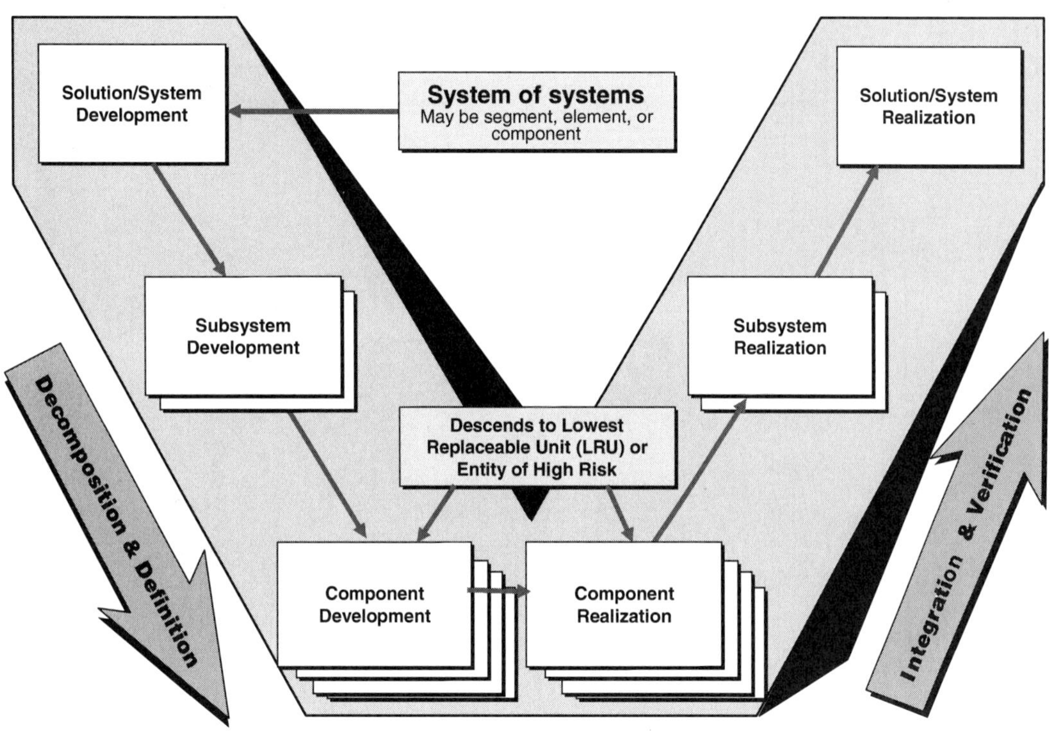

in the right Vee leg. The thickness of the Vee increases downward to reflect the increasing number of elements as a single system or solution is decomposed to its many individual subsystems and their lowest replaceable units (LRU).

At each decomposition level, there is a direct correlation between activities on the left and right sides of the Vee. This is deliberate. For example, the method of integration and verification to be used on the right must be determined on the left for each set of requirements and entities developed at each decomposition level (see Figure 23.2).

This minimizes the chances that requirements are specified in a way that cannot be measured or verified. It also forces the early consideration and preparation of the verification sequence, methods, facilities, and equipment required to meet the verification objectives as well as schedule and cost targets.

Verification facilities may become a task on the critical path requiring stakeholder approval. For example, mechanisms to be deployed in the weightlessness of space may require construction of a large float pool to demonstrate deployment using floatation devices to compensate for gravity. Similarly, a software system might require acquisition of special verification hardware, development, and loading of a verification database, or development of specialized verification drivers. Figure 23.3 illustrates verification planning at one level of

FIGURE 23.2. PROJECT CYCLE VEE+ MODEL WITH IV&V PLANNING.

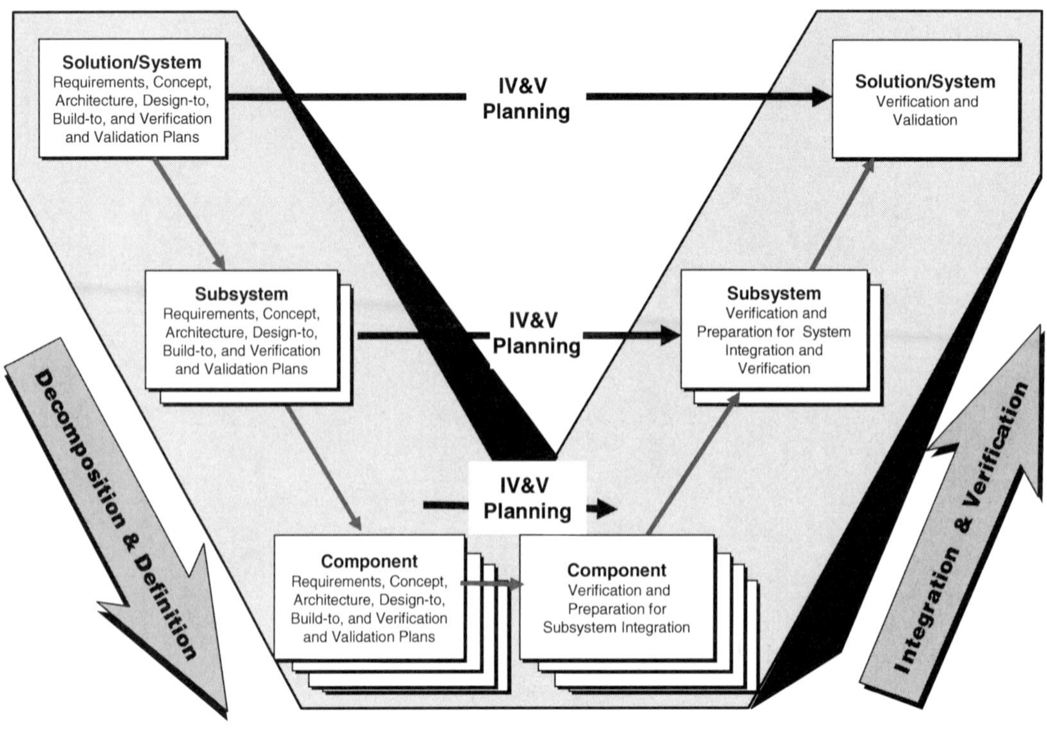

decomposition. This figure also illustrates the investigation of opportunities and their risks to whatever decomposition level is appropriate together with the affirming of the resultant baseline at the user level. However, these aspects of the Vee model are not relevant to this chapter and are not further explained.

There are four key steps in planning for integration, verification, and validation (see Figure 23.4) involving three user types (see Figure 23.5). The ultimate user is the end user of the system. The direct user is up one level in the decomposition from the item being validated. An associate user is any other user potentially impacted by the item being validated and usually exists at the same decomposition level. The three types are further clarified with examples later in this chapter.

- *Step 1.* Determine the integration sequence for combining the entities.
- *Step 2.* Determine how to prove that the solution when built is built right and satisfies both the design-to and build-to specifications.
- *Step 3.* Determine how to prove that the solution when verified is the right solution for both the direct user and the ultimate user.

FIGURE 23.3. INTEGRATION, VERIFICATION, AND VALIDATION PLANNING.

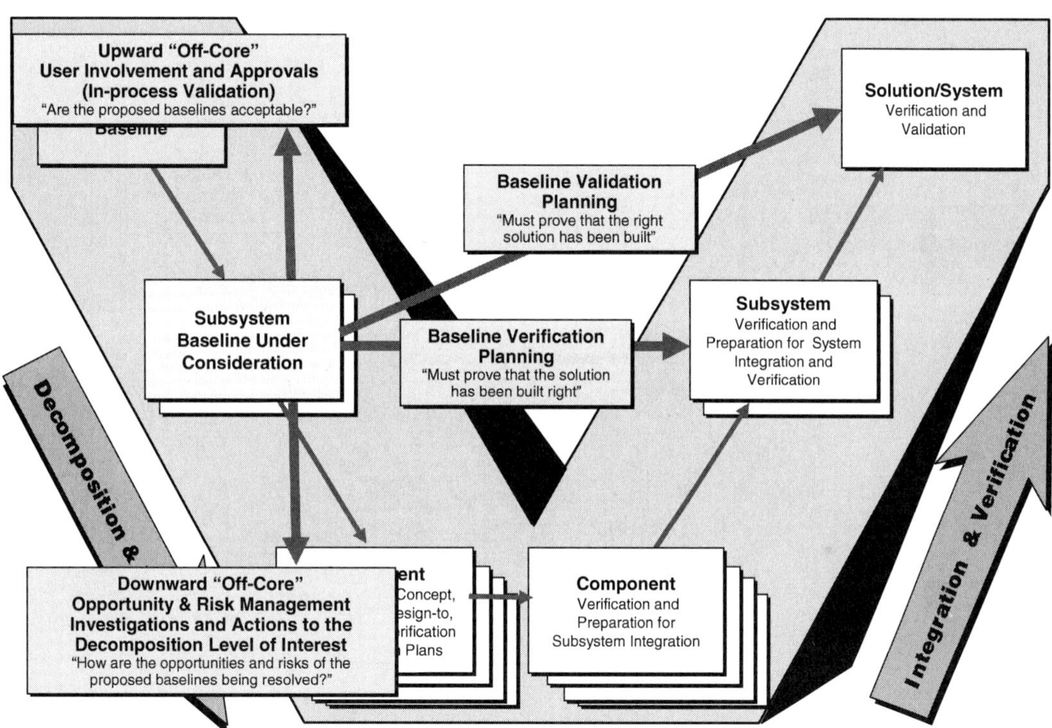

- *Step 4.* Determine if the concept as proposed and the associated proposed integration, verification, and validation approaches are acceptable to the associate, direct, and ultimate users of the solution (see Figure 23.5).

Risk: The Driver of Integration/Verification Thoroughness

Some projects are human-rated—that is, they must work flawlessly, as human lives are at stake. Some projects are quick-reaction attempts of a concept or an idea, and if they don't work, it is less serious compared to a human-rated project. Human-rated projects require extreme thoroughness, while the quick-and-dirty projects may be able to accept more risk. It is important to know the project risk philosophy as compared to the opportunity being pursued. This reward-to-risk ratio will then drive decisions regarding the rigor and thoroughness of integration and the many facets of verification and validation. There is no standard vocabulary for expressing the risk philosophy, but it is often expressed as "quick and dirty," or "no single point failure modes," or "must work," or "reliability is 0.9997,"

FIGURE 23.4. FOUR INTEGRATION, VERIFICATION, AND VALIDATION PLANNING STEPS.

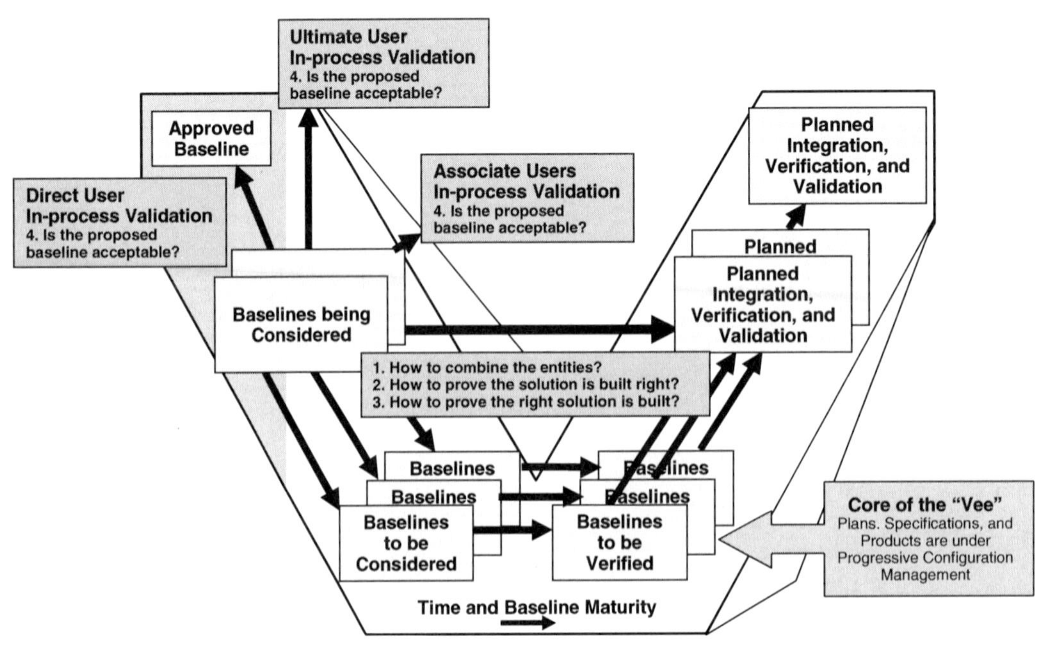

FIGURE 23.5. EXPANSION OF FIGURE 23.4 DETAILING THREE TYPES OF USERS.

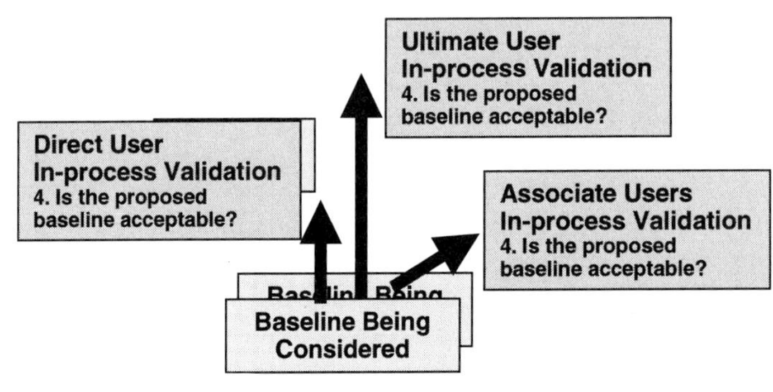

or some other expression or a combination of these. The risk philosophy will determine whether all or only a portion of the following will be implemented.

Integration

Preparation for integration, verification, and validation begins with planning for integration. The product breakdown structure (PBS) portion of the work breakdown structure (WBS) should reveal the integration approach but often does not. Integration planning must determine the approach so that the interfaces and intrafaces can be provided for, managed, and verified. Figure 23.6 illustrates four possible sequences to integrating four entities into the same higher-level combination.

Each approach reaches the same end result, but for each option, the interfaces are different and must be appropriately managed, followed by verification of both the interfaces and the entity performance before combining into higher-level combinations.

Interface management to facilitate integration and verification should be responsive to the following:

1. The product breakdown structure (PBS) portion of the work breakdown structure (WBS) should provide the roadmap for integration.
2. Integration will exist at every level in the product breakdown structure except at the most senior level.
3. Integration and verification activities should be represented by tasks within the work breakdown structure (see Figure 23.7).

FIGURE 23.6. FOUR INTEGRATION OPTIONS FOR A SYSTEM.

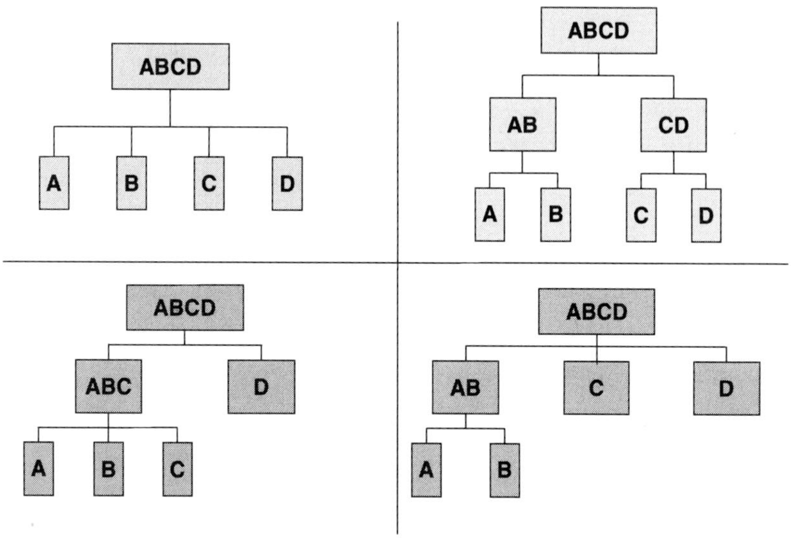

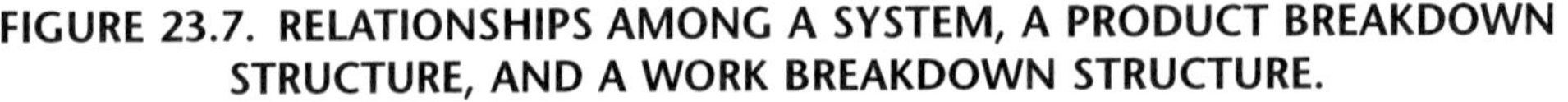

FIGURE 23.7. RELATIONSHIPS AMONG A SYSTEM, A PRODUCT BREAKDOWN STRUCTURE, AND A WORK BREAKDOWN STRUCTURE.

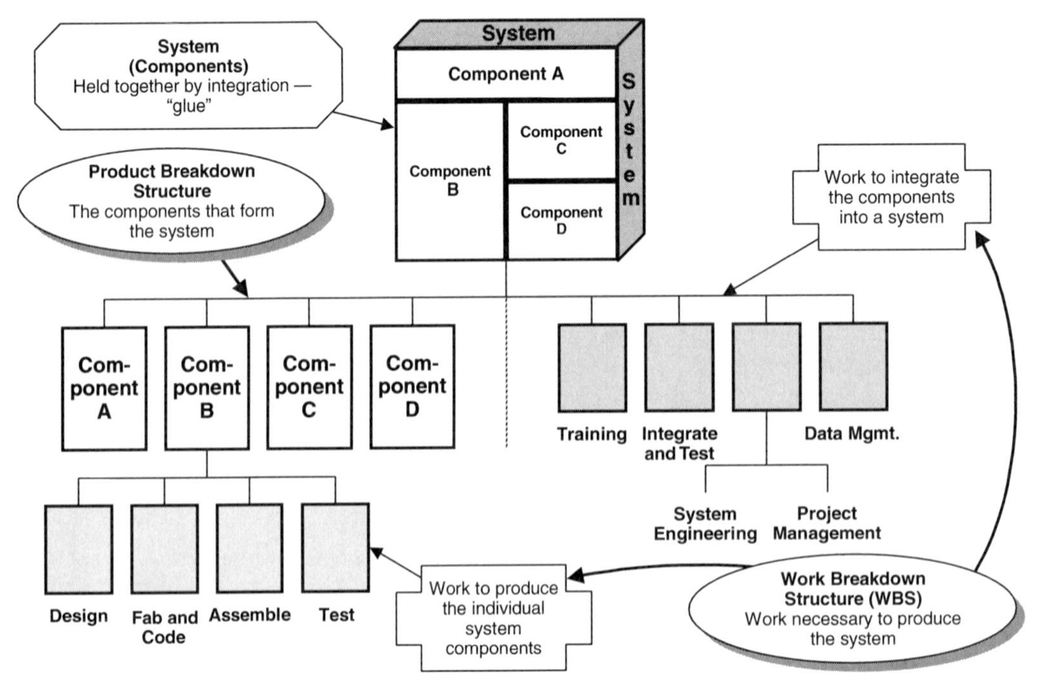

4. The work breakdown structure is not complete without the integration and verification tasks and the tasks to produce the intermediate work products (e.g., fixtures, models, drivers, databases) required to facilitate integration.
5. Interfaces should be designed to be as simple and foolproof as possible.
6. Interfaces should have mechanisms to prevent inadvertent incorrect coupling.
7. Interfaces should be verified by low-risk (benign) techniques before mating.
8. "OK to install" discipline should be invoked before all matings.
9. Peer review should provide consent-to authorization to proceed.
10. Haste without extra care should be avoided.

Integration and verification can be performed in a methodical incremental sequence by adding entities one at a time and proving the combined performance; or, all entities can be combined and then verified as a group in the "Big Bang" approach; or any combination between these two extremes can be used. In the sequential approach, anomalies are usually quickly resolved to the last entities; combined and corrective action can be swift. In the Big Bang approach, anomalies are more difficult to resolve, as there may be multiple causes working together to produce an undesired result. While the Big Bang approach, if it works, can result in substantial cost and time savings, it rarely works on newly developed systems that have not been adequately debugged. Other incremental variations, especially in software integration, though not limited to software, include the top-down, bottom-up, thread, and mixed approaches (see Figure 23.8). Note that prior to initiating any of these integration approaches each component or software module should have been verified against its specification.

Hindsight and lessons learned can be beneficial to the avoidance of future problems and to the development of improved methods. The following are valuable lessons learned related to the integration and verification of solution entities:

- Make sure names and identifiers are consistent and correct across entities being integrated.
- Ensure the correct versions of the entities are being integrated.
- Ensure no changes to external interfaces during integration.
- Be aware that logical integration problems are subtle. (They don't emit smoke.)
- Ensure that software and hardware baselines are compatible.
- Use peer reviews, software walk-throughs, and inspections to confirm compatibility.
- Verify software modules incrementally and resolve discovered anomalies.
- Use mechanical mock-ups to verify space, access, clearances and to practice the installation process.
- Use thermal models to confirm thermal predictions.
- Use an electrical/electronic simulator to verify functionality on both sides of the interface before mating.
- Enforce power off during connector mating.
- Ensure frame ground is common with power ground.
- Use connector keying and clocking to prevent incorrect mating.
- Use a single supplier for both halves of mating connectors.

FIGURE 23.8. INCREMENTAL INTEGRATION APPROACHES.

Technique	*Features*
Top - Down	• *Control logic testing first* • *Modules integrated one at a time* • *Emphasis on interface verification*
Bottom - Up	• *Early verification to prove feasibility and practicality* • *Modules integrated in clusters* • *Emphasis on module functionality and performance*
Thread	• *Top down or bottom up integration of a software function or capability*
Mixed	• *Working from both ends toward the middle* • *Choice of modules designated top-down vs. bottom-up is critical*

- Use "OK to install" discipline to make sure everything is perfect prior to each and every mating.
- Examine all connectors for debris, pushed and bent pins, and correct clocking.
- If it doesn't mate easily, STOP. Don't force it.
- Use "OK to power" discipline before applying power.
- Compare results to predictions; then identify and resolve discovered anomalies.

Validation and Validation Techniques

Validation is proof that the users are satisfied regardless of whether the specifications have been satisfied or not. Occasionally a product meets all specified requirements but is rejected by the users and does not validate. Famous examples are the Ford Edsel, IBM PC Junior, and more recently, Iridium and Globalstar. In each case the products were exactly as specified but the ultimate users rejected them, causing very significant business failures. Conversely, Post-It-Notes failed verification to the glue specification, but the sticky notes then catapulted into our lives because we all loved the failed result. The permanently temporary or temporarily permanent nature of the glue was just what we were looking for, but it hadn't been specified.

Traditionally, validation occurs at the project's end when the user finally gets to use the solution to determine the level of satisfaction. While this technique can work, it can also cause immense waste when a project is rejected at delivery. Too many projects have been relegated to scrap or a storage warehouse because of user rejection. Proper validation management can avoid this undesirable outcome. When considering the process of validation, recognize that except for the top product level having just the ultimate or end user, there are direct users, associate users, and ultimate users at each decomposition level and for each entity at that level, all of whom must be satisfied with the solution at that level. Starting at the highest system level, the ultimate user is also the direct user. At the outset, the ultimate users should reveal their plans for their own validation so that developers can plan for what the solution will be subjected to at delivery. A user validation plan is valuable in documenting and communicating the anticipated process.

Then within the decomposition process, as the solution concept and architecture is developed, the ultimate users should be consulted as to their satisfaction with the progression of proposed concepts. The approved concepts then become baselined for further decomposition and rejected concepts are replaced by better candidates. This process is called *in-process validation* and should continue in accordance with decomposition of the solution until the user decides the decisions being made are transparent to his or her interface and use of the system. This on-going process of user approval of the solution elaboration and maturation can reduce the probability of user dissatisfaction at the end to near zero. Consequently, this is a very valuable process to achieve and maintain user satisfaction throughout the development process and to have no surprise endings.

Within the decomposition process, validation management becomes more complex. At any level of decomposition, there are now multiple users. The ultimate user is the same. However, there is now a direct user that is different from the ultimate user, and there are associate users that must also be satisfied with any solution proposed at that level of decomposition. Consider, for instance, an electrical energy storage device that is required by the power system within the overall solution. The direct user is the power system manager, and associate users are the other disciplines that must interface with the storage device's potential solutions. If a chargeable battery is proposed, then the support structure system is a user, as is the thermodynamic system, among others. In software, a similar situation exists. Software objects have defined characteristics and perform certain specified functions on request, much like the battery in the prior example. When called, the software object provides its specified service just as the battery provides power when called. Associate users are any other element of the system that might need the specified service provided by the object.

All direct and ultimate users need to approve baseline elaboration concepts submitted for approval. This in-process validation should ensure the integration of mutually compatible elements of the system. In eXtreme and Agile programming processes, intense user collaboration is required throughout the development of the project to provide ongoing validation of project progress.

Ultimate user validation is usually conducted by the user in the actual user's environment, pressing the solution capability to the limit of user expectations. User validation may incorporate all of the verification techniques that follow. It is prudent for the solution developer to duplicate these conditions prior to delivery.

Verification and Verification Techniques

As stated at the outset, verification is proof of compliance with specifications. Verification may be determined by test, inspection, demonstration, or analysis. The following four techniques should be applied as appropriate to the verification objectives.

Verification by test. Direct measurement of specification performance relative to functional, electrical, mechanical, and environmental requirements. (Measured compliance with specified metrics).

Verification by inspection. Verification of compliance to specifications that are easily observed, such as construction features, workmanship, dimensions, configuration, and physical characteristics such as color, shape, software language, style, and documentation. (Compliance with drawings, configuration documents)

Verification by demonstration. Verification by witnessing an actual operation in the expected or simulated environment, without need for measurement data or post-demonstration analysis. (Observed compliance without metrics)

Verification by analysis. An assessment of performance using logical, mathematical, or graphical techniques, or for extrapolation of model tests to full scale. (Predicted compliance based on history)

Verification Objectives

The definition of verification calls for proof of specification performance. However, since specifications can require nominal performance, design margin, quality, reliability, life, and many other performance factors, the verification plan must be formulated to prove compliance within each of these requirement categories. Figure 23.9 illustrates the context of design margin.

To be conservative, engineers include design margins to ensure that their solution performs its function. Verification may then be designed to prove both nominal performance and a specified design margin with or without deliberately forcing the solution into failure.

The more common verification objectives are outlined in the following paragraphs:

Design Verification

Design verification proves that the solution's design performs as specified, or conversely, that there are identified design deficiencies requiring design corrective action. Design verification is usually carried out in nominal conditions unless the specification has design margins already built into the specified functional performance. Design verification usually includes the application of selected environmental conditions. Design verification should confirm positive events and the absence of negative events. That is, things that are supposed to happen happen, and things that are not supposed to happen do not. Software modules are often too complex to verify all possible combinations of events, leaving a residual risk within those that have not been deliberately verified.

eXtreme Programming and other Agile methods advocate thorough unit testing and builds (software integration) daily or even more frequently to verify design integrity in-process. Projects that are not a good match for an Agile methodology may still benefit from rigorous unit tests, frequent integrations, and automated regression testing during periods of evolving requirements and/or frequent changes.

Design Margin Verification: Qualification

Design margin verification, commonly called qualification, proves that the design is robust with designed-in margin, or, conversely, that the design is marginal and has the potential of failing when manufacturing variations and use variations are experienced. For instance, it is reasonable that a cell phone user will at some time drop the phone onto a concrete surface from about four or five feet. However, should the same cell phone be designed to

FIGURE 23.9. DESIGN MARGIN.

survive a drop by a high lift operator from, say, 20 feet? Qualification requirements should specify the margin desired.

Qualification should be performed on an exact replica of the solution to be delivered. The best choice is a unit within a group of production units. However, since this is usually too late in the project cycle to discover design deficiencies which would have to be retrofitted into the completed units, qualification is often performed on a first unit that is built under engineering surveillance to ensure that it is built exactly to print and as the designers intended. Qualification testing usually includes the application of environment levels and duration to expose the design to the limits that may be accumulated in total life cycle use. Qualification tests may be performed on specially built test articles that simulate only a portion of an entity. For instance, a structural test qualification unit does not have to include operational electronic units or software; inert mass simulators may be adequate. Similarly, electronic qualification tests do not need the actual supporting structure, since structural simulators with similar response characteristics may be used for testing.

The exposure durations and input levels should be designed to envelop the maximum that is expected to be experienced in worst-case operation. These should include acceptance testing (which is quality verification) environments, shipping environments, handling environments, deployment environments, and any expected repair and retesting environments that may occur during the life of an entity. Environments may include temperature, vacuum, humidity, water immersion, salt spray, random vibration, sine vibration, acoustic, shock, structural loads, radiation, and so on. For software, transaction peaks, electrical glitches, and database overloads are candidates.

The qualification margins beyond normal expected use are often set by the system level requirements or by the host system. Twenty-degree Fahrenheit margins on upper- and lower-temperature extremes are typical, and either three or six dB margins on vibration, acoustic, and shock environments are often applied. In some cases, safety codes establish the design and qualification margins, such as with pressure vessels and boiler codes. Software design margin is demonstrated by overtaxing the system with transaction rate, number of simultaneous operators, power interruptions, and the like. To qualify the new Harley-Davidson V Rod motorcycle for "Parade Duty," it was idled in a desert hot box at 100 degrees Fahrenheit for eight hours. In addition, the design was qualified for acid rain, fog, electronic radiation, sun, heat, structural strength, noise, and many other environments. Actual beyond specification field experience with an exact duplicate of a design is also admissible evidence to qualification if the experience is backed by certified metrics.

Once qualification has been established, it is beneficial to certify the design as being qualified to a prescribed set of conditions by *issuing a qualification certification* for the exact design configuration that was proven. This qualification certification can be of value to those that desire to apply this design configuration to other applications and must know the environments and conditions under which the design was proven successful.

Reliability Verification

Reliability verification proves that the design will yield a solution that over time will continue to meet specification requirements. Conversely, it may reveal that failure or frequency of repair is beyond that acceptable and anticipated. Reliability verification seeks to prove *mean time between failure* (MTBF) predictions. Reliability testing may include selected environments to replicate expected operations as much as possible. Reliability verification tends to be an evolutionary process of uncovering designs that cannot meet life or operational requirements over time and replacing them with designs that can. Harley-Davidson partnered with Porsche to ultimately achieve an engine that would survive 500 hours nonstop at 140 mph by conducting a series of evolutionary improvements.

Life testing is a form of reliability and qualification testing. Life testing seeks to determine the ultimate wear-out or failure conditions for a design so that the ultimate design margin is known and quantified. This is particularly important for designs that erode, ablate, disintegrate, change dimensions, and react chemically or electronically, over time and usage. In these instances the design is operated to failure while recording performance data. Life testing may require acceleration of the life process when real-time replication would take too long or would be too expensive. In these instances acceleration can be achieved by adjusting the testing environments to simulate what might be expected over the actual life time. For instance, if an operational temperature cycle is to occur once per day, forcing the transition to occur once per hour can accelerate the stress experience.

For software, fault tolerance is the reliability factor to be considered. If specified, the software must be tested against the types of faults specified and the software must demonstrate its tolerance by not failing. The inability of software to deal with unexpected inputs is sometimes referred to as "brittleness."

Quality Verification

In his book Quality is Free, Phillip Crosby defines quality as "conformance to requirements" and the "cost of quality" as the expense of fixing unwanted defects. In simple terms, is the product consistently satisfactory, or is there unwanted scrapping of defective parts? When multiple copies of a design are produced, it is often difficult to maintain consistent conformance to the design, as material suppliers and manufacturing practices stray from prescribed formulas or processes. To detect consistent and satisfactory quality—a product free of defects—verification methods are applied. First, process standards are imposed and ensured to be effective; second, automatic or human inspection should verify that process results are as expected; third, testing should prove that the ultimate performance is satisfactory. Variations of the process of quality verification include batch control, sampling theory and sample inspections, first article verification, and nth article verification. Quality testing often incorporates stressful environments to uncover latent defects. For instance, random vibration, sine sweep vibration, temperature, and thermal vacuum testing can all help force latent electronic and mechanical defects to the point of detection. Since it is difficult to apply all of these environments simultaneously, it is beneficial to expose the product to mechanical environments prior to thermal and vacuum environments where extended power-on testing can reveal intermittent malfunctions.

Software Quality Verification

The quality of a software product is highly influenced by the quality of the individual and organizational processes used to develop and maintain it. This premise implies a focus on the development process as well as on the product. Thus, the quality of software is verified by verifying that the development process includes a defined process based on known best practices and a commitment to use it; adequate training and time for those performing the process to do their work well; implementation of all the process activities, as specified; continuous measurement of the performance of the process and feedback to ensure continuous improvement; and meaningful management involvement. This is based on the quality management principles stated by W. Edwards Deming that "Quality equals process—and everything is process."

-ilities Verification

There are a host of -ilities that require verification. Figure 23.10 provides a list of common -illities.

Verification of -ilities requires careful thought and planning. Several can be accomplished by a combined inspection, demonstration, and/or test sequence. A verification map can prove to be useful in making certain that all required verifications are planned for and accomplished.

Certification

Certification means "to attest by a signed certificate or other proof to meeting a standard." Certification can be verification of another's performance based on an expert's assurance. In the United States, the U.S. Food and Drug Administration grades and approves our meat to be sold, and Consumer Reports provides a "Best Buy" stamp of approval to high value products.

Certification often applies to the following (see also Crawford's chapter):

- *The individual.* Has achieved a recognized level of proficiency
- *The product.* Has been verified as meeting/exceeding a specification
- *The process.* Has been verified as routinely providing predictable results

FIGURE 23.10. OTHER -ILITIES REQUIRING VERIFICATION.

Accessibility	Efficiency	Reusability
Adaptability	Hostility	Recyclability
Affordability	Integrity	Securability
Compatibility	Interoperability	Survivability
Compressability	Liability	Scalability
Dependability	Mobility	Testability
Degradeability	Manageability	Usability
Distributability	Producibility	Understandability
Durability	Portability	Variability

In all cases certification is usually by independent assessment or audit to a predefined standard. When material is "certified," it should arrive at the user's facility complete with a pedigree package documenting the life history of the contents and a signed certification with associated test or other verification results substantiating that the contents of the container are as represented.

Certification is becoming more and more popular as professional organizations promote organizations and individuals to improve their performance capability and to be recognized for it. Individual or organizational certification such as ISO (International Standards Organization; see Figure 23.11), Carnegie Mellon's Software Engineering Institute's CMM® (see Figure 23.12) or CMMI,® and the Project Management Institute's PMP® Project Management Professional (see Figure 23.13) designation are achieved by demonstrated compli-

FIGURE 23.11. ISO 9000 QUALITY STANDARD.

- Quality Policy Statement ☑
- Quality Organization ☑
- Management Quality Reviews ☑
- Quality System Procedures and Planning ☑
- Contract Review Procedures ☑
- Design Control Procedures ☑
- Document and Data Control System ☑
- Purchasing Control System ☑
- Control of Customer-Supplied Products ☑
- Product Identification and Traceability System ☑
- Process Controls ☑
- Inspection and Testing Procedures ☑
- Control of Inspection, Measuring, and Test Equipment ☑
- Inspection and Test Status System ☑
- Control of Nonconforming Products ☑
- Corrective and Preventive Action Procedures ☑
- Handling, Storage, Packaging, Presentation, and Delivery Procedures ☑
- Quality Record Control System ☑
- Internal Quality Audit Procedures ☑
- Training in Quality ☑
- Procedures for Servicing ☑
- Statistical Techniques ☑

90437

FIGURE 23.12. SEI CMM; CAPABILITY MATURITY MODEL.

Implications of Advancing Through CMM Levels.

	Level 1	Level 2	Level 3	Level 4	Level 5
Processes	Few stable processes exist or are used.	Documented and stable estimating, planning, and commitment processes are at the project level.	Integrated management and engineering processes are used across the organization.	Processes are quantitatively understood and stabilized.	Processes are continuously and systematically improved.
	"Just do it"	Problems are recognized and corrected as they occur.	Problems are anticipated and prevented, or their impacts are minimalized.	Sources of individual problems are understood and eliminated.	Common sources of problems are understood and eliminated.
People	Success depends on individual heroics.	Success depends on individuals; management system supports.	Project groups work together, perhaps as an integrated product team.	Strong sense of teamwork exists within each project.	Strong sense of teamwork exists across the organization
	"Firefighting" is a way of life.	Commitments are understood and managed.	Training is planned and provided according to roles.		Everyone is involved in process improvement.
	Relationships between disciplines are uncoordinated, perhaps even adversarial.	People are trained.			
Technology	Introduction of new technology is risky.	Technology supports established, stable activities.	New technologies are evaluated on a qualitative basis.	New technologies are evaluated on a quantitative basis.	New technologies are proactively pursued and deployed.
Measurement	Data collection and analysis is ad hoc.	Planning and management data used by individual projects.	Data are collected and used in all defined processes.	Data definition and collection are standardized across the organization.	Data are used to evaluate and select process improvements.
			Data are systematically shared across projects.	Data are used to understand the process quantitatively and stabilize it.	

95-357 drw 12B

FIGURE 23.13. PMI/PMP KNOWLEDGE AREAS.

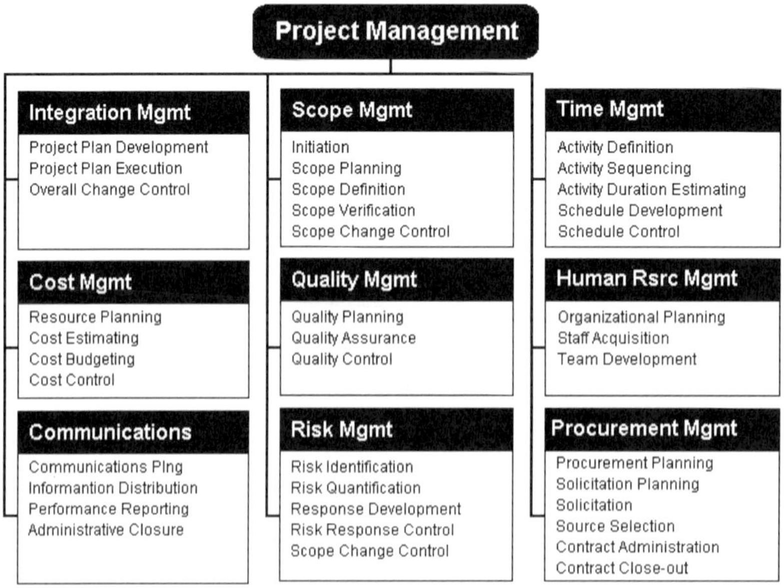

Project Management

Integration Mgmt	Scope Mgmt	Time Mgmt
Project Plan Development Project Plan Execution Overall Change Control	Initiation Scope Planning Scope Definition Scope Verification Scope Change Control	Activity Definition Activity Sequencing Activity Duration Estimating Schedule Development Schedule Control
Cost Mgmt	**Quality Mgmt**	**Human Rsrc Mgmt**
Resource Planning Cost Estimating Cost Budgeting Cost Control	Quality Planning Quality Assurance Quality Control	Organizational Planning Staff Acquisition Team Development
Communications	**Risk Mgmt**	**Procurement Mgmt**
Communications Plng Informantion Distribution Performance Reporting Administrative Closure	Risk Identification Risk Quantification Response Development Risk Response Control Scope Change Control	Procurement Planning Solicitation Planning Solicitation Source Selection Contract Administration Contract Close-out

ance to a recognized and controlled set of standards. It has become increasingly common for buyers of services to include these certifications in their buying decision criteria.

In some cases, certifications take the form of licenses to do business, such as a certified public accountant and for lawyers who are required to pass a bar exam to practice law.

Many other individual certifications support the project management discipline. Most are in the quality discipline, such as:

- Certified Quality Manager
- Certified Quality Engineer (CQE)
- Certified Quality Auditor (CQA)
- Certified Reliability Engineer (CRE)
- Certified Quality Technician (CQT)
- Certified Mechanical Inspector (CMI)
- Certified Software Quality Engineer (CSQE)

In addition, in 2004 the International Council on Systems Engineering initiated certification of systems engineers as Certified Systems Engineering Professionals.

The objective of organizational and personal certification is to ensure that the required level of individual and organizational competency exists throughout a project's internal and supplier organizations so as to achieve the project's objectives the first time and every time. Product and material certification is evidence that results are consistently being achieved at delivery.

The ultimate project certification is the system certification provided by the chief systems engineer that the solution provided to the customer will perform as expected. This testimonial is based on the summation of the verification history and the resolution of all anomalies. Figure 23.14 is an example certification by a chief systems engineer.

Verification and Anomaly Management

In the management of verification, it is important to keep latent biases removed from the process as much as possible. To achieve maximum objectivity, the verifiers should be independent of both the developers and the verification planners. Figure 23.15 illustrates a candidate organization structure.

Verification Management

The management of verification should be responsive to lessons learned from past experience. A few are offered for consideration:

1. A Requirements Traceability and Verification Matrix (RTVM) should map the top-down decomposition of requirements to their delivering entity and should also identify the integration level and method for the verification. For instance, while it is desirable to verify all requirements in an all-up systems test, there are many requirements that cannot be verified at that level. There may be stowed items at the system level that cannot and will not be deployed until the system is fielded. In these instances, verification of these entities must be provided at a lower level of integration. The RTVM should

FIGURE 23.14. CSE SYSTEM CERTIFICATION EXAMPLE.

Date: __________
I __________ certify that the _______system delivered on __________ will perform as specified. This certification is based on the satisfactory completion of all verification and qualification activities. All anomalies have been resolved to satisfactory conclusion except two that are not repeatable. The two remaining are:
1. ____
2. ____
All associated possible causes have been replaced and regression testing confirms specified performance. If either of these anomalies occurs during the operational mission there will not be any effect on the overall mission performance.

Signed __________
Chief Systems Engineer (CSE)

FIGURE 23.15. ORGANIZATION FOR VERIFICATION.

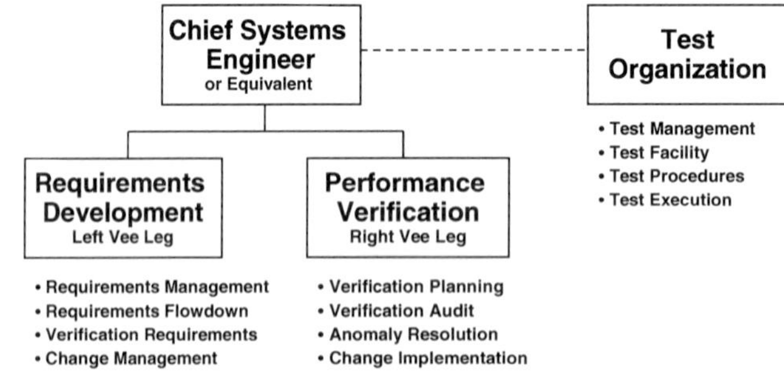

ensure that all required verification is planned for, including the equipment and faculties required to support verification at each level of integration.

2. The measurement units called out in verification procedures should match the units of the test equipment to be used. For example. considerable damage was done when thermal chambers were set to degrees Centigrade when the verification procedure called for degrees Fahrenheit. A perfectly good spacecraft was destroyed when the range safety officer, using the wrong flight path dimensions, destroyed it during ascent thinking it was off course. Unfortunately, there are many other examples that caused perfect systems to be damaged in error.
3. Red-line limits are "do not exceed" conditions, just as the red line on a car's tachometer is designed to protect the car's engine. Test procedures should contain two types of red-line limits. The first should be set at the predicted values so that if they are approached or exceeded the test can be halted and an investigation initiated to determine why the predictions and actuals don't correlate. The second set of red-line limits should be set at the safe limit of capability to prevent system failure or injury. If these limits are approached the test should be terminated and an investigation should determine the proper course of action. One of the world's largest wind tunnels was destroyed when the test procedures that were required to contain red-line limits did not. During system verification, the testers unknowingly violated engineering predictions by 25 times, taking the system to structural failure and total collapse.
4. A Test Readiness Review (TRR) should precede all testing to ensure readiness of personnel and equipment. This review should include all test participants and should dry-run the baselined verification procedure, including all required updates. Equipment used to measure verification performance should be confirmed to be "in calibration," projected through the full test duration including the data analysis period.
5. Formal testing should be witnessed by a "buyer" representative to officially certify and accept the results of the verification. Informal testing should precede formal testing to

FIGURE 23.16. REQUIREMENTS TRACEABILITY AND VERIFICATION MATRIX—BICYCLE EXAMPLE.

Level	Rev	ID	Name	Make or Buy	Requirement		Predecessor	Verification		Auditor	Date
0	0	0.0	Bicycle System	M	0.0.1	"Light Wt" - <105% of Competitor	"User Need" Doc ¶ 1	0.0.1	Assess Competition		
0	0	0.0	Bicycle System	M	0.0.2	"Fast" - Faster than any other bik	"User Need" Doc ¶ 2	0.0.2	Win Tour de France		
1	0	1.1	Bicycle	M	1.1.1	8.0 KG max weight	0.0.1, Marketing	1.1.1	Test (Weigh bike)		
1	0	1.1	Bicycle	M	1.1.2	85 cm high at seat	Racing rules ¶ 3.1	1.1.2	Test (Measure bike)		
1	0	1.1	Bicycle	M	1.1.3	66 cm wheel dia	Racing rules ¶ 4.2	--	*Verif at ass'y level*		
1	0	1.1	Bicycle	M	1.1.4	Carry one 90 KG rider	Racing rules ¶ 2.2	1.1.4	Demonstration		
1	0	1.1	Bicycle	M	1.1.5	Use advanced materials	Corporate strategy ¶ 6a	--	*Verif at ass'y level*		
1	0	1.1	Bicycle	M	1.1.6	Survive FIVE seasons	Corporate strategy ¶ 6b	1.1.6	Accelerated life test		
1	0	1.1	Bicycle	M	1.1.7	Go VERY fast (>130 kpm)	0.0.2	1.1.7	Test against benchmark		
1	0	1.1	Bicycle	M	1.1.8	Paint frame Red, shade 123	Marketing	1.1.8	Inspection		
1	0	1.2	Packaging	B	1.2.1	Packaged for Shipment	0.0.4, Marketing				
1	1	1.2	Packaging	B	1.2.1	Photo of "Hi Tech" Wheel on Box	0.0.4, Marketing				
1	0	1.2	Packaging	B	1.2.2	Survive 2 m drop	Industry std				
1	1	1.3	Documentation	M	1.3.1	Assembly Instructions	0.0.4				
1	1	1.3	Documentation	M	1.3.2	Owner's Manual	0.0.4				
2	0	2.1	Frame Assembly	B	2.1.1	Welded Titanium Tubing	1.1.5, 1.1.6				
2	0	2.1	Frame Assembly	B	2.1.2	Maximum weight 2.5 KG	1.1.1, allocation				
2	0	2.1	Frame Assembly	B	2.1.3	Demo 100 K cycle fatigue life	1.1.6				
2	0	2.1	Frame Assembly	B	2.1.4	Support 2 x 90 KG	1.1.4, 1.1.6				
			•			•					
			•			•					
			•			•					

discover and resolve all anomalies. Formal testing should be a predetermined success based on successful informal testing.

6. To ensure validity of the test results the responsible tester's or quality control's initials should accompany each data entry.
7. All anomalies must be explained including the associated corrective action. Uncorrected anomalies must be explained with the predicted impact to system performance.
8. Unrepeatable failures must be sufficiently characterized to determine if the customer/users can be comfortable with the risk should the anomaly occur following operations.

Anomaly Management

Anomalies are deviations from the expected. They may be failure symptoms or may just be un-thought-of nominal performance. In either case, they must be fully explained and understood. Anomalies that seriously alter system performance or that could cause unsafe conditions should be corrected. Any corrections or changes should be followed by regression testing to confirm that the deficiency has been corrected and that no new anomalies have been introduced.

The management of anomalies should be responsive to the past experience lessons learned. A few are offered for consideration:

1. Extreme care must be exercised to not destroy anomaly evidence during the investigation process. An effective approach is to convene the responsible individuals immediately on detecting an anomaly. The group should reach consensus on the approach to investigate the anomaly without compromising the evidence in the process. The approach should err on the side of care and precaution rather than jumping in with uncontrolled troubleshooting.
2. When there are a number of anomalies to pursue, they should be categorized and prioritized as Show Stopper, Mission Compromised, and Cosmetic. Show stoppers should be addressed first, followed by the less critical issues.
3. Once the anomaly has been characterized, a second review should determine how to best determine the root cause and the near- and long-term corrective actions. Near-term corrective action is designed to fix the system under verification. Long-term corrective action is designed to prevent the anomaly from ever occurring again in any future system.
4. For a one-time serious anomaly that cannot be repeated no matter how many attempts are made, consider the following:
   a. Change all the hardware and software that could have caused the anomaly.
   b. Repeat the testing with the new hardware and software to achieve confidence that the anomaly does not repeat.
   c. Add environmental stress to the testing conditions, such as temperature, vacuum, vibration, and so on.
   d. Characterize the anomaly and determine the mission effect should it recur during any phase of the operation. Meet with the customer to determine the risk tolerance

TABLE 23.1. IV&V ARTIFACTS.

Artifact	Purpose
System Engineering Management Plan	The technical strategy, including the overall approach to integration, verification, qualification, and validation.
Interface and Intraface Specifications	The requirements for entities to properly combine into higher assemblies.
Validation Plan	The approach to in-process and final validation.
Validation Procedures	The step-by-step actions required to accomplish validation.
Verification Plan	The approach to verification including environments imposed.
Verification Procedures	The step-by-step actions required to accomplish the various types of verification.
Verification Data	The raw data produced by verification activity.
Discrepancy Report	The characterization of an anomaly.
Failure Analysis Report	The results of failure analysis and the recommended corrective action.
Qualification Certificate	The summation of evidence and certification that a configuration item has survived a defined set of environmental and operational conditions.
Verification Report	A summation of the verification history and the verification results.
Requirements Traceability and Verification Matrix (RTVM)	A map of verification results against their requirements to ensure completeness and adequacy of verification.

of the using community and whether deployment with the risk, as quantified, is preferred over abandoning the project.

IV&V Artifacts

The integration, verification, and validation process is managed by an integrated set of artifacts. Table 23.1 summarizes the most popular artifacts and their purpose.

Summary

Following is a summation of the important points of this chapter:

Integration should be planned early and should be reflected in the product breakdown structure, the work breakdown structure, and the tactical project network.

Interfaces should be designed to be simple to facilitate integration and to simplify verification of those interfaces.

An "okay to install" discipline should be imposed for all integrations and matings.

At each level of decomposition, determine how the solution users will determine their satisfaction (validation).

Practice in-process validation with both the direct and ultimate users, being aware that, at some point in the decomposition, decisions may be transparent to the user, since they won't be impacted. For instance, the ultimate user probably won't be concerned if slotted or Phillips head screws are used as fasteners. The direct user who must apply the fasteners will care.

When new customers and users emerge, re-baseline their validation expectations.

One of the systems engineer's jobs is to reduce the expectations of the customer and users back to the approved baseline. Customers and users often expect more without adding funds or schedule.

Verification must prove performance of design, design margin, quality, reliability, and many other "-ilities."

Verification incorporates various combinations of testing, demonstration, inspection, and analysis.

Anomalies must be characterized and resolved without destroying the evidence.

For adequate qualification, all life cycle environments must be understood and planned for including the possibility of multiple environmental retests of a unit that has failed several times and has been repaired and retested each time. Certification demonstrates a level of capability and competency attested to by an authority.

Individual and organizational certification is available and expanding.

References

Ambler, S. W., and R. Jefferies. 2002. *Agile modeling: Effective practices for extreme programming and the unified process.* New York: Wiley.

Beck, K. 1999. *Extreme programming explained: Embrace change.* Reading, MA: Addison-Wesley.

Buede, D. M. 2000. *The engineering design of systems.* New York: Wiley.

Carnegie Mellon Software Engineering Institute Capability Maturity Model (CMM) and Capability Maturity Model Integrated (CMMI). www.sei.cmu.edu/cmm/cmms/cmms.integration.html.

Crosby, P. 1992. *Quality Is Free.* Denver: Mentor Books.

Cockburn, A. 2001. *Agile software development.* Reading, MA: Addison-Wesley.

Forsberg, K., H. Mooz, and Cotterman, H. 2001. *Visualizing project management.* New York: Wiley.

Forsberg, K. and H. Mooz. 2001a. Visual explanation of development methods and strategies including the Waterfall, Spiral, Vee, Vee+, and Vee++ models. INCOSE 2001 International Symposium.

Grady, J. O. 1998. *System validation and verification.* Boca Raton, FL: CRC Press.

International Organization for Standardization (ISO). www.iso.ch/iso/en/ISOOnline.openerpage.

Martin, R. C. 2002. *Agile software development: Principles, patterns, and practices.* Upper Saddle River, NJ: Prentice Hall.

Mooz, H., K. Forsberg, and H. Cotterman. 2002. *Communicating project management.* New York: Wiley.

Project Management Institute (PMI). www.pmi.org/info/default.asp.

International Council On Systems Engineering (INCOSE). www.incose.org/.

Walton, M., and W. E. Deming. 1988. *Deming management method.* New York: Perigee

Stevens, R., P. Brook, K. Jackson, and S. Arnold. 1998. *Systems engineering.* Upper Saddle River, NJ: Prentice Hall.

CHAPTER TWENTY-FOUR

MANAGING TECHNOLOGY: INNOVATION, LEARNING, AND MATURITY

Rodney Turner, Anne Keegan

In this chapter we take a slightly wider view of the management of technology than is usually taken, and in two ways. First, rather than viewing "technology" just as the engineering skills an organization uses to do its projects, we are going to focus on all the skills it brings to bear to do its projects. We suggest those skills exist on three levels:

1. The ability of an organization to manage projects in general. This is the organization's general competence or maturity at managing projects, stored in its collective wisdom, in standards, and elsewhere. Turner (1999) has described this as the "projectivity" of an organization, and it is related to the concept of maturity described by Cooke-Davies in his chapter later in this book.
2. The ability of an organization to manage a given project. This skill reflects the organization's ability to recognize the relevant success criteria and appropriate success factors for a given project, and to take its standards and develop a strategy for this project to deliver it successfully (see the chapter by Jamieson and Morris). It also reflects the organization's ability to learn what works and what does not work in given situations, and to manage the risks so that it does not keep on repeating the same mistakes. (See the chapters by Simister and by Ward and Chapman.)
3. The ability of an organization to use its technology ("engineering" skills) to build its assets as efficiently and as effectively as possible, and thereby obtain the best value (whatever that may mean for them).

Second, rather than just considering how an organization uses those skills to repeat previous performance, we focus on how an organization gets better at doing its projects. We describe how an organization learns from its previous experience to get better at what it does, and how it innovates to introduce new ideas to get better still. Surprisingly, project-based organizations often do not provide an environment supportive of innovation. This chapter looks at why that is and what can be done to overcome it. It also considers linear-rational and organic approaches to innovation management.

A Four-Step Process of Innovation and Learning

Turner, Keegan, and Crawford (2003) developed a four-step process of innovation and learning, based on earlier studies from management learning (Miner and Robinson, 1994), adapted for project-based organizations:

1. *Variation.* The organization experiments with new ways of delivering its projects.
2. *Selection.* It chooses those innovations that work and that it wants to adopt.
3. *Retention.* It stores the selected innovations in its collective wisdom.
4. *Distribution.* The organization then needs to distribute those new ideas to new projects and ensure they are used to improve project performance. This step is not needed in functional organizations; it is specific to project-based ways of working.

In a functional organization, only the first three steps are necessary, and all take place within the line organization. The function experiments with new ways of working, selects those it wishes to retain, and stores them in the functional organization where they are immediately accessible to people working in the function. In a project-based organization, variation occurs on one project, which comes to an end. Selection occurs in the post-completion review process, and ideas then need to be retained in a central project management function. But those ideas are not used there, so they need to be distributed to new projects about to start, and different project managers must be encouraged to use them. There are, however, two problems in the project-based context, the issues of deferral and attenuation:

Deferral

We have suggested (Keegan and Turner, 2001) the process can build in a delay at each step. The emphasis is on *post*-completion reviews. Assuming a two-year delay at each step, it can be eight years between a new idea being generated to its being used on a future project. We adopted the idea of viscosity of information. Some information oozes through the organization like treacle, with eight years from idea generation to its becoming widely adopted. To overcome this, people suggest the use of intranet-based technologies. Then information zips through the organization like gas in a vacuum. Yesterday's hearsay becomes today's perceived wisdom. Later in the chapter we discuss how to achieve a balance and discuss the use of internal project reviews to assess the technology used on a given project.

Attenuation

Cooke-Davies (2001) has measured the loss of information at each step. Of the worthwhile new ideas generated, about 70 percent are selected; about 70 percent of those get retained; about 70 percent of those get distributed to the organization; and about 70 percent of those get reused on future projects. The project-based organization gets to reuse about a quarter of the worthwhile new ideas it generates. (We are surprised the numbers are as high as they are.)

Creating an Environment Supportive of Innovation

Innovation is vital for technological development within organizations, developing new engineering and project management process skills to enhance project performance. It occurs at the first of the preceding four steps. There are many things an organization can do to create an environment supportive of variation and innovation (Keegan and Turner, 2003), including:

- Creating channels for formal and informal communication
- Blurring organizational boundaries, using integrating boundary spanners
- Creating flexible roles and multidisciplinary teams
- Allowing some stress and ambiguity
- Facilitating projects rather than rigidly controlling them

Information and Communication

Communication is essential, to ensure the right people have the right information. When establishing a project, the manager needs to ensure that everybody who may have knowledge or information or opinions about the design and specification of the end product is properly consulted, and that those people who need to know are informed. But just as important, he or she does not want to be overwhelmed with conflicting opinions from people who have no real input to make. Further, if everybody on the project is informed about every decision, those people who need to know may ignore the bit of information targeted at them.

Informal channels of communication are also important for innovation. New things happen and new ideas are generated because people who do not normally talk to each other make new contacts, and through those contacts generate the new ideas that underpin technological development (Keegan and Turner, 2003). The Nobel laureates Sir Alexander Fleming and Niels Bohr encouraged creative, informal contacts and communication (Larsson, 2001). Later in this section, when discussing organic approaches to innovation, we cover how to encourage new contact. Having people work in cross-discipline teams is also essential to that.

Management can often be viewed as a series of dilemmas, and information and communication on projects can be no different:

- Communication can be too formal, stifling cross-functional working, innovation, and people's ability to perform (see Figure 24.1).
- Or communication can be too informal, with too many conflicting opinions, too many siren voices, and too much hype about the possibilities for the project.
- The project team can talk to too many people, again with too many conflicting opinions and informing too many people about progress so that nobody bothers to listen.
- Or they can talk to too few, behaving in a cloak-and-dagger fashion, so that nobody knows what they are doing and nobody cares.

Information and communication on projects needs to be carefully managed to achieve the desired innovations.

Cross-Functional Working with Boundary Spanners

Innovation requires cross-functional working. It is about creating change, and change involves people working together in novel ways. We have just seen that people working across functions can lead to new ideas through new contacts. But it also needs people to be empowered to make progress together; a project must be a rugby scrum, not a relay race. The

FIGURE 24.1. EXCESSIVELY FORMAL COMMUNICATION.

Rodney Turner wrote a case study on an IS/IT project in the UK's Civilian Aviation Authority. One of the people interviewed had just joined from the private sector. He said he was working on a project involving people from other departments, but found it impossible to make progress. If he needed to write a memo to somebody in another department to make a decision, he had to write a draft of the memo to give to his boss to critique. He would then revise the memo, and it would go to his boss's boss to critique, and so on until it got to the lowest common boss, when it would be sent to the relevant person. The reply would come back the same way, and the whole process would take weeks. He said you cannot make progress on a cross-functional project working in this way.

You can understand why managers are doing this. If a plane crashes, the media will be on a witch hunt, and managers want their stamp on decisions they will be held accountable for. But in the temporary organization that is a project, you have to empower people to make decisions to make progress (see the chapter by Huemann, Turner, and Keegan).

The next person spoken to was a Royal Air Force Officer on secondment to the CAA. Asked if it was true, he said he was afraid it was. What he did was send the first draft of the memo to the person he was trying to communicate with and they got on with it, while the official memo did the rounds. You can see the military is used to empowering people. In the heat of battle you cannot refer decisions up the line; there isn't time. Projects are like the heat of battle; you need to empower people (within firm guidelines).

communication methods described in Figure 24.1 result in projects being artificially extended, as they go from one department to another. People must work in flexible, cross-discipline teams, with decentralized authority. People can be given roles to act as boundary spanners—people whose role is to bridge gaps between people and get them working together as a team. Figure 24.2 (adapted from Fong, 2003) illustrates the need, bridging boundaries between disciplines, hierarchies, and areas of expertise to achieve new thinking and innovation.

Stress and Ambiguity

Surprisingly, or perhaps not, some stress and ambiguity can lead to better innovation. It is well known that people work better under reasonable levels of stress than they do under no stress at all. But this is another dilemma. Too much stress can be deleterious. Likewise, ambiguity can encourage innovation. It is recognized by psychologists that it is not risk that people fear, but ambiguity and the chance of loss (Bernstein, 1998). People will respond to risk if they see it as a chance of gain. But if it is unclear whether the risk will result in loss of gain, they fear that. Thus, where ambiguity exists, people will try to eliminate it, leading

FIGURE 24.2. BOUNDARY SPANNERS.

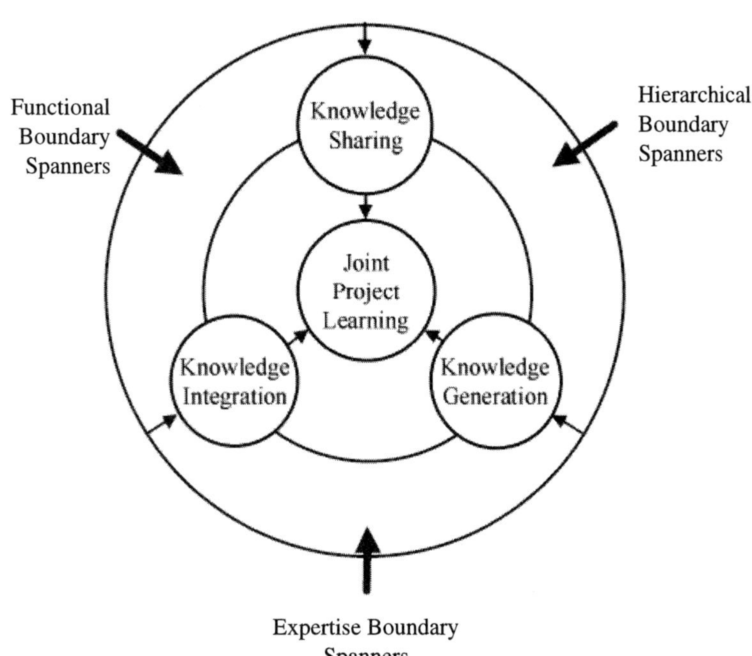

to innovative solutions to their problems. Reasonable levels of stress and ambiguity should be encouraged on projects to find the most innovative solutions.

Facilitation vs. Rigid Control

Projects involve stress and ambiguity, they involve cross-discipline teams with boundary spanners, and they involve communication among team members. They should therefore encourage innovation, right? Unfortunately not. The project environment often kills innovation through too-rigid control. On projects we can do all the other things but still kill innovation. We consider below the need to achieve a balance between the rigid linear-rational approach usually adopted on projects and the organic approach sometimes adopted. But let us first consider some of the problems of too-rigid control.

Competency Traps

Projects are more exposed to competency traps than routine operations. A competency trap is where there is an established way of working, which may not be optimal (Levitt and March, 1995). But people fear trying an alternative for risk of failure. In a routine environment it is easier to try something new. If it does not work on the first attempt, it may be improved next time, and if it doesn't work at all, the old way can be reinstated. It is easier to experiment in a routine environment. Projects (in their pure form) are only done once. If they do not work the first time, they do not work at all. There may be a way of doing a project that is twice as good as the preferred way, but with, say, a 20 percent chance of failure. So if it were done several times, on average it would be 60 percent better. In a routine environment, people can experiment, find the flaws, and get it right second time around. But on a project, if it is only done once, people prefer the certain, though less efficient, way. They are trapped in the inferior way of working.

Rigid Evaluation Criteria

Project management has developed evaluation criteria for assessing the value of projects and their contribution to corporate wealth, including techniques such as net present value (NPV) and internal rate of return (IRR) (Turner, 1995; Lock, 2000). However, these do not properly evaluate IS/IT and innovation projects (Akalu, 2003). For such projects a technique known as option pricing gives a better view, but unfortunately is more difficult to apply. Akalu (2003) showed that most organizations are reduced to applying qualitative assessment criteria to innovation projects and IS/IT projects. This can create problems in firms wanting to compare the IRR of all of their projects and applying strict hurdle rates that are not appropriate to all the projects they do. Artto and Dietrich, Archer and Ghasemzadeh, and Thiry describe benefits management and the evaluation of projects by linking their outcomes to business objectives in their chapters in this book.

Rigid Resource Utilization

Standard project management techniques also suggest tight assignment of resources, allocating the precise number required to do the job. This is not always appropriate for some projects. Projects are risky, and some flexibility is required to deal with uncertainty. But it is especially inappropriate for innovation projects. People with time to think develop much more innovative, creative solutions. Innovation projects require creativity, coupled with high uncertainty. That is stress enough, without adding additional stress by making the project team work to tight resource limits.

Rigid Control

Traditional project management also suggests tight control (see the chapters on control by Harpum and on changes by Cooper and Reichelt). On innovation projects this may be appropriate at the development stage, but not at the research stage. Innovation projects should still be managed, but at the research stage more organic approaches are appropriate, emphasizing facilitation and coordination of the people working on the project. At the development stage there can be more rigid deadlines—for instance, the new product needs to be delivered to market by a certain date, or the Web space needs to be online by a certain date. Thus, the appropriate form of control, organic or linear rational, depends very much on the type of project and the stage it is at.

Rigid Contract Management

Traditional contract management procedures can also be a block to innovation (see Figure 24.3). Innovative contracting techniques, such as partnering and alliancing (Scott, 2001) and appropriate sharing of risk on conventional contracts (Turner, 2003) are necessary to achieve

FIGURE 24.3. RIGID CONTRACT MANAGEMENT DISCOURAGING INNOVATION.

A client wanted a vessel to be shot-blasted and painted. They drew up a specification of how the job should be done, and asked potential contractors to bid competitively to do the job on a fixed-price basis. A contractor won the bid process, but as they were about to sign the contract, the contractor said they could do the job for half the price, but they would only tell the client how under two conditions:

- The client would not reopen the bid process.
- The contractor would earn the same absolute profit, not the same percentage profit.

You can see the contractor's concern. The client was so stuck in rigid contract management procedures that they could not award the job without compulsory competitive tendering, and they only wanted their contractors to make a certain profit margin; the client wouldn't share their increased profit with the contractor.

In this case the client agreed. They got the job done for 55% of what it would have cost them, and the contractor made their same absolute profit and a higher percentage profit.

innovation on projects. (See the chapters by Venkatarman, Langford and Murray, Lowe, and others in this book.)

Achieving Innovation in Project-Based Firms

In this section we consider how innovation can be achieved in a project context.

Encouraging Variation

"If you always do it the way you have always done it, you will always get what you have always got."

SIR MICHAEL LATHAM.

To improve the performance of their projects, organizations need to innovate and try things new. So how do people do that? The organic approaches we describe in the following help encourage the creation of new ideas. Other techniques that our research into innovation and learning have identified include the following.

Obtaining Senior Management Support

First we mention the importance of senior management support. This was identified as a key part of improving the management of information services (IS) projects in the Research and Development Department in SmithKline Beecham by Gibson and Pfautz (1999). Without senior management support, junior people will either fear making changes or not take the initiative.

1. A manager in IBM suggested junior people may avoid making honest reports in project reviews for fear of upsetting middle managers. Organizations must learn not to shoot the messenger, and the support of senior management helps junior people to make honest reports. The nature of the organization also has an impact here. If the organization has a blame culture, nobody will give honest reports, either for fear of attracting blame to themselves or through fear of damaging their immediate colleagues, particularly their immediate superior. A learning organization, on the other hand, will welcome honest reviews and treat them as opportunities to improve, rather than a basis for witch-hunts.
2. In many organizations in the construction industry, including government procurement departments, it is the "jobsworths" at junior levels who block the adoption of new contracting practices that would lead to improved effectiveness as suggested by Latham (1994). They fear that if they try something new and it goes wrong, they will get the blame. "The risk of that going wrong is more than my job's worth." In reality, this fear is often imagined and is an excuse for making their lives comfortable by doing what they have always done (and getting what they have always got). More than senior management support is needed here. Junior managers need to recognize it's more than their job's worth not to adopt new practices.

Involving Construction and Operations People at the Design Stage

We saw previously the importance to innovation of multidisciplinary teams and cross-functional working. On a project, one way of achieving that is to involve construction and maintenance people in the design process. Without being stimulated to think new thoughts, design people may apply their traditional ways of thinking. Figure 24.4 contains examples of obtaining improved value by bringing other experiences into the design review process. Thiry further describes in his chapter on value management how value can be improved by involving a broad range or project participants in value management workshops during the design stage.

Researching New Techniques at the Feasibility Stage

The project manager from the main contractor on one of the early alliance contracts in the Netherlands told us that he extensively researched alliance contracts as early as the bid stage. He assigned members of the potential project team to research different elements of alliance contracts. That helped the firm to learn a new approach to both contract management and in the area of project team building.

Managing Innovation Projects

There are two opposing approaches to managing innovation projects:

- *Linear-rational approach.* Emphasizes rigid evaluation criteria, rigid resource utilization, rigid control, and the following of a strict process
- *Organic approach.* Emphasizes more fluid and flexible approaches.

The linear-rational approach is about keeping to the straight and narrow, moving as briskly as possible to the final objectives. It suggests tight, rigid control. On the other hand, the

FIGURE 24.4. INVOLVING MAINTENANCE IN BUILDING DESIGN.

The British Airports' Authority, BAA, was planning an extension to one of its airports, and held workshops with its maintenance people during the design stage. In a building project, operation and maintenance costs are five times construction costs. Considerable improvements in value can be obtained by reducing operation and maintenance costs.

A window cleaner pointed out that if the windows were made of frosted glass they wouldn't need cleaning so often. And another member of maintenance staff pointed out that if gray grout was used in the bathrooms, the floors and walls would be easier to clean.

Both suggestion were adopted, to give considerable saving in maintenance costs at no change in capital cost organic approach is more pagan, following the seasons through cycles of development to the end objectives. It may be less efficient, but in the right circumstances is far more effective. Which is appropriate for a given innovation project depends on the nature of the project and the stage of development the product is at. If the project is in the research phase, organic approaches may be more appropriate, as it is here the new ideas must be generated. If it is in the final product development stage, more rigid control may be appropriate, as time-to-market is now significant (Turner, 1999).

The Linear-Rational Approach

The linear-rational approach is usually based on a version of the life cycle, and on toll-gates or stage-gates. Projects move through a series of go/no-go decisions, being evaluated against strict criteria at each stage-gate before being allowed to proceed. Wheelwright and Clark (1992) suggested a funnel as a metaphor, indicating that the surviving projects become fewer and fewer at each stage-gate. (See the chapters by Archer and Ghasemzadeh, by Milosevic, and by Cooper, Aouad, Lee, and Wu.) The PRINCE2 process, shown in Figure 24.5 (OGC, 2002), is in essence a Stage-Gate model, where the number of stage-gates can be varied to meet the needs of the project. At the completion of each stage, the project is evaluated against business and project control criteria.

Table 24.1 is the Stage-Gate model used by a financial services supplier to assess data products they supply over their own network and over the Web. They suggest that at each stage-gate, half the surviving projects fall. Thus, 64 product proposals need to start for one to be released to market. Even at the last stage-gate, one out of two products falls, two have

FIGURE 24.5. THE PRINCE2 PROCESS.

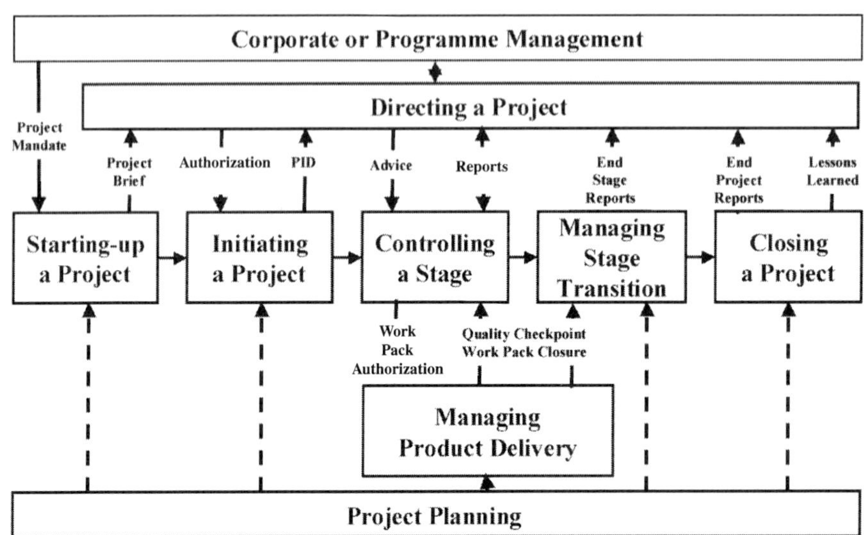

TABLE 24.1. THE STAGE-GATE MODEL FOR A FINANCIAL SERVICES SUPPLIER.

Stage	Time	Action
Identify market opportunity	One week	Sales prepares a brief description of the customer's requirement.
Initial product description	One month	Marketing appoints a product manager who develops a product description and initial project plan.
Project portfolio committee	Three months	An the resource requirements are quantified and the product portfolio committee prioritizes products for development-based resource availability.
Requirements definition		Functional and system requirements are defined.
Assess risks, and reconcile project success criteria		The project is planned in detail, a formal risk analysis is conducted, and success criteria and factors are evaluated.
Approve release		The product is ready for release to market. Likely maintenance costs and profitability are assessed.

made it all the way through the product development process, and only one is released to market.

At an innovation conference in Ireland a few years ago a speaker described the 63 projects that do not make it as "failures." They are not failures. To be an innovative organization, it is necessary to try out many things in order to have the one successful one. The challenge is to shut projects down as quickly as possible. Table 24.1 illustrates that seven-eighths of the projects should be shut down within three months, the time of the third stage-gate, and that is the role of the project portfolio committee (see the chapters by Archer and Ghasemzadeh, and by Arttos and Dietrich). The same speaker said that in his company, out of every four products released to market, two made a loss and only one made a profit. The financial services agency tried to stop that with their last stage-gate, so every product released at least breaks even.

Table 24.1 and Figure 24.5 illustrate that project reviews are held at the completion of each stage. Thus, they may be held at the completion of initiation, feasibility, and design. One purpose of project reviews is to ensure the right technology is being used in the form of the project management process and appropriate engineering skills, as well as checking that the project meets required business and operational criteria. This is discussed again at the end of the chapter. (See the chapters by Thiry on value management and by Huemann on quality reviews.)

The advantages of the linear-rational approach are as follows:

- It provides clear no-go decisions, encouraging the development of a business plan and allowing the project to be matched to company strategy.
- It allows ideas to be tried and tested; closure is not seen as a failure.

- It provides clear, strict control, through the stage-gates and through milestone planning.
- It helps manage risk; ideas are tested before progressing, and the stage-gates are clear review points.
- It creates a system that all employees are familiar with and provides a systematic methodology for evaluation across disciplines or functional areas.

The disadvantages of the linear-rational approach are as follows:

- It favors efficiency and control over creativity and effectiveness
- It can artificially extend projects if you insist that one stage-gate is passed before work on the next begins.

The Organic Approach

The organic approach overcomes these weaknesses, but at the expense of efficiency and control. It is still possible to provide vision and direction for the project, and coordinate the input of resources. But the organic approaches favor more flexible management. Through our research we have identified the following organic approaches to innovation on projects.

Deliberate Redundancy. Advertising agencies, when starting work on a new account, create three or four teams to work independently to develop ideas. They then sample those ideas, to come up with an overall proposal for the client. We wonder how many project-based organizations would consider having four teams work independently in the feasibility stage to think of different ideas. (BAES does.) Many high-tech companies often have two people do a job where strictly one will do. The advantage is two people working together come up with more creative solutions to the client's need, and when the job is over, two people have the new skill, improving learning in the organization. We call this the creation of Nellies. In traditional companies, the new recruit learns the work of the company by serving an apprenticeship, sitting next to an experienced person, "Nellie." There are no Nellies in high-tech companies because the technology is changing so fast, so firms create Nellies in the way described. Some people ask how can they afford this inefficiency. But if they obtain more creative solutions, with better learning for the organization, then the cost is repaid. Unisys, Intel, and Hewlett-Packard have all told us they adopt this approach. For innovation on their projects, organizations should consider having several people or teams work independently during the early stages and then choose the best solution. Those people can deliver more creative solutions and communicate those solutions to the rest of the organization. People might say they cannot afford the cost. Well, they need to compare the value of the creative solution to the additional resource input, and sometimes (not always) this approach provides better value.

Sampling. Having generated several different solutions to the problem, the best solution needs to be selected. That is what advertising agencies do. But the best solution may not be one of the proposals, but a mixture of them all. So the sampling needs to be more a blending process, where the best solution emerges as a mixture of the proposals. In the financial services company mentioned previously, the one product that emerged at the end was not one of the 64 that started but a mixture of them. This approach led to a tension between the product development personnel and senior management. Senior management wanted to shut half the projects down at each stage-gate, but the product development personnel wanted to keep them alive, as they could not know which ones would contribute to the final product. Somebody described it as being like blending whisky. Sampling from the different casks, they could not know they had the right blend, and what it would comprise, until they had it. The problem here is that the decision to keep projects going on the off chance they may be needed or that unanticipated synergies will emerge can be a cost-prohibitive move. The main reason for review gates within most organizations is cost control. But sometimes (not often) this approach can lead to better value solutions. Advertising agencies have a strictly limited time for the parallel working, and the financial services agency relied heavily on the intuition of the product development experts, and usually they were reliable.

Chance Encounters. We said previously that innovation comes through people making contacts that have not existed previously. New ideas come from old ideas reforming in new ways. An Irish advertising agency decided to try to reduce office rent and commuting costs for their employees by having them work from home. All their creative people were given laptops and an ISDN link at home. Creativity plummeted! Sir Alexander Fleming, who discovered penicillin, deliberately left petri dishes lying around to see if something unexpected happened, and it did (Larsson, 2001). We spoke above about the need for cross-discipline teams and boundary spanners. For innovation on projects, do not let the design people work in isolation.

Creative Communications. Chance encounters lead to creative communications and vice versa. The Danish hearing aid company Oticon encourages people to chat at the water fountain. At Henley Management College, morning and afternoon tea are something of a ritual, but creative communication occurs at them.

Creative Tensions. Difference between people, rather than being avoided, should be encouraged, as they too can lead to new ideas by reforming old ones in new ways. In advertising agencies, the tension between the suits and non-suits (businesspeople and creative people) is encouraged. On projects, rather than avoiding differences between engineers and marketing people, it should be encouraged to find the best solution (Graham, 2003).

The linear-rational and organic approaches are quite different, but not entirely incompatible. Clearly, deliberate redundancy is incompatible with efficiency, but not with effectiveness. Further, it is possible to adopt organic and linear-rational approaches at different stages of the life cycle. At the early stages organic approaches are best. Here the new ideas need to be generated, so cross-functional working with deliberate redundancy will be used through the first and second stage-gates in Table 24.1. Then ideas can be sampled at the second stage-gate, and then more strict management processes applied from then on, when costs begin to increase and time-to-market becomes significant (Turner, 1999).

Viewing Uncertainty as an Opportunity. Some project managers, in a desire to achieve certainty and strict control, try to squeeze risk and uncertainty out of their projects as quickly as possible. They follow Path A in Figure 24.6. However, in the process they lock themselves into high-cost solutions at an early stage. It has been suggested by Latham (1994) and Egan (1998) that in construction following Path B can lead to a 30 percent reduction in cost, by allowing more innovative, cheaper solutions to be found. It has even been suggested that following Path C can lead to a further 30 percent reduction in cost.

The problem with Path B, and more so with Path C, is that they are not compatible with predictability and certainty. They are not compatible with conventional project management thinking, which likes rigid control, following the straight and narrow, linear-rational approach. They are not compatible with normal relations between clients and contractors, based on compulsory competitive tendering for fixed-priced contracts and confrontational relationships.

However, there is now a growing body of case study evidence that Paths B and C can achieve what is promised of them, allowing options to be explored for a longer period and innovative, cheaper solutions to be found. The application of strategies B and C is based on at least two requirements, though:

1. The use of strict configuration management (Turner, 1999, 2002; see also the chapter by Kidd and Burgess) to manage the reduction in uncertainty and track the various options as they are explored, merged, or discarded
2. The use of modern contracting techniques that encourage collaborative working between clients and contractors, who view the project as an opportunity to work together toward

FIGURE 24.6. STRATEGIES FOR RISK REDUCTION.

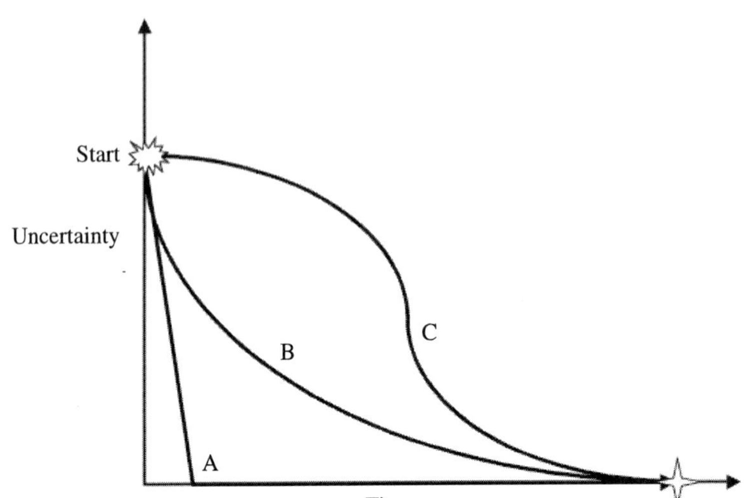

a common objective, with appropriate sharing of risk (Turner, 2003; Scott, 2001; see also the chapter by Langford and Murray).

Retaining and Using Technological Developments

Innovations occur at the first of the four steps we identified earlier in the in the section on variation. Having made new technological developments, the organization must decide which are worthwhile for further use (selection), record them so that people can draw on the knowledge (retention), and pass the knowledge on to people working on other projects so that they can use them (distribution). The organization must also ensure that the right technological solutions are being used on a given project. We describe here practices adopted for the selection, retention, and distribution of technological developments at the three levels described previously, and for checking the technological solutions on a given project. We have observed the use of four practices:

1. Systems and procedures
2. Project reviews
3. Benchmarking
4. Project management communities

We describe these four practices and then show how they support learning and maturity.

The Four Practices

Systems and Procedures. This is where the organization formally stores its technological knowledge, in "written" systems, procedures, and standards. They can take many forms:

- Procedures manuals
- Engineering standards
- Computer-based project management information systems
- Virtual project office in the intranet

An organization's competence can be described as its collective knowledge or wisdom. The systems and procedures are the concrete evidence of its collective wisdom. Systems, procedures, and standards are a key way organizations capture knowledge and experience. They are the collective representation of the firm's experiences.

The procedures and standards should be treated as flexible guidelines, tailored to the needs of each project. Every project is different, and so requires a unique procedure. Standard procedures represent captured experience and best practice, but they must be tailored project by project. Hopefully the tailoring is marginal, but it must done. It is part of a project manager's tacit knowledge that enables the manager to know how the procedures need to be tailored to individual projects. People who have the lack of maturity that makes them want to follow procedures to the letter are not yet ready to practice as project managers. The United Kingdom's Office of Government Commerce in its maturity model overtly states that part of maturity level 3 (of 5) is the ability to tailor procedures. A main contractor from the engineering construction industry reported that new project personnel are told to follow procedures strictly on their first project (when they are in a support role—sitting next to Nellie). On subsequent projects, they can reduce the amount they refer to the standards, as they internalize the firm's good practice. They are encouraged to adapt the procedures to individual projects as their experience grows.

Ericsson requires that its PROPS process should be used on all projects, although it is not mandatory. PROPS is designed to be tailored to the needs of individual projects. It represents good practice in Ericsson, but that good practice is flexible enough to be adapted to the size and type of project. PROPS is also continually updated to reflect new experiences, and the changing technology and nature of projects. The same is true for the PRINCE2 process produced by the United Kingdom's Office of Government Commerce (OCG, 2002). PRINCE2 certification is becoming mandatory to bid for many projects in both the public and private sector in the United Kingdom. In this way the government is contributing to the increasing competence of public sector projects, and to the increasing project management competence of the society. Organizations that have not captured their experience in project procedures are able to use industry-standard procedures, such as PRINCE2, ISO 10,006 (ISO, 1997), the PMI Guide to the Body of Knowledge (PMI, 2000), or other bodies of knowledge.

The emphasis on procedures, both as a learning medium and as a measure of maturity, does tend to emphasize process over outcome and intent (Levitt and March, 1995). However, both process and outcome should be emphasized on projects, and an emphasis on one is not mutually exclusive with an emphasis on the other (ISO, 1997). Project managers need to learn to emphasize both. The emphasis on procedure can lead to redundancy of experience and competency traps. However, the need to develop project-specific procedures for each project helps to ensure new processes are developed and tried. This encourages variation, although many project-based organizations do tend to be very conservative (Keegan and Turner, 2003).

Reviews. Reviews can fulfil two purposes:

1. They can be conducted internally throughout the project, to check that the project's requirements are properly defined (see the chapter by Davis, Hickey, and Zweig) and the right technologies (project management process and engineering skill) have been selected for the project
2. They can be conducted at the end of the project, so that the organization can learn how well it did and capture its success and learn from its failures.

Huemann describes both types of reviews in her chapter on quality reviews, though she reserves the word "review" for the first case and uses the word "audit" for the second.

Internal reviews may be conducted at the completion of project initiation, feasibility, design, and other project stage transitions. They were described previously in the discussion on project stage-gates.

Post completion reviews, or audits, play a vital part in capturing experience. PRINCE2 and ISO 10,006 suggest a review be conducted at the end of every project, and company procedures updated to reflect that learning. Pinto (1999) reported that one contributing factor for the failure of many IS projects was a failure of the organization to review its performance on previous projects and learn from experience. People working on failing projects are met with a strong sense of déjà vu: "We have been here before and can see we are locked on a path to failure." Many project-based organizations continually benchmark their procedures and processes, gathering data about project performance, storing that as historical data to help plan future projects, and thereby improving overall project performance.

However, many firms report less than satisfactory use of project audits. They find the practice difficult to enforce, and where it is enforced, it is a meaningless box-ticking exercise. An ICT contractor reported that post-completion reviews were an essential part of their quality assurance procedures, but there was no check on the quality of the outputs. Further, where reviews are conducted, it can be difficult to transmit the learning to the organization, because of the problems of deferral and attenuation identified earlier.

Benchmarking

The organization compares its performance on projects to projects elsewhere (Gareis and Huemann, 2003). It may compare its performance with the following:

- Earlier projects it undertook, to track performance improvement
- Projects undertaken by other parts of the same organization
- Projects undertaken by other organizations if it can access the data

Benchmarking is also essential to increasing project management performance, but is not something that is well done by many organizations. The European Construction Institute and the American Construction Industry Institute are benchmarking projects in the engineering industry in the two continents. There are also many benchmarking communities in Europe, the Far East, and Australasia.

Project Management Communities

The fourth learning practice adopted by many organizations is the maintenance of a project management community. The importance of the project management community is mentioned by many authors, for example, Gibson and Pfautz (1999) and Pinto (1999). The last step of the innovation and learning cycle is distribution, and project management communities help achieve that. Specific practices used by organizations through their project management communities to distribute innovations and technological knowledge are as follows:

- Regular (quarterly) seminars and conferences
- Mentoring of project management professionals
- Career committees, and support for individual competence and career development

- Overseas postings
- Centers of excellence
- The use of the intranet

Seminars and conferences: Many organizations, especially from high-technology industries, have regular, quarterly meetings of project management professionals, where they can network and share experiences. These can range form informal to extremely formal. The Dutch bank ABN-Amro has a quarterly meeting of its project managers from its information services (IS) department. This lasts a couple of hours, during which they have one or two lectures, followed by a *borrel.* In the Dutch army, the meeting lasts all day. Other organizations have a more formal conference one to four times a year.

As Huemann, Turner, and Keegan show in their chapter on human resource management (HRM) that most industries maintain industry-wide communities. These may be in the form of professional associations for individuals, such as the Project Management Institute or International Project Management Association, or professional institutes for companies, such as the European Construction Industry and Construction Industry Institute. Such associations hold seminars and conferences and provide other networking opportunities in the industry, rather than in individual companies.

Mentoring: Pinto (1999) reports that another contributing factor to the failure of IS projects in many organizations is a failure to mentor new project managers. They do not serve an apprenticeship; they do not spend time "sitting next to Nellie."

Career committees: These are described by Huemann, Turner, and Keegan in their chapter as an essential HRM practice to manage the learning and development of individual project managers, which in itself is critical to increasing performance in the organization.

Overseas postings: Moving people around the organization, through the spiral staircase career, is a way of spreading technological competence and learning throughout the organization. People take their learning with them and pick up new learning as well.

Centers of Excellence: Many project-based firms maintain centers of excellence for retaining learning and disseminating it throughout the company. They may maintain the company procedures and offer consultancy advice and training within the firm. The Office of Government Commerce is the UK government's Centre of Excellence in project management, maintaining the PRINCE2 process. It is also establishing satellite offices in all government departments. (The role of the PM office, which fulfills this role, is described in the chapters by Young and Powell, and by Gareis.)

The Intranet: Many firms are also using the Intranet to support organizational learning. However, experience is patchy, and the main risk is totally inviscid information. Yesterday's hearsay becomes entered in the system without being tested and proven, and becomes today's perceived wisdom. Some companies suggest the use of gatekeepers to monitor the entering of information, but then cannot afford the cost.

The Four Practices and Four Steps of Innovation and Learning

Table 24.2 shows how the four practices contribute to the development and distribution of technological competence throughout the organization through variation, selection, retention, and distribution. The four practices can also be related to organizational maturity and learning in organizations.

The Four Practices and Project Management Maturity

The purpose of innovating—of developing new technologies for project delivery, new project management processes, and engineering skills—is to improve project performance. The organization aims to get better at doing its projects, to increase its competence at project delivery. The jury is still out on what the project management competence of organizations should be called and how it should be measured. In the late 1980s, Turner labeled it *projectivity* (Turner, 1999). More recently it has been called *capability* and *maturity* (Fotis, 2002).

Cooke-Davies offers a challenging review of the application of the maturity concept to project management in his chapter later in this book but if we take the five levels of the Organizational Project Management Maturity Model, OPM3, we can see how the four practices contribute to increasing maturity, so defined.

Level 1—Initial. There is no guidance or consistency in the organization's approach to project management.

Level 2—Repeatable. The organization begins to pick off individual project management processes (scope, quality, cost, time, risk) and defines how those should be managed. It begins to write *procedures* for individual processes, the most often used, and begins to give minimum guidance to its project managers on how to use those through the embryonic *project management community*.

TABLE 24.2. THE ROLE OF THE FOUR PRACTICES IN INNOVATION AND LEARNING.

Learning Process	Contributing Themes
Variation	Project management communities Benchmarking
Selection	Benchmarking Reviews
Retention	Reviews Systems and procedures
Distribution	Systems and procedures Project management communities

TABLE 24.3. ORGANIZATIONAL PROJECT MANAGEMENT MATURITY MODEL, OPM3.

No.	Level	Theme	Attainment
1:	Initial	Procedures	Ad hoc processes
		Review	
		Benchmarking	
		Community	No guidance, no consistency
2:	Repeatable	Procedures	Individual processes for the most often used
		Review	
		Benchmarking	
		Community	Minimum guidance
3:	Defined	Procedures	Institutionalized processes across the board
		Review	
		Benchmarking	
		Community	Group support
4:	Managed	Procedures	Processes measured
		Review	Experiences collected
		Benchmarking	Metrics collected
		Community	
5:	Optimized	Procedures	Continuous improvement
		Review	Defects analyzed and patched
		Benchmarking	Data collected
		Community	Continuous improvement

Level 3—Defined. The organization begins to formalize the individual processes into a coherent, integrated *project management procedures.* It offers group support, through the *project management community*, mentoring apprentice project managers in the use of the company procedures

Level 4—Managed. *Review* and *benchmarking* become formalized, and the systems and *procedures* are measured as a basis for *benchmarking* and performance improvement.

Level 5—Optimized. The organization moves into continuous improvement. Data is collected, and defects are analyzed and patched to achieve that continuous improvement. *Procedures, reviews, benchmarking, and the project management community* are practiced to achieve *variation, selection, retention, and distribution* of new technological knowledge so the organization moves into a permanent state of innovation and performance improvement:

1. Improvement in the performance of its project management systems and procedures
2. Improvement in the performance individual projects
3. Improvement in the performance of its technological and engineering skills and in the efficiency and effectiveness of its products

The Four Practices and Organizational Learning

To increase its project management competence, in order to better use its technology, the organization needs to learn how to better use its project management processes and engineering skills. Learning is considered formally by Bredillet and by Morris in their chapters in this book. However, we wish to show here that the four practices described previously for selecting, retaining, and distributing technological knowledge do contribute to organizational learning using a model developed by Nonaka and Takeuchi (1995), two authors whose work is described more fully in the chapters on learning by Bredillet and by Morris. Figure 24.7 illustrates the organizational learning spiral postulated by Nonaka and Takeuchi (1995). It shows an organization and the people in it learning by cycling between explicit and tacit knowledge:

Explicit knowledge—Codified knowledge as reflected in the technological systems, project management procedures, and engineering standards used by the organization.

Tacit knowledge—Inherent knowledge reflected in the combined wisdom of the project management community.

Nonaka and Takeuchi suggest that organizations move clockwise through this cycle to improve explicit and tacit knowledge and so enhance organizational learning. We can see as

FIGURE 24.7. NONAKA AND TAKEUCHI'S LEARNING CYCLE.

		To	
		Tacit knowledge	Explicit knowledge
From	Tacit knowledge	*Socialization* Sharing-creating tacit knowledge through experience	*Externalization* Articulating tacit knowledge through reflection
	Explicit knowledge	*Internalization* Learning and acquiring new tacit knowledge in practice	*Combination* Systematizing explicit knowledge and information

an organization moves through this cycle, it follows the four-step process of variation, selection, retention, and distribution, using the four practices of standards, reviews, benchmarking, and community. This is best described starting at the selection (review) step, socialization:

Socialization. The *project management community* consolidates its tacit knowledge through reflection and *review*. It *selects* the tacit knowledge considered valuable for further use.

Externalization. Through further reflection, itarticulates that tacit knowledge and converts it into explicit knowledge. It decides what should be *retained* in its systems and *procedures*. It compares how it is doing by *benchmarking* its performance internally and externally.

Combination. Itsystematizes that explicit knowledge into systems and *procedures*, *retaining* it for further use. It can now be *distributed* to the organization through the *project management community*.

Internalization. The *project management community* can now use the explicit knowledge, and through use convert it into tacit knowledge. It can also try new ideas through a process of *variation* and thereby acquire new tacit knowledge.

Returning to socialization thus we see that the four practices suggested for selecting, retaining, and distributing technological knowledge contribute to increasing project management competence of organizations through a process of learning.

Summary

Organizations achieve superior project performance through the effective use of the technological knowledge available to them. However, they can either try to repeat past performance or they can try to improve their performance through the development and use of new technological knowledge. This chapter looked at how organizations can do that.

First the scope of technological knowledge was widened. Technological knowledge includes engineering skills, but also includes an organization's ability to manage projects—that is, its overall competence at project management, as well as its ability to manage specific projects through its ability to identify appropriate success criteria and key performance indicators, and to identify and manage risk effectively. A four-step process was introduced for the development and use of new technological knowledge: variation, selection, retention, and distribution. It was first shown what organizations can do to encourage innovation through variation and manage innovation and development projects effectively. Four practices were then introduced for the selection, retention and distribution of technological skills in project-based organizations. These practices are as follows:

- The use of systems, procedures, and standards
- Internal and cost completion reviews on projects
- Benchmarking project performance internally and externally
- The maintenance of project management communities

It was shown how these four practices contribute to increasing project management competence and maturity by supporting organizational learning.

References

Bernstein, P. L. 1998. *Against the gods: The remarkable story of risk.* New York: Wiley. ISBN: 0-471-29563-9.

Cooke-Davies, T. 2001. Project close-out management: More than just "good-bye" and move on. In *A project management odyssey: Proceedings of PMI Europe 2001,* ed. D. Hilson and T. M. Williams. London: Project Management Institute, UK Chapter.

Crawford, L. 2003. Assessing and developing the project management competence of individuals. In *People in project management,* ed J. R. Turner. Aldershot, UK: Gower. ISBN: 0-566-08530-5.

Egan, J. 1998. *Rethinking construction.* Construction Task Force report, London.

Fong, P. S. W. 2003. Knowledge creation in multidisciplinary project teams: An empirical study of the processes and their dynamic interrelationships. *International Journal of Project Management* 21(6). To appear Fotis, R. 2002. Maturity. *PM Network* 16(9):39–43.

Gareis, R., and M. Huemann. 2003. Project management competences in the project-oriented company. In *People in project management,* ed. J. R. Turner. Aldershot, UK: Gower.

Gibson, L. R., and S. Pfautz. 1999. Re-engineering IT project management in an R&D organization—a case study. In *Managing business by projects: Proceedings of the NORDNET Symposium,* ed K. A. Arrto, K. Kähkönen, and K. Koskinnen. Helsinki: Helsinki University of Technology.

Graham, R. G. 2000. Managing conflict, persuasion, and negotiation. In *People in project management,* ed. J. R. Turner. Aldershot, UK: Gower.

ISO. 1997. *ISO 10,006: Quality management: Guidelines to quality in project management.* Geneva: International Standards Organization.

Keegan, A. E., and J. R. Turner. 2001. Quantity versus quality in project based learning practices. *Management Learning* (special issue on project-based learning) 32(1):77–98.

———. 2003. The management of innovation in project based firms. *Long Range Planning* Larsson, U., ed. 2001. *Cultures of creativity: the Centennial Exhibition of the Nobel Prize.* Canton, MA: Science History Publications.

Latham, M. 1994. *Constructing the team: Final report of the government/industry review of procurement and contractual arrangements in the UK construction industry.* London: The Stationery Office.

Levitt, B. and J. G. March. 1995. Chester I Barnard and the intelligence of learning. In *Organization theory: From Chester Barnard to the present and beyond.* Exp. ed., Oliver E. Williamson. New York: Oxford University Press.

Lock, D. 2000. Project appraisal. In *The Gower handbook of project management.* 3rd ed, J. R. Turner and S. J. Simister. Aldershot, UK: Gower.

Miner, A., and D. Robinson. Organizational and population level learning as engines for career transition. *Journal of Organizational Behaviour* 15:345–364.

Nonaka, I., and H. Takeuchi. 1995. *The knowledge-creating company*. New York: Oxford University Press.

OGC. 2002. *Managing successful projects with PRINCE2*. 3rd ed. London: The Stationery Office.

Pinto, J. K. 1999. Managing information systems projects: regaining control of a runaway train. In *Managing business by projects: Proceedings of the NORDNET Symposium*, ed. K. A. Arrto, K. Kähkönen, and K. Koskinnen. Helsinki: Helsinki University of Technology.

Project Management Institute. 2000. *A guide to the Project Management Body of Knowledge*. Newtown Square, PA: Project Management Institute.

Scott, R., ed. 2001. *Partnering in Europe: Incentive based alliancing for projects*. London: Thomas Telford.

Turner, J. R., ed. 1995. *The commercial project manager*. London: McGraw-Hill.

Turner, J. R. 1999. *The handbook of project-based management*. 2nd ed. London: McGraw-Hill.

———. 2002. Configuration management. In *Project management pathways*, ed. M. Stevens. High Wycombe, UK: Association for Project Management.

———. 2003. Farsighted project contract management. *In Contracting for project management*, ed. J. R. Turner. Aldershot, UK: Gower.

Turner, J. R., A. E. Keegan, and L. Crawford. 2003. Delivering improved project management maturity through experiential learning. In *People in project management*. ed. J. R. Turner. Aldershot, UK: Gower.

Wheelwright, S. C., and K. B. Clarke. 1992. *Revolutionizing new product development*. New York: Free Press.

SECTION II.3

SUPPLY CHAIN MANAGEMENT AND PROCUREMENT

INTRODUCTION

An extremely important development in the management of projects over the past decade or so has been the manner in which logistics and concern for supply chain functions has impacted how we develop projects. We could reasonably argue whether these have ever been associated with project management, and yet, as more and more organizations adopt project management as a principal method for operating their primary activities, they are discovering that the traditional models of procurement—lowest-cost bidding, contract administration, supplier expediting, tracking, and so forth—once regarded as overridingly the concerns of the construction industry, have branched out and embraced most organizations managing projects today. Understanding and proactively managing the critical steps in a firm's supply chain have been proven to directly contribute to a company's bottom-line success. This phenomenon is particularly important in project-based industries. This section of the *Wiley Guide* takes us directly into the mainstream of supply chain logistics and procurement for the management of projects.

In Chapter 25 David Kirkpatrick, Steve McInally, and Daniela Pridie-Sale address integrated logistic support (ILS). The emphasis is on looking at whole-life operations; again, these need to be planned and managed from the very earliest stages of the project, however. They look at ILS largely from the defense sector's perspective, though with examples from civil manufacturing, medical equipment, and construction. ILS, though clearly centered within the acquisition process, involves significant interaction with the project's/program's technical functions, as can be seen, for example, in the discussion on logistic support analysis. They also show how ILS integrates with systems engineering and with private sector finance initiatives (PFI, etc.). They discuss CALS (Continuous Acquisition and Life Cycle Support), largely from the data-handling perspective (resonating with the parallel discussion in Chapter 21 on configuration management). They conclude, very usefully, with a review of the difficulties faced in implementing ILS: the quality of data available, the difficulties of forecasting over such long periods, changes in usage and organizational composition, managing stakeholders toward long-term objectives, and so on.

Ray Venkataraman turns to a less specialized topic in Chapter 26 with his overview of supply chain management (SCM). Having described the general critical issues in SCM—value optimizing the way customers, suppliers, design and operations, logistics, and inventory are effectively managed—he refines these in terms of the key challenges in projects. A three-stage project supply chain framework is proposed, covering procurement, conversion, and delivery, and then ways value can be enhanced through enhanced integration are discussed. A model (the Supply Chain Council's SOR model) is proposed for tracking supply chain performance. This operates at three levels: overall structure and performance targets based on best-in-class performance; supply chain configuration; and finally operational metrics such as performance, tools, processes, and practices. Ray concludes by describing the upcoming issues in project SCM as he sees them.

In Chapter 27 Mark Nissen bears down more specifically on project procurement. Mark takes a broad, process perspective, illustrating the customer-buying and vendor-selling activities and, in particular, the role of the project manager in optimizing key "hand-offs" (friction points) from his research in the U.S. high-tech sector. The project manager is presented with a dilemma Mark believes—rightly so—in being torn on the one hand to be tough and firm and on the other to be accommodating and build synergy through trust and cooperation. What advice has he therefore for the project manager? Mark proffers several tips: Don't tinker, manage the critical path, question the matrix, benchmark, and really watch IT and software.

Dave Langford and Mike Murray, in Chapter 28, make the discussion more specific with their analysis of procurement trends in the UK construction industry and elsewhere. They show that there have been some major shifts in project procurement practice since the early 1980s (the time of Cleland and King's *Project Management Handbook*). Again, they show how much of the key procurement activity happens in the early stages of the project—not just in acquisition planning but in the whole involvement of construction/manufacturing in the early design and definition stages (as can be seen, of course, in the chapters by Ray Venkataraman and Hans Thamhain, among others). These changes in procurement practices reflect a general trend away from simple transaction-based procurement to more long-lived, relational procurement where trust and value for money (whole life, see Ive, etc.) count for more than simply lower capital cost; trends accentuated by (a) the increase in technical and organizational complexity on many projects and (b) the increasing sophistication and active involvement of clients in the management of their projects. New forms of procurement have arisen and become increasingly dominant, the most significant being partnering and performance-based contracting.

George Steel, in Chapter 29, takes us through one of Mark Nissen's "friction points"—tender management—in considerable detail, showing how business benefit can be obtained by clearly following established processes and practices. The key essentially is to build a contracting and tender management strategy that reflects the organization's values and drivers; to recognize the difference between "hard money" and "soft" (actual costs versus estimated intangibles); and however tough the bidding process may have been, to build the supply chain synergy (team spirit, etc.) once the contract is awarded.

David Lowe in Chapter 30 similarly keeps us at a highly practical level in his expert review of contract management. This is a vitally important aspect in the management of any project that entails third-party contracts (or in-house ones for that matter). He uses the contract form of FIDIC as his reference. Though broadly construction-oriented, it is not exclusively so, and it is used in over 60 countries. David shows how contracts deal primarily with risk identification, apportionment and management, and relationship management. He believes a project manager should have a thorough understanding of the procurement process and post-tender negotiation, of the assumptions made by the purchaser and the supplier, of the purchaser's expectations of the service relationship, of the contract terms and conditions, and of the legal implications of the contract. To this end he takes us on a high-level tour of contractual issues; contract types and strategy; roles, relationships, and responsibilities; time, payment, and change provisions; remedies for breach; bonds, guarantees, and insurance; claims; and dispute resolution. He concludes with two lists of best-practice guidance drawn from the Association for Project Management.

Change management is one of the most important and delicate areas of project management. Poorly handled, it will lead to the project getting out of control. Sometimes, however, change can be for the project's benefit. Kenneth Cooper and Kimberly Sklar Reichelt in Chapter 31 examine this, calling on the results of interviews with dozens of managers and using simulation models to look at the potential disruptive impacts of changes. They show why it is better for changes to be handled quickly, before their effect can begin to become cumulative, and also show that disruption is reduced by less tight schedules. They conclude by looking at good practice in managing change claims and disputes.

About the Authors

David Kirkpatrick

Professor David Kirkpatrick was trained as an aeronautical engineer (and later an economist) and had a distinguished 33 year career in the scientific civil service of the UK Ministry of Defence (MoD). During this period he did research on aerodynamics and aircraft design at the Royal Aircraft Establishment Farnborough, military operational analysis for the chief scientist (RAF) cost analysis, and forecasting in support of defence equipment procurement. He also served for three years as an attaché in Washington, D.C., promoting UK/U.S. collaboration in defense technology. He retired from the MoD in 1995 and joined the Defence Engineering Group at University College, London to lecture in various aspects of defense equipment acquisition, and to undertake associated research consultancy work. He was appointed to a personal chair of Defence Analysis in 1999. In addition to many technical and official papers printed within the MoD, he has written for external publication over 60 papers on aerodynamics and aircraft design, the cost and effectiveness of defense equipment, defense economies, and military history. He is an independent member of the Defence Scientific Advisory Council and a specialist adviser to the House of Commons Defence Committee. He is a Fellow of the Royal Aeronautical Society and an Associated Fellow of the Royal United Services Institute.

Steve McInally

Dr. McInally initially trained as an electromechanical engineer, working in telecommunications industry in the mid-1970s and early 1980s. Between 1985 and 1994, he was employed by Philips Medical Systems in a variety of engineering roles, including installing, commissioning, and maintaining radiotherapy equipment, then later as a requirements elicitation specialist on radiotherapy system design projects. With Philips' sponsorship, he completed his BSc in Business and Industrial Systems at Leicester Polytechnic in 1992, and his PhD in Instrument System Design with UCL's Defence Engineering Group in 2001. He was appointed Research Fellow at the Defence Engineering Group in 2002. Recent publications include papers on requirements elicitation in the systems engineering process, the use of heuristics in systems engineering, and introspective learning models for advanced motorcycle riding. Dr McInally also designs online teaching modules for systems engineering education and acts as rapporteur for organizations such as Royal United Services Institute and the Royal Academy of Engineering in London.

Daniela Pridie-Sale

Daniela Pridie-Sale has studied languages, geography, and management. She commenced her career in the leisure industry; after a period of teaching, she went on to work in the oil and finance industries. More recently she was marketing manager for an international college before joining UCL's Defence Engineering Group. She provides teaching and research support in the application of management science within systems engineering and is currently studying for an MSc in international business at Birkbeck College, London.

Ray Venkataraman

Dr. Ray Venkataraman is an Associate Professor of Operations Management at the School of Business at Penn State university, Erie, PA. He received his PhD in Management Science from Illinois Institute of Technology. Dr. Venkataraman has published in *The International Journal of Production Research, Omega, International Journal of Operations and Production Management, Production and Operations management* (*POMS*), *Production and Inventory Management, Production Planning and Control, The International Journal of Quality and Reliability Management*, and other journals. His current research interests are in the areas of manufacturing planning and control systems, supply chain management, and project management. Dr.Venkataraman is also a member of The Decision Sciences Institute, Production and Operations Management Society (POMS,) and American Production and Inventory Control Society (APICS). He is currently on the editorial board of *IEEE Transactions on Engineering Management* and has served on the editorial review board of the *Production and Operations Management* (*POMS*) journal.

Mark Nissen

Mark Nissen is Associate Professor of Information Systems and Management at the Naval Postgraduate School and a Young Investigator. His research focuses on the study of knowledge and systems for innovation, and he approaches technology, work, and organizations as an integrated design problem. Mark's publications span the information systems, project management, and related fields, and he received the Menneken Faculty Award for Excellence in Scientific Research, the top research award available to faculty at the Naval Postgraduate School. Before his information systems doctoral work at the University of Southern California, he acquired over a dozen years' management experience in the aerospace and electronics industries, and he spent a few years as a direct-commissioned officer in the Naval Reserve.

David Langford

David Langford has published widely, and books to which he has contributed include *Construction Management in Practice, Direct Labour Organisations in Construction, Construction Management Vol. I and Vol. II, Strategic Management in Construction, Human Resource Management in Construction,* and *Managing Overseas Construction.* He has coedited a history of government interventions in the UK construction industry since the war. He has contributed to seminars on the field of construction management in all five continents. David Langford holds the Barr Chair of Construction in the Department of Architecture and Building Science at the University of Strathclyde in Glasgow. He has published widely in the field of construction management, coauthoring eight books and editing threevolumes on construction research. He is a regular visiting lecturer at universities around the world. His interests are travel, theatre, and cricket, and he plays golf with more enthusiasm than skill.

Michael Murray

Michael Murray is a Lecturer in construction management within the Department of Architecture and Building Science at the University of Strathclyde. He completed his PhD research in January 2003 and also holds a first-class honors degree and MSc in construction management. He has lectured at three Scottish universities (the Robert Gordon University, Heriot Watt, and currently Strathclyde) and has developed a pragmatic approach to both research and lecturing. He has delivered research papers to academics and practitioners at UK and overseas symposiums and workshops. He began his career in the construction industry with an apprenticeship in the building services sector and was later to lecture in this topic at several further education colleges. Mike is coeditor of two textbooks, *Construction Industry Reports 1944–1998* (2003) and *The RIBA Architects Handbook of Construction Project Management* (2003).

George Steel

George Steel is the founder and Managing Director of INDECO, a management consultancy specializing in project and contract management. He has personally led many international corporate value improvement initiatives and has been responsible for developing and negotiating many major contracts. Prior to founding INDECO, George was a partner of Booz Allen Hamilton, New York, where he worked with a number of international oil and gas companies on the development of their organization, and on the management of major development programs. Earlier in his career, George was a project manager with an international engineering and construction contractor designing and constructing oil refineries and Liquified Natural Gas (LNG) projects. He has an Honors degree in Engineering from the University of Edinburgh and is a Fellow of the Association for Project Management.

David Lowe

David Lowe is a Chartered Surveyor and a member of the Project Management, Construction and Dispute Resolution Faculties of the Royal Institute of Chartered Surveyors. He is a lecturer in project management at the Manchester Centre for Civil and Construction Engineering, UMIST, where he is Programme Director for the MSc in the Management of Projects. He is also joint program director for a distance-learning MSc in commercial management, a bespoke program for a blue-chip telecommunications company. Consultancy work includes benchmarking the engineering and project management provision of an international pharmaceutical company. His PhD, completed at UMIST, investigated the development of professional expertise through experiential learning. Further research projects include the growth and development of project management in the UK construction industry, an investigation of the cost of different procurement systems and the development of a predictive model, and a project to assist medium-sized construction companies develop strategic partnerships and diversify into new business opportunities offered by public and private sector clients. Dr. Lowe has over 30 refereed publications.

Kenneth Cooper

Kenneth G. Cooper is a member of the Management Group of PA Consulting and leads the practice of system dynamics within PA. His management consulting career spans 30 years, specializing in the development and application of computer simulation models to a variety of strategic business issues. Mr. Cooper has directed over 150 consulting engagements, among them analyses of 100 major commercial and defense development projects. His group's office is in Cambridge, Massachusetts. Mr. Cooper received his bachelor's and master's degrees from MIT and Boston University, respectively.

Kimberly Sklar Reichelt

Kimberly Sklar Reichelt is a managing consultant in PA Consulting Group's Decision Sciences Practice. For 15 years, she has specialized in building and using system dynamics models to aid management decision making. While her experience has been in a variety of industries, from sports to medical to financial, she has focused in particular on project management assignments for both commercial and defense contractors. Ms. Reichelt received her bachelor's and master's degrees in Management Science from the Massachusetts Institute of Technology.

CHAPTER TWENTY-FIVE

INTEGRATED LOGISTIC SUPPORT AND ALL THAT: A REVIEW OF THROUGH-LIFE PROJECT MANAGEMENT

David Kirkpatrick, Steve McInally, Daniela Pridie-Sale

Traditionally, project management has been associated with the activities of an organization creating new products and has therefore focused on the early phases of a project—from concept through design and development to production, up to the point of sale to a customer, who is most generally an end user. Relatively little attention has been given to later phases in the project's life, perhaps because the operation and support of a product in service require different skills from those used earlier in its design and production, and perhaps because the sale is seen to mark a significant transfer of responsibility from the supplier to the customer.

During the early phases of a project, a variety of pressures on the project manager tend to encourage a short-term approach—seeking to solve immediate problems without due regard for the consequences that solution will impose on later phases. For example, the development of the Tornado attack aircraft was truncated because of short-term budgetary constraints; consequently, the aircraft in service initially provided an unduly low level of availability for active duty and required many expensive design changes to rectify problems that should have been resolved in development (UK Parliamentary Select Committee on Defence, 2000).

In the private sector, a short-term approach is promoted by the need to maintain the organization's profitability and its share price, in order to satisfy shareholders' expectations. In the public sector, politicians and their officials face a chronic shortage of resources immediately available to meet limitless demands for public services, so they may be tempted by a policy that matches supply and demand in the short term but that might create problems some years ahead. In both sectors project decisions should ideally be guided by a process of investment appraisal that takes account of all the resulting costs and benefits through the life of a project, but in practice managers often pay more attention to immediate problems and neglect through-life issues.

An emphasis on the early phases of a project may be justified in those cases where the transfer of a product from supplier to customer is a purely financial transaction, involving virtually no exchange of information, or where the product's (short) life after its transfer to the end user absorbs only insignificant resources. However, an emphasis on the early phases is quite inappropriate in those cases where the costs of the later phases (operations, support, and disposal) constitute the larger fraction of the project's through-life costs. This latter category includes many defense equipment projects, so the UK Ministry of Defence (for example) has repeatedly exhorted its project managers to adopt a through-life approach (Ministry of Defence, 1998).

Furthermore, a good project manager should be aware that unsatisfactory performance of a product in service could damage the organization's reputation and its future sales; inadequate performance could even subject it to crippling litigation, and the manager to prosecution, if the product adversely affects the health and welfare of customers. In many countries an increasing body of legislation insists that a product being used for its designed purpose should not damage the environment and that it can later be safely recycled. For instance, in response to social and legislative pressures carpet fiber manufacturers DuPont Antron have developed a carpet reclamation initiative as part of a life cycle management methodology (DuPont Antron, 2002).

Thus, today's project managers must address all the phases in a project's life in an integrated manner, to ensure that all phases meet their targets of performance, timescale, and cost.

Life Cycle Phases of a Project

The term project can be applied in at least two ways depending on one's point of view. For simple products such as pencils and personal computers, the term project would usually refer to the creation process in the early part of the life cycle that brings about the new product, system, or equipment, as in a *design project.* This point of view is reflected in the definition that the *Oxford English Dictionary* provides for a project (*Oxford English Dictionary*, 2000):

> A co-operative enterprise, often with a social or scientific purpose.

For more complex and costly products and systems such as spacecraft, hospitals, and aircraft carriers, the term project is generally synonymous with the whole life cycle of those products and systems. Given that this chapter is primarily concerned with more complex and costly products, the second interpretation that a *project* is concerned with all life phases applies in this chapter.

Different industries use different nomenclatures to describe various life phases. Figure 25.1 describes the life phases from three different perspectives.

FIGURE 25.1. COMPARISON OF LIFE CYCLES FOR DIFFERENT PRODUCTS AND SYSTEMS.

NASA

Mission Feasibility	Mission definition	System definition	Preliminary Design	Final design	Fabrication and Integration	Prepare for deployment	Deployment and Operations verification	Mission Operations	Disposal

CADMID

Concept	Assessment	Demonstration	Manufacture	In-service	Disposal

BS7000

Trigger, Product Planning, and Feasibility	Development and Production	Installation, Commissioning, Operation, and Use	Disposal and Recycle

- The National Aeronautics and Space Administration (NASA) life cycle model (Shishko, 1995) reflects the complex nature of space flight projects. NASA retains responsibility and ownership throughout a spacecraft's life, so all life phases, from Mission Feasibility through to Disposal, are included.
- The CADMID cycle adopted by the United Kingdom's Ministry of Defence (MoD) is applied to military equipment in all shapes and sizes. Like NASA, but unlike manufacturers of simpler products, UK MoD retains responsibility throughout the equipment's life; hence, the whole life cycle is viewed as a project.
- The British Standard BS7000- 1:1999 Guide to Managing Innovation (BSI, 1999) life cycle is simpler and is intended primarily to provide guidance for the development of mass-produced products.

For the purposes of illustrating ILS, this chapter will follow the MoD's CADMID designation.

It should be noted that in the MoD the start of the Assessment and Demonstration phases of the CADMID cycle must be formally authorized by the allocation of appropriate funding, provided that the preceding phases have produced satisfactory results. Other organizations adopt less formal procedures and may allow some overlap in the timescale of the project's phases.

Responsibility for Project Phases: Civil Sectors

In the civil sector, the supplier and the customer take or share responsibility for the various phases of a project. The supplier would typically be responsible for project phases up to the point of sale—for instance, through Product Planning & Feasibility and Development & Production phases of BS 7000, with the customer taking ownership after the Point of Sale (BS 7000 Operation and Disposal phases). However, the allocation of responsibility can be different in different industries, as demonstrated in the following examples.

In many consumer goods industries, the early Concept, Assessment, Demonstration, and Manufacture phases are the exclusive responsibilities of the supplier, guided by market surveys and by related insights on latent customer demand. The supplier then transfers ownership to the user and, at the point of sale (often by a retailer or agent), provides only simple, if any, instructions. The later In-Service and Disposal phases are the exclusive responsibility (and sometimes irresponsibility) of the customer.

The suppliers of expensive, durable products recognize that their reputations, and their hope for future business, depend on the continued acceptable performance of those products; they therefore provide an extensive set of detailed instructions and a guarantee to repair or replace the product if it fails through ordinary use within a specific period. In the automobile industry, for example, the suppliers often seek to retain responsibility for repair and maintenance activities in the in-service phase to ensure, as far as possible, that their cars remain safe and reliable and their owners remain satisfied. Similarly, organizations supplying capital equipment to commercial customers often undertake a contractual responsibility to provide repair and maintenance through the equipment's service life.

In the civil engineering and building sectors, a customer organization may take an active role alongside one or more suppliers in the early Concept and Assessment phases, assign total responsibility for the Demonstration and Manufacturing/Construction to a chosen supplier under an agreed contract, and later take full responsibility for the project's operation and support. In some cases, however, the supplier may retain responsibility for rectifying problems that arise within an agreed period.

Under the UK government's policy for the provision of public services and infrastructure under the Private Finance Initiative (PFI; see the chapter by Ive), it is now usual for a prime contractor to undertake the construction of a school or a hospital, and later to undertake its operation and support to deliver over an agreed period an agreed volume of educational or medical services. Similarly, under PFI, the contractor responsible for building a motorway may assume responsibility for its repair and maintenance to a satisfactory standard for an agreed period, as well as for its design and construction. In these and similar cases, the customer is involved in the Concept phase, which captures the requirement and considers alternative options; thereafter the customer adopts a detached supervisory role, providing the funds agreed and monitoring the supplier's performance via an appointed regulator.

Responsibility for Project Phases: Defense Sector

In the special case of the defense sector, the Armed Forces have a unique and exclusive knowledge of the realities of military operations and of any developing shortfalls in the capabilities of their current equipment. Accordingly, they take a leading role in the Concept phase of a new defense equipment project, although potential suppliers may, in this phase, offer advice on alternative options that might provide the required increment in capability. As the project passes through the Assessment, Demonstration, and Manufacturing phases, the Armed Forces and the relevant MoD branches and agencies play a less active role, and the chosen prime contractor takes a progressively greater share of responsibility. As the new equipment enters service, the Armed Forces take full responsibility for the operation of front-line equipment but may assign to contractors the operation of equipment located in "benign" rear areas. Responsibility for in-service support may be allocated according to circumstances; in a nuclear submarine on extended patrol all repair and maintenance during that period must be done by the crew, but a squadron of aircraft operating near the contractors' facilities can easily draw on their expertise and stocks of spares. It follows that in some cases suppliers are involved in the day-to-day support of their equipment, but in other cases they are involved only in any major refit or refurbishment.

In former times it was customary for the Armed Forces to maintain and repair their own equipment whenever practicable. This activity gave Service personnel a greater knowledge of the equipment's strengths and weaknesses and a greater ability to repair battle damage or to improvise modifications in a crisis. Today, by contrast, support arrangements vary widely between different nations and between different classes of equipment. In the United Kingdom it is perceived to be more cost-effective to rely on contractors to provide wherever practicable the in-service support for peacetime training and for expeditionary operations, except in the front line. While on some projects the contract for support may be negotiated separately from the contract for procurement, it is increasingly common for the equipment supplier to be given a portmanteau contract covering design, production, and support within an agreed payment schedule.

In the Disposal phase, the Armed Forces have full responsibility of ensuring either that sale of the equipment to a third party does not breach national arms control policies or that the equipment is safely destroyed according to current environmental legislation.

The Need for Through-Life Management

These examples indicate that there are many sectors of industry where project managers must adopt a through-life approach, taking full account of their product's durability, reliability, maintainability, and repairability in the In-Service phase of its life cycle and of the need for safe and efficient Disposal.

Integrated Logistic Support (ILS)

The Need for ILS

Modern military equipment requires many inputs to keep it operational. It may need regular supplies of fuel and ordnance. It undoubtedly needs regular attention from skilled artificers to undertake scheduled maintenance and unscheduled repair, and these activities will require technical documentation describing the equipment and its potential faults in useful detail. Specialist tools and test equipment, spare components or assemblies to replace those found to be damaged or faulty, and a logistic chain designed to provide supplies when they are needed (or at worst soon afterwards) are also required. In many cases the cost of operating and supporting equipment through its service life equals or exceeds the cost of its procurement.

Furthermore, a modern expeditionary force (and its associated equipment) needs a large and consistent inflow of supplies to maintain its effectiveness in a remote theater of operations. For example, in the Gulf War of 1991, the UK armoured division required some 2000 tons of supplies per day before the period of active operations, and triple that volume during the land campaign (White, 1995). The provision of this large quantity of supplies demands considerable planning and resources, and hence must be organized efficiently.

What Is ILS?

Usually applied to defense systems, ILS is a disciplined, structured, and iterative approach to ensure that all the inputs required by each item of defense equipment are provided where and when they are required and that the cost of providing them is minimized. During the life cycle of a defense equipment project, the principal aims of the ILS process are to

- analyze the through-life requirement for logistical support;
- formulate plans to provide sufficient support resource;
- influence the equipment design; and
- deliver the support resources when required.

The U.S. Department of Defense provides a useful and succinct definition for ILS (U.S Department of Defense, 1983):

> ILS is a structured management approach aimed at influencing the design of the asset and ensuring that all the elements of design are fully integrated to meet the client's requirements and asset's operational and performance, including availability, reliability, durability, maintainability, and safety at minimum whole life cost.

One of the most important features of ILS practice is the notion that it should be closely integrated with procurement and development cycles.

> The basic management principle of the ILS process is that logistic support resources must be developed, acquired, tested, and deployed as an integral part of the materiel acquisition process."
>
> From the *US DoD Integrated Support Manager's Guide* (U.S. Army, 1998)

Throughout that process the overall objective of the ILS is to maximize the cost-effectiveness of the equipment by striking a balance between its logistic requirements for resupply, its reliability and maintainability, the scale and the organization of the support resources, and the equipment's life cycle cost (LCC). It must also satisfy the dual objectives of being economical in the peacetime environment, which is familiar and well understood, and also being effective in the strange and demanding environment of conflict.

ILS was originally developed and applied in the USA in the early 1980s with the introduction of US DoD directive 5000.39 (U.S. Department of Defense, 1983) and became a compulsory part of all UK MoD projects since the early 1990s, as embodied in Def. Stan 00-60 (UK Ministry of Defence, 2002). The MoD ILS process is defined as:

> Integrated Logistic Support (ILS) provides the disciplines for ensuring that supportability and cost factors are identified and considered during the design stage of an equipment so that they may influence the design, with the aim of optimizing the Whole Life Cost (WLC)."
>
> From the UK MoD ILS Guide (UK Ministry of Defence, 2001).

ILS in the CADMID cycle

Ideally the ILS process should be used initially as part of the Concept phase and should then be progressively updated during the later phases of the CADMID cycle, incorporating additional project data as it becomes available.

Figure 25.2 illustrates (in a general manner) the relationship between the project life cycle (CADMID) and the support system life cycle (within ILS). At the beginning of

FIGURE 25.2. PHASES IN THE ILS LIFE CYCLE SIMULTANEOUSLY EVOLVE WITH THE CADMID LIFE CYCLE.

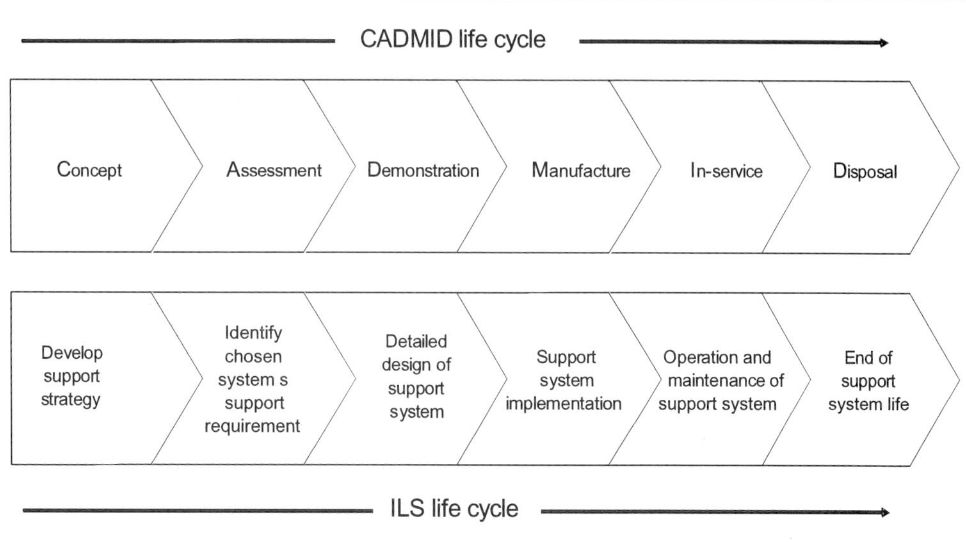

CADMID, there may be some general information available for the logistic requirements of systems "of this type," but there is little or no validated data for this specific system. As the CADMID cycle progresses, the ratio of vague/specific information decreases. Designers and procurers start to generate meaningful system design and performance data, and that data is communicated to those responsible for ILS analyses, specific guidance as to logistical requirements and constraints is fed back to system designers, and so on.

The interactive relationship between ILS and acquisition activities during the CADMID cycle, as illustrated in Figure 25.3, is fundamentally a learning process.

The majority of the expenditure in a project's life cycle is, by implication, committed by early design decisions, probably before the end of the assessment phase. Because the cost of in-service support of defense equipment is very high, sometimes higher than the cost of its procurement, this in-service stage has to be managed in a disciplined way and using the appropriate tools. It is therefore important that these early decisions are appropriately influenced by the results of ILS studies early in the project. In principle the early application of ILS can have a major influence on the project's through-life management plan, but in practice the ILS analysis is often constrained by predetermined (upper and/or lower) limits on the number of units to be deployed, or on the number of service personnel to be involved, or on the number of planned operating bases. Even constrained ILS studies can, however, favorably influence the initial design of equipment, provided that the study results are both timely and robust, and thus can significantly reduce the project life cycle cost.

FIGURE 25.3. AT EACH STAGE OF CADMID, THE VOLUME AND QUALITY OF ACQUISITION/DESIGN DATA INCREASES, ILS REQUIREMENTS BECOME CLEARER.

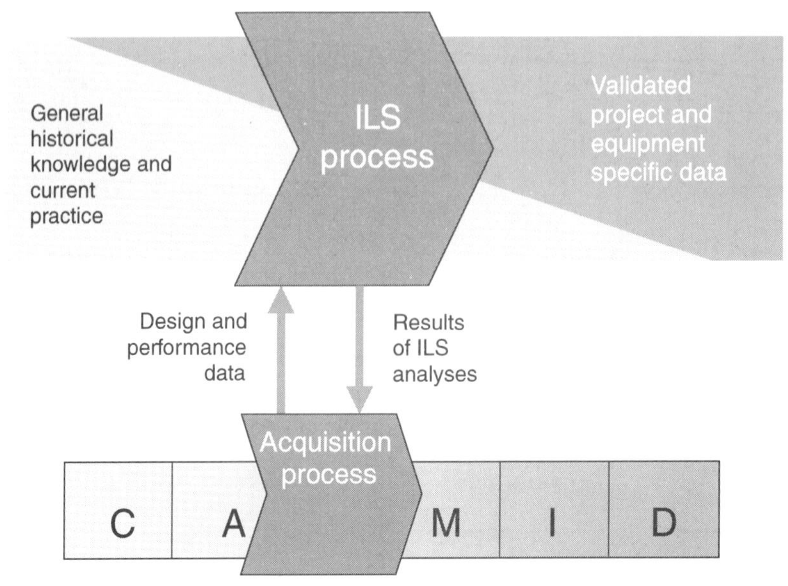

Components of ILS

ILS is a *framework* of tools and techniques; a *method* for prescribing the use of those tools and techniques; and in execution, a *process* whereby tools and techniques are systematically applied to a particular equipment life cycle. ILS starts from a proposed equipment design and a proposed support arrangement for the planned equipment and then uses a process of modeling and prediction to generate forecasts of equipment availability and life cycle costs. This provides a foundation for the comparative assessment of alternative design features and alternative support arrangements to identify the most cost-effective combination from a through-life perspective.

The ILS framework incorporates three principal activities (see also Figure 25.4):

- Logistic support analysis (LSA)
- Creation of technical documentation (TD)
- Formulating integrated supply support procedures (ISSP)

Logistic Support Analysis. The purpose of logistic support analysis is to identify the repair and maintenance tasks likely to be involved in the support of a new project and to plan how those tasks can most efficiently be accomplished. The results of this analysis can identify costs drivers in the proposed design and can stimulate trade-offs in which the design is refined to reduce its support costs without unacceptable penalties on performance, timescale, or procurement costs.

The LSA includes several discrete but integrated activities:

Failure modes effects and criticality analysis (*FMECA*) for each component in the proposed design determines how it might fail and the consequences of each failure for the equipment's safety and military capability. The FMECA results can guide decisions on component quality standards, duplication, and preventative maintenance.

Reliability-centered maintenance (*RCM*) considers alternative policies on inspection, preventative maintenance, and repair to establish the most cost-effective approach. Alternative policies include repairing or replacing items when they fail (with an appropriate level of servicing designed to delay failure); repairing or replacing items when electronic, visual, or other types of inspection reveal damage or deterioration approaching critical levels; and repairing or replacing items on a planned schedule linked to their durability (obtained from calculation or experiment) in order to avoid untimely failures. The optimal policy in each case depends on ease of inspection, the cost of preventative maintenance or repair or replacement, and the consequences of failure.

Maintenance Task Analysis considers the timescale and the resources of personnel and equipment required for each of the potential tasks. The personnel may require particular knowledge and skills and the equipment may include specialist tools and test facilities.

Level of Repair Analysis (*LORA*) is the process of determining the most efficient maintenance level for repairing items of equipment. Military organizations often have four levels of repair, with the *first line* in an operational unit, *second line* in a higher-level formation, *third line* in a base workshop, and *fourth line* at the contractor's factory. The

FIGURE 25.4. ILS INCORPORATING ITS THREE PRINCIPAL ACTIVITIES.

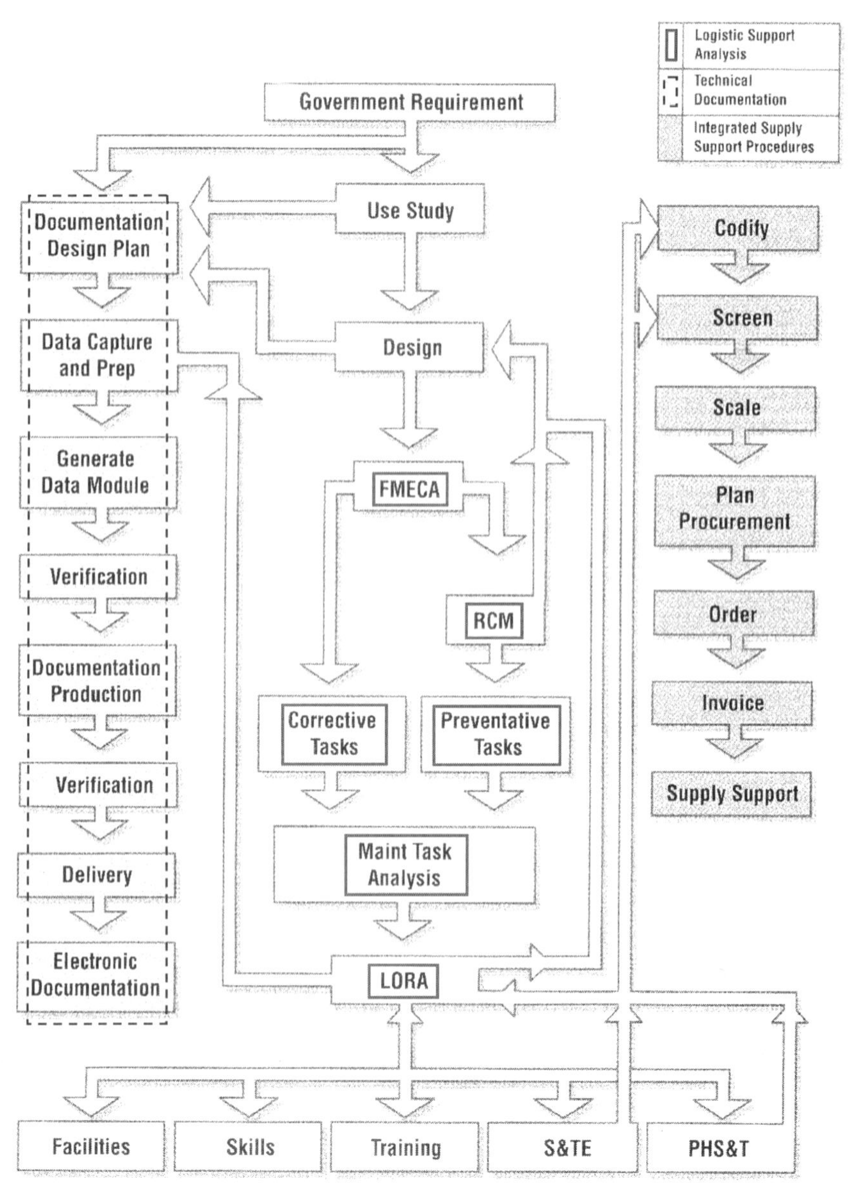

Source: Def Stan. 00-60.

number of levels and the arrangements within them varies between service environments and for different types of equipment. The LORA must balance the delay and resources required to transfer faulty equipment between the different levels of maintenance against the cost and risk involved in having skilled personnel, specialized test equipment, and spares holdings available in or close to operational units. The results of the LORA determine which types of spares should be held at each level of repair.

LSA coordinates these activities through five sets of tasks:

1. *Program planning and control* establishes the scale and scope of the analytical tasks and the procedures for ongoing management and review.
2. *Mission and support systems definition* considers how the equipment is to be operated and supported, and thus identifies design changes that would yield significant reductions in its support costs.
3. *Preparation and evaluation of alternatives* assesses detailed design trade-offs to determine the options yielding best value for money.
4. *Determination of logistic support requirements* quantifies the resources needed to support the equipment through its In-Service phase.
5. *Supportability assessment* reviews the effectiveness of the LSA and the lessons to be learned from it.

The data resulting from the LSA activities is assembled in a structured *logistic support analysis record (LSAR),* which can easily be used by the various government and commercial organizations involved in the project.

Technical Documentation (TD). Technical documentation contains all the information necessary to operate, service, repair, and support an equipment project through its service life and to dispose of it afterward. This information includes data on

- system description and operation
- illustrated parts data
- system servicing, maintenance, and repair
- diagnostic support equipment, and so on

and may be held as text or drawings on paper, fiche, text or drawings in electronic format, video, and data to support computer-aided design (CAD). For modern projects, electronic technical documentation (ETD), in the format established by Def. Stan. 00-60 part 10, is generally the most cost-effective option.

Integrated Supply Support Procedures (ISSP). ISSP cover the procurement of new spares, the repair and overhaul of defective items, and the administration of these processes. Given the multitude of spares of many types required by one equipment project, and of the overlapping sets of spares required by concurrent projects, it is vitally important that various service units and supporting contractors benefit from early, rapid, and unambiguous exchange of data, using electronic documentation.

The ISSP include the following:

- Codification that assigns to each item used by the Armed Services a unique identifier, using the NATO codification system
- Initial provisioning to provide adequate spares to support an initial period of operations (nominally two years in the UK) within which experience of reliability and maintainability yield definitive data
- Reprovisioning analysis that determines how many spares of each type should be held at each level of maintenance, when an order for replacement should be placed, and the economic size of the order
- Repair and overhaul plans that define how defective items, which cannot economically be replaced, can be restored to serviceability
- Procurement procedures that define how orders and invoices (ideally electronic) will be administered

Tailoring ILS

Although off-the-shelf procedures are widely used, every project is a unique enterprise, and therefore the ILS process should be tailored according to the realities of each and every project and program. Tailoring establishes which of the tasks and subtasks must be performed, when, and to what depth. A skillfully tailored ILS process can produce more saving than the use of off-the-shelf procedures, which are sometimes preferred by less experienced project managers because of their lower cost and their convenience.

All projects and programs have to accomplish certain core activities according to standards and regulations. When contracting for the ILS, the U.S. Army recommends that

> the ILS requirements will be tailored according to the acquisition strategy and included in the solicitation documents. The contractor will be required to define his approach to meeting the stated ILS requirements in the proposal developed in response to the solicitation.
>
> U.S. Department of the Army, 1999

Management of ILS

Within the ILS activities reviewed previously, it is necessary to take account of the following:

- Provision and upkeep of support and test equipment (S&TE)
- Test and evaluation (T&E) facilities
- Personnel and human factors
- Computing and IT resources
- Training and training equipment
- In-service monitoring
- Packaging, handling, storage, and transportation
- Safe and economical disposal

As described in the previous *Responsibility for Project Phases* sections, some of these tasks will be the sole responsibility of the ILS contractor; others will have to be carried out in conjunction with the customer or/and with other contractors.

The role of the ILS management process is to facilitate the development and integration of these elements. It is vital that ILS is integrated into the overall system development process in order to ensure the best balance between a system design, its operation, and its related support. The development of the ILS elements must be done in coordination with the system engineering process and with each other. When you are trying to achieve a system that fulfils all the desired criteria—performance, affordable, operable, supportable, sustainable, transportable, and environmentally sound—within the resources available, it is often necessary to have trade-offs between all these elements.

The ILS management process requires a demanding and rigorous approach to the development of a through-life management plan, requiring close attention to forecasts of the cost and duration of the successive phases of the project life cycle, and appropriate trade-offs of overall performance, cost, and timescale. The through-life management plan and detailed cost forecasts provide a good basis for a disciplined monitoring of the actual progress of the project. Furthermore, the logistic support analysis process supports a more precise forecast and assessment of the design costs and of the effect on costs of the changes that occur during the life of a project.

"ILS" in the Civil Sectors

Processes similar to ILS are used in the civil sectors of industry alongside a variety of techniques and methodologies that focus on identifying, analyzing, and optimizing with reference to issues that may emerge during the life of a product or system; for instance, systems engineering (SE), concurrent engineering (CE) and integrated product and process development (IPPD) are all through-life approaches. All these approaches, as well as certain proprietary life cycle management methodologies, have been successfully applied in the medical, automotive, nuclear, construction, and manufacturing industries for many years. All focus on analyzing and planning for a whole life cycle, and even for subsequent life cycles of replacement products. The motivation for developing and applying whole-life cost analysis and design techniques is born of a number of economic, environmental, and legislative factors.

Systems Engineering and ILS

The discipline of systems engineering was developed in response to the problems of managing complexity and reducing risk of failure in the design of large-scale, technology-driven systems such as information system, civil engineering, and aerospace development projects (see the chapters by Davis et al. on requirements management, Harpum on design management, and Mooz on verification). Systems engineering in its broadest interpretation includes a variety of concepts, models, techniques, and methods, including many or all of the concepts found in concurrent engineering, project management, integrated product and process development, as well as ILS (see the chapters by Thamhain, Cooper et al., and others). The central body for systems engineering, the International Council On Systems Engineering (INCOSE), defines a through-life approach as key to systems engineering as (INCOSE, 1999):

> Systems Engineering is an interdisciplinary approach and means to enable the realization of successful systems. It focuses on defining customer needs and required functionality early in the development cycle, documenting requirements, then proceeding with design synthesis and system validation while considering the complete problem: Operations; Performance; Test; Manufacturing; Cost & Schedule; Training & Support; Disposal. Systems Engineering integrates all the disciplines and specialty groups into a team effort forming a structured development process that proceeds from concept to production to operation. Systems Engineering considers both the business and the technical needs of all customers with the goal of providing a quality product that meets the user needs.

Although SE and ILS are two different concepts, in practice they are in some ways interdependent. SE is concerned with designing systems specifically with the emerging through-life considerations in mind, for instance, design for supportability. In his paper discussing the relationship between SE and ILS in the design of military aircraft, Strandberg describes logistic support analysis as the activity that bridges both ILS and SE (Bergen, 2000).

The concepts of SE and ILS are further integrated within international standard ISO 15288 Life Cycle Management—System Life Cycle Processes (International Organization for Standardization, 2002). The standard is intended to offer guidance for acquiring and supplying hardware, software systems, and services, but it also claims to offer a framework for the assessment and improvement of the project life cycle.

Although SE is concerned primarily with exploring and solving complex technical problems, it has a complementary relationship with project management. As Hambleton (Hambleton, 2000) puts it:

> You can't engineer a complex system without managing it properly and you can't manage a complex system without understanding its engineering. Systems Engineering and Project Management are two sides of the same coin . . ."

SE and ILS are different disciplines within project management. Both provide methodologies for the management of complexity to achieve specific organizational goals (such as optimal cost-effectiveness). SE is an overarching discipline integrating several other project management activities (such as requirement capture and equipment design) as well as ILS. Some of those other activities apply rigorous engineering methodologies, but ILS retains a more pragmatic approach with the methods and techniques applied being adapted to the project circumstances.

PPP, PFI, DBFO, and ILS

Public Private Partnership (PPP), Private Finance Initiative (PFI), and Design-Build-Finance-Operate (DBFO) initiatives have been key factors in the growth of interest in ILS and ILS-like methods The PPP initiative was introduced in the 1980s, the PFI launched in 1992, and more recently the DBFO initiative were all intended to bring the skills and resources of the public and private sectors together to improve the success of large-scale projects. (See the chapters by Ive and by Turner.) The shift toward a single contractor being responsible for the whole life of project has emphasized the need for contractors to adopt tools, methods, and techniques like ILS to reduce risk and cost in large projects.

For instance, the UK government has established a number of risk-sharing initiatives such as Contractor Logistic Support (CLS), financed by Private Finance Initiative (PFI) or Public Private Partnership (PPP) arrangements that will allow risk sharing for potentially expensive support services, which have traditionally been provided by the government. In order for nongovernmental organizations to provide these services, they need to understand fully how and why a system fails, what are the impacts of each failure, and what maintenance and resources would be required to carry out repairs. Under CLS initiatives, the UK government will no longer pay industry to perform the ILS activity of logistic support analyses and then use their own resources to carry out the work; rather, the industrial contractor will bid for the whole task at the Invitation to Tender (ITT) stage of a project.

According to the UK Confederation of British Industries (CBI), PPP, PFI, and DBFO initiatives have been a great success (Confederation of British Industries, 2002):

> Public Private Partnerships are a crucial element of delivering the government's commitments on improving public services. There is a vast range of PPP models and activities. Private Finance Initiative projects, for example, deliver public sector "capital and service package solutions", e.g. PFI prison service contracts where the private sector designs, builds and operates the prison for, say, 25 years. Over 400 PFI contracts had been signed to date. Investment in public services through the PFI is expected to increase from £1.5 billion to at least £3.5 billion by the end of the current spending round in 2003/4. The range of savings identified is considerable, ranging from less than 5% to over 20%.

In a report examining the value for money for the PFI-financed redevelopment of the West Middlesex Hospital, the National Audit Office (NAO) noted that

> the Trust considered that the unquantifiable benefits of doing this as a PFI deal outweighed the disbenefits (NAO,2001).

However, in a recent report from the UK Audit Commission, its chairman James Strachan indicated (BBC News, 2003)

> schools built by the Private Finance Initiative are "significantly worse" in terms of space, heating and lighting than new publicly-funded schools. . . . The early PFI schools have not been built cheaper, better, or quicker and learning from this early experience is critical.

Given that the application of PFI and PPP for many projects is still in the early stages, the benefits and implications of this type of financing are not yet fully understood. In June 2002 the Audit Commission for Scotland reported on the use of PFI contracts to finance the renewal of 12 schools projects in Scotland (Audit Scotland, 2000). It commented that

> we are at an early stage in the 25–30 year life-span of PFIs, so it is too early to judge their contribution to education. . . . it was not possible to draw overall conclusions on value for money as it is difficult to quantify the benefits associated with PFI. The Report notes that it is important to the whole integrity of the PFI process that councils as clients hold the providers to their contractual commitments.

The challenge for the project manager is that PPP financing significantly increases the scale and complexity of the management task. With PPP, PFI, and particularly with DBFO projects, the scope of the project cycle may well extend beyond the traditional handover point to many years into the future. This implies a need for "ILS" activity to support through-life management. Though it is unlikely that the original project manager would continue to be responsible throughout the complete life of the system being designed, the structure and processes of management must always address the whole life of the project and its associated costs, particularly at the early phases.

ILS in civil construction projects

In a study of the application of ILS techniques applied in the construction industry (El-Haram et al., 2001), researchers at the University of Dundee's Construction Management Research Unit noted a number of issues:

- PFI was a key motivator in adopting and applying ILS techniques.
- ILS needs to be broadly and thoroughly applied early in the development cycle in order to maximize benefits.
- In the absence of formal guidance as to the order and circumstances that the various ILS techniques and procedures should apply, participant organizations interpreted and adapted ILS to their own specific needs.
- Approximately one-third of the data used in ILS analyses was based on engineering intuition rather than recorded data.
- As ILS was relatively new to the organizations and their particular industry sector, co-ordination between stakeholders (designers, facility managers, manufacturers, and so on) was poor.

The study is part of the Construction Management Research Unit's ongoing research efforts, particularly in developing a framework for capturing and analyzing whole-life data for constructed facilities, and in developing guidance for which ILS techniques will be appropriate used in differing construction projects.

In recent years, the U.S. Department of Energy (DOE) has applied life cycle analysis methodologies to the nuclear industry in response to financial, social, and environmental pressures. In a recent report on the costs of managing nuclear waste, the DOE estimates that the total costs of radioactive waste management will be in excess of $49 billions (US

Department of Energy, 2001). In response to this, the DOE has published its own life cycle cost savings analysis methodology to assist the deployment of new technologies in the nuclear industry (U.S. Department of Energy, 1998), as part of DOE Order 430.1 (U.S. Department of Energy, 1998). Similarly, the Australian Federal Highway Administration (FHWA) has developed a life cycle cost analysis approach to support the choice of materials and design of major highway projects (Hicks and Epps, 2003).

Continuous Acquisition and Life Cycle Support (CALS) and ILS

The CALS acronym has come to take on various meanings since the term was first coined, for instance:

- Computer-Aided Logistic Support
- Computer-Aided Acquisition and Logistic Support
- Continuous Acquisition and Life Cycle Support
- Commerce at Light Speed

In general, though, all refer to the same fundamental objective: to acquire, store, manage, and distribute design data electronically. CALS is effectively the means by which ILS is implemented on acquisition and design projects.

CALS began life in the 1980s as a U.S. Department of Defense (DoD) initiative. The basic idea was that technical data should be exchanged between government and its contractors in electronic format rather than on paper; as the DoD puts it "a core strategy to share integrated digital product data for setting standards to achieve efficiencies in business and operational mission areas" (Taft, 1985).

In the United Kingdom the initiative was adopted by the Ministry of Defence, which, in 1990, developed its own strategy to implement CALS called CIRPLS (Computer Integration of Requirements, Procurement and Logistic Support), and in 1995 the use of CALS technologies became a common and obligatory strategy for organizations and governments in NATO member countries.

A key concept for CALS is "create data once, use many times." This idea was made feasible by the growth of computerized information networks with the subsequent increased connectivity between enterprises. The problem was that potentially useful technical data was being held in many locations on different systems in different organizations.

The aim of CALS was to allow any authorized individual, from any stake-holding organization, to access the body of data which grows and matures as a project develops. This would have the benefits of

- increasing the rate at which information was exchanged;
- reducing information management overhead costs; and
- allowing information to be reused through all stages of a product's life cycle.

The concept of sharing applied both to individuals and to collaborating organizations. Within a single organization, design engineers, manufacturing staff, and product support staff all need to share design and logistics data right from an early stage in the project, so a strategy that improved information sharing could lead to important gains, particularly in the reduction of product development and manufacturing costs, and in reduced lead times. Additionally, information shared between different organizations in partnering-style relationships reduces the burden of information systems development, populating, and maintenance.

According to the UK MoD's National Codification Bureau (NCB), the body responsible for ILS, CALS, and similar initiatives (Clarke, 2003), CALS and CALS-like strategies are being applied by many companies around the world in a variety of industries, from consumer goods to aircraft, petrochemical plants to building and maintaining a road network.

Since its debut, CALS has continued to evolve in response to political, industrial, and technological changes. According to the U.S. Department of Defense, the term CALS is starting to disappear (U.S. Department of Defense, 2003), not because of any inherent flaw in CALS, but rather by its success. The original concept of information sharing during the system acquisition and design process is evolving into strategies such as the Integrated Digital Environment (IDE), Interactive Electronic Technical Manuals (IETM), and a Common Operating Environment (COE).

The *Integrated Digital Environment* initiative is CALS-like in that it focuses on information sharing, particularly at the enterprise level, and at early project phases. The initiative aims to overcome the barriers to efficient communication caused by program-unique information environments. The aim is to create seamless collaborative digital business environments shared by stakeholders, allowing the right information to be acquired at the right time and leading to fewer formal reviews and the improved quality of analyses. The benefits are improved general visibility throughout the supply chain, online access to technical information, reduced need for a information management infrastructure investment, and reduced cycle time.

The concept of *Interactive Electronic Technical Manuals* has been evolving since the 1970s. The idea is straightforward enough. Shared electronic media replaces technical documentation such as books and manuals with the inherent problems of storage, distribution, and version management.

Technicians and managers are able to consult centrally stored electronic reference information, use that information, and provide immediate feedback if any amendments or updates are required. IETM also provides the opportunity for those who apply the information, and who are also experts on the documented procedures and methods to author new and additional procedures and methods.

The hope is that maintenance tasks can be accomplished quicker with fewer errors, with no opportunity to "lose" pages.

The concept of a *Common Operating Environment*, developed in the early 1990s, is that the various stakeholders involved in procurement and design processes benefit from economies of scale in the development of databases and communication system. The idea is that program cost and risk can be reduced by reusing proven solutions and by sharing common functionality. The benefits of COE are improvements in development times, technical obsolescence, training requirements, and life cycle costs.

The relationship between ILS and CALS is becoming ever more integrated, for instance, as seen in the NATO initiative, as described in detail in the NATO CALS Handbook

(NATO, 2000). The NATO CALS initiative funded by 11 of the 19 NATO member nations was formed in order to improve NATO's ability to exploit information and communications technology, in the acquisition and life cycle support of complex weapons systems. A key follow-on activity to this initiative will be to develop a new international standard based on ISO 10303 for industrial automation systems and integration—Product Data Representation and Exchange (International Organization for Standardization, 1994), also known as the Standard for the Exchange of Product Model Data (STEP), to cover in-service and disposal phases of a system's life cycle. STEP ensures that the information produced in digital form can be read by others, is not hardware- or software-dependent, and has a life cycle dependent on its value.

The continuing trend toward larger and more complex projects involving many organizations in many countries, and the concurrent complementary introduction of CALS and associated strategies and standards, has significant implications for project management in organizations large and small. Project management of large, technically advanced systems is becoming more complex, and project managers will have to access, use, and contribute to information in external as well as internal databases, and to manage the interfaces involved. The manager of projects involving many enterprises must operate a complex information interchange, in which problems may be compounded by language and cultural differences between the participants if the project spans several nations. Project managers in such situations must therefore learn to exploit initiatives like NATO CALS and to operate within standards like ISO 10303.

Medical systems Life Cycle Management

Medical equipment manufacturers face a number of commercial, technical, legal, and ethical challenges that force them to analyze and plan for a variety issues to emerge during the life of a product. Radiotherapy oncology systems and magnetic resonance imaging systems in particular have long in-service lives; are highly complex; require specialized technical staff to install, commission, operate, maintain, and dispose of; and are expensive to purchase and own. A single radiotherapy or magnetic resonance imaging suite costs a hospital around $10 million to install. Medical systems of this kind typically have design lives of 10 to 15 years and are often in service for 20 or more years. Although medical system manufacturers still compete largely on purchase price, there are increasing pressures from customers and purchasing authorities to identify and minimize costs of ownership.

In response to these pressures, companies such as Philips Electronics have developed and now apply a range of techniques in early project phases to optimize the design for many factors. The phases of Philips' proprietary life cycle model, the Product Creation Process (Sparidens, 2000), is illustrated in Figure 25.5.

The model applies to a wide range of product types from medical systems to manufacturing systems and consumer goods. Depending on the type of product, market and legislative pressures, a number of life cycle analyses and optimization strategies analogous to ILS will be applied, for instance, design for cost, usability, patient and operator safety, serviceability, environmental friendliness, and disposability.

FIGURE 25.5. KEY PHASES IN THE LIFE OF A PHILIPS PRODUCT.

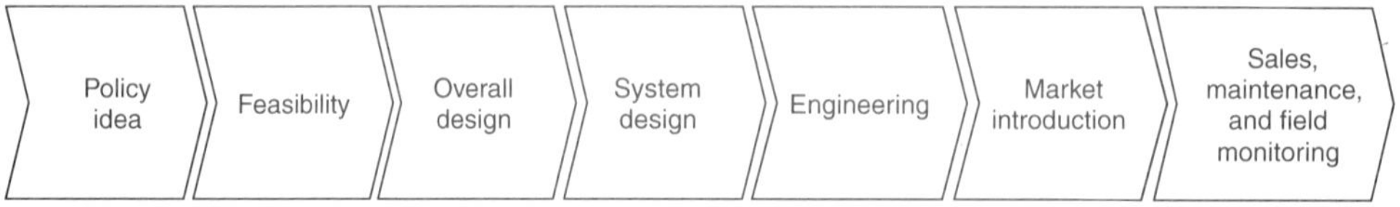

Though medical equipment manufacturers apply ILS-like methods and approaches, they are not always 100 percent successful. Equipment that incorporates leading-edge technologies makes prediction of life cycle costs and environmental impacts difficult. When Philips Medical Systems developed a lightweight solid-state digital replacement for its heavier glass tube image intensifier system, it was unable to predict all knock-on energy consumption effects in the supporting electronic control systems. Issues such as these seem obvious in hindsight, but at the time of development there was insufficient data on energy consumption and thermal radiation data with which to predict emergent properties. Philips was able to resolve the difficulties once operational data became available, but only with added costs, which seek to be recovered through sales and post sales revenue. Commercial medical equipment suppliers are obliged to seek other opportunities to recover research and development investment costs, through sales of service contracts, spare parts, user training, complementary products, and accessories.

Despite being fundamentally commercial products whose ownership transfers sometime shortly after being delivered to the customer, the manufacturer's responsibility for radiotherapy systems, as with other safety critical systems like aircraft and automobiles, does not cease at the point of sale.

Difficulties in Implementing ILS

While it is evident that many projects in the military and civil sectors can benefit from the application of ILS and related disciplines, there are some intractable difficulties in implementing ILS. These include a dearth of data on current systems, difficulties of forecasting accurately the characteristics of future systems, the sheer scale and complexity of the arrangements necessary for the logistic support of large projects, and the tendency of decision makers, in defiance of any existing ILS plans, to resolve urgent problems by solutions that are not cost-effective in the long-term.

Organizations have often failed to collect systematically data on the operation and support of equipment now in service. They may, for example, record the delivery of a batch of spare parts but take no account of when (and in what circumstances) these spares are used to repair existing equipment. They may record the delivery of fuel or utilities, but not identify the vehicles that consumed then. Not all organizations have yet been motivated to collect data that would be useful to ILS analyses. Company and service financial systems have been designed to monitor the various purchased inputs, rather than to facilitate input-output analysis linking such inputs to the organization's activities.

The problem of high-quality data varies between industries and product types. In the industries with particular concerns about safety, such as civil air transport and nuclear power generation, there is generally comprehensive data on all aspects of operation and support. In other industries having a large number of similar projects, such as civil engineering, it should be feasible to collect data of reasonable quality and volume. Data on the operation and maintenance of schools, hospitals, and prisons may be relatively easily acquired. Gathering good-quality data for, say, highly-innovative medical equipment, would be much more challenging.

Even when good data on the operation and support of current equipment has been collected, to provide a basis for forecasting the characteristics of future equipment and justifying the ILS policies chosen for it, the process of forecasting the operation and support of future equipment is extremely difficult. It is notorious that many forecasts of equipment reliability during its concept and the initial design stages have proved to be grossly inaccurate (Augustine, 1983), though in recent years a better understanding of the physics of failures has led to improvements in forecasting methodologies. However, when the new equipment incorporates unfamiliar technology or will be used in an unfamiliar environment, the initial estimates of equipment reliability and maintainability cannot be regarded as accurate until they have been confirmed by rigorous and realistic field trials.

In addition to doubts about the characteristics of future equipment, there are additional difficulties involved in forecasting the efficacy of some of the alternative arrangements for logistic support considered in the ILS process, particularly on those arrangements involving unfamiliar contractors and innovative contractual arrangements.

Even if the performance of the equipment itself (and of the organizations involved in its operation and support) could be forecast with confidence, there often remains considerable uncertainty about the employment of the equipment in service and the duration of its service life. Such uncertainty is greatest for military projects and for other capital equipment with long life cycles. The equipment's planned service life may be lengthened or shortened, according to the vagaries of military or corporate policy. It may or may not be subjected to mid-life upgrades or improvements. A military vehicle may be used for training in a benign peacetime environment or may be exposed to the rigors of warfare of various intensities and in different climates. A civil construction project may (during its lifetime) have to withstand more damaging levels of traffic or climatic conditions, or radical changes of use.

Because the future is inherently uncertain, any forecast of a project's life cycle cost is unlikely to be accurate, and hence should be accompanied by upper and lower confidence limits covering a substantiated range of uncertainty. Some project managers, accustomed to precise engineering calculations or auditable balance sheets (depending on their past experience), become demoralized by the distance between realistic confidence limits and cannot for that reason regard ILS as a really important influence on their management plan. Some of these managers may therefore be reluctant to allocate sufficient resources to ILS, when there are many urgent problems to engage the attention of their staff. In fact, many of the future uncertainties apply to all of the alternative design configurations and to all of the alternative logistic support arrangements; so it is it possible to select with confidence the most cost-effective designs or support arrangements based on their relative life cycle costs, even the where the absolute values of life cycle cost are very obscure.

Another inherent difficulty with ILS is the scale and complexity of some of the projects on which it must be used, the number of different organizations that must contribute, and the nature of the interfaces between these organizations. If these interfaces are blocked by mistrust or distorted by perverse contracting, the ILS process is unlikely to be completed satisfactorily. Furthermore, the proliferating multitude of interacting analyses, studies, and plans for the ILS of major projects encourages the growth of management procedures, bureaucracy, jargon, and acronyms, which together obscure the underlying principles of ILS and tend to insulate decision makers from operational realities.

Even when the ILS process has been satisfactorily completed and the most cost-effective strategy has been determined to manage the project through its entire life cycle, it remains difficult to ensure that the stakeholders are always guided by the best long-term policy. The politicians, government officials, and service officers directing military projects may be involved with the project for only a few years before at their respective career paths take them to other responsibilities. Business executives managing commercial projects may have personal goals (such as an annual bonus, stock options, or ambitions for promotion) whose attainment in the short term may not exactly correspond with the optimal policy for the project. In both cases, the stakeholders may take decisions that are attractive in the short term, but that in the long run can prove enormously expensive. The existence of an ILS plan can inhibit such decisions by highlighting and quantifying the scale of their adverse consequences, but it requires an appropriately forceful ILS manager to insist that the ILS management plan is widely understood and acknowledged as a significant factor in decision making.

Although a rational notion, there is a risk that ILS will lead to being unduly conservative in design. One criticism of PFI projects (a key driver for ILS) is that its application may lead to mediocrity of end product. Early consideration of later life cycle issues such as maintainability may stifle creativity and innovation, so that the end products may be maintainable but excessively dull as a result of the compromises made to make them so.

Summary

It is evident that integrated logistic support (or any similar process under another label) is an essential part of the development of a new product in the defense or civil sectors of industry. The ILS process specifies the facilities and supplying arrangements that are required to maintain and repair the products in service and to achieve the target level of availability. ILS is particularly necessary for large and complex projects that are expected to remain in service for many years, such as major capital items of defense equipment, investment goods, or infrastructure. ILS specifies the resources necessary for equipment support and hence defines their contribution to the equipment's life cycle cost, which is an essential input to its through-life project management plan (including budgeting).

There are many difficulties in implementing the ILS process, and these increase with the scale and complexity of the project considered. ILS involves many stakeholders who may have imperfect understanding of each other's problems and who may offer various levels of cooperation of the ILS process. The information available to support the ILS process is inevitably incomplete, particularly near the start of the product's life cycle, and the process itself is therefore prone to error and inaccuracy.

The ILS process should accordingly be tailored to match the information available and will help to identify critical areas of uncertainty. There are often inadequate resources (human and/or financial) and insufficient time to implement the ILS process as rigorously as would ideally be appropriate, since the project manager must always balance limited resources between ILS and various other activities required in creating a new product.

In poorly managed projects there is the risk that the ILS process is accomplished early in the life cycle only in order to obtain the funding necessary to launch the project but may subsequently be ignored during the Demonstration, Manufacturing, and In-service phases.

Despite these difficulties, ILS is a necessary activity since it provides vital inputs to through-life project management, except in those very rare cases where the supplier bears no accountability whatsoever for outcomes after the point of sale.

References

Audit Scotland. 2000. Taking the initiative: Using PFI contracts to renew council schools. Report to the Auditor General for Scotland in June 2002. Accounts Commission Scotland. www.audit-scotland.gov.uk/.

Augustine, N. R. 1983., *Augustine's laws.* p. 176. New York: AIAA.

BBC News. 2003, PFI schools criticised by report. BBC news report. Thursday, January 16, 2003, http://news.bbc.co.uk/1/hi/wales/2662999.stm.

Bergen, T. 2000. Supportability: A key to system effectiveness. Conference paper for Norwegian Systems Engineering Council (NORSEC) Annual Symposium, January 2000. www.incose.org/norsec/Dokumenter_og_nedlastbare_filer/NORSEC_moter/20000111/teknisk_referat20000111_2.pdf.

BSI. 1999. BS7000-1:1999. *Guide to managing innovation.* London: British Standards Institution.

Clarke, J. 2003. An Introduction to Codification, Statement by the Director of the National Codification Bureau, Glasgow, March 5. www.ncb.mod.uk/.

Confederation of British Industries. 2002. Making PFI / PPP work. Issue statement from the CBI Information Centre. September 23, 2002.

DuPont, A. 2002. Ensuring sustainability. Sustainability Brochure H93234, published by de Pont de Nemours and Company, United States, http://antron.dupont.com/pdf_files/literature/sustainability.pdf.

El-Haram, M., Marenjak. Horner. 2001. The use of ILS techniques in the construction industry. *Proceedings of the 11th MIRCE International Symposium.* Exeter, December.

Hambleton, K. 2000. Systems engineering: An educational challenge. *Ingenia* (November). Royal Academy of Engineering, London.

Hicks, R. and J. A. Epps. 2003. Life cycle costs analysis of asphalt rubber paving materials. Industry report. The Rubber Pavements Association, Tempe, AZ, May 1. www.rubberpavements.org/library/lcca_australia.

INCOSE. 1999. What is systems engineering? International Council on Systems Engineering. May 1. www.incose.org/whatis.html.

ISO. 2002. ISO/IEC 15288:2002(E). Systems engineering: System life cycle processes. Geneva: International Organization for Standardization/International Electrotechnical Commission.

ISO. 1994. ISO 10303-1:1994. Industrial automation systems and integration: Product data representation and exchange. Part 1: Overview and fundamental principles. Geneva: International Organization for Standardization.

NAO. 2001. The PFI contract for the redevelopment of West Middlesex University Hospital. National Audit Office Press Notice. National Audit Office. London: Stationary Office. ISBN:

NATO. 2000. *NATO CALS Handbook.* Version 2, June 2000. Available at www.dcnicn.com/ncmb/ (accessed May 1, 2003).

Oxford English Dictionary. 2000. Oxford, UK: Oxford University Press.

Shishko, R. 1995. *NASA systems engineering handbook.* SP-6105. Washington, D.C.: National Aeronautics and Space Administration.

Sparidens, H. 2000. Purchasing and supplier involvement is the product creation process. Philips Medical Systems corporate communication. Technische University Eindhoven, October 16. Available at www.tm.tue.nl/ipsd/educate/pms-2000.pdf (accessed May 1, 2003).

Taft, W. H. 1985. Computer Aided Logistics Support (CALS). Memorandum for Secretaries of the Military Departments, Defense Logistics Agency, U.S. Department of Defense. Report no. MIL-HDBK-59A. Washington, D. C.

UK Ministry of Defence. 2001. *MoD Guide to integrated logistics support.* Andover, UK MoD Corporate Technical Services. www.ams.mod.uk/ams/content/docs/ils/ils_web/ilsgdef.htm (accessed May 1, 2003).

———. 2002. *Integrated logistic support.* Defence Standard 00-60 Part 0, Issue 5, May. Glasgow: Directorate for Standardisation.

UK Parliamentary Select Committee on Defence. 1998. *The strategic defence review.* HC 138, Eighth Report. Vols. I–III.

———. 2000. *European security and defence* HC 264, Eighth Report. ISBN: 0-10-229400-3.

U.S. Army. 1998. *Integrated logistic support (ILS) manager's guide.* PAM 700-127. Washington, D.C.: United States Army Publishing Agency.

U.S. Department of the Army. 1999. Logistics, Integrated Logistic Support, Army regulation 700-127, November 10, p. 6. Washington DC Department of the Army.

U.S. Department of Defense. 1983. *Acquisition and management of integrated logistical support for systems and equipment.* Directive 5000, 39. November 17. Washington, D.C.: U.S. DoD Directives and Records Division.

———. 1983. Military Standard 1388-1A, Logistics support analysis.

———. 2003. *Integrated Digital Environment Initiative.* Integrated Digital Environment

———. 2003. Performance-centered learning module: IDE relation to CALS. January 13. www.acq.osd.mil/ide/learning_modules/ide/what_is_an_ide/ide_relation_to_cals.htm (accessed May 1, 2003).

U.S. Department of Energy. 1998. *Life cycle asset management.* U.S. Department of Energy order DOE 430.1 A.

———. 2001. *Analysis of the total system life cycle cost of the Civilian Radioactive Waste Management Program.* Report DOE/RW-0533 May 2001. Washington D.C.: U.S. Department of Energy.

———. 1998. *Standard life-cycle costs-savings analysis methodology for deployment of innovative technologies* Washington, D.C.: U.S. Federal Energy Technology Center.

White, M. 1995. *Gulf logistics from Blackadder's war.* London: Brasseys.

CHAPTER TWENTY-SIX

PROJECT SUPPLY CHAIN MANAGEMENT: OPTIMIZING VALUE: THE WAY WE MANAGE THE TOTAL SUPPLY CHAIN

Ray Venkataraman

During the 1990s, many organizations, both public and private, embraced the discipline of supply chain management (SCM). These organizations adopted several SCM-related concepts, techniques, and strategies such as efficient consumer response, continuous replenishment, cycle time reduction, vendor-managed inventory systems, and so on to help them a gain a significant competitive advantage in the marketplace. Companies that have effectively managed their total supply chain, as opposed to their individual firm, have experienced substantial reductions in inventory- and logistics-related costs, shorter cycle times, and improvements in customer service. For example, Procter & Gamble estimates that its supply chain initiatives resulted in $65 million savings for its retail customers. "According to Procter & Gamble, the essence of its approach lies in manufacturers and suppliers working closer together jointly creating business plans to eliminate the source of wasteful practices across the entire supply chain" (Cottrill, 1997).

While the adoption and implementation of total SCM-related strategies is quite prevalent in retail and the manufacturing industries and their benefits are well understood, project-based organizations have lagged behind in their acceptance and use of such strategies. For instance, the engineering and construction industry worldwide has been plagued by poor quality, low profit margins, and project cost and schedule overruns (Yeo and Ning, 2002). It is estimated that in the construction industry about 40 percent of the amount work constitutes non-value-adding activities such as time spent on waiting for approval or for materials to arrive on project site (Mohamed, 1996). The current project management practices of the construction industry in the areas of resource and materials scheduling would seem to be inefficient and lead to considerable waste. There is an urgent opportunity to adopt the practices of total supply chain management to reduce inefficiencies, improve profit margins, and optimize value. Sir John Egan, who headed a construction task force backed by the British government in 1997, strongly recommended in his report *Rethinking Construction* that the construction industry's performance would dramatically improve if it adopted the partnering approach in its supply chain (Watson, 2001).

Given that there are proven benefits in adopting total supply chain management-related strategies, the challenge then for project managers is to integrate these strategies into their management of projects.

What Is Supply Chain Management?

Supply chain management is a set of approaches utilized to efficiently and fully integrate the network of all organizations and their related activities in producing/completing and delivering a product, a service, or a project so that systemwide costs are minimized while maintaining or exceeding customer-service-level requirements. This definition implies that a supply chain is composed of a sequence of organizations, beginning with the basic suppliers of raw materials, and extends all the way up to the final customer. Supply chains are often referred to as *value chains*, as value is added to the product, service, or project as they progress through the various stages of the chain. Figure 26.1 illustrates typical supply chains for manufacturing and project organizations. Each organization in the supply chain has two components: an inbound and an outbound component (Stevenson, 2002). The inbound component for an organization may be composed of suppliers of basic raw materials and components, along with transportation links and warehouses, and it ends with the internal operations of the company. The outbound component begins where the organization delivers its output to its immediate customer. This portion of the supply chain may include wholesalers, retailers, distribution centers, and transportation companies, and it ends with the final consumer in the chain. The length of each component of the supply chain depends on the nature of the organization. For a traditional make-to-stock manufacturing company, the outbound or the demand component of the chain is longer than the inbound or supply component. On the other hand, for a project organization, the inbound component is typically longer than the outbound component. These concepts are illustrated in Figure 26.1.

The Need to Manage Supply Chains

Business organizations in the past have focused only on the performance and success of their individual firms. Such firm-focused approaches, however, will not help companies achieve a competitive edge in the current global business environment. Survival, let alone success, hinges on the ability of companies to manage their total supply chain. There are several reasons that make it necessary for companies to adopt supply chain management approaches.

First, businesses are encountering competition that is no longer regional or national; it is global. Competitive pressure from foreign competitors in both domestic and international markets is intense. Customers increasingly are seeking the best value for their money, and

FIGURE 26.1A. TYPICAL SUPPLY CHAIN FOR A MAKE-TO-STOCK MANUFACTURING COMPANY.

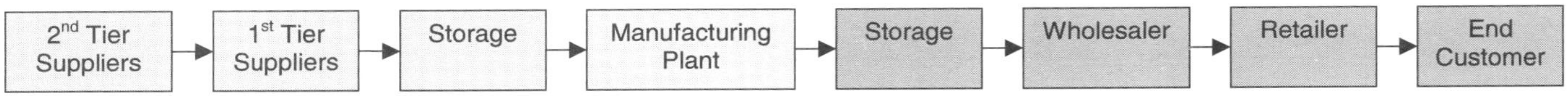

2nd Tier Suppliers → 1st Tier Suppliers → Storage → Manufacturing Plant → Storage → Wholesaler → Retailer → End Customer

FIGURE 26.1B. TYPICAL SUPPLY CHAIN FOR A PROJECT ORGANIZATION.

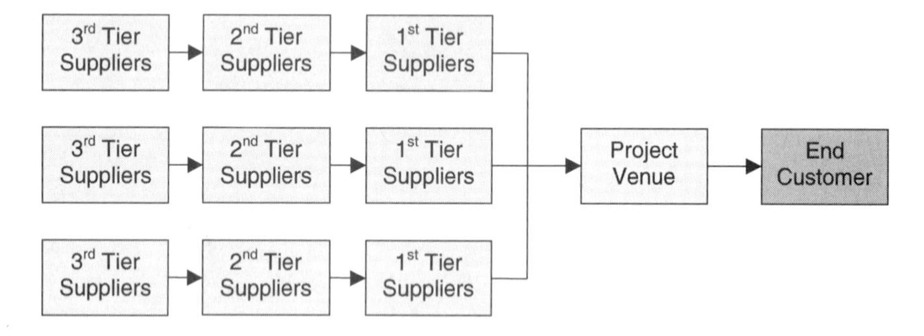

the advances in information technology and transportation have provided them the ability to buy from any company anywhere in the world that will provide that value. To win over these customers, business organizations need to reduce costs and add value, not just for their individual firm, but throughout their supply chain.

Second, inventory is a non-value-adding asset and is a significant cost element for businesses. The increasing variability in demand as we move up in the supply chain, known as the "bull-whip effect", can force some individual members of a supply chain to carry very high levels of inventory that can substantially increase the final cost of the product. Effective supply chain management approaches can enable a business to achieve a visible and seamless flow of inventory, thereby reducing inventory-related costs throughout the supply chain.

Third, the chain of organizations involved in producing and delivering a product or completing and delivering a project is becoming increasingly complex and is fraught with many inherent uncertainties. For example, inaccurate forecasts, late deliveries, equipment breakdowns, substandard raw material quality, scope creep, resource constraints, and so on can contribute to significant schedule and cost overruns for a project organization. The more complex the supply chain, the greater would be the degree of uncertainty and hence the more adverse the impact on the supply chain.

Supply chain management approaches such as partnering, information, and risk sharing can greatly reduce the impact of these uncertainties on the supply chain. Finally, management approaches such as lean production and TQM enabled many organizations to realize major gains by eliminating waste in terms of time and cost out of their systems. New opportunities for businesses to improve operations even further now rest largely in the supply chain areas of purchasing, distribution, and logistics (Stevenson, 2002). In the present-day global environment, because the competition is no longer between individual firms but between supply chains, companies need to better manage their supply chain to remain viable.

While several project-based organizations have adopted SCM-related strategies, evidence indicates their efforts to mitigate project schedule and cost overruns have fallen woefully short of expectations. The reason may be that project supply chain management is considerably more difficult as project supply chains are inherently more complex. For example, many projects typically involve a multitude of suppliers and experience considerable variability in supply delivery lead times and resource constraints, as well as frequent changes to the project scope. Such project supply chain complexities underscore the importance and need for project-based organizations to manage their total supply chain in a more formal and organized manner.

SCM Benefits

Companies that effectively manage their supply chain accrue a number of benefits. A recent study by Peter J. Metz of the MIT Center for eBusiness found that companies that manage their total supply chain from suppliers' supplier to customers' customer have achieved enormous payoffs, such as 50 percent reduction in inventories and 40 percent increase in on-time deliveries (Betts, 2001). Effective SCM has enabled Campbell Soup to double its inventory turnover rate, Hewlett-Packard reduced its printer supply costs by 75 percent, profits doubled and sales increased by 60 percent for Sport Obermeyer in two years, and National Bicycle achieved an increase in its market share from 5 percent to 29 percent (Stevenson, 2002; Fischer, 1997). Companies such as Wal-Mart that have better managed their supply chain have benefited from greater customer loyalty, higher profits, shorter lead times, lower costs, higher productivity, and higher market share. These benefits are not restricted to traditional manufacturing or retail businesses. Organizations that manage projects can enjoy similar benefits by effectively managing their supply chains.

Critical Areas of SCM

Effective supply chain management requires companies to focus on the following critical areas: Customers, suppliers, design and operations, logistics, and inventory.

Customers

Customers are the driving force behind supply chain management. Effective supply chain management, first and foremost, requires a thorough understanding of what the customers want. In a project environment, determining customer requirements and integrating the voice of the customer by working with the customer throughout the project will in all likelihood lead to a satisfied customer and project success. An important mechanism to achieve such a customer focus in projects is the integration of all project activities and participants into the larger framework of supply chain management. However, given that customer expectations and needs are ever-changing, determining these is tantamount to hitting a moving target. In recent years, customer value as opposed to the traditional measures of quality and customer satisfaction has become more important. "Customer value is the measure of a company's contribution to its customer, based on the entire range of products, services, and intangibles that constitute the company's offerings" (Simchi-Levi et al., 2003). Clearly, effective supply chain management is a fundamental prerequisite to satisfying customer needs and providing value. The challenge for project organizations, then, is to provide this customer value by managing the inevitable scope changes without incurring significant project schedule and cost overruns.

Suppliers

Suppliers constitute the back-end portion of the supply chain and play a key role in adding value to the chain. Their ability to provide quality raw materials and components when they are needed at reasonable cost can lead to shorter cycle times, reduction in inventory-related costs, and improvement in end-customer service levels. Traditionally, the relationship between suppliers and buyers in the supply chain has been adversarial, as each was interested in their own profits and made decisions with no regard to their impact on other partners in the chain. Supplier partnering is vital for effective supply chain management, and without the involvement, cooperation, and integration of upstream suppliers, value optimization in the total supply chain cannot be a reality. For project-based organizations, this issue is even more critical, as the supply or back-end portion of the chain is typically long, and without the total involvement of each and every supplier, value enhancement and project supply chain performance will be less than optimal. For example, in the case of highly technical projects, it is not atypical to have fifth or even sixth tier suppliers upstream in the project supply chain (Pinto and Rouhiainen, 2001). Managing the dynamic interrelationships and interactions that exist among these suppliers is considerably more complex and requires the effective integration of these supplier and their project activities into the larger framework of supply chain management.

Design and Operations

Design and operations play several critical roles in a supply chain. New product designs often seek new solutions to immensely challenging technical problems. In the face of uncertain customer demand, changes to the existing supply chain may have to be made to take advantage of these new designs. They require consideration of trade-offs between higher logistics- or inventory-related costs and shorter manufacturing lead times. The operations function creates value by converting the raw materials and components to a finished product. This function is present in every phase of the supply chain and is responsible for ensuring quality, reducing waste, and shortening process lead times.

Logistics

Logistics involves the transfer, storage, and handling of materials within a facility and of incoming and outgoing shipments of goods and materials. By ensuring that the right amounts of material arrive at the right place and at the right time, the logistics function makes a significant contribution to effective supply chain management. In project management, the logistics function requires a thorough understanding of customer requirements, reduces waste throughout the supply chain in order to reduce costs, and ensures timely completion and delivery of projects.

Inventory

Inventory control is an essential aspect of effective supply chain management for three reasons. First, inventories represent a substantial portion of the supply chain costs for many companies. Second, the level of inventories at various points in the supply chain will have a significant impact on customer service levels. Third, cost trade-off decisions in logistics, such as choosing a mode of transportation, depend on inventory levels and related costs. In project-based organizations, inventory-related costs can be substantial. It is obvious that effective inventory management can only be achieved through the joint collaboration of all members of the supply chain.

SCM Issues in Project Management

The benefits of utilizing the total supply chain management approach in the traditional make-to-stock manufacturing and retail environments have been well documented. Increasingly, project organizations and project managers are realizing that the integration of the total supply chain in managing projects, as opposed to a firm-focused approach, has the potential of reducing project schedule and cost overruns and the chances of project failure. However, as shown in Figure 26.1B, the typical chain for a project is considerably more complex. Problems associated with scope changes, resource constraints, technology, and numerous suppliers that may require global sourcing makes the total integration of the project supply chain risky and challenging. Consider, for example, the $200 billion Joint Fight Striker program, a mammoth and one of the most complex project management undertaking in history. "The principals of this project supply chain include

1. a consortium comprised of Lockheed Martin, Northrop Grumman and BAE Systems, overseeing design, engineering, construction, delivery and maintenance,
2. a matrix of partners, including Boeing, engine-makers Pratt & Whitney and Rolls-Royce and a handful of other subcontractors, all of which will lean on their own myriad suppliers for hundreds of thousands of components,
3. a multifaceted customer, the Pentagon, which is representing the U.S. Air Force, Navy and Marines, as well as the British Royal Navy and Air Force" (Preston, 2001).

Integrating and managing the total supply chain for this project is a Herculean task that will involve careful balancing of different vested interests and collaboration among all these partners to meet the stringent cost, quality, and delivery criteria set by the customer, the Pentagon. If the project's goal is focused only at the department or at the individual company level, instead of the total project supply chain, value optimization for the project cannot be achieved.

Projects in the construction industry are notorious for ill-managed supply chains. A recent research study of the UK construction sector found that fundamental mistrust and skepticism among subcontractors and other supply chain relationships was quite prevalent in this industry (Dainty et al., 2001). Such a lack of trust among supply chain partners will have a detrimental effect on the project delivery process. The key issue here is how to foster the necessary attitudinal changes throughout the project supply chain network to improve project performance.

Effective inventory management is yet another important supply chain issue in projects. In the airline industry, for example, enormous inventory inefficiencies such as duplication of distribution channels and excessive parts in storage have led to increasing costs for the total supply chain. In addition, a significant portion of every dollar invested in spare parts inventory constitutes holding and material management costs. Clearly, efficient inventory management throughout the total supply chain for projects in this industry has the potential for significant reduction in project life cycle costs.

Value optimization in projects cannot occur without the joint coordination of activities and communication among the various project participants. Consider, for example, a development project for an aluminum part to be delivered to an airline customer that was originally designed with a certain anodize process specified in the drawing. When the part designed is ready for production, the supply chain department of the project development group will then typically choose from a list of its favorite suppliers to get the lowest possible price. Often, these companies are rarely the ones that worked on the development hardware, and not surprisingly, they all will have different design changes that they would like to enforce on engineering to efficiently produce the part that will fit their particular set of processes. This can often lead to substantial increase in costs by way of engineering modification and requalification efforts. Furthermore, in the interest of price reduction, if the supply chain department later changes the design, without communicating or coordinating with the engineering department, to allow a supplier organization to use a different anodize process to fit its capabilities, then the part that will be delivered to the customer will be different from what the customer wanted, with colors that may be aesthetically displeasing when installed in the aircraft. Much time, money, and effort may have to be expended to rectify the situation with the irate customer. Project managers should be aware that without the joint collaboration of all project stakeholders working toward a common goal for the overall project, suboptimization will occur and the project is likely to fail.

Accurate, timely, and quality information on supply-chain-related issues is often not available to project managers and, as a result, causes them to make suboptimal decisions. Effective project supply chain management requires an infrastructure that can accelerate the velocity of information and will enable all project participants to collaborate throughout the project life cycle. For example, in a chemical plant construction project, the Global Project and Procurement Network uses the Internet to streamline and accelerate information flow that enables all supply chain participants to collaborate from plant design through operation and maintenance (Cottrill, 2001).

The terrorist attack of September 11, 2001, has heightened interest in security matters in the management of the total supply chain for many organizations. The challenge for many project organizations may range from designing facilities that are secure against outside intrusion to ensuring that the product can be protected from tampering till it reaches the end consumer. Ensuring security throughout the total supply chain is an enormous problem that will require project managers to provide unique and innovative solutions.

Value Drivers in Project Supply Chain Management

Value drivers in a project supply chain are those strategic factors that significantly add or enhance value and provide a distinct competitive advantage to the chain. The typical value drivers for a project supply chain are listed in Table 26.1.

The customer is the most important value driver in project supply chain management. In the context of project supply chains, the project client is the final recipient of the completed project. It is this customer's definition or perception that determines what constitutes value in a project. All other upstream supply chain activities in a project are triggered by this concept of customer value. If the customer values price, then all supply-chain-related activities of the project should focus on efficiency and eliminating waste throughout the total supply chain. On the other hand, if the customer values completion of the project on time or ahead of schedule, then all of the project supply chain activities should be geared toward achieving this goal. Thinking in terms of customer value requires project managers to have a clear understanding of customer preferences and needs, profit and revenue growth potential of the customer, and the type of supply chain required to serve the customer, and they must make sure the inevitable trade-offs that need to be made are indeed the correct ones (Simchi-Levi et. al, 2003).

The need to significantly lower or control project costs will also drive changes and improvements in the supply chain. In the retail industry, for example, the policy of everyday low prices required Wal-Mart to adopt the cross-docking strategy in its warehouses and distribution centers and strategic partnering with its suppliers. In the personal computer industry, Dell Computer Corporation uses the strategy of postponement (i.e., delaying final product assembly until after the receipt of the customer order) to lower its supply chain costs.

TABLE 26.1. PROJECT SUPPLY CHAIN VALUE DRIVERS.

Value Drivers	Definition
Customer	The final customer at the end of the project supply chain
Cost	Total cost incurred at the end of the project supply chain
Flexibility	The ability of the project supply chain to quickly recognize and respond to changing customer needs
Time	Refers to on-time delivery or delivery speed of completed projects to the end customer
Quality	The ability to deliver a completed project that meets or exceeds end customer expectations

Flexibility, the ability to respond quickly to changes in customer needs or project scope, is yet another important value driver in project supply chains. For example, the willingness of the project organization to provide the client the freedom to make significant design changes through development with the help of a strong and supportive engineering staff will enhance the value of the project supply chain (Pinto and Rouhiainen, 2001). Dell Computer Corporation is a classic example of a company that used flexibility to enhance customer perception of value. By allowing the customers to configure their own personal computer systems, Dell gained a significant competitive advantage in its industry.

The dimension of time has always been an important success factor in project management. Time in the form of project scheduling, in conjunction with cost and quality, represents the three most important constraints in projects. In event project management such as the Olympic Games, for instance, the dimension of time is of overriding importance, as the whole world is watching and the games must start on time. In other project-oriented situations, however, cost or quality can be the more important value drivers, and trade-offs in terms of time may have to be made in such projects. In any event, the ability to complete a project on time or ahead of schedule will certainly contribute to value in project supply chain management. In the retail industry, for instance, several time-based supply chain strategies such as continuous replenishment systems, quick response systems and efficient consumer response evolved as a direct result of the value-adding nature of time.

Quality, in a project context, is defined as achieving the project objectives that are "fit for purpose." "Fit for purpose means that the facility, when commissioned, produces a product which solves the problem, or exploits the opportunity intended, or better. It works for the purpose for which it was intended" (Turner, 1999). Project quality, simply defined, is that the project's product meets or exceeds customer expectations (Turner, 1999). Quality has several dimensions. For example, a person wanting to buy a Steinway grand piano for a price of $25,000 is more likely interested in the performance dimension of quality, whereas a person who wants to buy a Baldwin vertical piano for $5,000 is probably looking for a piano of consistent quality. Understanding the level of quality a customer wants in a project, ensuring the functionality of the project's product at that level of quality, and delivering the project at a reasonable price and time that will delight the customer should be the ultimate goal of every project manager. Meeting or exceeding the quality expectations of the customer adds value by fostering and sustaining customer loyalty and goodwill for years—long after the project is completed. Achieving this level of quality in projects, however, is easier said than done. It requires the total commitment to quality by each and every member of the total project supply chain and the integration of all their quality management activities.

Optimizing Value in Project Supply Chains

Choosing the Right Supply Chain

A fundamental prerequisite for value optimization in projects is the choice of the right supply chain for the project. More often than not, less-than-stellar supply chain performance is due to the mismatch between the nature of the product and the type of supply chain chosen to produce that product (Fischer, 1997). In the context of a project environment, this implies that first and foremost, the nature of the project, whether it is primarily functional or primarily innovative, should be clearly delineated. The next step is to choose the right supply chain for this project that will directly contribute to its core competencies and provide a distinct competitive advantage. Without having the right supply chain that is best for a particular project, value optimization in projects cannot be achieved.

Total Quality Management (TQM)

Quality, as defined by the customer, is the primary value driver in project supply chain management. Therefore, optimizing value in projects requires an unyielding commitment to total quality by all members of the project supply chain. A way to achieve this commitment is to adopt and integrate Total Quality Management (TQM) in project supply chains. "TQM is a holistic approach to continuously meeting customer needs and aims at continual increase in customer satisfaction at continually lower real cost. Total Quality is a total systems approach (not a separate area or program), and an integral part of high-level strategy. It works horizontally across functions and departments, involving all employees, top to bottom, and extends backwards and forwards to include the supply chain and customer chain" (Rampey and Roberts, 1992). The integration of the quality management activities of the project supply chain members through TQM is vital to complete a project in such a way that the multiple objectives of the customer in terms cost, quality, time, and safety can be met. The construction industry, for instance, is increasingly embracing TQM to solve its quality problems and ensure customer satisfaction. Quality assurance has always been difficult in this industry, as the products are one-off, the production processes are nonstandardized, and project design changes are frequent. Furthermore, the general contractor for a construction project is totally dependent on the goods and services of suppliers and other subcontractors to meet the quality requirements of the customer. The integration of quality management activities in a construction project supply chain through the application of TQM will enable the general contractors, suppliers, and other subcontractors to improve their own quality performance and will contribute toward optimizing customer value. For example, Shui On Construction Company in Hong Kong had successfully adopted TQM in 1993 and since then has been known for its good performance in building housing projects and has won the "Contractor of the Year" award three years in a row from 1995 to1997 (Wong and Fung, 1999).

Project Supply Chain Process Framework

The rest of the discussion in this section will be based on a simple framework of the project supply chain process that is presented in Figure 26.2. In this figure, the square box represents the procurement component of the chain. The oval-shaped box represents the conversion or fabrication phase of the project. It is in this component of the chain where the project's product is created. The rectangular box is the front-end portion of the project supply chain, delivery of the completed project to the customer.

FIGURE 26.2. PROJECT SUPPLY CHAIN PROCESS FRAMEWORK.

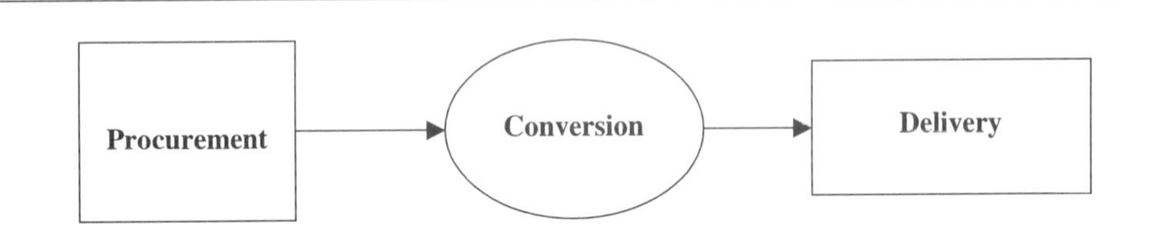

Procurement

The procurement portion of the project supply chain is typically long, and it is not uncommon to find fifth or even sixth tier suppliers upstream. The greatest opportunities for enhancing value of the total project supply chain exist in this area. Procurement involves all activities that are vital in acquiring goods or services that will enable an organization to produce the product or complete a project for its customer. The decision to buy from an outside vendor should be made only after a thorough "make or buy" analysis. In general, an organization should produce a product or component if it directly contributes to its set of core competencies. Otherwise, the product or components should be purchased from outside suppliers.

Procurement involves identifying and analyzing user requirements and type of purchase, selecting suppliers, negotiating contracts, acting as liaison between the supplier and the user, and evaluating and forging strategic alliances with suppliers. For many organizations, materials and components purchased from outside vendors represent a substantial portion of the cost of the end product, and hence effective procurement can significantly enhance the competitive advantage of an organization. Managing these suppliers and ensuring that parts and components of appropriate quality are delivered on time is a truly daunting challenge. In 1997, for example, Boeing, in its desire to respond to an unprecedented demand for new airplanes, attempted to double its production overnight without realizing the impact such a move would have on its supply chain. Parts and worker shortages at the assembly stage forced Boeing to close its 747 and 737 assembly lines and the company was hit with a $1.6 billion loss. Four years later, Boeing, through the use of lean manufacturing techniques, began to revamp its supply chain process and now requires tighter integration with suppliers and just-in-time delivery of their parts (Holmes, 2001). In the aerospace industry, companies such as Boeing and Rolls-Royce typically incur 60 percent of their project cost and 70 percent of their lead time because of purchased materials.

Hence, effective procurement strategies such as international sourcing, long-term supplier contracts, partnering with suppliers in project design, and risk and information sharing can maximize these companies' purchasing power, contribute to their business success, and significantly enhance the value of their supply chain. In 1993, Sikorsky Aircraft adopted the method of supplier Kaizen and realized the benefits of supplier long-term commitment and partnering for its future growth, declining prices and shorter lead times (Foreman and Vargas, 1999). "Supplier Kaizen is a method of bringing the suppliers to the same level of operations as the parent company, through training and improvement projects, to ensure superior performance and nurture the trust that is required for strong partnerships" (Foreman and Vargas, 1999).

The construction industry, as a whole, is characterized by significant distrust and antagonism within existing supply chain relationships. The key to future performance improvements in this industry is through the adoption of effective procurement strategies such as supplier selection and partnering, e-procurement, and supplier Kaizen. A recent Hong Kong-based study of factors affecting the performance of the construction industry has shown that the methods used for selecting the overall procurement system, contractors, and subcontractors are critical and the use of information technology/information systems can facilitate appropriate selection through all stages of the construction supply chain (Kumaraswamy et al., 2000). In the context of supplier selection, the series of international standards on quality management and quality assurance called ISO 9000 developed by the International Organization for Standardization (ISO) can be highly useful. For instance, companies that are ISO 9001 certified have demonstrated to an independent auditor that their systems and operations have met the rigorous international standards for quality and therefore can be included in the list of potential suppliers.

Supply Chain Relationships. Value optimization in projects cannot be achieved in the absence of close and trusting relationships among the project participants. Building trust and integrating the information systems among the supply chain members can lead to the elimination of certain redundant processes and simplification of sourcing, negotiating, and contracting procedures. Planning efficiency and project performance would improve, as suppliers are in a better position to provide valuable inputs to project planning because of the availability of timely and accurate information (Yeo and Ning, 2002). A recent study of two construction projects in the UK has shown that significant supply chain benefits and improvements can be realized through close partnerships and involvement of suppliers and subcontractors very early on in the project (Ballard and Cuckoo, 2001). Through partnering, all members of the supply chain are involved in translating the design concept into reality and ensuring that the appropriate cost criteria are met. The suppliers, along with other partners can be more innovative; problems can be resolved early, as there are more open channels of communication; and the end result will be a project that is completed on time, of higher quality, at a lower cost, and that gets the clients better value for their money. It is estimated that in the construction industry, supply chain partnering alone would lead to a 10 percent reduction in cost and time, similar increases in productivity and quality, and a 20 percent reduction in defects and accidents (Watson, 2001).

Supplier Development. Supplier development is yet another strategy that can add value to the procurement phase of the project supply chain. General Electric Company, as part of their global sourcing initiative, has a program for supplier development that involves providing extensive training by GE personnel to vendors in improving their own operations. Vendors who have improved their operations to the level of quality and efficiency that GE requires are awarded long-term contracts. In the final analysis, procurement in a project context requires extensive planning and coordination of project activities with suppliers. Strategies such as supplier Kaizen, partnering based on trust, vendor development, information and risk sharing, long-term strategic alliances with suppliers, and integrating quality management activities of suppliers through TQM will significantly reduce procurement and inventory costs, shorten lead times, and improve quality of purchased materials, and the value of the total project supply chain will be enhanced.

Conversion

The next phase of the project supply chain shown in Figure 26.2 that requires attention for value optimization is the conversion or fabrication phase of the chain. This is the project venue where the project's product is actually created, as in the case of new product development, creation of a new software package, or building an offshore oil-drilling vessel (Pinto and Rouhianen, 2001). The degree of success, in terms of value, that can be achieved in this area, to a large extent is dictated by the efficiency and effectiveness of the procurement phase of the project supply chain. As in the case procurement, the challenges encountered in this phase will depend on whether the project is relatively routine or highly complex. Regardless of the nature of the project, however, several strategies that have proven to be successful in the traditional manufacturing environment can be employed to enhance value in the conversion phase of the project supply chain. For instance, Boeing Corporation, in order to thwart the stiff competition from Airbus, is employing lean manufacturing practices to effect an innovative company-wide implementation of gigantic, moving assembly lines in its commercial aircraft division. For Boeing, such a technological advancement is reckoned to speed up production by 50 percent and increase its profit margins to double-digit levels on commercial aircraft sales (Holmes, 2001). The application of lean manufacturing techniques can add value in a project environment by eliminating waste and unnecessary inventories and by shortening process lead times.

Delivery

The final phase of the project supply chain process in Figure 26.2 is the delivery of the completed project to the customer. Normally, the transfer of the completed project to the client is relatively straightforward. In recent years, however, the project delivery process has undergone some significant changes, particularly in the case of clients from foreign countries. For example, in construction projects for large plants, some foreign countries require the project organization to operate the plant jointly with the foreign client for some extended period of time to mitigate potential start-up problems and to reduce the risk to the foreign client (Pinto and Rouhianen, 2001). While this increases the risk to the project organization, it also has the potential to enhance customer value. Clients are becoming increasingly more risk-averse, and the willingness of the project organization to assume additional project risks is certain to add more value and provide a distinct competitive advantage to the total project supply chain.

Integrating the Supply Chain

The obvious key to value optimization in projects is the total integration of the various components of the project supply chain. Several strategies can be implemented to achieve this goal. First, as shown by a recent study of two demonstration projects in the United Kingdom, development of "work clusters" and the application of concurrent engineering principles can lead to project supply chain integration, which in turn can improve value, eliminate inefficiencies, and reduce project costs (Nicolini et al., 2001). Second, project supply chain integration can be achieved through collaboration and standardization of business processes among the project supply chain partners. Such collaboration, however, requires understanding and managing the differences and interests of all the project supply chain members to create a common vision and work culture (Padhye, 2001). Third, accelerating information velocity by building an Internet-based supplier network for procurement purposes can further facilitate collaboration and integration in project supply chains (Cottrill, 2001). Building such a network also presupposes the presence of a viable IT/IS infrastructure among the project supply chain members. For example, the Joint Strike Fighter project discussed earlier in this chapter will require the various organizations involved in the design, engineering, manufacturing, logistics, finance and so on to collaborate over the internet to meet the stringent cost, quality, and time requirements set by their customer, the Pentagon.

Project supply chain integration and therefore value optimization requires the supply chain partners to change traditional thinking and practices. Effecting such a change requires the commitment and involvement of the people in each organization in the project supply chain. The impetus for achieving such commitment and involvement should come from the senior management of the project supply chain partners. Ultimately, it is the responsibility of the senior managers to prepare their organization for change, to overcome the cultural and organizational barriers to change, and to achieve cross-functional and cross-business unit cooperation (Burnell, 1999). Without the senior managers assuming the role of project champions, project supply chain integration and, hence, value optimization cannot be achieved.

In addition to the strategic initiatives discussed, the following specific steps can be undertaken to add value to a project (Hutchins, 2002):

1. *Flowchart the project supply chain processes before the project is initiated.* Such a flowchart will show the various links or steps involved in completing the project. Each step will potentially have a customer and a supplier. The flowchart can identify potential areas of redundancies, waste, or other non-value-adding activities in the chain and thus facilitate the use of lean management initiatives to eliminate them.
2. *Standardize processes.* Standardization of processes throughout the project supply chain by the use of methods such as simultaneous design, concurrent engineering, lean manufacturing, mistake proofing, total productivity maintenance, and collaborative teamwork will ensure consistency in the chain.
3. *Control process variation.* It is essential that processes across the total project supply chain are monitored and controlled for variation. For example, variability in lead times or

quality in materials and production processes should be controlled. Once the supply chain processes are stabilized, they can be improved.

4. *Prequalify suppliers through supplier certification.* Ensure that suppliers in each link of the project supply chain process are QS-9000 or ISO 9001 certified. Such certification guarantees a pool of quality suppliers.
5. *Audit the project supply chain processes and take corrective and preventive actions.* Processcs should be audited periodically for improvement and risk identification. Corrective action should be taken to eliminate the root causes of nonconformances and deficiencies that were uncovered through the audit. Preventive action ensures the recurrence of such problems.
6. *Measure project supply chain performance.* Measure project supply chain performance through the development and use of performance metrics and competitive benchmarking. Without the availability of specific quantifiable performance metrics, project supply chain performance in terms of both efficiency and customer satisfaction cannot be gauged. Such metrics will convey immediately how the project supply chain has been performing over time or in comparison with the best-in-class competition.

Performance Metrics in Project Supply Chain Management

Measuring project supply chain performance is a complex and challenging endeavor. Appropriate metrics should be carefully developed at the planning stage of the design of the total project supply chain. Involvement of all members of the project supply chain is critical to ensure that meaningful metrics are developed and will be used to monitor the performance of the total project supply chain. This will require reconciling differences and reaching consensus among the supply chain members on which metrics are appropriate for comparison with those of the best-in-class competition to measure success.

While there are a number of approaches to classify performance metrics, we will use the project process value drivers—time, cost, quality, and flexibility—to examine project supply chain performance. These performance metrics categories are presented in Table 26.2 (Coyle et al., 2003).

Time, in particular, project completion time, has always been considered an important measure of project performance. However, in addition to project completion time, this metric should capture other elements of time such as operational and start-up times, and procurement and manufacturing lead times. Furthermore, for routine projects the potential variability in these times should also be measured to track consistency and reliability of the project supply chain. For example, assume that historically the estimated completion time for routine construction projects has been 36 weeks. How frequently the project supply chain achieves this completion time is an indicator of consistency and reliability and can provide important insights for future improvements to the supply chain.

Some of the cost metrics noted in Table 26.2 are fairly straightforward. The important caveat here is that the emphasis should be on the cost incurred for the total project supply chain and not on just the cost incurred by the project organization. The total project supply chain cost is multidimensional and includes several elements, such as procurement and manufacturing cost of materials and goods, inventory costs, and so forth. Focusing on the

TABLE 26.2. PROJECT SUPPLY CHAIN PERFORMANCE METRICS CATEGORIES.

Performance Category	Performance Issues
Time	1. Was the project completed and delivered on time? 2. What is the potential variability in project completion times? 3. Was the completed project operationalized on time to the satisfaction of the customer? 4. Were the purchased materials and manufactured components delivered on time by upstream suppliers? 5. What is the potential variability in procurement lead times?
Cost	1. Was the completed project within budget for each of the project supply chain member? 2. What was the total project supply chain cost? • Procurement cost of purchased materials • Manufacturing cost • Inventory-related cost • Transportation cost • Project acceleration costs • Cost of liquidated damages • Other relevant costs: administrative, etc.
Quality	1. Did the project meet the technical specifications and does it provide the functionality desired by the customer? 2. Was the customer satisfied with the service provided during start-up, implementation, and final project transfer? 3. Were the purchased raw materials and manufactured components defect-free? 4. Was the completed project's product reliable and durable during its life cycle?
Flexibility	1. Was the customer accorded reasonable freedom within reasonable a time frame to make changes to the project scope, design, or specifications? 2. Were the upstream suppliers responsive to the reasonable needs of their downstream partners in terms of delivery time and quality issues?

total cost incurred will enable project participants to identify inefficiencies in the supply chain and facilitate coordination to devise ways to eliminate them. The ultimate goal is to optimize value by reducing waste and unnecessary cost throughout the supply chain.

Like cost, the quality metric also has several dimensions. In a project context, the most obvious ones are the dimensions of performance—that is, the functionality of the project's product and conformance to design or technical specifications. In addition to these dimensions, the level of service provided to the customer during the start-up and implementation phase and throughout the project's life cycle are also important quality measures. Ultimately, it is the customer perception of quality that matters, and the response of the project supply chain to meet this value perception should be the focus of this metric.

The last project supply chain performance metric category is flexibility. This metric measures the willingness and ability of the project supply chain to respond to reasonable changes in scope or design requested by the customer. Building an effective configuration and change control system that spans the total project supply chain can help achieve such flexibility and provide a distinct competitive edge to the value chain.

Project Supply Chain Metrics and the Supply Chain Operations Reference (SCOR) Model

Project supply chain metrics span the entire supply chain, with specific focus on common processes, and they should capture all aspects of supply chain performance. The SCOR model developed by the Supply Chain Council (SCC) provides the framework to track such performance and has been the basis for supply chain improvement for both global as well as site-specific projects (Yeo and Ning, 2002). By integrating the well-known concepts of business process reengineering, benchmarking, and process measurement, the model provides a cross-functional framework for improving supply chain performance. It spans all aspects and interactions of the supply chain, from the customer's customer all the way back to the supplier's supplier. "The SCOR model provides standard descriptions of relevant management processes; a framework of the relationships among the standard processes; standard process performance metrics; and standard alignment to features and functionality. The ultimate aim is to produce best-in-class supply chain performance" (Coyle et al., 2003). The model uses five key aspects of a supply chain—plan, source, make, deliver, and return—as building blocks to describe any supply chain. The SCOR model can be adapted to describe a project supply chain, as shown in Figure 26.3.

In a project supply chain context, the planning process in Figure 26.3 encompasses all aspects of planning, including the integration of the individual plans of all supply chain members, into an overall project supply chain plan. The planning phase essentially involves understanding customer needs and project scope; the best course of action to meet the sourcing, producing, and delivery requirements of the project; and developing the criteria to evaluate the total project supply chain performance. The sourcing phase focuses on all processes related to procurement, such as identifying, selecting and qualifying suppliers, contract negotiation, and inventory management, and so on.

The "make" process encompasses all aspects of creating the project's product, such as design and testing, and building and completing the project. It also includes systems and processes for quality and change control and performance reporting. The delivery process covers all aspects related to the final transfer of the completed project to the customer including installation, start-up, and so forth to the satisfaction of the customer. The return phase encompasses all activities that may range from addressing problems associated with the completed project's functionality at the customer site to return of raw materials to the vendor.

The SCOR model for a project supply chain is composed of three levels. At the top level, the scope and content of the model is defined and performance targets based on best-in-class competition are established. The next level focuses on the configuration of project's supply chain. The last level includes process elements such as performance metrics, systems

FIGURE 26.3. THE SCOR MODEL ADAPTED FOR A PROJECT SUPPLY CHAIN.

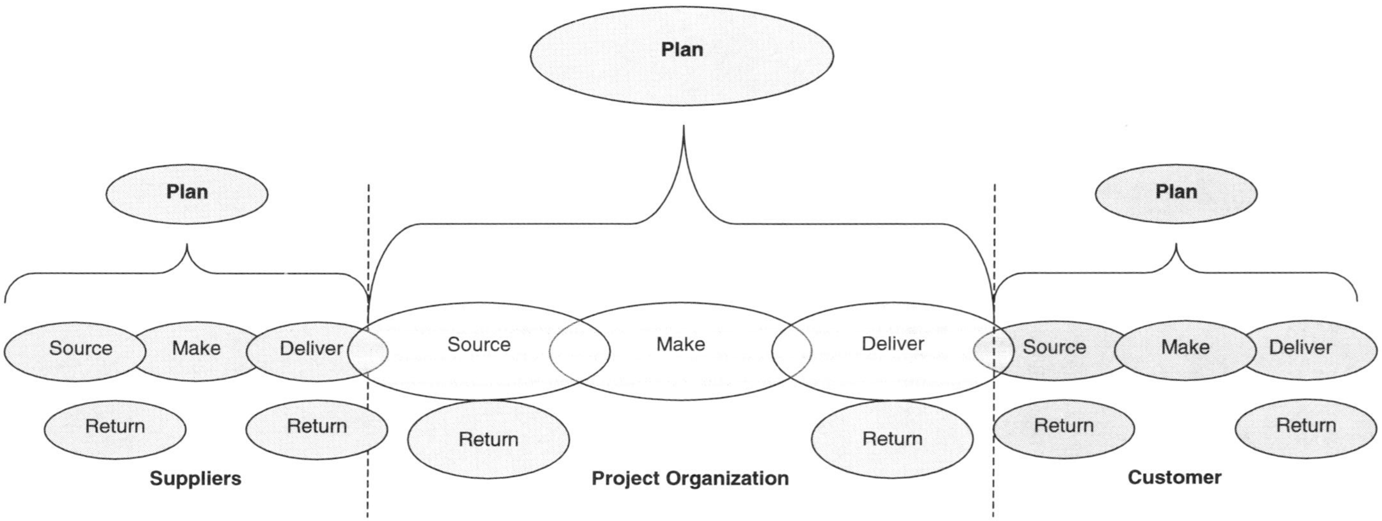

Source: Supply Chain Council Inc. www.supply-chain.org.

and tools, best practices, and the system capabilities to support them. As the SCOR model is based on standard processes and standard language, meaningful performance metrics for the project supply chain can be developed.

Future Issues in Project Supply Chain Management

Project supply chains in the twenty-first century will encounter a number of challenges, as well as opportunities for improvement. First, the availability and power of information technology will dramatically transform project supply chains. It will facilitate the virtual integration of project supply chains and thus provide the benefits that accrue from tight coordination, partnering, quick and efficient communication, focus, and specialization. The Internet and related e-commerce technologies can be exploited to overcome major systemic constraints. The challenge is to create and build a boundary-spanning information infrastructure that enables quick and efficient information sharing and communication.

Second, the trend toward globalization in project supply chain management will accelerate, as it has the potential to provide significant cost advantages. Boeing Corporation, as part of an initiative to reduce the number parts handled in its production lines, has partnered with European suppliers to procure higher-level assemblies (Sutton and Cook, 2001). Finally, businesses are increasingly concerned about the environment and are undertaking environmental projects to reduce costs, to reduce pollution and hazardous materials, to improve manufacturing performance and quality, to improve relationship with external stakeholders, and to proactively deal with environmental regulation. This trend toward environmental friendliness will require the supply chain for such projects to address issues such as recycling, reuse, asset recovery, minimization of waste, and handling and disposal of hazardous materials (Carter and Dressner, 2001).

To effectively respond to those challenges and exploit the opportunities, project supply chains need to adopt a comprehensive and integrated supply chain perspective. Such integrated supply chains will significantly enhance their value and enjoy a distinct competitive advantage.

Summary

The retail and the traditional manufacturing industries have enjoyed great success by adopting the principles and strategies of supply chain management. More recently, project-based organizations have also realized that the use of SCM-related strategies can significantly enhance value in projects.

Supply chain management is a set of approaches that can be used to integrate the network of all organizations and their activities in producing and delivering a product or undertaking and completing a project so that systemwide costs are minimized while meeting or exceeding customer requirements.

Given intense global competition, businesses have to embrace effective supply chain management approaches to remain viable and provide value to their customers. Significant benefits accrue to companies that effectively manage their supply chains. The benefits include lower costs, shorter lead times, increased productivity, greater customer satisfaction, and higher profits.

Customers, suppliers, design and operations, logistics, and inventory are critical areas of a supply chain. Value optimization in supply chains requires that these critical areas be effectively managed. While project supply chains encounter many of the same issues and challenges faced by supply chains in other industries—such as effective inventory management, supplier partnering, coordination of activities and effective communication among supply chain members, availability and sharing of information, security, and so on—the complexity of project supply chains makes management more difficult and challenging.

The important value drivers in project supply chains are customers, cost, flexibility, time, and quality. Strategic management of these factors by choosing the right suppliers, adopting TQM and lean management approaches, supplier partnering, and, above all, having a customer focus, will ensure value optimization.

Specific steps that can be taken for value optimization in project supply chains include process flowcharting, process standardization, process control, prequalifying suppliers, periodic supply chain audits to take preventive and corrective action, and measuring project supply chain performance. The four process value drivers of time, cost, quality, and flexibility constitute the major performance metrics categories for project supply chains and the Supply Chain Operations Reference (SCOR) model can provide an excellent framework for measuring and tracking performance in projects.

Increasing globalization, advances in information technology, and environmental concerns are some of the challenges that project supply chains will face in the future. Strategic thinking and an integrated approach to managing project supply chains can overcome these challenges and lead to success and value optimization.

References

Ballard, R., and H. J. Cuckow. 2001. Logistics in the UK construction industry. *Logistics and Transportation Focus*. 3(3):43–50.

Betts, M. 2001. Kinks in the chain. *Computerworld* 35(51):34–35.

Burnell, J. 1999. Change management is the key to supply chain management success. *Automatic I. D. News* 15(4):40–41.

Carter, C. R., and M. Dresner. 2001. Purchasing's role in environmental management: Cross-functional deployment of grounded theory *Journal of Supply Chain Management* 37(3): 12–26.

Cottrill, K. 1997. Reforging the supply chain. *Journal of Business Strategy*, 18(6):35–39.

———. 2001. Engineering a value chain. *Traffic World*. 265(9):21–22.

Coyle, J. J., E. J. Bardi, and C. J. Langley. 2003. *The management of business logistics: A supply chain perspective. 7th ed.* Cincinatti: South-Western.

Dainty, A. R. J., G. H. Brisco, and S. J. Millet. 2001. Subcontractor perspectives on supply chain alliances. *Construction Management and Economics* 19(8):841–848.

Fisher, M. L. 1997. What is the right supply chain for your product? *Harvard Business Review* 75(2): 105–116.

Foreman, C. R., and D. H. Vargas. 1999. Affecting the value chain through supplier Kaizen. *Hospital Materiel Management Quarterly* 20(3):21–27.

Holmes, S., (2001). Boeing Goes Lean. *Business Week.* (June 4): 94B–94F.

Hutchins, G. 2002. Supply chain management: A new opportunity. *Quality Progress* 35(4):111–113.

Kumaraswamy, M., E. Palaneeswaran, and P. Humphreys. 2000. Selection matters: In construction supply chain optimization. *International Journal of Physical Distribution and Logistics Management* 30(7/8): 661–669.

Mohamed, S. 1996. Options for applying BPR in the Australian construction industry. *International Journal of Project Management* 14(6):379–385.

Nicolini, D., R. Holti, and M. Smalley. 2001. Integrating project activities: The theory and practice of managing the supply chain through clusters. *Construction Management and Economics* 19(1):37–47.

Padhye, A. 2001. Apply leverage to ensure business process integration in your supply chain. *EBN* 1282:PGL38.

Pinto, J. K., and P. J. Rouhiainen. 2001. *Building customer-based project organizations.* New York: Wiley.

Preston, R. 2001. A glimpse into the future of supply chains. *Internet Week* 885:9–10.

Rampey, J., and H. V. Roberts. 1992. Perspectives on total quality. *Proceedings of Total Quality Forum IV.* Cincinnati.

Simchi-Levi, D., P. Kaminsky, and E. Simchi-Levi. 2003. *Designing and managing the supply chain: Concepts, strategies and case studies.* 2nd ed., p. 11. New York: McGraw-Hill/Irwin.

———. 2003. *Designing and managing the supply chain: Concepts, strategies and case studies.* 2nd ed. New York: McGraw-Hill/Irwin.

Stevenson, W. J. 2002. *Operations management.* 7th ed. New York: McGraw-Hill/Irwin.

Sutton, O., and N. Cook. 2001. Quest for the ideal supply chain. *Interavia,* 56(657):24–27.

Turner, J. R. 1999. *The handbook of project-based management,* p. 150. Marlow, UK: McGraw-Hill.

Watson, K. 2001. Building on shaky foundations. *Supply Management* 6(17):22–26.

Wong, A., and P. Fung. 1999. Total quality management in the construction industry in Hong Kong: A supply chain management perspective. *Total Quality Management* 10(2):199–208.

Yeo, K. T., and J. H. Ning. 2002. Integrating supply chain and critical chain concepts in engineer-procure-construct (EPC) projects. *International Journal of Project Management* 20:253–262.

CHAPTER TWENTY-SEVEN

PROCUREMENT: PROCESS OVERVIEW AND EMERGING PROJECT MANAGEMENT TECHNIQUES

Mark E. Nissen

Effective procurement is critical for effective project management. Depending upon the specific type of project being managed, over 50 percent of the total project cost can be attributed to parts, supplies, and services procured, and for many high-technology projects, this procurement fraction can approach 90 percent. Further, because of long lead times, procured items nearly always define the critical path through the project schedule network. And dependence upon markets for procured items can create project management difficulties in terms of agency (e.g., information asymmetries, incentives) and coordination (e.g., redundant project management organizations, interorganizational communications). In short, if a project manager is not managing procurement, then he or she is only managing 50 percent or less of the project as a whole.

Despite this critical role, however, the term "procurement" remains broadly defined and is used to describe a variety of entities (e.g., functions, organizations, systems, processes). The term is also evolving through time, as the activities associated with procurement have become increasingly important to enterprise success. For instance, procurement was once descriptive of the simple clerical activities associated with purchasing well-specified items, but it has evolved in some organizations to describe instead strategic partnering efforts made by senior executives.

In the former case, all that is required is buying an item that has already been specified, from a vendor that has already been selected, for a purpose that has already been determined (e.g., a part to be installed on an assembly line). But procurement also involves the activities associated with deciding whether an item will be made in-house or purchased from outside vendors (i.e., the make/buy analysis), and deciding from which vendor—or collection of vendors—to purchase an item is commonly included within the responsibilities assigned to procurement organizations. Procurement further represents a central activity in terms of supply chain management, which seeks to integrate the processes and activities of vendors, suppliers, producers, and customers, and procurement executives are routinely relied upon to shape enterprise strategy based on opportunities to form partnerships, alliances, and joint ventures with "vendors."

As implied, the project management professional has a number of different lenses through which to view procurement. For instance, procurement has long been referred to in functional terms, which depict a division of labor (e.g., buying items vs. making them), specific job tasks (e.g., market research, obtaining vendor quotations), and worker skills (e.g., contract interpretation, negotiation). As another instance, procurement is also referred to in organizational terms, which depict a specific department or other organizational entity in the enterprise, complete with its own managerial hierarchy, worker roles, and organizational responsibilities.

Procurement is further referred to in terms of a system, which involves examination of its inputs (e.g., requirements, information), outputs (e.g., purchase orders, received vendor items), transfer function (e.g., vendor selection, vendor management), and environment (e.g., corporation, industry). But I find it particularly useful to describe procurement in terms of a process—actually, a set of processes interlinking vendors, producers, and customers along the supply chain—with its attendant work activities (e.g., requirements determination, source selection), actors (e.g., market researchers, buyers), organizations (e.g., purchasing, contract management), and technologies (e.g., electronic catalogues, communication networks). Although every lens offers a unique perspective, the process view is inherently cross-functional, interorganizational, and systemic, and it enables one to focus on those aspects that are most important in terms of project management, even supporting analytical efforts that can effect dramatic performance improvement.

In this chapter, I adopt a process perspective to provide a focused overview of procurement, and I illustrate how it interacts with and interrelates to other key enterprise processes of importance to the project manager. Because this chapter is followed by more detailed treatments of supply chain management, procurement practice, bidding/tender management, and contract management, I refrain from delving into these areas in any depth. Instead, I leverage the process perspective and show how it can be used to improve procurement performance on the project. This performance improvement focus presumes you understand the basic elements of project management (e.g., project planning, staffing, coordination, monitoring, intervention) and discusses some emerging project management techniques from current research. In addition to providing an overview of procurement in the project management context, I hope to make a contribution in terms of these new techniques for the project manager's toolbox, and we critically review the current state of procurement practice to provide some guidance for effective management. The chapter closes with key conclusions pertaining to the process view and emerging techniques for the management of procurement in the context of project management.

Procurement Process

Nissen (2001) describes procurement in terms of complementary processes, which provides a useful perspective for understanding and management.

Complementary Processes

Two complementary processes are involved with procurement: (1) customer buying and (2) vendor selling. Although these customer and vendor processes can be viewed as separate intraorganizational activities within each of the respective buying and selling enterprises, a strong case can be made for viewing such activities *together*, as an integrated interorganizational procurement process. This reflects the integrated focus of prominent procurement models, such as the one used by Gebauer et al. (1998) to discuss the revolutionary potential of Internet- and Web-based procurement. Other models, such as the one proposed in Kambil (1997), include variations on a similar process flow and set of activities.

I find the General Commerce Model (Nissen, 2001) to be particularly useful for the task at hand, because it is specifically designed to highlight those aspects of project procurement that lend themselves to process integration. Process integration (i.e., managing all supply chain participants and activities as single coherent whole) represents an emerging project management technique for use in the modern enterprise. In particular, this model makes explicit the activities and media associated with *exchanges*, which tie together the various participants and activities.

Exchanges are those activities that constitute procurement proper, and they demarcate interorganizational process integration points; such integration points highlight avenues for a project manager to influence procurement process performance. The General Commerce Model diagram presented in Figure 27.1 depicts the process flow (from left to right) associated with a procurement transaction or relationship. A transaction generally occurs over a relatively short period of time, whereas a relationship is more enduring. Whether ephemeral or enduring, most transactions and relationships can be seen to progress through the steps along the process flow depicted in the model. Clearly, these steps represent procurement at a very high level.

Referring to the figure, from the buyer's perspective, note that the process begins with the identification of some need (B1) and proceeds through sourcing (B2) and purchasing

FIGURE 27.1. GENERAL COMMERCE MODEL.

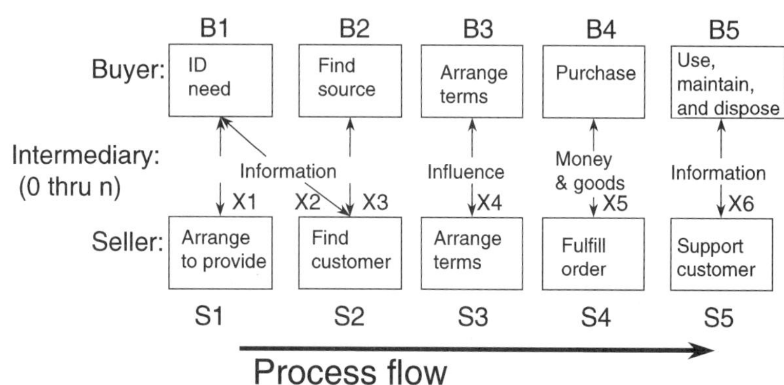

Source: Adapted from Nissen, 2001.

(B4) to the use, maintenance, and ultimate disposal of whatever product, service, or information is purchased (B5). The simple purchase by a price-taker (e.g., of commercial off-the-shelf, or COTS, products) will not generally include negotiation (B3) or involve significant terms other than price, whereas the exchange of influence through negotiation can become very involved in more complex procurements. The seller's process begins with some arrangement to provide (S1) a product, service, or information (e.g., through internal research and development, service process design, information acquisition) and proceeds through customer search (e.g., marketing and advertising; S2), pricing (S3), and order fulfillment (S4) to customer support (S5). The seller of COTS products may similarly not engage in negotiation.

The arrows connecting these high-level process steps are used to represent key items of exchange between buyer and seller. For instance, information is exchanged at several points along the process flow, as are money and goods (or services, information) and even "influence," as delineated at the negotiation stage. As depicted in the figure, zero or more levels of intermediaries (e.g., brokers, dealers, agents) can also participate in the process. Additional exchanges may also take place between such intermediaries and the buyers and sellers. In either case, by examining the activities through which information and other exchanges are made, one can acquire insight into those offering good potential for process integration, which can improve project performance dramatically.

Enterprise Supply Chain Illustration

Here, I draw from recent field research (Nissen, 2001) to describe an enterprise supply chain for illustration. As delineated in the General Commerce Model, we find this enterprise supply chain is actually composed of two complementary process instances: customer buying and vendor selling. Specifically, the buying part of the process pertains to work done by the supply department at a medium-size government facility in the United States. As a government institution, this facility is subject to a full complement of procurement policies and procedures, not unlike those that govern the purchasing activities of most large enterprises, in both the public and private sectors. The selling part of the process pertains to work done by a leading-edge U.S. technology development company. This firm is a leader in its COTS product market and maintains an active research and development activity that drives frequent product introductions, updates, and releases. This kind of rapid product evolution has been noted as problematic for procurement in major enterprises such as the large corporation and government agency (see Nissen et al., 1998).

Referring to Figure 27.2, note that this enterprise procurement process is used to instantiate the General Commerce Model. Notice that the process includes a supply department intermediary, which performs specialized purchasing activities on behalf of diverse users (e.g., engineering, manufacturing, or sales personnel) in the organization. The enterprise process depicted here includes a more detailed delineation of activities than the General Commerce Model in Figure 27.1 does. For instance, the "ID need" (B1) and "find source" activities (B2) from Figure 27.1 are decomposed and expanded to account for a number of activities and exchanges that take place between the user, contractor, and supply department—for instance, conduct market survey (X2), complete purchase request (PR) form (X2'), research sources, and issue requests for quotation (X3). Exchanges internal to the buyer

FIGURE 27.2. Enterprise Supply Chain Process.

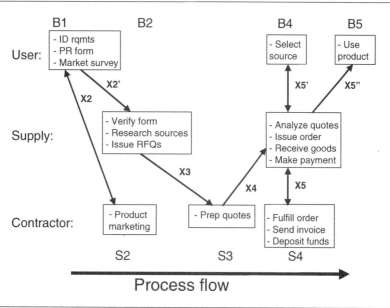

Source: Adapted from Nissen 2001.

organization are differentiated from their interorganizational counterparts by the prime symbol (e.g., X2', X5'). As another instance, the "purchase" activity (B4) from Figure 27.1 is similarly decomposed and expanded to account for more detailed activities and exchanges—for instance, analyze quotations, select source (X5'), issue order, receive goods, and make payment (X5).

As is appropriate for a general model, not all of the General Commerce Model activities are relevant to this particular enterprise supply chain process. For example, the buyer does not arrange terms (B3) for simple COTS purchases. With more complex procurements, arrangement of terms by the government buyer can be extensive. Alternatively, notice the seller effectively performs this activity (S3) for both parties when it prepares quotations. Such quotations will generally include information pertaining to product specifications, warranty, delivery, payment, and other terms set by the seller in addition to price. In one sense, this enterprise procurement process has evolved and specialized so that only the seller arranges terms. The other activities are relatively straightforward and map neatly to the General Commerce Model. From start to finish, we note a total of seven exchange points in the process between buyer, seller, and intermediary.

Exchange-Point Analysis

In this section, I build upon the procurement process perspective mentioned previously to discuss exchange-point analysis and outline how it can be used by the project manager to improve procurement process performance. As noted previously, exchange points highlight opportunities for process integration, which is noted for performance improvement opportunities (Nissen, 2001). The key is, exchanges between different organizations along the supply chain (e.g., buyer, seller, intermediary) are associated with process *friction*, a phenomenon that is known well for increasing project cost and schedule (Nissen, 1998a).

Friction occurs whenever information or other exchange items must pass through separate organizational hierarchies (e.g., for management approval) or cross-functional specialties (e.g., from engineering to procurement), or involve market-based contracting mechanisms (e.g., legal documents). In terms of procurement process innovation (Nissen 2000), such exchanges are expressly labeled as "handoffs" to highlight their associated performance degradation. To the extent that the project manager can decrease friction, project coordination costs can be decreased and time can be saved in terms of project schedule. The key to process integration is to convert key elements of procurement and the supply chain process from one based on laws and commercial norms pertaining to markets (e.g., treating one's vendor as an *external* corporate entity) to one that relies upon trust and goal alignment through the hierarchy (e.g., treating one's vendor as an *internal* project group). Williamson (1975) provides a thorough treatment of markets and hierarchies.

Even though a customer and vendor may be organized as distinct legal entities (e.g., corporation, governmental agency, military unit), when they are mutually engaged in a common project effort, project-focused subsets of both customer and vendor organizations can align their processes with the common goal of project performance (e.g., satisfying project requirements, meeting cost and schedule estimates). This requires a different mindset for the project manager, one that diverges from "us versus them" thinking and envisions the customer and all supplier tiers as one integrated virtual organization (Davidow and Malone, 1992). Consistently, current best practices in industry and government project management include trust-based operations between customers and vendors (STSC, 2000).

However, in procurement as with shoes, one size does not fit all, and trust-based procurement is not for every organization and situation. For instance, where items to be purchased represent fungible commodities (e.g., wheat, standard grade steel, computer cables) that are offered through standard terms, price generally represents the single decision criterion, and there may be little to gain through developing close relations with any single vendor. Alternatively, where criteria include factors other than price alone, trust-based relationships can be critical. For instance, if you are using just-in-time or similar techniques to minimize inventory costs, you are likely to depend upon vendors to keep your line running, and you may need them to make short-term production adjustments and rush deliveries to meet your unanticipated demand changes. Consider this heuristic: Where a vendor becomes critical to your success, it may be worthwhile to invest in trust-based relations.

RFP/RFQ Example

To help make these ideas concrete, consider, for example, the kinds of information exchanges that routinely occur between customers and prospective vendors associated with a request for proposal (RFP) or quotation (RFQ). Such exchanges are delineated as "X3" in Figures 27.1 and 27.2. A typical buying organization will spend considerable project time and effort developing specifications for an item to be procured, conducting market research to identify the kinds of product capabilities possessed by industry vendors and preparing often-lengthy quasi-contractual documents to request formal proposals (i.e., offers) or binding price quotations.

In turn, a typical selling organization will spend considerable project time and effort analyzing specifications, preparing cost and schedule estimates, and preparing often-lengthy contractual documents in the form of formal proposals or firm quotations. Because of their quasi-contractual and often binding nature, such documents generally must pass through one or more levels of management review—including legal analysis—before being officially transmitted through informational exchange. Although RFQs and price quotations can sometimes be prepared in a matter of minutes or hours for simple commercial and catalogue items, RFPs and proposals associated with large and/or complex projects often require weeks or months to prepare. This RFP/proposal information flow depicts a prime example of an exchange point that is ripe for process improvement through process integration.

For instance, the project manager can ask him or herself: How would such information be exchanged *within* our own organization? Would the same number and levels of management review be required? Would the lawyers be required to review all of the documents before transmittal? Would the documents require several days' processing time to wend their way through two organizations' internal mail systems and some third-party delivery enterprise (e.g., the post office)? Would the specifications require exact definition before being shared with product scheduling and manufacturing managers? If the answer is yes, then one's internal organization represents a rich area for process redesign. But if the answer is no, then exchange point analysis has identified several promising opportunities for process improvement, and such improvement can be made, in many cases, simply by treating the customer or vendor on a project as an "insider" as opposed to an external legal organization.

Why not empower project managers of both buyer and seller organizations to approve all project transactions, for example, with a caveat that lawyers from both sides must eventually (e.g., long after the items have been procured and used for the project) agree on a common set of terms and conditions? Some aspects of this suggestion are included under the rubric *alpha contracting* (Nissen, 1998b), but exchange point analysis enables the project manager to reach more deeply into the fundamental elements that join buyer and seller processes.

This specific approach (e.g., project manager empowerment, deferred legal review) is not so important as its ability to illustrate the kinds of procurement process changes that can be envisioned through exchange point analysis. Indeed, notice that this particular approach involves no technological innovation or even organizational change; only the respective attitudes (e.g., trust) between project customer and vendor require change, along with some intraorganizational rule changes (e.g., project manager authority) and process-work flow modifications (e.g., legal review).

Clearly, other changes such as employing network technology to electronically link customers and vendors, establishing electronic catalogues and virtual malls, and implementing intelligent software agents can further innovate the procurement process associated with a project, and organizational changes (e.g., forming project-oriented joint ventures) offer even more dramatic opportunities for buyer/seller process integration. Additionally, the other exchange points noted in the preceding figures (e.g., X1, X2, X4, X5) offer further opportunities for process integration and the associated performance improvement. But this example should convey the key ideas and illustrate how opportunities for procurement process improvement can be envisioned through exchange point analysis in the context of project management. The process view of procurement is central to envisioning such opportunities.

Procurement Guidance

The current state of procurement practice reflects dynamic interaction between two opposing forces, and as such it remains in considerable flux. On the one hand, strong drivers both internal (e.g., increased corporate accountability) and external (e.g., electronic commerce) to the project organization press for change. On the other hand, procurement has been practiced by organizations for hundreds if not thousands of years, and through a Darwinian process, the systems and techniques in practice today are proven in terms of efficacy. Indeed, most organizations have volumes of procedures describing the procurement process in considerable detail, and resources such as the PMBOK lay out all the basic process steps. So what should the project manager think about the state of procurement today? What advice can we give for keeping the best of the old without missing opportunities of the new? Based on continuing research into advancing the practice of procurement, I offer guidance in the following seven project management heuristics or rules of thumb:

1. *Don't tinker with procurement systems.* The colloquialism "if it ain't broke, don't fix it" is not an exemplar of good grammar, but it captures an important consideration in terms of project procurement: Most systems and procedures serve a useful purpose and are understood by the people in your organization. Unless there is compelling reason for change, the project will likely be better off if you refrain from tinkering with the procurement process. Many managers seem compelled to involve themselves in every detail of a project, and such detailed involvement is clearly warranted in many areas (e.g., maintaining political support for the project, assessing complex technical/financial tradeoffs). But system change is costly, as people's performance levels decrease reliably when they are required to learn new systems and procedures. Particularly in mature organizations with stable environments and established procedures, procurement systems is likely to be the last area requiring detailed project management attention.
2. *Do manage the critical path.* Notwithstanding the guidance in the preceding heuristic, I noted at the beginning of the chapter that procurement often lies on the critical path of a project, and we discussed exchange point analysis as an approach to systematically identifying opportunities for process enhancement. As such, any significant improvement in procurement efficiency or efficacy can effect direct improvement to the project as a whole, and the project manager sits in a prime position to view potential improvements. However, it's important for the project manager to balance the potential for performance gains through procurement enhancement with project risk. Many techniques for decreasing procurement cycle time (e.g., concurrent vendor development, close

customer-vendor interaction, increased trust) require additional coordination and increase project risk. Reducing the *most likely* project duration (e.g., by accelerating procurement schedules) only makes sense so long as management is also willing to accept the *worst-case* scenario that may emerge if coordination fails and vendor projects spin out of control.

3. *Question the matrix.* Most modern projects are organized using matrix management; that is, most people working on modern projects belong to two organizations: a functional group and a program or product group. In the case of procurement, the question is whether and how to integrate procurement people into the project organization. On the one hand, where standard commodities are required for a project, and specifications and schedules can be developed well in advance of their need, there is little advantage to having procurement specialists join the project team. Indeed, excluding procurement specialists from the project team reduces the number of people requiring direct supervision by and attention of the project manager. On the other hand, many project organizations depend critically upon certain key vendors, and understanding specific vendors' capabilities in detail can be central to project success. In such cases, procurement specialists on the project team may be indispensable.
4. *Balance efficiency with flexibility.* Efficiency is key to controlling costs and earning profits. But so is developing a product that sells and adapting to marketplace shifts. Efficiency is often obtained through standardization of procedures, specialization of labor, and buying in quantity with long lead time. However, standardized procedures are not generally flexible to change; specialized personnel require time and money to train for different tasks; and large-quantity, long-lead time contracts can be expensive to change or break. The project manager must maintain a balance between efficiency and flexibility, and such balance is likely to shift over the life cycle of a project, from one project to the next, and certainly across different technologies and industries.
5. *Benchmark.* Procurement systems and processes tend to be quite visible outside of organizations, and the astute project manager can periodically scan the environment to ascertain how various other organizations (e.g., competitors, partners, suppliers, customers) conduct their procurement processes (e.g., in terms of the four previous heuristics). Through such periodic looks, one's own procurement processes can be compared to those of other organizations to assess where one stands. Such assessment is commonly referred to as *benchmarking*, and it represents a relatively simple and inexpensive source of ideas for potential improvement—as well as confirmation that one's own processes are performing well (or at least adequately). Specialists and managers within the procurement organization are likely to perform benchmarking on a continuous basis, so procurement benchmarking information is often as close as a telephone call.
6. *Time carefully advancing information technology.* As noted, procurement is an exchange-oriented process, and information represents by far the principal object of exchange. Thus, advancing information technology offers omnipresent potential to enhance the procurement process, and because information technology advances so rapidly (e.g., annual doubling in terms of performance/price), last year's infeasible approach may very well develop into this year's competitive advantage and next year's crisis. Unfortunately, information technology improvements to a procurement process generally take

substantial time for implementation, so one must anticipate promising new technologies well in advance of their integration into the project organization. Further, information technology implementations often represent complex projects themselves, so the project manager can find him- or herself managing *two projects* (i.e., one concerning information technology and one concerning the organization's products) simultaneously. Careful timing is required here, as a project manager cannot generally afford to get too far ahead or too far behind in terms of information technology.

7. *Manage software procurements closely.* For over 50 years, software projects have been consistently underestimated in terms of cost, schedule, and complexity, and the success rate of large software projects in particular is dismally low. All projects clearly share many similarities, but it is important to note those aspects of software projects that make them unique (e.g., software is intangible, quality is difficult to evaluate, requirements are hard to specify and keep static) and manage them separately. Further, organizations rely increasingly upon vendors for software expertise, so the project manager depends upon the procurement system to select a capable vendor and establish means for effective coordination. As software becomes increasingly complex and products become increasingly software-intensive, managing the procurement of software can only become more critical to project success. There is no substitute for people with software (procurement) experience, and your project may benefit from one or more specialists if it meets the criteria above.

Summary

Effective procurement is critical for effective project management; if a project manager is not managing procurement, then he or she is only managing 50 percent or less of the project as a whole. The project management professional has a number of different lenses through which to view procurement, but I find it particularly useful to describe procurement in terms of a process, as it enables one to focus on those aspects that are most important in terms of project management, even supporting analytical efforts that can effect dramatic performance improvement.

In this chapter, I adopted a process perspective to provide a focused overview of procurement and illustrated how it interacts with and interrelates to other key enterprise processes of importance to the project manager. But, presuming you understand the basic elements of project management, I introduced the emerging project management technique from current research called exchange point analysis. In addition to providing an overview of procurement in the project management context, I hope to make a contribution in terms of this new technique for the project manager's toolbox, and we critically reviewed the current state of procurement practice to provide some guidance for effective management. In terms of project guidance, we developed and discussed seven heuristics for the project manager: (1) Don't tinker with procurement systems; (2) Do manage the critical path; (3) Question the matrix; (4) Balance efficiency with flexibility; (5) Benchmark; (6) Time carefully advancing information technology; and (7) Manage software procurements closely.

In closing, research to develop the kinds of emerging project management techniques and heuristics outlined in this chapter seeks to extend the state of the art in terms of both the theory and practice associated with project management. As such, techniques and heuristics have now emerged from the drawing board and laboratory to help inform practice. I feel they are appropriate for inclusion in a book such as this. Additionally, I remain very interested in tracking projects in which emerging techniques and heuristics are employed. Project tracking along these lines will help to assess and refine further the associated project management knowledge, and I hope to incorporate important lessons learned through practice into other techniques and heuristics that are still being conceived and developed. This represents the kind of partnership between the academic and practitioner that can decrease the friction often associated with exchanges between theory and practice, and such partnership is essential to enable the flow of project management knowledge from the laboratory to the field. I welcome the opportunity to continue our existing partnerships and engage in new ones.

References

Davidow, W. H., and M. S. Malone. 1992. *The virtual corporation.* New York: Harper Business Press.

Gebauer, J., C. Beam, and A. Segev, A. 1998. Impact of the Internet on procurement. *Acquisition Review Quarterly.* Special Issue on Managing Radical Change. 5(2):167–184.

Kambil, A., 1997. Doing business in the wired world. *Computer* 30 (5, May): 56–61.

Nissen, M. E. 2002. "An Extended Model of Knowledge-Flow Dynamics," *Communications of the Association for Information Systems* 8, pp. 251–266.

———. 2001. Agent-based supply chain integration. *Journal of Information Technology Management.* Special Issue, Electronic Commerce in Procurement and the Supply Chain. 2(3):289–312.

———. 2000. *Contracting process innovation.* Vienna, VA: National Contract Management Association.

———. 1998a. Redesigning Reengineering through Measurement-Driven Inference. *MIS Quarterly* 22(4):509–534.

———. 1998b. Alpha contracting JSOW style. *National Contract Management Journal* 29(1):15–32.

Nissen, M. E., and J. Espino. 2000. Knowledge process and system design for the Coast Guard. *Knowledge and Process Management Journal.* Special Issue: Into the "E" Era. 7(3):165–176.

Nissen, M. E., and E. Oxendine. 2001. Knowledge process and system design for the naval battlegroup. *Knowledge & Innovation: Journal of the KMCI* 1(3):89–109.

Nissen, M. E., K. F. Snider, and D. V. Lamm. 1998. Managing radical change in acquisition. *Acquisition Review Quarterly.* Special Issue on Managing Radical Change. 5(2):89–106.

STSC. 2000. *Guidelines for successful acquisition and management of software-intensive systems.* Version 3.0). Software Technology Support Center, Hill AFB.

Williamson, O. 1975. *Markets and hierarchies: Analysis and antitrust implications.* New York: Macmillan.

CHAPTER TWENTY-EIGHT

PROCUREMENT SYSTEMS

David Langford, Mike Murray

No single issue has dominated project management in certain industries more than procurement. Yet what is it? The *Oxford English Dictionary* defines procurement as the "act of obtaining by care or effort, acquiring or bringing about." The Association for Project Management (APM, 2000) describes procurement as "the process of acquiring new services or products. It covers the financial appraisal of the options available, development of the procurement or acquisition of suppliers, pricing, purchasing, and administration of contracts. It may also extend to storage, logistics, inspection, expediting, transportation, and handling of materials and supplies." Within this generic definition we can see that procurement in a project management sense has a wider definition; it is really about management on behalf of a client or user of a product that is delivered using a project process.

The APM definition of procurement suggests that this process is undertaken in a linear and sequential mode. However, much anecdotal evidence would suggest that the process is often less rational, conducted in an iterative mode, and often influenced by political game playing, groupthink, and even unethical or illegal decision making. Risk management procedures are rather ineffective at combating such behavior. Indeed, the generic definition of the term requires the service of a shoehorn to make it fit different industrial sectors and nationalities. This is evident if we consider the description of the procurement process given below. The definition is taken from a U.S. Web page (www.mgmtconcepts) advertising instructional courses for procurement managers. It clearly considers procurement to be undertaken by an in-house department involved in a buying-selling relationship with external suppliers. The course syllabus is broken down into seven areas that represent six key procurement processes: procurement planning, solicitation planning, solicitation, source selection, contract administration, and contract closeout.

1. *Procurement planning.* Factors in the decision, make-or-buy analysis, and contract type selection
2. *Outsourcing and partnering.* Reasons to outsource, global outsourcing market, buyer/seller relationships, the partnering process
3. *Solicitation planning.* Tools and techniques, specifications, procurement document contents, evaluation criteria
4. *Solicitation.* Developing qualified sellers lists, contacting prospective sellers, conducting a bidders conference
5. *Source selection.* Screening and weighting systems, proposal scoring, contract negotiation strategies, making the decision, elements of a contract.
6. *Contract administration.* Roles, responsibilities, and coordination; kickoff meetings, project performance responsibilities, change control procedures, contract administration, payment system
7. *Contract closeout.* Contract documentation, steps in the claims process, termination of contracts, lessons learned

This syllabus is particularly suited to organizations in the manufacturing and service industries that have procurement departments. However, although the generic principles of buying-selling are appropriate to all, the six key processes noted are less effective in describing the key milestone in procurement within the construction and engineering industries. Owners in this sector often rely entirely on external procurement advisers (public clients are likely to have internal procurement departments/procurement champions) who may be the lead consultant, and the importance of the brief-design-manufacture interface is critical.

Procurement involves selecting from a range of acquisition options. Buying is not necessarily the only, or even necessarily the preferred, one: Making, renting, or leasing may be equally valid options alongside buying. Such options can, however, lead to confusion in project-based industries. Manufacturers and purchasers of products may talk of acquisition, whereas in construction the term "procurement" is most often used. However, even within this sector, the different professionals use preferred terminology. An estimator working within a contracting organization may talk of buying or acquisition, whereas the construction manager will consider these activities to be defined within the boundary of procurement. Thus, as the APM definition makes clear, the term "procurement" constitutes the system/process in which subactivities such as acquisition take place.

This chapter focuses on the strategic issues surrounding the decision to procure anew. The concept of procurement can, however, be seen to differ within various industrial sectors. The procurement structures used by the construction, aerospace, and motor vehicle industries can be considered broadly similar in that an end product is produced via a distinctive design and construction process. Nevertheless, they can also be considered a service industry, and within these three sectors, significant differences in practice exist—the high levels of subcontracting (also referred to as outsourcing) used in construction, for example.

Other industrial sectors may commission special in-house projects as a means to launch a new product, but procurement may be less concerned with design and construction than on "procuring" (often termed "solicitation") the material resources that are used to manufacture their products.

While many of the illustrations used in this chapter are taken from UK construction practice, the broad sweep of the movement from traditional methods to performance-based procurement systems has been observed in many project-based industries around the world. As such, the illustrations are local but the concepts are universal. Indeed, the increasing interest in company benchmarking has led to much cross-fertilization of procurement practice between industry sectors.

Strategic Issues in Procurement: Examples from UK Construction

In an era of the "cult of the customer" where business relationships are perceived to be the dominant driver of business performance, there is increasing interest in the role of client satisfaction in respect of the project process. Government reports (in the United States and the United Kingdom) since the 1940s have complained that many project-based industries from construction and provision of IT services to defense have not served clients well (Morris, 1997; Murray and Langford, 2003) and that overruns on time, budget, and poor quality beset the project process.

Client Satisfaction

Naturally one of the most important client roles in the procurement system is the provision of a brief. This "statement of need" has to flow from the client's expectation in terms of the function of what is to be delivered, and performance in terms of time, cost, and quality. (See the chapter by Davis et al. on requirements management.) Clients will typically want an appropriate level of involvement in the whole procurement process; this involvement may be minimal, but client satisfaction needs to be defined early on, as it will have powerful implications for the selection of the procurement route. The issue of risk sharing between the client and the project delivery team is at the heart of such early decisions about procurement, and as Rowlinson (1999) says, the client's role is a strategic one—it sets the project objectives and the project contractor's role is to turn such objectives into a finished product that satisfies the client, the users, and society at large.

Procurement—and to a large extent therefore project management practice—varies between industries. But even within industries, practice (of both) varies.

In the construction industry, for example, procurement strategies of today still follow methods of procurement that were available to the medieval builder. The master mason designed and built the structure; he (for it invariably was) designed and managed the works by engaging with other craft guilds to undertake such packages such as woodwork, roofing, glazing, and so on. Since that time professional roles have emerged that have fragmented the design and construction process. (See the chapters by Morris on construction, and Cooper on process modeling.) Most government reports into project-based industries since World War II have complained of this fragmentation, and since the late 1960s, attempts have been made to integrate the design and of many of the professional roles used to deliver major projects. (This has paralleled other initiatives aimed at improving project management practice.) Let's look in a little detail at the various procurement options taking construction as an example.

Sequential Models

Traditional Approach

Here the key feature is the separation of design from construction/manufacture with the sequence of events being design-tender-build/manufacture, and the design phase being divided up into roles or specialist consultants. In this system the lead designer has a key role in leading the design function and managing the project. The project delivery company is likely to be selected by competitive tender, and the work of delivering the project will be subcontracted to specialist trades. The structure is as shown in Figure 28.1.

Accelerated Traditional

Because of dissatisfaction with the traditional elongated and often disputatious process, this model is often replaced with a fast-track model that overlaps design and manufacture/construction. With this model, project work starts as soon as sufficient design is available. The process is shown in Figure 28.2.

The design work is carried out with the same range of consultants, but the contractor will be selected by a negotiation or a pricing of a schedule of likely combinations of materials and processes. The roles and relationships will be as those in the traditional model.

Design-Build

The disadvantages of the traditional methods are that the organizations charged with design on the one hand and with realizing the design on the other are separated. Consequently there is a lack of single-point integration and responsibility, and fragmentation increases. To

FIGURE 28.1. TRADITIONAL PROCUREMENT.

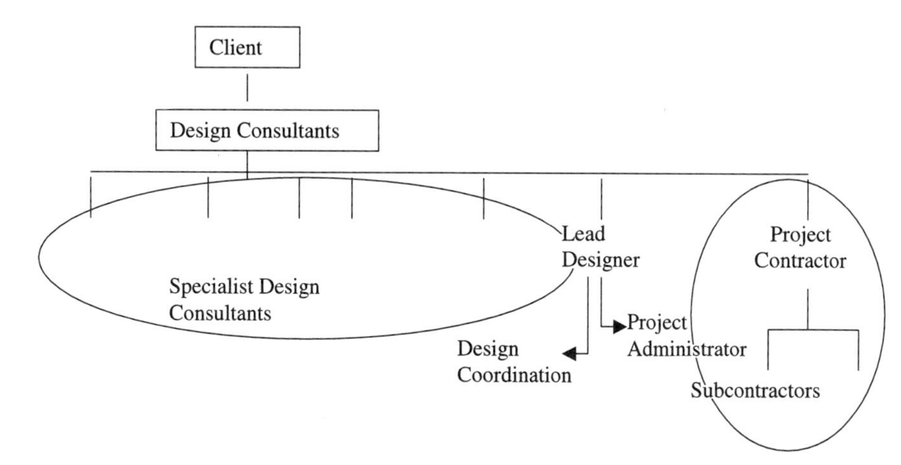

FIGURE 28.2. TRADITIONAL (ACCELERATED) PROCUREMENT.

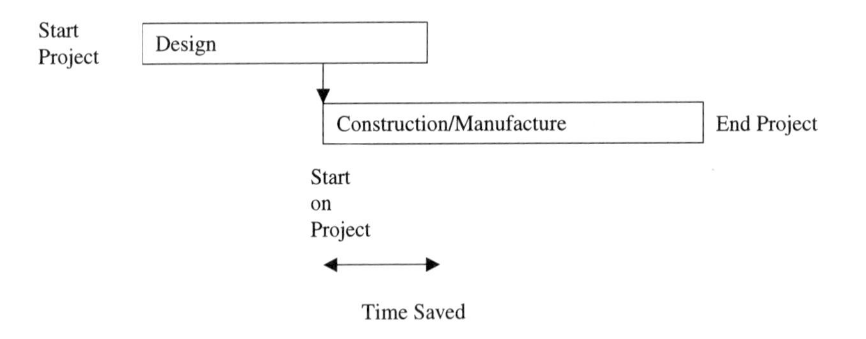

counter this, most project-based industries have a form of design-build model providing greater unification of responsibility (though in construction the traditional system still retains a strong presence throughout the world). In construction this method of procurement has evolved such that it has many variants. Broadly speaking, the design-build organization tenders/bids or negotiates with the client prior to commencing design work. Part of the tendering or negotiations will be schemas that demonstrate how the design-build organization is going to respond to the client's needs. In such circumstances, different costs will be associated with design solutions, and thus bid evaluation can be complex. Three main organizational configurations can be seen.

- *Integrated design-build.* This occurs when the design, costing, and implementation expertise lie within one organization and the contract for work is between the design and deliver company and the client.
- *Separated design-build.* This arrangement puts together a temporary organization comprising the necessary design and construction expertise. The project team is a consortium of independent practices for design and a construction or manufacturing firms, put together for a specific bid. It may well be that this arrangement has some semi-permanence in that every time a design-build job is imminent, a broadly similar set of firms configure themselves to present a design-build solution. Alternatively, bespoke arrangements are made to suit a particular kind of project.
- *Novated design-build.* Third, a more recent innovation is novated design build. The word "novated" is taken from the meaning of *novation,* which means substitution of a new obligation for the existing one. In this case the "new obligation" is for the project contractor to take over "existing" designs drawn by a lead designer. The system works by the client commissioning a design firm, often selected through an invited competition, to create the concept, key features of the design, and outline budget. When the conceptual work is agreed, then the designer's work is handed over to the project contractor who then "owns" the design and so develops it to a condition that enables the design to be realized. The method has the benefit of delivering solutions that have a schema design

developed without production issues dominating the designers' thinking, thus allowing design flair to flourish as the contractors develop the design with an eye to retaining the design concept while producing production drawings. This brings benefits in terms of assembly, function, and value. In short, the method seeks to maximize the design benefits of the traditional system and the production benefits of the design-build model.

The contractual relationships of all three variants are shown in Figure 28.3.

"Management" Systems

This model is characterized by the separation of the management system and the operational system required to deliver projects. The technical system—the design and manufacture/construction—are undertaken by specialist organizations, while a construction manager (or management contractor), or project management organization, provides the integrating management system.

The management organization is appointed early on in the project process, its role being to coordinate the design, procurement, and logistics to be used in common by all contractors and to provide the project control (overall scheduling and cost management). In the United Kingdom there are two principal variants to the construction form of this system: management contracting (MC) and construction management (CM), the differences being shaped by the relationship between the parties to the contract. The two forms are summarized in Figure 28.4.

As Tookey et al. (2001) have pointed out, such arrangements are often tailored for a particular project with payment systems, legal arrangements, and specialist contractor selection methods all being bespoke. This has an influence upon professional roles and relationships. Walker and Newcombe (2000), for example, see that the social system of a project is influential with the personality and power bases of the leaders of the various organizations engaged in the project shaping performance in practice. This theme will be explored in greater depth later.

FIGURE 28.3. VARIANTS OF DESIGN-BUILD.

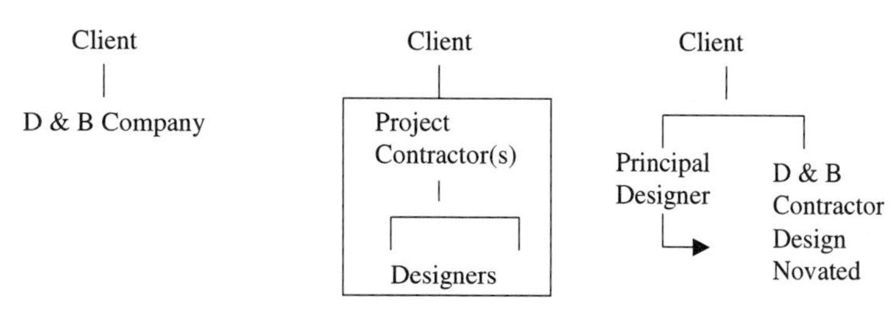

FIGURE 28.4. MANAGEMENT SYSTEMS.

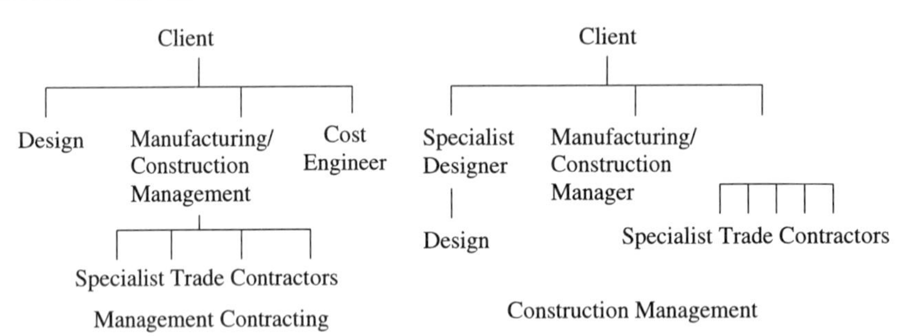

The advantages of the system are really drawn from Adam Smith's (1838) classical economic theory proposed in the *Wealth of Nations*, that of specialization producing economic benefits because of lower costs and more efficient production. The use of specialist trade contractors enables designs and installations to be rolled out as the project progresses, thus ensuring a fast-track approach. Moreover, the specialist packages can be tendered and so have the dubious benefit of work being let for the lowest cost. (This philosophy is waning in the face of "best value" rather than lowest-priced procurement strategies).

In construction, "project management"—as a procurement form rather than as a discipline—is similar to construction management, the principal difference being that the scope and authority of the project manager is likely to be greater that the CM. The PM organization will be appointed early and acts as a quasi-client. The PM may be asked to undertake on behalf of the client functions such as a site acquisitions, arranging the funding, obtaining necessary permissions—in effect undertaking all of the client-led feasibility studies. The project manager will commission the designs and constructors. In short, the PM acts in lieu of the client organization.

Explaining the Changes

The drive toward a more professional role-based procurement model has surpassed the more contractual arrangements of the past. Curtis et al. (1991) see the spectrum as shown in Figure 28.5.

These changes in construction procurement need to be understood in the context of organizational theories and social relations between the parties. In the following, the framework used draws upon the construction industry, although the argument applies in other project based industries.

Organizational Theories and Procurement

Of key importance is how the players in the project process make sense of the world of procurement. Hence, procurement is an "issue" that carries a definition of something that

FIGURE 28.5. ROLE-BASED PROCUREMENT.

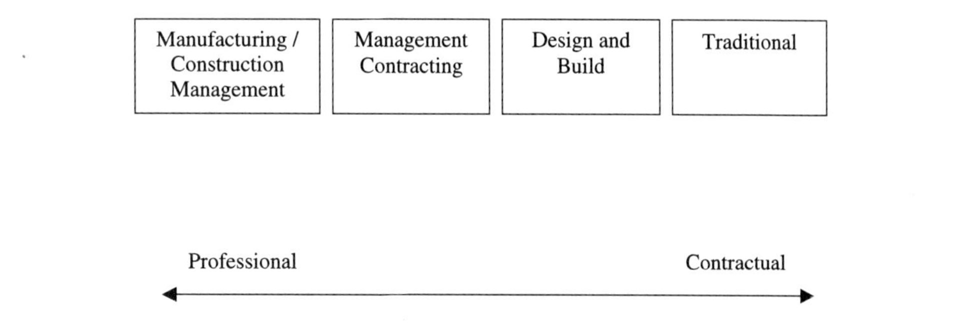

Source: After Curtis et al. (1991). Copyright CIRIA Publication SP81.

is "problematic and requiring action." In the context of procurement of projects, major clients, including government, have complained of poor performance of the project-based industries since World War II. Murray and Langford (2003) reviewing postwar government reports into the UK construction industry conclude that procurement is one of the "issues" for the industry to address. Similar evidence may be found in defense, shipbuilding, and other capital project industries.

The earlier models of traditional procurement have been found wanting. The emphasis upon "contractual" rather than "trust" relationships used to link the parties together has all too often, as the Latham report (1994) complained, ended in court. Such contractual arrangements relied upon a machine metaphor (Morgan, 1986) to explain how the procurement system worked. Each role in the process was carefully defined and well understood and was underpinned by contracts that spelled out the legal obligations and responsibilities of each party. Codes of practice for tendering and plans of work proceduralized the process. The parties in the contract have a role that may be replaced by others performing a similar role in the same way that a defective clutch on a car is replaced. The organization behaves in a hopefully predictable way.

As projects become more complex in technical and organizational terms, the machine metaphor begins to be no longer capable of containing the aspirations of different groups in the process. Clients become unsettled by delays and cost overruns and seek to assert authority over the project process. Dealing with a multiorganization and multicontractual environment expressed in the traditional paradigm becomes irksome. Hence, in a bid to work with "single-point responsibility," organizations encouraged the growth of design-build contracts. In Morgan's terms this would be the "organic" metaphor at work. The contractors who formerly were engaged as the principal builders of a project and had little to do with design organically evolve to fit the needs of the new environment. The client buys construction services from one organization, and the contractor, by integrating design with construction, reduces the physical and financial risks associated with converting someone else's design into reality. The shift in client expectations created preferences for changed procurement systems.

A third epoch may be detected in the development of different procurement routes: the move from design-build to management forms of procurement. Powerful clients and contractors looking at project based work in the United States in the early 1980s saw that practice there seemed able to deliver projects faster and cheaper than in the United Kingdom. The U.S. Construction Management form came into the UK environment as an alternate to management contracting and stimulated other forms of innovation procurement, not least project management. Clients and constructors no longer just polished the mirror to "tell it like it is." The challenge for such innovative organizations was to "tell it like it might become" and to unseat conventional assumptions about roles and relationships in the procurement process.

The metaphor here is of the organization as a "political" entity. Different interest groups have conflicting ideas of how to move forward. The contours of the professional roles start to change: Designers, for example, saw a diminished responsibility with fewer managerial duties. Cost advisors to the client saw a new future in project management as traditional ways of pricing projects began to fade away and new consultants such as value engineers, program managers, health and safety managers, and so on began to emerge. Inevitably in such paradigm shifts in roles and relationships, bitter arguments break out. The idea that "quality" refers to the product over long periods of time, it is said by some, is replaced by the concept that quality lies in the project process. This further diminishes the power of the design professionals.

In breakdown situations such as this, the old political order begins to fragment. Disputes then become more prevalent. Clients and government consequently have to seek new order. In the UK construction industry, for example, legislation has been introduced to restore order by making arbitration mandatory, ordering prompt payment to trade and specialist contractors. The phase of the political metaphor is complete.

This shift from traditional to management contract can be read as a reflection of the shifting nature of social hierarchies in society. In the traditional system the designer was "commissioned" as an independent artist and the contractor was "contracted" because legal constraints were needed to enforce performance. In this model the designer with extensive power and authority represents the ruling class and the contractor the working class. As society progresses, the class barriers are perceived to fall and political and social plurality is encouraged. The power exercised by the designer in relation to other professionals diminishes, and power and authority becomes more equally spread amongst the participants in the project process. In the new situation, the power of the client is emphasized and brought into greater relief. In the UK construction industry, the Egan report, "Rethinking Construction" (1998), legitimized the new authority of the client.

The new epoch is governed by the cultural metaphor. Here strong uniform norms of behavior are expected from all players in the project team. Epithets such as "teamwork," "singing from the same songsheet," and "pulling together" become watchwords of the cultural phase. In the UK construction sector, the trigger for this was the Egan report, which identified key issues in the procurement process that commanded reform. Central to these was the one that the procurement process should concentrate on customer focus with a strong emphasis upon measurement of a range of outcomes that could lead to improvement for all. In the United Kingdom the transformation of the industry culture is well under way and has shaped the procurement practices by the introduction of supply chain management to not only align the technical aspects of procurement but also to create project cultures that harmonize or at least make less diverse the cultural differences between participants in a project. (See the chapter by Venkataraman.)

The UK government sought to reshape government procurement by installing "best value" rather than "price" as the leading agent of contractor selection. Again, cultural values were seen as an important part of the "best value." Government agencies such as the UK Ministry of Defence and the Health Service refashioned their procurement practices to engage "prime contractors" who manage projects based upon functional and performance driven criteria. For example, the army may wish to procure a facility for keeping soldiers fit, a prime contractor is selected, and an early role will be to define whether the requirement is best delivered by the provision of a gym or an outdoor assault course. In this environment it is important for all those seeking to work with major clients to be compliant to this new culture. This means integration of software systems and capacity to engage in e-procurement. The journey is depicted by Kumarasawamy et al. (2003) in Figure 28.6, and we shall turn now to consideration of this new procurement paradigm.

Performance-Based Methods of Procurement

Performance-based procurement systems (PBPS) are being implemented within a wide range of project-based industries as a means to overcome problems associated with performance. The culture of soliciting suppliers on the basis of lowest bid price has been common in many industries, and particularly international construction. The low barrier to entry in this sector has exacerbated such conditions, and the dissatisfaction often experienced by the end user of the product/service is well documented. However, experienced owners who regularly commission projects are now seeking "best value" rather than simple capital cost reduction in a product/service.

Within project-based industries, best value procurement incorporates the following principles:

- Integrating value management and risk management techniques within normal project management
- Defining the project carefully to meet the user needs
- Taking account of whole life costing
- Adopting change control procedures
- Use of partnering arrangements
- Not appointing suppliers on lowest cost

The end to a reliance on formal contacts is also a core philosophy behind PBPS, whereby the relationship between buyer and sellers emphasizes partnership rather than distrust. Competition is based on clear targets for improvement, in terms of quality, timeliness, and cost.

The terminology used by different organizations to describe this relatively new model of procurement is evident if we consider NASA. Goldin (1999) refers to the implementation

FIGURE 28.6. EXAMPLE OF A FORCE FIELD "AGAINST" RELATIONAL INTEGRATION IN A CLIENT-CONTRACTOR RELATIONSHIP.

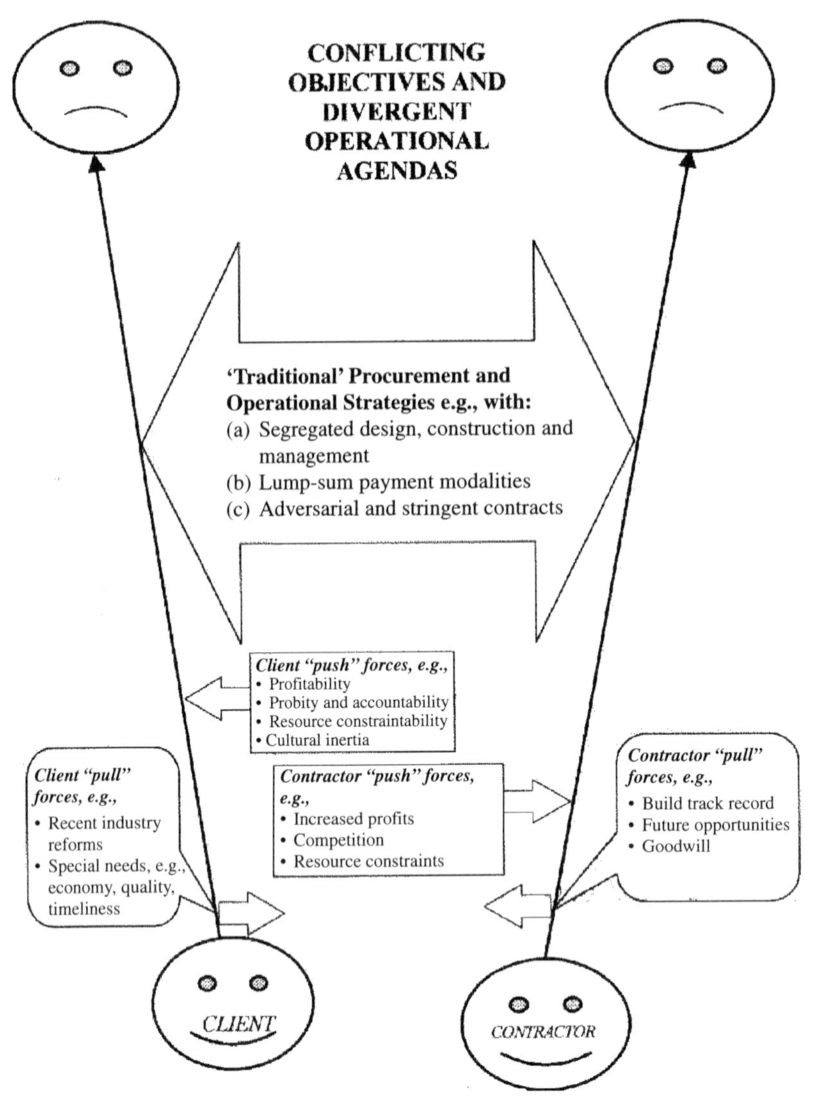

Source: Kumarasawamy et al. (2003).

of performance-based contracting (PBC), whereby all aspects of acquisition are structured around the purpose of the work to be performed as opposed to how the work is to be performed or broad and imprecise statements of work.

Kashiwagi (1997) notes that the success of PBPS rests in the supply of performance information and suggests that a facility owner use the supplier information in the list that follows. Although referring to the U.S. construction industry, any purchaser could use this performance information checklist.

- Expertise and experience
- Price
- Contractor margins, financial stability, and payment of subcontractors
- Previous size of jobs
- Previous types of contracts
- Completion rates on time and below budget
- Performance of previously constructed facilities of facility systems
- Personnel proposed for construction management.

The list should not be seen as exhaustive. Prequalification procedures are subject to evolution, and the current need to demonstrate both sustainability and ethical compliance are but two examples; both are increasingly added to such a list.

Kashiwagi notes the importance of information in facilitating a culture of trust between parties. The more information, the less distrust. It allows all parties to understand who and what all other parties bring to the partnership, including positive and negative characteristics. Figure 28.7 shows the PBPS proposed by Kashiwagi based on input from, inter alia, Motorola, Honeywell, IBM, McDonnell Douglas, Phelps Dodge, the State of Wyoming, and the U.S. Army command.

The transparency noted by Kashiwagi is also a core feature in the procurement of automotive parts at the Nissan UK plant in Sunderland. The "open-book" accounting allows Nissan to guarantee a fixed profit margin based on open-book accounting. If the price of raw materials rises, Nissan pays the supplier extra; if the raw material prices fall, Nissan expects to pay less. The Nissan factory also uses performance data to assess its suppliers. The benchmarking between the automotive and construction industry can be seen to exist (see Figure 28.8).

Partnering

The concept of partnership within business transactions has longevity within many cultures. Conventional wisdom suggests that Japan may offer excellent benchmarking potential. The work of W. Edwards Deming in the 1950s and his contribution to the managerial revolution—Total Quality Management—can be viewed as a catalyst for the interest in partnering today. Deming's help in rebuilding Japan's post WW II economy centered around continuous quality improvement, and today we consider this to be achievable through the use of

FIGURE 28.7. PERFORMANCE-BASED PROCUREMENT PROCESS.

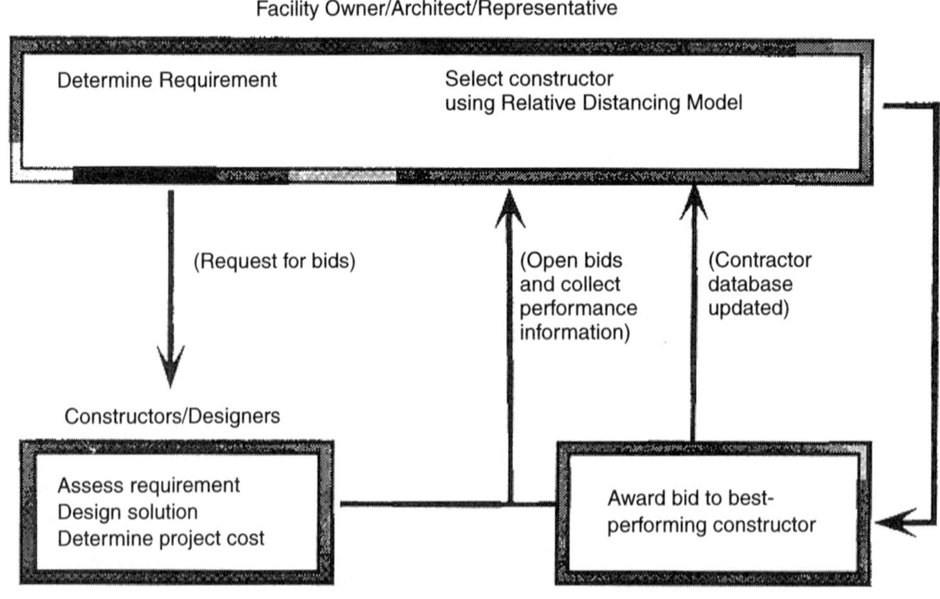

Source: Kashiwagi (1999).

integrated project teams who share mutual objectives and who can resolve disputes with a win-win rather than win-lose outcome.

Although the concept of partnering proliferated throughout much of Japan's industry, particularly car manufacturing, and although it has been subsequently adopted in Western manufacturing, process engineering, and defense-aerospace industries, it is not free from critical appraisal. Townsend (1996), for example, has doubts over the conventional belief that Japanese construction firms are more productive than Western counterparts.

Despite the potential for problems in partnering relationships, two UK construction industry reports, "Constructing the Team" (Latham, 1994) and "Rethinking Construction" (Egan, 1998) have both advocated the need for project teams to work together on serial contracts where a culture of continuous learning can assist in promoting project success. However, despite these laudable recommendations, the implementation of partnering in a project process is often confusing for all involved. Bresnen and Marshall (2000) have argued that partnering is an imprecise and inclusive concept capturing within it a wide range of behavior, attitudes, values, practices, tools, and techniques.

A recent survey of the benefits derived from partnering (Galliford, 1998) polled the views of over 500 managers and directors from organizations that directly appoint construction companies (see Table 28.1). The results show that clients regard partnering as providing

FIGURE 28.8. REPLACING CONTRACTS WITH PERFORMANCE MEASUREMENT.

Replacing Contracts with Performance Measurement

Nissan UK and Tallent Engineering Ltd have no formal contract beyond an annual negotiation of the cost and quality of rear axles that Tallent produce for Nissan's cars, and rigorous targets for improving performance. Each morning Tallent receives an order from Nissan detailing the precise mix of axles required by Nissan and five times a day Tallent delivers to Nissan's Sunderland Plant. If a problem were to occur with quality Tallent would send engineers to Nissan to fix it on the car production line. If a problem resulted in a significant loss of production, Nissan would expect to be compensated by Tallent for lost business or vice versa, but this has never happened and both sides work hard to ensure it cannot. Both Nissan and Tallent use similar non-contracts agreements with the firms delivering their construction projects.

Nissan's QCDDM supply chain management system is acknowledged to be among the most effective in the world. It measures all suppliers on Quality, Cost, Delivery, Design, and Management against negotiated continuous improvement targets. For each element, the supplier is marked on a range of product and process items which are aggregated on a weighted basis to give performance percentage for the element. Competition is created across the supply chain by collating the performance information every month and informing each supplier of its performance in relation to the others.

Source: Egan (1998).

TABLE 28.1. BENEFITS OF PARTNERING.

Which of the following statements do you think best describe the benefits of partnering?	%
1. Creates a sense of team-working on-site	56
2. Identifies mutual objectives	52
3. Solves problems faster	49
4. Provides the best rout to a successful outcome	39
5. Encourages the completion of projects within budget	30
6. Encourages a higher quality of job	23
7. Reduces lead in time	23
8. Saves time	19
9. Encourages completion of projects on time	17
10. Reduces program time	9
11. Saves money	7

Source: Galliford (1998).

an atmosphere where a team can develop trust and cooperation and focus on achieving a common goal. And there is evidence that such conditions have the potential of reducing waste, thus offering better value for money while contributing to greater client and end user satisfaction (Chartered Institute of Building, 2002).

Partnering can be "project-specific" or "strategic," where the partners work together on a series of projects that are consecutive, if not continuous (CIB, 1997). Monaghan (2000) uses an interesting analogy to emphasize this difference when he talks of "partnering for comfort" (project) and "partnering for improvement" (strategic). Although project partnering is often considered to be less effective, Bennett and Jayes (1995) found that in a study of partnering in the United States, project partnering provided benefits, even where there was no possibility of the client providing further work. They do, however, recognize that strategic partnering provides more long-term benefits, since it allows for continuous improvement.

What are the major barriers to successful partnering in construction projects? Larson and Drexler (1997) note that partnering represents a paradigm shift in how one approaches construction projects. Their study revealed that a general level of mistrust existed between owners and contractors and that this was engineered by years of viewing and treating each other as potential adversaries. Of particular importance was the failure to build a true relationship of trust and the reliance of legal loopholes in documents. Other significant difficulties include the inability of the project management structures to synchronize goals of the numerous subcontractors, a misunderstanding of partnering concepts by upper management, and a failure to "walk the talk." Another investigation into the decline of relationships (Drexler and Larson, 2000) examined owner-contractor relationship over time to identify the factors that contributed to decline or improvement in the relationship. The research was based on response from 276 members of PMI. Four categories of owner-contractor relationship were identified:

- *Adversarial.* Parties perceive themselves as adversaries, with each party pursuing their own concerns at the other party's expense.
- *Guarded adversarial.* Participants cooperate within the boundaries of the contract.
- *Informed partners.* Participants attempt to sustain a cooperative relationship that goes beyond the boundaries of the contract.
- *Project partners.* Participants trust each other as equal partners with a common set of goals and objectives

Fifty-eight percent of the projects experienced some fundamental change in working relationship, either positive or negative. Projects that began as formal partnerships were most stable, with two-thirds ending as they had begun. Of the relationships that changed, half regressed to an adversarial relationship, while half progressed into some form of partnership. Several common themes were seen to develop as to why relationships improved or declined. Relationships that had deteriorated were characterized by unclear contracts and resulting litigation, changes in scope and schedule, personnel failing to perform, lack of trust, and underbidding contracts. Relationships that improved were characterized by trust and positive relationships, shared goals, teamwork and communication, personnel changes, and the presence of a clear contract.

In the United Kingdom the use of pioneering partnering contracts has formalized much of the ideology inherent in partnering. The Project Partnering Contract 2000 (PPC200) published by the Association of Consultant Architects (ACA) and the Engineering and Construction Contract (partnering option document) published by the Institution of Civil Engineers are two examples. The philosophy underpinning these contracts is built on a team-based multiparty approach within a fully integrated design/supply/construction process. The use of co-located offices and a common information system (project Web site) also help to promote a common objective, and the application of value engineering and risk management are considered prerequisites.

Best-Practice Procurement

Three prominent owners who procure the services of the UK construction industry on a regular basis—British Airport Authorities (BAA), the Ministry of Defence (MoD), and the National Health Service (NHS)—have reformed their procurement practices and adopted a PBPS ideology.

Framework Agreements (The Case of BAA)

In 1996 BAA implemented an initiative know as 21st Century Airports with the aim of establishing itself as the most successful airport operator in the world. This continues to be an ambitious challenge for a formerly public client privatized in 1987. The ambition to be "world class" had implications for BAA's development program, given that construction costs (around £500 million annually) have a direct impact on profitability. For those readers outside the UK, the relationship with the then chief executive and the current agenda to "rethink construction" should be made apparent. Sir John Egan, the chief executive of BAA between 1990 and 1998, had previously spent 11 years as chairman of Jaguar, the car manufacturer. In October 1997 he was commissioned by the deputy prime minister to lead a Construction Task Force, the remit being to examine the performance of the UK construction industry. Egan was considered to be a tough and demanding task master who had previously revolutionized the production process at Jaguar cars and was the driving force behind the changes in procurement practices taking place at BAA. Indeed, on comparing construction costs with other airport operators, he discovered that UK projects were comparable to those in Europe. In the United States, a different story was revealed: "In America my eyes were opened and I kept wondering: Is he talking in dollars—Christ that can't be possible, and they were saying: Yeah and we spent too much. They would be over budget and still be half the cost" (cited in *New Builder*, 1994).

BAA's framework process is based on contractors tendering to become framework partners, and the successful bidders are then allocated work on BAA's construction program for five to ten years. Under the agreements, the contractors are selected on various criteria that include price, quality of the company and its staff and products, attitude, and the ability to work with BAA. The successful designers and contractors are then designated into specific delivery teams grouped around different product ranges: baggage, fit-out, shell and core, and infrastructure. These prequalification "tests" were seen to strain many contractors, who considered them to be overly bureaucratic (European Union procurement legislation) and time-consuming and who were used to competing in a "lowest price wins" market. In contrast, the philosophy behind the framework agreements is based on continued improvements vis-à-vis longer-term and more open relationships where collaboration rather than confrontation is king (see Cox and Townsend, 1998, for an account of BAA's reengineering of the construction supply chain).

MOD's Prime Contracting: Building Down Barriers

In the United Kingdom, the procurement of defense equipment such as tanks and aircraft has suffered major cost escalations and late delivery for many years, though recent moves to performance and partnering base procurement—"smart acquisition"—may now be reversing this trend (NAO, 2002a). "Building Down Barriers" (BDB) is a new procurement initiative pioneered by the MOD's Defence Estates (DE) department and supported by the government's Department of Environment Transport and Regions (DETR). Its purpose has been to create a learning mechanism for establishing the working principles of supply chain integration in construction. Two pilot projects established in 1997 involved the construction of recreation facilities and involved two "prime" contractors. In each project, the prime contractor was expected to integrate the supply chain into the building design, construction, and maintenance for a trial period of up to two years. The projects were monitored by the Tavistock Institute, who were also involved in establishing the project organization structures (see Holti, 1998; Holti et al., 2000).

The new organizational structure concept involved the concept of "work clusters." In each work cluster, the designers, subcontractors, and key suppliers were involved in a reasonably self-contained element of the building, undertaking a form of simultaneous engineering. Typical clusters included groundwork, frame, and envelope; swimming pools, mechanical, and electrical services; and internal finishes (see Figure 28.9). Within each work cluster the participants used techniques including value management, risk analysis, and risk management and contributed knowledge to anticipate the areas of interdependence that were likely to arise during detailed design and construction. The clusters were responsible not only for designing but also for delivering construction of their element of the building. Interdependencies that spanned the spheres of two or more clusters were resolved by taking them to an overall project team, where a cluster leader represented each cluster. Nicolini et al. (2001) note that the cluster process is in fact a collaborative, team-based approach to project decision making. The pilot projects are now completed and have shown positive benefits in labor productivity, material wastage, construction time, and whole-life costs.

NHS Estates ProCure21

NHS Estates is the property arm of the UK's National Health Service (NHS) and has a capital works program worth about £3 billion a year. The aim of ProCure21 is to promote

FIGURE 28.9. NEW INTER-ORGANIZATIONAL ARRANGEMENTS.

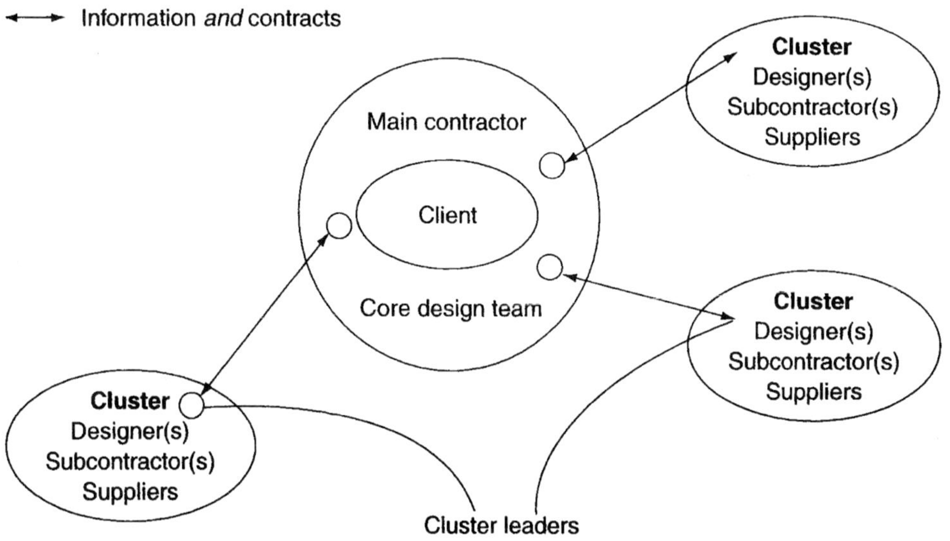

Source: Holti (1998). © Tavistock Institute of Human Relations.

better capital procurement in the NHS, the intention being to achieve this through delivering better-quality healthcare buildings and improved value for money by invoking a major cultural change. This reform involves the following:

- Establishing a partnering program for the NHS by developing long-term framework agreements with the private sector that will deliver better value for money and a better service for patients
- Enabling the NHS to be a best client
- Promoting the use of high-quality designs
- Monitoring performance through benchmarking and performance management

The need for such reform within NHS Estates is clear if we consider the disastrous procurement practices so prevalent in healthcare construction projects during the latter half of the twentieth century. Typically the gestation time to plan, design, and construct large healthcare projects lasted decades. During such a long process, the original design brief would "creep," with subsequent detrimental impacts on the original budget project program. Once completed, hospital projects were often found to be unsuitable for their end users, combined with excessive whole-life costs. Guys Hospital Phase 3 development in London is indicative of such conditions The National Audit Office (NAO 1998) investigated the project that was delivered three years late with an overspend of £68.7 million above its original budget. The key factors identified were as follows:

- Delay in putting design team in place
- Delay in resolving cost and funding problems
- Failure to freeze design and design changes
- Delay in designing the engineering services and producing drawings
- The insolvency of works package contractors
- Technical matters (defective copper pipework)
- Changes in statutory regulations (building regulations)
- Delays to the construction works
- Large number of claims associated with construction works
- Change to design team's fee rates
- Inflation

As with the Prime Contract initiative, ProCure21 means that NHS clients will work in partnership with substantially fewer suppliers. This will involve framework agreements with carefully selected companies from within the construction industry. The intention is to provide these companies with opportunities to undertake projects within an ongoing program of work at an appropriate level of profit. The current Procure21 pilot projects are said to be contributing to a saving of 4 percent in procurement costs. According to the initiative's program director, this can be contrasted with conventional procurement methods where an overspend of 8 percent is normal (Contract Journal, 2002d).

Principal Supply Chain Partners (PSCPs) who fulfill the selection criteria will be appointed onto a national framework. The PSCPs will take single-point responsibility for design and construction, and in the case of Private Finance Initiative (PFI) projects, facilities management and finance. (See the chapter by Ive on PFI.) Each PSCP has the responsibility for managing its integrated supply chain. Providers of key construction services appointed by the PSCPs are known as primary supply chain members (PSCMs). The PSCPs will also be responsible for appointing all other supply chain members. When a NHS client wants to construct a new scheme, it will select a PSCP, including its associated supply chain, from the framework agreement

Contemporary and Future Issues in Procurement

Technology and culture are two variables that have a large impact on the procurement process. Both are subject to constant and often unplanned pressure from both within and outside a project's boundary. The scientific management principles developed by the likes of Taylor and Galbraith (time and motion study) and adopted by Ford in their automotive plants were a product of this era. New project procurement enablers such as lean thinking and just-in-time are, however, steeped in process improvement. Thus, the procurement process could be said to be subject to management fads and fashions. This impacts on the various industrial sectors around the globe with varied degree and at different times.

The use of project intranets and electronic procurement has complemented the desire for integrated project teams and has permitted knowledge to be shared throughout project supply chains. Logistics management and its associated just-in-time philosophy are now a core ideology in some prestigious projects.

In addition, the recognition that adversarial "claims-ridden" procurement is destructive (notwithstanding the opportunities for construction lawyers!) has led to clients seeking less confrontational relationships where trust is the key motivator. This stakeholder view is becoming more apparent in society, and the topic of sustainability has encouraged clients to be more discerning about their supply chain, while contractors and consultants must demonstrate compliance with legislation. This has led to both external and internal forces moulding contemporary and future procurement strategies and to a post-reengineering world where strategic transparency and leanness are reflected in operational prefabrication, standardization, and mechanization However, one should not be deluded that risk and uncertainty have been removed from the procurement process. The sad toll of company insolvencies, accidents and deaths, litigation, and project failures continues to taint many project based businesses throughout the world.

E-Commerce

E-business is the term used to describe activities that involve the sharing of information through electronic networks, including companies being able to sell or order and pay for goods online, check availability, and get further information on products. It can also include using IT for project collaboration (Construction Confederation, 2002). The Construction Confederation describes two types of e-business: "process" e-business that helps to manage the flow of information within industry supply chains and "transaction" e-commerce that includes selling products and services, the latter being on a business-to-business (B2B) or business-to-consumer basis (B2C).

The benefits of adopting e-business are commonly cited as a reduction in transaction costs and increased project personnel collaboration. The business case for the adoption of e-commerce is made explicit by Walker and Rowlinson (2000), who argue that clients who are dependent upon intranets, e-mail, and electronic transfer of information are unlikely to be impressed by contractors, suppliers, and design teams who have yet to grasp the technology of e-communication and are not using such tools effectively. This conclusion is supported by research undertaken by the Construction Industry Institute in the United States. Voeller (2002) found that owners are leading the implementation of e-procurement and that many of them were experimenting with or using e-marketplaces and reverse auctions. (However the attempts to initiate industry-wide e-procurement practices in the United Kingdom have been far from successful. A service termed Arrideo to include buying, tendering, procurement, project collaboration, contractual agreements, delivery, and invoice information and payment facilities [Contract Journal, 2000] failed disastrously [Construction News, 2002b].)

Ethical Procurement

The concept of corporate social responsibility (CSR) and its associated drivers that include business ethics and transparency are becoming increasingly important in project-based industries. However, Loo (2002) notes that most books on project management fail to give this topic sufficient credibility and that this is repeated in both project management journals and conference proceedings. Such findings are perhaps worrying if we consider the findings of a survey conducted by the Ethics Resource Center in Washington. Nearly one-third of the 4,000 U.S. employees surveyed said that pressure had been put on them by their company to violate company policy in order to achieve business objectives (People Management, 1995).

Project-based industries, which are often characterized by interdisciplinary temporary teams, without a common objective, may perhaps provide an environment that facilitates enhanced conflict between business and society needs. The generic project procurement process has largely failed to account for unethical behavior or has perhaps unknowingly sanctioned such illegal activity! Risk management and value management are now considered a prerequisite in project-based industries, but how many organizations consider ethical management? One example that suggests improvements are being made is that of Lucas Aerospace UK. The dissemination of its ethics program throughout the company was seen to be crucial for its success following revelations that two of the company's North American divisions were found to have falsified tests on components supplied to the U.S. Navy (People Management 1995).

Summary

Procurement is a constantly evolving phenomenon that accommodates the complex nature of project-based industries. This chapter has lent heavily on case study examples from the construction industry. No apology is offered for this, as it is perhaps the oldest known project-based industry. Its fragmented nature with multi-disciplinary and multi-organizational players present procurement managers with often unique challenges in fulfilling a project's completion.

It was also made explicit that the one-size-fits-all description of procurement is unhelpful in explaining the peculiarities within the various project-based industries. The frameworks used consist of much more than an explanation of a buyer-seller relationship and whether to outsource or not. However, the importance of PBPS may be illusive for far too many who procure projects, and impact of practices such as of organizational (project) learning, knowledge management, benchmarking, lean construction, and supply chain management remain as yet unknown. This is also the case for e-commerce despite its commercial advantage being clear to many.

If we are to move from twentieth- to a twenty-first-century project-based practices, then the recommendations suggested by Cain (2001) are appropriate for revitalizing the procurement process and the relationships within. He argues that best-practice procurement differs from all other forms of traditional procurement in two ways: It involves the abandonment of lowest capital cost as a value comparator, and it includes involving specialist contractors and suppliers in design from the outset. He concludes that these two key differences can be broken down into six primary goals that are essential for best-practice procurement:

1. Finished products delivers maximum functionality, which includes delighted end users.
2. End users benefit from the lowest optimum cost of ownership.
3. Inefficiency and waste in the utilization of labor and materials is eliminated.
4. Specialist suppliers are involved in design from the outset to achieve integration and buildability.
5. Design and construction of the building is achieved through a single point of contract for the most effective coordination and clarity of responsibility.
6. Current performance and improvement achievements are established by measurement.

The next epoch in procurement will be to build social improvement practices into procurement. Best value will hopefully be judged beyond the current mantra of price; performance and value will be extended to incorporate agreed goals that have a special purpose. Here issues of gender and ethnicity balance in the project team will need to be considered when procurement decisions are being made. New businesses set up by groups formerly excluded from the current business culture will lead to new procurement frameworks. Thus, the policies of procurement operated by project-based industries have some way to travel before it may finally realize its business and social potential.

References

Association for Project Management. 2000. *Project Management: Body of Knowledge. 4th Ed.,* ed. Miles Dixon. High Wycombe, UK.

Bennet, J., and S. Jayes. 1995. *Trusting the team: The best practice guide to partnering in construction.* Centre for Strategic Studies in Construction. Reading, UK: Reading Construction Forum.

Bresnen, M., and N. Marshall. 2000. Partnering in construction: A critical review of issues, problems, and dilemmas. *Construction Management and Economics* 18:229–237

Cain, C. 2000. Cited in Was the Gain Worth the Pain? *Contract Journal* (June 1): 32–33.

Chartered Institute of Building. 2002. *Code of practice for project management: For construction and development.* 3rd ed. Oxford, UK: Blackwell Science.

Construction Confederation. 2002. An Introduction to E-Business in Construction. Construction House, London. www.theCC.org.uk.

Construction Industry Board. 1997. *Partnering in the team: A Report by Working Group 12.* London: Thomas Telford.

Construction News. 2002a. Fraud Probe at Navy Dock, p. 1. August 1.

Construction News. 2002b. Domain name firesale as the construction net dream dies. 1–2.

Contract Journal. 2000. Five contractors set to launch B2B venture. July 26.

———. 2002a. Up e-revolution. p. 9. July 19.

———. 2002b. Sharing e-efficiencies. 10–11. August 29.

———. 2002c. Investment takes off. 14–15. August 9.

———. 2002d. Procure21 shaving 4% off building costs. September 26.

Curtis, B., S. Ward, and C. Chapman. 1991. *Roles, responsibilities and risk in management contracting.* Special publication 81. London: CIRIA.

Chrichton, C. A. 1966. *Interdependence and uncertainty: A study of the building industry.* London: Tavistock Publications.

Confederation of Construction Clients. 2000. *The Clients' Charter Handbook.* London. www.clientssuccess.org.

Drexler, J., and E. W. Larson. 2000. Partnering: Why project owner-contractor relationships change. *Journal of Construction Engineering and Management* 126(4):293–297

Egan, J. 1998. Rethinking construction. The Report of the Construction Task Force to the Deputy Prime Minister, John Prescott, on the scope for improving the quality and efficiency of UK construction. London: The Stationery Office.

Galliford. 1998. *Partnering in the construction industry.* Leicestershire, UK: Galliford UK Limited.

Goldin, D. S. 1999. Performance based contracting. Taken from www.ksc.nasa.gov/procurements/nls/perfbase.htm (accessed June 24, 2003.)

Holti, R. 1998. The lost world: Virtual organisation in the building industry. 44–49. Discussion paper in the *Tavistock Institute Review* 1996/97.

Holti, R., D. Nicolini, and M. Smalley. 2000. The handbook of supply chain management. CIRIA Report C546. London: CIRIA

Kashiwagi, D. T. 1997. The development of the performance-based procurement system. 275–284. *ASC Proceedings of the 33rd Annual Conference.* University of Washington, Seattle, April 2–5.

Kumarasawamy, M., M. Rahman E. Palaneeswaran, S. Ng, and O. Ugawa. 2003. Relationally integrated value networks. *CIIFE Conference,* Loughborough, UK.

Larson, E., and Drexter, J. A. 1997. Barriers to project partnering: Report from the firing line. *Project Management Journal* (March): 46–52.

Latham, M. 1994. *Constructing the team: Final report of the government/industry review of procurement and contractual arrangements in the UK construction industry.* London: The Stationery Office.

Loo, R. 2002. Tackling ethical dilemmas in project management using vignettes. *International Journal of Project Management* 20: 489–495.

Monaghan, K. 2000. Trust me, I'm a contractor. *Building* 26 (May): 36–37.

Morgan, G. 1986. *Images of organisations.* Beverly Hills, CA: Sage Publications.

Morris, P. W. G. 1997. *The management of projects.* London: Thomas Telford.

Murray, M., and D. Langford. 2003. *Construction reports 1944–98.* Oxford, UK: Blackwell Science.

National Audit Office. 1998. Cost over-runs, funding problems, and delays on Guys' Hospital phase 3 development. Report by the Comptroller & Auditor General HC 761 1997/98.

———. 2002a. Ministry of defence: Major project report 2001, HC 330.

———. 2002b. The construction of nuclear submarine facilities at Devonport. Report by the Comptroller and Auditor General, HC 90, Session 2002–2003. London: The Stationery Office.

New Builder. 1994. World view, April 8, pp. 21–22 Nicolini, D., R. Holti, and M. Smalley. 2001. Integrating project activities: The theory and practice of managing the supply chain through clusters. *Construction Management and Economics* 19:37–47.

Nicolini, D., C. Tomkins, R. Holti, A. Oldman, and M. Smalley. 2000. Can target costing and whole life costing be applied in the construction industry?: Evidence from two case studies. *British Journal of Management:* 303–324

People Management. 1995. Business Ethics. pp. 22–34.

Smith, A. 1838. *The wealth of nations.* London: Longman.

Tookey, J. E., M. D. Murray, C. Hardcastle, and D. A. Langford. 2001. Construction procurement: Redefining the contours of organisational structures in procurement. *Engineering Construction and Architectural Management* 8 (1, February): 20–30.

Townsend, M. 1996. Is the Japanese way working? *Contract Journal* 21 (November): 22–23.

Voeller, J. 2002. E-commerce applications in construction. Research summary. Austin, TX: Construction Industry Institute, University of Texas at Austin.

Walker, A., and R. Newcombe. 2000. The positive use of power to facilitate the completion of a major construction project: A case study. *Construction Management and Economics* 18(1):37–44.

Walker, D. T, and S. M. Rowlinson. 2000. A construction industry perspective on the use of the Web for establishing a marketing presence. *International Journal of Construction Information Technology* 8(1):93–112.

www.mgmconcepts.com/scripts/mcicoursepage.asp?MCICourse=6126 (accessed June 24, 2003).

CHAPTER TWENTY-NINE

CONTRACT MANAGEMENT

David Lowe

Contract management has been defined as "... the process which ensures that all parties to a contract fully understand their respective obligations enabling these to be fulfilled as efficiently and effectively as possible to provide even better value for money" (CUP, 1997). This process commences with the identification of the purchaser's needs and concludes with the completion of the contract. Further, the process has two dominant characteristics:

- *Risk identification, apportionment, and management.* Related to contract performance
- *Relationship management.* Between the purchaser and supplier

A prudent project manager and contract management team, therefore, will have a thorough understanding of the following:

- Procurement process and post-tender (bid) negotiation
- Assumptions made by the purchaser and the supplier
- Purchaser's expectations of the service relationship
- Contract terms and conditions; for example:
  - Purchaser's duties and responsibilities under the contract
  - Supplier's obligations under the contract
  - Main cost determinants, how they relate to the outputs and quality standards, and how they will be measured
  - Certification and payment mechanism
  - Purchaser's and supplier's rights if things go wrong
  - Legal implications of the contract for which they are responsible (CUP, 1997)

There is, however, a competing view: To facilitate successful projects, contracts should be "left in the drawer." Latham (1994) is sympathetic to this viewpoint; he considers the function of a contact is to serve the contract process, not vice versa. However, Hughes and Maeda (2003) contend that this view is incorrect; planning for future events in the contract process could be very problematical without knowledge of the contract.

The aim of this chapter is to develop a critical understanding of the factors that influence commercial contract practice. The intention is to provide the project manager with an overview of contract provisions and procedures. The chapter seeks to explain contract management in terms of generic contract provisions: the key components of contracts. However, to illustrate these principles, reference will be made, where appropriate, to the Fédération Internationale des Ingénieurs-Conseils (FIDIC) Conditions of Contract for Construction (first edition, 1999), also known as the FIDIC Construction Contract. Essentially, this contract is intended for building and engineering works designed by the employer; however, the contract does allow for the inclusion of some elements of contractor-designed civil, electrical, mechanical, and/or construction works. Members of FIDIC come from over 60 countries worldwide, and the FIDIC Construction Contract is an internationally recognized standard form.

Contractual Issues

Definitions

Goods and services necessary for the completion of projects are procured through the use of contracts with suppliers. Likewise, main suppliers use contracts to procure from subcontractors; while contracts in their own right, they are generally referred to as subcontracts (denoting a relationship to the main contract).

Contracts

A *contract* is an agreement between two parties under which one party promises to do something for the other in return for a consideration, usually a payment. This places obligations on both parties to fulfil their part of the agreement. It is also the foundation for the relationship between the parties.

Elements of a valid contract vary depending on country of origin. Specifically:

- *Canadian law.* Intention to enter into the contract, consideration, capacity of the parties, offer and acceptance, and lawful object (Jergeas and Cooke, 1997).
- *English law.* Intention to be legally bound, consideration, capacity of the parties, and offer and acceptance.
- *French law.* Mutual assent, cause, capacity of the parties, and lawful object (French Civil Code, Article 1108).
- *German law.* Mutual assent, intent to confer a benefit, capacity of the parties, and lawful object (German Civil Code, §§ 518, 761, 780, 781).
- *United States law:* Mutual assent, consideration, capacity of the parties, and lawful object (American Law Institute, Restatement of the Law, Second: Contracts, §§ 3, 8, 12).

The parties to a contract are as follows:

- *The purchaser.* The party that acquires or obtains goods or services by payment or at some cost. Alternatively referred to as the buyer, client, customer, employer, owner, proposer, sponsor, user, and so on.
- *The supplier.* The provider of goods and services. Also referred to as the contractor, main supplier, main contractor, prime contractor, prime supplier, seller, vendor, and so on.

This text will refer to the purchaser and supplier unless reference is made to a specific form of contract.

Although not a party to the contract, most contracts make reference to the following:

- *A project manager.* The person who leads the purchaser's contract management team. Alternatively, the terms "architect," "contract manager," or "engineer" may be used. This text will refer to the project manager.
- *The subcontractor.* A supplier to the main supplier, main contractor, prime contractor, prime supplier, and so on.

While contracts can take the form of a single document, generally, commercial contracts comprise several documents. For example:

- *The contract agreement.* Itemizes the documents comprising the contract. It includes the identities of the parties and defines the scope of work, the contract price, and the schedule for its execution.
- *General specification and scope of work.* Describe the scope of work to be undertaken, the technical standards required, and the administrative procedures to control the implementation of the project.
- *General conditions of contract.* Normally a recognized standard form of contract. This details the obligations to produce and to pay, clarifies the offer and acceptance, allocates risks, describes the consequences of failure to pay or produce, and includes relevant issues, for example, insurances, bonds, safety, industrial relations, defects, and disputes, and so on.
- *Special conditions of contract.* These cover additions and amendments to the general conditions as required by the purchaser and specific circumstances of the project.
- *Administrative and coordination procedures.* Frequently the procedural aspects of a contract are covered separately as an appendix to the general conditions.

In practice, however, these documents will be interlinked, with some having greater importance than others. Because of the potential for conflicting information within these documents, an order of priority needs to be established prior to inviting bids.

Letters of Intent

Occasionally, letters of intent are used as an interim arrangement to permit a successful bidder to start work in advance of signing a contract and in the knowledge that they will ultimately be awarded the contract. To be operative, a letter of intent must have the properties of a contract and refer to those elements of the bid where agreement has been reached. As a minimum, a valid letter of intent should always state that the purchaser

- intends to place a contract with the supplier;
- wishes the supplier to begin work in advance of the contract; and
- authorizes the supplier to begin work in advance of the contract.

A letter of intent may not be legally enforceable and may be revoked by the issuer without any redress to the courts where it is held to be an announcement of one party's wish to do something. A further disadvantage arises from the purchaser's declaration of intent: It will reduce the ability of the client to satisfactorily negotiate the outstanding terms. HM Treasury, Central Unit on Procurement (CUP, 1989) recommends the use of a "start-up contract" with appropriate controls, limits, and safeguards, rather than a letter of intent.

Contracts and Orders

Legally there is no distinction between the terms "contract" and "order." Both refer to legally binding agreements for the supply of goods and/or services in return for some form of remuneration. Commonly, the term "contract" is used in relation to an agreement involving a longer time period and a greater outlay than a purchase order.

Subcontracts

In parallel to the contract between the purchaser and the supplier (the main contract), the supplier employs subcontractors and suppliers of materials, plant, equipment and services, and so on. The supplier is generally held responsible for any subcontracted work, and the purchaser, within the main contract, retains the right to vet subcontractors and limit the extent to which the work is subcontracted. The supplier may be free to choose the terms of subcontracts, or alternatively the terms may be required to correspond, "back-to-back," with those of the main contract.

Purchasers may also reserve the right to nominate subcontractors, requiring the supplier to enter into a subcontract with a subcontractor chosen by the purchaser or the project manager, usually to carry out specialist work. Contracts, therefore, require specific clauses to manage specific risks imposed by nomination, for example, in relation to the default of the nominated subcontractor, and to ensure that the supplier pays the nominated subcontractor.

Standard and Model Forms

Many industry sectors use standard or model conditions of contract as oppose to bespoke contracts. Wright (1994) contends that standard forms are used because they

- provide a recognized and predictable contractual basis;
- save time, both in writing and in negotiating the contract; and
- are familiar to the project/contract management teams, resulting in smoother-running projects, or at least in the avoidance of some mistakes that could disrupt progress.

Most organizations have their own standard conditions of contract such as standard sets of conditions of sale and purchase, for example, supplier's terms. Invariably, these contracts are biased, to a greater or lesser degree, in favor of the party that composed them. Alternatively, model conditions are used where the balance of power between the parties is approximately equal. Model conditions tend to be drawn up by an association including representatives of all parts of an industry and are, therefore, generally held, according to Wright, to represent a reasonable basis upon which organizations within that industry might be prepared to do business with each other.

Legal Interpretation of Contracts

Contracts operate within the framework of law. Globally, European legal systems are dominant. Basically, three systems prevail: Those based on English, Roman, and Russian law, the latter being a fusion of Roman and English legal principles. As a broad assertion, however, the general principles of commercial contract law are the same the world over, but the detail will vary. Therefore, specific legal advice should be sought on the implications of the interpretation of contract clauses with regard to a particular legal system.

International contracts need to state which country's law or other jurisdiction will apply. The choice of law has a significant effect on the administration of a contract. It determines how the contract is formed and establishes the underlying terms of the contract. It also provides a structure within which the parties function, for example, laws concerning trading standards and practices, safety, tax, and so on. Further, it tends to establish where and how disputes are resolved. Where a choice of legal systems exists, it is important to contemplate the consequence of that choice before entering into the contract.

Where the contract is produced in more than one language, the contract needs to determine the ruling language. Likewise, the contract needs to state the language for communication purposes. This is important because, in the event of a dispute occurring, the exact words used in the contract will be carefully interpreted in order to determine the precise agreement made between the parties. Immense care, therefore, is required when selecting the words used in a contract as they can have different interpretations.

Contract Terms

The terms of a contract are all the rights and obligations agreed between the parties, together with any terms implied by law.

Express and Implied Terms, Incorporation by Reference

- *Express terms.* Those terms that are stated (written) within a contract
- *Implied terms.* For example, within English law, those terms that form part of a contract but are not expressed. Implied terms include the following:

    - Conformity with statutes
    - Supplier's responsibility for their subcontractors
    - Fitness for purpose: provided the purpose has been communicated to the supplier and has not been overruled by an imposed specification
    - Furtherance of purpose, where both parties endeavor to perform the contract as best as they can
    - Duty to utilize competence and care
    - Supplier's liability to execute the work at a reasonable pace
- *Incorporation by reference.* For example, where the contract makes reference to terms contained within other documents.

Conditions and Warranties. The terms of a contract under English law can comprise conditions and warranties. A *condition* is a key term within the contract: a promise of considerable magnitude. Failure to fulfil the promise entitles the injured party to terminate the contract and to claim damages for failure to comply. A *warranty* is a less serious term within the contract. Failure to comply with such a term entitles the injured party to claim damages but not to terminate the contract.

Commercial Contracts

Complexity of Contracts. Commercial contracts are relatively complex documents; for example, construction projects generally require extensive contracts in order to express precisely the legal, financial, and technical facets of the project. As a result, one potential source of risk is the contract document. The contract conditions, therefore, according to Bubshait and Almohawis (1994), need to be assessed for clarity, conciseness, completeness, internal and external consistency, practicality, fairness, and effect on project performance—that is, on quality, cost, schedule, and safety. They present a simple and systematic instrument to evaluate these attributes.

Commercial Manager. The last 15 years has seen the emergence, primarily within large UK organizations, of the role of commercial management. A commercial manager has been defined as ". . . a person controlling or administering the financial transactions of an organisation with the primary aim of generating a profit generating whilst minimising associated risk" (Lowe et al., 1999). The function involves advising the organization on the use of contracts, formulating bespoke contracts, and in negotiating contracts.

Contract Strategy and Type

The provisions of a contract should do the following:

- *Define the responsibilities of the parties.* For example, define the project's objectives and priorities; project finance, innovation, development, design, quality, standards, procurement, scheduling, implementation, installation; project management, safety, inspection, testing, commissioning, and managing operating decisions.

- *Allocate risk.* For example, financial investment in the project, project definition, design, performance specification, subcontractor selection, subcontractors' defaults, site productivity, delays, mistakes, and insurances.
- *Determine effective payment terms.* For example, for development, design, demolition, construction, fabrication, implementation, management, and others services (Smith and Wearne, 1993; Wearne, 1999).

Contracts are usually classified in terms of strategy (procurement methodology or organizational choice)—for example, traditional, design and build, turnkey, and management contracts—or by type (allocation of risk and payment terms)—for example, lump-sum, remeasurement, and target cost contracts. Contract strategy and type should be planned concurrently.

Contract Strategy

For a discourse on the various procurement/contract strategies, see the chapter on procurement systems by Langford and Murray. The following are examples of contracts classified in terms of strategy.

Design Combined with Production

- *Design and build contracts.* FIDIC Conditions of Contract for Plant and Design-Build; JCT Standard Form of Building Contract with Contractor's Design (1998); AGC 415 Standard Form of Design-Build Agreement and General Conditions Between Owner and Design-Builder (Where the Basis of Payment is a Lump Sum Based on an Owner's Program Including Schematic Design Documents) 1999 Edition.
- *Turnkey contracts.* FIDIC Conditions of Contract for EPC (Engineering, Procurement and Construction)/Turnkey Projects

Design Separate from Production. Two alternative organizational structures exist where design is separate from production:

- Sequential contracts
  - *Conventional or traditional contracts.* FIDIC Conditions of Contract for Construction; JCT Standard Form of Building Contract (1998).
- Parallel contracts
  - *Management contracting.* JCT Standard Form of Management Contract (1998), NEC Engineering and Construction Contract: Management Contract (1995).
  - *Construction management.* JCT Construction Management Documentation (2002); AIA A101™/Cma Standard Form of Agreement Between Owner and Contractor—Stipulated Sum—Construction Manager—Adviser Edition/A201™/Cma General Conditions of Contract for Construction (1992); AGC 230 Standard Form of Agree-

ment and General Conditions Between Owner and Contractor (Where the Basis of Payment is the Cost of the Work with an Option for Preconstruction Services), 2000 Edition.

Parallel contracts are advantageous where an early project completion date is crucial, the design requirements are uncertain at the outset, supplier involvement in the design process is advisable, there is a requirement to maintain the operation of existing installations, the segmented work is of a specialist nature, and/or suppliers have a limited capability.

Alternative Organisational Arrangements

- *Term contracting.* Term contracting refers to a particular type of work to be executed over a given time period. It is commonly used for the provision of a service, for example, repair and maintenance work where the general nature of the work is known but the extent of it is not. Each individual order issued under the term contract becomes a discrete contract, and at this point the terms of the bid become binding. Example: JCT Standard Form of Measured Term Contract (1998).

Strategic Cooperative Arrangements. According to Smith et al. (1995), there are two major areas of operational difficulty in joint ventures, which have implications for both bid preparation and project implementation: conflict and culture. Likewise, Walker and Johannes (2001) found that equalization of power is crucial within joint venture partnerships, while the need to understand organizational cultural diversity was also seen to be pivotal.

Specific contract conditions are required, therefore, to ensure the establishment of a suitable organizational structure that will encourage the successful completion of the project and that will safeguard the purchaser in the event of the default or liquidation of one of the joint venture members. Example: AGC 299 Standard Form of Project Joint Venture Agreement Between Contractors, 2002 Edition.

Contract Type

Essentially, there are two categories of project contract payment terms: price-based and cost-based.

Price-Based

- *Fixed price.* Where the supplier is paid a fixed price or lump sum (a single tendered price) for the entire project. The terms "fixed" or "firm" usually indicate that the contract price will not be subject to escalation payments, whereas lump-sum contracts may. Additionally, fixed and firm contracts generally may not include variation clauses. However, the terms fixed and firm have no precise meaning. The payment terms included within a specific contract are the key factor.

  Examples include AIA A101™ Standard Form of Agreement Between Owner and Contractor—Stipulated Sum/A201™ General Conditions of Contract for Construction

(1997); AGC 200 Standard Form of Agreement and General Conditions Between Owner and Contractor (Where the Contract Price is a Lump Sum), 2000 Edition; IChemE Form of Contract for Use in the Process Industries: Lump-Sum Contract (The Red Book), 4th Edition.

- *Measurement.* Where a list of the items and quantities of the work to be executed under the contract (bill of quantities) is incorporated into the bid/contract documentation: The purchaser pays a standard rate based on agreed productivity rates and unit rates.

  Examples include JCT Standard Form of Building Contract Private With Quantities (1998); NEC Engineering and Construction Contract: Priced contract with bill of quantities (1995).
- *Remeasurement.* Where the actual work carried out by the supplier is measured on completion, as implemented, based upon either
  - *an approximate bill of quantities.* A list of the items and approximate quantities of the work to be executed under the contract; where the purchaser pays a standard rate based on agreed productivity rates and unit rates.

    Example: JCT Standard Form of Building Contract Private with Approximate Quantities (1998).
  - *a schedule of rates.* A list of potential items to be executed under the contract; where the purchaser reimburses the supplier using agreed unit rates.
  - *a bill of materials.* A list of the materials expected to be used, together with a unit of measurement; where the purchaser pays a standard rate based on a pre-agreed composite unit of measure.

Price-based contracts incentivize the supplier; by working efficiently, cost can be controlled and profit maximized. Likewise, the supplier will generally only supply goods and services that meet the absolute minimum required by the specification. With regard to risk, price-based contracts require the supplier to bear a comparatively high level of risk: They are required to perform all the necessary work to meet the specification within a specified timescale. From the purchaser's perspective, the major limitation of a price-based contract is that it establishes a relatively inflexible contract structure.

Cost-Based

- *Cost-plus.* Where the supplier is reimbursed all their entitled expenditure plus an agreed profit margin, which can either be a percentage of the final cost (cost plus percentage fee) or a fixed amount (cost plus fixed fee).

  Examples include NEC Engineering and Construction Contract: Cost reimbursable contract (1995); AGC 230 Standard Form of Agreement and General Conditions Between Owner and Contractor (*Where the Basis of Payment is the Cost of the Work*) 2000 Edition; IChemE Form of Contract for use in the process industries: Reimbursable Contract (The Green Book), 3rd Edition.

  Cost-based contracts have the benefit of being more collaborative, but they impose a much lower degree of control on the supplier, requiring more managerial effort by the purchaser. Compared with price-based contracts, the level of risk borne by the supplier will reduce, while that of the purchaser will rise; however, the contract will contend with high levels of change.

Incentives and Contract Type. Incentive provisions can be incorporated within fixed-price and cost-reimbursable contracts. Herten and Peeters (1986) describe and illustrate three specific types: cost incentives, schedule incentives, and performance incentives. They also refer to multiple-incentive contracts, where two or more of these incentives are combined, either dependently or independently, in the same contract. Bubshait (2003) puts forward a fourth type, safety incentives, although he found only limited support for its value.

Incentive contracts are not as extensively used as they might be. According to Ward and Chapman (1994), this is perhaps due to a lack of appreciation of the limitations of conventional fixed-price contracts and/or of the ability of incentive contracts to motivate suppliers. However, within industrial projects, Bubshait (2003) highlights the variation in the perception of purchasers and suppliers concerning incentive/disincentive (I/D) contracting. While he found a general agreement on the effectiveness of I/D contracting in encouraging supplier performance, few organizations incorporate I/D principals into their contracts. Moreover, penalty systems were used, rather than incentive systems, to penalize the supplier for late completion.

Examples of incentive contracts include NEC Engineering and Construction Contract: Target contract with bill of quantities (1995); NEC Engineering and Construction Contract: Target contract with activity schedule (1995); AGC 250 Standard Form of Agreement and General Conditions Between Owner and Contractor (*Where the Basis of Payment is a Guaranteed Maximum Price*), 2000 Edition.

Choice of Contract

The choice of contract type is one of the most significant strategic decisions, since it determines how the supplier is paid and how risk is allocated between the parties. As a general principle, contract type should aim to give the maximum likelihood of attaining the objectives of a project (Wang et al., 1996); they should be regarded as a means to an end.

Griffiths (1989) summarizes the advantages, problems, and resource requirements of the major contract type alternatives. In addition, based upon 93 R&D defense projects, Sadeh et al. (2000) found that contract type has a considerable impact on project success. Under increasing technological uncertainty, both parties to the contract benefit from cost-plus contracts, while fixed-price contracts generate more benefits when uncertainty is lower. They recommend two-stage projects. At the first stage, the preliminary design and feasibility study stage, where technological uncertainty is very high, they recommend the use of cost-plus contracts. At the second stage, the full-scale design and development stage, a fixed-price contract is preferable.

Likewise, Turner and Simister (2001) have demonstrated that, when using transaction cost analysis to indicate when alternative contract pricing terms should be adopted, it is uncertainty of the final product and not risk per se that determines the most appropriate type of contract. Further, they suggest that, if the purpose of a contract is to create a project organization based on a system of cooperation not conflict, then the requirement for goal alignment is more significant. This, they consider, requires that all parties to a contract should be properly incentivized, and that this is accomplished by incorporating contract pricing terms, as illustrated in Figure 29.1.

Turner and Simister (2001) conclude that the main criterion for selecting contract pricing terms is goal alignment, however, transaction costs are minimized *en passant.*

Roles, Relationships, and Responsibilities

Roles

A contract defines the roles of the two parties: the purchaser and the supplier. Additionally, the contract apportions roles: for example, project management, design execution and integrity, production supervision, and dispute determination.

The Role of the Parties to the Contract. Generally, the purchaser will be involved in the following:

- Defining exactly what services are to be provided
- Setting service levels
- Providing relevant and timely information to the project manager and supplier
- Informing the project manager of under-performance

The degree of empowerment of a supplier is dependent upon the procurement approach adopted and the terms of the contract. Generally, the supplier will be responsible for the following:

- Deciding how to provide the service
- Delivering the service to specification;

FIGURE 29.1. SELECTION OF CONTRACT TYPES.

<table>
<tr><td></td><td></td><td colspan="2">Uncertainty of the product</td><td></td><td></td></tr>
<tr><td></td><td></td><td>Low</td><td>High</td><td></td><td></td></tr>
<tr><td rowspan="2">Uncertainty of the process</td><td>High</td><td>Fixed-Price
Design and Build</td><td>Cost-Plus
Design and Build
Alliance</td><td>High</td><td rowspan="2">Complexity</td></tr>
<tr><td>Low</td><td>Remeasurement
Build Only</td><td>This situation was not researched</td><td>Low</td></tr>
<tr><td></td><td></td><td>Low</td><td>High</td><td></td><td></td></tr>
<tr><td></td><td></td><td colspan="2">Ability of the client to intervene</td><td></td><td></td></tr>
</table>

Source: Turner and Cochrane (1993).

- Deciding priorities to achieve the service
- Meeting purchaser requirements within the contract terms and budget
- Monitoring the service delivery performance
- Development and implementation of agreed procedures
- Providing information as required by the contract (CUP, 1997)

The contract may also define the role of other parties (agents of the purchaser), such as project manager, engineer, architect, landscape architect, interior designer, quantity surveyor, superintending officer, clerk of work, and so on, as well as the relationships with other suppliers, including subcontractors, nominated subcontractors, and nominated suppliers, and so on.

Relationships

Rules and Procedures. The establishment, from the outset, of transparent procedures will enhance contract management while reducing the disruption that problems may generate. While some will relate to the routine contract management activities, others will operate when needed.

Procedures will be required for the following:

- Performance/service management
- Risk assessment
- Contingency planning
- Payment submission, processing, and certification
- Budget review and control
- Change management—instigated by either the supplier or the purchaser
- Price adjustments
- Interrelationship of management and control
- Security
- Problem management
- Disputes resolution
- Compliance monitoring
- Termination requirements

Effective implementation of a project will be reliant upon the relationships between the parties, not necessarily on the contract or role definition. Further, it is crucial that these relationships be established at the outset, continually reviewed, and actively managed. Relationships need to be balanced between, on the one hand, flexibility and openness, and on the other, professionalism and businesslike behavior.

Risk and Responsibilities

Contracts set down the rights and obligations of the parties to the contract and describe the responsibilities and procedural roles of those named within the contract. For any project, achievement of its objectives is the principal risk; this is borne by the purchaser. Likewise, the purchaser bears the key risks of any project, for example, in deciding to instigate the project, defining the project's scope and specification, selecting a contract strategy, and choosing a supplier.

Other risks relate to the design, implementation, and delivery of the project; contracts seek to allocate these risks to the parties. However, both parties may be at risk irrespective of the contract, for example, where forces outside their control frustrate the work. Kangari (1995) summarizes the attitude of U.S. construction contractors towards the allocation and importance of risks within contracts. He also reviews trends in these perceptions.

Ideally, the allocation of risk between the parties should be based upon the following:

- *Managerial principals*. A satisfactory completion of a project is more likely to be achieved through effective planning and supervision rather than requiring guarantees and imposing rights to damages for default.
- *Commercial principles*: A risk should be borne by the party best able, economically, to control, manage, or insure against its consequences.
- *Legal principles*: Unfair contract terms, for example, penalties may not be enforceable (Wearne, 1999).

Ultimately, it is not in the purchaser's financial interest to ask a supplier to absorb all risks. The purchaser's objectives are more likely to be attained through the use of contract terms that motivate the supplier to perform on time, economically, and so on, and where the risks transferred are not so great as to be detrimental to either party in the short or long term (Barnes, 1983).

The Obligations and Entitlement of the Purchaser. The purchaser has three main obligations: to enable the supplier to complete the works/product/service, to pay the agreed price, and to accept the works/product on completion. Contracts also include entitlements, such as the right to appoint a project manager or engineer and the right to employ and pay others to complete work, if the contractor fails to perform in accordance with the contract.

The purchaser discharges their contractual responsibility by paying the contractor the accepted contract amount, amended where required under the contract. For a discussion of the role and responsibility of the purchaser under the FIDIC suite of contracts, see Van Houtte (1999).

The Obligations of the Supplier. Contracts generally contain numerous clauses that command the supplier to either comply with an instruction or do something; the FIDIC Construction Contract includes over 80. For example, the contractor shall "design, execute and complete the Works in accordance with the contract; comply with instructions given by the Engineer; remedy any defects in the Works; and institute a quality assurance system."

The supplier discharges their contractual responsibility by fulfilling their obligations under the contract. For example, under the FIDIC Construction Contract:

> The Contractor shall complete the whole of the Works . . . including achieving the passing of the Tests on Completion, and completing all work which is stated in the

Contract as being required for the Works or section to be considered to be completed for the purposes of taking over . . ." (Subclause 8.2).

Management and Supervision. Purchasers often delegate the functions of contract administration, which Wearne (1992) refers to as a *concierge de contract,* and project supervision to third parties via contracts. These functions are often combined, for example, in contract strategies where design is separate from production; the initial designer, architect, or engineer usually undertakes both roles. More uncommonly, contracts separate these functions, for example, in the NEC the project manager is responsible for contract administration and the supervisor for ensuring the works are implemented in accordance with the contract. (NEC, 1995).

The supervision of a supplier, in terms of what, how, and when to supervise, is dependent upon the risks inherent in the project, the contract terms, and the inclusion of incentives to encourage satisfactory performance. Normally, the supervisor has no authority to amend the contract or to relieve either party of any duties, obligations, or responsibilities under the contract.

Time, Payment, and Change Provisions

This section addresses the key areas of time issues, payment provisions, and change mechanisms within contracts.

Time Issues

Commencement. Contracts include a date for commencement of the project, usually determined by the purchaser or by negotiation between the parties. The commencement date should be set so that it enables the supplier to mobilize resources. Further, contracts should determine what happens in the event of the purchaser failing to provide access to a site or make available plant, service, or any other resources required under the contract, as such failure could frustrate the contract.

Schedule. Generally, contracts include statements regarding the progress of the project. For example, the FIDIC Construction Contract requires the supplier to submit a detailed program (schedule) within a stipulated time frame and to submit revised schedules whenever the previous schedule is inconsistent with actual progress or with the contractor's obligations.

Suspension. Contracts can include provisions that enable the purchaser to suspend the project. The consequences of suspension of the project, the entitlement to payment for plant and materials in event of suspension, the resumption of the project, and the possible termination of the contract as a result of prolonged suspension need also to be addressed within the contract.

Completion. Usually purchasers specify a completion date or alternatively the number of calendar or working days authorized for executing the work. Failure to complete a project within this stipulated time limit can be grounds for a significant dispute between the parties. Numerous contracts include two completion targets: substantial completion and final acceptance. Generally, the contract will stipulate the procedures to be used to determine substantial completion, that is, where the supplier has achieved substantial performance.

Payment Provisions

The contract states how, what, and when the supplier will be paid, for example, stage payments based on work completed at monthly intervals or milestone-based. Further, it will determine whether payment is incentivized and the level of retention. A purchaser, when planning a contract, should consider what payment terms are most likely to motivate the supplier to achieve the purchaser's objectives for the project.

Fixed-Price Terms of Payment. Fixed-price terms of payment are appropriate for projects that are fully specified prior to inviting potential suppliers to bid and where the completion date of the project is more important to the purchaser than the need to make changes to the specification or any contract terms.

Advance Payments. Advance payments, alternatively referred to as down payments or payment for preliminaries, are inducements to suppliers to commence work promptly; they also reduce the supplier's financing charges. A potential risk, however, is that the purchaser could lose the value of early payments if the supplier subsequently defaults. To avoid this, the supplier can be required to obtain a performance bond before receiving payment.

Milestone and Planned Progress Payments. The supplier can receive payments "on account" in a series of payments for achieving defined stages of progress. Two examples are "milestone" payments, where payment is made based upon progress in completing defined segments of the project, and "planned payment systems," where payment is activated upon achieving defined percentages of a supplier's schedule. Early payment systems should reduce the supplier's risks and financing charges.

Stage payments provide the supplier with an incentive to complete the work promptly. Incorporating additional "bonus" payments for attaining a milestone ahead of schedule can increase this incentive. However, it has the disadvantages that the contract and its management are more complex, and disputes may arise if the milestones or equivalents have not been adequately defined or their achievement proved. Additionally, the contract should state what happens when a stage is achieved ahead of schedule, and what payment is due if a stage is missed but the subsequent one is attained.

Payment Based upon Agreed Rates. In this provision, payment for work executed is based upon rates (unit prices) provided by the supplier when bidding for the contract, with the anticipated quantities of each item of work listed in a "bill of quantities" or "schedule of measured work." Unit rate terms of payment provide a basis for paying a supplier relative to the extent of work completed. The final contract price is calculated using fixed (pre-agreed) rates but is adjusted if the quantities change.

Alternatively, some contracts incorporate a "schedule of rates," where the rates are provided on the basis of indications of possible total quantities of each item of work in a defined period or within a limit of variation in quantity. Schedules of rates have the potential advantage of creating a basis for payment when the type of work is known but not the exact quantities. However, there is the potential disadvantage that supplier will incorporate uneconomical rates.

Contract Price Adjustment/Price Fluctuations/Variation of Price. Contracts can include terms for reimbursing suppliers for escalation of their costs as a result of inflation—for example, a clause that allows a contractor to raise prices in line with a pre-agreed index. Such terms have the potential advantage that suppliers have to attempt to forecast inflation rates, which may result in the submission of higher prices; however, a disadvantage for purchasers is that they generate uncertainty over the final project cost.

Where such terms are not incorporated into the contract, the supplier's prices are prone to be higher in periods of inflation; however, the final contract sum will be independent of inflation. Also, in periods of inflation the suppliers will have the incentive to complete their work more quickly to reduce the impact of increases in their costs.

Cost-Reimbursable Terms of Payment. Cost-based terms of payment can be referred to as cost-reimbursable, prime cost, dayworks, time and materials, and so on. There are two versions:

- *Cost plus fixed fee.* Where the purchaser reimburses all the supplier's reasonable costs: employees on the contract, materials, equipment, and payments to subcontractors, plus usually a fixed sum for financing, overheads, and profit.
- *Cost plus percentage fee.* As above, but the fee is added as a fixed percentage.

Cost-plus contracts can be let competitively where the supplier is required to provide their rates per hour or per day for categories of personnel, equipment, and other services. While this is also a unit rate system, it varies from those mentioned previously, as payment is for cost not performance.

With regard to the supplier, under cost-plus contracts their risks are limited at the expense of potential profit. For the purchaser, cost-plus contracts are appropriate if all the categories' potential resources can be predicted although the exact extent of the work initially remains uncertain.

Target-Incentive Contracts. Target-incentive contracts are a development of the reimbursable type of contract, where the purchaser and supplier agree at the beginning a probable cost for an as-yet-undefined project; however, they also agree to share any savings in cost relative to the target. However, if the target is exceeded, the supplier will be reimbursed less than cost-plus.

Cost plus incentive fee. Where the fee may fluctuate either up or down within set limits and in accordance with a formula linked to permissible actual costs. Veld and Peeters (1989) consider the most important aspect in cost incentives is the sharing factor. They note that the sharing formula can be nonlinear, it can vary between overrun and under run, and sometimes a neutral zone is introduced.

Al-Subhi Al-Harbi (1998) explains how purchasers and suppliers select the supplier's sharing fraction based upon a risk-averse, risk-taking, or risk-neutral perspective. Veld and Peeters (1989), Ward and Chapman (1994), Al-Subhi Al-Harbi (1998), and Berends (2000) provide examples of incentive fees expressed as equations.

Berends (2000) believes that in order to be effective, the incentive scheme must be aligned with the overall project objectives, not just the cost objective, of the owner. Further, it must provide a positive relationship between the supplier's performance and the supplier's profit margin. Subject to both parties having the ability to prepare realistic cost estimates, the contract negotiation process affords a means to deliver an effective incentive scheme.

Convertible Terms of Payment. A convertible contract incorporates an agreement that after any significant uncertainties have been decided it will be changed into a fixed-price or unit-rate-based contract. Potentially, such an agreement has the advantage of limiting the contract price once it is converted; however, there is little or no opportunity for competitive bidding.

Periodic Payments. Contracts may incorporate clauses that entitle the supplier to interim payments: payments on account. These payments are usually on a monthly basis, based on the estimated value of the project executed by the supplier in the preceding month and include any amount to be added or deducted under the terms of the contract, for example, retention.

Retention. The majority of large contracts incorporate a clause where a percentage of any payment due to the supplier, for example, as fixed price, milestone achievement, or value of work completed, is retained by the purchaser for a specified period (generally up to one year). The retained amount (retention) is paid (released) once the supplier has satisfactorily completed his or her obligations, for example, the rectification of any defective work.

Such a clause, for the purchaser, has the potential advantage of motivating the supplier to complete the project appropriately the first time, thereby, activating the release of the retention at the earliest opportunity.

Incorporating Change

A purchaser's needs may alter during the period of a contract, for example, in quality, quantity, and even character. Where a supplier provides an extra service or additional work, they are entitled to request extra payment. Contracts, therefore, should incorporate terms that facilitate the effective management of change. Where, in the course of a contract, major variations are expected, suppliers should be granted greater empowerment to initiate change, possibly in the form of value management. This can be of significant advantage to the purchaser where there is the potential to incorporate advances in new technologies or new techniques.

Contracts should also provide a mechanism for pricing any changes in the project deliverables, although, with regard to the purchaser, these mechanisms are usually less competitive than those in the original contract. Any changes need to be managed by both the purchaser and the supplier; inevitably this adds to the cost of the contract

Change management requires the following stages: identification of a potential change requirement, contemplation of the full impact of the change, production of a formal change order, and notification to the parties of the agreed change. Additionally, each stage of the process should be documented and the decision maker should possess the appropriate authority to agree the change.

Latham (1994) identified variations as one of the main problems confronting the UK construction industry, regarding them as probably being a significant cause of disruption, disputes and claims. (See also the chapter by Cooper and Reichelt elsewhere in this book.)

For example, under the FIDIC Construction Contract the engineer can instigate variations at any time before issuing the taking-over certificate for the works, either by issuing an instruction or by requesting that the contractor submit a proposal. The contractor is required to comply with a variation, although the contractor can object to a variation where the goods required for the variation cannot readily be obtained. Further, the contract prescribes how a variation is to be measured and evaluated. Generally, appropriate rates or prices specified in the contract will be used or form the basis to derive new rates or prices to value the work. Where no appropriate rates or prices exist, the work is to be valued based on the reasonable cost of carrying out the work, plus reasonable profit.

Remedies for Breach of Contract

Contracts need to incorporate provisions to manage the consequences of default by either party. This section introduces the concepts of performance indicators, liquidated damages, and termination.

Performance Indicators (PI)

The contract should contain appropriate and achievable performance indicators. While it is essential to include the general principles of performance indicators within the bid documentation, the actual indicators are often determined during the contract negotiation stage with the preferred supplier—that is, before the contract price is agreed and the contract signed.

Determining an effective performance measurement system requires the identification of the following:

- The services to be provided
- The critical success factors for a particular contract
- The key performance indicators, targets, and measures
- The components to be measured or assessed, both in terms of the outputs and outcomes of the service (CUP, 1997).

Liquidated Damages

Where it is possible to derive a preestimate of the loss to be suffered under certain circumstances, it is generally prudent to incorporate liquidated damages into the contract. Liquidated damages, however, should be a bona fide preestimate of the loss in the given situation.

"Liquidated damages" places a limited liability on the supplier to pay a specified sum for a defined breach in performance, for example, late delivery of a product or late completion of a project. The aim is to encourage suppliers to meet their contract obligations; however, their effectiveness may be limited, for example, where the cost of performing their obligations is more than the liquidated damages.

In addition to liquidated damages, contract terms can be used to motivate the supplier, for instance, by offering additional payments where the supplier completes on time or recovers time after a delay in meeting their contract obligations.

Termination

Under certain circumstances a contract may need to be terminated. As this action results in a lose-lose situation, it is generally only used as a last resort, due, usually, to the unacceptable performance of the supplier. However, other exceptional circumstances include a change in government machinery, a change of policy, or a change in user needs.

Termination of a contract requires contingency plans to ensure continuity of supply, construction, or completion of a service or product. This is essential when the service is critical to the business and the supplier can terminate the contract. These contingency plans may need to be incorporated into the contract; for example, to effect transition from one supplier to another, the existing supplier could be required to cooperate with the new suppliers. Also, special contract provisions may be needed to deal with intellectual property rights in software developed or improved during the contract.

For example, under the FIDIC Construction Contract the employer is entitled to terminate the contract if the contractor abandons the works, subcontracts the whole of the works or assigns the contract without the required agreement, becomes bankrupt or insolvent, or gives or offers to give a bribe (This list is indicative, not exhaustive.)

Likewise, the contractor is entitled to terminate the contract: if the employer fails to provide reasonable evidence of their financial arrangements, or an amount due under an interim certificate; the engineer fails to issue a payment certificate; or where the employer substantially fails to perform their obligations under the contract. (Again, this list is indicative, not exhaustive, and most clauses contain a time frame for compliance.)

Bonds, Guarantees, and Insurances

The Principles of Bonds and Guarantees

Financial guarantees are issued by banks, insurance companies, surety companies, or a parent company so that funds will be available should the purchaser have a legitimate claim against the supplier. Chaney (1987) classifies the following types of guarantee:

- *Tender (bid) guarantee (bond).* Typically required by purchasers to ensure that once a bid has been submitted, the supplier can be held to it, and, so far as possible, to exclude suppliers who lack the necessary financial resources to complete the contract. They will virtually always be conditional and have a time limit.
- *Repayment guarantee.* Generally incorporated where an advance payment is involved. They can either be "on-demand" or "conditional." An on-demand bond permits the purchaser to invoke the guarantee without having to establish default, while under a conditional guarantee, the purchaser can only invoke the guarantee once the supplier has admitted a breach of contract, or upon a ruling of a court or an arbitrator that the supplier is in default.
- *Performance guarantee.* Seeks to ensure that a purchaser can recover damages in the event of a supplier's failure to perform. Again, they may be classified as on-demand or conditional.
- *Retention guarantee.* An alternative to a retention clause, discussed earlier. While a purchaser may consent to this arrangement from the start of the contract, retention guarantees are more commonly used during the maintenance period, that is, upon attaining practical completion.
- *Surety bond.* The usual form of guarantee in North America. They vary significantly from the principle of indemnity: rather than merely paying the amount of the bond following a default, the emphasis is on the surety organizing the completion of project. A performance and payments bond is usually required on all publicly bid construction projects and may be required on some private projects (Carty, 1995).

The liability of the guarantor is therefore for costs incurred by the purchaser limited to the amount of the guarantee, while a surety has the added responsibility of arranging for the completion of the work by a third party. The guarantor does not insure the supplier; they merely provide a guarantee. Therefore, as a condition of issuing the guarantee, the supplier will be required to indemnify the guarantor to the extent that, if the guarantee is called in, the guarantor will only suffer financially if the supplier enters into liquidation. Further, the guarantor will generally require the supplier to provide details of their financial position and their capability and resources to undertake the contract. Depending upon who issues the guarantee, premiums can amount to 2 percent of the sum guaranteed per annum, a cost either directly or indirectly borne by the purchaser.

The Principles of Insurances

Contracts state what types of insurance are required, determine who is to be responsible for obtaining the insurance, and specify the particular terms of the policies and limits of coverage.

Claims

Smith and Wearne (1993) define a claim as a demand or request, usually for extra payment and/or time. Usually, a claim is an assertion of a right in a contract; but in others it is the converse: a submission outside the terms of a contract.

Suppliers

Contracts include provisions for the submission of claims for time and money where the supplier's work is likely to be disrupted or delayed. In such circumstances the supplier is generally required to give notice to the project manager of such disruption or delay. Further, the supplier will be required to give specific details of what caused the delay and to quantify the likely delay or disruption.

In the first instance, contracts usually empower the project manager to agree or determine any matter, by consulting with each party so as to reach an agreement. If there is a failure to reach an agreement, then the project manager is usually authorized to make a fair determination. If the project manager concludes that completion is or will be delayed, the supplier will be entitled to either an extension of time, an extension of time and payment of any cost incurred, an extension of time and payment of any cost incurred plus reasonable profit, or payment of any cost incurred plus reasonable profit.

Extension of Time. Events that would lead to the supplier having an entitlement to an extension of time, under the FIDIC Construction Contract, include the following, where completion is or will be delayed due to the following:

- A substantial change in the quantity of an item of work included in the contract
- Exceptionally adverse climatic conditions
- Any delay, impediment, or prevention caused by or attributable to the employer, the employer's personnel, or the employer's other contractors on the site
- Or, if the contractor is prevented from performing any of their obligations under the contract by force majeure and suffers delay and/or incurs cost due to force majeure. (This list is indicative, not exhaustive.)

Force majeure is defined as an exceptional event or circumstance, which is beyond a party's control; could not have been reasonably provided against before entering the contract; having arisen, could not have been reasonably avoided or overcome; and that is not substantially attributed to the other party (Subclause 19.4).

Extension of Time and Payment of Any Cost Incurred. Events that would lead to the supplier having an entitlement to an extension of time for completion and payment of any cost incurred, under the FIDIC Construction Contract, include the following:

- Encountering unforeseeable physical conditions
- Discovery of antiquities and the like found on the site
- Engineer's suspension of the works
- Suffering delay and/or incurring cost from rectifying any loss or damage to the works, goods, or contractor's documents due to the employer's risks. (This list is indicative, not exhaustive.)

Extension of Time and Payment of Any Cost Incurred Plus Reasonable Profit. Events that would lead to the supplier having an entitlement to an extension of time for completion and payment of any cost incurred plus reasonable profit, under the FIDIC Construction Contract, include the following:

- Failure of the engineer to issue any necessary drawing or instruction within reasonable time.
- Failure by the employer to give the contractor right of access to, or possession of, the site (within a specified time limit).
- Executing work that was necessitated by an error contained in the contract or notified by the engineer. (This list is indicative, not exhaustive)

Payment of Any Cost Incurred Plus Reasonable Profit. The supplier, under the FIDIC Construction Contract, is entitled to payment of any cost incurred plus reasonable profit if costs are incurred:

> . . . as a result of the employer taking over and/or using a part of the works, other than such use as is specified in the contract or agreed by the contractor (Subclause 10.2).

The use of words such as reasonable and unforeseen, which the FIDIC Construction Contract further defines as "not reasonably foreseeable by an experienced contractor by the date for submission of the tender," introduces ambiguity into contracts. The interpretation of words such as these has led to disputes. It is widely acknowledged that it does not mean what it says; rather, it is interpreted as what might have been foreseen and allowed for by an experienced contractor (Corbett, 2000).

Purchaser's Claims

Likewise, the purchaser can claim against the supplier, that is, include an amount as a deduction in the contract price or payment certificate. Typical claims against the supplier, according to Bubshait and Manzanera (1990), relate to the following:

- Use of nonspecified materials
- Defective work
- Damage to property
- Late completion by the supplier

Purchaser's claims can be broadly categorized as liquidated damages, claims explicitly provided for by the contract, and claims for damages for breach of contract by the contractor (Corbett, 2000).

For a discussion of force majeure (Van Dunne, 2002) and claims under the FIDIC Construction Contract, see Seppala (2000) and Corbett (2000).

Formal Procedures for Submitting a Claim

Generally, contracts require the project manager to assess and award extensions of time and/or expenses; however, prudently, most contracts place the onus for substantiation entirely on the claimant. Further, they state specific time frames for submission of details and assessment, and specify the course of action open if the claimant disagrees with the decision.

Kumarasawamy and Yogeswaran (2003) review the techniques and approaches available to substantiate claims for an extension of time and give recommendations on their use. These include global impact, net impact, time impact, snapshot, adjusted as-built critical path method, and isolated delay type techniques.

Dispute Resolution

Adversarial and Nonadversarial Dispute Resolution

Problems can arise when implementing contracts regardless of the relationship between the parties and the type of project. In addition to those mentioned earlier, problems can arise as a result of one or both of the parties having conflicting objectives; failing to anticipate significant risks or variations; failing to consult; making erroneous assumptions; and, at a basic level, making a mistake.

A major role of the project manager and, therefore, contract management is predicting potential difficulties in implementing the project to prevent problems turning into disputes. The project manager should, therefore, endeavor to reduce the effects of problems by instituting transparent procedures that are adhered to, by promoting collaboration, and by engendering a shared aspiration to resolve difficulties. The effect of problems can be reduced by the following:

- Establishing approved procedures
- Establishing and adhering to boundaries of delegated authority
- Making contingency plans
- Setting up regular reviews with both the purchaser and the supplier
- Instigating timely recognition and corrective action
- Implementing appropriate contract changes
- Escalating, where appropriate, the problem to senior management (CUP, 1997)

Traditionally, both in the United States and the United Kingdom, the preferred method of dispute resolution arising out of contracts was by litigation—a very slow and costly process. Contracts, therefore, generally incorporate methods of dispute resolution to avoid disrupting the implementation of the project and to resolve any dispute in a fair and timely manner. Examples include mediation, conciliation, dispute review boards, adjudication, and arbitration.

Fenn et al. (1997) distinguishes between conflict and disputes. Conflict, they assert, although an unavoidable fact of organizational life, has positive aspects to do with commercial risk-taking; alternatively, disputes afflict industry. They propose a conflict management/dispute resolution taxonomy (see Figure 29.2).

Elsewhere, Cheung (1999) presents a dispute resolution "stair-step" chart composed of the following steps:

- *Prevention.* Includes risk allocation, inceptive for cooperation, and partnering
- *Negotiation.* Includes direct and step negotiation
- *Standing neutral.* Includes dispute review board and dispute resolution adviser
- *Nonbinding resolution* Includes mediation, mini-trial, and adjudication
- *Binding resolution.* Arbitration
- *Litigation.* Judge

Antagonism and cost increase as one goes down the list (up the stairs). Moreover, resorting to arbitration or litigation is unlikely to improve the implementation of the project and should, therefore, only be used as a final measure. Furthermore, if they are used, the contract has effectively failed.

In the Hong Kong construction industry, Cheung found that when deciding upon a method of dispute resolution, the parties are mainly interested in the benefits, whether tangible or perceived, that may be gained. These benefits include prompt resolution, low cost, and preservation of the relationship between the parties.

Generally, governments are reluctant to legislate in regard to commercial contracts. However, in the case of construction contracts there are exceptions:

- In 1981 the State of California established mandatory arbitration for state agencies' construction contracts (Carty, 1995).
- In the United Kingdom, the Housing Grants, Construction and Regeneration Act 1996 established the right for a party to a construction contract to refer a dispute arising under the contract for adjudication, as outlined in the Scheme for Construction Contracts (England and Wales) Regulations 1998. The decision of the adjudicator will be binding until the dispute is finally decided by legal proceedings, by arbitration, or by agreement. For a detailed review of the implications of this act, see Paterson and Britton (2000).

Dispute Resolution under the FIDIC Construction Contract

Initially, under the FIDIC Construction Contract the engineer will proceed to agree or determine any matter. In doing so:

FIGURE 29.2. FENN ET AL.'S PROPOSED TAXONOMY.

Conflict Management

Non-binding

Dispute review boards

Dispute review advisors

Negotiation

Quality matters:

- Total quality management
- Co-ordinated project information
- Quality assurance

Procurement systems

Dispute resolution

Non-binding	*Binding*
Conciliation	Adjudication
Executive tribunal	Arbitration
Mediation	Expert determination
	Litigation
	Negotiation

> . . . the engineer shall consult each Party in an endeavour to reach agreement. If agreement is not achieved, the engineer shall make a fair determination, in accordance with the Contract, taking due regard of all relevant circumstances" (Subclause 3.5).

The contract also makes provision for the establishment of a Dispute Adjudication Board (Subclause 20.2), comprising either one or three suitably qualified persons, to adjudicate disputes that may arise among the parties. Where a dispute has been referred to the DAB and a decision, if any, has not become final and binding, the contract states that it shall ultimately be settled by international adjudication

For a discussion on the resolution of construction disputes by disputes review boards, see Shadbolt (1999) and specifically, under the FIDIC Construction Contract, see Seppala (2000).

Flexibility, Clarity, and Simplicity

Effective contracts require flexibility, clarity, and simplicity, and should stimulate good management. Standard forms have been criticized for failing to provide these attributes. For example, standard forms of contract within the UK construction industry have been criticized for lacking clarity and simplicity; criticisms that could be leveled at standard forms of contract in many industries and countries. The possible reasons for this, according to Broome and Hayes (1997), relate to their origin, being derived from very old precedents; their age, the language and phrasing being derived from English contracts of the late nineteenth century; development by committee; partisanship; lack of direction; and amendment, where specific users heavily amend and supplement the standard form.

Likewise, a review of procurement and contractual arrangements in the UK construction industry (Latham, 1994) criticized the existing standard forms of contract. Contracts, it suggests, should be fair, comprise simple phrasing, set transparent management procedures, and encourage teamwork.

The New Engineering and Construction Contract (NEC, 1995), designed to be used internationally, includes virtually all Latham's recommendations. The NEC suite of contracts has been written to form a manual of project management procedures rather than an agenda for litigation. However, a survey by Hughes and Maeda (2003) found their respondents to be ambivalent about the concept of a spirit of mutual trust; moreover, they held that authoritative contract management would improve performance. This finding is divergent from the principles behind current steps toward innovative working.

Other recently revised standard forms of contract include the FIDIC Suite of Contracts (First Edition 1999), the American Institute of Architects, AIA Contract Document, for example, the A201-1997 General Conditions of the Contract for Construction, and the recent Associated General Contractors of America AGC 200 Series of General Contracting Documents.

Summary

A contract, according to the Association for Project Management, Specific Interest Group on Contracts and Procurement (SIGCP, 1998), should be designed to be the basis for successful project management: being right in principle (contract strategy) and right in detail (contract terms). This chapter discussed the factors that influence commercial contract practice, providing the project manager with an overview of generic contract provisions and procedures.

To summarize, a good contract stipulates what, where, and when something is to be provided; identifies the supplier and purchaser; defines various roles, relationships, and responsibilities; sets standards; deals with issues of time, payment, and change; provides remedies for breach of contract; covers the issues of bonds, guarantees, and insurances; and contains mechanisms for the submission of claims and the resolution of disputes. Further, effective contracts require flexibility, clarity, and simplicity, and should stimulate good management.

Finally, the following best practice, in terms of planning the contract and contract management, is provided by the Association for Project Management, Specific Interest Group on Contracts and Procurement (SIGCP, 1998):

Planning the Contract

- Select suppliers that will best serve the interests of the project.
- Understand how a contract is formed and how it can be discharged.
- Choose the terms of a contract logically, taking into account the nature of the work, its certainty, its urgency, and the competence, objectives, and motivation of all parties.
- Consider how the contract will impact on other contracts and projects, and plan the coordination of the work carried out under the contract in relation to existing facilities and systems.
- Plan, before starting, how the contract will terminate.
- Envisage what can go wrong, utilize risk management, and allocate risks to best motivate their control.
- Define the obligations and rights of each party.
- Specify only what can be tested.
- Agree criteria for the assessment of satisfactory performance.
- Determine and incorporate effective payment terms.
- Ultimately, say what you mean and be clear about what you want.

Contract Management

- Control the contract through the appointment of a single manager, who has the authority to decide how to avoid problems and has experience of the potential conflicts of interest that can arise.
- Determine what power the suppliers' manager really has over resources.
- Scrutinize the contract and examine the obligations and rights of all parties.
- Recognize that, in the course of most contracts, objectives and priorities vary.

- Control variations and derive appropriate advantage from potential variations.
- Keep records and notes of reasons for decisions. Use routine headings.
- Distinguish between legal rights and project/commercial interests.
- Finally, realize that a contract should be a means to an end.

References

Al-Subhi Al-Harbi, K. M. 1998. "Sharing fractions in cost-plus-incentive-fee contracts," *International Journal of Project Management* 5(4):231–236.

American Institute of Architects (AIA). 2002. AIA contract documents. www.aia.org/documents. Washington, D.C.: The American Institute of Architects.

Associated General Contractors of America (AGC). 2002. *AGC contract documents at a glance.* Alexandria, VA: The Associated General Contractors of America.

Barnes, M. 1983. How to allocate risks in construction contracts. *International Journal of Project Management* 1(1):24–28.

Berends, T. C. 2000. Cost plus incentive fee contracting: Experience and structuring. *International Journal of Project Management* 18:165–171.

Broome, J. C., and R. W. Hayes. 1997. A comparison of the clarity of traditional construction contracts and the new engineering contract. *International Journal of Project Management* 15(4):255–261.

Bubshait, A. A. 2003. Incentive/disincentive contracts and its effects on industrial projects. *International Journal of Project Management* 21:63–70.

Bubshait, A., and S. A. Almohawis. 1994. Evaluating the general conditions of a construction contract. *International Journal of Project Management* 12(3):133–136.

Bubshait, K., and I. Manzanera. 1990. Claim management. *International Journal of Project Management* 8(4):222–228.

Carty, G. J. 1995. Construction. *Journal of Construction Engineering and Management* 121(3):319–328.

Central Unit on Purchasing (CUP). 1989. *Contracts and Contract Management for Construction Works.* CUP Guidance No. 12. London: HM Treasury.

———. 1997. *Contract Management.* CUP Guidance No. 61. London: HM Treasury.

Chaney, A. R. 1987. Financial guarantees. *International Journal of Project Management* 5(4):231–236.

Cheung, S. 1999. Critical factors affecting the use of alternative dispute resolution processes in construction. *International Journal of Project Management* 17(3):189–194.

Corbett, E. 2000. FIDIC's new rainbow 1st Edition: An advance. *The International Construction Law Review* 17(2):253–275.

Fenn, P., D. Lowe, and C. Speck. 1997. Conflict and dispute in construction. *Construction Management and Economics* 15:513–518.

The Fédération Internationale des Ingénieurs-Conseils (FIDIC). 1999. *Conditions of Contract for Construction (First Edition).* Lausanne, Switzerland: FIDIC.

Griffiths, F. 1989. Project contract strategy for 1992 and beyond. *International Journal of Project Management* 7(2):69–83.

Herten, H. J., and W. A. R. Peeters. 1986. Incentive contracting as a project management tool. *International Journal of Project Management* 4(1):34–39.

Hughes, W., and Y. Maeda, Y. 2003. Construction contract policy: Do we mean what we say? FiBRE—Findings in Built and Rural Environments, RICS Foundation, The Royal Institution of Chartered Surveyors, London. www.rics-foundation.org/publish/documents.aspx.

Institution of Chemical Engineers (IChemE). 2003. IChemE Forms of Contract. Rugby, UK: IChemE. www.icheme.org

The Institution of Civil Engineers (ICE), 1995. *The engineering and construction contract: Guidance notes.* London: Thomas Telford.

———. 1995. *The new engineering and construction contract (NEC).* 1995. The Institution of Civil Engineers. London: Thomas Telford.

Jergeas, G. F., and V. G. Cooke. 2000. Law of tender applied to request for proposal process. *Project Management Journal* 28(4):21–34.

Joint Contracts Tribunal (JCT), 2002. JCT contracts. London. The Joint Contracts Tribunal Ltd. www.jctltd.co.uk/contracts.htm.

Kangari, R. 1995. Risk management perceptions and trends of U.S. construction. *Journal of Construction Engineering and Management* 121(4):422–429.

Kumaraswamy, M. M., and K. Yogeswaran. 2003. Substantiation and assessment of claims for extensions of time. *International Journal of Project Management* 21:27–38.

Latham, M. 1994. Constructing the team: Final report of the government/industry review of procurement and contractual arrangements in the UK construction industry. London: The Stationery Office.

Lowe, D. J., P. Fenn, and S. Roberts. 1997. Commercial management: An investigation into the role of the commercial manager within the UK construction industry. CIOB Construction Papers, No. 81, 1–8.

Paterson, F. A., and P. Britton, eds. 2000. *The Construction Act: Time for review.* London: Centre of Construction Law & Management. King's College.

Sadeh, A., D. Dvir, and A. Shenhar. 2000. The role of contract type in the success of R&D defense projects under increasing uncertainty. *Project Management Journal* 31(3):14–22.

Seppala, C. R. 2000. FIDIC's new standard forms of contract: Force majeure, claims, disputes and other clauses. *The International Construction Law Review* 17(2):125–252.

Shadbolt, R. A. 1999. Resolution of construction disputes by dispute review boards. *The International Construction Law Review* 16(1):101–111.

Specific Interest Group on Contracts and Procurement (SIGCP). 1998. *Contract strategy for successful project management.* Norwich, UK: The Association for Project Management.

Smith, C., D. Topping, and C. Benjamin. 1995. Joint ventures. In *The Commercial Project Manager.* J. R. Turner, Maidenhead, UK: McGraw-Hill.

Smith, N. J., and S. H. Wearne. 1993. *Construction contract arrangements in EU countries.* Loughborough, UK: European Construction Institute.

Thomas, H. R., G. R. Smith, and D. J. Cummings. 1995. Have I reached substantial completion? *Journal of Construction Engineering and Management* 121(1):121–129.

Turner, J. R., and R. A. Cochrane. 1993. The goals and methods matrix: coping with projects with ill-defined goals and/or methods of achieving them. *International Journal of Project Management* 11(2): 93–102.

Turner, J. R., and S. J. Simister. 2001. Project contract management and a theory of organisation. *International Journal of Project Management* 19(8):457–464.

Van Dunne, J. 2002. The changing of the guard: Force majeure and frustration in construction contracts: The foreseeability requirement replaced by normative risk allocation. *The International Construction Law Review* 19(2):162–186.

Van Houtte, V. 1999. The role and responsibility of the owner. *The International Construction Law Review* 16(1):59–79.

Veld, J. in't, and W. A. Peeters. 1989. Keeping large projects under control: the importance of contract type selection. *International Journal of Project Management* 7(3):155–162.

Walker, D. H. T., and D. S. Johannes. 2001. Construction industry joint venture behaviour in Hong Kong: Designed for collaborative results?" *International Journal of Project Management* 21:39–49.

Wang, W., K. I. M. Hawwash, and J. G. Perry. 1996. Contract type selector (CTS): A KBS for training young engineers. *International Journal of Project Management* 14(2):95–102.

Ward, S., and C. Chapman. 1994. Choosing contractor payment terms. *International Journal of Project Management* 12(4):216–221.

Wearne, S. H. 1992. Contract administration and project risks. *International Journal of Project Management* 10(1):39–41.

Wearne, S. H. 1999. Contracts for goods and services. In *Project management for the process industries.* G. Lawson, S. Wearne, and P. Iles-Smith. Rugby, UK: Institution of Chemical Engineers.

Wright, D. 1994. A "fair" set of model conditions of contract: Tautology or impossibility? *International Construction Law Review* 11(4):549–555.

Recommended Further Reading

Specific Interest Group on Contracts and Procurement. 1998. *Contract strategy for successful project management.* Norwich, UK: The Association for Project Management,

Smith, N. J. 1995. Contract strategy. In *Engineering Project Management.* Oxford, UK: Blackwell Science.

Turner, J. R. 1995. *The Commercial Project Manager.* Maidenhead, UK: McGraw-Hill.

CHAPTER THIRTY

TENDER MANAGEMENT

George Steel

Most projects will, at one stage or another, need to go to the marketplace to procure the expertise, materials, equipment, and services required for development, implementation, or execution. The way this is done can have a profound impact on project success and profitability.

This chapter on tender management addresses the business imperative of the procurement function and the evolution of the way in which tenders for goods and services are structured and managed. It describes the tender management process in detail, starting at the early stages of contract development and finishing at contract award. By and large the chapter looks at the tender management function from the perspective of the owner organization awarding the contract. Tender management is also a critical function within supplier organizations, and this is also discussed.

The Business Imperative

The economics of top-class procurement are compelling. With the growth of outsourcing, third-party goods and services can account for between 40 percent and 80 percent of a project or company activity. The difference in value between the routine "three bids and a buy" approach by administrative staff going through a procurement process with little imagination and entrepreneurial drive, and the application of a world-class procurement process by an experienced and motivated team, can be from 10 percent to 30 percent or more. The impact on the bottom line is significant for major operating companies, as shown in the following example.

> **Case Study:** A European oil company spent €100 million per annum on the development and refurbishment of its gas stations throughout Europe. Implementing the kind of world-class tender management process described in this chapter resulted in 10 to 30 percent savings of third-party spend on materials equipment and construction services.

The impact on contracting companies can be even more dramatic. This is due to their relatively low profit margins and high levels of spend on subcontractors and other outsourced services. (See Figure 30.1.)

Evolution of the Procurement Function

Because of the potential impact on the bottom line, it is hardly surprising that the procurement function has evolved from a fairly routine administrative function into a major strategic lever of the project, or indeed the enterprise (Lassister, 1998; Latham, 1994). Initially, effective procurement meant casting user requirements in stone and squeezing individual suppliers till the "pips squeaked." This evolved into the "aggressive sourcing" methodologies

FIGURE 30.1 IMPACT OF PROCUREMENT SAVINGS ON THE BOTTOM LINE.

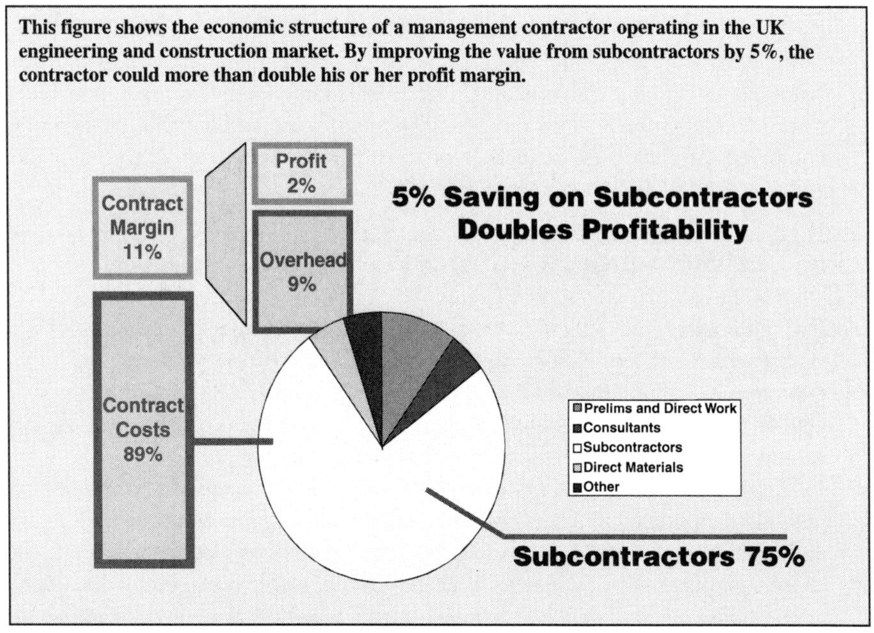
This figure shows the economic structure of a management contractor operating in the UK engineering and construction market. By improving the value from subcontractors by 5%, the contractor could more than double his or her profit margin.
Profit 2%
Contract Margin 11%
Overhead 9%
Contract Costs 89%
5% Saving on Subcontractors Doubles Profitability
Prelims and Direct Work
Consultants
Subcontractors
Direct Materials
Other
Subcontractors 75%

pioneered in the automotive industry, with exhaustive analysis of the supplier market, aggregation of demand, reduction of suppliers, and very aggressive negotiation. In recognition of the importance of organizational linkages, aggressive sourcing soon merged with the "Business Process Re-Engineering (BPR)" movement to improve the way the organization translated business objectives into commercial reality.

BPR was supported by the systematic application of "value management techniques (VMT)" to optimize user requirements. As recognition grew of the mutual dependencies between owners, contractors, and subcontractors, so the whole concept of "supply chain management (SCM)" became the new mantra, as the vehicle for obtaining sustainable improvement from the marketplace. In the 1990s "alliance contracting" and "partnering" forms of contracting became very fashionable, particularly in the UK engineering and construction industry, where initiatives by Sir John Egan and Sir Michael Latham (Egan, 1998; Latham, 1994) stressed the importance of collaboration across the project supply chain.

Finally, the emergence of online tendering systems is streamlining and accelerating the tendering process, and reverse-auction capability is unleashing the ultimate in competitive pressure. This is potentially very powerful in the project environment, where there is often the need to balance the rigor, and duration, of the procurement process with project deadlines. Over the last year or so, the use of online reverse auctions has increased significantly, and many companies are bidding significant construction contracts and contracts for professional services online. A leading pharmaceutical group, for example, has started to use online auctions for nonspecialized buildings such as warehouses and for professional building services.

The way in which the project "industry" balances the use of the intense competition generated by online bidding, with the advantages of supply chain alignment, will be one of the major issues in project procurement over the next few years. The name of the game, however, is improving the bottom line of *your* project; incorporating the procurement and tendering methodologies that integrate every single value improvement technique referred to previously within your project management process. This chapter describes how leading companies are using the tendering process to improve the profitability of their projects.

Overview of the Tendering Process

Figure 30.2 presents an overview of the procurement and tendering process covered in this chapter. The process starts with an evaluation of the project drivers and the statutory and corporate environment, along with the development of an execution strategy and contracting plan for the project. The process is complete when every package in the contracting plan has been awarded.

The process is generic and has been successfully applied to the procurement of commodities, complex equipment, and services of all kinds of projects—from IT systems to multimillion-dollar contracts for oil tankers and the engineering and construction of major process plants. Each of the steps indicated in the preceding figure is described in some detail in the following pages. The next section deals with strategic contracting issues at project

FIGURE 30.2. PROJECT AND PROCUREMENT STRATEGY.

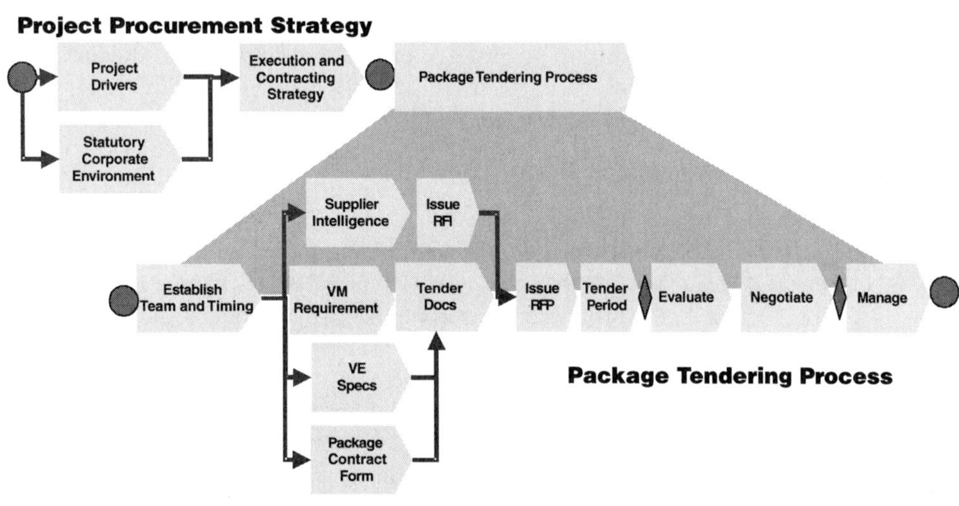

level. Subsequent sections deal with various aspects of procuring the specific work packages that form part of the overall project scope.

Project Level Issues

Understand Statutory and Corporate Compliance

It is absolutely essential that the tendering process reflects the statutory regulations and complies with corporate standards for the following reasons:

- Failure to comply with the statutory regulations that govern procurement in many countries, such as those in the European Union, may result in considerable embarrassment, abortion of the procurement process, and even expensive litigation.
- Similarly, failure to understand corporate standards for delegation of authority and contract approval, or the existence, may result in considerable delay prejudicial to the project.
- Many companies have framework contracts in place for certain materials, equipment, and categories of service. Framework contracts may save time and money for the project. The use of existing framework contracts may be an essential corporate policy.

The statutory and regulatory environment can have a profound impact on (a) the procurement process, (b) the time it takes to procure a given contract and (c) the composition of the team. This is especially true in most public sector procurement, where any minor noncompliance with process can result in elimination of the bidder or the need to retender the package. Most public institutions, and indeed many corporations, recognize the need for transparency in the procurement process and publish information on this on their Web sites.[1,2]

Understand Project Drivers and Priorities

In many cases, project business drivers will have a significant impact on contracting strategy and process. For example, in a project with a high net present value (NPV), or where *time to market* is critical, the need to get resources on board and working or ensuring the delivery of critical equipment may outweigh the economies, which would result from a more rigorous procurement process. These factors will determine the project execution and contracting strategy and the priorities of Package Procurement.

Before embarking upon the development of a project contracting strategy or a package tendering initiative, the project team must have a very clear understanding of the project business case, the risks that have been identified, and particularly the impact of time on project benefits.

Project Contracting Strategy

The development of an overall contracting strategy can be a complex subject in its own right, which goes beyond the intent of this chapter. The following points, however, will give some indication of the issues involved in developing and documenting an appropriate contracting strategy.

Package Breakdown and Basic Contractual Philosophy. The contracting strategy adopted for a project will have a significant impact upon its eventual success and profitability. The contracting strategy consists of (a) Identification of all the *work packages,* that will be undertaken by third-party organizations, (b) the scope of each work package, and (c) the basic nature of commercial arrangements governing that work package, including, but not limited to, the transfer of risk.

Competencies in Owner and Supplier Organizations. The optimal contracting strategy will be a balance between the competence of the owner organization to either undertake or to manage the work in question, and the competencies available of the supplier market. The point of departure is the definition of the minimum "core competencies" that the owner wishes to maintain within his or her own organization for strategic reasons. This will shape the services he or she wishes to buy from suppliers.

[1] Federal Acquisition Regulations 2001 S/N 922-006000008. provides a guide to the U.S. government procurement process.

[2] www.tendersdirect.co.uk provides a useful guide to EU procurement procedures.

Case Study: A European oil company requested suppliers to provide technical competence in electrical and instrumentation engineering as part of their offering. The company, however, regarded the electrical and instrumentation engineering as part of their core competence. This resulted in duplication of effort, confusion over roles and interfaces, and ultimately higher costs.

Similarly another oil company wished the engineering procurement and construction (EPC) contractor to provide overall project management of major refinery shutdowns. Unfortunately, the EPC contractor selected did not have this competence, and the oil company was required to take over the management of the turnaround project.

Major cost reductions were achieved in North Sea projects when owners stopped "man for man marking" and defined who was accountable for the result, and who was responsible for execution and reflecting this in their contracts.

Appropriate Apportionment of Risk. The contracting strategy will also define the way in which project risk is shared between the owner of the project and its suppliers and will also determine the shape of the owner project organization and the way the project is managed. The most appropriate apportionment of risk will depend on the following:

- Who has the financial resources to bear the consequences of the risk.
- Who has the best competencies to manage the risk.

Many owners will write contracts in which suppliers are asked to take risks that go way beyond their net asset value. More surprisingly, some suppliers will agree to accept such risks. Basically the owner is fooling him- or herself, because if the risk materializes, the contractor will be bankrupt and the owner will still suffer the consequences. Particularly in more complex contracts, it is probably in the interests of both the owner and contractor to ensure that risks are clearly defined and mutually understood as part of the contracting process and prior to contract award.

Risks are normally documented in a *risk register*, which lists the significant identified risks on the contract scope. In many cases the risk register is a key element in the contracting package. The risk register, shown in Figure 30.3, lists the risks that have been identified on the project, indicates the mitigating actions that can be taken, allocates responsibility for dealing with them, and specifies the contractual and commercial implications (Wideman, 1992).

Contract Scope Statements Based on Work Breakdown Structure

Many, if not most, contracting problems stem from failure of either the owner or contractor to fully understand the scope of the work of a particular contract package. Failure to appreciate scope can lead to unrealistic expectations on the part of the owner, underestimation of time and cost by the contractor, and problems of interface between the various contractors working on the project.

FIGURE 30.3. PROJECT RISK REGISTER.

Ref	Risk	Impact	Actions	Cost Impact	Schedule Impact	Resp.	Contractual
Loading Jetty							
10.1	Sea Bed Conditions Uncertain	Could seriously impact the piling specification, hence schedule and cost	Try to convince sponsor to accept floating point mooring system instead of jetty (see detailed report on relative economics). More detailed survey required as part of design process	\$1–3 Million	3–6 weeks on activity but non critical	EPC Contractor	EPC contractor to be reimbursed on unit rate basis according to the piling schedule

A very useful discipline is to base the *contract scope statement* on the project w*ork breakdown structure*. This technique ensures that all of the work necessary to complete the project is defined in one contract package or another. The technique also forces a very clear definition of the interfaces between contract packages. This is inevitably the area where problems will arise.

Project Contracting Plan

Once the overall contracting strategy has been developed, the next step is to produce an overall *project contracting plan*. This plan should be an integral part of the overall project plan and ensures that the deliverables expected from each tender are delivered when they are needed to meet project requirements. The contracting plan reflects the deliverables and resources that can be made available, particularly from the supporting disciplines such as engineering, legal, and the like. The project and the procurement team must be extremely clear on the priorities of the different packages. In most projects the award of one or more contract package will almost certainly be on the critical path of the project.

Figure 30.4 illustrates the relationship between a series of design, procurement, and execution activities that are all part of an overall project plan.

Procuring Project Packages

Once the overall contracting plan has been agreed at the appropriate level in the organization, the emphasis shifts from consideration of strategic issues to the systematic procurement of every package on the plan. The package tendering process is shown schematically in Figure 30.5.

FIGURE 30.4 BASIC TIMING OF CONTRACTING PLAN.

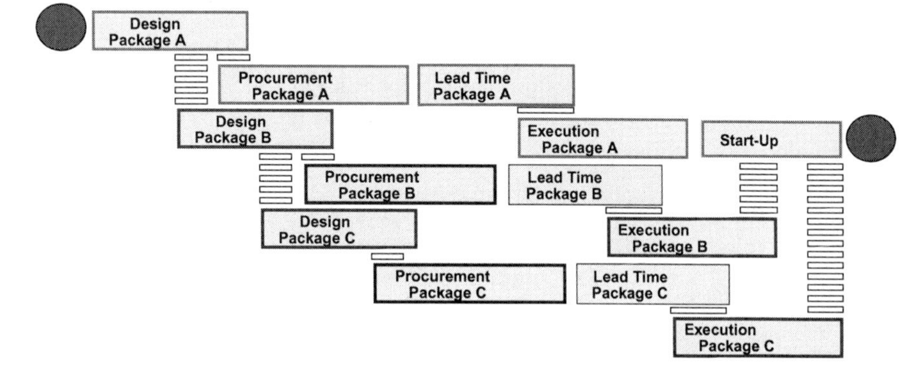

FIGURE 30.5. PACKAGE TENDERING PROCESS.

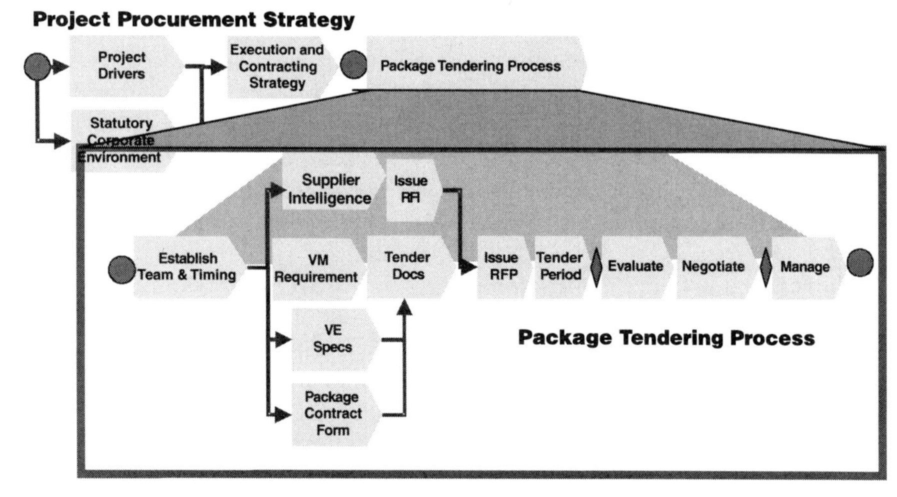

Forming the Team and Defining Roles

Many traditional organizations still work in their functional "silos," where the end user specifies operating requirements, the technical department documents technical specifications, and the procurement department defines commercial conditions and procures the goods and services from the traditional list of approved and customary suppliers. Indeed, in some companies the engineering department will decide on the supplier and ask the procurement department to regularize their decision by placing the purchase order or contract. This does not tend to produce the best value.

> **Case Study:** Traditionally the technical department in a large organization would specify the complex equipment required for product analysis and ask the purchasing department to regularise the situation by placing a purchase order. By procuring the equipment through the tendering process described savings of 30 to 40 percent were achieved. Similar levels of saving on other complex products and services achieved through the application of more rigorous procurement processes.

Most companies now recognize that procurement is a team effort and that the structure and composition of the procurement team are critical success factors for anything other than the simplest of commodities. The following distinct functions and competencies may provide a useful framework for defining roles and responsibilities within the project team. However, it is worth noting that in smaller organizations, it would be perfectly acceptable and possibly desirable for one person to wear more than one hat.

- *Project sponsor.* The ultimate decision maker. Takes P&L responsibility for the contract in question.
- *Project management.* Provides overall leadership, pulling the team together and driving them through the most appropriate procurement and tendering process.
- *Procurement.* Provides detailed knowledge of supplier markets and trends in the industry.
- *Technical.* Provides detailed input on technical specifications either directly or as a conduit to the technical resources within the company or external experts in the field.
- *Operations.* Represents the ultimate users of the goods or services to be purchased and ensures operational requirements are respected.
- *HSEQ (Health, Safety, Environment, and Quality standards).* Supports functions such as financial, legal, and PR, providing expertise and counsel within their area of competence.

In fact, it is probably worth conducting a formal RACI workshop to ensure that everyone is absolutely clear on the roles and responsibilities of the various participants. (RACI stands for Responsibility, Accountability, Consulted, and Informed. In a RACI workshop all the individuals and functions responsible for undertaking the activities that make up a business process debate who is *accountable* (who carries ultimate responsibility for the result), *responsible* (who actually does the work), *consulted* (who must be consulted as part of the activity), and *informed* (who is informed of the decision once it has been taken). The consensus is documented in the kind of RACI Chart shown in Figure 30.6.

The specific accountabilities and responsibilities will, of course, vary from organization to organization, but this chart may serve as a useful basis of a workshop to debate and agree

FIGURE 30.6. RACI CHART.

Activity	Accountable	Responsible	Consulted	Informed
Gaining budget authority	BU	BU	Finance	Procurement
Preparing contract/purchase requisition	BU	BU	HSEQ, Procurement	Engineering
Determining contract/purchase scope	BU	BU/ Procurement	HSEQ, Engineering	
Determining technical specifications	BU	Engineering	HSEQ	Procurement
Determining HSEQ requirements	BU	HSEQ	Engineering	Procurement
Selecting appropriate terms and conditions	Procurement	Procurement	Legal, Engineering	BU
Selecting form of remuneration	Procurement	Procurement	BU, Finance	Engineering
Preparing tender evaluation criteria—commercial	Procurement	Procurement	BU, Finance, Engineering	
Preparing tender evaluation criteria—technical	BU	Engineering	HSEQ, Procurement	
Choosing suppliers/contractors to receive ITT	Delegated Authority*	Delegated Authority*	Procurement Engineering	
Preparing invitation to tender	Procurement	Procurement	BU	Finance, Engineering
Issuing invitation to tender	Procurement	Procurement	BU	Finance, Engineering
Receiving and safekeeping of tenders	Procurement	Procurement		BU
Conducting tender opening	Procurement	Procurement		BU
Conducting technical evaluation of tenders	BU	Engineering	HSEQ	Procurement
Conducting commercial evaluation of tenders	Procurement	Procurement	BU	Engineering
Clarification/negotiation with suppliers	BU	Procurement	Engineering, Legal, Finance	Legal, HSEQ
Preparing evaluation report and recommendation	BU	Procurement	Engineering, HSEQ	
Reviewing recommendation and approval	Delegated Authority*	Delegated Authority*	Procurement BU	Engineering, Legal, HSEQ
Awarding contract	Delegated Authority*	Procurement	Legal, Finance	All Functions
Maintaining contract file/audit trail	Procurement	Procurement	BU, Finance	
Post-award operational supervision	BU	BU	Procurement, Finance	Finance, Engineering
Contractor audit	Procurement	Procurement	Engineering, Finance	

RESPONSIBLE	Individuals are the "doers."
ACCOUNTABLE	Individuals are ultimately answerable to the board.
CONSULTED	Individuals must be consulted and involved before a final decision or action is taken.
INFORMED	Individuals need to be informed once a decision or action has been taken.

on the specific responsibilities within your own organization or project. The ultimate effectiveness of the team will, however, ultimately depend on the level of domain expertise within each of the functions listed, along with the ability to work together creatively as a team and to communicate with the wider range of stakeholders who may be involved in or affected by the procurement initiative.

Agree on the Tender Event Schedule

Once the team has decided on and documented the process, the next step is to establish the tender event schedule. The overall time required will depend on the nature of the contract and can be anything from a few weeks to many months. Figure 30.7 shows a typical tender event schedule of 16 weeks for a fairly complex alliance contract.

Do Not Underestimate Resource Requirement

Working out a rigorous tendering process takes time. In the project environment there may be a large number of contracts to procure in parallel, and often the same people will be involved on different packages. Because the time frame for each package might appear quite relaxed, there is a tendency to overestimate the ability to handle a number of packages in parallel. This leads to an underestimation of the time, and hence resource, required. This

FIGURE 30.7. TENDER EVENT SCHEDULE.

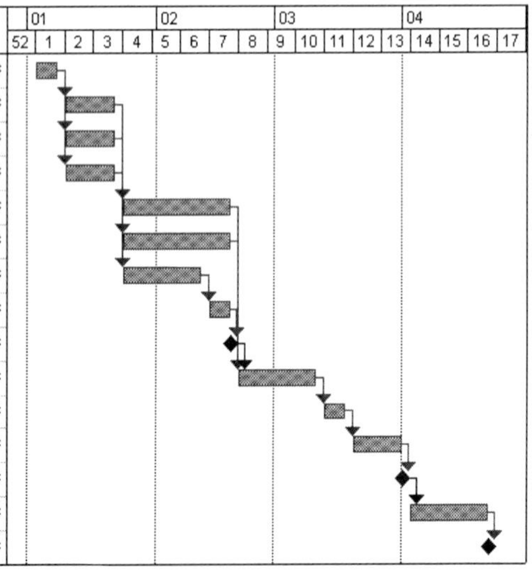

ID	Task Name	Duration
1	Agree Team and Timing	5 days
2	Develop Contractual Framework	10 days
3	Discussions with Contractors	10 days
4	Document Shutdown Programme	10 days
5	Provisional Workscope SD-1	20 days
6	Develop Commercial Conditions	20 days
7	Issue RFI	15 days
8	Initial Screening	5 days
9	Agree Long List	0 days
10	Issue RFP	15 days
11	Initial Sift	5 days
12	Alliancing Scorecard	10 days
13	Agree Short List	0 days
14	Post Tender Negotiations	15 days
15	Award Contract	0 days

is often particularly the case in the supporting functions without a dedicated project team member or in corporate functions such as the legal department who are offline and often "march to the beat of a different drum." Failure to assign adequate resources can often result in suboptimal procurement or schedule slippage or both.

One way around this problem is to insist on a fully resourced contracting schedule where the aggregate resource requirement is clearly demonstrated for all functions involved in the activity.

> **Case Study:** The critical path in the implementation of a petrochemical complex ran through the procurement of the main equipment. Selection of this was required to obtain the vendor design data required to undertake detailed design. This in turn paced the construction contracts. The contractor underestimated the level of effort required to meet the early procurement deadlines. A project delay of three months was incurred almost immediately which it was not possible to recover.

Determine Potential Impact of Package on Project

The relative importance of the various packages will undoubtedly have been discussed when the overall project contracting strategy was being developed. One of the first tasks of the procurement team is to really think this through to the next level of detail:

- What percentage of the total project spend does the package represent?
- How critical is timing?
- What are the issues and specific risks inherent in the scope of work?
- Can we access experience within the company, or in the professional community?
- What were the lessons learned?
- What are the implications of all of the above on package procurement?

> **Case Study:** One of the packages on the construction of an air separation unit was the insulation of the cold box. The actual work did not take place till late in the construction schedule, so there was plenty of time to develop and negotiate the insulation contract. This activity had always been on the critical path on previous projects so time was of the essence and 24-hour working was required. This was built into the tender conditions and evaluation criteria from day one. On previous jobs this had involved expensive variations during execution. A saving of 15 percent and a reduction in time of five days over previous jobs was attained.

Value Manage User Requirements

Value management is a set of techniques that can be used at project or package level to ensure that the equipment or service defined provides the best value for money for the owner. For example, by value managing the deliverables of the package, it may be possible to relax the functionality, timing, or operational constraints to reduce the price of the package without any prejudice to project profitability. Value management tends to apply to the scope, performance, and timing, rather than to optimization and technical specifications, which are covered under value engineering.

> **Case Study:** Many major projects are undertaken in remote locations. Typically the contractor will build a camp for his or her senior construction personnel that is demolished when the plant is completed. By advancing the construction of the operators' accommodation, the contractor could dispense with temporary accommodation for his or her project team and make a significant saving, which would be passed, in part, to the owner.

To encourage thinking, many companies invite participation outside of the project team in value management workshops, which are held at key points in the project life cycle to challenge the current definition. Many companies also use external facilitators with a particular experience in value management.

> **Case Study:** After a series of acquisitions a major bank required to restructure its branch network across Europe. Because of the costs involved, the board asked for an independent review. This review revealed that there was confusion over the bank strategy, which impacted on staffing levels. The policy on space allocation per full-time equivalent (FTE) was also ambiguous. In addition, architects were (a) allowing more space per FTE than necessary and (b) failing to design to need. By addressing these issues over a series of strategy and value management workshops, the scope, and hence potential cost, of the project was reduced by approximately 25 percent.

Value Engineer Technical Specifications

Normally, drawings and technical specifications will be produced by the in-house engineering department or external consulting engineers working under the direction of the engineering representative of the project team. Indeed, many companies have a complete suite of technical specifications that form an integral part of their tender documentation.

In seeking to optimize the return from project expenditure, the project team must examine these specifications critically. Many companies have found that replacing prescriptive technical specifications by performance specifications, which define the output required rather than the inputs demanded, enables significant advantage to be taken of supplier experience and lower-cost, more standardized components.

Often technical specifications have evolved over many years. The process of subjecting engineering designs and standards to critical examination is called *value engineering* (VE).[3] Responsibility for driving this process must rest with the technical representative on the project team. Many companies build in VE workshops as a key aspect of their design and procurement processes. VE, however, does require time, and this must be reflected in the tender event schedule for the package.

[3] A value engineering professional association called VE Today is at www.vetoday.com.

Case Study: An international oil company was procuring the contract to paint their refinery tank farm. Technical specifications called for three coats on a particular painting system. By discussing the requirements with suppliers one was found who would guarantee a two-coat system. This reduced the overall cost of materials and labor by around 25%.

Pick Supplier Brains

Before launching into the development of detailed package contract documentation, the owner organization must find out who the leading suppliers in the area are and pick their brains. Most suppliers are working with a wide range of clients, including competitors, and may be prepared to share some of the latest developments. It is advisable to try and do this before developing formal contract documentation. In this way it will be possible to incorporate these developments in the bidding scope and thus subject them to competitive pressure. If this is not done, the supplier may choose to offer them as an alternate. If these alternates are attractive, the owner may then be obliged to negotiate without the advantage of competitive pressure.

There is an ethical, and possibly legal, issue here about the incorporation of the ideas that emerge in this fashion into the competitive bidding process. In the international oil industry, this process is often formally structured as a design competition where bidders are paid to propose technical solutions. The owner thus obtains the intellectual property rights (IPR), which can be bid competitively. The winner of the design competition normally improves his chances of obtaining the contract.

Determine Precise Contract Form for Package

Contract Forms. While the basic contract form will have been established as part of the project contracting strategy, it has to be translated into specific and very precise contract documentation. Although it is always possible to develop a contract from first principles, in general it is advisable to use forms of contract in common use with an established body of case law.[4] Owners and contractors, and their lawyers, understand them. They can form more precise evaluation of their risk and will require less contingency as a result. Most major companies, however, adapt these basic contract forms to meet their own precise requirements based on their experience over the years.

Creating Win-Win Conditions. In some forms of contracts the interests of the owner and contractor are directly opposed. For many years, in the UK construction industry, contractors were subjected to intense competition and were often forced to bid under their actual estimate of cost to win the work. As a response to this, many contractors developed a

[4]Most of the forms of contract in general use have been developed by the professional institutions, including Institute of Chemical Engineers (www.icheme.org), International Federation of Consulting Engineers (www.fidic.org), American Institute of Architects (www.aia.org), and Associated General Contractors of America (www.asce.org).

specialist function to examine tenders for weaknesses, in either the definition of scope or in contract conditions which could be exploited later as the basis of claims and variations. These claims and variations were then negotiated during the contract execution at very high profit margins when the owner had become totally dependent on the contractor for completing the project. As a result, both owners' and contractors' management devoted considerable attention to structuring and opposing variations and claims, often to the detriment of the actual project.

Better results can often be obtained if the commercial interests of the owner and contractor can be aligned. For example, in a high NPV contract where time to market is critical and the execution of the package is on the critical path, a bonus can be awarded for compressing schedule. The same principles can be applied to reducing the overall cost of the project. Figure 30.8 illustrates a typical incentivized commercial arrangement.

In the contract form shown schematically in the figure, the owner and the main engineering procurement and construction (EPC) contractor agree to work on the basis of a fixed fee to cover project management, engineering, procurement, and construction management costs, including overhead and profit. The direct cost of materials and equipment and construction contracts were estimated by the contractor and a target *total installed cost* (TIC) established. If the contractor manages to procure the equipment and complete construction for less than the target cost, the saving is shared with the owner on a pre-agreed basis and the contractor is paid a "bonus." If, however, this target cost is exceeded, the contractor and the owner share the overrun and the contractor suffers a *malus.* It is normal to limit the extent of both bonus and malus. It should be noted that the these incentivized

FIGURE 30.8. INCENTIVISED CONTRACT ARRANGEMENT.

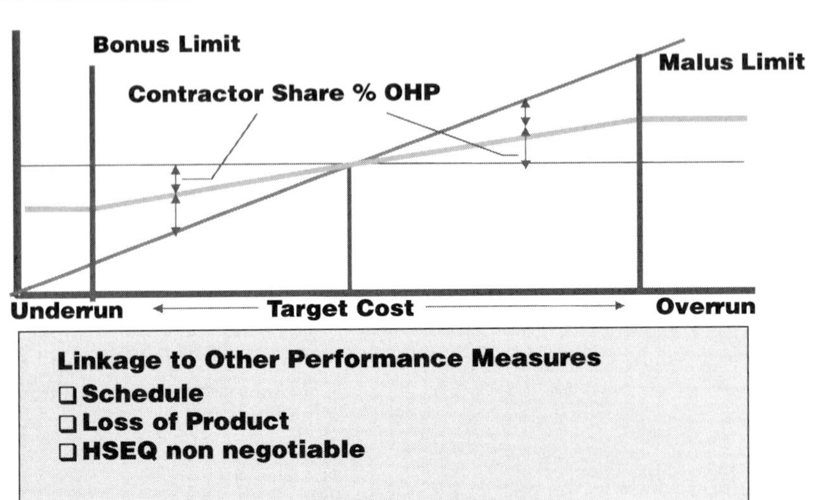

contracts are normally bid competitively where the contractors are required to specify both their fixed fee and the target they hope to achieve.

In the purest form of an "alliance open-book contract," only the fee for overhead and profit is fixed. All other costs are reimbursed "at cost" without any margin or profit. All costs incurred by the contractor are visible to the owner and subject to audit, hence the expression "open book." In theory these open-book contracts are volume-neutral so that all variations can be discussed on their technical merits.

> **Case Study:** An international oil company already had called for lump-sum turnkey (LSTK) bids for the engineering and construction of a major petrochemical complex. They were, however, concerned by the lack of flexibility to introduce changes to plant layout and specification to improve operating and maintenance. By moving to the incentivised target cost EPCm framework indicated previously, the owner could inject operating experience into the project design without incurring major variation charges. The EPCm scope was completed for 18 percent less than the original LSTK price, and the owner has a plant that is cost-effective to operate and exceeds performance targets.

While an incentivized or Win-Win contract framework can help promote alignment between the owner and contractor organizations, the real benefit comes from the ability of key individuals within the owner and contractor companies to collaborate professionally and fairly. The kind of incentivized and open-book contract forms referred to previously in fact may require a much more sophisticated management on the part of both the owner and the contractor to realize their theoretical advantages (ENR, 2001; Bennett and Baird).

> **Case Study:** A major refiner thought an alliance contract would be ideal for plant turnarounds, since the contractor could be engaged early to assist in detailed planning before scope was fully defined and then deal with changes on a commercially neutral basis. Neither the owner nor contractor management was used to working within such a framework, relative responsibilities were not well defined, and costs escalated out of control.

Request for Information—Widening the Supplier Network

Many companies tend to work, by default or by design, with a limited number of suppliers who have provided good service in the past. These suppliers understand their business and have established relationships, often at various levels, within owner organizations. Indeed, it is conventional management wisdom to reduce the number of suppliers, and it is fashionable, and even sensible, to consider tighter linkages throughout the supply chain. Furthermore, it takes considerable time and effort to search out and qualify new suppliers, and in certain areas, particularly service contracts, there is an increased risk in the level of performance that will be actually delivered by untried organizations.

However, a recent study by McKinsey concluded that productivity gains over the last decades were essentially driven by a combination of innovation and competition. In such circumstances, the continual search for new and emerging suppliers is a major value improvement driver that must be balanced with the need to limit the overall number of suppliers and build better relationships with them.

> **Case Study:** A European petrochemical producer opened up the maintenance of its facilities and grounds to contractors with no specific process industry experience. This enabled a number of smaller, local contractors with a lower cost structure to compete for the work. The track record of these contractors and the competence and motivation of their management were carefully evaluated. There were some problems as the new contractors came to terms with a more industrialized environment but costs were reduced by 30 to 40 percent.

The identification of new suppliers is, however, a time-consuming task that needs to be approached in a very systematic and structured fashion. It helps if this identification and qualification of suppliers can be done off-line without the pressures of project deadlines. Figure 30.9 shows the approach used by a major oil company to identify contractors to work on refinery shutdowns in various regions across Europe. The following are details:

- Over 1,000 contractors were identified from various trade registers and supplier databases.
- Each was sent an e-mail describing the nature of the work and the basic prequalifications required and asking whether they wished to be considered.

FIGURE 30.9. IDENTIFICATION AND EVALUATION PROCESS.

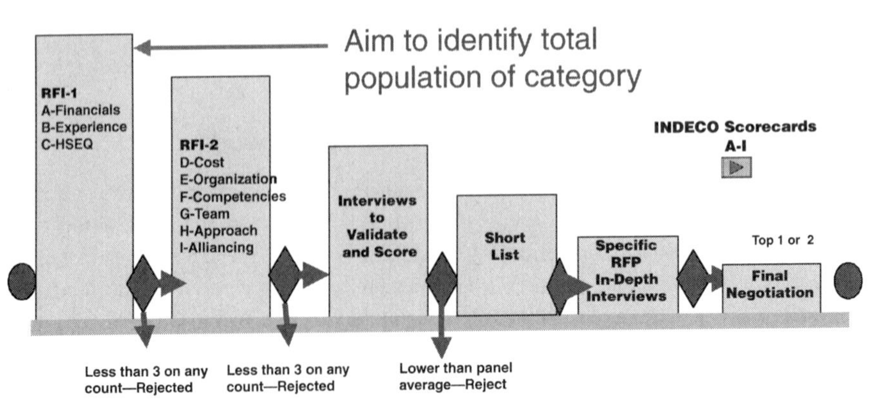

- The 300 who responded in the affirmative were sent a *request for information* (RFI) essentially asking for their financial statements, relevant experience, and HSEQ record.
- This information was used to identify about 50 potentially suitable contractors. These contractors were then sent a more detailed questionnaire soliciting information on their organization, competencies, and trade skills, including their approach to refinery shutdowns and alliancing.
- Short-listed contractors were then invited to interview to assess the quality of their management and validate their responses.
- Based on these interviews, a shortlist was created who were invited to bid for suitable projects.

A detailed set of information schedules and scorecards were established, as shown in Figure 30.9. The first RFI included Schedule A—Financial Structure, Schedule B—Relevant Experience, and Schedule C—HSEQ Credentials and Certification. RFI-2 sent to companies who passed the first cut included Schedule D—Cost Structure, Schedule E—Organization, Schedule F—Competencies, Schedule G—Project Team, Schedule H—Approach to the Specific Tender, and Schedule I—Experience and Approach to Alliancing. Companies who passed the next cut were invited to interview to validate the previous data and to present themselves and their ideas. Detailed interview guides were established that were agreed with and completed by the procurement team.

Tender Documentation

Clarity and Completeness of Tender Documentation is Critical. The structure and content of the tender documentation is at the core of the tender management process. The clarity and thoroughness of this documentation will determine the quality of the bids received and the prices tendered. It will also provide the basis for the execution and eventual cost of the contract. Any ambiguity will inevitably become the subject for debate, and possibly the source of contract variations and extra costs. The necessity of defining precisely what you want, what you are requesting the contractor to provide, and how you are paying for it cannot be overstated (Kennedy, 2000).

> **Case Study:** If you are in any doubt about the preceding, refer to the celebrated case of the British Museum where the client wanted a "Portland stone" masonry construction. The technical specification referred to the technical specification of Portland stone. The contractor constructed the building in French limestone, which apparently met the technical spec but which did not come from Portland. The resulting furore became a *cause celebre* in the British construction industry.

Structure of Tender Documentation. Although a number of variants are possible, the classic structure of a tender documentation package is described in the following. Most companies will standardize on this or something similar to ensure completeness and consistency and to speed up the development of the tender package within their own organization.

- *Invitation to tender.* This is a standard "one-page" letter inviting the contractors to submit a bid for the package in question under the conditions defined and documented in the package.
- *Instructions to tenderers.* This includes instructions to contractors defining the form, place, and time of tender submission and the duration of validity. Instruction to tenderers will reference a series of sections to be completed by the tenderer. These sections will ultimately form an integral part of the contract between the owner and contractor.
- *Form of Agreement.* The basis of the tender defining the contracting parties and referencing the various supporting schedules (see Figure 30.10). This will be signed by the contractor and form the basis of his offer. This form of agreement will be underpinned by a series of sections precisely documenting contract scope and conditions. The tenderer will be required at this stage to identify any exclusions—that is, any conditions or any other aspects of the documentation that are unacceptable.

The following sections are typical on many project related contracts, although they may vary according to sector:

- *General conditions of contract.* Most companies accustomed to awarding contracts will have these as standard articles or legal "boilerplate". Figure 30.11 shows a typical heads of agreement. Articles requiring particular attention are those covering "limits of liability," "variations to contract," and the "termination."
- *Contract scope.* This schedule defines matters such as the following:
  - Services or end result to be provided by the contractor
  - Operating conditions and environment
  - Delivery dates
  - Testing and handover requirements and protocols
- *Technical specifications and standards.* This section defines any performance or technical specifications and standards to be met by the contractor. Any drawings or technical data will normally be included and referenced in this schedule.
- *Remuneration.* This section defines how the contractor will be paid for the services provided under scope. Payment events, incentive arrangements and payment conditions, fiscal responsibilities, and currency issues are normally defined in this section. In addition, this section generally contains a pricing schedule. The pricing schedule has vital commercial implications. Some considerations on structuring this are developed in the *Pricing Schedule* section, coming later in the chapter.
- *HSEQ.* This section documents compliance requirements with Health, Safety, Environment, and Quality standards (HSEQ).
- *Provided by owner.* This section defines the services, information, and facilities to be provided to the contractor by the owner. This could cover such things as free issue materials, office and workshop space, access to owner computer systems, and the like.
- *Contract administration.* This section defines how the contract is to be administered, including the names of designated individuals to represent each party and the method of serving official notices and when they are required.

FIGURE 30.10. FORM OF AGREEMENT.

FORM OF AGREEMENT

This CONTRACT is made between the following parties:

xxxxxx a company having its registered office at xxxx, hereinafter called the COMPANY;

and

Contractor xxx having its main or registered office at xxxxxx, hereinafter called the CONT RACTOR.

WHEREAS:

1) the COMPANY wishes that certain WORK shall be carried out, all as described in the CONTRACT; and

2) the CONTRACTOR wishes to carry out the WORK in accordance with the terms of this CONTRACT.

NOW:

The parties hereby agree as follows :

1) in this CONTRACT all capitalised words and expressions shall have the meanings assigned to them in this FORM OF AGREEMENT or elsewhere in the CONTRACT.

2) the following Sections shall be deemed to form and be read and construed as part of the CONTRACT:

Section 1	Form of Agreement
Section 2	General Conditions of Contract
Section 3	Scope of Work
Section 4	Remuneration
Section 5	Health, Safety, Environment and Quality
Section 6	Items to be provided by the COMPANY
Section 7	Contract Administration

The Sections shall be read as one document, the contents of which, in the event of ambiguity or contradiction between Sections, shall be given precedence in the order listed.

3) In accordance with the terms and conditions of the CONTRACT, the CONTRACTOR shall perform and complete the WORK and the COMPANY shall pay the CONTRACT PRICE.

4) The terms and conditions of the CONTRACT shall apply from the date specified in Appendix 1 to this Section I - Form of Agreement which date shall be the EFFECTIVE DATE OF CO MMENCEMENT OF THE CONTRACT.

5) The duration of the CONTRACT shall be as set out in Appendix 1.1 to this Section I - Form of Agreement.

The authorised representatives of the parties have executed the CONTRACT in duplicate upon the dates indicated below:

For: COMPANY

Name:

Title:

Date:

For: CONTRACTOR NAME

Name:

Title:

Date:

FIGURE 30.11. GENERAL CONDITIONS OF CONTRACT.

General Conditions Table of Contents			
1.	Definitions	20.	Patents And Other Proprietary Rights
2	Interpretation	21.	Laws And Regulations
3	Company And Contractor Representatives	22.	Indemnities
4.	Contractor's General Obligations	23.	Insurance By Contractor
5.	Responsibility For Company P rovided Items	24.	Insurance By Company
6.	Contractor To Inform Itself	25.	Consequential Loss
7.	Contractor To Inform Company	26.	Confidentiality
8.	Assignment And Subcontracting	27.	Customs Procedures
9.	Contractor Personnel	28.	Completion
10.	Co-Operation With Others	29.	Defects Correction
11.	Programme	30.	Termination
12.	Technical Information	31.	Audit
13.	Inspection And Testing	32.	Liens
14.	Variations	33.	Business Ethics
15.	Force Majeure	34.	General Legal Provisions
16.	Suspension	35.	Liquidated Damages
17.	Terms Of Payment	36.	Limitations Of Liability
18.	Taxes And Tax Exemption Certificates	37.	Resolution Of Disputes
19.	Ownership		

Source: INDECO.

Under the structure of tender documentation described in the preceding text, the submitted tender forms the basis of the contract binding on "contractor."

Document Quality Assurance and Version Control Essential. Contract documentation is often assembled under considerable time pressure with input coming from many sources. Mistakes and omissions reflect badly upon the organization and may have extremely serious consequences on contract and project execution. It is therefore absolutely essential to operate a formal document management and quality assurance procedure. Ideally, each section should be checked and signed off by a competent person who is not the author.

Pricing Schedule

The *pricing schedule,* which defines how and how much the contractors will be paid for his services, is a critical part of the documentation and is worthy of special note. The pricing schedule should be carefully constructed to provide the basis for the detailed comparison between the eventual tenderers. Ideally, the pricing schedule should provide the owner with enough information to compare the offerings of the tenderers on a *whole-life cost* basis. (See the section *Whole-Life Cost Analysis* coming up in the chapter for a detailed explanation and example of whole-life cost analysis).

> **Case Study:** A distribution company was soliciting bids for the maintenance of their national retail distribution system, which was split between a number of contractors. A unit rate contract form and schedule was developed and bid by over 20 contractors. These pricing schedules were inspected and any anomalous rates removed. A schedule of rates was then developed using the lowest rate from the competing contractors. In a second round of bidding, the short-listed bidders were requested to bid this schedule of rates at a discount. This reduced maintenance costs by over 20 percent. The contract has now been running for about five years and has been very successful for both parties.

The detailed breakdown of bid prices can be an invaluable aid in the evaluation of tenders, allowing the owner to identify areas of anomaly, where perhaps the tenderer has either misunderstood the scope of work or has developed a creative approach to reduce costs. A detailed pricing schedule can also be an invaluable aid during post tender negotiation, allowing the owner to use a *cherry-picking approach* during post tender negotiation. In this approach the prices of the competing bidders are compared and negotiated at elemental level. This evaluation, which can be done in Microsoft's Excel or in a database application, highlights where certain contractors may have overestimated or "padded" the element. Pressure can be put on these elements to reduce unit, and hence total, costs. An extreme example of this is described in the case study later in the chapter.

In more complex cases it can be very effective to produce a *pricing wizard* using Excel. In this way the bidders are obliged to enter their information in a very structured way. More importantly, the buyer can easily manipulate and compare the data. The example shown in Figure 30.12 shows the front sheet of a pricing wizard for a contract to provide technical maintenance on over 300 elevators in many different buildings in three countries.

Manage the Tender Period

Control of information is absolutely critical during the tender period. During the tender period bidders actively seek to gain intelligence on their competitive situation. Even a chance

FIGURE 30.12. PRICING WIZARD.

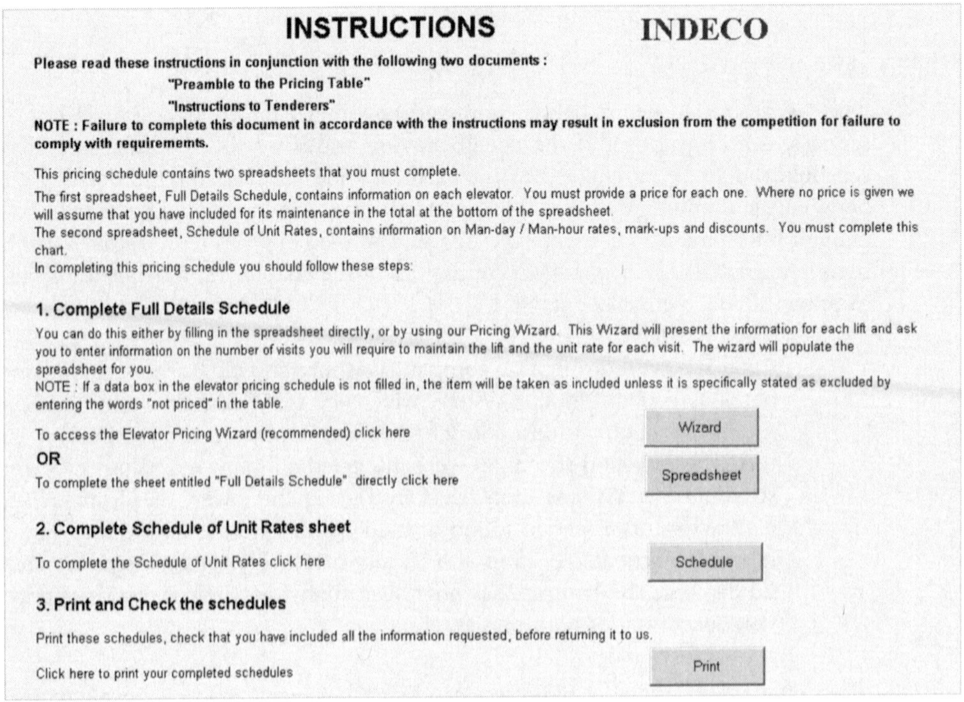

INSTRUCTIONS **INDECO**

Please read these instructions in conjunction with the following two documents :

"Preamble to the Pricing Table"

"Instructions to Tenderers"

NOTE : Failure to complete this document in accordance with the instructions may result in exclusion from the competition for failure to comply with requiremements.

This pricing schedule contains two spreadsheets that you must complete.

The first spreadsheet, Full Details Schedule, contains information on each elevator. You must provide a price for each one. Where no price is given we will assume that you have included for its maintenance in the total at the bottom of the spreadsheet.

The second spreadsheet, Schedule of Unit Rates, contains information on Man-day / Man-hour rates, mark-ups and discounts. You must complete this chart.

In completing this pricing schedule you should follow these steps:

1. Complete Full Details Schedule

You can do this either by filling in the spreadsheet directly, or by using our Pricing Wizard. This Wizard will present the information for each lift and ask you to enter information on the number of visits you will require to maintain the lift and the unit rate for each visit. The wizard will populate the spreadsheet for you.

NOTE : If a data box in the elevator pricing schedule is not filled in, the item will be taken as included unless it is specifically stated as excluded by entering the words "not priced" in the table.

To access the Elevator Pricing Wizard (recommended) click here [Wizard]

OR

To complete the sheet entitled "Full Details Schedule" directly click here [Spreadsheet]

2. Complete Schedule of Unit Rates sheet

To complete the Schedule of Unit Rates click here [Schedule]

3. Print and Check the schedules

Print these schedules, check that you have included all the information requested, before returning it to us.

Click here to print your completed schedules [Print]

remark on apparently noncommercial matters may encourage contractors to maintain a higher pricing level. Also, inconsistency of information supplied to bidders may cause considerable confusion and even potentially embarrassing and costly litigation.

For all of these reasons, one single point of contact in both the owner and contractor organization is highly desirable (see Figure 30.13). All other contacts should be actively discouraged. All questions arising should be formally documented, and identical responses answers should be sent to each bidder. Some companies run bidder conferences where bidders have the opportunity to ask questions in an open forum. The answers given during this tender period form an integral part of the contract documentation and conditions and so have to be very carefully considered and documented. Failure to manage the tender period in a manner that is absolutely and transparently equitable can, particularly in the public arena, result in a legal challenge from unsuccessful bidders.

Tender Evaluation

The object of the tender evaluation phase is to identify the proposal that offers the best value for money to the company. This can be more difficult than it sounds. Often the

FIGURE 30.13. SINGLE POINT OF CONTACT.

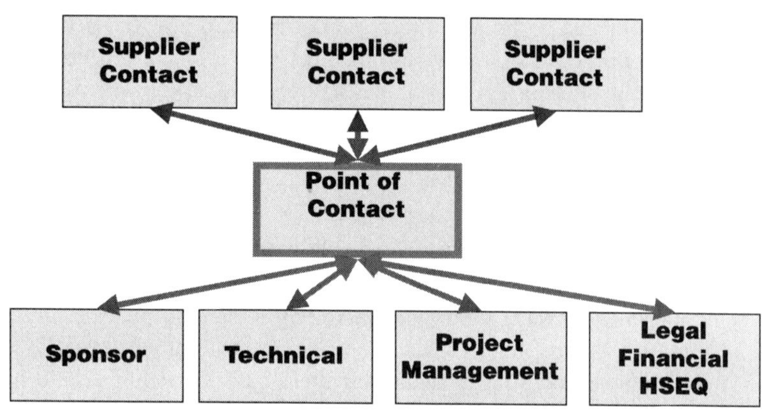

ultimate value for money to the company will depend on the quality of the contractor providing the service and the contractor's ability to meet performance targets and deadlines often over quite a long time frame. There may also be real or perceived different levels of risk associated with each tender. The more sophisticated tender evaluation systems described in the section *Whole-Life Cost Analysis* will take all of these factors into account.

Tender Evaluation: The Formal Approach. In public institutions and many large organizations, the procedures governing tender evaluation are by necessity very rigorous and formal (see Figure 30.14). In such cases there is a formal opening of tenders often in front of

FIGURE 30.14. FORMAL TENDER EVALUATION PROCESS.

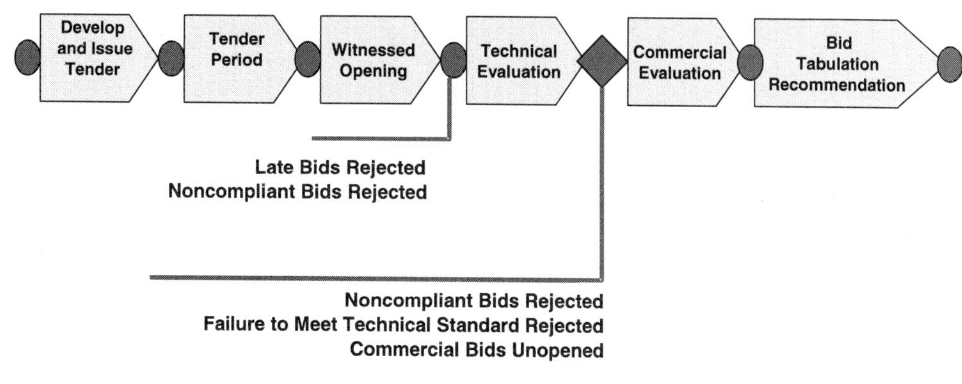

official witnesses. Each bid is divided into the technical and commercial package and the commercial package placed unopened in a secure locked filing cabinet or safe. The technical bids are then evaluated by a technical committee and graded according to the strength of the technical proposal. Bids that do not meet the technical requirements are rejected and the commercial offering left unopened. The commercial proposals of the bids still under consideration are then evaluated by a commercial committee and the overall cost to the owner determined. In many cases, particularly in the public sector, the lowest-cost bid of the proposals adjudges to be technically adequate is accepted.

This approach, while straightforward and transparent, does not, however, necessarily give the best value for money to the owner. Specifically:

- The offering of a technically adequate contractor may provide the lowest initial capital expenditure (capex), but this may be at the cost of higher operating expenditure (opex).
- There may well be a fairly significant spread in the technical and management quality of contractors who are all deemed to be technically adequate. Simply picking the technically adequate contractor with the lowest price may not necessarily provide the best value for money for the owner. There may also be a higher level of risk associated with delivering the required level of service between a contractor who is technically adequate and one who offers a higher level of management competence.

More sophisticated evaluation approaches, described in the following, can be used to take both these situations into account.

Whole-Life Cost Analysis. The structure of commercial proposal should permit the evaluation of the *whole-life cost* of the service. For example, in the case of an equipment package, such as an air-conditioning unit, the whole-life cost normally expressed as net present value would include the following:

- Original capital cost
- Associated freight and handling costs
- Power consumption
- Cost of operation
- Cost of spares, service and insurance.

For many equipment packages, the cost of spares and service over the operating life can be many multiples of the original capital cost. Unless the tender has been very precisely framed, many of these costs will be at best estimates, with little hard commercial underpinning. It is also evident that basing the evaluation on capex alone could be grossly suboptimal for the company over the life cycle of the project or asset. Figure 30.15 shows the comparison between two equipment packages where the lowest capex cost is definitely not the best buy over a ten-year life cycle for the project. This is, of course, highly relevant in the case of lump-sum turnkey bids, where, unless it is very carefully specified, the contractor will inevitably select the equipment package with the lowest initial cost.

FIGURE 30.15. WHOLE-LIFE COST.

VENDOR-A

Assumptions											
Discount Rate		8%									
000 Euros Year	WLC	1	2	3	4	5	6	7	8	9	10
Capex											
Capital Cost		250									
Insurance and Freight		10									
Installation		25									
Capex	**285**	285									
Opex											
Spares	110		10	10	15	10	15	10	15	10	15
Service	113		10	10	12	12	12	12	15	15	15
Consumables	100	10	10	10	10	10	10	10	10	10	10
Utilities	120	12	12	12	12	12	12	12	12	12	12
Opex	**443**	22	42	42	49	44	49	44	52	47	52
Total Cost Ownership	**728**	307	42	42	49	44	49	44	52	47	52
NPV	**€ 551.81**										

VENDOR-B

Assumptions											
Discount Rate		8%									
000 Euros Year	WLC	1	2	3	4	5	6	7	8	9	10
Capex											
Capital Cost		225									
Insurance and Freight		12									
Installation		15									
Capex	**252**	252									
Opex											
Spares	180		20	20	20	20	20	20	20	20	20
Service	113		10	10	12	12	12	12	15	15	15
Consumables	150	15	15	15	15	15	15	15	15	15	15
Utilities	120	12	12	12	12	12	12	12	12	12	12
Opex	**563**	27	57	57	59	59	59	59	62	62	62
Total Cost Ownership	**815**	279	57	57	59	59	59	59	62	62	62
NPV	**€ 600.81**										

Commercial Evaluation

Vendor-B	**12%**	**Lower Capex than Vendor-A**
Vendor-A	**8.15%**	**Lower Whole Life Cost than Vendor B**

Hard Money-Soft Money Evaluation. In some projects the managerial and technically quality of the contractor can significantly affect the outcome and business benefit which the owner receivers from the service. "Hard money-soft money" evaluation is a technique that attempts to make this assessment on an analytical basis. ("hard money" refers to the actual costs defined in the tender; "soft money" refers to the estimates made on the basis of the perceptions of the team evaluating the bid on the basis of their professional judgment.)

Structured *scorecards* are used to quantify the perception of the less tangible parameters affecting the technical quality of the bid and hence the ability of the bidder to add value to the project. These scorecards are completed by each member of the review team. Scores are then debated until an overall consensus reached. Figure 30.16 shows an example of a scorecard used to evaluate the team proposed by a tenderer to undertake a significant project. Scorecards help focus the mind and provide the basis of communication for the evaluation team. These scorecard systems can be quite complex. Scorecards obviously enable us to express a perception of quality in hard numbers.

In the hard money-soft money approach, this perception of quality is translated into measurable benefits for the owner. For example, in the tender to undertake the maintenance turnaround of an oil refinery or power station, the quality of the management team proposed by a contractor is obviously a key success factor. How then to compare the offering of one contractor, who has a more experienced team, with another who has an adequate team and a lower price?

In hard money-soft money evaluation the owner team translates their perception of quality of the management team into the additional margin they can really generate by completing the project earlier. Ideally, these benefits should be linked to incentive arrangements in the contract form. Figure 30.17 shows the hard money-soft money evaluation of a contract for pipework refurbishment during the revamp of a process unit.

Post Tender Negotiation

In virtually all public procurement, and in many companies, *post tender negotiation,* or negotiation of the price after tender submittal, is not permitted. Procurement professionals often debate whether post tender negotiation is in fact desirable. Theoretically, if bidders know there will be no further scope for negotiation, they will submit their best price in their original tender. In practice it doesn't seem to work like that. First of all, in most relatively complex proposals there is the need to debate the tender in great detail to ensure that both parties have a common understanding of the issues and the deliverables.

In fact, in the case of equipment procurement, the savings range is much higher because of the cost structure of the supplier organizations. In most multiround negotiating situations in a competitive environment, including the use of reverse auctions, prices tend to converge as expectations are progressively stripped away and suppliers cut to the absolute limit of their cost structure. Figure 30.18, taken from a real negotiation, shows how prices tend to converge in a typical multiround negotiation. Online reverse auctions also exhibit a similar pattern.

Negotiation is a very fine art where the invisible line between buyer and seller can be stretched to the limit but never broken. Experienced negotiators from both buyers and sellers

FIGURE 30.16. EVALUATION SCORECARD.

Scoring Guide Section G-Organisation and Project Team

Non existent-totally unacceptable-no appreciation of issues	0
Poor by Industry standards-Disappointing	1
Below sector average-Below expectations	2
Average for the sector-What we expect	3
Quite good-Better than average	4
Quite exceptional-the best we have seen	5

NOTE: *Weightings of sections agreed by Team Prior to Evaluation Interviews.*

Ref:	Key Questions	Overall Weight	Weight Within section	Points Earned	Score 0	1	2	3	4	5	Comment
G-1 Organisation											
	Linkage to Owner organisation appropriate?		20								
	All functions covered?		20								
	Structure coherent?		20								
	Resourcing adequate?		20								
	Delegations appropriate?		20								
	Other?										
	G-1 Sub-Total Organisation	20	100								
G-2 Profiles of Key Staff											
Appreciation of the experience motivation and flexibility of key staff											
	Business Unit Director										
	Experience		50								
	Motivation & flexibility		50								
	Overall Rating		100								
	Designated Project Manager										
	Experience		50								
	Motivation & flexibility		50								
	Overall Rating		100								
	Designated Design Manager										
	Experience		50								
	Motivation & flexibility		50								
	Overall Rating		100								
	Designated Procurement Manager										
	Experience		50								
	Motivation & flexibility		50								
	Overall Rating		100								
	Designated Construction Manager										
	Experience		50								
	Motivation & flexibility		50								
	Overall Rating		100								
	Designated Planning and Control Manager										
	Experience		50								
	Motivation & flexibility		50								
	Overall Rating		100								
	G-2 Sub-Total Project Team	80									
	Overall Rating on Experience										
	Overall Rating on Motivation & Flexibility										
Overall Score Organisation and Project Team		**100**									

FIGURE 30.17. HARD MONEY-SOFT MONEY EVALUATION.

Assumptions

Cost/Day € 50000

				Scope Creep Allowance				Soft Cost		
COMPANY	Manhour Estimate Fixed Scope	Price ?/MH	Fixed Price	Day Rates ?/Hr.	Estimated Extra Hours	Estimated Cost Dayworks	TOTAL HARD COST	Over-Run Risk Days	Soft Cost	TOTAL COST
Magyar Pipe										
Roma Pipe										
Gypipe										

From Manpower Planning

will work throughout the pre-tender and tender period to develop the climate of the negotiation to their own advantage. From the owner's perspective, it is important to manage the environment during the tender period to heighten the desire of the bidder for the contract and lower the bidder's expectation on selling price. In fact, there is clear evidence that the harder someone expects the negotiation to be, the lower the price they will settle for (Field, 2003). The actual negotiations will be carefully planned as the use of time is a major negotiating lever. The various members of the team must clearly understand their roles and "script."

FIGURE 30.18. PRICING CONVERGENCE.

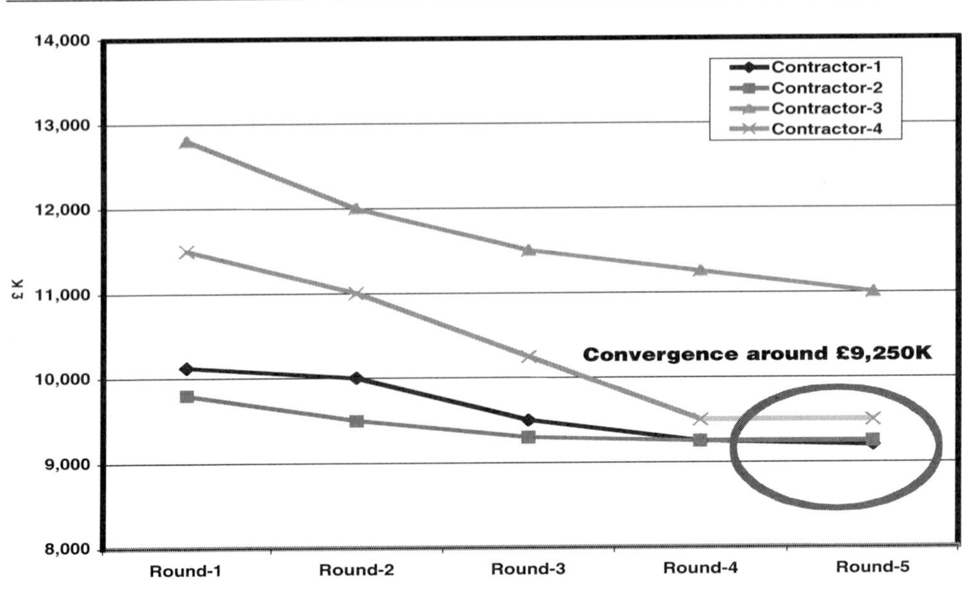

Remember that it is a vitally important time when often over a few hours or days, the ultimate profitability of the project or contract, involving years of work, can be made or broken. It is a time when the management teams from buyer and seller can develop a huge amount of mutual respect or loathing (P. Marsh).

Transition to Execution

Once the contract has been awarded, the buyers and successful vendors celebrate, and in many cases depart, leaving the operating organizations with the task of making the contract work—which can often be quite difficult, particularly in more complex contractual arrangements. Many of the understandings developed within the negotiating team may not be fully understood by others and may result in difficulties throughout the life of the contract. It is therefore advisable to recognize the need to manage the transition from negotiation into operations. This is true in all project situations, but in large complex projects managed by an integrated owner-contractor project team where many members of the project team are new to each other, it can be a key success factor. Such companies will spend considerable time constructing a complete integration program, including the following:

- Social events and expeditions to promote "bonding" across the project team.
- Contract workshops where the owner and contractor team that negotiated the contract explain it to the operational staff and jointly answer questions of detail from the execution team.
- RACI workshops where the owner and contractor staff can work out how best to manage day-to-day operations and develop management alignment, eliminating any confusion over relative responsibilities.

All of these events improve communication and generate a common understanding and language across the project team. They may even result in mutual appreciation of professionalism and fairness. Such common understanding and language can only help resolve problems, remove constraints, and ultimately deliver a better result and business benefit for both parties

Online Tendering Systems

The mechanistic aspect of the tendering process described previously lends itself to automation through the use of information systems. While the real value added in the tendering process is the development of creative contracting strategies and value managing requirements, the pure mechanics of issuing and producing tenders probably accounts for 80 percent of the level of effort and management time expended. This aspect can be really streamlined through the use of online tender management systems. Figure 30.19 outlines such a system.

The best of these systems support each step of the procurement process and provide online access as appropriate to everyone involved in the procurement process:

FIGURE 30.19. TENDER MANAGEMENT SYSTEM.

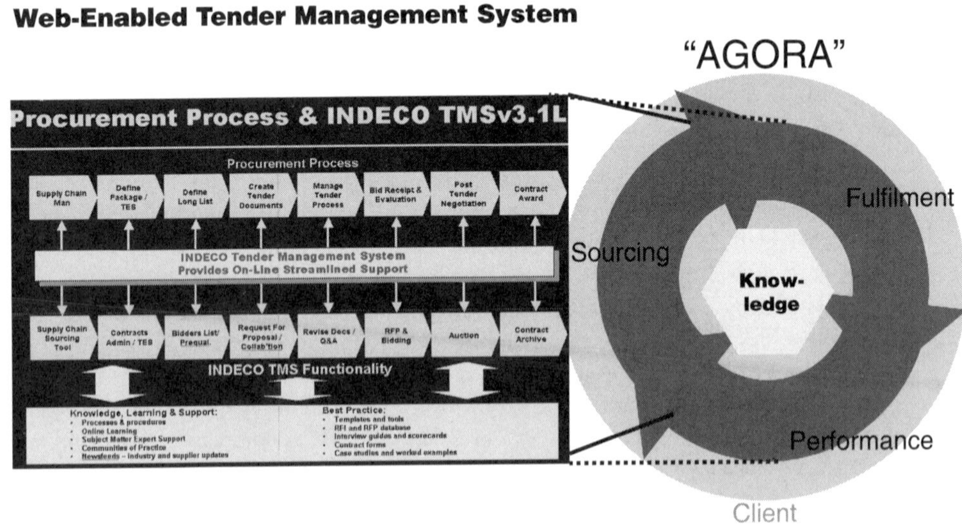

- Users can interrogate extensive supplier databases to identify potentially suitable suppliers.
- Request for Information can be published on the Web, completed, and processed online to produce a bidding list for a particular tender (see Figure 30.20).
- Questions and answers can be tabled and processed consistently online.
- Requests for proposals can be published and completed online, including all supporting documentation and drawings.
- Bids can be evaluated and a short list developed.
- Post tender negotiation can, where appropriate, be conducted through a reverse-auction process.

Tender management systems can readily be linked to a knowledge management database providing users access to case studies, templates, organizational documentation, and the like that is accessible at point of use. Such a system provides an extensive audit trail of every procurement transaction and can reduce the professional time to develop and conclude a typical tender by at least 25 percent.

The Bidders' Perspective

By and large this section has been written from the perspective of the organization issuing the tender. If effective tender management is important to the "owner" organization, it can

FIGURE 30.20. TRANSMITTING TENDER DOCUMENTS.

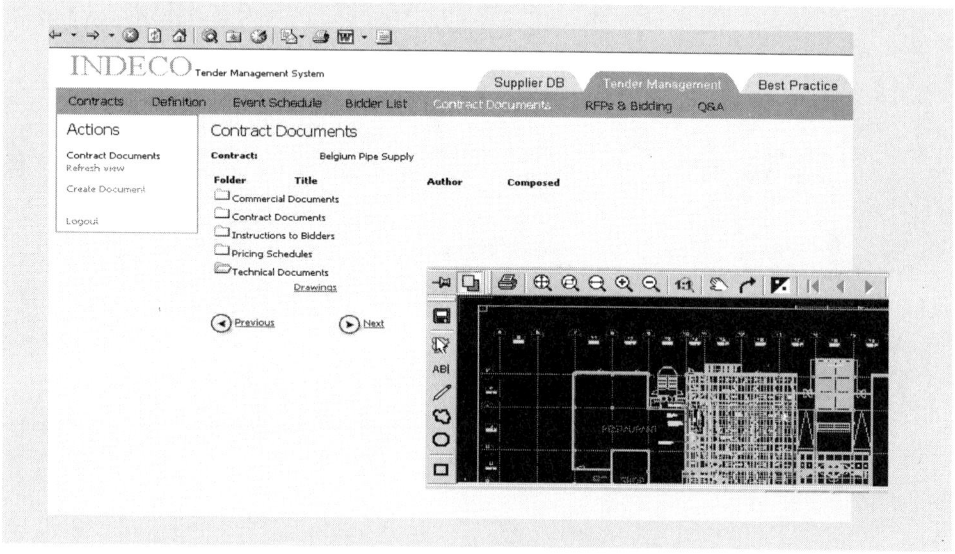

be even more so to the organizations responding to the tender. In many cases the risk a bidder assumes can be a very significant percentage of the bidder's net value. On the one hand a major contract can represent the opportunity for significant profit or strategic advantage. On the other, many companies have been bankrupted by contracts that went wrong. Figure 30.21 outlines the classic bidding process for a tendering company. Every company should have a formalized process for receiving, responding to, and submitting tenders. Tenders are the lifeblood of every contracting organization. They are mission-critical and must be treated as such.

Decision to Bid

The decision to bid or not to bid is critical. Tenders should never be taken lightly. Producing a tender costs money, often a great deal of money, and will tie up the energies and time of the most creative people. Many contractors have been bankrupted by taking on uneconomic contracts. Every contractor should ask the following questions for every tender being considered:

- Is this opportunity in line with the strategic objectives of the company?
- Is this a client the contractor wishes to work with?
- Does the contractor have the competence and capacity to undertake this work?

FIGURE 30.21. TENDER MANAGEMENT, BIDDERS' PERSPECTIVE.

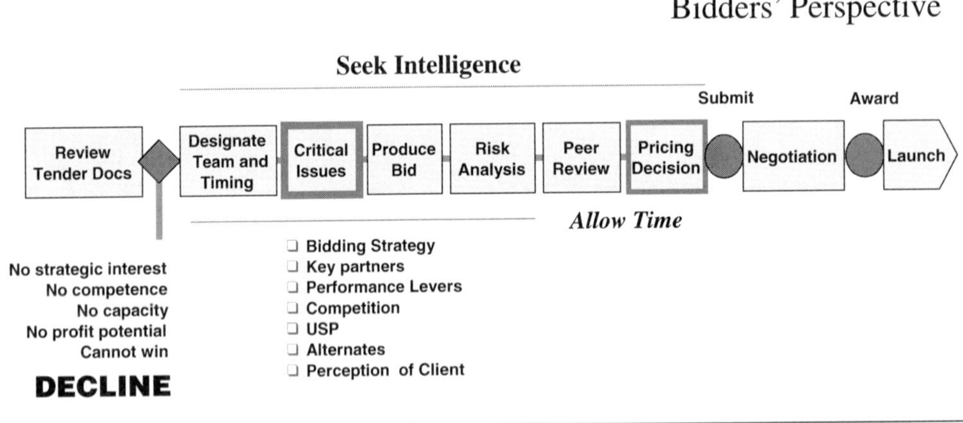

- Are there competitors the contractor cannot hope to beat?
- Does bidding this work make economic sense?

If the answer is no to *any* of these questions, the contractor should seriously consider whether it is advisable to bid. While it may be politically necessary to respond to a client with whom the contractor might wish to work on another project at another time, it is inadvisable to waste time and money on tenders the contractor does not really want to win. Figure 30.21 illustrates the typical tendering process we would expect to see within a typical reasonably sophisticated contracting organization.

Assuming the contractor wishes to tender, a bid manager is generally assigned with access to the appropriate in-house and external expertise and resources immediately. This bid manager is responsible for coordinating the company resources and producing a winning bid that will be profitable for the company.

The first key event in the bidding process is normally the *critical issues meeting.* This will be held as soon as possible and aims to table the collective wisdom of the contracting organization on the tender at hand. This meeting would seek to identify the following:

- The lessons learned from previous experience on similar work, or with the particular client.
- The performance levers in the scope of work that will generate real value for the owner organization.
- Any contacts, possibly within the client organization, who can provide intelligence on the tender at hand.
- Key suppliers who may provide a unique advantage and formulate ways to do this and deprive the competition of their services.

- The strengths and weaknesses of the competition and their likely response and pricing levels.
- The way we might differentiate ourselves from the competition and frame an offer the client cannot refuse.
- Any alternate approaches that might give a steep change in cost or any other client benefit.
- The major risks associated with the tender and how we can mitigate these risks.
- Any administrative, legal, or fiscal implications associated with the contract.

On the basis of this meeting a formal bidding strategy, budget, and schedule will generally be produced and responsibilities for the various aspects of the tender assigned. During the bidding period every attempt will be made to get as much intelligence on client thinking and the competitive position as possible. A formal risk analysis will be produced to support the pricing recommendation of the tender team. It is important to allow enough time for a formal peer review when the tender can be presented to senior staff, who have not been engaged on the tender, and subjected to challenge.

A formal memorandum will be tabled making the pricing recommendation to management. This memorandum will summarize tender scope, strategic implications, the key assumptions, the major risks, and mitigating strategies. The pricing of the tender should be recommended, including the basis of the negotiation strategy and the "walk away" price.

Summary

Tender management is a vitally important management function for buyers and sellers alike. Done well, it can significantly improve the profitability of both project owners and contractors. Done badly, it can seriously jeopardize the profitability of owner organizations and the very survival of contractors. It is undoubtedly an exciting space but not one for the faint-hearted.

This chapter is based on many years' experience of project and tender management in many different sectors, in many different countries, and from the perspective of both an owner and contractor. Unfortunately, many of the lessons learned have been very expensive!

While details will vary from case to case, I absolutely believe that the approach to tender management described in this chapter is applicable in most sectors of industry and that the eventual success of any procurement and tendering initiative depends on the following three fundamental success factors:

- The composition and competence of the team and the quality of teamwork
- The rigor of the procurement and tendering process
- The creation of an innovative management culture that encourages the whole project team to strive for higher levels of performance

Tender management is a crucial period in the life cycle of most projects. It can have a fundamental impact on the success and profitability of the project. If this chapter provides a little enlightenment, or even reassurance, to those concerned, it will have served its purpose.

References

Bennett, J., and A. Baird. *NEC and partnering: The guide to building winning teams*. London: Thomas Telford. ISBN: 0-7277-2955.

Egan, J. 1998. Rethinking construction. The Report of the Construction Task Force to the Deputy Prime Minister, John Prescott, on the scope for improving the quality and efficiency of UK construction. London: The Stationery Office.

ENR (2001) BP & Bovis boost relationship to remake oil giant facilities. ENR 2001.

Field, A. 2003. How to negotiate with a hard nosed adversary. March.

Kennedy, M. 2000. *Guardian*. December 6.

Lassiter, T. 1998. Balanced sourcing the Honda way. *Strategy & Business*. Booz Allen Hamilton, 4th quarter.

Latham, M. 1994. *Constructing the team: Final report of the government/industry review of procurement and contractual arrangements in the UK construction industry*. London: The Stationery Office.

McKinsey Report: Driving productivity and growth in the UK.

Marsh, P. Contract negotiation handbook: ASIN 0566-024039.

Value Engineering Professional Association. *www.vetoday.com*

Wideman R. R. M. 1992. Project and programme risk management. *PMI USA*.

CHAPTER THIRTY-ONE

PROJECT CHANGES: SOURCES, IMPACTS, MITIGATION, PRICING, LITIGATION, AND EXCELLENCE

Kenneth G. Cooper, Kimberly Sklar Reichelt

The project was slated to finish in another three years when the first of many change requests came in. The contractor added new staff to accommodate the changes and also worked 60-hour weeks trying to stay on schedule. The new staff needed extra supervision, which was in short supply. Some vendor-supplied design information was late, and the team implemented workarounds to keep things moving. Productivity suffered. Rip-out and rework became a routine "surprise" condition in the build effort. Staff morale worsened and key people left the project and the company. Problems snowballed until, in the end, the project was completed two years late and 50 percent over budget. No one understood how it had gone so wrong.

Was this (A) a naval shipbuilding project, (B) a big civil construction project, (C) a new military aircraft, or (D) the latest big upgrade of a software product? Or was it your latest project? Sadly, the answer could be "all of the above." But why? How is it that such diverse projects could share so consistently the same phenomena?

If there is one thing we should know to expect as project managers and customers, it is that our projects will change. The changes may be to the design of the product, changes from the expected construction conditions, changes in technology, schedule changes, or any of a host of other sources. So what do we do when all those carefully prepared project plans are established and the changes begin?

In this chapter we examine the beneficial and the challenging aspects of project changes. We review what experienced managers believe to be the most effective practices in managing changes and their project impacts. It is not a chapter for the faint of heart, for among the consequences we will examine is a set of impacts that have been responsible for ruining many a project and career. Study after study and survey after survey point to the disturbingly consistent pattern of projects failing to meet targeted objectives, whether they are failures in the product itself or in the cost and time required to get there. How much of a role do changes play in these oft-cited failures? In a word, huge. The ability to handle changes well is a rare and valuable corporate asset. The more common state is one in which changes lead to a persistent pattern of problem-ridden projects.

We have surveyed dozens of managers for their input on best practices and for their descriptions of the most troublesome aspects of managing changes. Among the proactive points of advice are many things that will seem straightforward. Yet they are the practices to which we seem to have the most difficulty adhering in the rush of project execution. In the absence of diligent implementation of these practices, we pay dearly for the consequences, in terms of cost and schedule overruns, lost profits, and ruined relationships. We will examine here how changes can cause projects to spiral out of control, through impacts that are many times the level expected. We will examine the dark side of change impacts, because, at their worst, they become the fodder for major disputes between customer and contractor. And we will describe how many firms have avoided those phenomena by rigorous adherence to a set of practices that have been successful in containing the adverse impacts on projects.

Project change management is arguably the most important under-addressed aspect in all of project management. Why? Do we simply not want to think that the carefully planned effort will be subject to changes (when nearly all of our experience is to the contrary)? Is it such a contentious aspect of project contracting that we are loathe to have the topic infect the euphoria of that exciting new effort (when avoiding it virtually ensures it will hurt the project)? Are we, as some cynics assert, purposefully obfuscating the issue at the outset, with contractors believing that they will "get well on changes" (when the clear experience is that this is rarely achieved)? Regardless of the motivation or mix thereof, one thing is clear: While changes occur with the intent of improving the project, most often we do not do a good job of managing those changes, to the peril of the project.

The situations in which these challenges manifest themselves most acutely are typically in the scenario of a "conventional" contracted project, with a business relation between customer and contractor for the purpose of designing and/or building a product—a building or plant, an electronic system, a ship or aircraft, software, or anything sufficiently complex as to require the coordination of design and build activities. The challenges of handling changes on such projects lie primarily in the hands of the contracted managers. This is not to minimize the importance of the customer's role, for, as we shall see, the customer is pivotal in determining the success or failure of change management and of the project itself. Nor is it to say that the principles described herein do not apply to all manner of projects, even if conducted internally in an organization without a formally contracted customer. We will, however, orient most of this discussion toward the contracted project, and within that, even more specifically to the contracted managers of such projects, on whose shoulders rests the primary burden of managing to typically aggressive cost and schedule targets in an environment so often riddled with change.

What Changes Are

"Change is any deviation from the way the work was planned, budgeted or scheduled."

THOMAS, 2002

Change is a necessary fact of project life. Projects are not just about meeting contractual requirements; they are about achieving the outcomes the end users need. In a world in which markets shift, technology advances, and requirements evolve, projects must be able to accommodate all of these types of changes. The result can be a more capable product that better meets the users' needs. Changes handled well generate long-term business relationships of trust and understanding; handled poorly, however, they can spiral into overruns and disputes. This, then, is the degree of leverage changes have on projects: from the surprisingly successful to the unforeseeably disastrous.

With that much at stake, first we need to know just what changes are. When we think of change, we may be tempted to define it narrowly; in terms of a house, it might be an extra window here, a higher ceiling there. However, we must think more broadly of change, as quoted previously, as "*any* deviation from the way the work was planned, budgeted or scheduled." With this more inclusive definition, changes may come in many forms, such as:

- Design changes
- Work-scope changes
- Late receipt of important technical information
- Excessive delays in design review and approval
- Diversion of key management and technical resources
- Unplanned site conditions
- Inadequately defined specifications or design "baseline"
- Changes in standards and regulations
- Late or inadequate subcontractor performance
- Schedule changes or acceleration
- Superior knowledge
- Technology advances

Whenever a change may have occurred, managers must review the change, plan its execution, consider mitigations, assess its impact, and importantly, communicate with the customer.

How Changes Impact a Project

To support an informed, proactive, ideally collaborative decision on changes, we need an accurate view of their impacts—and these may be as diverse as projects themselves. There are, however, highly consistent *categories* of impacts, which we describe in this section. No discussion of impacts can occur without first reminding ourselves that changes do not exist simply for the purpose of causing project management problems, overruns, and associated nightmares. Changes exist on nearly every project with an objective to enhance the product of the project, or to accommodate or correct conditions that would otherwise harm the project. While recognizing that these noble purposes exist, let us move on to the ways in which project impacts of the undesirable sort (cost, schedule) occur and what can be done about them.

Logically, analytically, contractually, procedurally, and practically, there are two fundamental categories of change impacts on projects: direct impacts and disruption impacts. Each category may have many alternate labels (direct/hard-core/primary . . . disruption/ripple/productivity loss/knock-on/impact/cumulative . . .). Basically, however, the practical distinction is this: The visible cost of directly implementing or accommodating a change is *direct* impact; the change's impact on the cost of executing unchanged work (or, indeed, even other changed work) is *disruption.* Among many dozens of management interviews, a universally cited observation is the much greater difficulty of dealing with the latter. Because of that, much of the discussion of change impacts herein will focus on the less well understood phenomenon of disruption. First, however, a brief look at direct impacts follows.

Direct Impacts

The direct impact of any given change is likely to have multiple dimensions. All impacts need to be translated eventually into one measure, of course: cost. But there are three dimensions that are helpful in achieving that translation. (Two excellent sources detailing the direct impacts of many types of changes are Cushman and Butler, 1994, and Hoffar and Tieder, 2002.)

Added Expenditures/Scope. This dimension of direct impact can be the most straightforward to document. "Build three, not two." "Buy another generator." "We need to install another two hundred feet of piping." The measures may be dollars of material and/or additional hours (and dollars) of labor.

Delays. Two types of delay can directly impact a project. Some sources of delay can be demonstrated to extend the period of performance on the project. For these delays there are accepted categories and formulas for added costs incurred (Hoffar and Tieder FedPubs, 2002).

Other sources of delay may have little or no direct impact on schedules and cost but may still have disruptive consequences. Examples could include late material delivery or delayed access to testing facilities. Even without critical path-delaying impact on the whole project, such conditions should be documented, as they can cause staff productivity impact (disruption).

Design Uncertainty. Another dimension of direct impact of changes is design uncertainty. Although not a direct cost source, one of the best early indicators of disruption impact is the degree of the design package affected by changes. The origin of this impact may be explicit design changes, or merely lack of resolution of needed design decisions (such as through delayed design change approval). Monitoring the percentage of the design so impacted, and the duration of the impact, is valuable. It is a strong leading indicator of the magnitude of disruption throughout the project, as it can cause design work to be done out of sequence, reduce productivity, increase rework, and thus cause construction productivity loss as well. This leads us, then, to the subject of disruption impact.

Disruption

Regardless of the label (cumulative impact, loss of productivity, knock-on effects, ripple, "death by a thousand cuts," secondary impacts, etc.), disruption is the change-induced additional cost of performing work not directly changed. Hence, it is necessarily the added cost (beyond what would have been required otherwise) from lowered productivity or increased rework on the unchanged work, as traceably caused by the change(s). In this section we seek to bring some added clarity to the phenomena to which we will collectively refer in this discussion as "disruption."

The Challenges of Disruption. Let us just acknowledge from the start that disruption is the most difficult and most abused aspect of contractual change pricing. It is precisely because it is the most difficult that it has been the most abused: some contractors use it as a blanket to hide their own problems and cover all cost overruns with "total cost" claims to customer; some customers have taken equally outrageous positions of denying the very existence of disruption and refusing any consideration or compensation for it. Each extreme position is equally absurd. Disruption happens. It is a fact of project life when changes occur. It is a legally recognized set of phenomena that have been established, upheld, and refined over multitudes of court and board decisions of every venue. (For an extensive discussion of the legal background and approaches to cumulative impacts, see Jones, 2001). So why is it so difficult to address, to describe, to analyze, and to quantify? The list of reasons is long.

The Seven Barriers to Rational Analysis of Disruption

1. Disruption can be widely separated in space and time from the precipitating event(s), but to be claimed successfully must be causally tied to their source: "Although a change order may directly add, subtract, or change the type of work being performed in one particular area of a construction project, it also may affect other areas of the work that are not addressed by the change order." (Jones, 2001, p. 2) A construction supervisor could be managing a set of impacted work (a) a year after engineering changes have (b) caused more overtime use that (c) lowered productivity, and thus (d) delayed needed work product from (e) a distant part of the organization. The construction supervisor is unlikely even to recognize that they are being impacted by change-induced disruption, let alone be able to quantify it!
2. Disruption impacts can be cumulative across large numbers of individual impacts.
3. Disruption is fundamentally about productivity and rework, which are hard to measure and thus are rarely measured well.

4. The ideal form of damage quantification is to define the amount of impact, including disruption, that would put the injured contractor in the condition the contractor would have been but for the damaging events—a challenging analytical task.
5. Disruption must screen out the effects of other concurrently occurring contributors (such as strikes, difficult labor markets, or mismanagement).
6. Contractor-customer discussions of project cost growth tend to be adversarial, even while the project continues, making efforts to quantify, explain, and mitigate disruption especially challenging.
7. Finally, with all of these difficulties, there is also, quite frankly, a poor track record of rigor in disruption quantification; it is far easier, and usually tempting, for both sides to put all blame on the other without rigorous analysis to back their claims; sloppy logic and analysis may drive assertion ("You caused all our problems!") and counter-assertion ("You mismanaged everything!") (Cooper, 2002).

Disruption Explained. Many hundreds of project managers have described to us various parts of disruption phenomena on hundreds of projects. When we assemble those comments into a cause-effect description, there is quite a high consistency of the kind of disruption phenomena, even as experienced by a wide range of project types. What follows is a description independently offered by managers of aerospace, construction, IT, shipbuilding, electronics systems, and many other projects and programs. Understandably, each has its own peculiarities and variations, but it is the similarity of the descriptions that is striking and that offers the prospect of usable guidelines for explaining project disruption.

The "+ & Δ" shown in Figure 31.1A represent additions and changes to the project's work scope. The direct consequence of changes and additions is to increase the work scope and reduce management's perceived progress on the project. With a lower progress estimate, management's expected hours at completion will grow, and they will increase their staffing requested.

Increasing the staffing request may have the short-term consequence of requiring the use of more overtime until the new hires are brought on board. Sustained high levels of overtime reduce the per hour productivity of staff (see Figure 31.1B). Hiring in a constrained labor market dilutes skills and experience and strains supervision, which further erode productivity and quality (see Figure 31.1C).

Later, the impact of the reduced quality will be felt. The errors created by the fatigued and less experienced staff will have propagated, as subsequent work products build off earlier faulty ones, and thus have been done at lower productivity and quality as well (see Figure 31.1D).

Later still, all the pressures of overrunning the budget and schedule, and finding more and more rework, lead to morale problems, furthering the decline in performance (see Figure 31.1E).

Problems early in the project propagate to downstream work. Change impacts originally isolated in the engineering phase end up affecting construction as well. There another similar set of dynamics is triggered, with the addition of some physical impacts as well, such as crowding (congestion, "trade-stacking"). All these dynamic effects in construction, as they

FIGURE 31.1A. DIRECT CONSEQUENCE OF CHANGES AND ADDITIONS.

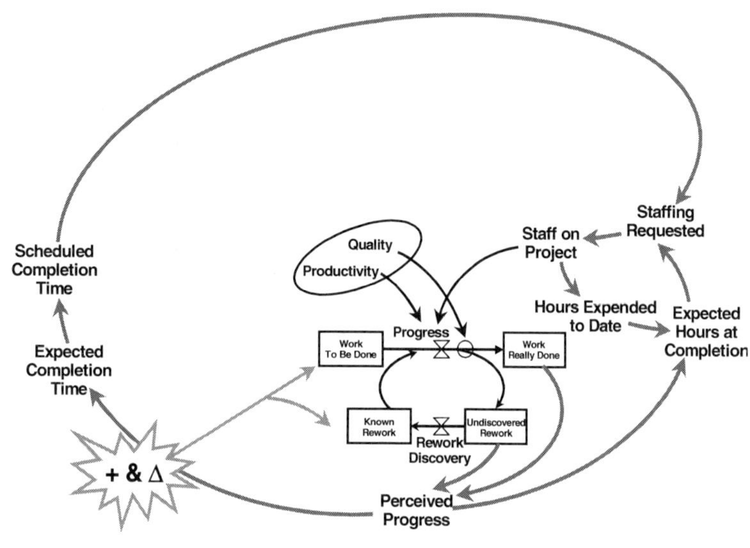

FIGURE 31.1B. INCREASED STAFFING NEEDS ARE MET WITH HIRING AND OVERTIME.

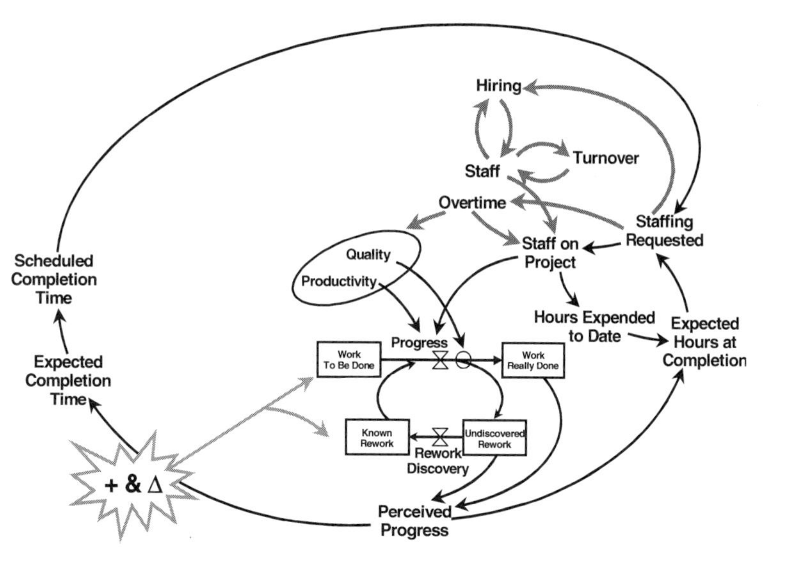

FIGURE 31.1C. HIRING DILUTES BOTH SUPERVISORY ATTENTION AND OVERALL SKILL LEVELS.

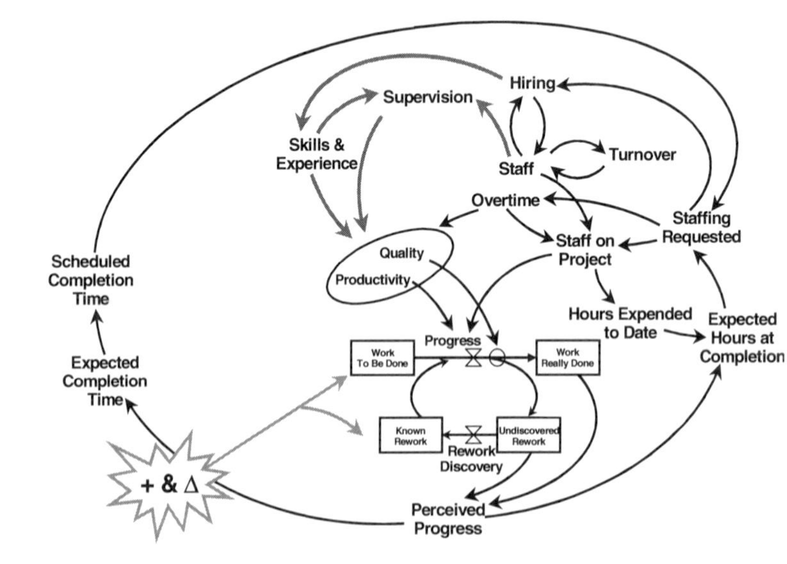

FIGURE 31.1D. EARLY QUALITY PROBLEMS AND SLOWER PROGRESS CAUSE PROBLEMS LATER.

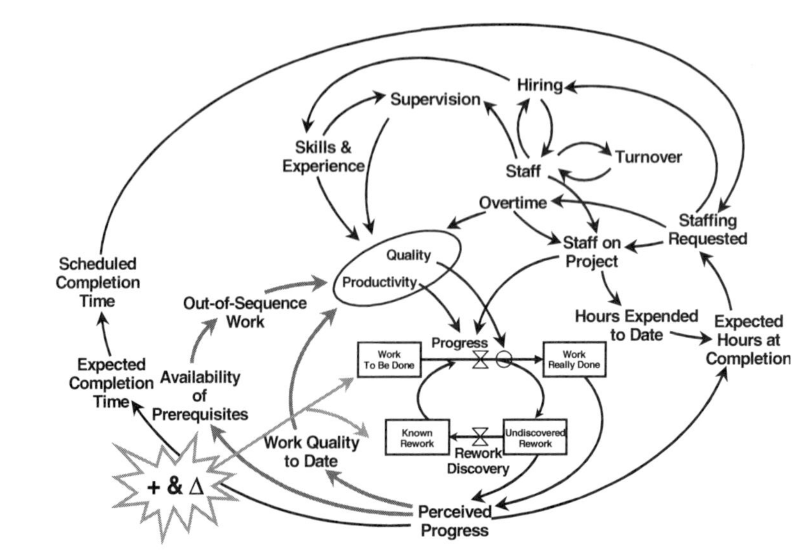

FIGURE 31.1E. BUILDING PRESSURES EVENTUALLY AFFECT PROGRAM MORALE.

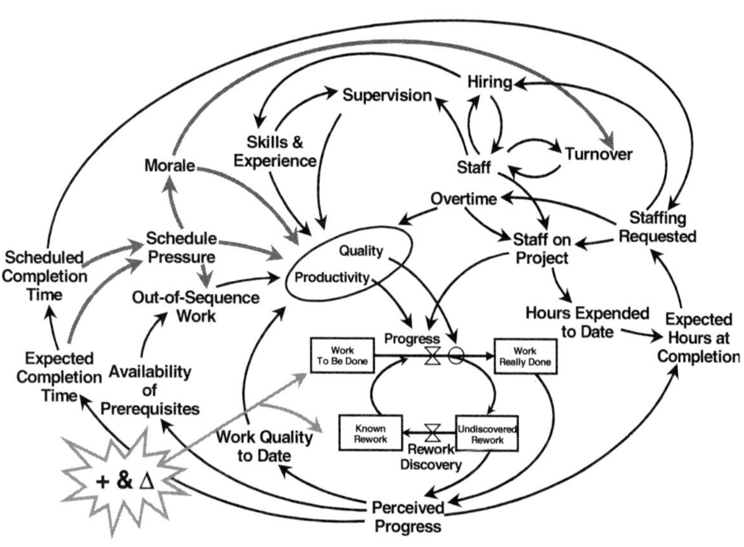

affect labor productivity, can amplify the magnitude of disruption impacts from changes (see Figure 31.1F).

With this causal description of project dynamics, it becomes obvious how problems early in a project can propagate through many stages—as ripple effects, knock-on, . . . Whatever phrase we choose, we really mean disruption.

Now, pause to look back on the assembled diagram (see Figure 31.1F) and note the degree of interconnected *loops* of cause and effect—those circular paths formed by the arrows. Most of these loops are self-reinforcing, and when they are interconnected as they are, the chance for disproportionate cost growth is high when there are many changes or substantial scope growth. Much more scope, for example, generates the need for a more overtime and hiring, diluting productivity and slowing progress versus plans. Work moves more out of sequence, and more rework is generated. Morale suffers, thus worsening productivity more. Staff turnover increases, and so even more overtime and hiring of new staff is needed and so on. It is the self-reinforcing character of those interconnected loops that generates the "cumulative impact" associated with many disruption cases.

With that qualitative description in mind, it is no wonder that the quantification (and responsibility allocation) of disruption remains a challenge. Of one thing we can be sure: When it comes to *anticipating* the additional disruption quantum caused by changes, we are almost universally guilty of underestimation. Some organizations are delighted to receive "extra" compensation of 10 to 50 percent on direct change costs, when in fact the disruption costs can be several *times* that.

FIGURE 31.1F. PROBLEMS IN ENGINEERING LATER FLOW DOWNSTREAM TO AFFECT CONSTRUCTION.

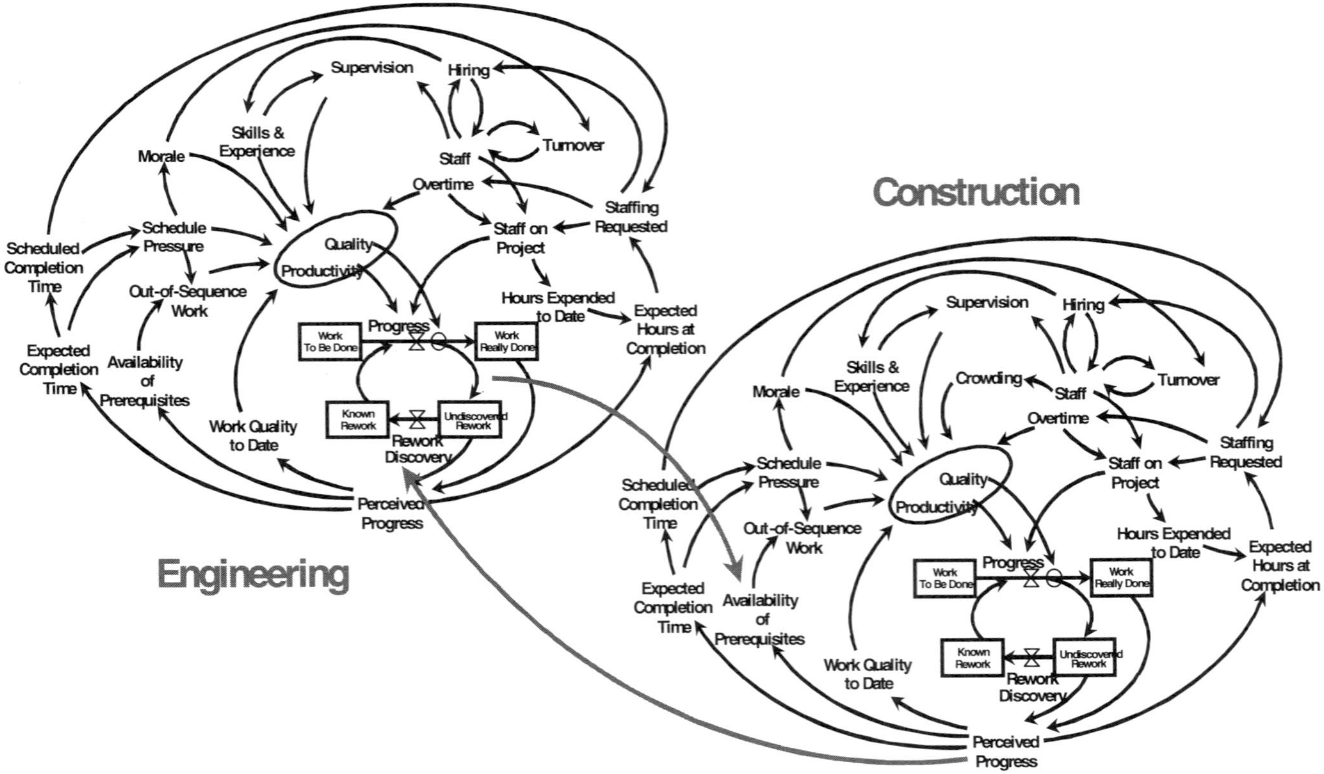

By way of quantitative illustration, we can examine results of several real project simulations. These simulations employ the cause-effect explanation described previously, coded in a model (Cooper, 1994) that is populated by numerical factors drawn from analyses of hundreds of real projects (and thus representative of many projects, but not a model of any single specific project). The model incorporates the most often cited sources of productivity impact from analyses and surveys of hundreds of projects (many excellent reviews and surveys describe sources of project productivity impact. See, for example Schwartzkopf, 1995 and Ibbs and Allen, 1995).

The Top Ten Causes of Project Productivity Loss. The following are the ten most prevalent causes of loss of project productivity, in no particular order:

- *Changing work sequence.* Workarounds are frequently necessary, especially when design changes are required and when tight schedules force moving ahead even when necessary information is unavailable, and tasks must be performed out of their ideal or planned sequence.
- *Skill dilution.* Higher staffing in a tight labor market usually means new, less experienced hires are brought into the program.
- *Supervision dilution.* Higher staffing levels, especially with less experienced personnel, can divert supervisors and dilute their effectiveness.
- *Overtime/second shifts.* Overtime or added shifts are frequently used to accelerate work progress; overtime is among the most researched effects on productivity.
- *Rework.* Productivity and rework creation suffer when downstream work products build off of flawed earlier work.
- *Congestion/crowding.* Especially when work is done in constrained spaces, higher staffing levels will hurt work productivity.
- *Late/changing engineering.* Changes or delays in the design will slow work progress and can cause rip-outs in build efforts.
- *Morale.* When program problems mount, the morale and productivity of the workforce may suffer and cause increases in absenteeism and staff turnover.
- *Tools/equipment/materials.* When necessary prerequisites or tools are late, the affected workforce suffers reduced productivity.
- *Schedule pressure.* If the program schedule begins to look difficult to achieve, pressure to regain schedule may cause a "haste makes waste" effect. While output may seem to be accelerated, it is frequently at the cost of increased rework.

Disruption Quantified. *A Disrupted Project.* We start with a look at the (simulated) project with no changes. It is a two-million-hour new-design and build project (650,000 hours in engineering, 1.35 million hours in construction direct and support). The construction starts one year after engineering starts and is scheduled for completion in three years (see Figure 31.2). Next we inject growth of scope and associated changes in both engineering and construction that has a *direct* impact of adding 15 percent of those hours to both stages. What additional cost is driven by *directly* adding nearly 300,000 hours to the project?

The graphical view of the impact is displayed in Figures 31.3A to 3C. Figure 31.3A displays the growth caused in engineering and construction staffing versus the no-impact

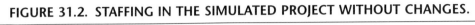

FIGURE 31.2. STAFFING IN THE SIMULATED PROJECT WITHOUT CHANGES.

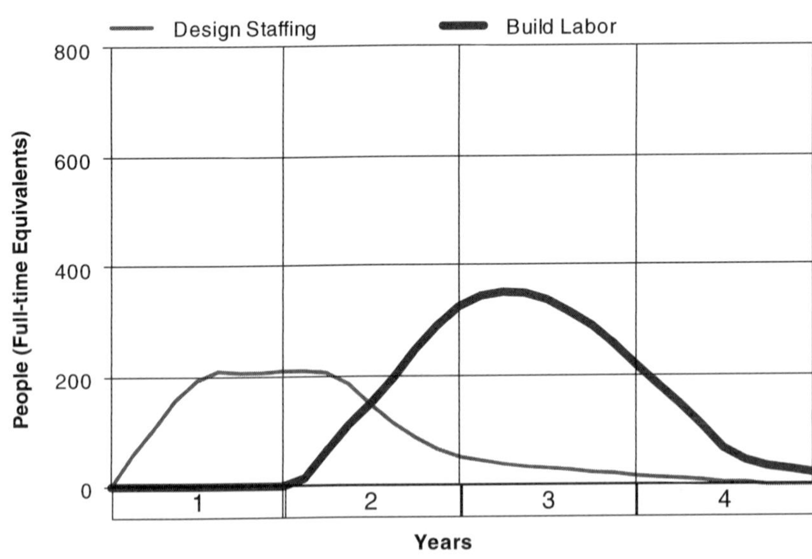

case. Why? Think back to the diagrams in the Figure 31.1 series. The changes lead to significantly higher (and later) engineering effort reworking designs (Figure 31.3B). Later and more engineering changes induce lower construction productivity and thus require more labor. Increased hiring and use of labor reduces skill levels, increases workplace congestion, and more, thus further reducing construction labor productivity (see Figure 31.3C). In the end the total amount of disruption caused through these phenomena is 3.3 times the directly added hours—near one million hours of disruption impact on this project.

Rule 1: Disruption impacts can appear years after the incident change event and can exceed the direct impact of the change.

The Cumulative Character. Even within a given project, the same change could have different amounts of impact if it is among the first of, or one of few, changes, versus being among the last of many. Some call it "death by a thousand cuts." *Cumulative impact* is the phenomenon of the impact of many changes being greater than the sum of the impact of the individual changes (as discussed previously). In the project simulation model, we tested the injection of multiple levels of changes in order to see the variation in impact. Figure 31.4 illustrates the nonlinearity of total impact, as more increments of exactly the same amount of extra work are successively added to the project. These plots of cumulative construction hours expended show how impacts can increase disproportionately as more and more changed work is added. As noted in Figure 31.5, with the first 5 percent of directly added work, the amount of disruption impact across this project is 2.2 times that of the direct impact. Adding another 5 percent takes that ratio to 2.8. Another 5 percent (the 15 percent-

FIGURE 31.3. IMPACT OF CHANGES ON SIMULATED PROJECT.

The Project with the Full Impact of Changes
(Direct Impact = 15% of original scope)

Figure 31.3A: Staffing

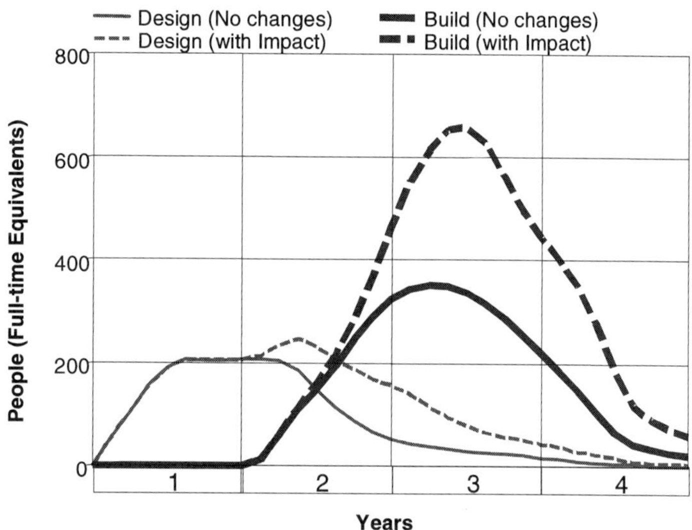

Figure 31.3B: Design Rework Effort

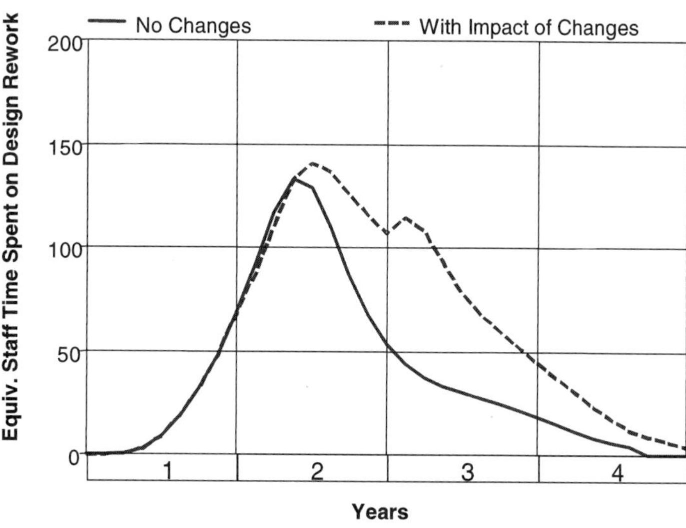

FIGURE 31.3. (*Continued*)

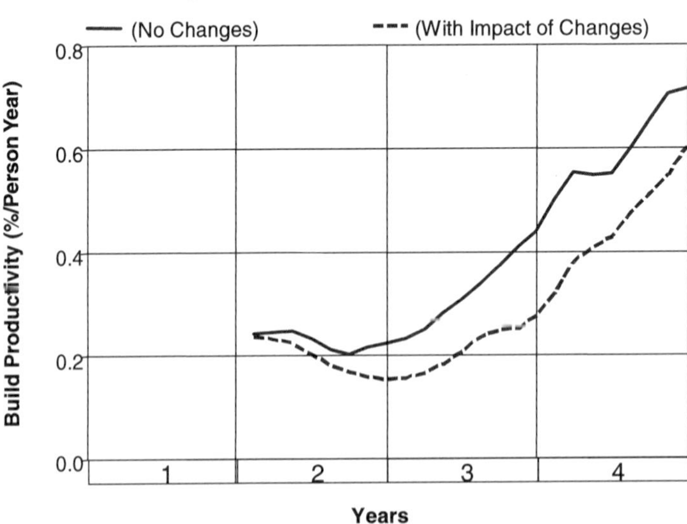

added-scope case we described previously) takes it to 3.3. And yet another 5 percent causes the total disruption impact to grow to 3.7 times the direct impact on this project. The phenomena at work driving these impacts are simply more and more of the self-reinforcing feedback effects described previously.

Rule 2: Disruption impacts grow disproportionately over more and more changes.

Timing Matters. One of the most consistently cited cautions in our management interviews is the need to resolve early and speedily the content of contemplated changes. Indeed, the timeliness of resolution can make a substantial difference in the impact of changes on project costs. How much? In our simulated project, resolving more rapidly the design issues associated with the changes could cut the disruptive impacts by as much as 40 percent. Figure 31.6A shows the significant improvement (reduction) in disruption impact, as less resolution time is required (moving right to left on the chart). The expanded view in Figure 31.6B shows the timing-induced disruption improvement (again, right to left) for the full range of direct impacts (i.e., from 5 percent to 20 percent of directly added work scope). And recasting these results to the form of display as in Figure 31.5, we can see the reduced range of disruption impact, for each amount of change, with more rapid issue resolution (see Figure 31.6C). The consistent pattern observable here is that cutting the resolution time in half reduces the amount of disruption by 10 to 20 percent, with the biggest percentage improvements in the most disrupted conditions.

FIGURE 31.4. IMPACT OF CHANGE GROWS WITH GREATER CHANGE MAGNITUDE.

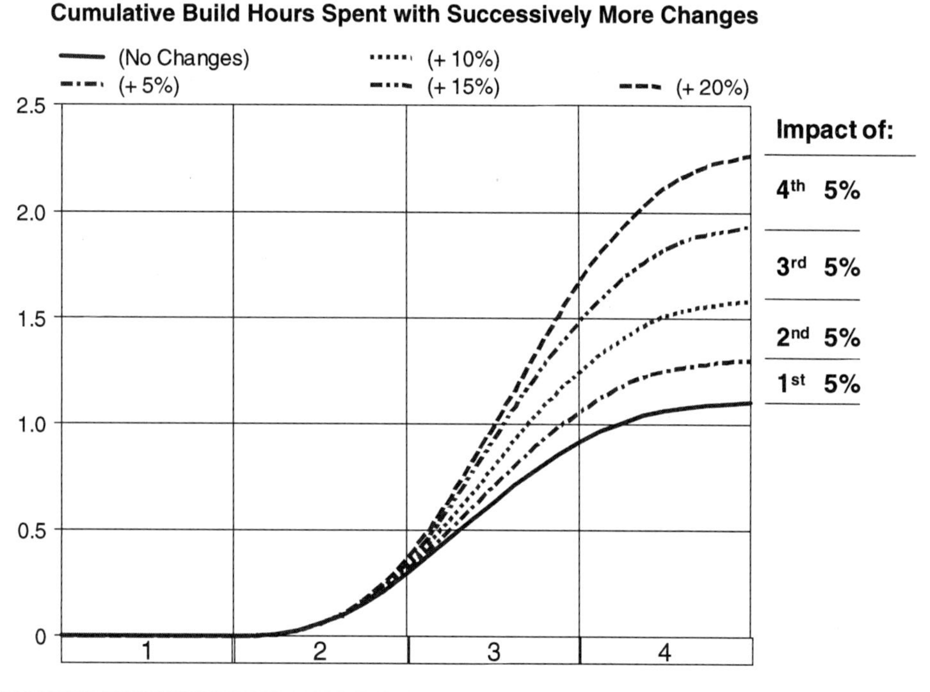

This degree of improvement cannot be surprising. In the extreme, if the changes were known at the project start and resolved instantly, they would not be changes at all; they would, in effect, become part of the project specifications and the known baseline.

Rule 3: Early issue resolution cuts disruption impacts significantly. Not All Projects Are the Same.

Surely we can't be saying that all projects are subject to disruption of two, three, or four times the direct impact of changes? Of course not; this is to explain how projects *can* be subject to this degree of impact, but not all projects share the same degree of sensitivity to the many factors active in this simulated project. For example, two of the most significant construction productivity-affecting factors here are (a) the quality of the design package and (b) the workforce skill levels. We tested the very same changes in the very same project, but with no susceptibility to these conditions—that is, if there were no significant design change impact on construction or no dependence on skilled labor (or at least plentiful supply thereof). Eliminating the dependence on skilled labor cuts the disruption impact of the changes from the base ratio of 3.3 (times the direct impact) to 2.2. Eliminating dependence

FIGURE 31.5. DISRUPTION RATIO INCREASES WITH MORE CHANGE.

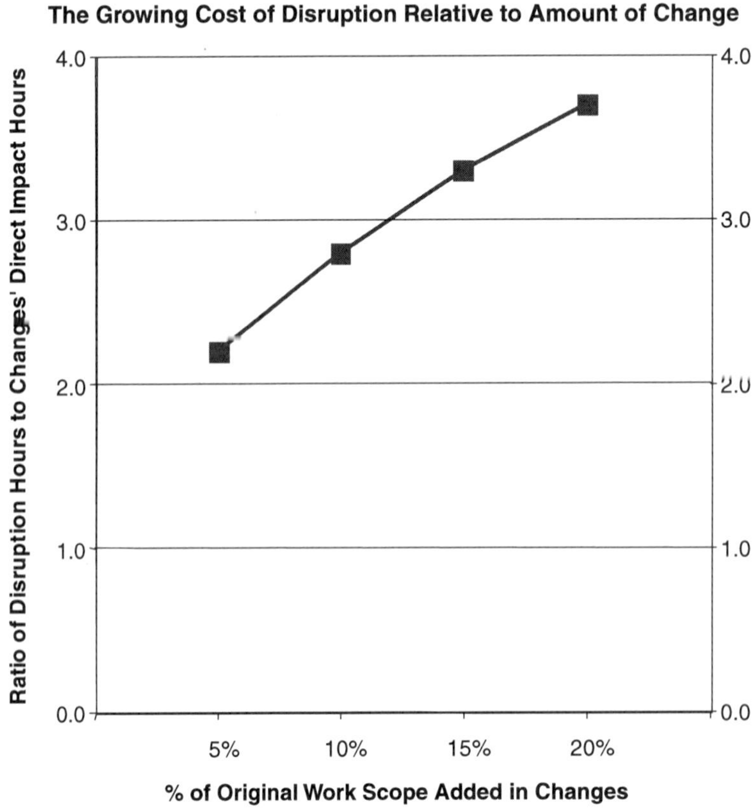

on design (as in the case, say, of building a completely standardized structure) cuts the ratio to 1.6! Eliminating *both* of these significant conditions reduces the disruption ratio down to "only" 1.2.

This illustrates yet another reason why "every case is different": Any variation in sensitivity to the different productivity-affecting factors on a project alters the amount of disruption caused by changes. Indeed, let's examine the full range of direct impacts tested previously (adding changes with direct impacts totaling 5 percent to 20 percent of original budgets), under the alternate skill and design conditions. Figure 31.7 shows the very different picture for disruption impacts in these scenarios. With no sensitivity to labor skill conditions, the remaining paths of impact yield disruption amounts that are substantially lower throughout all the different levels of change—with impact ratios of 1.4 to 2.7, versus the base conditions of 2.2 to 3.7. With no sensitivity to design conditions, the impact range is even lower—from 1.1 to 1.9. Eliminating sensitivity to *both* skill *and* design brings the range of

FIGURE 31.6. QUICK RESOLUTION REDUCES CHANGE IMPACT.

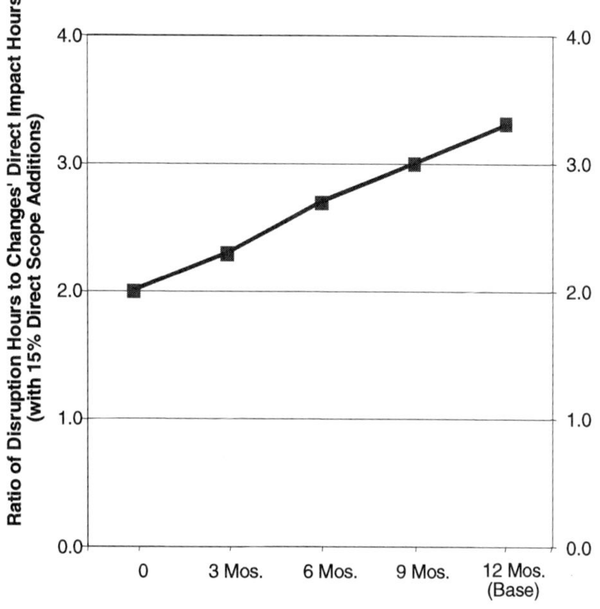

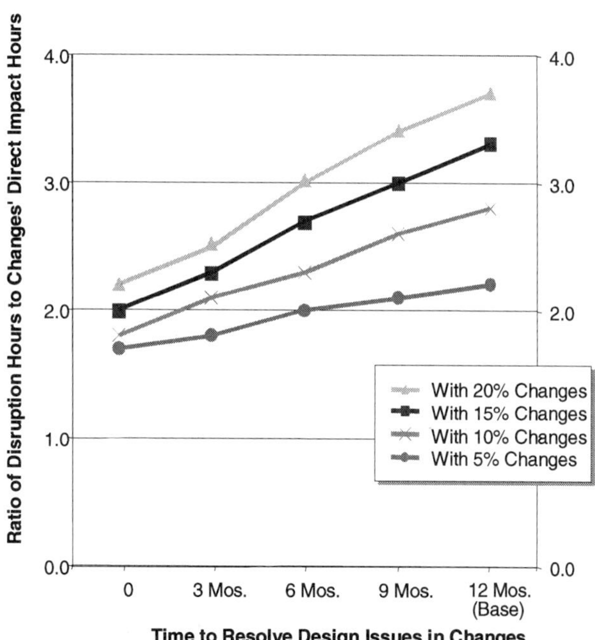

FIGURE 31.6. (*Continued*)

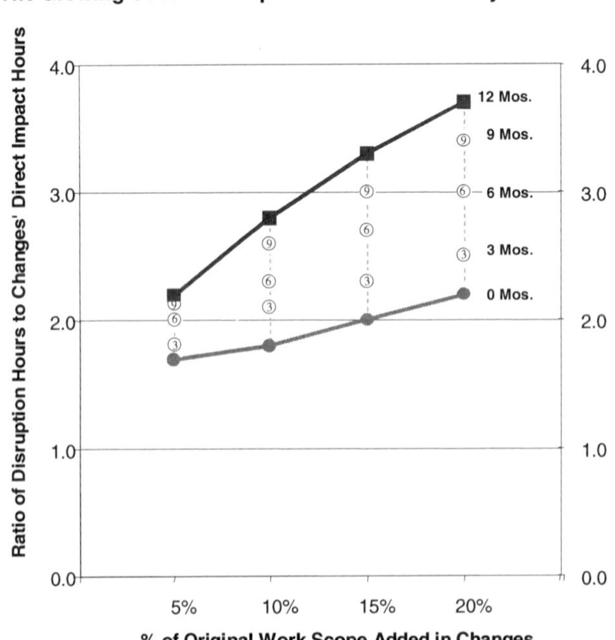

the disruption ratio down to a range of 0.8 to 1.0. These reduced ratios of disruption impact stem not only from the eliminated skill and design effects themselves but also because cumulative impacts through other feedback effects (less crowding, less overtime, and so on) also lessen.

Rule 4: Variations in project conditions drive significantly different disruption impacts.

Dominoes, Acceleration, and Mitigation. Of all the variations in conditions that can cause variation in changes' disruptive impacts, none are more important than the tightness of the schedule. Recall the cumulative cascade of phenomena that can hurt construction productivity—overtime fatigue, skill dilution, out-of-sequence work, congestion, . . . and, especially, late and changing engineering. Each of these phenomena can be significantly aggravated or ameliorated by the tightness of schedule conditions. And, since each can contribute to self-reinforcing cumulative impact, one should expect a significant difference in the domino effect under different schedule conditions. Indeed, this is just what we find in our project simulation analysis results. In this series of analyses, we set identical changes to occur in the

FIGURE 31.7. DISRUPTION COSTS VARY DEPENDING ON PROJECT CONDITIONS.

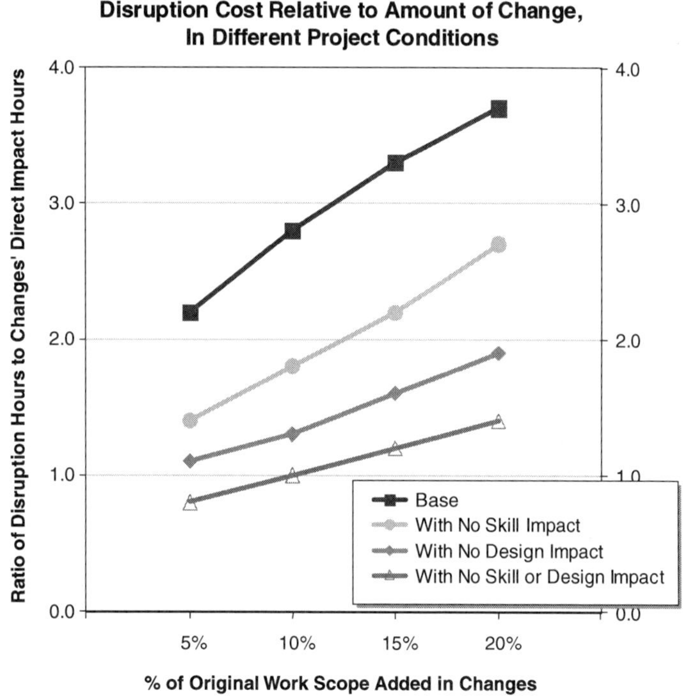

identical project, but under different schedule conditions, ranging from even tighter, with more design-construction overlap, to less tight. Figure 31.8 displays the resulting disruption impact. The even earlier construction start more than doubles the disruption ratio to 7.2 from the base value of 3.3, while the later start of build *drops* the impact ratio 30 percent to 2.3. Combine that later start with an extension in the scheduled completion date, and see a further reduction to 0.7 (i.e., disruption is less than the direct impact).

How tight is your project schedule? Some schedules are intended to be "optimal," a healthy balance between cost and time targets. Other project schedules are acknowledged to be excessively tight—"optimistic," "success-oriented . . ." (If your project schedule is acknowledged to be "loose," you are operating in a rare environment and a distinctly small minority among project managers.) In either case, the addition of changes to the project will tip the schedule toward "tighter" and increase the potential for higher amounts of disruption impact from those changes, as a result of increased hiring and skill dilution, overtime, crowding, and so on.

In a "schedule is supreme" world, it may be viewed initially as career-threatening to talk of schedule extensions. Nevertheless, the addition of work into a tight schedule constitutes a *de facto* acceleration and is likely to trigger the kind of disruptive impacts described

FIGURE 31.8. TIGHTLY SCHEDULED PROJECTS ARE MORE PRONE TO SUFFER DISRUPTION.

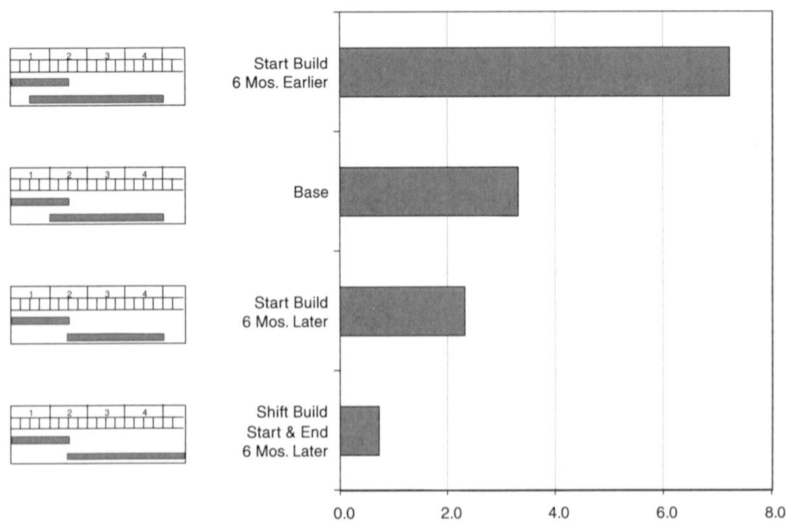

previously. Such impacts can grow costs significantly and may jeopardize that earlier schedule anyway.

Perhaps, then, it is only responsible to explore and to offer to the customer/contractor the mitigating option of *reduced* cost impacts with *extended* schedules. Of the cost-mitigating options available, none may be so highly leveraged as schedule relief. In business markets in which even small delays are very costly to the project customer, it may well be worth the much higher price to adhere to tighter schedules, but it is a rare market for which it is "at any price." In this simulated project, the labor disruption costs difference between starting construction six months earlier versus starting and ending six months later is two million hours . . . on a two-million-hour project budget! Indeed, *each successive construction schedule move (start or target completion) that is equal to the percent of direct impact (in this case 15 percent) cuts disruption cost impact successively by 30 to 70 percent.*

Cost-schedule trade-offs, therefore, should be a part of every discussion of significant change impacts.

Rule 5: Disruption impacts are significantly reduced by less tight project schedules.

Thus, we have seen five rules of changes' disruption impacts:

The Five Rules of Changes' Disruption Impacts

> *Rule 1: Disruption impacts can appear years after the incident change event and can exceed the direct impact of the change.*
>
> *Rule 2: Disruption impacts grow disproportionately over more and more changes.*
>
> *Rule 3: Early issue resolution cuts disruption impacts significantly.*
>
> *Rule 4: Variations in project conditions drive significantly different disruption impacts.*
>
> *Rule 5: Disruption impacts are significantly reduced by less tight project schedules.*

Bear in mind that each time "disruption impact" appears in these rules it means *cost.*

The "impact" means productivity reductions or rework increases that are added project cost, for which someone must pay. Best to resolve the full cost impacts of changes early, for if we don't, it can all go wrong.

When It All Goes Wrong: Disputes

If anyone should doubt the importance of dealing effectively with changes proactively, one only need talk with an executive or project manager who has been through a major contract dispute. For both customer and contractor (or subcontractor), a contract dispute or claim can be the most management-distracting, relationship-damaging, costly process of their careers. Whether ultimately resolved via negotiations, mediation, arbitration, or trial before a board or court of any jurisdiction, the contract dispute over unresolved costs of project change impacts can consume enormous staff and executive time, decimate customer-contractor goodwill, and cost millions of dollars.

The existence of and toll taken by such disputes is testament to the value of timely resolution of change impact costs during a project. When agreement on those costs cannot be resolved by routine processes, at least one party may choose to initiate a claim process. Processes and venues for such claims will have been spelled out in the contract terms.

Regardless of process or venue, the burden of proof is essentially the same: (1) the contractor must establish entitlement to recover for those "changes"; (2) causation of impact must be established; and (3) the full quantum of that impact must be demonstrated. Very much gone are the days of (successful) total cost claims. Between (a) universal disfavor among courts and (b) the use or threat of use of U.S. federal and state False Claims Acts, contractor claims that overreach (by attempting to recover all costs regardless of demonstrated causation) are likely doomed to failure. Indeed, *any* contractor claim that fails to provide adequate proof of liability, causation, and damage is so doomed. The characteristics of an analysis that are needed for a strong case are that it

- explain variations in productivity and rework;
- assess what would have happened under alternate conditions;
- describe the causation that ties effect to precipitating event(s);
- account for and explain cumulative impact of many individual claimed events;
- explain *why and how* productivity would have been affected;

- account for other concurrent events that influenced project performance;
- permit validation; and
- be testable and auditable.

Given the extreme challenge of explaining and quantifying the disruption elements, it is no wonder that many claims involving disruption are resolved at no more than 30 to 40 percent of the claimed value. This may be less a measure of truth and accuracy than an indicator of the difficulty of explaining and quantifying such claims.

Management of Change Process: Excellent Practices

With the many adverse impacts that can occur as consequences of change, it is crucial that the management of change be a high-priority aspect of managing projects. Actual practice, however, varies widely. Some project managers see change as something to be avoided. Others plan and behave as though change simply will not happen to them. Still others view change as an opportunity, a chance to recover on bidding mistakes. The best approach is to accept that change is normal and to be expected. On all but the most standard cookie-cutter projects, changes will happen. The key is not to avoid all changes but to anticipate them, be prepared for them, and react to them properly, with trained staff aided by clear guidelines and processes and systems.

In our research for this chapter, we interviewed dozens of top executives, project managers, attorneys, and consultants in the construction, software, and defense industries. The many interviews we conducted highlighted that change management practices vary significantly, as do the resulting project outcomes. Further, when it comes to change management, the experience of the manager is not a good predictor of success; success requires having the right processes in place and following them.

The process areas that are keys to success in handling change are as follows:

- Managing the contract
- Managing the project
- Managing the contractor-customer relationship

Managing the Contract

"Most mistakes happen before the shovel hits the ground."

LANG, 2002

Change planning is critical during the drafting of the contract documents. Contractor and customer should have an open discussion about change, including how much change is reasonable to expect and how, together, changes will be handled.

Central Contract Review. One key element to successful management of contracts is to have central contract review. Firms that do a good job at managing project changes (in terms of pricing, mitigating impacts, recovering damages, and avoiding disputes) have a strong central coordination, if not control, of the process from beginning to end. Starting from before the project begins, *all* contracts should be reviewed centrally by the legal department. While it is fine to craft custom change clauses as needed, central review ensures compliance with key principles (e.g., reservation of rights, consequential damages, damage-delay clauses).

Equally important is knowing when to walk away. Thorough contract review enables parties to make informed decisions regarding the acceptability of a contract and the risks it would bring.

As part of this whole process, the legal department should develop a standard checklist of all the elements that should appear in the contract. Central review will uncover deviations from the norm and enable informed decision about variances. Indeed, the contractor should assign an attorney to each contract and have that attorney support the contract beginning to end. This attorney should be responsible for all the legal aspects of the contract and stay involved throughout, ensuring familiarity with the project when issues arise. In this way, the attorney can serve as an ongoing resource to the project management team, who should themselves be familiar with the contract terms and conditions.

Specify Planned Conditions. One the first challenges in dealing with change is getting both sides to agree whether a change has occurred. For this reason, it is important to specify clearly in the contract what the planned conditions and dependencies are (Hatem, 2002). With the baseline conditions clearly specified, the emergence of changed conditions is easier to identify.

Consider in Advance How to Deal with Disputes. There are a variety of constructive approaches to resolving disagreements proactively. If both sides can agree in advance on a process with which they are comfortable, much of the tension surrounding change can be alleviated.

- Consider a project "neutral" to review changes. A neutral party who is familiar with the project and knowledgeable about the industry can be invaluable. An unbiased third party can quickly and fairly review proposed changes and their impacts (Shumway, 2002; Sink, 1999).
- Agree how disputes will be escalated. Have an agreed process in place by which disputes are rapidly escalated to the necessary levels of decision-making authority. This prevents issues from lingering too long, which can both breed resentment from both sides and also worsen the impact of the very problem under dispute (Cady, 2002). See the *Disruption* section earlier in the chapter.
- Agree in the contract on whatever alternative dispute resolution (ADR) methods will be employed. In some settings, these can shorten resolution times significantly (Bird, 2002). The two most commonly used methods are mediation (a nonbinding approach that attempts to bring the parties to a settlement with the help of a facilitator) and arbitration (in which a neutral third-party hears their case and provides a ruling that is binding on the parties). ADR can be used to resolve disagreements more quickly and cost-effectively than litigation if both parties are committed. An excellent way to foster this commitment is to establish during the contractual stage a partnering agreement, a pact between con-

tractor and customer to work together proactively. The partnering process is typically kicked off with a contractor-customer retreat and culminates in an agreement describing how the parties will work together, including establishing a fast-track approach to dealing with issues that may arise. These agreements can be effective in fostering proactive communication and setting the right stage for a cooperative working relationship between contractor and customer (Ness, 2002). Consider among the various ADR methods establishing a dispute review board (DRB). If possible, the contractor and customer should pick the DRB as a team. The more traditional we-pick-one-you-pick-one-they-pick-one approach tends to be more adversarial. Instead, work together to pick a team of people who really understand the industry. Involve the team from cradle to grave on the project, and have them invest enough of their time to understand the project thoroughly.

The importance of legal matters continues beyond the contracting. Continue monitoring the legal aspects throughout project execution: Contractors should adhere to notice provisions, be careful about what they sign away, and be clear about what should and should not be in writing.

Finally, ensure that legal talent is seen as part of the team. Project managers often distance themselves from attorneys, viewing their presence as a sign of weakness, that the managers were unable to resolve problems themselves. Instead, Legal should be actively integrated into the project team, and the attorney involved with the project should be viewed as a resource, not as a last resort (Kieve, 2002).

Managing the Project

"Know thy project."

CHIERICHELLA, 2002

Provide Training. Companies that are successful in managing changes make sure the team understands the project and how to identify change. They publish explicit guidelines and standard processes and teach them throughout the company. They train the management team in documentation, letter writing, and handling constructive changes. They provide basic legal education for the project management team. They teach the standard systems that will be employed on each project, evaluate individual lists of responsibilities, and reallocate as deemed appropriate. They provide training on the specifics of the contract and conduct role-playing of possible events (Goff, 2002). They teach key staff not to take verbal instruction from customer representatives without treating it as a possible change (Mountcastle-Walsh, 2002).

Exercise Discipline. Most of the managers and attorneys interviewed emphasized the importance of discipline. Central standards and guidelines may sound bureaucratic and, frankly, boring, but they are helpful in ensuring projects run smoothly. As one executive noted, "new ideas are not needed as much as enforcement of basic 'blocking and tackling' around project change" (Stafford, 2002).

Develop a standard set of processes for projects, train people in the processes, and enforce guidelines to ensure the processes are actually followed. An important element to these processes is to make compliance with the processes as easy as possible. Forms for logging changes, for example, while covering the essentials, should require as little input as possible.

Before the project begins, have a transition session to impart key information from the proposal team to the project team. The proposal manager and project manager should review together all the assumptions used in generating the proposal, including a review of project risks and early warning signs that those risks are developing. Best of all is for the project manager to have been involved in the proposal effort, so the person responsible for executing the project will have been involved in defining it.

Every piece of paper exchanged between customer and contractor should be reviewed for change content. However, even a thorough review of paperwork may miss many changes on a project. Therefore, it is important to monitor constantly for change. If a contractor has confidence in the estimating process employed, then the very fact of overruns in various cost cells might signal that a change occurred. The overrun implies either that there was a misestimation or that something changed—perhaps there were verbal instructions from the customer that never made it into writing, or perhaps the specifications were too vague. Investigate these overruns. Whatever their source, whether changes initiated by the customer, misestimation, or unexpected productivity losses, the contractor should ensure proper diagnosis (Chierichella, 2002).

Find and Process Changes Promptly. Contractors need to provide supervisors in the field the right incentive to identify changes. One successful approach is to create a work task category such as "Pending Items," items that have been identified and for which the customer will be or has been notified, but which have not yet been approved. In earned-value progress reporting, supervisors would get credit for these items, such that their performance measures are not penalized as a result of changes. Contractors should provide the customer with timely notice and then ensure they do not execute any work that is neither authorized as a change nor pending (Newman, 2002).

Assign dedicated personnel for handling the change process. Particularly on those jobs with tight budgets (i.e., those most likely to end up with disputes), there is a tendency to cut back on overhead, but it is exactly these projects that are most likely to end up overrunning and have disputes. In the event of a dispute, daily reporting, schedule updating, and disciplined documentation become especially important and can, in fact, promote the resolution of disputes.

Document Thoroughly

"One reason that you have disputes is that the parties aren't working with the same factual description of what happened. Having the documentation increases the likelihood of reaching a resolution that both sides can be happy with."

KRAFTSON, 2002

To increase the likelihood that both sides are working with the same facts, the following are helpful practices (see Hoffar and Tieder FedPubs, 2002, pp. 1–5, and Currie and Sweeney 1994 for more detailed review of documentation practices):

- Prepare daily time and material sheets, and submit reports to the customer every day. Daily reporting accelerates communication and ensures all are well informed on project status (Currie and Sweeney, 1994, p. 218).
- Log time and cost for every change in a separate cost code. Contemporaneous tracking of costs is far better than a re-creation done later. While even real-time reporting is certain to miss some of the impacts of change, it at least provides a more complete picture of the direct costs.
- Document efficiency losses (e.g., when workers are waiting on material, log their time idle). Comment on inefficiencies at job meetings, and record these in meeting minutes. Contemporaneous tracking of productivity problems will ensure that everyone is aware of all relevant problems (perhaps enabling some mitigation ideas), and it provides a good resource for computing disruption.
- Take pictures of the job site and date them. Photographs (or videos) often can relay the story far more clearly than written documentation. For example, a photograph showing workers crowded into a site is better documentation than a report including comments that congestion has been an issue (Currie and Sweeney, 1994, p. 220).
- The contractor should document the customer's performance as well as their own. It is not only the contractor whose performance can drive cost and schedule. Delayed approvals from a customer or late information can impact contractor performance and should be documented and included in reports to the customer (Goff, 2002).

Managing the Contractor-Customer Relationship

"If you trust the person on the other side of the table, you can work through just about any problem."

DEAN, 2002

Communicate Effectively. A key to success is open, honest, and direct communication between the customer and contractor. It is important for the contractor to invite active customer involvement early in the project. Some contractors, for example, require an open exchange with the customer before the project starts to discuss priorities and processes, including the handling of changes.

Reports from contractor to customer should be regular and frequent, include both the good and the bad, and be complete and fair in reporting the true status of the project. It is important that customers be confident that they are hearing about all issues and that the contractor will alert them quickly and be ready to discuss any problem that might arise. In the same spirit of openness, the contractor might consider keeping the regularly updated schedule for the project open to all, including customer and subcontractors. This openness is an important part of developing a partnership.

When a change is identified, the contractor should communicate with the customer quickly. First, the contractor must inform the customer they believe a change has occurred, or is about to. By providing the customer notice (rather than just acting on the change), the customer is afforded the option to avoid the change. If the customer opts to implement the change, the contractor needs to provide a quick, accurate assessment of the impact (and, for changes that involve significant cost or schedule impact, provide mitigation options as well). Indeed, the contractor and customer might assign staff to review changes together before change orders are written and submitted.

Nearly every manager interviewed noted that speed is essential in managing change—speed in identifying changes, estimating direct and, when possible, productivity impacts, (see the *Disruption* section below earlier in the chapter) defining explicitly the cost-schedule trade-offs for the customer, and obtaining customer concurrence. By moving quickly on change issues, much of the impact of the uncertainty that accompanies change can be mitigated (see the *Disruption* section).

Price Completely

"Add a door, I can tell you what that costs. Change the door three times, and it's hard to quantify."

LAX, 2002

When changes are small and uncomplicated, it is usually straightforward to develop an accurate assessment of the total cost of a change. When changes are numerous, large, and/ or complex, it is much more difficult. While it is important to resolve issues as quickly as possible, it is wiser to provide an informed estimate than to present a number that may increase dramatically later (Grimes, 1989).

Know the project conditions, and understand how the change fits in before trying to estimate the impact (Schwartzkopf, 1995). It is essential to understand the impacted site, as it is in this environment in which the change will take place (Sanford, 2002). Is labor already crowded into the site where the change will add more people? Has the affected work already started? Are people working in a comfortable environment or are they welding in mid-summer heat?

When providing an estimate, contractors must be realistic, practical, and not overstate the costs. Estimate as much of the costs as possible, minimizing the number of reservations (Allen, 2002). In the estimate, consider additional rework that may occur, as well as any anticipated lost productivity (see the *Disruption* section earlier in the chapter). Industry-standard measures arc available, such as MCAA and CII studies, to provide some reasonable benchmarks for estimation. (There have been numerous studies on the impact of change orders and project conditions on labor productivity. See, for example, MCAA, 1994; Thomas and Rayner, 1988; Ibbs and Allen, 1995; Hanna, 1999 et al.; Adrian, 2001). Be cautious, however, as it can be difficult to anticipate the many ways in which a project may be impacted or the degree to which each type of impact will be caused by the changes.

The contractor and customer should review the reasoning for the estimated additional costs. In addition, this is the time to look for offsets that might provide a trade-off in the project scope, or cost-schedule trade-offs that might enable containment of impact costs (see below). With an understanding of the full impact of a change, the parties may even decide to forego it.

The Importance of Getting it Right

Project changes are inevitable. They are essential in order to improve products, adopt new technologies, adapt to market conditions, and accommodate changed or unexpected circumstances. If we as project managers, executives, and customers know to expect changes, we must learn to manage them far better. We need to execute the basic "blocking and tackling" of change planning and monitoring. We need to evaluate alternatives and trade-offs and review them openly between customer and contractor. We need to anticipate and communicate more completely and candidly the likelihood and nature of disruption impacts, and thus the ultimate cost effects on the project. Companies who do so see substantial performance improvement. The failure to do so results in project overruns and failures, damaged customer-contractor relations, resource-consuming disputes, ruined careers, and lost profits and shareholder value.

Acknowledgments

Many executives, consultants, and attorneys generously gave us their time and experience-honed ideas in the dozens of interviews conducted in support of this chapter. Although only a few have been quoted and cited, many others helped shape the ideas and content herein. To all of them we owe our gratitude. In addition, without the contribution of colleagues in our firm and our clients over many years, the insights into and analyses of change impacts summarized here would not have been possible. Of most recent significant help have been Tom Kelly and Sharon Els, who helped us conduct so many of the interviews; Hua Yang, who conducted all of the simulation analyses reported herein; Sheri Dreier and Jane Hemingway, who prepared the graphics for this chapter; and Doris Walsh, our patient and ever-productive assistant, who (many times over) converted bits of notes into the assembled text and charts for the manuscript.

References

Adrian, J. 2001. *Jim Adrian's Construction Productivity Newsletter.* see, for example, Change orders: How they affect time and Cost (Vol. 18, No. 2); The impact of temperature on productivity" (Vol. 19,

No. 5),; and The impact of the loss of learning on construction productivity (Vol. 19, No. 6); among many others.

Allen, S. 2002. Stephen Allen, Washington Group; Boise, Idaho. Personal communication. June 18, 2002.

Bird, K. 2002. Karl Bird, USAF, Wright-Patterson Air Force Base. Personal communication. July 29, 2002

Cady, J. 2002. Jim Cady, Granite Construction, Inc.; Watsonville, California. Personal communication. July 3, 2002

Chierichella, J. 2002. John Chierichella, Fried, Frank, Harris, Shriver & Jacobson; Washington, D.C. Personal communication. August 7, 2002

Cooper, K. 1994 The $2000 hour: How managers influence project performance through the rework cycle. *Project Management Journal.* (March 1994).

Cooper, K.G. and K. S. Reichelt (2002) Quantifying project disruption with simulation. San Antonio: Project Management Institute

Currie, O. A., and N. J. Sweeney. 1994. Prelitigation advice. In *Construction Change Order Claims*, Chap. 12, 215–237, ed. R. F. Cushman and. S. D. Butler. New York: Wiley.

Cushman, R. F. and S. Butler, eds. 1994. *Construction change order claims.* New York: Wiley.

Dean, W. 2002. William Dean, The Clark Construction Group; Bethesda, Maryland. Personal communication. October 3, 2002

Goff, C. M. 2002. Colleen Mullen Goff, Zachry Construction Corporation; San Antonio, TX. Personal communication. July 23, 2002.

Grimes, J. E. 1989. *Construction paperwork: An efficient management system*, 159. Kingston, MA: R. S. Means Company, Inc.

Hanna, A. S., J. S. Russell, E. V. Nordheim, and M. J. Bruggink. 1999. Impact of change orders on labor efficiency for electrical construction. *Journal of Construction Engineering and Management—ASCE* 125(4):224–232

Hatem, D. 2002. David Hatem, Donovan Hatem, Boston. Personal communication. July 24, 2002.

Hoffar, Julian F. and Tieder, John B. 2002. *Proving construction contract damages.* Federal Publications, Inc. Washington, D.C.

Ibbs, C. W. and W. E. Allen. 1995. Quantitative impacts of project change. Source Document 108. Construction Industry Institute; Austin, TX pages 1–46.

Jones, R. M. 2002. Lost productivity: Claims for the cumulative impact of multiple change orders. *Public Contract Law Journal* 31: (Fall, 1).

Kieve, L. 2002. Loren Kieve, Quinn Emmanuel; San Francisco. Personal communication. September 5, 2002.

Kraftson, D. J. 2002. Daniel J. Kraftson, Jenkens & Gilchrist; Washington D.C. Personal communication. July 15, 2002.

Lang, R. 2002. Roger Lang, Turner Construction; New York. Personal communication. June 14, 2002.

Lax, P. 2002. Paul Lax, Lax & Stevens; Los Angeles. Personal communication. June 18, 2002.

MCAA. 1994. Change orders, overtime, and productivity. Mechanical Contractors Association of America; Rockville, MD. Publication M3.

Mountcastle-Walsh, H. 2002. Harriet Mountcastle-Walsh, Honeywell International; Columbia, MD. Personal communication. July 18, 2002.

Myers, J. M. 1994 Changes resulting from delays. In *Construction Change Order Claims*, ed. R. F. Cushman and S. D. Butler. 215–237. New York: Wiley.

Ness, A. 2002. Andy Ness, Thelen Reid & Priest; Washington D.C. Personal communication. August 8, 2002.

Newman, J. 2002. Joe Newman, Bechtel Corporation; San Francisco. Personal communication. September 26, 2002.

Sanford, J. 2002. Jim Sanford, Northrop Corporation; Los Angeles. Personal communication. July 25, 2002

Schwartzkopf, W. 1995. Calculating lost labor productivity in construction claims," 125–130. Frederick, MD: Aspen Law and Business.

Shumway, R. 2002. Ron Shumway, KPMG; San Francisco. Personal communication. August 7, 2002.

Sink, C. M. 1999. Ten ways to improve the contract claims process. *Water Environment & Technology* (April).

Stafford, T. 2002. Trevor Stafford, Fluor Corporation; Aliso Viejo, CA. Personal communication. July 29, 2002

Thomas, H. R., and K. A. Rayner. 1988. *The effects of scheduled overtime and shift schedule on construction craft productivity*. Austin, TX: Construction Industry Institute, Source Document 98.

Thomas, M. E. 2002. Mary Edith Thomas, Harris Corporation; Melbourne, FL. Personal communication. July 9, 2002

SECTION II.4

CONTROL

INTRODUCTION

Project control represents more than the simple evaluation of project performance. In fact, the control cycle—as described in Chapter 1 by Pete Harpum—is typically set up around a four-stage, recurring process that first challenges us to establish targets, then measure performance, compare actual results with intended goals, and make necessary adjustments.

While this model offers a useful conceptual backdrop, in practice we find ourselves dealing with some aspects of project management that are seemingly as old as the hills (somewhere between 50 and 5,000 years anyway!) but also some newer concepts, like quality management or value management. Thus, readers will find that embedded in these chapters are not only well understood concepts but also important, cutting-edge work that is having a profound impact on the manner in which many organizations are managing projects today.

Chapter 32 begins conventionally enough. Asbjørn Rolstadås takes us, with some rigor, through the principles of project time and cost control. Gantt charts and critical path are explained together with the use of contingencies in project estimating.

Larry Leach, in Chapter 33, extends Asbjørn's text by offering a detailed review of critical chain project management. Beginning with Goldratt's "theory of constraints" (Dettmer 1997)—which, as he points out, isn't really a theory as such—Larry explains how one of the limitations of critical path planning, resource leveling, can be treated as a constraint, and how this constraint can be exploited by more sophisticated statistical modeling of activity times. This is done through project buffering. The implications of this methodology are then reviewed, including performance measurement (see Chapter 34), resourcing, and decision making. Multiproject critical chains and organization-wide critical chain management are then reviewed, with particular emphasis on the behavior changes that will be required by managers.

Reporting performance is obviously a key part of project control, but as anyone who has tried to do this soon realizes, it is not easy. The essential challenge is what measures to use and how to report on these in an integrated way. Dan Brandon tackles this head-on in Chapter 34. Beginning, like Peter Harpum in Chapter 1, from the control cycle view, Dan covers basic schedule and cost reporting before moving on to earned value. He then broadens the discussion to look again at the question of project success (previously covered in Chapter 5) and how to measure this concept. As we've seen before (George Steel, for example, in Chapter 29), the project values and drivers need to dictate the measures that will be employed. Dan ends by discussing the balanced scorecard approach as a framework for project performance measurement, integrating earned value and project success measures within this.

Risk management has already been expertly reviewed by Steve Simister in Chapter 2. Stephen Ward and Chris Chapman, two of the leading scholars in the field of risk management, take us on a second, more critical discussion of the topic, however, in Chapter 35, suggesting in the process that we probably should be thinking of it now rather as uncertainty management than risk management. In doing so they touch on many of the issues we have recently been discussing: estimating business benefit, design and technology risk, statistical probability, and the selection of performance measures, among others.

Value management (VM), the process of formally optimizing the overall approach to the project (including whether or not it should be done) could quite legitimately have been placed in the strategy section (II.1) but insofar as technology and procurement, among many other things, need addressing before VM can really be brought to bear, we decided to delay it till these topics had been discussed. Michel Thiry discusses VM in Chapter 36. He begins by discussing what is meant by value and by defining the various terms used in VM (value engineering, value analysis, etc.). VM is positioned as a strategic process comprising sensemaking, ideation, elaboration, choice, and mastery. Techniques within each of these, such as function(al) analysis, are described. Overall, Michel takes an ambitious view of VM, positioning it as "the method of choice to deal with the ambiguity of stakeholders' needs and expectations and the complexity of changing business environments at program level and project initiation."

Quality management (QM) is a subject that bears both on strategic planning and operational control. Martina Hueman, in Chapter 37, traces the evolution of quality management from quality control to Total Quality Management before looking at quality management in project management. This she does under the headings of certification, accreditation, the quality "Excellence" model, benchmarking, audits and reviews, coaching and consulting, and evaluation. Quality standards for projects and programs are then discussed in detail, with reference to engineering, construction, and IT/IS. As she points out, typically much of the project management focus tends to end up being about product quality; there is, however, opportunity for it to be applied to project management processes and practices—and people—as she shows, not least in discussion of QM on organizational change projects. Martina's discussion is particularly valuable on the role of management audits and reviews, showing how these can lead to improvements in project management competency. The chapter ends by noting the role of the project management office, the topic of the next chapter, in supporting this process.

James Young and Martin Powell review the project management support office in Chapter 38. Initially the project office was often seen as a project status reporting unit. Increasingly, however, it has become a "home" for project management and a center for project management excellence. The PMSO may be found at three levels in an organization: at the corporate/enterprise level, at the business unit level, and at the project/program level. It has particular applicability in organizations where there is considerable virtual working. Its activities cover portfolio, program, and project support; enterprise-wide project management support (resource planning, communications, benchmarking, performance measurement); competency development; and support in tools and techniques. James and Martin walk the reader through examples of how the PMSO might provide support in each of these areas. They conclude the chapter with a discussion of how PMSO effectiveness can be measured.

About the Authors

Asbjørn Rolstadås

Asbjørn Rolstadås is professor of production and quality engineering at the Norwegian University of Science and Technology. His research covers topics such as numerical control of machine tools, computer-aided manufacturing systems, productivity measurement and development, computer-aided production planning and control systems, and project management methods and systems. He is a member of the Royal Norwegian Society of Sciences, the Norwegian Academy of Technical Sciences, and the Royal Swedish Academy of Engineering Sciences. He serves on the editorial board of a number of journals, and is the founding editor of the International Journal of Production Planning and Control. He is past president of IFIP (International Federation for Information Processing). He is also past president of the Norwegian Computer Society. He is currently the head of the Norwegian Centre for Project Management. He has done studies of project execution of major governmental projects, mainly within development of oil and gas in the North Sea, research on risk analyses and contingency planning in cost estimates, and developed training courses in project planning and control using e-learning.

Larry Leach

Larry Leach is the president of the Advanced Projects Institute (API), a management consulting firm. API specializes in project management, including leading the implementation of the new critical chain method of project management. Larry supports many companies large and small with diverse projects, ranging from R&D to construction. Larry developed and operated the project management office for the American National Insurance Company (ANICO) in Galveston, Texas, performing large IT projects. He has worked at the vice president level in several Fortune 500 companies, where he managed programs of large and small projects of many types. Prior to that, Larry successfully managed dozens of projects, ranging up to one billion dollars. Larry has master's degrees in both Business Management from the University of Idaho and in Mechanical Engineering from the University of Connecticut. He was awarded membership in Tau Beta Pi, the Engineering honorary society, while earning his undergraduate degree in Engineering at Stevens Institute of Technology. Larry is a member of the Project Management Institute (PMI) and a certified Project Management Professional. His has published many papers on related topics, including a *PMI Journal* article on critical chain in June 1999 and a pair of papers published in *PM Network* in Spring 2001. He presents seminars for PMI and authored the recently published book *Critical Chain Project Management.* Larry also serves as faculty for the University of Phoenix, facilitating courses in business management.

Dan Brandon

Dr. Daniel Brandon is a Professor and the Department Chairperson of the Information Technology Management (ITM) Department at Christian Brothers University (CBU) in Memphis, Tennessee. His education includes a BS in Engineering from Case Western University, an MS in Engineering from the University of Connecticut, and a PhD from the University of Connecticut, specializing in computer control and simulation. He also holds a PMP (Project Management Professional) certification. His research interest is focused on software development, both on the technical side (analysis, design, and programming) and on the management side. In addition to his eight years at CBU, Dr. Brandon has over 20 years' experience in the information systems industry, including experience in general management, project management, operations, research, and development. He was the Director of Information Systems for the Prime Technical Contractor at the NASA Stennis Space Center for six years, MIS manager for Film Transit Corporation in Memphis for ten years, and affiliated with Control Data Corporation in Minneapolis for six years in several positions, including Manager of Applications Development. He is also an independent consultant and software developer for several industries, including finance, transportation/logistics, medical, law, and entertainment.

Stephen Ward

Stephen Ward is Professor of Risk Management at the School of Management, University of Southampton, UK. He holds a BSc in Mathematics and Physics (Nottingham), an MSc in Management Science (Imperial College, London), and a PhD in developing effective models in the practice of operational research (Southampton). He is a member of the PMI and a Fellow of the UK Institute of Risk Management. Before joining Southampton University, he worked in the Operational Researchgroup at NatWest Bank. At Southampton he was responsible for setting up the school's MBA program and he is now director of the school's master's program in Risk Management. He founded and edited for ten years the Operational Research Society's quarterly publication *OR Insight*, which continues to publish articles on the application of management science. Professor Ward's teaching interests cover a wide range of management topics, including decision analysis, managerial decision processes, insurance, operational and project risk management, and strategic management. His research and consulting activities relate to project risk management systems and the management of uncertainty. He has published a range of papers on risk management and coauthored two books (with Chris Chapman): *Project Risk Management* (Wiley, 2nd ed., 2003) and *Managing Project Risk and Uncertainty* (Wiley, 2002). The latter text provides a case based treatment of key issues in uncertainty management using constructively simple forms of analysis. His latest book *Risk Management Organization and Context* (Witherby, 2004) takes an organization-wide approaches to integrated risk management, building on emergent issues in project risk management.

Chris Chapman

Chris Chapman has been Professor of Management Science, University of Southampton, since 1986. He is a former Head of the Department of Accounting and Management Science (1984–91) and Director of the School of Management (1995–98). He was founding Chair, Project Risk Management Specific Interest Group, Association for Project Management 1986–91, President of the Operational Research Society (1992–93), and a panel member, Business and Management Studies, HEFCE Research Assessment Exercise, 1992 and 1996. He was elected Honorary Fellow of the Institute of Actuaries, 1999. His consulting and research have focussed on risk management since 1975. He undertook seminal work as a consultant to BP International, developing project planning and costing procedures for their North Sea operations, 1976–82, adopted worldwide by BP. The new ideas associated with this work were developed and generalized on a range of assignments in the United States, Canada, and the United Kingdom for many major clients. Publications include 55 books (4 of which were coauthored), 15 book chapters (7 of which were coauthored), and 40 refereed academic journal papers (32 of which were coauthored), and one paper, published in 1985, received the ORS President's medal in JORS. His most recent book is *Project Risk Management: Processes, Techniques and Insights*, with Stephen Ward (Wiley, 1997).

Michel Thiry

Michel Thiry is managing partner of Valeuse Ltd, a European-based organizational consultancy. He has 30 years' experience in project management in North America and Europe and has worked in Canada, the United States, Australia, the United Kingdom, and continental Europe. He is currently adjunct professor for the Lille Graduate School of Management (France) and seminar leader for PMI Seminars World. He is also visiting professor at UIS (Australia) and external lecturer at Reading University (UK). He holds a MSc in Organizational Behaviour from the School of Management and Organizational Behaviour, University of London and is currently reading for a PhD at University College London. He regularly speaks and publishes at the international level. He has also provided value and project or program management expertise to major organizations, in various fields, including construction, pharmaceutical, IT and IS, telecom, water treatment, transportation (air and rail), and others. He has authored the book *Value Management Practice* and coauthored the "Managing Programmes of Projects" chapter in the *Gower Handbook of Project Management,*

3rd Edition. He also authored the program management and value management chapters in *Project Management Pathways*, published by the Association for Project Management (UK) in early 2003. In addition, he regularly writes and reviews for *the International Journal of Project Management*. Mr. Thiry is also past Director of the Project Management Institute's Montreal and UK chapters and past President of the European Governing Board for Value Management Certification and Training, based in Paris.

Martina Huemann

Dr. Martina Huemann holds a doctorate in project management from the Vienna University of Economics and Business Administration. She also studied business administration and economics at the University Lund, Sweden and the Economic University Prague, Czech Republic. Currently she is assistant professor in the Project Management Group of the Vienna University of Economics and Business Administration. There she teaches project management to graduate and postgraduate students. In research she focuses on individual and organizational competencies in project-oriented organizations and project-oriented societies. She is visiting fellow of the University of Technology, Sydney. Martina has project management experience in organizational development, research, and marketing projects. She is a certified project manager in accordance with International Project Management Association (IPMA) certification program. Martina organizes the annual PM Days research conference and the annual PM Days student paper award to promote project management research. She contributed to the development of the pm baseline—the Austrian project management body of knowledge—and is a board member of Project Management Austria.. Further, she is assessor of the IPMA Award and trainer of the IPMA advanced courses. Martina is trainer and consultant of ROLAND GAREIS CONSULTING. She has experience with project-oriented organizations of different industries and the public sector. Martina is specialized in management audits and reviews of projects and programs, and human resource management issues like project management assessment centers for project and program managers.

James Young

James Young is a senior consultant with INDECO Ltd. He is highly experienced in the development of organizational and individual project management competencies, as well as advising companies on strategies for the successful delivery of portfolios, programs, and projects. He has undertaken a number of high-profile initiatives for major blue-chip organizations. He brings with him experience from work undertaken across Europe, Scandinavia, South America, and the United States. Earlier in his career James worked in the UK construction industry. James has authored a number of published articles on various aspects of project management. His research has been presented at the World Congress for Project Management, the Project Management Institute's European Conferences, and the International Project Managers Association Conference. James has a first-class degree from UMIST.

Martin Powell

Martin Powell is a Managing Director of Cambridge Management and Research, a consultancy specializing in project management and performance improvement practices. Prior to this Martin was a management consultant at INDECO International Management Consultants. He is responsible for the delivery of assignments in both the pharmaceutical and oil and gas sectors. He has been working with a major pharmaceutical company developing a global implementation strategy for a project management support office, as well as providing strategic advice to a number of companies looking to develop one. He leads teams in the development of tailored guidelines and also supports wider initiatives in knowledge management and communications. Martin has also worked as a Project Engineer for Impresa Federici S.p.a., an Italian Engineering Company in Rome, and as a Project Manager for Ove Arup & Partners, an engineering design consultancy, where he was responsible for the delivery of several high-profile projects in Italy, Spain, and Asia. He also worked with many of leading architects, such as Richard Rogers, Renzo Piano, and Zaha Hadid, supporting various master planning submissions for design competitions. Martin has a degree in Civil Engineering from the University of Dundee. He also took electives in Spanish at St. Andrews University. He also obtained a scholarship to study at Stevens Institute of Technology in Hoboken, New Jersey.

CHAPTER THIRTY-TWO

TIME AND COST

Asbjørn Rolstadås

Time and cost are two important planning and control variables in a project. They are interdependent. For example, an acceleration of a schedule may lead to reduced productivity on the work carried out and thus to increased costs or it may require resources that only are available at extra costs (such as overtime). A delay or a prolongation of a project may also involve extra costs to carry the project management and administration for the extra time.

In this chapter scheduling will first be discussed without taking resource constraints into account. The traditional network scheduling techniques such as CPM (critical path method) and PERT (program evaluation and review technique) will be explained by use of examples. Then techniques for handling resource constraints will be briefly discussed.

Scheduling Representation

All projects will have to comply with a deadline for their finish. For a project owner, the deadline may be set to comply with the needs of the results of the project. Quite often the deadline is important in the project's overall profitability analysis. An extension to the deadline may in the worst case turn a profitable project into a nonprofitable one. Deadlines may also be set by external conditions such as weather. As an example, offshore oil platforms for the North Sea are fabricated on shore and can only be towed to field during the summer because of the risk of bad weather in the winter time. For a contractor the deadline is usually a part of the contractual conditions and may involve penalties if not met.

This section of the chapter discusses how it is decided when each single activity of the project should be executed in order to meet a predetermined deadline or when a project can be expected to be finished given that the duration of each activity is known. This is referred to as scheduling and represents the process of determining when an activity should start and when it should finish.

There is a rich literature available on the scheduling of projects. Almost all discuss bar charts and networks. The difference is how deep into network techniques they go and whether or not they consider resource constraints. Some general references are given at the end of the chapter.

In scheduling a distinction is made between the following:

- Activities
- Events

An *activity* is defined as a number of job assignments that requires resources to be accomplished. An *event* is a point in time where an activity starts or ends. Hence, there is a duality between activities and events. For any activity, there are two associated events (start and stop), and for any two adjacent events, there is one, and only one, activity connecting them.

Milestones are a special type of events. A *milestone* represents an event against which achievement is measured. Milestones are used to control progress at a high level. They are selected as major completion points or decision points. A milestone that represents a decision point is also referred to as a *decision gate* or a *tollgate*. The idea is that before the project is allowed to proceed, the gate must be passed. The gatekeeper checks that necessary documentation exists and that the required preceding tasks are satisfactory completed.

The most widespread scheduling tool is the Gantt chart (or bar chart). Henry Gantt, who tried to schedule logistics for the war front in World War I, invented it. The Gantt chart is very simple. It is a rectangular diagram with time on the horizontal axis and activities on the vertical axis. In the diagram, bars show the timing and duration of activities. An example is given in Figure 32.1.

The Gantt chart is unsurpassed in its efficient way of communicating a schedule. Any reader can understand and read the diagram without any prior training or knowledge. However, there are a few drawbacks with the diagram:

- It does not include information on resource level for the activity. It just states when the activity starts and when it finishes.
- It does not include any information on precedence relationships.

For example, with reference to Figure 32.1, if the activity "deck mating" should start earlier, the diagram does not show that "GBS fabrication" and "deck fabrication" must also be moved forward. However, there is a revised version of the original Gantt chart, referred to as a *linked Gantt chart*, that includes this information.

The Gantt chart can also be used for monitoring schedule progress, as shown in Figure 32.1. The open bars indicate unfinished activities. As the activity progresses, the bar is filled. A vertical (dashed) line indicates the present day. A quick look at the diagram reveals crucial progress information. In the figure, it is easily detected that "GBS fabrication" is seriously behind schedule.

FIGURE 32.1. EXAMPLE OF A GANTT CHART.

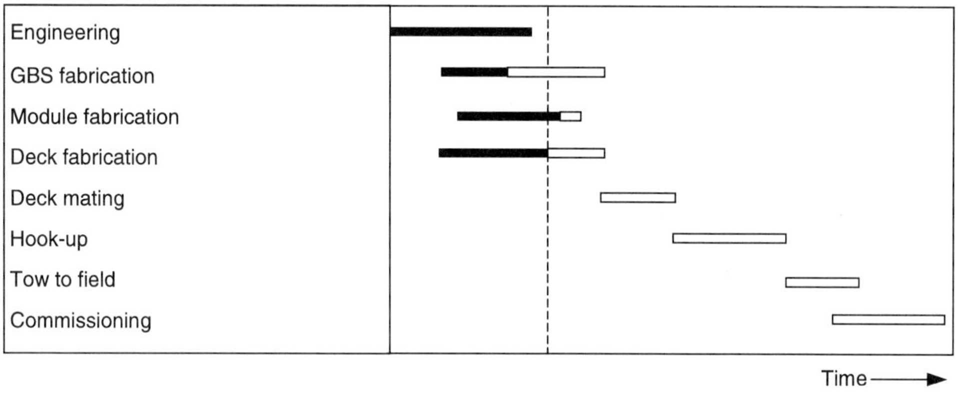

Engineering
GBS fabrication
Module fabrication
Deck fabrication
Deck mating
Hook-up
Tow to field
Commissioning

Time ⟶

Network Representation

Even though a linked Gantt chart can include precedence relationships between activities, it is an unsuitable tool if these relationships grow beyond a certain complexity. In such situations, the scheduling should be carried out by a network technique. However, the presentation of the schedule to the project participants may still be done in a Gantt chart format. Most project control packages available offer this flexibility.

There are two types of network representations:

- *Activity on arrow* (AOA)
- *Activity on node* (AON)

Figure 32.2 shows an example of an AOA representation. The arrows of the graph represent the activities. The arrows also define the precedence relationships, for example, "D and E must be finished before H may start." The nodes in the AOA network represent events.

In AON representation the very same project is shown in Figure 32.3. The nodes of the graph represent the activities. The arrows give only precedence relationships. Events are not shown in AON networks.

In order to design a network, basically three types of information are needed:

- A list of all the activities
- A list of all precedence relationships
- An estimated duration of each activity

The list of activities and the precedence relationships define the structure of the network. Table 32.1 gives an example of these data for the network shown in Figures 32.2 and 32.3. The precedence relationship is given by defining the immediate preceding activities. For

FIGURE 32.2. EXAMPLE OF AN AOA NETWORK.

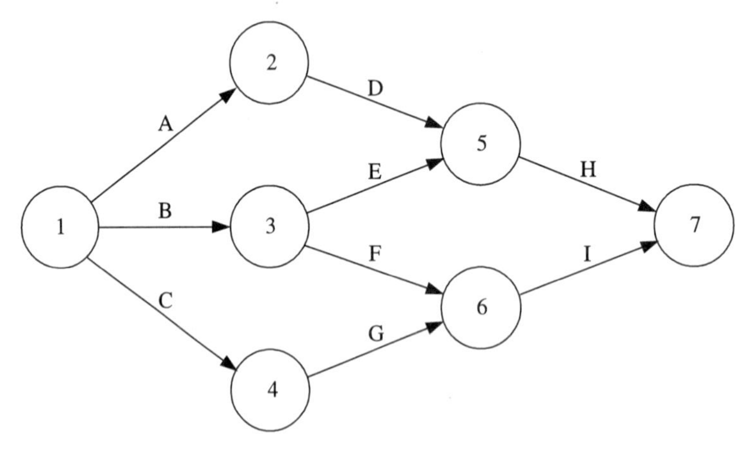

FIGURE 32.3. EXAMPLE OF AN AON NETWORK.

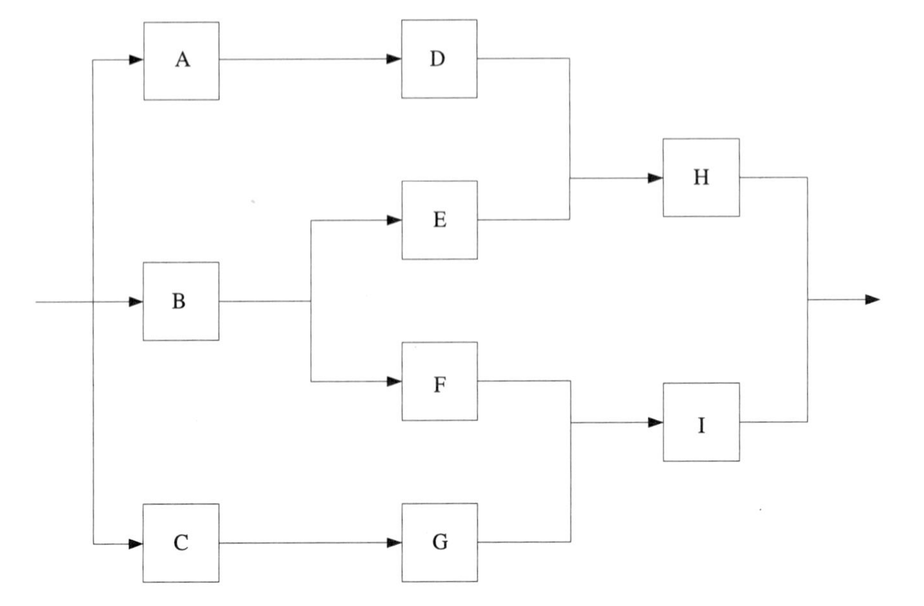

TABLE 32.1. DATA DEFINING STRUCTURE OF NETWORK IN FIGURES 32.2 AND 32.3.

Activity	A	B	C	D	E	F	G	H	I
Preceding activity	—	—	—	A	B	B	C	D, E	F, G

example, for activity H, the preceding activities are D and E. This means that D and E must be finished before H can start.

Constructing the Network

During project scoping, a work breakdown structure (WBS) is usually developed. During scheduling, each work package normally represents one activity in the network. The schedule serves as a baseline for time control and is usually referred to as a *control schedule*. The one that is approved for execution start is referred to as *the master control schedule* (MCS).

When the project execution starts, invariably there will soon be a need for changing and updating the schedule. Since the control schedule is a reference document for monitoring and control, it should at any time be as realistic as possible. This requires that it be kept updated. Usually it is updated on a periodic basis, depending on variability of the schedule and the need for up-to-date control information. The reference schedule at any time is referred to as the *current control schedule* (CCS). Sometimes a number is included to indicate the update number; for example, CCS (3) is the third update of the schedule.

Since the schedule is directly connected to the WBS, aggregate schedules may be obtained by aggregation to a higher WBS level than the work package. For example if the WBS levels are

$$\text{Project} \rightarrow \text{Contract package} \rightarrow \text{Work package}$$

a master schedule with contract packages as activities may be derived from the CCS by aggregation.

The purpose of a network scheduling is twofold:

- To determine when the project will finish (its duration)
- To determine which activities directly influence the project duration

Network scheduling was developed during the 1950s. Two techniques were developed in parallel:

- Program evaluation and review technique (PERT)
- Critical path method (CPM)

The main difference between the two techniques is how the duration of the activity is estimated. In PERT networks the duration is a stochastic (uncertain) variable following a statistical distribution. In CPM, the duration is deterministic (certain). PERT is therefore able to handle uncertainty and may give answers to questions like "What is the probability of finishing the project by March 12" or "What date should be given for milestone X if there should be more than 90 percent probability of meeting it." CPM does not handle uncertainty. It assumes that the duration is a fixed number.

CPM Networks

CPM calculates a network in several steps:

- For each event:
  - *Earliest possible time.* The earliest possible time the event can occur
  - *Latest possible time.* The latest possible time the event can occur
- For each activity:
  - *Early start* (*ES*). The earliest possible start time for the activity
  - *Early finish* (*EF*). The earliest possible finish time for the activity
  - *Late start* (*LS*). The latest possible start time for the activity
  - *Late finish* (*LF*). The latest possible finish time for the activity
  - *Float* (*FL*). The amount of time an activity may be delayed compared to early start without jeopardizing the project deadline

The calculation of events is done in a forward (for early times) and a backward (for late times) "pass." In the forward pass all possible activities leading into an event are examined and an event time is calculated along each of them. The maximum value is chosen. Early event time for event i can be computed as:

$$e_i = \max_j(e_j + t_k)$$

$$j \in PE_i, \; k \in PA_i$$

where:

e_i is early event time for event i.
t_k is duration of activity k.
PE_i is set of directly preceding events to event i.
PA_i is set of activities having event i as finish event.

For the backward pass, the late event times are determined in a similar way:

$$l_i = \min_j(l_j - t_k)$$

$$j \in SE_i, \; k \in SA_i$$

where:

l_{Ii} is late event time for event i.
SE_i is set of all directly succeeding events to event i.
SA_I is set of all activities having event i as its start event.

Figure 32.4 shows the results of the calculations for the network in Figure 32.2. The activity durations are taken from Table 32.2 and are shown on the network in parentheses after the activity name. How durations can be calculated is shown in the *Resource Constraint* section later in the chapter.

For example, for the calculation of early event time of event 6, it is realized that there are two preceding events (3 and 4) with early event time of 7 and 12, respectively. Following activity F from event 3 to 6 would give an early event time of 7 + 13 = 20 for event 6. Following G from event 4 to 6 would likewise give 12 + 16 = 28. The larger number of 20 and 28 (28) is then selected as the early event time for event 6, since the event can occur only when both activities F and G are finished.

For the activities the following calculations are done:

FIGURE 32.4. CPM CALCULATIONS FOR AN AOA NETWORK.

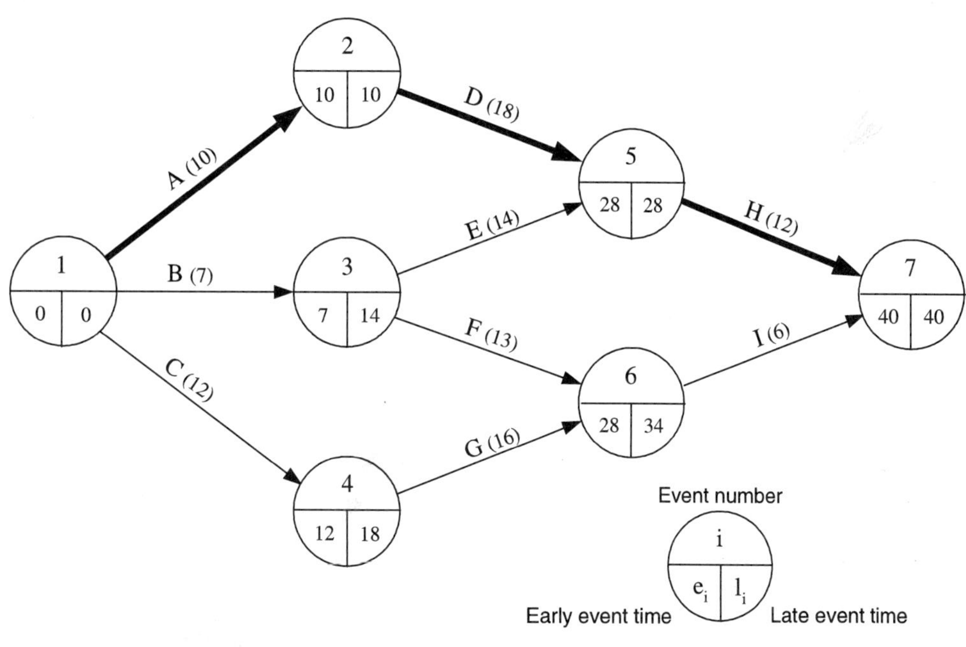

TABLE 32.2. ACTIVITY DURATIONS FOR NETWORK IN FIGURE 32.4.

Activity	A	B	C	D	E	F	G	H	I
Duration (weeks)	10	7	12	18	14	13	16	12	6

$$ES_i = e_e$$

$$EF_i = ES_i + t_i$$

$$LF_i = l_l$$

$$LS_i = LF_i - t_i$$

$$FL_i = LF_i - EF_i$$

where:

e_e is earliest possible time for start of event for activity i.
l_l is latest possible time for finish of event for activity i.

The calculations for the network in Figure 32.4 are shown in Table 32.3.

The term "float" may need some further explanation. With reference to Figure 32.4, there is a time window for execution of activity G limited by the early time of event 4 (12), and the late time of event 6 (34). The time window spans a total of 34 − 12 = 22 days. The duration of G is 16 days, which leaves a surplus of 22 − 16 = 6 days. These 6 days represent a freedom in scheduling of activity G. The start of G may be delayed by up to

TABLE 32.3. CALCULATION OF ACTIVITY TIMES FOR NETWORK IN FIGURE 32.4.

Activity	Duration	ES	EF	LS	LF	FL
A	10	0	10	0	10	0
B	7	0	7	7	14	7
C	12	0	12	6	18	6
D	18	10	28	10	28	0
E	14	7	21	14	28	7
F	13	7	20	21	34	14
G	16	12	28	18	34	6
H	12	28	40	28	40	0
I	6	27	33	34	40	6

six days without affecting the finish time of the project. Activities with no such freedom (float) are denoted critical activities. A chain of critical activities from start to finish in the network is called the *critical path.* In Figure 32.4 the critical path is A-D-H.

The concept of critical activities is quite interesting. It draws the attention of the project manager to the activities that need the closest monitoring. Any delay of a critical activity leads to an equivalent delay of the total project.

Usually activities are scheduled for start as early as possible (ES). If this is the case, the entire float can be taken up by a delay without affecting the finish time of the project. However, it must be kept in mind that the float is valid for a chain and not a single activity. Again with reference to Figure 32.4, the chain of activities C and G together have a float of 6 days. That means that the sum of delays for C and G may run up to 6 days without affecting the project finish time.

Analysis of float is a particularly neat tool for calculating consequences of schedule variance. An example (with reference to Figure 32.4 and Table 32.3) may help to understand this. Assume the following field report with respect to schedule:

- B will be delayed by 4 days.
- D will be delayed by 1 day.
- E will be delayed by 5 days.
- G will be delayed by 3 days.

It is recognized that D is critical. Hence, a delay of at least 1 day to the overall project is unavoidable. Activity G has a float of 6 days. Since no other activity on that chain has a delay, the float will accommodate the 3-day delay of G, and this delay will therefore not influence the project finish date. Further, B and E are both on the same chain. The float along this chain is 7 days, and the total delay is 4 + 5 = 9 days. This means a two-day delay of the project. In conclusion the project will be delayed by 2 days and B-E-H will be the new critical path. A-D will have a float of 1, and C-G a float of 4.

The CPM calculation of an AON network is similar to that of an AOA network, but the calculation of events is omitted. Figure 32.5 shows these calculations for the same network (first shown in Figure 32.3).

The calculations are done in two passes as for event calculation of an AOA network. In the forward pass, all early start and finish times are determined:

$$ES_i = \max_j(ES_j + t_j)\ j \in PA_i$$

$$EF_i = ES_i + t_i$$

where PA_i now is defined as the set of all directly preceding activities of activity i.

For the backward pass the following calculations are done:

FIGURE 32.5. CPM CALCULATIONS FOR AN AON NETWORK.

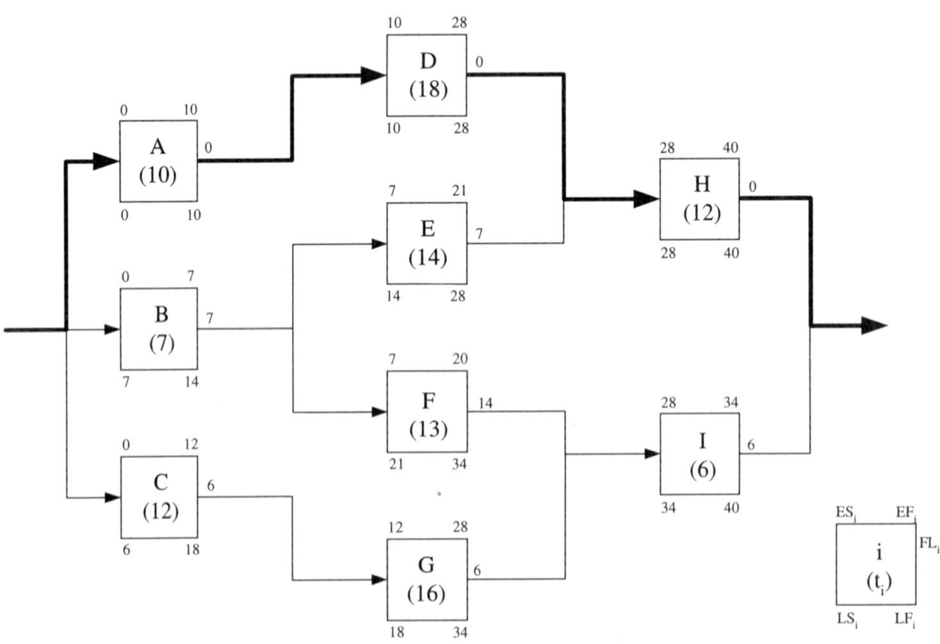

$$LF_i = \min_j(LF_j - t_j)\ j \in SA_i$$

$$LS_i = LF_i - t_i$$

where again SA_i now denotes the set of all directly succeeding activities to activity i.

Float is calculated as before.

Precedence Networks

In traditional network theory only one type of precedence relationship applies: "Activity B may start as soon as activity A is finished." This is called a "finish to start" precedence relationship. The finish of one activity is related to the start of some other activity.

Naturally, there may in principle be four different types of precedence relationships, as indicated in Figure 32.6:

- Finish to start (FS)
- Start to start (SS)
- Finish to finish (FF)
- Start to finish (SF)

FIGURE 32.6. PRECEDENCE RELATIONSHIPS BETWEEN TWO ACTIVITIES.

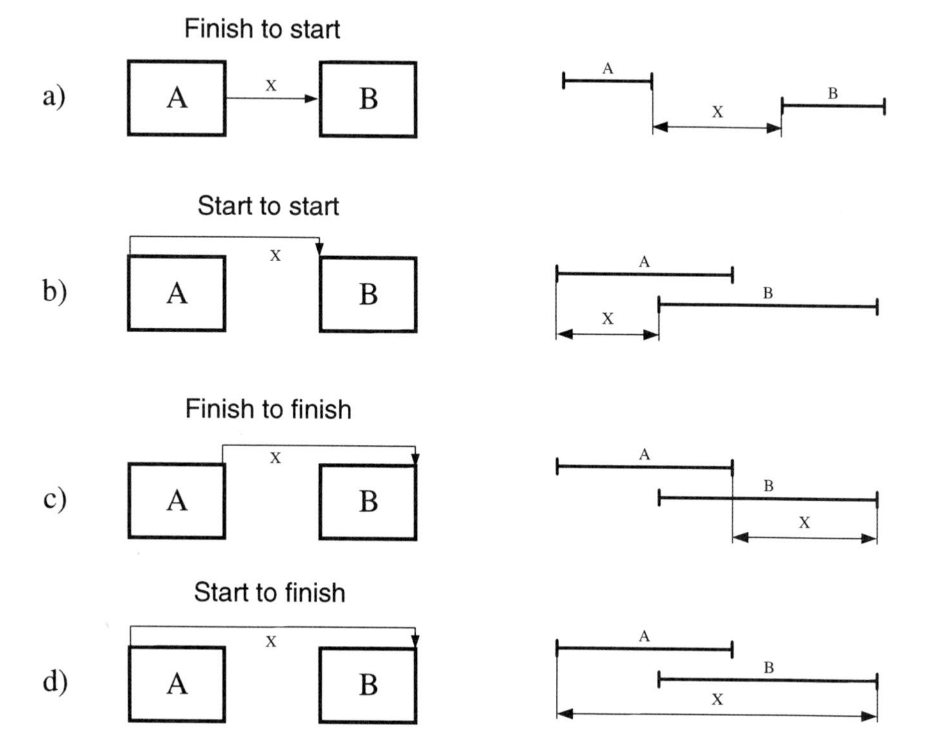

Further, a time delay of x units may be added. An FS_x relationship between A and B means that B can start x time units after the finish of A. For an SS relationship, the start of A determines the start of B. A and B may run in parallel with a time phase of x time units. The same is true for an FF relationship with the exception that it is the finish of the two activities that are time phased rather than the start.

The SF relationship, on the other hand, defines a total time window of x time units for the activities. As an illustration, assume that activity A is "engineering design of a petrochemical process" and that activity B is "procurement of equipment items." Obviously, in practice one may allow these activities to overlap. However, they cannot overlap fully. To procure equipment, a certain amount of engineering design must be completed, say, 30 percent representing a duration of 8 weeks. This means that there is a start-to-start relationship between A and B with an 8-week (30 percent) time lag (SF_8).

Also, in this example, the procurement needs some time after the finish of engineering design to do the last purchasing, say, about 3 weeks. This means that there is also an FF_3 relationship between activities A and B.

Any of these activities may be combined with each other. This provides a good flexibility to define a network with different overlaps of activities. Many say this is more realistic.

However, there is also a danger to start adapting the overlap conditions to meet a desired project due date. In this way the project manager may be fooled into overlap conditions that are unrealistic and that will show during project execution.

Precedence networks are calculated very much in the same way as traditional networks. It is beyond the scope of this chapter to go into this, and readers are encouraged to reference other literature on this topic.

PERT Networks

PERT allows activity duration with uncertainty. It assumes a statistical distribution for the duration of each single activity.

The most commonly applied distribution is a β-distribution. The β-distribution is finite between *a* and *b* and has a mode (most probable value) of *m*. *a* and *b* are interpreted as the most optimistic and pessimistic estimate of the duration. *m* is the most likely duration. By selection of the parameters *a*, *b*, and *m*, almost any skewed distribution may be constructed. Usually the distribution is skewed to the right, indicating that a delay is more likely than a finish ahead of scheduled due date.

The PERT method assumes that all activity durations are stochastically independent. This means that there is no underlying connection leading to simultaneous duration variations of two or more activities. If activity X is delayed, this does not lead to a similar delay of activity Y. This is, of course, a questionable assumption. Quite often there are activities that are connected (covariance). For example, a delay of an activity may be caused by a shortage of labor. In this case it is reasonable to believe that this is the case also for other activities performed in the same region.

PERT calculates expected durations for all activities and then does an ordinary CPM calculation of the network using these expected values as durations. For the β-distribution, the expected value and variance may be calculated as:

$$E(t_i) = \frac{1}{6}(a_i + 4m_i + b_i)$$

$$Var(t_i) = \frac{1}{36}(b_i - a_i)^2$$

where:

$E(t_i)$ is expected value of activity *i* duration.
$Var(t_i)$ is variance of activity i duration.
a_I is optimistic estimate of activity *i* duration.
b_I is pessimistic estimate of activity *i* duration.
m_I is realistic estimate of activity *i* duration.

The variance is a measure of the uncertainty of the duration. The larger variance, the larger is the uncertainty.

If p is a path (a chain of activities) in the network, the total duration along that chain is

$$T_p = \sum_{i \in p} t_i$$

This is, of course, trivial, since it just says that the duration of the path is the sum of the durations of the activities along the path. If the path is the critical path, π, the duration, T_π, equals the project duration.

With the assumptions made, the expected value and variance for T_p can be computed as follows:

$$E(T_p) = \sum_{i \in p} E(t_i)$$

$$Var(T_i) = \sum_{i \in p} Var(t_i)$$

Even though each t_i follows a β-distribution, T_p, may be approximated with a normal distribution. This is convenient, since the normal distribution is readily available in a table format.

An example may clarify the use of PERT networks. Assume the same network as earlier (see Figure 32.5). Table 32.4 gives the three estimates (a, b, m) for all the activities. In practice, these estimates have to be provided, either based on experience data or an assessment of the uncertainty of the activity. In the table, expected values are calculated for all the activities and the variance is calculated for the critical activities. Note that the most likely estimate (m) is the same as the duration used for the CPM calculation.

The network is now calculated using the expected values as shown in Figure 32.7. The new total duration is 44 days instead of the 40 days that the CPM calculation gave. The 4-day difference is a measure of the risk associated with each activity's duration. Since for most activities a delay is more possible than a finish ahead of schedule, the expected value

TABLE 32.4. DURATION ESTIMATES FOR A PERT NETWORK.

Activity	m	a	b	E(t)	Var(t)
A	10	9	17	11	1.78
B	7	5	9	7	0.44
C	12	10	20	13	2.78
D	18	16	32	20	7.11
E	14	13	21	15	1.78
F	13	10	16	13	1.00
G	16	15	23	17	1.78
H	12	11	19	13	1.78
I	6	5	7	6	0.11

FIGURE 32.7. EXAMPLE OF A PERT NETWORK CALCULATION.

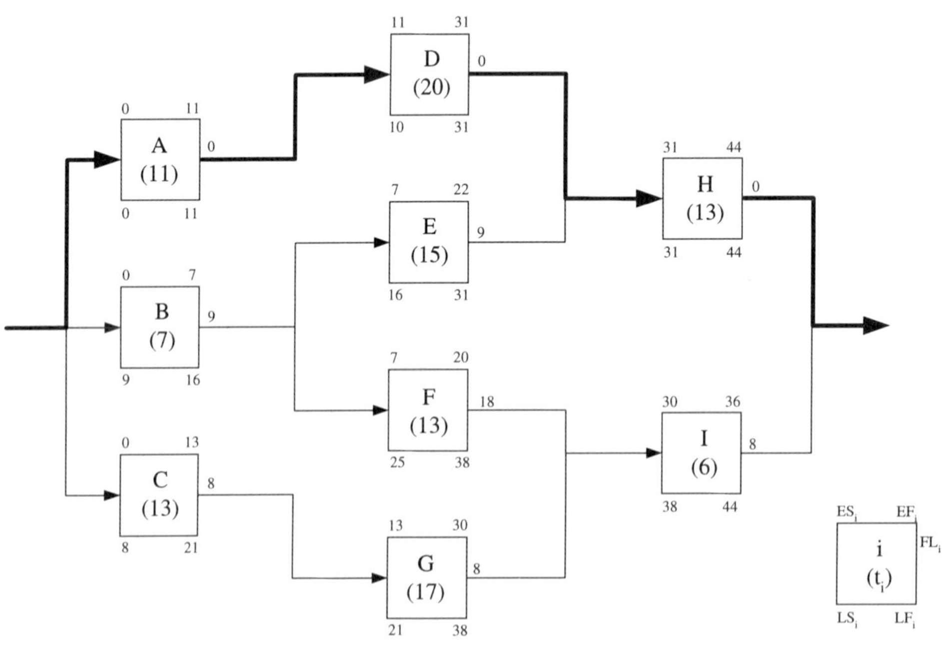

is somewhat higher than the most likely value. For example, for activity A the most likely value is 10 days, but the expected value is 11 days. The one extra day compensates for the risk of overrun.

The critical path is as before A-D-H. The duration of the project is

$$T_{\pi} = t_A + t_D + t_H$$

and expected value and variance is

$$E(T_{\pi}) = E(t_A) + E(t_D) + E(t_H) = 11 + 20 + 13 = 44$$

$$Var(T_{\pi}) = Var(t_A) + Var(t_D) + Var(t_H) = 1.78 + 7.11 + 1.78 = 10.67$$

The variance as such does not provide much practically useful information about uncertainty. It has to be transferred to a decision parameter that the project manager can understand and use. One way to do this is to calculate the probability of finishing the project on or before a given date or the probability of meeting a milestone that has been set. For example, the project manager would like to know what the probability is of reaching the 40-days deadline calculated by the CPM method. The calculations are as follows:

$$\Pr(T_\pi \leq 40) = \Phi\left[\frac{40 - E(T_\pi)}{\sqrt{Var(T_\pi)}}\right] = \Phi\left(\frac{40 - 44}{\sqrt{10.67}}\right) = \Phi(-1.23) = 0.109$$

where Φ is the normal distribution (0.1).

As can be seen, the chance of meeting this deadline is not very large, only 10.9 percent. The project manager can rephrase the question and ask what deadline he or she should give if he or she wants a 90 percent probability of finishing on or before the deadline. If this deadline is D, the calculations are as follows:

$$\Pr(T_\pi \leq D) = \Phi\left(\frac{D - E(T_\pi)}{\sqrt{Var(T_\pi)}}\right) = \Phi\left(\frac{D - 44}{\sqrt{10.67}}\right) = 0.90 = \Phi(1.28)$$

$$\frac{D - 44}{\sqrt{10.67}} = 1.28 \Rightarrow D = 48.2$$

So in order to have this at least 90 percent guarantee against a delay, the deadline quoted should be 49 days.

In the calculations $\Phi(x)$ is taken from a table of the normal distribution. This may be found in any statistical tables book.

Earlier, the assumption that all activity durations are stochastically independent was discussed. It was said that this assumption might be unrealistic. There is another assumption with the PERT method that is also quite unrealistic. Assume that there is a path other than the critical path that is nearly critical. For example, it may have a float of 1 of 2 days. Since durations may vary stochastically, there may be a situation where one of the activities on the critical path becomes very short at the same time as one activity on the other path becomes very long. This may cause the critical path to change from the original one to one that was nearly critical. PERT does not handle such situations, since all calculations are done under the assumption that the critical path does not change.

This may be one reason why the PERT method is not frequently applied in practice. Another reason is, of course, the problem of making the three estimates for activity duration (a, b, and m). In any case, a computer will do calculations. Then it may be just as easy to do a Monte Carlo simulation.

Monte Carlo Simulation

A Monte Carlo simulation will use the same input data as PERT—that is, statistical distribution for the duration of each activity. Then it will draw a set of durations for all activities from its statistical distribution. Then the network is calculated as a usual CPM network with these durations. The procedure is repeated a large number of times (for example, 1,000), and in this way an empirical statistical distribution is achieved. This may again be used to answer questions of what is the probability of meeting a milestone or a due date.

If a Monte Carlo simulation is done, there is a very interesting by-product. By counting the number of times an activity is critical, the probability of an activity being critical can be estimated. This is referred to as the *critical index* of the activity, and it may be a powerful decision parameter for the project manager. CPM only says whether an activity is critical or not. For example, A may be critical, but B is not. If A has a critical index of 51percent and B a critical index of 49 percent, it is seen that both A and B may be critical with almost the same probability (51percent versus 49 percent).

If a project scheduling needs to take uncertainty into account, a Monte Carlo simulation is probably the best approach. There are commercial products available for doing such analysis.

Resource Constraints

In the preceding discussions, resource constraints have not been taken into account. In practice, resource constraints, of course, apply in a number of ways:

- To determine the duration of an activity
- As a trade-off between time and cost
- As a limited resource during scheduling

Each of these three cases is briefly discussed in the succeeding paragraphs.

When the project is being scoped, an estimate of work volume is done, for example, as a number of person-hours for an activity (work package). If the activity is labor-intensive, it is common to apply a buildup and a rundown period to and from peak manning of the activity. This may be referred to as *mobilization* and *demobilization*. In this case the resource profile looks like a trapezoid as shown in Figure 32.8. It is assumed that the work scope is known and that the peak manning level and the transfer mobilization and demobilization are given. Realizing that the area of the trapezoid equals the work scope, the activity duration, t, can be calculated as:

FIGURE 32.8 THE TRAPEZOIDAL METHOD.

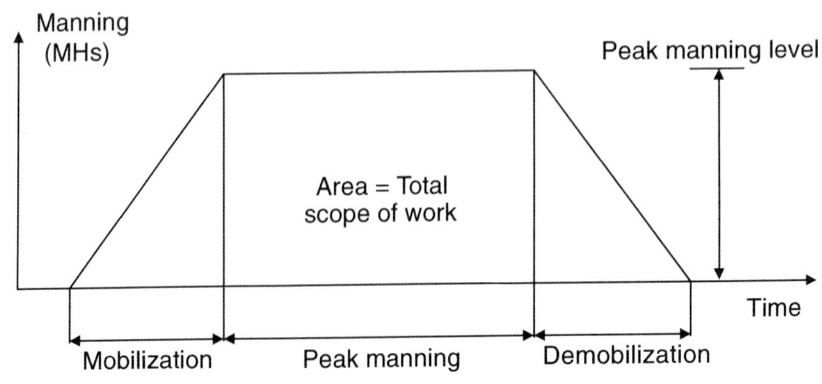

$$t = t_m + t_p + t_d$$

where:

t is activity duration.
t_m is mobilization time.
t_p is time with peak manning level.
t_d is demobilization time.

Then the following calculations can be done:

$$S = c \cdot P_{\max} \frac{t_m + t_d + 2 \cdot t_p}{2}$$

where:

S is scope of work (person-hours).
c is net person-hours per person per week.
P_{max} is peak manning level.

By solving this equation with respect to t_p and inserting in the expression for t, the trapezoidal formula is derived:

$$t = \frac{S}{c \cdot P_{\max}} + \frac{1}{2}(t_m + t_d)$$

This formula gives the duration as a fixed number. In practice, the duration may always be influenced. For example by the use of overtime or extra manpower, the activity may be accelerated. This would then involve a higher cost. This again allows for a trade-off between time and cost, as shown in Figure 32.9.

The curve shows the normal duration and its associated cost. It further indicates that the activity may be accelerated (duration shortened) until a limit referred to as "crash duration." The corresponding costs are called "crash costs."

The time/cost relationship allows formulating the selection of duration as an optimization problem. There will therefore be an optimal duration. The mathematical approach to this is called *operations research* and has found wide application among researchers but is hardly used in practical project planning and control. There may be various reasons for this; an important one may be the complexity of the problem and that the model the optimization is based on most often is quite unrealistic.

In a practical situation one would start with a schedule based on normal durations. Then one would use a heuristic strategy to reduce this as much as is required. The most common strategy is to calculate the cost gradient for each activity:

FIGURE 32.9. TIME/COST TRADE-OFF.

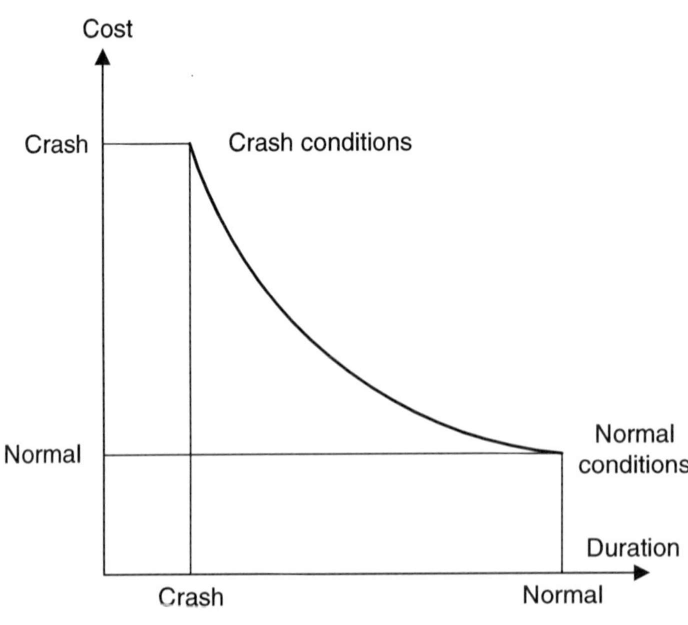

$$c' = \frac{c_D - c_d}{d - D}$$

where:

c' is cost gradient.
c_d is normal cost.
c_D is crash cost.
d is normal duration.
D is crash duration.

Then one would look at the critical path activities and reduce the one with the lowest cost gradient as much as possible. If this is not sufficient, one carries on with other activities in increasing sequence of the cost gradient value. Of course, it would make little sense to crash noncritical activities, because that would only buy extra float in the network.

If the resources are limited, the scheduling problem is referred to as "scheduling under resource constraints." In this connection there are really two variables that are considered:

- The project duration
- The resource level

Trying to fix both may create a problem with no solution. For example, if two welding activities running in parallel requires 2 and 3 welders, respectively, and the total number of available welders is 4, the only two solutions are either to obtain another welder (increase resource level) or to prolong the duration of one of the activities. Depending on the amount of float on that activity, the prolongation may involve longer project duration. Therefore, a standard approach is first to calculate the schedule without resource considerations and then adjust the schedule under resource limitations, keeping either the resource level or the project duration fixed and allowing the other to vary.

Scheduling with resource constraints is difficult and is a combinatorial problem of a magnitude that makes any exhaustive search (check all possible alternatives) impossible, even with the largest computers. However, a number of heuristic algorithms have been developed. It is beyond the scope of this chapter to discuss such algorithms, and you are referred to specialized books on scheduling or operations research.

Cost Management

In this section planning and control of costs are discussed. The cost estimate and its contingencies are introduced, and some basic rough estimating techniques are briefly explained. Then estimate updating and cost control is discussed. For a more complete picture of cost control including performance measurement by the earned value method, you are to the chapters by Harpum and by Brandon.

The Cost Estimate

The cost estimate is a forecast of the final project cost broken down on work packages and specified according to a code of account. This means that every estimate carries a certain amount of uncertainty. An estimate shall serve as a baseline for cost control. A fair estimate will have equal probability of overrun and underrun.

Any estimate will carry a contingency. There are two types on contingencies:

- *Contingency allowance.* Money to cover, normal, expected variances
- *Contingency reserve.* Money to cover unexpected variances

The *contingency allowance* should cover costs that are likely to occur but cannot at the present time be identified. The contingency allowance is meant to be spent in a normal project. It is not intended to cover unexpected events. A contingency to cover unexpected events is usually referred to as a *contingency reserve* and should cover costs that are unlikely to occur. It is intended not to be spent.

If each cost item has an associated uncertainty, the contingency allowance may be calculated. The uncertainty of the cost item can be estimated in the same way as for activity duration—that is, by giving optimistic, realistic, and pessimistic estimates and assuming a β-distribution.

Since the distributions normally will be skewed, the expected value will be higher than the most likely value. The sum of the most likely value for each cost item is referred to as a *base estimate.* Figure 32.10 shows the estimate value as a function of the probability. The amount that has to be added to the estimate to bring it up to a 50 percent probability may be interpreted as the contingency allowance. The contingency reserve is the additional amount needed to bring it up to, for example, 70 percent.

Usually, there are several estimates in a project. A distinction is made between estimates made prior to the start of project execution and estimates made during project execution.

Estimates made prior to the project start of execution should be classified according to at which project phase it has been made. This will allow an identification of estimate with respect to the following:

- The phase at which the estimate was made
- The documentation it is based upon
- The uncertainty the estimate carries

After project execution has started, the estimate will be updated to include approved variances at regular intervals. As with the schedule, the first estimate is denoted *master control estimate* (MCE) and the latest update is called *current control estimate* (CCE).

FIGURE 32.10. CALCULATION OF ESTIMATE CONTINGENCIES.

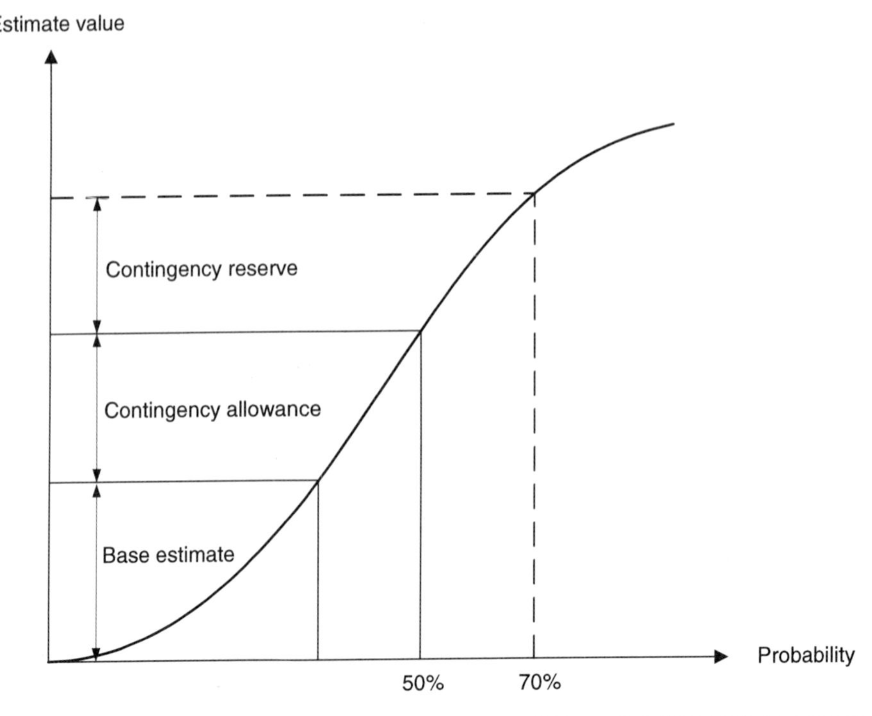

Estimating Techniques

Cost estimates are made at different stages in a project. At an early stage an estimate has to be provided before all technical details or design are developed. At a later stage more accurate estimates are made based on detailed design or bill of quantities. Developing the estimates at different stages will require different estimating techniques.

Basically there are two sets of estimating techniques:

- Synthetic
- Analytic

The synthetic techniques are used to obtain rough estimates. A project or work breakdown structure does not exist. The analytic techniques are based on a known WBS. Detailed estimates are made for each work package in accordance with a code of accounts. The analytic techniques are used for detailed estimates and are usually brought down to an uncertainty level of ± 10 percent. The synthetic techniques may be brought down to approximately $\pm 20\%$ uncertainty. They fall in two categories:

- Relational estimation
- Factor estimation

Relational estimation takes the cost of a finished similar project and corrects this for the following:

- Capacity of the facilities
- Time
- Location where the work is done

Factor estimation assumes that the percentage distribution of all cost items is known. Then one category is estimated in detail—usually the equipment—and the others are determined using the percentage distribution.

Both factor estimation and relation estimation are frequently used in practice. Their main application is selecting projects at a very early stage. Then the selected project will need an analytical estimate to be used as reference.

Cost Control

During project execution, costs need to be monitored and controlled (in addition to scope of work and time). The purpose of a cost control is to

- make project management aware of possible cost overrun at an early stage;
- inform all project team members of the current control estimate and the budget frame the work against; and
- establish cost consciousness amongst all project team members.

A full cost control in a project comprises the six steps indicated in Table 32.5. The first three are the traditional ones that are used in monitoring cost. The last three serve the purpose of really managing costs by analyzing the development and implementing corrective actions.

To implement the first three steps of the approach outlined in Table 32.5, a form like the one presented in Table 32.6 may be applied. Contingency allowance is used to cover variations, while contingency reserve is used to cover changes in scope of work.

The master control estimate (MCE) is updated with approved scope changes and variations to the current control estimate (CCE). To monitor the costs, both accrued expenditures and future commitments are registered. In addition, an independent forecast to complete is made. The sum of expenditures, commitments, and forecast to complete represents a revised forecast. This will be further analyzed through the Steps 4 to 6 in the approach outlined in Table 32.5. If the changes involved are approved, they will be included next period in the next update of the CCE.

Summary

In this chapter time and cost has been discussed. Together with the scope of work they make up the main control variables of a project. They are represented as two plan documents:

- Project schedule
- Project cost estimate

Together with the scope definition these two documents make up the project control baseline.

The purpose of these two documents is to clarify when the activities of a project should be executed and what the budget limit is for each of them. In addition, they serve as the reference for schedule progress and cost control of the project.

TABLE 32.5. THE SIX STEPS OF COST CONTROL.

Type of Activity	Step	Activity	Purpose
Cost monitoring	1	Know control estimate.	Know what to do.
	2	Keep account of commitments.	Know what has been done.
	3	Estimate costs to complete.	Know what remains.
Cost management	4	Analyze cost deviations.	Identify problem and its cause.
	5	Take corrective actions.	Minimize cost overrun.
	6	Develop revised forecast.	Estimate effect of actions.

TABLE 32.6. EXAMPLE OF A COST CONTROL FORM.

Cost Item	MCE	Scope Changes	Variations	CCE	Expenditures	Commitments	Forecast to Compl.	Revised Forecast
Contract A	2500	300	100	2900	400	900	1800	3100
Contract B	4000		500	4500	2000	300	2200	4500
Contract C	3500	100		3600	800	1500	1200	3500
Cont. allow.	2000	−400		1600				1500
Total estimate	12000		600	12600				12600
Cont. reserve	1500		−600	900				900
Total budget	13500			13500				13500

References

Clark, F. D., and A. B. Lorenzoni. 1997. *Applied cost engineering,* 3rd ed. New York: Marcel Dekker.

Cotterell, M., and R. Hughes. 2002. *Software project management.* 3rd ed. London: McGraw-Hill.

Elmaghraby, S. E. 1977. *Activity networks.* New York: Wiley.

Granli, O., P. W. Hetland, and A. Rolstadås. 1986. *Applied project management.* Trondheim, Norway: Tapir Forlag.

Guthrie, W. 1977. *Managing capital expenditures for construction projects.* Carlsbad, CA: Craftsman Book Company.

Jessen, S. A. 2002. *Business by projects.* Oslo: Universitetsforlaget, 2002.

Lockyer, K. 1984. *Critical path analysis and other project network techniques.* 4th ed. London: Pitman.

Morris, P. W. G. 1994. *The management of projects.* London: Thomas Telford.

Turner, J. R. 1999. *The handbook of project-based management.* New York: McGraw-Hill

Wysocki, K. W., R. Beck, and D. B. Crane. 1995. *Effective project management: How to plan, manage, and deliver projects on time and within budget.* New York: Wiley.

CHAPTER THIRTY-THREE

CRITICAL CHAIN PROJECT MANAGEMENT

Lawrence P. Leach

Critical chain project management (CCPM) is a relatively new entry to the Project Management Body of Knowledge (PMBOK), first reaching the broad project audience with a presentation I gave at the Project Management Institute (PMI) international conference in Long Beach, California, in 1998. This followed the 1997 publication of the book *Critical Chain* (1987) by Eliyahu M. Goldratt, best known for his much earlier production management classic *The Goal* (1984). *Critical Chain Project Management* (Leach, 2000) in February of 2000 integrated the schedule preparation and management method of Goldratt's critical chain approach with the rest of PMI's PMBOK.

The excitement about critical chain stems from it being the first *new thing* to project planning since CPM (critical path method), PERT (program evaluation and review technique), and Monte Carlo simulations came on the scene in the 1960s. It also results from the reports by some early adopters of CCPM quoting statistics like the following:

- Project success rates (triple constraint): Up from < 10 percent to near 100 percent.
- Projects complete in half the previous time, or less.
- Much less stress on project managers and team members.

This early excitement was tempered in the PMI community by charges that CCPM really is nothing new; that everything it includes was already available in the PMI's PMBOK®. As companies beyond the early adopters sought the benefits promised by CCPM, some found that it required changes in management behavior they were not ready for. Consequently, a growing number of CCPM implementations did not get the rapid benefits promised by the early adopters. CCPM is now becoming a standard tool for project planning and control. This chapter explains why.

Any Project Worth Doing Is Worth Doing Fast!

Projects we are concerned with all have a purpose that can be expressed in terms of return on investment. The return need not be monetary; for example, it may be lives saved by a humanitarian project. Project investment can usually be expressed in monetary terms; but even if you choose to express it in terms of staff-hours or some other measure of resource consumption, the following applies. Most projects provide no return (or at least very little) until they are done. Thus, return on investment, no matter what the units of measure, is negative until the project completes, and then begins to pay back and hopefully eventually becomes positive. Because the investment is roughly the same whether you complete a project slowly or fast, the return on investment accelerates in time if the project completes sooner, with no additional investment. (Worse, projects that take longer usually cost more for the same final result.) Further, if the project is a new product development, the overall profitability could be impacted by hundreds of percent by being first to market. Therefore, any project worth doing is worth doing fast.

The basic idea underlying CCPM derives from the theory of constraints (TOC), developed by Dr. Eliyahu Goldratt (Dettmer, 1997). There is no consensus on the basic statement of this theory. It starts as a system theory, recognizing that people build systems to achieve a goal. Usually businesses can best measure progress toward the goal as throughput of the business. TOC asserts that goal achievement for any system is limited by a constraint. It then follows that if you do not know what the constraint is, most of the work you might do to improve the system will improve things that are not the constraint and thus lead to no improvement of the system in terms of its goal (which, for profit-making companies, is to make money now and in the future). Goldratt had great success taking this simple idea into the world of production with five focusing steps for system improvement:

- *Identify* the constraint.
- *Exploit* the constraint (i.e., do whatever is necessary to ensure the constraint works at full capacity on quality input to produce quality output and pass on the work result as soon as possible).
- *Subordinate* everything else to the constraint (i.e., eliminate interferences with exploiting the constraint to achieve system throughput). Often this includes eliminating subsystem efficiency measures and policies.
- *Elevate* the constraint (i.e., get more of the constraint, be it machines or people).
- Do not let *inertia* keep you from doing the cycle again (as a new constraint always arises).

TOC accounts for statistical fluctuations in process step time and the dependent flow of material through process steps in a production line. Both of these factors apply to projects as well.

Single Project Critical Chain Project Management

Identify the Constraint

Applying TOC to project management immediately confronts one with the triple constraint (scope, time, budget). Although called the triple constraint, the idea of limited resources always lurks nearby. TOC cannot abide a triple constraint, much less a quadruple constraint. The first step is to identify the constraint of most projects. Keep in mind the opening remarks; it is the constraint to the goal, and the constraint to getting more of the goal sooner, that is of interest.

One finds much discussion of the critical path in the project literature . . . a concept we need not redefine in this chapter. Figure 33.1 illustrates a sample critical path schedule as a point of departure to illustrate the single project constraint, and how critical chain differs from the critical path method (CPM).

Two things about the critical path are interesting from a TOC perspective. First, most projects have finite resources. Second, activity performance times always show statistical fluctuations. CPM does not take either of these factors into account. The project literature suggests that you should resource-level your CPM plan. Figure 33.1 illustrates a plan without resource leveling, assuming one resource of each type shown on the plan. (Many project managers do not even identify the resources needed in the project plan. Identifying the required resources is called "resource loading" the plan.) Note that tasks 3, 9, and 14, which are planned in parallel in Figure 33.1, all require an engineer. If there is only one engineer and these tasks assume the engineer is dedicated to the task, these tasks must overrun by at least a factor of three.

Try it and you will find that the overall project duration is significantly longer than the sum of the critical path durations. Figure 33.2 illustrates the resource-leveled version of Figure 33.1. Note that the plan now only requires one resource of each type at a given time. The plan is substantially longer. Thus, the critical path is not the constraint to getting the project done sooner. As such, it is questionable if it is the critical path. Further, as illustrated in Figure 33.2 (the space between task 3 and 14), the critical path usually has gaps in it after resource leveling—that is, apparent float. The PMBOK® Guide (PMI, 2000) defines the critical path as the path with zero float. Thus, after resource leveling, no path may qualify as the critical path.

Critical chain corrects this logic problem by considering equally both the resource constraint and the activity logic when defining the critical chain. The critical chain is the longest path through the project . . . it is the real resource-leveled critical path. It jumps the logic chains whenever a resource contention requires it to do so. Except for rare exceptions, the critical chain is not the same sequence of activities as the critical path. Figure 33.3 illustrates the critical chain for the example project. This chapter addresses the changed activity duration and the buffers.

Exploit the Constraint

The second TOC focusing step is to exploit the constraint, which is the critical chain of the project. Exploit means to get more out of the system with no more input. For the reasons discussed previously, this means complete projects quicker. Goldratt relied on his production experience to suggest a direction for the solution to this step, taking into account the statistical fluctuations in project activity times and the dependent nature of the project logic network.

FIGURE 33.1. EXAMPLE CRITICAL PATH SCHEDULE.

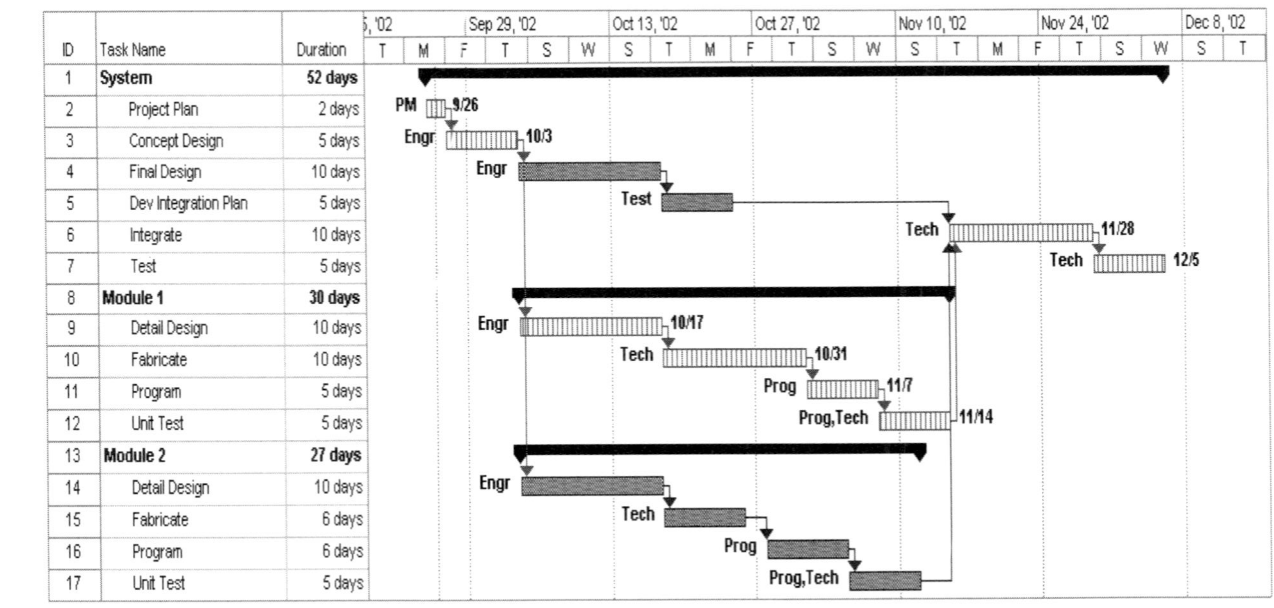

FIGURE 33.2. RESOURCE-LEVELED CRITICAL PATH.

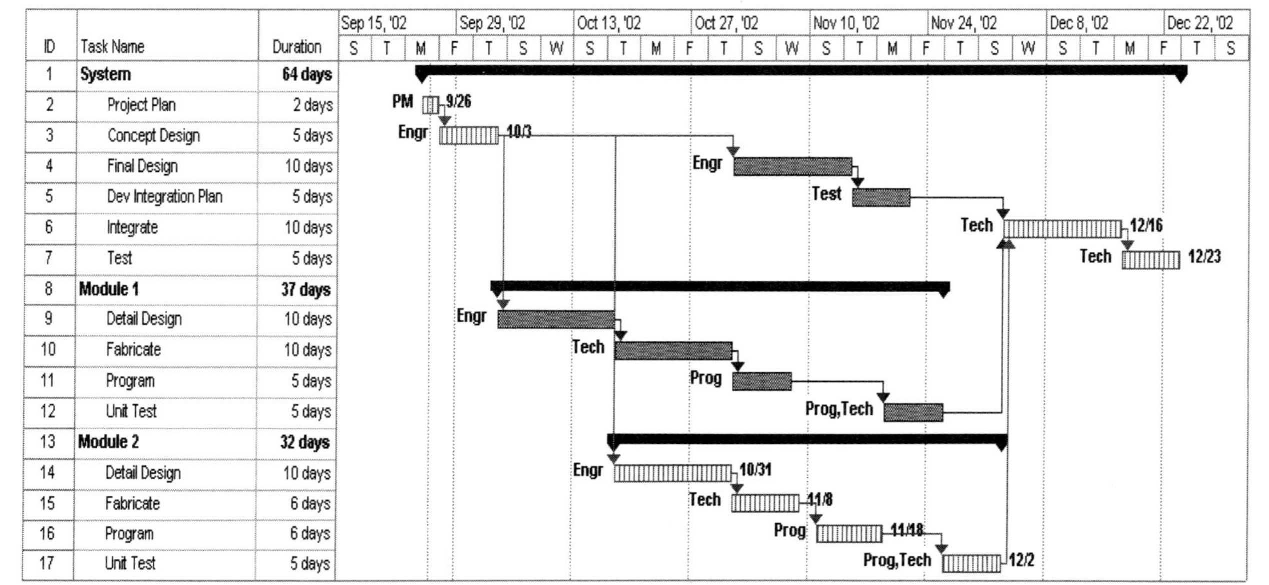

ID	Task Name	Duration
1	**System**	**64 days**
2	Project Plan	2 days
3	Concept Design	5 days
4	Final Design	10 days
5	Dev Integration Plan	5 days
6	Integrate	10 days
7	Test	5 days
8	**Module 1**	**37 days**
9	Detail Design	10 days
10	Fabricate	10 days
11	Program	5 days
12	Unit Test	5 days
13	**Module 2**	**32 days**
14	Detail Design	10 days
15	Fabricate	6 days
16	Program	6 days
17	Unit Test	5 days

FIGURE 33.3. CRITICAL CHAIN FOR EXAMPLE PROJECT.

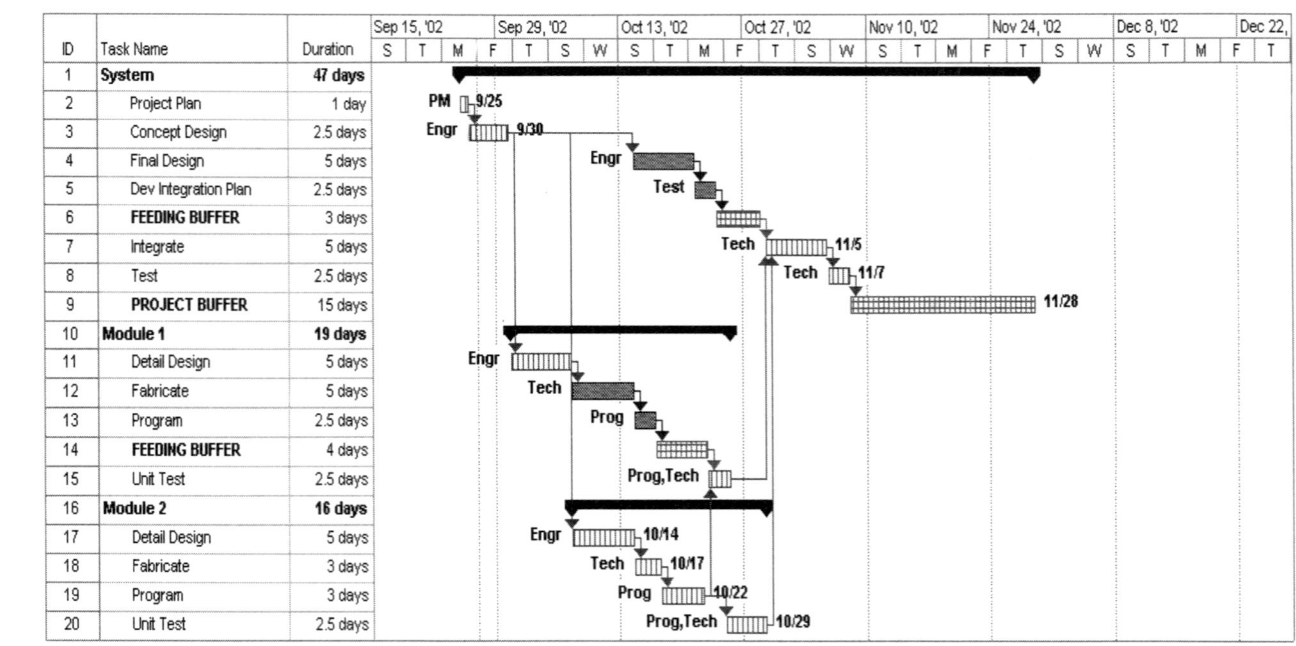

Mean Estimates. Most project work is known to have a completion time probability density function something like that shown in Figure 33.4. Functions that approximate this shape are the mainstay of probabilistic project planning methods such as PERT and Monte Carlo simulation.

There are actually at least two distributions that have characteristics similar to Figure 33.4. One results from simply doing the same activity over and over and measuring how long it takes. That is a measure of the actual variation in activity performance duration. Numbers that you find in databases for estimating activity performance—for instance, the probability distribution of bricks laid per hour—show such a distribution. Many project activities seem unique or first of a kind. One of the primary tools of project management is to use the work breakdown structure to decompose such activities down until you have activities on which you have an experience base to make an estimate.

The second contribution important to project schedules is the variation in estimation. Ultimately, project schedulers are interested in the comparison of actual duration to the estimated duration. If different estimating techniques are used, one can get different results from the same data. For example, even the most basic construction activities allow for adjustments for worker productivity by region or for whether one is working on ground level or on a scaffold.

Consider Figure 33.4 to understand another point about activity estimates. The probability associated with any point on the x-axis—that is, with any specific duration estimate—

FIGURE 33.4. ACTIVITY TIME PROBABILITY DISTRIBUTION.

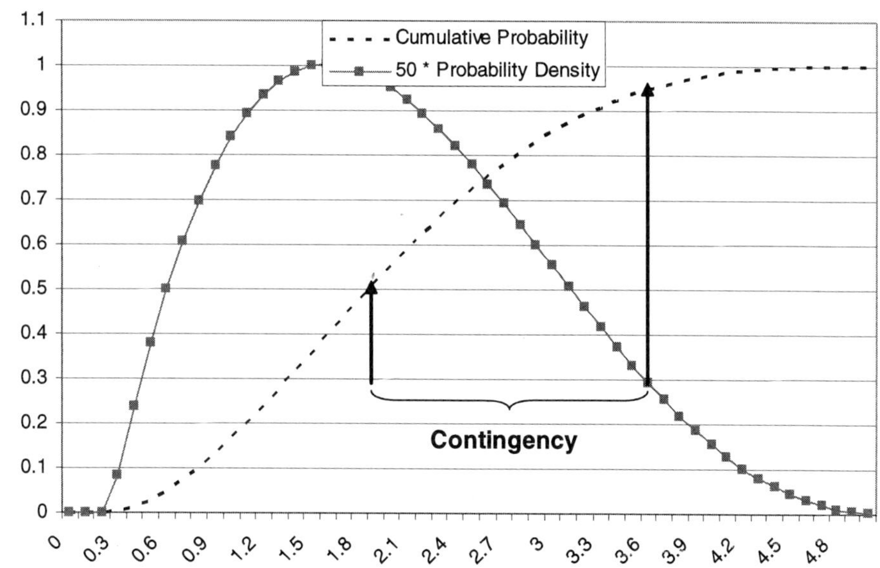

is exactly the same. It is zero. In a conventional schedule with single deterministic schedule duration estimates, all of the individual activity start and stop dates (and with CPM usually both early and late start and stop dates) thus have a probability of zero . . . they are a complete fiction. Yet people get driven to those specific dates.

One only gets a probability by integrating between two points. Usually when one gives a single point estimate, the meaning is "within that time or less"; in other words, the associated probability is the integration from zero up to that duration. Figure 33.4 shows that as the dashed curve. This is why statisticians insist that point number estimates, without some indication of the associated uncertainty, are meaningless.

Project managers face the problem of what to do about such distributions. The PMBOK® Guide encourages estimating the uncertainty for all project estimates. Most current software tools include the PERT capability. Despite these facts, most project schedules do not address uncertainty. Most schedules use a single activity duration estimate and focus on calculated start and finish dates for each activity.

PERT attempts to model the curve of Figure 33.4 by making three estimates (time or cost): an optimistic, most likely, and a pessimistic estimate. It uses these estimates and an assumed distribution (usually a version of the beta distribution with specific parameters, or a triangular distribution, which never exists in reality) to estimate the mean value and variance for each activity along the critical path. Critical chain uses essentially the same approach, but often with only two estimates, and sometimes with no explicit estimate of the individual activity variation. PERT usually limits this thinking to the critical path, while critical chain applies the same thinking to all activities in the schedule. Monte Carlo simulations attempt to do the same thing and usually allow many more degrees of freedom on the statistical distributions applied. This is an area where the claim that critical chain is "nothing new" is valid . . . people have understood the element of uncertainty from the beginning of project management and have been making attempts to use the knowledge. CCPM takes the same knowledge and builds a different approach on how to use it through integrating other elements of the process, such as measurement and control.

While some project literature (correctly) urges using mean times for the deterministic activity duration estimate, the culture that persists around the world does not support using mean estimates. It is correct to use the mean because individual mean estimates along a path do add linearly, whereas other central tendency statistics (e.g., mode, median) do not. But everywhere you look, people are judged as good if they get done sooner than an activity estimate and bad if they are late. Thus, people are behaviorally reinforced to make estimates that are much higher probability than the mean, perhaps as high as a cumulative probability of 90 to 95 percent. For most activity distributions like Figure 33.4, the estimated duration must therefore be two to three times the mean.

I find much of the project literature emotionalizes the reality of statistical fluctuations. They talk about activity time "padding" as if there were some absolute reference to pad against (the mean, mode, median?). Some suggest adding an activity at the end of the critical path to account for these fluctuations . . . but quickly caution to call it something that will not draw attention to it, as though if you do, management will take it away. This thinking illustrates poor understanding of statistical fluctuations. Such poor understanding leads to mismanagement of uncertainty. A key aspect of CCPM is to recognize and utilize the fact of statistical fluctuations. To do that, you need a tool.

Project Buffer. The project buffer is the primary tool to manage statistical fluctuations. Using the mean duration for all project activities, the probability distribution of the end of chain of activities will approach a normal distribution due to the central limit theorem of statistics. This means there is only a 50 percent chance of completing the project at that time or less. Thus, it is necessary to add the project buffer as an allowance to the end of the critical chain to bring up the probability of delivering the project on time (or earlier) to some desired level; usually $>$ 95 percent. Figure 33.3 illustrates a project buffer as activity 9.

Project Buffer Sizing. Properly sizing the project buffer has been a matter of much discussion. Goldratt initially suggested starting with activity estimates as people had always made them, cutting the individual estimates in half and adding back in a project buffer equal to one half of the resulting critical chain. Although the subject of much criticism, this method has shown its merits many times.

I believe that the best project companies will eventually come to use control charts to size project buffers. Tracking a control chart of the difference between scheduled and actual completion time for projects reveals the amount of adjustment necessary to the mean project duration to achieve any desired level of predictability.

In the meantime, many people have suggested a method that is similar to PERT: sizing the buffer as the square root of the sum of the squares (SSQ) of the differences between low-risk and most-likely or optimistic activity duration estimates (Newbold, 1998, p. 94). This method has the same advantages and disadvantages of the PERT approach. Leach (2003) suggests modifying this method to account for sources of bias in project performance to plan, with a minimum project buffer limit of 25 percent of the length of the critical chain. The sources of bias (that is, a tendency for projects to overrun the schedule and not underrun) include the following:

- Oversights (necessary activities not included in the plan).
- Errors.
- Merging project activity chains (more than five or six, or nearly equal in length and uncertainty).
- Queuing effect of multiple projects demanding the same resource.
- Overconfidence as to the variation of activity performance. This is institutionalized in PERT, where there is an inherent assumption that the range people estimate is plus-or-minus three sigma. Analytical evidence demonstrates people's estimates usually range more on the order of plus-or-minus one to two sigma (Kahneman1982).
- Multitasking.
- Student syndrome: Delaying the start of activities until the need seems really urgent, such as students do with term papers.
- Date-driven behavior: Failure to turn in the results of activities that complete early; spending the time "polishing the apple" or sitting in an out box until the due date.
- Failure to report rework: Passing on work with known defects to meet the scheduled due date. The rework gets discovered later, where it has a much larger impact on completing the project.

The effects of the first two items on this list are evident, and this chapter also discusses merging and multitasking.

One may use Monte Carlo simulations to size the project buffer, but they are subject to most of the preceding biases that the PERT and SSQ approach overlook. An exception is that some Monte Carlo schedule analysis tools may take path merging into account.

Subordinate to the Constraint

Feeding Buffers. Most project networks have multiple parallel chains. All chains must tie in by the end of the project. You should always have a milestone at the end of the project schedule for project completion. Many chains tie in to the critical chain before the end of the project, usually into assembly, test, or transition to operation activities. Any of these merging chains can make the project critical chain late, as the successor activities require all input activities to be complete before they can start. In CCPM, these merging chains are called *feeding chains*, as they feed the critical chain.

Conventional project management seeks to use float to manage feeding chains. *Float* is the result of a backward pass calculation on the network and reveals how much later than the start of the project activities can start and still not make the project late. By definition, the critical path or critical chain has zero float. The assertion is that the float allows for uncertainty in the chains that feed the critical path.

Unfortunately, float is a very poor tool to account for uncertainty in merging chains. Float has nothing to do with the uncertainty of the duration of the activities in the feeding chain. Float results from only network logic construction. In general, longer feeding chains have less float. Thus, as Figure 33.1 illustrates, the amount of float available is inversely proportionate to the amount needed. (Compare the float for activity 5 to that for activity 17.) A chain just short of the critical chain, such as the lower one in Figure 33.1, has nearly zero float. Yet, as the next longest chain in the project, it needs the most float.

CCPM inserts feeding buffers at the point feeding chains join the critical chain. They connect the last activity on the feeding chain to the successor on the critical chain. Figure 33.3 illustrates two feeding buffers for the sample project. The scheduler sizes feeding buffers based on the uncertainty of the duration of the activities in the feeding chain, applying the same method used to size the project buffer. In addition to providing assurance that the successor critical chain activity will have the necessary input from its predecessor on the critical chain and on the feeding chain, feeding buffers also enable measuring progress on the feeding chains.

A simple illustration serves to show the importance of feeding buffers to the overall schedule. If each chain of activities has a reasonable number of activities, and if mean estimates are used, the probability distribution for completing the chain as scheduled would have a 50 percent probability of completing at the merge point. Thus, the chances of having both are only 25 percent, the product of 0.5 × 0.5. If a third path joins at the same point, the probability of having all three reduces to just 12.5 percent. And so on. Thus, it is difficult to protect a highly parallel schedule. Analysis demonstrates that sizing the feeding buffers using the same method as recommended for project buffers is effective for a modest number of feeding chains: up to five or six. More merging chains require larger feeding buffers, or an addition to the project buffer.

Late-Start Feeding Chains. Most project software defaults to early-start scheduling. This may be for two reasons:

1. The belief that the float thus introduced helps manage project activity time variation
2. No alternative but late-start

This in turn can create several problems for project execution:

1. The project manager is confronted with a bewildering array of activity chains to start while at the same time needing to align all of the project stakeholders.
2. Early-start can cause resource contention not otherwise present, leading to bad multitasking (see *Resource Decisions* coming up in the chapter for a more thorough description of bad multitasking.)
3. If one is using earned value (EV) against an early-start schedule, the project will look behind from the beginning, even if it is doing fine. This can cause the project manager to take unnecessary actions, leading to increased variation in project performance.

CCPM recommends late-starting feeding chains. Keep in mind that each feeding chain has a feeding buffer activity where it joins the critical chain, meaning that the start of the chains will not be delayed as much as in CPM or PERT late-finish schedules. It is applying the just-in-time principle to project performance, while using the feeding buffer to account for the reality of activity duration variation.

Earned Value and CCPM. One may apply earned value (EV) to CCPM projects. Because the schedule variance (SV) and Schedule Performance Index (SPI) do not reflect the critical chain (or path), and because they can provide misleading information on actual project status and even encourage inappropriate behavior (e.g., performing expensive noncritical tasks before inexpensive critical chain tasks), I recommend not using SV or SPI for project control or forecasting. Buffer management provides the schedule forecasting and control functions for CCPM projects.

When applying earned value to CCPM projects, you must consider the statistical nature of the estimates: The task estimates are the mean values, and the buffers statistically sum the variances. When providing a high-probability estimate of project cost and schedule completion, project completion is at the end of the project buffer. The baseline budget at completion (BAC) must include an allowance for the cost of working into the project buffer. This cost allowance is commonly called a *contingency* or *management reserve*, but is usually not integrated with the schedule impact, as it must be in CCPM. The project baseline completion date and the baseline BAC for the project must include these buffers. The planned value (PV) for the project extends to the end of the buffers in both time and money. Figure 33.5 illustrates this compared to the standard project S curve (i.e., the accumulated project

FIGURE 33.5. CCPM EARNED VALUE.

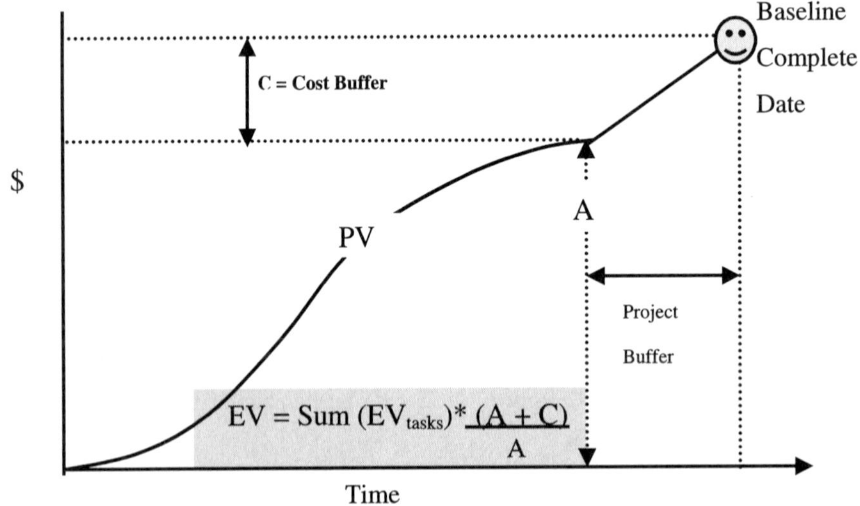

progress and cost vs. time first raising slowly, then rapidly, and then slowly to final project completion).

The sum of the activity estimates is the mean time and cost to complete the project. This must be adjusted to predict how the project will complete relative to the baseline. This is best accomplished by setting EV* = EV (activities) × R, where R = ratio of the BAC* (with contingency) to the BAC = sum of the activity (mean) estimates. With this adjustment, other EV calculations will work normally—for instance, EAC = BAC/CPI, where AC = actual cost, CPI = Cost Performance Index, and CPI = EV*/AC.

For example, assume a project where the sum of the estimated tasks (BAC) is $100K and the cost buffer = $20K. Thus, BAC* = $120K. Halfway into the project (i.e., half the tasks completed), EV = $50K. Let's assume that the project is right on cost: AC = $50K. Then, EV* = 50 * (120/100) = 60. Then CPI = 60/50 = 1.2. Thus, EAC = 120/1.2 = $100K.

Buffer Management

Buffer management provides the CCPM cost and schedule control system. (CCPM projects need effective quality management, just like any project.) Buffer management starts with weekly updates of the schedule by asking all activity performers, "How many working days until you will be done?" This forecast of activity completion is used to project the completion of the project activity chains. Keeping the feeding and project buffers fixed in time allows determining the penetration, incursion, or use of the buffer. Incursion into the project buffer forecasts project completion. Figure 33.6 illustrates tracking buffer incursion.

FIGURE 33.6. TRACKING BUFFER INCURSION VS. TIME.

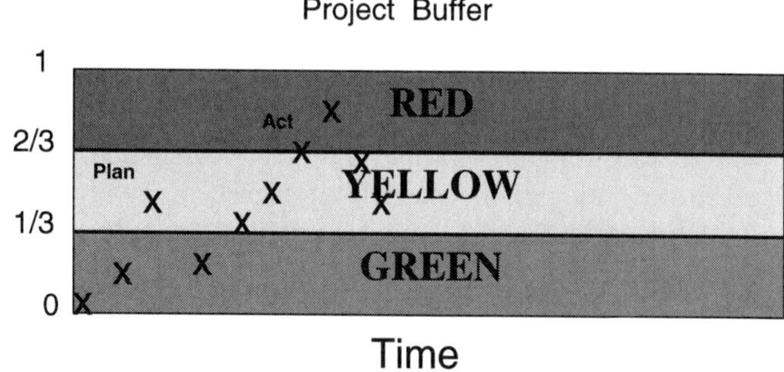

The specific question "How many working days until you will be done?" is important. Experience demonstrates that estimates of percent complete tend to underestimate the time necessary to complete an activity. Causing activity performers to think about what remains to be done, rather than what they have completed, leads to more realistic estimates.

The entire project team uses buffer management to make decisions about the project. The two primary decisions are those made by resources, who must decide which activity to work on next, and by the project manager, who must decide when to take action on the project to keep it in control.

Resource Decisions. A cardinal rule of CCPM is that all resources should work on only one project activity at a time and turn in their result as soon as it is completed. This is called roadrunner or relay-runner activity performance. The reason is that dilution of work time amongst different project activities makes them all take longer and ultimately makes all projects take longer. This is what CCPM calls bad multitasking. An insidious aspect of bad multitasking is that if one keeps a database of activity durations in a multitasking environment, the database will reflect this poor practice. This combined with current practice in many organizations to turn in work when due, but not earlier, leads to a self-fulfilling prophecy of protracted activity duration.

To avoid bad multitasking, each project resource must have a tool that specifies which activity to work on next. The buffer report provides this tool. Each resource should work on the activities in their queue (following the project logic first, of course), in accordance with the following run rules:

1. Critical chain activities take priority over non-critical chain activities.
2. If two activities are both on or both off the critical chain, the resource should work on the one with the greatest relative buffer penetration.
3. Nonproject work is lower priority than project work.

Project resources like to know when activities are coming up to be worked. Some suggest using resource flags or buffers along the critical chain. The resource buffer does not consume time in the project; it is an "alarm clock" lead time to let resources know when a critical chain activity is coming up to be worked—usually with a fixed lead time. Most critical chain implementations do not require this added formality. Project status meetings and communication of project schedules often fulfill this function.

Some resources will have difficulty with not knowing a fixed date to start and complete activities. In reality, they did not have fixed dates before, only estimates. In time, they will come to understand this. Where you must use fixed dates, be sure to precede them with an appropriately sized buffer.

Project Manager Decisions. The project manager can use buffer reporting as a control chart to decide when to take action on the project. As described in two *PM Network* articles (Leach, 2001), it is necessary to discriminate between common-cause (i.e., within the natural variation of the process that produces the result, in this case an actual project task duration) and special-cause variation (i.e., an identifiable cause of variation beyond the natural process limits) before taking action on a project. Mistaking one for the other will increase overall project variation. Control charts set limits on variables such that variation within the control limits represents common-cause variation, and should not be acted upon. Variation outside these limits is special-cause variation, and calls for action by the project manager to bring the project back into control.

Figure 33.6 illustrates a buffer control chart. Limits are preset for planning to take action and for the initiation of action. Variation below the planning threshold requires no action by the project manager. Variation beyond the normal range, but not yet to the trigger point, signals the project manager to develop mitigation plans. Variation beyond the action trigger causes the preplanned action to take place.

One may use buffer management for all of the buffers in the project. In practice, feeding buffers are often used up very rapidly, and tracking on a chart is frequently not necessary. Some suggest that the buffer planning and action thresholds should vary over the planned duration of the project, with tighter limits at the project beginning. I have not found that extra sophistication to have significant value, nor have I found it to be detrimental.

What Is Different about CCPM?

The schedule is different from most CPM plans.

- All CCPM schedules are resource-loaded and -leveled.
- The critical chain jumps logic chains where necessary.
- Mean activity durations are used.
- Project and feeding buffers are used.
- Project start times are sequenced to project priority for access to an organization drum resource.

- A capacity constraint buffer delays project start times.

Task performance differs from many projects.

- Bad multitasking is eliminated.
- Activities do not have start and stop dates; they are performed as fast as possible in sequence as the resource is available.
- Early activity completions are passed on as soon as possible to the nest activity.

Measurement and control is different.

- Status asks "how many days to complete?" working activities.
- The buffer report forecasts project completion.
- The project manager uses buffer thresholds to take pro-active project action.
- Resources and resource managers use the buffer report to decide which activity to focus on next.

Multiple-Project Critical Chain

CCPM defines a multiproject environment as one in which resources are shared across projects. Informal surveys I have conducted confirm that nearly all project environments qualify. The measure for the multiproject environment is completing the most projects in the shortest possible time to support the business goal. Applying the TOC five focusing steps identifies a different constraint than the critical chain of a single project. One of the resources shared across all of the projects controls the overall throughput of project results. CCPM calls this resource the "drum," with the image of the drummer on an ancient galleon setting the pace for the whole organization.

Figure 33.7 illustrates why it is so important to identify and control using the constraining resource. The upper three projects share represent multitasking of the resource supply over the three projects. In the lower three projects, the resources are sequenced, thus reducing the duration of each activity proportionately.

However, the upper case could represent one person shifting back and forth between the three projects, or three resources, one on each project. Usually having more than one resource helps lighten the load—that is, two people get the work done in less than half the time of one.

I use these simple project constructs in a training simulation, and the result is always the same. If one performs like most organizations do and starts two or more projects at the same time (e.g., the first of the fiscal year), it always takes much longer to complete each project than if the projects are staggered to enable the drum (i.e., the shared resource used most across all projects: the system constraint) resource to focus on one project at a time. More remarkably, both projects complete sooner than either project did in the multitasking case. In real organizations with dozens of projects, the results are much more impressive.

FIGURE 33.7. COMPARISON OF MULTITASKING AND PROJECT SEQUENCING.

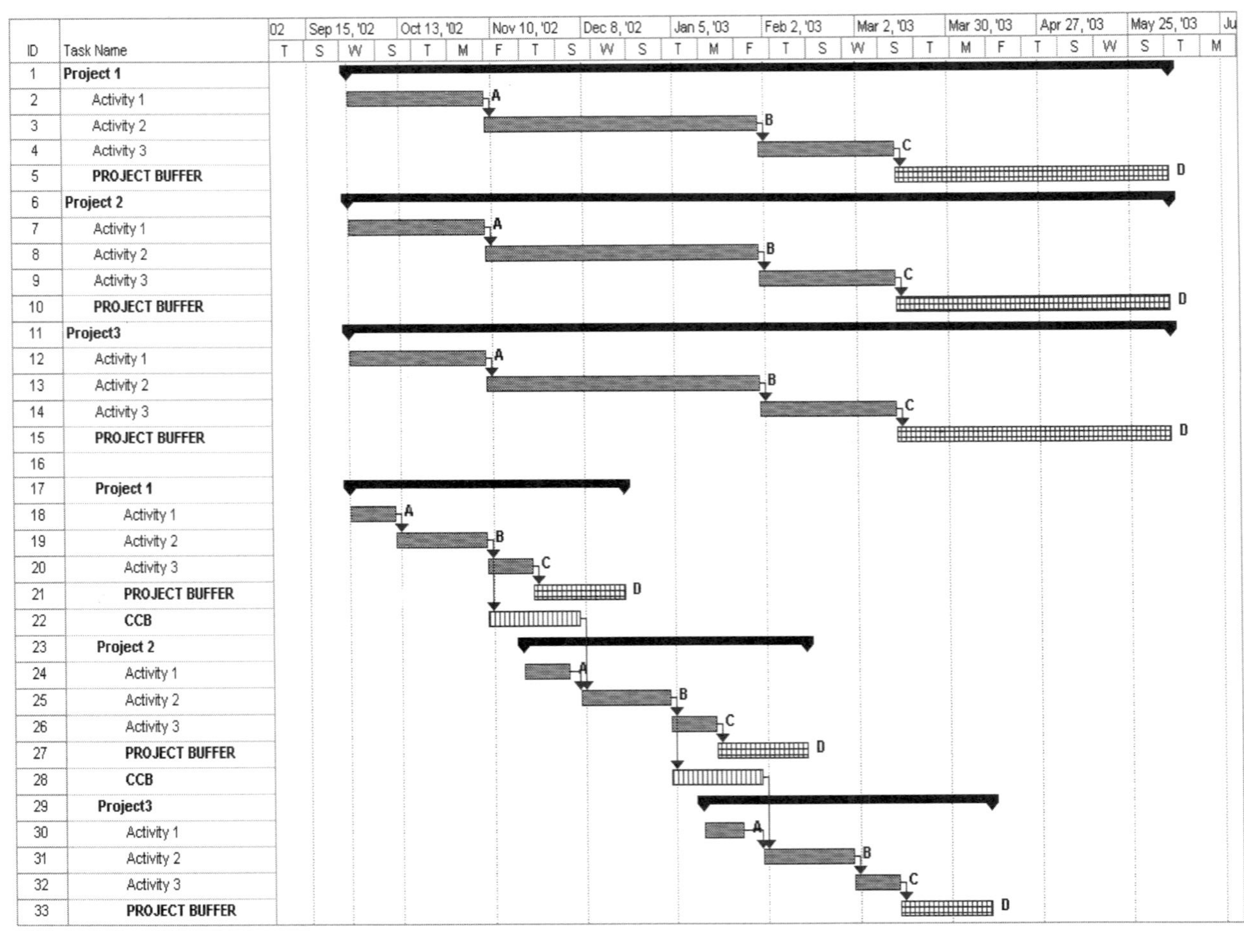

Note that in the lower case, even the last project finishes before any of the projects in the upper case.

Figure 33.7 is conservative, as it assumes no switching loss for multitasking in the upper three projects. Research indicates the multitasking switching loss can be on the order of 40 percent. Multitasking also does not allow for the learning curve to help later projects.

Identify the Constraint. The first focusing step is to identify the constraint—the drum resource. Our purpose in identifying the constraint is to use the information to sequence the start of projects. It has nothing to do with how important the resource is. However, if a resource can be increased relatively quickly and inexpensively, it is not a good choice to remain as your company project constraint. The value of the constraint far outweighs its cost, as it limits your entire project throughput. So it should also be a resource that is used relatively heavily by your projects.

Although identifying the drum is usually not difficult; it can sometimes cause frustration for your project managers. Some prefer to take an analytical approach and add up the resource demands from all of the extant project plans. This usually does not work because resource identification is not sufficiently precise to do so, and the set of resource loaded project plans is often a poor reflection of the actual projects that the organization will perform.

Selection of the wrong resource as the drum is not a fatal mistake. You will get information to correct the error. More importantly, selection of any resource and using it to stagger the start of projects will improve your organization performance. Selecting the right resource will improve it the most, but often the difference of staggering projects to any resource far outweighs the importance of having the right constraint to start with.

Project Priority. The next step is to prioritize all projects that will be performed. The priority is for access to the drum resource and for scheduling the start of projects. It has little to do with the absolute value of the project, although if your information is good enough to do so, you should place higher priority on the projects that deliver the greatest throughput per use of the drum resource.

Project prioritization must become an ongoing process. All new projects must be placed in the priority list before they are scheduled. You can adopt any priority rules you wish, but you must stick to them on an ongoing basis.

In parallel with setting the project priority, plan each project using the planning tool to create a resource loaded critical chain schedule, just as in the single project case. Each project should assume access to the maximum amount of each resource that it can use effectively, as if it was the only project the company was doing. Each project should assume 100 percent dedication of the resources over the duration of the activity.

Subordinate to the Constraint. Resource B is the constraint (drum) in the Figure 33.7 case. A final step necessary to get the start and completion of the projects is to place a properly sized capacity constraint buffer (CCB) between the use of the drum resource in one project and its use in the next. Keep in mind that the drum schedule uses mean activity times. Figure 33.7 shows the placement of the CCB between the lower projects. It does not actually appear in a project schedule, but it does set the start date for the schedule by establishing a start-no-earlier-than constraint on the first drum using activity in the downstream project.

The manager of the identified drum resource can then take all of the project plans and their priorities and create a schedule for the work of the drum resource. The schedule for the drum resource on each project will then determine the start and completion date of each project. (This discussion can be an oversimplification if projects use the drum resource for many tasks. Describing the process for those cases goes beyond the scope of this essay, but the illustration here adequately describes the function of the process.)

Multiple-project CCPM does not resolve all resource contentions across all projects. That is one of the great simplifications provided by the TOC approach. Experience demonstrates it is not possible to maintain information integrity across all projects on a timescale comparable with the rate at which things change. Since all non-drum resources have excess capacity, on a statistical basis they cannot control project output. Because of statistical fluctuations, there will be times that the demand for nonconstraint resources exceeds capacity. The buffer management rules tell the resources the sequence of tasks to work on. The buffers absorb the impact of the delay of some tasks.

Multiple-Project Buffer Management. Multiple-project buffer management adds one rule to the resource decisions. Project priority comes first. Then, the remaining buffer management rules apply.

Organization-Wide Implementation of CCPM

Any individual can productively use the principles of TOC and CCPM. For example, you can eliminate bad multitasking in everything you do. You can use the five focusing steps at home and at work to focus your energy where it matters most. Most project managers can use mean estimates, buffers, the critical chain instead of CPM, CCPM status determination and reporting, and buffer-threshold-based decision making without disturbing the rest of your company. But if you are in a multiproject environment (as most of us are), to really supercharge your company's project results, you are going to have to lead a company-wide change.

Geoffrey Moore wrote two books about bringing a new product into the mainstream. You can think of CCPM (or any significant new management process) as a new product, and the managers and employees of your company as the market. *Crossing the Chasm* (Moore, 1991) describes how the original new product adoption life cycle traces through a series of customers, starting with Innovators and proceeding through Early Adopters, Early Majority, Late Majority, and Laggards. He added the concept of the chasm between the Early Adopters and Early Majority. Most new products die in this chasm. Most new management processes (which some call fads; e.g., excellence, TQM, just-in-time, teams, reengineering) die in this chasm as well.

To succeed, you must plan and manage your implementation like a new product introduction. Although new products such as TQM worked very well for the early adopters, they fell into the chasm in later companies for a variety of reasons. Moore suggests, "The winning strategy does not just change as we move from stage to stage, it actually reverses the prior strategy . . . [it is] the need to abandon each one in succession and embrace its opposite that proves challenging.

Usually the ones interested in CCPM at first are the techies . . . new project managers, schedulers, project control people, and in some cases software people. Your best project managers may not feel they need it, because they have learned how to succeed against all odds and aren't necessarily interested in making it easier on their internal competitors. Early adopters are visionaries who will look to the promised benefits and give it a try. You may find a senior management sponsor willing to take on this role. If so, you can make it to the chasm. To cross it, you have to bring along the Early Majority. They are the pragmatists. These "show me" people are generally more interested in evolution than revolution. Moore characterizes them as people who "look to adopt innovations only after a proven track record or useful productivity improvement, including strong references from people they trust." This means you are going to have to run some credible pilot projects in your company to convince these people and bring them along.

Before CCPM becomes "the way we do projects around here," you need to bring the Late Majority on board. They are the real conservatives. They are pessimists about anything new and are only going to change what they are doing under duress. The key to getting them to follow is to make it simply easier to do it the new way, and perhaps to close the door to doing it the old way.

You can safely almost forget about the Laggards. They will not change. They are skeptics, and not worth the investment it will take to bring them along. You simply have to keep them from blocking your way.

Implementing CCPM requires changing certain behaviors. For example, in many companies the ability to multitask is considered a virtue. When you are deploying CCPM, it is a vice for any project resources to multitask while working on a project activity. Usually multitasking behavior is a result of feedback from management that encourages it and often due to the inability of management teams to agree on a company priority. Functions tend to suboptimize and do what is best for their function, instead of what is best is for the company. Frequently this is at least in part because they have no way of knowing what is best for the company. Tables 33.1 through 33.6 list potential behavior changes that you may face. Not all companies reflect the present behavior listed so you may not have to address all of the behaviors as a change from a contrary behavior. You do have to succeed to create the CCPM behaviors.

Because these behaviors have been in place a long time, it will be difficult to change them. It usually requires focusing on changes at the individual, social group, and management level to achieve the full benefits of CCPM. Hellriegel et al. (1998) poses an organization behavior framework that addresses individual processes, group and interpersonal processes, organizational processes, and organizational issues. They clarify that you must consider this model in light of environmental influences that affect behavior in each element of the model. They note that "the management of change involves adapting an organization to the demands of the environment and modifying the actual behavior of employees."

The following sections outline the actual behavior of employees' changes that many companies face. Each company has to develop its own plan for how to cause the change.

As a common saying goes, "If you always do what you've what you've always done, you always get what you always got." You can't expect to double throughput without changing what you do.

Some people adapt to change better than others—they are the Early Adopters and your initial customers. Most people follow the crowd; your strategy is to gradually enlist

TABLE 33.1. BEHAVIOR CHANGES SENIOR MANAGEMENT MUST MAKE.

Change	Present Behavior	Future Behavior
Only commit to feasible delivery dates.	Sometimes commit to arbitrary delivery dates: determined without consideration of system capability to deliver.	Only commit to delivery dates with a critical chain plan and (if multiple projects), after sequencing through the drum schedule.
Eliminate interruptions.	Insert special requests to the system with no assessment of system capability to respond. Sometimes place demands for routine administrative work above project work (e.g., salary reviews).	Prioritize all requests using buffer report.
Set project priority.	No clear project priority, or changing project priorities,	Set project priorities; including the priority of new projects relative to ongoing projects.
Select drum resource.	No consideration of system constraint.	Select the drum resource to be used for sequencing the start of projects and creating the drum schedule.
Select drum manager and approve project sequencing.	Start each project independently as funding is available.	Drum manager creates drum schedule. Senior management approves. Project managers schedule projects to the drum.
Project Status.	Looking over shoulders.	Buffer report

TABLE 33.2. BEHAVIOR CHANGES RESOURCE MANAGERS MUST MAKE.

Change	Present Behavior	Future Behavior
Resource priority.	Assign resources on a first come, first served priority; or attempt to meet all needs by multitasking.	Assign resources using the buffer report.
Resource planning.	Plan resources by name and activity.	Plan resources by type, and assign to activities as they come up using the buffer report priority.
Early completion.	Turn in activities on due date.	Turn in activities as soon as they are complete.
Eliminate multitasking.	Ensure resource efficiency by assigning to multiple activities at the same time.	Ensure resource effectiveness by eliminating bad multitasking.
Resource buffers.	Resources planned far ahead and not available when needed.	Use resource buffers and buffer report to dynamically assign resources to activities.

TABLE 33.3. BEHAVIOR CHANGES PROJECT MANAGERS MUST MAKE.

Change	Present Behavior	Future Behavior
Mean activity duration estimates.	Project managers send a message that they expect due dates to be met.	Project managers first get low-risk activity duration, and then get "average" duration; using activity uncertainty to size buffers.
Roadrunner/relay racer activity performance.	Provide start and finish dates for each activity, and monitor progress to finish dates.	Provide start dates only for chains of activities, and completion date only on the project buffer.
Feedback on activity duration overruns.	Management provides negative feedback when activities overrun due dates.	Management provides positive feedback and help if resources perform to roadrunner paradigm.
Project status.	Varies. Often use earned value as the schedule measure.	Buffer report (including a cost buffer).
Project changes.	Varies. Often submitted to minimize minor variances.	When triggered by buffer report.
Response to management demands for shorter schedule.	Arbitrary activity duration cuts.	Add resources or make process changes to get a feasible and immune schedule.
Early start.	Start activities as early as possible.	Start activity chains as late as possible, buffered by feeding buffers.
Sequence projects.	Start project as soon as funding is available.	Schedule project start using drum schedule.
Assign resources dynamically according to critical chain priority and buffer report.	Get resources as soon as project funding is available, and hold resources until they can't possibly be used anymore on the project.	Get resources only when needed, and release as soon as activity is complete.

TABLE 33.4. BEHAVIOR CHANGES PERFORMING RESOURCES MUST MAKE.

Change	Present Behavior	Future Behavior
Perform 100% on one activity at a time.	Multi-activity to satisfy all management demands.	Eliminate bad multitasking (i.e., multitasking that lengthens project activity duration.)
Turn in early completion.	Turn in on due date.	Turn in as soon as "good enough."
Eliminate (bad) early start.	Start as soon as funding is available.	Start chains as scheduled.
50-50 duration estimates.	Provide protected (low-risk) activity duration estimates.	Provide 50/50 duration estimates.

TABLE 33.5. BEHAVIOR CHANGES SUBCONTRACTORS MUST MAKE.

Change	Present Behavior	Future Behavior
Deliver to lead times.	Deliver to due dates.	Deliver to lead times.
Shorten lead times.	Deliver to due dates.	Shorten lead times.

those groups be demonstrating success with the change leaders. The people you enlist will provide the testimonials to bring the later crowd along.

Moore's second book, *Inside the Tornado* (1995), deals with the phases of product growth after the chasm. In the case of CCPM implementation, crossing the chasm means you have had a few successful pilot projects. Once you have crossed it, you can have unprecedented success bringing this innovation to your company . . . rapid growth that Moore equates to the chaos of a tornado. As the leader of this innovation, it will tax your resources mightily. But if you do not lead it properly, your innovation will fall by the wayside along with the other management fads your company has tried. To quote Moore:

> The market is not yet ready to buy in as a whole. It still has too much invested in the old paradigm and will drag its feet for some time to come. If you try a broad frontal assault now, you will only consume your resources in advance of the real opportunity. Instead, it is time to focus on winning niches of the marketplace, made up of customers who are marginalized under the old paradigm, not well served by it, and who find themselves under pressure to reengineer their businesses to become more competitive.

The market and customers we are addressing are the other project managers and resource managers in your company.

Eliyahu Goldratt (1996) posed a model for organizational change called the "layers of resistance model." This model focuses on change from the perspective of answering concerns

TABLE 33.6. BEHAVIOR CHANGES PROJECT CUSTOMERS SHOULD MAKE.

Change	Present Behavior	Future Behavior
Eliminate project scope changes.	Customers spend little time initially establishing requirements, and then introduce late changes	Establish requirements as part of the project work plan; change as little as possible with formal change control.
Support using project buffer.	Customers interpret contingency as "fat."	Customers understand the need for buffers to reduce project lead time and ensure project success.
Eliminate arbitrary date milestones.	Demand arbitrary date milestones.	Use plan to set milestones.

of individuals. Two of these concerns are obstacles within the organization that might prevent deploying CCPM and potential unintended consequences if the changes succeed. It helps to reduce resistance if you solicit and address these concerns as part of your change plan. I believe that these considerations are necessary, but not sufficient to bring about the behavior changes needed for complete success with CCPM.

Software

Four commercial software packages are currently available for deploying CCPM, and more are expected. Eventually we expect to see it as an option in all schedule software.

It is feasible and reasonable to plan and manage critical chain projects with primitive project planning and control tools. Some very large construction projects ($250 million) did it using colored pieces of paper to represent resources. The following are three software developers currently providing CCPM software.

CCPM+

The latest offering on the market, developed under the direction of the author, CCPM+ provides an add-in to plan and execute critical chain projects. It allows using more features of MS project than other offerings (e.g. task priority and a variety of relationships) and employs an innovative algorithm to help resources determine which task to work on next. More information on CCPM+ is available at http://www.advanced-projects.com/CCPM+.htm.

Concerto

Concerto is an innovative solution to deploying CCPM in a multi-project environment. It employs many features helpful to larger organizations, including a WEB-based interface to update progress and display a variety of task and resource information in useful formats. It is the most expensive tool to implement CCPM. More information is available at www.realization.com/.

Sciforma PS8

Sciforma PS8 is a full-function tool that rivals Microsoft Project in all ways and has the inherent ability to deploy CCPM. It is intuitive and easy for anyone skilled in other mainline project software (e.g., MS Project, Primavera) to adapt to. It is very flexible in reporting and well suited to multiple projects of any size. More information is available at www.sciforma.com/products/ps_suite/ps_suite.htm.

Any critical path scheduling tool can aid planning and controlling critical chain projects. Most such tools have the ability to resource-load and -level the plan. You can put in buffers as dummy activities. You can put in the resource constraint using the activity-linking capability of the software. Although this will become daunting if you have more than about 50 activities in your plan, it provides some control advantages over using off-the-shelf software that usually contains "undocumented features" that can influence your plan. The specific procedures that you must develop depend on the software you use.

Summary

CCPM integrates the critical chain approach to scheduling and schedule control with the rest of the PMBOK (PMI, 2000). It is one step in the process of ongoing improvement to the project delivery system. But it is only that. Projects often go awry because of causes that critical chain does not address at all, such as failure to get and maintain stakeholder alignment, poor scope definition, or ineffective quality assurance or change management. Critical chain only addresses common-cause variation impacts on the project. CCPM adds conventional deterministic project risk management to address special cause variation (Leach, 2001).

When the other PMBOK® necessary conditions are in place, CCPM has the potential to reduce overall project duration for single projects by up to one-half. Overall, project duration reduction in the multiple-project environments can be significantly larger, depending on the amount of bad multitasking the organization engages in. Much of this acceleration comes about because the information provided by critical chain schedules and buffer management enables resources to avoid bad multitasking.

Critical chain enforces the discipline of developing effective schedule logic, resource loading, and resource leveling (Elton, 1998). For many project environments, taking these steps without critical chain would resolve much of the project chaos and improve schedule predictability.

Most importantly, critical chain uses the reality of variation in activity performance duration and uncertainty in estimates. It integrates the knowledge of variation in a simple way that makes it unnecessary for project managers to become experts in simulation or statistics. It connects the method to address uncertainty in developing the schedule with the measurement and control system (buffer management) in a way missed by previous approaches to analyze uncertainty.

References

Dettmer, W. H. 1997. *Goldratt's theory of constraints: A systems approach to continuous improvement.* Milwaukee: ASQC Quality Press.

Elton, J., and J. Roe. 1998. *Bringing discipline to project management. Harvard Business Review* 76 (2, March–April): 78–83.

Goldratt, E. M. 1985. *The goal.* Great Barrington, MA: North River Press.

———. 1996. *My saga to improve production.* New Haven, CT: Avraham Y. Goldratt Institute.

———. 1997. *Critical chain.* Great Barrington, MA: North River Press.

Hellriegel, D., J. W. Slocum, Jr. and R. W. Woodman. 1998. *Organizational behavior.* 8th ed. Cincinnati: South-Western.

Kahneman, D., P. Slovic, and A. Tversky. 1982. *Judgment under uncertainty: Heuristics and biases.* New York: Cambridge University Press.

Leach, L. P. 2000. *Critical chain project management.* Boston: Artech House.

———. 2001. *Putting quality into project management.* Parts I and II *Per Network.* Newtown Square, PA: Project Management Institute.

———. 2003. *Schedule and cost buffer sizing: How to account for the bias between project performance and your model. Project Management Journal* 34 (2, June): 34 ff.

Moore, G. A. 1991. *Crossing the chasm.* New York: HarperCollins.

———. 1995. *Into the tornado.* New York: HarperCollins

Newbold, R. 1998. *Project management in the fast lane.* Boca Raton, FL: St. Lucie Press.

Project Management Institute. 2000. *A guide to the Project Management Body of Knowledge.* Newtown Square, PA: Project Management Institute.

CHAPTER THIRTY-FOUR

PROJECT PERFORMANCE MEASUREMENT

Daniel M. Brandon, Jr.

This chapter concerns project performance both from the classical shorter-term "tactical" perspective and the longer-term "strategic" perspective. Despite many innovations in project management and performance methods, many projects fail; in some industries, such as information technology (IT), most projects still fail. A Standish Group International study found that only 16 percent of all IT projects come in on time and within budget (Cafasso, 1994; Johnson, 1995). Field's study discovered 40 percent of information services (IS) projects were canceled before completion (Field, 1997). The problem is so widespread and typical that many IT professionals accept project failure as inevitable (Cale, 1987; Hildebrand, 1998).

This continued failure in project performance can be attributed to a large degree to lack of long-term successful strategies and accompanying metrics. Earned value analysis (EVA) has proven successful for accurate performance measurement and improvement for time and cost on a single project (within a constant scope and quality). And while time and cost values (and accurate estimates of those values at project completion) are very important, there are other important success criteria for project-based work including employee attitudes, the satisfaction of all stakeholders, and leaving a legacy of relevant "stuff" for future projects. Clear identification of these success criteria (and the underlying critical factors leading to success for each criteria) is essential for overall project success. For longer-term success, a system of methods and metrics has not yet been widely adopted for the project management discipline. The balanced scorecard (BSC) is a strategic long-term management approach that has had considerable success in the last decade. Currently, EVA and BSC are two of the most powerful and successful management tools for measuring and increasing performance. However, the two techniques have not been used together. This chapter explains and illustrates these approaches (EVA, BSC, and critical success factors) and also presents a method to integrate the approaches together to achieve both short-term tactical success and long-term strategic project management success.

Project Performance and Feedback Control Systems

The Project Management Institute defines "process groups" (initiation, planning, performing, controlling, and closing) and "knowledge areas" that are related, as shown in Figure 34.1. Later when the strategic issues in project management are addressed, this "knowledge matrix" may require extension. In this chapter our primary focus is on execution and control; however, while the focus is on just those two process groups, there are key activities within process planning and other groups that are key to both short-term and long-term successful performance improvement.

The basic control process used in project management is the same process used in most engineering and business systems. It is based on the definition and establishment of key measures, then the comparison of those measurements to some desired values or standards to formulate algebraic formulas usually called metrics. If the difference between the measurement and the desired value exceeds some threshold, then corrective action (feedback) of some type is invoked and the degree of corrective action may be a function of the size of the difference (and/or the integration [accumulation] or differentiation [rate] thereof). The measurements may be of process outputs or of the process itself, and the measurement level may be tactical (generally how things are being done) or strategic (generally what things are being done). This is illustrated in Figure 34.2.

What to Measure and Control

Once the decision is made to measure and control performance, the next questions involve "what to measure" and "what to control." Project performance typically involves a trade-off of several dimensions, specifically, what is done (scope and quality) versus the resources used to do the work (time and cost). Scope involves both stated needs (requirements or deliverables) and unstated needs (expectations). Quality involves both built-in quality (internal and/or less readily noticeable aspects) and inspected quality (visible and/or noticeable aspects).

In terms of resources, the consumption of those resources is usually measured. But also important are other factors that might affect the behavior of those resources, including the elements of risk and satisfaction. So a list of things that could possibly be measured include scope (percent complete), time (calendar), cost, risk, quality, and the satisfaction and attitudes of the following:

- Project team, performing organization, and line management
- Benefiting organization ("customer")
- Project sponsors and other "stakeholders" (suppliers, collaborators, etc.)

FIGURE 34.1. PMI PROCESS GROUP AND KNOWLEDGE AREAS.

	Initiation	Planning	Executing	Controlling	Closing
Integration		Project Plan Development	Project Plan Execution	Overall Change Control	
Scope	Initiation	Scope Planning	Scope Verification	Scope Change Control	Scope Verification
		Scope Definition			
Time		Activity Definition		Schedule Control	
		Activity Sequencing			
		Activity Duraiton Estimation			
		Schedule Development			
Cost		Resource Planning		Cost Control	
		Cost Estimating			
		Cost Budgeting			
Quality		Quality Planning	Quality Assurance	Quality Control	
Human Resources		Organizational Planning	Staff Acquisition	Team Development	
Communications		Communications Planning	Information Distribution	Performance Reporting	Administrative Closure
Risk	Risk Identification	Risk Identification		Risk Response Control	
		Risk Quantification			
		Risk Response Development			
Procurement		Procurement Planning	Solicitation	Contract Administration	Contract Closeout
		Soliciation Planning	Source Selection		
			Contract Administration		

FIGURE 34.2. THE GENERIC CONTROL CYCLE.

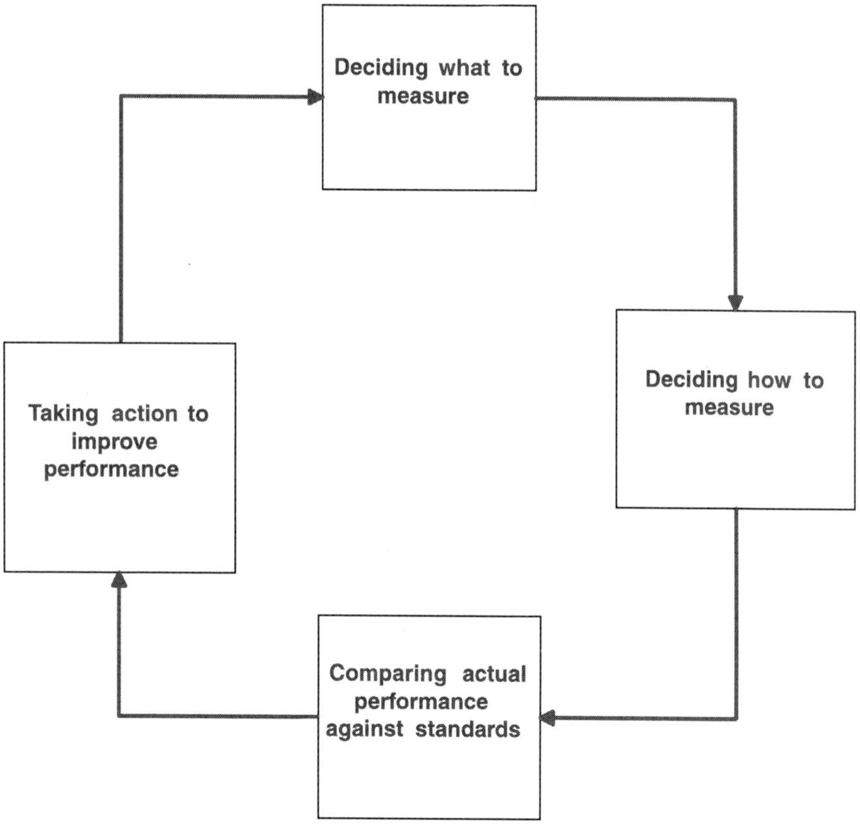

What can and should be measured varies with the type of project and the perspective of the organization managing the project. This is further discussed and quantified later in this chapter under the discussion of success criteria.

Measurement of all relevant variables is important for both management information and also for the specification of "what kind" and "how much" corrective action is necessary. Corrective actions are management prerogatives that are available to a manager based upon the type of organization (functional/hierarchical, projectized, matrix), the position of the project manager, the organization culture, and the governing laws of the state or nation. Examples of corrective action include fast tracking, "crashing," adding additional resources, (people, money, time, etc.), scope reduction, compromising quality, increasing risk, employee/contractor disciplinary actions (negative reinforcement), employee/contractor incentives (positive reinforcement), pep talks and other motivation, and so on. Some corrective actions tend to be more tactical, and some more strategic.

Earned Value Analysis (EVA)—Time and Cost Control within Scope

Traditional project measurement at the tactical level involved only "cost versus budget" and qualitative percent activity complete (Gantt chart). Today, improved quantitative methods are coming into general use such as EVA. PMI now uses the term EVM (earned value management) to describe not only the analysis portion of earned value but the overall management approach of using earned value; both of these topics are included in this discussion. Earned value can be used to more accurately evaluate the performance of a project both in terms of cost deviation and schedule deviation. It also provides a quantitative basis for estimating actual completion time and actual cost at completion. It is one of the most underused cost management tools available to project managers (Fleming, 1994). Refer to Brandon (1998 and 1999) for a more complete discussion of EVA and the effective implementation thereof.

The earned value concept has been around in several forms for many years dating back to types of cost variances defined in the 1950s. In the early 1960s PERT (program evaluation review technique) was extended to include cost variances and the basic concept of earned value was adopted therein. PERT did not survive, but the basic earned value concept did. For many U.S. government contracts, the nature of the contract was such that the government assumed most of the burden for cost overruns. To minimize those overruns, the government was in search of project performance measurements to better control cost. Thus, the earned value concept was a key element in the 1967 DoD (Department of Defense) policy called Cost/Schedule Control Systems Criteria (C/SCSC). Early implementations of C/SCSC met with numerous problems—the most common of which was "overimplementation" because of excessive checklists, data acquisition requirements and other paperwork, specialist acronyms, and overly complicated methods and tools. However, C/SCSC has been refined over the years and is now very effective. The government has accumulated many years of statistical evidence supporting it, and earned value has now met the test of time for nearly three decades on major government projects.

Earned value is basically the value (usually expressed in dollars) of the work accomplished up to a point in time *based upon the planned (or budgeted) value* for that work. The U.S. government's term for earned value is "budgeted cost of work performed" (BCWP). Note that EVA analysis is based upon a predefined scope of work and also a predefined specification for the degree of "quality" built into the resulting product.

Typically when a schedule is being formulated, the work to be done is broken down into tasks or work packages that are organized into a logical pattern usually called a *work breakdown structure* (WBS). The WBS is usually formulated in a hierarchical manner that may follow methodology established for a particular industry or organization. The amount and type of cost-to-complete each work packet is then estimated and resources to perform the work are identified, either generically or specifically. The estimated cost is typically a function of the amount and type of resources; dependent tasks are identified (a list of tasks that must be completed before starting this task).

These tasks are then typically input to a scheduling program that produces a time phasing of task start and end dates based upon the project start date, task resource needs, resource availabilities, and task interdependencies. When these tasks are rolled-up the WBS hierarchy, the total cost plan is derived, as shown in Figure 34.3. The U.S. government's term for this planned cost curve is "budgeted cost of work scheduled" (BCWS).

As the project progresses, actual cost are incurred by the effort expended in each work package and the total actual cost can be plotted as shown in Figure 34.4. Also, the relative amount of the things needed to be accomplished within the work packet that have actually been completed (% complete) can be determined or at least estimated. For example, if an activity had an estimated total cost of $10,000, and if the things to be done in the activity were 70 percent complete (or 70 percent of the elements were complete), then the earned value would be $7,000. Since percent complete and earned value can be estimated for each work packet, the total project earned value at a point in time can be determined by a WBS roll-up of the values.

The earned value is a point on the planned cost (BCWS) curve. This is illustrated in Figure 34.5, which shows the planned cost and actual cost curves for a project analysis. Variances between the three values BCWS (planned cost), BCWP (earned value), and actual cost (ACWP) yield the earned value metrics. There are earned value metrics available for both cost and schedule variances. The cost metrics are as follows:

- Cost variance (dollars) = BCWP − ACWP
- Cost variance (percent) = (BCWP − ACWP) × 100%/BCWP

FIGURE 34.3. THE PROJECT COST CURVE.

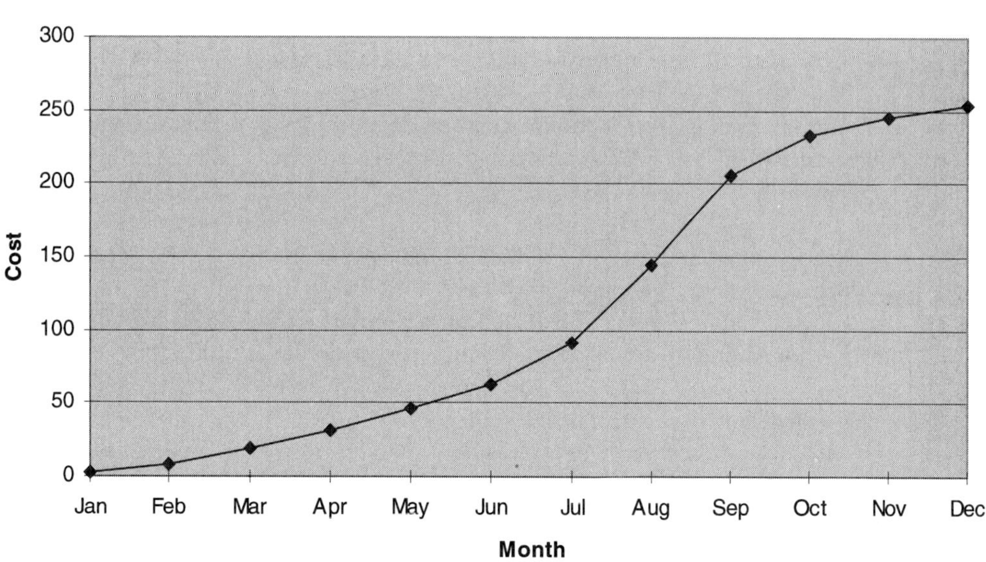

FIGURE 34.4. COMPARISON OF ACTUAL (TRIANGLE) VERSUS PLANNED COSTS (DIAMOND).

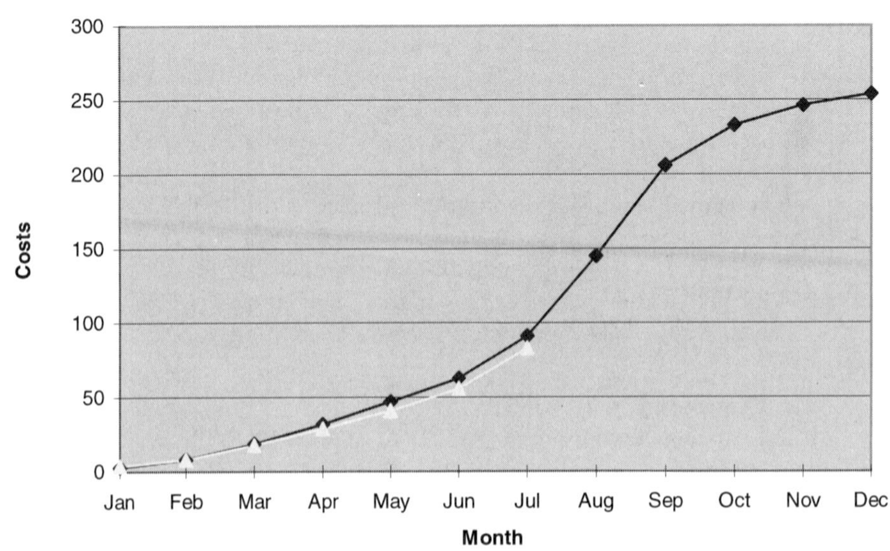

- Cost efficiency factor or Cost Performance Index (CPI) = BCWP/ACWP
- Estimated cost at completion (EAC) = Budget at completion/(CPI)

There are several other EAC formulas, and the most appropriate depends upon project type and when the EAC is calculated (Christensen, 1995). The schedule metrics are as follows:

- Schedule variance (dollars) = BCWP − BCWS
- Schedule variance (months) = (BCWP − BCWS)/(Planned cost for month)
- Schedule efficiency factor or Schedule Performance Index (SPI) = BCWP/BCWS
- Estimated time to complete = (Planned completion in months)/(SPI)

The schedule variance in time (S) is shown in Figure 34.5 along the time axis.

Usually when project progress is reported, two types of information are presented: schedule data and cost data. Schedule data is typically shown in a Gantt or similar type chart, as shown in Figure 34.6 for an example project. Cost data is typically reported as actual costs versus planned costs at some upper level of the WBS. The cost variance is often just reported at the total level as total actual cost incurred versus budget.

FIGURE 34.5. GRAPHICAL DEPICTION OF EARNED VALUE.

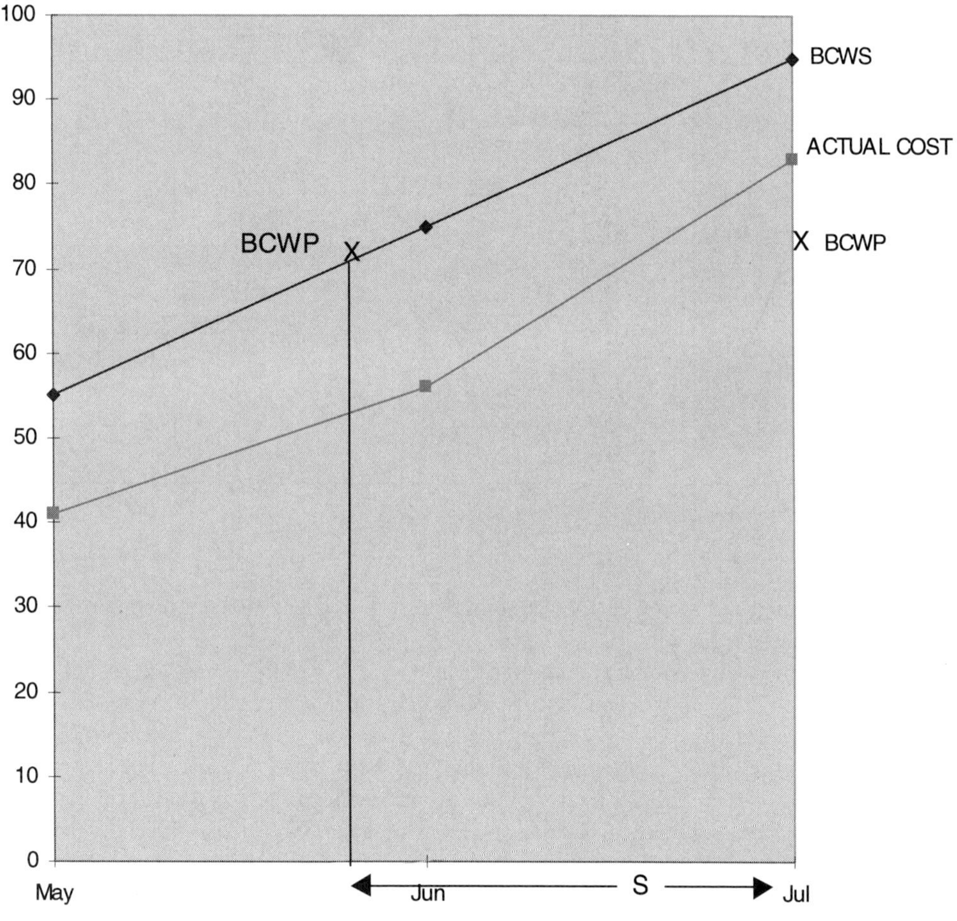

The problem with these usual methods is that they do not provide a clear quantitative picture of the true project status, nor do they provide a means for extrapolating project cost to complete or completion date. For example, when we look at the Gantt chart (which also shows the task % complete as dark bars stripes inside the bars) and we then see that we are not over budget (say, an actual cost of 83,000 versus planned cost of 91,000), it is hard to determine how much we are behind schedule and it *appears we are not overspending.*

However, on this project *we are well behind schedule and are overspending.* An earned value analysis would show that the schedule variance is 0.67 months (behind schedule) and the cost variance is $10,000 (overspent). The estimated time to complete is 15 months instead of the 12 months planned, and the estimated cost to complete is 289,000 instead of 254,000.

FIGURE 34.6. A GANTT CHART SHOWING SCHEDULE DATA.

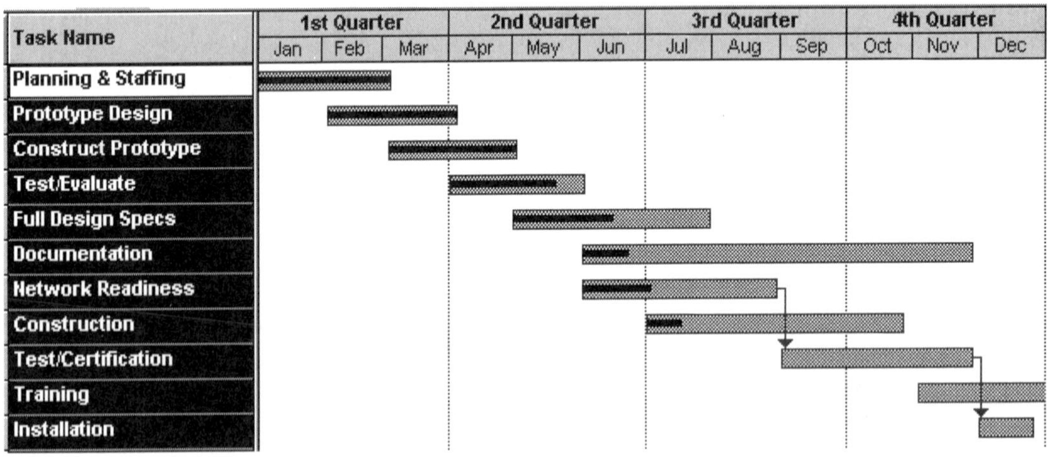

Task Name
1st Quarter: Jan Feb Mar
2nd Quarter: Apr May Jun
3rd Quarter: Jul Aug Sep
4th Quarter: Oct Nov Dec
Planning & Staffing
Prototype Design
Construct Prototype
Test/Evaluate
Full Design Specs
Documentation
Network Readiness
Construction
Test/Certification
Training
Installation

The exact calculation of these numbers and their illustration in a spreadsheet is shown in Brandon (1998).

Often, *effective* implementation of earned value is difficult in some organizations. A common key problem involves data acquisition of "percent complete" data. Data on task completion must be obtained at regular intervals, but the effort involved in gathering the information must not be burdensome. It is best to set up this data acquisition as a byproduct of some other required reporting mechanism such as weekly time card reporting.

Another common problem concerns the accuracy of the estimates versus the time required to calculate the estimate. As long as the project WBS is developed to point where each task at the lowest level of the WBS is only a person-week or so effort, then approximation techniques are sufficiently accurate and quick. There are a number of approximation techniques described in the EVA literature: One of the simplest uses 0 percent for a task that has not begun, 50 percent for a task that has started, and 100 percent for a task that is complete.

Another potential EVA problem area concerns employee resistance and the honesty of completion percentage reporting. EVA should not be used directly for employee evaluations; doing so will certainly compromise the main purpose of the system: project performance measurement. Computer systems to do EVA analysis are available, and an organization can always program their own (or set it up with spreadsheets). For example, a computer EVA system can look for employee reporting problems such as showing the same % complete on a task from week to week where hours have been charged to that task. Note that most off-the-shelf project scheduling software will not offer these EVA management analyses. These issues are also addressed in Brandon (1998).

Earned value methods have another advantage over current reporting techniques (Gantt charts and cost versus budget). Since earned values are quantitative numbers expressed in dollars or person-hours (for both cost and schedule deviations), these numbers can be rolled up, along an OBS for example, to give a picture of how all projects of varying sizes are performing in an organization. Since underspent projects do not necessarily help overspent projects (in either time or dollars), often the positive variations are set to zero and estimate at completion (EAC) is unchanged. This is illustrated in Figure 34.7. If spreadsheet models are used for earned value, then these are easily interfaced with most executive information systems.

Success Criteria and Success Factors

EVA methodology is based upon time and cost metrics for a given scope and quality. Cost, time, and quality (often referred to as the "iron triangle") have formed the prime basis for measuring project success for the last 50 years (Atkinson, 1999). However, a number of authors in more recent years (Morris and Hough, 1987; Pinto and Slevin, 1988; DeLone and McLean, 1992; Lim and Mohamed, 1999; and Atkinson, 1999) have suggested that other criteria for success are also important. (See also the chapter by Cooke-Davies.) Some of these other criteria may be less quantitative and more difficult to measure. Also some of these criteria may be temporary in that their values may be much more important at some points in the project, but less important at other points like the end of the project.

Many of these authors raise the question of "what is a successful project." Different stakeholders involved with the same project may have different opinions about a project's success. One example given concerns the construction of a shopping center that is eventually

FIGURE 34.7. AN EARNED VALUE SPREADSHEET FOR MULTIPLE PROJECTS.

PROJECT	BCWS	BCWP	- TIME VAR ($)	VAR +	ACWP	- COST VAR ($)	VAR +	PLAN	EAC
Project 1	91	73	18	18	83	10	10	254	289
Project 2	130	135	-5	0	125	-10	0	302	302
Project 3	65	60	5	5	75	15	15	127	159
Project 4	25	23	2	2	27	4	4	48	56
Project 5	84	82	2	2	81	-1	0	180	180
Project 6	53	47	6	6	48	1	1	110	112
Project 7	102	103	-1	0	110	7	7	190	203
Project 8	35	37	-2	0	40	3	3	78	84
	585	560			589				1385

Total Schedule Overage	33	**Total Cost Overage**	40
Relative Schedule Overage	5.64%	**Relative Cost Overage**	6.84%
Schedule Overage (Months)	0.68	**Cost at Completions**	1385

completed to the quality standard—however, with significant cost and time overruns. Some stakeholders are very unhappy, and that depends upon the type of contracts involved and which party(s) contractually bear the burden of the cost overruns. Other stakeholders, such as the public using the mall and the merchants renting space in the mall and the government getting tax revenue from the mall, are pleased with the results and see the project as a great success. They define two perspectives: the macro perspective, which involves all the stakeholders, and the micro perspective, which involves only the construction parties, such as the developer and contractors. (See the chapter by Winch on stakeholder management.) The macro perspective is relevant for all phases of a project from conceptualization through construction and then operation. The micro perspective is most relevant for the construction phase. This is a theme explored extensively later in this book.

Lim and Mohamed (1999) define two types of success criteria: completion and satisfaction. Completion criteria include contract-related items such as cost, time, scope, and quality. Satisfaction criteria include utility (fitness for purpose) and operation (ease of use, ease of learning, ease of maintenance, etc.). The macro perspective involves both completion and satisfaction criteria; the micro perspective only involves completion perspectives. This is illustrated in Figure 34.8.

Lim and Mohamed draw a clear distinction between "success criteria" and "success factors." The criteria are "a principle or standard by which anything is or can be judged"; factors are "any circumstance, fact, or influence which contributes to a result". Figure 34.9 from their article illustrates this point. Factors for the completion criteria would typically include resource variables (cost, availability, skill, motivation, etc.), management variables (project manager skill, line management support, etc.), and risk variables (weather, economy, technology, etc.). Factors for the satisfaction criteria would be those things that drive the satisfaction of the stakeholders.

Success criteria tend to be relatively independent of the type of project being measured. However, the factors are very dependent on the type of thing being built (or accomplished). For example, a factor for the satisfaction type criteria of utility (using the shopping center model) might be "ample parking"; a factor for the operation factor of the satisfaction criteria might be "ease of parking".

DeLone and McLean (1992) developed a taxonomy of success criteria and factors for information systems. The six major criteria are system quality, information quality, use

FIGURE 34.8. A GENERALIZATION OF LIM AND MOHAMED'S TWO TYPES OF SUCCESS CRITERIA.

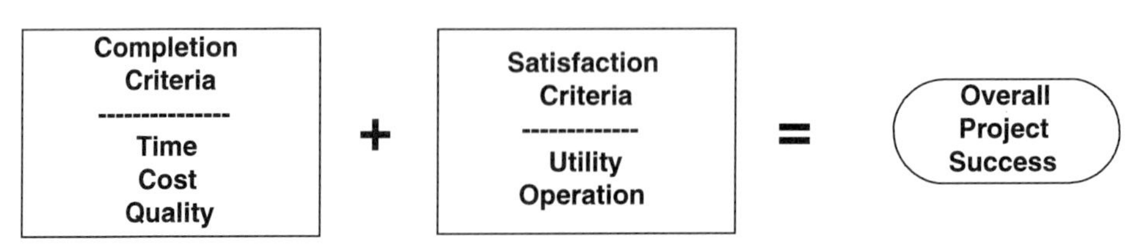

FIGURE 34.9. LIM AND MOHAMED'S RELATIONSHIP BETWEEN FACTORS AND CRITERIA.

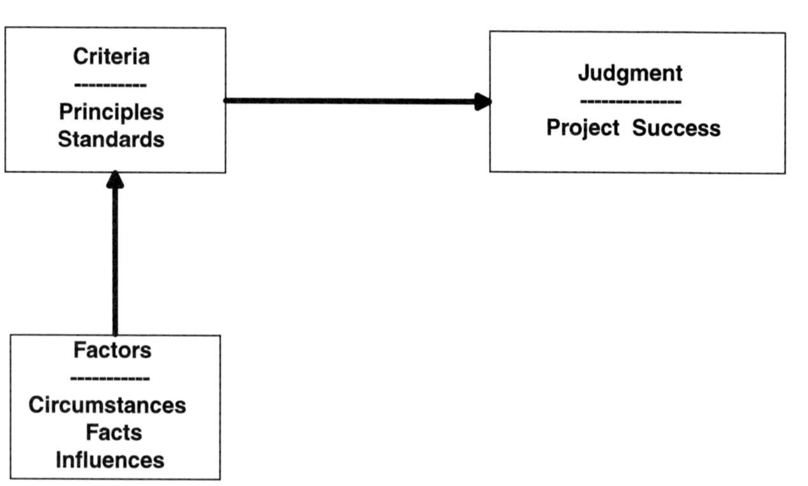

(amount), user satisfaction, individual impact, and organizational impact. Figure 34.10 illustrates this taxonomy: system and information quality foster use and satisfaction, which determine individual impact, which then determines organizational impact. They did not divide these into completion and satisfaction types, but the first two would likely be more completion-oriented, and the last four more satisfaction-oriented. Their taxonomy listed a number of factors (which they called measures) for each criterion. Molla and Licker (2001) recently extended DeLone and McLean's taxonomy to e-commerce systems. Their success criteria are system quality, content quality, use, trust, and support.

In terms of project performance measurement, both success criteria and factors are important and useful. Even the less quantitative factors can be used as an early-warning system to potential problem areas. We need to first determine our concern and perspective: macro or micro. Then we need to list our success criteria. Next the success factors for each criterion need to be determined. Then we need to set up feedback systems using our understanding of the cause-and-effect relationship between the criteria/factors and our management decision rights over adjustable items (resource type, compensation, etc.). For completion criteria (time, cost, quality, scope, performance, safety), EVA is an excellent performance tool. For the less quantitative factors (satisfaction, utility, etc.), some other metrics have to be employed. Many of these less quantitative success criteria and factors are also considered in PMI's PMBOK. PMBOK processes are defined to manage these factors, particularly in the area of communications, human resources, and risk. Metrics for these factors may involve formal or informal surveys, and these are discussed later in the *Balanced Scorecard (BSC) Approach* section coming up. Consideration of both the completion

FIGURE 34.10. DELONE AND MCLEAN'S TAXONOMY OF SUCCESS CRITERIA AND FACTORS.

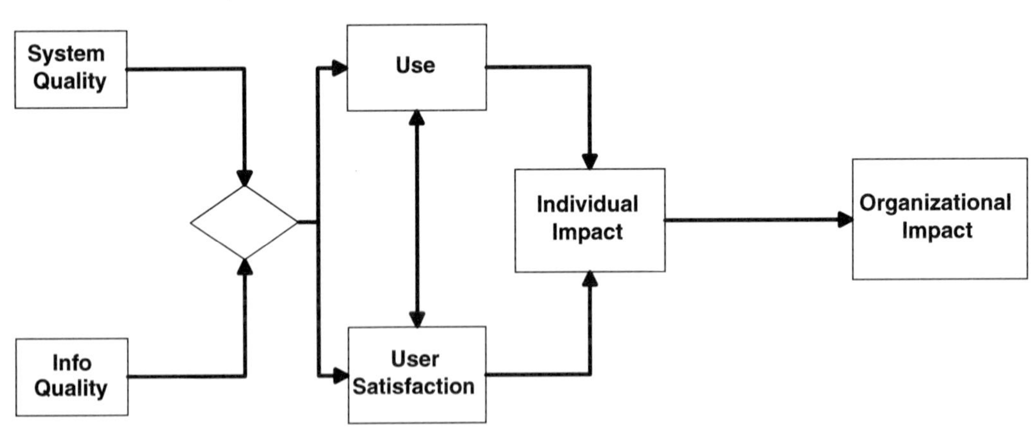

criteria and the satisfaction criteria leads to a more balanced approach toward project success and performance measurement.

Balancing Time/Cost with Satisfaction

EVA methods offer key management information to take corrective action primarily at the short-term tactical level for a particular project. When EVA metrics from all projects in a program are combined, management has a greater visibility to examine issues across multiple projects and take actions—some of which may be strategic as well as tactical. Project management is often criticized as being too shortsighted and limited to tactics that may not be in the best interest of the long-term success of an organization. Certainly, we understand (and most of us have seen cases) where employees are pushed too hard for success of one project only to become demoralized for performance on future projects.

Organizations should consider implementing a "corporate project management strategy". Implementing a strategy involves the development and/or adoption of some type of strategic foundation upon which to translate corporate vision and goals into winning strategies with effective tactics. Long-term winning strategies must address both the quality and satisfaction dimensions, as well as the financial and temporal dimensions. (This idea is examined in some detail in the chapters by Artto and Dietrich, Jamieson and Morris, Cleland, Thiry, and others in this book.)

A number of strategic foundations have been used by organizations in the past, but lately the balanced scorecard approach seems most successful and relevant in today's global economic setting. We know that today's economic setting is often typified by the phrase "better-cheaper-faster." We also realized that companies compete is a global economy today, and this is becoming even more so with e-commerce and the very rapid progression in technology.

However, the two main factors in the increased need for consideration of the quality and satisfaction dimensions in overall performance are the fact that traditional project management techniques are mostly tactical, not strategic, and the fact that the corporate value of intangible versus tangible assets has been shifting considerably in the last 20 years. Note that in disciplines that have always had problems with project performance (such as information technology), there were always a high percentage of intangible assets.

Balanced Scorecard (BSC) Approach

The BSC is an approach to strategic management that was originally proposed in 1992 by Kaplan and Norton (Kaplan, 1992). They recognized some of the problems with classical management approaches including, traditionally, an overemphasis on metrics that were strictly financial based and that looked mostly at the results of management decisions in the past. This rear-facing approach was becoming obsolete in today's fast-moving economic and technology-based economy. The balanced scorecard approach provides definitive procedures as to what companies should measure in order to balance this traditional prime focus of solely a financial perspective.

While the processes of modern business have changed dramatically over the past several decades (particularly in the last five years with the growth of the Internet), the methods of performance measurement have stayed much the same. Past measurement methods were well suited to asset-based, slow-changing manufacturing organizations. But past performance measurement systems are no longer as relevant to capture the value-creating mechanisms of today's modern business organizations (Kaplan, 1996). Today intangible assets such as employee knowledge, customer base and relations, supplier base and relations, and access to innovation are the key to creating value; Figure 34.11 shows this shift for U.S. companies. When we view financial statements (profit/loss, balance sheet, cash flow), Gantt charts, or even EVA analysis, we are seeing the results of actions and decisions that occurred in the past ("lagging indicators") (Eickelmann, 2003). Some of these past actions and decisions that determined an organization's present financial state may have taken place a month ago, a year ago, or a decade ago.

The balanced scorecard does not ignore financial matters but changes the perspective from one of reaction to one of proactive involvement. The balanced scorecard approach takes a look at the key management actions (including metrics) that will most likely affect a company's financial state *in the future* ("leading indicators") (Eickelmann, 2003). It has organizations first clearly and quantitatively define their vision and strategy, and then turn them into measurable actions. It provides feedback around both the internal business processes and external outcomes to continuously improve overall performance. When

FIGURE 34.11. THE GROWTH IN INTANGIBLE ASSETS IN U.S. COMPANIES SINCE THE 1980S.

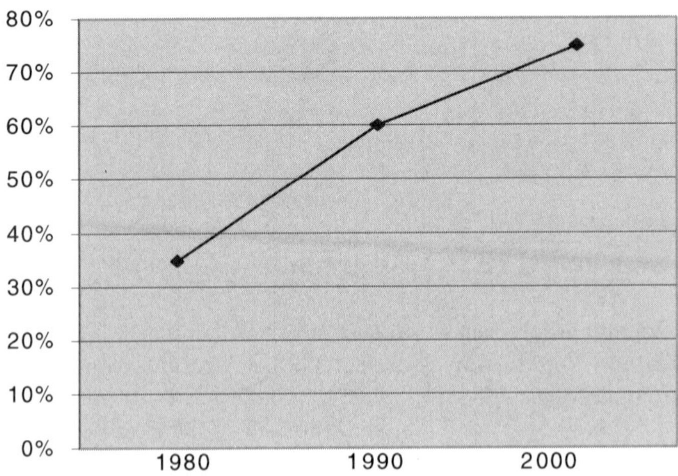

appropriately implemented, a "balanced scorecard approach transforms strategic planning from an academic exercise into the primary control mechanism" of an organization (Niven, 2002; Averson, 2002).

In the language of the founders of BSC (Kaplan and Norton):

> The balanced scorecard retains traditional financial measures. But financial measures tell the story of past events, an adequate story for industrial age companies for which investments in long-term capabilities and customer relationships were not critical for success. These financial measures are inadequate, however, for guiding and evaluating the journey that information age companies must make to create future value through investment in customers, suppliers, employees (Kaplan, 1996)."

Today BCS has been implemented successfully in thousands of organizations around the world, both "for-profit" and "nonprofit." Reengineering success ratios using BSC are reported as (Averson, 2002):

- Nonmeasurement-managed organizations (55 percent)
- Measurement-managed organizations (97 percent)

The BSC approach defines four "perspectives" from which an organization is viewed. Metrics are developed, then data collected and analyzed relative for each of these perspectives. This philosophy is illustrated in Figure 34.12.

The first perspective is Learning and Growth. This involves the investment in human capital through activities like positive feedback, motivational techniques, setting up mentors, communications facilitation, employee training, and development of a "company culture" supporting quality. In today's knowledge worker organization, employees (the main repository of know-how) are the main resource. In the current climate of rapid technological change, it is becoming necessary for knowledge workers to always be learning—lifelong education. Organizations often find themselves unable to hire new technical people and at the same time have reduced training of existing employees. This is a leading indicator of brain drain that can ultimately destroy a company (Averson, 2002). Metrics can be put into

FIGURE 34.12. THE FOUR MEASUREMENT PERSPECTIVES OF THE BSC APPROACH.

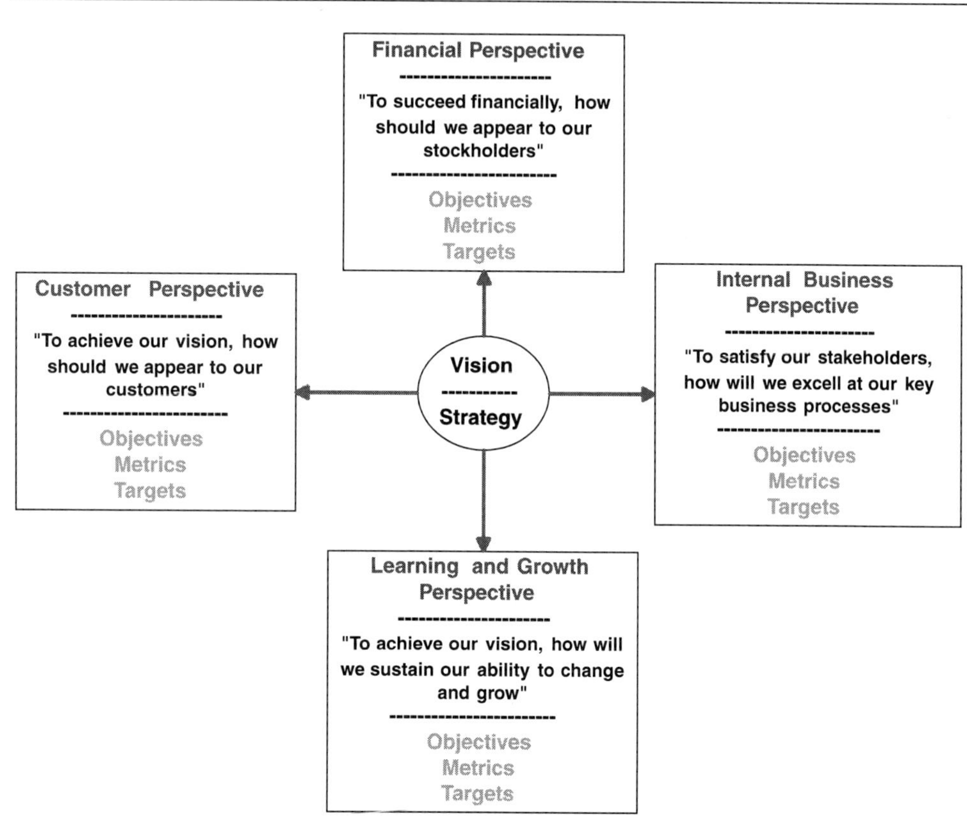

place to guide managers in focusing personnel development funds where they can help the most. This perspective is the foundation for long-term success of an organization.

The next perspective is the Business (or Internal) Process. This relates to the internal methods used to produce the goods and/or deliver the services. Metrics based on this perspective provide managers with knowledge on how well their business is running and is fine-tuned, and whether the products and services conform to customer-stated requirements and unstated expectations. These metrics have to be carefully designed by those who know these processes best. There are two kinds of business processes in an organization's value chain: primary (mission-oriented) processes, which are the unique functions of an organization, and support processes (accounting, legal, procurement, HR, etc.), which are more repetitive and common and hence easier to measure and benchmark using classical metrics.

The next perspective is the Customer. Recent management philosophy and the popularity of IT products as CRM (customer relations management systems) and SFA (sales force automation systems) have demonstrated an increasing appreciation of the importance of customer focus and customer satisfaction in any business. If customers are not satisfied, they will eventually find other companies that will meet their needs. Poor performance from this perspective is also a leading indicator of future decline, even though the current financial picture may look rosy. From a project management perspective, some customer relations are included in the PMI process groups and knowledge areas, particularly for external projects.

The last perspective is Financial. Kaplan and Norton do not ignore the traditional importance of financial data. Timely and accurate accounting data will always be important. Their point is that the current emphasis on financials leads to the unbalanced situation with regard to other perspectives. Today there is also a need to include additional financial-related data, such as risk assessment and cost-benefit data, in this category. Again from a project management perspective, financial data is a key metric, including EVA.

Figure 34.13 shows the perspectives in a waterfall, long-term, strategic cause-and-effect scenario. Motivation, skills, and satisfaction of employees are the foundation for all improvements. Motivated, skilled, and empowered employees will improve the ways they work and also improve the work processes. Improved work processes will lead to improved products and services that will mean increased customer satisfaction. Increased customer satisfaction will lead to long-term improved financial performance.

(The BSC methodology incorporates some key notions of Total Quality Management (TQM) such as customer-defined quality, continuous improvement, employee empowerment, and measurement-based management and feedback. In the early industrial revolution, quality control and zero defects were big management buzzwords. To prevent the customer from receiving poor-quality products and services, stringent methods and efforts were put into inspection and testing at the end of the production line. The problem with this approach, as pointed out by Deming (Walton, 1991), is that the true causes of defects can never be identified, and there will always be inefficiencies because of the rejection of defects. What Deming saw was that variation is created at each point in a production line, and the causes of defects need to be identified and corrected. Deming emphasized that all business

FIGURE 34.13. THE WATERFALL RELATION BETWEEN THE BCS MEASURES.

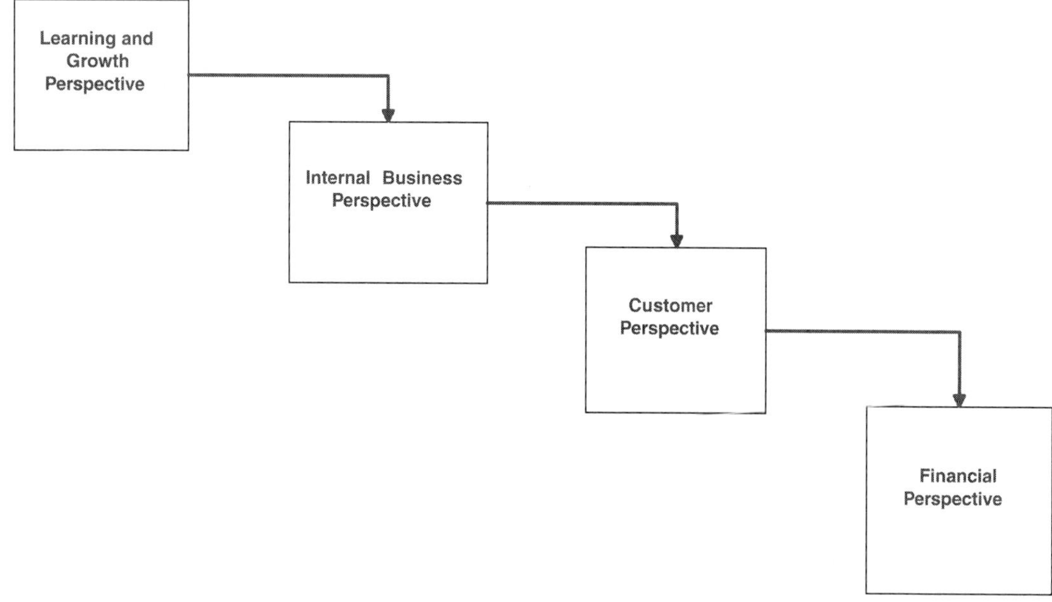

processes should be part of a system with feedback loops. The feedback data should be examined by managers to determine the causes of defects (the processes with significant problems), and then attention can be focused on fixing those particular problem processes. BSC emphasizes feedback control from each of the four perspectives in each major business process; that is, each perspective for each key process should have objectives, metrics, and targets for proper feedback control.)

Integrating EVA and Success Factors into BSC to Maximize Performance

How would one integrate the successful concepts of BSC in a project management context (Phillips et al., 2002)? Considering a project as a key business project, one needs to establish metrics for each of the four perspectives at the individual project level and also at the PMO (project management office) level. (See the chapter by Young and Powell for a discussion of the PMO.) EVA would be a key tool for the BSC financial perspective. Success factors from the "completion" perspective could be used in the BSC internal perspectives. Success factors from the "satisfaction" perspective could be used in the BSC customer perspective. This is illustrated in Figure 34.14.

FIGURE 34.14. THE POSITION OF EVA AND SUCCESS FACTORS IN THE BALANCED SCORECARD.

A scorecard defines the key activities for each perspective and the metrics that are going to be used to monitor the performance of that activity. One possible scorecard is shown in Figure 34.15 for the project level and for the PMO level. For example, at the BSC customer perspective, we would measure not only the customer's satisfaction with the system but also how involved our project team was with customer personnel. These metrics could be determined both at project completion and also at project milestones for early warning of potential problems.

The integration of BSC concepts into PM could also eventually lead to extensions of the PMI process groups and knowledge areas of Figure 34.1—that is, PMBOK.

Summary

EVA, critical success factors, and BSC are some of the most powerful and successful management tools for measuring and increasing performance. However, these techniques are so far rarely if ever used together, and proponents from one camp often criticize the methods of the other, probably because of a lack of full understanding of the other's approach.

FIGURE 34.15. BALANCED SCORECARD EXAMPLE.

Project Management BSC Metrics.

Perspective	Activities	Metrics	PMI Knowledge Area
Learning and Growth	Effective Communications Team Building and Motivation Employee Development Mentor Program	Communication Problems Employee Feedback Training Funding Mentor Reviews	Communications Human Resources Human Resources Human Resources
Customer	Customer Relations Customer Involvement	Customer Satisfaction Customer Interaction Logs	Communications Scope
Internal	Planning Sizing "Built-in Quality" Testing and Inspection Scope Control	EVA Time, SPI Estimation Accuracy Standards Methodology Specifications, QA Change Orders	Time and Risk Time Scope Risk Quality Scope
Financial	Use of Resources Contracting/Procurement	EVA Cost, CPI Procurement Standards	Cost Procurement

PMO BSC Metrics.

Perspective	Activities	Metrics
Learning and Growth	Effective Communications Management Team Building and Motivation Project Manager Development Motivation PM Forum	Communication Problems PM and Employee Feedback Training Funding PM Incentive Packages Sharing of "Lessons Learned"
Customer	Customer Relations Customer Involvement	Customer Satisfaction Customer Feedback (about PM)
Internal	Planning PM Support (from line management)	Roll-up of EVA Time PM Feedback
Financial	Use of Resources Contracting/Procurement	Roll-up of EVA Cost Procurement Problems

EVA, critical success factors, and BSC need not be opposing management techniques. Tactical techniques like EVA and critical success factors can be used as some of the key metrics for the strategic BSC approach.

This chapter has presented one way that the approaches can be fully integrated into the managing of projects.

References

Atkinson, R. 1999. Project management: Cost, time, and quality. *International Journal of Project Management* 17(6):337–342

Averson, P. 2002. The Balanced Scorecard Institute. www.balancedscorecard.org.

Brandon, D. 1998. Implementing earned value easily and effectively. *Project Management Journal* 29(2).

———. 1999. Implementing earned value easily and effectively. In *Essentials of project control,* 113 ff. Newtown Square, PA: Project Management Institute.

Cale, E. G., J. R. Curley, and K. F. Curley. 1987. Measuring implementation outcome: Beyond success and failure. *Information and Management* 3(1):245–253.

Cafasso, R. 1994. Few IS projects come in on time, on budget. *Computerworld* 28(50):20.

Christensen, D., and D. Ferens. 1995. Using earned value for performance measurement on software projects. *Acquisition Review Quarterly* (Spring): 155–171

Christensen, D. S., et al. 1995. A review of estimate at completion research. *Journal of Cost Analysis* (Spring).

DeLone, W., E. McLean. 1992. Information systems success: The quest for the dependent variable. *Information Systems Research* 3(1):6–95.

Eickelmann, N. 2003. Achieving organizational IT goals through Integrating the balanced scorecard and software measurement frameworks. Chapter II in *Technologies and Methodologies for Evaluating Information Technology in Business,* ed. C. K. Davis. Hershey, PA: IRM Press.

Field, T. 1997. When bad things happen to good projects. *CIO* 11(2):54–62

Fleming, Q., and J. K. Koppelman. 1994. The essence of evolution of earned value. *Cost Engineering* 36(11):21–27

———. 1996. *Earned value project management.* Newtown Square, PA: Project Management Institute.

———. 1998. Earned value project management: A powerful tool for software projects. *CROSSTALK* (July).

Hewitt, L., and M. O'Connor. 1993. Applying earned value to government in-house activities. *Army Research, Development & Acquisition Bulletin.* (January–February): 8–10.

Hildebrand, C. If at first you don't succeed. *CIO Enterprise.* Section 2:4–15.

Horan, R., and D. McNichols. 1990. Project management for large systems. *Business Communications Review* 20 (September): 15–24.

Kaplan, R., and D. Norton. 1996. *The balanced scorecard.* Cambridge, MA: Harvard Business School Press.

———. 1992. The balanced scorecard: Measures that drive performance. *Harvard Business Review* (January).

———. 1993. Putting the balanced scorecard to work. *Harvard Business Review* (September).

Kiewel, B. 1998. Measuring progress in software development. *PM Network* (January): 29–32.

Lewis, J. 1995. *Project planning, scheduling, and control.* Chap. 10. Homewood, IL: Irwin.

Lim, C. S., and Z. Mohamed. 1999. Criteria of project success: An exploratory re-examination. *International Journal of Project Management* 17(4):243–248.

Molla, A., and P. Licker. 2001. E-commerce systems success: An attempt to extend and respect the DeLone and Maclean model of IS success. *Journal of Electronic Commerce Research.* 2(4):131–141.

Morris, P. W. G., and G. H. Hough. 1987. *The anatomy of major projects.* Chichester: Wiley.

Niven, P. 2002. *Balanced scorecard: Step by Step.* New York: Wiley.

Phillips, J. J., T. W. Bothell, and L. Snead. 2002. *The project management scorecard: Measuring the success of project management solutions.* Oxford, UK: Butterworth-Heinemann.

Pinto, J., and D. Slevin. 1998. Critical success factors across the project lifecycle. *Project Management Journal* XIX:67–75.

U.S. Department of Commerce. 1999. Guide to balanced scorecard performance management methodology (July). Available at http://oamweb.osec.doc.gov/bsc/guide.htm.

Walton, M., and W. E. Deming. 1991. *Deming management at work.* New York: Perigee, 1991.

Yeates, D., and J. Cadle. 1996. *Project management for information systems*, London: Pitman.

CHAPTER THIRTY-FIVE

MAKING RISK MANAGEMENT MORE EFFECTIVE

Stephen Ward, Chris Chapman

The chapter on risk management by Steve Simister provided an introduction to project risk management and described a generic process for managing risks in projects. This chapter builds on these fundamental ideas to consider how risk management can be made more effective in a given project context. This implies maximizing the benefits obtained from the risk management process for any given level of cost—designing and using processes that are cost-effective.

In designing a risk management application, the project team must address the basic questions associated with the "six Ws" of the risk management process:

- ***W****ho* wants risk analysis to support risk management, and **w**ho is to undertake it?
- ***W****hy* is analysis being undertaken?
- ***W****hat* is the scope of risks to be included in the analysis?
- ***W****hichway* should the analysis be carried out?
- ***W****herewithal*—what resources are required?
- ***W****hen* should analysis be undertaken?

These "six W" questions provide a convenient framework for discussing generic risk management process design principles, although for expository convenience, the ***w****ho* question will be considered last. As will become apparent, there are significant interdependencies in the answers to each of these questions, and the order of discussion is somewhat arbitrary when general principles are addressed. For example, several central benefits of risk analysis (***w****hy* issues) depend on how one defines risk, and are not obtainable unless risks are quantified (a ***w****hichway* issue).

Why Is Analysis Being Undertaken?

Many people who are not familiar with effective formal risk management assume quite limited roles for risk management, such as simple measurement of risk or identification of things that might go wrong. With this perspective, the benefits of risk management are seen as showing how risky a particular venture is, or protecting project performance from adverse impacts. If this is the limit of the rationale for undertaking risk analysis, then the opportunity to benefit from risk management will be largely wasted. Further, it will prove ineffective in relation to these limited purposes because it will be seen as a counterproductive "add-on" instead of a highly productive "add-in." Effective risk management is not just about measuring or protecting project performance. It is also about assessing and modifying project objectives, base plans, and contingency plans in ways that enhance the prospects for good project performance.

A central reason for employing formal risk management should be to guide and inform the search for favorable alternative courses of action. Central to achieving this is the concept of "risk efficiency," which is concerned with the trade-offs between expected performance and risk that must be made in selecting one course of action or investment strategy over another. Risk efficiency is a core concept in the literature of economics and finance, with a prominence underlined by the award of a Nobel prize for economics to Markowitz (1959) for his portrayal of it in terms of mean-variance portfolio selection models, but the concept is not widely understood by many project managers with engineering backgrounds.

Risk efficiency has to address cost, time, quality, and other measures of performance, but assume, for the moment, that achieved performance can be measured solely in terms of cost outturn, and that achieved success can be measured solely in terms of realized cost relative to some approved cost commitment. In this context risk can be defined in terms of the threat to success posed by a given plan in terms of the size of possible cost overruns and their likelihood. More formally, when assessing a particular project plan in relation to alternatives, we can consider the expected cost of the project as one basic measure of anticipated performance (what should happen on average), and the only other relevant measure of performance as associated cost-risk in terms of downside variability relative to expected cost.

In these terms, some ways of carrying out a project will involve less expected cost and less cost-risk than others—they will be better in both respects, and relatively more efficient. The most efficient plan for any given level of expected cost will involve the minimum feasible level of cost-risk. The most efficient plan for any given level of cost-risk will involve the minimum feasible level of expected cost. Risk efficiency in this sense defines a set of what economists call *Pareto optimal plans*. In choosing between this set of risk-efficient plans, expected cost can only be reduced by adopting a more risky plan, and cost-risk can only be reduced by adopting a plan that increases the expected cost. This concept is most easily pictured using a graph like Figure 35.1.

Consider a set of feasible project plans portrayed in relation to expected cost and cost-risk as indicated in Figure 35.1. The feasible set has an upper and lower bound for both expected cost and cost-risk because there are limits to how good or bad plans can be in both these dimensions.

FIGURE 35.1. RISK-EFFICIENT OPTIONS.

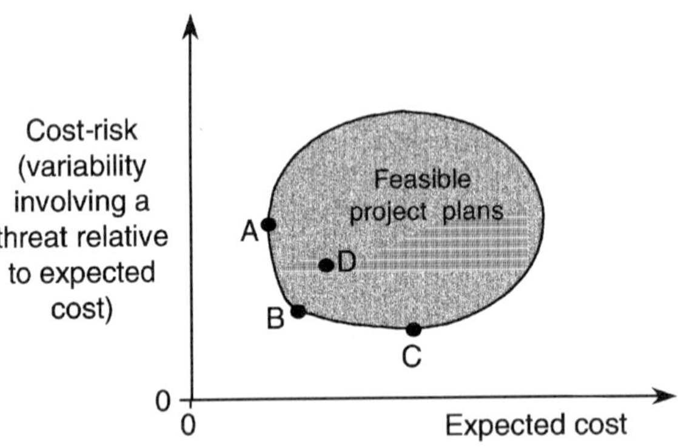

The "risk-efficient boundary" portrayed by the line through points A–B–C defines that set of feasible project plans that provides a minimum level of cost-risk for any given level of expected cost, and the minimum level of expected cost for any given level of cost-risk. Any point off the boundary, like D, represents an inefficient plan, which can be improved upon with respect to both expected cost and cost-risk—moving to B, for example. If a project base plan and associated contingency plans are risk-efficient, any change to these plans that reduces the associated risk will increase the expected cost, and any change that reduces the expected cost will increase the risk.

Diagnosis of potential changes to base or contingency plans to improve risk efficiency is the central purpose of effective project risk management. We can never be sure our plans are risk-efficient. However, we need to search systematically for improvements in risk efficiency, and we need to understand what we are looking for; otherwise, we will never find them. In addition to identifying risk-efficient courses of action, there is a need to consider choices. This involves considering preferred trade-offs between expected cost and level of risk. In relation to Figure 35.1, point A represents the minimum expected cost project plan, with a high level of cost-risk despite its risk efficiency. Choosing A involves not spending money on proactive risk management and a calculated gamble that may not pay off. Point C represents the minimum cost-risk project plan, with a high level of expected cost despite its risk efficiency. Choosing C involves spending money on proactive risk management with the possibility that, in the event of good luck, this expenditure will have been unnecessary.

If an organization can afford to take the risk, A is the preferred solution. Plan A may be the only viable option if an organization's competitors all operate at this minimum expected cost point. Organizations that do not take calculated gambles reduce their average profitability, and this may guarantee going out of business eventually. We have seen programs specifically designed to demonstrate the need for such calculated gambles in organizations that are spending too much on avoiding gambles, the equivalent of persistent overinsurance.

If the risk associated with A is too great, it must be reduced by moving towards C. In general, successive movements will prove less and less cost effective in terms of reducing threat intensity, larger increases in expected cost being required to achieve the same reduction in absolute or relative risk. In practice, an intermediate point like B usually needs to be sought, providing a cost-effective balance between risk and expected cost, the exact point depending upon the organization's ability to take risk.

The scale of the project relative to the organization in question is a key issue in terms of the relative desirability of plans A, B, or C. If the project is one of hundreds, none of which could threaten the organization, plan A may be a sensible choice. If the organization is a one-project organization and failure of the project could lead to failure of the organization, a more prudent stance may be appropriate, adopting a risk-efficient plan closer to C than A. This in turn implies it is very worthwhile defining a level of potential threat below which the organization can ignore cost-risk and above which cost-risk needs to be considered and managed.

In relation to risk-efficient plans and trade-offs between risk and expected performance, risk analysis can help to

- diagnose alternative risk-efficient plans;
- demonstrate the implications of such alternatives; and
- inform choices between alternative risk-efficient plans.

In this way risk management can produce very much more substantial improvements in project performance than a limited focus on merely "keeping things on track." It can also encourage a recognition of the role of considered trade-off decisions involving calculated gambles, so that managers are not blamed for bad luck (and not always held in esteem for experiencing simple good luck).

Taking this perspective, if risk analysis and management is effectively applied to a stream of individual projects over time, a number of important corporate benefits will accrue. Consider, for example, a contracting organization that undertakes risk management prior to and after tendering on individual contracts. For such an organization a number of interrelated benefits can accrue, all driving up profitability, through lower-level benefits like the following:

- *Losing more contracts that ought to be lost.* Better appreciation of uncertainty, which enables more realistic pricing and the avoidance of potential loss making "disaster" contracts where uncertainty is too great.
- *Winning more contracts that ought to be won.* Keener pricing, better design, and stronger risk management abilities providing competitive advantage in terms of improved chances of winning contracts.
- *Lower project costs.* Ability to manage risks to lower project costs with direct profit implications.
- *Lower tendering costs.* Reduce tendering costs through more efficient and effective risk analyses contributing to higher profits directly.

- *More business with higher prices.* Even in the face of substantially lower competitive bids, plausible bids demonstrating effective associated risk management will win when discriminating clients are involved, and these are the most desirable clients.

Risk efficiency, which addresses cost, time, quality, safety, and other measures of performance, may take an aggressive approach to cost-risk, but a conservative approach to safety risk, for example. In any event, the trade-offs between performance criteria and associated uncertainty will need careful management (Klein, 1993). Thoughtful attention to trade-offs and risk efficiency issues can amplify the benefits discussed previously many times over. Conversely, a failure to manage risk efficiency in terms of *all* relevant performance measures can make performance incentives seriously perverse. This is especially important if a contractor is the guardian of issues like quality in the context of incentive contracts (Chapman and Ward, 2002).

In addition to addressing risk efficiency, introducing formal risk analysis and management can lead to valuable cultural changes, treating uncertainty as a source of opportunities rather than as something to be avoided. Formal processes also encourage a proactive approach in which uncertainty is addressed in advance, in a calm and creative way, while there is time to work around problems and exploit opportunities. Reactive crisis management is not eliminated, but it is reduced to a tolerable level. Frustration is reduced, and morale is improved. Such changes can make an organization more exciting to work for and make going to work more enjoyable. This in turn can lead to higher-quality staff wanting to join (and stay with) the organization, with obvious general benefits. All of these benefits should not simply be allowed to happen or not; they should be encouraged by designing them into the risk management process.

It follows that all the preceding benefits are more likely to accrue if organizations develop a corporate capability in (project) risk management, with corporate guidelines and appropriate support, rather than simply encourage the use of risk management in an *ad hoc* way on individual projects (Office of Government Commerce, 2002). It is also the case that organizations with an established infrastructure for supporting risk management will be able to deploy risk management on individual projects much more rapidly and efficiently, and to much greater effect.

What Is the Scope of Risks to Be Included in the Analysis?

A key issue for achieving effective risk management is determining the scope of risks to be included in analysis and subsequent management. Partly this is a question of understanding what is meant by risk in general, and partly it is about delineating the scope of risks that are considered to be project-related. As noted earlier, a process that is limited to a threat perspective on risk management will have limited value because it fails to consider the management of opportunities, in the sense of potential welcome effects on project performance.

In any given decision situation, both threats and opportunities are usually involved, and both should be managed. A focus on one should never be allowed to eliminate concern for the other. While opportunities and threats can sometimes be treated separately, they are seldom independent—just as two sides of the same coin can be examined one at a time, but they are not independent when it comes to tossing the coin. Just as it is inadvisable to pursue opportunities without regard for the associated threats, it is rarely advisable to concentrate on reducing threats without considering associated opportunities. Courses of action are often available that reduce or neutralize potential threats and simultaneously offer opportunities for positive improvements in performance.

To emphasize the desirability of a balanced approach to opportunity and threat management, the term *uncertainty management* is increasingly used in preference to *risk management* and *opportunity management.* However, uncertainty management is not just about managing perceived threats, opportunities, and their implications. It is also about identifying and managing all the sources of uncertainty that give rise to and shape our perceptions of threats and opportunities. It involves exploring and understanding the origins of project uncertainty before seeking to manage it, with no preconceptions about what is desirable or undesirable. Key concerns are understanding where and why uncertainty is important in a given project context, and where it is not. This is a significant change in emphasis compared with most project risk management processes.

Uncertainty in the plain-English sense of lack of certainty is in part about *variability* in relation to performance measures like cost, duration, or quality. It is also about *ambiguity* associated with lack of clarity because of the behavior of relevant project players, lack of data, lack of detail, lack of structure to consider issues, restrictive working and framing assumptions used to consider the issues, known and unknown sources of bias, and ignorance about how much effort it is worth expending to clarify the situation. These aspects of uncertainty can be present throughout the project life cycle, but they are particularly evident during conception, design, and planning, and they contribute to uncertainty in five areas:

- The variability associated with estimates of project parameters
- The basis of estimates of project parameters
- Design and logistics
- Objectives and priorities
- Relationships between project parties

All these areas of uncertainty are important, but generally areas are more fundamental to project performance as we go down the list. Potential for variability is the obvious focus of the first area, but ambiguity rather than variability becomes the more dominant underlying issue in latter areas. Uncertainty about variability associated with estimates is often driven by the other four areas, each of them in turn involving dependencies on later areas in this list.

Variability Associated with Estimates

An obvious area of uncertainty is the size of project parameters such as time, cost, and quality related to particular activities. For example, we may not know how much time and effort will be required to complete a particular activity. The causes of this uncertainty might include one or more of the following:

- Lack of a clear specification of what is required
- Novelty in terms of a lack of experience of this particular activity
- Complexity in terms of the number of influencing factors and interdependencies between these factors
- Limited analysis of the processes involved in the activity
- Possible occurrence of particular events or conditions that could have some (uncertain) effect on the activity

Only the last of these items really relates to specific events or conditions that might be thought of as threats or opportunities. The other sources of uncertainty arise from a lack of understanding of what is involved. Because they are less obviously described as threats or opportunities, they may be missed unless a broad view of uncertainty management is adopted.

Uncertainty about Assumptions Underlying Estimates

A particularly important source of uncertainty is the nature of assumptions underpinning estimates. The need to note assumptions about resources, choices, and methods of working is well understood if not always fully operationalized. However, a large proportion of those using probabilistic project risk management processes often fail to address the conditional nature of probabilities and associated measures used for decision making and control. Key outputs of estimation and evaluation phases of the risk management process (see the chapter by Simister) are estimates of expected values for project parameters and measures of plausible variations on the high and low side. Interpretation of expected values, or plausible extremes like a 95 percent confidence value, have to be conditional on the assumptions made to estimate these values. For example, a sales estimate may be conditional on a whole set of assumed trading conditions, such as a particular promotion campaign and no new competitors. Invariably, estimates ignore, or assume away, the existence of uncertainty, which relates to three basic sources: known unknowns, unknown unknowns, and bias:

- *Known unknowns* are of two types: explicit, extreme events (triple Es), and scope adjustment provisions (SAPs). Triple Es are *force majeure* events, like a change in legislation that would influence an oil company's pipeline design criteria in a fundamental way. SAPs are conditions or assumptions that may not hold and that are explicit, like the assumed operating pressure and flow value for an oil pipeline, given the assumed oil recovery rate.
- Unknown unknowns are the unidentified triple Es or SAPs that should be factored in to the risk management process. We know that the realization of some unknown unknowns is usually inevitable. They do not include issues like "the world may end tomorrow," because it is sensible for most practical decision making to assume we will still be here tomorrow, but the boundary between this extreme and what should be included is usually ill-defined.

- Bias may be conscious or unconscious, pessimistic or optimistic (McCray, Purvis and McCray, 2002). It is usually difficult to identify, and clues or data may be available or not.

All three of these sources of uncertainty can have a very substantial impact on estimates, and this needs to be recognized and managed.

Uncertainty about Design and Logistics

In the conception stage of the project life cycle the nature of the project deliverable and the process for producing it are fundamental uncertainties. In principle, most of this uncertainty should be removed in pre-execution stages of the life cycle by specifying what is to be done; how, when, and by whom; and at what cost. In practice, a significant amount of this uncertainty may remain unresolved through much of the project life cycle (see, for example, Drummond, 1999). The nature of design and logistics assumptions and associated uncertainty may drive some of the uncertainty about the basis of estimates.

Uncertainty about Objectives and Priorities

Major difficulties arise in projects if there is uncertainty about project objectives, the relative priorities between objectives, and acceptable trade-offs. These difficulties are compounded if this uncertainty extends to the objectives and motives of the different project parties, and the trade-offs parties are prepared to make between their objectives. A key question is "Do all parties understand their responsibilities and the expectations of other parties in clearly defined terms that link objectives to planned activities?" Value management has been introduced to encompass this concern (Kelly and Male, 1993; also see the chapter by Thiry). The need to do so is perhaps indicative of a perceived failure of risk management practices. However approached, *risk management and value management need joint integration into project management.*

Uncertainty about Fundamental Relationships between Project Parties

Uncertainty about objectives and priorities is compounded by any uncertainty about the identity of project parties, their respective roles, and their relationships with one another. The relationships between the various parties may be complex. They may, or may not, involve formal contracts. The involvement of multiple parties in a project introduces uncertainty arising from ambiguity with respect to the following:

- Specification of responsibilities
- Perceptions of roles and responsibilities
- Communication across interfaces
- The capability of parties
- Contractual conditions and their effects
- Mechanisms for coordination and control

Ambiguity about roles and responsibilities for bearing and managing project-related uncertainty may be involved here. For example, interpretations of risk apportionment implied by standard contract clauses may differ between contracting parties (Hartman and Snelgrove, 1996; Hartman, Snelgrove, and Ashfrati, 1997). This ambiguity ought to be addressed systematically in any project, not just in those involving formal contracts between different organizations. Contractor organizations are often more aware of this source of ambiguity than their clients, although the full scope of the risks and opportunities that this ambiguity generates for each party in any contract (via claims, for example) may not always be fully appreciated until rather late in the day (Ackermann et al., 1997; Cooper, 1980; Williams et al., 1995).

Whichway Should Analysis Be Carried Out

An important aspect of developing cost-effective approaches to risk management involves consideration of the appropriate structure and level of detail in the analysis to be undertaken, and the choice of models and techniques. Choices here will be strongly influenced by the who?, why?, and what? questions discussed previously.

To address uncertainty in both variability and ambiguity terms, we need to modify and augment existing risk management processes and adopt a more explicit focus on uncertainty management. An obvious first step is to replace terminology involving the word *risk* with the word *uncertainty* and avoid using purely threat-orientated descriptors in identification exercises. Other steps involve modifications to risk management processes to address each of the sources of uncertainty outlined in the previous section. What these modifications involve is outlined in the following text. More detailed explanations and commentary can be found in Chapman and Ward (2002).

Expose and Investigate Variability

Difficulty in estimating time or effort required to complete a particular activity may arise from a lack of knowledge of what is involved rather than from the uncertain consequences of potential threats or opportunities. Attempting to address this difficulty in conventional risk management terms is not appropriate. What is needed is action to improve knowledge of organizational capabilities and reduce variability in the performance of particular project-related tasks. For example, uncertainty about the time and cost needed to complete design or fabrication in a project may not be readily attributable to particular sources of risk, but to variability in efficiency and effectiveness of working practices. An uncertainty management perspective would seek an understanding of why this variability arises, with a view to managing it. This may require going beyond addressing uncertainty associated with a specific project, to trigger studies of operations that provide an input into a range of projects, as illustrated by this example.

Further investigation of variability, and consideration of risk-efficient alternative courses of action in base plans or contingency plans, requires quantification of the perceived variability. Single-point estimates of a particular parameter are of limited value for uncertainty management purposes without some indication of the potential variability in the size of the parameter. For example, a best estimate of the cost of a particular activity is of limited value without some indication of the range or probability distribution of possible costs. Quantifying variability (and uncertainty about this variability) forces management to articulate beliefs about uncertainty and related assumptions. In addition, quantification obliges managers to clarify the significance of differences between *targets, expected values* and *commitments*, with respect to costs, durations, and other performance measures. This in turn highlights the need to clarify the distinction between *provisions* and *contingency allowances*.

In cost terms, expected values are our best estimate of what costs should be realized on average. Setting aside a contingency fund to meet costs that may arise in excess of the expected cost defines a probability of being able to meet the commitment. The contingency allowance provides an uplift from the expected value, which is not required on average if it is properly determined. Determining this probability of being able to meet a commitment ought to involve an assessment of all related downside variations and the extent to which these may be covered by a contingency fund, together with an assessment of the implications of both over- and underachievement in relation to the commitment.

Targets, set at a level below expected cost, with provisions accounting for the difference, need to reflect the opportunity aspect of risk. Targets need to be realistic to be credible, but they also need to be lean, to stretch people. If optimistic targets are not aimed for, expected costs will not be achieved on average, and contingency funds will be used more often than anticipated.

Organizations that do not quantify risks have no real basis for distinguishing these three very different kinds of estimates. As a consequence, single values attempt to serve all three purposes, usually with obviously disastrous results, not to mention costly and unnecessary dysfunctional organizational behavior. The *cost estimate*, the *completion date*, or the *promised performance* become less and less plausible, there is a crisis of confidence when they are moved, and then the process starts all over again. Sometimes differences between targets, expectations and commitments are kept confidential, or left implicit. Effective risk management requires these differences to be explicit, and a clear rationale for the difference needs to be understood by all, leading to an effective process of managing the evolution from targets to realized values. The ability to manage the gaps between targets, expected values, and contingency levels, and setting those values appropriately in the first place, is a central concern of risk management.

Quantifying uncertainty about levels of performance in terms of targets, expectations, and commitments is useful if concern is with aggregate performance on a single criterion such as cost or time. However, this approach is less helpful if applied to quantifying uncertainty about each activity individually in a chain or collection of activities. For example, setting commitment levels for the duration of each task in a chain of tasks may be counterproductive. In *Critical Chain* Goldratt (1997) describes the problem in the following terms: "we are accustomed to believing that the only way to protect the whole is through protecting the completion date of each step"; as a result, "we pad each step with a lot of safety time." Goldratt argues that this threat protection perspective induces three behaviors that, when combined, waste most of the safety time:

- The student syndrome (leaving things to the last minute)
- Multitasking (chopping and changing between different jobs)
- Delays accumulate, advances do not (good luck is not passed on)

The challenge, and opportunity, is to manage the *uncertainty* about performance for each link in the chain in a way that avoids these effects and ensures that good luck is not only captured but shared for the benefit of the whole project. Joint management of the good luck, efficiency, and effectiveness of each project-related activity is needed. This implies some form of incentive agreement between the project manager and those carrying out project activities that encourages the generation and delivery of good luck, efficiency, and effectiveness, with the minimum of uncertainty. This agreement needs to recognize interdependency between performance measures. Duration, cost, quality, and time are not independent, and all four are functions of the motivation and priorities of those involved. For example, uncertainty about the duration of an activity is usually driven by ambiguity about quality, cost, and inefficient working practices. This needs to be managed, to reduce uncertainty and to capture the benefits of managing good luck, in the sense of the scope for low duration, high quality, and low cost.

To illustrate these ideas briefly, consider a procurement project involving significant design work before the next stage of the project can proceed. If an internal design department is involved, do we need to set a commitment date for design completion? The simple answer is we do not. Rather, we need a target duration plus an expected duration that becomes firm as early as possible, in order to manage the good luck as well as the bad luck associated with variations in the duration of the design activity. We also need an agreement with the design department that recognizes that the design department can make and share their luck to a significant extent if they are motivated to do so. As the design department is part of the project-owning organization, a legal contract is not appropriate. However, a contract is still needed in the form of a memorandum of understanding, to formalize the agreement. Failure to formalize an internal contract in a context like this implies psychological contracts between project parties that are unlikely to be effective. The ambiguity inherent in such contracts can only generate uncertainty, which is highly divisive and quite unnecessary.

Most internal design departments have a *cost per design hour* rate based on an historic accounting cost. A *design hours* estimate times this rate yields a design cost estimate. Design actual cost is based on realized design hours and the internal contract is *cost plus*. The duration agreed to by the design department is a "commitment" date with a low chance of being exceeded, as noted earlier. To address the problems that this arrangement induces, some form of incentive agreement is required that recognizes the potential for trade-offs between different measures of performance. What is needed is a *fixed nominal cost* based on the appropriate expected number of design hours, with a premium payment scale for completion earlier than an appropriate *trigger duration*, and a penalty deduction scale for later completion. Additionally, a premium could be introduced for the correct prediction of the design completion date to facilitate the more efficient preparation of following activities. The trigger duration might be something like an 80-percentile value, comparable to a commitment duration. The target should be very ambitious, reflecting a plausible date if all goes as well as possible. Other performance objectives can be treated in the same way, with premium and penalty payments relative to cost and quality level triggers as possible options. However, premiums and penalties need to be designed to ensure that appropriate trade-offs are encouraged.

Such *trigger contracts* can have beneficial effects on performance that go beyond the immediate project. For example, carrying on with the preceding example, if a trigger-based incentive contract with an internal design department is used, fewer hours may be required because the incentive structure will reduce multitasking. If multitasking is reduced, the efficiency of the design department might improve enough to eliminate rumors of selling off the design function (outsourcing all design), which might be underpinning risks related to loss of staff and low morale. If these downside risks are eliminated, improved morale, low turnover of good staff, and easier hiring may follow leading to further gains in efficiency and effectiveness in a virtuous circle.

Clarify Assumptions Underlying Estimates

The previous section noted that estimates of project parameters invariably ignore, or assume away, the existence of uncertainty that relates to three basic sources: known unknowns, unknown unknowns, and bias. A full description of how best to address these sources of uncertainty is complex. An outline is provided elsewhere (Chapman and Ward, 2002), a summary here.

The starting point is recognizing that all estimates are *conditional*, in the sense that assumptions they depend upon have been made. The second step is understanding that all estimates are *subjective*, in the sense that truly objective estimates that are fit for purpose do not exist. The third step is recognizing that judging the quality of someone else's estimates necessarily involves judging the quality of the process they used to arrive at that estimate, as well as the nature of their explicit and implicit conditions.

The impact of these three basic sources of uncertainty can be considered via the use of three scaling factors: F_k known unknowns, F_u unknown unknowns, and F_b bias. For example, an F_k scaling factor might be defined by the user of an estimate in relation to an estimated expected cost provided by a provider of the estimate in the form:

F_k	Probability of F_k
F_k = 1.0	Probability of F_k = 0.1
1.1	0.7
1.2	0.1
1.3	0.1

This example involves a mean $F_k = 1.0 \times 0.1 + 1.1 \times 0.7 + 1.2 \times 0.1 + 1.3 \times 0.1 = 1.12$. The subjective probabilities might be based on the user's view of the impact of conditions noted by the provider, of the type "normal market conditions will be involved."

In simple, crude terms, the example numbers would imply that an uplift in the estimated cost of the order of 30 percent is plausible for a pessimistic scenario value, like a 95 percentile, and an uplift of 12 percent is an appropriate expectation, because of the potential impact of an abnormal market or the violation of other conditions noted. In practice, F_k values could be much higher than in this example. A very careful risk-management-driven estimation process resulting in a $F_k = 1$ might suggest an $F_u = 1$ and an $F_b = 1$, but much higher values may be appropriate.

Combining these three scaling factors provides a single *cube factor* (short for kuuub from **kn**own **u**nknowns**,** **u**known **u**knowns, and **b**ias), designated F^3, and defined by $F^3 = F_k \times F_u \times F_b$, which is then applied as scaling factor to conditional estimates. This cube factor, F^3, can be estimated in probability terms directly or via these three components to clarify the conditional nature of the output of any quantitative risk analysis. This avoids the very difficult mental gymnastics associated with trying to interpret a quantitative risk analysis result that is conditional on exclusions and scope assumptions (which may be explicit or implicit), and no bias, without underestimating the importance of the conditions.

The key value of explicit quantification of F^3 is forcing those involved to think about the implications of the factors that drive the expected size and variability of F^3. Such factors may be far more important than the factors captured in the prior conventional quantitative risk analysis. There is a natural tendency to forget about conditions and assumptions and focus on the numbers. Attempting to explicitly size F^3 makes it possible to avoid this. Even if different parties emerge with different views of an appropriate F^3, the process of discussion is beneficial. If an organization refuses to estimate F^3 explicitly, the issues involved do not go away; they simply become unmanaged risks. Many of them will be betting certainties. Variability, that needs to be managed in a risk management context, must embrace cube factors explicitly, and "variability" is defined here in this sense.

An important source of ambiguity concerns the extent to which different project parties need to be concerned about particular cube factors. For example, suppose a project manager decides to contract out design work. In estimating design costs, the design contractor will not scale its estimates to allow for known unknowns if the contractor can negotiate a contract to avoid bearing any risk associated with known unknowns. Similarly, it will not be appropriate for the project manager to scale the project design budget to incorporate an allowance for these known unknowns, unless they are wholly under the control of the project manager. For example, certain scope adjustments may come in this category. The potential impact of other unknown unknowns needs to be recognized at some organizational level above the project manager, where there is an ability to bear the consequences of any unknown unknown occurring, and an appropriate cube factor estimated. A similar cube factor would need estimation if the project company's own design department undertook the work, but the risk allocation issues would be more complicated. Much post project litigation arises because of a failure to appreciate or acknowledge exposure to cube factors, and a failure to resolve ambiguity about responsibility for cube factors earlier in the project.

Address Uncertainty about Fundamental Relationships, as well as Design and Logistics

Careful attention to formal risk management is usually motivated by the large-scale use of new and untried technology while executing major projects, where there are likely to be significant threats to achieving objectives. A threat perspective encourages a focus on these initial motivating risks. However, key issues are often unrelated to the motivating risks and are usually related to sources of ambiguity introduced by the existence of multiple parties and the project management infrastructure. Such issues need to be addressed very early in the project and throughout the project life cycle, and should be informed by a broad appreciation of the underlying roots of uncertainty. A decade ago most project management professionals would have seen this in terms of a suitable high-level activity structure summary and a related cost item structure. It is now clear that an activity structure is only one of six aspects of a project that need consideration (Chapman and Ward, 2003). To review, these are as follows:

- Who (parties or players involved)
- Why (motives, aims, and objectives of the parties)
- What (design of the deliverable)
- Whichway (activities to achieve the deliverable)
- Wherewithal (resources required)
- When (time frame involved)

Understanding the sources of uncertainty associated with each of these aspects is fundamental to effective identification and management of both threats and opportunities. Use of this "six Ws" framework for the project from the earliest stages of the PLC could usefully inform development of project design and logistics by clarifying key sources of uncertainty. However, it is important not to treat all these sources of uncertainty as independent—an assumption that is implicitly encouraged by summary risk registers (Williams, 2000). In practice it is better to assume dependency and seek to understand its nature, recognizing that dependencies can be complex, involving chains of knock-on effects and undesirable self-reinforcing feedback loops (see, for example, Drummond, 1999; Eden et al., 2000). Such dependencies can be effectively represented using influence diagrams, causal maps, and systems dynamics models (Ashley and Avots, 1984; Eden et al., 2000; Howick and Eden, 2001).

Address Uncertainty about Objectives and Priorities

As part of the process of understanding the relationships between the six Ws of the project, project owners need to

1. identify pertinent performance criteria;
2. develop a measure of the level of performance for each criterion;
3. identify the most preferred (optimum) feasible combination of performance levels on each criterion;
4. identify alternative combinations of performance levels on each criterion that would be acceptable instead of the "optimum"; and
5. identify the trade-offs between performance criteria implied by these preferences.

These steps should be undertaken by any project owner, particularly in the early conception and design stages of a project. Adopting an iterative process may be the most effective way to complete these steps. The process of identifying and considering possible trade-offs between performance criteria is an opportunity to improve performance, as noted earlier. It should enable a degree of optimization with respect to each performance measure, and it is an opportunity that needs to be seized. In particular, the information gathered from these steps can be used to formulate appropriate incentive contracts by selecting some or all of the performance criteria for inclusion in the contract, developing payment scales that reflect the acceptable trade-offs, and with the supplier, negotiating acceptable risk-sharing ratios for each contract performance criterion. A detailed discussion on the formulation of such contracts is given in Chapman and Ward (2002, Chapter 5).

Constructively Simple Estimating

To facilitate insight and learning in a cost-effective process, uncertainty in all the various forms discussed previously needs to be explored using an iterative process, with process objectives that change on successive passes. An iterative approach is essential to optimize the use of time and other resources during the uncertainty management process, because initially we do not know where uncertainty lies, whether or not it matters, or how best to respond to it. To begin with, a first pass is usually about sizing variability, to see if it might matter, and to reflect potential variability in an unbiased estimate of the expected outcome. If a very simple first-pass conservative estimate suggests expected values and variability do not matter, we should be able to ignore variability without further concern or further effort. If the first pass raises concerns, further passes are necessary in order to effectively manage what matters. Final passes may be concerned with convincing others that what matters is being properly managed. The way successive iterations are used needs to be addressed in a systematic manner. A simple one-shot, linear approach is hopelessly inefficient.

A common first-pass approach to estimation and evaluation employs a probability-impact matrix (PIM). The PIM approach typically defines low, medium, and high bands for possible probabilities and impacts associated with identified sources of uncertainty (usually risks involving adverse impacts). These bands may be defined as quantified ranges or left wholly subjective. *The PIM approach offers a rapid first-pass assessment of the relative importance of identified sources of uncertainty, but otherwise delivers very little useful information* (Ward, 1999a).

Even with the availability of proprietary software products such as Risk for quantifying, displaying, and combining uncertain parameters, use of PIM has persisted (further encouraged by PIM software). This is surprising, but it suggests a gap between simple direct prioritization of sources of risk and quantification requiring the use of specialist software. To address this gap, Chapman and Ward (2000) describe a 'minimalist' first-pass approach to estimation and evaluation of uncertainty. This approach defines uncertainty ranges for probability and impact associated with each source of uncertainty. Subsequent calculations preserve expected value and measures of variability, while explicitly managing associated optimistic bias.

The minimalist approach involves the following steps in a first-pass attempt to estimate and evaluate uncertainty:

1. Identify the parameters to be quantified.
2. Estimate crude but credible ranges for probability of occurrence and impact.
3. Calculate expected values and ranges for composite parameters.
4. Present results graphically (optional).
5. Summarize results.

In step 1 a clear distinction is made between sources of uncertainty that are useful to quantify and sources that are best treated as possible scenarios. For example, suppose an oil company project team wants to estimate the duration and the cost of the design of an offshore pipeline using the organization's own design department. Current common best practice would require a list of sources of uncertainty (a risk list or risk log), which might include entries like "change of route," "demand for design effort from other projects," "loss of staff," and "morale problems". "Changes of route" would probably be regarded as a source best treated as a condition by the project manager and by the head of the design department. Subsequent steps apply only to those sources of uncertainty that are usefully quantified.

In step 2 the probability of a threat occurring is associated with an approximate order of magnitude minimum and maximum plausible probability, assuming a uniform distribution (and a midpoint expected value). This captures the user's feel for a low, medium, or high probability class in a flexible manner, captures information about uncertainty associated with the probability, and yields a conservative (pessimistic) expected value. For trained users it should be easier than designing appropriate standard classes for all risks and putting a tick in an appropriate box.

Similarly, in step 3 the impact of a threat that occurs is associated with an approximate order of magnitude minimum and maximum plausible value (duration and cost), also assuming a uniform duration. This captures the user's feel for a low, medium, or high impact class in a flexible manner, captures information about the uncertainty associated with the impact, and yields a conservative (pessimistic) expected value. For trained users it should be easier than designing appropriate standard classes for all risks and putting a tick in an appropriate box.

In step 4 the expected values and associated uncertainties for all quantified sources of uncertainty are shown graphically in a way that displays the contribution of each to the total, in expected value and range terms, clearly indicating what matters and what does not, as a basis for managing subsequent passes of the risk management process in terms of data acquisition to confirm important probability and impact assessment, refinement of response strategies, and key decision choices.

Although simple, the minimalist approach is sophisticated in the sense that it builds in pessimistic bias to minimize the risk of dismissing as unimportant risks that more information might reveal as important. Also, it is set in the context of an iterative approach that leads to more refined estimates wherever potentially important risks are revealed. Sophistication does not require complexity. It requires *constructive simplicity*, increasing complexity only when it is useful to do so.

The concern of the minimalist first-pass approach is not a defensible quantitative assessment. The concern is to develop a clear understanding of what seems to matter based on the views of those able to throw some light on the issues. This is an attempt to resolve the ambiguity associated with the size of uncertainty about the impact of risk events and the size of uncertainty about the probability of risk events occurring, the latter often dominating the former. A first pass may lead to the conclusion there is no significant uncertainty, and no need for further effort. This is one of the reasons why the approach must have a conservative bias. Another reason is the need to manage expectations, with subsequent refinements of estimates indicating less uncertainty/more uncertainty providing an explicit indication that the earlier process failed. An estimator should be confident that more work on refining the analysis is at least as likely to decrease the expected value estimate as to increase it. A tendency for cost estimates to drift upward as more analysis is undertaken indicates a failure of earlier analysis. The minimalist approach is designed to help manage the expectations of those the estimator reports to in terms of expected values. Preserving credibility should be an important concern.

Readers used to single-pass approaches that attempt considerable precision may feel uncomfortable with the deliberate lack of precision incorporated in the minimalist approach. However, more precise modeling is frequently accompanied by questionable underlying assumptions such as independence between parameters and lack of attention to uncertainty in original estimates. The minimalist approach forces explicit consideration of these issues.

What Resources (Wherewithal) Are Required?

Just as resources for the project require explicit consideration, so too do resources for effective risk analysis and management, the process ***wherewithal*** question. In a given project context, there may be specific constraints on cost and time. Resource questions are likely to revolve around the availability and quality of human resources, including the availability of key project personnel, and the availability of information processing facilities.

If one of our clients asks "How long will it take to assess my project's risk?," the quite truthful response "How long is a piece of string?" will not do. A more useful response is "How long have we got?" (the process ***when***), in conjunction with "How much effort can be made available?" (the process ***wherewithal***), "Who wants it?", (the process ***who***), and "What do you want it for?" (the process ***why***). The answer to this latter question often drives the process ***what*** and the ***whichway***. It is important to understand the interdependence of these considerations. Six months or more may be an appropriate duration for the initial, detailed project risk analyses of a major project, but six hours can be put to very effective use if the question of the time available is addressed effectively in relation to the other process *W*s. Even a few minutes may prove useful for small projects. In these circumstances, it is essential to adopt an iterative approach with a minimalist first pass as outlined in the previous section.

Computing power is no longer a significant constraint for most project planning, with or without consideration of uncertainty. Even very small projects can afford access to powerful personal computers. However, software can be a significant constraint, even for very large projects. It is important to select software that is efficient and effective for an appropriate model and method. It is also important to prevent preselected software from unduly shaping the form of the analysis.

In the early stages of the risk management process, the risk analysis team may be seen as the project planning player doing most of the risk management running. However, it is vital that all the other players see themselves as part of the team and push the development of the risk management process as a vehicle serving their needs. This implies commitment and a willingness to spend time providing input to the risk analysis and exploring the implications of its output.

When Should Analysis Be Undertaken?

The nature of projects undertaken is likely to be a primary influence on the scope, level of detail, perspective, and extent of quantification that is appropriate. For example, a combination of substantial investment with high levels of uncertainty warrants serious management attention, greatly facilitated by a comprehensive, systematic risk management process. However, if projects are of a routine, low-risk nature, project managers hardly need sophisticated risk management systems.

In general, risk management could be usefully applied on separate and different bases in each stage of the project life cycle without the necessity for risk management in any previous or subsequent stages. For example, risk analysis could form part of an "evaluation" step in any stage of the life cycle. Alternatively, risk analysis might be used to guide initial progress in each life cycle stage. In these circumstances, the focus of risk analysis is likely to reflect immediate project management concerns in the associated project stage. For example, risk analysis might be undertaken as part of the Plan stage primarily to consider the feasibility and development of the work schedule for project execution. There might be no expectation that such risk analysis would or should influence the design, although it might be perceived as a potential influence on the subsequent work allocation decisions. In practice, many risk analyses are intentionally limited in scope, as in individual studies to determine the reliability of available equipment, to determine the likely outcome of a particular course of action, or to evaluate alternative decision options within a particular project life cycle stage. This can be unfortunate if it implies a limited, *ad hoc* bolted-on or optional extra approach to risk management, rather than undertaking risk management as an integral built-in part of project management.

As a general rule, the earlier risk management can be carried out in the project life cycle, the better. Implementing risk management earlier than the planning stage can be difficult because the project is more fluid and less well defined, with more degrees of freedom and more alternatives to consider. A less well defined project also means appropriate documentation is harder to come by and alternative interpretations of what is involved may not be resolvable (Uher and Toakley, 1999). That said, implementing risk management earlier in the life cycle is usually much more useful than if it is first attempted later on. Risk management earlier in the life cycle is usually less quantitative, less formal, less tactical, more strategic, more creative, and more concerned with the identification and capture of opportunities. There is scope for much more fundamental improvements in project plans, perhaps including initial design or redesign of the product of the project. Also, it can be particularly useful to be very clear about project objectives as early as possible, in the limit-decomposing-projects objectives and formally mapping their relationships with project activities, because preemptive responses to threats need to facilitate lateral thinking that addresses entirely new ways of achieving objectives.

Implementing risk management later in the project life cycle gives rise to somewhat different difficulties, without any compensating benefits. Contracts are in place, equipment has been purchased, commitments are in place, reputations are on the line, and managing change is comparatively difficult and unrewarding. Risk management can and should encompass routine reappraisal of a project's viability. In this context, early warnings are preferable to late recognition that targets are incompatible or unachievable, but better late than never.

Who Wants the Risk Analysis, and Who Is to Undertake It?

In any given project context, an important initial step in scoping the project risk management process is to clarify ***who*** is undertaking risk analysis for whom, and how the reporting process will be managed. The key players should be as follows:

- Senior managers, to empower the process, to ensure the risk analysis effort reflects the needs and concerns of senior managers, and to ensure it contains the relevant judgments and expertise of senior managers
- All other relevant managers, to make it part of the total management process and ensure that it services the whole project management process
- All relevant technical experts, to ensure it captures all relevant expertise for communication to all relevant users of that expertise in an appropriate manner
- A risk analyst or risk analysis team, to provide facilitation/elicitation skills, modeling and method design skills, computation skills, and teaching skills that get the relevant messages to all other members of the organization and allow the risk analysis function to develop and evolve in a way that suits the organization.

Some organizations refer to risk analysts as risk *managers*. This usually implies a confusion of roles. Risk is a pervasive aspect of a project that can be delegated in terms of analysis, but not in terms of management. Proper integration of project risk management and project management requires that the project manager takes personal responsibility for all risk not explicitly delegated to managers of components of the project. Further, ownership of the risk management process should not be portioned off. It needs to be embedded in the thought processes and actions of the team as a whole.

Drivers of Participant Performance

As with any task, the effectiveness of a project participant in undertaking risk management is driven by both the working environment and characteristics of the participant (Ward, 1999b). From a risk management perspective, the working environment could be characterized by factors such as location of the risk management effort in relation to the project organization; the nature of the organizational structure; the quality of supporting information systems; the availability of information; and the existence of an organization culture with respect to attitude to risk and risk management, resources available, and time available to undertake risk management. These factors either have to be managed to facilitate risk management or else taken into account in determining the form of risk management process that is feasible.

In terms of effectively undertaking risk management, relevant characteristics of a project participant comprise the participant's perception of their responsibilities for undertaking risk management in the given project, the participant's capability and experience in risk management, and his or her motivation to undertake risk management.

Perceived Responsibilities

The need for clearly specified responsibilities has long been recognized as a central requirement for effective performance in organizations, reflected in the widespread use of "management by objectives" and related techniques. In a project context this translates into a need for responsibility for project tasks to be clearly allocated to one or more parties, and for these parties to have a clear idea of what is expected of them. In terms of risk management, this translates into a need for sources of uncertainty to be identified and clearly allocated to appropriate parties. For example, a client organization may seek to manage risks by implicitly transferring them to a contractor via a firm fixed-price contract, but this is no guarantee that the contractor will identify these risks and manage them in the client's interest. (See the chapters by Langford and Murray, and by Lowe.)

Capability and Experience

In any project, a participant's perception of his or her responsibilities will depend on the nature of the project being undertaken and the extent of their capability and experience. This applies to project tasks in general and risk management in particular. For example, contractual requirements that a contractor shall be responsible for certain project risks, or undertake risk management, are no guarantee that effective risk management will actually occur. Selecting contractors with appropriate capability and experience in risk management will go some way to ensuring that such contractual obligations are met to the satisfaction of the client.

Similar considerations apply to project parties within the project owner's own organization. In any given project, the risk analysis team must be seen by the project team as an effective support function. If this is not the case, the cooperation necessary to do the job will not be forthcoming and the risk management process will flounder. However, it is equally important that the risk analysis team be seen by the project owning organization as unbiased, with a demonstrable record for telling it as it is, as providers of an honest broker external review as part of the process. If this is not the case, the risk management process will sink without trace.

Undertaking risk management is a high-risk project in itself, especially if embedding effective risk management in the organization as well as in the project in question is the objective. Often the project planning team provide a high-risk environment for risk analysis because, for example, project management is ineffective, or project team members

- are not familiar with effective project risk management processes;
- are familiar with inappropriate risk management processes;
- come from very difficult cultures;
- come from competing organizations or departments.

If the quality of the project management process or staff is a serious issue, this can be the biggest threat to the risk management process, as well as to the project itself. If this is the case, it deserves careful management, for obvious reasons.

Recognition of these issues by top management usually results in initiatives to enhance *corporate* capability in project risk management. Hillson (1997) has characterized corporate capability for risk management in terms of four levels of risk maturity:

1. The level of commitment to risk management
2. The degree of formality in risk management processes
3. The level of in-house expertise and training in risk management skills
4. The extent to which risk management tools and methods are applied to the organization's activities

Once an organization has assessed its level of risk maturity in terms of these four attributes, Hillson identifies a number of actions that the organization then needs to undertake in order to raise its level of risk maturity.

Motivation

With respect to risk management, project participants need to be convinced that risk management activity will help them meet their own objectives. For powerful customers, a requirement that all tenders include a risk management plan may be sufficient inducement for offering contractors to comply at some level. However, if the customer ignores the needs of the contractor and insists on inappropriate allocation of project risks, then contractors will be motivated to use risk management in their own interest rather than in the interests of the customer. A simple example is claims-seeking behavior by contractors who think they are going to lose money on an onerous fixed-price contract. Enlightened customers expecting risk management from contractors will take pains to demonstrate how contractors can benefit directly by improved cost estimation, greater efficiency, improved project control, and, ultimately, higher profitability.

Similar motivation issues arise in persuading different units in the same organization to undertake risk management. A risk management process that is treated by organization units as just more bureaucracy is unlikely to be fully effective in bringing about proactive management of project risks. In relation to all key players, the risk management process must be seen as immediately useful and valuable, in the sense that it more than justifies the demands made upon them. Furthermore, if the risk management process threatens any of the players, there must be a balance of power in favor of meeting that threat rather than avoiding it.

Summary

Like Simister's, this chapter is concerned with general principles and processes in project risk management that are applicable in all project contexts. These principles can be applied either to *de novo* applications of risk management in particular projects, or, more usefully, to the design of corporate-based formal processes. In the latter case, process design is "strategic" in the sense that an overview of what is appropriate for a given organization is the first consideration. In addition to taking on board the principles discussed under the questions ***who***?, ***why***?, and ***what***?, developing corporate guidelines for deployment of risk management will also involve consideration of ***when***? issues in terms of the range of projects that will be subject to risk management processes. In general, comprehensive risk management will tend to be most useful when projects involve one or more of the following:

- Significant novelty (technological, geographical, environmental, or organizational)
- Significant complexity (in terms of technology, organizational structure, political issues, etc.)
- Significant size (in terms of cost, total resources, committed in-house resources, etc.)
- High cost of failure (financially or in terms of reputations)
- Long planning horizons

In time, organizations institutionalizing project risk management may apply different guidelines in terms of the six W questions to different projects, depending on the extent to which the preceding factors are present. However, such sophistication needs to wait on the development of experience with comprehensive risk management processes on selected projects. Failing to consider this issue would be rather like operating a car hire firm that always offers a Rolls-Royce, or a Mini, regardless of the potential customer's wallet or needs. It is difficult to over-emphasize this point because the systematic nature of risk management processes can easily seduce those who ought to know better into the adoption of a single analytical approach for all projects. "If the only tool in your toolbox is a hammer, every problem looks like a nail" is a situation to be avoided.

Even with such sophistication, there will still be considerable need to modify the scope and approach of analysis for individual projects. In effect, the risk management aspect of a parent project needs to be regarded as a project in its own right, requiring a specific design phase prior to execution of risk analysis, and periodic, ongoing development of the process as necessary. This design phase needs to address the six W questions: ***who***?, ***why***?, ***what***?, ***whichway***?, ***wherewithal***?, and ***when***? Addressing the first three of these six W questions refines the scope of analysis and subsequent management actions within the framework of any corporate guidelines. Addressing the latter three questions leads to more detailed, project-specific "tactical" planning of the analysis to be undertaken. However, there are significant interdependencies between these six Ws. For example, in a particular context, ***why?*** analysis being undertaken may be closely related to who wants the analysis and the scope of risks being included in the analysis. Limiting time and resources may constrain ***whichway*** the analysis is carried out. Recognition of these interdependencies is important in determining the most appropriate approach to risk management for a given project and organizational context.

References

Ackermann, F., C. Eden, and T. Williams, T. 1997. Modelling for litigation: Mixing qualitative and quantitative approaches. *Interfaces* 27:48–65.

Ashley, D. B., and I. Avots. 1984. Influence diagramming for analysis of project risk. *Project Management Journal* 15(1):56–62.

Chapman, C. B., and S. C. Ward. 2003. *Project risk management: Processes, techniques and insights*, 2nd ed. Chichester, UK: Wiley.

———. 2002. *Managing project risk and uncertainty: A constructively simple approach to decision making.* Chichester, UK: Wiley.

———. 2000. Estimation and evaluation of uncertainty: A minimalist, first pass approach. *International Journal of Project Management* 18:369–383.

Cooper, K. G. 1980. Naval ship production: A claim settled and a framework built. *Interfaces* 10(6): 20–36.

Drummond, H. 1999. Are we any closer to the end? Escalation and the case of Taurus. *International Journal of Project Management,* 17(1):11–16.

Eden, C., T. Williams, F. Ackermann, and S. Howick. 2000. The role of feedback dynamics in disruption and delay on the nature of disruption and delay (D&D) in major projects. *Journal of the Operational Research Society* 51, 291–300.

Goldratt, E. M. 1997. *Critical chain.* Great Barrington, MA: North River Press.

Hartman F. and P. Snelgrove. 1996. Risk allocation in lump sum contracts: Concept of latent dispute. *Journal of Construction Engineering and Management* (September): 291–296.

Hartman F., P. Snelgrove, and R. Ashrafi. 1997. Effective wording to improve risk allocation in lump sum contracts. *Journal of Construction Engineering and Management.* (December): 379–387.

Hillson, David A. 1997. Towards a risk maturity model. *The International Journal of Project and Business Risk Management.* (Spring): 35–45.

Howick, S., and C. Eden. 2001. The impact of disruption and delay when compressing large projects: Going for incentives? *Journal of the Operational Research Society* 52:26–34.

Kelly, J., and S. Male. 1993, *Value management in design and construction: The economic management of projects.* London: Spon.

Klein, J. H. 1993. Modelling risk trade-off, *Journal of the Operational Research Society.* 44(5):445–460.

Markowitz, H. 1959. *Portfolio selection: Efficient diversification of investments.* New York: Wiley.

McCray, G. E., R. L. Purvis, and C. G. McCray. 2002. Project management under uncertainty: The impact of heuristics and biases. *Project Management Journal* 33(1):49–57.

Office of Government Commerce. 2002. *Management of risk: Guidance for practitioners.* London: The Stationery Office.

Uher, T. E., and A. R. Toakley. 1999. Risk management in the conceptual phase of a project. *International Journal of Project Management* 17(3):161–169.

Ward, S. C. 1999a. Assessing and managing important risks. *International Journal of Project Management* 17(6):331–336.

———. 1999b. Requirements for an effective project risk management process. *Project Management Journal* 30(3):37–43.

Williams, T. M. 2000. Systemic project risk management. *International Journal of Risk Assessment and Management* 1:149–159.

Williams, T., C. Eden., F. Ackermann, and A. Tait. 1995. The effects of design changes and delays on project costs. *Journal of the Operational Research Society* 46:809–818.

CHAPTER THIRTY-SIX

VALUE MANAGEMENT:

A Group Decision-Making Process to Achieve Stakeholders' Needs and Expectations in the Most Resource-Effective Ways

Michel Thiry

Value management (VM) is not a "new fad," but a proven methodology formally developed in the late 1940s that evolved from "a problem-solving system to deliver products with appropriate performance and cost" (Miles, 1972) to "a style of management . . . with the aim of maximizing the overall performance of an organization" (BSI, 2000).

From Value Analysis to Value Management

Historical Background

From the late 1940s to the early 1960s, Lawrence D. Miles, who is today considered the father of value methodologies, developed the concept of value analysis at General Electric. He specifically identified three elements that are at the core of all value methodologies: the concept of function, the multidisciplinary team workshops, and a structured "job plan." During that period, value analysis (VA) and value engineering (VE), considered synonymous, were mainly applied to manufactured products.

In the early 1960s, Charles W. Bytheway (1965) developed the concept of the Function Analysis Systems Technique (FAST) diagram, and VA and VE started to be applied in construction and new product development. In the early 1980s, the Society of American Value Engineers developed the 40-hour (5-day) workshop in the United States, as part of their certification system. (A few issues limit the usefulness of the 40-hour workshop: it is generally carried out by external experts, which greatly reduces buy-in from the project team and implementation of proposals, and it is mostly focused on cost reduction, which limits its use in early strategic situations.) In the late 1980s and early 1990s, in Europe and

Canada, a number of practitioners started to apply VA and VE in earlier stages of projects and to integrate them into the project process; it is also the period when the first applications to organizational processes were attempted.

Although the term value management was first used in the 1980s, it was only in the late 1990s that value management emerged as a discipline distinct from VA/VE, drawing on management techniques and fully integrating it in the project life cycle as a "collaborative group-learning approach" (Barton, 2000). Today, VM is used in a number of new fields, like strategic planning, process reengineering, organizational management, change management, concurrent engineering, and others; it is also integrated with known processes like organizational effectiveness, quality management, design to objectives (DTO), and risk management.

Associations and Standards

In late 1958, a first group of value "engineers" associated under the acronym of SAVE (Society of American Value Engineers), now renamed SAVE International. Soon after SAVE was founded, VE associations started to spread around the world. Value societies were founded in Japan in 1965, the United Kingdom in 1966, Germany in 1974, France in 1978, and a number of other countries since; today more than 20 countries are represented by value associations, among them the United States and Canada, eight EU countries that have formed the European Governing Board of the VM Training and Certification System (EGB), Hungary, the Czech Republic, Kuwait, Brazil, Japan, Hong Kong, South Korea, Australia and New Zealand, India, and South Africa.

The U.S. Department of Defense established the first governmental value program in 1954. The method was to be applied at the engineering stage, which brought about a change in name from value analysis to *value engineering*. In the early 1970s, GSA asked SAVE to develop a certification program for value practitioners. The status of Certified Value Specialist (CVS) was established by SAVE as a standard (SAVE, 1993), recognizing competence in the field of value engineering. Today, many such programs exist around the world; specifically a European System of Training and Certification in Value Management was set up in 1997 as a result of discussions between eight European National Value Associations.

Germany developed the first value standard: DIN 69 910 on "Wertanalyse" was created in 1973. From 1985 to the early 1990s, AFNOR (French Standards Body) developed a number of value analysis standards in France. In 1987, the Bureau of Indian Standards set up standard IS:11810-1986 on VE. Australia and New Zealand published Standard AS/NZS 4183:1994 "Value Management"—*as an analytical process*—in 1994. The American Society for Testing Materials (ASTM, 1995) released a "VE in Construction" standard in December of 1995 and the European Committee for Standardization (CEN) issued the first standard on value management as a management approach in 2000 (BSI, 2000).

Current Definitions

As early as 1731, Daniel Bernoulli, presenting a paper at the Imperial Academy of Sciences in St. Petersburg, stated that "the value of an item must not be based on its price, but rather on the utility which it yields" (Bernstein, 1996). This concept of the relationship between value and function is at the heart of all value disciplines.

In 2003 and 2004, SAVE International and the EGB agreed to develop a framework for certification reciprocity. As part of this framework, definitions were developed for common value terms; they are recognized as a common basis for value practice. The definitions are as follows

- "*Value* has been defined as a ratio between 'Quality & Cost', 'Function & Cost', 'Worth & Cost', 'Performance & Resources', 'Satisfaction of needs & Use of resources' and 'Benefits & Investment'. All those ratios and others are acceptable, as long as, on one side, lies the satisfaction of an explicit or implicit need and on the other, the resources invested to achieve it." In the United States and U.S.-influenced countries like English Canada, the Middle East, Japan, the focus is on cost, whereas Europe and countries like Australia, French Canada, and Hong Kong have developed a broader view of resources.
- "*Value Management* (VM) consists of the combined application of value methodologies and other methodologies at organizational level (from strategic to operational) in order to improve organizational effectiveness." (In the U.S.-influence zone it is an umbrella term for VA and VE, whereas in Europe and other countries it is considered a management approach, as described in EN-12973:2000 European Standard).
- "*Value Analysis* (VA) and *Value Engineering* (VE) are specific value methodologies aimed at improving existing products or developing new products. Products can include both goods and services."
- "*Value methodologies* [no acronym] include all processes, tools and techniques derived from the work of Lawrence Miles. They specifically include the concept of functions and function analysis (and its derivatives), the concept of cross-functional teamwork and the concept of a structured process based on creative thinking (alternative left/right brain use).

  "There are a number of specific methodologies that have been developed worldwide on the basis of these, including, but not limited to: FAST Diagramming (US and Canada), 40 hour workshop with job plan (US & Canada), Split workshops (UK), Function Tree (Europe & Canada), Function Cost (US & Europe), Functional Performance Specification (France, French Canada & Europe), Soft VM (Australia, UK) etc. The use of one methodology or another should not limit the value practitioner, whose first aim is to improve value within the limits of their intervention."

There are three key concepts that, combined, underlie all value methodologies:

- The *function*, which is the expression of needs in terms of purpose, independent from any solution
- The *cross-functional team*, which enables a broad view and an increased knowledge of a situation
- The *structured process*, based on creative thinking; the alternate use of creativity and analysis, or lateral and vertical thinking (de Bono, 1990)

The Concept of Value

As outlined previously, there are many definitions of value. For the purpose of this chapter, I will concentrate of the one developed for the European Standard: "Value may be described, in the context of VM, as the relationship between the satisfaction of need and the resources used in achieving that satisfaction". Although there are other definitions of value, this one is the most appropriate to the practice of VM.

The European Standard (BSI, 2000) uses the representation shown in Figure 36.1, with the following note: "The symbol α signifies that the relationship between the satisfaction of need and the resources is only a representation. They must be traded off one against the other in order to obtain the most beneficial balance" (BSI, 2000).

Using a more specifically strategic approach, I have developed a more elaborate diagram for value, shown in Figure 36.2, where the Benefits Variance (BV) = Offered Benefits (OB) − Expected Benefits (EB) and BV must be ≥ 0. The assumption is that *Benefits* is directly related to the *Satisfaction of needs*. In addition, Resource Variance (RV) = Available Resources (AR)–Required Resources (RR) and RV must be ≥ 0.

This means that the value manager is not only concerned with balancing resources with satisfaction of needs (benefits) but also with making sure that both for benefits and resources, capacity—including capability—is matched with intent.

Distinctions between Value Analysis, Value Engineering, and Value Management

Value analysis was the term first used by Miles to define the process he developed; it was later changed to value engineering when he started working with the military. Today there is a consensus to define VA and VE as specific methodologies aimed at improving or developing better products. A common distinction is to consider VA as the value-based analysis of existing products to improve them and VE as the application of value techniques to develop new products. Officially, VE and VA are considered synonymous in both the SAVE (1997) Standard and in the European Standard (BSI, 2000). In their 1997 Standard, SAVE uses the term value methodology to encompass both VE and VA and defines it as "the

FIGURE 36.1. THE EUROPEAN STANDARD DIAGRAM DESCRIBING VALUE.

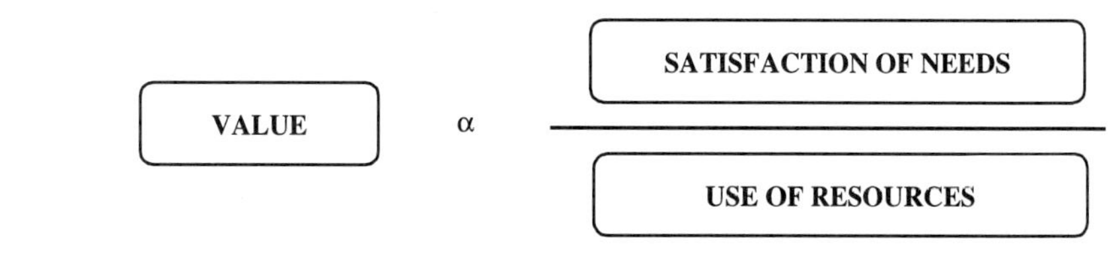

FIGURE 36.2. THE VALUE RATIO.

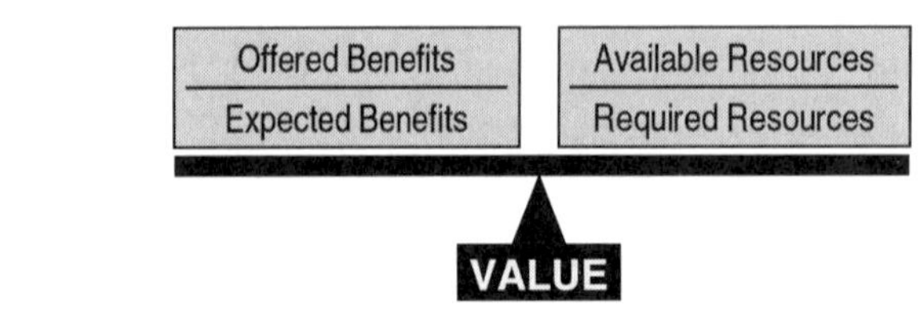

systematic application of recognized techniques which identify the functions of the product or service, establish the worth of those functions, and provide the necessary functions to meet the required performance at the lowest overall cost".

SAVE International has traditionally used the term value management to define the management of value proposals generated through the use of VE and VA (SAVE, 1997). Considering the recent reciprocity agreement between SAVE International and the EGB, as well as recent publications (Green, 1995; Kaufman, 1997; Barton, 2000; Brun and Constantineau, 2001; Woodhead and Downs, 2001). We can now consider that the definition of value management given previously and detailed in the European Standard (BSI, 2000) is widely accepted. In Australia (Barton, 2000) and in the United Kingdom (Green, 1995), researchers, using developments in systems engineering (Checkland and Scholes, 1994), have distinguished "soft" VM from "hard" VM; the former being applied at strategic level or in the early stages of projects to identify stakeholders and define needs and expectations, and the latter, based on VA and VE techniques, being applied in later stages of projects, when customers and problems can be clearly identified, to improve delivery and outcomes.

Value Management Today: A Style of Management

At a management level, VM is based on three root principles:

1. A focus on objectives and targets: stakeholders' needs and critical success factors
2. A focus on the function: purpose of the product, system, or service
3. A continuous awareness of value: measures for improvement and innovativeness

Three key concepts underlie it:

(a) *A transverse approach, translated in practice by cross-functional teamwork.* The VM methodology is based on a multidisciplinary team approach in a workshop environment; it enables the team to broaden perspective, assess every angle, share opinions, and reach consensus.

(b) *A structured decision process leading to better business decisions.* The VM process is a systematic approach, based on creative thinking, which uses time in the most efficient way to ensure that the scope of the VM study is covered in the best sequence.
(c) *The use of functions, increasing effectiveness and enhancing competitiveness.* VM relies on functions, rather than on predefined solutions, to identify the expected outcome of a project, thus allowing evaluation of a broader range of possible options.

A Management Approach

The application of VM at the organizational level is a source of competitive advantage. The combined use of a cross-functional group process, the concept of functions and of creative thinking principles combined to foster learning and enhance innovation, creates an undeniable combination of differentiation and implicit knowledge, leading toward competitive advantage.

Organizational decisions are often characterized by a high degree of uncertainty (lack of information), ambiguity (conflicting information), and stress (lack of time). Group decision making (GDM) enhances the ratio of facts versus assumptions, the quality (truthfulness) of information, consensus and objective sharing, and group commitment to decisions (also a relief of stress). VM is a GDM process that through sensemaking (Thiry, 2001) reduces ambiguity and enhances quality of information, and builds consensus and buy-in from participants in the process. At the elaboration phase, it uses scope definition, feasibility, and risk management concepts to increase the level of information and reduce uncertainty, as well as the possibility to develop more resource-effective options. If the players are well chosen among the key stakeholders, VM will increase decision support and implementation success through its participative group process.

The concept of functions is central to all value methodologies; applied at the strategic level, the traditional concept of functions can be defined as expected benefits in general and critical success factors (CSFs) in particular. CSFs have been defined in many studies as key issues to be addressed in every project, in which case they are generic and apply to all projects (Pinto and Rouhiainen, 2001). In a VM perspective, CSFs are specific expected benefits that are identified as critical for success and prioritized for each project by the key stakeholders, which are qualitative; these are defined in quantitative measurable terms through key performance indicators (KPIs).

Recently, an emergent perspective has linked VM to a learning paradigm (Thiry, 2000, 2002). This concept aligns VM with developments in organizational effectiveness like balanced scorecard (Kaplan and Norton, 1986), the EFQM Excellence Model (EFQM, 2000), and ISO-9000:2000 (ISO, 2000). When used iteratively in a systemic perspective, at program level and at project gateways, this approach ensures that strategic benefits will not only be explicitly defined at a strategic level but also effectively measured and delivered.

All the above would not be possible without a robust and structured process. Each value association around the world has developed their own "job plan," which can vary from five (SAVE, 1997) to seven (AFNOR, 1985) or even ten (BIS, 1987) steps. All those "plans" originate from the concepts developed by Miles for VA. They are all based on creative thinking, which has been detailed by de Bono (1990) as the alternate use of lateral and vertical thinking. Miles described a standard problem-solving process, where the decision-making authority lay outside the process, to which he added cross-disciplinary participation and function identification.

Today, VM is more associated with a decision-making process, where the decision makers are actively participating in the process and have authority over the resources required to implement the decision. In complex situations and environments, standard problem-solving or decision-making techniques are not applicable to define the situation; one needs to work with soft systems methodologies (Checkland and Scholes, 1994; Green, 1994; Neal, 1995) or more constructivist approaches (Weick, 1995; Guba and Lincoln, 1989; Quinn, 1996) like sensemaking.

The VM Process

For the purpose of this chapter, we shall treat VM as a strategic decision-making and control process in the context of program and project management. Figure 36.3 represents this view.

The VM process comprises the following subprocesses, which are typically carried out as facilitated workshops or meetings where the key stakeholders (at least) participate actively. Some tasks can be carried on individually, but research (Vennix, 1996; Woodhead and Downs, 2001; Fong, 2002) has demonstrated that facilitation is a key aspect of successful

FIGURE 36.3. VALUE MANAGEMENT IN CONTEXT.

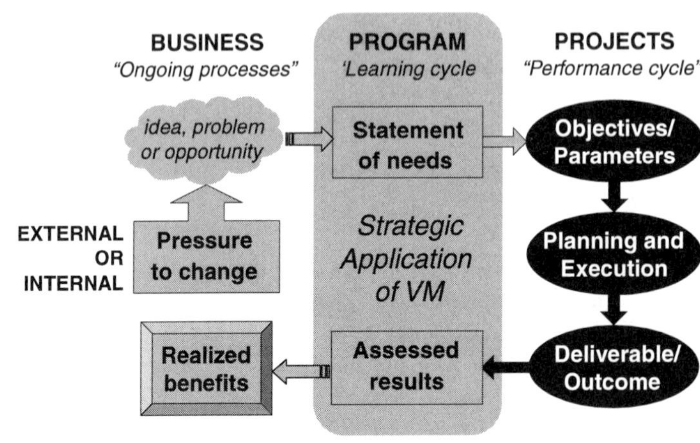

group decision processes in general and VM in particular. In this section, the "team" will mean those participating in the VM process through workshops or individual work.

- *Sensemaking*, which includes function(al) analysis and can use a variety of techniques like scenario planning, soft systems analysis, gap analysis, and others, is used to understand the situation and come to a shared agreement about the critical success factors (qualitative-level expected benefits) and key performance indicators (quantitative measures).
- *Ideation* is the creative generation of alternatives that enables the process to be truly innovative.
- *Elaboration* consists of the evaluation of alternatives in terms of their achievability and contribution to expected benefits, and their combination/modification to develop viable options. The definition given by Spradlin (1997) will be used for this chapter: "An option is an alternative that permits a future decision following revelation of information. All options are alternatives, but not all alternatives are options."
- *Choice* is the action of selecting/prioritizing the best options, in regards of the critical success factors.
- *Mastery* is a formative evaluation and control process based on improvement rather than on a baseline.

Focus on Results

The main success factor of value methodologies has always been their focus on tangible results. It starts with the analysis of stakeholders' needs and expectations and their translation into measurable objectives, which are then addressed to identify the most profitable options. Following the identification of the best options, a "formative evaluation" (Guba and Lincoln, 1989) process enables the team to deliver results that are in line with the expected benefits. The key aspect of VM is the direct link that is established between needs and results, through functions.

A case study follows, which will be used to demonstrate application of the VM techniques outlined in this section.

Case Study: Outline of Situation

Consider a medium-size company that develops and implements projects for external clients. The company has been in this business since 1985 and has built a good reputation with its clients.

The market is growing, and in the last year the company has hired 50 new personnel, of which 15 are project managers. About 20 percent of the personnel are project managers; the others are technicians, operations and support staff, and product developers. There are five program managers.

Recently, a number of clients have complained about project performance and it has come to the ears of members of the board. A significant number of projects are either running late or over budget.

The MD calls one of the program managers and asks her to take care of the problem. The mandate calls for quick results on the most significant projects and general improvement of the situation within six months.

Among the new people that have been hired, many have good experience but have not yet fully integrated in the company; others have little practical experience, although they show good potential.

The company has always relied on a few experienced PMs to "run the show," but those are the ones that are also the most resistant to change; they know their business and do not accept criticism of their methods easily: "I have always delivered what I was asked; don't come and tell me what to do."

For a few months now, the HR Department has talked about induction courses and PM courses. The Quality Management Department has put in a budget for the standardization of PM processes and procedures; it would involve hiring an external consultant.

Most people work remotely at client sites.

Sensemaking

The first step in the VM process is to understand the situation that has created the need for change and, from that understanding, identify tangible benefits expected by the key stakeholders—the *functions* that the program or project must provide. The traditional *information* and *function analysis* phases of VA/VE are not adequate to deal with complex situations; "[. . .] whereas in a well-defined situation, with cohesive groups it may be possible to shorten the sensemaking process to a simple 'information phase', in complex, ambiguous, multileveled situations it is necessary to allow and foster a sensemaking interaction . . ." (Thiry, 2001).

A well-managed sensemaking process enables participants to construct a shared view of a complex situation and model their expected benefits through a function diagram. This model, which has been called function breakdown structure (FBS) (Thiry, 1997) enables the group to define CSFs and KPIs, which are the key to the successful achievement of expected benefits.

1. The first step of sensemaking in VM is to perform a *stakeholder analysis*, which encompasses the identification of the stakeholders, their classification, and their ranking.
2. The second step is to carry out a *functional analysis*, which consists of determining stakeholders' needs and expectations, translating them into expected benefits using a verb-noun semantic, identifying any additional benefits required, and organizing all these into an FBS.
3. The next step lies in the identification of those expected benefits that are *critical success factors* of the program or project and their prioritization, which will support decision making throughout the delivery process.

4. Finally, those functions that have been identified as CSFs are *characterized* (BSI, 2000) through the definition of key performance indicators (KPIs), using the concept of a criterion of measure, an expected level of performance, and an accepted range of flexibility. (This method was standardized by the French Value Analysis Association and is now included in the VM Standard.) The definition of KPIs enables the stakeholders to move from qualitative to quantitative measures of success and to be able to assess benefits on clearly quantifiable terms.

Stakeholder Analysis

To define needs and expectations of stakeholders, the team must first identify the stakeholders. They will typically use intuition and historical data to list a number of possible stakeholders without attempting to categorize them or classify them—a lateral thinking process (de Bono, 1990). Once stakeholders have been identified, the team can group them in a hierarchical structure by creating groups and subgroups, not unlike a WBS; it is usually represented as a *Mind Map*™ (Buzan, 1974). This is a vertical thinking process, the purpose of which is to complete the list, understand it, and simplify it; *it is not a hierarchical representation of the structure of the organization.*

The second step is to categorize the stakeholders—a vertical thinking process—to measure their potential influence on the program or project process and their outcome, in order to identify the key, or significant, stakeholders. There are many ways to do this; they can be classified by power level (preponderant to affected party), area of interest (financial, technical, regulatory, etc.), or structural layer (regardless of direct influence). A simple, effective way to quickly map the stakeholders is to develop an influence diagram. Based on the level of power—of all types, as described in the chapter by Magenau and Pinto—they could exercise on the program or project and their level of interest (whichever the area of interest), which can be positive or negative.

FIGURE 36.4. THE STAKEHOLDER INFLUENCE DIAGRAM.

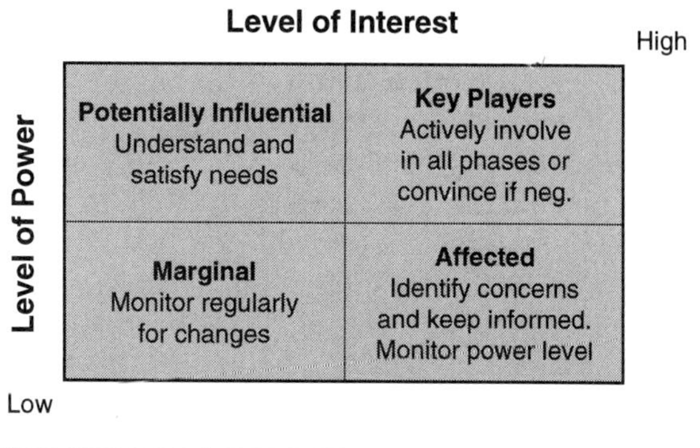

The influence diagram can be divided into four major areas (as shown in Figure 36.4), or each axis can be graded from 1 to 10 to create a more detailed picture of stakeholders' influence. Level of interest could also be divided into negative and positive, each with a gradation. When doing this, though, the team should not forget that a high level of interest, whether positive or negative, should be seen as similarly influential, but with a different potential impact.

Case Study: Stakeholder Influence Grid

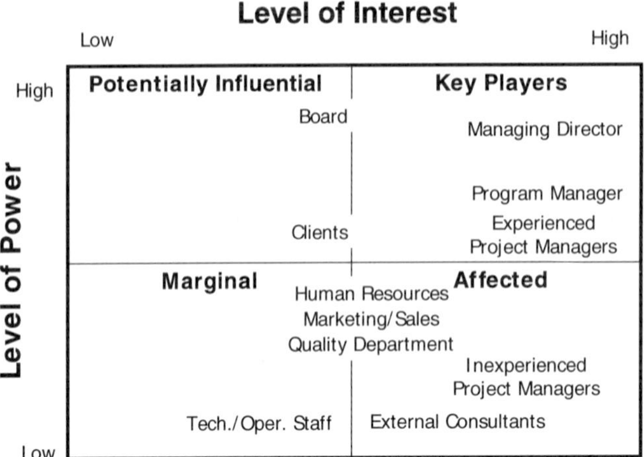

Function(al) Analysis

The PMBOK Guide (PMI, 2000, p.18) states that "finding appropriate resolution to [stakeholders] differences can be one of the major challenges of project management." The VM Standard (BSI, 2000) claims that "stakeholders . . . may all hold differing views of what represents value. The aim of VM is to reconcile these differences. . . ." To reconcile the stakeholders' differences, the first step is to identify stakeholders' needs and expectations; traditionally, project management has associated needs with requirements and expectations are considered undefined requirements. More precisely, Dorothy Kirk has identified expectations as "beliefs or assumptions about the future . . . usually set unintentionally" (Kirk, 2000). Conversely, VM has defined needs as "what is necessary for, or desired by the user. A need can be declared or undeclared, it can be an existing or potential one" (BSI, 1997). For VM, needs and expectations should be considered as one and the same, but it identifies needs and wants in order to distinguish what is absolutely necessary from what is not.

Identification of Needs and Expected Benefits. Although identification of needs is technically part of functional analysis, stakeholders' analysis and functional analysis are usually a seamless process. Before the workshop, the team will be required to gather information about the subject to be addressed through expert judgment and historical information. The team may also be required to carry out interviews of stakeholders that have been identified as key and cannot, or will not, attend.

Functional analysis requires clarifying expectations, rendering these explicit by asking questions to make stakeholders express them openly (Barton, 2000), and making any remaining assumptions agreed and well documented, so stakeholders can examine and clarify them. Once expectations are explicit, they need to be managed, which means comparing available versus required resources to fulfill them and resolving any gaps. This can be achieved only by communicating openly and early in the program or project. Finally, measurable criteria will be identified to ensure that the achievement of these expectations can be followed through.

Generally, needs identification is carried out using intuitive techniques like brainstorming or simple discussion. The European VM Standard (BSI, 2000) defines a more thorough method called Method of Interaction with the External Environment, or more simply, Interactors' Analysis. It consists of identifying the different interactive agents of the environment of the program or project (stakeholders as well as physical parameters and constraints) and identifying both direct expectations and interactions regarding the product or process, as well as functions created between interactors through the product or process. Although most functions can be identified using intuitive methods, it is worthwhile, if resources are available, to complete the study with an interactors' analysis.

Example:

In our case study, a direct expectation for the Managing Director toward the improvement of project management delivery may be to standardize PM processes; an interaction function may be the need to choose between different standards, which will be identified through the use of interactive agents like existing PM standards.

When identifying stakeholders' needs, one should consider both direct, or tangible, needs—usually translated into hard benefits (economic, technical, operational)—and indirect, or intangible, needs—usually translated into soft benefits (power, politics, communications). In addition, one should also identify contradictions and assess their consequences, prioritize values, and manage trade-offs. This process should be iterated during decision implementation.

Case Study: Needs/Expected Benefits

Board:

- *Maintain/increase revenue*
- *Increase market share*
- *Stop clients' complaints*

Managing Director:

- *Demonstrate control*
- *Deliver within set parameters*
- *Maintain profits*

Clients:

- *Deliver within agreed parameters*
- *Maintain quality*
- *Improve current processes*

Experienced Project Managers:

- *Maintain freedom*
- *Get clear objectives/deliverables*
- *Understand company culture (new)*

Inexperienced Project Managers:

- *Understand company culture*
- *Get clear instructions (processes and procedures)*
- *Know where to find information (mentor?)*

Organization of Expected Benefits: The Function Breakdown Structure (FBS). Once benefits have been identified, it is time to start organizing them. This organization is accomplished through a hierarchical structure, based on needs, which has been labeled function breakdown structure for its similarity to the WBS (except it is, by convention, oriented from left to right, instead of top-down). In the United States, diagrams based on the same basic concepts are called FAST diagrams (Function Analysis Systems Technique). FAST diagrams have stricter rules than FBS and are often too constraining for program or project preinitiation situations; they are well adapted to project-product improvement or development. In Europe true FAST is little practiced and VM practitioners use what they call function trees or functional diagrams, which are similar to the FBS. The FBS process and the development of high-level WBS for projects are an iterative process, and the FBS must be reviewed regularly to continue representing the stakeholders' views.

The foundation rule of any function diagram is to create a hierarchy of functions that goes from the more abstract needs (the why) to concrete actions or tasks/activities (the how). The objective for the FBS is to model the expected benefits of the stakeholders in a way that represents their shared view of the situation. At the higher levels, these will be labeled vision, mission, objectives, and goals; at the lower levels, they will be called functions, tasks, and activities. If used to define programs or projects, the FBS should be a seamless continuum from the objectives and goals to the actions (tasks and activities) required to satisfy that purpose; at that level it could be labeled a task-oriented WBS and therefore clearly links project definitions to the stakeholders' expected benefits in terms of their contribution. Like a WBS, it helps define the scope of the needs, by identifying gaps and verifying completeness and by eliminating unnecessary functions. It also shows specific relationships by identifying the main purpose, critical success factors, and prioritizing needs and wants. It is an effective way to model complex situations objectively, when ambiguity is high and agreement low.

To build the FBS, the team will start with a minimal number of functions and relate them to each other using a how?-why? logic, as shown in Figure 36.5, to develop the basic structure. More functions will then be added in relationship with those already in place. Eventually, the completeness of the FBS will be verified by adding a when? relationship vertically; the when? constitutes a sequence of functions/tasks that, if achieved, will fulfill the higher-level function/benefit.

Identification of Critical Success Factors (CSF). Once the team is relatively satisfied that the FBS represents their shared view of the situation and their expected benefits, it must identify the CSFs. CSFs must be of a high enough level to be manageable and of a low enough level to be easily measured; they are generally qualitative. There are few rules as to what constitutes a CSF, except that the team identifies them as significant measures of the program or project's success. CSFs can be identified at different levels of the same FBS but should cover the complete range of needs. This means that if the team decides to go down one level, they must take all the subsidiary functions of that function; also, they cannot identify CSFs in the same branch at two different levels. Breaking these rules would create a gap or a redundancy.

FIGURE 34.5. THE HOW?-WHY? LOGIC.

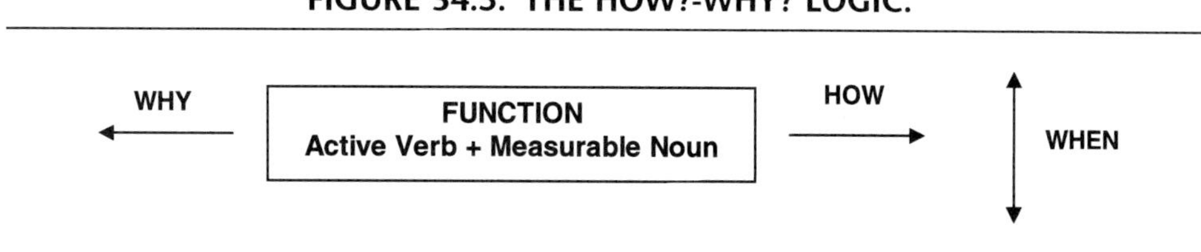

Case Study: Function Breakdown Structure with CSFs

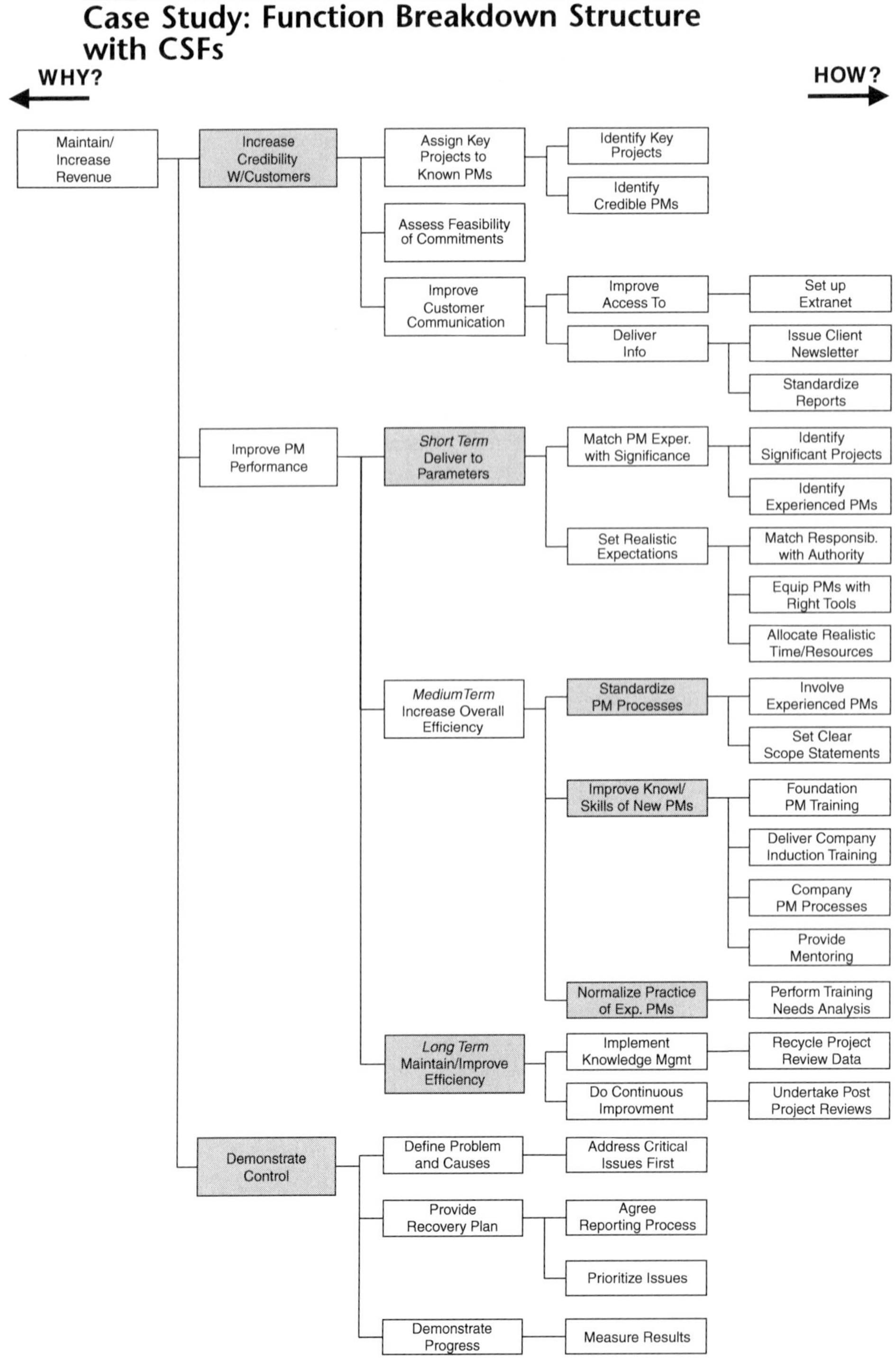

Note: High-level goal of program on left; critical success factors in gray; specific actions on right.

Eight to twelve CSFs is generally a good manageable number for most programs and projects, although large programs may require more. Once the CSFs have been identified and agreed by the team, they will become the baseline value criteria for the evaluation of all options in decision making and change management in the program or project. To increase the effectiveness of this process, CSFs need to be prioritized. Paired comparison is a fast and objective way to achieve this. Generally, 5 points will be allocated between each pair of CSFs. Combinations can therefore be 5-0, 4-1, 3-2, and vice versa, as shown in Figure 36.6. Once this is done, the scores will be brought back onto a 100 percent scale. Prioritization should be reviewed regularly to confirm ongoing validity, mostly in programs that are more long term and can therefore be more susceptible to changes in priorities.

Case Study: Weighting of CSFs using paired comparison:

A—Increase credibility with customers
B—Deliver to parameters (Short term)
C—Standardize PM processes
D—Improve knowledge and skills of new PMs
E—Normalize practice of experienced PMs
F—Maintain/improve efficiency
G—Demonstrate control

	A	B	C	D	E	F	G	Total
A		3	5	4	4	4	3	**28**
B	2		4	3	3	4	4	**24**
C	0	1		2	2	2	1	**11**
D	1	2	3		3	2	2	**17**
E	1	2	3	2		1	1	**14**
F	1	1	3	3	4		3	**15**
G	2	1	4	3	4	2		**16**

RANKING OF CSFS	SCORE	WEIGHT
1. Increase credibility with customers.	28	22
2. Deliver to parameters (short term).	24	19
3. Improve knowledge and skills of new PMs.	17	14
4. Demonstrate control.	16	13
5. Normalize practice of experienced PMs.	15	12
6. Maintain/improve efficiency.	14	11
7. Standardize PM processes.	11	09
Total:	125	100%

Characterization of Critical Success Factors. Although traditional VM calls for characterization of all functions of lower level, experience has shown that it is more resource-effective in programs and projects to characterize only the CSFs. Characterization, as explained earlier, requires identifying key performance indicators. The French have developed and standardized a characterization process that works very well in programs and projects (AFNOR, 1991; Thiry, 1997). It basically consists of three steps:

1. For each CSF, identify one or more measurable criteria.
2. Set the expected level of performance.
3. Identify an acceptable range or tolerance.

This process is repeated for each CSF in order to create a baseline related to the needs of the stakeholders.

Example: Deliver to Parameters (Short Term)

Criteria could be as follows:

- Actual time/cost vs. baseline in percentage of difference
- Time to achieve objective in weeks

Level could be as follows:

- 5% maximum
- 8 weeks

Flexibility could be:

- −10% or +5%
- +2 weeks

In general, sensemaking processes are performed in a team environment. It is possible, if stakeholders cannot commit the necessary time, to perform some of the tasks in smaller groups, as long as the results are offered to the whole team for discussion and final approval.

Ideation

VM practitioners use creativity techniques in a number of VM processes, but if there is one area where creativity becomes essential, it is the generation of alternatives. Although the functional analysis has already generated a number of actions (how?), ideation will broaden the scope of possibilities from which to develop options and therefore increase the quality of those options from which the team will choose solutions. It will also generate a bank of possible solutions that could be exploited and developed further if circumstances or stakeholders' expectations changed, thereby saving important redevelopment time and resources. Ideation is the traditional *creativity phase* of VA/VE; it consists of identifying as many alternatives as possible for the fulfillment of one or more benefits. It is an area where lateral thinking cannot be mixed with vertical thinking. The team will start with CSFs, or alternatively, low-level functions, and creatively identify possible ways of achieving them.

There are a number of techniques available to the team to generate alternatives. The best known are brainstorming (Osborn: where the subject is clearly defined, or Gordon: where the subject is known only to the facilitator), forced comparison, synectics, interviews, stepladder technique (Rogelberg et al., 1992), or Delphi technique. In a program or project context, forced comparison and synectics are usually not focused enough and competent facilitators in those techniques are not easily available; Delphi is time-consuming and generally costly; interviews are a good alternative, but robustness of alternatives given by the group process is lost. Osborn technique brainstorm, with an experienced facilitator, is an easily applicable and valuable technique for small groups (five to eight), and stepladder would be the best choice where large groups are concerned, as it combines the quality of exchanges of the group process with the participation and creativity of every individual. As there is an evolution toward more stakeholder involvement, VM is slowly moving from using traditional brainstorm for creativity toward techniques akin of stepladder.

In the *Osborn brainstorm technique*, the group must first agree on the definition of the problem and make a clear statement of it; this is usually achieved through the functional analysis. The group, under the guidance of a facilitator, then identifies as many alternatives as possible to resolve the problem; or in this case, the team brainstorms on the CSFs or lower-level functions to offer the expected benefits. There are typically ten rules for an effective brainstorming session:

1. Write down all ideas and comments.
2. Target quantity rather than quality.
3. Exclude criticism; assume that each idea will work.
4. Hold judgment until the evaluation phase.
5. Eliminate the word impossible from your vocabulary.
6. Let your imagination roam free (the craziest ideas are often the most important).
7. Use piggybacking (build on other ideas and comments).
8. Cross-fertilize ideas (associate or modify ideas and comments).
9. Let everybody talk.
10. Do not interrupt!

Research has demonstrated that there are some problems associated with traditional brainstorming, like social inhibition, created by shyness or dominant individuals; social loafing, which distracts from objectives; and production blocking, which increases inhibition of individuals in a group situation (Rogelberg et al., 1992; West, 1994).

The *stepladder technique* directly addresses some of the problems identified for brainstorming and particularly a strong criticism of group brainstorming, that "quantity and often quality of ideas produces by individuals working separately are consistently superior to those produced by a group working together." (West, 1994). The principle of stepladder is that each individual works independently on the problem before joining the group, which also fosters transformational learning through *critical reflection* (Cranton, 1996). Ideas are then shared with the group in turn as each new individual joins (to avoid being influenced), and discussion can start only when all ideas have been expressed. The first part of the discussion should follow the ten rules of brainstorming, as the objective is to develop group alternatives. Only when the group is satisfied that they have explored all possible alternatives will they switch into vertical thinking mode.

A short form of the stepladder technique consists of having individuals work first independently and then discuss their ideas in pairs and in fours before sharing all ideas with the group. In this case there is a series of short, lateral–vertical thinking processes, with the first part of every discussion lateral, before the final evaluation and development of options.

In Australia, Roy Barton (2000) used a variation of stepladder for large VM workshops (20 to 30 or more participants) at strategic planning level for the New South Wales Public Sector. He divided the group in smaller working units of five to seven people, who work independently from each other before sharing findings with the whole group. A survey of 200 participants in more than 20 of his workshops showed that 90 people felt their views were adequately considered, and in more general terms, 76 percent felt that VM was very important to their project.

Case Study: Alternatives for Three Top CSFs

CSF 1: Increase Credibility with Customers.

- *Assign key projects to known/best PMs.*
- Improve customer communications.
- *Assess feasibility of commitments.*
- Communicate success.
- *Deliver within parameters.*
- Acknowledge past failures to customers.
- Involve customers in problem identification and solving.
- Create focus groups.
- Give performance guaranties and metrics.
- *Review customer account history to ensure good fit.*
- Involve Business Unit Managers with clients.
- *Realign dates and estimates to be realistic.*
- *Validate sales commitments before commitment to customer.*

CSF 2: Deliver to Parameters

- *Match PM experience with significant projects.*
- *Set realistic expectations.*

- *Provide strong incentives to deliver.*
- Renegotiate expectations with customers, based on achievability.
- *Provide PMs with required tools.*
- *Create team incentives to meet objectives.*
- Implement overtime incentives.
- *Re-scope deliverables in regards of resources.*

CSF 3: Improve Knowledge and Skills of New PMs

- Provide general training.
- Send new PMs to PMI seminars.
- *Create a mentoring program.*
- *Share lessons learned.*
- *Start new PMs on small, less risky projects.*
- Have them develop their own teaching material on PM methodology.
- Provide regular feedback.
- *Reward good performance.*
- Spot recognition program.
- *Set-up incentive plans.*

Alternatives in italic are part of chosen option (see next section).

In a project of program context, it is important to keep the team focused during ideation; although creativity requires the group to accept all ideas, it is also easy to lose sight of the initial objective. While the loose-rein approach is valid for purely creative processes, the management of programs and projects require some focus because of the limited resources available. An experienced facilitator should be able to manage the process to allow enough freedom to foster creativity without losing sight of the objectives. It is also noticeable that fewer ideas will be generated at higher levels (strategic or tactical) than at lower levels (technical or operational) because they are more abstract and cover more ground.

Elaboration

This phase of the VM process combines two traditional VA/VE phases: *evaluation* and *development.* It is the assessment of those alternatives that have been generated in the previous phase and the development of viable and profitable options. It uses *vertical thinking* concepts like feasibility, cost-benefit analysis, weighted matrices, risk analysis, estimating, and so on.

The first step of elaboration is to eliminate all nonviable alternatives; alternatives could be deemed nonviable because they are clearly unachievable or deliver none of the expected benefits, or more simply, because they will not be accepted by the stakeholders. Other alternatives will be clearly viable because of their intrinsic value in terms of achievability and benefits, or their wide acceptance by stakeholders. Some alternatives will fall into a gray zone where acceptance or viability will not be clearly established and rejection is not clearly an option; these alternatives will be looked at more closely to establish if they can be combined or improved to make them clearly viable.

The second step is to combine, modify, or develop the alternatives further to generate options. The development of options includes the gathering of additional data and a degree of analysis that enables the team to make a well-documented decision and persuade sponsors to support them. A practical way to achieve this is to identify "champions" for each potential option and to ask these champions to develop their options up to a point where they feel comfortable to make a recommendation that outlines benefits, states advantages and disadvantages (risks), and draws contingency plans.

Case Study: Elaboration and Comparison of 3 Options

Option 1: *Develop a Comprehensive Staffing Plan*

- Addresses both short term performance objectives and long term strategic goals.
- Cost invested brings return.
- Plan addresses both achievability and benefits.

Option 2: *Develop a Performance Incentive Plan*

- Could be costly (escalation and equity).
- Does not address right PM for right project issue.
- May not increase credibility.

Option 3: *Develop a Customer Communication Plan*

- Does not really address overall problem.
- Could backfire if delivery does not follow.
- Only addresses short term.

The third step is to compare a number of options between them based on their relative capacity to deliver CSFs, in order to prioritize resource expenditure. Decision factors may consider aspects like available resources, quick wins, ease of implementation, or, simply, benefits.

The benefits variance (BV) concept ranks and weighs CSFs—using paired comparison, for example. The team then determines expected benefits (EBs) by defining a minimum acceptable score (note that, if weighting adds up to 100, as shown in example, and scores are set on a scale of 0 to 10, the maximum possible score will be 1,000). Options are evaluated against each of those CSFs to generate, for each, a combined score that corresponds to their offered benefits (OBs). Any option that is below the minimum score is rejected; others are ranked. Thus, an order of priority is determined.

Case Study: Evaluation of Options

WEIGHTED MATRIX SCORING OF OPTIONS ON BENEFITS VARIANCE.

	Increase credibility with customers		Deliver to parameters (Short term)		Standardize PM processes		Improve knowledge and skills of new PMs		Normalize practice of experienced PMs		Maintain/improve efficiency		Demonstrate control		
Options CSF/Weight	**A**	**22**	**B**	**19**	**C**	**14**	**D**	**13**	**E**	**12**	**F**	**11**	**G**	**9**	**Total**
1—Develop a comprehensive staffing plan	7	154	6	114	7	98	9	117	9	108	7	77	9	81	**749**
2—Develop a performance incentive plan	3	66	7	133	2	28	1	13	5	60	8	88	5	45	**433**
3—Develop a customer communication plan	6	132	2	38	0	0	0	0	0	0	0	0	7	63	**233**
Maximum Possible Score	10	220	10	190	10	140	10	130	10	120	10	110	10	90	**1000**
Minimum Acceptable Score	7	154	7	133	6	84	6	78	5	60	5	55	4	36	**600**

Resource variance (RV) is estimated, considering a concept of achievability based on the organization's capabilities, or available resources (ARs) against program or project requirements, or required resources (RRs), in terms of:

- *Financial factors*: Capital cost, cash flow, life cycle costs
- *Parameters and constraints*: Size/scope, cost, type of work, etc.
- *Human resources*: Expertise, spread, external versus internal, etc.
- *People factors*: Availability, competence (skills and expertise), customer perception
- *Complexity*: Innovativeness, interdependencies, stakeholders, etc.

To be accepted, an option must be clearly achievable and RV must be superior to the required minimum.

If a number of options are contemplated, the RV result is added to the BV and the combined score is used to prioritize resources against benefits; it is possible to establish a minimum acceptable score if achievability elements are quantified. This means that the value manager is not only concerned with balancing resources with satisfaction of needs (benefits) but also with making sure that both for benefits and resources, capacity is matched with intent.

Once this is done, alternatives that have been judged viable are offered for final decision/prioritization. To complete the process, each major option will be assessed against the KPIs and risks will be identified and analyzed; then risk responses should be included in the final options.

Choice

The decision itself is the last step of the process. Traditionally, in VA/VE this was the *recommendation* phase; in VM, the team is expected to "own" the power over resources and therefore make the actual decision rather than recommend options. This is a major change with traditional value methodologies, as the value practitioner has, until recently, mostly acted as an external consultant and did not have decision-making power. Under more recent developments, where VM is considered a group decision support process, decision making is part of the VM process, since the decision makers are on the team.

A number of authors (Vennix, 1996; Checkland and Scholes, 1994; Guba and Lincoln, 1989; Waring, 1989) point out that in complex situations, decision making is subjective and intuitive rather than objective and rational. These findings outline the importance of aiming toward consensus, which, as a number of research studies have demonstrated (Vennix, 1996), will foster support and commitment to the implementation of the decision and increase its intrinsic quality. Group decision making is influenced by the significance of issues for the group, the shared understanding among group members, and participants' representation in the process (Eden, 1992; Deetz, 1995). VM, through its sensemaking, ideation, and elaboration processes, fosters all those elements and makes consensus easier to achieve. Still, there is a gradation in decision making from the most to the least consensual: consensus, compromise following discussion, vote and consensual discussion, vote following discussion, simple vote, leader decision following discussion, and finally, leader decision. If there is no other choice than to have a leader decision, the team must still be given a fair opportunity to discuss and try and reach consensus before the final decision is made.

Interestingly, a study by Holloman and Hendrick in 1972 demonstrated that consensus after majority vote is the most effective method in terms of time, quality, and satisfaction with the decision (Vennix, 1996). In this case the team takes a majority vote to choose the options and subsequently tries to arrive at a consensual decision. Experience shows that, when using VM, consensus is fairly easy to achieve, as it is a logical conclusion to the process, and when vote needs to be taken, there is usually little discussion or conflict.

The following case study example shows a vote with discussion, where a vote was taken to choose option 1 (unanimously). Following agreement, a discussion was initiated to decide which elements it should include. The final choice is slightly different from the initial option presented for evaluation and now includes incentives, which were part of option 2.

Case Study: Choice

Option 1 Components:

- Assess competency levels of all staff.
- Assess PM relationship/experience with different customers.
- Empower PMs, give authority to make decisions.
- Identify required tools and provide them.
- Tie incentives to restructuring.
- Tie incentives to company goals.

- Gain comprehensive knowledge of workload versus demand.
- Assign lesser tasks to new PMs.
- Team up new PMs with experienced PMs across organization.

Mastery—Benefits Management

If VM is to be considered a style of management, gatekeeping becomes an essential part of the process to ensure that value is delivered. With power over resource prioritization, VM can really achieve management of value over time if it is used for program appraisal and at project "gateways," which correspond to milestones. These gateways generally relate to deliverables, allowing stakeholders to monitor benefits methodically. As the program or project progresses toward its outcome, the focus of VM will evolve from strategic to tactical and technical/operational level, and expected benefits and context may evolve along the way. Hence, VM must be an iterative process.

Change management is an essential part of the management of value; the VM process is also applied to change, especially regarding evaluation of results and integration with overall needs and expectations. When VM is applied to the change control process, it ensures that the "real" issues are addressed and that changes are made in line with the CSFs and other expected benefits. This leads us to the concept of benefits management, which has recently been identified as a significant aspect of the management of programs. VM provides a clear link between identified expected benefits (functions) at different levels of the organization and results. As an iterative process, VM regularly reassesses stakeholders' needs and expectations and alerts program and project managers early enough to identify the most resource-effective alternatives and evaluate them on a rational basis.

There are a few elements that must be put in place to support this iterative process:

- Gateways for approval of deliverables need to be defined (typically part of program management) and a VM process (sensemaking–ideation–elaboration) used to review the deliverables.
- Regular reviews of stakeholders' needs and expectations should be planned, specifically at project gateways and program appraisals; CSFs and other expected benefits adjusted in consequence.
- Value criteria (CSFs and KPIs) are to be the basis for change request evaluation; again the VM process should be applied to the management of change.

To achieve this, the value team must develop concepts of responsive and "formative" evaluation, as opposed to strictly baseline and "summative" evaluation. Whereas summative evaluation is based on preset standards (the baseline), where the actual situation is compared with these standards, formative evaluation is the iterated negotiation of program or project evaluation criteria and priorities, more appropriate to complex situations. VM, applied to program appraisal, project gateways, and more generally, change management enables the team to structure this process around a robust framework.

Organizations that apply the preceding principles often closely link the VM team with finance or resource management and follow projects on an ongoing basis. Project initiation is based on VM; reviews are carried out at gateways by the VM team and include the project team. Outputs of this process include resource reprioritization, contingencies and risk response management (reallocation of unused funds or resources to other projects on an ongoing basis), revalidation of CSFs, and other success factors to improve value.

Summary

The PMBOK Guide (PMI, 2000) states: "Finding appropriate resolution to [stakeholders] differences can be one of the major challenges of project management." On the other hand, the European Value Management Standard (BSI, 2000) says: "Stakeholders . . . may all hold differing views of what represents value. The aim of VM is to reconcile these differences and enable an organisation to achieve the greatest progress towards its stated goals with the use of minimum resources." This chapter has defined the methodology, tools, and techniques that enable the achievement of the latter statement.

To be most effective, value management must be linked to strategy, success factors, programs, and prioritization, as well as to change management. VM is a group decision management (GDM) process, which enables groups of stakeholders to make sensible and well-grounded strategic decisions, based on needs; define and prioritize expected benefits; and quantify them. It also enables program managers to select and prioritize projects and other actions, based on the expected benefits that have been defined at a strategic level and the most effective use of resources. Value management, if it is applied at the project gateways and program appraisal phase, becomes a change management methodology that ensures a choice of options directly related to the expected benefits.

Over its history of more than 50 years, value methodologies have spread across the world, and although VM, VE, and VA are widely practiced, there are still differences in practice depending on regions and value association allegiance. However, although methods and techniques may be different from region to region, the same basic principles apply everywhere. The grouping of European value associations under a common Standard and Training and Certification System, and the recent discussions toward certification agreement between SAVE International (United States), the EGB (European Governing Board), and preliminary talks between the EGB and the AIVM (Australia), are all signs of the desire to unify the knowledge and practice of VM. The SAVE-EGB Certification agreement proposal identifies, among others, the two following medium-term objectives:

- Develop universally recognized value knowledge and practice standards.
- Extend common understanding and respect for regional differences in the concepts and practice of value methodologies.

In conclusion, VM is the method of choice to deal with the ambiguity of stakeholders' needs and expectations and the complexity of changing business environment at program level and project initiation. It will bring structure and objectivity to what has often been a highly subjective and intuitive process and provide a framework for decision making throughout the delivery process. The VM process requires involvement of the whole program/project team, at different levels and times, but to be effective, decision makers—with authority over resources—must be involved at all stages of the process.

References

AFNOR, Commission de normalisation. 1985. *Analyse de la valeur, recommandations pour sa mise en oeuvre,* norme NF X 50-153, AFNOR, Paris.

———. 1991. *Analyse de la valeur, Analyse fonctionnelle, Expression fonctionnelle du besoin et cahier des charges fonctionnel,* norme NF X 50-151, AFNOR, Paris.

ASTM. 1995. Subcommittee E-06.81 on Building Economics. *Standard Practice for Performing Value Analysis (VA) of Buildings and Building Systems.* Standard Designation: E 1699-95. Philadelphia: American Society for Testing Materials.

Barton, R. 2000. Soft value management methodology for use in project initiation: A learning journey. *Journal of Construction Research* 2(1):109–122.

BSI. 2000. *Value management.* Standard BS EN 12973:2000. European Committee for Standardization (CEN) Technical Committee CEN/TC 279–British Standards Institute Technical Committee DS/1, Chelsea, UK.

———. 1997. *Value management, value analysis, functional analysis vocabulary.* Standard BS EN 1325-1:1997. European Committee for Standardization (CEN) Technical Committee CEN/TC 279–British Standards Institute (BSI) Technical Committee DS/1, Chelsea, UK.

Brun, G., and F. Constantineau. 2001. *Le management par la valeur: Un nouveau style de management.* AFNOR, Paris, FR.

BIS. 1987. *Guidelines to Establish a Value Engineering Activity.* Management and Productivity Sectional-Committee, EC 9., IS: 11810-1986. New Delhi: Bureau of Indian Standards.

Buzan, T. 1974. *Use your head.* London: BBC Consumer Publishing.

Bytheway, C. W. 1965. FAST diagramming. *SAVE Proceedings.* SAVE, Northbrook, IL.

de Bono, E. 1990. *Lateral thinking: A textbook of creativity.* 3rd ed. Harmondsworth, UK Penguin.

Checkland, P., and J. Scholes. 1994. *Soft systems methodology in action.* Chichester, UK: Wiley.

Cranton, P. 1996. *Professional development as transformative learning.* San Francisco: Jossey-Bass Publishers.

Deetz. S. 1995. *Transforming communication, transforming business.* Creskill, NJ: Hampton Press.

EFQM 2000. *The EFQM Excellence Model.* Brussels: European Foundation for Quality Management. www.efqm.org/.

Fong, P. S. W. 2002. Effective facilitation in value management workshops. *Proceedings of the 5th PMI-Europe Conference.* Cannes, June.

Green. S. D. 1994. Beyond value engineering: SMART value management for building projects. *International Journal of Project Management* 12(1):49–56.

Guba, E.G., and Y.S. Lincoln. 1989.. *Fourth generation evaluation.* Newbury Park, CA: Sage Publications.

ISO 2000. *ISO-9000:2000 Quality management systems.* Geneva: International Organization for Standardization.

Kaplan, R. S., and D. P. Norton. 1996. Using the balanced scorecard as a strategic management system. *Harvard Business Review.* (January–February): 75–85

Kaufman, J. 1997. *Value management: Creating competitive advantage.* Menlo Park, CA: Crisp Publications Inc.

Kirk, D. 2000. Managing expectations. *PM Network.* (August).

Miles, L. D. 1972. *Techniques of value analysis and engineering.* 3rd ed. New York, McGraw-Hill.

Neal. R. A. 1995. Project definition: The soft systems approach. *International Journal of Project Management* 13(1):5–9.

Project Management Institute. 2000. *A guide to the Project Management Body of Knowledge.* Newtown Square, PA: Project Management Institute.

Quinn, J. J. 1996.. The role of "good conversation" in strategic control. *Journal of Management Studies* 33(3):381–394.

Rogelberg, S. G., J. L. Barnes-Farrell, and C. A. Lower. 1992. The stepladder technique: An alternative group structure facilitating effective group decision-making. *Journal of Applied Psychology* 77: 730–737.

SAVE. 1993. *Certification examination study guide.* Northbrook, IL: Society of American Value Engineers.

———. 1997. *Value methodology standard.* Northbrook, IL: Society of American Value Engineers.

Spradlin T. 1997., A lexicon of decision making. *Decision Analysis Society Web site*: http://faculty.fuqua.duke.edu/daweb/lexicon.htm.

Thiry, M. 1997. *Value management practice.* Sylva, NC: Project Management Institute..

———. 2000. A learning loop for successful program management. *Proceedings of the 31st PMI Seminars and Symposium.* Newtown Square, PA: Project Management Institute.

———. 2001. Sensemaking in value management practice. *International Journal of Project Management.* Oxford, UK: Elseveir Science.

———. 2002. Combining value and project management into an effective programme management model. *International Journal of Project Management.* Oxford, UK: Elseveir Science. Also appears in 2001 *Proceedings of the 4th Annual PMI-Europe Conference.* London.

Vennix, J. A. M. 1996. *Group model building: Facilitating team learning using system dynamics.* Chichester, UK: Wiley.

Waring, A. 1989. *Systems methods for managers: A practical guide.* Oxford, UK: Blackwell Science.

Weick, K. E. 1995. *Sensemaking in organizations.* London: Sage Publications.

West, M. A. 1994. *Effective teamwork.* Leicester, UK: The British Psychological Society.

Woodhead, R., and C. Downs. 2001. *Value management: Improving capabilities.* London: Thomas Telford.

CHAPTER THIRTY-SEVEN

IMPROVING QUALITY IN PROJECTS AND PROGRAMS

Martina Huemann

In this chapter quality management in projects and programs is described. It begins with a brief history on quality management and an overview on different quality management concepts. The application of general quality concepts such as certification, excellence models, reviews and audits, benchmarking, and accreditation are described in a projects context. In the chapter I differentiate between quality of the contents processes and their outputs and the management processes of the project or program and their outputs. I also make the point that the management quality can be assessed.

When looking in the literature on quality management in projects and programs, we can find a lot for engineering, software, or construction and product development projects. But what quality means in "softer areas"—for example, organizational development—is hardly ever treated. The chapter addresses these projects too, aiming to reflect quality issues on all types of projects.

Project management audit and reviews are described and introduced as learning instruments. Quality management methods—which can be applied to all types of projects, like the project excellence model based on the EFQM Excellence Model—are then described. The chapter concludes with a summary view on quality management and the role of the PM office in the project-oriented company.

From Quality Control to Continuous Quality Improvement

Quality can be defined as the totality of features and characteristics of an entity that bear on its ability to satisfy stated or implied needs (ISO 9000:2000), where an *entity* can be a product, a component, a service, or a process. The need for quality management derives from mass production at the beginning of the twentieth century. Quality management has developed from product-related quality control to company-related Total Quality Management, aiming for continuous process improvement (Seaver, 2003).

Quality Control Based on Statistics

In the early part of the last century, quality control was established in the manufacturing industry to control the quality of parts and products. Quality control became necessary because of the reorganization of the working process introduced by Frederick W. Taylor and the shift to mass production. Regular inspections had to be carried out to find the defective parts and sort them out. When the products became more complex, the work of the inspectors, who were responsible for the quality control, increased. To reduce the quality control costs statistical methods were introduced, based on the assumption that 100 percent quality control is not needed as long as the inspections were done according to statistical parameters. This quality control method is called *sampling inspections.* The objective of quality control was to fulfill an "average outgoing quality level." At that point in time no preventive actions were taken.

During World War II, statistical methods were widely used for mass production in the defense industry in the United States and in Great Britain. Walter Shewhart recognized the need for process control and introduced the first control charts to visualize the variations of the output in a graphical way. Control charts are still used in quality management to observe whether the variations observed are normal process variations or if the process is getting out of control.

Six Sigma, which has gained some attraction in the project management community, is a more recent offshoot of one of the early quality initiatives based on statistics. The Greek letter sigma, σ, is the symbol for standard deviation. It is a measure of variance. The goal of Six Sigma is to reduce process output variation so there is no more than $+/-$ six standard deviations (Six Sigma) between the mean and the nearest specification limit. When a process is operating at Six Sigma, no more than 3.4 "defects" per million opportunities will be produced (Tennant, 2002; Anbari, 2003).

The Deming Approach

Juran, Feigenbaum, Crosby, and Deming are generally considered as the founders of the quality movement. Juran, together with Deming, made a significant contribution to the Japanese quality revolution, published the *Quality Control Handbook* (Juran, 1950, 1986). Feigenbaum, who worked for General Electric, contributed a publication on "Total Quality Control" that already included a lot of features of Total Quality Management (Feigenbaum, 1991). Crosby defined quality as "conformance to requirements" and proved that an organization can get quality for "free" (Crosby, 1979). A detailed presentation of all their approaches is not possible in this chapter. To set the ground, I will concentrate on the approach of Deming (1992).

Deming showed that quality and productivity do not contradict each other but correlate positively, if production is perceived as horizontal process in which the customer relation and feedback for improvement are important features. He describes this as a *chain reaction*. Improvement in quality leads to lower production costs because of less rework, fewer defects, and increased usage of machinery and material. The lower production costs lead to improvement of productivity and make better quality for lower costs possible, which results in an extended market share. This extended market share protects the continued existence of the factory and maintains job.

Later Deming introduced the Deming Circle with the focus on defect correction as well as defect prevention. The Deming Circle, shown in Figure 37.1, consists of the steps "plan, do, check, and act" and is therefore also referred to as the PDCA cycle.

The cycle establishes the base for continuous improvement of the (production) process. If, based on data quality, deficits are detected, a change in the process is planned and tried out on a small scale. The results of this change are checked. If the data improves, the change in the process is introduced. If the improvement did not happen, the cycle is started anew with fresh planning.

Japanese Approach to Quality

In Japan the ideas of Deming and Juran were pursued and further developed by Ishikawa and Taguchi. Ishikawa postulated that all company units and all staff members are responsible for quality and that quality is defined by the customer. He used the word customer in

FIGURE 37.1. PDCA CYCLE BY DEMING.

4. Act: Implement change in large scale or try again.
1. Plan: Plan a change (do a test) to improve quality.
Act
Plan
Check
DO
3. Check: Study the results.
2. Plan: Implement the change in small scale.

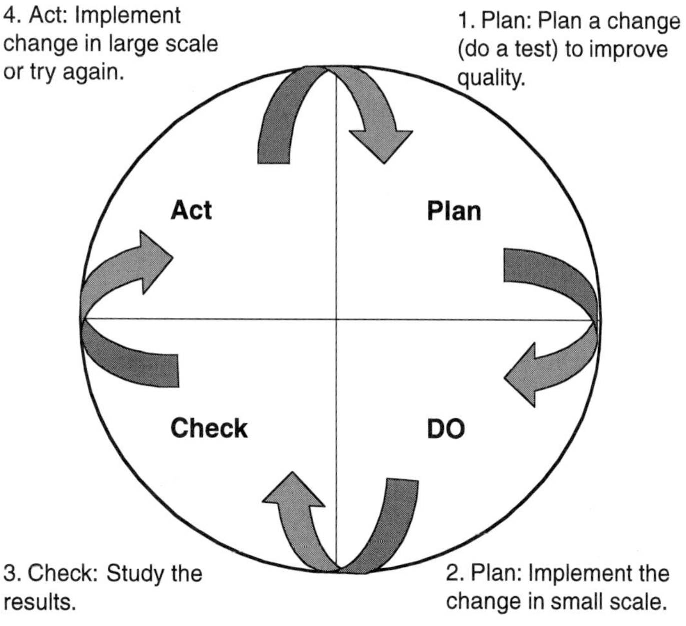

a broader sense. Customer does not only mean the end customer who pays for the product, but always the next person in the process. Each staff member is both a customer and a supplier at the same time. Further, Ishikawa advocated the statistical methods and introduced the cause-and-effect diagram, also known as a fishbone diagram because of its appearance, which separates the process logically into branches. It is used to visualize the flow of work to determine the cause and effect of problems that are encountered. Quality control circles were established to analyze and to solve quality problems based on Deming's PDCA Cycle (Ishikawa and Lu, 1985).

Taguchi stated that quality is an issue for the whole organization, where quality means conformance to requirements. Statistical process control has been a feature of quality assurance methods for many years, but the application of statistical methods to the design process was his invention. He introduced the concept of "robustness" in design; by that he means the ability of a design to tolerate deviations without its performance being affected (Logothetis and Wynn, 1989; Roy, 1990).

The Japanese word for continuous improvement is Kaizen, a quality management technique that aims for constantly looking for opportunities to improve the process. This has become one of the main features of Total Quality Management. Even if the processes are operating without problems, they are objects for quality improvement. Thus, the quality of a process improves in small increments on a continuous basis (Imai, 1986).

Total Quality Management and Continuous Quality Improvement

Total Quality Management, which was founded by Deming and further developed by the Japanese quality movements, can be summarized in seven principles, which are interrelated (Bounds, Dobbins, and Fowler, 1995).

1. *Customer orientation.* The customer buys a product or service because of the benefits to be gained from it. Therefore, the needs and requirements of the customer have to be known. But "the customer" is not only the external customer; the internal customer in the process is also relevant. Customer orientation is the main principle of Total Quality Management.
2. *Continuous improvement of systems and processes.* This requires also that the quality standards and procedures have to be continuously checked and further developed.
3. *Process management.* Performance is based in the competencies of the staff members to fulfill the business processes in the company. Most of the defects have their origin in inadequate processes. Blame the system instead of blaming the worker. Table 37.1 compares the attitude of traditional managers with process managers (Harrington, 1991, p. 5).
4. *Search for the true reason.* The basic assumption is that any problem is only a symptom and has a true origin in the system. Only by digging into the process can the true reason be found and a repetition of the mistake be prevented.
5. *Data collection and analysis.* Data has to be collected as a basis for improvement. For quantitative data collection and analysis, statistical methods such as process control carts, cause-and-effective diagrams, and Pareto charts can be used. Additional, more qualitative methods are audits, reviews, and benchmarking.

TABLE 37.1. TRADITIONAL MANAGERS IN COMPARISON WITH PROCESS MANAGERS.

Traditional Managers	Process Managers
• The staff member is the problem.	• The process is the problem.
• Do the own job.	• Support the others to do the job.
• Know own job.	• Know own job within the whole process and context.
• Evaluate staff members.	• Evaluate process.
• Change staff members.	• Change process.
• Search for a better staff member.	• The process can always be improved.
• Control staff members.	• Train and further develop staff members.
• Find the guilty one to blame for a mistake.	• Find the reason for the mistake.
• The quantitative result is important.	• Customer orientation is important.

6. *People-orientation.* A continuous improvement is only possible if the competencies and skills of the staff members are also continuously further developed. The employees have to be educated, trained, and empowered. In high-performance organizations, people are *enabled* to do their best work. They have the adequate tools, standards, policies, and procedures.
7. *Team-orientation.* Team orientation is part of the culture and structure of the organization. Members of one team also communicate and cooperate with members of other teams to fulfill a process. That makes the traditional department-oriented thinking obsolete and changes it into a process-oriented, trans-sectional way of thinking.

Overview on Quality Management Methods in the Project-Oriented Company

In project-oriented companies that apply modern quality management based on Total Quality Management and Continuous Improvement, different quality management methods—often in combination—are applied. These include the following:

- Certification
- Accreditation
- Excellence model
- Benchmarking
- Audit and review
- Evaluation
- Coaching and consulting

Certification

Certification is a procedure in which a neutral third party certifies that a product, process or a service meets the specified standards. An important—but by no means the only—certification process is ISO Certification according to the International Standards for Quality (ISO). The ISO is a network of national standards institutes from 147 countries working in partnership with international organizations, governments, industry, business, and consumer representatives. ISO has developed over 13,000 International Standards on a variety of subjects (www.iso.ch/).

One advantage of ISO certification is that it is international. For example, the ISO 9000:2000 series of standards are far more process-based than ISO 9000:1994, which were mainly based on procedures. This was a shortcoming, as ISO auditors ended up ticking boxes whether a document was there or not, without considering any further quality dimensions. The ISO 9000:1994 became obsolete by December 2003. The new ISO 9000:2000 also considers the quality of results and is therefore compatible with excellence models and accreditation schemes. Therefore, the new series will fit better to the process philosophy of modern quality management in project-oriented companies.

For projects and programs, the new ISO 10006:2003, "Quality Management Systems: Guidelines for Quality Management in Projects," exists, which provides a structured approach for the optimal management of all processes involved in the development of any type of project. (See also the chapter by Crawford for a discussion of ISO 10006 and standards in general.)

Other certifications relevant in the context of project management and project-oriented organizations are certifications done on the PRINCE2 project and MSP program management methodology (OGC, 2002) and professional certification by IPMA, IPMA member organizations, and PMI for individual project management personnel. (Again, see the chapter by Crawford.)

Accreditation

An accreditation is an external evaluation based on defined and public known standards. Accreditation was originally established to support customer protection. Consumers can be protected by certification, inspection, and testing of products and by manufacturing under certified quality systems. Consumers need confidence in the certification, inspection, and testing work carried out on their behalf, but that they cannot check for themselves. The certifiers of systems and products as well as testing and calibration laboratories need to demonstrate their competence. They do this by being accredited by a nationally recognized accreditation body. Accreditation delivers confidence in certificates and reports by implementing widely accepted criteria set by, for instance, the European Committee for Standardization (CEN) or international (ISO) standardization bodies. The standards address issues such as impartiality, competence, and reliability; leading to confidence in the comparability of certificates and reports across national borders. (See www.european-accreditation.org/).

Accreditation is a commonly used quality management method, for instance, in the healthcare sector, in which a lot of projects and programs are carried out. There the accreditation was established to protect staff members and patients from faulty organizational processes. Participating in an accreditation program is voluntary. The applying organization does a standardized self-assessment. The results of the self-assessment are the basis for a site visit, where the surveyor uses documentation analysis, observations, and interviews for information gathering. Results of the site visit are summarized in a report. The applicant gives feedback to the report. This feedback discussion can be organized in the form of a workshop. Then the final result of the accreditation—which can be numerical or a descriptive like "substantial compliance, partial compliance, minimal compliance, or noncompliance"—is provided. Accreditations have to be renewed every couple of years.

In the project management context, accreditation is done, for example, by PMI for project management education and training programs. The degree and nondegree programs are accredited for their content and progress compliance with the standards set by the Global Accreditation Center for Project Management. (see www.pmi.org)

Excellence Model

Excellence models are nonnormative models that provide a framework to assess an organization in it's degree of excellence in the application of practices. All excellence models differentiate between enabler criteria and result criteria as a basis for the assessment. The most important excellence models have been developed in the frame of regional quality programs, which award organizations for outstanding quality improvement. I mention here the following:

- Deming Prize
- Malcom Baldridge National Quality Award
- European Quality Award
- International Project Management Award

Deming Prize. The Union of Japanese Scientists and Engineers (JUSE) created a prize to commemorate Deming's contribution to quality management and to promote the development of quality management in Japan. The prize was established in 1950 and annual awards are still given each year (See www.deming.org/demingprize/)

Malcom Baldridge National Quality Award. The Baldridge Award is given by the President of the United States to businesses and to education and healthcare organizations that apply and are judged to be outstanding in seven areas: leadership, strategic planning, customer and market focus, information and analysis, human resource focus, process management, and business results. Congress established the award program in 1987 to recognize U.S. organizations for their achievements in quality (see www.quality.nist.gov)

European Quality Award. The European Foundation for Quality Management (EFQM) was founded in 1988 with the endorsement of the European Commission. It is the European framework for quality improvement along the lines of the Malcolm Baldridgc Model in the United States and the Deming Prize in Japan. The European Model for Business Excellence—now called the EFQM Excellence Model—was introduced in 1991 as the framework for organizational self-assessment and as the basis for judging entrants to the European Quality Award, which was awarded for the first time in 1992 (see www.efqm.org/).

International Project Management Award. An excellence model in the context of projects is the project excellence model, which is based on the European Model for Business Excellence. The project excellence model is described in detail in a later section in this chapter (see www.gpm-ipma.de).

Benchmarking

By the end of the 1970s companies like Xerox in the United States were suffering from competition from Japan. They had to find out what their competitors were doing. Benchmarking as a tool to compare the performance and practices of one company with other companies derived from the work of Robert Camp (1989). The aim is to understand the reason for the differences in performance by examining the process in question in detail. Benchmarking is a tool for improving performance by learning from best practices and understanding the processes by which they are achieved. Application of benchmarking involves following basic steps:

1. First, understand in detail your own processes.
2. Next, analyze the processes of others.
3. Then compare your own performance with that of others analyzed. Comparison can be done within one's own organization or with other organizations from the same industry or different industries.
4. Finally, implement steps necessary to close the performance gap.

A number of benchmarking models and processes have been developed and are applied for a wide range of subject areas. In the context of project-oriented companies, maturity models are often used as a basis for the benchmarking exercise. Most of the models are used for benchmarking on the company level and not on the single-project level. (See the chapter by Cooke-Davis on project management maturity models.) Also, specific benchmarking communities for project and project management benchmarking exist. (See, for example, the chapter by Ibbs, Reginato, and Hoon Kwak). Most of these benchmarking activities are industry-specific.

Audit and Review

ISO 19011:2002 defines auditing as a "systematic, independent and documented process for obtaining audit evidence and evaluating it objectively to determine the extent to which the audit criteria are fulfilled." The audit criteria are a set of policies, procedures, or requirements. Reasons for audits are, for instance, certification, internal audits and review, and contract compliance. Reviews are considered to be less formal than audits.

Audits and reviews are applied to ensure quality in projects and programs. Peer reviews for projects and programs are done by peer professionals such as program managers, project managers, or other experts who are not part of the project or program under consideration. Audits and reviews are not meant to be a replacement for other exchange of experience activities, like coaching or experience exchange workshops. Audits and reviews as methods of quality assurance and quality improvement in the context of projects and programs are further described later in this chapter.

Evaluation

In general, evaluation is referred to as a systematic inquiry of the worth or merit of an object. Evaluations are applied for projects and programs. While audits and reviews are performed during the project or program, evaluations are carried out when the project or program is finished. Objects of evaluation are the management processes, the technical processes, and performance criteria.

Coaching and Consulting

Coaching and consulting might not be considered as part of quality management at first sight. But management consulting on projects and programs as well as management coaching of project and program managers are definitely quality management methods to ensure management quality. These methods are applied in many advanced project-oriented companies. While in management coaching the client is an individual—for example, the project manager or program manager—the object of consideration in consulting is the project or program. Coaching is often considered as a method to further develop personnel. If a new project management approach has been implemented in a company, coaching may be provided to project managers to support the implementation. In this way the quality of the management process is ensured.

A typical situation for management consulting is a program start-up. The situation is typically rather complex but very important for the success of the program, as in the start-up process, the quality for the management of the program is set. Another typical situation for management consulting is a project or program crisis. Then the consultant helps to manage the discontinuity. Consulting activities can also support the program to implement the corrective and preventive actions, which have been agreed on after a management audit of the program. Large project-oriented companies often provide these services through their PM office and have internal PM coaches and management consultants for projects and programs.

Application of Quality Management Methods in the Project-Oriented Company

All quality methods described in the preceding text are applied in the project-oriented company. They are often combined based on a continuous-improvement philosophy. Some of the methods described are commonly applied, such as certification, benchmarking, and design audits, while others are rather new, such as PM audits, PM coaching, and PM consulting. In general, for project-oriented companies quality management is relevant for the following:

- The project-oriented company as such
- The company units (e.g., human resources department, engineering department, PM office, finance department)
- The single program
- The single project

In this chapter I concentrate on the quality management issue of projects and programs and look at some of the quality management methods in more detail. Still, the quality management of the project-oriented company as such is an important context to the quality management in projects and programs, as the company should have a quality system in place that provides quality standards and procedures for the projects and programs.

Quality Standards for Projects and Programs

Product and process quality in projects and programs

Good quality in the context of projects and programs is defined (Turner, 2000) as being to

- meet the customer requirement
- meet the specifications
- solve the problem
- fit the purpose
- satisfy or delight the customer.

Possible objects of considerations regarding quality in projects and programs are as follows:

- The product (services)—that is, *the material results* of the project or program, which could range from a marketing concept, a feasibility study, a conference, a new product line, or a power plant
- The project or program *content processes* to create the project or program results (e.g., designing, engineering, implementing, and testing)
- The *management processes* in the project and program: These include project management processes such as project start-up, project coordination, project controlling, and project closedown.

For each of these objects of consideration, standards need to be provided in the project-oriented company to set the basis for quality management. We can think of this in knowledge and learning terms. Organizations have the capability to gather knowledge and experience and to store such knowledge in a "collective mind" (Senge, 1994). The organizational knowledge is hidden in the systems of organizational principles, which are often anonymous and autonomous; these define the way the organization works. Through these, one can discover the organization's knowledge and experience, operation procedures, description of work processes, role descriptions, recipes, routines, and so on.

Standards as Basis for Quality in Projects and Programs

In general, standards can be

- generic—therefore applicable to all types of projects and programs in all project-oriented companies of any industry; or
- specific—includes specifics for a certain industry, company, project, product, or customer.

Further, standards can be normative or non-normative. There are different types of standards relevant for projects and programs, which are

- product standards
- project and program standards
- project management standards
- other standards, such as procurement standards, health and safety regulations, and so on.

Product Standards. Product standards can either be based on regulations or common shared standards or specific agreements with the customer. These product specifications covers issues like the following:

- The required functionality of the product or service
- Design standards
- Cost of the product or service and time at which is should be delivered
- Availability, reliability, maintainability, and adaptability

Project specifications are agreed on with the customer and (internal) project sponsors. Some companies have company-internal specific product quality and safety regulations, which then also apply to their product development projects. In some industries, like the pharmaceutical industry, there are very strict regulations by authorities, which of course then also apply to project work. For repetitive projects, standardized product breakdown structures may be applied.

Project and Program Standards. Project standards are normally project-type-specific and provide a standard procedure for the content and management processes of the project or program, as in the following examples:

- Phases for a construction project: concept, definition, design, supply, construction, commissioning
- Phases for a software development project: Requirement definition, requirement specification, prototyping, design specification, system development, test system, install system, pilot application

Project-oriented companies may wish to further standardize repetitive projects by providing detailed project process descriptions. Project work breakdown structures and even work packages specifications may be standardized. This is very often the case in industries with a lot of repetitive projects such as in the engineering industry.

Figures 37.2 and 37.3 show an example for standardized processes and their relation to project standards. The company is part of an international concern and acts as supplier for standardized intermediates and high-quality fine chemicals for the life science industry (pharma, agro, food). In this project-oriented company, different processes have been defined, of which parts are carried out in the form of projects. Different types of projects are carried out, of which R&D projects as well as investment projects have a repetitive character (Stummer and Huemann, 2002).

For all repetitive projects, the company provides standard project plans, with the following objectives:

- To ensure the quality of processes and results
- Therefore to optimize project and program costs and duration
- To enable the use of existing knowledge instead of permanently reinventing the wheel

FIGURE 37.2. TYPES OF PROJECTS IN A CHEMICAL COMPANY.

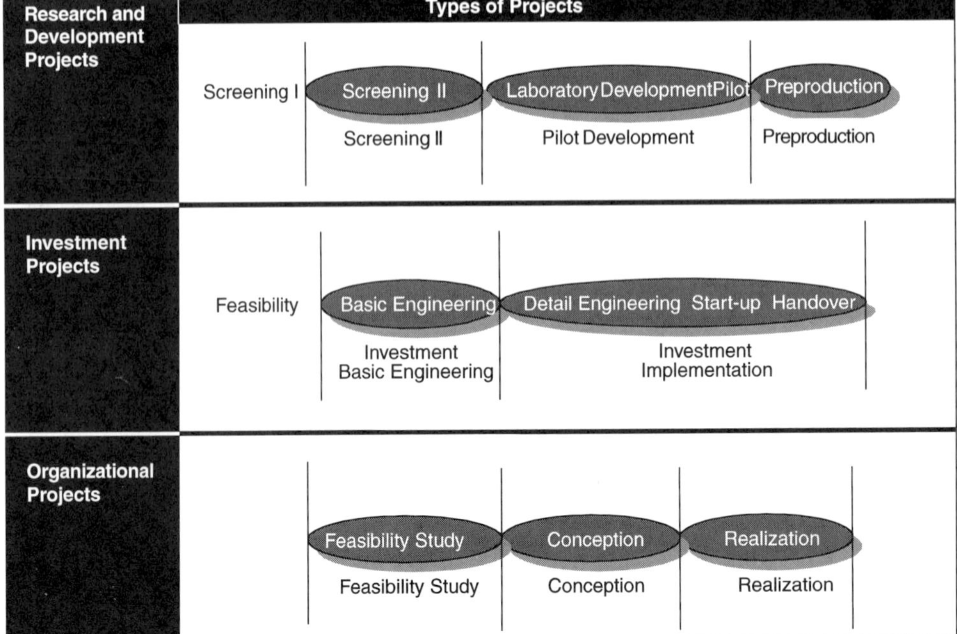

FIGURE 37.3. EXAMPLE: STANDARD PROJECT PLAN: WBS LABORATORY DEVELOPMENT.

Laboratory Development Pilot

1.1 Project Management	1.2 Plan Laboratory Development	1.3 Perform Lab. Development	1.4 Reporting and Process Handbook	1.5 Decision, Plan Preproduction Project
1.1.1 Project Start	1.2.1 Analyze Client Requirements	1.3.1 Performance experiments	1.4.1 Laboratory Reports	1.5.1 Preparation Presentation
1.1.2 Project Co-ordination	1.2.2 Analyze Product Definition	1.3.2 Test material	1.4.2 Process Handbook	1.5.2 Presentation
1.1.3 Project Controlling	1.2.3 Plan Laboratory Tests	1.3.3 Test Process and Safety	1.4.3 Up date Client Requirements	1.5.3 Decision
1.1.4 Project Closedown	1.2.4 Literature Research	1.3.4 Test Technology	1.4.4 Update Market Potential	1.5.4 Plan Pre-production Project
	1.2.5 Plan Technology	1.3.4 Pilot Sample	1.2.4	
	1.2.6 Plan Material	1.3.4 Test Pilot Sample		

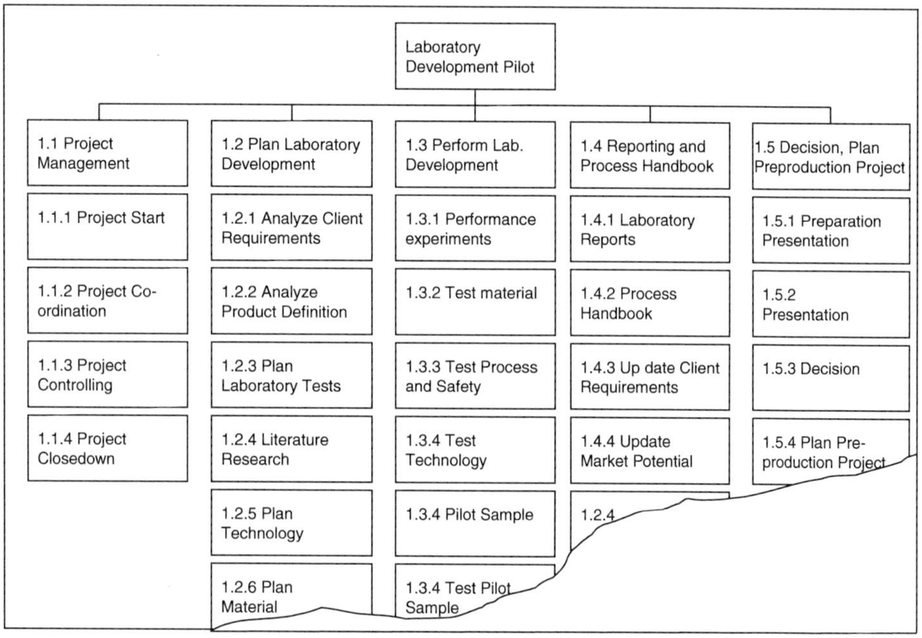

In this company there are standard project plans such as project objectives plan, list of objects of consideration, work breakdown structures, work package specifications, project milestone list, project organization, and so on. The standard work breakdown structure of a laboratory development project is shown in Figure 37.3.

Project Management Standards. Generic PM standards are standards that are applicable to all types of projects; examples for generic project management standards are PRINCE2 (OGC, 2002) the PMI's PMBOK (PMI, 2002), and pm baseline (Gareis, 2002). For a detailed discussion, see the chapter by Crawford.

Many project-oriented companies have implemented their own project management guidelines, often based on one of the international project management standards. Table 37.2 shows a content structure of guidelines for the management of projects and programs based on the pm baseline. These PM guidelines limit the variety of different management approaches and methods to a common company standard and provide means of support and help to ensure the program and project management quality in the single program or project. The guidelines prevent every project manager having to invent the same processes again and again.

TABLE 37.2. GUIDELINES FOR THE MANAGEMENT OF PROJECTS AND PROGRAMS.

Guidelines for the Management of Projects and Programs

Table of Contents

1 Introduction
- 1.1. Purpose and Contents of the Guidelines
- 1.2. Updating the Guidelines

2 Projects and Programs
- 2.1 Definitions: Project, Small Project, Program
- 2.2 Types of Projects in the Company X

3 The Processes of Project and Program Management
- 3.1. The Project Management Process
- 3.2. The Project Management Subprocesses: Project Start, Project Controlling, Project Coordination, Dealing with Project Discontinuities, Project Closedown
- 3.3. The Program Management Process

4 The Application of Project and Program Management Methods

5 The Organization of Projects and Programs
- 5.1. Organization Charts of Projects and Programs
- 5.2. Roles in Projects and Programs

6 Tools for Project and Program Management (Checklists, Forms)

Checklists and standard project handbooks may also provide the structure for the project manager and his or her team to manage the project.

Quality Management in Projects and Programs

PMI's PMBOK (PMI, 2002) states that project quality management includes the processes required to ensure that the project will satisfy the needs for which it was undertaken. Hence, says PMBOK, quality management includes the following:

- *Quality planning.* To identify all the quality standards relevant for the project and plan how to satisfy them
- *Quality assurance.* To evaluate the project to ensure that the relevant quality standards will be met
- *Quality control.* To monitor, to compare with the relevant quality standards, and to correct the product (components, their configuration, the facility) and the processes

Traditional quality management approaches for projects and programs, says PMBOK, concentrate more on the product quality as such and on quality control by statistical means like inspections, control charts, Pareto diagrams, statistical sampling, and so on. This perception is rather shortsighted, as projects need more than manufacturing quality management approaches. Quality assurance by using reviews and audits becomes more important in projects. The processes need to be checked rather early to ensure the quality of the project deliverables, as only sound processes lead to good products and solutions.

Quality of Processes as Basis for Product Quality

An example for quality planning and assurance in a program is the Austrian Railway revitalization initiatives. Within this program 40 railway stations were revitalized. The program work breakdown structure is shown in Figure 37.4 (below). To plan the quality management of the program, a standardization project was performed to develop process standards, to introduce a documentation management system and to collect all relevant standards to be applied within the program. Within the program, three types of repetitive projects existed, namely, conception projects, planning projects, and realization projects. For these repetitive projects, standard project plans including work package specifications were developed. The project outputs were also standardized. For instance, for each of the railway stations, a feasibility study was carried out. The structure of how this study should look was standardized, so the feasibility studies were comparable.

Quality Management in Engineering, Construction, and IT/IS Projects and Programs

Solution quality and process quality in engineering, construction, and IT/IS projects and programs are typically assured as follows (Turner, 2000):

FIGURE 37.4. WORK BREAKDOWN STRUCTURE OF AN INFRASTRUCTURE PROGRAM.

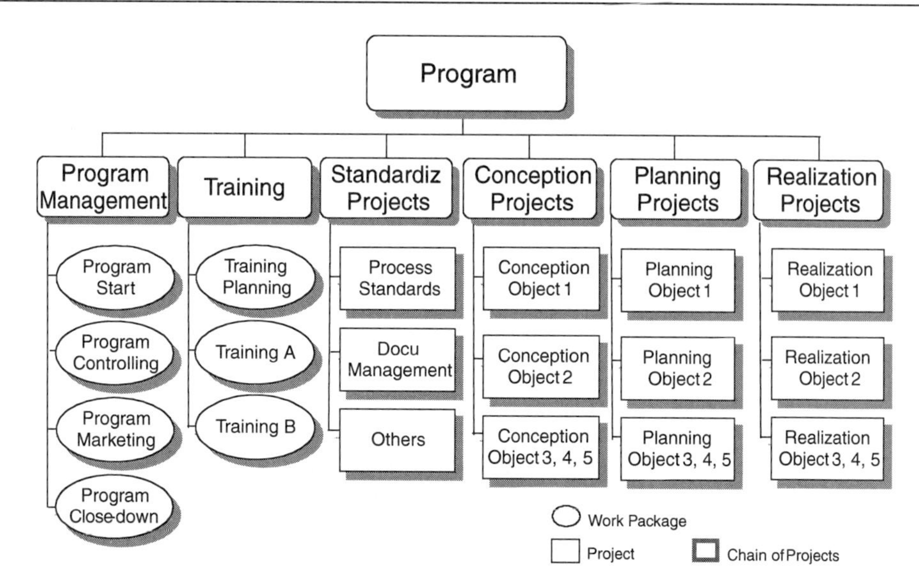

- *Competent project personnel.* Previous experience of the project team and the project-oriented organization in creating the facility
- *Well-defined specifications.* These include required functionality of the facility and its components, design standards it is required to meet, the time and cost at which it should be delivered, availability, reliability, maintainability, and adaptability.
- *Standards.* These include standards the solution has to meet and process standards the project should follow.
- *Audits and reviews of the project deliverables.* These include, for example, design reviews (Webb, 2000) of the project processes.
- *Change control and configuration management.* As described in the chapters by Kidd and Burgess, and by Cooper and Sklar Reichelt.

Audits and reviews to check the project processes and results are the most common methods for quality assurance in projects and programs. They are applied at the end of phases. For example, the gate model for process improvement of an international engineering company—applicable for their integrated solution delivery programs—shows the phases: concept, design, development, implementation, and benefits delivery. Reviews are carried out to evaluate the deliverables produced during the phases. The reviews are assessments of the solution under development, which include the following:

- *Concept phase review(s).* To assess the completeness of the design concepts, including consideration of alternative designs.
- *Design phase review(s).* To assess the completeness of the design phase work, which include, for instance, process design and system requirements, logical design, operations plan, and test plan.
- *Detailed design review.* To do a complete technical assessment of the detail design before beginning extensive coding or purchasing of software.
- *Pilot readiness review.* To assess whether the solution is ready to pilot.
- *Implementation readiness review.* To assess the readiness of implementing the solution to its planned full extent.
- *Implementation reviews.* To assess the implementation on each site that implements the new solution. It includes validation of implementation measurements, system performance, site adjustments, planning adjustments, implementation logistics, budget, and schedule.

The reviews are linked to the gates, which are go/no-go decision points. Only successful reviews allow a project to schedule a gate meeting. Beyond these reviews, they are tested by third parties to ensure that the solution is in accordance to specifications. (See also the chapter by Roulston for an illustration of this in defense contracting.) In concurrent engineering the gates are difficult to specify, leading to Cooper's notion of fuzzy gates and other flexible project checks (see the chapter by Thamhain).

These quality assurance activities are an inherent part of the technical content processes and have to be visualized in the work breakdown structure, the Gantt chart, and the cost plan of such a project. To have such reviews and to improve the solution on the way is part of best-practice quality management for these kinds of projects. But what about other types of projects and programs?

Quality Management is an Issue in All Types of Projects and Programs

The approaches to quality management in projects and programs typically quoted reflect those types of projects that result in a facility or system, such as a power plant, a new product, or an IT system. While some of the quality management issues mentioned previously, like competent project personnel, are relevant for other projects like organizational development, event management, marketing, or indeed any kind of project, quality issues like design reviews or product standards or configuration management may not be. Either they are not applied at all or different terminology is used.

In organizational development projects and programs, the best solution is no good if it is not accepted by the organization. So quality is very much linked to acceptance. This means that the process has to be designed so that people can get involved and contribute to the solution (Frieβ, 1999). Instead of reviews to ensure the quality of the solution in organizational development projects, working forums like workshops or presentations are often more relevant. In workshops, solutions are commonly created and evaluated. Cyclic working, doing first a draft then presenting it, getting feedback, and including the feedback improves the quality. Typically, no standard program plans for such programs are available in the organization, as they rarely have repetitive character. Like the railway station program shown earlier, program standards can be developed within the program to ensure the quality of the content's processes and of the deliverables. Project and program management is responsible for the adequate design and the management of the content's processes. Thus, high-quality program and project management is crucial for these kind of projects and programs of high social complexity.

Management Quality of Projects and Programs

The quality of the deliverable that is created in the project—whether it is a marketing concept, a new organization for a company, a newly implemented IT system, or an international congress for 500 PM experts—depends always on the processes of creating it. If there are no standard project processes, it is the task of project management to design these content processes and the adequate organization by providing a process-oriented work breakdown structure, appropriate schedules, adequate organizational roles and responsibilities, and so on. It is also a project management task to manage these processes by controlling the status, adapting plans if necessary, agreeing on new responsibilities, and so forth.

The less repetitive a project, the more parties are involved in the project; therefore, the more socially complex, the more important is the management process and its quality. The quality of the project management process can be measured by looking at the design of the management process and at its results (Gareis and Huemann, 2003).

- Design elements of the management process are, for example, participants taking part in the management, working forms applied, infrastructure used—there is no question that there is a difference in quality of the project plans if they are only prepared by the project manager behind his or her computer or if they are the result of a two-day start-up workshop for the project team, including customer and supplier representatives.
- Results of the start-up process for a project or program will include, for example, adequate program plans like a work breakdown structure, objective plan, schedules, cost

plans, and so forth; adequate consideration of the program context; adequate program organization and communication structures; agreed on responsibilities; an established program management team; an established program office; an appropriate program culture; and so on.

The project organization contributes to the quality of the solution delivered. High-quality results can be promoted by building an integrated project organization with suppliers and customers working in one project organization together instead of building parallel investor, contractors', and subcontractors' project organizations. If this is based on a common contract, it is called *partnering* (Scott, 2001; and see the chapters by Langdon and Murray, Roulston, and Morris). This form of project organization saves time, reduces costs, and improves quality.

The principle of partnering is rather simple: Performance is improved by buyers and suppliers working together closely on a long-term Total Quality base. Aims are matched, resources pooled, teamwork promoted, and planning integrated (Morris, 1999). One documented case is the Ruhr Oel case (van Wieren, 2002), an alliance of Ruhr Oel, the client; Fluor Daniel, the engineering contractor; and Strabag, Fabricon, Ponticelli, the construction contractors. All partners had their own quality management system; the Fluor Daniel system formed its basis. They had common project execution plans and alignment meetings to share information. An alliance board, which consisted of senior managers of each company and the project managers from Ruhr Oel and Fluor Daniel met monthly to monitor the project and check that all parties remained aligned. The best-qualified person was nominated for each key position. Strong cooperation between and integration of all team members from all companies was supported. As no claims were possible according to their alliance contract, the only way to solve conflicts was through communication. The alliance managed to complete the EPC work for two refineries in one year and seven months. Although despite a 25 percent increase in work due to scope development during the EPC phase, the project was completed on time and 9 percent under the target price. The plants were operating as specified and meeting quality standards, and all authority requirements were fulfilled. By the way, the project won the International Project Management Award in 2002.

Management Audits and Reviews of Projects and Programs

A project audit is a systematic and independent investigation to check if the project is performing correctly with respect to project and or project management standards. A *project review* is defined as a formal examination of the project by persons with authority in order to see whether improvement or correction is needed. Another word for project review is project health check (Wateridge, 2000). A special form of the project review is the peer review: Here the review is carried out by experienced peer project managers to give feedback and advice to the project.

In the previous section the review of the deliverables was introduced as one method of quality assurance commonly applied in construction, engineering, and product development projects and programs. In this section I concentrate on management audits and reviews of projects and programs that are applied in all kinds of projects and programs. Management audit and reviews of projects and programs assess the management competencies of the projects or program, namely, the organizational, team, and individual competencies to perform the management processes. Thus, the project or program management process and its results are reviewed. Results of the project start process could be, for instance, that adequate project plans exist and the project team has been established. The benefits of management audits and reviews of projects are on the one hand to provide a learning opportunity to the single project to improve its project management quality. On the other hand, by evaluating the results of several management audits, patterns can be found. For instance, if a lot of the projects have a low-quality cost plan or do not apply a stakeholder analysis, this shows that these issues are general subjects for improvement in the project-oriented company (Huemann and Hayes, 2003).

Differences between Management Audit and Review

Table 37.3 gives an overview of the differences between management audits and reviews. The main differences can be seen in terms of obligation and formalization. We consider reviews to be less formal and the obligation to implement the review results is medium in comparison with audits. Management reviews serve the purpose of learning, feedback, and quality assurance, while management audits are also used for problem identification and controlling.

We differentiate between the initiator and the owner of the management audit or review. The management audit or review owner is the project or program owner, as he or she has to provide the resources. The audit can also be initiated by a profit center, the PM office, or the customer. The management review is often initiated by the project manager or program manager. The client is always the project or the program and not the single project manager or program manager. Management audits and reviews are always carried out by persons who are external to the project. The auditors or reviewers are never part of the project or program organization.

PM Audit Criteria Depend on the PM Approach Used

The PM audit criteria depend on the PM approach used. The basis for the PM audit is the PM procedures and standards. If the PM audit is based on a traditional PM approach, the audit criteria are limited to the PM methods regarding scope, schedule, and costs. Additional PM objects of consideration like the project organization, the project culture, and the project context become only PM audit criteria, as, for example, if PRINCE2 is used. If project management is considered as a business process consisting of the subprocesses' project start, project controlling, project coordination, management of project discontinuities, and project closedown, the design of the PM process becomes an audit criterion. The PM approach used in a PM audit has to be agreed on.

TABLE 37.3. DIFFERENCES AND COMMONALTIES OF MANAGEMENT AUDIT AND REVIEW.

	Management Audit	Management Review
Initiator	• Project or program owner • Profit center • PM office • Customer	• Project manager or • Program manager
Owner	• Project owner or • Program owner	• Project owner or • Program owner
Client	• Project or • Program	• Project or • Program
Purpose	• Learning • Feedback • Controlling • Problem identification • Quality assurance	• Learning • Feedback • Quality assurance
Obligation	• High	• Medium
Formalization	• High	• Medium
Methods	• To be agreed on • All possible	• To be agreed on • All possible
Object of consideration	• Management process(es) and results • Organizational, team, and individual project or program management competence	• Management process(es) and results • Organizational, team, and individual project or program management competence
Homebase of auditors/reviewers	• Company external • Company internal	• Company internal (peer review)
Number of auditors/reviewers	• 1–3	• 1–3
Duration	• 1–2 weeks	• 2 days–1 weeks
Resources	• Depending on scope of the project or program and methods agreed on: 8–12 days	• Depending on scope of the project or program and methods agreed on: 2–8 days

Adequate Times for PM Audits

PM audits can be done randomly, regularly, or because of a specific reason. They are still very often carried out if somebody in the line organization has a bad feeling about the project. Then the method is used for problem identification and controlling, not so much for learning purposes and quality assurance. Nevertheless, the ideal point in time to do a project management audit is in a relatively early phase of the project—for instance, after the project or program start has been accomplished. That gives the project the chance to further develop its management competence. Further PM audits/reviews later in the project are possible to give further feedback but also to verify if the recommendations agreed on in earlier PM audits were taken care of by the project or program.

PM Audit Process

An audit needs a structured and transparent approach (Corbin, Cox, Hamerly and Knight, 2001). A PM audit process established in a project-oriented company in accordance with the ISO (19,011: 2002) includes following steps.

- Situation analysis
- Planning PM audit
- Preparation PM audit
- Performance of analysis
- Generation of PM audit report
- Performance of PM audit presentation
- Termination of PM audit

The follow-up of the PM audit, thus verifying if the corrective and preventive actions recommended by the PM auditors have been implemented in the project, is not part of the PM audit process. There is the need for an agreement between representatives of the project and the project owner as to which of the actions recommended by the PM auditors have to be implemented. A PM audit follow-up agreement form is shown in Figure 37.5.

Methods of PM Audits

In the PM audit, a multimethod approach is used. The following methods can be applied for gathering information:

FIGURE 37.5. PM AUDIT FOLLOW-UP AGREEMENT FORM.

PM Audit Follow-up Agreement		
Name of project audited:		Name of project manager:
PM audit completed at:		Name of project owner:
Name of auditors:		Date of follow-up agreement:
Corrective Actions Action	Responsibility	Deadline
Preventive Actions Action	Responsibility	Deadline
.... Project Manager		.... Project Owner

- Documentation analysis
- Interview
- Observation
- Self-assessment

For presenting the PM audit findings, the following methods can be used:

- PM audit report
- PM audit presentation
- PM audit workshop

Documentation Analysis. In a documentation analysis, the organizational PM competence of a project can be observed. Documents to be considered are PM documents like, for example, project work breakdown structure, project bar chart, project environmental analysis, project organization chart, project progress reports, and minutes of project meetings. The auditors can check whether or not the required PM documents exist. This might end up in ticking boxes without checking whether the contents of the document make sense. A further step is to audit also the quality of the single PM document, as well as the constancy between the single PM documents. Figure 37.6 shows an example of a checklist for assessing the quality of a work breakdown structure. Criteria are as follows:

- Completeness
- Structure
- Visualization
- Formal criteria

FIGURE 37.6. AN EXAMPLE FOR A PM AUDIT CHECKLIST.

PM Audit: Work Breakdown Structure			
Project:	Company:		
Criteria:	Weight	Result	Weighted Result
Completness			
Structure			
Visualisation			
Formal Criteria			
Total			
PM Auditor:	Date:	Page:	

The criteria used for checking the quality very much depend on the project management standards of the company. If they use a project management approach that promotes the use of the PM plans as communication instruments in the project team, then the criteria "visualization" becomes important. The criteria might even be weighted. So, for instance, to fulfill the "formal criteria" is less important than having a complete project plan.

Interview. Interviews are conducted to obtain more detailed information based on questions that arise from the documentation analysis. One can differentiate between group interviews and interviews conducted with a single person. In the case of group interviews, the PM auditor has the opportunity to see how the interview partners interact with each other and react to different opinions. In management audits of projects, interviews can be held with representatives of the project organization, the project manager, the project sponsor, the project team members, as well as with representatives of relevant environments like, for example, the customer, suppliers, and so on. When doing interviews with the customer, which might be quite sensitive, the PM audit result is based on information provided from different angles. That approach can be compared with a 360-degree feedback approach. The customer and suppliers can give another perspective from outside, which might be very different to the perspective the project team has.

Observation. In the observation, the PM auditors collect further information about the project management competence in the project by using observation criteria. Project owner meetings, project team meetings, and project subteam meetings can be observed.

Self-Assessments. Within the PM audit, self-assessments of the individual PM competence of representatives of the project organization— for example, project manager, project owner, and project team member—can be applied (Huemann, 2002). In some project-oriented companies, such assessments are applied on a regular basis as part of the human resource management to further develop the PM personnel. Then the self-assessment of the individual PM competence is not part of the PM audit.

Further, a self-assessment of the PM competence of the project team can be carried out in a PM audit. The PM competence of the project team can be described as the knowledge and the experience of the project team to develop commitment in the project team, to create a common "big project picture," to use the synergies in the project team, to solve conflicts, to learn in the project team, and to jointly design the PM process. These self-assessment activities very much can make the PM audit a learning experience for the project or program team.

PM Audit Reporting. The objective of the PM audit report is to summarize the findings of the PM audit and give recommendations regarding the further development of the PM of the project. It also includes recommendations for the further development of the project management in the project-oriented company. The PM audit report is the basis for the follow-up agreement between the project and the PM audit owner on which actions have to be taken. Table 37.4 shows the structure of a PM Audit report as an example. The PM auditors are not responsible for checking whether their recommendations are followed. How-

TABLE 37.4. STRUCTURE OF THE PM AUDIT REPORT.

1.	Executive Summary
2.	Situation Analysis, Context, and Description of the PM Audit Process of Project XY
3.	Brief Description of the Project XY
4.	Analysis of the PM Competence of Project Management of Project XY
4.1	Analysis of the Project Start
4.2	Analysis of the Project Coordination
4.3	Analysis of the Project Controlling
5.	Further Development of Project Management of Project XY
6.	Further Development of Project Management in General
7.	Enclosures

ever, if there is a PM audit at a later point in time, the PM auditor will also have a look at the previous PM audit report.

PM Audit Presentation. The PM audit reporting will always be in written form. Often there is also a PM audit presentation, before the written report is handed in by the auditors. Participants in the PM audit presentation are the PM audit owner, project manager, and further representatives of the project. Further representatives of relevant environments of the projects—for example, representatives of the client, supplier, and so on—can be invited. In many project-oriented companies, the PM audit presentation is considered as important. The objectives of the PM audit presentation, sometimes even organized as workshops, is to understand the PM audit results. That leads to more acceptance of the PM audit results and provides the chance to the project to become a learning organization as defined by Peter Senge in *The Fifth Discipline Fieldbook* (1994).

Application of Adequate PM Audit Methods

Which methods for information gathering are used depend on the specific case and on the agreement between the PM audit owner and the project manager of the project to be audited. The PM audit should at a minimum (ISO 19011) include documentation analysis and interviews at least with the project manager, the project owner, and representatives of the project team. In the case of a program, interviews with the program owner, the project managers, and project sponsors of the different projects are required. In the case of projects, further interviews with representatives of relevant environments like client and suppliers are important. By observing meetings, the PM auditors get an insight in the PM team competence. The self-assessments help achieve a holistic picture. Self-assessments provide the individuals and the project team the possibility to reflect and very much add to the learning perspective of the PM audit. If the individual PM competence of PM personnel is assessed by the project-oriented company, this self-assessment will not be part of the PM audit. The quality of the results of the PM audit depends very much on the scope of the methods and the professional application of these.

PM Audit Organization and Roles

In the PM audit system, the roles PM audit owner, PM auditor, representatives of the project, and representatives of relevant environments can be differentiated. The PM audit owner is responsible for the assignment of the PM audit and for agreeing the scope and timing of the PM audit with representatives of the project. Further, the PM auditor has to ensure resources for the PM audit. Often the PM audit is performed by two to three auditors. Then one of the PM auditors takes over the role of the lead auditor. The PM auditors analyze the PM competence of the project and give recommendations regarding the further development of the project management of the project. The PM auditor needs not only profound PM competencies but also audit competencies like designing the PM audit process or performing an interview professionally. Thus, social competence and emotional intelligence are important.

The role of the representative of the project is taken over by the project manager of the project audited. The objective of this role is to contribute information for the PM audit and to invest resources. Tasks of the project manager in a PM audit are, for example:

- Contribution to clarify the situation in the project
- Feedback to the PM audit plan
- Agreeing scope and methods of the PM audit
- Interview partner in the PM audit
- Provision of PM documents of the project for the documentation analysis.

PM Audit Values and Limits

The communication policy should be agreed on between the PM auditor and project manager at the beginning of the PM audit. From a learning perspective, the PM audit should be done in a cooperative and not a hostile way. This would also mean that the project manager of the project that is audited should be kept informed by the PM auditors. Circumstances that should lead to a cancellation of the PM audit and the consequences of a cancellation should also be agreed on at the start. One major challenge is that the audit result is not perceived as a feedback to the single project manager who then will be blamed for mismanagement. This requires a certain culture of openness in the project-oriented

FIGURE 37.7. PM AUDIT SYSTEM

company. A good example is CMG, a global consultancy company that has a long tradition in performing management audits and reviews of projects and programs. In 2002 about 7 percent of their annual turnover was spent on quality management. About half of it was spent for management audits of projects and programs, which also include the training of the auditors in doing their job correctly.

The International Project Management Award

Project Excellence Model

The International Project Management Award is based on the project excellence model. The project excellence model was developed by the German Project Management Association (GPM; *www.gpm-ipma.de*) for the IPMA and is based on the EFQM Model. The project excellence model is applicable to any project type. There is no specific consideration of programs. The model, as shown in Figure 37.8, altogether assesses nine criteria divided into two sections: Project Management and Project Results: The Project Management section

FIGURE 37.8. PROJECT EXCELLENCE MODEL.

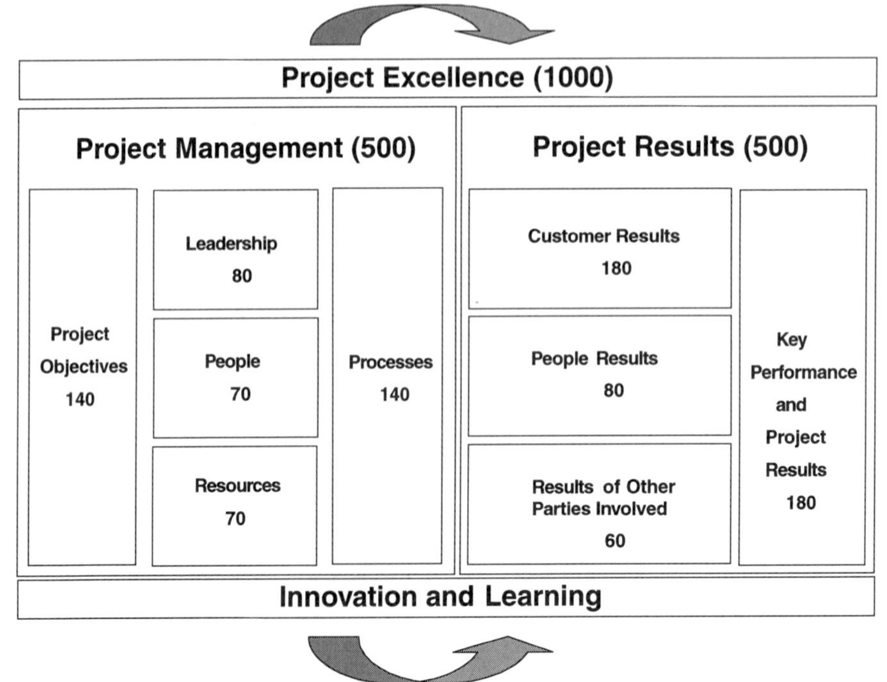

evaluates how far the enabler processes are excellent, while the Project Results section evaluates the degree of excellence of the project results.

The Assessment Criteria

The criteria for the assessment of project management include the following:

- *Project objectives*. How the project formulates, develops, checks, and realizes its objectives
- *Leadership*. How the behavior of all leaders within the project inspires, supports, and promotes project excellence
- *People*. How project team members are involved and how their potential is seen and utilized
- *Resources*. How existing resources are used effectively and efficiently
- *Processes*. How important project processes (content and management processes) are identified, checked, and changed, if necessary

The criteria for the assessment of project results include the following:

- *Customer results*. What the project achieves regarding customer expectations and satisfaction
- *People results*. What the project achieves concerning expectations and satisfaction of the employees involved
- *Results of other parties involved*. What the project achieves concerning expectations and satisfaction of other stakeholders involved
- *Key performance and project results*. What the project achieves regarding the expected project results

All criteria are further described. The criteria for project objectives of the section project management is shown in Table 37.5 as an example.

Table 37.6 shows how the processes regarding project management can be evaluated as excellent. A similar table is used for evaluating how far the project results are excellent.

Process of Assessment and Methods

The project excellence model may be applied in project reviews and evaluation and is quite commonly applied, for instance, in Germany for internal use as a self-assessment and pos-

TABLE 37.5. CRITERIA: PROJECT OBJECTIVES.

Criteria: Project Objectives (140 points)
How the project formulates, develops, checks, and realizes its objectives: 1.1. Application and demands of parties involved are identified. 1.2. Project objectives are developed, as well as how competitive interests are integrated. 1.3. Project objectives are imparted, realized, checked, and adapted.

TABLE 37.6. ASSESSMENT TABLE FOR PROJECT MANAGEMENT.

Sound Process	Systems and Preventions	Checking	Sophistication and Improvement of Business Effectiveness	Integration into the Normal Project Work and Planning	Model for Other Projects	Evaluation
Clear and extensive proof	Clear and extensive proof	Frequently and regularly checked	Clear and extensive proof	Perfectly integrated	Could be an example	**100%**
Clear proof	Clear proof	Frequently checked	Clear proof	Very well integrated	—	**75%**
Proof	Proof	Occasionally checked	Proof	Well-integrated	—	**50%**
Some proof	Some proof	Rarely checked	Some proof	Partly integrated	—	**25%**
		No proof				**0%**

sibility for reflection by the project team. Based on the self-assessment results, steps for further improvement are taken.

If a project team applies for the International Project Management Award, the process is as such:

1. *Application.* A project team applies for with a written statement explaining how it fulfils the criteria of the model for "Project Excellence." The statement is a self-assessment that helps the project team to understand how to achieve success and to identify and use their strengths and improvement potential.
2. *Assessment.* The assessment of the application is done by a team of at least four assessors with a high amount of project management competence. The assessor teams have different nationalities and different project backgrounds. In the first step the assessors do their assessments on their own. The lead assessor collects the single results. The criteria where the assessors have different opinions have to be discussed in a consensus meeting. The assessor has to find a consensus on the results and give recommendation to the jury regarding whether the project should enter the second stage of the assessment. The jury has the final decision on which of the candidate projects gets a site visit.
3. *Site visit.* The site visit takes one to two days, where the assessors conduct interviews with project representatives, as well as observations and documentation analysis. The assessors conclude their assessment report and again give a recommendation to the jury.
4. *Jury decision and award ceremony.* There are three categories of prizes. All the projects who got a site visit are finalists. Among the finalists, the jury decides the Prize Winners; one of these can be selected as Award Winner. Not every year an Award Winner is selected.
5. *Assessment report.* Finally, all candidate projects get their assessment report, which shows them strengths and areas of improvement.

Limits of the Project Excellence Model

There is no doubt that the project excellence model is a good project management quality management tool, as it provides a clear link between the quality of processes and the quality of results. Nevertheless, there are limitations to the model. The assessment has to stay at a high level, since it is a generic model applicable to any type of project. But for the project management section, it would be possible to go one step further and include the choice of a project management approach as the basis for the assessment. Currently, no specific PM method has to be applied by the candidate project. For instance, whether or not the stakeholder analysis is used may not matter as long as the project can prove that some kind of structured method is used to analyze their stakeholders. This kind of criticism is, however, inherent in excellence models, because they are non-normative. Nevertheless, the project excellence model has proven to be a useful project management quality management method.

Quality Management in the Project-Oriented Company

For all types of projects and programs, one can summarize that quality management has to be an integral part of the contents and management processes. The quality management of

TABLE 37.7A. OVERVIEW OF QUALITY MANAGEMENT METHODS IN THE POC.

	Objective	Standards	Methods	Roles
Project audit and review	To assess • Project the quality of deliverables and/or • The contents process	• Process and product standards	• Different combinations possible	• Representatives of the project or program • Auditor/reviewer • Audit/review owner
Management audits and reviews	To assess • the management competence of the project	• Project management and program management standard • Standards for Management audits and reviews	• Different combinations possible: • Interviews, documentation analysis, observations, self-assessments	• Representatives of the project or program • PM auditor/reviewer • PM audit/review owner
Project Excellence Model	To assess • the excellence of the project management process and the content processes • the excellence of the project results	• The model is the standard. • Reference to ICB, but no further standards.	• Self assessment or • Assessment for award • Combination with benchmarking possible	• Project or program team or • Project or program team • Assessors • Jury
Project and program benchmarking	To assess and compare • the project performance and or • the content processes and or • the project management process and it results	• Maturity models • Best practices • Performance criteria • Process models • Project management standards	• Different step models possible	• Benchmarking institution • Benchmarking partners
Project and program consulting	To support • The project or program in a complex situation	• Project management and program management standard	• Different combinations possible: Interviews, documentation analysis, observations, self-assessments • Facilitation of workshops	• Representatives of the project or program • PM consultant

TABLE 37.7B. OVERVIEW OF QUALITY MANAGEMENT METHODS IN THE POC.

	Objective	Standards	Methods	Roles
Project manager or program manager coaching	To support • the project manager or program manager in a complex situation	• Project management and program management standard	• Different combinations possible: Interviews, documentation analysis, observations, self-assessments	• Project manager or program manager • PM coach
Project or program evaluation	To assess the • project performance and the management process • after the project or program is completed	• Project performance criteria • Project management and program management standards	• Different combinations possible: Interviews, documentation analysis, observations, self-assessments	• Evaluator • Project team
ISO certification	To assess and certify • The quality of the processes of a company or company unit • By a third party	• ISO standards	• According to ISO, e.g., ISO audits	• Certification body • Auditors • Candidate
PM Certification	To assess and certify • the PM competence of project management personnel • By a third party	• IPMA Certification: ICB, NCB • PMI Certification: PMBOK	• According to PM certification system	• PM certification body • Assessor(s) • Candidate
Accreditation	To assess and accredit • a product, methods or processes • by a third party	• Accreditation standards	• According to accreditation procedure	• Accreditation body • Surveyor(s) • Candidate

the project or program further depends on the quality management of the project-oriented company. Project-oriented companies often have a quality management system based on a combination of ISO certification and excellence models and use different quality management methods to continuously improve their performance. The challenge is to integrate process management, quality management, and project management as specified in the new ISO 9000:2000 standards.

Table 37.7 summarizes the quality management methods applied in the project-oriented company. ISO Certification and accreditation is mainly used at the level of the project-oriented company or the line unit. They are both relevant for the project and program; for instance, if the company is ISO-certified, its projects and programs have to run according to ISO processes.

Within the project-oriented company, the PM office has an important role regarding quality management in projects and programs. Many PM offices provide quality assurance services like management audits and reviews, management consulting of project and programs, coaching of project managers and program managers, and project and program evaluation. The PM office is responsible for the management processes of project and programs and the standards and guidelines to ensure the quality. (See the chapter by Powell and Young.) Often the PM office supports PM certification of PM personnel and is responsible for other human resource functions, as competent PM personnel is also an issue of quality management (see the chapter on HR by Huemann, Turner, and Keegan). The reason for making the PM office responsible for project and program management quality assurance in the project-oriented company is that improvements can be implemented and communicated faster than when the task is left to the single project or program (Brucero, 2003).

One major thing I learned from the project-oriented company Transsystem, situated in the middle of nowhere, as they stated themselves: You can be professional and deliver high-quality solutions no matter where your company is located (Stroka, 2002).

References and Further Reading

Anbari, F. T. 2003. An integrated view of the six sigma management method and project management. In *Project Oriented Business and Society* 17th IPMA World Congress on Project Management. June 4–2, Moscow.

Bounds, G. M., G. H. Dobbins, and O. S. Fowler. 1995. *Management: A total quality perspective.* Cincinnati: South-Western.

Bucero, A. 2003. Implementing the project office: Case study. In *Creating the project office: A manager's guide to leading organizational change.* L. Randall, R. J. Graham, and P. C. Dinsmore. New York: Wiley.

Champ, R. C. 1989. *Benchmarking: The search for the industry best practices that lead to superior performance.* Milwaukee, WI: ASQC Quality Press.

Corbin, D., R. Cox, R. Hamerly, and K. Knight. 2001. Project management of project reviews. *PM Network* (March)

Crosby, P.B. 1979. *Quality is free.* New York: McGraw-Hill.

Davenport, T. 1993. *Process innovation.* Boston: Harvard Business Press.

Deming, W. E. 1992. *Out of the crisis.* Cambridge, MA: MIT

Feigenbaum, A. V. 1991. *Total quality control.* 3rd ed. New York: McGraw-Hill.

Frieβ, P. M.,(1999. *Projekt Management für den tiefgreifenden organisatorischen Wandel mittelgroβer Einheiten.* Bremer Schriften zu Betriebstechnik und Arbeitswissenschaften, Band 25, Verlag Mainz, Wissenschaftsverlag Aachen.

Gareis, R., and M. Huemann. 2003. Project management competences in the project-oriented company. In *People in Project Management.*, ed. J. R. Turner. Aldershot, UK: Gower.

Gareis, R. ed. 2002. *pm basline: Knowledge elements for project and programme management and for the management of project-oriented organisations.* Projekt Management Austria, www.p-m-a.at/publikationen.htm.

Huemann, M. 2002. *Individuelle Projektmanagement Kompetenzen in Projektorientierten Unternehmen.* Europäische Hochschulschriften, Peter Lang Verlag, Frankfurt-am-Main.

Huemann, M., and R. Hayes. 2003. Management audits of projects and programs a learning instrument. In *Project oriented business and society.* 17th IPMA World Congress on Project Management. June 4–6, Moscow.

Imai, M. 1986. *Kaizen: The key to Japan's competitive success.* New York: McGraw Hill.

Ishikawa, K., and D. Lu. 1985. *What is quality control? The Japanese way.* Englewood Cliffs, NJ: Prentice Hall.

ISO. 2000. *ISO 9,000: Quality management systems—Fundamentals and vocabulary.* Geneva: International Standards Organization.

———. 2000. *ISO 9,001: Quality management systems—Requirements.* Geneva: International Standards Organization.

———. 2000. *ISO 9,004: Quality management systems—Guidelines for performance improvement.* Geneva: International Standards Organization.

———. 2002. *ISO 19,011: Guidelines for quality and/or environmental management systems auditing.* Geneva: International Standards Organization.

———. 2003. *ISO 10,006: Quality management—Guidelines to quality management in projects.* Geneva: International Standards Organization.

Juran, J. M. 1950. *Quality control handbook,* New York: McGraw-Hill.

———. J. M. 1986. *Out of the crisis. Cambridge,* MA: MIT.

Logothetis, N., and H. P. Wynn. 1989. *Quality through design: Experimental design, off-line quality control and Taguchi's contributions.* Oxford, UK: Oxford University Press.

Morris, P. W. G. 1999. Key issues in project management. In *The Project Management Institute project management handbook,* ed. J. K. Pinto. San Francisco: Jossey- Bass.

OGC. 2002. *Managing successful projects with PRINCE2.* 3rd ed. London: The Stationery Office.

Pharro, R. 2002. Processes and procedures. In *The Gower handbook of project manage*ment, ed. J. R. Turner and S. J. Simister. Aldershot, UK: Gower.

Pinto, J. K. 1999. Managing information systems projects: Regaining control of a runaway train. In *Managing business by projects. Proceedings of the NORDNET Symposium,* ed. K. A. Arrto, K. Kähkönen, and K. Koskinnen. Helsinki: Helsinki University of Technology.

PMI 2000. *A guide to the project management body of knowledge.* Newtown Square, PA: Project Management Institute.

Roy, R. K. 1990. *A primer on the Taguchi method.* New York: Van Nostrand Reinhold.

Scott, B., ed. 2001. *Partnering in Europe: Incentive based alliancing for projects* London: Thomas Telford.

Seaver, M., ed. 2003. *Gower handbook of quality management.* Aldershot, UK: Gower.

Senge, P. 1994., *The fifth discipline fieldbook: Strategies and tools for building a learning organization.* New York: Doubleday.

Sroka, S. 2002. Reorganisation of transsystem to a POC for the re-inforcement of the customer focused market. In *Making the vision work. 16th IPMA World Congress on Project Management.* Berlin, June 4–6.

Stummer, M., and M. Huemann. 2002. Development of competences as a project-oriented company: A case study in the chemical industry. In *Making the vision work.* 16th IPMA World Congress on Project Management. June 4–6, Berlin

Tennant, G. 2002. *Design for Six Sigma, Launching New Products and Services without Failure,* Aldershot, UK: Gower.

Turner, J. R. 1999. *The handbook of project based management. 2nd ed.* London: McGraw-Hill.

———. 2000. Managing quality. In *Gower handbook of project management,* ed. J. R. Turner and S. J. Simister. Aldershot, UK: Gower.

———. 2003. Farsighted project contract management. In *Contracting for project management,* ed. J. R. Turner. Aldershot, UK: Gower.

Van Wieren, H. D. 2002. Alliance, an excellent solution to meet project execution challenges. In *Making the vision work.* 16th IPMA World Congress on Project Management June 4–6, Berlin.

Wateridge J. 2000. Project health checks. In *The Gower Handbook of Project Management,* ed. J. R. Turner and S. J. Simister. Aldershot, UK: Gower.

Webb, A. 2000. *Project management for successful product innovation.* 2nd ed. Aldershot, UK: Gower.

CHAPTER THIRTY-EIGHT

THE PROJECT MANAGEMENT SUPPORT OFFICE

Martin Powell, James Young

As the practice of project management has grown, so has the demand for a systematic method of implementation. In the late 1970s, the establishment of a project support office was seen as the vehicle to achieve this and was traditionally responsible for status reporting. It was staffed by project management professionals serving the organization's needs through the provision of support. From this project office or project support office stemmed the project management support office (PMSO), combining the duties of supporting projects and reporting project status with a Project Management Centre of Excellence.

The PMSO is a central organizational unit that is responsible for ensuring the portfolio of projects performs optimally. It does this through the provision of support to portfolio, program, and project managers; the creation of a project management function (this is often "virtual," i.e., it has no functional mandate but provides a "home" for project management); the development of management competencies appropriate for the management of the portfolio; and the provision of effective tools.

For project team members, the PMSO should be an invaluable source of support. The PMSO can supply team members with mentors, facilitators, and just-in-time support. It offers guidance on project management methodology, standards, and processes and can provide team members with the tools they need to do their jobs effectively. It can maintain a library of previous project management solutions that might be reusable on new projects, be the owner of project knowledge, be responsible for communication, be a source of topic expertise, provide audit and "health check" support, and be a "home" for the project management community.

The roles that a PMSO performs are broad ranging—they need to reflect the needs of the organization, its structure, the type of projects it undertakes, and, in particular, its culture. A PMSO often comes into its own in a "virtual" organization—one where the company's operations and projects are conducted across a number of locations—where communication is stretched, standardization is difficult to achieve, and support is difficult to deliver. What is also true is that implementing a PMSO in a virtual organization presents a far greater challenge, both organizationally and culturally.

The Need for a PMSO

Every day, organizations commit enormous quantities of resources to projects. Not all organizations can be certain that they have sufficient data at hand to ensure that they are investing in the right projects, that the business case for the projects they have chosen to invest in remain valid, or that their current portfolio of projects will deliver the required product within the time and cost resources specified. The role of the PMSO is to support management by validating that the information that is received is accurate and that those they put in charge of delivering their portfolio are properly equipped to do so.

Not everyone embraces the idea that a PMSO adds value to the project management capabilities of an organization. Many may view it as yet another layer of bureaucracy that reduces the agility of an organization's project management function. Others see it as an attempt to stifle innovation, expose them to audit, and diminish their ability to manage their projects the way that they see fit and put them under the watchful eye of a centralized control group.

Organizations that wish to establish a PMSO should realize that they might encounter substantial skepticism and resistance to their efforts. It is important at the outset that well-defined steps are taken in educating individuals about the need for a PMSO and that the way it will function and its deliverables are clearly communicated. To be successful, a PMSO must be supported by the project management community; the community must feel as though the PMSO is there to provide support and guidance it when required.

A good starting point is to identify what questions might be raised about a PMSO. A PMSO must be able to define its value to projects, define what it is going to produce and why it is producing it, and be able to understand the impact of its work.

The aim of a PMSO must ultimately be to improve the performance of projects in an organization. How this is measured depends on the context in which the organization operates—but it always has to deliver greater "benefit."

Purpose

The objectives of a PMSO are a key factor in determining its structure, responsibilities, and organizational "location." Usually the principle objective of establishing a PMSO is to improve the overall performance of projects within an organization. But what does this mean? In some companies this is viewed as ensuring projects deliver effectively within the classic time, cost, and quality criteria; in others it can mean improving the performance of the portfolio and its impact on the business.

The objectives often reflect the sector or "orientation" of the organization; contracting companies—for example, construction or IT service organizations—often seek to use the

PMSO to improve the competence and quality of their project managers and thereby the service they offer to their clients. In contrast, product development companies—for example, software vendors or pharmaceuticals companies—often mandate the PMSO with improving the benefit that the portfolio of projects delivers back to the business in terms of return on investment.

The principal differences between these two models are the scope of the PMSO's responsibilities, the mandate that they are handed, and its location within the organization.

Structure

The structure of the PMSO depends upon the scope of its responsibilities, its location within the organization (does it provide support at project, program, and/or portfolio), and the size and structure of the organization it serves. The PMSO can exists at three levels (Crawford, 2002):

- *The strategic PMSO.* Responsible for setting corporate policies and standards for projects, for coordinating their implementation down through the organization, and for supporting portfolio managers in aligning the project portfolio with business strategy. The strategic PMSO is also responsible for ensuring effective communication between business unit PMSOs to facilitate the sharing of best practice and the transfer of knowledge.
- *The business unit PMSO.* A dedicated project office responsible for implementing the standards and policies established by the strategic PMSO within a specific business function or unit. Often corporate standards need interpreting to the specific needs of a business unit (BU) or function. While the principles remain the same, the specific processes, templates, and control standards may vary depending on the types of projects under way. The BU PMSO might also be responsible for managing project resources and for ensuring the effective transfer of knowledge between projects within the BU.
- *The project office.* Large, complex projects might warrant the existence of a dedicated project office organization. This office is responsible for ensuring the standards and policies established by the strategic PMSO and BU PMSO are applied. It is also often responsible for the implementation of information management standards, as well as the hands-on management of the project control function.

In fact, these offices do not need to be separate entities—in some organizations one PMSO might undertake the roles of strategic PMSO, BU PMSO, and project office.

Mandate

The roles and responsibilities of the PMSO must be clearly established from the very outset. This includes direction on which of the PMSO's services, standards, and policies are to be "imposed" on projects—mandatory—and which should be made available to projects should they be required.

FIGURE 38.1. LOCATIONS OF THE PMSO.

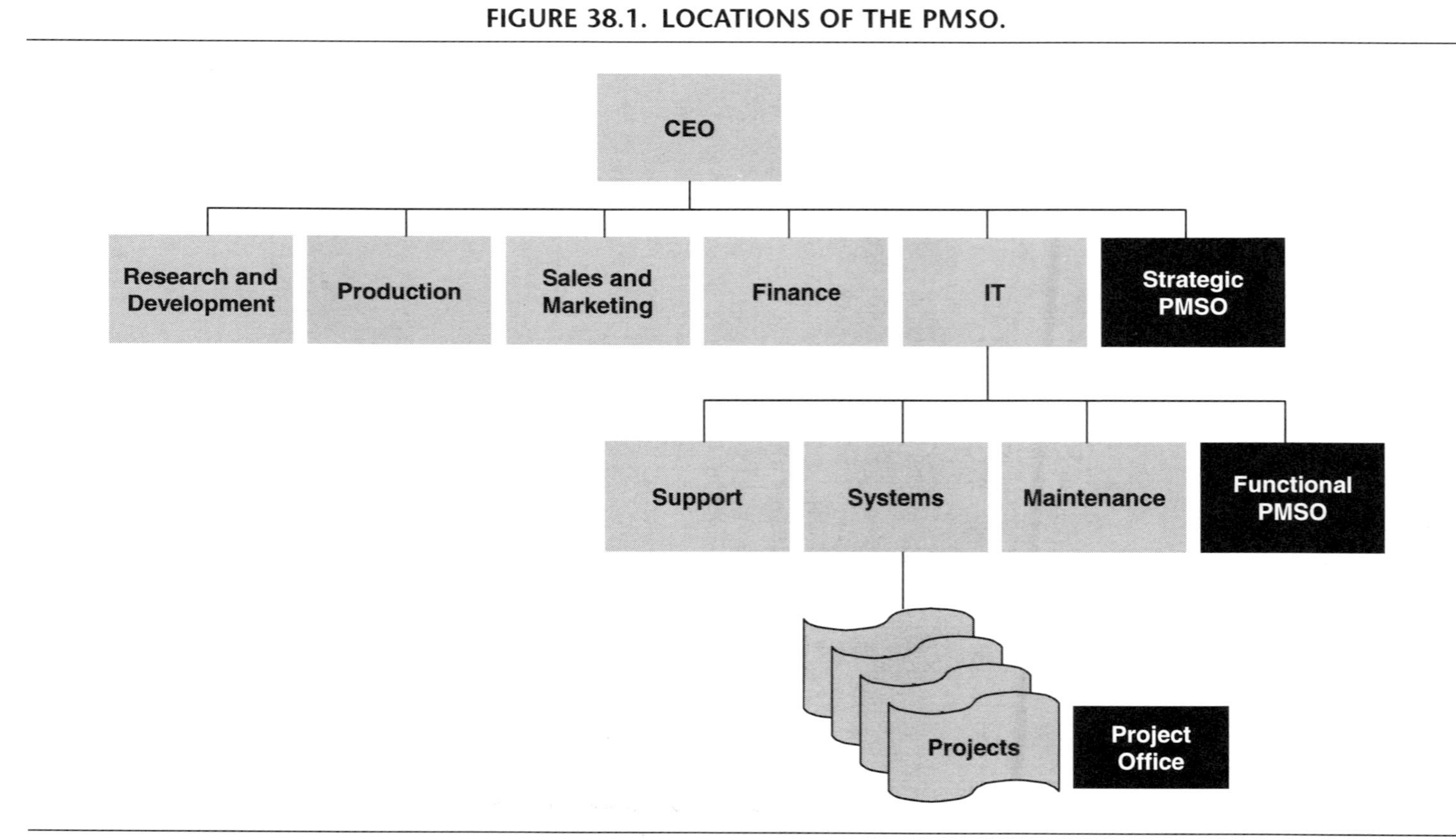

Services that are mandatory must be determined by the sponsor of the PMSO but must reflect the structure and culture of the organization—a devolved organization operating in autonomous business units often rejects "interference" from a central organization, while centralized organizations tend to be far more accepting of company-wide standards.

There are no hard-and-fast rules on which standards, policies, and services should be mandatory; however, those that are mandated typically tend to focus on the use of the PMSO to provide reviews of projects at key milestones and standard structuring, recording, and reporting of project information—in order that it can be easily aggregated to program and portfolio level.

Another service that can be mandated is the use of the PMSO to audit projects. Where this is done, care should be taken to ensure that the role of the PMSO, to provide objective support to project managers, is not compromised by its role in reviewing their performance.

Sponsorship

During PMSO implementation, a critical factor is the appointment of a sponsor at an appropriately senior level. The role of the sponsor is to ensure that the deliverables, in this case the services of the PMSO, remain in line with the needs of the organization and to provide it with strategic direction.

The sponsor of the PMSO has some basic roles: to support, promote and provide strategic direction to the PMSO, as well as determining its responsibilities, structure, and mandate. The sponsor for each of the three levels of the PMSO can be different, but the basic roles remain the same.

Strategic PMSO

The sponsor must be a senior manager within the organization who has a vested interest in the PMSO making an impact on the performance of the business. Because the strategic PMSO often lies across several business units or functions, the sponsor's role may often be undertaken by a sponsoring board. The board will be made up of senior representatives from each of the functions or business units covered by the PMSO.

Business Unit/Functional PMSO

The sponsor is usually a senior manager, often the head of the business unit or function, who requires an improvement in the performance of projects and greater certainty of outcome of projects within their organization.

Project Office

The sponsor is usually the project manager who requires a project control function with added responsibility for ensuring that corporate standards and policies are being adhered to. Whoever the sponsor is, they must be willing to actively communicate the roles and responsibilities of the PMSO (at any level) and to champion its use throughout the organization.

Charging for PMSO Services

Charging projects for the use of the PMSO's services is often a delicate matter; a balance needs to be struck between encouraging the project community to engage with the PMSO in order to embed best practice across all projects and ensuring that the PMSO continues to provide tangible value-added services to projects.

Studies have indicated that there are mixed feelings about charging for any PMSO services, although generally it is accepted that charging might be appropriate for "consultancy"-type services—for example, project ramp-up support, workshop facilitation, training, and project "health checks."

One such study (Morris et al., 2000) found that charging for these types of service provided some benefit to the PMSO:

- It helps prevent the businesses and projects from using the PMSO for trivial, nonmandatory, or non-critical tasks.
- It encourages the PMSO to meet some of its budget through the sale of its services to projects. This encourages the PMSO to ensure that the services it provides are cost-effective and relevant to the business.
- It enables the PMSO to raise its profile across the organization, helping to attract high-caliber project personnel to work within it.

Responsibilities of the PMSO

The responsibilities of the PMSO are split between the different levels described earlier in this chapter. The chart in Figure 38.2 identifies the responsibilities that each of the offices might reasonably take.

The strategic PMSO role focuses on supporting the portfolio management role, defining best practice, facilitating communication across the project community, and collecting and reporting the status of all projects within the portfolio. BU PMSOs focus on the implementation of best practice in the context of their specific organization of function, while ensuring that project information is reported consistently in accordance with standards set by the strategic PMSO. The project office is concerned with applying the policies and standards at a micro level on specific projects.

The responsibilities of the PMSO fall into four categories:

- *Portfolio, Program, and Project Management Support.* Where the PMSO provides support directly to portfolio managers, program managers, and project teams
- *Organization.* Concerned with the creation and management of the program and project management community
- *Competence.* Identification, structuring, and delivery of knowledge, including training and communities of practice
- *Systems and Tools.* Designing, training, and supporting the tools and systems used to support portfolio, program, and project managers

FIGURE 38.2. RESPONSIBILITIES OF THE THREE LEVELS OF PMSO.

Responsibilities	Strategic PMSO	Business Unit PMSO	Project Office
Portfolio, Program, and Project Support			
Portfolio management support	✓	✓	
Project health checks and audits		✓	
Project governance and reviews	✓	✓	
Reporting	✓	✓	✓
Project ramp-up support and project rescue		✓	✓
Project Management within the Organization			
Resource planning		✓	✓
Communications	✓	✓	
Benchmarking	✓	✓	✓
Performance measures and metrics	✓	✓	
Competency			
Best-practice guidance	✓	✓	
Tailored methodologies	✓	✓	
Communities of practice	✓	✓	
Training and building capability	✓	✓	
Coaches and mentors	✓	✓	
Knowledge management	✓	✓	
Systems and Tools			
Project management systems design	✓	✓	
Software reviews and recommendations	✓	✓	
Design of standard templates and documents	✓	✓	✓
Design of procedures	✓	✓	✓

Portfolio, Program, and Project Support

The PMSO is responsible for providing support to all levels of the business on the status of projects:

- Strategic PMSO provides support to the portfolio management function on the current status of its cadre of programs and projects, while informing BU PMSOs of the standards that it must adhere to.
- BU PMSOs provide support to the strategic PMSO and the function or business unit in which they operate in determining the current status of programs and projects, as well as establishing and communicating the standards that programs, projects and project offices must adhere to.

- Project offices provide support to program and project managers to determine the current status of the work that they are responsible for, as well as structuring data for aggregation by the BU PMSO and/or strategic PMSO.

Portfolio Management Support

The strategic PMSO can provide support to the portfolio management function: its primary function in this area is the acquisition, structuring, and reporting of portfolio-level information from programs and projects. Key to this process is the establishment of data standards—defining which information is required (this should include an agreed definition of terms such as "accrual" and how often it is required.) The strategic PMSO should define these standards and ensure that they are communicated to the functional PMSOs and the program and project managers.

The strategic PMSO should ensure that the data in the portfolio system is regularly updated in accordance with the process agreed by portfolio managers and that the data submitted is accurate. The information stored in the portfolio system must be the same as that in the project control system, although it might not be updated as frequently. Portfolio reports should be automatically generated from the portfolio management system.

In addition to this, the strategic PMSO might also provide assistance in applying different scenarios into the portfolio system to demonstrate the impact on the performance of the portfolio of changes to project cost, time-to-market, and project revenues.

Project Health Checks and Audits

The provision of project health checks takes on two forms: audit and self-assessment. Auditing is typically seen as a policing role deemed mandatory for projects. An audit consists of a formal review of the project against a set of quality criteria. It focuses on compliance as well as providing a snapshot status review of the project for the benefit of the sponsor. Often the competency of the staff is also reviewed and reported to the sponsor.

The other form of health check is self-assessment, where the PMSO provides the project manager with the tools and training to enable him or her to assess the performance and compliance of his or her own project. Self-assessment of projects is rarely deemed to be mandatory, but it is used to encourage project managers to revisit their projects prior to major gates or milestones.

The PMSO can face a dilemma if project audits are mandated; a balance must be struck between the PMSO's role as the provider of knowledge and support—demand-driven services—and audit. Projects might be less inclined to discuss issues, problems, failures, or their support requirements if they feel that their performance is to be judged by that same organization during an audit. This is particularly true if the outcome of audits have any bearing on the way project managers are assessed for career progression.

Project Governance and Reviews

The PMSO (strategic or BU) can assist the businesses manage their projects through the governance process. The role of governance is to monitor project progress against budget and milestones; to ensure that the business case remains valid; to provide strategic direction on issues and changes; to ensure that corporate standards, processes, and procedures are being properly applied; and to facilitate the removal of organizational blockages to successful project completion.

A project's governance board is made up of project stakeholders—people with a vested interest in the project's execution and delivery. A PMSO often has representation on the governance board with a specific focus on the application of good project management and to ensure that progress information provided is accurate. The PMSO can provide advice to project managers regarding how to address shortfalls in the application of project processes and where support may be sought.

Reporting

The PMSO is often responsible for generating status reports. The first stage in this process is to gather, structure, and validate progress data. The data collected and reports generated vary by level in the organization:

- The strategic PMSO collects portfolio-level progress information, typically, completion date, forecast cost at completion, and updated business case. It then report this to investment committees and business planning.
- BU PMSOs collect status information on programs, projects, and the BU portfolio typically performance against budget and baseline, as well as updated business case. Reports are generated for the business unit managers, and the data is structured to allow it to be aggregated by the strategic PMSO into its portfolio management systems.
- Project offices are responsible for acquiring updated information on the progress of the program or project that they serve. Reports are generated for the program/project manager, the project sponsor, and any other party that requires project information. In addition, the project office structures its information in such a way as to allow it to be aggregated by the BU PMSO and the strategic PMSO.

Project Ramp-up Support and Project Rescue

The PMSO often provides the traditional project office role of project control. The PMSO can either provide a resource to fulfil the project control role or provide the project control service to the project as a clearly defined set of deliverables.

PMSOs might retain the services of some high-performing project managers that can be deployed rapidly on projects, temporarily, either to rescue projects that are failing or to ensure that projects are structured correctly from the outset.

Where the PMSO is called in to rescue a project, its first task is to assess the situation and to provide a clear strategy on what actions must be taken to rescue it, including the implications to time, cost, and so on. Where a project has encountered significant problems or appears to be badly failing, a judgment on whether to continue the project must be made with the sponsor.

The PMSO's services for supporting the set up of projects are often used on particularly complex or large projects where highly competent and experienced project personnel are required to ensure the project is correctly structured, or where the project is unable to staff itself with suitable resources in the immediate term. The PMSO is responsible for ensuring that the project is established in accordance with the best practices that it has established.

Where these types of services are provided, the PMSO often charges the project for its project managers' time on a consultancy basis.

Project Management within the Organization

The following section deals with the responsibilities of the PMSO on an organization-wide basis. It addresses the role undertaken in planning project management resources across business units and functions and how these resources are managed. This section also addresses how the PMSO communicates with the organization and how benchmarking should be undertaken to ensure that project practices are effective.

Resource Planning

The PMSO can take on a number of roles with regard to the management of the project resource pool. It can have full responsibility for the day-to-day management of project managers at one end of the spectrum, and at the other it can have no responsibility at all. It can also take on a myriad of roles between these two scenarios.

In organizations where project management is a central function, the PMSO undertakes the same roles as any other corporate function. As well as the traditional "hiring and firing" roles, it might also include the identification of skill requirements (competencies); forecasting the number and types of resource that are needed to manage the forthcoming portfolio of projects; monitoring actual resource use; providing guidance, mentoring, and training to project managers; conducting performance reviews; and monitoring career progression.

Where the PMSO is responsible for forecasting resource requirements it must liase closely with the portfolio management function to plan the number of project managers required in the future and to ensure that the right caliber and seniority of project manager is available to manage the up-and-coming portfolio of projects. Coordination between the business unit PMSOs is often required to meet peaks in demand across the organization.

The PMSO might be responsible for managing the whole project management resource pool or might retain a few high-caliber individuals in a central pool. In the latter case, these project managers might be deployed to manage high-profile, complex, or very large projects, or might be used to undertake project audits, sit on project review boards, provide project ramp-up support, or step in to rescue projects that are in trouble.

A *project management competency framework* is often used in mature PMSOs to ascertain the project management skill sets across the organization. The competency framework is also useful in identifying skill gaps across the organization and allows the PMSO to focus its training to ensure that the skill needs are met across the organization and resource can be deployed cross-functionally if required.

Where the PMSO is not responsible for the management of the project management resources pool it often maintains a database of project managers and their skill sets in order that project managers can be redeployed across the organization where required.

Communications

The PMSO should also establish and execute a strategy for communicating with the project management community. This strategy should determine what messages it intends to broadcast—usually updates on services offered, news, access to information, knowledge, training and contact details, and its target audience, including program managers, project managers, project team members, project sponsors, business managers, and the media that it intends to use to communicate.

In recent years Internet technologies (Internet, intranet, and extranet) have provided a highly effective medium for supporting communication strategies. More advanced technologies can be deployed to aggregate information on progress, status, performance, or any other measure, as well as reference and support information. This is discussed later in this chapter under *Portals*.

To support its role as "competency developer" the PMSO should play a role in advertising project management courses within the organization and generating course attendance. The PMSO also has a major responsibility to monitor the quality of the training efforts and to provide the guidance needed to deliver first-rate course offerings.

The methods of communication can vary widely depending upon the functions of the PMSO. Many organizations communicate through e-mail notifications, but other forms of media such as posters, calendars, summary sheets, and intranet sites promoting the products and services are also used.

Benchmarking

For a company to assess its performance, it must compare itself against a reference point. This process is known as *benchmarking*, and it is critical in the management of any organization. The PMSO should actively lead or support the process for benchmarking projects and project management.

The PMSO must begin with an assessment of the current condition of the organization with regard to its project management maturity, and then it must establish a baseline. While no industry standard yet exists for baselining the capabilities of an organization's project management functions, several models exist designed to measure project management maturity. (See the chapters by Cooke-Davies and the chapter by Ibbs and Reginato.) If the maturity can be assessed, it provides the future ability to quantify the value of project management against the original baseline.

Benchmarking is undertaken for a range of reasons:

- To establish performance in comparison to competitors—to determine whether a company's project delivery capability is as effective as its rivals'.

- To determine whether the performance of the organization is improving—monitoring a metric over a period of time to determine whether improvements are being achieved.
- To establish performance by comparison with companies that have a similar focus—for example, measuring the competency of project management staff. This is often done where no industry standard is available for the metric.
- To monitor the effect of an initiative—for example, training—on the performance of projects.

Any aspect of an organization can be benchmarked—the return on investment can be compared with competitors or the competency of project managers can be compared with other companies in different sectors. Crucial to the benchmarking process is that the information being benchmarked is accurate and the population against which comparison is being made is known.

Performance Measures and Metrics

A company's portfolio of projects is the vehicle for developing new products and services or for delivering products and services to clients. Whichever it is, the company needs to know how well its portfolio is performing.

The PMSO should be involved in determining which measures to put in place, and once established, then it should be responsible for ensuring that they are effectively implemented, understood, and continuously updated on projects. The PMSO should be active in monitoring the performance of projects against these metrics, reviewing the metrics at regular intervals to ensure that projects remain on track, and helping to communicate what the metrics mean to project sponsors. The PMSO should be identifying where the metrics indicate that a project is not performing and, ideally, providing advice on corrective actions.

The measurement and monitoring of specific areas of performance allows managers to determine whether desired performance levels are being achieved or whether improvements are being made. Crucially, it also allows failures in performance to be identified and trends towards nonperformance to be identified and management actions to be taken. These measures are commonly known as "metrics."

Figure 38.3 illustrates that metrics can either be empirical—for example, *number of concurrent projects*—or subjective—for example, *client satisfaction*. They can also be classified as *input* or *output* metrics. Input metrics indicate that projects should perform well, for example, *highly competent project managers*, while output measures indicate that projects are performing well, for example, *return on investment.*

Metrics that have an empirical basis are generally more reliable (although every project manager will tell you that the numbers can be massaged) and easy to update, as the data can generally be derived from project control, finance, and HR systems. While metrics do not have to have an empirical basis, efforts must be made to ensure that subjective measures are described in sufficient detail to allow them to be applied consistently every time.

Metrics do not necessarily provide any valuable knowledge on their own. To have real meaning, they must be benchmarked. The metrics used vary according to different levels within the company: Strategic PMSOs help business and portfolio managers to focus on strategic metrics, BU PMSOs help functional and business unit managers to focus on or-

FIGURE 38.3. EXAMPLES OF METRICS BY CLASSIFICATION.

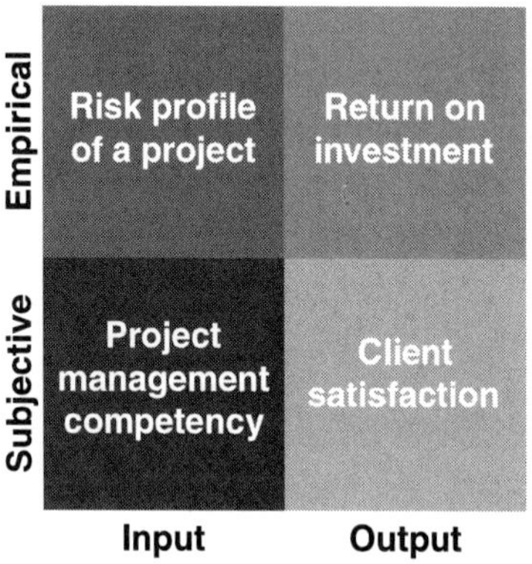

	Input	Output
Empirical	Risk profile of a project	Return on investment
Subjective	Project management competency	Client satisfaction

ganizational metrics, and project offices support program and project managers with project performance metrics.

The actual metrics used depends on a number of factors:

- *Whether the project organization is focused on delivering projects internally or externally.* Companies focusing on the delivery of internal projects focus more on return on investment and portfolio performance, while those delivering projects externally focus more on resource usage and margin.
- *The project management maturity of the organization.* Companies that are relatively immature find it difficult to identify information to support the calculation and update of metrics.
- *The sector that the company operates in.* The sector influences the metrics used at the program and project level to measure performance. This generally reflects industry cultures and data standards; for example, system availability is commonly used for measuring performance in the ICT sector, while earned value calculations are more common on military and construction projects.

Whichever metrics are used, perhaps the most important factor in their effectiveness is whether they are aligned to the strategic and tactical needs of the company—that is, do they tell management what they need to know? There is no correct set of metrics that should be used, but those that are implemented must be current, accurate, and repeatable.

Competency

The PMSO needs to be an ambassador for project management within the organization. It must be seen as the "voice" of how projects should be managed. To do this successfully, it needs to be the following:

- Holder of best practice guidance
- Owner of a set of methodologies
- Key driver in knowledge management initiatives
- Catalyst for communities of practice
- Locator of subject matter experts
- Key provider of training
- Possessor of up-to-date literature and institutional knowledge
- Assessor of the maturity of the organization and measure improvement
- Capable to carry out health checks and support to projects either through facilitation or general project management training
- Provider of tools to facilitate practice and also to benefit portfolio reporting

Figure 38.4 shows the PMSO as the "owner" of project or project management knowledge. This knowledge is distributed through a series of sources—guidelines, processes, procedures, and so on. Key to maintaining knowledge is for projects to implement and feedback new knowledge in the form of lessons learned or process improvements through the use of subject matter experts and communities of practice.

Best-Practice Guidance

The PMSO is the owner of best practice. It needs to hold and communicate a clear message of what project management is and how it should be carried out within the organization to optimize effectiveness. A significant factor in the success of the PMSO is its ability to effectively articulate, disseminate, and maintain a set of practices and processes that reflect how the organization expects projects to be managed. It should also identify weaknesses or areas where improvements are required and set forth a vision of where it should be developing in these areas. To achieve this success, these standards need to be organizationally specific, contextualized, concise, and, above all, practical.

The PMSO serves as a central library for these standards and is the center for expertise on their deployment. The PMSO also incorporates lessons learned on projects in project management methodologies. Clearly, lessons learned can be very specific and the methodologies should have some method of arranging learning in degrees of generality and specificity.

In a well-developed PMSO, there is usually a separate individual responsible for process development and maintenance to ensure that knowledge is used effectively. His or her role is to leverage the existing networks within the organization, or create new networks of subject matter experts who can maintain and develop specific process. The individual can also facilitate the formation of communities of practice, as discussed more fully later.

Bodies of Knowledge. As the owner of best practice, the PMSO needs to decide whether the best practice is based upon industry standards, internal company standards, or proprietary standards bought off the shelf.

FIGURE 38.4. THE PROJECT MANAGEMENT KNOWLEDGE TRANSFER.

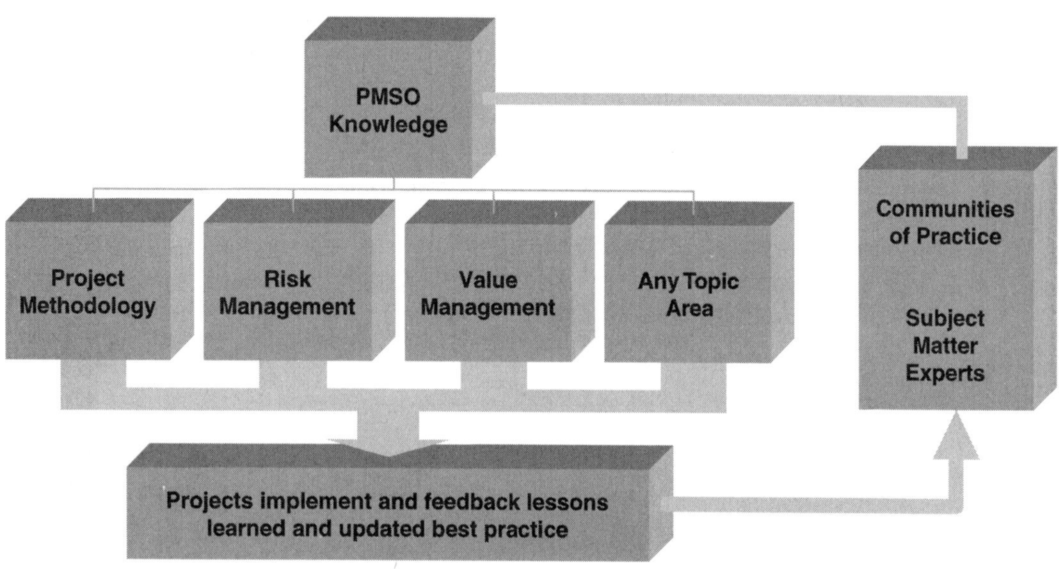

The two most significant standard guides to project management that are generally available to the public are those produced by the Project Management Institute (PMI) and the Association for Project Management (APM). (The two chapters by Crawford address these, and other project management institutions and their bodies of knowledge.)

PMI's Project Management Body of Knowledge (PMBOK) provides an overall framework for thinking about project management and details nine "knowledge" areas that PMI considers are the areas "unique" to project management (PMI, 2000).

The *APM* Body of Knowledge looks at areas that research has shown can contribute to projects being successes or failures, and that therefore need to be managed. There are a total of 42 topics grouped into seven areas. (Dixon, 2000)

Generic Methodologies

In the development of a set of project management methodologies, general project management practice has to be reviewed in terms of the organization's project management needs. Topics within the established "Bodies of Knowledge—PMBOK and APM—can be evaluated in terms of an organization's life cycle or its key activities in developing its products or services. Relevant topics can then be selected to be included within the set of methodologies. More importantly, the business unit PMSOs can select topics that apply to their business area. In large organizations governance boards may mandate certain processes to be incorporated into the methodologies that the business units have selected.

A project management methodology spells out the steps to be followed for the development and implementation of a project. A sample sequence of a project management methodology starting after project approval is shown in Figure 38.5.

As the owner of these methodologies, the PMSO also maintains the templates, forms, and checklists developed to assist project managers in the delivery of their projects. In many cases the templates are an integral part of the methodologies themselves. The project management methodology may require, for example, a risk plan, and project managers are helped considerably if they can see a sample of a risk plan, including instructions for completing one. There may be several templates such as:

- The project charter
- Communication plan
- Risk log
- Issues log
- Schedule template
- Reporting template
- Resource allocation and leveling samples and matrices

These templates should be standardized across the organization, but appropriate flexibility should be built into the templates to ensure they do not stifle the team's ability to innovate. Standardized templates should facilitate the production, aggregation, and review of projects.

While the PMSO should be the owners of the methodologies, parts or all of the ownership may also exist within:

FIGURE 38.5. A SAMPLE SEQUENCE OF A PROJECT MANAGEMENT METHODOLOGY.

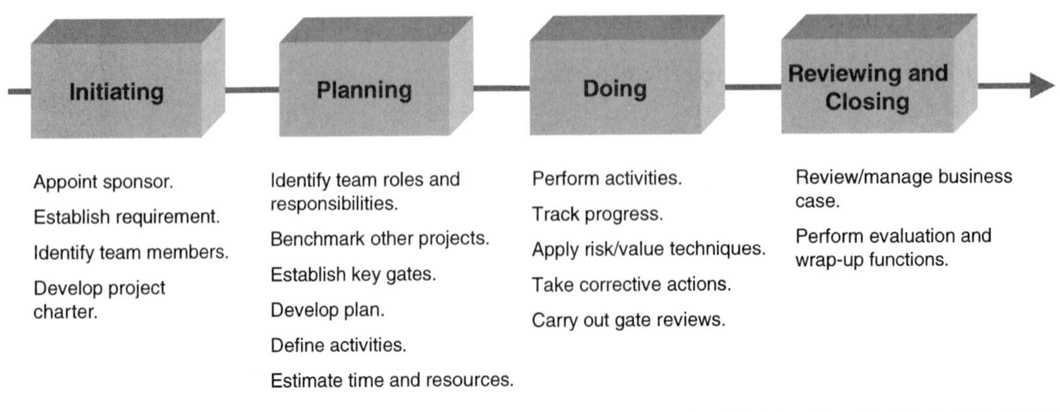

- Specialist groups or functions
- Functional PSOs
- Formal networks of practitioners
- Informal networks of practitioners
- Organizational functions (i.e., Human Resources or Finance)

This is very dependent on the organizational position of the PMSO.

Tailored Methodologies

A tailored project management methodology ensures that people and systems can speak a common language across a multiple-project enterprise setting. It must ensure that the way in which projects are carried out fits the context and the culture of the organization. It is created, and therefore "owned," by the organization, and it focuses on its specific needs—sector, culture, size, structure, and so on.

In the development of project management methodologies, a structured approach involving a series of interviews with key stakeholders and practitioners, along with a series of reviews, is necessary in order to obtain organization-wide buy-in to the proposed approach.

In large organizations this needs to be carefully planned to ensure all parts of the business are represented. To be seen as the keeper of best practice, people need to accept this proposed approach and have to make the transition from what they do now to what they need to be doing.

Figure 38.6 shows the components that make up a methodology:

- *Practices*. These are a definition of what you are trying to do.
- *Methods*. These describe how to carry out the practices.
- *Processes*. These define the sequence to how something is done.
- *Procedures*. These offer how-to detail about particular processes, methods, and practices. The methodology summarizes this detail.
- *Rules*. These are the constraints to the methodology.

These components are disseminated and presented to the organization through tools, templates, case studies, principles, lessons, and standards or standard operating procedures. The success of how this is presented to the organization is through the understanding of the needs of the people who are to apply these methodologies. This is fundamental.

Methodologies vary widely from industry to industry and from company to company. For example, Japanese product development methodologies are sharply different from U.S. software development methodologies (Rad and Levin, 2002). Global companies like IBM that do a wide range of projects, like product development, network services, manufacturing, and information technology systems integration, face the challenge of creating a common methodology that meets the needs of all project users. Once a methodology is in place, it has to be documented in the form of procedures. Procedures are the detailed how-to instructions describing the steps in the methodology, and these might need to be tailored to the needs of individual needs of different business units and functions.

FIGURE 38.6. A METHODOLOGY TRANSLATED INTO AN ORGANIZATIONAL CONTEXT.

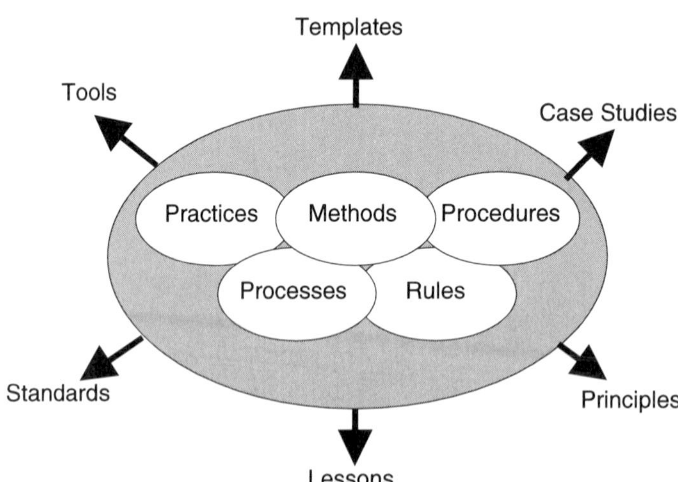

Guidelines. A set of project management guidelines offers a compromise to the methodologies. They describe at a higher level how to do project management and offer people a conceptual and more holistic view of what best practice should look like and how each project management topic should be undertaken. Where the target project management community is working across several business units, a set of guidelines would potentially achieve greater buy-in than a more detailed methodology.

Many organizations incorporate within both the guidelines or methodologies detailed information such as decision gates, toll gates, milestones, standard reporting times, portfolio management reporting, go/no-go decision criteria, and so on. This may be applied in a stepwise fashion once buy-in has been obtained to the more holistic and high-level approach.

Communities of Practice

Communities of practice are important resources, along with trainers, coaches, facilitators, mentors, line managers, and others, in ensuring that project management practice is clearly and appropriately articulated, disseminated, practiced, and kept up-to-date and relevant. *Communities of practice* are networks of interested parties who can contribute to the development and maintenance of project management best practice (Wenger, 1998). Communities can form, and operate, powerfully in organizations when focused around practice areas. In this way, groups develop a shared understanding of meaning and get improved buy-in. The PMSO may need to incentivize membership of the communities, particularly in the first one to two years, before a more steady-state community is established and membership is reliant on enthusiasm and belief.

In particular, they can be valuable in a virtual organization in using resources from the project management community (rather than relying just on a large, central PMSO) to develop guidelines and disseminate knowledge as well as facilitating the sharing of best practice and networking.

Figure 38.7 shows the communities representing a project management topic or methodology and having membership across all of the functions. Where project management is a function in the organization, this model is still applied.

Communities of practice often include people recognized as leaders or experts in one or more aspects of the practice area—for example, risk management or scheduling. These people are commonly referred to as "subject matter experts." They have an important role in articulating these project management practice areas—and in ensuring that practice is kept up-to-date and relevant. They can also

FIGURE 38.7. COMMUNITIES OF PRACTICE.

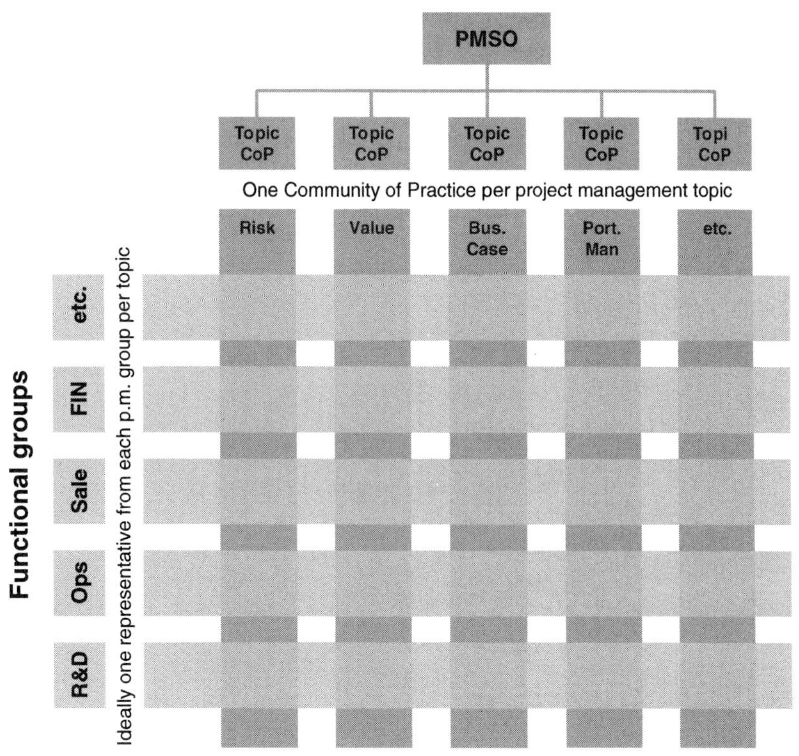

- provide help in answering questions;
- champion the topic; and
- assist in training, facilitation, coaching, or mentoring.

Communities of practice should develop as self-nominating groups as well as formally nominated groups, but as mentioned earlier, they may require incentivization at inception. Senior managers should become involved in the communities. These communities ensure that best practice remains current and promote and develop project management capability within the organization on behalf of the PMSO. The crucial task of the PMSO is to help these communities grow and to work with them so that their work is harnessed around the development of best practice for the benefit of the organization as a whole.

The level of commitment required as an individual within a community for follow-up activities, such as reviewing training material, meeting with colleagues to present practices, and acting as workshop facilitators or coaches to people taking training in the topic areas, varies depending on the following:

- Nature of the topic
- Number of people wanting assistance
- Level of effort that individuals are prepared to put in

The communities should report directly into the PMSO. This ensures that any significant changes to the current best practice can be coordinated across the entire project management community as well as across other communities of practice that may be affected by the change. There should be no limit to the size or membership of the community itself, and cross-functional or cross-business-unit participation can only strengthen the purpose of its very existence.

Training and Building Capability

The PMSO is the central focus for ensuring appropriate project management training. It identifies competencies needed by high-performing project managers, for effective executive awareness and team member participation. The PMSO participates, and typically leads, in helping tailor standardized courses around the methodologies that apply specifically to the organization.

Initially the PMSO must identify the set of capabilities that the organization needs in order to improve the performance of its project management. These capabilities can be expressed in terms of the "individual competencies"—the knowledge, skills, and behaviors needed by the project managers—and "organizational competencies"—the structure, size, geographical location(s), facilities, systems, lines of reporting, and culture in which the projects are undertaken.

The first step in improving individual competencies is to identify the knowledge and skills and behaviors that an organization's project managers should possess. This is usually done by crafting an appropriate body of knowledge framework and set of methodologies.

Once this has been done, the PMSO can then undertake a process of assessing where it actually is—that is, establishing what are the competencies and capabilities of the existing project managers. This is often done through a series of structured interviews or by a questionnaire designed to assess the project management knowledge and skill levels in the organization and also how well these are being applied.

This process allows the PMSO to identify the "gap" between where they are and where they want to be, and to set out a roadmap of how to address the shortfalls. Training courses can then be developed to address specific areas of weakness and for different maturities of project management knowledge. This form of training needs assessment can also be applied to individuals so they can determine their own training needs.

Although the training department within HR is usually the coordinator of corporate training, the PMSO may provide subject matter expertise in project management. The PMSO identifies the appropriate training that is required and may participate in the selection of the trainers. The PMSO also helps identify the required levels of knowledge and competency and the required segments of training that are necessary in order to achieve maximum performance. Thus, the PMSO is the focal point for establishing the means to measure project manager competency. Competency building can be provided in a variety of ways, and the range of offerings should be determined from the questionnaire.

A learning styles assessment can also help in ascertaining how people learn. These offerings include the following:

- *Face-to-face (F2F) classroom training.* This is typically offered in a lecture or workshop style and is designed to encompass as much of the methodology topic areas as possible. This is an ideal environment to "expose" the project management community to a wide range of topics.
- *Blended learning.* This is a combination of e-learning being supported and led by a trainer. This can be applied across different geographical locations through. The trainer controls the progress of the material.
- *e-learning.* This can be designed as pre-F2F learning or to reinforce the classroom training. As a stand-alone training package, e-learning tends to be of greater benefit where knowledge needs to be acquired rather than a skill developed. As pre-reading, it helps make the F2F training more efficient, and as a refresher to the classroom training, it helps greatly in the retention of knowledge.
- *Decision support tools.* These provide an interactive version of online learning to be viewed and used just-in-time, thus allowing practitioners to reference the training on an as-needed basis. They are designed to provide practitioners with possible solutions to real problems.
- *Facilitation (and JIT).* Perhaps the most effective form of competency development because it links the training to a real environment, this combination of applying best practice to a practical situation provides the maximum benefit and training effectiveness; people learn on the job.
- *Coaching or mentoring.* These roles promote best practice and target all practitioners on an individual basis. Strictly, coaching is more one-to-one training; mentoring is more general

advice, often of a career development nature. While they may not be structured training, they provide a means of leveraging the wealth of knowledge that exists within the organization.

The PMSO plays an important role in offering project management training to employees in the organization. The PMSO can carry out this responsibility in several ways. For example, it can work closely with the training department to develop courses that would be offered through the training department. PMSO staff could offer courses themselves, or they could identify and select outside vendors who would develop and deliver the course material. The PMSO, in every case, should play a key role in the design of the training or suite of training courses, and the selection of attendees—in order to maximize the effectiveness of the training and exploit any policy around mandatory training and optional training.

A crucial element in developing training material is how well it fits the organizational context; does it reflect the environment, culture, processes, and industry that the project managers and project team members operate in? Other than at a pretty basic level, the same training course would struggle to engage, and therefore impact, project managers working on an oil facility and those on a new financial product.

As organizations devote more resources and energy to conducting their work on a project basis, the need for project management training grows. Project staff need this training to strengthen their ability to organize and implement their work. Workers from the functional areas (such as accounting, design, marketing, purchasing, and engineering) might also need some measure of project management training to make sure their efforts mesh with the organization's project focus.

The types of training that should be offered can vary:

Project Management Basics. A course on the project management basics is geared toward project management novices or to people who just need an appreciation of projects so they can work with them more effectively. It describes the methodology and its fit with the business, people's roles and responsibilities, and some of the useful tools and techniques typically available to assist in enhancing the management function. It introduces core topics such as time, cost, and people management; the project life cycle; project management players; project politics; control; and evaluation. It would usually run from two to five days. Its principal objective is to develop an understanding of what project management is, what it does, and how it fits into the organization.

Advanced Project Management. Advanced project management may be a single course or a series of courses geared toward developing specific project management competencies, such as scheduling, cost management, or resource allocation skills. Unlike project management basics, students are expected to participate in exercises, case studies, and role-playing. The number of training days associated with an advanced project management curriculum typically ranges from 5 to 15 days.

Preparation for the Certification Examination. Many organizations accept the Project Management Institute's Project Management Body of Knowledge (PMBOK) as a useful initial basis for accreditation on project management standards. However, the APM and other country-based associations provide accreditation programs, and some companies also work with consulting companies and universities to acquire additional certification programs. There are also many public methodologies that have training offerings. The length of training depends on the standard of certificate and the overlap with other forms of training being offered.

Specialist Topics. Beyond studying the project management topics in areas like cost control and scheduling, project workers may benefit from investigating more specialized topics such as risk management, value management, and contracting and procurement (or any of the methodology topics developed by a PMSO). Not every project worker may need take these courses, which are aimed primarily toward those who wish to assume higher levels of project management responsibility. Each of these courses can be handled as a one- or two-day offering or offered as a combination of methodology topics tailored to demand, or developed in specialist cases into longer programs. These specialist topics may be undertaken as facilitated workshops; for example, value engineering may require three to five days of team effort at a key stage gate, or risk management may require three to five days over a period of weeks or months.

General Business Management. To enable project managers and other project staff to develop general business management skills, they should be encouraged to take courses that cover topics such as finance, marketing, information systems, and organizational behavior. The material can be bundled into one or two business basics courses, or it can be offered through separate specialized classes.

The need to sustain a good project management operation with training is obvious. The question that arises is this: What role can the PMSO play in developing and delivering the training material and in coordinating the overall training effort? The real issue is whether the PMSO should provide the lead role, or whether it should support the training activities through the training department. The response in most organizations seems to be that the PMSO should do some of both. On specialized topics, the PMSO often plays the lead role. On more generalist training, it plays a support role.

The PMSO plays a lead role in course development, delivery, and coordination. it must assume leadership in identifying the curriculum, because through competency assessment, as owners of best-practice and gap analyses, the PMSO staff are in the best position to identify the organization's project management training needs.

Organizations with Centralized Training Departments. It makes sense to have these departments play the lead role in developing and delivering project management courses, just as they do with other courses. Because of their centralized location, they know what the overall training program is for the entire organization and can fit the project management curriculum into the organization's broader training portfolio. This may lead to a more cohesive training effort and yield economies of scale in the delivery of training.

In this case, the PMSO gives the training staff the technical information it needs to assemble a good program. For example, if the centralized training department decides to develop its own training material, the PMSO supplies the experts who create course content.

In a sense, they are "contracted" by the training department to develop material. The training department plays the role of client, the PMSO the role of developer.

The PMSO could coordinate the supply of the instructors, since the organization's project management expertise resides there. The PMSO may assign one of its staff to work full-time with the training department throughout that period or at least until the trainers have been trained.

Organizations That Lack Centralized Training Departments. It may make sense here to have the PMSO assume a strong leadership role in developing and delivering project management courses. The PMSO must work closely with the training department in this effort. However, the roles have been reversed: Now the PMSO plays the lead role while the training office plays a support role. Support from the training office might include the following:

- Guidance on good curriculum development practice
- Historical training needs and learning styles data
- Review of course material to determine its "teachability"
- Development of standard formats for producing course material
- Assistance in the production of course handouts
- Provision of teaching technology, such as overhead projectors, LCD projectors, flip charts
- Assistance in the administration of training—organizing venues, delegates, materials, and so on.

Coaches and Mentors

Individuals assigned to look after the training needs of others (i.e., coaches and mentors) can work with the PMSO to serve a just-in-time training function. For example, when design engineers need help in learning how to capture their activities in a work breakdown structure, they can contact a coach who provides them with on-the-spot instruction on building one.

The PMSO should develop a network of coaches through the communities of practice or through interest from individuals who have undergone training. The PMSO should also ensure that mentors are sufficiently briefed about the project management offerings and needs of their staff.

The PMSO plays the lead role in supplying just-in-time training. The training department is generally not equipped to handle this kind of spontaneous response to the temporary training needs of the organization's employees. In some cases the PMSO might provide this type of support, training, facilitation, or ramp-up support in a consultancy capacity, charging for its time.

Knowledge Management

Building a world-class set of methodologies involves taking advantage of the lessons learned by project managers and the project team. An archive of lessons learned is one of the PMSO's key contributions to standardizing methodology across the organization. (See the chapter by Morris on best practice in this area.) This library of information and data is assembled from past projects, for example:

- What worked
- What didn't work
- How it can be reapplied more effectively to benefit other projects

The PMSO may also serve as the quality audit and continuous improvement function for the project management community, since it understands what should be done in terms of methodology and training and can audit against whether or not it is being done—and if it is, whether or not it is showing value and productivity. (See the chapter by Huemann.)

The embedded knowledge and understanding of the coach, mentor or trainer, and facilitator also impacts the effectiveness of the methodologies and training. The ability for both the material and trainer to deliver both knowledge and practical solutions for implementing their practices back in the workplace has a profound impact on the value and effectiveness of training.

Furthermore, while a candidate can develop new skills during training, he or she can be lost if sufficient support is not provided back in the workplace. Learning and knowledge retention are separate activities; often just six months later, competencies are lost, despite being learned during training (Baldwin and Ford, 1988). The ability for candidates to refer back to training and knowledge once in the workplace again is key to retaining knowledge, and to the ongoing contribution of the training intervention toward improving the performance of the organization.

Systems and Tools

One of the key responsibilities of the PMSO is the definition of the tool set required to support portfolio, program and project managers, as well as to support its own activities. The term "tools" not only refers to large integrated systems; spreadsheets, Web sites, templates, and search engines are also a fundamental part of the project manager's tool set. The tools used by today's project managers can be categorized into two groups:

- Project management applications
- Project management resources, guidance and support

Project Management Applications. By project management applications we mean the systems and tools that allow project managers to collect, structure, manipulate, and interrogate information to support them in the day-to-day management of their project. Typically, these tools provide functionality to assist the project manager in the following:

- Estimating and managing cost
- Generating and distributing reports
- Recording the status of documents

- Capturing, structuring, and controlling requirements
- Recording and monitoring risks
- Tracking actions
- Communicating across the project team

For these systems to be truly effective, they need to be deployed in a consistent manner across the organization; data captured in an ad hoc manner is very difficult to structure or analyze retrospectively, and information gathered and categorized using different conventions make comparison and benchmarking difficult to perform in the future.

For this reason, many companies choose to mandate certain elements of a standard tool for use across the organization. When these tools are developed or integrated across an organization, they allow structured information to be accessed in a consistent and meaningful manner at project, program, and portfolio level. Unfortunately, this structuring often limits the flexibility with which the project manager can use the systems.

The key to developing these tool sets is not in the way that the systems interact with one another, but how intuitive the user interface is and how well the structure that the information stored in the system suits the context. For example, an oil company would need to establish different project work breakdown structures for its exploration, refining, distribution, and service station divisions, as well as corporate functions such as IT, R&D, HR, marketing, and finance. The way that project information would be structured, as well as the different terminology used, would make the sensible structuring of information in a consistent manner across the organization almost impossible, although the basics must, and can, be harmonized.

Project Management Resources, Guidance, and Support. The second category of tools covers those that provide support to the project manager in ensuring that the correct processes are applied, that proper advice or knowledge is provided at the right time, and that communication across and between projects (including the development of a virtual project management community) is facilitated.

Typically these tools provide the project manager with the ability to access the following:

- Guidelines and up-to-date good practice
- Lessons learned on previous projects—what went well and what went badly;
- The ability to contact communities of practice and subject matter experts for advice
- Access to worked examples
- Magazines and project features
- Access to the output of studies, benchmarking data, and examples from other sectors and industries

These tools allow the PMSO, as owner of best practice and guidance for project management, to deliver its knowledge and content to the project community. By providing this information in an electronic format, the PMSO can create a virtual presence across the organization without the overhead of having a physical presence across multiple locations.

This can be extremely attractive to geographically dispersed organizations. It also allows the PMSO to create and position itself at the center of a virtual project management community where issues can be discussed, problems shared, news broadcast, and information traded.

While the PMSO might not actually "own" these tools, one of its key responsibilities is to participate in the specification of the tool set functionality, selection of software, and the implementation of tools on projects.

Definition of Systems and Tools

Any organization, whether concerned with the management of projects or not, needs tools to support the exchange of information, to facilitate communication, and to assist with the collection and dissemination of knowledge. The tool set that a company deploys needs to reflect and support its people and processes—tools do not perform a function on their own, they merely support people in applying process (and in some cases force the user to comply with process).

In determining the tool set that should be deployed, the PMSO should first establish what the tools should achieve for the organization. Then the functionality of the tool set can be determined, followed by the level of integration with other systems.

Data Standards and Templates

For meaningful information to be collected and aggregated up through an organization, clear guidance must be given on what each piece of data means and how it should be calculated. The PMSO must establish these standards and ensure that they are understood and implemented vertically and horizontally throughout the organization.

A glossary of data definitions should be produced giving details on the following:

- *Data description.* What is the data and what should it be used for
- *Data definition.* How is the data identified or calculated
- *Data format.* Describe any additional information associated with the data, e.g. currency, date of production, language, and so on.
- *Frequency of update.* How often should the data be updated to reflect current status

Clear processes should also be established for the exchange of data.

Systems and Tools Functionality

Typically a project tool set deployed does some, but not necessarily all, of the following:

Provide Access to Information and Knowledge:

- Support the delivery of learning and training to users in a just-in-time manner.
- Facilitate the capture and disseminate knowledge in structured manner.

- Provide access to subject matter experts and communities of practice.
- Provide a communication platform across the project management community.
- Provide easy access to data and information (both project and functional information).

Support the Basic Program/Project Management Functions:

- Estimating and scheduling.
- Resource management.
- Cost control.
- Risk management.
- Document management.
- Action tracking.
- Communications.
- Simplify administrative tasks, such as time-sheeting, expenses, and so on.

Support the Data Capture and Reporting Processes:

- Support the capture of project information (progress, issues, risks, and so on) in a structured manner.
- Facilitate portfolio, program and project reporting plus the production of ad-hoc reporting as required.

All this is often to be achieved in an environment where the project management community is virtual—that is, geographically dispersed—and where the systems in place are designed to meet functional requirements, not project requirements.

Whether an organization implements a corporate tool set depends largely on the standardization that is mandated. The PMSO's role in defining the tool set must reflect this. The PMSO has five principle responsibilities for tools:

- Assessing the need for a tool
- Determining the functionality (requirements) of the tools, including the need for any integration with each other or with corporate systems such as HR or Finance
- Providing guidance on nonstandard tools, for example, making recommendations on which risk tools to use where none is mandated
- Supporting the effective implementation of the tool
- Providing training

Portals

One of the challenges facing project managers is being able to access the disparate range of information systems and tools that they need to support them in their everyday work. Similarly, the PMSO faces a challenge in delivering the knowledge, policies, and standards to the project community at the point of use, in a consistent manner, and in a way that can be easily accessed and referenced.

To achieve this, companies are starting to develop project "portals." As shown in Figure 38.8, portals sit across the top of disparate systems, aggregating their content to a single point. This means that from a single point, a project manager can have access to:

- Project management applications:
  - Project control
  - Risk management
  - Document management
  - Time-sheeting
  - Etc.
- Project resources, guidance, and support:
  - Knowledge
  - Guidance and best practice
  - e-learning
  - Training material and courses
  - Communities of practice and subject matter experts
  - Etc.
- Other applications:
  - E-mail
  - News
  - Functional knowledge repositories
  - Etc

Furthermore, these systems can be tailored to the individual users, so that each project manager can decide what information he or she sees depending on the project that he or she is working on, geographical location, function, or any other category. This provides a powerful tool for project managers—a single point of access to any information that they need to perform their role. At the same time it provides a vehicle for the PMSO to deliver its content and knowledge directly to project managers at their point of work. A win-win situation.

Measuring Success of the PMSO

An organization must measure the impact that the PMSO has on the performance of projects. During the start-up of the PMSO, a measure of output based on the number of guidelines or methodologies produced and the demand for training based on this output provides the PMSO with a measure of its "impact" on the organization. The PMSO can take this organizational "appetite" as a sign of success. Over time, maybe several years, the training demand can be continually assessed against the same output. The correlation can then be made against the demand for the different "levels" of training, thus providing a measure of the capability or maturity of the organization.

The impact of the PMSO may also culminate in the release of manpower in other areas. The BU PMSOs may have been duplicating efforts and no longer need to sustain

FIGURE 38.8. PORTALS INTEGRATE DISPARATE SYSTEMS AND TOOLS.

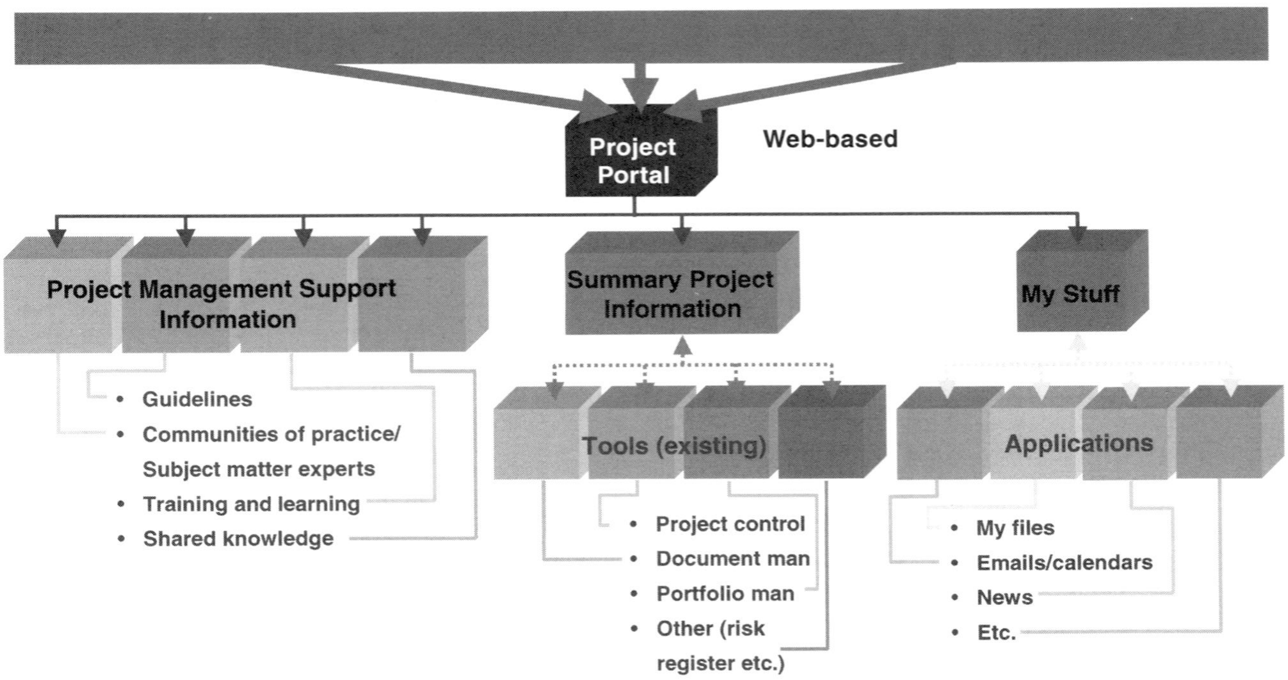

Source: INDECO.

their level of support to their function. It can be very difficult to attribute these changes solely to the PMSO, particularly when other organizational changes are occurring simultaneously.

Perhaps the most difficult measure of success to quantify is the direct impact that the PMSO has on the delivery of projects (i.e., reduction in project delivery timescales, costs, and so on). The PMSO should aim to demonstrate compliance to best practice across the projects and the effects it has on meeting key criteria—that is, resource utilization, meeting milestones, scope, and so on. This can be achieved be carrying out a project health check.

Finally, surveys can provide information about the perceived value added by the PMSO and the range of services that it offers. Caution should be exercised when using surveys to ensure that the data captured accurately reflects the subjective view of those responding to the survey.

Measures and Metrics

One of the ways that the performance of the PMSO can be measured is through the use of measures and metrics. These focus on specific, measurable areas that can be consistently tracked to determine whether the PMSO is meeting its targets and its impact on the organization. Commonly used metrics are as follows:

- *Indicative.* Simple measures used both during the setting up of the PMSO to monitor its productivity and as part of business-as-usual activity. These measures might include the number of guidelines produced, the number of people trained, and the number of requests for PMSO assistance.
- *Resource.* These focus on the amount of resource freed up by establishment of the PMSO by reducing the amount of rework and duplication across the organization in developing guidelines, identifying and assessing tools, and designing organizational structures.
- *Quantitative.* The impact that the PMSO has on the overall portfolio in terms of reduced time to market, reduced cost overruns, and so on. The problem with these metrics is that it is difficult to extract exactly what impact the PMSO has had in isolation of any other initiative within the organization.
- *Feedback.* "Client" satisfaction surveys to determine the perception of the impact of the PMSO. This is a highly subjective measure, and responses tend to be highly anecdotal.
- *Health.* Measuring the overall "health" of the project portfolio to determine whether projects are generally more effectively managed, whether they are hitting targets more frequently, and whether there is greater certainty in outcome.

Whichever measures and metrics are used, they must be baselined, measured, and monitored. Where possible, they should also be benchmarked to determine whether the PMSO is achieving its targets in line with industry practice.

Summary

Today's PMSO can take many forms, working at all levels of an organization, with many different service offerings and charging mechanisms. What remains true across all PMSOs is that their ultimate function is to improve the performance of projects, programs, and portfolios.

The PMSO can assume the role of mentor, auditor, supporter, facilitator, and trainer to the project management community, while its responsibilities cover the development of organizational project management competency; the hands-on support of portfolios, programs, and projects; as well as supporting the provision of the systems and tools required by project managers to effectively carry out their responsibilities.

The implementation of a PMSO can be organizationally challenging—cultural, structural, and geographical issues must all be addressed, as must the perennial barriers to change of individual inertia and organizational latency. But the benefits that a PMSO brings can be significant: Ensuring that organizations not only do projects right but also do the right projects can have a dramatic impact on overall business performance.

Whichever approach an organization takes to their implementation and ongoing management, the key to their successful establishment is to align their mandate with the strategic needs of the organization to get strong organizational support, to focus on areas of weakness and to identify itself as the natural "home" for projects and project people.

References

Baldwin, T., and J. Ford. 1988. Transfer of training: A review and directions for future research. *Personnel Psychology* 41:63–105.

Bernstein, S. 2000. Project offices in practice, *Project Management Journal* 31(4):4.

Block, T. and J. Davidson Frame. 1998. *The project office.* Menlo Park, CA: Crisp Publications Inc.

Davidson Frame J. and Block, Thomas R., 1998. *The project office* Menlo Park, CA: Crisp Publications Inc.

Franz, K-F. 2002. Crystal: Novartis' framework for IT related projects. (PMI Europe Conference, Cannes).

Dai, C. X. 2001. The role of the project management office in achieving project success. Doctoral dissertation. Washington, D.C.: George Washington University.

Dinsmore, P. C. 1999. *Winning in Business with enterprise project management.* New York: AMACOM.

Lave, J., and E. Wenger. 1991. *Situated learning.* Cambridge, UK: Cambridge University Press.

Levin, G., and P. F. Rad. 2002. *The advanced project management office.* Boca Raton, FL: CRC Press.

Ibbs, C. W., and Y. H. Kwak. 2000. Assessing project management maturity. *Project Management Journal* 31(1):32–43.

Ibbs, C. W., and J. Reginato. 2002. *Quantifying the value of project management.* Newtown Square, PA: Project Management Institute.

Kent Crawford, J., 2002. *The strategic project office.* New York: Marcel Dekker, Inc.

Kent Crawford, J., and J. Pennypacker. 2002. Put an end to project mismanagement. *Optimise* 12.

Kwak, Y.H., and C. W. Ibbs. 2000. Calculating project management's return on investment. *Project Management Journal* 31(2):38–47.

Kwak, Y. H. and C. W. Ibbs. 2002. Project management process maturity model. *Journal of Management in Engineering* 18(3):150–155.

Morris. P. W. G., and J. D. Young. 2001. Building long-term project management competencies by aligning personal competencies with organizational requirements (PMI Europe Conference, Berlin).

Morris, P. W. G., H. Khatau, J. Young, I. Keates, and J. A. Wright. 2000. Benchmarking best practice in the project support office. IPMA World Congress for Project Management, London.

Phillips J. J., T. W. Bothell, and G. Lynne Snead. 2002. *The project management scorecard: Measuring the success of project management solutions.* Oxford, UK: Butterworth-Heinemann.

Rad, P., and R. Asok. 2000. *Establishing an organizational project office.* pp. 13.1–13.9. Morgantown, WVA: AACE International Transactions, Morgantown,

Wells, Jr., W. G., and C. X. Dai. 2001. Project management offices: Organizational use is on the rise. *ESI Horizons.* (July).

Wenger, E., 1998. *Communities of practice.* Cambridge, UK: Cambridge University Press.

Young, J. D., and P. W. G. Morris. 2002. How companies are using the Internet to support their project management functions. (PMI Europe Conference, Cannes).

SECTION II.5

COMPETENCE DEVELOPMENT

INTRODUCTION

At bottom, project management represents a "people" challenge; that is, the ability to successfully manage a project from inception through delivery is predicated on our ability to appreciate and effectively employ the competencies of all those who are associated with the project development and delivery process. This section builds off Chapters 3 and 4 to cover a range of organizational and people-based topics that are occupying the project management world today. Foremost amongst these are issues to do with knowledge, learning, and maturity. But first is a discussion of teams, leadership, power and negotiation, human resources, and competencies.

Connie DeLisle, in Chapter 39, brings the reader up-to-date with her chapter on contemporary views on shaping, developing, and managing teams. Connie begins in a characteristically arresting manner: "We have knowledge and wisdom to change, but why do we not do so, or even act in ways contradictory to successful team building and management?" She looks at team working in three sections: first, the forces shaping teams (crisis management, globalization, donated time, organizational anorexia, senior exec priorities); then team development (team characteristics, key responsibilities, knowledge competencies, personalities, resource negotiation, development itself, nature/nuture, and virtual working); and finally team management, where her focus is particularly on addressing team pathologies.

Peg Thoms and John Kerwin look in Chapter 40 at leadership in projects. They begin by reviewing some of the key concepts, particularly around the so-called charismatic and transformational schools of leadership theory. They then focus on the importance of vision creation as a key activity in leadership—not just having one but being able to communicate it (imaging). Visioning goes in part with personality but also in part reflects one's future-looking orientation. Techniques are presented for helping create a vision, and time orientation—past, present, and future—is discussed in terms of project leadership. Finally, these ideas are illustrated by two cases: a highway project and a film production. Peg and John end by summarizing: "we can learn a great deal from leadership theories, but even more by observing effective leaders. Paying attention to how problems are solved, how innovative strategies are developed, and how great project leaders communicate and motivate are the best ways to improve our leadership ability."

Few project managers have much formal power. Much is informal and has to be negotiated. Influencing skills are very important. In Chapter 41, Jeff Pinto and John Magenau review the different sources of power and forms of influence. They then look at developing influencing and negotiating skills and how to prepare for, and conduct, negotiations.

Martina Huemann, Rodney Turner, and Anne Keegan broaden the discussion in Chapter 42, on human resource management challenges in project-based organizations. Beginning with a review of the specific challenges posed by projects—essentially, the lack of certainty over future work and career development—they move on to a central discussion of the importance of competences in project management, something that will now occupy much of the rest of this section.

In Chapter 43, Andrew Gale looks in depth at organizational and personal competencies. He goes over our definitions of competencies, noting the basic normative idea that they are what are required to fulfill a specific role and concluding that "competence is concerned with the capacity to undertake specific types of action and can be considered as an holistic concept involving the integration of attitudes, skills, knowledge, performance, and quality of application." Becoming, and even remaining, competent clearly involves learning, and learning thus is important in any discussion of competency (hence, Chapters 44 and 45). Andrew then looks at the idea of project management (professional) competency frameworks, which he sees as predominantly normative, reductionist, deterministic, and restricting. He looks at alternative models (for example, actor-network theory) and at studies on organizational competencies. He concludes by looking at experience in trying to measure competency: No one has yet found it possible to generate hard measures.

Knowledge management (KM) and organizational learning (OL) became subjects of considerable interest from the late 1990s onward. In Chapter 44 Christophe Bredillet offers a stimulating *tour d'horizon* of the field applied to the management of projects. All the key ideas in both fields, KM and OL, are presented before the author goes on to look at projects, which, in his view, are specially good places for learning to happen: "learning at the edge of organization."

In Chapter 45 Peter Morris poses the question, how do we know what's what in project management—how valid are the rules, practices, insights, and other knowledge that we may wish to pass on to, or even impose on, others? First, we have to define our frame of reference; this tells us at least what the ballpark looks like. Undoubtedly, at some level there are generalizable insights. The question is this: At what point do they become inapplicable or unuseful? Much of the most valuable project management knowledge is process-based rather than substantive in a context-specific way. Risk management is a good example: The process rules are relatively straightforward and the substantive judgments often quite difficult. Best-practice rules on project-based learning suggest that while it ought to be possible to support real role-specific learning (at a price), general competency standards and accreditation is more questionable.

Lynn Crawford, in Chapter 46's comprehensive review of global project management knowledge and standards, probably agrees. Lynn surveys with great thoroughness all the international project management professional standards, as well as related national and international standards. She addresses first what she calls project-based standards—essentially the four major bodies of knowledge (PMI, APM, IPMA, PMCC)—then people-based standards (the Australian, UK, and South African competency standards), and finally organizational standards—basically the maturity models and methodologies (OPM3, PRINCE2, etc.). Lynn then goes on to compare the contents of the different bodies of knowledge before revisiting the issue of competency assessment (Chapter 43), noting, quite rightly, the lack of much evidence on the attitudes and behaviors that several project management competency assessing bodies understandably rate so importantly, and at project manager qualifications. Touching briefly on international initiatives to form a global perspective on a project management body of knowledge and standards, Lynn concludes on a cautionary note, pointing to the difficulties many feel in (a) translating "hard" engineering-based project management to "soft" organizational projects and (b) the differences between program and project management. Also, "the process for standards development, which is largely a process of making explicit and codifying through consensus the tacit knowledge of experienced practitioners, ensures that standards will remain conservative and will lag behind the cutting edge of both research and practice".

In Chapter 47 David Frame brings us back to the world of real projects with his discussion of project evaluation (pre-, mid-, and post-project) and in doing so touches on many of the issues addressed previously. David covers the basic principles of evaluation and what's needed to convert learning into action. He deals in particular with structured walk-throughs, EISA, and customer acceptance tests. He concludes by tying all this back to the learning organization and HR management.

Bill Ibbs, Justin Reginato, and Young Hoon Kwak in Chapter 48 extend the discussion of how the organization can get value out of project management, by discussing the program of research they have been doing for several years on the return on investment of project management. Their method is heavily based around the concept of organizational maturity: Data on project management practices is benchmarked against other organizations' and the different levels of maturity assessed against the organizations' abilities to achieve projects "on time and to cost."

Organizations are indeed now looking closely at how they can leverage value from project management, and to many, the maturity concept is an appealing way to tackle this challenge. Terry Cooke-Davies in Chapter 49 carefully explains the difficulties with the concept, however. He traces the origins of maturity work from the quality movement through software (SEI's CMM) into project management. He reviews the OPM3 and PMMM models before moving on to "untangle the vocabulary and distinguish the relevant concepts." As in several of the previous chapters, he notes the importance of what is being measured and the challenge of distinguishing between the maturity of processes and practices (in various different types of projects the organization might undertake) and its overall ability to perform. Project management maturity models make a lot of assumptions here. Like Lynn Crawford, Andy Gale, and Peter Morris, Terry is skeptical that uniform approaches of assessment will work for all project situations. In short, the approach needs applying with care, tailored to the organization's specific circumstances.

About the Authors

Connie Delisle

Connie Delisle completed a PhD in the Department of Civil Engineering in the Project Management Specialization, a MSc in resource management at the University of Calgary, and two bachelors' degrees from the University of Victoria. The study of the psychology of human behavior and teamwork has been an integral part of her educational studies for the past 15 years. Her recent practical work experience is in the area of managing sustainable development and results-based delivery of public services' projects within the context of federal government policy, programs, and operations. Earlier roles include course design and delivery of on-line learning to executive MBAs working in virtual teams, and working as an environmental advisor and project team leader for local government in British Columbia and for senior oil and gas companies in Alberta. Experience as a Canadian national team rower and triathlete provide Connie with a solid understanding of the principles and practical aspects of creating winning team experiences. She has published over 35 articles in the area of success, teams, and communication, as well as a recent joint publication with Dr. Janice Thomas and Dr. Kam Jugdev that represents a three-year international research effort in the area of selling project management to senior executives.

Connie recently moved to Ottawa, Canada and is currently employed as a Principal Consultant with Consulting and Audit Canada, a cost-recovery-based branch of Public Works and Government Services Canada.

Peg Thoms

Peg Thoms earned her PhD in Organizational Behavior from the Fisher School of Business at Ohio State University. She is currently an Associate Professor of Management and the Director of the MBA Program at Penn State, Erie, The Behrend College. She also recently served as a visiting professor of project management at Umeå University in Umeå, Sweden. In addition, Dr. Thoms has 16 years of business management experience. She conducts research and has published in the areas of leadership vision, time orientation, leadership development, and self-managed work teams. She has published articles in a number of journals, including *Human Resource Development Quarterly, Journal of Organizational Behavior, and Journal of Management Inquiry*, as well as a chapter on the motivation of project teams in the *Project Management Handbook* (edited by Jeffrey Pinto). She coauthored a book entitled *Project Leadership from Theory to Practice.* Her new book, *Driven by Time: A Guide to Time Orientation and Leadership*, is published by Greenwood/Praeger (2003). She has consulted with various organizations in the manufacturing, healthcare, insurance, and banking industries. She has also worked with many not-for-profit groups. Dr. Thoms has won numerous teaching and research awards, including the Walter F. Ulmer, Jr. Applied Research Award from the Center for Creative Leadership for her work on leadership vision and *Project Management Journal's* Paper of the Year for her article "Project Leadership: A Question of Timing." She teaches management, human resources, and leadership.

John Kerwin

John Kerwin received his master's degree from the Edward R. Morrow School of Communications at Washington State University. Previously, he was producer, director, and writer at his production company in Los Angeles, California, Kerwin Communications, Inc. In that capacity, his company produced programs and commercials for NBC, CBS, ABC, and ESPN, and he had his own show on ESPN for four years. Kerwin started in the production business with NBC and worked on Super Bowls, World Series, Wimbledon tennis, the NBC Nightly Newspolitical campaign coverage, elections, and prime time news specials, one of which was honored with an Emmy. His company also worked on movies and theatrical productions. Kerwin's area of research is the application of visual production technology to creative and practical concepts in communications. He is currently assistant professor of communications at Penn State University and has produced and published several works for the marketing and public awareness of services and resources in the educational and corporate sectors. His latest work will be "The First Year Experience," a 30-minute production that will be distributed internationally on the subject of the freshman experience in higher education.

Jeff Pinto

Dr. Jeffrey K. Pinto is the Samuel A. and Elizabeth B. Breene Professor of Management in the Sam and Irene Black School of Business at Penn State, Erie. His major research focus has been in the areas of project management, the implementation of new technologies, and the diffusion of innovations in organizations. Professor Pinto is the author or editor of 17 books and over 120 scientific papers that have appeared in a variety of academic and practitioner journals, books, conference proceedings, and technical reports. Dr. Pinto's work has been translated into French, Dutch, German, Finnish, Russian, and Spanish, among other languages. He is also a frequent presenter at national and international conferences and has served as keynote speaker and as a member of organizing committees for a number of international conferences. Dr. Pinto served as Editor of the *Project Management Journal* from 1990 to 1996 and is a two-time recipient of the Project Management Institute's Distinguished Contribution Award. He has consulted widely with a number of firms, both domestic and international, on a variety of topics, including project management, new product development, information system implementation, organization development, leadership, and conflict resolution. A recent book, *Building Customer-Based Project Organizations*, was published in 2001 by Wiley. He is also the co-developer of SimProject, a project management simulation for classroom instruction.

John Magenau

John Magenau holds a PhD from the State University of New York-Buffalo (1981). His research and teaching interests are in the areas of labor-management relations and negotiation. He is director of the Sam and Irene Black School of Business and serves as a trustee on the board of Lake Erie College of Osteopathic Medicine, as president of the board of the Enterprise Development Center of Erie County, and as secretary of board of directors for the Center for eBusiness and Advanced Information Technology.

Martina Huemann

Dr. Martina Huemann holds a doctorate in project management from the Vienna University of Economics and Business Administration. She also studied business administration and economics at the University Lund, Sweden, and the Economic University Prague, Czech Republic. Currently she is assistant professor in the Project Management Group of the Vienna University of Economics and Business Administration. There she teaches project management to graduate and postgraduate students. In research she focuses on individual and organizational competencies in project-oriented organizations and project-oriented societies. She is visiting fellow of the University of Technology, Sydney. Martina has project management experience in organizational development, research, and marketing projects. She is a certified project manager in accordance with the International Project Management Association (IPMA) certification program. Martina organizes the annual PM Days research conference and the annual PM Days student paper award to promote project management research. She contributed to the development of the pm baseline—the Austrian project management body of knowledge—and is board member of Project Management Austria. Further, she is assessor of the IPMA Award and trainer of the IPMA advanced courses. Martina is trainer and consultant of ROLAND GAREIS CONSULTING. She has experience with project-oriented organizations of different industries and the public sector. Martina is specialized in management audits and reviews of projects and programs, and human resource management issues like project management assessment centers for project and program managers.

Rodney Turner

Rodney Turner is Professor of Project Management at The Lille Graduate School of Management and at Erasmus University, Rotterdam, in the Faculty of Economics. He is also an Adjunct Professor at the University of Technology Sydney, and Visiting Professor at Henley Management College. He studied engineering at Auckland University and did his doctorate at Oxford University, where he was also for two years a post-doctoral research fellow. He worked for six years for ICI as a mechanical engineer and project manager on the design, construction, and maintenance of heavy process plant, and for three years with Coopers and Lybrand as a management consultant. He joined Henley in 1989, Erasmus in 1997 and Lille in 2004. Rodney Turner is the author or editor of seven books, including *The Handbook of Project-Based Management*, the best-selling book published by McGraw-Hill, and the *Gower Handbook of Project Management*. He is editor of *The International Journal of Project Management* and has written articles for journals, conferences, and magazines. He lectures on and teaches project management worldwide. From 1999 and 2000 he was President of the International Project Management Association, and Chairman for 2001 to 2002. He has also helped establish the Benelux Region of the European Construction Institute as foundation Operations Director. In addition, he is a Fellow of the Institution of Mechanical Engineers and the Association for Project Management.

Anne Keegan

Anne Keegan is a University Lecturer in the Department of Marketing and Organisation, Rotterdam School of Economics, Erasmus University, Rotterdam. She delivers courses in Human Resource Management, Organisation Theory, and Behavioural Science in undergraduate, postgraduate, and executive-level courses. She has been a member of ERIM (Erasmus Research Institute for Management) since 2002. In addition, she undertakes research into the project-based organization and is a partner in a European-wide study into the versatile project-based organization. Her other research interests include human resource management in knowledge-intensive firms, new forms of organizing, and critical management theory. Dr. Keegan has published in *Long Range Planning* and *Management Learning* and is a reviewer for journals such as the *Journal of Management Studies* and the *International Journal of Project Management*. She is a member of the American Academy of Management, the European Group for Organisation Studies (EGOS), and the Dutch HRM Network. Dr. Keegan studied management and business at the Department of Business Studies, Trinity College, Dublin, and did her doctorate there on the topic of management practices in knowledge-intensive firms. Following three years postdoctoral research, she now works as a university lecturer and researcher. Dr. Keegan has also worked as a consultant in the areas of human resource management and organizational change to firms in the computer, food, export, and voluntary sectors in Ireland and the Netherlands.

Andrew W. Gale

Andrew Gale is Senior Lecturer in Project Management and Program Director for the MSc Project Management Professional Development Program in the Manchester Centre for Civil & Construction Engineering UMIST. He is a Chartered Civil Engineer and Chartered Builder specializing in construction project management. He teaches project management with emphasis on people and culture, equality and diversity, and group and team process. He is actively involved in joint collaborative teaching of art and civil engineering students. He is leading the introduction of project management curriculum in the new British University in Dubai. He has managed many research and consultancy grants since 1990 and published over 90 papers and articles. He has extensive experience in working with Russian construction firms and academic institutions and led development training programs, distance learning, and curriculum development in Russia, funded by UK and EU government agencies. He has over 18 years' experience in research on construction organization and project culture, with specific interests in diversity, equality, and inclusion. Currently he is developing a collaborative research program (with the Art and Design Faculty at the Manchester Metropolitan University) investigating how civil engineers and artists can learn from each other in the context of project management. He is an active member of the European subgroup of the ICE International Policy Committee with special responsibility for Russia.

Christophe Bredillet

Professor Christophe N. Bredillet, PgD in Project Management, MBA, DrSc, MSc Eng EC Lille, Certificated Program Director IPMA Level A, CMP, CCE, has 18 years of experience in project and program management with several industries (banking, sporting goods, and IT). For the past ten years, he has been the program director of the MS, MBA and Doctorate in Project & Program Management at ISGI-Groupe ESC, Lille. He was appointed Professor of Project Management at UTS (University of Technology, Sydney) in 2001. The field of his research in project management is the design of learning and knowledge management systems, the development of standards, and the epistemology and evolution of the field. He is strongly involved in development of project management professional associations worldwide. He is Editor of the *Project Management Journal*. He was till recently PMI Global Accreditation Committee Member and has been for the past three years Research MAG, and in the co-lead team member of two standards development projects for PMI (Project Managers Competences Framework and Organizational Project Management Maturity Model). He is a steering committee member of the Global Project Management Forum and of the international nonaligned think tank group named OLCI. He is Vice-President International Affairs, a member of the Board of Directors, and Regional Manager, North of France, for AFITEP (the French professional project management Association). He is founder and President of the PMI chapter Hauts-de-France.

Peter Morris

Peter Morris is Professor of Construction and Project Management at University College, London, Visiting Professor of Engineering Project Management at UMIST, and Director of the UCL/UMIST-based Centre for Research in the Management of Projects. He is also Executive Director of INDECO Ltd, an international projects-oriented management consultancy. He is a past Chairman and Vice President of the UK Association for Project Management and Deputy Chairman of the International Project Management Association. His research has focused significantly around knowledge management and organizational learning in projects, and in design management. Dr. Morris consults with many major companies on developing enterprise-wide project management competency. Prior to joining INDECO, he was a Main Board Director of Bovis Limited, the holding company of the Bovis Construction Group. Between 1984 and 1989 he was a Research Fellow at the University of Oxford and Executive Director of the Major Projects Association. Prior to his work at Oxford, he was with Arthur D. Little in Cambridge, Massachusetts, and previously with Booz Allen Hamilton in New York and with Sir Robert McAlpine in London. He has written approximately 100 papers on project management, as well as the books *The Anatomy of Major Projects* (Wiley, 1988), *The Management of Projects* (Thomas Telford, 1997) and *Translating Corporate Strategy into Project Strategy* (PMI, 2004). He is a Fellow of the Association for Project Management, Institution of Civil Engineers, and Chartered Institute of Building and has a PhD, MSc and BSc, all from UMIST.

Lynn Crawford

With a diverse background as architect, project manager, regional planner, and policy adviser, and with qualifications in human resource management and business administration, Dr. Lynn Crawford has experience both as a project manager and an adviser to project-based organizations on human resources, strategic and business planning, and development issues. Ongoing research and practice includes working with leading organizations that are developing their organizational project management competence by sharing and developing knowledge and best practices as members of a global system of project management knowledge networks. A particular area of research and expertise is the assessment and development of individual and corporate project and program management capability. Lynn was until recently Director of the postgraduate Project Management Program at the University of Technology, Sydney. She has been Project Director for several major research projects funded by the Australian Research Council in areas of project management competence and the management of multiple, interdependent, and soft projects and continues to conduct research in these areas, working with industry partners. Lynn was a member of the steering committee for the development of Australian National Competency Standards in Project Management and is currently leading initiatives aimed at development of global standards for project management. These involve all major project management professional associations, recognized leaders in project management, and representatives of global corporations.

J. Davidson Frame

Dr. Frame is Dean at the University of Management and Technology. Prior to joining UMT in 1998, he was Professor of Management Science at the George Washington University, where he served on the faculty from 1979 until 1998. At GWU, he was chairman of the Management Science Department, Director of the Program on Science, Technology, and Innovation, and founder of the project management master's degree program. Between 1973 and 1979, Dr. Frame was vice president at Computer Horizons, Inc. While at CHI, he headed the Washington office and directed some 30 software development and research projects.

David Frame has written eight books, including the business best seller, *Managing Projects in Organizations* (3rd ed., 2003). His most recent book is *Managing Risk in Organizations*, published in July 2003. He has also written some 40 scholarly articles in the area of technology management, the management of intellectual property, and project management. Dr. Frame is active in the project management professional arena. He served as Director of Certification at the Project Management Institute from 1990 to 1996. He then served as PMI's Director of Educational Services in 1997 to 1998. He was elected to PMI's international Board of Directors and served in that capacity from January 2000 until December 2002. He was awarded PMI's Distinguished Contribution Award in 1993 and its Person of the Year Award in 1995. He is on the editorial board of *The International Journal of Project Management.*

William Ibbs

William Ibbs is the founding principal of The Ibbs Consulting Group. Dr. Ibbs is also Professor of Project Management at the University of California at Berkeley. He has authored more than 150 scholarly papers and received various awards, including PMI's Presidential Citation and the National Science Foundation's Presidential Investigator Award. Active in the Project Management Institute, he has served in a number of positions including Research Director. PMI sponsored his research, which led to the Berkeley Project Management Process Model and new ways to measure project management's return on investment (the PM/ROI concept). Two of his books resulted from that work and were published by PMI. Current research is investigating role of project management for managing innovation. This includes distinguishing characteristics and management styles of different project categories: derivative, platform, and breakthrough. Dr. Ibbs holds BS and MS degrees from Carnegie Mellon University and a PhD from the University of California at Berkeley, all in civil engineering.

Justin Reginato

Justin Reginato is currently a candidate for a PhD degree in Engineering and Project Management, emphasizing in the Management of Technology, at the University of California, Berkeley. The focus of his research is determining project management's role in the development of innovative and research-intense projects and its impact on the market value of corporations. Additionally, Mr. Reginato is an adjunct professor of Engineering at Santa Clara University, where he teaches courses on project management. Justin has over five years of project management experience with URS Corporation, as well as consulting experience with several high-technology companies. He is coauthor, with C. William Ibbs, of *Quantifying the Value of Project Management*, published in 2002 by PMI.

Young Hoon Kwak

Dr. Young Hoon Kwak is a faculty member of the project management program at the management science department at George Washington University, Washington, D.C. He received his BS (1991) in civil engineering from Yonsei University in Seoul, Korea, and his MS (1992) and PhD (1997) in engineering and project management from the University of California, Berkeley. Before joining GWU, he taught at Florida International University in Miami and was a postdoctoral scholar at the Massachusetts Institute of Technology. Dr. Kwak published and presented numerous papers on engineering in project management, construction management, and technology management at peer-reviewed journals and conferences. He is serving as a member of the Editorial Review Board for *Project Management Journal*. Dr. Kwak has also consulted for U.S. Naval Facilities Engineering Command, DAEWOO Engineering and Construction, Construction and Economy Research Institute of Korea (CERIK), and other Fortune 500 companies. He was the coprincipal investigator of Project Management Institute's nationwide research Benefits of Project Management: Financial and Organizational Rewards to Corporations. Dr. Kwak's major interests include project management and control, project risk management, construction management, technology management, and international project management.

Terry Cooke-Davies

Terry Cooke-Davies has been a practitioner of both general and project management continuously since the end of the 1960s. He is the Managing Director of Human Systems Limited, which he founded in 1985 to provide services to organizations in support of their innovation projects and ventures. Through the family of project management knowledge networks created and supported by Human Systems, he is in close touch with the best project management practices of more than 70 leading organizations globally. The methods developed in support of the networks are soundly based in theory, as well as having practical application to members, and this was recognized by the award of a PhD to Terry by Leeds Metropolitan University in 2000 for a thesis entitled, "Towards Improved Project Management Practice: Uncovering the Evidence for Effective Practices through Empirical Research." He is now an Adjunct Professor of Project Management at the University of Technology, Sydney and an Honorary Research Fellow at University College, London. Terry is a regular speaker at international project management conferences in Europe, North America, Australia, and Asia and has published more than 30 book chapters, journal and magazine articles, and research papers. He has a bachelor's degree in Theology, and qualifications in electrical engineering, management accounting and counseling in addition to his doctor's degree in Project Management.

CHAPTER THIRTY-NINE

CONTEMPORARY VIEWS ON SHAPING, DEVELOPING, AND MANAGING TEAMS

Connie L. Delisle

What defines a project team? What has really changed in nearly 20 years in team research and practice? How have Western scientific views influenced our understanding of team development and management? We have knowledge and wisdom to change, but why do we not do so, or even act in ways contradictory to successful team building and management? Why push technical solutions in a global business "game" where all the players more or less understand the rules? To help address these really quite deep questions, this chapter begins to challenge the very assumptions about behaviors and attitudes that dominate our thinking about teams.

The chapter employs a deductive approach in examining teams and team building in order to take the broader business context into consideration. Part I takes a high-level look at the business context that impacts teams in a variety of potential ways. Part II concentrates on the more micro-level aspects of team development to consider when growing teams in ways that can be creative as well as efficient. Specifically:

- It covers the identification of human resources and the understanding of competencies, and personalities, as primary considerations in selecting team members. Once a team is selected, it needs to "grow" or be developed;
- It provides a brief discussion of evolutionary models of team development, as well as current thinking about how teams perform over time. In doing so, it challenges some long-held beliefs about how team building interacts with core environmental (external) and behavioral factors that influence how teams are developed. Although many environmental factors exist, team composition, demographics, and how teams view their progress in time seem to have a significant influence on team development. From a behavioral

point of view, mental toughness, cohesion, and motivation appear as key forces that shape team development;

- It introduces virtual project teams as a context that reflects how team development has shifted since its inception, adding to our understanding of how teams develop.

Part III takes a brief look at team management, specifically at the most common team pathologies. Suggested strategies to mitigate or avoid pathologies are also provided for consideration.

Part I: Teams and Forces Shaping Teams

Team by Definition or Characteristics

The acronym TEAM means, to some people, "Together Everyone Achieves More," but what does "team" really mean? There seems to be an ever-increasing number of conflicting definitions where it once appeared that this simple term was commonly understood. Team members, according to the PMBOK Guide (PMI, 2000), are those who report to the project manager or leader; nothing is mentioned about who does the work. In contrast, the Association of Project Management (UK) (Dixon, 2000) presents a brief discussion on teams in Section 7, focusing on describing characteristics of effectiveness rather than being prescriptive on what a team does.

What about the term "project"? According to the PMBOK Guide, a *project* is a temporary endeavor whose purpose is in producing a unique product or services (PMI, 2000). Consider process teams (e.g., manufacturing), which are supposed to act in predictable, repetitive, standardized ways with project teams (e.g., on a software project), which will be working within a defined start and end period, where the work is unique or difficult to standardize and that may take a long time to conclude. Rather than focusing on the correctness of the definition, effort may be better spent in finding a set of core characteristics that reflect what is happening in the real world.

Trends Impacting Teams

The division of labor creates and is in turn impacted by trends concerning the selection, development, and management of project teams. Five of the most palpable trends are briefly introduced and are then followed by a few practical responses to consider. These trends do not fit into categories labeled "drivers and barriers" (Wilemon and Baker, 1988; Miller, 1988), because changes in constraints over time will determine in part whether a driver becomes a barrier or vice versa. For example, crisis management may be considered a barrier to organization learning or a driver because it may trigger a recognition concerning the need to change.

Crisis Management. Crisis management seems to be the norm rather than the exception. The self-perpetuating problem stems from organizations losing their ability to plan and prioritize because they practice crisis management to handle dramatically increased workloads, which in turn further increases workload (Duxbury and Higgens, 2001). Thomas, Delisle, and Jugdev (2002) identify that senior executives often react to the time pressure when in organizational crises by maintaining the status quo (deny the severity of the crisis) or purchase a solution from outside their organization. To maintain the status quo, they may appoint internal project members who do not have the qualifications or expertise to fix the problems, resulting in the creation of "accidental project managers" who often make the situation worse. When the crunch is on, people either "hide out" or engage in dysfunctional behaviors marked by withdrawing or banding together in nonproductive cliques or preparing themselves for exiting the organization (DeMarco and Lister, 1987). This limits communication and the flow of information even further.

The management level response is to do the following:

- Treat organizational crises as a symptom of a more serious and systemic internal problems.
- Avoid buying into fear campaigns and reflexively reacting to crises. Focus on creating adaptive triggers or situations where employees can provide input, ideas and ask deep questions about the organization's values and beliefs and the way it conducts business.
- Notify employees of changes early, be honest and communicate often (Blount and Janicik, 2001).

Globalization and the Highly Skilled Mobile Workforce. The blurring of lines between temporal, geographical, and organizational boundaries impacts teams in many organizations. This has not been well reflected in the literature on developing and managing teams over the past two decades. Different fads, tips, and strategies tend to recycle the same principles in more attractive packages. Doing the right things right, at the right time, continues to be the fundamental challenge regardless of the traditional or virtual nature of a project team. Like any resource, accelerated rates of change increase the complexity of addressing the challenge. Grazier (2002) identifies three major trends in the global workforce. First, high-intensity collaboration that cuts through functional power levels in an organization and its networks. Second, shifts in leadership competencies from technical to human skills (this might also be interpreted as requirements for technical *and* human skills). And third, enhanced/expanded role for frontline workers so that they operate more like independent project contractors within a team context moving from team to team on the basis of need rather than job security. (See also the chapter by Huemann, Turner, and Keegan.)

Workers have responded to these pressures by becoming multiskilled and mobile. Upgrading education entails time off without pay to engage in full-time, part-time, or evening study, or simultaneous work and study through (often virtual) post-secondary institutions. Workers focus on developing and maintaining current referral networks to allow them to quickly move on to other opportunities, refocusing and fully exploiting their skills, experience, and education at the first hint of crises related to merger or market downturns. The most pronounced change related to organizational expectations of teamwork relates to the need for a vigorous justification of the cost of training based on return on investment (ROI) and organizational impact (Brown, 2002).

The management level response here is to do the following:

- Recognize that individuals being trained often have more depth of knowledge and/or experience than the trainer or senior levels of management. Capitalize on their expertise (strategic minds) rather than seeing them as bodies charged with carrying out the work.
- Co-manage team members' careers by assisting then in acquiring the right skills and current education, and move them to positions that test their mettle.
- Fully account for the cost of training and return on investment to the organization for each team member. Take this into account in hiring and firing decisions.

Donated Time. The phenomenon of donating work time does not appear to be novel, but it is finally validated in research. Duxbury and Higgens (2001) find that over half of study respondents report working 3.5 to 5 unpaid overtime hours per month, equating to between 40 to 60 unpaid days of overtime per year. The bad news—as with Newton's second law of thermodynamics—is that for every action there is an equal and opposite reaction. The flip side of overtime is undertime. Undertime represents the compensatory days involved in workers catching up on their lives. DeMarco and Lister (1987) give a general rule of thumb that of a 1:1 ratio for over and undertime. Duxbury and Higgins (2001) conclude that the link between hours of work, role overload, work-life conflict, and health problems makes this practice nonsustainable over the long run (see Case Study 1).

Case Study 1

> The project team of experienced information technology (IT) professionals raced to meet the timeline for completion of a highly complex product in response to a major scope change nearly 3/4 of the way into the first test cycle. The team worked between 70 and 85 hours a week for the last 3 weeks of the project. They did not "count" the hours worked out of the office at home because being on "salary" implicitly meant "getting the job done" and being paid overtime was not a company policy. After the project was delivered, some team members took time off for various reasons including vacation and health. Other members sat in their offices and played "catch up," and others justified personal appointments as something that the company owed them.

The management level response here is to do the following:

- Investigate and make explicit the quantitative link between days off from work, health claims, and so on.
- Assess levels of job satisfaction, and open the door to discussion about unpaid overtime. Be honest about its impact on the organization and don't expect employees to suck it up!
- Aim for doable, not perfect, solutions by considering resource constraints and needs, not "wants" of clients.
- Triple time estimates for complex projects (rule of thumb), make sure the resources are in place, or be prepared to say "no" to unreasonable client demands late in the project life cycle.

Organizational Anorexia. Strassmann (1995) notes that outsourcing to cut labor costs shares attributes of the psychological disorder anorexia nervosa, where people refuse to eat to the point of starvation and have distorted self-images of being fat even when emaciated. Strassmann (1995) and later Duxbury and Higgins (2001) find that organizations with anorexia chose downsizing as a preferred method for restoring competitiveness even when they know they have too few resources and exponentially increasing workloads. The evidence is in a longer workweek, up from 37.5 hours to 50 hours in Canada, and higher job stress up from 13 percent to 35 percent from ten years ago (Duxbury and Higgins, 2001). A recent U.S. study of 750 employees and 250 employers shows that 44 percent of employees have more job stress than a year ago (Bureau of Labor Statistics, 1999–2000). Conditions appear similar across the European Union where nearly 30 percent of 21,500 workers interviewed cited stress as the second most serious problem (The European Foundation for the Improvement of Living and Working Conditions, 2000). Sixty percent of 19,000 Australian workers surveyed in a study by the Department of Industrial Relations stated having more stress than one year ago (Beder, 2001).

Why work so hard? In Canada, two incomes are necessary just to keep from losing ground and maintaining a family's standard of living (Statistics Canada, 2000). Employers blame global competition as the need to extend work hours to allow work across time zones, and compete and keep costs down by limiting the number of employees it deems feasible to hire (Duxbury and Higgens, 2001). The underlying fear of job layoff for refusal to work unpaid overtime continues to make branding an effective method. Johnson, Lero, and Rooney (2001) state that "job angst"—fear of losing one's job and being unable to find a comparable one—cropped up strongly in the 1990s, but it still seems to have plenty of momentum.

The management level response here is to do the following:

- Formally assess "perceived" stress levels related to work hours to achieve a baseline before any intervention: Talk to your employees and really listen to what they are saying.
- Establish a policy to have budgeted and actual working hours recorded. Examine the trends, use results to effect change, and reassess stress levels. An organization cannot afford not to know, and it cannot afford to penalize its employees for speaking the truth.

Senior Executive Priorities. CEO turnover is on the rise in all three major world markets. Drake, Beam, and Morin (2000) studied 476 public and private business organizations representing over 50 different industries in 25 countries to find that in just the past five years, nearly two-thirds of all major companies replaced their CEO. Almost half of CEO replacements occur because of mergers and acquisitions. Thomas, Delisle, and Jugdev (2002) also report that nearly 60 percent of international senior executives, project managers, and team member respondents were with their current employer for five years. In short, implementation of plans ends up with shorter time frames that focus on short-term business results (Drake, Beam, and Morin, 2000). Despite self-preservation priorities, corporate debt reduction, business ethics, and overreaching concerns for national security, attracting and retaining high-caliber employees is said to be second only to profitability as the most important priority (Meyer, 1988; Stephenson, 2000).

The management-level response here is to do the following:

- Educate top executives; they have to know the people behind the organization.
- Be involved in creating value-driven arguments for hiring that can be tied to quantitative outcomes.
- Reframe people information in executive terms; people are human resources that can deliver on value propositions related to saving the organization time, money, and so on.
- Do not leave the hiring to the human resources department needs; it is important to understand that the organization is hiring "minds" to make a strategic contribution, not just bodies to carry out the work.

Part II: Team Development

Understanding Teams

Not much has changed in the discussion about team selection and building from influential books such as the 1988 *Handbook on Project Management* (Clelend and King). What are some of the reasons why this should be so?

One reason may be that the concept of teams has shifted over time, but our understanding has not. Thus, what once looked liked a team may more aptly act as a group or committee. Table 39.1 presents a list of team-based literature. Rather than being exhaustive, the table serves as a starting point to help teams identify their dominant characteristics.

To add some structure, Miller (1988) suggests placing each characteristic in three larger categories to demonstrate their link to "task, result, or people factors" (p. 826). Keep in mind that results-oriented characteristics as described by Miller (1988) tend to look like success factors as described in project management literature (i.e., on-time performance). Decisions about what characteristics are most important in delivery of successful projects can assist a project manager in figuring out what kind of team can be built. However, this is certainly just a starting point! The tough work lies ahead in creating the conditions necessary for successful teamwork.

Another reason may be that the pervasive Western view has narrowed thinking about building a better team to finding ways to be more efficient and effective at tasks. Consider this line of wisdom in simple mechanical terms: To build a better or higher-performing machine, the wheels simply need to be tightened. In terms of people, simply make the same processes more efficient by tightening timelines can only improve efficiency a certain degree. Consider that over time, even a highly performing team may grind to a halt. Doing things differently means loosening the bolts on the "machine," letting the pieces fall, and reorganizing in a way that allows teams to do different things and to do things differently.

Once the team concept is more fully understood with respect to which characteristics are linked to successful teams, the business of identifying team members and the relationship between competencies and personalities may be more clearly understood.

Identify Key Resources

Miller (1988) comments that staffing the project is the "first milestone during the project formation phase" (p. 834). Project managers/leaders need to be involved in identifying and

TABLE 39.1. KEY CHARACTERISTICS of TEAMS.

Some Key Characteristics	Research Examples	Practice Examples
Celebrate, have fun, recognition of effort	Hoffman (2000); Delisle (2001)	Forsberg, Mooz, and Cotterman (2000; Hartman (2001); NASA (2002)
Common goals/cooperative team focus	Hoffman (2000); Chen and Barshes (2000); Delisle (2001)	Forsberg, Mooz, and Cotterman (2000); Kerzner (2001); McGannon (2001); NASA (2002)
Empowered	Hoffman (2000)	NASA 2002
Structured (establish clear boundaries around roles, conduct)	Hoffman (2000)	Forsberg, Mooz, and Cotterman (2000); Nasa (2002)
Cohesive (tribal culture)	Hoffman (2000); Delisle (2001)	Hartman (2000); Nasa (2002)
Interdependency	Hoffman (2000); Chen and Barshes (2000)	Forsberg, Mooz, and Cotterman (2000); Kerzner, (2001); NASA (2002)
Communicative (open communication and strong skills)	Chen and Barshes (2000); Hoffman (2000); Delisle (2001)	Miller (1988); Hartman (2000); Kerzner (2001); Nasa (2002)
Committed (to team and job)	Hoffman (2000); Delisle, (2001);	Nasa (2002); Kerzner (2001); Hartman (2000)
Diversity	Hoffman (2000)	Nasa (200)
Results-oriented—Shared milestones/ performance criteria	Hoffman (2000)	Miller (1988); McGannon (2001); Kerzner, (2001); Nasa (2002); Hartman (2000)
Trust	Crisp and Jarvanpaa (2000); Delisle (2001)	Miller (1988); Hartman (2000); Kerzner (2001)
Highly skilled/competent	Mills, Tyson, and Finn (2000); Dyrenfurth, (2000)	McGannon (2001); Kerzner (2001)
High spirit and *energy*	Next Step (2002); European Commission (2002)	Miller (1988); Forsberg, Mooz, and Cotterman (2000); Kerzner (2001)

negotiating for key resources to increase the probability of obtaining the necessary skilled resources to match the job requirements. Recruiting good people for projects is too important a job to be left to the human resources department (Meyer, 1998). In reality, the project manager/leader often has little influence on who will be a member of the project team. However, in an interview with Madigan (1998), Pinto suggests actively finding people to lessen the risk of having problem employees dumped on the project by department heads. Team leaders or self-directed teams need to communicate with decision makers about desired personal attributes, areas of competence (things a person will be required to do), and personality traits/behaviors that will likely result in a strong project team.

Knowing Competencies—Job-Related and Personal

Assessing skills and abilities and knowing personality traits and key behaviors of individuals increase the chances of choosing a team that has the potential to succeed (Sugarman, 1999). What are competencies and how do they fit in? Competencies differ from KSA (knowledge, skills, and aptitudes/abilities) in that competencies are based on the individual and their capability rather than on the job and its associated tasks (Kierstead, 1998). (See also the chapter by Gale.) Competencies generally fall into broad categories of intellectual, management, relationship, and personal (Kierstead, 1998). Key resources for project teams are those who have strong relationship competency (ability to build and maintain interpersonal relationships and communication ability) and personal competency (behavior flexibility or ability to be responsive to change, workload, transitions) (Kierstead, 1998).

Understanding competency as person-based rather than as job-based appears to be particularly important in a management-by-project environment, where jobs and associated tasks may not be narrowly defined. (Though Crawford, in her chapter, makes the case for professionally based competencies.) Decision makers can reduce the risk of making the wrong resource selection by identifying the areas of competence for the actual job or role descriptions. In an ideal situation, team member selection follows by assessing person-based competencies of each potential candidate and matching them against the job or role requirements (see Case Study 2).

Case Study 2

The project entailed developing a new book store as part of a national retail chain. The members of the team were brought together through human resources from different functional divisions within the same company. The project manager was inexperienced and did not have the influence or know-how to get involved in the team selection process.

Although the talent did exist within the organization, the human resources department did not clearly match the skill set of each person to the job requirements for the project. The mismatch was partly due to the project manager not communicating the requirements (roles and level of responsibility) needed for each position to HR.

The team therefore did not contain the competencies needed to get the project completed. In frustration, the project owner fired the project manager. The new project manager was more cognizant of the need to match skill level with the project requirements and thus was able to pull in a few replacements to strengthen the team and eventually complete the work.

Understanding Personalities

Personality tends to be enduring and, some argue, unchangeable. Thus, principles presented in Hill and Summers' (1988) chapter in the *Project Management Handbook* provide a solid grounding in personality and conflict dynamics, and these topics need not be redressed in detail. Social conflicts tend to be relational (between people), but triggers that ignite conflict reside in the individual (Hill and Summers, 1988). Thus, assessment of the way individuals gather and process information instruments such as the Myers-Briggs Type Indicator provide useful information about the strengths and weaknesses of individuals, as well as the potential triggers that might blindside the team. In addition, Hill and Summers (1988) make a strong case for assessing the emotional dynamics to gain insight into how individuals act in interpersonal relationships.

Frankel's (1999) work serves as a unique foundation from which to explore the role of emotions versus logical thinking on project teams. He asserts that an individual's cognition does play a role in elaborating and refining a repertoire of coping skills to help individuals adapt to their environment. However, emotions essentially govern how individuals continuously estimate how to avoid risks and exploit opportunities (Frankel, 1999). Although a simplified explanation, people tend to act as utility theorists by first seeking to reduce pain and then increase pleasure. Personality then is partly driven by the intensity of the positive or negative feelings as well as the valuation of the risk and opportunity, and not by pure rational thinking. Individuals practice what Frankel (1999) calls "emotional economics" to make sense of their world. If estimates are out of proportion with the situation (too low or high), a problem may erupt. When individuals estimate emotions accurately in a specific situation, responses are within the realm of their adaptive cognitive skill set. Thus, asking questions that test the adaptive coping skills of individuals on a cognitive as well as emotional level may help in the long run in avoiding intense conflicts.

Negotiate for Human Resources

Line managers or members of self-directed project teams need to know what problems senior executives face—what keeps them up at night. Knowing executive's top challenges makes it possible to negotiate for the resources that will provide the best fit, allowing the project team to deliver the level of value added needed to justify their costs. Step three and four of Belgard, Fisher, and Rayner's (2000) process provides the most opportunity for a project leader to influence in the negotiation process:

1. Size up the work to be done.
2. Establish selection criteria.
3. Find possible candidates.
4. Evaluate candidates and make selection.
5. Formulate the offer.
6. Orient the new member.

Once the offer is on the table, it is more difficult to do any negotiation. From the negotiator's perspective, backward planning might be a useful process. First, know what you need the outcome of the negotiation to be. Next, decide where you will draw the line—how far are you willing to go to get a resource? How willing is the candidate to join the team? (Miller, 1988) Finally, work out a "best" alterative to the negotiated agreement in the event the process goes further than you are willing to go.

Developing or Growing a Team

An obvious yet often overlooked step is to identify the purpose of the team. Traditionally, literature on project teams points to the attainment of a *common goal* as the hallmark of a high-performance team. However, consider that individuals make up the team and carry out the component tasks charged to the project team. Individuals attain goals that result in the team delivering a product or service. DeMarco and Lister's (1978) argument still makes sense: Align individual goals as a means of achieving a common goal. From a management point of view, it matters less that the team has the exact same concerns and more that the project team carefully focuses its energy on those concerns that matter to the project to make sure it will contribute to the overall business success.

Project managers and senior executives often make the naive assumption that teams accept the employer's goals as a condition of employment. Individuals tend to have many different reasons for involving themselves in the organizational or project goals, so it is naive to assume that simply refocusing the project team on an arbitrary company goal will bring success. DeMarco and Lister (1987) suggest that this may actually hinder the team's ability to gel. Project teams tend to gel much more readily around the challenge of achieving a goal that is related to the team itself. For example, Hartman (2001) describes a "tie tribe" (pp. 272 to 273). Deliverables are written on each tie stripe, starting with the last one at the top. As the team successfully achieves a deliverable, the tie is trimmed and the highest contributor wears the tie.

Evolutionary Model. How can we grow or develop a team once we know its purpose? Many practitioners and researchers still utilize a nearly 40-year-old model by Tuchman (1965). The model describes five sequential stages of team development (see Table 39.2, left). The traditional team-building model is grounded in the assumption that over time, a team moves toward better performance.

Tuchman's (1965) team building model assumes that teams develop sequentially. Once a stage has been mastered, a team moves to the next stage, ultimately leading to better performance. The leader gives direction, and the team carries out the work in a relatively predictable way. Social dynamics are dealt with in the early stages to minimize interference with the general task's focus. McGrath's (1990) model (see Table 39.2, right) is predicated on the same evolutionary principles, and even the labels for each stage appear similar.

TABLE 39.2. TEAM DEVELOPMENT MODELS.

Sequential Team Development Stages	Potential Activities by Combination
Model by Tuchman (1965) **Forming** Learn about members, learn about the environment, and size up members.	Model by McGrath (1990) **I. Inception** *Production focus:* Select goals, plan, and brainstorm.
Norming Gain consensus about acceptable individual and group behaviors; clarify and align expectations; learn about roles, responsibilities; and set guidelines for operation.	*Social focus:* Ensure inclusion, establish parameters for participation, and flesh out roles. **II. Problem Solving** *Production focus:* Identify problems (technical and task-related), provide alternatives, and solve problems.
Storming Express anxiety, defensiveness, engage in power struggles and interpersonal conflict, learn to deal with conflict and find resolution.	*Social focus:* Address power, status issues; clarify roles, expertise, and competencies. **III. Conflict Resolution** *Production focus:* Identify and resolve conflicts (actual or anticipated) related to the project methodology, processes, etc.
Performing Achieve commitment and full attention of members, use of each other as resources, respect of boundaries; express cohesion (gel, bonding).	*Social focus:* Address interpersonal relationships, styles of learning, cultural differences (organizational and ethnic). **IV. Execution** *Task focus:* Perform task and continually scan environment for barriers/opportunities.
	Social focus: Revisit participation, accountability, and communication competency.

McGrath's original model shows a team's activities as a set of three interlocking functions: production (task) function, member support functions, and group well-being function. Durarte and Tennant Synder (2001) combine the member support functions and group well-being functions and label it "social" focus which is reflected in figure 39.2 (right). They use the same labels for each activity and apply the model to virtual project teams, although McGrath (1990) does not make this distinction.

McGrath's (1990) model aptly reflects how teams grow and change, creating "patterns" that suit the needs of the situation (see Figure 39.1). As constraints change, the growth pattern of the team may change. For example, the introduction of new communication technology may shift the path and/or the balance of energy spent on social rather than task functions.

Pattern A indicates that teams move through stages or activities I to IV sequentially, most closely resembling Tuchman's model (teams may not spend the same amount of time in each phase). Pattern B (focus on activity I and IV) often occurs if teams are given a project to complete under extreme time pressure. They tend to skip social and production functions concerning problem solving and conflict activities, focusing on production functions in relation to the project's execution. This describes teams that end up stamping out crisis fires instead of making use of the midpoint change in direction to improve strategy (McGrath, 1990).

Pattern C (focus on activity I, II, IV) does not suggest that teams avoid conflict. However, teams tend to address problem solving at the task level that may not challenge earlier established strategy, thus avoiding personal types of conflict (McGrath, 1990). Pattern D (focus on activity I, III, IV) reflects project situations where teams receive direction rather than work together to solve problems (i.e., task force teams that are brought together to solve predetermined problems). Patterns B to D appear most susceptible to neglect of social functions, which increases the risk that collaboration will suffer over time and hamper the desire to work together again (McGrath, 1990).

Revolutionary Model. Evolutionary models, however, do not adequately explain why team performance varies over time in relationship to its evolution. A common observation is that teams, regardless of the stage or phase, tend to go through at least one major transformation about halfway through their growth cycle (Gersick and Hackman, 1990; McGrath, 1990; Duarte and Tennant Synder, 2001). Gersick and Hackman (1990) note it is during this

FIGURE 39.1. POTENTIAL TEAM GROWTH PATTERNS.

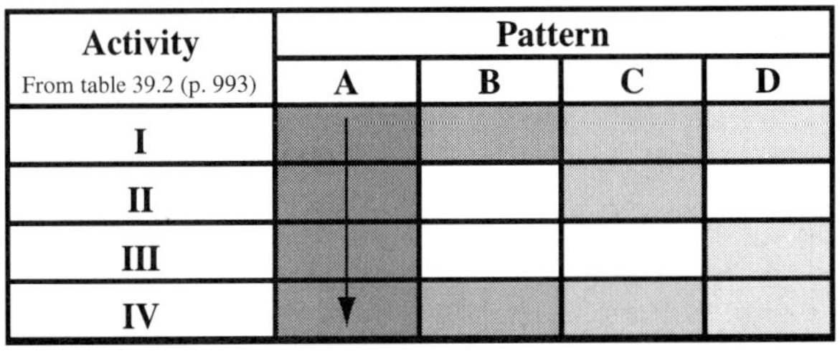

Activity	Pattern			
From table 39.2 (p. 993)	A	B	C	D
I				
II				
III				
IV				

process that teams "punctuate their equilibrium." That is, after midway into any length of a project, efficiency drops as teams enter a more revolutionary period of change. McGrath (1990) notes that this action does not support the kind of sequential linear patterning explained by Tuchman's development model. Instead, McGrath's model works more readily with a cyclical concept of time.

Reworking of the original plan, a sense of urgency to finish, and closer contact with sponsors and clients may trigger efforts to punctuate equilibrium, but the underlying cause-effect relationship is not known for certain (Duarte and Tennant Synder, 2001). Perhaps a more internal, fundamental process such as trust plays a role in the midpoint transition of a team's revolutionary development. Crisp and Jarvenpaa's (2000) study lends support to this idea, finding that the level of team trust drops sometime near the middle of a project. Few changes in trust levels occur until the end of the project. Their work shows that by the midpoint of team development, members have more substantive information about the project and thus, greater insight into the "real" personality and behavior of team members.

Team members tend to engage in relationship building early on in the formation to gain a sense of which person to trust. This will ultimately be measured by how much weight each appears to be pulling to get the project work completed. Project team members may be closest to their *true* nature when nearing or engaging in critical questioning about other's performance and contribution (see Case Study 3).

Case Study 3

An external consultant was hired to create a post-project review process for a major oil and gas company. The project team was chosen by careful attention to personality traits, experience, and team work skills. Its members had agreed to work cooperatively together to produce a high-quality project. About a third of the way through the project when the first set of deliverables were due, internal grumbling began about having to pick up the slack because one member did not follow thorough on task or team processes as agreed. Not until midway through the project did two of the five members openly address this problem in a group discussion. The initial level of trust based on early agreements for performance had broken down at a critical point in the project, forcing the team to rework the plan, reorganize the workload, and shift roles and responsibilities to enable the successful completion of the project.

In consideration of time and its relationship to team development, no one pattern of growth is "right or wrong." Rather, team awareness of the features and benefits of each pattern may allow its members to bring production functions and social dynamics into sharper focus as the project moves through its life cycle. Overall, teams appear to mature by visiting and revisting the interlocking sets of tasks and social activities, with recognition that midpoint in any project may serve as an important platform to rethink current processes and practices. As such, resolutions may be partial, simultaneous, or even parallel in some or all stages of team development.

Nature or Nurture: What Influences Team Dynamics?

How can you make a team gel so they work through development more easily? DeMarco and Lister (1987) astutely note that nothing can make a team gel, only improve the odds of growing a team that *may* gel. A grow rather than build model recognizes that the process is inherently nonlinear and prone to chaotic processes. Thus, the initial point of becoming a "team" cannot be known in enough detail to predict its evolution (Duarte and Tennant Snyder, 2001). Sommerville and Dalziel (1988) find evidence that teams also don't know how they will handle their end point and often engage in a period of "mourning" after performing. In reality, it may take teams until the end of the cycle to gel, or they may never gel. A technique called "inversion" developed by DeBono (1977) has been used with success to increase the odds that they will gel. Inversion simply refers to thinking of ways to achieve the opposite of a team—to make team formation impossible. DeMarco and Lister (1987) point to phony deadlines, defensive management, and fragmentation of people's time to be surefire ways of failing, or committing "teamicide."

Environmental Factors Affecting Team Dynamics. Many factors act to influence whether a project team gels or not. Team composition, demographics, and temporal influences appear to be important dynamics that influence team development and performance. Each dynamic is briefly introduced in the following subsections as a starting point for further investigation.

Team Composition. Decision makers need to consider the size and level of expertise on the project team. A team of approximately ten core or central members seems to work effectively in many project-based organizations (Delisle, 2001). Considering that projects utilize a blend of physical and virtual contributors, determining the actual size of the team may be extremely difficult. Deciding who belongs to the core, mid-layer, and outer stakeholder group may in itself be an important exercise in establishing desired role and authority relationships. More important, how many teams should any one individual be part of at one time?

Team Demographics. Not enough can be said here about the changes to the labor force and cultural composition to justify its importance to understanding the dynamics of a modern-day project team. More women than men are entering the labor force in both the United States and Canada. The Bureau of Labor Statistics (1999 to 2000) in the United States projects that the share of the workforce by each gender will be almost equally divided by the year 2008. Statistics Canada Labor Force Survey shows the same trends, with women accounting for 46 percent of the workforce in 2001 (Ministry of Industry, 2002). In terms of race, projected growth rates of over 3 percent are expected for the Asian and Hispanic labor force by 2010, resulting in the strongest presence of ethnic groups in the U.S. labor market (Fullerton and Toosi, 2001). In Canada, immigration is expected to account for almost all of the net growth in the Canadian labor force by the year 2011 (HRDC, 2002).

How are team dynamics impacted? Individuals are members of multiple cultures that include national/ethnic origin, profession, function (within organizational departments), and business or corporate cultures. Individuals are also expected to become part of or establish a team's culture when selected for a project. Project leaders and managers need to consider the range of diversity that enables the team to manage differences as the project progresses. Understanding the extent of each team member's cultural memberships offers a starting point or common ground from which to grow a strong yet diverse team.

Temporal Orientation. Time or "temporal" orientations of the organization, project leader, and team members may also influence a team's dynamics and play a significant role in project performance (Thoms, forthcoming; Thoms and Pinto, 1999, and Thoms and Greenberger, 1995). Project leaders and member's *thoughts* may be aligned mostly in the past, present, or future (Thoms, forthcoming; Thoms and Greenburger, 1995). For example, past-oriented individuals may think that strategic planning about the future as a waste of time. Future-oriented individuals may have a difficult time in conducting project reviews because they don't see the value in going over what has already been accomplished. Present-oriented individuals may be so caught up in day-to-day operations that they cannot see the big picture and how past lessons can be applied to avoid future pitfalls.

As well, individuals' *behaviors* may be described in terms of the "temporal skills" they use to make sense of the past, present, and future orientation (Thoms and Pinto, 1999). Thoms and Greenburger (1995) propose that a new class of management skills based on temporal consideration needs to be taken into consideration, because overall adaptability is tied to the individual's time orientation (past, present, future). For example, a temporal skill such as "chunking" involves the creation of time units (i.e., workweek = 5 days) to allow for the team to attack the project in "doable" units rather than as one time frame (Thoms and Greenburger, 1995). Furthermore, "time warping" by constantly reaffirming that the team is moving toward the project milestones is a skill that helps team members stay motivated.

Thus, *attunement* or match between the temporal skill and thought pattern alignment in the past, present, or future orientation seems largely controlled by an individual's choice of and prowess in using a temporal skill (Thoms, 2003). Consider that as individuals become attuned, they may be drawn to members that have a similar resonance (i.e., they seem to "click") in relation to the flow of the work. McGrath (1990) finds evidence that by imposing an initial tight project deadline, teams switch to a high task focused work at a fast rate, and deliver poorer quality. This carries over to subsequent projects even when they do not have the same restricted time deadlines. Teams become "entrained" or work in synch much like the cardiac rhythm of two individuals synchronizing when their chest cavities touch, for example.

Behavioral Factors Affecting Team Dynamics. Although arguments can be made about many significant team behaviors, three internal factors identified by Sugerman (1999) make sense in terms of distinguishing high from poorly performing teams. A brief discussion about mental toughness, cohesion, and motivation follows.

Mental Toughness. Mental toughness refers to the ability to bounce back positively from an adverse situation or event (Sugerman, 1999). The majority of study and research has focused on sport and military teams. These lessons are sometimes taken and applied to the business world. Mentally tough team members, according to Brown (2002b) are those that

- seek feedback about performance, even when it is not guaranteed to be positive;
- choose tasks that are challenging but add value rather than taking the path of least resistance;
- trust gut feel and ability instead of procrastinating when they need to make critical decisions;
- face conflict by quickly taking action, engaging it as required or resolving it quickly;
- focus on solutions rather than reacting negatively and defensively to perceived threats; and
- contribute ideas that could make a solid impact on the company's bottom line even if fearful of failure or rejection.

From a personal trait point of view, a major barrier that individuals on a project team face relates to learning how to respond by technique and not emotion in the face of adversity. This does not mean becoming emotionally hardened or cold. Rather, detaching oneself from taking an issue personally helps to avoid destructive blame tendencies that erode a sense of shared loyalty and trust.

Cohesion. Cohesion refers to the closeness or "glue" that holds a team together. Popular literature is rife with examples of fire, police, and sport teams pulling together to beat the odds. Sugerman (1999) asserts that building cohesion is a matter of assessing and aligning expectations, establishing open communication, and engaging in direct assessments of team members. However, does cohesion need to occur for top performance? Does cohesion also have a flip side?

Cohesive teams may or may not visibly show signs of intense social commitment such as "high fives" often seen in professional hockey games. Professional sport teams spend a great deal of time together on and off the playing field, and socialization is part of the culture. In contrast, the Royal Canadian Mounted Police Emergency Response Team (ERT) does not spend a great deal of time together on or off duty. The team consists of a two highly trained and specialized tactical units. Daily activity is mostly individually planned, although tactical and technical training sessions are group-based. Training and conditioning is primarily done on an individual basis, unlike highly organized group training of elite teams in the armed forces. As well, individuals do not spend much "social" time together, such as eating meals. However, when called into action, they work seamlessly together on each project they undertake. How can both types of teams be highly cohesive?

At one end of the continuum, an underlying force driving cohesion of ERT stems from overcoming threats to physical safety by cooperatively uniting as a team. At the other end of the continuum, elite sport teams understand the feelings associated with overcoming the "odds" and reaching self-actualized goals through cooperative game play, creating strong cohesion. Consider the paradox of individual needs of ERT versus group needs of an elite sport team. The ERT appears driven to satisfy individual physiological *self-interests* related to safety, whereas the elite sport team is driven to satisfy higher-order or *socially related interests*. Thus, cohesion looks different to different teams.

Motivation. Sugerman (1999) characterizes motivation as the intensity of behavior (arousal) or psychological force that pushes a person to perform. Theories of motivation range widely, including instinct theory (we are genetically predetermined), drive theory (seek to reduce tension, gain balance); arousal theory (seek to calm down or psych up), and incentive theory (perform for external reward). Perhaps the whole question of whether motivation is extrinsic (learned in response to reward) or intrinsic (because you love something) needs to be reframed as subquestions that we can answer. An important subset of the motivation question relates to perceived lack of control. Psychology studies consistently find a relationship between lack of control over self, events, and situations; increased burden of responsibility; and stress associated with lack of motivation.

Teams members may experience intensified feelings of loss of control if they receive little or no acknowledgment of or responses to their contributions. This may even be more intense in a virtual environment because of the lack of physical cues. Eventual withdraw from participation in online collaboration may be taken as a lack of motivation when the problem stems from loss of awareness or incomplete awareness of other's activities in the project or inability to extract salient information from an overwhelming mass of information (Conklin et al., 1998). Whether virtual or not, the leader's task is to ensure team members control their effort levels and not hold them responsible and accountable for external events that they *cannot control.* Consider as well that, as utility theory, people are engaging in what Frankel (1999) calls "emotional economics." They calculate the "net interaction value" partly as a function of personal value less personal cost (Conklin et al., 1988). Cost and value differ by individual—depending on whether they consider and how they weight related emotional, intellectual, and social impacts.

Another important subset of the motivation question relates to an individual's perception of themselves within the team. Swann, Polzer, Seyle, and Ko's (2002) research challenges commonsense thinking about the way people's self-views influence their responses to motivating feedback. Their study finds that participants with positive self-views worked hardest if they believed they would receive positive feedback, and participants with negative self-views worked *least* when they received positive feedback. As predicted by self-verification theory, people with negative self-views withdrew effort when they thought that they would be receiving positive feedback because they felt undeserving (Swann et al., 2002). If the group verifies this view, the member feels detached and may quit. If the group does not verify this view, the member feels committed to the team, yet alienated. These findings point to the need for personal transformation where individuals on a team first get in touch with their own views to enable a clearer view of the group and the overall organization. Signs that point to negative self-views include perfectionist tendencies, taking on too much work, and taking blame for events beyond one's control.

Influence of Virtual Teams

The evolutionary and revolutionary models help us to understand the changes in team composition and dynamics over time. However, shifts in division of labor over time essentially helps to explain the bigger picture of how teams relate to their environment, especially as we continue to hear discussion about challenges of working in virtual teams—those teams who work primarily using telecommunications devices and who may seldom or never meet face to face. Figure 39.2 depicts the division of labor by organization type, showing a rough time frame for each division (Delisle, 2001).

A review of the organizational and management literature shows that organizations tend to move through cycles of more formal controls to periods of loose connections (Delisle, 2001). Lipnack and Stamps (1996) do not suggest that any one organizational structure disappears; rather, its dominance shifts over time. For example, pre-1900s was marked by the division of labor into small tribal groups that relied on nomadic wisdom or storytelling. Moving to the right in the figure, cottage industries roughly emerged in the 1940s, borne out of the shift to an agriculture focus.

This agriculture era transitioned into bureaucracies where gains in labor use came from the mass production of goods and services in the 1970s. In the 1980s packets or nodes of workers typically organized by groups or teams began to emerge in workplace. Nodes provided more autonomy, although control remained centralized and expectations were to follow organizational norms and rules (Delisle, 2001). The continuing surge of communication technology noticeably changed the division of labor in the 1990s, providing choices such as distributed work through telecommuting and hot desking (sharing desks between office workers).

Most recently, the literature notes the emergence of virtual teams or "tribes" composed of a more highly educated and mobile workforce driven by the need for organizations to increase profits/cut costs and fulfill individual needs of workplace autonomy. Virtual workers may act independently as knowledge contractors, as well as be part-time members of one or more traditional organizational teams. Early indications are that the workforce of the

FIGURE 39.2. DIVISION OF LABOR OVER TIME.

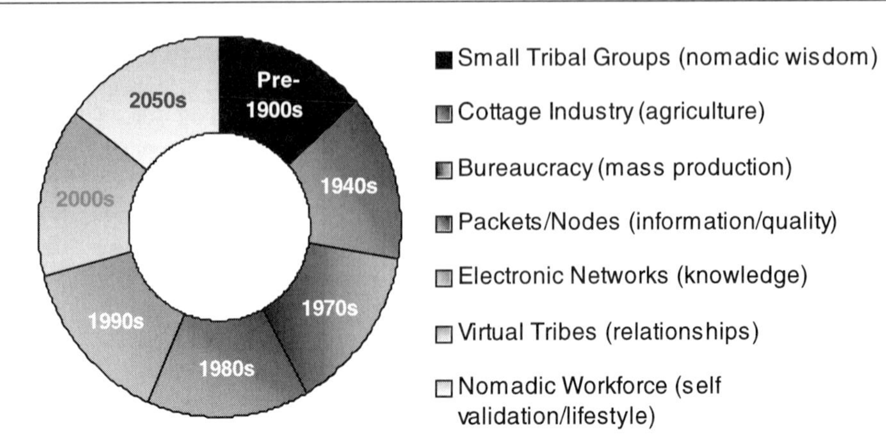

Adapted from Delisle, 2001.

future may become even more fragmented and driven by independent subject matter "nomads" who value their lifestyle sometimes ahead of work. Distinctions between the division of labor may be primarily around differentiating levels of technology sophistication, more fluid leadership sharing arrangements, and unique linkages. The linkages refer to interconnections between people, activities, and knowledge, rather than formal control (Lipnack and Stamps, 1996).

Challenges in Virtual Teams. What specific challenges do we face in building functional virtual teams? Relationships networks are becoming increasingly important to enabling virtual project teams. Thus, workers outside of large city centers may have difficulty gaining access to professional networks to allow them to select work. Therefore, organizations are beginning to focus more on building relationships with competitors and external suppliers that normally fall outside the commonly accepted "definition" of a project team (Delisle, 2001). On a micro level, virtual teams face specific challenges, of which two key ones are briefly touched on in this section. These challenges are in defining roles and responsibilities whose clarity is needed to assist in managing communications.

Given that virtual project teams emerge to solve a problem (task-driven), it might at first not seem necessary to define roles the way traditional teams do (Ott and Nastansky, 1997). To make matters more complex, traditional rules-based team design does not fit well with the reality of a virtual team. The most important issue is that responsibility tends to shift by task, *regardless of role definition.* Delisle, Thomas, Jugdev, and Buckle (2002) note that team members take on the tasks they believe they feel most skilled to do rather than the ones they feel obligated to do.

The Types of Work Index (TOW) by Margerison and McCann (1994) serves as a useful guide to helping virtual teams explore and understand their own internal capabilities (see Table 39.3). If the time is not afforded from this activity, the team can become frustrated (i.e., when timelines tighten), and they are not able to quickly identify who to draw from in solving problems related to the work and team function.

In terms of the TOW Index, a member may engage in one task (i.e., organizing) more often than another because he or she has had previous successes with the preferred behav-

TABLE 39.3. TYPES OF WORK INDEX.

Task	Description
Advising	Gathering and reporting information
Innovating	Creating and experimenting with ideas
Promoting	Exploring and presenting opportunities
Developing	Assessing and testing the applicability of new approaches
Organizing	Establishing and implementing ways of making things
Producing	Concluding and delivering outputs
Inspecting	Controlling and auditing the working of systems
Maintaining	Upholding and safeguarding standards and processes

iors. This, in turn, reinforces the likelihood of similar future behavior (Delisle, 2001). One area that appears to be consistently weak in virtual teams is the role of "maintaining," because unlike face-to-face teams, the administrative function is not tied to any specific role. Thus, meeting notes, coordinating work to be done, and following up on action items often falls off the team's radar screen unless someone volunteers. This essentially shifts a burden to one or two individuals, which can create resentment over time because of the added workload.

A closely related issue is of responsibility, because it tends to shift throughout a task depending on the work and workload demands. The use of a tool such as the RACI (Responsibility, Accountability, Consult and Inform) chart seems to elevate confusion specifically between who has a stake in contributing to the work (responsible) versus who has to deliver the work at the end of the day (accountable). As well, this tool when used consistently seems to resolve problems around what consult means (active involvement) versus inform (rubber-stamping near-final draft), which seem to be particularly problematic in virtual teams.

Part III: Team Management

Common Team Pathologies

Part III takes a practical look at ways to overcome common team pathologies. Research literature, popular books, magazines, and Web articles are filled with tips, tools, practices, and methods for managing teams. Rather than trying to cover off every tip or practice invented, this section focuses on a few key areas to focus efforts to do things differently and do different things in managing teams.

Arguably, teams go through some type of dysfunction or cycle of pathology at one point or another over the project life cycle. Pathologies are typically tied to dysfunctional behaviors. Lack of mental toughness, cohesion, and motivation as discussed appear to underlay many team dysfunctions. This section provides a brief introduction to common team pathologies in these three areas and suggests possible courses of action and tips to help overcome these difficulties.

Lack of Mental Toughness. Project manager/leaders often face the challenge of handling personalities that play a major role in shaping a team. Lack of mental toughness may be expressed in many forms, particularly tied to situations were people "give up" or are take the situation personally.

Two simple strategies can be used to increase metal toughness. First, all review and feedback from the project manager to team members should be couched in terms of "I" talk not "you" talk, especially concerning emotionally sensitive topics. For example, "I feel that your performance is not as strong as it was last month—I would like to spend some time hearing about what you think and finding ways that I can help." This is directly contrasting typical communication that begins with "you are not working to your potential; what are you going to do about it?"

Second, have the team agree on a code of conduct/guidelines that includes an agreement not to take things personally. This simple step will help the team go a long way in keeping blame from crippling the team's effectiveness.

From a behavioral point of view, Seligman's research in the mid-1960s on "learned helplessness" sheds light on mental toughness. For example, an individual fails at a task and believes he is at fault. Over time, the belief that he is incapable of doing anything in order to improve performance spills over to similar tasks. Perhaps the most damaging aspect is that this belief interferes with the ability to learn in a new situation where avoidance or alternate solutions are possible. For example, if conditions in the workplace change, the perceived lack of controllability holds employees in a motivational paralysis, and job burnout may occur over time (Potter, 1998). As a team, the first step is to become aware of learned helplessness and then work to create an environment to help members relearn adaptive behaviors.

Ineffective Cohesion. How does one go about holding together teams that work in the regular business world when cohesion is not based on fulfilling basic survival needs or achieving dizzying heights of self-actualization? Ironically, in the business context, there is greater incentive to *compete* than cooperate as a team. The practice of individual reward and recognition tends to reinforce individualistic tendencies.

One strategy may be in revisiting the message that is being sent: Does it reinforce team or individual behavior? Are the team member's choices driven by team member loyalty friendships rather than what is in the best interest of getting the work completed? An example from the popular television show *Survivor* where members try to outwit, outlast, and outplay their counterparts over the course of 39 days serves as a useful example. The team's survival (safety, food, shelter) is not ultimately threatened; thus, team member selection is mostly based on relationships formed among those "conforming" to the majority-driven social norms that emerge as the game evolves. Members who do not follow emergent social norms and form loyalty alliances are often quickly voted off the team. So are those who show innovation, strength, or high levels of motivation because they are seen as a "threat." In a corporate setting this behavior may be expressed as "groupthink," a condition where team members make decisions on the basis of social relationships or preserving self-esteem, not on business outcomes.

Strategies to reward and recognize team behavior are effective when tied to performance and are peer-driven (not manager-determined). Monetary rewards are not always the most effective; often a simple and informal recognition venue or events will suffice.

Lack of Motivation. The propertics of the team itself in increasing or decreasing motivation deserve mention. Lack of motivation may be expressed symptomatically in many ways, including showing up late for work, producing poor quality, leaving work unfinished, complaining about other team members or policy, and so on.

However, root causes such as "social loafing," identified by Ringleham in 1913, are also critical to understand. Ringleham's psychological research identified that the greater the number of people working together on a task, the less effort each person decides to contribute (Ingham, Levinger, Graves, and Peckham, 1974). Studies of physical strength in rope-pulling contests reveal an individual's behavior is also affected by group dynamics. In group situations, the same individuals tended to pull less vigorously. Ultimately, the pressure to perform well is shared by the group, perhaps at an unconscious level. Awareness of social loafing is important in improving the quality of the team, because members are able to identify the signs and work on prevention rather than cure. Specifically, steps can be taken to identify individuals' inputs, evaluate inputs, involve all members in the task, and diffuse responsibility among members.

Another strategy is a modified "brainstorming" called the nominal group technique (Gersick and Hackman, 1990). Individuals first tackle the problem or question individually on paper. Contributions are collected and recorded on a flip chart like conventional brainstorming. Discussion continues until the alternatives are exhausted, and then individuals are asked to prioritize alternatives. This information is recorded by a facilitator and displayed to the group for further discussion. However, creation of a psychologically safe environment is as important as picking a method that will stimulate the team's thinking. Consider that if the body itself on a purely physiological level is far less responsive to any exercise routine beyond a six- to eight-week cycle, the mind is also likely to become bored by repeated use of any one method.

Misunderstanding of Success. Studies about success tend to talk about project- or business-level success without often considering the link between the two. Do businesses succeed when they make a profit? Do teams succeed when they deliver a successful product or service? Research shows that perceptions about success differ greatly, depending on who is asked and what context they are judging success. Success is a whole lot more complex than just meeting "priority triangle" criteria related to cost, time, and quality. Overall, project managers need to identify what role their team has in creating value.

First, the project teams need to establish what success means. Much confusion exists in practice and in the literature about what critical success factors (CFS) really mean. Initial research into success in the 1980s lumped success in one box and popularized CSF (Pinto and Slevin, 1988). To me, the CSF represents an overarching concept that includes both critical success indicators (CSIs) and critical success criteria (CSCs). Critical success indicators refers to the "softer" processes and markers organizations/teams agree to heed as a way to increase their chances of delivering successful projects (i.e., top management support). They are not often measured during or at the end of the project (Delisle, 2001). Critical success criteria (CSC) refers to "harder or more quantitative dimensions" that a project will be judged against as being successful (Delisle, 2001). Both CSIs and CSCs are often talked about in reference to the success of project management process. Outcome success then refers to the actual product/service itself that may be measured in relation to how well the project met the business goals/objectives (Delisle, 2001). (See the chapters by Brandon and by Cooke-Davies on this topic.)

Next, project teams need to determine what is most critical to their success. Part 1 of a study by Delisle (2001) asked participants what indicators of success they considered most critical. Respondents chose six CSIs from a list of 29 compiled from an extensive literature review (see Figure 39.3). As shown by the frequency of responses, participants chose open communication, communication skills, fun, trust, commitment, and culture. Crisp and Jarvenpaa's (2000) research on virtual project teams reveal a positive correlation between early team communication and increasing trust levels.

FIGURE 39.3. TEAM-LEVEL SUCCESS.

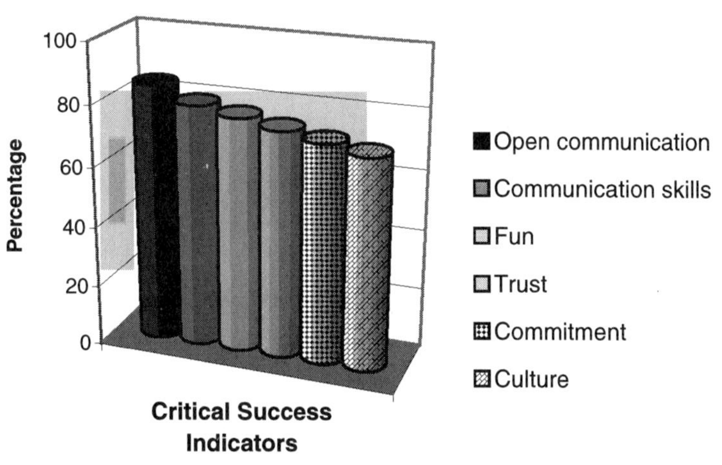

For example, if teammates communicated early on in the project, members of the team were perceived as more competent and worth trusting. As well, teams that communicated actively in the project tended to keep communicating more readily than those that did not as the project progressed (Crisp and Jarvenpaa, 2000).

These points underscore the importance of establishing predictable communication patterns early on in a project. Hartman (2000) found trust, fun, and open communication as three of seven elements for effective teams, *regardless* of whether they are virtual or not. Reaching team agreement about what could be done, and aligning this with what can be done in light of resource and project constraints, will help to identify CSIs that make sense for the project team rather than identifying one "right" list. Managing team success as an ongoing process will encourage the team to make adjustments to the CSIs and CSCs (add or drop) during the project life cycle as constraints change and the overall outcome success shifts.

Finally, project teams need to determine what they will pay attention to and what they will measure their success by during the project. Delisle's (2000) research identifies that only 45 percent of respondents report identifying CSIs or CSCs, and only 20 percent of these respondents report that their teams define and measure both. There is no magic formula that will ensure a successful team experience or guarantee a successful project. From a practical point of view, it may be more difficult for a team to agree whether they have successfully arrived at their destination if they have not taken the time to identify what signposts they expect to pass along the road to a successful project.

Misaligned Learning Styles. People differ in the way they take in and process information. For example, a team member may learn by listening (auditory), seeing a demonstration (visual), or touching or interacting with a model or real product (tactile). More likely, people learn by using a combination of learning styles. Many theories try to account for how people learn, although this has generated little agreement on the actual process. Learning style inventories simply identify the preferred way a person encodes information and changes it to make sense of a problem or situation.

Often, individuals are not even aware of which style they rely on and under pressure default to patterns that may not be as effective as possible to get the job done. Gardner (1993) advanced thinking about learning styles to accommodate multisensory abilities or "multiple intelligence" (MI). The term MI should not be confused with IQ. Individuals have varying levels of skill and can enhance their ability to learn by becoming more self-aware in terms of their strengths and weaknesses in relation to each of the MIs shown in Figure 39.4. For example, individuals who learn best through *words* use verbal/linguistic intelligence, those who learn best though *questioning* use logical/mathematical intelligence, and individuals who learn best through *pictures and images* rely on visual and spatial intelligence. Intuition as a way to learn seems to be guided by instinctive gut feel or "vibes."

Consider using the learning wheel to assess team member's MI at the start of the project. Sharing the results among the team may help avoid frustration and misunderstanding that often results from assuming people will readily adopt standard communication protocols (see Case Study 4).

FIGURE 39.4. MULTIPLE SENSORY LEARNING.

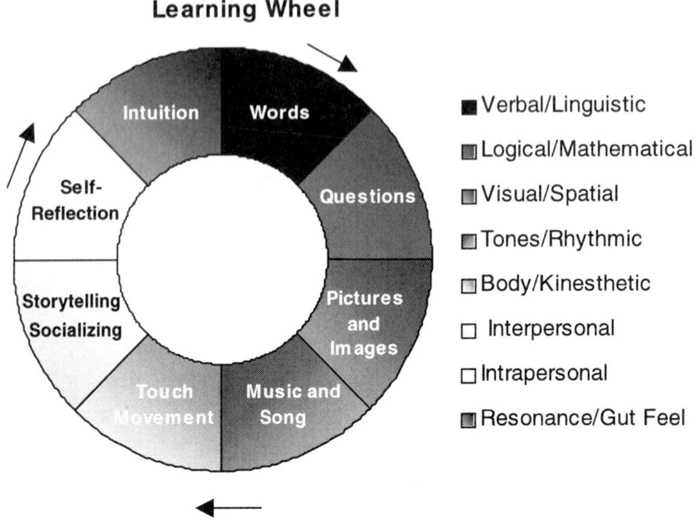

Case Study 4

> An instructional design project team that produced interactive online learning tools met to brainstorm ideas to solve a technical problem. The team leader and the senior designer were engaged in a discussion that was going in circles. The leader kept saying "can you see what I am saying?" in response to the designer saying "I hear you, but I need you to answer these three questions." Another team member was fiddling with the actual tool, which created additional frustration because other members thought she was not paying attention. Had team members known what their natural preferences were for learning, their may have spent less time being right and moved on to solving the problem they could deal with.

Looking at One Side of the Equation. To breathe more life into project team that has lost its ingenuity, reintroduce small amounts of disorder. Why bother, who has time anyway? Research shows the importance of learning how to recognize the patterns of interaction that undermine learning because patterns of team interaction quickly become routinized (Conklin, 2001; Cha and Edmondson, 2003). Consider that what we may value as "good" also has an equal and "opposite" side, given the nature of the laws of thermodynamics. Cha and Edmondson (2003) identify individual vigilance, maximizing efficiency, and empowerment as three "good" patterns that can also *inhibit* learning.

First, *individual vigilance* encourages people to take personal responsibility to solve problems as they arise. Counterintuitively, this may create barriers to team success because it encourages independence. Next, business culture most often rewards and expects people to take *personal responsibility*. Consider the flip side. Cha and Edmondson (2003) report that 70 percent of nurses believe their managers expect them to work through the "daily disruptions" on their own. Speaking up about a problem or asking for help is perceived as incompetence. Finally, *empowerment* is often touted as a solution to productivity problems. Empowerment allows for the removal of mid-level managers or use of self-directed (leaderless) teams, leaving team members on their own to resolve problems that may stem from parts of the organization. Often, quick-fix solutions are applied, making a project look balanced. However, problems may resurface elsewhere, because the balance is artificial. The consequences then are even worse than before the problem was "solved."

Consider examining the lessons learned from a previous or similar project; look for patterns that do not work that once did make sense. Take steps to feed this information back into initiation phase of a new project. As well, a simple yet often overlooked step is to meet and ask for "permission" (at a social level, not in terms of authority) to change a process—don't assume that members all think the same as you do. Above all, keep an eye out for shifting constraints that may wreak havoc on things if monitoring and follow-up does not become part and parcel of actions that encourage learning.

Lack of a Common Vocabulary. Education and experience has predominantly prepared project teams to solve routine or "tame" problems. Teams set out to complete projects better and faster, yet projects failure rates remain high. Fragmentation of effort may be one cause linked to difficulties teams have in identifying and solving complex problems (Conklin, 2001). How do you know if fragmentation of effort is hurting your project? Blame is one visible symptom. Conklin (2001) characterizes blame as a kind of persistent white noise that becomes normal over time. High team personnel turnover is another symptom.

Developing a common vocabulary may be one antidote to reduce blame and turnover. On a micro level, developing a shared glossary of terms before the project starts often heads off problems as the project progresses, particularly if the project team training is multidisciplinary. Assuming that terminology in documents to assist in managing projects such the Project Management Institute's (PMI) *A Guide to the Project Management Body of Knowledge* is *generally accepted* and will be commonly understood appears to be a faulty assumption.

Herein lies the wicked problem: The term "generally accepted" according to PMI "means that knowledge and practices exists that apply to most projects most of the time, and that widespread consensus endures about their value and usefulness" (PMI, 2000, p. 3). Studies show that experts in any profession can only ever share parts of terminology and conceptual systems (Gaines and Shaw, 1989). Practical evidence is in the number of entries in a popular online project management glossary by Max Wideman soaring to over 5,400 terms/definitions from 100 sources, containing over 50,000 internal cross-referencing links (Delisle and Olson, 2004). The hundreds of project management glossaries filled with conflicting terms and definitions have paradoxically increased fragmentation instead of helping to create a shared social understanding. (See also the chapter by Crawford on standards in this regard.)

The first step lies in raising awareness about the extent of the problem. With this aim, a recent study asked 51 participants who on average had just over 15 years of project and business experience to define 20 "common" project management terms without using outside sources (Olson, 2001). The results are presented in Figure 39.5.

Only 37 percent consensus was achieved as an aggregate for the 20 common terms, disconfirming the belief that people-managing projects naturally have a high level of shared understanding. The problem is not easy to address. For example, 47 percent defined the term "competency" differently (high level of correspondence). Thirty-seven percent of respondents defined "project management maturity" totally differently. As well, they used different labels to describe it (i.e., age, experience), pointing to a high level of correspondence.

I have subsequently validated these findings by repeating the exact exercise using the same type of participants in the same conditions over the past two years. Consensus has never reached beyond 40 percent for all 20 terms. What level of shared understanding can we realistically achieve? Cha and Edmonson's (2003) study provides some insight into the why it is difficult to achieve a shared understanding (see Case Study 5).

Case Study 5

The CEO of a young marketing company strongly espoused and clearly articulated his vision of the core company values (i.e., one was unpretentiousness). Employees initially embraced his *sent* or *stated* values, but over time as the company grew, employees made increasingly negative attributions about the CEO. The way that employees made sense of the CEO's stated values did not map exactly onto the

FIGURE 39.5. DIFFERENCE IN SHARED UNDERSTANDING OF 20 TERMS.

Terminology \ Definitions	Same	Different
Same	**37% Consensus** Use terminology and definitions in the same or very similar way.	**20% Correspondence** Use different definitions for the same terminology.
Different	**28% Conflict** Use the same terminology but define concepts differently.	**13% Contrast** Totally differ in the use of terminology and definition.

> actual words, nor did their meaning fit exactly. The employee's *elaborated* values were broader and more ideological than the CEO's sent values. For example, the CEO's sent value of unpretentiousness (meaning being informal, not egotistical) was elaborated on by employees to encompass "no hierarchy and elimination of rank" and labeled "equality." When they finally raised concerns about the decision to name shareholders within the company, it was limited to asking the CEO for change and not in seeking alternate explanations for the problem or findings solutions. In short, employees harshly judged the CEO as hypocritical because they believed the company's "shared" values were violated.

In the initial honeymoon or appraisal period, positive or negative attributions are made about a leader's actions (Cha and Edmonson, 2003). If subsequent actions are inconsistent with the espoused values, people engage in "sensemaking" where they critically analyze and judge the situation. On a biological level, physical stress increases and results in more simplified responses if actions are deemed to be inconsistent (Weick, 1995). On a cognitive level, people seek to blame someone that is highly visible for the problem, disregarding any role that they many had or have in order to preserve self-esteem (Cha and Edmonson, 2003).

What can a project team do about creating a shared understanding?

- Avoid assuming anything and seek clarification.
- Create a safe environment that invites questions and clarification without judgment.
- Raise awareness and acknowledge differences and similarities among member's training.

- Identify how large a gap exists in shared social understanding on the team up front in the project.
- Capture the output using a display system (mind map or equivalent) to make the issues and ideas "visible" and reusable.
- Engage the team in a process of articulating issues/ideas as a way of getting to a reasonable level of consensus on terms that everyone needs to understand. This is a process of making knowledge "explicit" and visible so that it can be reused by the team (including stakeholders).
- Aim to identify and agree on the key characteristics of complex concepts rather than trying to nail down one catchall definition.
- Understand that meaning is socially constructed; having the project leader impose a list of definitions or state his or her values with no further discussion does not mean complete buy-in.

Above all, project managers/leaders need to invite discussion and questions and promptly follow up if gaps occur between their stated values and those values elaborated on by team members.

Summary

This chapter on teams and effective team building may serve as a reminder to some and act as a source of information to others about the processes, conditions, and behaviors that influences teams. The chapter provided a snapshot of some of the key areas to consider when building and developing teams, as well as identified common team pathologies that may arise over the course of a typical project life cycle.

Following good team practices to achieve successful projects is arguably not a cause-and-effect relationship. What is more certain is that if the points brought up in this chapter are disregarded, the job of identifying, growing, and developing teams may be that much more difficult.

References

Beder, S. 2000. Working long hours. *Engineers Australia.* (March): 42.

Belgard, Fisher, Rayner, Inc. 2000. Building a collaborative team environment. Team Tools Module 14: 20811 NW Cornell Road, Suite 100, Hillsboro, OR 97124.

Blount, S., and J. Gregory. 2001. When plans change: Examining how people evaluate timing changes in work organizations. *Academy of Management Review* 26:566–585.

Brown, S. M. 2002. Changing times and changing methods of evaluating training. Lesley College, Cambridge, MA.

Brown, L. 2002b. The secret to mental toughness at work. www.lisabrown.ca/mttforwork.html (accessed November 17, 2002).

Bureau of Labor Statistics. 1999–2000. Women's share of labor force to edge higher by 2008. *Occupational Outlook Quarterly* (Winter): 33–38.

Cha, S., and A. Edmondson. 2003, How promoting shared values can backfire: Leader action and employee attributions in a young, idealistic organization. Harvard Business School. Division of Research, paper 03-013.

Chen, X., and W. Barshes. 2000. To team or not to team. *The China Business Review.*

Cleland, D. I., and W. R. King. 1988. *Handbook on project management.* New York: Wiley.

Conklin, J. E. 2001. Wicked problems and fragmentation. Cognexus Institute. This paper is Chapter 1 in the forthcoming book *Dialogue mapping: Defragmenting projects through shared understanding.* CogNexus Institute. 2003. http://cognexus.org.

Conklin, J., C. Ellis, L. Offermann, S. Poltrock, A. Selvin, and J. Grudin. 1998. Towards an ecological theory of sustainable knowledge networks. Touchstone Consulting Group White Papers. 1920 N Street NW, Suite 600, Washington, D.C. 20036.

Crisp, B. C., and S. K. Jarvenpaa. 2000. Trust over time in global virtual teams. Presented at Organizational Communication & Information Systems Division of the Academy of Management Meeting.

DeBono, E. 1977. *Lateral thinking: Creativity step by step.* New York: Harper & Row.

Delisle, C. 2001. Success and communication in virtual project teams. PhD diss. Dept. of Civil Engineering, Project Management Specialization. The University Of Calgary, Calgary, Alberta.

Delisle, C., J. Thomas, K. Jugdev, and P. Buckle. 2001. Virtual project teaming to bridge the distance: A case study. PMI seminar and symposium. Nashville, TN.

Delisle, C., and D. Olson. 2004. Would the real project management language please stand up? *International Project Management Journal.* p. 327–337.

Delisle, C., and J. Thomas. 2002. Defining success to get traction in a turbulent business climate. *Proceedings of the 2nd Annual PMI Research Conference.* Seattle, WA.

DeMarco, T., and T. Lister. 1987. *Peopleware: Productive projects and teams.* New York: Dorset House Publishing Co.

Dixon, M. 2000. *Project Management Body of Knowledge.* 4th ed. High Wycombe, UK: The Association for Project Management.

Drake, B., and Morin. 2000. *CEO turnover and job security: Research highlights from a worldwide survey.* 101 Huntington Avenue. Boston, MA 02199. Toll-free 800 DBM-2242.

Duarte, D., and N. T. Snyder, 2001. *Mastering virtual teams.* 2nd ed. San Francisco Jossey-Bass:.

Duxbury, L., and C. Higgins. 2001. Work-life balance in the new millennium: Where are we? Where do we need to go? Discussion Paper No. W|12. Canadian Policy Research Networks. 250 Albert Street, Suite 600. Ottawa, ON K1P 6M1.

Dyrenfurth, M. J. 2000. Trends in industrial skill competency demands as evidenced by business and industry. 1–18. *Proceedings from the International Conference of Scholars on Technology Education. European Commission.*

European Foundation for the Improvement of Living and Working Conditions. 2000. Third European survey on working conditions. EU0101292F. http://217.141.24.196/2001/11/study/tn0111109s.html.

Forsberg, K., H. Mooz, and H. Cotterman. 2000. *Visualizing project management.* New York: Wiley.

Frankel, C. 1999. *Emotions and emotionsand economic knowledge: Drivers of self-regulatory information processing.* Pacific Graduate School of Psychology

Howard, N F., Jr., and T. Mitra. 2001. Nov. labor force projections to 2010: Steady growth and changing composition. *Monthly Labor Review* 124(11):21–38.

Gaines, B. R., and M. Shaw. 1989. Comparing conceptual structures: Consensus, conflict, correspondence and contrast. *Knowledge Acquisition* 1(4):341–363.

Gardner, H. 1993. *Frames of mind: The theory of multiple intelligences.* New York: Basic Books.

Gersick, C., and J. Hackman. 1990. Habitual routines in task-performing groups. *Organizational Behavior and Human Decision Processes* 47(1):65–97.

Grazier, P. 2002. Work in the 21st century will recognize human potential. Telephone: (610) 358-1961.

Hartman, F. T. 2000. *Don't park your brain outside: A practical guide to improving shareholder value with SMART management.* Newtown Square, PA: Project Management Institute.

Hill, R. E., and T. L. Summers. 1988. Project teams and the human group. In *Project management handbook.* 2nd ed. D. Cleland and W. R. King. New York: Wiley.

Hoffman, E. J. 2000. Developing superior project teams: A study of the characteristics of high performance in project teams. NASA's Academy of Program and Project Leadership. *Conference Proceedings.* PMI International Research Conference, Paris.

Human Resources Development Canada (HRDC). 2001. Recent immigrants have experienced unusual economic difficulties. *Applied Research Bulletin* 7 (1, Winter/Spring).

Ingham, A. G., G. Levinger, J. Graves, and V. Peckham. 1974. The Ringlemann effect: Studies of group size and group performance. *Journal of Experimental Social Psychology* 10:371–384.

Johnson, K. L., D. S. Lero, and J. A. Rooney. 2001. Work-life compendium. 150 Canadian statistics on work, family and well-being. Centre for Families, Work and Well-Being, University of Guelph.

Kerzner, H. 2001. Project management: A systems approach to planning, scheduling, and controlling. 7th ed. New York: Wiley.

Kierstead, J. 1998. Competencies and KSAO's. Public Service Commission of Canada. www.psc-cfp.gc.ca/resaerch/personnel/comp_ksao_e.htm (accessed October 15, 2002).

Lipnack, J., and J. Stamps. 1996. *Virtual teams: Reaching across space, time and organizations with technology.* 2nd ed. New York: Wiley.

Madigan, C. O. 1998. Perfecting Project Management Skills: Part One. *Business Finance Magazine* (December): 28.

Margerison, C. J., and D. J. McCann. 1994. The types of work index: A measure of team tasks. *The Occupational Psychologist* (23):24–31.

McGannon, R. 2001. Will your project team get the job done? *ESI Horizons Newsletter.* (March).

McGrath, J. E. 1990. Time matters in groups. In *Intellectual teamwork: Social and technological foundations of cooperative work,* ed. J. Gallegher, R. E. Kraut, and C. Egido. 23–61. Hillsdale, NJ: Erlbaum Press.

Meyer, P. 1998. Trouble finding good people? Stop trying to hire them. *The Business and Economic Review.*

Miller, T. E. 1988. Teamwork: Keys to managing change. In *Project Management Handbook.* D. I. Cleland and W. R. King. 2nd ed. New York: Wiley.

Mills, T., S. Tyson, and R. Finn. 2000. The development of a generic team competency model. *Competency and Emotional Intelligence* 7(4):1–6.

Ministry of Industry. 2002. Women in Canada: Work chapter updates. Statistics Canada Housing, Family and Social Statistics Division. Catalogue no. 89F0133XIE.

NASA. 2002. *Characteristics measured by TeamMates: NASA's Project Team Development Survey.* NASA Academy of Program and Project Leadership.

Next Step. 2002. Case study: London & Manchester Assurance Company Limited. Customer Service Teams. www.nextstepltd.co.uk/case_studies/archive/london_manchester.html.

Olson, D. 2001. Is a common vocabulary lacking in project management? Information Technology Project Management. Executive MBA program at Athabasca University.

Ott, M., and L. Nastansky. 1997. Modeling organizational forms of virtual enterprises, University of Paderborn, Business Computing 2. Warburger Str. 100, D-33098 Paderborn, Germany.

Pinto, J., and D. Slevin. 1988. Project success: Definitions and measurement techniques. *Project Management Journal.* (February): 67–71.

Project Management Institute. 2000. *A guide to the Project Management Body of Knowledge.* Newtown Square, PA: Project Management Institute.

Potter, B. 1998. *Overcoming job burnout: How to renew enthusiasm for work.* Berkeley, CA: Ronin Publishing.

Sommerville, J., and S. Dalziel. 1998. Project teambuilding: The applicability of Belbin's team role self perception inventory. *International Journal of Project Management* 16(3):165–171.

Statistics Canada. 2000. Income in Canada, 1998. Labour force and participation rates. Ottawa: Statistics Canada, Catalogue 75-202XIE.

Statistics Canada. 2001. Canadian Statistics: Labour force and participation rates. (Online table). Women in Canada. Ottawa: Statistics Canada, Catalogue 89-503-XPE. www.statcan.ca (accessed November 22, 2002).

Stephenson, C. M. 2000. Innovation through people: One CEO's experience. Lucent Technologies Canada Inc. Presentation to the York Technology Association.

Strassmann, P. A. 1995. Outsourcing: A game for losers. *Computerworld* (August 21): 75.

Stark, M. 2001. Five keys to successful teams. PricewaterhouseCoopers. www.pwcglobal.com/Extweb/NewCoAtWork.nsf/docid/ 5D9D4B372EC8367485256C6100771C85 (accessed November 10, 2001).

Sugarman, K. 1999. *Winning the mental way.* Buligame, CA: Step Up Publishing.

Swann, W. B., J. T. Polzery, D. C. Seyle, and Sei Jin Ko. 2002. Finding value in diversity: Verification of personal and social self-views in diverse groups. National Institutes of Mental Health MH57455 Department of Psychology, University of Texas at Austin. Division of Research at Harvard Business School.

Thomas, Janice, Delisle Connie, and Kam Jugdeo. 2002. Selling project management to senior executives—framing moves that matter. Newtown Square, PA: Project Management Institute.

Thoms, P. Forthcoming. *Driven by time: A leader's guide to time orientation.* Portsmouth, NH: Praeger/Greenwood.

Thoms, P., and D. B. Greenberger. 1995. The relationship between leadership and time orientation. *Journal of Management Inquiry* 4(3):272–292.

Thoms. P, and J. K. Pinto, J. 1999. Project leadership: A question of timing. *Project Management Journal* 30 (1).

Tucker A. L., and A. C. Edmondson. 2002. Why hospitals don't learn from failures: Organizational and psychological dynamics that inhibit system change. Boston: Harvard Business School.

Tuchman, B. W. 1965. Developmental sequence in small groups. *Psychological Bulletin* 63(6):384–389.

Walsh, M. W. 2001. Luring the best in unsettled times. www.nytimes.com/library/financial/01working-wals.html

Weick, K. E. 1995. *Sensemaking in organizations.* Thousand Oaks, CA: Sage Publications.

Wideman, R. M. 2001, June. Project Management Glossary. www.maxwideman.com/pmglossary/PMG_E00.htm (accessed March 2002).

Wilemon, D. L., and B. N. Baker. 1988. Some major research findings Regarding the human element in project management. In *Project Management Handbook.* 2nd ed. D. I. Cleland and W. R. King. New York: Wiley.

CHAPTER FORTY

LEADERSHIP OF PROJECT TEAMS

Peg Thoms, John J. Kerwin

Leadership is difficult under any circumstances, but the leadership of project teams is particularly challenging for a number of reasons. First, projects almost always involve initiating changes. In addition, projects are typically assigned to project leaders by a person or people who have their own visions of what the end product should be. The project leader must understand and achieve those visions while creating his or her own image of the project outcomes and, particularly, of the process necessary to complete the project. With the management of large projects, there may be numerous stakeholders who have their own ideas about the end product. The project leader must please everyone, even when there are contradictory opinions. Finally, unlike the day-to-day organizational management of a particular operation, projects typically have a limited duration and utilize team members who move from project to project. Project team members must be self-directed and goal-oriented, and must take on leadership roles themselves in many situations. The challenges for effective project leaders are successfully bringing about change, satisfying their stakeholders, and leading teams of contributors who may be involved on a short-term basis.

This chapter explains several of the best-known theories of leadership that have implications for project leadership. In particular, readers will be introduced to transformational leadership theory and how it applies to project teams. Readers will learn to develop and articulate a project vision. In addition, the chapter focuses on how leaders can ensure that the project vision is being implemented. We also introduce the concept of time orientation and how leaders can utilize their own and others' time perspectives to make projects more successful. Finally, we examine two complex projects as examples of how a vision can drive project decisions and inspire project teams.

Leadership Theories

Leadership is defined in different ways by different experts. For purposes of project leadership, we will define leadership as the process of influencing others to understand what needs to be done and how it can be done, coordinating and motivating the work of various individuals and subcontractors, and delivering a successful product in the context of a project. Because project leaders often take someone else's vision and execute it, it is essential that they are able to grasp the basic concept that is being developed. In other cases, project leaders develop the vision and sell their idea to the funding sources. The "how" portion of project leadership requires technical expertise that the leader brings to the project through personal knowledge and experience or through recruiting appropriate team members who have the expertise. For example, a number of subcontractors work on large projects, which makes it necessary for the leader to organize and schedule the work of various team members while providing a motivating work environment. The leader must be constantly aware of criteria that should and will be used to measure project success. Continuous assessment and monitoring of performance are a critical component of the delivery of the finished product.

Most leadership theories fall into one of two categories: contingency or universal. Contingency leadership theories suggest that different times, tasks, and organizations may require different types of leaders or leadership behavior. These models imply that leaders can and do change their behavior as the needs change. Universal theories suggest that an effective leader is an effective leader regardless of the situation. These theories describe traits and behaviors that should work with any organization. Here are some examples of each type of theory that relate to project leadership.

The *Situational Leadership Model* (Hersey & Blanchard, 1977) suggests that leaders need to use more relationship-oriented behavior in some situations and more task-oriented behavior in others. Specifically, they tell leaders that when a team member is capable, but lacks motivation, a participative style is best. When a team member is capable and motivated, it is best to delegate authority. When a team member is inexperienced but motivated, the leader should provide guidance, explain decisions, and clarify procedures. When the team member is inexperienced and lacks motivation, the leader should dictate tasks and closely monitor the work. This theory focuses on the follower's level of maturity and how leaders should interact with them.

Path-Goal Theory (House, 1971) explains how the behavior of a leader influences the feelings of satisfaction and performance of subordinates. Essentially, this theory suggests that leaders should increase the personal payoffs to team members for work-goal attainment and make the path to these payoffs easier by clarifying work direction, eliminating roadblocks, and increasing opportunities for personal satisfaction on the way to meeting the goal. According to this theory:

1. Leaders must clearly communicate the desired outcomes and any necessary steps or requirements along the way.
2. Leaders must ensure that team members are consistently rewarded.
3. The rewards must fit the needs and interests of the individual team member.

The Normative Decision Model (Vroom & Yetton, 1973; Vroom & Jago, 1988) provides a framework to help leaders determine the optimal level of participation of team members needed for effective decision making. Levels of participation are as follows:

AI. The leader makes the decision alone using only information available at the time.

AII. The leader obtains information from team members and makes the decision alone.

CI. The leader shares the problem with team members individually, gets input, and makes the decision alone.

CII. The leader shares the problem with team members in a group setting, gets input, and makes the decision alone.

GII. The leader shares the problem with team members as a group, and the group generates alternatives, reaches agreement on a solution, and reaches a decision.

The model provides questions that help the leader determine which level of participation is best for each situation. In general, the following guidelines can be used by project leaders:

1. The more important the decision, the more participation is needed.
2. The less information the leader has, the more participation is needed.
3. The more information team members have, the more participation is needed.
4. The more important it is that team members accept the decision, the more participation is needed.
5. The fewer rules, procedures, and policies, the more participation is needed.

Leadership Substitutes Theory (Kerr & Jermier, 1978) is one of the most useful for project leaders because they are likely to be working with experienced professionals on their teams. This theory suggests that certain characteristics of a project team may actually substitute for leadership. These substitutes include team members with a professional orientation, experience, ability, and training. Other substitutes for leadership include structured and routine project tasks, intrinsically satisfying work, and feedback that comes directly from the work, like automatically generated progress reports. Finally, other substitutes include a cohesive team and established formal roles and policies. This theory helps leaders understand that choosing the right team members and setting up efficient work systems make the job of a leader much easier. Individuals who enjoy what they do and receive feedback about their work through the system do not need the level of interaction and supervision required in other situations.

In addition to the lessons we can learn from the theories described. Yukl provides some other suggestions relevant to project leaders that are based on the research done on contingent leadership theories (Yukl, 2002).

1. Spend more time planning for long complex tasks.
2. Consult with team members who have relevant knowledge.
3. Provide more direction to team members with interdependent roles.
4. Monitor critical tasks and unreliable team members or subcontractors more closely.
5. Provide more coaching to inexperienced team members and to those who have stressful tasks.

Universal theories of leadership are based on the belief that a good leader is a good leader, regardless of the situation, because of personal traits and patterns of behavior that always work. The *Great Man Theory* was one of the earliest theories of leadership. This theory tells us that great leaders are born and are born to be effective. When this theory evolved, leaders were typically male and inherited their positions of power. Although few people give this theory much credibility today, other universal theories do not differ much from the idea that some leaders "have it" and others do not.

Charismatic Leadership Theory was introduced in 1947 and refined later (Weber, 1947). Charisma is a Greek word that means "divinely inspired gift." It refers to the ability to perform miracles or predict the future. Modern theorists use the term to describe the ability of a leader to solve an immediate crisis, articulate a vision, attract followers, and achieve parts of the vision. A recent version of the theory suggests that charisma is attributed to the leader by followers (House, 1977). Followers are more likely to see a leader as charismatic if the leader:

1. Advocates a vision that is different from the status quo but socially acceptable.
2. Acts in unconventional ways to achieve the vision.
3. Makes personal sacrifices, takes risks, and pays a cost in order to achieve the vision.
4. Appears to be confident.
5. Uses persuasive appeals rather than using an authoritative approach with followers.

Charismatic Leadership Theory suggests that these types of leaders influence others because followers identify with and want to please and imitate the leader. Followers measure their own success by gaining approval from the leader. Leaders who praise and recognize followers who perform well reinforce desired behavior. In addition, charismatic leaders introduce and develop new values and beliefs. For project leaders, the implications are a bit more complicated, since these theories suggest that one is either charismatic or not. Charismatic leaders are most apt to emerge during a crisis. For example, if a project is in serious trouble, there is an opportunity for a project manager to demonstrate the behaviors typical of charismatic leaders. These behaviors may occur naturally, or it might take hard work on the part of the leader to change and inspire followers.

Transformational Leadership Theory (Bass, 1985) is the most widely studied theory in the past 20 years. This theory distinguishes between transformational leaders and transactional leaders. Transformational leaders motivate followers by making them aware of the importance of their work, convincing them to sacrifice self-interest for the sake of the group, and encouraging them to achieve higher-order needs like belonging to a group and achieving important goals that help others. Transactional leaders simply exchange rewards for task performance, but do not inspire followers or build commitment to the group or organization. Bass suggested that transformational leaders engage in specific behavior that allows them to affect followers' motivation. These behaviors are as follows:

1. *Idealized influence.* Behavior that brings about strong emotions among followers who identify with the leader
2. *Individualized consideration.* Providing support and encouragement to followers and coaching them to bring about improved performance
3. *Inspirational motivation.* Communicating a positive future vision, using symbols to focus work effort, and modeling desired behavior

4. *Intellectual stimulation.* Increasing follower awareness of problems and influencing them to see problems from a different perspective

A number of other leadership experts have explored the behaviors that seem to make transformational leaders so effective in bringing about organizational change. Kouzes and Posner have studied this behavior for many years and suggest that there are five critical behaviors common to transformational leaders (Kouzes & Posner, 1995):

1. *Challenging the process.* Continuously seeking new options and exploring others. Transformational leaders want to make things better even when the current system is broken.
2. *Inspire a shared vision.* Creating a vision that appeals to the values, goals, and interests of followers. Transformational leaders have an uplifting and rewarding image of the future that they effectively communicate to followers.
3. *Enable others to act.* Empowering followers to make decisions and act. Transformational leaders educate followers, allow them to make important contacts, and encourage them to create and try new things.
4. *Model the way.* Modeling the behavior that they value and want others to practice. Transformational leaders set an example for the level of performance they expect from others. They are dramatic, take their work personally, and tell stories which reinforce their values.
5. *Encourage the heart.* Rewarding and encouraging followers. Transformational leaders reward and recognize followers publicly. They provide feedback to followers and create winners by helping people become successful, even when it seems unlikely.

Project Vision

Vision is the common thread between the various theories of transformational leadership. According to these theories, it is necessary to be able to create and communicate a positive vision of the future in order to bring about major changes in organizations. Vision is not a new concept. There are legends and stories about visionary leaders that trace their origins to prehistoric times. Vision is a concept that cuts across cultures as well. Most peoples on this planet have known of the importance of vision for millennia. Over 1,000 articles and books have been written on vision in the academic press alone.

Despite the evidence of its importance, vision is an overused and widely misunderstood term in organizations. The problem is that few people really understand vision and fewer still have it. In most organizations and groups, the vision is nothing more than a long general statement that appears on a plaque on the wall and makes it to the second page of the annual report. These vision statements all look alike and provide little if any guidance to followers regarding the direction of the organization. A vision that will drive project and organizational performance is much different, and we define it this way: *A vision is a cognitive image of an organization or project that is positive enough to followers to provide motivation and elaborate enough to provide direction for planning and goal setting.* It may exist in the mind of the leader or, if well communicated, be understood by all constituents of the organization.

It is critical that a vision be idealistic. Visionary leaders are idealistic. Just as challenging goals lead to higher levels of performance, idealistic visions lead project teams to higher levels of accomplishment. If a project leader envisions a project where there are no cost overruns and personally behaves in a manner consistent with that, communicates that vision to team members, and is persistent whenever even a slight overrun occurs, the project will come in closer to budget. Part of the project planning will include specific strategies to ensure no cost overruns. An idealistic vision drives the planning process in ways that a "realistic" vision (or no vision at all) will never do. Visionary leaders always start with the best possible outcome in mind.

Based on recent research (Thoms & Blasko, 1999), the ability to create a vision is related to other personality characteristics. One study found that high visioning ability is related to an optimistic style toward life and situations. People who are positive and optimistic about their chances for success are more likely to imagine a positive future for themselves and their organizations. In addition, people who have higher future time perspective are more apt to have high visioning ability. People who are present- and past-oriented are less likely to think about the future and to create images of what the future could be. Nonetheless, most leaders can learn to create and communicate a vision.

Having a vision is not enough for a transformational leader, however. The vision must drive the leader's and the organization's behavior. Imaging is an important concept in management, and we use imaging training to improve sales, athletic, and task performance. We teach salespeople to picture a meeting with a customer and themselves overcoming objections. We teach athletes to imagine a ball going into a hoop. We know that creating mental images of ourselves being successful leads to successful performance. Vision works the same way, except that a vision is often broader than just one individual's performance. We create a positive image that drives our behavior, and, in turn, our behavior influences followers (Thoms & Govekar, 1997). At the project level, the behavior most likely to lead to achievement of our project vision would include the following:

1. *Planning.* Identifying the potential barriers to achieving the vision and finding ways around them.
2. *Influencing.* Convincing followers to stay committed to the vision. Particularly when problems occur, leaders must focus attention on achieving the vision and not compromising quality for an easier route.
3. *Selecting project team members.* Recruiting and hiring people who share similar values, goals, and interests who will be able to commit to the vision and have the skill to perform at the appropriate level.
4. *Providing feedback.* Executing a vision requires continuous monitoring of progress and adherence to the vision. Followers must be given accurate feedback regarding their performance and the performance of the team. When performance is inconsistent with the vision, people must know so that they can change their behavior.
5. *Choosing the appropriate use of the leader's time.* Deciding from day-to-day how to spend the time. If one crew is not on target and the vision includes meeting all timelines, the leader will have to spend part of that day finding out what is wrong and helping to find solutions. Other tasks that do not influence the success of the vision should be set aside.

Often, project leaders let tedious repetitive tasks, some of which could be delegated, take up their time and fail to focus on the aspects that will determine the success of the project.

6. *Goal setting.* Establishing the goals for each aspect of the project and for each team member. Transformational project leaders set challenging goals, communicate them clearly, and monitor team members' progress toward their goals. The goals should answer the question, "How will the vision be achieved?" The vision answers the question, "Why do we have these goals?"
7. *Communicating the vision.* Articulating the vision or parts of the vision to the team members who are responsible for the execution of each aspect of the project. Communicating vision is done in a variety of ways on a project. For example, blueprints and storyboards are clear pictures of the leader's vision. Explaining how time frames will be followed or why using a particular type of material or a specific location is important may not be so simple to understand. Effective project leaders must translate the image in their minds into words and talk with the appropriate individuals, explain their vision, influence team members to commit to it, and provide direction. Every team member does not have to understand the entire vision for a complex project, but they had better understand what the leader is trying to achieve as it relates to them and their work.

Communicating vision takes place every day in everything that the leader does. We typically think of speeches made by leaders when we consider communication of vision, but it goes well beyond that and includes modeling appropriate behavior, relating performance feedback, and sharing success stories about people working on the project. When problems occur and it appears that the vision will not be achieved, the leader must decide whether to stick with the vision. When project leaders consistently sacrifice their vision for cheaper, faster, easier solutions to problems that occur, they send a message to followers that the vision doesn't matter.

As mentioned at the beginning of the chapter, project leaders often find themselves having to execute the vision of others who hire them. For example, the project engineer of a highway construction project is paid to make sure that the highway is built according to the blueprints and plans drawn by engineers and officials. Or, the director of a film may be executing the vision of a producer or a screenwriter who came up with the original idea. In these cases, the project leader has to understand the vision of the primary stakeholder. Nonetheless, the project leader can and should create a vision of the process that will be used to achieve the vision. Visions do not just encompass end products like a well-built highway or a successful film. The vision must also include the process. This means that the project vision may include the following: the project will fall within its budget, the workers will follow all laws, the employees will feel that they were treated fairly, and the community will be satisfied with the progress made. The ideal process is the vision that the project leader, who is charged with executing someone else's idea, must create.

Creating a Project Vision

Visionary leaders create visions automatically. Images of the future appear in their minds as naturally as the rest of us eat breakfast. That doesn't mean that they can achieve their visions, only that they probably have excellent imaginations, are future-oriented, and have an optimistic outlook toward situations. The rest of us have to focus a bit more in order to create a project vision and will also have to work hard to achieve it. One approach to vision creation that has been empirically tested involves three steps and utilizes several creativity development methods (Thoms & Greenberger, 1998; Pinto et al., 1998).

Step 1: Mapping

The first step is designed to help the project leader identify aspects of the project and elements that must be considered in order for the project to be successful. To begin, put the name of the project in the center of a sheet of paper. Identify each aspect of the project. For example, for a highway project, you would list budget, subcontractors, environmental issues, location, legal requirements, and existing traffic patterns, among other things. Distribute these aspects around the center, and consider each element of the project that relates to that aspect. For example, around environmental issues, you would list local laws, flora, fauna, landscaping, noise levels, and potential chemical pollutants, among other things. You may be able to break this down even further. For example, next to fauna, you may have listed breeding areas, endangered species, and nesting locations. You would continue the mapping, creating a detailed web or cluster of associations that reveal every aspect of the project that the vision should encompass. Figure 40.1 provides an example of what this might look like. More detail would be added by an experienced project leader who understands this type of project.

Step 2: Generating "Wouldn't It Be Great If . . . ?" Statements

The second step begins by returning to each aspect of the project identified in the first step and creating as many statements that begin with the phrase, "Wouldn't it be great if . . . ?" as appropriate to capture what you would consider to be the best possible outcomes. These statements capture the ideal outcomes for the project. Do not worry about being realistic at this point. By creating challenging goals, we achieve more in the long run even if every goal is not fully achieved. The same is true for a challenging vision. Going back to the example of the highway construction project, Figure 40.2 lists some possible statements that a project leader might generate for a highway construction project. Depending on the complexity of the project, there could be hundreds of these statements.

Step 3: Creating the Move in Your Mind

The third and final step is to pull all of the statements together and visualize the project at the end. Because this step occurs in the mind of the leader, it is difficult to impossible for others to monitor it. It requires the leader to spend a significant portion of time imaging how the project will progress and how each stage of the project will look. The leader would imagine the work proceeding, how team members are interacting, how the work site sounds and smells, the condition of the equipment, the attitude of the workers, the cooperation of the various subcontractors, and so on. In essence, the leader pictures every aspect of the

FIGURE 40.1. EXAMPLE OF THE DEVELOPMENT OF A MAP.

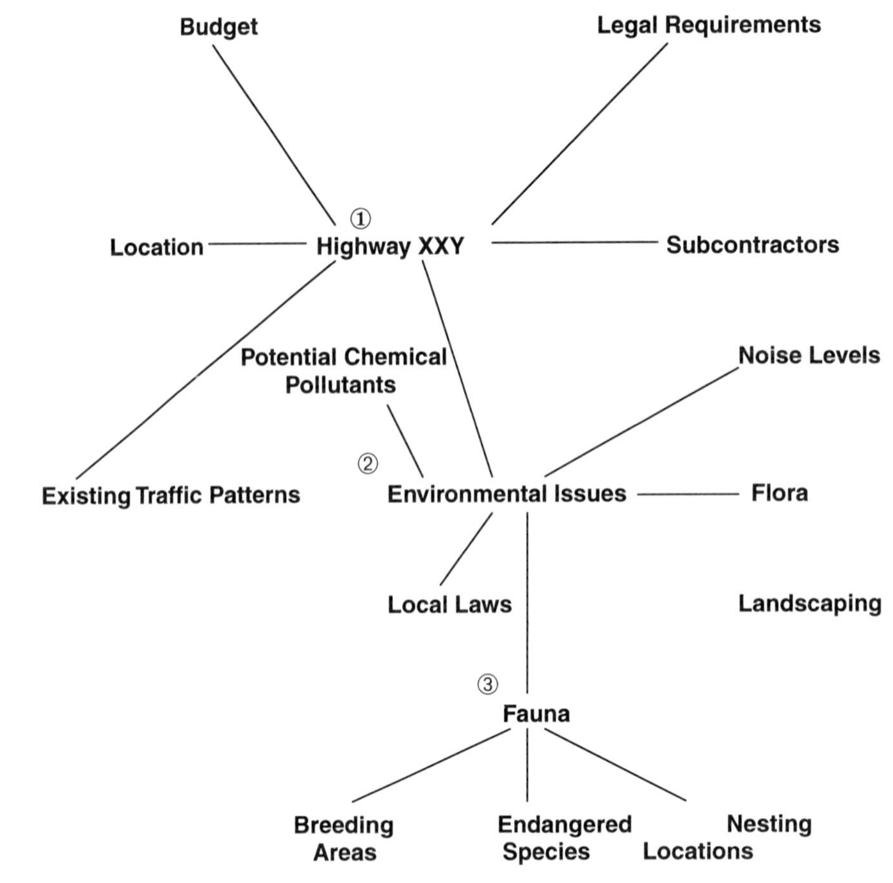

project before it occurs using the map and the statements generated in previous steps. This becomes the vision that drives the project. Visionary leaders do this automatically. Those of us who are not born visionary leaders must force ourselves to imagine successful projects and to mentally rehearse the behavior it will take to be successful. Mentally rehearsing difficult situations and specific challenges is a technique that has been used successfully for many years by the best athletes, salespeople, actors, and leaders.

Visionary leaders may not imagine every aspect of a project at the same time. It would be impossible for most humans to hold a clear image of every aspect of a complex project in one's mind at once. Most likely, leaders think about their projects in pieces and envision the best-case scenario for each aspect. At any point in time, the overall vision may be symbolized by positive feelings and the drawing, blueprint, model, or storyboards of the completed project. The images of the various pieces of the project come into focus only as each becomes important. As they emerge, the leader imagines that aspect of the project and

FIGURE 40.2. EXAMPLES OF "WOULDN'T IT BE GREAT IF . . . ?" STATEMENTS FOR A HIGHWAY CONSTRUCTION PROJECT.

Depending on the complexity of the project, there would be hundreds of these statements.

Wouldn't it be great if we had a complete list of all legal requirements with a list of the regulatory bodies and a contact person for each?

Wouldn't it be great if we had software that would track all legal issues as the project proceeds and alert us to any potential problems?

Wouldn't it be great if inspectors from each regulatory body made visits to the construction site at the appropriate time?

Wouldn't it be great if the project was completed with no labor stoppages?

Wouldn't it be great if all overtime expenses could be eliminated?

Wouldn't it be great if every worker had perfect attendance during the length of the project?

Wouldn't it be great if no OSHA inspections were requested during this project?

communicates it to team members. The leader of a construction project may move continuously around a work site, causing different images to come to mind as the focus shifts from excavating to framing to meeting with subcontractors. The key is that the leader is constantly comparing the work in progress against the vision of the ideal. It is critical to monitor progress toward the vision, looking for discrepancies between performance and the vision, and correcting the errors.

In the leadership literature, there is a lot of discussion about groups developing a vision. Although developing a vision is an activity in which many project teams engage, someone must commit to the vision in order for it to drive behavior and impact performance. Visioning or imaging only works if it is in the mind of the person who must execute it. It is highly unlikely that an entire team will imagine the same things and commit equally to a vision. Leaders do not have to use a participative style when they create a vision, but they do have to sell it to others. It can be useful, however, to involve other leaders in the process in situations where you do not have the necessary expertise or need to generate many ideas. Just don't assume that everyone on your team will see the vision the same way that you do or at the same time.

Project Leadership and Time Orientation

Recent research suggests that every person has a unique temporal alignment (Thoms & Pinto, 1999). *Temporal alignment* is made up of a variety of psychological constructs or biases that relate to time. Essentially, this means that we are all oriented primarily toward the past, the present, or the future. Each time orientation has its own strengths and weaknesses. In addition, different types of tasks may require unique time orientations. This suggests that different projects and different stages of projects require different temporal alignments. In other words, sometimes it is better to be past-oriented than future-oriented. Other times it is better to be present-oriented than future-oriented. It depends on the situation. When the leader has the temporal alignment that the situation demands, we call it *attunement.*

Effective leaders must occasionally behave in ways that fall outside of their temporal comfort zone in order to achieve goals. Most people understand that sometimes even shy people must be outgoing and friendly and that extroverted people must sometimes work alone. Because the concept of temporal alignment is relatively unknown, few people understand that sometimes we must be present-oriented when we would rather be working on an innovative project that won't be completed for ten years. Understanding temporal alignment and adapting to various situations can make leaders more effective.

The Future-Oriented Project Leader

Future-oriented project leaders are particularly effective when a project environment is dynamic and changing. They are the most likely of the three types of leaders to imagine a positive vision of the future. Envisioning the future is automatic behavior for most future-oriented leaders. Because most people are not future-oriented, visionary leaders must use their visions to motivate followers—to pull them into the future with them. Future-oriented project leaders tend to be very good at gathering information about what other organizations are doing to improve their effectiveness. This information gathering is done with the future in mind. As they read an article about another company trying something new, the future-oriented project leader is considering how the approach might work on their project. Future-oriented project leaders are constantly looking for opportunities—which may even be new projects. They are also apt to challenge current systems, seeking ways to improve. This creates tremendous frustration for followers who have done the work to make things flow smoothly and do not think about the future. Real problems can occur when future-oriented project leaders make changes without waiting for team members to catch up. One example of this would be a product development project leader who begins to market a product the company is adding to its line before the engineers on the team finish the design and production begins. Future-oriented project leaders must force themselves to slow down to accommodate the operational stages of change. This is difficult and will require effort depending on the strength of their future orientation.

The biggest problem that we see with future-oriented project leaders is their failure to recognize past accomplishments. This isn't just true for followers, but also for the leaders themselves. The past matters less to them. Most of the rest of us expect to be recognized for our performance. Some future-oriented project leaders may be less good at performance appraisals. They do not focus on the past year, may not remember what each individual reporting to them did or did not do, and tend not to value past performance. At best, feedback is usually delivered when the performance takes place. These leaders must use specific methods in order to do performance appraisals well. One example would be making notes when a follower does something out of the ordinary and putting the note in a file. Leaders who have administrative staff may ask them to keep track of specific information that can be easily accessed when an appraisal must be done. At the end of the evaluation period, the notes can be organized and used to develop feedback on performance.

Sometimes, future-oriented leaders are viewed as "off in the clouds," "daydreaming," "out there," or "in their own world." Most people are present-oriented, and they deal with life one day at a time. People who spend a great deal of time imagining the future are not mentally in the present much of the time and may even be seen as strange, depending on how well they can hide it. In meetings, future-oriented project leaders may not hear everything that is said or focus on the subject of the meeting. They are thinking about the future and are not available to the team, cognitively. This creates anxiety for team members who must get tasks done and need guidance or permission from leaders. Not hearing information provided in meetings can create critical problems, especially given that future-oriented people rarely read minutes of meetings.

Future-oriented project leaders are especially comfortable and effective in dynamic organizations. Because their orientation lends itself to taking advantage of opportunities and trying to make the future better, they work best on flexible projects with open management styles. They thrive in situations where creativity is rewarded, where they have access to decision makers, where leaders empower followers to make decisions, and where intuition or hunches are encouraged.

The Present-Oriented Project Leader

Leaders who live for the present and focus their attention on what is happening today are probably the most common of the three types. In a leadership role, present-oriented project leaders are very effective at dealing with day-to-day issues. This is true for several reasons. First, they care about the present. Problems are addressed one day at a time instead of speculating on how each aspect of the business may someday cause a problem. Their project teams tend to produce high-quality work when the leader communicates well, is accessible to constituents, and provides worthwhile feedback to team members.

Second, they are cognitively available on a daily basis. They will be able to help team members address concerns and develop methods to accomplish specific tasks. It would be unlikely to find this leader not paying attention or staring off into space when help is needed. The present-oriented project leader is likely to be circulating around the work site talking, observing, anticipating concerns, and answering questions.

Present-oriented project leaders are a good fit when leading projects similar to others they have done in the past. When they find themselves in dynamic situations, they will manage the production areas well, providing a critical anchor for those who are not future-oriented. Present-oriented project leaders are very good at solving problems. Because, they are not as reluctant to talk about the past as future-oriented leaders are and not as reluctant to talk about the future as past-oriented leaders are, they are willing to research the cause of the problems and help develop solutions for the future. When they have good communication skills, they are effective at providing information to team members and can be charged with this responsibility. They will deal better than other types of leaders with problem project workers. They will be more apt to go through the procedures in place, follow disciplinary guidelines, forget about problems that occurred five years ago that are not relevant to the current situation, and focus on solving the problem one day at a time.

The downside of present-oriented project leaders is that they may not make major contributions to the planning process. They may be too busy and may even sabotage the process if it will disrupt their operations. When they sabotage planning, they often kill it. They may also have a tendency to micromanage details and overlook the big picture. Vision is something they often lack, and they will need to make sure they have tangible things like blueprints or models to use to drive their behavior.

The Past-Oriented Project Leader

Past-oriented project leaders are very good at remembering and using the history of their project, department, and organization. This is often valuable. For example, it is important when a problem occurs and it is necessary to trace past meetings, decisions, and behavior. In this case, it is critical that information is preserved and that an oral history of the events leading to the problem can be re-created. Another contribution of past-oriented project leaders is their ability to trace patterns in their industry and to identify trends that may recur. These trends and patterns are very useful in predicting what may happen in a particular project or industry. Past-oriented project leaders with extensive experience will also remember the past behavior of specific leaders in other organizations, problems with certain subcontractors, and political issues that occurred on previous projects.

Past-oriented project leaders remember and value the contributions team members have made in the past. Future- and present-oriented project leaders tend to focus on "What have you done for me lately?" Past-oriented project leaders are more likely to forgive recent performance gaps if they know one is capable of good work based on past behavior. The past-oriented project leader believes that what followers did in the past matters. And, of course, it does. Projects with high turnover pay a price in terms of flexibility (bringing in new team members actually slows down processes and the ability to make changes), loyalty (workers without job security are also planning their next career move), and continuity of service to customers.

The downside to this orientation toward the past is that sometimes past-oriented project leaders overlook current performance problems. Previous performance can become a halo over the team member. The leader may simply think of the team member as a good employee without noting more recent problems or an inability to respond to current performance demands. The biggest problem with past-oriented project leadership is that the future is sometimes viewed as something that is going to happen to the organization, rather than something the organization will create. Past-oriented project leaders tend to respond well to trends but are less likely to set them. Maintaining the status quo is the result. Sometimes this works and sometimes it does not. These leaders are hard to engage in planning. They are often too busy, distract others involved in planning meetings, and fail to provide critical information. Once the strategic plan is complete, it is common for past-oriented project leaders to ignore the plan. An additional problem with past-oriented project leaders is a frequent lack of attention to day-to-day operations.

Applications to Project Leadership

A project that must deal with rapidly changing conditions in a dynamic industry needs future-oriented project leaders. If that is not your natural orientation, you will have to adapt to the situation. Highly future-oriented leaders must adjust when their projects are in the operational stages. Instead of spending their time thinking about the next project, they must focus on day-to-day activities, monitor progress, and provide feedback to team members. This may also mean that leaders who understand their own time orientation will select other team members who complement theirs.

In other words, if the project involves problem solving, a future-oriented project leader must recruit others who are more past- and present-oriented and use their abilities to do the research into the past and to explore the causes and solutions to the problems. A present-oriented project leader may use a future-oriented team member to attend strategy sessions, report back, and respond to issues related to the plan. Past-oriented project leaders are very good at remembering the local inspectors from previous projects and the issues they raised which can help the team anticipate problems. Attunement does not necessarily mean that a leader has to change in every situation, but it does mean that the project leader must find some way of understanding and getting the required work done so that every issue is adequately addressed. Because projects are often very complex, a variety of time orientations will be required during the course of each.

Developing Team Members into Leaders

Project leaders must rely on team members to carry out most of the tasks involved in large and complex projects. The most important part of delegating authority or empowering team members is making sure that they understand the vision—the vision of the constituents and the vision of the project leader. If the vision is understood, the team member must gain experience by managing small aspects of the project and moving up to larger and larger amounts of responsibility. This is a good opportunity to develop team members in order to complement the skills, expertise, and personality of the project leader.

For example, on a highway construction project, a promising team member may be assigned to research and develop the protocol for dealing with environmental issues. The project leader may need to monitor the team member's progress carefully the first time. After the team member has started to work successfully, the project leader can loosen the supervisory strings and allow the team member to make a few decisions. After each decision is carried out, the leader and the team member should debrief and talk about what should be done differently in the future. Then, the team member can be assigned a more complicated aspect of the project. As the team member develops more skill and confidence, the project leader can turn over more responsibility. This should be going on with several team members simultaneously.

Research tells us that most managers choose to hire, develop, and mentor people who are like them in terms of work experience, educational background, and personality. We do this because we are most comfortable with people who are like us. However, effective leaders specifically choose to work with people who are different because they bring skills, abilities, and personality traits that we do not have. They complement us and give our project teams and organizations balance. A heterogeneous project team will have more conflict and will take longer to make decisions, but it will also be more creative in developing new ideas, seek more and better solutions to problems, and find ways to make a leader's vision a reality despite barriers.

Effective project leaders also encourage and empower team members to take a leadership role on their own. In many organizations, the chain of command is strictly followed. Project teams tend to be different because we select experts to work on our teams. Expertise is a tremendous source of power, and project leaders need to give tested team members the authority and freedom they need to make decisions and follow through. This authority and freedom should be earned either through past project experience or a gradual building of levels of responsibility on the current project.

Examples of Visionary Project Leaders

To illustrate how visionary project leaders work, we have provided two examples. The first example is of a highway construction project. The second example is of a film production. Although film production is a relatively unexplored area in the project management literature, movie making shares the characteristics of typical projects utilizing a director as a project leader and crews of technical experts. Many of the problems of project teams result when various technical experts disagree on how different aspects of the project should be done. The role of a project leader is explored in these illustrations.

Example 1: Highway Construction Project. The East Side Access Highway took years to develop. After local and state officials interacted with the federal government during a ten-year planning period, ground was broken and this $130 million vision was a "done deal" by the time project engineers Reggie Jannetti and Dan Pellegrini assumed their leadership roles. They had to take the vision for this 6.2-mile highway link from Interstate 90 into downtown Erie, Pennsylvania, and convert the master plan from the blueprints of highway and bridge designers into a well-built, safe roadway. Currently, the East Side Access Highway is under construction. Although the project leaders were given the blueprints for the highway, they had to envision the construction process and make sure that the plans were executed in a safe, legal, timely, and cost-effective way.

Dick Corporation, a general contractor from Pittsburgh with extensive experience in road and bridge construction, was awarded the initial contract for $45 million to build two phases of the five-phase highway. Before ground was broken, the right-of-way officials had to buy land and properties in the highway's path so that the designer's vision could become a reality. In the initial stages, there was no one leader or engineer who could steer this huge project before the earth-movers arrived. Instead, teams of professionals, each with its own leader, had to utilize its area of expertise. To be successful, they would have to understand the vision of the constituents and the designers and the complexities of the job while sharing the overall project goal of completing a modern-day highway on budget and on schedule.

The Project Leaders: Project Engineers and Crew Chiefs. Reggie Jannetti is the project engineer in charge of a one mile phase of construction. A number of different crews and teams are involved and have to be managed during this phase of the construction. There are crews working for PennDOT, Jannetti's employer. Then there is a team of 15 inspectors who were hired by a consulting firm selected by PennDOT. Dick Corporation is the prime contractor responsible for hiring numerous subcontractors. One subcontractor paves the blacktop, another does curbs and gutters, one installs drains, and another removes and clears trees. Of course, various materials including concrete, stone, gravel, and paint come from suppliers who are critical players on the team. There are approximately 50 suppliers. Direct supervision of different crews could not come from one individual. Instead, leadership must come from the project engineers. Their directions are filtered through inspectors, then down to crew chiefs who see that the work gets done. It's a day-by-day, situation-by-situation challenge. For example, if ten trees stand in the way of the construction of an $18 million bridge essential to the highway, a project engineer must decide what to do.

Innovation and Creative Options: Key Ingredients in Leadership. Both Jannetti and Pellegrini agree that project crew chiefs are put to the test every day. There are numerous challenges in highway construction, but one of the most common is field conditions that are not on the blueprints but must be dealt with in order for the highway to be built. Jannetti faced a field condition that had the potential for stopping his phase of construction. There were ten trees whose height exceeded the maximum allowed and were located under the construction site of the bridge that was part of the highway. Chopping the trees was an easy option, except that they had been marked as untouchable for ecological concerns. It was time for an innovative solution. Jannetti and his staff, after consultation with all parties involved, topped off those trees so construction could proceed. In return, they planted over 50 new trees on surrounding Penn State University property. It was a trade-off that came out of Jannetti's experience in the field coupled with a talent for innovative solutions—a prime component of effective project leadership.

Example 2: Project Leader of a Film Production. Pete Jones of Chicago had a vision for a movie that he converted into a film script called *The Stolen Summer.* Jones spent one year writing his story of the struggles of two adolescent boys, one Jewish and one Catholic, growing up in Chicago, and submitted his screenplay to Miramax Studios in a nationwide competition. To his surprise, Jones, a first-time writer/director, won the competition. After the shock wore off, he realized that he would now assume the leadership responsibility of numerous crews of Hollywood professionals in order to carry out his creative vision. He had won the coveted green light to begin *The Stolen Summer.*

The Vision of The Stolen Summer. The Stolen Summer was not like most film projects that are pitched to Hollywood studios in a steady stream—industry estimates are that there is only one acceptance given per hundred submissions that are pitched each month. Jones considered himself lucky to get the go-ahead for his film. On the other hand, Jones knew that Miramax was a bottom-line-conscious studio whose box-office successes such as *The Matrix* and *Hamlet* were as creative as they were cost-effective film productions. Jones knew that he would have to be as frugal as he was creative. In addition, Ben Affleck and Matt Damon would be looking over his shoulder as executive producers of *The Stolen Summer*. Jones had gone from selling insurance to Hollywood director. He would now test his ability to communicate his creative concept, scene-by-scene, while infusing his actors and crew chiefs alike with the enthusiasm and direction needed during a nonstop, three-month production schedule. In short, he had to become an overnight leader.

The Storyboard—A Filmmaker's Blueprint. During the first phase of preproduction, Jones had to sell and convince studio executives that his screenplay was not only creative but also doable for the allotted budget. After considerable negotiations, he passed his first test of leadership by convincing key players that his project was financially viable. Then he proceeded to the most difficult phase of production: converting his vision and script into a scene-by-scene storyboard. This visual blueprint for his movie had to translate into a clear, concise schedule. *The Stolen Summer* began as a 132-page screenplay consisting of actor's dialogue, camera shots, scene and mood descriptions, and lighting effects. This was translated into a scene-by-scene detailed storyboard that was drawn by an artist and would serve as the creative backbone of the film. Using the storyboards, Jones had to lead his entire production crew from conceptualization to the actual shooting of the film, scene by scene.

His first task was to cast the actors for his movie by working with a casting director and his staff. At the same time, he had to meet with the costume designer and her staff in order to create the wardrobe look of the seventies in Chicago. Then he had to bring the line producer on board. Collectively, they had to connect the frame-by-frame pictures on the movie's storyboard into specific filming locations around Chicago. The second week of production began with intensive meetings with his director of photography, who was responsible for the cinematic look of the film. Then the production designer, who functions like an interior designer, joined the team in order to create sets and add props. By the end of the week, the lighting director, chief sound technician, and camera operator joined the planning process as Jones was ready to leave the safe confines of his offices and lead his crews on location in Chicago. There, they would be joined by electricians, carpenters, grips, production assistants, security personnel, caterers, and other logistical personnel. Jones was now the leader of a dozen separate film production crews. The total number of personnel on location was 52 people each day. Despite the enormity of this first-time task, Jones assumed the role of a film director who must be an effective and dynamic project leader in order to be successful.

A Day on the Lake: Leadership in Action. As in the case on most film shoots, circumstances arise that cause changes or alterations of the vision on the storyboard. The final call is the director's, who must lead the crews accordingly.

As a first-time venture for Jones, *The Stolen Summer* was experiencing more than its share of production problems and resultant financial crises. His idea for shooting in Chicago was true to his vision, but his producers, actors, and crew chiefs were looking for ways to cut the location filming short and film several scenes in Hollywood in order to save time and money. Jones had to be strong when suggestions for changes were made, especially when a critical scene in the story needed to be shot on the shores of Lake Michigan.

His line producers insisted that the scene could be shot in a more controlled environment in Los Angeles. The director of photography had numerous problems conceptualizing the shots, especially with scenes where the crew would be required to shoot in the water on a makeshift floating scaffold in order to get an overhead shot. The camera crew joined in the criticism, questioning the significance of the shot versus the time required to shoot it. In addition, the water was ice cold. The result was an across-the-board questioning of Jones' insistence on this shot for just a few seconds of a usable scene in the finished film. In light of all this, Jones' challenge was formidable—he could be dogmatic and demand that the shot be done here on the shores of Lake Michigan; or he could take the time to communicate the importance of these scenes to each of the concerned crews and convince them of the visual importance of the scenes to the finished product. In his mind, where else was he going to get the gorgeous sunset over the horizon of Lake Michigan edited with cutaways of the Chicago skyline? In Jones' vision, these shots were essential to the film, and he had to lead the crew through a long, cold day on the lake in order to put these shots in the can.

First, he had to convince his three producers. If they said no, everybody would pack up their gear and head back to Hollywood. He was persuasive and won their approval. Then he had to describe the look and camera angle of the scenes so that his director of photography and the camera crew would be on the same visual page with him. The audio technician and his crew were next as they all waded into the chill of the lake to record sound. Makeup and costume personnel had their hands full trying to cope with windy conditions and freezing actors. Jones had to rally his crew. Although it was a far cry from the demands of selling insurance, which was what he did before writing the screenplay, his sense of commitment and dedication to his movie were greater than the bone-chilling temperatures of Lake Michigan.

Jones led his crew that day to the windy shores and into the cold waters of Lake Michigan. In the final cut of *The Stolen Summer*, the overhead shots from the floating scaffold were not used and his child actor got so cold that he was unable to complete the filming. However, the shots of the boy walking along the shore were brilliant and added a great deal of emotional impact to the film. That made everyone happy, and the shoot was considered a success.

The ability to lead during the production of a film requires a clear vision of the project, well-honed people skills, incredible patience, a thorough knowledge of each crew's function and responsibilities, and a burning passion for the film that translates into enthusiasm and commitment. As a first-time director, Jones had the vision and the passion. As far as the future is concerned for Jones, public acceptance at the box office will be the ultimate indicator whether or not he will have future projects that get the green light.

Summary

Clearly, there is a considerable difference between *The Stolen Summer* and the East Side Access Highway project. But despite these differences, there was a vision that had to be translated to understandable terms, teams of technical experts that had to be managed, and dozens of workers who had to be motivated. Pete Jones, Reggie Jannetti, and Dan Pellegrini share some common leadership traits even though filming on the shores of Lake Michigan in three months is a world apart from building a six-mile roadway in three years. But the same project commitment, innovative solutions to problems, and the ability to keep crews enthusiastic about their work assignments are essential to their successful leadership.

Effective project leaders share many common characteristics and exhibit similar behaviors. We can learn a great deal from leadership theories, but even more by observing effective leaders. Paying attention to how problems are solved, how innovative strategies are developed, and how great project leaders communicate and motivate are the best ways to improve our leadership ability. In addition, volunteering for challenging assignments, learning from our missteps, and using our vision to drive our day-to-day behavior will help us develop into great leaders.

References

Bass, B. M. 1985. *Leadership and performance beyond expectations*. New York: Free Press.

Conger, J. A., and R. Kanungo, R. 1998. *Charismatic leadership in organizations*. Thousand Oaks, CA: Sage Publications.

Hersey, P., and K. H. Blanchard. 1977. *The management of organizational behavior*. 3rd ed. Englewood Cliffs, NJ: Prentice Hall.

House, R. J. 1971. A path-goal theory of leader effectiveness. *Administrative Science Quarterly* 16:321–339.

———. 1977. A 1976 theory of charismatic leadership. In *Leadership: The cutting edge*, ed. J. G. Hunt and L. L. Larson. 189–207. Carbondale: Southern Illinois University Press.

Kerr, S., and J. M. Jermier. 1978. Substitutes for leadership: Their meaning and measurement. *Organizational Behavior and Human Performance* 22:375–403.

Kouzes, J. M., and B. Z. Posner. 1995. *The leadership challenge: How to keep getting extraordinary things done in organizations*. 2nd ed.. San Francisco: Jossey-Bass.

Pinto, J. K., P. Thoms, J. Trailer, T. Palmer, and M. Govekar. 1998. *Project leadership from theory to practice*. Newtown Square, PA: Project Management Institute.

Thoms, P. Forthcoming. *Driven by time: Leadership and time orientation*. New York: Praeger.

Thoms, P., and D. Blasko. 1999. Preliminary validation of a visioning ability scale. *Psychological Reports* 85:105–113.

Thoms, P., and M. A. Govekar. 1997. Vision is in the eyes of the leader: A control theory model explaining organizational vision. *OD Practitioner* 29:15–24.

Thoms, P., and D. B. Greenberger. 1998. A test of vision training and potential antecedents to leaders' visioning ability. *Human Resource Development Quarterly* 9:3–19.

Thoms, P., and J. K. Pinto. 1999. Project leadership: A question of timing. *Project Management Journal* 30:19–26.

Vroom, V. H., and A. G. Jago. 1988. *The new leadership: Managing participation in organizations*. Englewood Cliffs, NJ: Prentice Hall.

Vroom, V. H., and P. W. Yetton. 1973. *Leadership and decision making*. Pittsburgh: University of Pittsburgh Press.

Weber, M. 1947. *The theory of social and economic organizations*. Translated by T. Parsons. New York: Free Press.

Yukl, G. 2002. *Leadership in organizations*. Upper Saddle River, NJ: Prentice Hall.

CHAPTER FORTY-ONE

POWER, INFLUENCE, AND NEGOTIATION IN PROJECT MANAGEMENT

John M. Magenau, Jeffrey K. Pinto

When we speak of power in project management settings, it is important to understand the meaning of the term and why it is so vital to achieving project goals. So much has been miscommunicated about the importance of power that it is easy for many of us to be mistrustful or wary of ever using power in organizations. The truth, however, is that successful project managers must become both adept and comfortable at using power all the time. Adept because they have to recognize how to use it well in order to avoid its misuse. Comfortable because it is the principal means by which major decisions are made, resources allocated, teams governed, and stakeholders satisfied. Successful project managers understand the constructive uses of power, as must each one of us.

Power is simply the ability to get activities or objectives accomplished in an organization the way one wants them to be done. Among the key activities that power enables us to achieve are the ability of affect decisions and control resources (Dubrin, 1995). Some definitions of power set it up as a confrontational issue; for example, the belief that power implies forcing someone to do something that they would not ordinarily do otherwise. Other definitions are more benign, stating that power is simply the mechanism to positively affect a project team's natural inertia. Underlying all these definitions of power is the belief that power "enables" some members of the organization to pursue objectives. Whether those objectives are for the good of the organization as a whole or are purely self-centered is another issue. Nevertheless, as part of any discussion of the project management process, it is important to understand the nature of power in our organizations, its various bases, how one gets and holds of power, and its potential effects on power holders.

Sources of Power

There are many different ways in which individuals can acquire power within organizations. The types of power individuals choose to exercise depend, in large degree, upon the types of people they are, the opportunities available to them, and the companies they work for. Sometimes, gaining power is merely the result of luck or fortunate circumstances. To understand how each of us may acquire power, it is necessary to consider some of the more common sources of power and the reasons why they exist. Table 41.1 gives a list of some of the more common types of power routinely found in organizations.

Positional Power. Positional power typically refers to the power an individual gains from occupying a position with the organization. For example, departmental managers would have a higher degree of positional power than their subordinates. Another common term for positional power is authority, stemming from the right to direct the behavior of others because of the higher position one may occupy in the organizational hierarchy. Positional power offers some important subcomponents of power, all based on positional authority. They include (1) legitimate power, (2) reward power, and (3) coercive power.

1. *Legitimate power.* Legitimate power is the power that is granted to people because of the position they occupy within the organization's chain of command. The higher up the corporate ladder an individual sits, the more legitimate power they are able to exercise in performing their duties. One benefit to a project team of enlisting the active support of senior executives is their ability to directly affect change or impact on the project's development.
2. *Reward power.* Often coupled with legitimate power is the power of distributing rewards to members of the organization as a reward for their performance or compliance with directives. This power to provide rewards often serves as the motivational "carrot" to

TABLE 41.1. TYPES OF POWER.

1. *Positional power.* Power that derives to an individual from the position they occupy in an organizational hierarchy
2. *Personal power.* Power stemming from personal qualities or personality characteristics within an individual
3. *Resource power.* Power that derives from one individual's control over critical resources needed for the project
4. *Dependency power.* The power that one individual or organizational unit acquires when others depend upon them, their output, or resources they can provide
5. *Centrality power.* The power an individual or unit receives from being closely linked to the primary activities within an organization
6. *Nonsubstitutability power.* The power that comes from the perception that an individual or group possesses a competency or skill that cannot be replicated by others within the firm
7. *Coping with uncertainty.* The power that derives from the ability to effectively cope with environmental uncertainty

induce project teams to perform to optimal levels. It is important to note, however, that reward power is only viable when the manager actually can employ it in meaningful ways. For example, in many organizations where employees are unionized, all monetary rewards are collectively bargained in advance, greatly limiting a manager's ability to provide additional rewards for performance. As a result, they may have to employ creative, nonmonetary rewards for superior performance.

3. *Coercive power.* In addition to the power to reward compliant or appropriate behavior, power that enables the leader to punish noncompliance is referred to as coercive power. Coercive power is best understood as the power that derives from the fear of punishment. Less powerful individuals are inclined to follow the direction of the leader who wields coercive power because of their wish to avoid punishment. Research suggests, however, that coercive power is only of limited usefulness as a motivator and tends to work best in the short term.

Within the arena of project management, the whole issue of positional power becomes much more problematic. Project managers in many organizations operate outside the standard functional hierarchy. While that position allows them a certain freedom of action without direct oversight, it has some important concomitant disadvantages, particularly as they pertain to positional power. First, because cross-functional relationships between the project manager and other functional departments can be ill-defined, project managers discover rather quickly that they have little or no legitimate power to simply force their decisions through the organizational system. Functional departments usually do not have to recognize the rights of project managers to interfere with functional responsibilities; consequently, novice project managers hoping to rely on positional power to implement their projects are quickly disabused.

As a second problem with the use of positional power, in many organizations, project managers have minimal authority to reward team members who, because they are temporary subordinates, maintain direct ties and loyalties to their functional departments. In fact, project managers may not even have the opportunity to complete a performance evaluation on these temporary team members. Likewise, for similar reasons, project managers may have minimal authority to punish inappropriate behavior. Therefore, they may discover that they have the ability to neither offer the carrot nor threaten the stick. As a result, in addition to positional power, it is often necessary that effective project managers seek to develop their *personal* power bases.

Personal Power. Another set of "power bases" relate to an individual's personal power, or the power that derives from characteristics or qualities within an individual. In the case of personal power, the source of power is not the result of an organizational mandate or position in the hierarchy; it is due to the traits that individuals possess. As a result, whereas positional power tends to collect within the standard authority structure of a company and is most obviously seen in higher levels, personal power can be used by people regardless of their formal position in the organization. When we think of the informal leader within a social group or the most respected members of a research and development team, we are observing that some individuals possess or use personal power to a greater degree than their peers. Just as positional power could be broken down into subcategories, so too can we look at various forms of personal power, including: (1) referent power, (2) expert power, (3) information power, and (4) connection power.

1. *Referent power.* Referent power simply refers to the situation that other organizational members like someone and want to be like them; that is, the power holder acts as a reference point for others. Referent power is an extremely significant form of power, as advertisers are well aware. Their use of star athletes to endorse products is their acknowledgment of the fact that a large percentage of people in our society will be swayed by the opinions of those they hold in high regard. Adolph Hitler's charismatic presence and oratorical abilities gained him a measure of personal power long before he became first chancellor and finally dictator of Nazi Germany. Likewise, within organizational circles, fine examples of referent power can be found at all levels of the organization. When members of a loading dock gang gravitate toward a friendly or physically large coworker, we find evidence of referent power. When a junior manager willfully conflicts with a superior and is lionized by coworkers for having the guts to stand up to the boss, he is experiencing a level of referent power. In all of these examples, the power holder is one who has the ability to sway the opinions of others through the dynamism of personal power; power that is evidenced by the regard with which the power holder is held by other members of the organization.
2. *Expert power.* A second form of personal power is called expert power and refers to the follower's belief that the power holder has some expertise or knowledge that the others need to perform their jobs. A common example of expert power can be found within the R&D departments of most organizations. Those individuals who are generally regarded by their peers as having expert knowledge will typically wield far more *real* power in the laboratory than the designated lab manager, particularly if that manager is not perceived to have an acceptable level of expertise.
3. *Information power.* Information power is another form of personal power and has strong ties to expert power. It is defined as the belief by followers that the power holder either possesses or has access to information that is necessary to perform a job. Some managers serve as the conduit for all forms of organizational information. Whether that information is conveyed in memos, gossip and hallway rumors, or direct access to upper management, these individuals hold a form of power because they possess this information. Other organizational members are willing to defer to them precisely to the degree that they perceive that this information, whether gossip or activity-based, is relevant and useful for their own work and organizational survival.
4. *Connection power.* The final type of personal power is connection power. The power of connections is well known in all organizational settings. Some individuals, regardless of their formal position, possess tremendous power solely because of their connections to other, powerful people. The classic adage, "It isn't *what* you know, it's *who* you know" succinctly illustrates the importance of connections as a source of power, not only with peers but, potentially, with superiors as well.

Note that each of the personal power bases illustrates a similar feature: The "power" found in each derives directly from human relationships. By definition, unless we are willing to explore relationships with peers and superiors in organizations, we cannot reasonably expect to develop any form of personal power. Additionally, these personal power bases offer project managers a wide range of options. Not everyone is blessed with a magnetic personality making it easy to cultivate and maintain referent power. Likewise, some people are adept at creating a network of powerful connections, while others, either through personality or external circumstances, do not have similar opportunities. On the other hand, acquiring information or developing expertise is within most managers' control and should be explored as alternative bases of power (Pinto, 1996).

Resource Power. Some individuals and organizational units have more power than others because they have the ability to control important resources. For example, when a project manager needs a critical resource, such as a lead programmer, for a project and must get the department head's approval to recruit such an individual, the department head is able to wield resource power over the project manager. Somewhat facetiously, it has been observed that the "real" golden rule states, "Whoever has the gold makes the rules." With resource power, there is a degree of truth to this adage. To the degree I can control your access to resources you require, I have power over you.

Dependency Power. Closely related to resource power, dependency power suggests that when one person or department possesses something another organizational unit needs, they have dependency power. It is very common for organizations to set up many operations so that dependencies exist between different departments. Before a construction project can begin, materials must be purchased. Prior to the start of engineering and prototype development, a corporation's design department must complete their work assignments. In both of these examples, one department operates with dependency power over another in that the follow-on unit "depends" upon the first department completing their work before the next can initiate their operations. Following this logic, we can expect that the more a department depends upon other units, the lower its relative power is within the organization.

Centrality Power. An important opportunity to acquire power within organizations occurs when individuals or departments have high visibility or are perceived as being closely linked to the primary activities within the organization. For example, a Senior Vice President for Projects within a construction organization will possess a significant amount of power because of the centrality of his or her role in the overall organization. Because project management is key to organizational success, the head of projects is likely to be very visible and central to firm operations (Daft, 2001).

Nonsubstitutability Power. A final common form of power is referred to as nonsubstitutability power. This form of power is best understood as the power an individual or department has when their activities or expertise are unique and not easily substituted. If, for example, a project organization has only one person with the important contacts or personal relationships with a potential client, that person has power because she cannot be easily substituted. In the 1970s, when computers were not as widely used as today, programmers had tremendous power within many companies because no one else could replicate their skills. As programming has become more widely disseminated throughout society, the nonsubstitutability power of computer programmers has diminished (Pettigrew, 1973).

Power versus Influence

Influence and influencers are pervasive in our society. Television and radio advertisements, televangelists, and salespeople represent examples of some of the most common types of influence we experience on a daily basis. Note that in each case, the influencer cannot *force* your compliance. Each of us has the power to simply change the channel or leave the store if we are offended or threatened by the influencer's message. How, then, do we explain the success that these people have in raising money and gaining sales? In a different context, how can we explain the success that some of our peers have in gaining compliance from other organizational members even though they have no direct authority over them? The answer to these questions typically focuses on the greater influencing ability that some of us possess relative to others.

Because influence is a key component in organizational life, it is important to distinguish between influence and power. Many managers define power in terms of influence as a convenient shorthand. Even textbooks make the same definitional link—suggesting, for example, that power is the capacity to influence others. More appropriately, we can define *influence* as one person's ability to get another to do something they want when there are no gross power differences between the two parties. That is, the influencer has no formal ability to "force" the other person to seek some goal or perform some task. From this definition, it is clear that there are some important similarities between power and influence: Both are used to change another's behavior. We will see, however, that the two concepts are very different in some important ways while demonstrating that a thorough knowledge of influence tactics can be an important tool for better managing within the organization's political climate.

Table 41.2 demonstrates some of the key differences between power and influence. We can classify these differences under three headings: scope and generality, strength of foundation, and tenure. *Scope and generality* refers to the nature of how one is able to use influence versus power. Typically, influence, in order to be successful, is situation-specific; that is, one who is adept at influencing others knows intuitively when and under what circumstances to attempt to change the other's behavior. Good influencers do not misuse or overuse their abilities, because they know that the more often they employ them, the more likely that coworkers will begin to refuse to comply with their wishes.

In addition, good influencers rely almost exclusively on face-to-face meetings, or at least opportunities in which one side can view the other. There are two primary reasons for preferring to meet head-to-head rather than using telephones or other media. First, it is

TABLE 41.2. POWER VERSUS INFLUENCE.

	Power	Influence
1. Scope and generality.	Cuts across situations and relationships.	Situation-specific and usually face-to-face.
2. Strength of foundation.	Strong base. Does not have to be done well to work.	Weak base. Must be used well or will not work.
3. Tenure.	Long-term.	Short-term.

much harder for the average person to refuse another during direct contact. Memos or telephone calls offer an impersonal approach that makes it easier to refuse an influencer's requests. The second reason is that good influencers are often adept at reading body language and other forms of nonverbal responses from their "target." When we observe good influencers in action, we see them constantly altering the "angle of attack" or promotional pitch as they perceive that one line of argument is either likely to be accepted or rejected. This sensitivity to the other individual's reactions is simply not possible while using other media.

Power does not accept the constraints of situation and approach. When some individuals have power over others, they are in the position to operate without regard to concerns about scope and generality. The "boss" or power holder is in the position to force compliance regardless of the situation and via whatever means they choose. Having power enables me to communicate through multiple means, any of which is just as effective for me as face-to-face meetings.

Another distinction between influence and power lies in *strength of foundation.* This concept refers to the fact that influence, in order to be effective, must be used well; that is, because my base (or foundation) of actual power over another may be weak, I must substitute effective planning, preparation, and role-playing in influencing another. Simply put, influence must be done "well" in order to work. Power, on the other hand, gives a manager a much stronger base from which to operate. The manager with power over another does not have to be constrained to exercising that power in clever or situation-specific ways. He or she simply tells another to do something and that subordinate is bound to comply. Further, because the power base is strong, power holders do not have to be particularly sensitive in using their authority. "Do it because I say so!" may be all the information that the power holder is required to convey.

A final difference between power and influence refers to its tenure. Power is much more long-lived than influence. As noted initially, just as influence is situation-specific, so too is it used sparingly. To overuse influence, particularly with any one individual, is often to lose influence. On the other hand, power lasts. As long as some functional managers occupy higher positions in the organizational hierarchy than others, they will have legitimate power over them.

Forms of Influence

What are some of the more common methods by which one individual can seek to influence the behavior of another? Because, as noted, influence is highly personal as well as being situation-specific, there are several tactics one can employ in trying to apply influence. Note, however, the underlying characteristics of the forms we will discuss in the following; they work best in face-to-face settings, they are situation-specific, and they implicitly assume that the influencer cannot directly force his or her will on the other person. Among the common methods are the following.

Persuasion. Persuasion is a tactic in which one person attempts to influence another simply by arguing the merits of their position. Persuasion suggests that if the target individual will simply give a fair hearing to the influencer, he or she will be won over on the strength of the argument. Persuasion is a sound influence tactic to employ when the influencer perceives that their arguments or evidence supporting them are strong. For example, in a project on which two organizations are cooperating, technical disagreements between the firms can often be resolved through persuasive argument in which detailed discussion, data analysis, and objective assessment are used as tools by one of the project partners.

Ingratiation. Ingratiation is the art of flattery, cajolery, or a search for common ground to win favor and gain another's willingness to cooperate. Ingratiation, as an influence tactic, offers the simple argument that it is easier to catch flies with honey than vinegar. For example, a project manager working for Chrysler Corporation made a point of saving all organization notices (transfers, promotions, awards of advanced degrees, and so forth). He filed them away, and prior to calling on another manager or executive for the first time, he always read up on any relevant information concerning that manager. Thus, he knew what the person's educational background was, where he or she was from, and something about the person's work history. This tactic offered the project manager a valuable source of common ground that made contacts cooperative and supportive of his needs.

Pressure. Pressure is a form of influence that applies external considerations as a supplement to the message itself; for example, time constraints or cost issues. Pressure tactics frequently seek to limit the target's freedom of choice or movement in order to gain compliance. A salesperson who claims that a product will only be on sale for a limited time is hoping that the added pressure of the time constraint will reinforce the influential message. Pressure, of course, can backfire. In fact, research in advertising has found that message content that the listener perceives to be too pressure-inducing will often have a boomerang effect, as the target individual resents the too-obvious display of pressure tactics.

Multiple Pressure. This alternative is an extension of simple pressure, in which the target is receiving the combined attention of two or more additional individuals. In employing this approach, other members of the project team "gang up" on the recalcitrant member or target for influence. Research suggests that pressure from multiple sources can be a very effective influence tactic, as it is often extremely difficult for most of us to resist the combined influence of several others.

Guilt. Guilt can be a powerful but sometimes overlooked form of influence; guilt implies a relationship based on obligation between the two parties. As a result, one party may attempt to sway the other by an appeal to his sense of duty or obligation, regardless of whether or not obligation actually exists. As with research on the effects of pressure tactics, recent research on guilt appeals in advertising demonstrate that in order to be effective, guilt must be used moderately. Too blatant examples of guilt-inducing appeals actually lower a person's willingness to cooperate and instead promote active resistance and resentment.

The preceding are just some of the more common influence tactics that project managers may use. Note the common feature of each of these tactics, regardless of how seemingly powerful or effective: They are only successful when employed correctly, used sparingly, and limited to exchanges that allow for immediate feedback (such as face-to-face meetings). Influence, though not the same thing as power, can still be an excellent method for inducing compliance when used effectively. Its effects are particularly apparent when we examine the importance of influence tactics in both political behavior and project negotiations.

Developing Influence Skills

How does a project manager succeed in establishing the sort of sustained influence throughout the organization that is useful in the pursuit of project-related goals? A recent article (Keys and Case, 1990) highlights five methods managers can use for enhancing their level of influence with superiors, clients, team members, and other stakeholders (see Table 41.3). They suggest that one powerful method for creating a base of influence is to first establish a reputation as an expert in the project that is being undertaken. This finding was borne out in research on project manager influence styles. A project manager who is widely perceived as lacking any sort of technical skill or competency cannot command the same ability to use influence as a power mechanism to secure the support of other important stakeholders or be perceived as a true "leader" of the project team. One important caveat to bear in mind about this point, however, is that the label of "expert" is typically a perceptual one. That is, it may or may not be based in actual fact. If one party perceives that another is an expert, that belief is often sufficient grounds for the leader to maintain influence over others.

A second technique for establishing greater influence is to make a distinction between the types of relationships encountered on the job. Specifically, managers need to make

TABLE 41.3. FIVE KEYS TO ESTABLISHING SUSTAINED INFLUENCE.

- Develop a reputation as an expert.
- Prioritize social relationships on the basis of work needs rather than on the basis of habit or social preference.
- Develop a network of other experts or resource persons who can be called upon for assistance.
- Choose the correct combination of influence tactics for the objective and the target to be influenced.
- Influence with sensitivity, flexibility, and solid communication.

conscious decisions to prioritize their relationships in terms of establishing close ties and contacts with those around the company who can help them accomplish their goals, rather than on the basis of social preference. Specifically, from the perspective of seeking to broaden their influence ability, project managers need to break the ties of habit and expand their social networks, particularly with regard to those who can be of future material aid to the project.

The third tactic for enhancing influence is to network. Networking consists of the practice of establishing a series of social and professional relationships with as large a set of individuals as possible. As part of creating a wider social set composed of organizational members with the power or status to aid in the project's development, canny project managers will also establish ties to acknowledged experts or those with the ability to provide scarce resources the project may need during times of crisis. It is always helpful to have experts or resource providers handy during times of munificence.

A fourth technique for expanding influence process suggests that for influence to succeed, project managers seeking to use influence on others must carefully select the tactic they intend to employ. For example, many people who consider themselves adept at influencing others prefer face-to-face settings rather than using the telephone or leaving messages to request support. They know intuitively that it is easier to gain another's compliance when the request is direct and personal. If the tactics that have been selected are not appropriate to the individual and the situation, influence will not work.

Finally, and closely related to the fourth point, successful influencers are socially sensitive, articulate, and very flexible in their tactics. For example, in attempting to influence another manager through a face-to-face meeting, a clever influencer will recognize how best to balance the alternative methods for attaining the other manager's cooperation and help. Through reading the body language and reactions of the other manager, the person attempting to use influence may instinctively shift the approach in order to find the argument or influence style that appears to have the best chance of succeeding. Whether the approach selected employs pure persuasion, flattery, and cajolery, or the use of guilt appeals, successful influencers are often those people who can articulate their arguments well, read the nonverbal signals given off by the other person, and tailor their arguments and influence style appropriately to take best advantage of the situation.

Block's Framework

Peter Block (1987) proposed a framework for developing strategies as a part of the influence process that often occurs between managers. For example, if Bob, the Project Manager, is interested in gaining support for a proposed technical change to a current project, he may have to first gain the cooperation of other significant project stakeholders, including Anne—Head of Technical Services, Nancy—V.P. for Engineering, and Tom—Lead Designer. How Bob chooses to attempt to influence these individuals to gain their cooperation with his plans is an important step and needs to be carefully considered.

Figure 41.1 shows Block's framework for developing influence strategies. There are some important points to note about the framework in general. First, the model makes clear

FIGURE 41.1. BLOCK'S FRAMEWORK.

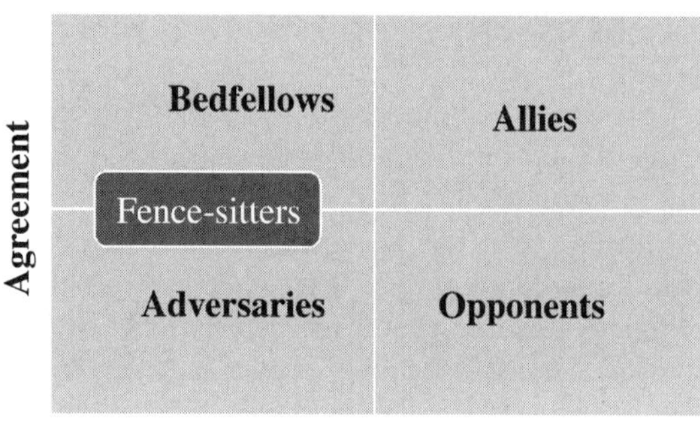

that influence must be done on an individual-by-individual basis; that is, there is little reason to attempt to develop a blanket influence tactic. By definition, each individual is different, and the steps taken to attempt to influence that person have to acknowledge this fact. Each person to be influenced merits an individualized strategy. Second, the model identifies two important features that weigh heavily on our ability to influence others: agreement and trust. Agreement refers to short-term or issue-relevant assessments. In other words, do you and I agree substantially on the current issue at hand? The second dimension, trust, is based on longer-term, relationship assessment. That is, regardless of our level of agreement on the current issue, if we have a positive working relationship, we may continue to trust each other and each other's intentions, even in the face of short-term disagreements. Note that agreement and trust do not have to coincide. I can disagree with your position without sacrificing the trust relationship we have developed. Likewise, it is possible for us to work together based on an issue of mutual agreement without ever trusting each other or each other's motives.

Figure 41.1 illustrates five groups that we are likely to confront at different times as we attempt to use influence effectively: (1) bedfellows, (2) allies, (3) opponents, (4) adversaries, and (5) fence-sitters. For each of these potential groups, we can formulate an approach that is most likely to gain us their compliance or, at least, minimize their potential negative impact on our goals.

1. *Bedfellows.* Bedfellows are characterized by circumstances in which we (they and ourselves) have an immediate issue of mutual concern or interest, although there is no longer-term basis of trust between us. For example, during World War II, the Western allies, headed by Winston Churchill and Franklin Roosevelt, were able to make common cause with Joseph Stalin, dictator of the former Soviet Union in their joint desire to defeat Nazi Germany. There was little trust between the Western leaders and Stalin, but circumstances and immediate concerns enabled them to put aside differences and

collaborate in dealing with the general menace of Nazism. Block argues that the best influence strategy when you are attempting to influence bedfellows is to do the following:

- *Reaffirm agreement.* Ensure that both parties are in agreement on the key issue of mutual concern.
- *Acknowledge the caution that exists.* Because the trust factor is low, there is little point in pretending otherwise; hence, it is best to acknowledge this caution immediately and directly.
- *Be clear what we want from bedfellows and what we intend to do.* Create a joint "map" of our plan for cooperation with the other party, including our commitments and the commitments we expect from the bedfellow.
- *Try to reach agreement on how we intend to work together.* Establish a shared protocol for how we can together achieve our goals, even in the absence of mutual trust. The less the trust between partners, the more important it is that a formal agreement of some sort is established.

2. *Allies.* Allies are other parties or individuals with whom we share both immediate agreements on current issues as well as long-term trust. For example, in almost all significant foreign policy issues in recent decades, the United States and Great Britain have worked hard to present a united front based on both the long-term quality of our relationship as well as the desire to maintain agreement on how to approach the specific issue at hand. Allies are our best source of support in influence situations because they are fundamentally in favor of our views and choices of action. The best influence strategy when dealing with allies includes the following:
   - *Reaffirm the agreement on the project or vision.* First, take steps to ensure that we are in substantive agreement with our allies on the issue at hand.
   - *Reaffirm the quality of the relationship.* It is also key to take time to reinforce the long-term nature of the trust relationship that exists between us and our allies.
   - *Acknowledge the doubts and vulnerability that we have regarding the project or other issues at hand.* With allies, it is not necessary to hide any difficulties or concerns that we may face in achieving our goals for the project. Because we have a relationship that is both trusting and affirmative, we can work with allies to seek answers to any lingering questions.
   - *Ask for advice and support.* Allies should be used to strengthen our position regarding the specific project issues we face. As a source of support and information, we can depend on allies to work with us rather than at cross-purposes.
3. *Opponents.* Opponents are those within the project stakeholders who disagree with us regarding the specific issue at hand, even while the long-term, trust-based relationship remains healthy. For example, a company may have a strong, long-term relationship with its suppliers, although disagreeing with them on an issue of immediate concern. The key to working with opponents is to always remember the importance of maintaining the level of trust that exists, regardless of short-term disagreements. Among the steps to employ are the following:
   - *Reaffirm the quality of the relationship.* The keystone of our relationship with opponents is trust. The first step in dealing with them is to reemphasize that trust remains the cornerstone of our interactions.

- *State our position.* We need to make a clear case for our position, recognizing that disagreements have arisen. Our goal is to influence through persuasion, not emotion.
- *State in a neutral way what we think their position is.* As objectively as possible, we need to demonstrate that we understand their principal objections with our decisions and make them aware that we appreciate their views. Because trust is key, it is vital that our attempts to persuade opponents cannot be misinterpreted into them thinking we are deliberately misleading them.
- *Engage in problem solving.* The key to resolving disagreements with opponents is to treat these issues as mutual problems requiring joint discussion and solution. Even if we are unsuccessful at persuading opponents to cooperate, the steps we take to influence them must be perceived as objective and free from deception.

4. *Adversaries.* In any circumstance involving the need to influence others, there will be adversaries who oppose our views both from an immediate "disagreement" perspective, as well as from a long-term lack of trust. Adversaries differ from other groups in that it is doubtful that we can successfully influence them; that is, turn them around to our point of view. The successful approach, therefore, is usually to find a method for minimizing their potential for negatively affecting our position and project. Among the most appropriate steps in dealing with adversaries are the following:
   - *State our position, including vision for the project.* Be clear as to our goals and directions for the project when speaking with adversaries.
   - *State in a neutral way what we think their position is.* It is important to demonstrate our own objectivity when dealing with adversaries. While this approach may not positively influence their perceptions, it ensures that we have dealt with our adversaries in a direct and honest manner. It is also helpful to demonstrate that while we disagree with adversaries, we understand their views and the means by which they came to them.
   - *End the meeting with our plans and no demands.* In adversarial relationships, it is unlikely that either side is going to effectively influence the other party or alter that person's perceptions and way of thinking. The best approach, as a result, is usually to maintain clear lines of communication, make our position unambiguous, and proceed with our plans.
5. *Fence-sitters.* Fence-sitters are usually described as wavering between agreement and disagreement with our position. Because no long-term trust has been established, fence-sitters are interested exclusively in the current issue at hand. Our strategy for influencing fence-sitters must be carefully considered because they often represent an important group in any situation in which influence is being applied. The key steps to appealing to fence-sitters include the following:
   - *State our position on the issue at hand.* Be clear and direct as to how we view the problem or issue, what the various options are that we are presented with, and why we have chosen the present course. The key is to be as comprehensive as possible to demonstrate that we have considered the issue from multiple perspectives.
   - *Ask where the fence-sitters are on the issue.* Attempt to collect as much information as possible about the views of these individuals. Find out, for example, their principal objections to your position, misunderstandings, or discomfort. It is important to gain

as much useful information as possible because it will allow us to formulate a plan for best alleviating their concerns.

- *Apply gentle pressure.* Following the process of collecting information, we are now able to develop the best approach to influencing fence-sitters. First, consider the reason for their concerns. If it was due to a misunderstanding of our position, we can now clearly explain our views and why they are appropriate. If their concern was due to some other substantive reason, we are in a position to take steps to minimize this concern. Applying gentle pressure can be coupled with a plan for correcting their misapprehensions about our approach to the problem.
- *Encourage fence-sitters to consider the issue and what it would take to gain their support.* We are attempting to influence through bargaining behavior. Rather than exerting excessive pressure or threats that can actually turn a fence-sitter into an adversary, we are looking to find the key that will make them bedfellows. In this attempt, it is important to acknowledge any concerns or issues they might have and, if possible, create a circumstance in which we can engage in a trade: their cooperation for our cooperation on some important issue of theirs.

Block's framework is a useful model for recognizing the various types of individuals we are likely to face in using influence, ranging from the trusting and cooperative all the way to the hostile and suspicious. Creating different strategies for dealing with each of these possible constituencies in a project-related problem will make our task much easier as we seek to use influence tactics most effectively.

Negotiation Skills

Project managers need good negotiation skills. Earlier in this chapter we stated that most project managers lack the positional power or authority to order people to do things. As a consequence, they must rely on less formal types of power such as personal power and use influence tactics and negotiation to accomplish their objectives.

Negotiation is ubiquitous in project management. For example, there are negotiations with department managers for the services of the members of the project team. There are negotiations with suppliers to obtain the materials and contractors to obtain the services needed to keep projects on schedule and within budget. There also are negotiations among team members and with upper management about project specifications, team member responsibilities, task completion dates, and budgets. And there are negotiations with customers about costs, project completion dates, and project change orders.

Effective negotiation requires careful preparation, the choice of a negotiation strategy appropriate to the situation, and the skillful use of power and influence to implement the chosen strategy. For example, a project manager may be very aggressive in a negotiation with a supplier if there are several sources that can provide equally good materials or services. On the other hand, if there is only one good supplier, thus giving that supplier nonsubstitutability power, a more conciliatory strategy may be required. Block's framework suggests problem solving or compromise may be the best choice if maintaining the trust of team members or customers is important to future project success or repeat business. In this section we discuss the characteristics of negotiation, the basic steps in preparation for negotiations, negotiating strategies, and considerations in choosing a negotiation strategy.

What Is Negotiation?

Negotiation is a process involving two or more people who start with apparently conflicting positions and attempt to come to an agreement by revising their original positions or by inventing new proposals that reconcile the interests underlying them. Elaboration of the parts of this definition will help explain what negotiation is and when it is used.

Negotiation is a process. Negotiation involves an exchange of proposals and counterproposals that take place over time. The time might be a few minutes, or it could be several months or even years.

Negotiation involves two or more people. The simplest negotiations involve two people. However, many negotiations involve more than two individuals. For example, a project may entail multiparty negotiations between several companies. Each company may field a team of negotiators who represent the interests of others (constituents such as higher-level management) not present at the negotiation table. When several actors are involved, the social structure of negotiation can become very complicated. Negotiations take place not only between the representatives of the various organizations but also within a negotiating team and between team members and their constituents.

Negotiators have apparently conflicting positions. There would be no need to negotiate if the people involved did not perceive differences between their initial positions. Sometimes further discussion reveals real differences between positions, but at other times further discussion may lead to the discovery of solutions that reconcile or integrate what initially appeared to be conflicting interests.

Negotiators attempt to reach agreement. Although it may not always be accomplished, usually the purpose of negotiation is to reach voluntary agreement about something. None of the negotiators have enough of a power advantage (or if they have it, they decide not to use it) vis-à-vis the other to make decisions unilaterally. Both sides believe they can improve their outcomes through negotiation. Negotiators will continue trying to reach agreement as long as they believe an agreement is possible and preferable to the alternative of not reaching agreement.

Negotiators revise their initial positions. During negotiation the parties usually change their initial proposals. Often this requires both sides to make concessions toward some middle-ground position. Consequently, negotiators have an expectation of give-and-take during the negotiation process. Sometimes negotiators are able to invent or discover options that reconcile or integrate their underlying interests in such a way that a high degree of satisfaction is achieved by all concerned. This allows both parties to satisfy their interests to a greater extent than would have been possible if they had simply conceded to some middle-ground position.

Preparation for Negotiation

Although the time and money a negotiator invests in preparation depends upon the importance of a particular negotiation, successful negotiation requires thorough preparation. Without careful preparation, negotiators are likely to find themselves reacting to events rather than influencing them. In highly competitive situations, a well-prepared opponent can put an unprepared negotiator on the defensive. The chances of concluding an agreement that satisfies the reasons for negotiating in the first place are reduced when thoughtful preparation does not occur. An opponent's confidence and commitment to a position may increase if the other side is unprepared. In this section a summary of the basic steps of preparation for negotiation is presented. A more detailed description of the steps involved in preparation for negotiation can be found in Lewicki, Saunders, and Minton (1999, pp. 29–69).

Consult with Constituents and Other Project Stakeholders. Negotiators need to consult with the various groups or individuals they represent and those who will be affected by the negotiations. For example, negotiators should learn about the requirements upper management has with regard to project costs, schedules, liability, intellectual property, and so on. In addition, negotiators need to consult with engineers, production managers, subcontractors, and suppliers, who will be critical to implementing any agreement that is reached. Consultation with constituents and others allows the negotiator to identify his or her side's interests and to frame, prioritize, and develop positions on the issues, along with arguments to support them.

Identify Interests. Next, the negotiator needs to think about the reasons for negotiating with the other side. What are the basic wants or concerns that an agreement should satisfy? A project manager negotiating a change order may need to complete the project within budget and on schedule. Although there surely will be an interest in avoiding the additional cost of making unnecessary changes, there also may be an interest in keeping clients happy in order to increase the chances of future business.

Interests are different from the specific concrete proposals that will be discussed at the negotiating table. They are the concerns underlying the proposals, and proposals are intended to satisfy them. There may be more than one proposal that satisfies an interest equally well, and alternative proposals can be evaluated in terms of how well they satisfy interests. The success of the negotiations also can be evaluated in terms of how well interests have been satisfied by any agreements reached. Interests are not limited to the obvious or tangible substantive matters such as project cost or completion date. Other interests may relate to the process of negotiation, the relationship with the other side, and preserving important principles (Lax and Sebenius, 1993). Interests also can be classified as instrumental or intrinsic for each of these three categories.

Interests are *instrumental* if favorable terms are valued because of their impact on future dealings. Interests are *intrinsic* if favorable terms are valued regardless of their effect on future dealings. A concern about possible future business is an instrumental interest, as are concerns about the impact of the completion date for a production facility on the ability of a company to honor a possible contract with one of its clients. On the other hand, staying within a budget is an intrinsic interest in a negotiation over a change order for a project.

Aside from the outcomes of negotiation, the parties may also have an interest in the *process* of negotiation. Some may prefer to conduct business in a smooth, harmonious fashion characterized by honesty and openness. Others may enjoy a more aggressive and competitive process and the feeling that they have extracted as many concessions as possible from the other side. When it comes to *relationships*, most negotiators want to be treated with respect by the other side. Having the other's respect makes the negotiator feel better while negotiating (intrinsic interest), and it also may have important implications for the negotiator's treatment in future negotiations (instrumental interest).

Negotiators also may have interests in *principles* that they wish to establish or have honored when negotiating. These principles may assume great importance because they can have precedent-setting implications in future negotiations.

Thinking about different types of interests is useful because it reminds negotiators that they should consider more than substantive interests when preparing for negotiation. There may be nonobvious but important immediate and longer-term interests at stake in a negotiation related to process, relationship, and principles.

Learn about the Other Side. Negotiators should not only think about their own interests, they also should try to learn about the other's interests, likely issues, sources of power, probable negotiation strategies and tactics, alternatives to reaching agreement, as well as any other information relevant to the negotiation. Any pertinent information they can learn about the other side prior to negotiation will help them anticipate their behavior during the negotiations.

For example, in advance of negotiation, a project manager might learn though a company sales representative that the other side may be willing to spend more money for a project if a firm completion date can be guaranteed. Or the other side may have plans for several other projects over the next several years. Alternatively, a project manager may learn from industry sources that the other company is experiencing financial difficulty and needs several future construction projects to remain profitable. These are valuable items of information for negotiators to have as they prepare for negotiation.

There are several sources of information about the other negotiator. Archival data may be used to learn more about the other party (Lewicki et al., 1999, p. 64) and may come from publicly available sources such as Dun & Bradstreet, financial statements, newspaper articles, stock reports, and legal judgments. Other ways of obtaining useful information about the other side may be visiting the other side's place of business, having preliminary discussions with them about what they would like to achieve in the forthcoming negotiations, trying to anticipate their interests through role-playing, and talking to people who know or who have previously negotiated with them. Previous negotiations with the other side may provide notes and experiences that are useful. If other information is unknown prior to negotiation, many times it can be obtained through careful observation and listening as negotiations progress.

Negotiators form perceptions of the other side based on the information they collect. These perceptions can be very useful in anticipating the other's behavior but it is important to remember that such information may be incomplete and the specific circumstances of the upcoming negotiation may lead to different behavior. Therefore, it is important not to act hastily on the basis of untested assumptions in ways that may create obstacles to reaching agreement and to remain receptive to new information as the negotiations progress.

Identify issues. An issue is a continuum on which negotiators can take different positions. The interests of the two sides define the issues to be negotiated. For example, one negotiator may have an interest in minimizing project costs and the time required to complete the project, in being treated fairly by the other side, and in establishing principles for allocating the costs of changes on this and future projects. These interests might lead to the following issues on which the two sides will take different positions: the total cost of the project, the project completion date, penalties for failure to complete the project on schedule, the ground rules for conducting the negotiations, and the criteria to be considered or procedures to be used in making changes during the project. On each of these issues each side can take different positions. For example, one side will probably favor lower project costs and an earlier completion date, while the other may propose higher cost and a later completion date.

Although having many issues to negotiate can complicate and lengthen negotiations because there are more things to negotiate, multiple issues also can facilitate negotiations by creating the possibility of trade-offs. For example, one side might agree to a penalty clause if it does not complete the project on time if the other agrees to pay 100 percent of the cost of overtime associated with any project changes.

Prioritize Issues. Negotiation very often requires making a concession on one issue in order to gain on another. To maximize their outcomes, negotiators should make larger and more frequent concessions on less important issues than they should make on more important ones. For example, if the completion date for a project is more important than the total cost of the project, a negotiator may be very inflexible about the completion date but willing to pay more to ensure that the project is completed on schedule. For these reasons, negotiators need to evaluate the relative importance of the issues.

Develop Proposals. Once issues and underlying interests have been identified, it is time to think about proposals. A proposal should usually encompass the entire set of bargaining issues that need to be resolved in order to reach agreement. It is beneficial to think about at least three alternative proposals prior to starting negotiation. The first is an *initial offer* that will be presented to the other side at the start of negotiation, the second is a *target point* or proposal that represents the negotiator's preferred outcome, and the third is the *resistance point*, or the least favorable proposal a negotiator is willing to accept for the foreseeable future.

Because negotiation involves give-and-take, initial offers should be more favorable to the negotiator than a target point and a target point should be more favorable to a negotiator than a resistance point. It can be advantageous to have the other side present their initial offer first. This allows the negotiator the opportunity to gauge the other side's position on various issues and avoids the possibility of offering the other side a proposal that is more favorable than necessary.

Sometimes a resistance point is based on what Fisher, Ury, and Patton (1991) call a BATNA. A BATNA, or *b*est *a*lternative *t*o a *n*egotiated *a*greement, represents a negotiator's best option outside the current relationship. In a case of negotiating with suppliers, a BATNA might be what another vendor would charge for a product or service. Or the best alternative might be mediation, arbitration, or a court settlement. Establishing a BATNA before negotiation can help prevent agreeing to less favorable terms than are available elsewhere or rejecting a proposal that is more favorable than the alternatives. In addition, a negotiator can strengthen a negotiation position by finding a better BATNA.

When the negotiator is establishing an initial offer, target point, and resistance point, it is very useful to consider objective standards or standards of fairness that may apply to the situation. These will help in supporting proposals and make them seem more realistic and supportable to the other side. There are a number of principles that negotiators can use to support their demands in a negotiation (Magenau and Pruitt, 1979).

Perhaps there is an industry-standard practice or custom that one or both project managers might use to support their position regarding the change orders. If there is no industry norm to rely on, perhaps the project managers might adopt an *equality* norm that would support a 50-50 division of the costs and benefits. An *equity* norm would favor allocation of project revenue according to the amount of investment or risk each of the parties have at stake in the project. Under a *needs* rule, the greatest portion of the cost would be assumed by the party with the greatest capacity to pay (least needy). In other situations, market value, scientific judgment, efficiency, costs, what a neutral third party would decide, or moral standards may be used as possible objective criteria (Fisher et al., 1991).

It is likely that a negotiator will be asked to explain how the specifics of a proposal were established. Use of objective standards will help ensure that a plausible explanation is available and that proposals appear realistic. Proposals without any supporting rationale may be dismissed without serious consideration and may lead to loss of credibility for the negotiator. Serious negotiation is unlikely to occur until a more realistic position is adopted, or the other side may become convinced that further negotiation is a waste of time.

What if there are several standards that could be applied in a negotiation? Assuming that some are more favorable to a negotiator than others, the most favorable could be used to establish an initial offer. A target point could be set according to a standard less favorable to the negotiator but that perhaps is more compelling than the standard underlying an initial offer. Other objective standards, if available, could be used to support proposals intermediate between an initial offer and target point or a resistance point.

Linking an initial offer, target point, and resistance point to principles can be a double-edged sword and should be done with caution. On the one hand, this tactic may make proposals more rational, defensible, and likely to prevail. On the other hand, adopting positions linked to principles may lead to rigidity that makes reaching agreement more difficult (Pruitt and Carnevale, 1993). That is, opposing negotiators may adopt different positions favoring their own interests, each supported by competing principles. If this occurs, there is a danger that one or both negotiators may become so committed to their respective positions that they are psychologically unable to make the concessions that are necessary to reach agreement.

To avoid the psychological trap of becoming too committed to positions based on principles, Fisher, Ury, and Patton (1991, p. 88) advise negotiators to "frame each issue as a joint search for objective criteria," to "reason and be open to reason," and to "never yield to pressure, only to principle." In other words, negotiators should be firm about reaching an agreement based on principles of fairness but at the same time be flexible enough to consider different principles that may legitimately apply to a given situation.

Negotiators should prepare for negotiation by developing a series of proposals that are supportable by objective criteria. At the same time they should be flexible enough to acknowledge the validity of other legitimate principles that may apply to the situation. This flexibility may motivate a search of innovative solutions that reconcile or integrate various principles previously thought to be contradictory. Or it may lead to a compromise solution that incorporates some elements of various principles. One side may even adopt the position of the other side if, after discussion, it proves to be fairer and easier to implement.

Team Organization

In important complex negotiations, typically each side is represented by a team instead of a single individual. Teams should be composed of members who bring the necessary negotiating skill and specialized technical expertise to the negotiating table, and team members should be briefed on their responsibilities prior to negotiations. It is difficult for a single individual to talk, listen, think, write, observe, and plan simultaneously. For this reason, Kennedy, Benson, and McMillan (1982) recommend that negotiation teams include three generic roles: leader, summarizer, and recorder.

The *leader* is usually the senior member of the bargaining team and serves as the team coordinator and spokesperson, and generally leads the team's effort. The leader usually does most of the talking at the negotiation table but may call upon others to speak when needed. Channeling communication through a leader allows a team to control the information provided to the other side and avoid having its negotiating position undermined by revealing sensitive information or within-team disagreements to the other side.

The *summarizer*' listens carefully to the arguments of the other side and gives thinking time to the leader by intervening when appropriate. This often can be done by asking the other side for clarification of a point or by summarizing the other side's position or arguments on some issue. The summarizer should be someone who is trusted and who can work closely with the leader.

The *recorder* remains silent during the negotiation unless called upon to speak about a particular issue related to his or her expertise. There could be several recorders who attend all or some of the negotiations sessions depending on the requirements of a particular meeting. These might include lawyers, engineers, architects, cost estimators, environmental experts, programmers, production schedulers, or other technical specialists knowledgeable of some critical aspects of the project. The responsibilities of this position also include watching the other team for verbal and nonverbal cues that might reveal something about their interests, priorities, or points of internal disagreement. Recorders may be called upon to report their observations to the team during private meetings away from the negotiating table.

Negotiating Strategies

Negotiation strategies can be classified into five basic categories (Pruitt and Carnevale, 1993; Lewicki et al., 1994):

- Concession-making
- Contending
- Compromising
- Problem solving
- Inaction/withdrawal

Concession-making involves changing a proposal so that it provides less benefit to you but more benefit to the other side. Concession-making may be your response to the other side's successful use of power and influence.

Contending involves trying to persuade the other side to make a proposal more favorable to you but less favorable to them. Contending may involve the use of coercive power communicated in the form of threats or influence tactics such as strong persuasive arguments, pressure, multiple pressure, guilt, and making statements about your unwillingness to make any additional concessions.

Compromising represents a strategy that is intermediate between concession-making and contending. A middle ground is sought that involves an exchange of concessions and some degree of sacrifice for both sides. A compromise strategy utilizes a "carrot and stick" approach as compared with the heavier reliance on the "stick" used with a contending strategy. Compromising involves a more positive approach to the other side than contending and therefore is probably better implemented with more positive forms of power such as reward (e.g., reciprocating concessions), information, and expert power (e.g., using objective information or expert opinion to support a compromise agreement) or referent power (e.g., increasing your attractiveness or prestige in the eyes of the other side). Influence tactics such as ingratiation and moderate persuasion also would be consistent with a compromise strategy. However, the use of coercive power and the influence tactics employed in a contending strategy may be needed to motivate concessions if the other is unwilling to move to a middle-ground position.

Problem solving involves efforts to find agreements that are highly favorable to both parties. If successful, both sides can avoid the deep concessions that may be necessary under a concession-making or compromising strategy. Problem solving often involves a joint effort in which the two sides work together by openly sharing information about interests and priorities and by discussing the pros and cons of various proposals.

Problem solving is more likely to occur when there is a positive long-term relationship and where trust exists between negotiators that allows for open exchange of information and frank discussions. It is a negotiation strategy that is generally incompatible with the use of coercive power and aggressive influence tactics used with a contending strategy. On the other hand, the use of reward power, referent power, expert power, and information power are consistent with problem-solving strategy.

Inaction and withdrawal are similar. Withdrawal involves terminating negotiations for the foreseeable future without an agreement. Inaction involves attempts to delay or avoid engaging in serious negotiations. If they chose withdrawal, both parties might adopt their BATNAs. An inaction strategy might involve ignoring the other's requests to negotiate change orders, refusing to discuss changes, or diverting the conversation to other topics when the other side proposes negotiation. Inaction is likely to be favored by those with advantages they wish to preserve and who view negotiation with others as an attempt to change the status quo. Inaction and withdrawal may be effectively employed by those who have resource power, dependency power, or nonsubstitutability to convince the other side that concessions are necessary to start or resume negotiation.

Choosing a Strategy

Pruitt (1983) proposed the dual concern model to explain how negotiators choose a negotiation strategy. What the model says about how negotiators actually choose a negotiation strategy provides a good analytical framework for considering strategic alternatives. According to the dual concern model, the choice of a negotiating strategy is primarily influenced by two key concerns. First is the negotiator's level of concern for his or her own outcomes, and second is the level of concern for the outcomes of the other side. A third consideration is the feasibility of the strategy under consideration. The relationship between the first two variables, or the dual concerns, and the choice of a strategy are shown in Figure 41.2.

The model indicates that negotiators choose contending or problem solving strategies when they have a high level of concern for their own outcomes. This makes sense because most people are less willing to sacrifice their own interests when they place great value on them. Thus, we are less likely to engage in concession-making or compromise when important interests are at stake, and we are unlikely to choose inaction or withdrawal if we believe these interests can only be satisfied by the other side.

The choice between contending and problem solving depends on a negotiator's level of concern with the other party's outcomes. If the negotiator has a high level of concern for the outcomes of the other side, he or she will choose problem solving because this is a strategy designed to satisfy the needs of both sides. If there is little concern for the other side's outcomes, contending is the strategy of choice because it seeks to maximize the negotiator's own interests at the expense of the other side. If the negotiator's concern for his or her own outcomes is low but the negotiator's concern for the other's outcomes is high, the negotiator is likely to choose concession-making in order to satisfy the other side. When both concerns are low, there is little interest in negotiation and inaction and withdrawal are probable strategy choices.

The importance negotiators place on these concerns and the feasibility of various strategies can change over the course of negotiations and across various issues under discussion. As a result, the choice of a strategy can vary over the course of negotiation and with different

FIGURE 41.2. THE DUAL CONCERN MODEL.

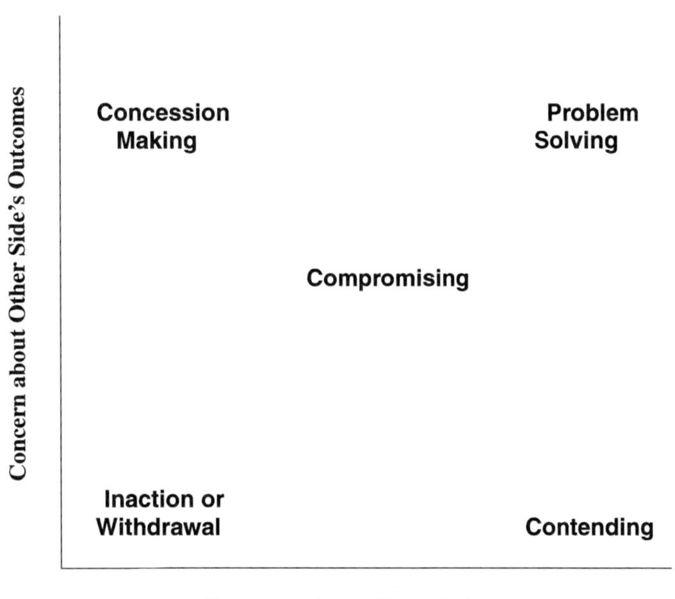

Source: Adapted from D. G. Pruitt (1983). Strategic choice in negotiation. *American Behavioral Scientist* 27:167–194.

issues. The strategies are not mutually exclusive in the sense that different strategies might be employed at different stages of the negotiation or on different issues. For example, the negotiator might begin by contending because he or she believes the other is weak and will concede in the face of our pressure tactics and persuasive arguments but shift to a compromising strategy after hearing the other side's arguments and learning that the other side is more resistant to change than anticipated. In addition, some issues in the negotiation may be perceived by the two sides as win-lose or zero sum. They may attempt to resolve these by contending, concession-making, or compromising, while other issues may be viewed as having problem-solving potential.

The dual concern model can be used as a guide to preparation for negotiation if negotiators ask themselves the following questions:

- What is my level of concern for my own outcomes?
- What is my level of concern for the other side's outcomes?
- What is the feasibility of the strategy suggested by my level of the preceding two concerns?

Concern for Own Outcomes. In thinking about their level of concern for their own outcomes, negotiators should consider the importance of the interests that will be affected by the outcomes of negotiation, the opportunity costs associated with the negotiation, and the extent to which they are willing to risk conflict with the other side in order to achieve their desired outcomes. The importance that constituents place on the outcomes the negotiator achieves also will influence the calculation, especially if constituents can hold the negotiator accountable for results. For example, I am likely to rate the importance of my side's outcomes as high if the company's profitability is significantly affected by the outcome of the negotiation, the current contract is larger than other contracts my company has, I think the other side has few alternatives to signing a contract with me, or my boss informs me that my future with the company hinges on negotiating contract terms favorable to the company.

Concern for the Other Side's Outcomes. Concern for the other side's outcomes will be greater if there are interpersonal bonds of friendship with the other side that exist or that the negotiator wishes to establish. Concern for the other side's outcomes also will be greater if the negotiator is dependent on the other side to achieve desired outcomes or if the negotiator will be dependent on them in future negotiations. It is important to keep such future dependencies in mind because negotiators sometimes become so absorbed in the current negotiation that they forget about the impact of their behavior on their long-term relationship with the other side. For example, we would expect a negotiator's concern for the other side to be greater if the other negotiator is an old and good friend, needs an agreement with the other side to complete a project before a fast approaching deadline, and wants to do business with the other again in the future.

In terms of Block's (1987) framework, in a negotiation between allies and opponents, concern for the other's outcomes would be highest where maintaining a positive long-term relationship is a key consideration. Concern for the other side's outcomes may also be a consideration when negotiating with bedfellows and fence-sitters, but it is clearly more of a short-term consideration and probably not as strong as it is for allies and opponents. There is likely to be low concern with the other's outcomes when negotiating with adversaries.

The Feasibility of Problem Solving

If the negotiator has a high level of concern with his or her own outcomes as well as a high level of concern for the other side's outcomes, the dual concern model suggests a problem-solving strategy. However, the model suggests that the choice of problem solving also depends on its feasibility.

Problem solving becomes more feasible as the perception of common ground on the part of the negotiators increases (Pruitt and Rubin, 1986). This occurs because greater common ground makes the likelihood of finding alternatives that satisfy the interests of both parties seem more promising. Common ground on an issue increases as the amount of overlap between the negotiator's resistance points becomes greater and to the extent that the parties believe that alternatives favorable to both sides exist or can be invented. The concept of common ground is obviously closely aligned with the agreement dimension in the Block (1987) model for choosing influence strategies.

Negotiators also are more likely to adopt a problem-solving strategy when they are confident in their own problem-solving ability, they have a track record of solving problems with the other negotiator, they have access to third parties who can assist them with locating or devising mutually beneficial alternatives, and they believe that the other side is inclined to problem-solve too. A negotiator's belief that the other side is ready to problem-solve depends on trust. Trust in the other in turn depends on a negotiator's belief that the other is concerned about his or her interests. Our trust in the other side is likely to increase when the other has a positive attitude toward us, is similar to us, is dependent on us, or has been helpful to us in the past, especially if the help has been voluntary.

The relationship between trust and problem solving is likely to occur as long as trust is coupled with the belief that the other side is strongly committed to their own outcomes. If the other side appears to have a weak commitment to their outcomes, trust can encourage contending rather than problem solving. This can occur because when we believe someone is concerned about us, but at the same time has a weak commitment to his or her own goals, we are likely to conclude that that person will concede easily.

There are several things a negotiator can do to encourage problem solving. One is to explore the possibility of common ground. Many negotiators have a "fixed-pie" bias (Bazerman, 1983). They assume, many times incorrectly, that there is a fixed amount of what is being negotiated and that one side can benefit only if the other side loses. Instead of making this assumption, you should approach negotiations with an open mind and try to explore areas of common ground.

Negotiators also can enhance their own confidence in their problem-solving ability by improving their communication skills. They can try to build the other side's confidence in their problem-solving ability by verbally reinforcing the other's problem-solving efforts or by encouraging them by clearly stating their desire to find a mutually beneficial agreement. If discussion of easier-to-solve issues is scheduled first on the agenda, success in solving them may lead to the belief and that more difficult issues can be solved later on in negotiations. Confidence may be increased if trusted and respected neutral third parties who can be relied upon to facilitate problem solving are identified. Finally, an indication of the other's readiness for problem solving can be gained through careful observation of clues in their behavior, by direct questioning, or by analyzing the factors relevant to their dual concerns.

The Feasibility of Contending

Contending is more feasible when (1) the other side has low concern for their own outcomes and therefore little resistance to conceding, (2) a negotiator has the power and the skill to effectively employ influence tactics and the other side lacks the power and skill to counter the tactics used, and (3) there is little risk of alienating the other side by using such tactics. Therefore, before deciding to use contentious tactics, a negotiator should attempt to evaluate the other side's level of concern for its own outcomes. The negotiator should consider the sources of power and influence tactics he or she has available as well as those available to the other side, and evaluate the risks of alienating the other side by deploying the tactics under consideration.

The Feasibility of Inaction and Withdrawing

The feasibility of inaction depends on the time pressure a negotiator is experiencing. There are two sources of time pressure: the cost of continued negotiations and the necessity of meeting deadlines. Time pressure caused by the continued cost of negotiation might result from lost business opportunities incurred while busy negotiating, the cost of supporting a negotiation team away from home, or loss in value of the object of negotiation as in the case of perishable items. Deadlines represent times in the future when significant costs will be incurred if there is no agreement.

In considering the feasibility of inaction as a strategy, negotiators should ask themselves what costs are associated with delaying the negotiation and what significant costs will result from failure to meet specific deadlines. For example, a supplier might try to use inaction for a time to postpone negotiation on price reductions but later realize that negotiation is necessary to avoid the loss of business.

The feasibility of withdrawing depends on the attractiveness of a negotiator's BATNA. Negotiators who have very attractive alternatives outside of their relationship with the other side are more likely to consider withdrawal as a feasible alternative.

The Feasibility of Concession Making

Concession making is a feasible strategy as long as one is able to make concessions. Because this strategy provides a way of reaching agreement quickly, it is favored when time pressure is high. The capacity to concede depends on how close the negotiator is to his or her resistance point. The closer the negotiator's current position is to the resistance point, the more difficult each additional concession becomes. For example, if my current proposal is for $500,000 to complete the work on a project, I am more willing to agree to a $50,000 reduction in my asking price if my resistance point is $350,000 than if it is $425,000.

The Feasibility of Compromising

Compromising can be thought of as a less intense form of concession making or problem solving (Pruitt and Rubin, 1986). The difference between compromising and pure concession making is that under a compromising strategy, negotiators attempt to coordinate the exchange of concessions with the other negotiator toward some middle ground rather than concede unilaterally. It differs from problem solving in that the compromise under consideration involves more sacrifice for the negotiators and the agreements reached tend to provide lower joint benefit.

The conditions making a compromise strategy feasible are similar to those underlying the feasibility of concession making and problem solving. That is, one must have the ability to concede to the proposed compromise alternative. And, as with problem solving, the perceived feasibility of a compromise solution is increased when there is a perception of common ground. Also, as in the case of problem solving, compromising is more likely to occur if negotiators previously have successfully exchanged concessions, believe the other side is ready to exchange concessions, and third parties are available who can help coordinate concessions if needed.

The use of conditional language is strongly recommended when attempting to coordinate concessions toward a compromise solution. For example, an negotiator might say, "If you agree to my proposed project completion date, I will agree to your proposed cost" (Kennedy, Benson, and McMillan, 1982, p. 88). Conditional proposals let the other side know precisely what you want in return for your concession. It also is clear that the concession is offered only if the other side reciprocates. If the other side does not agree to the proposed terms, the concession is withdrawn.

Summary

The nature of their work makes skill in the use power, influence, and negotiation indispensable to successful project managers. Power and influence are commonly used in negotiation, and understanding their use is valuable in preparation for and in the conduct of negotiation. In this chapter we defined power, influence, and negotiation. We discussed the sources of power and how to acquire it. Various types of influence were also described, as well as when they can be used most effectively.

Thorough preparation is essential to successful negotiation, and preparation involves several recommended steps. As with the use of power and influence, successful negotiation depends on the use of a strategy appropriate to the situation. It is important to remember that the choice of a strategy is contingent upon the dual concerns of the negotiators and the feasibility of the strategies under consideration. Different strategies are recommended under different circumstances. Further, negotiators should be prepared to change strategies over the course of a negotiation as the circumstances require.

References

Allen, R. W., D. L. Madison, L. W. Porter, P. A. Renwick, and B. Y. Moyes. 1979. Organizational politics: Tactics and characteristics of actors. *California Management Review* 22(1):78.

Bazerman, M. H., 1983. Negotiator judgment: A critical look at the rationality assumption. *American Behavioral Scientist* 27:211–228.

Beeman, D. R., and T. W. Sharkey. 1987. The use and abuse of corporate politics. *Business Horizons* 36(2):26–30.

Block, P. 1987. *The empowered manager.* San Francisco, CA: Jossey-Bass.

Daft, R. L., 1999. *Leadership: Theory and practice.* Fort Worth, TX: Dryden Press.

———. 2001. *Organization theory and design.* 7th ed. Cincinnati: South-Western.

Fisher, R., W. Ury, and B. Patton. 1991. *Getting to yes: Negotiating agreement without giving in.* New York: Penguin.

French, J. R. P., and B. Raven. 1959. The bases of social power. In *Studies in Social Power*, ed. D. Cartwright, 150–167. Ann Arbor, MI: Institute for Social Research.,

Gandz, J., and V. V. Murray. 1980. Experience of workplace politics. *Academy of Management Journal* 23:237–251.

Goodman, R. M., 1967. Ambiguous authority definition in project management. *Academy of Management Journal* 10:395–407.

Hickson, D. J., C. R. Hinings, C. A. Lee, R. E. Schneck, and J. M. Pennings 1971. A strategic contingencies theory Of intraorganizational power. *Administrative Sciences Quarterly* 16:216–229.

Kennedy, G., J. Benson, and J. McMillan. 1983. *Managing negotiations*. Rev. ed. Englewood Cliffs, NJ: Prentice Hall.

Keys, B., and T. Case. 1990. How to become an influential manager. *Academy of Management Executive* IV(4):38–51.

Lawrence, P. R., and J. W. Lorsch. 1967. Differentiation and integration in complex organizations. *Adminstrative Science Quarterly* 11:1–47.

———. 1969. *Organization and environment*. Homewood, IL: Irwin.

Lax, D. A., and J. K. Sebenius. 1993. *Interests: The measure of negotiation*. In *Negotiation: Readings, Exercises and Cases*, ed. R. J. Lewicki, J. A. Litterer, D. M. Saunders, and J. W. Minton. 2nd ed. Barr Ridge, IL: Irwin.

Lewicki, R. J., D. M. Saunders, and J. W. Minton. 1999. *Negotiation* 2nd ed. Boston: Irwin/McGraw-Hill.

Magenau, J. M., and D. G. Pruitt. 1979. The social psychology of bargaining: A theoretical synthesis. In *Industrial relations: A social psychological approach*, ed. G. Stephenson and C. Brotherton. Chichester, UK: Wiley.

March, J. G., and H. A. Simon. 1958. *Organizations*. Hoboken, NJ: Wiley.

Markus, M. L. 1983. Power, politics, and MIS implementation. *Communications of the ACM* 19:321–342.

Mayes, B. T., and R. W. Allen. 1977. Toward a definition of organizational politics. *Academy of Management Review* 2:675.

Mintzberg, H. 1983. *Power in and around organizations*. Englewood Cliffs, NJ: Prentice Hall.

Payne, H. J. 1993 Introducing formal project management into a traditionally structured organization. *International Journal of Project Management*. 11:239–243.

Pettigrew, A. M. 1973. *The politics of organizational decision-making*. London: Tavistock.

Pfeffer, J. 1981. *Power in organizations*. Marshfield, MA: Pitman.

Pinto, J. K., and O. P. Kharbanda. 1995. Project management and conflict resolution. *Project Management Journal* 26(4):45–54.

———. 1995. *Successful project managers: Leading your team to success*. New York: Van Nostrand Reinhold.

Pruitt, D. G. 1983. Strategic choice in negotiation. *American Behavioral Scientist*, 27:167–194.

Pruitt, D. G., and P. J. Carnevale. 1993. *Negotiation in social conflict*. Pacific Grove, CA: Brooks-Cole.

Pruitt, D. G., and J. Z. Rubin. 1986. *Social conflict: Escalation, stalemate and settlement*. New York: McGraw-Hill.

Slevin, D. P. 1989, *The whole manager*. New York: AMACOM.

Thamhain, H. J., and G. Gemmill. 1974. Influence styles of project managers: Some project performance correlates. *Academy of Management Journal* 17:216–224.

CHAPTER FORTY-TWO

MANAGING HUMAN RESOURCES IN THE PROJECT-ORIENTED COMPANY

Martina Huemann, Rodney Turner, and Anne Keegan

In this chapter we describe the characteristics of human resource management (HRM) in the project-oriented organization. Human resource management is a specific and strategically important process in the project-oriented organization. It includes recruitment, disposition and development, leadership, retention, and release of project management personnel.

The contents of this chapter are based on recent research into the HRM in the project-oriented organization and project-oriented society. First we describe the changing nature of HRM in the project-oriented society and consider the impact on project management personnel and their careers. We then consider the different types of project personnel who need to be managed in the project-oriented organization and describe the HRM processes in the project-oriented organization. We end by briefly describing the role of the PM office in managing project management personnel.

Human Resource Management in the Context of the Project-Oriented Society

A change toward a project-oriented society is observable. Gareis and Huemann (2001) define a project-oriented society as one that does the following:

- Considers projects and programs as an important form of (temporary) organization for achieving strategic and change objectives
- Supports a relatively high number of project-oriented organizations

- Has specific competencies for managing of projects, programs, and project portfolios
- Has structures to further develop these management competencies

The fact that there are increasingly more projects performed in society is explained by the evolutionary demand for projects (Lundin and Söderholm, 1998). Not just traditional industries, but many others, including the public sector, perceive temporary organizations such as projects and programs as appropriate to perform business processes of medium to large scope. Beside traditional contracting projects, other types, such as in marketing, product development, and organizational development, have gained in importance. Projects and project management are applied in new social areas, such as local municipalities, associations, schools, and even families. "Management by projects" becomes a macroeconomic strategy of the society, to cope with complexity and dynamics and to ensure quality of the project results (Gareis, 2002). Further, project management is being established as a profession. The Project Management Institute estimates that there are about 16 million people worldwide who consider project management as their profession (Gedansky, 2002).

Individuals Work More Often in Temporary Organizations

In project-oriented societies, there is a trend for individuals to get temporary assignments as they work on successive projects and programs. Project participants move from one project to another, often from one company to another, and even from one country to another. This creates a picture in our minds of "project nomads," whom we might think of as having an adventurous life. However, the personnel manager of an international engineering company pointed out that these nomads have to move from one place to the other because the country is too poor in which to settle down permanently. Similar pictures are drawn by Drucker (1994) when he describes the knowledge workers and Handy (2002) when he describes the life of a the self-employed "flea." Handy (1988) previously described such people as being like freelancers, literally mercenaries at the time of the crusades, who were not part of the regular army. Temporary employment and self-employment is increasing. Lifetime employment and permanent careers become rare. Acquiring project management competencies, keeping them state-of-the-art, and getting them certified becomes an issue, even for those project management personnel who belong (permanently) to a project-oriented organization. An individual has to take on the responsibility for the acquisition of the competencies demanded and of his or her professional development to keep employable.

Characteristics of HRM in Project-Oriented Organizations

What are the features of project-oriented firms that influence the nature of employment within them? Projects are temporary organizations undertaken to bring about change (Turner and Müller, 2003; Lundin and Söderholm, 1998). Some, primarily functional, organizations undertake occasional projects to enact specific changes. They can adopt classical human resource management practices and assign resources to projects from within the functions as necessary. But for project-oriented organizations, projects are their business; the majority of the work they do is project-based. Turner and Keegan (2003) showed that they need a different approach to human resource management than the classical approach adopted by functionally oriented organizations.

As temporary organizations, projects are unique, often novel, and transient. Being unique, the organization has never done exactly this before. They often require novel processes and have novel resource requirements. Being unique and novel, the method of delivery can be uncertain. The consequences on human resource management requirements are as follows:

- The present and future resource requirements of the organization are uncertain.
- People follow careers other than climbing the ladder up the functional silo.
- People may not have a functional home to belong to.

Uncertain Requirements

In the classically managed, functional organization, resource requirements are assumed to be well determined. The jobs to be done are well known from past experience. A job description is written for a job, defining what is to be done, the levels of management responsibility required, and the competence required, including levels of education and training and past experience. Somebody is recruited in accordance with that specification. There is a saying, "You grade the job and not the person." The requirements of the job are defined, and the best match is found to those requirements.

That level of certainty often does not exist in the project-oriented organization:

1. Projects are unique and transient, with high uncertainty. It is often not possible to define precisely the requirements of the current job. You need to recruit people known to work well on projects and, to an extent, let them define the job around themselves. (Though this is true of many other management positions as well, of course.)
2. Contract organizations often cannot precisely predict the levels of resource requirements into the immediate future. They may have several jobs at the moment, with one coming to the end, and several bids out. For instance, consider that they have five bids out, with a normal success rate of winning one bid in five. If they achieve that, they will have one job to replace the one coming to an end. If they are successful with none, their workload will fall; if they win two, they may just cope; if they win three, they will be overloaded. Keegan and Turner (2003) report that the only way project-oriented organizations cope with this uncertainty is by employing between 20 percent and 40 percent contract staff. They report one organization employing up to 80 percent contract staff. This is essential to cope with fluctuating and uncertain workloads.
3. As for forecasting future resource requirements, if it is not possible to predict resource requirements one month out, how can anyone predict them one year out? Organizations can assume they will carry on doing the same types of projects, and they will try to use economic forecasts to predict future numbers of projects in the industry. However, it is much less certain than in a functional organization.

The Spiral Staircase Career

The consequence for people's careers is good news and bad news. The bad news is they do not have the comfortable certainty of a clear career path where they can climb the ladder up the functional silo. The good news is they have much more varied and interesting careers. Projects, being transient, cannot provide careers, but each project can be a learning opportunity in a career. Projects provide an opportunity for a broad sweep of learning experiences. Keegan and Turner (2003) coined the phrase "the spiral staircase career" to reflect that people will move through a series of varied and wide-ranging jobs. They might spend time in the design function, time as lead designers on a project, and time as project managers. Rather than each move being a whole step up the ladder, moves can be half or even a quarter of a step sideways and upwards. People can also avoid the Peter Principle—namely, being promoted to the level of their incompetence. If they find themselves in a job that does not suit them, they can take a move sideways, which does not carry any stigma, compared to taking a step down the ladder of the functional silo.

No Home Syndrome

Coupled with varied career is the "no home syndrome" (Keegan and Turner, 2003). People spend their working lives moving from one project to another. They generally do not have a permanent home, or a permanent sense of belonging. They work on one project for 9 to 18 months; then that team breaks up and they move to a new team. This creates the nomadic life mentioned previously, but it also increases the need for team building on projects to create a sense of belonging to the project (Reid, 2003). A practice adopted by many project-oriented firms is the creation of the PM office, or an expert pool of project managers. This can provide workers a "home base" between projects and a place to continue to belong to and seek support while working on projects. Sometimes the PM office may be virtual but still satisfy these needs.

Project Management Personnel

In project-oriented organizations, we can differentiate several different types of resources, including line management, technical experts, and project management personnel. Project management personnel are those human resources who need to draw on project management knowledge and experience to fulfil their roles. They include project managers but also include people in other project roles. The HRM practices we discuss in this chapter apply to project management personnel in the first instance. The project-oriented organization may apply similar processes, or conventional ones, to people working in line management or as technical experts. Project management personnel include people working in temporary structures such as projects and programs, and people working in permanent structures such as a project management office, a project portfolio group, or an expert pool. The former group includes the following:

1. Project personnel, such as:
   - The project owner, project sponsor, or project champion
   - The project manager, project leader, or project director
   - The project management assistant
   - The project controller
   - Project team members and project contributor
2. Program personnel, such as:
   - The program owner or program director
   - The program manager, project director, or program coordinator
   - The program assistant or program controller
   - Program office members

People working in permanent structures include the following:

- Project management office personnel such as the office leader and office members. They are the process owners for the project management process within the project-oriented company. Further functions of the project management office are described later in this chapter.
- Project portfolio group members who take the responsibility to manage the project portfolio from a strategic perspective. Usually these members of the project portfolio group are managers of those business units of the permanent organization, which are frequently involved in projects and programs.
- Quality management personnel such as project or project management auditors and reviewers, project or project management coaches, and project or project management consultants.
- Expert pool personnel such as the leader of the project expert pool and the members of the project expert pool. From these expert pools the project personnel is drawn.

The project portfolio office leader, project portfolio group members, and project expert pool leaders are often labeled as "project executives." Employees in the project-oriented company often have more than one role and can therefore belong to different groupings of project management personnel. For example, one person can be a program manager for one project and at the same time work as project coach for a different project.

Competences of Project Management Personnel

As part of their HRM policies and practices, project-oriented firms need to define competence requirements for all these project management personnel. (Competence development is described in Gale's chapter). Competence is the knowledge, skills, and behaviors (experience) a person needs to fulfill his or her role (Huemann, 2002). Project management personnel need a set of several competencies covering not just the management of projects but also the following:

- *Project management.* Knowledge and experience about project and program management including methods and processes
- *Organization.* Knowledge and experience about the project-oriented organization at its specific processes like portfolio management, assignment of projects and programs, and so on
- *Business.* Social networks, product, industry, and so on
- *Technical.* Technical, marketing, engineering, and so on
- *Cultural and ethical awareness.* As in the case of international projects.

How these competencies are described is specific to the company and the project management approach used. It may be traditional, emphasizing scope, cost, and time, as in PMI's PMBOK (2002); it may be more holistic, emphasizing process orientation, as in PRINCE2 (OGC 2002); or it may emphasize also project context and organization, as proposed by Gareis (2002) and Morris (1997) or the Association for Project Management (APM, 2000). There is always a lot of discussion on how much technical competencies the project managers need to manage a project. The range goes from nontechnical competencies to being a technical expert as well as a project manager. The more project management is considered a profession in the organization the less technical competencies may be asked for. Figure 42.1 illustrates minimum competence requirements of a senior project manager in an engineering company. The competence profile required very much depends on the size of the project, its type, and the industry. The competencies will be developed through the individual's career, which we discuss next. However, there is a different emphasis in the project-

FIGURE 42.1. MINIMUM COMPETENCE REQUIREMENTS OF A SENIOR PROJECT MANAGER.

Competences	Knowledge					Experience				
	5 very much	4 much	3 average	2 low	1 none	1 none	2 low	3 average	4 much	5 very much
Project and Program Management	■								■	
Management of the Project-Oriented Company		■						■		
Business		■						■		
Project Contents				■			■			

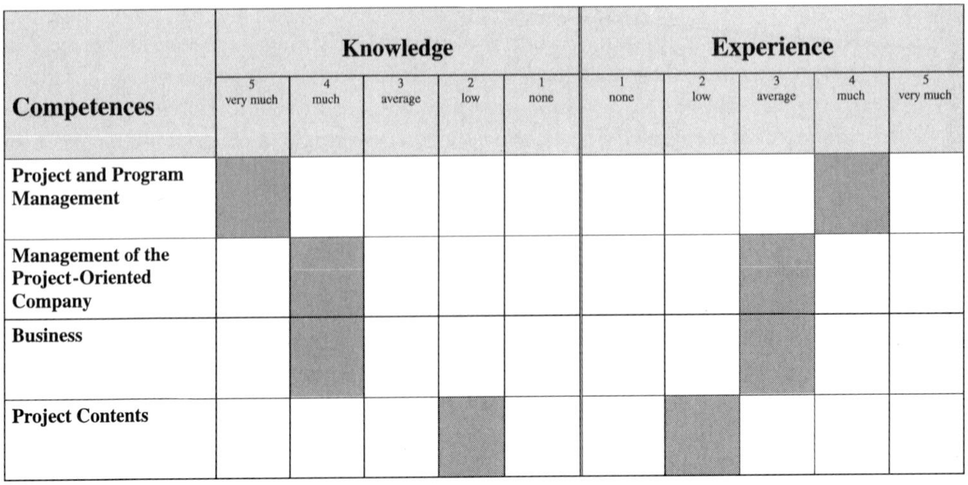

oriented firm. In the traditional organization, the individual gains his or her knowledge and experience working within one function, in the stable organization. In the project-oriented firm, the person gains knowledge and experience through a series of projects, fulfilling different roles on those projects, following the spiral staircase career. Such careers need careful management.

Career Development

We have already seen how project-based ways of working fundamentally change the careers of individuals. Rather than climbing the ladder up the silo, they follow a spiral staircase career, with wide and varied career experiences. There are several practices project-based organizations adopt to support project management careers, including the following:

- A defined project management career
- Measuring "up" in novel ways
- Career committees
- Project management communities
- Individual responsibility

A Defined Project Management Career. Many project-based organizations from both the engineering and high-technology industries recognize project management as a defined career path. Table 42.1 shows a typical seven-step career for many high-technology companies. Many organizations support the career path with professional certification for the Project Management Institute or International Project Management Association, and with formal education programs. IBM, for instance, requires its personnel to take PMI PMP certification followed by a master's degree.

Some organizations have parallel career paths. One high-technology company profiled by Keegan and Turner has a career structure like an upside-down table, with four recognized careers in the company:

- Project management
- Line management
- Technology management
- Sales and marketing

The four careers followed a common structure up to stage 3 in Table 42.1. This enabled people to follow a spiral staircase, sampling different possible careers, until they reached the start of stage 4. Then they were expected to specialize, climbing the ladder up one of the four legs for the remainder of their career.

Measuring "Up". Discussing career movement in terms of climbing a ladder or a spiral staircase implies there is some measure of "up"—some measure of increasing seniority and responsibility. Traditionally, the way to measure up was by the number of people managed

TABLE 42.1. SEVEN-STEP CAREER MODEL.

Stage	1	2	3	4	5	6	7
Name	New start	Team member	Team leader	Junior project manager	Project manager	Senior project manager	Program director
Responsibility			Single function	Several functions	Several companies	Complex projects	Many complex projects
IPMA certification			Level D		Level C	Level B	Level A
PMI certification				PMP			
Education				Certificate	Diploma	MBA, MPM MSc (PM)	

and the budget of the individual's department. This idea was widely discredited early in the history of HRM but is still applied by many organizations.

In spite of this, many organizations continue to reward people according to the number of their direct subordinates right up to the present day. Recently one of the authors was interviewing a man from a company that made electronic equipment. The company wanted to projectize their business and wanted to know what might stand in the way of that. The reward structure was seen as a potential barrier: The company still rewarded people according to the number of direct subordinates. That meant the manager of an engineering department with 1,000 engineers would be scaled as very senior. On the other hand, the manager of a project of £5 million, with a profit margin of £500,000 and of critical importance to the UK's defense, would have very few direct subordinates and so would be scaled as very junior.

Many project-oriented organizations measure "up" in other ways. A practice common in high-technology and engineering companies is to measure "up" by control of risk (which is related to impact on profit). In organizations in both industries, the head of a function or department may not be the most senior person in that department. For example, as part of the spiral staircase career, a potential project manager may return to manage the design function for a while and while doing this might not be the most senior member of the team. Management of the design function carries a certain level of risk, while being a senior designer may carry more.

We interviewed somebody from the engineering industry who had gone from being projects director on the company's board to director of a $1.5 billion project for a major client. The project was considered to be of such high risk that the director of that project was a more critical role than a company board member.

Career Committees. The career development process in Table 42.1 does not happen on its own. Many organizations have committees of senior project or program managers, managing the development of project management professionals. In the engineering industry these committees tend to be fairly ad hoc. The process is managed, but in an informal way. In high-technology firms, the process tends to be more formally managed, linked to career development process of the organization as a whole. This may also be the role of the PM office, as discussed later in the chapter.

Project Management Communities. Project management communities are often used to aid organizational learning. These are networks of project managers to support the development of individual project managers, through mentoring or via events where project management professionals can meet and exchange experiences. Developing the competence or maturity of an organization and its people are closely linked. Project management communities can be company-internal or company-external. Project management associations, such as the Project Management Institute and the International Project Management Association, fulfill the role of external communities. The European Construction Institute in Europe and the Construction Industry Institute in the United States provide an external community for people from the engineering construction industry. Internal communities are maintained in large, high-technology companies, such as IBM, Telekom Austria, and the information services (IS) department of the Dutch bank ABN AMRO. In large companies the project management community also helps project managers to meet, which would otherwise be difficult. It can also help promote the profession and facilitate knowledge management.

Individual Responsibility. Although companies maintain committees to manage careers and communities to support development, individual project management professionals are expected to take responsibility for their own career development. It is easy to get lost on the spiral staircase, both to lose your way and for people to stop noticing you. Turner, Keegan, and Crawford (2003) report that in the engineering industry personal ambition is a key criterion for identifying someone as a potential project manager. Many organizations through their career committees and project management communities provide people with guidance on setting annual objectives and development plans, including training and certification. But individuals must take personal responsibility for achieving their plan.

A key issue that frequently occurs when a person has a development objective and a new project comes along that provides that opportunity. In this case, is the person made to finish his or her current project and be denied the opportunity, or can they be switched mid-project? Enlightened companies switch people mid-project, as long as the current project is not in start-up or closeout. It is better for the company that the individual gets the development opportunity, and individuals will be more loyal to companies that provide them with appropriate development opportunities.

Project Management Profession

The establishment of a project management career path contributed to the development of the project management profession in the project-oriented society. Many project-oriented organizations require potential project management personnel to seek certification. Such certificates prove that the person has a certain level of project management knowledge and experience. Project management certification is offered by global project management associations such as the International Project Management Association (www.ipma.ch), the Project Management Institute (www.pmi.org), or the many national project management associations. Figure 42.2 shows the IPMA project management certification as offered by Project Management Austria (PMA). This also illustrates the changing competence requirements of project management personnel at different levels. This structure and the associated levels have been adopted by many project-oriented companies in Austria, such as Unisys and Telekom Austria. The certification structures are associated to the project management career structure in these companies, and there is a link to the reward system. Certification is perceived as an external quality check for the project management personnel, and in many cases the customers ask for certified project managers.

Processes of Human Resource Management

In this section we describe the processes of human resource management, their specific characteristics in the project-oriented organization, and the methods applied. These processes include the following (Schein, 1987; Keegan, 2002):

FIGURE 42.2. PROJECT MANAGEMENT AUSTRIA 4 LEVEL CERTIFICATION BASED ON IPMA CERTIFICATION.

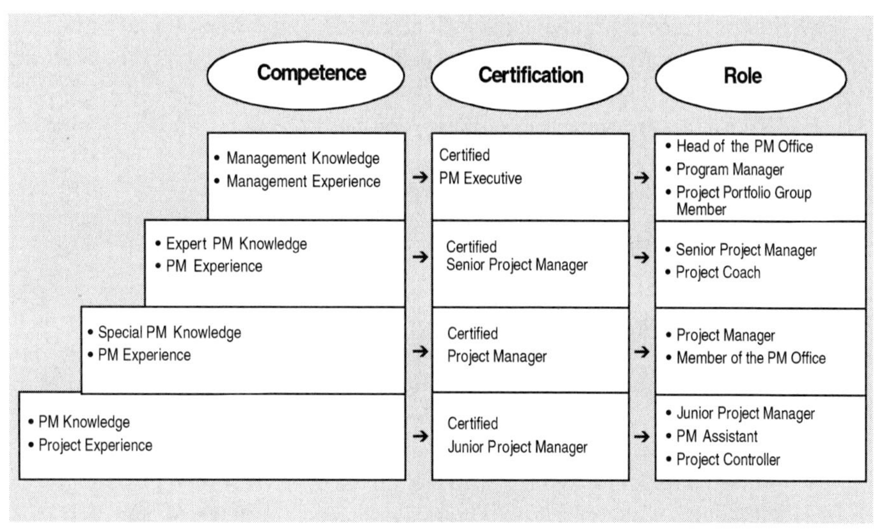

- Recruitment
- Disposition
- Development
- Leadership
- Retention
- Release

Recruitment

Recruitment comprises the search for competent personnel to meet current or future resource requirements. Search and selection can be done for the company in general or for a specific project or program. One can differentiate between company-internal and company-external recruiting. Company-internal recruiting for a specific project or program draws on expert pools within the company. Another possibility is to recruit project management personnel from outside the company from project management networks. Freelance personnel often appreciate the advantages of being part of a network. For example, the platform www.myfreelancer.at offers a marketplace for freelancers in the IT sector. Networks increase the flexibility of the company by enabling it to build relationships with cooperationpartners and experts and to have access to them in case of demand. We saw previously that the only way many project-oriented organizations cope with fluctuating demand is through the use of contract staff, with levels typically ranging between 20 percent and 40 percent. Often the recruiting is performed by a single project or program, the project portfolio office, the project management pool, or expert pool and not by the human resource management department.

Methods for Search and Selection. The traditional, formal manner of recruiting people to a post, whether recruiting them internally or externally, is to do the following:

1. Write a job description, describing what has to be done, and the competence (knowledge, skills and experience) required.
2. Identify candidates for the job, often through advertising.
3. Find the best match between the job description and the candidates.

The person is matched to the job. That process does not work so well in the project-oriented organization, especially for the people working in temporary structures. Jobs cannot be defined with precision. You need to find people with competencies required to work on projects and let them define the jobs that need to be done in the circumstances. In project organizations much less formal recruiting practices are adopted. People are often initially recruited to work on an individual project, perhaps on a freelance or contract basis. Because of the large number of contract workers, it is easy to take someone on initially as a peripheral worker. Then if that person performs well, or is a good fit, the person can be offered permanent employment and even be placed on the project management career development track.

Maintaining Networks. To recruit people in this way, it is essential to maintain networks in the industry, with clients, competitors, and suppliers, and with universities and professional associations. We mentioned previously that some organizations maintain a project management community external to the organization, belonging, for instance, to a professional association or network. As well as providing development opportunities for existing project management personnel, they can also be a source for new personnel. Companies also often maintain strong links with universities, offering students temporary employment during the summer vacation, to see if they are a good fit, and offer them employment if they are.

Assessment Centers. Crawford (2003) describes the use of assessment centers for competence assessment and development. Assessment centers (Woodruffe, 1990) use a process lasting from two to five days, during which candidates are put into a simulated project environment to see how they perform. An assessment center consists of a standardized evaluation of behavior based on multiple inputs. Judgments are made from specifically developed assessment simulations. These judgments are pooled in discussions among the assessors, and the participants themselves have an opportunity to give and receive feedback on the instruments and measures used. The discussion results in evaluations of the performance of the candidates on the competencies or other variables that the assessment center is designed to measure. The effectiveness of assessment centers depends on the involvement of senior personnel from the project management community to observe the candidates and give informed feedback. Their dedication and willingness to be involved often depends on their being convinced that the center is as much a development opportunity for them as for the organization. Assessment centers are highly resource-intensive. It therefore makes sense to assess the threshold competencies of the majority of project personnel through other means and to reserve assessment centers for the assessment of the performance of candidates for promotion to senior project management positions. The exercises used in project management assessment centers have to be specific to reflect the typical processes of the project-oriented organization. Table 42.2 shows some examples.

Disposition

As project management personnel work on several projects and programs at the same time, the disposition of project management personnel to the different projects and programs is a critical issue in the project-oriented organization (Eskerod, 1998). Coordination and disposition through organizational unit, which is of permanent nature, is required. This function may be carried through the PM office. Disposition comprises the following:

- Allocation of PM personnel to projects and programs
- Optimization of allocation of resources in case of multiproject engagement
- Organization and support of the transition of PM personnel from one project to another

In many organizations disposition is closely linked to the coordination of the project portfolio. The project portfolio database is often linked to the management of personnel resources.

Development

The objective of personnel development is to improve the competence of project management personnel by offering the possibility of gaining knowledge and experience. Development activities are carried out either on the job, in a project or program assignment, or in general outside project assignments. Development can be limited to project management personnel employed or extended to include freelancers in the network. Responsibility for

TABLE 42.2. PROJECT-MANAGEMENT-RELATED EXERCISES IN A PROJECT MANAGEMENT ASSESSMENT CENTER.

AC Method	PM-Related Exercises
Presentation	Project start, project controlling, project closedown situation
One on one	Project controlling, feedback
Group discussion	Nearly all PM topics
Role-playing	Nearly all PM situations, e.g., meeting of project manager and project owner in a crisis situation
Analysis	Interpretation of a portfolio report, PM audit result interpretation, planning of the project start workshop

the development of individuals is generally taken on by the career committee and the project management community as described previously, although as we saw, the individual is also often required to take responsibility for his or her own development.

Methods adopted for the development of project management personnel include the following:

- Education and training
- PM competence assessment
- Assessment centers for development
- Coaching
- Feedback
- Training on the project
- Job rotation (within the project or program, between projects, and to other organizations)
- Support networks and communities
- Career development committees

Turner and Huemann (2000, 2001) describe education programs offered in several project-oriented societies globally. Project management training is often organized by the project office, in cooperation with the HR department, using either internal or external trainers. What often happens is that firms train their junior project management personnel internally. They are coached in the organization's ways of working. With middle- to senior-level personnel, levels 5, 6, and 7 in Table 42.1, training is conducted externally. There are several reasons for this:

- There are fewer people to train.
- They need general management skills in addition to project management skills.
- They can network in the industry on the courses.

The project management community also plays an essential role in developing individuals, providing coaching and mentoring, as well as the opportunity for people to network internally and develop personal competence through assimilation. Coaching and mentoring in the project management community is also part of leadership in the project context.

Project management competence assessments are either used in combination with the project management training or included in assessment centers (for development as well as for selection). Figure 42.3 shows some results of a project management competence assessment of project managers, which was used to get an overview on the status of the project management personnel in a company. Based on the single results, tailored activities for the further development of the project management knowledge and competence were taken.

Feedback is a method that is not very often used. Feedback methods are introduced to project managers in project management courses at the University of Economics and Business Administration Vienna. The course participants, who are experienced project managers, give feedback to each other after they have been working together on some training projects. The discussion that springs up from this is often that project managers are lacking feedback from the project owners as well as from project team members. A special form of

FIGURE 42.3. RESULTS OF A PROJECT MANAGEMENT COMPETENCE ASSESSMENT.

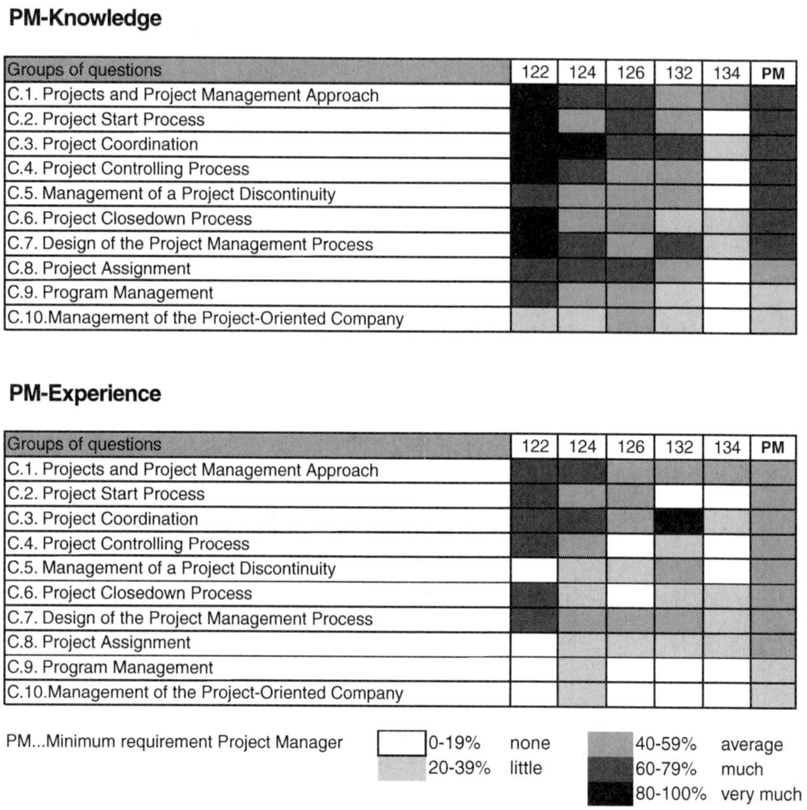

PM-Knowledge

Groups of questions	122	124	126	132	134	PM
C.1. Projects and Project Management Approach						
C.2. Project Start Process						
C.3. Project Coordination						
C.4. Project Controlling Process						
C.5. Management of a Project Discontinuity						
C.6. Project Closedown Process						
C.7. Design of the Project Management Process						
C.8. Project Assignment						
C.9. Program Management						
C.10.Management of the Project-Oriented Company						

PM-Experience

Groups of questions	122	124	126	132	134	PM
C.1. Projects and Project Management Approach						
C.2. Project Start Process						
C.3. Project Coordination						
C.4. Project Controlling Process						
C.5. Management of a Project Discontinuity						
C.6. Project Closedown Process						
C.7. Design of the Project Management Process						
C.8. Project Assignment						
C.9. Program Management						
C.10.Management of the Project-Oriented Company						

PM...Minimum requirement Project Manager

0-19% none
20-39% little
40-59% average
60-79% much
80-100% very much

feedback is the 360-degree review, shown in Figure 42.4. (Philips is one company that uses it for the project managers.) Here the project manager does a self-assessment based on a questionnaire. Other environments like project owner, project team members, suppliers, and customers give feedback to the project manager based on the same questionnaire.

Leadership

Leadership and team building are topics to which individual chapters are devoted elsewhere in this book (Thoms & Kerwin on leadership, and DeLisle on teams). However, they are both essential HRM processes, and for completeness, we discuss some core and new topics here.

In general, leadership is needed

FIGURE 42.4. CONCEPT OF THE 360-DEGREE FEEDBACK.

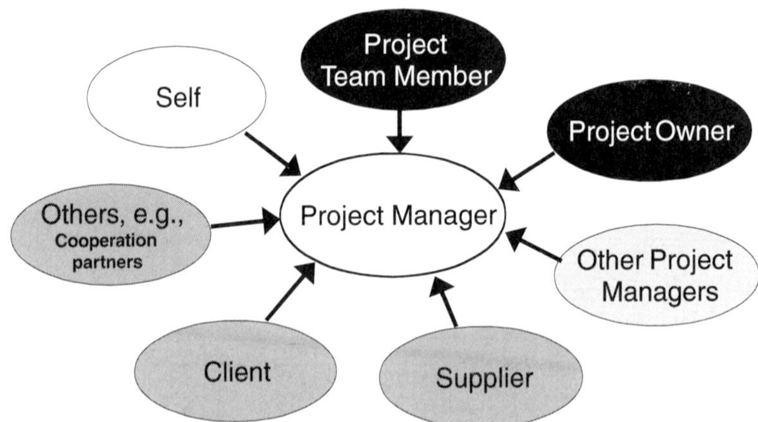

- at the level of the project-oriented company, with (inter alia) the company providing leadership to individual projects, programs, and project participants; and
- at the level of individual projects and programs, with project leaders (the project owner, the project manager, and the subteam leaders) providing leadership to other project participants (see Table 42.3).

Empowerment as a Key Value in the Project-Oriented Company. In high-performance organizations, people are *enabled* to do their best work. They have the adequate tools, standards, policies, and procedures. They are well trained and trusted. Empowerment is about goal setting, providing frameworks and limits within which subordinates can operate but

TABLE 42.3. LEADERSHIP IN THE PROJECT-ORIENTED COMPANY.

Level	Leader	Others
Project	Project sponsor Project manager Project subteam leader	Project manager Project team Project subteam Project team members
Program	Program sponsors Program manager	Project managers Program team Program team members
Project-oriented company	Portfolio group PM office Line manager	Program managers Project managers Project members PM office members

allowing freedom within those limits. Both van Fenema (2002) and Müller (2003) deal with the issue of empowerment and show that it leads to better project performance. It is needed by the following:

- The project itself
- The project team
- The single project manager
- The single project team member

Projects, as we have seen, are temporary organizations used to deliver change in organizations (Turner and Müller, 2003). It is a less efficient form of organization than the line organization, but it is more effective at delivering change, as it is more responsive and has lower inertia. However, if the change is to be achieved, the project must be removed from the line organization, and so empowerment of the project is essential. Empowerment here means to reduce the interventions of the line organization to a minimum and let the project work. The quality is ensured by providing adequate project management tools, standards, and guidelines.

Empowerment of the project creates the issue of the principal-agent relationship between the project sponsor and the project team, and so effective communication mechanisms must be put in place for empowerment to work (Müller, 2003). The project manager and project team must be made aware of the client's requirements, and the client needs to be made aware of progress. With effective communication, empowerment is possible; and with empowerment, effective project management is possible. An example of a symbolic act to empower the project team is to let them all sign the project charter. It is standard practice in some project-oriented companies that at the end of the project start workshop, the project owner and all the project team members (including the project manager) sign the project charter on a flip chart.

In researching communication between project sponsors and project managers, Müller (2003) found that high-performing projects were correlated with high collaboration between project managers and sponsors, and medium levels of structure. Collaboration was related to clearness of objectives and relational norms, and structure to clearness of work methods and "mechanicity". In Müller's sample, medium levels of structure and high levels of collaboration (empowerment) were necessary conditions for project success. Empowerment means that the project sponsor should set clear objectives, relational norms (high collaboration), and defined boundaries, but leave the project manager freedom to find the best solution within those boundaries (medium structure). Empowerment is not tight structure (no freedom), but equally it is not laissez-faire management (no objectives and boundaries).

Figure 42.5 illustrates three paths to falling collaboration. On failing projects, or projects with unclear objectives, tight structure and control tends to be adopted. Where there is remote working or infrequent reporting (van Fenema 2002), there is low collaboration with medium levels of structure. Where the clients and project managers objectives are misaligned, or reporting is informal, collaboration falls and anarchy reigns.

Empowerment of project managers includes clear agreements on the role and agreements on frequent communication structures and decision making with the project sponsor.

FIGURE 42.5. COLLABORATION AND STRUCTURE ON HIGH PERFORMING PROJECTS.

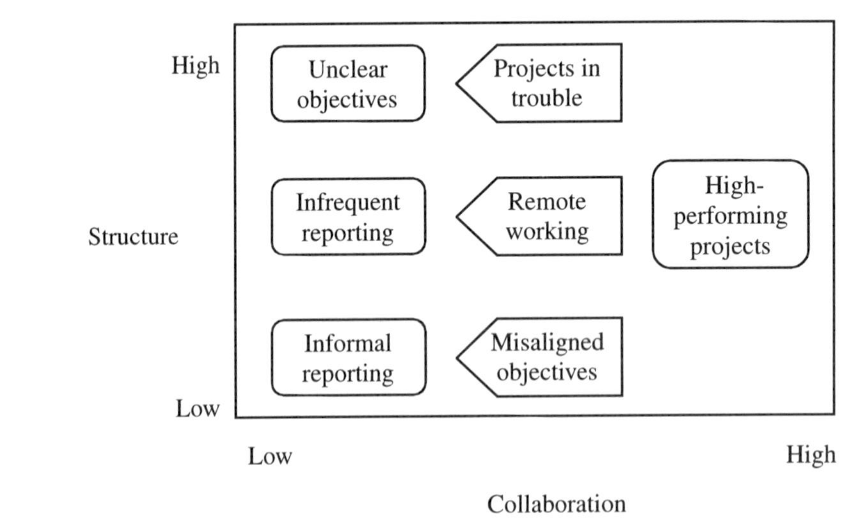

The role of the project manager should be described in relation to the role of the project sponsor. While the project sponsor has to take care of the company's interest and is responsible for strategic decisions and providing context information to the project, the project manager takes care of the project interest, is responsible for operative decisions and contributes to strategic decisions, and provides project information to the project sponsor.

Just as the project manager needs to be empowered by the project sponsor, so too does the project participant need to be empowered by the project manager, which leads us to the leadership role of the project manager.

Project Management Includes Project Leadership. Project management involves both people leadership and task management. Turner and Müller (2003) liken the role of the project manager to that of the chief executive of the temporary organization that is the project, and quote the classic text by Barnard (1938), who said the role of the chief executive is

> to formulate purposes, objectives and ends of the organization . . . This function of formulating grand purposes and providing for their redefinition is one which needs systems of communication, experience, imagination, interpretation and delegation of responsibility.

That sounds like the role of the project manager, who has to delegate, guide project team members, motivate, set goals, provide information, make decisions, and give feedback. But the project manager is not the only one taking on a leadership function. Within the project, leadership has to be provided by the project sponsor, the project subteam managers, and the project manager. The project organization chart in Figure 42.6 shows an empowered project organization. In the figure a differentiation is made between the leading of single individuals and the leading of teams.

Moderation functions have to be distinguished from leadership functions. While facilitation of workshops and meetings includes preparation of meeting, moderation of decision processes, structuring communication processes, and such is normally directionally neutral, while leadership sets interventions to steer into a direction. Recognizing this difference can allow the project manager to understand that in a workshop situations like project start-up, project crisis, and project closedown, he or she has to take over two roles: the role of the leader and the role of the neutral process facilitator. That might be difficult. Sometimes project and program managers support each other in such situations by bringing in someone else, for example, a manager from another project, to facilitate in the workshop. This helps the project manager to concentrate on the leadership function.

Leadership functions can be further described by looking at the subprocesses of project management and the subprocesses of team development

Leadership in the Project Start. Project start is the most important subprocess of project management. From a leadership point of view, during the project start the project team has to be established and the "big project picture" has to be commonly defined. In a project start workshop, not only do the project plans need to be created but the rules and values

FIGURE 42.6. LEADERSHIP IN THE EMPOWERED PROJECT ORGANIZATION.

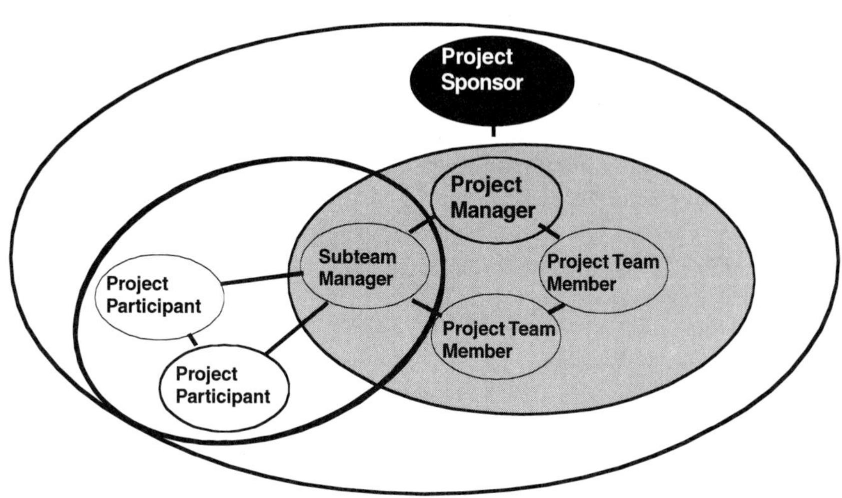

for working together have to be established and agreed on. Benchmarking of project management processes (Gareis and Huemann, 2003) suggested that traditional phases of team formation and maintenance (forming, storming, norming, performing) should be reconsidered, and instead the team actively work on the establishment of a project and team culture. That includes agreeing project rules and understanding roles in the project. Doing the forming and norming of the project team together in the project start situation reduces the storming and leads earlier to the performing of the project team. Further elements in establishing a project culture and identity are as follows:

- Project name
- Project motif and logo
- Newsletters

The name is important and agreeing on a project name might be quite sensitive as "nomen est omen." Guess what happened to a huge IT project by the name of Atlantis? It sank like Atlantis.

As part of the start-up process, it is important to build psychological attachment among project team members, especially for teams working remotely. Flying people to a central location for the start-up meeting may seem a great expense, but it can be invaluable for team formation. Once established, the team newsletter will help maintain psychological attachment.

Leadership in the Project Coordination and Project Controlling. In the project coordination, leadership functions of the project manager comprise delegating work packages, setting objectives, and giving feedback to project participants. As a communication structure, regular team meetings and meetings of single individuals are appropriate.

What leadership means in the project controlling process very much depends on the project management approach used. Project organization, project culture, team, and communication structures have to be questioned regarding whether they are still appropriate. If the project roles and rules are not adequate, they have to be adapted and agreed on within the team. Methods used here are reflection and feedback. Gareis and Huemann (2003) describe the project management competence of project team, which has to be built up and maintained during the project. A project team has to be able do the following:

- Together create the "big project picture."
- Create commitment.
- Use synergies.
- Managing conflicts.
- Learn.
- Together create the PM process.

One method to foster these competencies is to do have the team do a self-assessment of their team project management competence. An example question of such a self-assessment is given in Figure 43.7. Thus, one might go so far and say that the project team as such has a leadership function.

Leadership in the Project Closedown. While during project start the team is established, in project closedown the leadership function is to support the resolution of the project team, which includes making agreements for the final work, but it should also involve an emotional closedown by reflecting the process of working together and giving each other feedback. Very often, because of time pressure caused by another project waiting in the wings, this closedown is neglected.

Retention

Turner, Keegan, and Crawford (2003) reported that project managers tend to stay longer with one organization than other project participants. They feel their commitment to one firm as part of their career development. But commitment is a two-way street. Project management is a core competency to project-oriented organizations, and so they should make an effort to retain their project managers. One way to do that is make a commitment to their development as project managers, which means helping to manage their careers through the spiral staircase, giving them development opportunities as they arise. Sponsoring them through certification and master's programs demonstrates a clear commitment to their development. Building a psychological contract in this way can engender the commitment of individuals to the company.

Incentive systems and motivation are closely linked. Different incentive systems are possible in the project-oriented organization. We can look at incentives for project managers

FIGURE 42.7. SELF-ASSESSMENT OF THE PROJECT TEAM COMPETENCE.

Creation of "Big Project Picture" in Team		
1= none, 2= little, 3= average, 4= much, 5= very much	Knowledge	Experience
Common performance of workshops and meetings		
Use of project plans for communication		
Context orientation		
Holistic view		
Interpretation		

or for the whole project. Most incentives are monetary. Often there are difficulties in distributing the reward at the end of the project. The incentive system of the company is very much linked to the culture of the company. Again, size of budget may distort real priorities. Rewards linked to the budget of the project can lead to competition amongst the project managers to manage projects with the highest budget, which are often not the ones that are the most complex and the most strategically important. Creative incentive systems are more personalized. For example, after a very busy period in a project, a project manager gets as an incentive a week-end trip with his family. Recognition through little things can have a huge impact on motivation.

Release

Finally we consider release, which encompasses both release of the project and release of the temporary workers from the organization. There are two key elements of the release process, applicable to both: organizational learning and individual review and feedback. Organizational learning is covered in the chapters by Bredillet and by Morris. Feedback and review were discussed earlier in the chapter. With the release of freelance workers from the project and from the organization, it is also important to remain in contact to maintain the organization's network and to make future cooperation possible.

We have discussed several processes of human resource management in the project-oriented organization, including recruitment, disposition, development, leadership, retention, and release. These are summarized in Table 4.

The Role of the PM Office in HRM

So far we have discussed the practices and processes of human resource management in the project-oriented company. In this section we consider the role of the PM office as the unit that in cooperation with the central HR department is responsible for managing project management personnel. The PM office is a permanent function within the project-oriented company (Knutson, 2001; Rad and Levin, 2002; see also the chapter by Powell and Young).

An organization chart of a PM-office, which could be virtual, is shown in Figure 42.8.

The objectives of a PM-office, which in some companies is called the project management center of excellence, are to

- ensure a ready supply of professional project and program managers;
- provide management support to projects and programs, often by providing project managers and program managers to projects and programs;
- develop individual and organizational competencies in the project-oriented company; and
- manage project portfolios and related services, which will not be discussed further here (but see the chapter by Gareis and by Archer and Ghamazadeh).

Home for Project Management and Project Management Personnel

In many cases the PM office provides a pool of project and program managers. The PM office is therefore often seen as the home for the project management personnel. In addition,

TABLE 42.4. OVERVIEW: METHODS USED IN THE DIFFERENT HR SUBPROCESSES.

Recruiting	Disposition	Development	Leadership	Retention	Release
Assessment center	Portfolio database	Assessment center	Decisions	Incentive system	Documentation of learning
Informal section	Resource database	Feedback	Feedback	Reward system	Feedback
Liaison with universities	Reflection	Reflection	Reflection		
	Education and training	Information providing			
		Delegation Empowerment			

FIGURE 42.8. ORGANIZATION CHART OF A PM OFFICE.

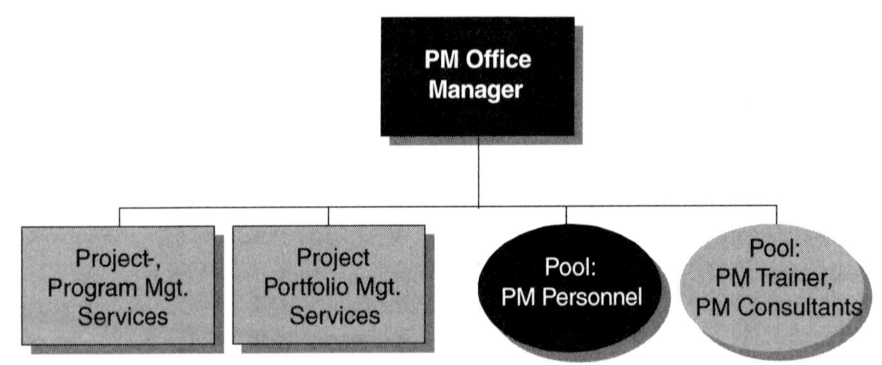

the PM office may provide HR services to the project management personnel, including the following:

- Provision of internal project management training and/or the organization of project management training with external training providers in accordance with the project management approach of the company
- Provision of coaching and mentoring of project management personnel, supporting the career committee
- Establishment and maintenance of the project management community as described previously
- Maintenance of the link with project management freelancers, by maintaining a database and inviting them to in-house networking activities and conferences organized by the project management community

Services to Empower Projects and Programs

As discussed, empowerment of projects and programs is one of the key values of the project-oriented company. To make empowerment possible and to ensure quality of the project management processes, the PM office provides and further develops the following:

- Project management guidelines and procedures, standard project plans for repetitive projects, and standard project management forms
- Project management infrastructure such as project management software software, project management portals, and collaboration platforms
- Management consulting services to projects and programs
- Management audits and (peer) reviews for projects and programs

Promoter of the PM Profession

The PM office acts as promoter of the project management profession within the company. In many project-oriented companies, the PM office is responsible for the following:

- The establishment of the project management career path
- The running of incentive and reward system suitable for project-oriented companies
- The holding of in-house PM conferences and support for project managers to attend project management conferences
- Cooperation with universities to have access to new theories and to well-educated project personnel for recruitment
- Cooperation with external project management communities and professional institutions to have access to best practices

Summary

In this chapter we showed human resourse processes and practices applied in the project-oriented organization. We argued that as individuals work more often in temporary organizations, such as projects and programs, and move from one project to the other like modern nomads. We identified specifics like the spiral staircase career and the no-home syndrome of project workers which leads to a specific view on human resource management in the project-oriented organization. We especially concentrated on project management personnel and defined the term and described the competencies which are developed through the individual's career.

We showed examples of career paths and how these are linked to certification offered by IPMA and PMI. We further discussed career committees and project management communities which support the establishment of project management as a profession.

We described specific human resource processes like recruitment disposition, development, leadership, retention and release, and the practices applied in these processes. Table 42.4 provides a compact overview on the methods used in different human resource processes in the project-oriented organization. We finally ended with the role of the PM office, which takes on human resouce management functions.

References and Further Reading

Association for Project Management (2000) *Body of Knowledge 4th ed.* www.apm.org.uk.

Barnard, C. I. 1938. *The functions of the executive.* Cambridge, MA: Harvard University Press.

Crawford, L. 2003. Assessing and developing the project management competence of individuals. In *People in project management,* ed. J. R. Turner., Aldershot, UK: Gower.

Drucker, P. F. 1994. *Post capitalist society.* New York: Harper Business.

Eskerod, P. 1998. The human resource allocation process when organizing by projects. In *Projects as arenas for renewal and learning processes,* ed. R. A. Lundin and C. Midler. Boston: Kluwer Academic Publishers.

Gareis, R. 2002. Project management for everybody: A visionary dimension of the project-oriented society. In *Proceedings of IRNOP V, the Fifth Biennial Conference of the International Research Network on Organizing by Projects*, ed. J. R. Turner. Renesse, Netherlands, May 29–31.

Gareis, R., and Huemann, M. 2001. Assessing and benchmarking project-oriented societies. *Project Management: International Project Management Journal, Finland* 7 (1, Summer): 14–25.

———. 2003. Project management competences in the project-oriented company. In *People in project management*, ed. J. R. Turner. Aldershot, UK: Gower.

Gedansky, L. 2002. Inspiring the direction of the profession. *Project Management Journal* 33(1).

Handy, C. B. 1988. *The Future of Work: a guide to changing society*. Oxford, UK: Blackwell.

———. 2002. *The Elephant and the flea: Reflections of a reluctant capitalist*. Cambridge, MA: Harvard Business Press.

Huemann, M. 2002. *Individuelle projektmanagement Kompetenzen in projektorientierten Unternehmen.*. Europäische Hochschulschriften. Frankfurt-am-Main: Peter Lang.

Keegan, A. E. 2002. Human resource management. In *Project management pathways*, ed. M. Stevens. High Wycombe, UK: Association for Project Management.

Keegan, A. E., and J. R. Turner. 2003. Managing human resources in the project-based organization. In *People in project management*, ed. J. R. Turner. Aldershot, UK: Gower.

Knutson, J. 2001. *Succeeding in project-driven organizations: People, processes and politics*. New York: Wiley.

Lundin, R. A., and Söderholm, A. 1998. Conceptualizing a projectified society: Discussion of an eco-institutional approach to a theory on temporary organizations. In *Projects as arenas for renewal and learning processes*, ed. R. A. Lundin and C. Midler. Boston: Kluwer Academic Publishers..

Morris, P. W. G. 1997. *The management of projects*, London: Thomas Telford.

Müller, R. (2003). *Communication of information technology sponsors and managers in a buyer-seller relationship*. DBA thesis, Henley Management College, Henley-on-Thames.

PMI, (2001). The PMI Project Management Fact Book, Secons Edition, Project Management Institute: Pennsylvania.

Rad, P. F., and G. Levin. 2002. *The advanced project management office: A comprehensive look at function and implementation*. Boca Raton, FL: St. Lucie Press.

Reid, A. 2003. Managing teams: The reality of life. In *People in project management*, ed. J. R. Turner. Aldershot, UK: Gower.

Schein, E. H. 1987. Increasing organizational effectiveness through better human resource planning and development. In *The art of human resources*, ed. E. H. Schein. Oxford, UK: Oxford University Press.

Turner, J. R. 1999. *The Handbook of Project Based Management, 2nd edition*, McGraw-Hill, London.

Turner, J. R., and M. Huemann. 2000. Current and future trends in the education of project managers. *Project Management: International Project Management Journal, Finland* 6 (1, Summer): 20–26.

———. 2001. The maturity of project management education in the project oriented society. *Project Management: International Project Management Journal, Finland* 7 (1, Summer): 7–13.

Turner, J. R., and R. Müller, R. 2003. On the nature of the project as a temporary organization. *International Journal of Project Management* 21(1).

Turner, J. R., A. E. Keegan, and L. Crawford. 2003. Delivering improved project management maturity through experiential learning. In *People in project management*, ed. J. R. Turner. Aldershot, UK: Gower.

van Fenema, P. C. 2002. *Coordination and control of globally distributed software projects*. PhD. thesis, Erasmus Research Institute of Management, Erasmus University Rotterdam. ISBN: 90-5892-030-5.

Woodruffe, C. 1990. *Assessment centers*. London: Institute of Personnel and Development.

CHAPTER FORTY-THREE

COMPETENCIES: ORGANIZATIONAL AND PERSONAL

Andrew Gale

Morris (1999) argues that the "rapidly changing climate of management enablement" has three sources of change: organization, IT, and people. He said recently an important way in which the organizational context of managing projects has changed is a new focus on competence at both organizational and personal levels.

Competence

The simple meaning of the word "competence" is the ability to do something well or successfully. However, the concept of competencies, as known and understood today, developed because of dissatisfaction with the so-called intelligence tests used during the 1970s. The cause of dissatisfaction, as argued by McClelland (1973), was that the intelligence tests were discriminatory, and this not only favored certain ethnic and socioeconomic groups but also rendered the tests largely invalid. Based on his research, McClelland suggested the adoption of criterion sampling, where a sample job and required skills are tested against the performance on that job. This concept of testing performance forms the basis of the majority of competency approaches. Competence related to both occupation and social and interpersonal skills.

Boyatzis (1982, p. 21) defines competency as an "underlying characteristic of a person in that it may be a motive, trait, skill, aspect of one's self-image or social role or a body of knowledge which he or she uses." He developed this definition from that of Klemp (1980), who stated competency was "an underlying characteristic of a person, which results in effective and/or superior performance in a job." After Boyatzis, many definitions were proposed. Some have suggested that competency could be defined as the knowledge, skills, and qualities of effective managers, and pointed to the ability to perform effectively the functions associated with management in the work situation. Hogg (1993) states that competencies are the characteristics of a manager that lead to the demonstration of skills and abilities and result in effective performance within an occupational area.

Competence is linked then with individual behavior and job performance. Regarding the effective performance in a job, Boyatzis (1982, p. 2) states that the "effective performance of a job is the attainment of specific results (i.e. outcomes) required by the job through specific actions while maintaining or being consistent with policies, procedures and conditions of the organizational environment".

Figure 43.1 illustrates a model for effective job performance showing the relationship between competence of an individual, task requirements, and the environment.

Particular importance in this model goes to the competence of the individual, because each individual must demonstrate the ability or characteristics to perform specific actions for a particular job to produce the desired results. The desired results are considered as the requirements of the task, and the environment is the culture and tradition, physical, economic and technical resources available, plus organizational constraints. The likelihood of effective performance increases when two of these components are congruent. However, the effective performance will occur only when all three components fit in this model.

Since the capability of performing the job effectively may or may not be known to the individual, these capabilities can be characterized as conscious as well as unconscious aspects

FIGURE 43.1. EFFECTIVE JOB PERFORMANCE.

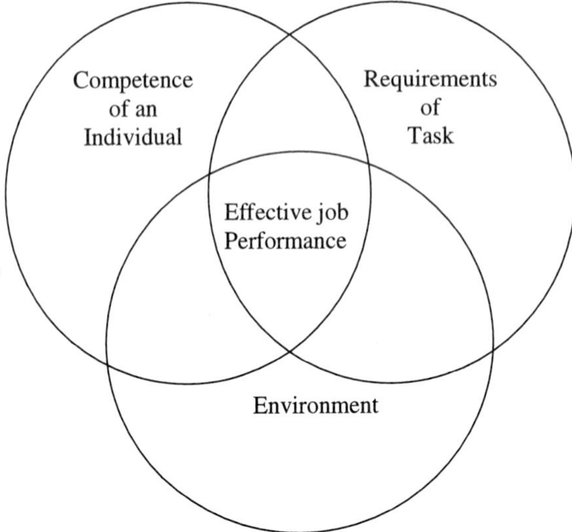

of the person (Boyatzis,1982). Describing the relationship between competence and effective performance, Boyatzis argues that to define competence, one must determine the required actions, their place in the system, and their sequence of behavior. One must also determine the results or effects and the margins or intent of the actions and results.

The definition of the concept of competence continues to cause confusion, and it may be that is it has become overdefined. Competence is a normative concept rather than a descriptive one. Competence can be delineated from performance in that competence is the ability to perform effectively in different ways in various contexts. Competence implies the integration of many aspects of practice and as such can be regarded as a psychological construct. However, performance may be measured for competence in relation to specific behaviors. Competence certainly includes the possession of particular knowledge and skills. Competence is concerned with the capacity to undertake specific types of action, and it can be considered a holistic concept involving the integration of: attitudes, skills, knowledge, performance, and quality of application.

Projects, their environments, and contexts require project managers and team members to be competent to address the challenges of their roles. Projects management bodies—for instance, Association for Project Management (APM), International Project Management Association (IPMA), Project Management Institute (PMI)—have for some time been concerned with the question of competencies, and through the development of bodies of knowledge (BOKs), they have begun to establish a frame of reference within which to relate the implications of competence so that they can be understood.

Learning Skills

Eraut (2002) argues that skills (or how things are done) must be viewed in the context of knowledge and learning, in which he distinguishes between three types of knowledge. These help to differentiate what is being learned and determine how learning occurs.

Eraut's three knowledge types are (1) codified, (2) cultural, and (3) personal. The BOKs are a form of codified knowledge. Cultural knowledge, argues Eraut (2002, p. 1), covers "much more than codified knowledge and introducing a totally different perspective. Its emphasis is on knowledge created, shared and used by groups of people working together, networking or socially interacting with each other." Finally, personal knowledge relates to the totality of a person's knowledge and background knowledge (Eraut, 2002).

Codified knowledge and cultural knowledge can often appear contradictory. Also, personal knowledge tends to be augmented by cultural knowledge. Therefore, there is much to consider in relation to the fragility of competence frameworks that do not take these issues into account.

Competence Frameworks

Whiddett and Hollyforde (2000) explain the structural properties of a useful competence framework. They say it should be clear and easy to understand, relevant, take account of expected changes, contain discrete elements (e.g., behavioral indicators) that do not overlap, and appear fair and balanced to all those affected by the use of the framework. Typically, a framework constitutes competency clusters that are broken down into different competencies with different levels. Behavioral indicators are stated for each competence listed by level (see Figure 43.2).

Frame (1999) has produced a major work on project management competence. He discussed the significance of competence and argued that in the case of project management it must be addressed at the level of (1) the individual, (2) the teams, and (3) the organization. He argues that there is a "competence dilemma" because of the conflict between the the-

FIGURE 43.2. TYPICAL COMPETENCE FRAME MODEL HIERARCHY.

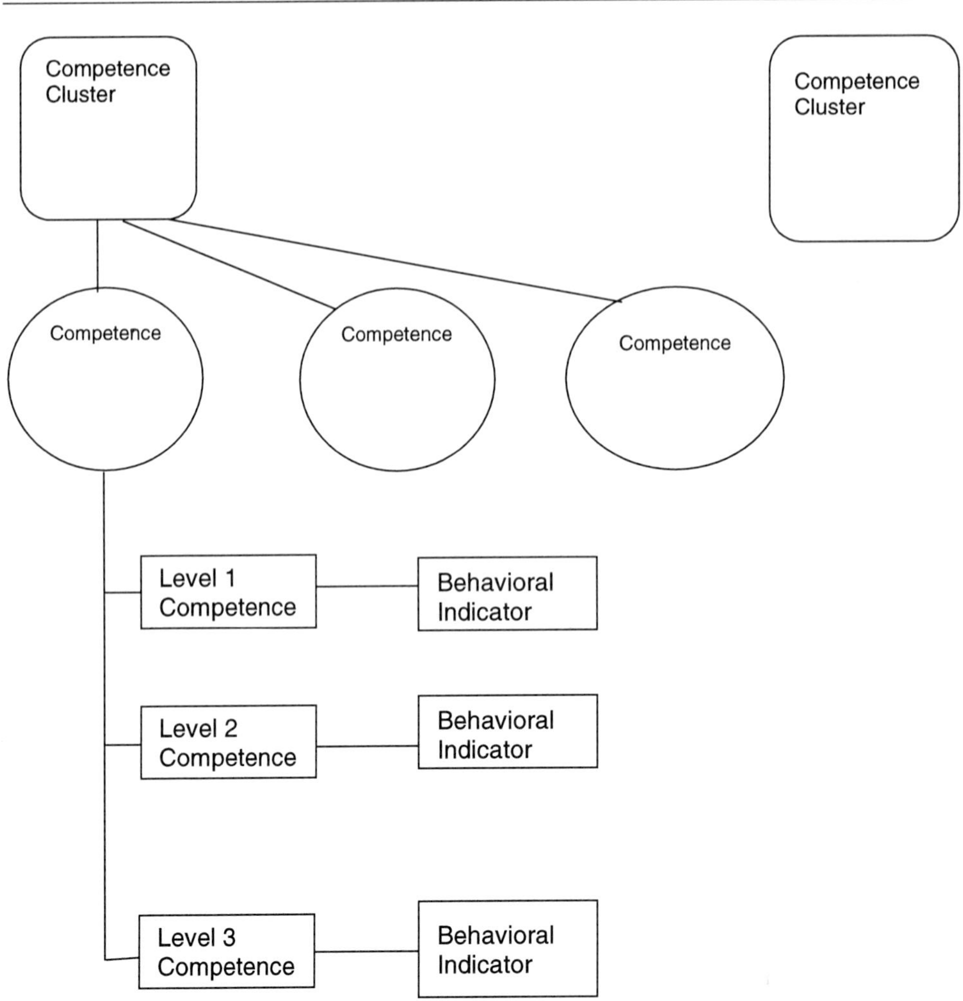

Source: Whiddett and Hollyforde, Figure 2:14.

oretical notion that all people are competent if given proper support and the working situation involving major differences in individual capabilities. He is also concerned with the microeconomic aspects of competence and the benefits brought to a team, project, and organization by a competent individual. Finally, he recognizes that an important connection must exist between a viable and healthy organization and the importance that should be attached to organizational culture and governance.

Some Recent Research

What Project Managers Do

Blackburn (2002) undertook a qualitative study into what project managers do and how project managers understand and talk about what they do. This study did not use a competence framework; rather, it focused on the network of relationships between project managers, along with environmental and technological factors. Actor-network theory (ANT) was used to interpret the project managers' stories about their work and project management techniques in the context of a variety of projects.

ANT is a school of sociological scientific knowledge in which networks constitute human actors, technological and natural agents, organisms, and human inventions. Project organizations can be viewed as generated in diverse, patterned networks but not structures with elements. The project network is more like a play with a script in which all the actors have their agendas and who seek to direct the play toward their own goals and interests. A dynamic, interactive process occurs in which both network and actors influence each other (Blackburn, 2002). The research found that project management processes enable project managers to interest and enroll team members and stakeholders and to mobilize the support of sponsors and other powerful players in a quasi-political process.

Blackburn (2002) argues there are two different prevailing schools of thought on the subject of project management: (1) methods, tools, and techniques, and (2) the heroic position, regarding the project manager as the key person responsible for bringing about project success. However, there is a third way in the form of ANT that is concerned with process not structures. Project management methodologies focus on "the right structures" and related processes linked with appropriate competencies on the part of project managers and project team members. Thus, these structures and processes form the agenda for the competence frameworks adopted and applied by project management experts. ANT is concerned with human and nonhuman actors and alliances used by people to make project organizations function. Rather than adding a "competent" project manager to a project team to ensure success, based on the prevalent perspectives on project management involving structures and processes, the ANT position is to understand project managers as competent through their effect on the actor-network of all the human and nonhuman elements that constitute the project.

Professional Competence. Cheetham and Chivers (1998) went in search of a comprehensive holistic model to address the concept of professional competence. Their research involved interviews with 80 practitioners in 20 professions. They drew on the theoretical position of the reflective practitioner, citing Schon (1983 and 1987) and challenging the techno-rational perspective on professional practice. They were concerned with what can be called "tacit knowledge" or "knowing in action," referred to also by Eraut (2002). Professional workers are able to draw on a set of solutions that fit the needs of a particular situation. Professional workers use their "artistry" to identify the need for a particular approach and make the judgment of fit, thus taking a professional approach to a situation all based on reflection—the crucial professional competence. This reflection can occur during an action or afterward and is the basis of continuous improvement. Cheetham and Chivers (1998) discuss the relationship between functional competence and personal competence. In the United States personal (or behavioral) competencies are a common focus for researchers, covering elements such as (1) confidence, (2) emotional control, and (3) interpersonal skills. Arguably, these personal competencies form a good basis upon which to predict a person's potential. However, there is no way of telling whether or not a person will be able to apply his or her personal competence to meet the demands of particular occupational requirements. Therefore, personal competence should be considered as complementary to functional competence.

Cheetham and Chivers (1998) also discuss the importance of competence in acquiring skills such as communication, problem solving, and critical analysis. These researchers further turned their attention to ethical considerations, arguing that ethics and values are usually ignored but should form an important element in professional competence. The Cheetham and Chivers (1998) model of professional competence argues that there is an over-arching set of meta-competencies:

- Problem solving
- Learning/self-development
- Mental agility
- Analysis
- Reflection

Under this are four interrelated components:

- Knowledge and cognitive competence
- Functional competence
- Personal and behavioral competence
- Values and ethics competence

The preceding all feed into professional competence, leading to outcomes that can be reflected upon in action and subsequent to the outcomes and feeding back into knowledge and cognitive competence, and values and ethics competence.

Organizational Competence

The literature reports some diverse recent findings with respect to organizational competence. Although none of these directly relate to project management, elements of these findings could be useful in the context of the project management environment, and project organization.

Stuart et al. (1995) reported the results of a research study on the importance and difficulties involved in translating existing competence frameworks into workable frameworks of specific contexts or organizations. While his work reports on small and medium enterprises (SMEs), it has general relevance to those interested in competence frameworks. A ten-step process was developed out of this research for translating a competence framework. The process enables the identification of individual team and organization development issues. From this management development, programs can be planned based on the competence-based management development initiatives. This is relevant to the interpretation and practical application of project management.

Research on organizational competence undertaken in Austria (Fischer and Schuch, 1994) identified four types of subcontractor organization. It was found that subcontractors unable to satisfy the demands of the client and who were poor in providing flexibility in production, just-in-time delivery, and quality control were unprofitable and ranked low in the hierarchy of subcontracting organizations. This has implications for the selection of suppliers and subcontractors, organization and specification of work packages, and supply chain integration in the context of complex projects.

Very (1993) reported a study in which the relationship between operational and strategic "relatedness" and the success or failure of seven large French industrial companies was investigated. No direct relationship was found between operational relatedness and success. However, the ability to exploit or strengthen competitive advantage through operational relatedness is linked with the success or failure of diversifications. These findings contribute to an understanding of the importance of "relatedness" in strategic diversification. This has a bearing on approaches to strategy.

Henderson and Cockburn (1994) discuss resource-based theory of the firm and focus attention on the role of heterogeneous organizational "competence" in competition. They seek to measure the importance of organizational competence. They distinguish between "component" and "architectural" competence by using data from ten firms, showing that the two forms of competence appear to explain variations in productivity across the firm. This has relevance to resource management in projects.

Sanchez (2002) proposes a five-mode competence-based management model. He argues that the whole field of competence has added benefits to modern management thinking. Confusion in the literature concerning the essential aspects of organizations' competence is attributable to the three reasons in the list that follows. Terminology is frequently differentially defined by different authors. The most obvious of these situations is the definition of competence itself. Different writers refer to different levels of activity within organizations, and there is often a tendency to consider competence as in some way static and unchanging in organizations. According to Sanchez (2002), different forms of competence arise from different levels of activity in organizations. He lists three levels of activity:

1. An organization's capability to create and produce products and services
2. The ability to organize, coordinate, and innovate in effective ways
3. The capability of senior management to imagine strategies for creating value in the market

The five organizational competence modes postulated by Sanchez (2002) are as follows:

1. *Cognitive flexibility to image alternative strategic logics.* The source of this mode is the "collective corporate imagination" of an organization's managers in perceiving feasible market opportunities in which the organization can create value.
2. *Cognitive flexibility to image alternative management processes.* This cognitive flexibility is concerned with managers' ability to conceive of alternative processes for implementing strategies.
3. *Coordinative flexibility to identify, configure, and deploy resources.* This depends on managers' ability to acquire or access, configure, and implement resources to achieve strategic goals.
4. *Resource flexibility to be used in alternative operations.* This relates to the organization's inherent level of flexibility. For example, it covers the extent to which an organization's resource can be described by a range of uses and the time frame within which this typically happens.
5. *Operating flexibility in applying skills and capabilities to available resources.* This relates to an organization's ability to apply resource flexibility to a range of operating conditions.

Sanchez argues that these five modes need careful interrelated management. Also, the type of competitive environment dictates the critical competence mode associated with success in that context. In stable competitive environments, Mode 5 is critical; for evolving environments, Modes 1 to 4 are critical, but led by Modes 3 and 4; and for dynamic environments, all five modes are critical, but led by Modes 1 and 2. Regarding success, Sanchez cites Bove et al., (2000): ". . . organizational competence does not depend simply on achieving excellence in one or two key success factors . . ."

The applicability of these five modes of competence to project management is easy to see. Some relate directly to key roles. This is particularly the case with respect to resource acquisition. Project managers, as change managers, need to be strategic as well as able to envisage and create linkage between many components in the project team and dynamic context.

Project Management Competencies, Standards, and Bodies of Knowledge

As practitioners of project management and the broader management community have increasingly focused on the importance of competence and standards, the leading professional bodies associated with project management have begun to address these issue of competence in a systematic manner. I have chosen, in spite of several national and regional competency models extant, to concentrate discussion around the two principal ones: in Europe and in North America.

European Bodies of Knowledge

Project management competencies have become the subject of much literature and debate. The International Project Management Association (IPMA) has produced an "International Competence Baseline" (IPMA, 1999). The baseline was produced from a study of the Association for Project Management (APM) UK Body of Knowledge (UK BoK), Swiss Assessment Structure (VZPM), German Projecktmanagement-Kanon (PM-ZERT), and the French assessment Criteria (AFITEP). The Baseline contains 42 elements on knowledge and experience (28 core and 14 additional) and a total of 18 attitudinal aspects.

The argument for improving the competence of employees is strong. The IPMA undertook a benchmarking study through all the members of the IPMA to produce the IPMA International Competency Baseline (ICB) and National Competency Baselines (NCBs) for each member country. These show the fields of project management qualification and competence. The IPMA defines competence as:

Knowledge + Experience + Personal Attitude

The IPMA classifies project management under four levels: A, B, C, and D. Knowledge and experience relate to function, and attitude relates to behaviors. Project management competence relate to the capability to manage projects professionally, by applying so-called best practices regarding the design of the project management process and the application of project management methods. Project management competencies require knowledge and experience in the subject, which enable objectives and deadlines to be met (Gareis and Huemann, 1999).

A. *Projects director.* Competent to "direct all projects of a company or branch or all projects of a program."
B. *Project Manager.* "able to manage complex projects". IPMA defines complex projects as having several interrelated subprojects and involving several companies or organizations across different disciplines and phases. These projects typically utilize many project management methods, techniques, and tools.
C. *Project management professional.* Defined as "able to manage noncomplex projects and/or assist the manager of a complex project in all elements and aspects of project management."
D. *Project management practitioner.* Said to "have project management knowledge in all elements and aspects."

Standards relating to project management competence fall into two main areas: Those relating to what project managers are expected to know and those relating to what project managers are expected to be able to do. The latter primarily takes the form of performance-based or occupational competence standards. For example, the principal competencies of a project manager in project management, as stated by Morris (1994), are as follows:

- Skills in project management methods and tools
- Team and people skills
- Basic business and management skills
- Knowledge of project sponsor role
- Knowledge and awareness of project environment
- Technical knowledge (specialized discipline skills)
- Integrative abilities of the preceding skills and knowledge.

Project management professionals working in projects where technical issues are important must have the competency to deal with them. Project managers must be able to recognize issues and be confident that appropriate action has been taken to deal with them. Professional project management competencies are achieved by the combination of education and the knowledge acquired during training, the skills developed through experience, and application of such acquired knowledge and experience. McCaffer et al. (2000, p. 113) lists the related field in which project manager should have a good working knowledge (see Figure 43.3).

To provide formal recognition that a project manager has reached a level of higher project management competence, the Association for Project Management, Body of Knowledge (1996) identified eight principal characteristics:

1. Attitude
2. Common sense
3. Open-mindedness
4. Adaptability
5. Inventiveness
6. Prudent risk taking
7. Fairness
8. Commitment

Key Project Management Competencies. The Association for Project Management Body of Knowledge (1996) identified 40 key competencies which are appropriate for the project management and are divided into four parts. The key competencies are as follows:

FIGURE 43.3. FIELDS IN WHICH PROJECT MANAGERS KNOWLEDGE IS CLASSIFIED.

Integration	**Human resource**	**Quality**
Time	**Scope**	**Risk**
Cost	**Procurement**	**Communications**

Source: McCaffer and Edum-Fotwe (2000, p. 113).

Part one: Project management

Part two: Organization and people

Part three: Process and procedures

Part four: General management

The preceding competencies are addressed fully by Crawford in her chapter.

UK Competence Standards. Occupational standards (e.g., UK National Vocational Qualifications) are based on functional competence in which job-specific outcomes are recognized. It has been recognized that learning is a continuous process that is not limited only to classrooms. In the world of competition, in order to survive and achieve success, everyone needs to raise and maintain his or her skills. The requirement of good-quality qualifications that are recognized and valued by individuals and employers is increasing. As the demand for skills and knowledge increases, individuals and industries are bound to improve competitiveness and productivity.

The UK government created the work-related National Vocational Qualification (NVQ). It is flexible and widely recognized by industry as evidence of performance standards, describing competence for a particular job (occupation). NVQs are classified from level one (routine and predictable activities) to level five (professional and managerial activities).

The assumption behind the development and use of project management standards is that the standards describe the requirement for effective performance of project management in the workplace. This is a controversial and deterministic approach and may have only limited value in the field of project management. However, it is acknowledged that for functional aspects of project management application, the use of standards may have a very useful role to play.

American Body of Knowledge. The Project Management Institute (PMI) in the United States has developed a Body of Knowledge (PMBOK) that has become their basis for knowledge testing. The PMI has recently introduced "The Project Manager Competency Development Framework" (PMI, 2002) called the PMCD Framework. This document claims to clearly identify the interdependencies between job knowledge, skills, and behavior. The PMCD Framework is founded on the following sources: *A Guide to the Project Management Body of Knowledge* (PMBOK, 2000), *Project Management Experience and Knowledge Self-Assessment Manual,* and *Project Management Professional* (PMP Role Delineation Study). The PMCD takes a performance-based approach that presumes a causal relationship between skills, attitudes, and behaviors, and job performance (Crawford (1997). According to the PMCD Framework, there are three dimensions of project management competency:

1. Knowledge competence
2. Performance competence
3. Personal competency

These dimensions are broken down into units of competence at various levels, as is typical of the framework approach: clusters, element, and performance criteria (Figure 43.4).

The PMCD Framework dedicates a page of definition for each "unit of competence," incorporating examples of assessment guidelines. An example of such a unit of competence is "Project Cost Management" (PMI, 2002, pp. 30–31). Under this unit of competence are "competency clusters" containing processes (e.g. "Planning"). The "Planning" cluster contains three "elements," which are cross-referenced to the PMBOK Guide:

- Conduct Resources Planning (PMBOK Guide 7.1) with 12 performance criteria (e.g., "No.6. Develop resource histograms.")
- Conduct Cost Planning (PMBOK GUIDE 7.2) with ten performance criteria (e.g., "No. 5. Evaluate inputs to the cost baseline development process.")
- Conduct Cost Budgeting (PMBOK Guide 7.3) with 3 performance criteria (e.g., "No. 3. Develop a cost baseline to determine cost performance.")

Under "Examples of Assessment Guidelines" for "Knowledge Competencies" is listed "Demonstrate a knowledge and understanding of the tools and techniques utilized for planning of resources and the compilation of cost estimates and budgets," and for "Performance Competencies," "Demonstrate an ability to develop/use Cost Baseline."

The PMCD Framework states that there are five steps in the competence development methodology:

1. *Determination of applicable elements and performance criteria.* This step involves the individual or organization identifying elements and performance criteria contained in the PMCD Framework.
2. *Determination of the desired level of proficiency.* This step is concerned with determining the desired level of proficiency under each performance criterion section in the PMCD Framework.
3. *Assessment.* This stage is an assessment of the project managers to assess their strengths and weaknesses against the elements and performance criteria in order to determine any "gaps."

FIGURE 43.4. PMCD FRAMEWORK.

Project Management Competency		
Knowledge Competency	**Performance Competency**	**Personal Competency**
9 units	9 units	6 units
5 clusters per unit	5 clusters per unit	2–4 clusters per unit
Elements for each cluster	Elements for each cluster	Elements for each cluster
Performance criteria for elements	Performance criteria for elements	Performance criteria for elements

Source: PMI, (2002, p. 12).

4. *Dealing with gaps in competence identified.* Once gaps have been identified, these should be addressed to enhance performance in specific areas. Reassessment must occur after action to establish if more action is needed.
5. *Progression towards competence.* By progressively addressing gaps, individual project managers can achieve competence in each dimension included in the assessment.

The whole thrust of the American approach is performance-based with a strong component of personal competencies. It appears reductionist.

Measuring Competence

There are a number of indicators commonly used in the measurement of competence: continuous education, examinations, portfolios, self-assessment, interviews, outcome and performance measures, direct observations, and peer review (assessment). Performance management, a distinct and large subject, is inextricably related to the question of measurement of competence.

According to Gratton (1989), there are three broad techniques for measuring competence: the checklist approach, observational method, and framework approach. Checklists and observational methods are fairly low-level techniques and appear to be appropriately engaged in the assessment of performance. Performance underlies competence (Hager, 1993). The framework approach is likely to be more related to an integrated perspective of competence—a relatively holistic perspective.

The generic skills approach, while not having the large number of specific competencies of the behaviorist approach, but rather a smaller set of generic competencies, seems to lack a plausible rationale.

The integrated or task attribute approach (Hager, 1993) argues that competence may be inferred from performance. Typically a manageable number of key competencies are developed and used in a framework. Arguably the Body of Knowledge philosophy of the project management professional bodies such as the Association for Project Management and the Project Management Institute are rather reductionist, tending toward the atomistic end of the continuum.

Olney (1999), in a short paper entitled "Measuring Project Management Competence," acknowledges that interrelatedness is axiomatic of projects. She describes project management as an "integrative function." The processes required in the management of projects are complex, and coaching and mentoring are useful ways of improving outcomes. She argues that a core competency model is "individual statements—measurable and observable skill, behaviours traits as applied to best in class performers." This concept of "best in class" is related to the concept of so-called best practice. While experts can perhaps judge what constitutes "best practice," there is no escaping that "practice" has sociocultural determinants. There is then cultural bias woven into the fabric of the measurement process.

Many methods are adopted to measure competence. These include specific techniques such as so-called 360-degree feedback. This is relevant when dealing with the concept of competence, as the type of information included in 360-degree feedback can certainly be regarded as constituting the competence of an individual—namely, knowledge (of the job, organization and industry), skills (task proficiency and efficiency), and behaviors (energy and general approach).

The concept of 360-degrees is based on the practice of obtaining data from an individual's peers, direct reports, internal customers, and line manager. The technique usually involves using questionnaires (sometimes software-based) to obtain the evaluation of an individual from many sources.

Olney (1999) goes on to explain that an organization can assess the efficacy of project management coaching and professional development, in terms of improved performance, through the application of 360-degree assessment. The 360-degree technique is said to have the advantage of obtaining more objective, triangulated data, and the results are thus better received by the individual as being helpful in determining personal and professional development needs. In other words, through this technique, the individual is able to understand how others see him or her. Teams are said to benefit from increased communication among members and enhanced involvement in the development process. Company benefits are argued to be enhanced development of employees, increased internal promotion, inclusion of customers through training involvement, and a powerful driver for training and professional development.

A Case Study of Project Management Competence Measurement

Rolls-Royce plc uses an electronic questionnaire instrument as an aid to employees, their managers, and business units to assess competencies, knowledge, skills, and experience for project management roles. The questionnaire is designed to take a person no more than 20 minutes to complete. If there is discussion time with a manager, this will obviously take somewhat longer.

The questionnaire instrument is an aid to appraising a person's knowledge and experience in the topics that make up the Association for Project Management's (APM) Body of Knowledge (BoK). It is used in conjunction with a list of "Core Competencies, Knowledge and Experience for Programme Management Roles," which gives guidance on the level of knowledge and experience for each of the four job titles: A: Project Director, B: Project Manager, C: Project Management Professional, and D: Project Management Practitioner; based on IPMA certification levels. Figure 43.5 shows the program management core competencies identified by Rolls-Royce for Project Directors: Level A and for Project Managers: Level B.

Individuals and/or their managers complete a questionnaire and are given feedback on the level of knowledge and experience against topics that are considered to be central to the project management task. As well as recording an individual's level of knowledge and experience against each of the APM BoK topics, the levels of knowledge or experience can be averaged across the questionnaire and overall average of knowledge and experience can be reported across all topics. To enhance feedback, an individual's average at Level B can be broken down into B− (slightly above C), B, and B+ (almost an A).

This aid may be used by a manager in a 180-degree appraisal, by an individual in self-appraisal, or a combination of these. Details at the personal level remain confidential between a manager and/or individual and are not retained centrally. To provide useful

FIGURE 43.5. CORE COMPETENCIES OR PROJECT DIRECTORS AND PROJECT MANAGERS.

Project Director: Level A	Project Manager: Level B
Managing vision and purpose	
Business acumen	
Customer focus	Customer focus
Priority setting	Priority setting
Directing others	Directing others
Leading from the front	Leading from the front
Drive from the results	
Dealing with ambiguity	
Composure	
Comfort around higher management	Comfort around higher management
Negotiating	Negotiating
Building effective teams	Building effective teams
	Conflict management
	Timely decision making
	Motivating others
	Organizing

Source: Brown (2003).

information at a central level for training need analysis, business provide feedback to a central database.

Project complexity is difficult to define in terms applicable to all businesses. Nevertheless, project complexity is often referred to in the preceding aid when a person's experience is assessed. Following are some indicators of project complexity, on a scale of 1 (Low) to 5 (High):

1. Multidiscipline
2. Multidiscipline, multicompany
3. Multidiscipline, multicompany, multinational up to £30m in value
4. Multidiscipline, multicompany, multinational £30 to £75m in value, complex funding
5. Multidiscipline, multicompany, multinational £75m plus, complex funding, BOT/BOOT or TCP included in scope

These five descriptions are only general indicators of project complexity, and each business unit considers what factors contribute to project complexity in its business area.

Return on Investment from Education and Training

There are some theoretical considerations relating to the content, delivery, and pedagogical strategy adopted for the program. These are all linked through a common theme of return on investment (ROI). Commercial organizations spend resources on educating and training their staff in order that the following intangible benefits occur: increased job satisfaction and increased organizational commitment. Brown (2003) prefers the concept of benefits metrics.

Some would argue that there is no valid and reliable relationship between indirect measures of the intangible benefits, mentioned previously, and the quantitative bottom-line indicators. There may be arguments based on anecdotal evidence, but these are not statistically valid associations. Rowe (1994) argues that it is not possible to make measurements that enable return on investment to be assessed or evaluated with respect to improved teamwork, improved customer service, reduced complaints, reduced conflict, or improved communication. These are measured indirectly using a number of methods. On the other hand, direct measures evaluating the success of an organization are quantitative and internationally understood (e.g., return on capital employed, or ROCE, profitability, turnover, and market share). Organizations invest in people through education and training to improve their bottom-line performance. Figure 43.6 makes linkages between investment in education and training and ROCE, relating competence, change, and the measurements necessary to close this loop.

Quantitative Approaches

Phillips (1997), on the other hand, writes convincingly on the relationship between competence and ROI. He is a devotee of developing algorithms for the calculation of real financial benefit in order to be compared with training, education, and development investment costs. He argues that organizations have moved from training for activity to training with a focus on bottom-line results. He insists that ROI methodologies must demonstrate simplicity, credibility, and soundness. The three most common measures are (1) cost-benefit ratio, (2) return on investment, and (3) payback period.

COST-BENEFIT RATIO

CBR = Program benefits/Program costs;
expressed as a ratio x: 1.

RETURN ON INVESTMENT FORMULAE

ROI= (Net program benefit/Program cost) × 100;
expressed as a percentage.

PAYBACK PERIOD

PP = Total investment/Amount saved;
expressed as x years.

FIGURE 43.6. RETURN ON INVESTMENT CYCLE.

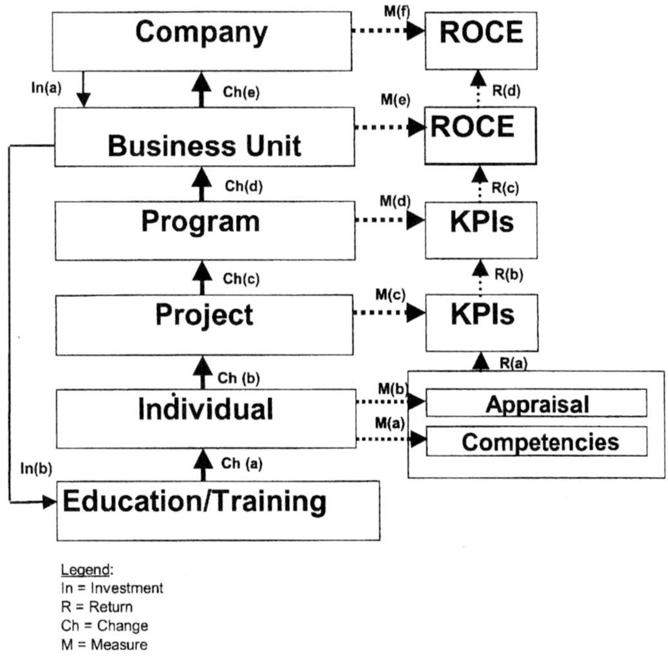

Other methods are (1) discounted cash flow (DCF), (2) internal rate of return (IRR), and (3) utility analysis.

UTILITY ANALYSIS

$$DU = T \times N \times dt \times Sd - N \times C$$

Where: DU = Monetary value of the program
T = Duration of the program
N = Number of employees on the program
dt = True difference in performance: trained cf. untrained
Sd = Standard deviation in performance of untrained
C = Cost of training per employee

Phillips acknowledges the importance and difficulty of quantifying benefits of training and development programs. There is no doubt that some types of program lend themselves to this quantitative approach. However, it is very difficult to see how project management development programs with complex inputs and slow-to-develop intangible benefits can be fitted into this style of approach.

When one considers the emerging body of research and debate on competence in project management, the strongest argument for an association between measurable bottom-line indicators and intangible benefits from the activities of education, training, management development programs, and continuing professional development may have to do with competence (Humphreys, 2001; Rowe, 1994; Seppänen, 2002; Skulmoski, 2000; Hartman 1999).

Rowe (1994) reported the experience of BAE SYSTEMS (formerly British Aerospace) in evaluation and effectiveness of open training. Trainees were asked to keep learning logbooks to record their experiences. This included how they learned, insights gained, benefits obtained, and other relevant matters. The reality was that very few people actually made entries in their logbooks. Regarding measurement of learning experience through end-of-course questionnaires; Rowe (1994) states that ". . . issuing assessment questionnaires at the end of a course neither measures nor evaluates a training event; it simply monitors it. But it does not tell us how effective the training was in terms of meeting the business needs."

Evidence for return on investment and individual learning gain remain unrecorded. Rowe points out that in his opinion it is not possible to effectively link the effects of training to improvements in organizational performance.

Rowe (1994) stresses that there is a contradiction between what he argues to be the two fundamental goals of training for an organization. One is the need to develop and maintain a *learning organization*, and the other is to *meet business needs*. The former is associated with the development and encouragement of open-ended stimulation, while the latter is concerned with things that we need to know, involving a relatively convergent perspective.

Summary

There seems little point in yet more adventures intended to define competence as such. It seems far more worthwhile addressing the linkage between learning and acquisition of knowledge and experience. An important issue to consider is the fact that the field of organizational competence continues to grow in importance. The relationship between project objectives and organizational objectives represents another area in this field of research. Cultural and diversity dimensions have been neglected by researchers and should be a major consideration for the future. Western values underpin the bodies of knowledge, and their use in developing competence frameworks in many parts of the world may be inappropriate because of differences in cultural contexts.

The field of return on investment or benefits metrics is certainly an important area to research. The focus should be on the relationships depicted in Figure 43.6. However, a serious effort is needed to move forward on the qualitative front and to get away from the evermore statistically based quantitative obsessions so often associated with this subject. The long-term view should be developed in which continuing professional development and communities of practice are identified as both outcomes in terms of professional competence development and sources of rich qualitative data for return on investment research.

Websites

Individuals

www.hrscope.com/project_management_competencies.htm
www.pmforum.org/pmwt01/duncomp.htm
www.pmforum.org/library/papers/cbwhitepaper.htm

Books and Publications

www.pm-prepare.com/BIBLIOGRAPHY.htm
www.cbponline.com/bookstore/project_management.htm
www.majorprojects.org/cgi-bin/pub_cont.cgi?range=az

Organizations

www.apm.org.uk/Default.htm
www.ipma.ch/index.htm
www.pmi.org/info/default.asp
www.aipm.com.au/html/
www.birminghamnow.com

References and Bibliography

Alic, J. A. 1995. Organizational competence: Know-how and skills in economic development. *Technology in Society* 17(4):429–436.

Andersen, P. H., N. Cook, and J. Marceauet. Forthcoming. Dynamic innovation strategies and stable networks in the construction industry: Implanting solar energy projects in the Sydney Olympic Village. *Journal of Business Research.*

Association for Project Management, 1996. *Body of knowledge,* 3rd ed. High Wycombe, UK: Association for Project Management.

Atkinson, S. 1999. Reflections: Personal development for managers: Getting the process right. *Journal of Managerial Psychology* 14(6):502.

Australian Institute of Project Management. 1996. *National competency standards for project management.* 1st approved ed. Sydney: Australian Institute of Project Management (www.aipm.com.au/html/ncspm.cfm).

Barnes, J. K. 1999. UK education's support for its aerospace industry. *Aircraft Engineering and Aerospace Technology* 71(2):136–142.

Baxendale, T., and O. Jones. 2000. Construction design and management safety regulations in practice: Progress on implementation. *International Journal of Project Management* 18(1):33–40.

Bergenhenegouwen, G. J. 1996. Competence development: A challenge for HRM professionals: Core competences of organizations as guidelines for the development of employees. *Journal of European Industrial Training* 20(9):29.

Birchall, D. 1993. Case study B: Senior managers and competence. *Management Development Review* 6(3): 13.

Binnersley, S., and C. Rowe. 1992. British Aerospace takes off with higher education in management development. *Executive Development* 5(1):10–13.

Blackburn, S. 2002. The project manager and the project-network. *International Journal of Project Management* 20(3):199–204.

Boam, R., and P. Sparrow. 1992. *Designing and achieving competency: A competency-based approach to developing people and organization.* London: McGraw-Hill.

Bove, K., H. Harmsen., and K. G. Grunert. 2000. The link between competencies and company success in Danish manufacturing companies. Vol. 6 in *Advances in applied business strategy: Implementing competence-based strategies,* ed. R. Sanchez and A. Heene. 287–312. Greenwich, CT: Jai Press.

Boyatzis, R. E. 1982. *The competent manager: A model for effective performance.* New York: Wiley.

———. 1993. Beyond competence: The choice to be a leader. *Human Resource Management Review* 3(1): 1–14.

Brown, M. R. 2003. Rolls-Royce plc. Personal communications.

Burnes, B. 1996. No such thing as . . . a "one best way" to manage organizational change." *Management Decision* 34(10):11–18.

Cannon, M. D., and A. C. Edmondson. 2001. Confronting failure: Antecedents and consequences of shared beliefs about failure in organizational work groups. *Journal of Organizational Behavior* 22(2): 161–177.

Cassells, E. 1999. Building a learning organization in the offshore oil industry. *Long Range Planning* 32(2):245–252.

Chandler, G. N., and E. Jansen. 1992. The founder's self-assessed competence and venture performance. *Journal of Business Venturing* 7(3):223–236.

Chastain, T., and A. Elliott. 2000. Cultivating design competence: Online support for beginning design studio. *Automation in Construction* 9(1):83–91.

Chaston, I., B. Badger, et al. 1999. Organizational learning: research issues and application in SME sector firms. *International Journal of Entrepreneurial Behaviour and Research* 5(4):191–203.

Chaston, I., and T. Mangles. 2000. Business networks: Assisting knowledge management and competence acquisition within UK manufacturing firms. *Journal of Small Business and Enterprise Development* 7(2):160–170.

Cheetham, G., and G. Chivers. 1998. The reflective (and competent) practitioner: A model of professional competence which seeks to harmonise the reflective practitioner and competence-based approaches. *Journal of European Industrial Training* 22(7):267–276.

Cleland, D. I. 1999. *Project management: Strategic design and implementation.* 3rd ed. New York: McGraw-Hill.

Crawford, L. H. 1997. A global approach to project management competence. *Proceedings of the 1997 AIPM National Conference,* 220–228. Gold Coast, Brisbane: AIPM.

———. 2000. Profiling the competent project manager. *Project Management Research at the Turn of the Millennium: Proceedings of PMI Research Conference.* 3–15. Newtown Square, PA: Project Management Institute.

CRMP 1999. *Guide to the project management, Body of knowledge.* Manchester: CRMP, UMIST.

Curtis, B., W. E. Hefley, et al. 1997. Developing organizational competence. *Computer* 30(3):122–124.

Dunphy, D., D. Turner, et al. 1997. Organizational learning as the creation of corporate competencies. *The Journal of Management Development* 16(4):232–244.

Eliasson, G. 1996. Spillovers, integrated production and the theory of the firm. *Journal of Evolutionary Economics* 6(2):125–140.

Ellström, P. E. 2001. Integrating learning and work: Problems and prospects. *Human Resource Development Quarterly* 12(4):421–435.

———. 2002. *Learning challenges for knowledge-based organizations.* Sussex: University of Sussex Institute of Education.

Fairtlough, G. 1994. Organizing for innovation: Compartments, competences, and networks. *Long Range Planning* 27(3):88–97.

Fischer, M. M., and K. Schuch 1994. Technological and organizational competence of Austrian subcontractors: An empirical study based on the machine construction, steel, electrical and electronic industries. *Mitteilungen Der Osterreichischen Geographischen Gesellschaft* 136:179–202.

Frame, J. D. 1999. Project management competence: Building key skills for individuals, teams, and organizations. San Francisco: Jossey-Bass.

Fraser, C. 1999. A non-results-based effectiveness index for construction site managers. *Construction Management and Economics* 17(6):789–798.

———. 2000. The influence of personal characteristics on effectiveness of construction site managers. *Construction Management and Economics* 18(1):29–36.

Frey, R. S. 2001. Knowledge management, proposal development, and small businesses. *The Journal of Management Development* 20(1):38–54.

Gareis, R., and M. Huemman. 1999. Specific competencies in the project-oriented society. Project Management Days '99: Projects and Competencies, Vienna: Austria, November 18–19.

General, M., and S. G. Genega 1997. Leadership: Essential to managing success. *Journal of Management in Engineering* 13(4):22–23.

Godbout, A. J. 2000. Managing core competencies: The impact of knowledge management on human resources practices in leading-edge organizations. *Knowledge and Process Management* 7(2):76–86.

Gongal, K. 2000. An investigation into the relationship between project management competencies and project success. MS, diss, UMIST, Manchester.

Gottshall, W. L. 2000. Competence or collapse. *Civil Engineering—ASCE* 70(10):74–75.

Gratton, L 1998. Work of the manager. In *Assessment and selection of organizations: Methods and practices for recruiting and appraisal,* ed. P. Herriot, 511–528. New York: Wiley.

Gray, C. J. 1998. A strategic investment in training and development by the UK steel construction industry. *Journal of Constructional Steel Research* 46(1–3):281.

Gronhaug, K., and O. Nordhaug 1992. Strategy and competence in firms. *European Management Journal* 10(4):438–444.

Hager, P. 1993. Conceptions of competence. *Proceedings of the Forty-Ninth Annual Meeting of the Philosophy of Education Society,* University of Technology Sydney, Philosophy of Education, www.ed.uiuc.edu/EPS/PES-yearbook/93_docs/HAGER.HTM.

Hartman, F, and G. Skulmoski, G. 1999. Quest for team competence. *Project Management* 5(1):10–15.

Haydock, W. 1995. Management development: A personal competency approach. *Training & Management Development Methods* 9(4):7–13.

Henderson, R., and I. Cockburn. 1994. Measuring competence: Exploring firm effects in pharmaceutical research. *Strategic Management Journal* 15:63–84.

Henriksen, L. B. 2001. Knowledge management and engineering practices: The case of knowledge management, problem solving and engineering practices. *Technovation* 21(9):595–603.

Hodgson, G. M. 1998. Competence and contract in the theory of the firm. *Journal of Economic Behavior & Organization* 35(2):179–201.

Hogg, B. A 1993. European managerial competences. *European Business Review* 93(2):21–26.

Holmberg, S. C. 2001. Systemic research on competence and competence development in SMEs. *Systems Research and Behavioral Science* 18(2):101–102.

Humphreys, P. 2001. Designing a management development programme for procurement executives. *Journal of Management Development* 20(7):604–623.

IPMA. 1999. www.ipma.ch/document /+CB20DL.pdf (accessed March 18, 2002).

Jang, Y., and J. Lee 1998. Factors influencing the success of management consulting projects. *International Journal of Project Management* 16(2):67–72.

Javidan, M. 1998. *Core competence: What does it mean in practice?* Long Range Planning 31(1):60–71.

Johnston, N. M. 1991. How to create a competitive workforce. *Industrial and Commercial Training* 23(2): 4–7.

Jones, N., and N. Fear. 1994. Continuing professional development: Perspectives from human resource professionals. *Personnel Review* 23(8):49–60.

Jones, N., and G. Robinson. 1997. Do organizations manage continuing professional development. *Journal of Management Development* 16(3):197–207.

Jurie, J. D. 2000. Building capacity: Organizational competence and critical theory. *Journal of Organizational Change Management.* 13(3):264–274.

Kersten, A. 2000. Diversity management. *Journal of Organizational Change Management* 13(3):235–248.

King, A. W., and C. P. Zeithaml. 2001. Competencies and firm performance: Examining the causal ambiguity paradox. *Strategic Management Journal* 22(1):75–99.

Klemp, G.O. 1980. *The assessment of occupational competence.* Report to the National Institute of Education, Washington, D.C.

Kræmmergaard, P., and J. Rose 2002. Managerial competences for ERP Journeys. *Information Systems Frontiers* 4(2):199–211.

Lampel, J. 2001. The core competencies of effective project execution: The challenge of diversity. *International Journal of Project Management* 19:471–483.

Larsen, H. H. 1997. Do high-flyer programmes facilitate organizational learning? *Journal of Managerial Psychology* 12(1):48–59.

———. 1997. Do high-flyer programmes facilitate organizational learning? From individual skills building to development of organizational competence. *Journal of European Industrial Training.* 21(9):310–317.

Lewis, D. 1998. Competence-based management and corporate culture: Two theories with common flaws? *Long Range Planning* 31(6):937–943.

Lewis, M. A. 2001. Success, failure and organizational competence: A case study of the new product development process. Journal of Engineering and Technology Management 18(2):185–206.

Lindsay, P. R., and R. Stuart 1997. Reconstructing competence. *Journal of European Industrial Training* 21(9):326–332.

Löfstedt, U. 2001. Competence development and learning organizations: a critical analysis of practical guidelines and methods. *Systems Research and Behavioural Science* 18(2):115–125.

Lloyd, C., and A. Cook. 1993. *Implementing standards of competence: Practical strategies for industry.* London: Kogan Page.

Lysaght, R. M., and J. W. Altschuld. 2000. Beyond initial certification: The assessment and maintenance of competency in professions. *Evaluation and Program Planning* 23(1):95–104.

Maister, D. 1982. Balancing the professional service firm. *Sloan Business Review.* 24(1):15–29.

Margerison, C. 2001. Team competencies. *Team Performance Management* 7(7/8):117–122.

Martin, G., and H. Staines. 1994. Managerial competences in small firms. *The Journal of Management Development* 13(7):23.

Mathiassen, L., and P. A. Nielsen. 1990. *Surfacing organizational competence. Soft systems and hard Contradictions.* Bjerknes, G. (Ed) Organizational competence in system development. Lund: Studentlitteratur.

McCaffer, R., and F. T. Edum-Fotwe. 2000. Development project management competency: Perspectives from the construction industry. *International Journal of Project Management* 18(2):111–124.

McCain, B. 1996. Multicultural team learning: An approach towards communication competency. *Management Decision* 34(6):65–68.

McClelland, D. C. 1973. Testing for competence rather than "intelligence." *American Psychologist* 28(1): 1–14.

McCreery, J. K. Forthcoming. Assessing the value of a project management simulation training exercise. Uncorrected proof. *International Journal of Project Management.*

McGrath, R. G., I. C. MacMillan, and S. Venkataraman. 1995. Defining and developing competence: A strategic process paradigm. *Strategic Management Journal* 16(4):251–275.

Merali, Y. 2000. Individual and collective congruence in the knowledge management process. *Journal of Strategic Information Systems.* 9(2–3):213–234.

Meyer, T., and P. Semark. 1996. A framework for the use of competencies for achieving competitive advantage. *South African Journal of Business Management* 17(4):96–103.

Morden, T. 1997a. Leadership as competence. *Management Decision.* 35(7):519–526.

———. 1997b. Leadership as vision. *Management Decision* 35(9):668–676.

Morris, P. W. G. 1994. The management of projects. London: Thomas Telford.

———. 1999. Project management in the twenty-first century: Trends across the millenium. IPMA/ SOVNET International Project Management Congress, June 28.

———. 2001. "Updating the Project Management Bodies of Knowledge", *Project Management Journal.* 32(3):21-30.

Mothe, C., and B. Quelin 2000. Creating competencies through collaboration: The case of EUREKA R&D consortia. *European Management Journal* 18(6):590–604.

Mulder, L. 1997. The importance of a common project management method in the corporate environment. *R&D Management* 27(3):189–196.

Murphy, L. 1995. A qualitative approach to researching management competences. *Executive Development* 8(6):32–34.

Murphy, S. E. 1988. Organizational competence: It depends on your staff. *Training and Development Journal* 42(1):34–36.

National Vocational Qualification. www.dfee.gov.uk/nvq/what.html.

The APM Body of Knowledge: The 40 key competencies. www.apmgroup.co.uk/the 40 key.html (accessed March 2002).

Nair, K. U. 2001. Adaptation to creation: progress of organizational learning and increasing complexity of learning systems. *Systems Research and Behavioral Science* 18(6):505–521.

Newman, V. 1997. Redefining knowledge management to deliver competitive advantage. *Journal of Knowledge Management* 1(2):123–128.

Nkado, R, and T. Meyer. 2001. Competencies of professional quantity surveyors: A South African perspective. *Construction Management and Economics* 19:481–491.

Olney, J. 1999. Measuring project manager competence. *PM Network.* (October).

Österlund, J. 2001. The forgotten revenue of product development: learning new competence. *Systems Research and Behavioral Science* 18(2):159–170.

Otala, L. 1994. Industry-university partnership: Implementing lifelong learning. *Journal of European Industrial Training* 18(8):13–18.

Overmeer, W. 1997. Business integration in a learning organization: The role of management development. *The Journal of Management Development* 16(4):245–261.

Owen, K., R. Mundy, W. Guild, and R. Guild. 2001. Creating and sustaining the high performance organization. *Managing Service Quality* 11(1):10–21.

Phillips, J. J. 1997. *Return on investment in training and performance improvement programs: Improving human performance series.* Houston: Gulf Publishing Co.

Project Management Association 2000. *Project Management Institute Body of Knowledge.* Newtown Square, PA: Project Management Institute.

———. 2002. *Project manager competency development framework*, Newtown Square, PA: Project Management Institute

Quality Assurance Agency for Higher Education, *Handbook for academic review*, Annex D, www.qaa.org.uk (accessed March 2002).

Raynaud, D. 2001. Competences et expertise professionnelle de l'architecte dans le travail de conception: The competence and expertise of architects during the phase of design. *Sociologie du Travail* 43(4):451–469.

Ritter, T., and H. G. Gemunden 2004. The impact of a company's business strategy on its technological competence, network competence and innovation success. *Journal of Business Research* 57(5): 548–556.

Rolls-Royce 2000. Personal communications with Change Management Department.

Rowe, C. 1994. Assessing the effectiveness of open learning: The British Aerospace experience. *Industrial and Commercial Training* 26(4):22–27.

Seppanen, V. 2002. Evolution of competence in software subcontracting projects. *International Journal of Project Management* 20:155–164.

Sanchez, R. 2001. *Knowledge management and organizational competence.* Oxford: Oxford University Press.

———. 2002. Understanding competence-based management; Identifying and managing five modes of competence. *Journal of Business Research* 57(5):518–532.

Schön, D. 1983. *The reflective practitioner: How professionals think in action.* London: Maraca Temple Smith.

———. 1987. *Educating the effective practitioner.* San Fransisco: Jossey-Bass.

Seppänen, V. 2002. Evolution of competence in software subcontracting projects. *International Journal of Project Management* 20(2):155–164.

Simkoko, E. 1992. Managing international construction projects for competence development within local firms. *International Journal of Project Management* 10(1):12–22.

Skulmoski, G., F. Hartman, and R. DeMaere. 2000. Superior and threshold project competencies. *Project Management* 6(1):10–15.

Stuart, R., J. E. Thompson, and J. Harrison. 1995. Translation: From generalizable to organization-specific competence frameworks. *The Journal of Management Development* 14(1):67–80.

Stuart, R., and P. Lindsay 1997. Beyond the frame of management competenc(i)es: Towards a contextually embedded framework of managerial competence in organizations. *Journal of European Industrial Training* 21(1):26–33.

Sundberg, L. 2001. A holistic approach to competence development. *Systems Research and Behavioral Science* 18(2):103–114.

Tovey, L. 1993. Competency assessment: A strategic approach—Part I. *Executive Development* 6(5):26–28.

———. 1994. Competency assessment : A strategic approach—Part II. *Executive Development* 7(1):16–19.

Very, P. 1993. Success in diversification: Building on core competences. Long *Range Planning* 26(5):80–92.

von Zedtwitz, M. 2002. Organizational learning through post-project reviews in R&D. *R&D Management* 32(3):255–268.

Wang, C. K. 2001. Organizational competence analysis: Experience of a Japanese multinational competitive intelligence review 12(3):3–9.

Whiddett, S., and S. Hollyforde. 2000. *The competencies handbook.* London: CIPD.

Williamson, O. E. 1999. Strategy research: governance and competence perspectives. *Strategic Management Journal* 20(12):1087–1108.

Yan, Y., T. Kuphal, and J. Bode. 2000. Application of multiagent systems in project management. *International Journal of Production Economics* 68(2):185–197.

CHAPTER FORTY-FOUR

PROJECTS: LEARNING AT THE EDGE OF ORGANIZATION

Christophe N. Bredillet

For the past 40 years project management has become a well-accepted way to manage organizations. The field of project management has evolved from operational research techniques and tools to a discipline of management (Cleland, 1994; Bredillet,1999).

Introduction: Some Conceptual Issues to Knowledge and Learning in Project Management

Management of/by Projects for Implementing Strategy

Many authors emphasize this evolution in the way of managing projects. Referring to his book *The Management of Projects*, Morris writes, "This book traces the development of the discipline of project management" (Morris, 1997). Project management becomes the way to implement corporate strategy (Turner, 1993; Frame, 1994) and to manage a company; ". . . value is added by systematically implementing new projects—projects of all types, across the organization" (Dinsmore, 1999, p. ix). *Management of Projects*, which discusses the way to manage projects within the same organization (Morris, 1997), and *Management by Projects*, which describes projects as a way to organize the whole organization (Gareis, 1990; Dinsmore, 1999), are both good examples of that tendency. Projects are a form of organization that positions a company in relation to its environment. As projects are the vectors of the strategy (Grundy, 1998), project management is a way to deal with the characteristics of the whole environment: complexity (Arcade, 1998), change (Voropajev, 1998), globalization, time, and competitiveness (Hauc, 1998). Thus, with the help of project management, strategic management becomes really the management of irreversibility (Declerck et al.,

1997), concentrating on the ecosystem's project/organization/context, operation/organization/context and their integrative management (Declerck et al., 1983).

Competencies, Sources of Competitive Advantage and the Creation of Value

Projects, as strategic processes, modify the conditions of the firm in its environment. Through them, resources and competencies are mobilized to create competitive advantage and other sources of value. As resources are easily shared by many organizations, the organization's competencies are the most important relevant driver. Thus, through the organization's processes or projects, past action is actualized as experience; present action reveals and proves competencies; future action generates and tries out new competencies (Lorino and Tarondeau, 1998). Competencies (both individual and organizational) are at the source of competitive advantage and the creation of value.

The Link with Performance

Recent research is being done on the assumption that the more competent the project managers, teams, or organizations (maturity), the more efficiently they will perform, the more effective will be the performance of the projects, and the more successful will be the organization (Crawford, 1998; PMI Project Manager Competency Development Framework, 2002). Such research, and indeed the development of professional certification programs in general, seems to contradict former findings. For example, Pinto and Prescott (1988) concluded that the "personnel factor," even if designated in theoretical literature as a crucial factor in project efficiency, is a marginal variable for project success at any of the four project life cycle phases considered (for a criticism of their findings, see Belout, 1998). A working paper (Turner, 1998) shows the influence of the project managers' competencies on value of shares of a company. But performance also comes from the maturity of the organization's ability to deal with projects. And in respect of maturity, learning is especially significant. The OPM3 research program (PMI Standards Committee) and others papers (for example, Remy, 1997; Saures, 1998; and Fincher et al., 1997) explore the relations between maturity of the organizations and success of the projects.

Knowledge and Competence

To develop competencies, an individual needs knowledge. Two main views of competence development may be considered. One traditional view is that it involves applying a body of knowledge to known situations in order to produce rational solutions to problems (what I call the "have" or "quantitative" perspective). However, in a rapidly changing world and information-based society, practitioners and organizations increasingly need to respond intelligently to unknown situations and go beyond established knowledge to create unique interpretations and outcomes (Schön, 1971; Ackoff, 1974; Toffler, 1980, 1990; Reich, 1991) (what I call the "be" or "quality" perspective). As a result, it is no longer adequate to base professional development just on transmitting existing knowledge and developing a predefined range of competences on the basis that one problem equals one solution. Instead, practitioners need to be able to construct and reconstruct the knowledge they need and continually evolve their practice (Schön, 1987:35–6), thereby leading to a systemic and dynamic development of their competencies. (For a review of the link between knowledge, personal, and performance-based dimensions of competence see Crawford, 1998). These alternative approaches of going beyond traditional models of production and knowledge use while recognizing its validity in some areas are based more on the reflecting, questioning, and creating processes.

An Epistemological Perspective

The term "epistemology" refers to the study of the nature and grounds of knowledge, including how we define or recognize it. Most of the works on organizational learning, learning organizations, knowledge management, knowledge-creating organizations, and so on are based on a traditional understanding of the nature of knowledge. We could name this understanding the "positivist epistemology" perspective, since it treats knowledge as something people have. But this perspective does not reflect the knowing found in individual and team practice, knowing as an "intelligent" action, "ingenium,": This mental faculty that makes possible connecting in a fast, suitable, and satisfying way of "the separate things" (as stated by Lemoigne (1995), quoting Giambattista Vico (1708) in calling for a "constructivist epistemology" perspective). The "positivist epistemology" tends to promote explicit over tacit knowledge (see *Tacit vs. Explicit Knowledge* coming up in the chapter), and individual knowledge over team or organizational knowledge.

This integrative epistemological approach for project management suggests that organizations will be better understood if explicit, tacit, individual, and team/organizational knowledge are treated as four distinct forms of knowledge (each doing work the others cannot), and if knowledge and knowing (intelligent action) are seen as inseparable and mutually enabling. Thus, knowledge may be seen as an input of knowing, and knowing as an aspect of our interaction with the social and physical world, and therefore the dynamic interaction of knowledge and knowing can generate new knowledge and new ways of knowing.

Knowledge Management: Overview, Key Issues

A Brief History of Knowledge Management (KM): Different Cultural Perspectives

KM offers a unique concept considered by many in the industry as simultaneously progressive yet soft and difficult in application., One may suggest that this is primarily because of the intangible elements of knowledge. However, the increased topicality—if not to say pervasiveness—of the term through the writings of such well-known and recognized authors as Drucker (1993), Wheatley (2001), De Geus (1997), and Senge (1999) strongly suggest that KM is becoming accepted as a credible concept.

Although the study of knowledge dates back at least to Plato and Aristotle, entertaining the management of knowledge throughout a corporation first gained visibility by Polanyi in

1958. O'Dell and Grayson (1998) state "Polanyi's work served as a basis for the much-acclaimed knowledge management theories and books by the Japanese organizational learning guru, Ikujiro Nonaka" (p. 3).

Polanyi (1958) presented knowledge as something that can have intrinsic value placed on it; he also outlined two types of knowledge: tacit and explicit (defined in the *Tacit vs. Explicit Knowledge* section, coming up). Nonaka (1991) confirmed Polanyi's two knowledge-level concepts and introduced what he named "the knowledge-creating company" (p. 22). Nonaka proposed that organizations were not so much like machines as living organisms. That insightful biological analogy created a logical link between knowledge and organizations and started a paradigm shift in the need to pay attention to the collective thoughts of the people within the organization as knowledge contributors. The focus was on knowledge sharing and knowledge transfer. The average life span of major corporations is 40 to 50 years, roughly half that of humans. The need for development of KM is potentially the answer to expand the life expectancy of organizations and in doing so improve the overall health of the organization within the process.

Prior to Nonaka's (1991) research, Westerners (predominantly in the United States) viewed organizations as "a machine for information processing. According to this view, the only useful knowledge is formal and systematic hard data, codified procedures, universal principles" (p. 23). Wheatley (2001) maintains that even today the Japanese approach the field of KM differently than Westerners. She explains that we in the West still focus on explicit knowledge, while our Japanese counterparts find most gains in the areas of tacit knowledge ("be" side). For instance, Davenport and Prusak (1997) focus on knowledge acquisition, providing a market perspective on organizational knowledge creation (the "have" side). Peng and Akutsu (2001) propose that "there are two fundamentally different mentalities for dealing with new knowledge: linear thinking and dialectical thinking" (p. 107). Linear thinking is defined as "distaste for ambiguity and contradiction and preference for consistency and certainty" (p. 108). They differentiate dialectical thinking in "synthesizing dialectical thinking," aiming at identifying contradiction and resolving it by means of synthesis or integration, from "compromising dialectical thinking," focusing on tolerating contradiction. They come to the conclusion that mentality is a major factor in understanding how people are behaving in front of new knowledge. They found in their research that the Japanese are more dialectical than Americans.

What Is Knowledge? What Is Knowledge Management?

Part of the interest in the possible value of corporate knowledge comes from the information age. The advent of sophisticated information systems, the World Wide Web, networks, e-mail, and instantaneous sharing of information led to the realization that knowledge (and its sharing) was the fundamental element behind an organization's activities. This is not to say that information is knowledge. Deming (1993) accurately states that "information, no matter how complete and speedy, is not knowledge" (p. 106). Once information is embodied in time and gets a temporal value, it then becomes knowledge. To differentiate knowledge from data and information, I would say that the bad thing about knowledge is that many people, experts and lay people alike, treat it as some sort of higher-level information: extended, synthetic, advanced, tacit, complex, and so on, but still as information. Although information is an enhanced form of data, knowledge is not an enhanced form of information. It is quite clear, even on an intuitive level, that knowledge is not and cannot be the same thing as information, not even a form of information. It cannot be handled as information, does not have the same uses, and will resist any simplistic and expedient methodological transfers from information systems to "knowledge systems." Having information is not the same as knowing: Not every reader of a cookbook is a great chef. It is therefore very important to define knowledge in a distinct, appealing, and operational way. Simply calling more complex forms of information "knowledge" will not make them knowledge, even if repeated for years. "Knowledge" of information can be demonstrated through a statement, recall, or display. Knowledge itself can only be demonstrated through action. What is knowledge? Knowledge is purposeful coordination of action (intelligent action, or "ingenium").

Knowledge management is "the art of creating value from an organization's Intangible Assets" (Sveiby, 1999). With Sveiby (2001), we can define knowledge management by looking at what people in this field are doing: "Both among KM-researchers and consultants and KM-users there seem to be two tracks of activities, and two levels."

The two tracks of *activities* are as follows:

1. *Management of information.* Researchers and practitioners in this field tend to have their education in computer and/or information science (Hayes-Roth et al., 1983). They are involved in the construction of information management systems, Artificial Intelligence (AI), reengineering, groupware, and so on. To them, knowledge equals objects that can be identified and handled in information systems. This track is new and is growing very fast at the moment, assisted by new developments in IT.
2. *Management of people.* Researchers and practitioners in this field tend to have their education in epistemology, philosophy (Kuhn, 1970), psychology, sociology (Polanyi, 1958, 1966), or business/management/economics (Silberston, 1967). They are primarily involved in assessing, changing, and improving human individual skills and/or behavior. To them knowledge equals processes, a complex set of dynamic skills, know-how, and so on that is constantly changing. They are traditionally involved in learning and in managing these competencies individually—like psychologists—or on an organizational level—like philosophers, sociologists, or organizational theorists. This track is very old and is not growing so fast.

The two *levels* defined by Sveiby are as follows:

1. *Individual perspective.* The focus in research and practice is on the individual (AI specialists, psychologists).
2. *Organizational perspective.* The focus in research and practice is on the organization (reengineering, organization theorists, etc.) (Sveiby, 2001).

Crossing these two dimensions, we can capture one essential issue: "There are paradigmatic differences in our understanding of what knowledge is" (Sveiby, 2001). "The researchers and practitioners in the "Knowledge = Object" column tend to rely on concepts from Information Theory in their understanding of Knowledge. The researchers and practitioners in the column "Knowledge = Process" tend to take their concepts from philosophy or psychology or sociology.

Another approach of KM schools can be founded in Earl (2001). Seven knowledge management schools grouped into categories are introduced. For each of them, attributes (focus, aim, unit critical success factors, principal IT contribution, and "philosophy") are proposed. See Table 44.1.

Useful Concepts in Knowledge Management

Some key concepts emerging from the KM movement and from other disciplines can be summarized as follows.

Tacit vs. Explicit Knowledge. This idea finds its origin in Polanyi (1966) but has been applied to business and knowledge management by Nonaka (1995). It suggests that there are two types of knowledge: tacit, which is embedded in the human brain and cannot be expressed easily, and explicit (Brooking, 1999), which can be easily codified. Both types of knowledge are important, but Western organizations have focused largely on managing explicit knowledge.

Codification vs. Personalization. This distinction is related to the tacit vs. explicit concept. It involves an organization's primary approach to knowledge transfer (Hansen and Ali, 1999). Organizations using codification approaches rely primarily on repositories of explicit knowledge ("have"). Personalization approaches imply that the primary mode of knowledge transfer is direct interaction among people ("be"). Both are necessary in most organizations, but an increased focus on one approach or the other at any given time within a specific organization may be appropriate.

Knowledge Processes. Knowledge processing may be seen as a social system positioned in the value chain of an organization. Two sides may be considered as shown in the Knowledge Life Cycle exposed by McElroy (2002, p. 6):

TABLE 44.1. KM SCHOOLS ACCORDING TO EARL (2001).

Groups	Schools	"Focus"
TECHNOCRATIC	Systems	Technology
	Cartographic	Maps
	Engineering	Processes
ECONOMIC	Commercial	Income
BEHAVIORAL	Organizational	Networks
	Spatial	Space
	Strategic	Mind-set

1. Demand side with knowledge production; also called first-generation KM
2. Supply side with knowledge integration; called second-generation KM.

The overall process can be generically represented as three subprocesses:

- Knowledge production and capture
- Knowledge integration/codification
- Knowledge transfer/use

Knowledge production and capture includes all the processes involved in the acquisition and development of knowledge. Knowledge integration/codification involves the conversion of knowledge into accessible and applicable formats. Knowledge transfer/use includes the movement of knowledge from its point of generation or codified form to the point of use. One of the reasons that knowledge is such a difficult concept is that this process is systemic and often discontinuous. Many cycles are concurrently occurring in businesses. These cycles feed on each other. Knowledge interacts with information to increase the space of possibilities and provide new information, which can then facilitate generation of new knowledge.

Knowledge Markets. This concept focuses on the interest that individuals have in holding onto the knowledge they possess. To part with it, they need to receive something in exchange (Davenport and Prusak, 1997). Any organization is a knowledge market in which knowledge is exchanged for other things of value—money, respect, promotions, or other knowledge.

Communities of Practice. This idea, developed in the "organizational learning" movement, states that knowledge flows best through networks of people who may not be in the same part of the organization but do have the same work interests (Brown and Duguid, 1991; Wenger, 1998, 2002). Some organizations have attempted to formalize these communities, although theorists argue that they should emerge in a self-organizing fashion without any relationship to formal organizational structures: "learning happens, design or not design" (Wenger, 1998, p. 225).

Intangible Assets. Many observers have recently pointed out that formal accounting systems do not measure the valuable knowledge, intellectual capital, and other "intangible" assets of a corporation (Sveiby, 1997). Some analysts have even argued that accounting systems should change to incorporate intangible assets and that knowledge capital should be reflected on the balance sheet. However, the esoteric and subjective nature of knowledge makes it impossible to assign a fixed and permanent value to knowledge. Intangible assets have, however, always been integrated in strategic analysis as a source of competitive advantage.

Knowledge Management in action

Since KM is a relatively new philosophy, Davenport (1999) argues that it is "not yet tied to strategy and performance in practice" (p. 2-1). The use of the term "knowledge" in business strategy is rampant; however, Davenport (1999) believes that only a very small number of companies use a knowledge strategy as part of their business strategy.

Some authors have provided principles that facilitate the implementation of KM. Among them some are offering a humanistic perspective: Wheatley's (2001) six principles that facilitate KM are more a testimonial of Nonaka's original principle of an organization being analogous to an organism. The six principles are (1) knowledge is created by human beings; (2) it is natural for people to create and share knowledge; (3) everybody is a knowledge worker; (4) people choose to share their knowledge; (5) knowledge management is not about technology; and (6) knowledge is born in chaotic processes that take time. De Geus (1997) comes within the scope of the humanistic side of an organization by not using the term "learning organization" and provided the exploration of the possibilities of a "living company." If one subscribes to the six elements that have been provided by Wheatley (2001) to facilitate KM, the next step would be one of capture, integration/codification, and transfer/use of knowledge. These functions would be limited to explicit knowledge and not applicable for tacit knowledge at this time. It should be noted, however, that as organizations create processes to facilitate the conversion from tacit to explicit knowledge, applicability increases. Although Wheatley (2001) states "knowledge management is not about technology" in her fifth of six elements, capture, integration/codification, and transfer/use of knowledge are accomplished through technology.

With a more "information technology" perspective, Zack (1999) discusses the management of codified (explicit) knowledge and the use of four primary resources to manage knowledge. They include "repositories of explicit knowledge, refineries for accumulating, refining, managing and distributing knowledge, organization roles to execute and manage the refining process, and information technologies to support the repositories and processes" (p. 47).

As stated previously, it is not my intent to say that information technology creates knowledge. Augier and Morten (1999) state "technologies manifest themselves as representers of knowledge" (p. 253). Huang, Lee, and Wang (2001) show that "technology and systems, however, are used as facilitators in the production, storage, and use of organizational knowledge" (p. 4). Wiig (1999) outlines 16 building blocks that should be considered for introduction of KM: "1. Obtain management buy-in; 2. Survey and map the knowledge landscape; 3. Plan the knowledge strategy; 4. Create and define knowledge-related alternatives and potential initiatives; 5. Portray benefit expectations for knowledge management initiatives; 6. Set knowledge management priorities; 7. Determine key knowledge requirements; 8. Acquire key knowledge; 9. Create integrated knowledge transfer programs; 10. Transform, distribute, and apply knowledge assets; 11. Establish and update KM infrastructure; 12. Manage knowledge assets; 13. Construct incentive programs; 14. Coordinate KM activities and functions enterprise-wide; 15. Facilitate knowledge-focused management; 16. Monitor knowledge management: Provide feedback on progress and performance of KM program and activities." (p. 3–6)

Thus, Wheatley (2001) provides the principles in facilitation of KM, Zack (1999) provides the primary resources to manage codified knowledge, and Wiig (1999) provides the thought process necessary to build upon (in sequential order) to have a successful KM program.

It was mentioned earlier that according to the Japanese perspective, tacit knowledge represents the fundamental element to enable knowledge. Nonaka has stated that to facilitate true "learning organizations," tacit knowledge must be the cornerstone of future investigation. Krogh, Ichijo, and Nonaka (2000) provide five knowledge enablers as a means to develop the power of tacit knowledge: "1—instill a knowledge vision, 2—manage conversations, 3—mobilize knowledge activists, 4—create the right context, and 5—globalize local knowledge" (p. 5). They state that it is through these "knowledge enablers" that sharing of knowledge can occur and that true organizational improvement will happen. They also identify KM in its current form as a "constricting paradigm" that has three pitfalls: (1) knowledge management relies upon easily detectable, quantifiable information; (2) knowledge management is devoted to the manufacture of tools; and (3) knowledge management depends on a knowledge officer (pp. 26–28). It is through "knowledge enabling" that they propose to avoid these pitfalls.

Clearly, the field of KM is extensive, complex, and still in its preparadigmatic stage (Kuhn, 1970), though with significant application.

Organizational Learning: Mapping the Domain, Considerations

Origins and Definitions

The first publications on organizational learning (OL) appeared in the 1960s (Cangelosi and Dill, 1965), but research on learning organizations principally gained impetus with the publication *of The Fifth Discipline* by Peter Senge (1990a) and the special edition on organizational learning in *Organization Science* (1991). An overview of the development of the field can be found in Dierkes et al. *Handbook for Organizational Learning & Knowledge* (2001, pp. 926–927).

The concept of a "learning organization" or "learning by organizations" has actually been taken from the psychological concept of "individual learning" (Weick, 1991). Almost all definitions of organizational learning are based on this analogy.

One can differentiate normative and descriptive definitions. The normative definition refers to some requirements that an organization must satisfy in order to be known as a learning organization (Garvin, 1993; Hayes et al., 1988; Bomers, 1989; Senge, 1990a). Various other authors propose a more descriptive definition (Kim, 1993; Levinthal and March, 1993). Kim argues that all organizations learn, whether consciously or not. Some organizations try to encourage learning; others abandon such efforts and in doing so, obtain habits that finally reduce their learning capability. However, in both situations there are, in one way or another, learning processes taking place.

Individual and Organizational Learning

Many authors emphasize the paradoxical nature of the relationship between individual and organizational learning (e.g., Argyris and Schön, 1978; Huber, 1991; Bomers, 1989). One can observe that an organization consists of individuals, and individual learning is consequently a necessary condition of organizational learning. In contrast, the organization is capable of learning independently of each single individual but not independently of all individuals (Argyris and Schön, 1978).

An organization learns through its individual members and is thus directly or indirectly influenced by individual learning. Therefore, it is not surprising that most theories about learning organizations are based primarily on observations of learning individuals, particularly in experimental situations (Sterman, 1989; Huber, 1991; Kim, 1993).

Hedberg (1981) makes a comparison between the brains of individuals and organizations as information processing systems. Organizations have cognitive systems and memories, through which certain modes of behavior, mental models, norms, and values are retained. For that reason, organizations are not only influenced by individual learning processes, but organizations influence the learning of individual members and store that which has been learned. This may take the form of manuals, procedures, symbols, rituals, and myths. Though the individual is the only entity able of learning, he or she must be seen as being part of a larger learning system in which individual knowledge is exchanged and transformed.

Single-Loop and Double-Loop Learning

Most authors refer to two kinds of learning processes: single-loop and double-loop learning (Argyris and Schön, 1978; Bomers, 1989; Duncan and Weiss, 1979; Fiol and Lyles, 1985; Pedler et al., 1991). Single-loop learning involves processes in which errors are tracked down and corrected within the existing set of rules and norms. According to Fiol and Lyles (1985), single-loop learning is the result of repetition and routine. Examples of the result of these sorts of learning processes are successful programs and decision-making rules (Cyert and March, 1963). These are particularly important in situations in which the organization controls its environment (Duncan and Weiss, 1979). The main characteristics of single-loop learning can be described as based on repetition and routine, and within existing structures. It mainly concentrates on a specific activity or direct effect, within a simple context. The expected results may be change of behavior or performance level, and problem-solving capacity.

Double-loop learning, in contrast, involves changes in the fundamental rules and norms underlying action and behavior (Argyris and Schön, 1978). Double-loop learning generally has long-term effects with consequences for the whole organization. Crisis situations often provide opportunities for double-loop learning. Argyris and Schön (1978) defined double-loop learning as a process in which errors are tracked down and corrected with the result that underlying norms, ideas, and objectives become the objects of discussion and, where necessary, change. To summarize, the main characteristics of double-loop learning can be described as based on cognitive processes and understanding the nonroutine, and aimed at changing rules and structures. It occurs within a complex context. The results are changes of mental frameworks, development of frames of reference, and interpretation on the basis of which decisions can be made, as well as the development of new myths, stories, and cultures.

Different Approaches to Organizational Learning

Organizational learning may be seen under many different perspectives. For instance, the *Handbook of Organizational Learning and Knowledge* (Dierkes et al. 2001) proposes in Part I, entitled "insights from major social disciplines", seven perspectives: psychological, sociological, management science, economic theories, anthropology, political science and historic. Easterby-Smith (1997) describes six academic perspectives that have made significant contributions to understanding about organizational learning: psychology and organization development, management science, strategy, production management, sociology, and cultural anthropology. Argyris (1999), introducing the evolving field of organizational learning (p. 1), suggests the following "subfields": sociotechnical systems, organizational strategy, production, economic development, systems dynamic, human resources, and organizational culture.

Here I can propose four different approaches to organizational learning: contingency theory, psychology, information theory, and system dynamics (Romme and Dillen, 1997). As such, these approaches seem to constitute the main alternative frameworks for thinking about organizational learning. Each approach is presented as an ideal type, although in practice there is, of course, some overlap.

Contingency Theory. The classic interpretation of the concept of organizational learning is based on contingency theory, which views organizations as open systems, which continually adapt themselves to their environment. Thus, the learning process in organizations is seen primarily as an adaptation process (Cangelosi and Dill, 1965; Cyert and March, 1963; Meyer, 1982; Hutchins, 1991).

Psychology. Organizational learning can also be considered from a psychological perspective. Here the fundamental assumption is that organizations translate their internal and external environment in terms of their own frames of reference. One of the best-known ideas produced by this perspective is Weick's enactment principle (Weick, 1979), which implies that members of organizations develop collective perceptions of the organizational environment. These sets of beliefs are, to a large extent, unique for an organization and lead to a collective language through which agreement on experiences and insights can be reached (see also Argyris and Schön, 1978; Argyris, 1982, 1990, 1991, 1992).

Information Theory. The two previous approaches do not indicate how learning processes take place and where frames of reference come from. The approach based on information theory gives some attention to these kinds of questions. Here, organizations are primarily considered as processes of acquisition, distribution, interpretation, and storage of information. Organizational learning is a continually evolving process that results in the expansion and improvement of knowledge. This knowledge can only be labeled as organizational knowledge if it is exchanged and accepted among the participants. To shape the structure and dissemination of organizational knowledge, format and informal learning systems can be institutionalized: examples of these systems include strategic planning systems, management information systems, informal information channels, and communication networks

(Duncan and Weiss, 1979; Walsh and Ungson, 1991; Ulrich et al., 1993; Huber, 1991; Nonaka, 1991; Boisot, 1998).

System Dynamics. The system dynamics approach uses principles and concepts from system dynamics and systems thinking to understand learning processes in organizations (Morgan, 1986; Senge, 1990a). Here, the main assumption is that human organizations are characterized by "dynamic complexity," which makes models with simple cause-effect relationships no longer applicable (Senge, 1990a). Consequently, principles of system dynamics, such as positive and negative feedback, are used to show that social reality consists of circles of causality. Organizational learning must first be understood as a cohesive, holistic process before more detailed theorizing can take place (Morgan, 1986). Authors from the Sloan School of Management (MIT) have contributed significantly to the development and dissemination of concepts of system dynamics (e.g., De Geus, 1988; Kim, 1993; Stata, 1989; Senge, 1990a).

Contingency and psychological theories approaches have to some extent been integrated in more recent publications from the perspective of information theory (Walsh and Ungson, 1991; Huber, 1991) and system dynamics (Senge, 1990a; Kim, 1993).

Learning from a System Dynamics Perspective

Daft and Weick (1984), March and Olsen (1975), and Kolb (1984) have tried to formulate integrated models of learning processes in organizations. These attempts included the crucial link between individual and organizational learning. However, the model developed by Kim (1993) incorporates the key concepts and dimensions of earlier models, and in addition, describes the interactions between individual and collective mental models as a transfer mechanism between individual and organizational learning.

At the level of individual learning, Kim distinguishes between two levels: operational learning, what is learned ("know-how") and conceptual learning, the understanding and use of this knowledge ("know what"). Operational and conceptual learning are linked to two aspects of the individual mental model. Operational learning is learning at a procedural level through which the individual learns the steps required for accomplishing certain tasks. This knowledge is rooted in routines. Routines and operational learning influence each other. Conceptual learning involves thinking about the underlying causes of required actions, through which conditions, procedures, and concepts are discussed and new frames of reference are created. Then, the individual learning model consists of a cycle of conceptual and operational learning that is fed by individual mental models. Individual learning cycles—that is, the processes through which individual learning and its results are stored in the mental models of individuals—influence the learning process at the team or organizational level through their influence on collective mental models. So, the organization can only learn through its members, but in doing so, it does not depend on every single member. On the other hand, individuals are able of learning without the organization.

Changes in the frames of reference of one or more individuals can lead to conceptual learning at the organizational level, in the form of changes in collective frames of reference, or *Weltanschauung*, which in turn can lead to changes in the frames of reference of other individuals in the organization. In general, collective frames of reference evolve very slowly. Changes in individual routines can lead to operational learning at the organizational level, in the form of adapted or new organizational routines, such as standard procedures. These organizational routines in turn influence the development of (other) individual routines. The extent to which individual mental models can influence collective mental models depends on the influence that certain individuals or groups can exert. In general, top management tends to be one of the most influential groups.

Kim's model also incorporates the concept of single- and double-loop learning. This characteristic applies at the individual level as well as the organizational level. Individual single-loop learning takes place if the learning cycle leads to changing behavior of the individual. Individual double-loop learning concerns the process by which the individual learning cycle influences the individual mental model and vice versa. Single-loop learning at the collective level takes place when individual actions lead to the intended changes in collective actions. Double-loop learning at a collective level happens when individual mental models are transformed into collective mental models.

The literature describes seven types of learning disturbances possibly occurring in the learning process, which can be mapped with Kim's model:

1. *Role-constrained learning.* This occurs if individual learning processes do not have an effect on individual actions (March and Olsen, 1975).
2. *Audience learning.* The problem arises from the link between individual and organization actions (March and Olsen, 1975).
3. *Superstitious learning.* The problem in this disturbance is the link between individual and collective actions and the response of the environment to these actions (March and Olsen, 1975).
4. *Learning under ambiguity.* In the case of learning under ambiguous conditions, the causal links between events and the environment are no longer obvious (March and Olsen, 1975; Levinthal and March, 1993).
5. *Situational learning.* This takes place if the individual does not secure his knowledge or forgets to code it for later use (Kim, 1993).
6. *Fragmented learning.* The link between individual mental models and the collective model is poorly maintained (Cunningham, 1994; Kim, 1993).
7. *Opportunistic learning.* This happens when collective actions are based on the initiative and vision of only one person or a small group of individuals. According to Kim, this disturbance occurs in situations in which individuals consciously try to by-pass the prevailing *Weltanschauung* and organizational routines, because the old way of doing things is considered inadequate. The link between collective mental models and collective actions is deliberately broken in order to create new opportunities without the whole organization having to change.

The identification of learning disturbances is important. Methods and tools, mainly resulting from training and consultancy work (e.g., Argyris, 1991, 1992; Cunningham, 1994; Senge, 1990a; Swieringa and Wierdsma, 1990), have been developed to overcome learning disturbances: system archetypes and learning laboratories, team learning and the role of dialogue, learning sets and contracts, and circular organizational structures (Romme and Dillen, 1997).

In general, the literature on this subject still has a strongly theoretical nature. The information-theoretical perspective principally appears to provide a broad analytical framework for describing and understanding organizational learning. An integrative framework with interesting practical implications has been developed in the field of system dynamics. Tools and methods with practical relevance have been developed mainly as part of training and consultancy work.

From Knowledge Management and Organizational Learning to Learning Organization

The focus now shifts to look at the applied area of organizational learning, which is normally associated with the label of the learning organization. In this area KM and OL are intimately and inextricably linked.

Finding a Common Ground

"Knowledge development constitutes learning"

WEICK, *1991, p. 122*

Similarities between knowledge management (KM) and organizational learning (OL) begin with ambiguity of definition. Garvin (2000) provides no less than seven definitions of learning organizations. He synthesizes various definitions to form one that appears broad enough to be considered as a template of what a learning organization is: "A learning organization is an organization skilled at creating, acquiring, interpreting, transferring, and retaining knowledge, and at purposefully modifying its behaviour to reflect new knowledge and insights" (p. 11).

Peter Senge (1990b, 1999) outlines the mechanisms to achieve a learning organization working on the basis of systems theory. In fact, the open environment needed to facilitate KM is the same environment that facilitates organizational learning. "Knowledge management is not a stand alone process. It is closely bound up with the inputs of organizational learning and strategy that govern its nature and scope" (Rostogi, 2000).

The information era has generated the need for focus on knowledge workers versus blue-collar labor. The new knowledge workers are only successful if they operate in particular environments. The environment proposed in learning organizations appears to facilitate KM. KM and OL reflect the collective focus of minds toward meeting common interactive organizational objectives. It can be plausibly deduced that, given the research provided on KM, they become inextricably linked to learning organizational development. The title, and the subjects, of Dierkes et al., *Handbook for Organizational Learning & Knowledge*, (2001) provides a good example. To further illustrate the interconnections that exist between KM and learning organizations, Kofman, Senge, Kanter, and Handy (1995) state "the learning organization is built upon an assumption of competence that is supported by four characteristics: curiosity, forgiveness, trust, and togetherness" (p. 47). These characteristics are also those providing the proper environment for the process to share explicit and tacit knowledge.

Two Learning Organization Models.

A typology of learning organization models has been proposed in Easterby-Smith (1997), as we have seen. Dierkes et al. (2001, p. 930) identify three main learning models: ". . . based on feedback loops between the organization and its environment," ". . . portrayed learning in term of steps or phases," and the spiral model (Nonaka and Takeuchi, 1995).

Among this diversity I will introduce two major models. My choice is driven by the fact these models are widely recognized (Nonaka et al.) and/or are bringing new perspectives (Boisot). Nonaka, Toyama, and Byosiere (in Dierkes et al. 2001, pp. 491-517), after a critical review of existing studies of OL, state that "creating knowledge is a continuous . . . not a special kind of learning at one point in time." They propose the basic concepts of knowledge creation process, management of the process, and the organizational structure for knowledge creation. The knowledge creation process is composed of three layers: SECI (four modes of knowledge conversion: socialization-externalization-combination-internalization processes), *Ba* (platforms for the knowledge creation process), and knowledge assets (inputs, outputs of SECI, and moderator of the knowledge creation process between *Ba* and SECI). The foundation of SECI process is *Ba*, four types of *Ba* being considered: originating *Ba* (socialization/face-to-face), dialoguing *Ba* (externalization/peer-to-peer), systemizing *Ba* (combination/collaboration), and exercising *Ba* (internalization/on site). Knowledge assets, inputs, and outputs of the knowledge creation process form the basis of organizational knowledge creation. They also influence how *Ba* works. The four types of knowledge assets (KA) are as follows: experiential KA (tacit knowledge shared through common experience), conceptual KA (explicit knowledge articulated through images, symbols, and language), systemic KA (systemized and packaged explicit knowledge), and routine KA (tacit knowledge routinized and embedded in actions and practices). An organization, building on its existing knowledge assets, creates new knowledge through the SECI process that takes place in *Ba*, the knowledge created becoming then part of the knowledge assets of the organization and the basis of a new cycle of knowledge creation.

The proper management of the knowledge creation process requires a new model called a "middle-up-down" model. Top and middle management have a leadership role by working on the three layers of the knowledge creation process, providing knowledge vision, developing and promoting the sharing of knowledge assets, create and energizing *Ba*, and enabling and promoting the continuous spiral of knowledge creation. But for effective knowledge creation a supportive organizational structure is needed. The proposed new organizational structure is called "hypertext" organization, articulating a knowledge-base layer (accumulation and sharing of knowledge) bottommost layer, a business-system layer (utilization of knowledge), and a project-team layer topmost layer (creation of knowledge) through a dynamic knowledge cycle, continuously creating, exploiting, and accumulating organizational knowledge.

FIGURE 44.1. KNOWLEDGE CREATION PROCESS.

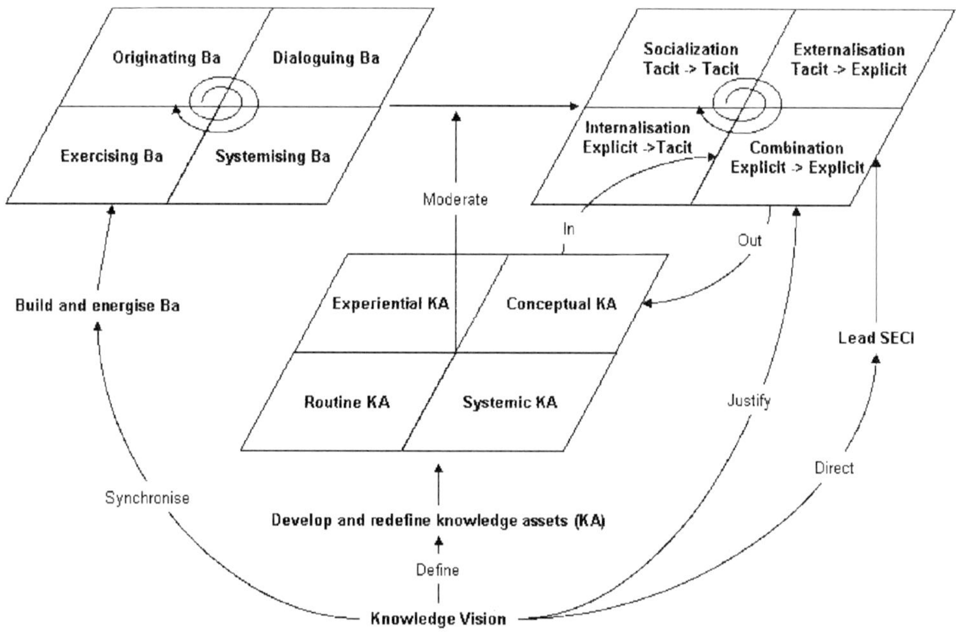

Source: Adapted from Nonaka, Toyama, and Byosiere (in Dierkes et al., 2001, p. 507). By permission of Oxford University Press.

Boisot (1998) proposes a different model grounded on an information perspective and complexity science, a set of theories describing how complex adaptive systems work. For him (p. 34), knowledge assets emerge as a result of a two-step process, constituting the two distinct phases of the evolutionary production function: creating knowledge ("process of extracting information from data") and applying knowledge ("testing the insights created in a variety of situations that allow for the gradual accumulation of experiential data"). He defines an information space (I-Space) according to three dimensions: codification (information codified/uncodified), abstraction (abstract/concrete), and diffusion (diffused/undiffused). The creation and diffusion of new knowledge occurs in a particular sequence (social learning cycle, or SLC, p. 59): scanning, problem solving, abstraction, diffusion, absorption, and impacting. Two distinct theories of learning, although not mutually exclusive, are introduced as part of identification of two distinct strategic orientations for dealing with the paradox of value (i.e., "maximising the utility of knowledge assets compromises their scarcity, and maximising their scarcity make it difficult to develop and exploit their utility," p. 90). In neoclassical learning (N-Learning), knowledge is considered cumulative. Learning becomes a stabilizing process. This approach may lead to excessive inertia and fossilization of the knowledge assets. In Schumpeterian learning (S-Learning), change is the natural order of things. Abstraction and codification are incomplete. "Knowledge may be progressive in the sense that successive approximation may give a better grasp of the underlying structures of reality, but it is not necessarily cumulative" (p. 99). S-Learning is more complex than N-Learning because it integrates both certainties and uncertainties, and it requires an "edge of chaos" culture (p. 116).

Knowledge management and organizational learning are thus linked in a systemic way through action and the quest of developing learning organizations. The diversity of epistemologies, scientific fields, theories, perspectives, and resulting models, not mutually exclusive, exemplify the plurality of approaches.

Project Management, Knowledge Management, and Organizational Learning: A Systemic Relationship

My purpose here is to introduce and illustrate the specificity of the project environment with respect to a learning organization perspective. Most of these developments are the results of research undertaken as part of the CIMAP Research Centre–Groupe ESC Lille and are grounded on the former works of the founders Decleck and Debourse (1983, 1997). (See also "Projects as Arenas for Renewal and Learning Processes," Lundin and Midler, 1998).

Projects vs. Operations: The Nature of Projects

Every organization acts according to two fundamentals modes: an operational mode, aiming at the exploitation of competitive advantage and current position on the market and providing profits and renewal or increase of resources; and an entrepreneurial mode, or project mode, focusing on the research of new position and new competitive advantage, consuming money and resources. To ensure their sustainability and development, all organizations need to combine both modes. (Declerck in Ansoff, Declerck and Hayes, 1976).

Thus, we have to face two types of activities, and I wish to propose the dichotomy operations/projects. Table 44.2 emphasizes the main characteristics of these activities. I focus here on these two types, although in reality, activities may be a blend of these two pure types.

With this differentiation in mind, you can now look at the characteristics of a project team not only in charge of project activities but, to some extent, of "operations" activities as well, and you can view the project team as a learning organization.

Project Team and Learning Dynamic

I would like firstly to compare some characteristics of groups and teams. Wenger and Snyder (2000, p. 142) draw a comparison between several forms of team organizations: community of practice, formal work group, informal network, and project team. To this it seems important to add the concept of *Ba*, as described previously. There are some fundamentals differences between project team, community of practice, and *Ba*, as summarized Table 44.3.

To understand the specificity created by the project environment and project team as far as learning is concerned, let us synthesize some of the key perspectives (see Table 44.4).

TABLE 44.2. OPERATIONS VS. PROJECTS.

Operations	Projects*
Ongoing and repetitive activities, being prone to influence of numerous factors. The factors of influence are mainly internal (endogenous) rather than environmental, and they can be manipulated by the operation manager. The environmental factors explain only a low part of the fluctuation of outputs. The inputs present random variations. It is possible to measure and to estimate the probabilities associated to these variations. The variation of inputs can be made statically stable. Future effects can be predicted with a specified margin of error. Nonusual variations coming from perturbations external to the operation lead to slight penalizing and never to disaster. Operations are reversible processes: perturbations can be detected, the nature of these causes can be identified, and these causes can be eradicated. The reversibility of operations can occur within economically acceptable limits. Operations may interact with the actions of the observer.	Nonrepetitive activities, one-shot. Decisions are irreversible. Projects are subjects to multiple influences. The main influences come from environment (exogenous) and may vary considerably. The decision maker cannot usually handle an important number of variables (exogenous variables). It is very tough to measure the effects or these influences. Project is generally not in statistical stability and it is not possible to associate probabilities to the effects one try to measure. A "bad" decision and/or a non controllable influence of a major event may lead to catastrophic result.
To summarize, operations involve	*Projects involve*
Planned actions	Creative actions
Masked actors	Unmasked actors
Process	Praxis
Cooperation	Confrontation
And they are	*And they are*
Rational	Para-rational
Algorithmic	Mosaic
Anhistoric	Historic
Stable and making one feel secure	Rich, ambiguous, unstable

*Entrepreneurial activities are assumed here to be managed using the project "form."

From the table, it is clear that projects as such are learning organizations or learning places. Projects, through the way the project team acts (praxis), are a privileged place for learning: Such project-based learning needs to integrate the two perspectives ("have" and "be" or "operations" and "projects" acting modes), as there is a need for a blend of creative or exploratory learning and application or exploitative learning (Boisot 1998, p. 116). Having in mind the need for efficiency and effectiveness, a project team acts as a temporary structure, generating first information and creating knowledge (adding complexity) with many degrees of freedom, and then applying it (reduction of complexity) in the former stage of a

TABLE 44.3. PUTTING IN PERSPECTIVE PROJECT TEAM, COMMUNITY OF PRACTICE, AND BA.

Community of Practice	*Ba*	Project Team
Members learn by participating in the community and practicing their jobs.	Members learn by participating in the Ba and practicing their jobs.	Members practice their jobs and learn by participating in the project team.
Place where members learn knowledge that is embedded in the community.	Place where knowledge is created.	Place where knowledge is created, where members learn knowledge that is embedded, and where knowledge is utilized.
Learning occurs in any community of practice.	Need of energy in order to become active.	Need of energy (forming the team) and then learning occurs.
Boundary is firmly set by the task, culture, and history of the community.	Boundary is set by its participants and can be changed easily.	Boundary is set by the task and the project.
Membership rather stable. New members need time to learn and fully participate.	Membership not fixed. Participants come and go.	Membership fixed for the project duration (temporary nature). May vary depending the phases of the project.
Participants belong to the community.	Participants relate to the *Ba*.	Participants may relate/belong to the project team for the duration of the project but may belong/relate to the operational/functional organization (department, contractors, suppliers, etc.).

project. Of course, the level of knowledge being created will depend of the nature of system project/organization/environment. Some construction projects require a little amount of creativity, while others in a different context will require a lot.

On a larger issue, the notion of knowledge management is so fascinating within projects precisely because all new project teams must solve a unique conundrum: to what degree is the information/knowledge available to complete the project based on past experience, replicable historical processes, and so on, and to what degree must all knowledge and learning be acquired or "emergent" as a result of the unique nature of the project tasks. I am thinking, for example, of the development of the Concorde, when the technical challenges were so new and nonhistorical that new cost estimating methods had to be developed, such as parametric estimation, precisely because all knowledge had to be emergent in regard to these activities.

The consequence at the knowledge management level is twofold. On the one hand, focusing on the "have" side, there is a need of for some form of knowledge—guidance, best

TABLE 44.4. SYNTHESIS OF TWO PERSPECTIVES REGARDING KM, OL, AND LEARNING ORGANIZATIONS.

Epistemology	Positivist—"Have"	Constructivist—"Be"
Main Acting Mode	*Operations*	*Projects*
Knowledge management	Western approach. Codification. Explicit knowledge. Linear thinking. Knowledge market.	"Japanese" approach (and actually French one). Personalization. Tacit knowledge. Dialectical thinking: "synthesizing dialectical thinking," aiming at identifying contradiction and resolving it by means of synthesis or integration, from "compromising dialectical thinking," focusing on tolerating contradiction.
Organizational learning	Single-loop learning. Information theory (knowledge as formal and systematic hard data, codified procedures, universal principles).	Double-loop learning. Information theory (Nonaka 1991, Boisot 1998). System dynamics theory (Senge 1990a, Kim 1993).
Learning organization	Neoclassical learning (N-Learning), knowledge is considered cumulative (Boisot, 1998).	SECI cycle, *Ba*, knowledge assets, needs for a supportive organization (Nonaka, 1991). Schumpeterian learning (S-Learning), change is the natural order of things (Boisot, 1998).

practice, standards, and so on—at the individual, team, and organizational level. The development of professional certification programs, as well as maturity models, are important in this. It's important to recognize that such standards have to be seen as largely social constructs, developed to facilitate communication and trust among those who are adopting them, but their evolution in line with the experiences gained by the users or because of new developments or practices is a vital to avoid any fossilization (Bredillet, 2002). On the other hand, on the "be" side, the need of more creative competence (for example, some professional certifications are incorporating personal characteristics), flexible frameworks (for example, use of meta-rules), and organizational structure to enable the sharing of experience is fundamental.

Consider now the organization of learning and the necessary supporting structures. Each organization running projects has its own characteristics. Each has to build its own learning organization system. Buying some off-the-shelf software or training methods is unlikely to be sufficient, even though they might form the backbone of a learning organization architecture. Being conscious of the specificity of projects, and being clear on the underlying assumptions of the concepts, methods, tools, and techniques available, should, however, certainly help in the design of an appropriate system. In this regard, promising research results have been presented by Morris (2002).

Summary

Projects being the way the organizations implement their strategy and knowledge being the ultimate source of competitive advantage, it seems logical to try to understand how knowledge and learning interact within a project organization. The knowledge management field and the organizational learning field, although relatively young, are nevertheless already well developed in term of concepts, methods, and tools. Thus, they can provide inspiration to organizations aiming at organizational performance and at creating a sustainable competitive position.

To design an effective project learning organization, you must understand the underlying theories and assumptions on which each discipline is grounded. The first choice is thus the choice of a conscious perspective regarding the organizational culture and strategy. Starting from this, it is possible to build a coherent, relevant structure, adapting the methods and tools as appropriate. This construction has to be contextualized, integrating the relevant available explicit knowledge, and providing the necessary and sufficient support for creating, combining, and applying explicit and tacit knowledge.

Project organizations are a privileged place for learning, as there is a need for combining creative and exploitative learning to manage projects efficiently and effectively. There is unlikely to be any panacea or new "one best way" to develop knowledge and learning within any kind of organization, however.

But over all, I strongly believe that at the beginning of any creation of a conscious learning organization, an act of faith is needed: the belief in knowledge as a vector of value for people, organization, and society.

Ordo ab chaos.

References

Ackoff, R. L. 1974. *Redesigning the future: A systems approach to societal problems.* New Jersey: Wiley.

Ansoff H. I. 1975. Managing strategic surprise by response to weak signals. *California Management Review* 18(2):21–33.

Ansoff, H. I., R. Declerck, and R. Hayes, eds. 1976. *From strategic planning to strategic management.* New Jersey: Wiley.

Arcade, J. 1998. Articuler prospective et Stratégie: Parcours du stratège dans la complexité. *Travaux et Recherches de Prospective* 8 (Mai): 1–88.

Argyris, C. 1982. *Reasoning, learning and action: Individual and organizational.* San Francisco: Jossey-Bass.

———. 1990. *Overcoming organizational defenses: Facilitating organizational learning.* Englewood Cliffs, NJ:: Prentice Hall.

———. 1991. Teaching smart people how to learn. *Harvard Business Review* 69 (May–June): 99–109.

———. 1992. *On organizational learning.* Oxford, UK: Blackwell Science.

———. 1999. *On organizational learning.* 2nd ed. Oxford, UK: Blackwell Science.

Argyris, C., and D. Schön. 1978. *Organizational learning: A theory of action perspective.* Reading, MA: Addison-Wesley.

Augier, M., and T. Morten. 1999. Networks, cognition, and management of tacit knowledge. *Journal of Knowledge Management* 3:252–261.

Belout, A. 1998. Effects of human resource management on project effectiveness and success: Toward a new conceptual framework. *International Journal of Project Management* 16(1):21–26.

Boisot, M. H. 1998. *Knowledge Assets: securing competitive advantage in the information economy.* New York: Oxford University Press.

Bomers, G. B. J. 1989. *De lerende organisatie.* Breukelen, Netherlands: University of Nijenrode

Bredillet, C. 1999. Essai de définition du champ disciplinaire du management de projet et de sa dynamique d'évolution. *Revue Internationale en Gestion et Management de Projets* 4(2):6–29.

———. 2002. Genesis and role of standards: Theoretical foundations and socio-economical model for the construction and use of standards. *Proceedings of IRNOP V.* Renesse, Netherlands, May 28–31.

Brooking, A. 1999. *Corporate memory: Strategies for knowledge management.* New York: International Thomson.

Brown, J., and P. Duguid. 1991. Organisational Learning and Communities of Practice. *Organisation Science* (March): 40–57.

Cangelosi, V.E., and W. R. Dill. 1965. Organizational learning: Observation toward a theory. *Administrative Science Quarterly* 10:175–203.

Cleland, D. I. 1994. *Project management: Strategic design and implementation.* New York: McGraw-Hill.

Crawford, L. 1998. Project Management Competence for Strategy Realisation. *Proceedings of the 14th World Congress on Project Management.* Vol. 1. pp. 12–14. Ljubljana, Slovenia, June 10–13.

Cunningham, l. 1994. *The Wisdom of Strategic Learning.* London: McGraw-Hill.

Cyert, R. M., and J. C. March. 1963. *A behavioral theory of the firm.* Englewood Cliffs, NJ: Prentice Hall.

Daft, R. L., and K. E. Weick. 1984. Toward a model of organizations as interpretation systems. *Academy of Management Review* 9:284–295.

Davenport, T. H., 1999. Knowledge management and the broader firm: strategy, advantage, and performance. In *Knowledge management handbook,* ed. J. Liebowitz. Boca Raton, FL: CRC Press.

Davenport, T. H., and L. Prusak. 1997. *Working knowledge: How organizations manage what they know.* Boston: Harvard Business School Press.

De Geus, A. P. 1988. Planning as learning. *Harvard Business Review* 66 (March–April): 70–75.

———. 1997. *The living company: Habits for survival in a turbulent business environment.* Boston: Harvard Business School Press.

Declerck, R. P., J. P. Debourse, and C. Navarre. 1983. *La méthode de direction générale: le management stratégique.* Paris: Hommes et Techniques.

Declerck, R. P., J. P. Debourse, and J. C. Declerck. 1997. *Le management stratégique: Contrôle de l'irréversibilité.* Lille, France: Les éditions ESC Lille.

Deming, W. E. 1993. *The new economics.* Cambridge, MA: Center for Advanced Engineering Technology.

Dierkes, M., A. Berthoin Antal, J. Child, and I Nonaka. 2001. *Handbook of organizational learning and Knowledge.* New York: Oxford University Press.

Dinsmore, P. C. 1999. *Winning in business with enterprise project management.* New York: AMACOM.

Duncan, R. D., and A. Weiss. 1979. Organizational learning: Implications for organizational design. In *Research in Organizational Behaviour,* ed. B. Shaw et al. 75–123. Greenwich, CT: JAI Press.

Drucker, R. E. 1993 *The post capitalist society.* Oxford, UK: Butterworth-Heinemann.

Earl, M. 2001. Knowledge management strategies: Toward a taxonomy. *Journal of Management Information Systems* 18(1):215–233.

Easterby-Smith, M. 1997. Disciplines of organizational learning: Contributions and critiques. *Human Relations* 50(9):1085–1113.

Fincher, A., and G. Levin. 1997. Project management maturity model. *Project Management Institute 28th Annual Seminars/Symposium.* Chicago, September 29–October 1.

Fiol, C. M., and M. A. Lyles. 1985. Organizational learning. *Academy of Management Review* 10: 803–813.

Frame, J. D. 1994. *The new project management: tools for an age of rapid change, corporate reengineering, and other business realities.* San Francisco: Jossey-Bass.

Gareis, R. 1990. Management by projects: The management strategy of the "new project-oriented company. In *Handbook of Management by Projects,* ed. R. Gareis. Vienna: MANZ.

Garvin, D. A. 2000. *Learning in action: A guide to putting the learning organization to work.* Boston: Harvard Business School Press.

———. 1993. Building a learning organization. *Harvard Business Review* 71:78–91.

Grundy, T. 1998. Strategy implementation and project management. *International Journal of Project Management* 16(1):43–50.

Guénon, R. 1986. *Initiation et réalisation spirituelle.* Paris: Editions Traditionnelles.

Hansen, M., N. Nohria, and T. Tierney. 1999. What's your strategy for managing knowledge? *Harvard Business Review* 77(2):106–116.

Hauc, A. 1998. Projects and strategies as management tools for increased competitiveness. *Proceedings of the 14th World Congress on Project Management.* pp. 1–4. Ljubljana, Slovenia, June 10–13..

Hayes, R. H., S. Wheelwright, and K. B. Clark. 1988. *Dynamic manufacturing: Creating the learning organization.* London: MacMillan.

Hayes-Roth, F., D. A. Waterman, and D. B. Lenat. 1983. An overview of expert systems. In *Building Expert Systems,* ed. F. Hayes-Roth, D. A. Waterman, and D. B. Lenat, 3–29. Reading, MA: Addison-Wesley,

Hedberg, B. L. T. 1981. How organizations learn and unlearn. In *Handbook of Organizational Design,* ed. P. C. Nyström and W. H. Starbuck, 3–27. Oxford, UK: Oxford University Press,.

Huang, K. T., Y. W. Lee, and R. Y. Wang, 1999. *Quality information and knowledge.* Upper Saddle River, NJ: Prentice Hall.

Huber, G. P. 1991. Organizational learning: the contributing processes and literatures. *Organization Science* 3:88–115.

Hutchins, E. 1991. Organizing work by adaptation. *Organization Science* 2:14–39.

Kim, D. H. 1993. The link between individual and organizational learning. *Sloan Management Review* (Fall): 37–50.

Kofman, F., P. Senge, R. M. Kanter, and C. Handy. 1995. *Learning organizations: Developing cultures for tomorrow's workplace.* Portland, OR: Productivity Press.

Kolb, D. A. 1984. *Experiential learning.* Englewood Cliffs, NJ: Prentice Hall.

Krogh, G. V., K. Ichijo, and I. Nonaka. 2000. *Enabling knowledge creation: How to unlock the mystery of tacit knowledge and release the power of innovation.* New York: Oxford University Press.

Kuhn, T. 1970. *The structure of scientific revolutions,* Chicago: University of Chicago Press.

Legay, J. M. 1996. *L'expérience et le modèle: Un discours sur la méthode.* Paris: INRA Editions.

Lemoigne, J. L. 1995. *Les épistémologies constructivistes.* Paris: PUF.

Levinthal, D. A., and J. G. March. 1993. The myopia of learning. *Strategic Management Journal* 14:95–112.

Lorino, P., and J. C. Tarondeau. 1998. De la stratégie aux processus stratégiques. *Revue Française de Gestion* 117 (Janvier–Février): 5–17.

Lundin R. A., and C. Midler. 1998. *Projects as arenas for renewal and learning processes.* Boston: Kluwer Academic Publishers.

March, J. G., and J. P. Olsen 1975. The uncertainty of the past: organizational learning under ambiguity. *European Journal of Political Research* 3:147–171.

McElroy, M. W. 2002. Corporate epistemology and the new knowledge management. *IV Conference, Institute for the Study of Coherence and Emergence.* Fort Meyers, FL, December 8.

Meyer, A. 1982. Adapting to environmental jolts. *Administrative Science Quarterly* 27:515–537.

Morgan, G. 1986. *Images of organization.* Beverly Hills: Sage Publications.

Morris, P. W. G. 1997. *The management of projects.* London: Thomas Telford.

———. 2002. Managing project management knowledge for organizational effectiveness. pp. *77–87. Proceedings of PMI Research Conference.* Seattle, July 14–17.

Nonaka, I. 1991. The knowledge-creating company. *Harvard Business Review* 69 (November–December): 96–104.

Nonaka, I. and H. Takeuchi 1995. *The knowledge creating company: How Japanese companies create the dynamics of innovation.* New York: Oxford University Press.

O'Dell, C., and C. J. Grayson, Jr. 1998. *The transfer of internal knowledge and best practice: If only we knew what we know.* New York: Free Press.

Pedler, M., J. Burgoyne, and T. Boydell. 1991. *The learning company.* London: McGraw-Hill.

Peng, K., and S. Akutsu. 2001. A mentality theory knowledge creation and transfer: Why some smart people resist new ideas and some don't. In *Managing industrial knowledge: Creation, transfer and utilization,* ed. I. Nonaka and D. J. Teece, 105–123. London: Sage Publications.

Pinto, J. K., and Prescott, J. 1998. Variations in success factors over the stages in the project life cycle. *Journal of Management* 14(1):5–18.

PMI Standards Committee. 2002. *Project manager competency development (PMCD) framework.* Newtown Square, PA: Project Management Institute.

Polanyi, M. 1958. *Personal knowledge.* Chicago: University of Chicago Press

———. 1966. *The tacit dimension.* London: Routledge and Kegan Paul.

Reich, R. B. 1991. *The work of nations.* London: Simon & Schuster.

Remy, R. 1997. Adding focus to improvement efforts with PM3. *PM Network* (July): 43–78.

Romme, G., and R. Dillen, R. 1997. Mapping the landscape of organizational learning. *European Management Journal.* 15(1):68–78.

Rostogi, P. 2000. Knowledge management and intellectual capital: The new virtuous reality of competitiveness. *Human Systems Management* 19:39–49.

Saures, I. 1998. *A real world look at achieving project management maturity.* Project Management Institute 29th Annual Seminars/Symposium, Long Beach, California, October 9–15.

Schön, D. A. 1971. *Beyond the stable state.* New York: Norton.

———. 1987. *Educating the reflective practitioner.* London: Jossey-Bass.

Senge, P. M. 1990a. *The fifth discipline: The art and practice of the learning organization.* New York: Doubleday Currency.

———. 1990b. The leader's new work: Building learning organizations. *Sloan Management Review* (Fall): 7–23.

Senge, P., A. Kleiner, C. Roberts, R. Ross, G. Roth, and B. Smith. 1999. *The dance of change: The challenges to sustaining momentum in learning organizations.* New York: Doubleday/Currency

Silberston, A. 1967. The patent system. *Lloyds Bank Review* 84 (April): 32–44.

Stata, R. 1989. Organizational learning: The key to management innovation. *Sloan Management Review* 30(3):63–74

Sterman, J. D. 1989. Modeling managerial behavior: Misperceptions of feedback in a dynamic decision-making experiment. *Management Science* 35:321–339.

Sveiby, K. E. 1997. *The new organizational wealth: Managing and measuring knowledge based assets.* San Francisco: Berrett-Koehler.

———. 1999. *The invisible balance sheet: Key indicators for accounting, control and valuation of know-how companies.* Stockholm, Sweden: Konrad Group

———. 1998. Measuring intangibles and intellectual capital: An emerging first standard. Internet version. Updated Aug 5, 1998. www.sveiby.com/articles/EmergingStandard.html.

———. 2001. What is knowledge management? Updated May 17, 2003. www.sveiby.com/articles/KnowledgeManagement.html.

Swieringa, J. and A. F. M. Wierdsma. 1990. *Op Weg Naar Een Lerende Organisatie.* Croningen, Netherlands: Wolters Noordhoff.

Toffler, A. 1980. *The third wave.* London: Collins.

———. 1990. *Power shift.* London: Bantam Press.

Turner, J. R. 1993. *The handbook of project-based management.* London: McGraw-Hill.

———. 1998. Projects for shareholder value: The influence of project managers. 283–291. *Proceedings of IRNOP III, "The nature and role of projects in the next 20 years: Research issues and problems.* Calgary, Alberta, July 6–8.

Ulrich, D., M. A. Von Glinow, and T. Jick. 1993. High-impact learning: Building and diffusing learning capability. *Organizational Dynamic* 22 (Autumn): 52–66.

Voropajev, V. 1998. Change management: A key integrative function of PM in transition economies. *International Journal of Project Management* 16(1):15–19.

Walsh, J. P., and G. Ungson. 1991. Organizational memory. *Academy of Management Review* 16: 57–91.

Weick, K. E. 1979. *The social psychology of organizing.* New York: Random House.

———. 1991. The nontraditional quality of organizational learning. *Organization Science* 2:116–123.

Wenger, E. 1998. *Communities of practice: Learning, meaning, and identity.* New York: Cambridge University Press.

Wenger, E., R. McDermott, and W. M. Snyder. 2002. *Cultivating communities of practice: A guide to managing knowledge.* Boston: Harvard Business School Press.

Wenger E. C., and W. M. Snyder. 2002. Communities of practice: The organizational frontier. *Harvard Business Review* (January–February): 139–145

Wheatley, M. 2001. The real work of knowledge management. *Human Resources Information Management Journal,* 5(2):30–37.

Wiig, K. M. 1999. Introducing knowledge management into the enterprise. In *Knowledge management handbook,* ed. J. Liebowitz. Boca Raton, FL: CRC Press.

Zack, M. 1999. Managing codified knowledge: A framework for aligning organizational and technical resources and capabilities to leverage explicit knowledge and expertise. *Sloan Management Review* 40: 45–59.

CHAPTER FORTY-FIVE

THE VALIDITY OF KNOWLEDGE IN PROJECT MANAGEMENT AND THE CHALLENGE OF LEARNING AND COMPETENCY DEVELOPMENT

Peter W. G. Morris

I was recently invited to speak at two seminars that sought to challenge the prevailing views on what we know about the management of projects. The first stated that it was time to move project management on from its perhaps rather tired and dated positivist, if not to say quasi-normative, origins stemming from its engineering-based background, to reflect a much more complex reality, as characterized, for example, by organizational change-type projects, where more interpretive views of the subject were appropriate. Fair enough, but the implication seemed to be that some of the knowledge we believe we have so carefully been trying to build up over the last 40 or more years since the discipline's emergence in its contemporary form was somehow no longer relevant. This was a position that seemed worth challenging.

The second was a seminar organized on "Managing Projects in the Pharmaceutical Industry" and a research institute's—Fenix's, in Göthenburg—work on product development contrasted against my own research, not least current work on strategy, concurrent engineering, and project-based learning (Morris and Jamieson, 2003; Miranda Lopez and Morris, 2003; Morris, 2002; Morris, Lampel, Jha, and Loch, 2003). Adler, of Fenix, for example, takes three management frameworks for the product development challenge of dealing with uncertainty (Adler, 1999): new product development,; organization theory, and project management. For Fenix, the "dominant model" of project management is control: planning, risk analysis, and monitoring to reduce uncertainty. But who says this is "the dominant model?" Certainly our own research over the years reflects a model of project management that is far broader than project control—what we in fact refer to as "the management of projects." But then again, who says we are right? What *is* "the model" of project management? How generalizable, as well as valid in the first seminar sense, is our knowledge of project management?

Why should there be so much doubt about how best to manage projects?

This chapter argues that there *are* such things as established good practices for managing projects successfully. That there are models, or rules, of project management (i.e., having a predictive ability) that more or less work. But there are indeed problems: Management *is* contextual and there are quite severe limits to our predictive capacity. And within this context, learning is nevertheless possible, though again there are quite significant challenges and limitations.

What Is Project Management?

Project management now occupies an established place in most business bookshops, with dozens of texts telling the reader how to manage projects. There have been literally thousands of books and papers on project management published over the last 40 years, with several hundred being added to the list each year. Most, it is true, talk about planning and control and organization and managing teams (these latter aspects are not included in Adlers' "standard model," incidentally). Increasingly, management of technical issues is also being included—particularly when catering to the systems market, where issues such as requirements management and configuration management figure strongly (e.g., Lientz and Rea, 1999; Forsberg, Mooz, and Cotterman, 1996; OGC, 2002). And if the literature reflects a construction, new product development or defense/aerospace background, procurement will likely figure strongly as well (e.g., DoD, 1996; Marsh, 2001), as the chapters by Venkataraman, Roulston, Milosevic, Morris, and several others in this book attest. All these "popular" writings are trying in effect to suggest norms or good practices of managing projects. But how valid is the knowledge they are seeking to impart? What indeed should be included in the definition of project management? Should the paradigm be an implementation, execution "on time, in budget, to scope" one, or should we be taking a broader view and be including the setting up of the project and the delivering of it to achieve stakeholder satisfaction?

What, first, should the model of project management be? And within this model, how valid is our knowledge?

There are probably at least something like 140,000 or more people—the membership of the Project Management Institute (PMI)—who might be expected to say that project management is as defined in PMI's *A Guide to the Project Management Body of Knowledge* (PMI, 2000).

Though widely accepted, many practitioners, academics, and others, however, believe this model to have serious shortcomings. It contains nothing detailed on project strategy, nothing on project definition, little on value management, nothing on technology management, and little on the linkage with programs and portfolios. There is nothing on leadership and minimal material on team-based development. One begins to suspect, in fact, that it represents an old-fashioned view of project management as tool-based, ignoring the broader context and treating strategy and technology as a given, with people essentially as an interchangeable commodity. All these shortcomings derive largely from its intellectual perspective of project management as primarily an *execution* discipline: of delivering a project "on time, in budget, to scope."

There are many people for whom this perspective is quite adequate. Many firms still separate the project execution end of projects from the project definition and development, labeling the former as project management and the latter as something else (e.g., project development). For others, however, from both a practical and theoretical view, it is not. It misses the whole area of setting up the project objectives (defining scope, budget, and schedule), as well as the linkages with business performance (through portfolio and program management, project strategy, and value management). For many, this really is *the* important area in the successful accomplishment of projects. And for them, seeing the management of the project as a whole brings added benefits of optimization. Project management has to be about delivering business benefit through projects, and this necessarily involves managing the project definition as well as downstream implementation.

Though this is obviously more an owner-oriented view of managing projects, even many contractors need a model of project management that involves managing the front end (commercial, procurement and logistical strategy, technology and design management, etc.), as, for example, in design-build type contracts.

This debate about the intellectual model of project management raises the first and probably the most important philosophical question about project management: What is its remit?

Addressing this question may be approached by looking at the research on project success and failure, as described, for example, in the chapter by Cooke-Davies in this book. The question of success and failure is important because it defines the goals, and hence the activities, of the discipline.

The work by Baker, Murphy, and Fisher (1974, 1988); Pinto and Slevin (1988); and Lechler (1998) assesses success both in terms of project outcome and project management process—did the project meet its mission, were the parent and client groups, and the project team, satisfied with the outcome? For Morris and Hough (1987), however, the measures of success are more explicitly outcome-oriented: What use is it following all the project management processes if the project is not a success? Success measures include both the traditional ones of "on time, in budget, to scope" delivery, but also those most relevant to the different stakeholders. The stakeholders may be many, but the two principal ones are the sponsor—the person funding the project who is responsible for its business success—and the suppliers. Hence, two of Slevin and Pinto's ten key success factors are concerned with consulting with, and selling to, "the client," who is treated almost as an external party, while Morris and Hough consider the client as integral to the project—perhaps even dominant.

Further, Morris and Hough found that the process of *defining* the project was critical to the chances of subsequently delivering the project successfully, and that often external factors such as business-driven changes, geophysical or socioeconomic factors, or technical or supply chain problems, impacted the likelihood of project success more than the control and organization ones typified by the PMBOK-type model. (Admittedly their study was of "major" projects, but interestingly both Pinto and Slevin and Lechler have the management of technology among the top ten factors, while PMBOK ignores this topic altogether.)

The net effect of the Morris and Hough work, I would like to suggest, is to emphasize more the importance of managing the project for a successful outcome and less on the internal processes of project management for their own sake. (Cooke-Davies takes this argument further in his chapter on maturity models where he talks about processes and practices as competency inputs to an organization's project management maturity, and the ability to use these to achieve output performance as capabilities.)

Indeed, it is ultimately perhaps more useful to recognize that what really distinguishes projects from nonprojects is primarily their development cycle, as shown in Figure 45.1. The sequence of going from concept into definition, development/execution and into completion/delivery (sometimes including operation, leading to closeout)—or some variant of these words. All projects go through the same sequence, no matter how trivial or complex (only the wording will change). Nonprojects—steady-state operations—do not follow this development cycle. Really, therefore, what we are talking about is *the management of projects.*

Making this recognition leads to a paradigm shift. For the "dominant model" is now far from just project control. It is beyond even PMBOK, but not the APM BoK (APM, 2000), shown in Figure 45.2, or even the IPMA "Competence Baseline" or Japanese ENAA P2M model (Caupin et al., 1998; Engineering Advancement Association of Japan, 2001). It is the management discipline of how one initiates, develops, and implements projects for stakeholder success and includes portfolio and program management, project strategy, technology management, and commercial management as much as the traditional areas of project control and organization. (Crawford concludes on the same point in her chapter on project management standards.)

How we define the paradigm within which we work is of the utmost importance, as Kuhn so clearly showed (Kuhn, 1970). It sets the terms of enquiry; stimulates the areas of research and of professional discourse; and, in a very practical way, as I shall be describing, outlines the challenge of the way we organize, learn, and improve.

Unfortunately, this broader paradigm involves not just a vastly enlarged intellectual framework but, insofar particularly as it requires knowledge leading to successful outcomes, a more contextual understanding where tacit knowledge and judgment are called upon, as much as observation of quasi-normative, positivist PMBOK-type rules.

Defining Our Knowledge about Managing Projects

If there really is a discipline of managing projects that can be described generally, then there needs to be some knowledge about it that can be articulated with a reasonable degree

FIGURE 45.1. THE GENERIC PROJECT DEVELOPMENT CYCLE.

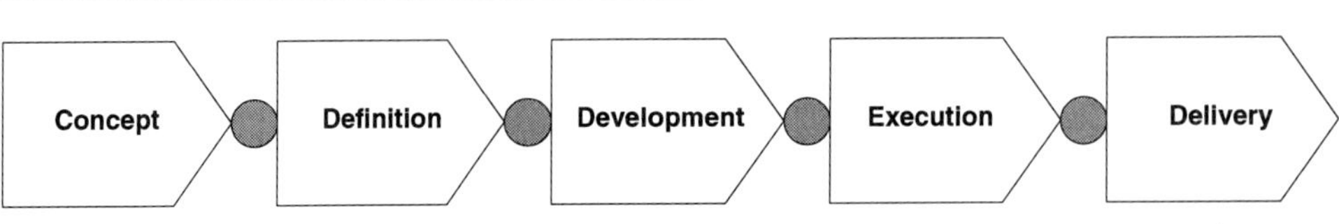

FIGURE 45.2. THE APM PROJECT MANAGEMENT BODY OF KNOWLEDGE (4TH EDITION, 2000).

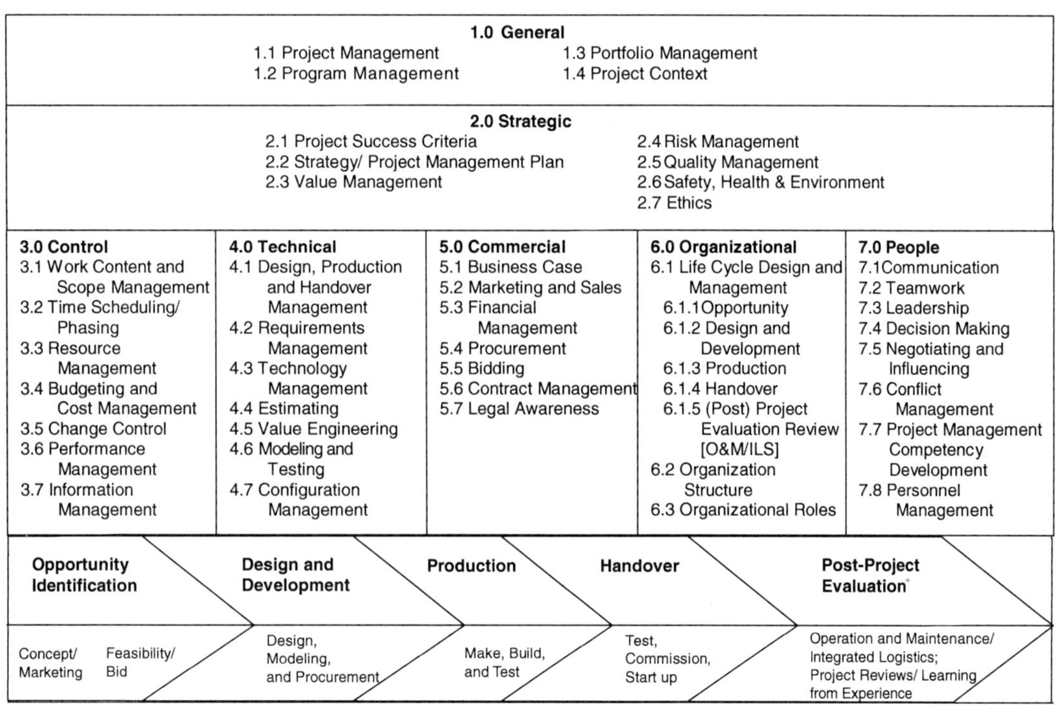

of robustness. One of the challenges implicit in the first seminar was the suggestion that much of the knowledge previously built up in this area is of questionable validity. How valid itself is this suggestion?

All management, as Griseri among many others has argued (Griseri, 2002), is contextual, and hence there will always be limitations to the generalizability of our knowledge of how to manage projects. Yct there have been very serious efforts over the years to document "lessons learned," and indeed I contend that there are indeed some "best practice" guidances that, if applied, should lead to improved chances of successfully managed projects than would be the case were they gone against.

There is nothing wrong in approaches such as ethnography or poststructuralism looking at organizations with fresh eyes, so to speak, to try to elucidate knowledge about how best to manage projects, but there *is* surely knowledge already extant that does have empirical as well as critical validity.

Projects are particularly goal-oriented forms of organization. Indeed, project management is close to an "instrumental rationality" view of the world—"rational action oriented to practical goals" (Weber, 1949). There is every reason why normative rules should apply in some instances; and therefore there is no reason why normative guidance on how to define and deliver projects should not be given (particularly in the more straightforward project management "on schedule, in budget, to scope" model of the discipline).

Thus, for example, I might say that at a general level, good project management will involve the following:

- Aligning the development of the project strategy with the sponsor's (and other stakeholders') business strategy, including reviewing the project strategy at formal review (investment) gates
- Defining the requirements (in a testable manner) so that these lead to specifications and solutions being designed and developed
- Managing design and technology so that innovations are thoroughly examined before proceeding to full project commitment
- Defining and managing the project scope, schedule, resource requirements, and budget (ensuring this represents optimal financing), including limiting changes once the design has been agreed (design freeze); integrating cost, schedule, and scope measurement; and conducting trend analyses on anticipated outturn performance
- Procuring/inducting resources into the project in as positive, cost-effective/value-creating manner as possible
- Building effective project teams
- Exercising leadership
- Ensuring effective decision and efficient communications
- Reviewing lessons-learned after and during the project and feeding other knowledge into the project through formal peer review sessions

The trouble, of course, is determining at what point such knowledge becomes so generalized that it is of limited value, and at what point is it so specific that it is no longer generalizable. Normative rules must inevitably be linked to concrete or generally agreed-to definitions of project management to be useful. Until a definitional basis is established, rules will simply generate more exceptions than use. Further, the more context-specific a discipline, the less normative rules can be generally advanced.

There are at least two problems here. One is the critical realism perspective that reality is stratified (Bhaskar, 1978, 1998; Outhwaite, 1987): There may indeed be causal relationships (laws, event sequences, etc.) discernable at a level of observation, but these are just subsets of what can be observed, and what can be observed is itself a subset of what, at a deeper level of reality, in fact exists. It is inevitable, particularly in the social sciences but in many of the physical ones too, that our knowledge of reality is incomplete.

The other issue is the nature of the knowledge being called upon when talking about how to manage projects. If we accept a framework such as the Morris and Hough/APM Body of Knowledge one, then in order to manage projects successfully, the project management practitioner will need to draw on knowledge, skills, and behaviors in strategic, technical, and commercial management areas, as well as ones in control, organization, and behaviors. The nature of our knowledge in these different areas differs.

Substantively it differs in several ways. The nature of our knowledge about finance, for example, is different from the nature of our knowledge about leadership; similarly production management is different from strategy; and so on. Project management, certainly in the broader paradigm of "the management of projects," covers many branches of management, and to an extent this even holds for the more limited PMBOK model (integration versus time management; quality versus scope).

Even within a topic area, of course, there will be epistemological variations: Our certainty of knowledge about scheduling can be quite robust normatively as regards critical path but more uncertain when it comes to critical chain (insofar as much of the power of critical chain lies in the aggregation of contingencies and in the benefit that management can make of this in motivating teams and juggling grouped contingencies).

Substantively, in fact, as I have already pointed out, it becomes much harder to capture and codify rules relating to the likelihood of certain outcomes emerging. But this is not necessarily the case from a process point of view. Risk management is a good example. The process steps for carrying out a risk analysis can be described easily. But making substantive judgments about which risks are a greater priority, and how best to deal with these, often requires considerable judgment.

In reality, much of our knowledge of how best to manage projects is process-based—how to manage changes, do a risk or value management exercise, prepare a quality plan, form a team, bid a contract, do a design review, prepare a schedule, and so on. This process-based knowledge, however, is then set in a context: how it is done in "our industry," "our company," "our business unit," or even "our project."

It is the process-based knowledge that largely forms the base of project management's generalizable rules. It would be naïve to maintain that these are to a degree also not context-dependent, but they are more generalizable than the substantive knowledge of the context-specific situations that sit above these basic process "good practices." The difference is not quite one of normative, positivist rules versus constructivist insights, but rather one of stratification.

Project Learning

This hypothesis about project management knowledge raises questions about how we can learn the discipline of project management and what relevance this knowledge has in delivering successful projects. Most work on learning in project-based organizations in recent years has, *pace* the preceding, recommended a number of process "good practices"—which themselves begin to resemble valid process rules (laws or event sequences). For example, project performance should be improved through

- the systematic collection of learning on projects (Dixon, 2000);
- periodic project reviews by peer teams (peer reviews and peer assists) (Collison and Parcell, 2001);

- distinguishing between tacit and explicit knowledge (Morris, 2002; Fernie et al., 2003; Nonaka and Takeuchi, 1995);
- identification of key persons as repositories of tacit knowledge and as "owners" of subject matter areas (Ayas and Zeniuk, 2001; Wenger, 1998);
- the use of information management tools to aid in the capture, storing, processing, archiving, looking up, retrieving, and presenting of information (Morris, 2003; Currie, 2003);
- a discipline of accessing knowledge—using checklists or other "look-up" guides by the project teams before beginning a new project task (Brander-Löf, Hilger, and André, 2000);
- having a definition in some way of the knowledge in a particular area: the "Body of Knowledge" (Morris, 2001; Wenger, 1998);
- having a knowledge management program in place (Turner, Keegan, and Crawford, 2000)
- distinguishing between individual, team, and organizational learning (Popper and Lipshitz, 2001);
- a program or programs for using the knowledge/learnings identified (Cross and Baird, 2000; Schindler and Eppler, 2003);
- applying metrics for the usage made of the knowledge and learning;
- implementing a competency development program for updating the knowledge.

But how far does this process view of project management help us in learning how to manage projects better? It is clearly helpful for the more process-oriented type of management challenge—as I have characterized the PMBOK-type model to be—but less so where people are looking for a combination of substantive and process knowledge, where context and judgment is important, and experience and tacit knowledge particularly valuable (Bresner et al., 2003; Fernie et al., 2003). In fact, even at the process level, context is unavoidable.

Many researchers have identified the need for contextual knowledge and have emphasized the importance of tacit knowledge (Dierkes, Antal, Child, and Nonaka, 2001; LeRoy, 2002; Scarbrough et al. 2002; etc.). Loch (2000) and Ferlie and Loch (2001) contend that the process of learning is mediated by both the "content" and the "context" of the process. Boisot (1998) has elaborated a similar idea in terms of his I-Space (information space), where learning takes the form of a spiral (scanning, problem solving, abstraction, diffusion, absorption, impacting) and where the learning process is affected by the degree to which the knowledge is

- codified or uncodified—the business context may be relatively codified (KPIs and strategy) or vague ("our preferred way of working");
- abstract or concrete—for example, a risk management support guide is generally less abstract than an overall project management methodology;
- diffuse or undiffused—the knowledge may be "generally available" or have high company specificity (IPR).

In short, different learning approaches will be required for different types of knowledge. Further, not only are there different learning requirements for different types of (project management) topics, it is generally recognized that the more strategic and cognitive type of learning (more prevalent in the broader model) is more difficult than the more routine and process-oriented learning (more prevalent in the traditional project management model). (This relates to the incrementalist, single-loop theory of learning (e.g., Levinthal and March, 1993; Miner, 1990) compared with the more strategic, cognitive—double-loop—perspective (Argyris and Schön, 1978; Fiol and Lyles, 1985).)

These findings are now being reinforced in research on the rollout of project management best-practice guidance, which suggests that different enabling mechanisms may be appropriate for different project management practices. The thrust of this work, which combines both the model of the spiral of learning of Nonaka et al. (Nonaka and Takeuchi, 1995) and the work on enabling mechanisms of Von Krogh (Von Krogh, Ichijo, and Nonaka, 2000), is that different enabling mechanisms are more relevant in the different modes of knowledge conversion (Socialization-Externalisation-Combination-Internalisation, per Figure 45.3) posited by Nonaka. Further, that within the modes of knowledge conversion

FIGURE 45.3. LEARNING AND SUPPORT MECHANISMS APPROPRIATE FOR DIFFERING STAGES OF THE KNOWLEDGE CREATION SPIRAL.

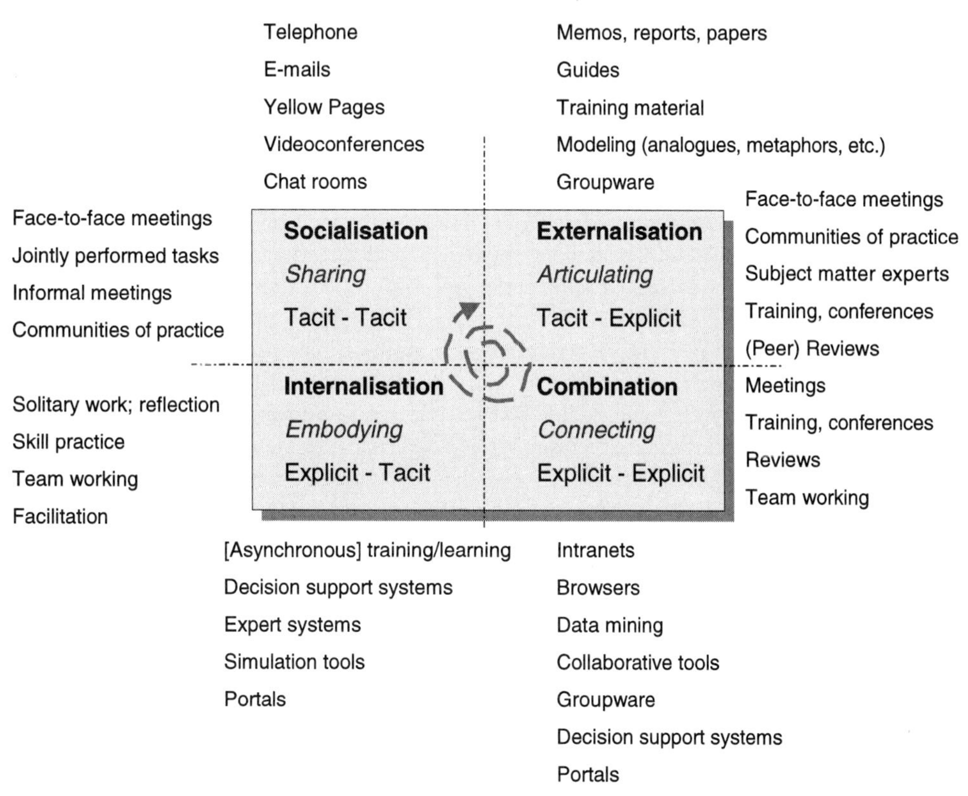

Source: After Nonaka, von Krogh, et al.

there could be a skew in the four quadrants. It may be skewed at different times in the organization, and at different stages of the project (Morris, Loch, Lampel, and Jha, 2003; LeRoy, 2000). For example, the socialization mode appears more dominant where distillation (stories/narratives/word-of-mouth) is the dominant strategic mode (Lampel, Morris, Loch, and Jha, 2003).

And though von Krogh suggests learning/knowledge creation happens best when all quadrants of the spiral are balanced, I have observed that, in a project context at least, this situation may not occur. In fact, all the organizations studied in this research experienced significant learning constraints, even though they might know the theory quite well, due for example to

- strategic intent (perhaps a conscious intent to focus on people rather than tools for knowledge sharing, or from being so preoccupied with creating customer solutions and doing the business that trans-project learning suffers);
- organizational constraints (lack of project orientation, the silo effect);
- virtual working (thus, a focus on explicit ways of working) or the attention it receives or does not receive (wrong priorities and incentives; lack of personal interest).

Implications for Project Management Competence Development

So what does this imply for the development of project management competence? Is there, for example, a place for a generic discipline-based certification program?

Competence is generally taken to be role-specific and to comprise the knowledge, skills, and behaviors needed to perform the role (Boyzatis, 1982). The role specificity of the concept allows us to focus and emphasize the contextual and tacit dimensions of the learning of the knowledge, skills, and behaviors needed to perform the role properly (which, of course, is also one of the reasons why "experience" requirements are often tagged onto definitions of competences.)

The outlook at a more general level is more problematic, however. If the noncontextual, generalized type of knowledge is more process-oriented (more PMBOK-like), then while being of value in helping people learn useful rules and routines, it is less valuable in helping deal with difficult contextual situations requiring judgment. More contextual learning will be more appropriate (such as case studies and team exercises). Professional accreditation or certification is hence at best an inadequate guarantee of professional performance (nice but not necessarily very useful)—which, I would suggest, just about accords with what most practicing project professionals believe to be the case.

Summary

It would be a mistake to ignore the "good practice" guidance and rules that have been built up and formalized as quasi-normative knowledge on project management over the last 40 or more years. Much of this knowledge relates to the processes of managing projects effectively. But even such process knowledge soon requires an understanding of the context in which management operates. This is particularly the case where more strategic and judgmental decisions are being made. The larger "management of projects" framework calls for more of this type of knowledge than the more circumscribed PMBOK-type model of project management.

This has implications for the way we learn about project management and for the value of project management accreditation programs. Incremental-type learning, of rules and practices, should be distinguished from the more challenging strategic and cognitive learning.

References

Adler, N. 1999. *Managing complex product development.* Stockholm: Stockholm School of Economics, EFI.

Argyris, C., and D. Schön. 1974. *Theory in practice: Increasing professional effectiveness.* San Francisco: Jossey-Bass.

Association for Project Management. 2000. *Body of knowledge.* 4th ed. *www.apm.org.uk.*

Ayas, K., and N. Zeniuk. 2001. Project-based learning: Building communities of reflective practitioners. *Management Learning* 32(1):61–76.

Baker, B. N., D. C. Murphy, and D. Fisher. 1974. *Determinants of project success.* NGR 22-03-028.

National Aeronautics and Space Administration. 1988. Factors affecting project success. In *Project Management Handbook.* 2nd ed., ed. D. I. Cleland and W. R. King. 902–919. Hoboken, New Jersey: Wiley.

Bhaskar, R. 1978. A realist theory of science. York, UK: Leeds Books

———. 1998. *The possibility of naturalism.* Atlantic Heights, NJ: Humanities Press.

Boisot, Max. 1998. *Knowledge assets.* Oxford, UK: Oxford University Press.

Boyatzis, R E. 1982. *The competent manager: A model for effective performance.* Hoboken, New Jersey: Wiley.

Brander-Löf, I., J-W Hilger, and C. André. 2000. How to learn from projects: The work improvement review. *IPMA World Congress 2000.* International Project Management Association, Zurich.

Bresner M., L. Edelman, S. Newell, H. Scarbrough,and J. Swan. 2003. Socialpractices and the management of knowledge in project environments. *International Journal of Project Management* 21(3):157–166.

Caupin, G., H. Knöpfel, P. W. G. Morris, E. Motzel, and O. Pannenbäcker. 1998. ICB IPMA competence baseline. International Project Management Association, Zurich. www.ipma.ch.

Collison, C., and G. Parcell. 2001. *Learning to fly.* Oxford, UK: Capstone Publishing Ltd.

Cooke-Davies, T. J. 2000. The "real" success factors on projects. *International Journal of Project Management* 20(3):185–190.

Cross, R., and L. Baird. 2000. Technology is not enough: Improving performance by building organizational memory. *Sloan Management Review* 14 (3, Spring)

Currie, W. L. 2003. A knowledge-based risk assessment framework for evaluating Web-enabled application outsourcing projects. *International Journal of Project Management* 21(3):207–218

Dierkes, M., A. B. Antal, J. Child, and I. Nonaka, eds. 2001. *Handbook of Organisational Learning and Knowledge.* Oxford, UK: Oxford University Press.

Dixon, N. 2000. *The organizational learning cycle: How we can learn collectively.* New York: McGraw-Hill.

DoD. 1996. *Mandatory procedures for major defense acquisition programs and major automated information systems.* Directive 5000.2-R. Washington, D.C.: Department of Defense,

Engineering Advancement Association of Japan. 2001. *P2M: Project and program management for enterprise innovation.* www.enaa.or.op.jp

Ferlie, E., and I. Loch. 2001. Change management and organisational learning in primary care. *Report to NHS Executive R and D (SE Region)*. London: The Management School, Imperial College.

Fernie, S., S. D. Green, S. J. Welleer, and R. Newcomber. 2003. Knowledge sharing: context, confusion and controversy. *International Journal of Project Management* 21(3):177–188

Forsberg, K., H. Mooz, and H. Cotterman. 1996. *Visualizing project management.* New York: Wiley.

Griseri, P. 2002. *Management knowledge: A critical view.* London: Palgrave.

Kuhn, T. S. 1970. *The structure of scientific revolutions.* Chicago: The University of Chicago Press.

Lampel, J., P. W. G. Morris, P. Jha, and I. Loch. 2003. Projects and the organisation: A strategic learning interface. *Organizational Knowledge and Learning Conference.* Barcelona.

Lechler, T. 1998. When it comes to project management, it's the people that matter: An empirical analysis of project management in Germany. IRNOP III. The Nature and Role of Projects in the Next 20 Years: Research Issues and Problems. Calgary: University of Calgary.

Leintz, B. P. and K. P. Rea. 1999. *Breakthrough technology management.* London: Academic Press.

LeRoy, D. 2002. Knowledge management and projects' capitalization: A systemic approach. *Proceedings of PMI Research Conference.*

Loch, I. 2000. Learning, change, and professional paradigms in primary care. Abstract in *British Academy of Management 2000 Conference Proceedings.* University of Edinburgh.

Lopez Miranda, A., and P. W. G. Morris. Forthcoming. A conceptual framework for maximizing the benefits from implementing concurrent engineering with project management. *International Journal of Agile Manufacturing.*

Marsh, P. 2001. *Contracting for engineering and construction projects.* Aldershot, UK: Gower.

Miner, A. S. (1990) 'Structural Evolution through Idiosyncratic Jobs: the Potential for Unplanned Learning', *Organization Science.* pp. 195-210.

Morris, P. W. G.. 2001. Updating the Project Management Bodies of Knowledge. *Project Management Journal* 32(3):21–30.

———. 2002. Managing project management knowledge for organisational effectiveness. *Proceedings of PMI Research Conference 2002, Seattle.* Newtown Square, PA: Project Management Institute.

———. 1992, 1997. *The Management of Projects.* London: Thomas Telford.

Morris P. W. G., P. M. Deason, T. M. S. Ehal, R. Milburn, and M. B. Patel. 2003. The role of IT in capturing and managing knowledge in designer and contractor briefing. *IT in Architecture, Engineering and Construction* 1(1):1–18.

Morris, P. W. G., and G. H. Hough. 1987. *The anatomy of major projects.* Chichester, UK: Wiley.

Morris P. W. G., J. Lampel, P. Jha, and I. Loch. 2003. Organisational learning and knowledge creation interfaces in project-based organisations. *EURAN 2003.* Milan, European Academy of Management.

Morris, P. W. G. and H. A. Jamieson. 2004. Moving from Corporate Strategy to project strategy. Newtown Square, PA: Project Management Insitute.

Morris, P. W. G., I. Loch, J. Lampel, and P. P. Jha. 2003. A construct for project-based learning: The PROBOL model. *OLK 5 Conference,* Lancaster University Management School, June.

Nonaka, I, and H. Takeuchi. 1995. *The knowledge-creating company.* New York: OUP.

Office of Government Commerce. 2002. *Managing successful projects with PRINCE2.* London: The Stationery Office.

Outhwaite, W. 1987. *New philosophies of social science: realism, hermeneutics, and critical theory.* New York: Macmillan.

Pinto, J. K., and D. P. Slevin. 1988. Critical success factors across the project life cycle. *Project Management Journal* 19(3):67–75.

Popper, M., and R. Lipshitz.. 2001. Organizational learning: Mechanisms, culture and feasibility. *Management Learning* 31(2).

Project Management Institute. 2000. *A guide to the Project Management Body of Knowledge,* Newtown Square, PA: Project Management Institute.

Scarbrough, H., M. Bresnen, L. Edelman, J. Swan, S. Laurent, and S. Newell. 2002. Cross-sector research on knowledge management practices for project-based learning. *European Academy of Management 2nd Annual Conference.* Stockholm.

Schindler, M., and M. J. Eppler., 2003 Harvesting project knowledge: A review of project learning methods and success factors. *International Journal of Project Management* 21(3):219–228.

Shenhar, A. J., O. Levy, and D. Dvir. 1997. Mapping the dimensions of project success. *Project Management Journal* 28(2):5–13.

Turner, J. R., A. Keegan, and L. Crawford. 2000. Learning by experience in the project-based organization. *Proceedings of PMI Research Conference 2000, Paris.* Newtown Square, PA: Project Management Institute.

Von Krogh, G., K. Ichijo, and I. Nonaka. 2000. *Enabling knowledge creation.* New York: Oxford University Press.

Weber, M. 1949. *The methodology of social sciences.* New York: Free Press.

Wenger, E. 1998. *Communities of practice.* Cambridge, UK: Cambridge University Press.

CHAPTER FORTY-SIX

GLOBAL BODY OF PROJECT MANAGEMENT KNOWLEDGE AND STANDARDS

Lynn Crawford

The definition of a body of knowledge and the development of standards for project management have been significant features of the growing interest in this field of practice as it aspires to recognition as a profession. There has been much debate within the field of project management as to whether it satisfies criteria for status as a profession (Zwerman and Thomas, 2001). One view, however, is that professions can be considered to begin either with the recognition by people that they are regularly doing something that is not covered by other professions or with the formation of professional associations (Abbott, 1988). The impetus behind the formation of professional associations is considered by Eraut (1994, p. 165) to be "derived from the perceived need of a relevant group to occupy and defend for its exclusive use a particular area of competence territory" and that, politically, their interest may lie in claiming that only their members are competent within that territory.

Definition of a distinct body of knowledge and of standards based on that body of knowledge are ways of marking professional territory (Morris et al., 2000; Berry and Oakley, 1994). Assessment and award of qualifications provides a process whereby professionals are recognized as meeting the standards of a profession by demonstrating mastery of the body of knowledge and either minimum or graduated levels of proficiency or competence (Dean, 1997). A body of knowledge, standards, and related assessment and qualification processes can therefore be seen as essential building blocks in the formation and recognition of a profession (see Figure 46.1).

Development and recognition of a distinct profession of project management has certainly been a strong driver in the development of standards. Other related factors that have been significant:

FIGURE 46.1. BUILDING BLOCKS OF A PROFESSION.

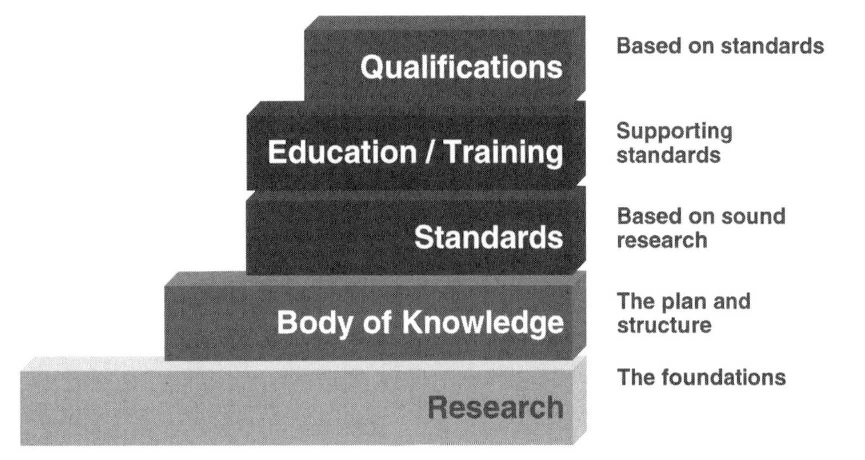

- The need to identify the role and tasks of project managers in an emerging field of practice where neither the project managers, their clients, nor their employers necessarily had a clear understanding of the role
- The need for common terminology
- The need for a common basis for employment and deployment of project personnel, working collaboratively, across functions in multidisciplinary teams; across organizations in strategic alliances and joint ventures; and across continents, in global projects

Project personnel are actively seeking sound guidance for the identification of project management competencies and credentials that will enhance their careers. As project based work takes over from position-based work and careers are defined less by companies and more by professions (Stewart, 1995), project personnel are keen to achieve professional status and independent recognition of their project management competence. If people are to be evaluated, not by rank and status but flexibly according to competence (Stewart, 1995), then evidence of this competence becomes extremely important to individuals as well as to organizations. A distinct body of project management knowledge and standards providing a baseline for project management competence are important building blocks in professional recognition for individuals and for an emerging profession.

This chapter presents an overview of the current principal project management standards and guides for project management knowledge and performance, including a comparison of their content and coverage and an indication of their use in assessment and as a basis for qualifications. Current developments and potential future directions are also reviewed.

Project Management Standards and Guides

Overview

A *standard* is a measure, devised by general consent, as a basis for comparison against which judgments might be made as to levels of acceptability. Standards, to have effect, do not need to be officially endorsed. They can be voluntarily accepted. In the field of project management, there is a very strong link between the definition of a project management body of knowledge and the development of standards, with a number of guides to aspects of the project management body of knowledge being treated as standards for what project management practitioners are expected to know.

It is important to note that although the general territory or coverage of a project management body of knowledge may be defined, the entire body of knowledge, which encompasses tacit and explicit knowledge embodied in published and unpublished material and in the established and emerging practices of practitioners, cannot be captured in any single document. Therefore, any documentation of that knowledge must be considered as a guide to one view or aspect of the project management body of knowledge at a point in time. This distinction is made very clear in the introduction to the Project Management Institute's *A Guide to the Project Management Body of Knowledge* (PMBOK® Guide, Project Management Institute, 2000, p. 3), which is the most widely distributed of a number of body of knowledge guides and is also recognized as a standard by the American National Standards Institute.

Guides and standards have been developed for project management for various purposes, which can generally be classified as the following (Duncan, 1998):

- *Projects.* Knowledge and practices for management of individual projects
- *Organizations.* Enterprise project management knowledge and practices
- *People.* Development, assessment, and registration/certification of people

The most widely known, distributed, and used guides and standards for project management are presented in Figure 46.2, indicating their general focus: projects, organizations, or people. They can be further classified as either focusing on knowledge or on description of practices, the latter being primarily in the form of performance-based competency standards or frameworks intended specifically for assessment and development of project management practice in the workplace. There are a number of standards for aspects of project management that are provided in languages other than English, but these are not included here, since their application tends to be limited to their country of origin. The Japanese P2M is an exception, as an English translation of a significant part of the standard has been widely distributed.

Projects:

Those standards and guides that focus primarily on what project management practitioners need to know (knowledge guides) are also those dealing essentially with management of individual projects. The Japanese P2M stands out as the one exception, specifically extending the focus beyond the management of single projects to management of programs of projects

FIGURE 46.2. SUMMARY OF STANDARDS AVAILABLE THAT FOCUS ON PROJECTS, PEOPLE AND ORGANIZATIONS.

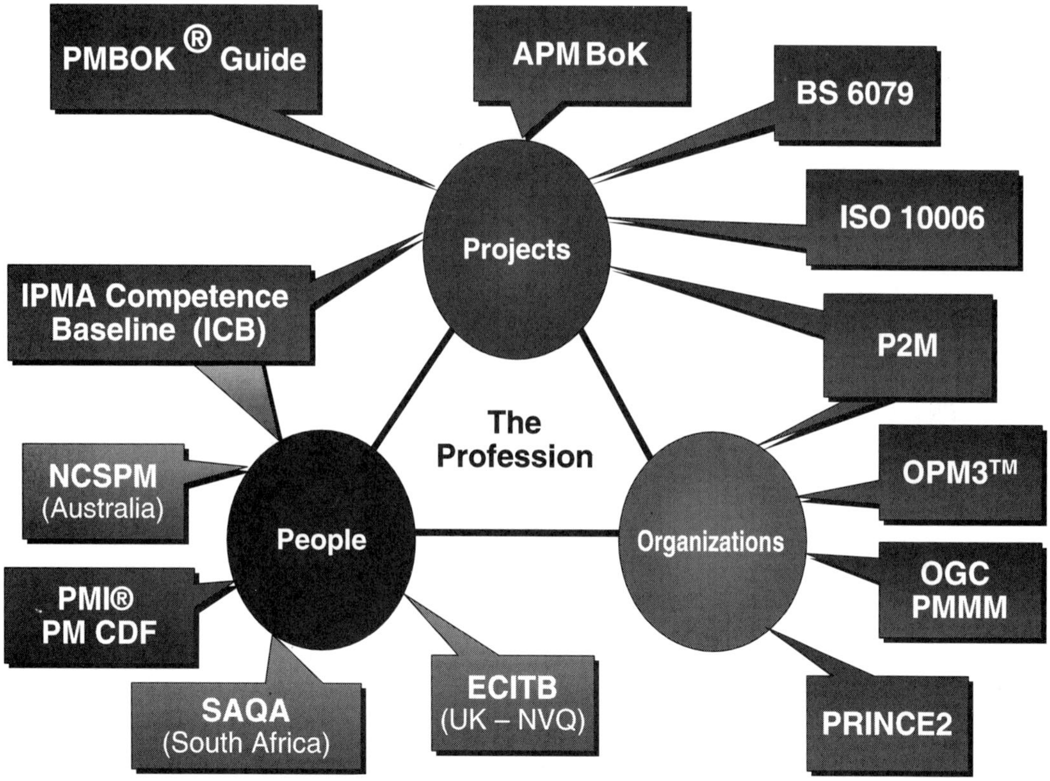

in the context of corporate strategy implementation and enterprise innovation and management. The standards and guides shown in Figure 46.2 as focusing on knowledge and relating primarily to management of individual projects are as follows:

- Project Management Institute's *A Guide to the Project Management Body of Knowledge* (Project Management Institute, 2000)
- *Association of Project Management Body of Knowledge* (APM BoK, UK; Dixon, 2000)
- BS6079 *Guide to Project Management* (British Standards Board, 1996)
- ISO 10006 *Guidelines to Quality in Project Management* (ISO, 1997)
- *ICB: IPMA Competence Baseline* (Caupin et al., 1999)
- *P2M: A Guidebook of Project and Program Management for Enterprise Innovation* (Engineering Advancement Association of Japan Project Management Development Committee; ENAA, 2002)

All of the documents listed are considered to be standards, either formally or informally. All of the documents, except BS6079 and ISO 10006, are used as the knowledge base or standard for professional certification programs. ISO 10006 provides guidelines to quality in project management and is certainly a standard. It is, however, primarily a quality management rather than a project management standard. There has been discussion of the potential for development of an ISO standard for project management, but to date no such standard has been produced.

The P2M was developed by the Project Management Development Committee of the Engineering Advancement Association of Japan with funding from the Japanese government through the Ministry of Economy, Trade, and Industry (METI). The P2M provides the basis for a certification program, and the Project Management Professionals Certification Center (PMCC) was established in May 2002 to manage this certification program and maintain the P2M. All of the other documents have been developed by and are maintained by project management professional associations.

PMBOK Guide. The Project Management Institute (PMI) had been working on defining or mapping what constituted the body of knowledge of project management since the mid-1980s (Wideman, 1986), and its first project management standard was published in 1983 as part of the PMQ Special Report on Ethics, Standards and Accreditation (Project Management Institute, 2002a). The standards portion of the report identified six major project management functions: human resources management, cost management, time management, communications management, scope management, and quality management. The *Project Management Body of Knowledge of the Project Management Institute* was first published in *PM Network* in 1987 and included the six functions identified in the 1983 document as well as two further project management knowledge functions: risk management and procurement management.

The publication, in 1996, of *A Guide to the Project Management Body of Knowledge* marked a major milestone in the development of project management as a field of practice and aspiring profession. It is this document that has been widely accepted throughout the world as a standard for project management knowledge and has been an important factor in the growth of interest in project management, its dissemination benefiting from rapidly increasing use of the World Wide Web. When first published in 1996, the document was made freely available for download from the PMI Web site. It was also made available to all PMI members through PMI chapters, through the PMI online bookstore, and through publication and dissemination by other project management professional associations such as the Australian Institute of Project Management, under generous cooperative agreements. Trainers and consultants were also able to publish and jointly badge the document.

Through this generous promotional program and as a result of the inherent quality of the document, dissemination was rapid, with over 570,000 copies distributed by the end of 2000. There are now over 1 million copies of the PMI Standard in circulation, and all new members receive a copy of the PMBOK Guide on CD-ROM when they join the Institute.

The PMBOK Guide, published in 1996, changed the functions to knowledge areas, adding one more (Integration), and introduced a process focus (see Figure 46.3). The 1996 and 2000 Editions of the PMBOK Guide include nine project management knowledge areas: Integration, Scope, Time, Cost, Quality, Human Resources, Communications, Risk,

FIGURE 46.3. STRUCTURE OF PMBOK GUIDE.

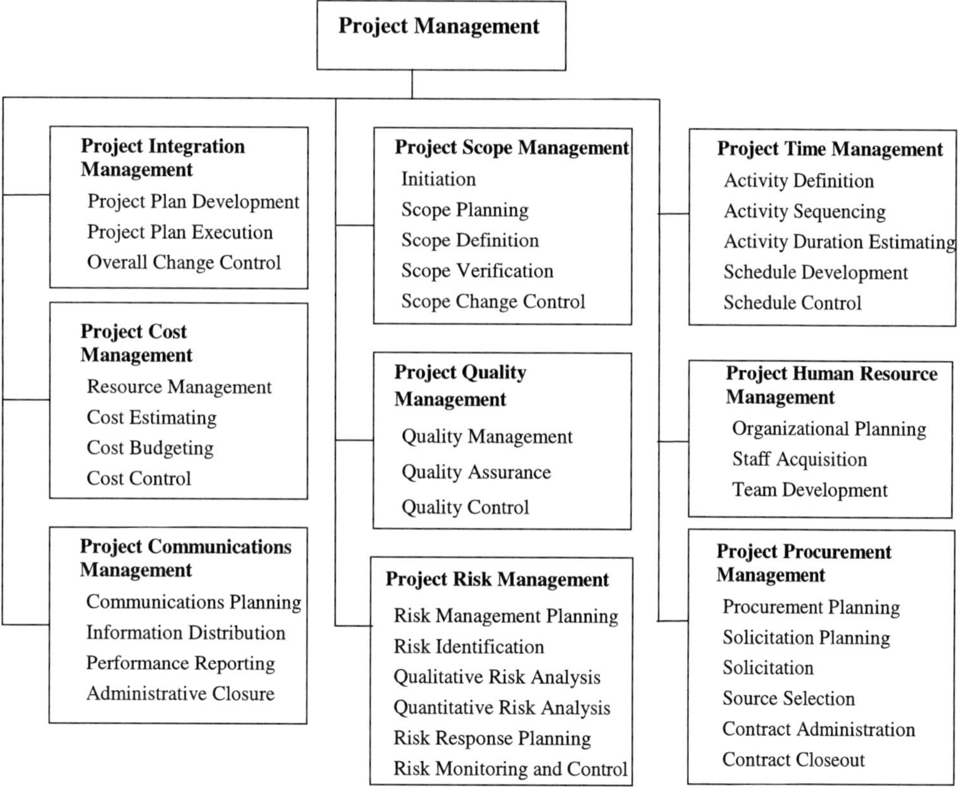

and Procurement, plus a section dealing with project management context and processes, identifying five process groups. A glossary was also added. In the 2000 Edition of the PMBOK Guide, there was "no fundamental revision of the structure or philosophy" (Morris, 2001, p. 23). The major change was a revision and extension of the section relating to risk management. Both editions clearly acknowledge that no single document can embody the whole body of knowledge relevant to project management.

The introduction to the PMBOK® Guide (2000 Edition, p. 3) states that it provides a guide to "that subset of the PMBOK that is generally accepted" in terms of knowledge and practices "applicable to most projects most of the time." The document is not intended to be either comprehensive or all-inclusive. It is used by the PMI to provide a consistent structure for its professional development programs, including but not limited to the following:

- Certification of Project Management Professionals (PMPs®)
- Accreditation of degree-granting education programs in project management

It is intended that *A Guide to the Project Management Body of Knowledge* will be a living document, subject to ongoing review. At the beginning of 2003, the Guide was available in eight languages: Mandarin Chinese, French, German, Italian, Japanese, Korean, Brazilian Portuguese, and Spanish.

In early 1999 PMI was accredited as a Standards Development Organization by the American National Standards Institute (ANSI), and in September 1999, the PMBOK® Guide was approved as an American National Standard (ANSI/PMI 99-001-1999) (Holtzman, 1999). In March 2000, the PMBOK® Guide, 2000 Edition replaced the 1996 Edition as PMI's American National Standard (ANSI/PMI 99-001-2000). The PMBOK® Guide has also been adopted as an IEEE Standard (1490-1998): *IEEE Guide to the Project Management Body of Knowledge* (IEEE, 2000).

Since publishing the PMBOK® Guide, 2000 Edition, PMI has published three other project management standards and guides: *PMI Practice Standard for Work Breakdown Structures*, the *Government Extension to the PMBOK® Guide*, 2000 Edition, and the *Project Manager Competency Development Framework*. PMI has a number of other standards and guides under development (refer to www.pmi.org).

APM BoK. Discussion concerning development of a body of knowledge reference document as a basis for an APM (initially the Association of Project Managers and now the Association for Project Management) certification program began in 1986, led by the Professional Standards Group of the APM (Morris, 2001). The first version of what was referred to as the *APM Body of Knowledge* was published in 1992, updated in 1994 (Second Edition) and 1996 (Third Edition), and significantly revised in 2000 (Fourth Edition). In undertaking what they refer to as a "fundamental revision" (Dixon, 2000, p. 7), the APM sought the assistance of UMIST's Centre for Research into the Management of Projects (CRMP).

The APM Body of Knowledge (Third Edition; APM, 1996) identified 40 key areas that the Association for Project Management considered people involved in project management should have both knowledge of and experience in. The document also listed eight principal personality characteristics that a Certificated Project Manager should display. This edition of the *APM Body of Knowledge* was clearly intended as the reference underlying a certification program, as it included several references to the knowledge, experience, and personality characteristics that would be expected of a Certificated Project Manager. Guidance was given for those seeking certification.

The 40 key areas of knowledge and experience were grouped under four headings:

- *Part 1.* Project Management
- *Part 2.* Organization and People
- *Part 3.* Processes and Procedures
- *Part 4.* General Management

The significantly revised Fourth Edition of the APM BoK has 42 topics listed under 7 headings (see Figure 46.4).

As mentioned earlier, the Fourth Edition of the APM BoK was primarily based on research cosponsored by APM and the industry and conducted at UMIST's Centre for Research into the Management of Projects (CRMP), under the direction of Professor Peter

FIGURE 46.4. APM BOK (FOURTH EDITION, 2000)—SECTIONS AND TOPICS.

1 General 1. Project Management 2. Programme Management 3. Project Context **2 Strategic** 4. Project Success Criteria 5. Strategy / Project Management Plan 6. Value Management 7. Risk Management 8. Quality Management 9. Health, Safety and Environment **3 Control** 10. Work Content and Scope Management 11. Time Scheduling / Phasing 12. Resource Management 13. Budgeting and Cost Management 14. Change Control 15. Earned Value Management 16. Information Management **4 Technical** 17. Design, Implementation and Hand-Over Management 18. Requirements Management 19. Estimating 20. Technology Management 21. Value Engineering 22. Modelling and Testing 23. Configuration Management	**5 Commercial** 24. Business Case 25. Marketing and Sales 26. Financial Management 27. Procurement 28. Legal Awareness **6 Organisational** 29. Life Cycle Design and Management 30. Opportunity 31. Design and Development 32. Implementation 33. Hand-Over 34. (Post) Project Evaluation Review (O&M/ILS) 35. Organisation Structure 36. Organisational Roles **7 People** 37. Communication 38. Teamwork 39. Leadership 40. Conflict Management 41. Negotiation 42. People Management

Morris. The aims of the CRMP research (Morris et al., 2000; Morris, 2000) included the following:

- Identifying the topics that project management professionals consider need to be known and understood by anyone claiming to be competent in project management
- Defining what is meant by those topics at a generically useful level.
- Interviewing and collecting data in over 117 companies over a 14-month period

According to Morris (2000, p. 5), the CRMP research "endorsed the breadth of topics in the APM BoK [Third Edition]" and suggested "an even broader scope of topics". Morris (2000, p. 5) points out that the original APM BoK was "strongly influenced by research then being carried out into the issue of what it takes to deliver successful" projects, including work by Baker, Murphy, and Fisher (1988) and undoubtedly by Morris' own work on the subject (Morris and Hough, 1987). Morris et al. (2000) contend that the PMBOK® Guide does not cover all the factors that must be managed to deliver successful projects, claiming that there are other important issues, included in the Fourth Edition of the APM BoK, such as technology, design management, environmental, external, business, and commercial issues. The authors of the PMBOK® Guide would respond by referring to the disclaimer, printed in the Guide, that it is not intended to be either comprehensive or all-inclusive, but to present that subset of the PMBOK that is generally accepted (Project Management Institute, 2000).

The purpose of the Fourth Edition of the APM BoK (2000) has a similar clarity of definition. It claims to be "a practical document, defining the broad range of knowledge that the discipline of project management encompasses" representing those "topics in which practitioners and experts consider professionals in project management should be knowledgeable and competent" (Dixon, 2000, p. 9). It does, however, clearly state that it is not "a set of competencies: (p. 7), although it is suggested that it might be used, in organizations, as "the basis of the project management element of a general competencies framework" (p. 9).

Topics included are those considered to be potentially applicable to all project management situations. They are described at a high level, and readers are referred to texts and other sources for further detail.

A separate *Syllabus for the APMP Examination* (Hougham, 2000) has been published by APM, based on the APM BoK (Fourth Edition). It defines the topics that candidates for the APMP Examination (APM's baseline professional qualification) are expected to know, learning objectives, and a glossary of key terms.

Following the release of the Fourth Edition of the APM BoK, the APM has produced an edited book titled *Project Management Pathways* (Stevens, 2002), which they claim draws upon "current and forward thinking concerning the theory and practice involved in the art and science of project and programme management throughout the world." (APM, 2002, p. 2) and is required reading for the APMP Examination and for all levels of APM professional certification.

ICB (IPMA Competence Baseline). The IPMA (International Project Management Association) has developed the *ICB: IPMA Competence Baseline* (also referred to as the "Sunflower"), which it considers to be a global standard (Pannenbäcker et al., 2002). Work on the ICB was initiated in 1993 and a First Version, in English, French, and German, was presented in June 1998. The primary purpose of the ICB is to provide a basis for certification of project managers. Another important role of the ICB was to bring together or "harmonise" a number of European national project management body of knowledge documents.

The following national project management body of knowledge guides formed the basis of the ICB:

English. Project Management Body of Knowledge, APM, Version 3

German. The Swiss Beurteilungsstruktur, VZPM, Ausgabe 1.0 and The German PM-KANON, PM-Zert, Version 1.0

French. The French Matrice des Projects (AFITEP)

There are 28 core elements and 14 additional elements of project management knowledge and experience (a total of 42 elements) identified from an analysis of the four national documents (see Figure 46.5).

The 28 core elements are presented as a "sunflower" to overcome the difficulties of achieving agreement on a knowledge structure (see Figure 46.6). As Morris et al. (2000) point out, the project management profession has less difficulty agreeing on content of a body of knowledge than they do on a structure for that knowledge.

FIGURE 46.5. ICB: IPMA COMPETENCE BASELINE: CORE ELEMENTS OF PROJECT MANAGEMENT.

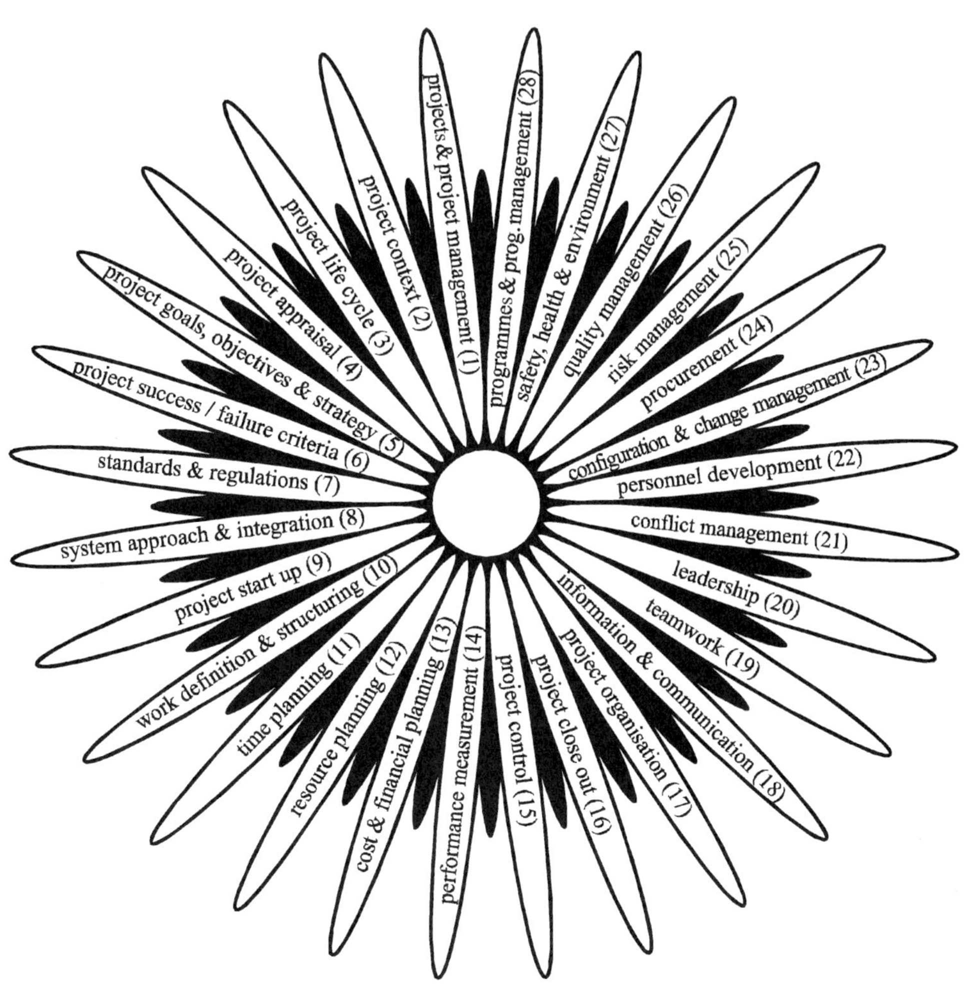

Source: Caupin et al., 1999.

FIGURE 46.6. ICB: IPMA COMPETENCE BASELINE: PERSONAL PROFILE.

Attitude	Inventiveness
Common sense	Prudent risk taker
Open-mindedness	Fairness
Adaptability	Commitment

The ICB also includes a section on the expected personality characteristics for a Certificated Project Manager: These are the same as appear in the APM Body of Knowledge (Version 3.0). They were developed in a series of practitioner workshops or meetings conducted by the APM. It is understood that they have no empirical basis.

The *ICB: IPMA Competence Baseline* is intended as the basis of the IPMA Validated Four-Level Certification Program (Pännenbacker, K et al., 1998).The IPMA is a federation of national project management professional associations, and it encourages the national associations to develop their own National Competence Baselines (NCB). The *IPMA Competence Baseline* (ICB) provides the "reference basis" (Pannenbäcker et al., 2002, p. 15) for National Competence Baselines. Each NCB is required, by IPMA, to include all 28 of the ICB core elements and at least 6 additional elements chosen by the nation plus the 8 aspects of personal attitudes and 10 elements of general impression. Up to eight of the additional elements can be eliminated or replaced by new elements, allowing each nation, in developing their NCB, to take into account local requirements or emerging developments in project management. The NCB is used as the basis for national certification programs validated by the IPMA Certification Validation Management Board.

At the end of 2001 there were 21 published NCBs and 5 in preparation. The majority of NCBs have been developed by European countries. An *Egyptian Competence Baseline* was published in 1999 and an Arabic version was in preparation in 2002. The Chinese NCB, also referred to as the C-PMBOK, was published in Standard Modern Chinese in 2001. An Indian NCB was in preparation in 2002.

P2M: A Guidebook for Project and Program Management for Enterprise Innovation. The P2M, *A Guidebook for Project and Program Management for Enterprise Innovation*, was released, in Japanese, in November 2001. An English-language summary version of the P2M (ENAA, 2002) is available covering the total of Parts 1, 2, and 3 and an overview of Part 4, which deals with 11 project segment areas. The Project Management Professionals Certification Center (PMCC) of Japan, a nonprofit organization established in May 2002, has published this summary version. The PMCC is responsible for maintenance of the P2M, for promotion of project management, and for the Certification System for Project Professionals, based on the P2M document.

The Japanese version of the P2M, published in November 2001, is a 420-page document developed following extensive worldwide research, over two and a half years, by the Innovative Project Management Development Committee of the Engineering Advancement Association (ENAA) under the leadership of Professor S. Ohara. The development of the P2M and the associated certification process received support from the Japanese government, primarily through the Ministry of Economics, Trade, and Industry (METI), in recognition that effective project and program management had potential to assist in revival of the Japanese economy.

The focus of the P2M is on "value creation to enterprises, either commercial or public, and a consistent chain from a mission, through strategies to embody the mission, a program(s) to implement strategies, to projects comprising a program" (ENAA, 2002, Preface). Therefore, the document has application to individual projects and also to programs of projects and the wider organizational context. The document is intended to provide a Capability Building Baseline (CBB) for project management and mission-performer professionals, where mission-performer professionals are described as "integration-oriented professionals who perceive complex problems and issues from a high perspective and realize right and optimal solutions" (p. 3). The authors of the P2M align it with PM knowledge guides and standards, claiming that although P2M is "considerably more extensive than existing project management bodies of knowledge or project management competency standards, it does not try to explore every detail of the topics discussed" (p. 2). Rather, it is intended that capability should be developed and expanded through professional experience, as well as related disciplines of science and technology in the context of continuing professional development.

The P2M uses what it calls a Project Management Tower to represent its coverage (see Figure 46.7). Part I: Entry describes how to make a first step as a professional; Part II: Project Management explains the basic definitions and framework of project management; Part III: Program Management introduces management of programs of multiple projects; Part IV: Project Segment Management, considered similar to "Knowledge Areas of Project Management," includes 11 "segments" or knowledge areas of project management that can be used in a "standalone or combined manner for individual tasks and challenges of project management and program management" (ENAA, 2002, p. 17).

People

All of the guides and standards linked in Figure 46.2 with 'People' are primarily concerned with management of individual or stand-alone projects, rather than programs of projects or enterprise project management. With the exception of the ICB (*IPMA Competence Baseline*), which is primarily a knowledge guide, they are generally in the form of performance-based competency standards. Performance-based competency standards are specifically designed for assessment and recognition of current competence, independent of how that competence has been achieved. They also encourage self-assessment, reflection, and personal development in order to provide evidence of competence against the specified performance criteria.

Standards referenced under People in Figure 46.2 that are used as a basis for assessment of competence of project management practitioners and are prepared and/or endorsed by national governments as follows:

FIGURE 46.7. P2M—PROJECT MANAGEMENT TOWER.

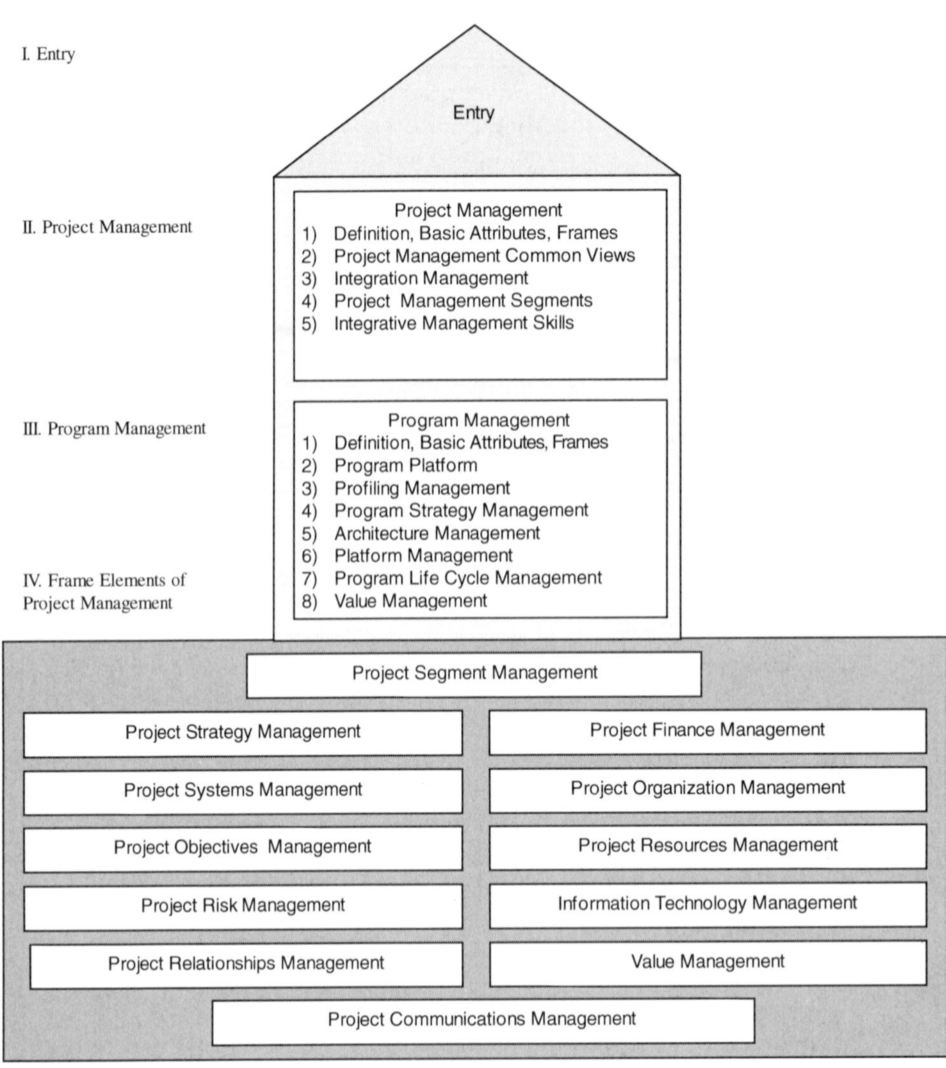

Source: ENAA, 2002.

- National Competency Standards for Project Management (NCSPM) (AIPM [sponsor], 1996) (Australian National Training Authority, or ANTA)
- National Occupational Standards for Project Management (ECITB, 2002) (Qualifications and Curriculum Authority, or QCA)
- National Certificate in Project Management—NQF Level 4 (SAQA) (South African Qualifications Authority, 2001)

All of these standards are formally recognized and provide the basis for award of qualifications within national qualifications frameworks. The National Competency Standards for Project Management also form the basis for award of professional qualifications by the Australian Institute of Project Management.

The Project Management Institute's Project Manager Competency Development Framework (PMI PMCDF; Project Management Institute, 2002c) is intended as a guide to self-assessment and development and is specifically *not* intended as a basis for award of qualifications. The purpose of this document is to describe the "competencies likely to lead to effective project manager performance across contexts," and it is intended for "use in professional development of project managers rather than for use in selection or performance evaluation." Part of the Framework, however, is in a form similar to the performance-based competency standards used by the government-endorsed standards of Australia, the United Kingdom, and South Africa.

The *ICB: IPMA Competence Baseline* (Caupin et al., 1999) is indicated as having application to people, as it is specifically intended to provide a basis for the International Project Management Association's (IPMA) certification program for project personnel. It purports to provide a basis for assessment of knowledge, experience, and personal attitude. The *IPMA Competence Baseline*, as a basis for assessment, includes a guide to what is considered to be the required knowledge and experience for an effective project manager, a taxonomy for identifying the extent of knowledge and experience required at each level of competence and a profile of the personality characteristics and attitudes that are expected in a project manager.

As the ICB has already been discussed under standards relating specifically to single projects, only the following people-focused standards will be discussed in further detail here.

Australian National Competency Standards for Project Management. The first performance-based competency standards for project management were the Australian National Competency Standards for Project Management that were developed through the efforts of the Australian Institute of Project Management and endorsed by the Australian Government on 1st July 1996.

At the commencement of 2003, there were 12 levels in the Australian Qualifications Framework (from school level to Doctorate), and the National Competency Standards for Project Management have been developed at Levels 4 (Certificate IV), 5 (Diploma), and 6 (Advanced Diploma). The standards have been adopted by the Australian Institute of Project Management for their 3 level professional registration process (see Figure 46.8).

The Australian National Competency Standards for Project Management were developed over a three-year period, commencing in 1993 and culminating in the endorsement

FIGURE 46.8. AIPM REGISTERED PROJECT MANAGER PROGRAM LEVELS.

AQF Level	AIPM Recognition Descriptors		Level of Experience Required for Assessment of Competence at This Level
Certificate IV	QPP	Qualified Project Professional	Specialist or Team Member
Diploma	RegPM	Registered Project Manager	Project Manager of well-defined or less complex projects Section Leader of complex projects
Advanced Diploma	MPD	Master Project Director	Project Manager of complex projects Project Manager or Director of multiple projects

of the standards by the Australian Government in 1996. Development was carried out by a consultant working under the guidance of a Steering Committee and Reference Group representing over 50 Australian organizations.

The standards development process is well documented (Gonczi et al., 1990; Heywood et al., 1992) and requires the examination of existing information about the occupation and "analysis of the purpose and functions of the profession and the roles and activities of its members" (Heywood et al., 1992, p. 46) in order to derive the units and elements of competency that provide the structure for the standards. In developing the standards, the team decided to follow the structure of the PMBOK® Guide and, at the same time, recognize the PMBOK® Guide as a contributing knowledge base to the standards.

There are nine units in the standards, described at Levels 4 (Certificate IV/QPP), 5 (Diploma/RegPM), and 6 (Advanced Diploma/MPD), as shown in Figure 46.9.

Level 4 of the standards does not include the unit relating to Integrative Processes. The only specific reference to personal characteristics or attributes in the standards is on page 16 of the Guidelines (AIPM [sponsor], 1996), which, in describing the nature of the project manager, suggests that "technical know-how alone is not sufficient to bring a project to successful completion" and "desirable attributes of a project manager include:

Leadership ability

The ability to anticipate problems

Operational flexibility

Ability to get things done

An ability to negotiate and persuade

An understanding of the environment within which the project is being managed

The ability to review, monitor, and control

The ability to manage within an environment of constant change

By following the structure of the PMBOK® Guide, which is by far the most widely distributed and recognized of the project management body of knowledge guides, the developers

FIGURE 46.9. UNITS IN THE AUSTRALIAN NATIONAL COMPETENCY STANDARDS FOR PROJECT MANAGEMENT.

	LEVEL 4	LEVEL 5	LEVEL 6
UNIT 1	Not applicable at Level 4	Guide Application of Project Integrative Processes	Manage Project Integration
UNIT 2	Apply Skills in Scope Management	Guide Application of Scope Management	Manage Scope
UNIT 3	Apply Skills in Time Management	Guide Application of Time Management	Manage Time
UNIT 4	Apply Skills in Cost Management	Guide Application of Cost Management	Manage Cost
UNIT 5	Apply Skills in Quality Management	Guide Application of Quality Management	Manage Quality
UNIT 6	Apply Skills in Human Resources Management	Guide Application of Human Resources Management	Manage Human Resources
UNIT 7	Apply Skills in Communications Management	Guide Application of Communications Management	Manage Communications
UNIT 8	Apply Skills in Risk Management	Guide Application of Risk Management	Manage Risk
UNIT 9	Apply Skills in Procurement Management	Guide Application of Procurement Management	Manage Procurement

of the Australian National Competency Standards for Project Management ensured that the standards would attract interest worldwide.

The Australian National Competency Standards for Project Management, as endorsed standards within the Australian Qualifications Framework, are the responsibility of Business Services Training Australia. With funding from the Australian National Training Authority (ANTA), Business Services Training Australia commenced a review of the Australian National Competency Standards for Project Management in 2001 and revised standards were finalized in mid-2004. Discussion has been based on the Australian National Competency Standards for Project Management as developed and endorsed in 1996. Information on the current status of these standards is available through the Australian Institute of Project Management Web Site (www.aipm.com.au).

Although the New Zealand Qualifications Authority (NZQA) has a qualification framework, they do not yet have project management standards, and it currently seems unlikely that they will develop their own standards. Mutual recognition arrangements between Australia and New Zealand are more likely to be utilized.

National Occupational Standards for Project Management (UK). In the United Kingdom, the Occupational Standards Council for Engineering produced standards for Project Controls (OSCEng, 1996), which were endorsed in December, 1996, and for Project Management (OSCEng, 1997), which were endorsed in early 1997. The Construction Industry Standing Conference (CISC), the Management Charter Initiative (MCI), and what was then called the Engineering Services Standing Conference (ESSC), now the Occupational Standards for Engineering (OSCEng), together developed Level 5 NVQ/SVQ competency standards for Construction Project Management. A section of the Management Charter Initiative Management Standards, titled "Manage Projects" (MCI, 1997), provided a further set of competency standards for project management, but in this case, within the general management framework.

As with the Australian National Competency Standards for Project Management, the OSCEng NVQ/SVQs in project management and project controls were developed and tested in conjunction with experienced project management practitioners (OSCEng, 1996, p. 3) from over 50 employer organizations. The standards claimed to be generic and applicable:

- *At Level 4.* To all those who take responsibility for managing projects at the operational level
- *At Level 5.* To those with a strategic role in project management

NVQ/SVQ Levels 4 and 5 do not equate directly to the AQF Levels 4 and 5, as there are 5 levels in the NVQ/SVQ Framework and 12 levels in the Australian Qualifications Framework. For general purposes of comparison, NVQ/SVQ Level 4 equates to AQF Level 5 (Diploma), and NVQ/SVQ Level 5 equates to AQF Level 6 (Advanced Diploma).

Review of the UK OSCEng Standards commenced in mid-2001, and the revised standards were endorsed by the regulatory authorities in August 2002 (ECITB, 2002). These reviewed standards are the responsibility of the ECITB (Engineering Construction Industry

Training Board) and are specifically intended as cross-industry standards. The standards have been written as 51 separate units of competence, each relating to a distinct functional area covering both strategic and operational project management functions. Revision of the standards is claimed to incorporate content of the 4th Edition of the APM Body of Knowledge (Dixon, 2000). Definition of terms within the document is taken from BS6079. Two project management qualifications have been designed based on the 51 units. The Level 4 qualification is essentially for operational project management, and the Level 5 qualification may be considered as strategic project management. Each of these qualifications comprises a total of 20 units that must be completed, of which 11 are mandatory and 9 can be drawn from a selection of options. This allows for variation in the context (range indicators) in which the profession/occupation is performed.

National Certificate in Generic Project Management (Project Administration and Co-ordination) at NQF Level 4 (PMSGB/SAQA). The South African NQF Level 4 standards were developed using a process similar to that followed in both Australia and the United Kingdom. This is the first set of standards and associated qualifications to be produced for project management in South Africa, and work is proceeding on development of standards at other levels. The NQF Level 4 standards are at approximately the same level as the Australian AQF Level 4 (Certificate IV).

The National Certificate in Generic Project Management (Project Administration and Co-ordination) at NQF Level 4 is available with electives offering specialization in one of the following:

- Supervising a project team of a developmental project to deliver project objectives
- Supervising a project team of a technical project to deliver project objectives
- Supervising a project team of a business project to deliver project objectives
- Supporting project environment and activities to deliver project objectives

The qualification is intended to provide recognition of basic project management skills in the execution of small simple projects or providing assistance to a project manager of large projects. The focus is primarily on skills as a project team member but includes working as a leader on a small project/subproject involving few resources and having a limited impact on stakeholders.

There are 15 core titles or units and 4 elective titles that relate to differing contexts or specializations as indicated previously (e.g., developmental or technical projects). The structure and number of units bears more similarity to the revised UK standards (ECITB, 2002) than to the Australian standards.

Organizations

The focus of attention in development of project management standards has been on the management of individual projects. Even those standards that are intended for assessment of the project management competence of individuals are concerned primarily with their ability to manage individual projects and to some extent, at the strategic level, consider the need to manage and report on multiple projects or programs of projects. Management of multiple projects, program management and aspects of enterprise management that foster the effective management of projects have not received the same level of attention as the management of single projects.

P2M: A Guidebook for Project and Program Management for Enterprise Innovation. As mentioned previously, the P2M (ENAA, 2002) can be linked to enterprise or organizational project management through its attention to integration across projects, management of programs of projects, and its declared focus on project and program management for enterprise innovation. The focus, however, is on the development of individuals rather organizational assessment and development.

OPM3. The OPM3, or Organizational Project Management Maturity Model, is a standards development project of the Project Management Institute that has been active since 1998 through a globally representative team of volunteers. The declared purpose of the OPM3™ project is "to develop a global standard for organizational project management," and the vision is "to create a widely and enthusiastically endorsed maturity model that is recognized worldwide as the standard for developing and assessing project management capabilities within any organization" (Project Management Institute, 2002b). At the end of 2003 the Project Management Institute released the OPM3™ as a book and interactive CD-ROM. The OPM3™ has three elements. A Knowledge element presents "Best Practices" and provides guidance on how to use the accompanying material and CD-ROM. An Assessment element is an interactive database on the CD-ROM that enables organizations to evaluate current practices and identify areas for improvement, for which guidance is provided in the third, Improvement, element (Project Management Institute, 2004).

During the long gestation period of the OPM3™ (1998 to 2004), a significant number of proprietary models designed for assessment of organizational project management capability and maturity were developed by commercial organizations, in many cases drawing on either or both of the PMBOK® Guide and the SEI Capability Maturity Model (CMM) (Software Engineering Institute, 1999). In their work on the OPM3™, the team reviewed more than 30 maturity models, including those that could be considered as derived from business excellence and quality models from the SEI's CMM, and others individually or corporately developed (Schlichter, 2002).

OGC PMMM. PMMM is the Project Management Maturity Model owned as a Crown Copyright product by the Office of Government Commerce (UK) (OGC, 2002a). The OGC PMMM was developed in response to requests from both public and private sector organizations for a benchmark or standard against which to assess and demonstrate their corporate project management capability. Its development base is similar to that of the OPM3™, drawing on the approach of the SEI Capability Maturity Model and recognized project management standards and bodies of knowledge as well as OGC experience in providing support for project management improvement. It is intended that organizations can seek assessment against the OGC PMMM by accredited assessors.

PRINCE2. PRINCE, which stands for "PRojects IN Controlled Environments," is a project management methodology or approach to the management of projects that was first developed in 1989 by the Central Computer and Telecommunications Agency (CCTA), now part of the UK Office of Government Commerce (OGC). When first developed, it was intended as a UK government standard for IT project management. PRINCE2 is a development of the original methodology that is intended as a generic approach applicable to management of all types of projects (Office of Government Commerce (OGC, 2002b). There are many project management methodologies commercially available, but PRINCE2 is in the public domain and was developed by a UK government agency, with the specific intention of providing a standard approach to management of projects in organizations. Training in the use of PRINCE2 is available through accredited training organizations, and there is a quality-assured process of assessment and certification in use of the methodology. Training and certification are available in many parts of the world.

PRINCE2 could have been included under "Projects," as it is essentially a methodology for use in management of individual projects. It has been included under the heading "Organizations" because it is a methodology for use within organizations for management of projects and because it relates the management of projects directly back to the business case or organizational justification for each project with a strong concern for project and corporate governance.

Managing Successful Programmes. Program (programme) management, defined by the CCTA as "the co-ordinated management of a portfolio of projects that change organizations to achieve benefits that are of strategic importance" (CCTA, 1999, p. 2) is considered under the heading of "Organizations," as its focus is beyond that of the single project. It is a term widely used in practice and has been the subject of a number of conference papers in recent years. However, program management has so far attracted very little interest in terms of development of guides or standards from professional associations. This may be due to an unstated assumption that project and program management are interchangeable terms or that the practices of project management are equally applicable to the management of programs of projects.

The UK Central Computer and Telecommunications Agency (CCTA), the developers of PRINCE, have provided a considerable amount of material on the management of programs, including "An Introduction to Programme Management" published in 1993 (CCTA, 1993). *Managing Successful Programmes*, now published by the Office of Government Commerce (OGC) (CCTA, 1999), describes the framework and strategies of program management and the delivery of business benefits from a set of related projects. Reference is made in the document to PRINCE2. Although *Managing Successful Programmes* is presented by its authors as a standard, it is not. It does not yet form the basis for an assessment process or qualification (although the OGC has developed a "Successful Delivery Skills Programme"; Office of Government Commerce, 2002b). Therefore, *Managing Successful Programmes* cannot really be considered as a formal standard, although it arguably provides a widely disseminated guideline for the management of programs.

Conclusion

Development of standards and guides for management of single projects and for assessment of the competence of individuals primarily relating to management of single projects has clearly attracted more attention and effort than the development of guides and standards for multiple project, program, organizational, or enterprise project management. It is easy to assume that this is merely a reflection of time and professional maturity. However, it may also be influenced by the nature of boundaries relating to professional territory. Effort to establish a professional territory for project management has understandably begun at the level of the single project, where territorial claims can be won without a great deal of competition, especially in an environment of change.

One way in which new professions are formed is by successfully claiming territory (knowledge and expertise) previously occupied by other professions (Abbott, 1988). Management of single projects by a new profession of project management is effectively an appropriation of territory previously held by a number of professions or disciplines including engineering, architecture, and general management. The cross-disciplinary nature of projects creates confusion between roles of traditional professions, enhancing the opportunities for project management to establish professional claims. As the concerns of project management extend upward in organizations toward programs and enterprise project management, competition is primarily with general management, making it increasingly difficult to define their professional boundaries, especially as general management, itself, is actively fighting for professional recognition.

This section has provided an overview of project management standards and guides relating to projects, people, and organizations. Underlying the development of these standards is agreement on what is rightly included in a project management body of knowledge. The following section focuses on project management knowledge standards, often referred to as project management bodies of knowledge, providing a review of coverage and intent. This is done on the basis that there is general agreement on what constitutes the body of project management knowledge, and that knowledge standards and guides focus on parts of that body of knowledge, each from a slightly different perspective.

Project Management Knowledge Standards

Project management knowledge standards such as the PMBOK Guide, ICB, APM BoK, and P2M identify what is considered as the minimum knowledge coverage required relative to the purpose of each standard. Although they differ in scope of topic coverage, depth of treatment, structure, format, and terminology used, they share a nucleus of core content. This shared coverage, although arranged in different structures in each of these knowledge guides, essentially represents the nine project management knowledge areas outlined in the PMBOK® Guide (Integration, Scope, Time, Cost, Quality, Human Resources, Communications, Risk, Procurement). Further, although the majority of content is common across all guides, there are areas that are not dealt with in some guides or where the degree of emphasis and detail of coverage varies considerably across the guides.

Much of the difference in coverage can be attributed to the difference of purpose of each of the guides, although they do share one purpose: to provide a basis for an assessment and certification program for project management practitioners. It is useful, in considering the difference in content of the guides, to recognize the difference in declared purpose of each of these guides. The purpose to some extent influences the length of the document. Both the APM BoK and the ICB are considerably shorter than the PMBOK Guide and the P2M, providing less detail as they include references to other documents. It is important to note that the content of the P2M, reviewed here, represents only part (Parts 1, 2, and 3, and an overview of Part 4 which deals with 11 project segment areas) of a much larger document. The number of words in each document is therefore included in Figure 46.10 as an indicator of the size and therefore the potential for detailed coverage. Both the APM BoK and the ICB are less than half the size of the PMBOK® Guide and the P2M in terms of length in words. The PMBOK® Guide is the most substantial by some margin, except that only the English Summary Translation of the P2M is included in this analysis. In its complete form (420 pages), the P2M would be significantly more substantial that the PMBOK® Guide.

Structure is another issue that needs to be considered in comparing the content of these four documents. Although the content of documents may be similar, it may appear to be different as a result of the way in which the content is structured. Issues of content are not limited to whether or not a topic is included or excluded but also the degree of emphasis given to particular topics in each document.

Content and coverage of project management knowledge guides/standards and performance-based competency standards is reviewed in the section titled *Content and Coverage of Knowledge Guides and Performance-Based Competency Standards* in this chapter.

Performance-Based Competency Standards for Project Management

Performance-based competency standards describe what people can be expected to do in their working roles, as well as the knowledge and understanding of their occupation that is needed to underpin these roles at a specific level of competence. A valuable aspect of such standards is that they are specifically designed for assessment purposes. At the same time they are developmental in their approach, with assessment being undertaken by registered Workplace Assessors, within a well-defined quality assurance process.

Such standards have been developed within the context of government-endorsed standards and qualifications frameworks in Australia (ANTA),[1] New Zealand (NZQA),[2] South Africa (SAQA),[3] and the United Kingdom (QCA).[4] Although there are some differences in the aims of these national frameworks, common themes are as follows:

[1] ANTA—Australian National Training Authority.
[2] NZQA—New Zealand Qualifications Authority.
[3] SAQA—South African Qualifications Authority.
[4] QCA—Qualifications and Curriculum Authority.

FIGURE 46.10. SUMMARY OF STATED PURPOSES AND WORD LENGTH OF KNOWLEDGE GUIDES REVIEWED.

Guide	Purpose	Approx. no of words
APMBoK	• To provide an overall guide to the topics that professionals in project management consider are essential for a suitable understanding of the discipline. • Scope includes not only specific project management topics, such as planning and control tools and techniques, but also broader topics found to have a significant influence on the success of projects such as social and environmental context, technology, economics and finance, organization, procurement, people, and general management • Topics are described at a high level of generality on the basis that detailed description of the topics can be sourced elsewhere. • Topics included are those that are considered generically applicable.	13,000
ICB	• Guide to knowledge, skills and personal attitudes expected in project managers and project personnel. • Contains basic terms, tasks, practices, skills, functions, management processes, methods, techniques, and tools that are commonly used in project mangement as will as specialist knowledge, innovative, and advanced practices used in more limited situations. • Not intended as a textbook or cookbook. • Based on four existing PM body of knowledge guides in three languages and presented in English, French, and German. • Intended as reference basis for National Competence Baselines of IPMA member countries and eferencing of project mangement competence in theory and practice (Caupin et al., 1999, p. 9).	10,000
PMBOK□ *Guide*	• To identify and describe that subset of the Project Management Body of Knowledge (PMBOK□) that is generally accepted (i.e., knowledge and practices applicable to most projects most of the time.) • To provide a common lexicon within the profession and practice, for talking or writing about project management. • To provide a basic reference that is neither comprehensive nor all-inclusive. Further sources of information are listed and application area extensions are either already available or in production.	56,000
P2M	• Specifically intended to encompass management not only of individual projects but of multiple projects, and programs of projects and the wider organizational context. • Intended as a guide. • Although P2M is *considerably more extensive than existing PMBoKs or PM competency standards, it does not try to explore every detail of the topics discussed* (ENAA, 2002, p. 2).	36,000

- To provide an integrated and consistent system for recognition of learning achievements and qualifications
- To facilitate access to and mobility of progression within education, training, and career paths
- To enhance the quality of education, training and outcomes in the form of employability and productivity
- To contribute to and encourage personal development and career progression through learning

The standards developed within these national frameworks in Australia, New Zealand, South Africa and the United Kingdom can therefore be used for many different purposes. Such purposes include both development and assessment, as well as the provision of a basis for recognition of current competence, regardless of how that competence has been achieved.

Performance-based inference of competence is concerned with demonstration of the ability to do something at a standard considered acceptable in the workplace, with an emphasis on threshold rather than high performance or differentiating competencies. Threshold competencies are units of behavior that are essential to do a job but that are not causally related to superior job performance (Boyatzis, 1982). "Performance based models are concerned with *results* (or "outcomes") in the workplace *rather than potential* competence as indicated by tests of attributes. Even when the underlying competence being tested is not itself readily observable—for example, the ability to solve problems—performance and results in the workplace are still observable and the underlying competence they reflect can be inferred readily. Performance-based models of competence should specify *what* people have to be able to do, the "*level of performance required* and *the circumstances in which that level of performance is to be demonstrated*" (Heywood et al., 1992, p. 23).

The definition of competency, within the context of performance-based or occupational competency standards, is considered as addressing two questions:

- What is usually done in the workplace in this particular occupation/profession/role?
- What standard of performance is normally required?

The answers to these questions are written in a particular format.

Units of Competency

Development of performance-based competency standards begins with an overview of the competency of the overall profession or occupation with an emphasis on the competency levels of particular interest. The overall competency of a profession or occupation is then subdivided into manageable components that are meaningful to practitioners and will be observable in the performance of individuals in the workplace. This first subdivision reflects significant functions of the profession and is generally referred to as a "unit." Each unit of competency describes a broad area of professional or occupational performance.

Elements or Specific Outcomes

As a unit is likely to be too large to be practically demonstrable or assessable for the purposes of recognition of competence of individuals in the workplace, units are usually further subdivided into what are referred to in the Australian system as *elements of competency* and in the South African system as *specific outcomes*. Elements of competency or specific outcomes constitute the building blocks of each *unit of competency*, describing in more detail what is expected to be done in the workplace for each unit.

Performance or Assessment Criteria

While units and elements or specific outcomes describe what is done in the workplace, *performance criteria* (Australia) and *assessment criteria* (South Africa) describe the standard of performance that is required. Performance criteria/assessment criteria specify the type of performance in the workplace that would constitute evidence that the required standard has been achieved. They describe what a competent practitioner would do, expressed in terms of observable results and/or behavior in the workplace. Performance or assessment criteria also specify the evidence, in the form of documentation, from which competent performance in an element of competency or specific outcome would be inferred.

Range Statements

Range Statements (Australia and South Africa) describe the circumstances or context in which competent performance is expected. They add definition to the unit by elaborating critical or significant aspects of the performance requirements of the unit. The Range Statement establishes the range of indicative meanings or applications of these requirements in different operating contexts and conditions. The term "scope" is used in the recently endorsed UK ECITB standards to refer to the same concept.

Underpinning Knowledge and Understanding (UKU)

The Australian standards recognize the need for performance to be underpinned by relevant knowledge and understanding (i.e., underpinning knowledge and understanding, often abbreviated to UKU). This is referred to in the South African standards as "embedded knowledge." The UK ECITB standards refer to both underpinning knowledge and understanding and "specific knowledge" required for each unit.

Key Competencies and Critical Cross-Field Outcomes

The South African standards include *critical cross-field outcomes*. These are outcomes that are useful for, and result from, all teaching and learning.

In the Australian context, standards include reference to *key competencies*, defined as generic skills or competencies considered essential for people to participate effectively in the workforce. Key competencies apply to work generally, rather than being specific to work in a particular occupation or industry. The Finn Report (1991) identified six key areas of competence that were subsequently developed by the Mayer committee (1992) into seven key competencies: collecting, analyzing, and organizing information; communicating ideas and information; planning and organizing activities; working with others and in teams; using mathematical ideas and techniques; solving problems; and using technology (see www.anta.gov.au/gloftol.asp).

Structure and Terminology

A summary of the equivalent structural units of each of the government-endorsed performance-based competency standards for project management introduced in section entitled *People*, earlier in this chapter is presented in Figure 46.11.

Levels

The concept of levels, relating to different roles and levels of responsibility in the workplace, as well as to different levels of academic or workplace achievement, is fundamental to government endorsed performance-based competency standards and qualifications frameworks. Competency Standards are written at a number of different levels corresponding to the demands of occupational roles and/or educational requirements. In National Qualification Frameworks, these levels start at the equivalent of secondary school and move through to postgraduate qualifications, reflecting roles from entry level to chief executive officer. As an example, the Australian National Competency Standards for Project Management have been written at three levels, generally corresponding to the following job roles:

Project Team Member

Project Manager

Project Director/Program Manager

FIGURE 46.11. COMPARATIVE STRUCTURE AND TERMINOLOGY OF GOVERNMENT-ENDORSED PERFORMANCE-BASED COMPETENCY STANDARDS FOR PROJECT MANAGEMENT.

NCSPM (Australia)	ECITB (UK)	PMSGB/SAQA (South Africa)
Unit		Unit
Element	Unit*	Specific outcome
Performance criteria		Assessment criteria
Range statement	Scope	Range statement
Underpinning knowledge and understanding (UKU)	Specific knowledge required for this unit	Embedded knowledge
Key competencies		Critical cross-field outcomes

Note: The level of detail of the units of the ECITB standards is nearer to that of the elements/specific outcomes in the other standards and guides.

Figure 46.12 provides a summary of the approximate equivalency of levels or roles addressed by standards introduced in the *People* section, earlier in this chapter.

Content and Coverage of Knowledge Guides and Performance-Based Competency Standards

Review and comparison of the content and coverage of knowledge guides and performance-based competency standards has been undertaken by the Project Management Research team at the University of Technology, Sydney, to provide the platform for initiatives aimed at development of a framework of Global Performance Based Standards for Project Management Personnel, described in the section of the same name, later in chapter.

In conducting this review and comparison, two main research approaches were taken. First, the content of the documents was mapped at the topic, heading, unit, element, or specific outcome level (high-level structure) to identify those concepts or topics that are covered in all or some of the documents. Second, detailed text analysis was carried out on the full text of the documents under review, using Wordsmith Tools Version 3.0 (Scott, 1999). In this way, it was possible to identify concepts/topics that were not identified at the topic level of knowledge standards or at the unit, element, or specific outcome level of the performance-based standards but were significantly represented throughout the text of the documents. Collocation of words was carefully reviewed to identify the use of words in context. Searches were made on words with similar meaning to counteract the possibility that a concept or topic would be identified as absent from a text because of differences in use of terminology.

FIGURE 46.12. EQUIVALENT LEVELS OF GOVERNMENT-ENDORSED PERFORMANCE-BASED COMPETENCY STANDARDS FOR PROJECT MANAGEMENT.

Standard / Guide	Project Team Member	Project Manager	Project / Program Director	Status
NCSPM (Australia)	AQF Level 4 (Graduate Certificate)	AQF Level 5 (Diploma)	AQF Level 6 (Advanced Diploma)	Govt.-endorsed standard
ECITB (UK)		NVQ Level 4	NVQ Level 5	Govt.-endorsed standard
PMSGB/SAQA (South Africa)	NQF Level 4			Govt.-endorsed standard

It is important to note that in any mapping exercise, it is necessary to select a "spine" to map against. The selection of any one of the existing standards as the "spine" for mapping purposes can be seen as privileging the selected standard. To overcome this difficulty, several mapping exercises were undertaken using the higher-level structure of the different documents as the spine. From this activity, lists of concepts or topics were derived. These lists of concepts or topics were then compared with a list of 44 topics compiled by Themistocleous and Wearne (2000) from the "various systems of Body of Knowledge 'elements' used by the APM, the International Project management Association (IPMA) and the US Project Management Institute (PMI), plus pilot testing on a steering committee" (p. 7). As a result of this comparison, some modifications were made to the concept/topic lists derived from the analysis of the knowledge and performance-based competency standards. These lists, and the Themistocleous and Wearne list, are shown in Figure 46.13. Some differences between the lists have been retained for specific reasons.

First, there were some concepts that were noticeably absent from the Themistocleous and Wearne (2000) list but present in the documents reviewed—or vice versa. Specifically, two additional concepts/topics have been included in the performance-based standards list but do not appear in either the Themistocleous and Wearne list or the knowledge guides. These are *Documentation Management* and *Reporting*. Although these might legitimately be categorized under *Information/Communication Management*, they have a level of significance in their own right in the performance-based standards that warranted their separate listing. For instance, while *Documentation Management* and *Reporting* are well represented in the PMSGB/SAQA Level 4 standards, other aspects of *Information/Communication Management*, which feature strongly in the other standards, are markedly absent. *Documentation* and *Reporting* are clearly more significant at the level of *practice* than they are as areas of *knowledge*, and it seemed appropriate that this should be highlighted.

Benefits Management was included because, although not represented in the Themistocleous and Wearne list and not well represented either at concept or text level in the government-endorsed performance-based competency standards reviewed, it is a potentially emerging concept and practice. Benefits management was identified as a topic in another research study I did (Crawford, 2002b), which included a wider range of standards and guides, including the OGC *Successful Delivery Skills Framework* (Office of Government Commerce, 2002), where it figures prominently.

Although *Configuration Management* and *Change Control* are included together in the Themistocleous and Wearne list, they are listed separately in the other lists. The characteristics of appearance of *Configuration Management* and *Change Control* are quite different in the performance-based standards. While *Change Control* is strongly represented in the high-level structures and throughout the text of the standards, *Configuration Management* receives very little mention in the government-endorsed performance-based standards. The term *Configuration* appears only three times, and then in only two of the performance-based competency standards reviewed. In the knowledge guides, *Configuration Management* is included at topic level in the APM BoK but is only mentioned once in the ICB and three times in the PMBOK Guide, although *Configurations and Changes* is a topic in the ICB.

Estimating is included as a separate item in the knowledge and performance-based standards lists, as it is quite strongly represented in these documents. Its application is not strictly

FIGURE 46.13. LISTS OF CONCEPTS/TOPICS AS PRESENTED BY THEMISTOCLEOUS AND WEARNE (2000) AND DERIVED FROM MAPPING AND TEXT ANALYSIS OF SELECTED KNOWLEDGE AND PERFORMANCE-BASED STANDARDS AND GUIDES.

	Themistocleous and Wearne (2000)	Knowledge Standards/ Guides	Performance-Based Standards/Guides	
1		Benefits management	Benefits management	1
2	Business need and case	Business case	Business case	2
3	Configuration management and change control	Change control	Change control	3
4		Configuration management	Configuration management	4
5	Conflict management	Conflict management	Conflict management	5
6	Cost management	Cost management	Cost management	6
7	Design management	Design management	Design management	7
8			Document management	8
9		Estimating	Estimating	9
10	Financial management	Financial management	Financial management	10
11	Goals, objectives, and strategies	Goals, objectives, and strategies	Goals, objectives, and strategies	11
12	Industrial relations			
13	Information management	Information/communication management	Information/communication management	12
14	Integrative management	Integration management	Integration management	13
15	Leadership	Leadership	Leadership	14
16	Legal awareness	Legal issues	Legal issues	15
17	Marketing and sales	Marketing	Marketing	16
18		Negotiation	Negotiation	17
19		Organizational learning	Organisational learning (inc. lessons)	18
20	Performance measurement	Performance measurement (inc. EVM)	Performance measurement (inc. EVM)	19
21	Personnel management	Personnel/human resource management	Personnel/human resource management	20
22	(Post-) Project evaluation review	(Post-) Project evaluation review	(Post-) Project evaluation review	21
23		Problem solving	Problem solving	22
24	Procurement	Procurement	Procurement	23
25	• Contract planning and administration			
26	• Purchasing			
27	Program management	Program/programme management	Program/programme management	24
28	Project appraisal	Project appraisal (Options/Modeling Investment/evaluation/analysis)	Project appraisal (Options/Modeling Investment/evaluation/analysis)	25
29	Project closeout	Project closeout/finalization	Project closeout/finalization	26
30	Project context	Project context/environment	Project context/environment	27
31	Project launch	Project initiation/start-up	Project initiation/start-up	28
32	Project life cycles	Project life cycle/project phases	Project life cycle/project phases	29
33	Project management			
34	Project management plan	Project planning	Project planning	30
35	Project monitoring and control	Project monitoring and control	Project monitoring and control	31
36	Project organization	Project organization	Project organization	32
37	Quality management	Quality management	Quality management	33
38		Regulations	Regulations	34

FIGURE 46.13. (*Continued*)

	Themistocleous and Wearne (2000)	Knowledge Standards/ Guides	Performance-Based Standards/Guides	
39			Reporting	35
40	Requirements management	Requirements management	Requirements management	36
41	Resources management	Resource management	Resource management	37
42	Risk management	Risk management	Risk management	38
43	Safety, health, and environment	Safety, health, and environment	Safety, health, and environment	39
44	Schedule management	Time management/scheduling/phasing	Time management/scheduling/phasing	40
45		Stakeholder/relationship management	Stakeholder/relationship management	41
46	Strategic implementation plan	Strategic alignment	Strategic alignment	42
47	Stress management			
48	Success criteria	Success	Success	43
49	Supply chain management			
50	Systems management			
51	Teamwork	Team building/development/teamwork	Team building/development/teamwork	44
52	Testing, commissioning, and handover/acceptance	Testing, commissioning, and handover/acceptance	Testing, commissioning, and handover/acceptance	45
53		Technology management	Technology management	46
54	Value improvement	Value management	Value management	47
55	Work management	Work content and scope management	Work content and scope management	48

limited to any of the other concepts such as Cost, Time, or Resources, and it was therefore considered inappropriate for it to be hidden within other concepts.

From analysis of the content of the documents reviewed, inclusion of separate topics of *Contract Planning and Administration* and *Purchasing*, used in the Themistocleous and Wearne list, did not appear necessary, as they were consistently represented in the context of procurement.

Knowledge Standards/Guides

In Figure 46.13, those concepts/topics that are shaded are those that are represented in the higher level structures of ALL of the standards reviewed. They may be considered CORE concepts or topics. Of these, *Project Environment/Context, Resource Management, Risk Management, Information/Communications Management*, and *Project Life Cycle/Project Phases* are the only topics that are present at this level in all four knowledge guides. *Scope, Time, Cost, Quality*, and *Procurement* are represented at the topic level in all knowledge guides except the P2M, where *Scope, Time, Cost* and *Quality* are included together in *Project Objectives Management* and *Procurement* within *Project Resources Management.*

Overall, topics considered core to the management of individual projects and that are extensively covered in the PMBOK® Guide receive little coverage in the Summary Translation of the P2M. The developers of the P2M gave full recognition to existing knowledge guides, and this may represent an acknowledgment that these topics are well covered in the PMBOK® Guide. They are also more fully covered in Part 4 of the P2M, which has not been translated into English and is not included in this analysis. In any case, the developers of the P2M specifically aimed to provide a knowledge guide that went beyond "delivery-focused traditional project management models and to develop a guide to allow the integration of project business strategy elements and utilization of valuable knowledge created through projects and programs" (ENAA, 2002, p. 1).

Programme/program management is understandably well covered in the P2M, for which the full title is *A Guidebook of Project and Program Management for Enterprise Innovation. Programme/program management* also receives topic-level coverage in the APM BoK but receives only passing mention in the PMBOK® Guide and ICB. Similarly, *Strategy* is covered at topic level in both the APM BoK and P2M, receiving only minor mention at text level in the PMBOK® Guide and ICB. The APM BoK is the only knowledge guide to specifically deal with the *Business Case.* Knowledge relating directly to interpersonal skills of project personnel, including *leadership, problem solving, conflict management,* and *negotiation* are only covered as discrete topics in the APM BoK and ICB.

Contextual issues such as *regulations, environment, health,* and *safety* are specifically covered in the APM BoK and ICB but receive very little mention, even in passing, in the PMBOK® Guide and P2M. Marketing is also covered, at topic level, in the APM BoK and ICB but is barely mentioned in the PMBOK® Guide and P2M. *Legal aspects* are covered at topic level in the APM BoK and ICB, and although not specified at topic level in the PMBOK® Guide and P2M, they are covered in the body of these texts.

The PMBOK® Guide is the only knowledge guide reviewed that mentions *life cycle costing,* while *modeling and testing* are only specifically included in the APM BoK. *Post project evaluation review* is mentioned only in the APM BoK and *reviews* in general receive little attention in any of the knowledge guides.

There are a number of aspects of management that are only covered at topic level in the APM BoK and receive little or no attention in the other knowledge guides. These include *Design, Technology* and *Requirements Management. Requirements Management.* The latter is particularly interesting, as this was one of the handful of topics on which there was less than 50 percent agreement concerning its importance as a knowledge area for competent project managers in the survey undertaken by the CRMP under the leadership of Professor Peter Morris in the review of the APM BoK (Morris et al., 2000). Morris expresses surprise that there is less than 50 percent agreement on the importance of *project success* criteria as a knowledge area for project personnel, yet success criteria are only covered at topic level in the APM BoK and ICB, and are not specifically dealt with in the PMBOK® Guide or P2M.

Value Management is covered only in the APM BoK and the P2M, with a single mention in the ICB and no mention at all even at detailed text level in the PMBOK® Guide, although *Value Engineering* is mentioned in the PMBOK® Guide.

Relationship Management is interesting, as it is a topic area in the P2M but does not appear in any form in the other three knowledge guides. This may be considered as another way of referring to *Stakeholder Management,* but although stakeholders are mentioned at text level in the other guides, and at topic level as Project Stakeholders in the PMBOK® Guide, *Stakeholder Management* is not specifically covered in any of the guides.

In summary, although the APM BoK and the ICB are the shorter of the four documents, they have the widest coverage. The APM BoK has slightly wider coverage than the ICB, but this is understandable, as the APM is a member of IPMA and the APM BoK is effectively the APM's National Competence Baseline (NCB). The PMBOK® Guide is very clearly focused on management of single projects, and the content of the P2M is directed more toward enterprise project management and the role of project management in value creation.

Performance-Based Competency Standards

A number of concepts/topics, not identified as core in Figure 46.13, that do not necessarily appear in the high-level structures and in some cases appear only in Range Statements, Underpinning Knowledge, and Understanding or other parts of the standards, are strongly or clearly represented at a text level in all the documents. These concepts/topics are shown in Figure 46.14.

Competency Models and Personal Competence

Performance-based competency standards represent an approach to competence that is strictly defined within the National Qualification Frameworks of countries such as Australia, New Zealand, South Africa, and the United Kingdom and is essentially concerned with threshold competence or minimum standards of performance required in the workplace. This approach to competence has attracted little interest in the United States, although it is worthwhile noting that Mexico has a National Qualifications Framework based on a performance-based approach to competence that includes a standard of competence for project management. When the term "competence" is used outside the context of National Qualifications Frameworks, it is often used in the sense of the Competency Model or Attribute-Based approach to competence, which is concerned with identification and definition of high-performing or differentiating competencies that contribute to superior performance.

The Competency Model approach is largely derived from the work of McClelland and McBer in the United States, beginning in the 1970s and reported by Boyatzis in the early 1980s (Boyatzis, 1982). Followers of this approach define a competency as an "underlying characteristic of an individual that is causally related to criterion-referenced effective and/or superior performance in a job or situation" (Spencer and Spencer, 1993). Five competency characteristics were defined by Spencer and Spencer (1993). Two of these competency characteristics—knowledge (the information a person has in specific content areas) and skill (the ability to perform a particular physical or mental task)—are considered to be surface competencies and the most readily developed and assessed through training and experience. Three core personality characteristics—motives, traits, and self-concept—are considered difficult to assess and develop.

FIGURE 46.14. CONCEPTS/TOPICS WELL REPRESENTED IN SOME DOCUMENTS ONLY AND/OR BELOW THE ELEMENT/SPECIFIC OUTCOME LEVEL.

Concept / Topic	Comment
Financial management	This concept is distinct from cost management. Cost management focuses on management of costs *within* the project, generally assuming that the necessary funds are available. Financial management is more concerned with interfaces between the project and its context and includes the processes involved in securing funds for the project and ensuring return on investment.
Goals, objectives and strategies	Although the term "goal" is rarely used, "objectives" is one of the most frequently occurring words across all four documents. These words are all used in the sense of goals, objectives, and strategies for the project (internally focused). Relationship to organizational strategy (externally focused) is treated as a separate concept (strategic alignment) and is far less strongly represented.
Legal issues	Legal issues are referred to in all documents but primarily in range statements and underpinning knowledge and understanding requirements.
Organizational learning	This includes the capture and sharing of knowledge between projects.
Performance measurement	Performance measurement appears in a number of ways including reference to 'assessment of performance' and is closely associated with organizational learning. Although earned value management falls within the ambit of performance measurement, it is only mentioned in the range statement at Level 6 of the Australian standards.
Personnel/human resource management	Human resource management is a unit in the Australian standards and is present as personnel management in the UK standards. It is outside the role addressed in the South African Level 4 standards.
Project appraisal	This is a concept applicable in various ways throughout a project and includes options, modeling, evaluation, and analysis, which are not represented in higher-level structures of the standards but appear in association with other concepts.
Requirements management	This is used primarily in the sense of information, procurement, quality, legal, resource, and other project requirements. In some cases (e.g., Australian standards AQF L6) it is used in the sense of quality and end product/stakeholder requirements. It can also be traced through the use of other words such as "needs."

The Competency Model approach, used extensively as the basis for numerous corporate competency development programs worldwide, sees competencies as clusters of knowledge, attitudes, skills, and in some cases personality traits, values and styles that affect an individual's ability to perform. While knowledge standards and performance-based standards as described in this chapter are available for project management, there are no such standards available for the behavioral competencies or personal competencies associated with project management.

A brief listing of attitudes and behaviors that are expected in a project manager are included in both the Association of Project Management Body of Knowledge (APM BoK) (Dixon, 2000) and the International Project Management Association's Competency Baseline (*ICB: IPMA Competence Baseline*) (Caupin et al., 1999) but do not constitute a standard. The Project Management Institute has included Personal Competencies in their Project Manager Competency Development Framework, but this document is very clearly intended as a guide to professional development rather than a standard that may be used as a basis for assessment.

Neither the listing of attitudes and behaviors in the APM BoK and the ICB, nor the Personal Competencies in the Project Management Institute's Project Manager Competency Development Framework, have a strong foundation in research. The Personal Competencies in the Project Manager Competency Development Framework are based on general management competencies presented in the work of Spencer and Spencer (1993). In some cases using methods similar to those presented in the work of Boyatzis (1982) and Spencer and Spencer (1993), a number of organizations have developed corporate competency models for project management, identifying the behaviors that are considered desirable and associated with superior performance within a specific corporate context. The NASA Competency Development Framework, incorporated in the NASA Project Management Development Process (PMDP) (www.nasaappl.com/ilearning/pmdp/pmdp.htm) includes both performance-based and behavioral or personal competencies and was developed by an expert group of NASA personnel.

Gadeken's work (Gadeken and Cullen, 1990; Gadeken, 1991; Gadeken, 1994) remains the most important work on behavioral competencies of project managers, but the results should be addressed with some caution because of the focus on both acquisition and the armed forces. Based on critical incident interviews with 60 US and 15 UK project managers from Army, Navy, and Air Force acquisition commands, the study identifies six behavioral competencies that distinguished outstanding program/project managers from their peers:

- Sense of mission
- Political awareness
- Relationship development
- Strategic influence
- Interpersonal assessment
- Action orientation

A further five behavioral competencies were demonstrated to distinguish outstanding program/project managers at a slightly lower level of significance:

- Assertiveness
- Critical inquiry
- Long-term perspective
- Focus on excellence
- Initiative

Project Management Qualifications

Each of the knowledge standards and guides and the performance-based competency standards reviewed previously are the basis for assessment for award of a project management qualification.

The APM *Body of Knowledge* and the *ICB: IPMA Competence Baseline* and associated National Competency Baselines (NCBs) form the basis for the IPMA's four-level certification program (Pännenbacker, et al., 1998). The entry level to this certification program, Level D, is a knowledge test based on the ICB or National Competency Baseline, which in the case of the United Kingdom is the APM Body of Knowledge.

The PMBOK® Guide is the project management knowledge guide of the Project Management Institute. The PMBOK® Guide is the basis for the Institute's single-level, Project Management Professional (PMP®) certification, which includes a multiple-choice knowledge exam plus project management experience. The PMI Certificate of Added Qualification (CAQ) recognizes knowledge, skills and experience in project management in specific industries. It is intended that Certificates of Added Qualification will be offered in a range of industries, with a CAQ relating to the automotive industry being the first to be made available (Project Management Institute, 2001).

The P2M is the basis for a three-level project management certification program that includes interviews, essay tests and project management experience. Qualifications are conferred by the Project Management Professionals Certification Center (PMCC), which was founded in April 2002.

The qualifications and assessment processes based on the knowledge standards and guides are summarized in Figure 46.15.

The performance-based competency standards discussed in the *Performance-Based Competency Standards for Project Management* section earlier in this chapter have been developed in the context of National Qualifications Frameworks and are therefore specifically designed to provide the basis for assessment and award of qualifications.

Assessment against government-endorsed performance-based competency standards is undertaken by Registered Workplace Assessors. Candidates are required to gather evidence of use of practices in accordance with Performance Criteria specified in the Competency Standards. It is part of the role of a Registered Workplace Assessor to work with candidates, advising and assisting them in achieving recognition of competence. Candidates are assessed either as competent at a particular level against the standards or "not yet competent." If assessed as competent against a set of performance-based competency standards at a particular level, a candidate may be awarded a qualification that is recognized within a government endorsed-qualifications framework.

FIGURE 46.15. EXAMPLES OF KNOWLEDGE-BASED PROJECT MANAGEMENT STANDARDS, ASSESSMENT PROCESSES, AND QUALIFICATIONS.

Standard or Guide	Level	Description	Form(s) of assessment
PMBOK® Guide *(Project Management Institute)*	PMP	Project Management Professional	• Multiple choice exam • Record of experience • Record of education
Industry specific extensions to the *PMBOK® Guide*	CAQ	Certificate of Added Qualification	• Must hold current PMP Certification • Record of industry-specific experience • Examination demonstrating industry-specific knowledge and skills
ICB: IPMA Competence Baseline and National Competence Baselines *(International Project Management Association, and member National Associations, e.g., AFITEP, APM)*	Level A	Programme or Projects Director	• Self-assessment, project proposal • Project report • Interview
	Level B	Project Manager	• Self-assessment, project proposal • Project report • Interview
	Level C	Project Management Professional	• Evidence of experience, self-assessment • Formal examination with direct questions and intellectual tasks • Interview
	Level D	Project Management Practitioner	• Formal examination, direct questions, and open essays
P2M *(Project Management Professionals Certification Center)*	PMA	Program Management Architect	• Interview and essay tests • Experience of at least three projects required
	PMR	Project Manager Registered	• Interview and essay tests • Experience of at least one project required
	PMS	Project Management Specialist	• Written examination

The Australian Institute of Project Management has a professional registration process that is aligned with the Australian National Competency Standards for Project Management and the Australian Qualifications Framework. Requirements for this registration process are available from the web site of the Australian Institute of Project Management (www.aipm.com.au). The equivalent project management role and professional and government recognized qualifications for Australia are shown in Figure 46.16.

In South Africa, the Project Management Standards Generating Body continues to work on development of standards for a range of project management roles. At time of writing, the only standards available for project management were those intended as the basis for award of a National Certificate in Generic Project Management (Project Administration and Co-ordination) at NQF Level 4.

The qualification is intended to provide recognition of basic project management skills in the execution of small, simple projects or providing assistance to a project manager of large projects. The focus is primarily on skills as a project team member but includes working

FIGURE 46.16. EQUIVALENT PROJECT MANAGEMENT ROLE AND PROFESSIONAL AND GOVERNMENT RECOGNIZED QUALIFICATIONS IN AUSTRALIA.

PM Role	Australian National Training Authority Qualification	Australian Institute of Project Management Award Title	Post Nominals
Project Team Member	Certificate IV	Qualified Project Professional	QPP
Project Manager	Diploma	Registered Project Manager	RegPM
Project Director/ Program Manager	Advanced Diploma	Master Project Director	MPD

as a leader on a small project/subproject involving few resources and having a limited impact on stakeholders.

The UK ECITB standards form the basis for project management qualifications recognized in the UK National Vocational Qualifications (NVQ) framework, namely:

- *At NVQ Level 4.* All those who take responsibility for managing projects at the operational level
- *At NVQ Level 5.* Those with a strategic role in project management

Demand for Global Project Management Standards

The possibility of achieving globally recognized project management standards and associated qualifications was a primary topic of interest amongst representatives of 29 countries at a Global Project Management Forum in October 1995 (Pennypacker, 1996). The idea of global standards and certification has continued as a topic for discussion at subsequent Global Forums held in association with major project management conferences, including the biannual World Congress of the International Project Management Association.

The International Project Management Association (IPMA) initiated a series of Global Working Parties at a meeting in East Horsley, England in February 1999. The working parties addressed Standards, Education, Certification, Accreditation/Credentialing, Research, and the continuation of the Global Forum process.

Following the inaugural meeting in February 1999, the Global Working Group: Standards subsequently met independently and in association with Global Forums, with participation by representatives of over 20 countries. The working group identified a number of global standards initiatives. One of these, aimed at development of a global body of project management knowledge, had already been commenced in late 1998, arising from an initiative of the Project Management Institute's Standards Committee. The other project, concerned with development of global performance standards for project management personnel (already mentioned in this chapter) was initiated by the Global Working Group: Standards. A further interest of the Global Working Group: Standards was the possibility of development of an ISO standard for project management, and this has been the subject of a watching brief rather than specific action.

The following are the primary reasons for interest in a global approach to the project management body of knowledge, standards and qualifications:

- Demand by corporations for standards and qualifications that are applicable throughout global operations as a basis for project management methodologies and for selection, development, and deployment of project management personnel
- Demand from practitioners for global recognition and transportability of professional and academic qualifications in project management
- Concern for international competitiveness by nations, corporations, and individuals
- Potential fragmentation of an emergent project management profession because of competition rather than cooperation in development and promotion of project management standards and qualification

Essentially, much energy and investment is wasted by individuals and organizations forced to make choices between competing project management standards and qualifications. A global approach would help to strengthen the image of project management and offer a more effective use of resources.

A brief review of the two primary initiatives relating to global approaches to the project management body of knowledge and standards will be presented here. These two initiatives are as follows:

- Towards a Global Body of Project Management Knowledge (OLCI)
- Global Performance-Based Standards for Project Management Personnel

Towards a Global Body of Project Management Knowledge (OLCI)

Definition of a body of project management knowledge is an essential step in development of a profession, and during the 1990s the key project management associations were all actively engaged in contributing to this development.

In a special edition of the *International Journal of Project Management*, focusing on project management bodies of knowledge, Wirth and Tryloff (1995) claimed that, at that time, at least ten national or international organizations appeared to be writing their own project management body of knowledge documents, while others were waiting to see what happened before attempting to write their own or adopting someone else's.

Although many separate efforts directed toward development of the project management body of knowledge were indicative of a very healthy and growing interest in project management, each organization, having developed its own guide to the body of project management knowledge, tended to become protective of its own content coverage, terminology, structure, and associated qualifications. The result was confusion in the minds of project management practitioners and their employers, particularly those involved in globally distributed operations when faced with decisions concerning which project management "body of knowledge" they should accept as a guide to their practice and which qualifications to support.

In October 1998, a small group of people actively involved in development and review of the PMBOK® Guide, the ICB, and the APM BoK took the initiative of holding an exploratory meeting to discuss ways in which they might work together in the interests of global cooperation. Those present at the meeting, although representing strong vested interests in existing project management body of knowledge guides, agreed to put aside those interests and work together to develop a global framework for project management knowledge. The initiative has come to be known as the OLCI or OLC Initiative, reflecting the initial name of the group that met in Long Beach, California, as the Operational Level Coordination Committee, a subcommittee, at that time, of the Project Management Institute's Standards Committee. The Project Management Institute's Standards Committee was disbanded at the end of 1998 because of a change of governance of the Institute. The Global Body of Project Management Knowledge initiative or OLCI has continued, but it has no formal status and has no affiliation with any project management association or other organization. All material produced by the group is in the public domain.

The work began at the first meeting in Long Beach, California, in October 1998 and was continued in a two-day workshop, hosted by NASA and attended by over 30 globally representative and recognized opinion leaders in Norfolk, Virginia, in June 1999 (Crawford and Pannenbäcker, 1999).

The group realized that they could not start with any of the existing documents but would have to start from a neutral base. Project management texts were nominated by the group in advance of the workshop and the indices of these texts scanned and analyzed. A resulting list of 700 words was reviewed and culled and then increased, at the workshop, to over 1,000 words. The initial list of 700 words and the results of two working parties that made a first attempt at structuring of the words are available in a full report on this workshop at www.aipm.com/OLC.

Prior to the workshop in June 1999, Max Wideman had begun a Glossary of Project Management terms and he has subsequently added the terms agreed at the June 1999 workshop as having a place in the project management body of knowledge. The Glossary continues to be updated and is available on the Web (www.pmforum.org).

Since the first two-day workshop in 1999, further meetings have been held annually, hosted by organizations that recognize the value of a global initiative to advance the development of the body of project management knowledge. The 2000 workshop was held in Norway and hosted by Telenor; the 2001 workshop was held in Lille, France, hosted by ESC Lille; and the 2002 workshop was held in Tokyo, hosted by the Project Management Professionals Certification Center (PMCC) and Japan Project Management Forum. A 2003 workshop is planned for Washington, D.C., hosted by NASA, and in 2004 the Project Management Research Committee has offered to host a workshop in China.

When this initiative was commenced in 1998, it is reasonable to assume that those involved generally envisaged that the outcome would be some form of document that represented a globally agreed body of project management knowledge. At the first two-day workshop in 1999 agreement was reached on the overall range of content of the body of project management knowledge. At this and subsequent workshops, there was realization that

- The existing body of knowledge documents (PMBOK® Guide, APM BoK, ICB, and more recently the P2M) are guides to specific parts of the overall body of project management knowledge.
- If a guide to the "global" body of knowledge were to be produced, it should not be in the form of a document but in a flexible and interactive form that enables stakeholders to view those parts of the project management of knowledge that are relevant to them at a point in time—a Web-based knowledge repository was proposed.
- Although structure is important, many structures are possible and different structures will address different needs and purposes.

The initial purpose of the OLCI, to work toward a global body of project management knowledge, has largely been achieved through shared recognition that the body of project management knowledge exists independently of the various guides representing views of parts of its content. An important role of the work of the OLCI has been to place each of the existing guides or standards in context, not as competing representations of the body of project management knowledge but as legitimately different and enriching interpretations of selected aspects of the same body of knowledge. The OLCI has therefore matured into a high-level and independent global "think-tank" and forum for advancement of the project management body of knowledge.

Global Performance-Based Standards for Project Management Personnel

A project for development of a framework of Global Performance-Based Standards for Project Management Personnel was initiated in mid-2000 at a meeting of representatives of project management and cost engineering professional associations, national standards and qualifications bodies, academic institutions and corporations. A history of this initiative is available at www.globalPMstandards.org.

The initiative responds to a growing recognition that standards for project management need to extend beyond knowledge to the application of that knowledge in the workplace. It provides an opportunity to respond to the needs of industry and project management practitioners interested in transferability of standards and associated qualifications across national boundaries. Governments are concerned with ensuring an internationally competitive workforce and in mutual recognition and transferability of qualifications.

The purpose of this initiative is therefore to develop an agreed framework for Global Performance Based Standards for Project Management Personnel that can be used by organizations, academic institutions, professional associations and government standards and qualifications bodies globally. It is proposed that these standards form a basis for review, development and recognition of local standards that map to a global framework.

In support of this initiative, the Project Management Research Team at the University of Technology, Sydney, conducted a detailed comparative review of the government-endorsed performance-based competency standards for project management of Australia, South Africa, and the United Kingdom, and of the Project Management Institute's Project Manager Competency Development Framework. The results of this research were referred to earlier in this chapter.

This research formed the background material for a working session of this initiative that was held in France in February 2003 and resulted in the development of a first draft of a framework of performance-based standards for the project manager role to which the content of existing standards can be mapped. This first draft was issued for wide stakeholder review in March 2003 and is available from the Global Performance Based Standards for Project Management Personnel Web site (www.globalPMstandards.org) (Global Performance Based Standards for Project Management Personnel). This report also indicates those individuals present and organizations represented at the Working Session.

Based on the 48 concepts identified in Figure 46.13, attendees at the Working Session identified 13 units of activities considered to be applicable to most project managers in most contexts and 4 units of activities considered to be applicable either in a role above that of project manager or only to some project managers in some contexts. The grouping of the 48 concepts into these units is presented in Figure 46.17. Keep in mind that what is presented here is a work in progress and may be expected to change significantly in response to review, stakeholder input, and further work. Further working sessions were held in Sydney, Australia in October, 2003 and in Cape Town, South Africa in May 2004. A further draft of performance-based standards has been developed, extended to cover two levels of project management activity. This and ongoing developments will be made progressively available on the Global Performance Based Standards for Project Management Personnel Web site. In the following Figure 46.17, Units considered applicable only to some project managers in some contexts are shown shaded.

SUMMARY

Standards development in project management over the last decade has made a significant contribution to codification of knowledge and practices as a means of establishing the territory of a project management profession, raising its profile, and making it accessible to a wide constituency. While the most widely distributed standards—the PMBOK® Guide, APM BoK, and ICB—have been instrumental in the growth of interest in project management, they can also be seen as limiting its development and sphere of influence. These standards, and current performance-based competency standards, are essentially reductionist and deterministic in their approach and focus on knowledge and practices applicable to management of single projects. Standards and guides for project management in organizations such as Japan's P2M, the emergent OPM3® and the work of the UK Office of Government Commerce (OGC) such as PRINCE2, *Managing Successful Programmes* and the *Successful Delivery Skills Framework* (OGC, 2002) are beginning to provide some guidance for

FIGURE 46.17. UNITS DEVELOPED FROM 48 CONCEPTS/TOPICS.

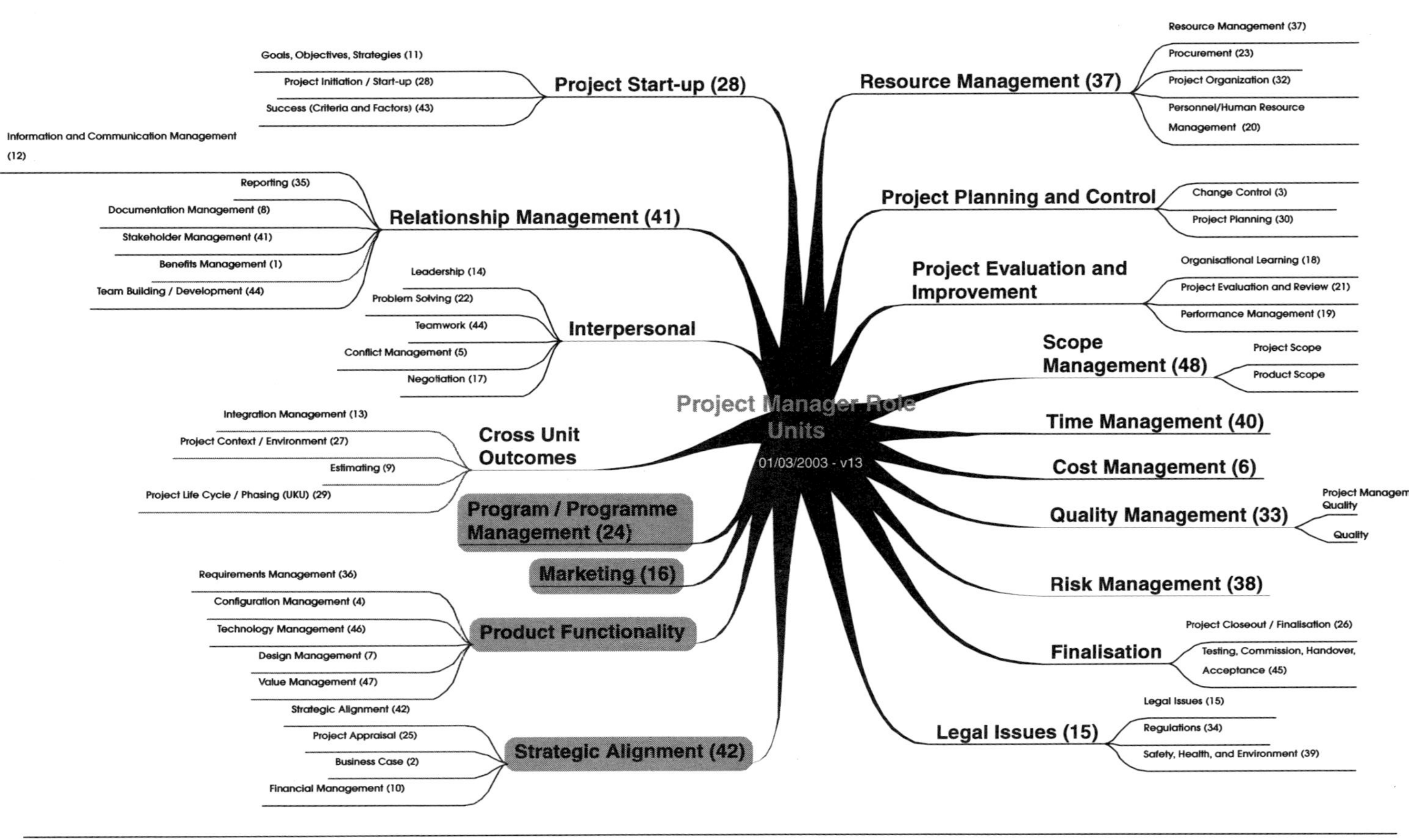

management of multiple projects and programs of projects, the role of the project sponsor, enterprise project management, and provision of corporate support for effective management of projects and programs. These standards and guides, however, enjoy less widespread recognition and appear to have less influence on professional formation than the knowledge and practice standards, largely focused on delivery of stand-alone projects, that form the basis for assessment and award of project management qualifications.

The most recent guide to the project management body of knowledge, Japan's P2M, is the most significant advance toward genuine integration and acceptance of the role of project and program management at the enterprise level. This is the first of the guides, or standards, intended as the basis for assessment and award of project management qualifications, that does the following:

- Develops an approach to enterprise project and program management that starts afresh from the viewpoint of the enterprise rather than drawing on project paradigms developed in the context of large, single, physical projects as the day-to-day business of project-based organizations.
- Directly addresses program management (rather than focusing only on single projects).
- Recognizes and responds to the complexities of fast-moving multistakeholder environments.
- Recognizes and addresses the systemic nature of projects and programs and specifically includes project systems management with reference to both hard and soft systems approaches.
- Addresses integration across programs and portfolios of projects at enterprise level (when other guides and standards mention "integration," in the P2M it is used in reference to integration *within* a single project).

Ensuring that project management develops as a field of practice and aspiring profession that has a strategic influence on the way organizations are managed and results are delivered is hampered by the image of project management that has been defined and presented by the project management knowledge standards and guides developed during the last decade. These standards and guides describe practices and a profession that is firmly lodged at the middle management level of organizations, and research (Crawford, 2002a; Crawford, 2002c) indicates that in the majority of organizations, senior management expects project managers to concentrate on what may be described as first-generation project management, managing efficient delivery of results within established time, cost, and quality restraints (Ohara, 2001, p. 4).

Results of two separate research studies indicate that on one hand less than 40 percent of senior managers surveyed considered strategy, systems, integration, and information management essential knowledge for project managers (Morris et al., 2000), and on the other, increasing levels of use of practices in these areas are associated with decreasing likelihood of being rated by senior management as a top project management performer (Crawford, 2001). This evidence suggests that general managers don't welcome encroachment by "project managers" on what they consider to be their territory. They expect those in project management roles to focus on delivery of their projects to specification, on time and within budget, which is hardly surprising, as this is the image of project management offered by the most widely distributed project management standards.

There is considerable rhetoric, however, promoting the potential contribution of project and program management to corporate strategy implementation (Pellegrinelli and Bowman, 1994), and there is increasing evidence to suggest that the project management approaches enshrined in current standards developed in the context of essentially 'hard' projects in the construction, engineering, defense, and aerospace industries are not uniformly successful when applied to 'soft' projects such as organizational change. Problems in applying hard project management practice to soft projects have prompted rethinking of standards and practices, with a number of writers and researchers turning to systems theory for possible enlightenment (Neal, 1995; Rodrigues and Bowers, 1996; Costello et al., 2002). There is potential for projects in complex, multistakeholder environments to be approached and managed as systemic interventions that engage stakeholders, enable environmental responsiveness, recognize the validity of different viewpoints, and facilitate organizational and individual learning.

Much of this is envisaged in the P2M, but it seems unlikely that it will be reflected in other knowledge and practice guides and standards in the near future. The process for standards development, which is largely a process of making explicit and codifying through consensus the tacit knowledge of experienced practitioners, ensures that standards will remain conservative and will lag behind the cutting edge of both research and practice.

References

Abbott, A. 1988. *The system of professions.* Chicago: The University of Chicago Press.

AIPM (Sponsor). 1996. *National competency standards for project management.* Sydney: Australian Institute of Project Management. www.aipm.com.au/html/ncspm.cfm.

APM. 1996. *Body of Knowledge* (*Version 3*). High Wycombe, UK: Association of Project Managers.

APM. 2002. APM's new book. www.apm.org.uk/pub/Pathways%20Flyer%20Final.pdf (accessed February 12, 2003).

Baker, B. N., D. C. Murphy, and D. Fisher. 1988. Factors affecting project success. In *Project management handbook,* 2nd ed., ed. D. J. Cleland, and W. R. King 902–919. New York: Van Nostrand Reinhold.

Berry, A., and K. Oakley, K. 1994. Consultancies: Agents of organizational development: Part II. *Leadership and Organization Development Journal* 15(1):13–21.

Boyatzis, R. E. 1982. *The competent manager: A model for effective performance.* New York: Wiley.

British Standards Board. 1996. *Guide to project management: BS6079: 1996,* London: British Standards Board.

Caupin, G., H. Knopfel, P. Morris, E. Motzel, E. and O. Pännenbacker. 1999. *ICB: IPMA Competence Baseline,* Version 2. Germany: International Project Management Association.

CCTA. 1993. *An introduction to programme management.* London: The Stationery Office.

CCTA. 1999. *Managing successful programmes.* London: The Stationery Office.

Costello, K., L. Crawford, L. Bentley, L., et al. 2002. Connecting soft systems thinking with project management practice: An organizational change case study. In *Systems theory and practice in the knowledge age,* ed. G. Ragsdell. New York: Kluwer Academic/Plenum Publishers.

Crawford, L., and Pännenbacker, O., eds. 1999. Towards a global body of project management knowledge. www.aipm.com/OLC/(accessed March 28, 2000).

Crawford, L. H. 2001. *Project management competence: The value of standards.* DBA thesis. Henley-on-Thames: Henley Management College/Brunel University.

———. 2002a. Senior management perceptions of project management competence. *Proceedings IRNOP V: Zeeland,* ed. J. R. Turner. Rotterdam: Erasmus University.

———. 2002b. *Project management qualification research study.* High Wycombe, UK: APM Group Limited.

———. 2002c. Profiling the competent project manager. In *The frontiers of project management research,* ed. D. P. Slevin, D. I. Cleland, and J. K. Pinto. Newtown Square, PA: Project Management Institute.

Dean, P. J. 1997. Examining the profession and the practice of business ethics. *Journal of Business Ethics* 16(15):1637–1649.

Dixon, M. 2000. *APM Project Management Body of Knowledge,* 4th ed. Peterborough, UK: Association for Project Management.

Duncan, W. R. 1998. Presentation to Council of Chapter Presidents. In *PMI Annual Symposium.* October 10, Long Beach, CA.

ECITB. 2002. *National occupational standards for project management: Pre-launch version September 200.,* Kings Langley: Engineering Construction Industry Training Board.

ENAA. 2002. *P2M: A guidebook of project and program management for enterprise innovation: Summary translation.* Revision 1. Tokyo: Project Management Professionals Certification Center (PMCC).

Eraut, M. 1994. *Developing professional knowledge and competence.* London: The Falmer Press.

Finn, B. 1991. *Young people's participation in post-compulsory education and training.* Canberra, Australia: AGPS.

Gadeken, D. O. C. 1991. Competencies of Project Managers in the NMOD Procurement Executive. Royal Military College of Science.

Gadeken, D. O. C., and B. J. Cullen. 1990. *A competency model of program managers in the DoD acquisition process.* Fort Belvoir, VA: Defense Systems Management College.

Gadeken, D. O. C. 1994. Project managers as leaders: Competencies of top performers. In *12th INTERNET (IPMA) World Congress on Project Management.* pp. 14–25. Oslo, Norway, IPMA.

Global Performance Based Standards for Project Management Personnel. Working Paper No. 1: Report from Working Session 24-26 February, 2003, Lille, France. Sydney: University of Technology, Sydney.

Gonczi, A., P. Hager, and L. Oliver. 1990. *Establishing competency standards in the professions.* Canberra, Australia: Australian Government Publishing Service.

Heywood, L., A. Gonczi, and P. Hager. 1992. *A guide to development of competency standards for professions.* Canberra, Australia: Australian Government Publishing Service.

Holtzman, J. 1999. Getting up to standard. *PM Network* 13(12)(December): 44-46

Hougham, M. 2000. *Syllabus for the APMP Examination.* 2nd ed. High Wycombe, UK: Association of Project Management.

IEEE. 2000. 1490-1998 *IEEE Guide to the Project Management Body of Knowledge.* Adoption of PMI Standard. http://standards.ieee.org/catalog/software2.html#1490-1998 (accessed April 30, 2000).

ISO 1997. *ISO 10006: 1997: Quality management: Guidelines to quality in project management.* Geneva: International Organization for Standardization.

Mayer Committee 1992. *Key competencies: Report of the Committee to advise the Australian Education Council and Ministers of Vocational Education, Employment and Training on employment-related key competencies for postcompulsory education and training.* Melbourne: Australian Education Council and Ministers of Vocational Education, Employment and Training.

MCI 1997. *Manage projects: Management standards—Key Role G.* London: Management Charter Initiative.

Morris, P. W. G. 2000. Benchmarking project management bodies of knowledge. In *IRNOP IV Conference—Paradoxes of Project Collaboration in the Global Economy: Interdependence, Complexity and Ambiguity,* ed. L. Crawford and C. F. Clarke. Sydney, Australia: University of Technology, Sydney.

———. 2001. Updating the project management bodies of knowledge. *Project Management Journal* 32(3): 21–30.

Morris, P. W. G., and G. H. Hough. 1987. *The anatomy of major projects*, Chichester, UK: Wiley.

Morris, P. W. G., M. B. Patel, and S. H. Wearne. 2000. Research into revising the APM project management body of knowledge. *International Journal of Project Management* 18(3):155–164.

Neal, R. A. 1995. Project definition: The soft-systems approach. *International Journal of Project Management* 13(1):5–9.

Office of Government Commerce (OGC). 2002a. Project Management Maturity Model (PMMM): OGC Release Version 5.0, London: The Stationery Office

———. 2002b. *Successful delivery skills framework, Version 1.0.* www.ogc.gov.uk.

Ohara, P.S. 2001. Project management and qualification system in Japan: Expectation for P2M and challenges. *Proceedings of the International Project Management Congress 2001: Project Management Development in the Asia-Pacific Region in the New Century.* November 16–21. Tokyo. Tokyo: Engineering Management Association of Japan (ENAA) and Japan Project Management Forum (JPMF).

OSCEng. 1996. *OSCEng Level 4: NVQ/SVQ in project controls.* London:: Occupational Standards Council for Engineering.

———. 1997. *OSCEng Levels 4 and 5: NVQ/SVQ in (generic) project management.* London: Occupational Standards Council for Engineering.

Pannenbäcker, K., H. Knopfel, and G. Caupin. 1998. *PMA and its validated four-level certification programmes.* Version 1.00. Nijkerk, Netherlands: International Project Management Association.

Pannenbäcker, O., H. Knoepfel, and J. Communier. 2002. *IPMA Certification Yearbook 2001.* Nijkerk, Netherlands: International Project Management Association.

Pellegrinelli, S., and Bowman, C. 1994. Implementing strategy through projects. *Long Range Planning* 27(4):125–132.

Pennypacker, J. S. 1996. *The Global Status of the Project Management Profession.* Newtown Square, PA: Project Management Institute.

Project Management Institute. 2000. *A guide to the Project Management Body of Knowledge..* Newtown Square, PA: Project Management Institute.

———. 2001.Certificate of Added Qualification. www.pmi.org/certification/CAQ/caq.htmwww.pmi.org/certification/CAQ/caq.htm accessed January 17, 2002.

———. 2002a. Introduction to the Project Management Institute (PMI). www.pmi.org/prod/groups/public/documents/info/ap_introoverview.aspgroups/public/documents/info/ap_introoverview.asp (accessed December 29, 2002).

———. 2002b. PMI Standards Open Working Session October 2002. www.pmi.org/info/PP_OWS02.pdf (accessed January 1, 2003).

———. 2002c. *Project Manager Competency Development Framework.* Newtown Square, PA: Project Management Institute.

———. 2004 Organizational Project Management Maturity Model (OPM3) www.pmi.org/info/PP_OPM3ExecGuide.pdf (accessed May 4, 2004].

Rodrigues, A., and J. Bowers. 1996. The role of system dynamics in project management. *International Journal of Project Management* 14(4):213–220.

Schlichter, John. 2002. Organizational Project Management Maturity Model: Emerging Standards. *www.pmi.org* (accessed January 1, 2002).

Scott, M. 1999. *Wordsmith Tools Version 3.* Oxford, UK: Oxford University Press.

Software Engineering Institute. 1999.SW-CMM Capability Maturity Model SM for Software. www.sei.cmu.edu/cmm/cmm.html (accessed February 14, 1999).

South African Qualifications Authority. 2001. General Notice No. 1206 of 2001: Notice of publication of unit standards-based qualifications for public comment: National Certificate in Project Management—NQF Level 4. *Government Gazette* 437 (22846, November 21).

Spencer, L. M. J., and S. M. Spencer. 1993. *Competence at work: Models for superior performance.* New York: Wiley.

Stevens, M. 2002. *Project management pathways.* High Wycombe, UK: Association for Project Management.

Stewart, T. A. 1995. Planning a career in a world without managers. *Fortune* 131 (5, March 20): 72–80.

Themistocleous, G., and S. H. Wearne. 2000. Project management topic coverage in journals. *International Journal of Project Management* 18(1):7–11.

Wideman, R. M. 1986. The PMBOK report. *Project Management Journal* 15 (Special Summer Issue): 102.

Wirth, I., and D. E. Tryloff. 1995a. Preliminary comparison of six efforts to document the project-management body of knowledge. *International Journal of Project Management* 13(2):109–118.

———. 1995b. Preliminary comparison of six efforts to document the project-management body of knowledge. *International Journal of Project Management* 13(2):109–118.

Zwerman, B., and J. Thomas. 2001. Barriers on the road to professionalization. *PM Network* 15 (4, April): 50–62.

CHAPTER FORTY-SEVEN

LESSONS LEARNED: PROJECT EVALUATION

J. Davidson Frame

The basic function of project evaluation is to engage in big-picture stock taking, where the most fundamental goals of a project are identified and the extent to which they are being achieved is determined. The implementation of effective evaluations on projects is important for at least three reasons:

- Evaluations force organizations to determine explicitly what it is that they are trying to achieve on their projects. That is, they require managers to identify the core objectives of projects (Locke and Latham, 1990).
- Evaluations supply feedback on project performance, enabling project staff to determine the degree to which the project is on target. This feedback may show that performance is on track (or even "ahead") or that performance objectives are not being attained. Without such information, managers have little or no idea of whether their projects are doing well or are headed toward failure. When evaluations are carried out this way, they are called *summative evaluations*: They summarize actual performance against established performance objectives (Farbey, 1999).
- Evaluations enable organizations to learn what works and what does not. Based on insights gained through evaluations, managers can adjust the organization's processes to improve organizational performance. Thus, evaluations are a core element of *organizational learning* (Symons, 1990). Evaluations that are carried out with a view of providing guidance on future behavior are called *formative evaluations* (Walsham, 1999; Remenyi, 1997).

This chapter focuses on post-project evaluations that are carried out as part of the project closeout process. It examines the function of post-project evaluation, identifies where it fits in the larger closeout effort, and describes how it can be conducted.

The Process of Project Evaluation

Post-project evaluations do not occur in a vacuum. They are in fact the final evaluative action in a process that commences before a project is even selected. There are three distinct types of evaluation that are carried out on projects: pre-project evaluation, mid-project evaluation, and post-project evaluation. (Farbey, et al., 1999). Taken together, they constitute overall project evaluation. The relationships among them are pictured in Figure 47.1.

Pre-project evaluation is also called *project screening, project selection,* and *project appraisal.* With pre-project evaluation, potential candidates for support are examined in respect to project selection criteria (Buss, 1984). Typical criteria include the following:

- *Financial.* What will the project cost? What are the anticipated financial returns?
- *Technical.* What technologies are needed to carry out the project? What technologies will the project create? Do you have the technical capabilities to carry out the project? Will the project enable you to strengthen your technical capabilities?
- *Marketing.* How can you sell your project idea to potential customers? What competitors do you need to contend with? How does the project relate to marketing's Four Ps (i.e., product, price, place, and promotion)?
- *Operational.* What demands will the project place on your operations? Do you have the technical and administrative infrastructure to implement the project smoothly?

FIGURE 47.1. THE PROCESS OF PROJECT EVALUATION.

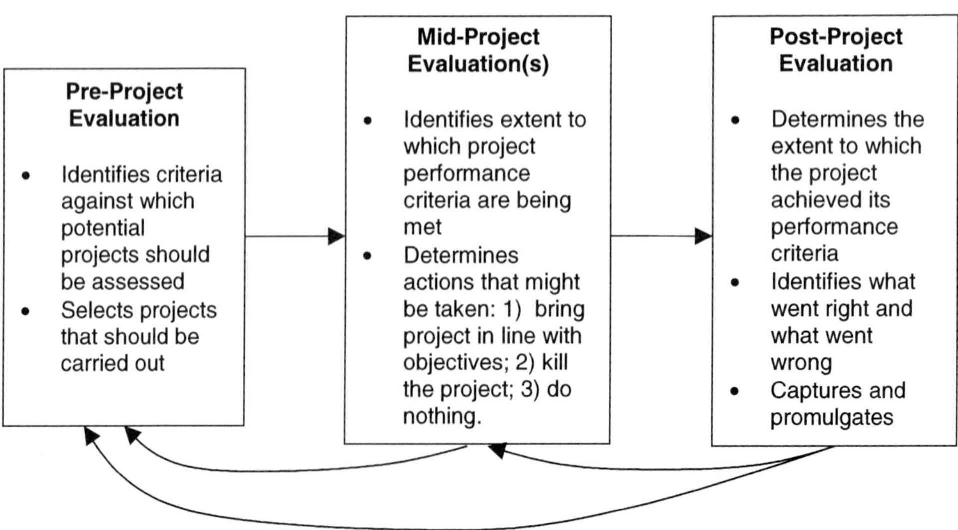

- *Strategic.* To what extent does the project address and support the organization's strategic goals? Does it complement the existing portfolio of projects?
- *Corporate culture.* Does the project align with the organization's overarching culture and goals?

In addressing the questions raised here, performance standards for the project effort emerge. For example, a preliminary budget of $2.3 million may be established for the project. Estimated revenues of $2.6 million may be projected. The need to hire two software designers to serve the project team may be identified. An advertising budget of $55 thousand may be established. Ultimately, these performance standards serve as the basis for determining whether or not the project effort is on target. Note that to answer basic questions raised at the project screening phase, analysts need good forecasting and estimation skills (Frame, 2002). If the projections are off target, then the wrong projects may be chosen.

The importance of effective pre-project evaluation is enormous. When due diligence is not followed in selecting projects, an organization will find that it has committed its resources to pursuing a loser. In this case, project failure has been hardwired into the project before any work has begun.

Mid-project evaluation is carried out periodically during the life of the project. (See also the chapter by Huemann.) Specific evaluative efforts should be established as milestones in the project schedule, requiring the project team to review performance in accordance with an established evaluation schedule. For example, the project plan may suggest that the preliminary design of a database system should be completed by June 12. A formal evaluation of the design effort can be scheduled to be conducted on June 13 and would include a review of the achievement of both technical and business objectives.

Mid-project evaluations address a number of standard questions, including the following:

- Is the project achieving its objectives as planned?
- If not, to what extent is it missing the objectives?
- Are the objectives still worth pursuing, knowing what you know today?
- If the project is not achieving its objectives, should you take corrective action to get it back on track?
- If you decide to bring the project back on track, what actions do you need to take?

A review of these questions shows that mid-project evaluations serve a cybernetic function: They seek *feedback* information for purposes of project *control.* (Harpum touches on this too in his chapter on project control.)

Mid-project evaluations assume a number of different forms, including the following:

- *Technical evaluations.* These evaluations address the technical performance of the deliverable: Is it meeting the specifications? Technical evaluations usually occur concurrently with system tests in order to assess whether the deliverable is achieving prescribed requirements. On software projects, a popular form of technical evaluation is the structured

walk-through. With structured walk-throughs, the project team members step through their work before a panel of outside evaluators whom they have chosen. Because project team members control the whole evaluation process with structured walk-throughs, the level of threat associated with the evaluation is reduced and the team members have little reason to fear surfacing the problems they are encountering. Other commonly encountered technical evaluations include preliminary design reviews and critical design reviews. (See Harpum's chapter on design management.)
- *Performance appraisal reviews.* These evaluations examine the performance of individual employees at predetermined times (e.g., every six months). The principal question they address is this: Are they achieving their performance goals effectively? Individual performance is measured against performance goals established earlier in the project life cycle.
- *Audits.* Mid-project audits are surprise evaluations conducted with little or no advanced warning. Their objective is to see how the project team is functioning at a given moment in time. (Again, see Huemann's chapter.) Through audits, evaluators can see how team members "really" function when they do not expect to be inspected. One reason audits are carried out is to keep project team members on their toes.
- *Managerial reviews.* One of the best-known evaluation methodologies is called *management by objectives* (MBO). Its principal proponent back in the 1950s was the management guru Peter Drucker (Drucker, 1985). With MBO, project teams and their managers negotiate objectives that the team members should achieve at different points in time. Once these points of time are reached, the performance of the team is reviewed to see the extent to which the objectives have been achieved. Unlike audits, MBO reviews eschew surprises—the team members know far in advance what is expected of them during the evaluative review.

Post-project evaluations are conducted after the project effort is completed. The objectives of post-project evaluation are different from those of mid-project evaluation. Clearly, once the project is done, the issue of taking action to get the project back on track is moot. The concerns at this time are as follows:

- At the end of the day, did you do what you said you would do, and did you achieve what you set out to achieve? (Morgan and Tang, 1993).
- What lessons can you learn from the project experience?
- What good practices and results did you encounter that you should attempt to replicate on future projects?
- What troubles did you encounter that you should strive to avoid in the future?

Given the results of the post-project evaluation, managers may determine that organizational processes need to be changed in order to build on strengths and avoid problems in the future. A significant challenge of post-project evaluation is to avoid the "pitfalls of hindsight" (Fox, 1984). That is, for the evaluation to meaningful, it must reconstruct the conditions facing the project as it was implemented in order to avoid critiquing the project on abstract principles that are not linked to reality.

Basic Principles Governing Evaluative Efforts

Effective mid-project and post-project evaluations strive to follow four basic principles: objectivity, internal consistency, replicability, and fairness. Each of these will be discussed briefly.

Objectivity

Evaluations must strive to be as objective as possible. Traditionally, the need for objectivity has been addressed by having evaluators come from outside the group being evaluated. The theory is that by using outsiders, you can avoid conflicts of interest that might arise when you have team members evaluate their own efforts. In other words, you strive to avoid having foxes guard the chicken coop.

While the rationale for objectivity may be solid, experience shows that in practice, relying on outsiders to conduct evaluations can lead to problems. Teams being evaluated under such circumstances often express serious concerns about the process, and their concerns have merit. Commonly encountered problems include the following (Frame, 2002):

- *The outside evaluators are unfamiliar with the circumstances the project team is addressing in its work.* The evaluators may not fully understand where the project team stands technically, what specific instructions it has received from its clients/bosses, how the organization works, what impediments team members have encountered while carrying out their tasks, and so on. Certainly, the evaluators can get up to speed on many of these details, but the effort can be quite disruptive. As the project team is responding to the information-gathering queries of the evaluators, project work may grind to a halt, jeopardizing the team's ability to achieve its goals. This may cause team members to feel hostile toward the evaluative effort.
- *The project team may be suspicious about who selected the evaluators and what instructions the evaluators have received.* Although the point of using outsiders is to maintain objectivity in the evaluation process, it is clear that the process is still susceptible to subjective influences. For example, if the outside evaluators hold a particular ideological perspective, their objectivity is questionable, particularly if their ideology runs counter to the perspective followed by the project at its outset. The assumption of objectivity of the outsiders is also doubtful if their marching orders have been given to them by an executive who is antipathetic to the project's raison d'etre. Project teams understandably want answers to the following questions: Who chose the evaluators? What instructions were they given? What obvious biases—if any—do they hold?
- *The outside evaluators feel compelled to find problems.* An important objective of evaluation is to uncover problems. The earlier problems can be identified, the easier it is to deal with them. If problems are surfaced after they have had time to fester, they can be enormously disruptive. Consequently, you carry out evaluations to discover problems in their infancy, so you can fix them as soon as possible. There is an important point here: Evaluations seek out problems so they can be fixed. Their fundamental rationale should *not* be to identify guilty parties who should be punished for their failings. As soon as evaluations

become associated with punishment, they become exercises in team demoralization and discourage honest reporting of problems. Evaluators should certainly be on the lookout for problems. That is a large part of their job. But they should guard against behaving like traffic police who have a quota of tickets to issue each day.

- *The outside evaluators are not competent.* It occasionally happens that the outside evaluators are not competent to perform their tasks. They may possess skills in the wrong areas or simply do not know what they are doing or what they are talking about. This can be a frightening situation for the project team that is being evaluated, because team members recognize their future lies in the hands of unqualified people.

In recent years, the view that employment of a team of outsiders satisfies the principle of objectivity has become passé. The pitfalls of using "objective" evaluation teams are well recognized, and consequently a number of approaches have arisen to rectify their inadequacies. Two approaches stand out. One is the employment of structured walk-throughs. The other is use of the EISA approach (defined later in the chapter). Key features of these approaches are discussed later.

Internal Consistency

Effective evaluations are conducted in a systematic, logical way. Procedures must be established and followed. Conclusions must map closely to the facts.

Without the employment of internally consistent evaluation procedures, the results of the evaluative effort can be viewed to be arbitrary. This is the predictable result of ad hoc evaluations, where evaluation team members make up the rules as they go along. For example, in tracking budget performance during an evaluation, the evaluation team members may use readily available financial data that, unbeknownst to them, only examines direct costs and leaves out indirect costs. From their review of the incomplete data, they conclude that the project is on target from the perspective of budget performance, when in fact it is experiencing a serious cost overrun. An evaluation team that employs a well-developed evaluation process would not fall into this trap. They would specify what kind of budget data they need to review and would conduct analyses *only* with the proper figures.

Replicability

Ultimately, employment of a consistent, systematic evaluation procedure contributes to the *replicability* of results. A fundamental principle of good science is that results that are achieved through scientific inquiry must be replicable (Garfield, 1987; Merton, 1996). If a scientist makes a seemingly great breakthrough, yet no one can replicate the findings, then the scientific community rejects them (Kuhn, 1996). Results are held to be reliable only after they can be replicated by others.

The same basic principle applies in the arena of evaluation. A properly conducted evaluation is one where the results of the evaluative effort are the same, whether the evaluation is conducted by outside Team A or Team B or Team C. If each evaluation team comes up with dramatically different conclusions, then the evaluative effort is flawed. The people being evaluated justifiably feel that the evaluation results they experience are tied to the luck of the draw—for example, being evaluated by a friendly group rather than an unfriendly one, or by a disciplined group rather than one that follows ad hoc procedures. Project teams that receive poor evaluations owing to the poor conduct of an evaluative effort can become demoralized when they question the fairness of the overall process.

Fairness

In general, healthy people who are in touch with their capabilities have a good sense of when they are doing a good job or bad job. If they are doing a good job, they expect to be praised for their efforts. If they are doing a poor job, they are prepared to experience a measure of criticism. The important thing is for them to believe that the evaluations they experience are conducted fairly. If they feel that they have been unfairly criticized—and if this criticism jeopardizes their job security or bonus status—it can cause them to be very unhappy about their jobs (Adams, 1963; Vroom, 1964).

Being fair can be tricky. For example, if a team is unable to achieve its performance goals on time, within budget, and according to specifications, it appears at first blush that its members deserve a low score on their evaluations. However, it may be that this team inherited a loser project—one that was underresourced and where unrealistic promises were made to the client. In fact, through their heroic efforts, the team may keep cost overruns and schedule slippages modest, as opposed to disastrous. Yet the fact remains that the team has encountered cost overruns and schedule slippages, and there is a danger that they may be punished for this. If they are punished, they may feel justifiable anger at the unfairness of the verdict.

Converting Lessons Learned into Action

It is not enough to go through the motions of conducting a lessons-learned exercise. When lessons are gathered and documented, there is a danger that they will be put on a shelf where they remain unread and gather dust. For the lessons to be truly learned, they must be formulated in such a way that ultimately leads to action. There are a number of approaches people take to promulgate lessons, several of which are described here (Frame, 2002):

- *Share lessons in informal meetings.* In some organizations, project staff meet informally from time to time to exchange project experiences. For example, employees in some companies hold monthly brown-bag lunch meetings where one or two attendees describe recent project experiences. All the people who are present at this meeting are invited to explore what went right and what went wrong on the described projects.
- *Maintain a case study library of project experiences.* Some organizations maintain files describing corporate project management experiences in a case study format. Each case provides a history of a project, from inception to closeout. It focuses attention on special issues and

challenges that arose during the life of the project and the project team's responses to them. It also contains a lessons learned section at the end of the case. A library of such cases can provide employees with valuable insights into the organization's project experiences, enabling them to develop realistic expectations about what they might encounter on future projects and possibly suggesting steps they can take to avoid problems.

- *Change organizational procedures to reflect lessons learned.* The surest way to make certain that the lessons learned are converted into action is to have them incorporated into organizational procedures. For example, a review of problems associated with the execution of a project might lead to the conclusion that a major source of friction in dealing with clients is that project staff take too long to respond to their queries. The conclusion derived from this review might be that project staff must deal with customer queries as quickly as possible. Procedures might be adopted requiring staff to touch base with clients who have questions before the close of the business day in which the query was generated. If this approach is taken, it is important to make sure that the list of procedures is kept lean—each time a new procedure is added, old ones should be examined with a view of throwing out procedures that no longer provide value.
- *Employ captured performance data to establish baseline measures.* A leading cause of struggles on projects is the absence of good estimates that form the basis of project plans. Often, cost, schedule, and other performance estimates tend to be optimistic in order to gain support to move ahead with a project. This is particularly true when trying to win a contract award. Sales staff may promise potential clients that their organizations can deliver incredible deliverables at bargain prices and according to phenomenal schedules. Once a contract is awarded, however, performance often falls far short of the promises, since the promises were based on wishful thinking and not on fact. A factual basis for making estimates can be created if actual performance data is used to develop realistic baseline performance measures. For example, if experience shows that it takes 2.5 days to install a piece of equipment, and that the cost of installation is typically $1,100, then these figures can be employed for schedule and cost-estimating purposes. The trick here is to establish and *enforce* procedures for archiving cost, schedule, resource, and performance experiences.
- *Employ information from the lessons-learned analyses into risk assessments.* In conducting lessons-learned exercises, you will find that most lessons you surface are mundane. For example, you may determine that in order to get reports to clients quickly and reliably, you should ship them using commercial overnight delivery services instead of the national postal service. Or you may find that copies of all correspondence sent to clients should be cc'd to the project manager. Occasionally, you come across a high-impact lesson, and it may be appropriate to embed the lesson into your organization's risk assessment efforts. For example, you may discover that the technical team rarely implements suggestions that arise during the critical design review and that this is a major source of customer unhappiness. Your risk assessment process can be adjusted to audit team performance a week after the critical design review sessions to make sure team members are following up on suggestions.

Conducting Friendly Post-Project Evaluations

Earlier, it was stated that effective evaluations follow four basic principles: objectivity, internal consistency, replicability, and fairness. It was also pointed out that it is often difficult to follow these four principles. For example, in the search for objectivity, you may hire outside evaluators who are ignorant of conditions the project team is facing. Or in a drive to be consistent in your evaluative efforts, you may be unfair in the verdicts you deliver to project teams because you did not take into account the extenuating circumstances facing the team members. To the extent that evaluations are perceived to be unfair and threatening by the people who are being evaluated, they will likely be less than honest in working with the evaluation team. Without honest feedback from the people being evaluated, the whole evaluation effort becomes suspect.

Two "friendly" approaches to evaluation have emerged over the years. What makes them friendly is that the people being evaluated are given a measure of control over the process. Consequently, they develop a sense of trust in the evaluative effort. One approach, developed by IBM in the 1960s, is called the *structured walk-through.* The second is associated with evaluative assessments for such initiatives as ISO 9000 and the Software Engineering Institute's Capability Maturity Model and is called the *EISA approach* (Wilson and Pearson, 1995). Each approach will be discussed in turn.

Structured Walk-Throughs

As mentioned, the structured walk-through approach to evaluation was developed and promoted by IBM in the 1960s (Yourdon, 1988). Managers at IBM recognized that when evaluations are perceived to be unfriendly, it is difficult to gain cooperation from the people being evaluated. So in order to make evaluations friendlier, the people being evaluated should be empowered to run the evaluation effort.

Originally, structured walk-throughs were employed for the purpose of reviewing software code that was being developed. The programming team would walk a panel of evaluators through their work and gain feedback from the evaluators on what they were doing right and what they were doing wrong. Over the years, it grew apparent that the structured walk-through methodology could be expanded to cover a much broader range of evaluations, including design reviews, document reviews, and proposal reviews. This section describes how the structured walk-through can be an important vehicle to conduct post-project evaluation reviews.

As originally conceived, the structured walk-through entails following four rules:

- *Rule 1.* The team being evaluated selects the evaluators.
- *Rule 2.* The team being evaluated sets the evaluation agenda.
- *Rule 3.* The team being evaluated runs the evaluation meeting.
- *Rule 4.* No senior managers are permitted to attend the evaluation session.

Each of these rules will be described briefly.

Rule 1. The Team Being Evaluated Selects the Evaluators. One of the great complaints of people being evaluated is that they are unhappy with the evaluators assigned to judge them. This unhappiness has several roots that were discussed earlier. For example, the team may be concerned about *who* selected the evaluators. Was it someone friendly to the project? Someone unfriendly? Another example: Do the evaluators have any idea of what project team members are doing, or do the team members need to take valuable time getting them up to speed? Still another example: How can project team members deal with evaluators who are fundamentally not qualified to sit in judgment of the project effort?

By selecting their evaluators, the project team members can make sure that they choose people who are sympathetic to their efforts, who are up to speed on what the team is doing, and who are fundamentally competent. An obvious question raised here is this: What's to keep the project team from "rigging" the jury. That is, isn't it likely that they will select friends who will engage in mutual back-scratching? Clearly, this is a possibility. At least two approaches have arisen to deal with it. In one, team members get to choose the evaluators. Then their list of prospects is reviewed by an independent outside panel that will examine the qualifications of the proposed evaluators. In the second approach, the organization maintains a pool of people available to conduct evaluations. The team to be evaluated then selects evaluators from the pool.

Rule 2. The Team Being Evaluated Sets the Evaluation Agenda. With traditional evaluations, the outside evaluators establish the evaluation agenda. They determine what the evaluation will focus on. They define the rules for conducting the evaluation. They even set the time and date for the evaluation review sessions. In this environment, the team finds itself operating according to the vagaries of the evaluators, with minimum input into the process. One commonly heard complaint is that because the evaluators do not fully understand the environment in which the project is being executed, they often focus on the wrong issues and do not ask the right questions. Another concern is that during the evaluation effort, the outside evaluators may try to trick the project team into revealing problems and catching them with a smoking gun. Project team members understandably are concerned about dealing with surprises dealt them by the outside evaluators. A final, more mundane complaint is that activities associated with the evaluation are invariably scheduled at the convenience of the evaluators and not the team members. In fact, it may happen that poorly scheduled evaluative activities actually cause schedule delays on projects.

The structured walk-through empowers project team members to establish the evaluation agenda. They select the topics to be covered. They determine the order in which the topics are treated and who gets to speak on them. They schedule the evaluation at *their* convenience to minimize disruptions to the project effort. Through a process like this, the project team can develop a sense of control over their fate. Anxieties about the relevance of the evaluation and surprise attacks by evaluators disappear.

As with Rule 1, concerns may be raised that by empowering project team members to conduct the evaluation, they may structure the agenda to avoid dealing with real problems. They may do this consciously and cynically keep known problems off the agenda. Or they may do it unconsciously, because they are too close to the work to identify objectively what the evaluation should address. To deal with the possibility of skewed agendas, it is a good idea to let the project team members establish it to the best of their abilities, and then to have it reviewed by an objective outside panel. By following such a process, the team feels empowered while a measure of objectivity can be maintained in establishing an evaluation agenda.

Rule 3. The Team Being Evaluated Runs the Evaluation Meeting. In the spirit of providing project team members with a feeling that they are empowered to run their projects, the structured walk-through has them running the evaluation meetings. If a meeting drifts from the agenda and addresses irrelevant topics, they have the power to bring it back to the agenda. If people speak out of turn, they have the power to insist that speakers stick to the proper protocol.

The principal problem associated with implementing Rule 3 is that project team members seldom have good meeting facilitation skills. In practice, they do not know how to keep the meeting focused on the agenda or how to make sure that speakers stick to their allotted time allocations. If the meeting becomes disordered, the evaluation effort can lose much of its value.

The problem of poor meeting facilitation can be handled by selecting a professional facilitator to run the evaluation session. Professional facilitators have the needed skills to keep the meeting moving forward and focused on the important issues. Because they are objective outsiders, they will not be not be cowed by some of the political dynamics that may arise during the evaluation. It is important when hiring facilitators to be clear that their job is to serve the project team, *not* to serve the organization in some abstract sense. They are like defense attorneys whose responsibility is to defend their clients to the extent possible, not to see that justice is achieved in the abstract. Facilitators *must* view project team members as their clients, in order to maintain the team members' trust in the evaluation process. If the facilitator is presented as a fair arbiter whose task is to serve the organization, it is likely that the project team members will not be completely forthcoming in their participation in the evaluation.

Rule 4. No Senior Managers Are Permitted to Attend the Evaluation Session. Rule 4 is the best-known rule of structured walk-throughs. Its rationale is obvious. How honest will project team members be in describing the problems they are encountering if the people who determine their salaries and career development are sitting in the room? Interestingly, many organizations have extended Rule 4 to cover customers as well. That is, they stipulate that customers should not attend structured walk-through sessions. How honest will project team members be in describing problems if customers are sitting in the room?

Ultimately, senior managers and customers need to be brought into the loop if they are going to have the information they need to function properly. In dealing with updating senior management on project issues, some organizations carry out two walk-throughs. The first—conducted with no senior managers present—entails a tough, honest review of problems. At the end of the session, the project team spends time determining how to deal with problems. Once they have developed solutions to the surfaced problems, a second walk-through can be conducted. Senior managers attend this session. Problems can be discussed frankly. What is good is that in this second walk-through, the team can also present solutions to the problems.

Customers can be brought into the loop in many ways—for example, through progress reports, customer walk-throughs, and customer partnering arrangements. It is important that they not be kept in the dark. However, customers and senior managers should recognize that the driving rationale of structured walk-throughs is to make sure that project team members feel no constraints to being honest.

Employing Structured Walk-Throughs in Post-Project Evaluations

Structured walk-throughs can be employed effectively in post-project evaluations. There is a lot to be gained by empowering project team members to conduct this final project review. Because they had daily exposure to the project, they know better than anyone what worked and what did not. With the structured walk-through, their personal insights can be tempered by the expert views of outsiders. If project team members are asked to select the outside reviewers (subject, of course, to approval by senior managers), you have some assurance that the evaluation team will be composed of qualified people. Furthermore, if the project team establishes the evaluation agenda (subject, again, to approval by senior managers), you have some assurance that the right topics will be addressed.

EISA Approach

EISA is an acronym for External, Internal, Self-Assessment (Wilson and Pearson, 1995). The EISA approach to evaluation has become standard when undertaking major assessments, such as those conducted under the auspices of ISO 9000 reviews, Capability Maturity Model reviews, and Baldridge Award reviews (Software Engineering Institute, 1993a and 1993b). As with structured walk-throughs, the EISA approach combines objective external assessments with major inputs from the group being evaluated.

With the EISA approach, the first round of review is conducted in the form of a self-assessment by the group being evaluated. The group identifies the performance goals it should be addressing in its work, then determines the extent to which these goals are being achieved. Once the self-assessment is complete, independent evaluators from within the organization conduct an internal evaluation. They examine the self-assessment, review the group's performance independently, then make a final judgment on the group's effort. Based on this assessment, deficiencies in the group's performance may be remedied. Finally, outside evaluators are hired to engage in a completely independent assessment of the group's performance.

Employing the EISA Approach in Post-Project Evaluations

As with structured walk-throughs, the EISA approach can be employed effectively in post-project evaluations. Its strength lies in the fact that it relies on input both from the people performing the work that is being evaluated and from independent assessors. Consequently, it likely will yield more meaningful evaluations than those carried out by one or the other party alone.

Customer Acceptance Tests: Built-in Post-Project Evaluation

A standard practice implemented on contracted projects is to have customers conduct a final inspection of the deliverable before accepting it and making final payments. The process of undertaking a final review of the deliverable is called a *customer acceptance test* (CAT). (In the information technology community, the term user acceptance test (UAT) is frequently used.)

The CAT is a form of post-project evaluation. Clearly, it is an important evaluation because it is being undertaken by the players for whom the deliverable is being developed. If they are unhappy with it, then the project has not met the important objective of customer satisfaction, which is central to all projects these days. They may not accept what has been produced and may insist that modifications be made to it.

CATs focus on the functional and technical performance of the deliverable. Is it achieving the functional requirements satisfactorily? For example, will the new data entry forms cut data entry errors in half, as required? Is the deliverable meeting technical requirements? Does the circuit board layout on actual circuit boards correspond 100 percent with layouts drawn on paper?

If a deliverable passes muster with a CAT review, then customers sign a statement that they are satisfied that the deliverable has met project requirements. If the CAT fails to satisfy the customers, things can get dicey. When there is an obvious deficiency in the deliverable, the contractor has a responsibility to fix it. However, the source of customer unhappiness might be tied to an interpretation of requirements that is different from the contractor's. In this case, the contractor and customers need to work out a solution that is satisfactory to both parties.

To close out the contract, a responsible member from the customer organization must attest that the contractor has met its obligations satisfactorily. The customer contract manager will also conduct a final review of the deliverable and payments to date to make sure that the organization has received what it has paid for. When the contract officer signs an acceptance document, final payments are made and the project is officially closed out.

Increasingly, well-managed organizations are adopting the CAT approach on noncontracted, internal projects—for instance, a project carried out by the IT department to update the organization's Web servers. To do this, they need to identify internal "customers" whose needs and wants should be satisfied through the project. This is not always easy to do, since projects always have multiple customers and the customers usually have contending needs and wants. With these internal projects, conducting a CAT is usually the last step taken before the project is closed out.

Post-Project Evaluation and the Learning Organization Perspective

Peter Senge's best-selling management book titled *The Fifth Discipline* stimulated substantial discussion on one of the topics covered in the book: the nurturing of a learning organization

(Senge, 1990). (See the chapters by Bredillet and by Morris.) The learning organization perspective is an extension of a basic cybernetic principle, which holds that for systems to survive they must continually use feedback data to help them adjust to changing conditions in their environments. Thus, learning organizations engage in a process of constantly learning lessons from their environmental experiences and adjusting their behaviors accordingly. A little reflection shows that this is precisely the central concern of post-project evaluation.

Through post-project evaluations, organizations can determine what works and what does not work when their employees execute projects. To the extent that these evaluations cover a broad range of issues—including technical, financial, operational, organizational, and legal—they provide insights that will enable organizations to survive and even thrive in highly competitive business environments. Through a learning processes rooted in the conduct of post-project evaluations, they function as learning organizations.

Organizations that continue to conduct business in the same old way—that live by the motto "If it ain't broke, don't fix it"—are likely to find themselves unable to deal with the surprises thrown at them from out of their ever-changing environment. They are not learning from their experiences, and ultimately this will translate into weak business performance, or worse.

Post-Project Evaluation and the Human Resource Management Perspective

Post-project evaluations can have significant human resource management implications. (See the chapter by Huemann, Turner, and Keegan.) Two are examined here: Post-project evaluations can be tied to an organization's reward system, and they can indicate the adequacy of the organization's staffing efforts. Each will be discussed briefly.

Post-Project Evaluations and Reward Systems

An important element of good management is *accountability* (Frame, 1999). People must be held accountable for their actions. High performers should be recognized for their achievements, while low performers should be alerted to their poor performance and provided guidance on how to improve it. Post-project evaluations can play a significant role in an organization's reward system. For example, they can identify highly successful projects and explain the reasons for success. Individuals who contributed to the success can be recognized. Letters of commendation can be written on behalf of these people and included in the performance appraisal review process. By the same token, they can identify troubled projects and the causes of problems. If the problems are closely tied to the behavior of specific individuals, this information can be incorporated into their performance appraisal reviews.

While rewards are often used to motivate project workers, their shadow side should be recognized. Even as they may stimulate employees to do their best, they may also lead to unhappiness among those who are not recognized (Kohn, 1993).

Post-Project Evaluations and Guiding the Organization's Staffing Efforts

In today's fast-paced world, staffing requirements are undergoing continual change. In developing our information systems, yesterday we needed Cobol programmers, today Oracle database experts, and who knows who we will need tomorrow? A major challenge facing human resource management specialists today is determining whether current staffing arrangements work and predicting future staffing requirements. Information on the adequacy of staffing can be gleaned from post-project evaluations. For example, the evaluation might suggest that a project's performance was hampered by the lack of software testing personnel. If this is deemed an important problem, then the personnel department can work to remedy it by hiring new resources or training existing personnel. To gain the maximum benefit from post-project evaluations, organizations should make sure that human resource management personnel are aware of evaluation findings that have staffing implications.

The Bottom Line: Dealing with the Realities of Post-Project Evaluation

Most organizations do not conduct effective post-project evaluations (Kumar, 1990). A problem they encounter is that these evaluations require the commitment of time, money, and expertise that many organizations are not willing to provide for what they perceive to be an overhead activity. The fact that post-project evaluations are carried out at the end of the project life cycle adds to the problem. By the time most projects end, project funds have already been expended. In fact, managers running projects that face cost overruns are not likely to serve as rigorous advocates for additional funding for post-project evaluations. In their desire to save money, they eschew "nonessential" expenditures, such as expenditures on post-project evaluations.

Beyond the matter of funding, at project's end, its momentum is gone and enthusiasm for project work is often low. The principal item of concern of team members at this time is future job assignments. Higher-level managers usually want to put the finished project behind them and focus attention on new revenue-generating prospects. No one is arguing that effective post-project evaluations must be carried out for the long-term good of the organization.

Because the realities of project termination work against the conduct of effective project closeout, proponents of evaluation need to be willing to take a graded approach. They must recognize that they can't have it all. The graded approach identifies which evaluation activities *must* be conducted and which are nonessential, although nice to have. The following is a list of some *must haves* and *might haves* (and the list compares well with Turner's listing at the end of his chapter on managing technology):

- Processes *must* be in place to capture information on project activities as they are being carried out. Included here is data on actual cost of work effort, actual task durations, actual amount of labor employed, and actual milestones achieved. By archiving cost, schedule, resource, and performance data, evaluators are able to examine project performance on these items retrospectively. The retrospective review enables them to identify exactly when cost overruns and schedule slippages occurred. Or when milestones were achieved early and at what cost. Without archived data, the post-project evaluation is based on anecdote and conjecture and its reliability and validity are suspect.

- Checklists of standard project closeout items *must* be developed. They should address questions such as: Have you protected your deliverables, documents, and processes from an intellectual property perspective (e.g., patents, trademarks, copyrights, trade secrets)? Has equipment been properly reassigned for use on other projects? Have people been reassigned to other work efforts? Have all the items in your statement of work been achieved? Have contractors submitted pertinent deliverables, documentation, equipment, and materials as required by the contract (Frame, 2003)?
- A bare-bones, post-project analysis *must* be carried out and its findings written up in a lessons-learned document. At a minimum, the analysis should describe the project experience in narrative format. Quantitative data on cost, schedule, and resource performance should be included as well. At the end of the document, a statement of key lessons learned should be offered and recommendations made for future projects.
- Procedures for capturing and implementing lessons learned *must* be established. For example, procedures might be developed where formal reviews are conducted biannually on lessons learned from recent project experiences. Then the key lessons might be incorporated into the organization's project and business processes. Another example: The organization might create a library of lessons-learned documents that makes real project experiences accessible to all project staff in the organization.
- Highly structured evaluations employing structured walk-through or EISA processes *might* be adopted. These structured evaluations are important for large, complex, high-impact projects. The hassles associated with conducting them do not usually make them worthwhile for smaller projects.
- Detailed and thorough root cause analyses of sources of project problems *might* be conducted. The goal of these analyses is to understand the underlying causes of problems on a given project. The information gleaned can then be used to improve performance on future projects. These analyses will likely be expensive: Senior managers and key project players need to be interviewed; data needs to be carefully analyzed; a detailed study with well-conceived recommendations must be written up and distributed.
- Steps *might* be taken toward creating a learning organization environment, where lessons learned are continually being incorporated into the organization's business processes and obsolete items are being jettisoned.

In the final analysis, effective post-project evaluation requires sensitivity to organizational realities. However, while it is argued here that evaluators must be willing to live with compromises in the evaluation effort, I am not suggesting that extreme shortcuts be taken or that organizations can afford to abandon post-project evaluations entirely. As noted at the outset of this chapter, post-project evaluations are important for a range of reasons and must be implemented so that organizations can develop an understanding of how their projects are doing, can gather information to enable them to see what works and what does not, and can employ this information to adjust their business processes to ensure better performance.

References

Adams, J. S. 1963. Toward an understanding of inequity. *Journal of Abnormal and Social Psychology* 67: 422–436.

Buss, M. D. J. 1984. How to rank computer projects. *Harvard Business Review* 61:118–125.

Drucker, P. F. 1985. *Management: Tasks, responsibilities, practices.* New York: Harper Business.

Ezingeard, J.-N., Z. Irani, and P. Race. 1999. Assessing the value and cost implications of manufacturing information and data systems: An empirical study. *European Journal of Information Science* 7(4): 252–260.

Farbey, B., F. F. Land, and D. Targett. 1993. *How to assess your IT investment: A study of methods and practice.* Oxford, UK: Butterworth-Heinemann.

———. 1999. Moving IS evaluation forward: Learning themes and research issues. *Journal of Information Technology* 8:189–207.

Fox, J. R. 1984. Evaluating management of large complex projects. *Technology in Society* 16(6):129–139.

Frame, J. D. 1999. *Project management competence.* San Francisco: Jossey-Bass.

———. 2002. *The new project management.* San Francisco: Jossey-Bass.

———. 2003. *Managing risk in organizations.* San Francisco: Jossey-Bass.

Garfield, E. December 1987. Is there room in science for self-promotion? *The Scientist* 1(27):9, 14, 187–188.

Kohn, A. 1993. Why incentive plans cannot work. *Harvard Business Review* 74(5):54–61.

Kuhn, T. 1996. *The structure of scientific revolutions.* 3rd ed. Chicago: University of Chicago Press.

Kumar, K. 1990. Post-implementation evaluation of computer-based IS: Current practice. *Communication of the ACM* 33(2):203–212.

Locke, E. A., and G. P. Latham. 1990. *A theory of goal setting and task performance.* Englewood Cliffs, NJ: Prentice Hall.

Merton, R. K. 1996. *On social structure and science.* Chicago: University of Chicago Press.

Morgan, E. J., and Y. L. Tang. 1993. Post-implementation reviews of investment: Evidence from a two-stage study. *Journal of Production Economics* 30(3):477–488.

Remenyi, D., M. Sherwood-Smith, and T. White. 1997. *Achieving maximum value from information systems: A process approach.* Chichester, UK: Wiley.

Senge, P. M. 1990. *The fifth discipline: The art and practice of the learning organization.* New York: Doubleday & Co.

Software Engineering Institute. 1993a. *Capability maturity model for software, Version 1.1, CMU/SEI-93-TR-24.* Pittsburgh: Carnegie Mellon University.

———. 1993b. *Key practices for the Capability Maturity Model, Version 1.1, CMU/SEI-93-TR-25.* Pittsburgh: Carnegie Mellon University.

Symons, V. 1990. Evaluation of information systems: IS development in the processing company. *Journal of information Technology* 5:194–204.

Vroom, V. H. 1964. *Work and motivation.* New York: Wiley.

Walsham, G., 1990. Interpretive evaluation design for information systems. Chapter 12. *In Beyond the IT Productivity Paradox,* ed. L. P. Wilcox and S. Lester. Chichester, UK: Wiley.

Wilson, P. F., and R. D. Pearson. 1995. *Performance-based assessments: External, international, and self-assessment tools for Total Quality Management.* Milwaukee: American Society for Quality Control Press.

Yourdon, E. 1988. *Structured Walkthroughs,* Englewood Cliffs, NJ: Prentice Hall Professional Technical Reference.

CHAPTER FORTY-EIGHT

DEVELOPING PROJECT MANAGEMENT CAPABILITY: BENCHMARKING, MATURITY, MODELING, GAP ANALYSES, AND ROI STUDIES

C. William Ibbs, Justin M. Reginato, Young Hoon Kwak

How good are your organization's project management practices? How well do your practices compare with those of your peers in the business world? Are you making the appropriate investments in new project management systems, processes, and practices? These are the questions that few firms can answer directly and accurately. Yet their answers can unlock the gate to superior business performance.

The first step in understanding an organization's project management effectiveness is to determine its Project Management Maturity (PMM). By having a grasp of where a company lies on the PMM spectrum, management can determine its project management strengths and weaknesses, which is enormous value in today's highly competitive, project-oriented marketplace.

Stated simply, a company's PMM is a measure of its current project management sophistication and capability. Knowledge about the most sophisticated project management tools does not necessarily mean that those complicated tools will be used on every project. Rather, appropriate knowledge means that the firm and its managers understand which tool is appropriate for the demands of the project. PMM helps gauge such management wisdom.

Once the PMM is known, it can be used to both understand the company's current standing and to develop a roadmap for future improvements in project management processes and practices. Once on the path to such enlightenment, companies can craft their capabilities and strategy to enhance competitive advantage and wealth creation.

The purpose of this chapter is to describe how PMM benchmarking can help organizations develop that roadmap. We do this by first highlighting the importance of PMM in today's competitive marketplace. Second, we exhibit techniques for determining current levels of PMM and defining a course for PMM improvement. Last, we demonstrate methods to enumerate the value of project management improvement to ensure that investments in project management are reaping the desired returns.

This chapter presents a number of quantitative findings, some of which are cast in a statistical manner. Keep in mind, though, that the subject of this chapter is classical statistical methodologies for "real-world management"—that is, t-tests, levels of significance, and other sophisticated statistical concepts—are not directly pertinent to, or reasonable for, such applications because the real world of management is very complex. These statistical relationships should therefore be seen as general tendencies and not treated as precise correlations and cause-effect relationships.

The Importance Of PMM in Today's Marketplace

Businesses are becoming increasingly projectized. Examples of current successful businesses that are organized around projects include Microsoft (operating and networking environments), Boeing (large commercial and defense aerospace ventures), and Amgen (biotechnology R&D). They are successful because they continually grow revenues and profits and have achieved remarkable stock market capitalization. The vast majority of their revenue and profit sources are from projects.

These firms share common characteristics such as devolved power, strong emphasis on intellectual property, powerful brand identification, as well as a premium on project-driven services and products. They are also very much bottom-line focused, managing themselves in a manner that creates increased shareholder value. These companies also share another trait: As they evolve organizationally, their abilities to deliver projects that advance their corporate strategies also evolve.

As projects become the currency for improved business performance, making project management a core capability of successful organizations in turn becomes paramount. No longer are being on time and on budget the only benefits or goals of strong project management. Additional core benefits of a project-centric focus are sophisticated project management tools that improve organizational effectiveness, meeting quality standards, and fulfilling customer satisfaction (Al-Sedairy, 1994; Boznak, 1988; Bu-Bushait, 1989; CII, 1990; Deutsch, 1991; Gross and Price, 1990; Ziomek and Meneghin, 1984).

But to demonstrate a true competence, project management success cannot be an occasional event. Performance that is good, on average, is not sufficient. Repeatability and relentless improvement must be the standard. Project performance that is, on average, good but erratic is not sufficient because one wayward project ripples through and affects the company's portfolio of projects. When management attention is diverted to the errant project, the loss of attention on all other projects can ripple through to other concurrent or subsequent projects, hurting them. (The issues associated with the concept of project success are discussed in Cooke-Davies' chapter. See also Cooke-Davies' other chapter on maturity.)

The first step in determining PMM and appropriate directions for future improvements to project management is to evaluate the benefits of project management in quantified terms. Project management cost, schedule and cost performance, PMM, and the financial return on project management investments (what we refer to as PM/ROI[SM]) are all important quantifiable metrics in evaluating the benefits of project management. Research shows that companies with high levels of PMM can leverage a range of quantified benefits (Ibbs and Reginato, 2002):

- *Companies with more mature project management practices have better project performance.* For example, companies with more mature practices deliver projects on time and on budget, whereas less mature companies may miss their schedule targets by 40 percent and their cost targets by 20 percent.
- *PMM is strongly correlated with more predictable project management schedule and cost performance.* The project portfolios of more mature companies, for instance, have lower standard deviations for schedule performance (0.08) and cost performance (0.11) than companies with lower PMM scores (corresponding values of 0.16 for both schedule and cost).
- *Good PM companies have lower direct project costs than poor project management companies.* Highly mature companies have project management costs in the 6 to 7 percent range, while their counterparts average 11 percent (and in some cases reach 20 percent). Note this is just the direct cost spent on project management. Organizations with low PMM risk other undesirable events such as increased indirect costs, late project deliveries, missed market opportunities, and dissatisfied customers.

Each of these points will be discussed in greater detail later in this chapter. The key point being made is that increased levels of PMM correspond to better project performance and, in turn, can be a key driver for corporate success.

PMM: Concepts and Quantification

So how does an organization measure PMM? The following section demonstrates techniques for determining current levels of PMM, as well as measuring it numerically so that it can be quantifiably valued.

Definition

PMM is the sophistication level of an organization's current project management practices and process (Kwak and Ibbs, 2002). Project management techniques serve businesses in planning, controlling, and integrating time and resource-intensive endeavors. Until recently, there was little quantifiable data, suitable methodologies, or well-defined processes that impartially measure PM practices. PMM models were developed to fill that void.

In addition to measuring the internal level of PMM, corporate executives needed a method to validate the investments they were making in project management. PMM can be used in conjunction with valuation techniques such as financial return on investment because it is a quantifiable value. That is, what profit return will a company realize for every $1 it invests in some new project management system, process, or procedure? Also, as many PMM models become standardized (which is a goal of the Project Management Institute's OPM3 initiative), PMM becomes a yardstick from which companies can compare their internal measures of PMM externally with other peer organizations.

Models

As PMM has been thrust to the forefront of building project management competence, several models for measuring maturity have evolved. One such model is the Berkeley Project Management Process Maturity Model. This model has been developed over a seven-year research period and has been successfully implemented in several industries, nonprofits, and government agencies (Ibbs and Reginato, 2001; Kwak and Ibbs, 2002).

The five-step Berkeley Project Management Process Maturity Model is used to establish an organization's current PMM level. This model demonstrates sequential steps that map an organization's incremental improvement of its project management processes. It is schematically illustrated in Figure 48.1.

The model progresses from functionally driven organizational practices to project-driven organizations that incorporate continuous project management learning. An organization's position within the model can be used to determine its position relative to the other companies in the same industry class or otherwise that have been assessed.

Level 1: Ad Hoc. At the Ad Hoc stage, there are no formal corporate procedures or plans to execute the project. The project activities are poorly defined, and cost estimates are inferior. Project management-related data collection and analyses are not conducted in a systematic manner. Processes are unpredictable and poorly controlled. There are no formal

FIGURE 48.1. BERKELEY PROJECT MANAGEMENT PROCESS MATURITY MODEL.

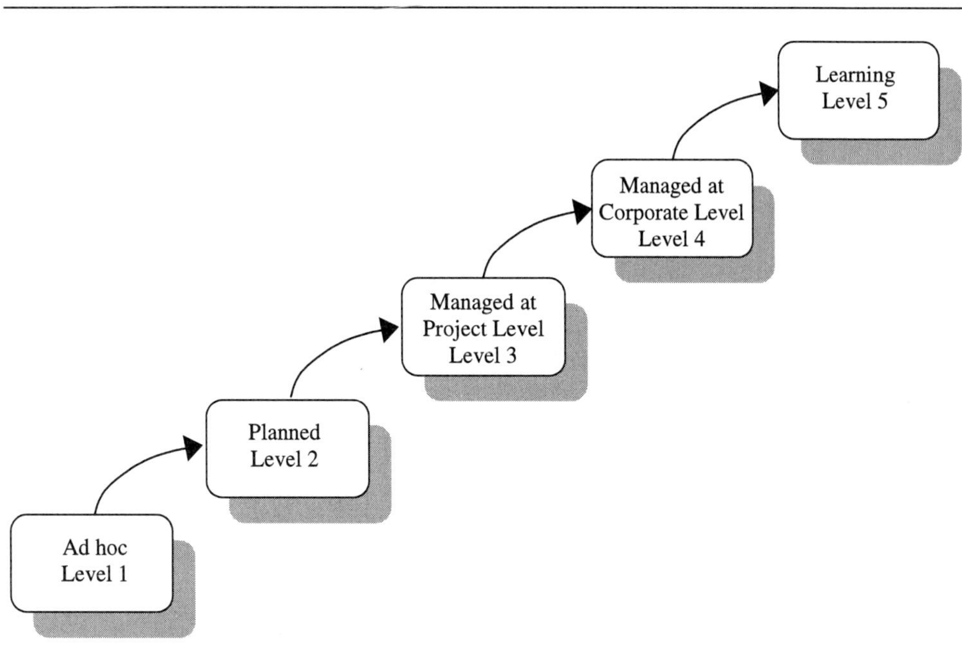

steps or guidelines to ensure continuity of project management processes and practices. As a result, utilization of project management tools and techniques is inconsistent and applied irregularly, if at all, even though the individual project manager may be very competent (Ibbs and Kwak, 2000).

Level 2: Planned. At the Planned level, informal and incomplete processes are used to plan, but not control, a project. Some project management problems are identified, but they are generally not documented or corrected in a systematic manner. Project management-related data collection and analysis are informally conducted but not documented. Project management processes are partially recognized and used by project managers. Nevertheless, planning and management of projects depends largely on individuals.

An organization at Level 2 is more team-oriented than at Level 1. The project team understands the project's basic commitments. This organization possesses strength in doing similar and repeatable work. However, when the organization is presented with new or unfamiliar projects, it likely experiences chaos in managing and controlling the project. Level 2 project management processes are efficient for individual project planning, but not for controlling the project, let alone any portfolio of projects (Ibbs and Kwak, 1997).

Level 3: Managed at the Project Level. At Level 3, PM exhibits systematic planning and control systems that are implemented for individual projects. Project management processes become more robust and demonstrate both systematic planning and control characteristics. The project management team typically works together in an informal setting. For the purposes of project control, most of the challenges regarding project management are identified and informally documented for each project. Various types of analyzed trend data are shared by the project team to help it work together as an integrated unit throughout the duration of the project. This type of organization works hard to integrate cross-functional teams to form a project team.

Level 4: Managed at the Corporate Level. For projects managed at Level 4, management processes are formal, while information and processes are documented informally. The Level 4 organization is fully integrated: It can plan, manage, and control multiple projects efficiently across an organization's project portfolio. A project management process model is probably well defined, with project requirement systems that are in place but not necessarily regularly used. Project-related data and records are formally and systematically collected, reviewed, and distributed to the appropriate parties but are not formally organized. Also, data is collected and analyzed to anticipate and prevent adverse productivity and quality impacts or other trends detrimental to project success. This allows an organization to establish a foundation for fact-based decision making.

In addition to effectively conducting project planning and control for multiple projects, the organization exhibits a strong sense of teamwork within each project and across projects. Project management training is available when needed and is provided to the entire organization, according to the respective role of project team members.

Level 5: Learning. The key characteristic of companies that operate at the Learning stage is that they continuously improve their project management processes and practices. Each project team member spends considerable effort to maintain and sustain the project-driven environment. Training is formally available when needed, presenting lessons learned and other techniques to improve project management on an ongoing basis. Project team members are typically together throughout the entire project duration, and their individual roles are defined based by their strengths and experience. Problems associated with applying project management are fully understood and addressed on an ongoing basis to ensure project success. Project management data are collected automatically to benchmark project management strengths and identify the weakest process elements. These data are then rigorously analyzed and evaluated to select and improve the management processes. Innovative ideas are also vigorously pursued, tested, and organized to improve processes.

Formal comprehensive requirement systems exist and are used regularly. A project management process model is formally defined, distributed, and discussed by all of the team members, using previous project experience as a guideline. Additionally, a project management consulting group is probably created and chartered, and its existence is communicated throughout the organization.

Each level within the Berkeley Project Management Process Maturity Model includes an assessment of PM processes and practices based upon six PM processes and nine knowledge areas, as shown in Figure 48.2. When each organization is assessed along these boundaries, PM strengths and weaknesses are determined. This assessment allows companies to prudently invest in areas to improve upon their weaknesses.

FIGURE 48.2. BENCHMARKED PROJECT MANAGEMENT PHASES AND KNOWLEDGE AREAS.

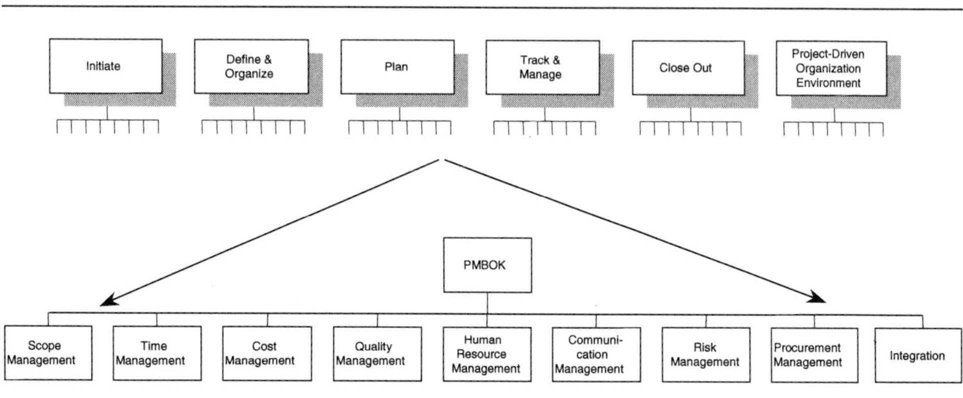

Ideally, an organization evolves smoothly and thoroughly from a less project management-sophisticated organization to a "learning" project-centered organization. However, in practice it is rare that when a company moves to maturity Level N+1, it has implemented *all* the characteristics of Level N. Rather, at Level N+1, an organization has the capability to choose the proper and relevant project management practices or processes that are suitable for a given project.

To illustrate, suppose that the scheduling techniques available to a company range from simple bar charts at the low end to complex simulation for resource optimization at the high end. An organization that has a high level of PMM does not always have to employ the most sophisticated techniques available to them, which in this case would be the simulation. Rather, they enjoy the ability to apply the most appropriate techniques based upon the complexity of the project. This allows for the construction of a broad-ranging PM toolkit as maturity increases.

In addition to the Berkeley Model are other maturity models developed by consultants and practitioners. Among them is the Center for Business Practices Model, Kerzner's Project Management Maturity Model, ESI International's Project Framework, and SEI's Capability Maturity Model Integration. PMI is striving to develop some commonality and consistency among these models through its OPM3 endeavors. For a good summary of maturity models, see Foti's (2002) article and Cooke-Davies' chapter.

Measuring PMM

Measuring PMM involves quantifying the internal level of maturity and then comparing it externally to peer organizations. Two interconnected methods—benchmarking and gap analysis—are discussed in the following sections.

Benchmarking. Benchmarking is a process that allows organizations to compare different aspects of current practices against best practices. The basic premise is to improve and learn tools and techniques from other organizations. The purpose of benchmarking is to analyze the internal operation, understand the competition and industry leaders, incorporate best practices, and gain a superior foothold in competitive markets (Camp, 1995).

To assess PMM between different organizations or functional groups within an organization, a rigorous and comprehensive benchmarking methodology must be developed. The methodology adopted by the Berkeley Model involves a detailed, three-part questionnaire for data collection. Part I involves collecting general data regarding each organization, including the size of the organization, personnel structures, and how much it spends on PM per year.

Part II consists of 162 multiple-choice questions. Its intent is to measure the maturity of the organization's standard project management processes. Examples of some such questions are displayed in Figure 48.3.

To calculate the overall PMM, the average score for all 162 questions and their standard deviation are computed for each organization. All questions are weighted equally, so un-

FIGURE 48.3. SAMPLE BENCHMARKING QUESTIONS.

65. Critical path identified

No critical path calculation done. Each subproject identifies critical tasks independently and sets work priorities 1

Critical path based on committed milestone dates. No CPM calculation performed, or CPM used on individual subprojects 2

Key critical tasks identified through nonquantifiable means and used to drive the critical path calculation 3

Critical path calculated through integrated schedule, but only key milestone dates communicated back to subprojects. 4

All critical tasks identified and indicated in each individual subproject schedule. Critical path determined through integrated schedule. 5

106. Quality management (QA/QC) system is utilized

No quality management system 1

Informal quality management system, not used 2

Informal quality management system, hardly used 3

Formal quality management system, occasionally used 4

Formal quality management system, intensely used 5

128. Project deliverables list reviewed and cross-checked against actual deliveries

No project deliverables list available 1

Deliverables list available, but not reviewed 2

Some informal review of original, approved deliverables list 3

Formal review of approved deliverables list, but with only informal comparison to actual deliverables 4

Formal review of approved deliverables list with point-by-point comparison to actual deliverables 5

---

derlying the assessment is the assumption that all questions are equal indicators to an organization's PMM. Because of industrial and organizational competition, situations arise where some questions are more relevant to an organization than others. However, neglecting such factors allows for achievement of nonbiased circumstances to specific variables.

Part III of the assessment tool collects project-specific data, such as cost and schedule performance, as well as metrics regarding scope and quality attainment. The PMM analysis of Part II can be compared with the project performance data collected in Part III and evaluated as to how project performance improves with corresponding improvements in maturity. It is important that efforts in improving PMM result in increased project performance—improving PMM for project management's sake is not likely to benefit the overall organization's market performance.

As a rough guide based on our experience, we suggest that at least five people partake in any assessment process, with 10 to 15 yielding better results. Less than five lends little to statistical evaluation, and more than 15 becomes difficult to manage. The people participating in the assessment process should be project management professionals that represent typical project managers for the organization. Again, it is important that the people partaking in the assessment process be *typical* to the organization to ensure that representative data is being collected.

Process efficacy is further ensured if a manager or organizational decision maker, preferably from the vice president level or higher, is involved as a champion of the assessment process. A person at this level can expedite the assessment process, foster buy-in, and help to ensure assessment data quality. Conducting post-assessment interviews with assessment participants can further ensure score stability. Such interviews are particularly helpful when assessment scores exhibit seemingly arbitrary results.

Data collection is highly dependent on the type and size of the organization being assessed. The overarching goal is to collect data that is representative of the entire organization or division being assessed. Obviously, if the group being assessed for PMM is multinational and consists of thousands of employees, then many people from multiple geographic locales should partake in the assessment process. Smaller organizations can, generally speaking, accurately assess their organizations with far fewer participants to the assessment process.

Gap Analysis. Working hand-in-hand with benchmarking is gap analysis. Gap analyses are characterized by the comparison of an organization's current state to its desired state. The current state is defined by current practices, and the desired state is represented by industry best practices (Camp, 1995). The gap between current and best practices serves as the basis for preferred improvement.

Industry best practices are determined by comparing the project management processes of multiple peer organizations. Understanding the project management processes of multiple comparative organizations can be achieved several ways. One common method consists of attending discussions and symposia where other companies discuss their practices and processes. Another common way is to partner with or hire an organization that conducts assessments for multiple organizations and hence has access to copious amounts of industry-specific data. The Berkeley database, for example, has project management process data from over 60 organizations in five industries. Other organizations have similar industry-specific best-practice data as well.

Once best practices are understood, the organization attempting to improve its PMM should critically examine which best practices it wishes to adopt. It is important to note that organizations undergoing PMM improvement initiatives should initially select a few areas in which to focus improvement. We suggest "picking the low-hanging fruit" first—that is, choosing the easiest practices to improve at the onset. These practices will provide the least expensive and easiest processes to improve and can serve as a springboard to further process improvement.

Also, once best practices have been determined, implementation should include input from customers, suppliers, subcontractors, operators, and so on. These parties will add further enhancement of the gap analysis by providing insights not attainable by benchmarking alone. Benchmarking is an excellent tool for illuminating best practices, but it is the ability to creatively implement (as opposed to copying) best practices that will allow an organization to become the best-of-class within the project management world.

Maturity's Role in Other Fields

For those readers who are in the software industry, much of the preceding may sound familiar. The Software Engineering Institute (SEI) at Carnegie Mellon University has several models that focus on maturity in the software industry. Research conducted by SEI is widely regarded, and many large organizations require that their software vendors meet certain levels of software capability maturity, usually stipulating that vendors provide continuous improvement over the life of the service contract. This continuous improvement stipulation is not unlike the gains that can be made by systematically improving PMM as shown by the Berkeley PM Process Maturity Model.

PMM'S Relationship to Business Results

Determining organizational PMM should not be an exercise in measurement for measurement's sake. Detailed goals should be outlined, with most, if not all, of those goals tied to business objectives. The following is a discussion of how improvements to PMM can improve overall business processes.

Measurable Maturity Benefits

As previously discussed, the assessment process begins by determining the overall maturity for a group of peer organizations. This step is highlighted in Figure 48.4.

The data presented in the figure were generated by research conducted at the University of California at Berkeley. In this figure companies that were involved in engineering and construction projects are labeled "EC," telecommunications and information management & movement companies are labeled "IMM," information service organizations are "IS," and high technology manufacturing companies are "HTM."

Knowing how an organization stacks up against others is important because the managers of that firm can then ascertain a relative projection of how PM is providing a competitive advantage or stands as a competitive barriers. However, to be most helpful the PMM must be assessed on a "subatomic" level. For example, consider company EC11 in Figure 48.4, as represented by the white bar. Its overall PMM is about 3.70, which puts it in the better half of its peers in terms of PMM.

FIGURE 48.4. OVERALL MATURITY FOR 44 ORGANIZATIONS.

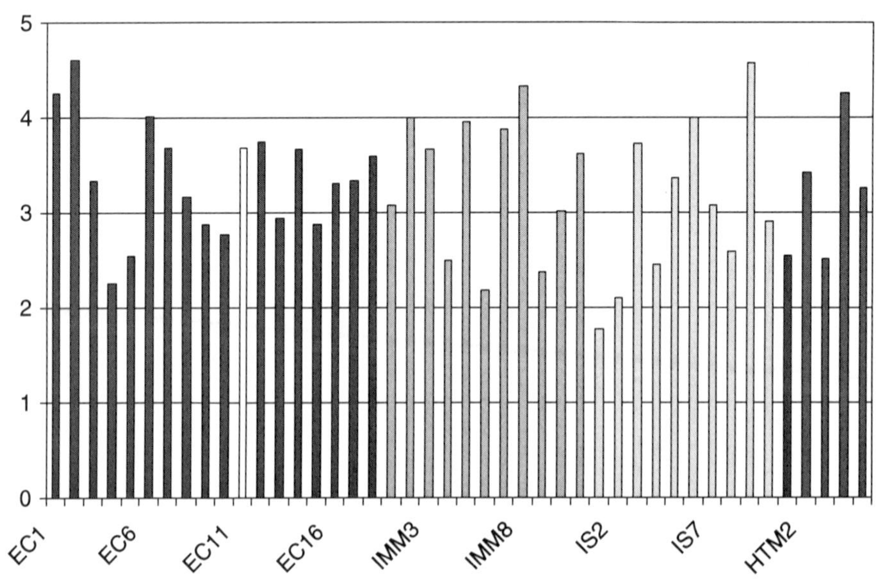

If the managers of EC11 wished to improve the company's PMM, where would they look? A gap analysis would point out specific areas in which EC11 could improve vastly with respect to its peers. For example, EC11 lags well behind its peers in terms of initiating projects, as shown by the white bar in Figure 48.5.

The gap analysis illuminates that EC11, while ahead of most of its peers in terms of overall PMM, can improve substantially by improving the process by which it initiates projects. This analysis allows EC11 to target the most appropriate areas of improvement rather than a hit-or-miss approach that may not improve (or even decrease) overall maturity.

As obvious as this seems, few companies conduct such a diagnostic test on their project management processes and teams before undertaking major project management improvement efforts that they *think* will help their companies. In terms of an analogy, most ailing people would be skeptical of a physician who prescribes a certain treatment before running a full battery of tests. Yet those same rational people routinely spend enormous sums of money, time, and effort on new project management software systems or training without first pinpointing where those improvement efforts would be most helpful.

This methodology can be extended to each of the project management phases and knowledge areas that are represented in Figure 48.2. Table 48.1 displays how a gap analysis can highlight en masse the areas where an organization leads or lags its peers.

FIGURE 48.5. DRILLING DOWN PMM TO THE INITIATING PROJECT'S LEVEL.

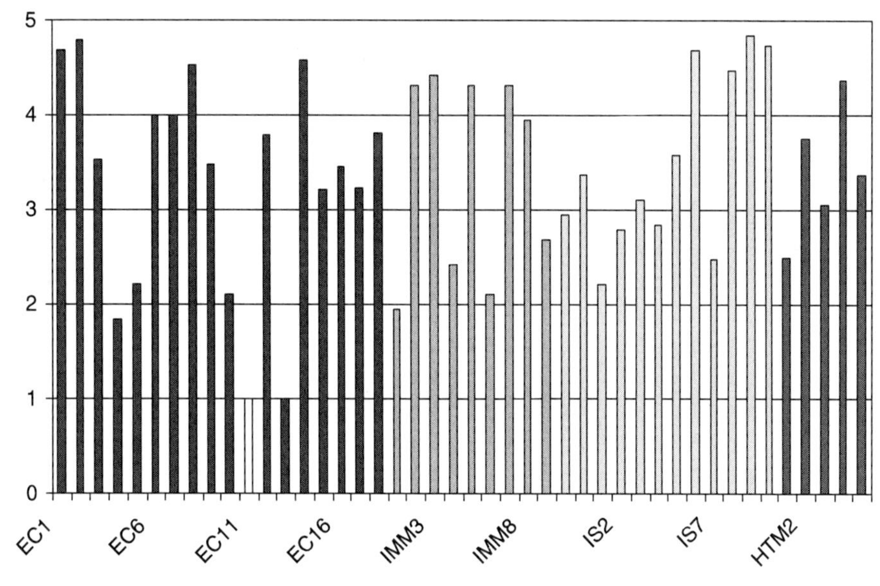

TABLE 48.1. GAP ANALYSIS FOR EC11 AND ITS PEERS.

Process Area	All Companies	Peers	EC11	EC11—Peers
Scope	3.42	4.15	3.84	−0.31
Time	3.37	3.86	3.85	−0.01
Cost	3.47	4.28	4.26	−0.02
Quality	3.00	3.15	3.93	0.78
Risk	2.97	3.35	3.97	0.62
Communication	3.53	3.81	4.10	0.29
Human resources	3.11	3.44	3.88	0.44
Procurement	3.15	4.40	3.97	−0.43
Integration	3.61	3.67	3.86	0.19
Initiate	3.35	4.27	1.00	−3.27
Define/organize	3.65	4.13	3.87	−0.26
Plan	3.21	3.54	3.79	0.25
Track and manage	3.32	3.65	4.40	0.75
Closeout	3.27	3.54	3.33	−0.21
Project-driven organization	2.97	3.63	3.45	−0.18
Overall	3.30	3.73	3.70	−0.03

In the EC11—Peers column, positive numbers represent areas where EC11 surpasses its peers and negative figures show areas of needed improvement. To target areas of improvement, EC11 can simply rank the negative numbers in order of highest to lowest. The largest discrepancies between EC11 and its peers are the areas in which EC11 should focus.

While the table demonstrates that, on average, EC 11 is similar to its peer organizations in terms of *overall PMM*, there are several areas in which it lags behind its peers. The process of benchmarking identifies these areas and allows EC 11 to concentrate and target specific areas for improvement, such as procurement management and project initiation. If EC 11 takes this targeted approach to PMM improvement, then it can more easily adopt best practices and improve its overall PMM above those of its peers.

As mentioned earlier, our initial benchmarking analysis treats each of the 162 questions as being equally important. At any stage of the analysis, though, some questions can be given more importance if the individual company so wishes. Experience shows this must be done with careful forethought, however; otherwise, the managers of the subject company may obtain an inaccurate analysis. For instance, they may think that quality management issues are paramount for their business and ask that such questions be super-weighted, whereas in point of fact their competitors are emphasizing some other aspect of project management.

Project Performance Improvement Benefits

Improving PMM will lead to improvements in project management processes and practices. However, the real goal of project management is not to improve processes and practices per se but to deliver projects more successfully. As discussed in the following section, there is a correlation between improved PMM and improved project performance.

Schedule and Cost Performance. What do companies get for their investments in project management? Our analysis of detailed assessments reveals that companies with higher PMM tend to deliver more of their projects on time and on budget. See Figures 48.6 and 48.7

These figures contrast Schedule Performance Index (SPI) and Cost Performance Index (CPI) against PMM for companies that we have benchmarked over the past six years. SPI and CPI are defined as the ratios of total original authorized duration or budget versus total final project duration or cost, respectfully. That is:

$$\text{Cost Performance Index} = \text{CPI} = \frac{\text{Planned budget}}{\text{Final costs}}$$

$$\text{Schedule Performance Index} = \text{SPI} = \frac{\text{Planned duration}}{\text{Final duration}}$$

It should be noted that our use of the terms CPI and SPI vary from that of common project management vernacular. We use the terms CPI and SPI because it is important to understand that as PMM improves, so do cost and schedule performance. However, the CPI and

FIGURE 48.6. SPI VS. PMM.

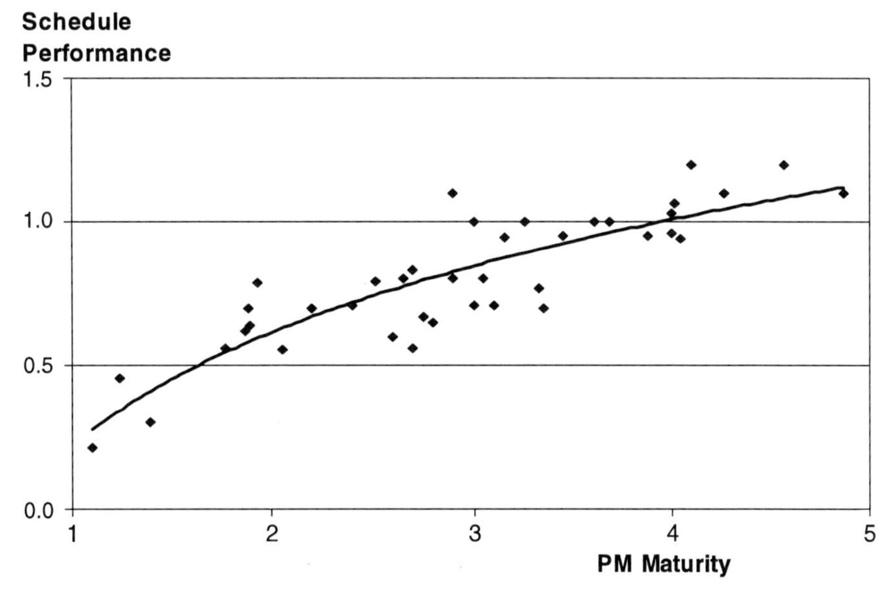

FIGURE 48.7. CPI VS. PMM.

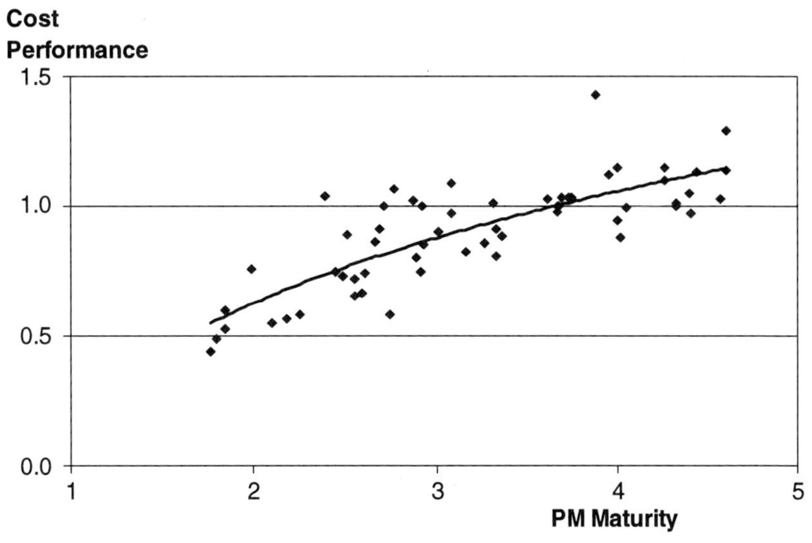

SPI ratios that we have listed are not the same as the ratios by the same name as applied to earned value analysis.

The value to organizations is apparent. In terms of schedule, as PMM increases so does the ability to complete projects on time. An SPI ratio of 1.00 equates to finishing projects in exactly the time that was originally estimated; a number < 1.00 indicates late completion. The ability to accurately forecast the time necessary to complete a project affords senior executives in the firm a powerful tool in meeting time-to-market windows.

Like SPI, CPI increases with higher PMM levels. Also similar to SPI, a CPI value approaching 1.00 signifies accuracy in estimating and delivering projects on budget. Increasing CPI is good because accurate cost forecasts allow companies to confidently and accurately allocate capital.

The R^2 value in these figures is called the *correlation coefficient.* A value $= 1.00$ would mean that the computed regression lines depicted in these figures are correlated perfectly with the actual data. Since the R^2 value for CPI is lower than that for SPI, we can say that cost estimating and control seems to be more erratically performed than schedule planning and control. One possible explanation for this is that companies are more schedule-driven in their projects than cost-driven and therefore are willing in actuality to overspend their projects to meet time commitments.

Schedule and Cost Reliability. At least as important as good SPI and CPI ratios is the reliability of such cost and schedule performance. That is, a PM organization that erratically delivers projects with SPI or CPI $= 1.00$ is not as trustworthy to top management as a team that delivers such reliably.

Our work shows that companies with higher PMM deliver projects with more predictable schedule and cost results. As companies improve their PMM, their individual SPI results tend to deviate less from the overall SPI average. This is seen in Figure 48.8 by examining the standard deviation of SPI over a portfolio of projects.

As also seen with these data, CPI standard deviation decreases as PMM improves. That is, companies with high PMM are less likely to have projects where the budgets escalate out of control. See Figure 48.9.

Budget accuracy is important because it reduces fiduciary risk. For capital-intensive projects, this can lead to a reduction in the cost of capital and large savings for companies that borrow money for project budgets or higher financial ratings for companies that obtain project financing from the capital markets.

A subtle though crucial point that many people overlook is the reliability of SPI and CPI metrics. Many people think that a SPI or CPI that averages more than 1.00 is good, but this is not necessarily the case. It is of little help to a company in estimating project durations if half of its projects have an SPI of 1.25 and the other half 0.75. Such a large variation thwarts effective planning and management of multiple projects.

Similarly, a company that has an average SPI and CPI substantially over 1.00 is being too conservative in its estimates. It may be "leaving money on the table" and not undertaking as many projects as it could with more realistic forecasts.

The data for Figures 48.8 and 48.9 come from a relatively few number of companies, 7 and 7, respectively, but a large number of projects, 46 and 41. In statistical terms this means there is a reliable number of degrees of freedom.

FIGURE 48.8. SPI STANDARD DEVIATION VS. PMM.

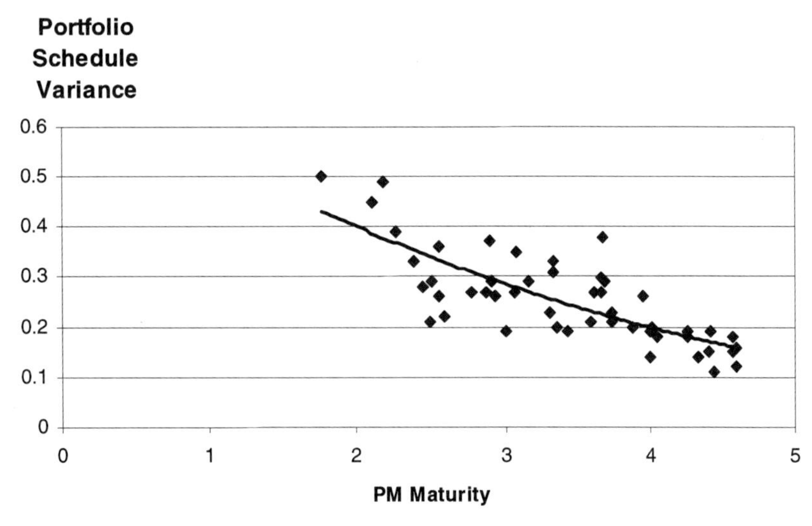

FIGURE 48.9. CPI STANDARD DEVIATION VS. PMM.

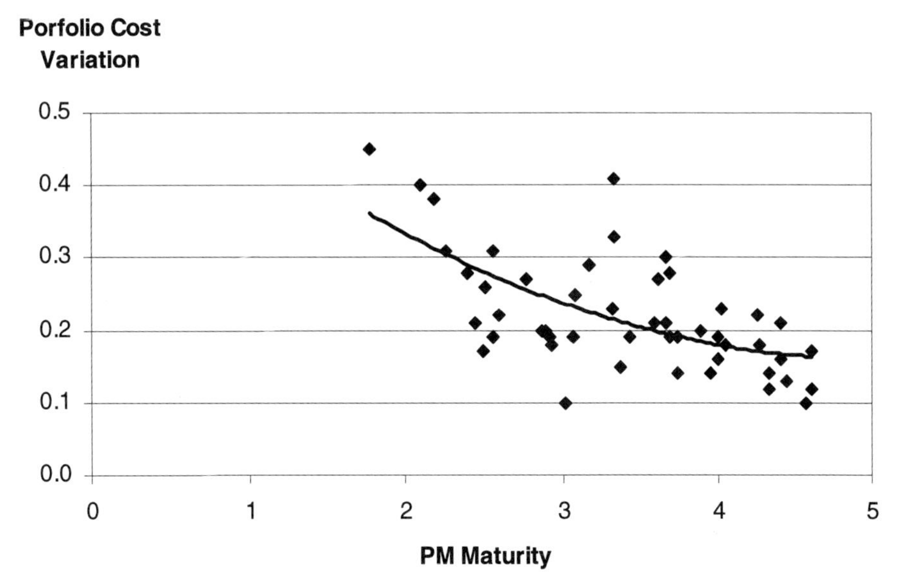

PMM and Project Management Cost Ratio. *Project management cost* entails summing all of the costs incurred by project management to deliver a project. It includes labor and burden costs for direct and indirect project management personnel; hardware, software and communications costs; and training costs; as well as those costs associated with consultants and subcontractors.

In companies that we have studied, this *project management cost ratio* is usually computed by annualizing and dividing all direct project management costs incurred by the total value of the projects executed during that same time frame (see Figure 48.10).

As the regression line in Figure 48.10 displays for N = 32 companies, the project management cost ratio increases until approximately PMM Level 3. From there the project management cost ratio steadily decreases with increasing PMM. This means that companies investing in project management will initially see their investment costs outstrip benefits. Eventually, however, the investments pay off, since mature companies actually pay less, as a percentage of project management costs, to improve their PMM. Economies of scale do hold rewards in project management, just like most other aspects of business.

Bringing It to Closure: The Virtuous Cycle of Project Management

Based on case study interviews and data collected from companies assessed during our research, we have created Figure 48.11 to illustrate what we call the Virtuous Cycle of

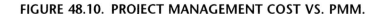

FIGURE 48.10. PROJECT MANAGEMENT COST VS. PMM.

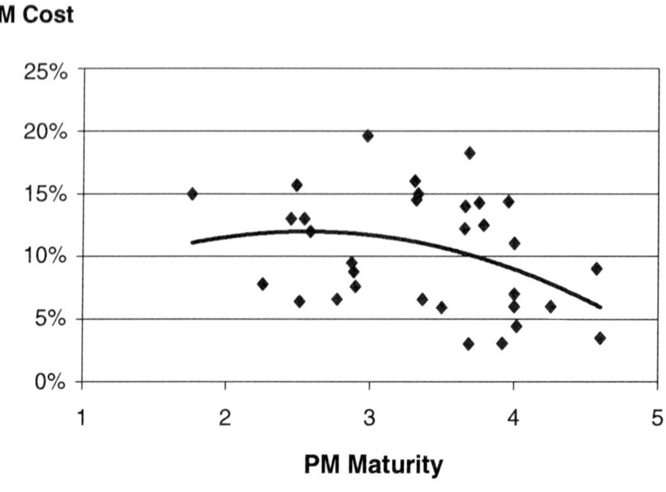

FIGURE 48.11. THE VIRTUOUS CYCLE OF PM.

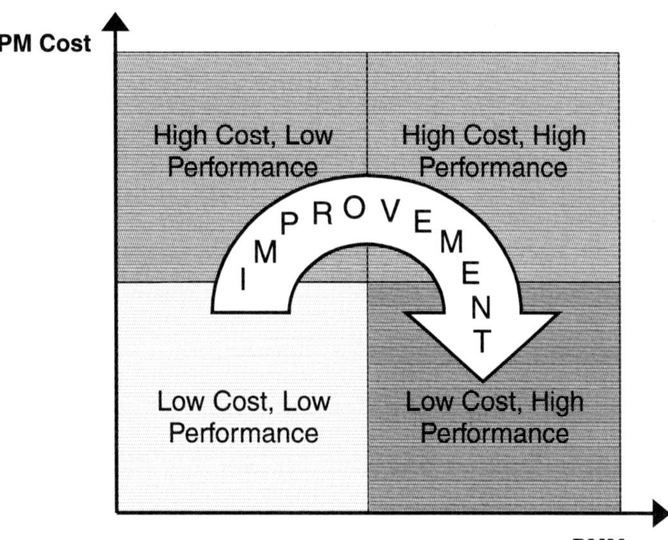

Project Management. This schematic allows organizations to map their PMM and investments in project management to plan and ensure a logical and sustainable progression along the project management cost-effectiveness journey.

The axes of the matrix pit PMM against the project management cost ratio. This diagram was developed by in-depth analysis of project management operations of 24 companies. Quadrants were divided so that each is represented by an equal number of companies to ensure no bias toward any one particular classification. The boundary between "good" and "bad" project management cost ratio % (the vertical axis) is approximately 5 percent to 7 percent; the PMM division is approximately 3.30 to 3.35. Project management growth should, we hope, go through the clockwise cycle depicted on this diagram, though the speed will vary depending on variables such as management commitment, industry circumstances, and the size and dispersion of the company.

Companies in the lower left-hand quadrant are underinvested in project management and are earning low returns, if any. Without adequate investment, both PMM and project performance will continue to lag behind peer organizations.

Continuing to the upper left-hand quadrant is the next step of the progression, and an area where no organization should reside for any prolonged duration. In this quadrant, PM investments have begun to increase, but benefits are not yet being proportionately realized. Organizations whose PM practices exist in this region for extended periods are in danger of paying a steep price for relatively poor PM performance.

In the upper right-hand quadrant, the benefits of improving PM are starting to be realized within the organization, but the cost of those improvements is still relatively steep. While having a high PMM is commendable, the victory is somewhat bittersweet in that PM is still costly for these organizations.

The lower right-hand quadrant is the ideal locale for company-wide PM practices. Companies in this category have best-of-class PMM, low PM cost and very high PM/ROI℠. These companies are in "PM nirvana" mainly because they have the highest throughput of projects with respect to their PM investment. For organizations in this arena, PM is a strong organizational competence and even, in some circles, regarded with competitive envy.

Since companies that have high-level PMM are, by definition, companies with high levels of PM learning (see Figure 48.1), they can be self-sustaining and self-improving. This allows them to become pioneering and agile organizations that grow and adapt to changing marketplace challenges, thus offering more value over time.

Summary

Project management can be a key lever for delivering projects. Many companies are interested in, and actively pursuing, initiatives to improve upon their project management processes. Understanding PMM can dramatically aid in the improvement of project management sophistication.

Improving PMM can be efficiently managed with the use of benchmarking and gap analyses. These tools allow for determining an organization's overall level of PM ability, as well as industry best practices. Most important, organizations can utilize benchmarking and gap analyses to make pointed and focused improvements in their PM processes.

Our research has shown that improvements in PMM help deliver three significant benefits:

1. Improvements in cost and schedule performance
2. Improvements in cost and schedule reliability
3. Lower overall project management cost in delivering projects

Large organizations certainly can apply statistical methods to PMM assessment methodologies, but we want to stress that in the dynamic corporate world, such analyses are as much art as science. We have presented methodologies that allow for the combination of statistical methods with experience and judgment to create a methodology that many companies can readily apply within their organizations to improve their PMM.

References

Al-Sedairy, S. T. 1994. Project management practices in public sector construction: Saudi Arabia. *Project Management Journal* (December): 37–44.

Boznak, R. G. 1988. Project management: Today's solution for complex project engineering. *IEEE Proceedings.*

Bu-Bushait, K. A. 1989. The application of project management techniques to construction and R&D projects. *Project Management Journal* (June): 17–22.

CII. 1990. *Assessment of owner project management practices and performance.* Special CII Publication (April).

Camp, R. C. 1995. *Business process benchmarking.* Milwaukee: ASQC Quality Press.

Deutsch, M. S. An exploratory analysis relating the software project management process to project success. *IEEE Transactions on Engineering Management* 38 (4, November).

Foti, R.. 2002. Maturity. *PM Network* 15(9):38–43.

Gross, R. L., and D. Price.1990. Common project management problems and how they can be avoided through the use of self managing teams. *1990 IEEE International Engineering Management Conference.* Santa Clara, CA.

Ibbs, C. W., and Kwak, Y. H. 1997. *The benefits of project management: Financial and organizational rewards to corporations.* Newtown Square, PA: Project Management Institute.

———. 2000. Calculating project management's return on investment. *Project Management Journal* 31 (2, June): 38–47.

Ibbs, C. W., P. W. G. Morris, and J. M. Reginato. 2001. Calculating the value of project management. *Project Management Institute's Annual Seminars & Symposium.* Newtown Square, PA: Project Management Institute.

Ibbs, C. W., and J. M. Reginato. 2002. *Quantifying the value of project management.* Newtown Square, PA: Project Management Institute.

Kwak, Y. H., and C. W. Ibbs. 2002. Project Management Process Maturity Model. *Journal of Management in Engineering* (July): 150–155.

Ziomek, N. L., and G. R. Meneghin. 1984. Training: A key element in implementing project management. *Project Management Journal* (August): 76–83.

CHAPTER FORTY-NINE

PROJECT MANAGEMENT MATURITY MODELS

Terry Cooke-Davies

A glance through the contents of this book provides ample evidence that project management is no longer seen as simply being concerned with the skillful and competent management of a single project. There is much more involved in it than that. Organizations undertake many projects, and so require a set of processes and capabilities, of systems and structures, to allow the right projects to be undertaken and supported and to achieve consistent project success. As this recognition has evolved, so has the desire on the part of organizations to assess these systems, structures, processes, and capabilities, and many have turned for help to so-called project management maturity models.

There is no shortage of them; more than 30 were considered as a part of the research leading up to the Project Management Institute's own draft standard OPM3 (Cooke-Davies et al., 2001), and they are supported by claims that an increase in maturity brings organizational benefits (e.g., Kwak and Ibbs, 2000; Pennypacker and Grant, 2003).

The reason for this upsurge in interest is not difficult to understand. As project management has expanded from its origins in the engineering, construction, and defense industries, IS/IT has played an increasingly prominent role in shaping the debate about project management. Well-publicized failures (e.g., Standish Group, 1994) have been accompanied by an increasing focus on developing robust software development and systems engineering processes, as well as improving the management of both software development and business change projects. A significant factor in this development has been the family of Capability Maturity Models (CMM) developed under the leadership of Watts Humphrey by the Software Engineering Institute of Carnegie Mellon University (Paulk et al., 1996).

The principle behind the original CMM is simple: If organizations wish to develop predictability and repeatability in their IS/IT production processes, they need to develop a number of capability areas, each of which consists of families of related processes. In turn, each of these processes needs to develop through a series of stages of maturity from informal at the lower end of the scale to highly routinized and with continuous improvement embedded at the higher end. To prevent the model from becoming excessively complex to understand, the capability areas and process maturity measures are combined into a series of five levels of organizational maturity, into one of which any organization can be categorized.

More and more organizations, in more and more countries, are using the software CMM, and procurers of software are increasingly specifying the level of maturity that must be achieved by would-be suppliers. As a consequence, the general level of maturity of software development organizations has shown significant improvements since the early 1990s (Software Engineering Institute, 2003). The model itself, originally for software development, has since spawned a number of other versions covering such fields as systems engineering, human resources, and, most recently, systems engineering, software development, integrated product and process development, and supplier sourcing in a model known as CMM-I, where the "I" stands for integration.

Since software is developed through projects, it is natural that the concept of organizational maturity would migrate from software development processes to project management, and this has been reflected in an interest in applying the concept of "maturity" to software project management (Morris, 2000; Cooke-Davies et al., 2001). Possibly as a result of this, a number of project management maturity models appeared during the mid-1990s that were more heavily influenced by the thinking of project management consultants and practitioners.

Against this background and recognizing the growing interest in the field, the Project Management Institute in May 1998 initiated a program known as the Organizational Project Management Maturity Model (Schlichter, 2001; Friedrich et al., 2003) to develop a standard for organizational project management processes that would complement the ubiquitous PMBOK Guide, its widely applied standard for the management of individual projects. The first draft of this standard was launched in December 2003.

As often happens in the early days of the development of new concepts, however, the field of maturity models is characterized by a tangle of confused concepts and unclear vocabulary. Several factors contribute to this confusion. In particular:

- There is no universal agreement as to the extent of the practices and processes that are necessary for the successful management of projects.
- Practices and processes are interwoven at many levels simultaneously within the field of "project management," and so it is by no means clear how and to what extent the concepts of "process control," "process maturity," and "capability" can be applied to the whole field.
- "Capability" and "maturity" are words that carry a multiplicity of meanings, some of them technically precise and others more broadly based in common usage. There is no general agreement on how such words, and the concepts that they signify, apply to the general field of project management.

But does this outpouring of creative activity add value to the field of project management, or simply confuse it by offering yet more silver bullets that will ultimately turn out to be illusory? That is the question this chapter addresses.

Before hazarding an answer to this question, however, we must shed light on confused concepts and clarify ambiguous vocabulary. But first of all it is appropriate to review briefly the literature on project management maturity (such as it is) and to describe in a little more detail the most recently developed maturity models that seek to offer themselves as serious candidates to become widely accepted standards.

A Brief Survey of the Literature on Project Management Maturity

This section will inevitably live up to its billing as brief, since there has been comparatively little written about such a new topic. The roots of the concept of process maturity seem to lie deep within the Quality movement and can be clearly traced in the copious writings of its gurus such as Walter Shewhart and W. Edwards Deming. The principle is simple: "a stable process . . . is said to be *in statistical control*. . . . A system that is in statistical control has a definable identity and a definable capability" (Deming, 1986, p. 321, italics in original text). Thus, the efforts of quality improvement are first of all to make the process stable and thus bring it under statistical control, and then to work on improving the capability of the process. A process can thus be said to mature as it passes through the stages from unstable, to stable, and then to enjoying improved capability. The effects can be clearly seen in charts such as Figure 49.1.

FIGURE 49.1. HOW THE PROCESS OF HITTING A GOLF BALL IMPROVES WITH LESSONS.

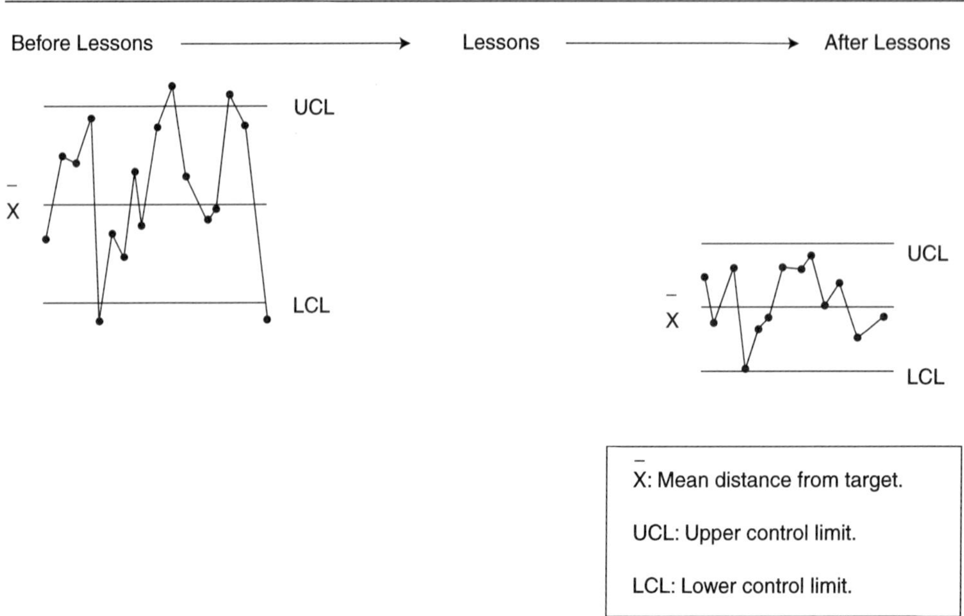

Source: J. Edwards Deming (1982). *Out of the Crisis*. Cambridge: Cambridge University Press, p. 252. Reprinted with the permission of Cambridge University Press.

These principles have been clearly embodied in the Capability Maturity Model for software that was developed by the Software Engineering Institute of Carnegie Mellon University between 1986 and 1993. Integral to the model is the concept that organizations advance through a series of five stages to maturity: initial level, repeatable level, defined level, managed level, and optimizing level. "These five maturity levels define an ordinal scale for measuring the maturity of an organization's software process and for evaluating its software process capability. The levels also help an organization prioritize its improvement efforts." (Paulk et al., 1996, p. 7). The prize for advancing through these stages is an increasing "software process capability," which results in improved software productivity.

As might be expected in a field that is so heavily dominated by practitioners, the literature on project management maturity models is concerned primarily with their practical application rather than with an exploration of the theoretical validity of the concept, or with empirical research to demonstrate their value.

Of the maturity models that have been described in the project management literature, a significant number (e.g., Couture and Russett, 1998; Ibbs and Kwak, 1997; Pennypacker, 2002) show their dependence on marrying two concepts together: "project management" as described in the PMBOK® Guide and "maturity level" as described in CMM (see Figure 49.2). Others, on the other hand, show signs of rethinking the concept of how an organization matures in its ability to manage projects (e.g., Gareis, 2001; Hillson, 2001; Kerzner, 2001). Not surprisingly, each of these develops its own unique description of the path to maturity and its own scope of the practices and processes that are to be assessed.

FIGURE 49.2. FIVE "STAGES" OF MATURITY OF THE SOFTWARE DEVELOPMENT PROCESS.

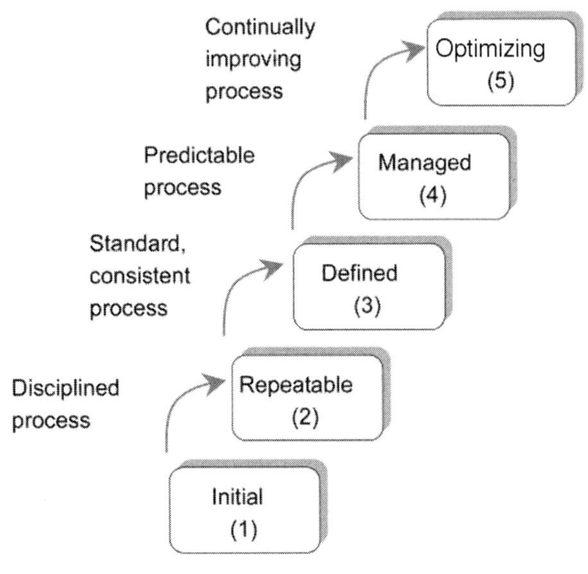

The Project Management Institute's forthcoming OPM3 (Fahrenkrog et al., 2003) has involved at one time or another more than 800 volunteers (Friedrich et al., 2003) drawn from the world's community of project managers, so it is perhaps not surprising that it has begun to influence the development of other custom in-house maturity models, such as that developed for BNY Clearing Services Inc. (Rosenstock et al., 2000).

In addition to descriptions of maturity models, conferences have heard tales of the improvements that can be obtained by individual corporations through their application (e.g., Suares, 1998; Peterson, 2000; Rosenstock et al., 2000). The models have been used in attempts both to assess the state of the art in project management (e.g., Pennypacker and Grant, 2003; Mullaly, 1998) and also (less successfully) to demonstrate the organizational benefits of project management (Ibbs and Kwak, 1997; Ibbs and Reginato, 2002).

Other models, however, are being used within organizations to assess project management maturity as a part of an overall assessment of the quality of business practices (Rosenstock et al., 2000; Cooke-Davies et al., 2001), using models such as the Baldridge National Quality Award (www.quality.nist.gov) or the European Forum for Quality Management's "Business Excellence" model (www.wfqm.org/imodel/model1.htm).

The award-winning article by Jugdev and Thomas (2002) is a refreshing exception to the complaint that little attention has been paid to questioning the fundamental relevance of maturity models to the total scope of managing projects in organizations. The paper examines maturity models (MMs, in the language of the article) from the viewpoint of four different resource-based models in order to assess whether or not the possession of a higher maturity level in project management confers competitive advantage on an organization. The article concludes that MMs possess some but not all of the characteristics of a strategic asset and thus cannot in and of themselves confer competitive advantage. It also asserts that although "MMs are a component of project management[; they are] not a holistic representation of the discipline" (p. 11).

This assertion implies an answer to this chapter's own question about the value of maturity models, about which more will be said later.

Before that, however, it is appropriate to examine two of the most recent additions to the field, each of which could conceivably become, for different reasons, a broadly used model and a widely accepted standard.

OGC's PMMM and PMI's OPM3

Each of these two models has some attributes in common with other models mentioned earlier in this chapter, but it is not the purpose of this chapter to conduct a detailed comparison between individual maturity models. The two models have been selected for special mention for two reasons: First, they are entering the field after sufficient time has elapsed to allow experiences with other models to have been taken into account, and second, each of them is backed by an organization that has demonstrated its ability to establish widely adopted standards—the Office of Government Commerce in the United Kingdom, which has produced both PRINCE2 and *Managing Successful Programmes*, and the Project Management Institute in the United States, publisher of the PMBOK® Guide.

At this writing, neither has yet been formally issued, so the possibility exists of significant change, but the general lines of development of each of them is sufficiently far advanced that their potential contribution to the field can be assessed. The information on PMMM has been based on Version 5.0 of the model, which is available on the OGC's Web site. The information on OPM3 has been taken from papers presented at the PMI Global Assembly 2003—Europe (Fahrenkrog et al., 2003; Friedrich et al., 2003).

PMMM

In the introduction to the Project Management Maturity Model (PMMM), the OGC observes that it has been developed because SEI's experience in the arena of software development between 1986 and 1991 indicated that maturity questionnaires provide a simple tool for identifying areas where an organizations' processes may need improvement. The model is descriptive, with the express intention of providing organizations with guidance to support their process improvement initiatives, and the document describing the model is at pains to point out that the model itself is not to be confused with any questionnaire that may be used to establish an organization's current maturity level.

Each stage is characterized by a discrete set of processes that are definitive of the stage of maturity (see Figure 49.3).

The description of each process includes the process goals and functional achievement, the approach laid down, the deployment that is to be expected, the method of review that

FIGURE 49.3. PROCESSES INCLUDED IN EACH STAGE OF MATURITY—PMMM.

Level 1: Initial Process	Level 2: Repeatable Process	Level 3: Defined Process	Level 4: Managed Process	Level 5: Optimized Process
1.1 Project Definition	2.1 Project Establishment 2.2 Requirements Management 2.3 Risk Management 2.4 Project Planning 2.5 Project Monitoring and Control 2.6 Management of Suppliers and External Parties 2.7 Project Quality Control 2.8 Configuration Definition and Control	3.1 Organizational Focus 3.2 Project Management Success 3.3 Project Training 3.4 Integrated Management 3.5 Lifecycle control 3.6 Interteam Coordination 3.7 Quality Assurance	4.1 Project Metrics 4.2 Organizational Quality Management	5.1 Proactive Problem Management 5.2 Technology Management 5.3 Continuous Process Improvement

is recommended, the way the organization should perceive the process, and the performance measures that should be used. Inherent in the idea of a mature organization is the existence of an organization-wide capability to manage projects based on a set of clearly defined common processes that can be tailored to meet the needs of individual projects. The introduction to the model includes a description of the two extreme states of maturity: immature and mature.

The model can be used for either or both of two purposes:

- To understand the key practices that are part of an effective organizational process to manage projects
- To understand the key practices that need to be embedded within the organization to achieve the next level of maturity

It could be used by any organization wanting to improve its capability to manage projects effectively, by governance bodies and consultancies for the purpose of developing maturity questionnaires, or by accredited service providers in assisting teams to perform project management process assessments or capability evaluations.

OPM3

The Project Management Institute has announced its firm intention to launch OPM3 as a draft standard before the end of 2003. It differs from many of the other models mentioned in this chapter in that it introduces a structure that owes little to the structure of CMM, but rather one that relates explicitly to the PMBOK® Guide (although it is dramatically different in scope); that covers the three domains of portfolio management, program management, and project management; and that explicitly relates the management of projects to organizational strategy.

The basic "building blocks" at the heart of OPM3 are five different kinds of entity:

1. "Best practices" that are associated with organizational project management
2. "Capabilities" that are prerequisite or that aggregate to each "best practice"
3. The observable "outcomes" that attest to the existence of a given "capability" in the organization
4. Key performance indicators (KPIs) and metrics that provide the means of measuring the "outcome"
5. Pathways that identify the capabilities aggregating to the "best practices" being reviewed

The relationships among these are shown in a simplified manner in Figure 49.4.

An example that has been given of a best practice is "Use Teamwork—Cross-functional teams carry out the organization's activities." The four associated capabilities leading up to this are "develop cross-functional training opportunities," "organize project work by functional area," "develop cross-functional teams," and "develop integrated program and project teams." Additional dependencies have been identified between specific "best practices" when

FIGURE 49.4. THE RELATIONSHIP BETWEEN FUNDAMENTAL ELEMENTS IN OPM3.

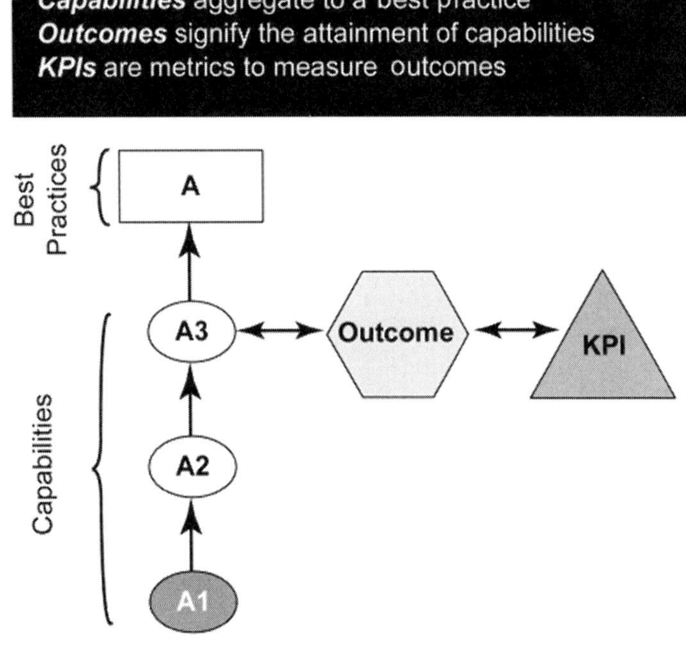
Capabilities aggregate to a best practice
Outcomes signify the attainment of capabilities
KPIs are metrics to measure outcomes
Best Practices
A
Capabilities
A3
A2
A1
Outcome
KPI

Source: © Project Management Institute: reproduced by permission.

dependencies exist between one or more capabilities that aggregate to different "best practices."

These basic building blocks are used within a framework of organizational project management processes that identifies five "process groups" (initiating, planning, executing, controlling, and closing processes) in each of three "domains" (portfolio management, program management, and project management).

Each process group in each domain is seen to progress through four stages of "process improvement" (standardized, measured, controlled, and continuously improved) to give an overall framework for the model. (See Figure 49.5.) Each of the "best practices" is mapped onto at least one location within this three-dimensional model, so that OPM3 will tell the user where a "best practice" falls within a "process group," "domain," and stage of "process improvement."

The scope of the model is vast: There are more than 600 "best practices," more than 3,000 "capabilities," and more than 4,000 relationships between capabilities. The finished

FIGURE 49.5. THE PROCESS CONSTRUCT OF OPM3.

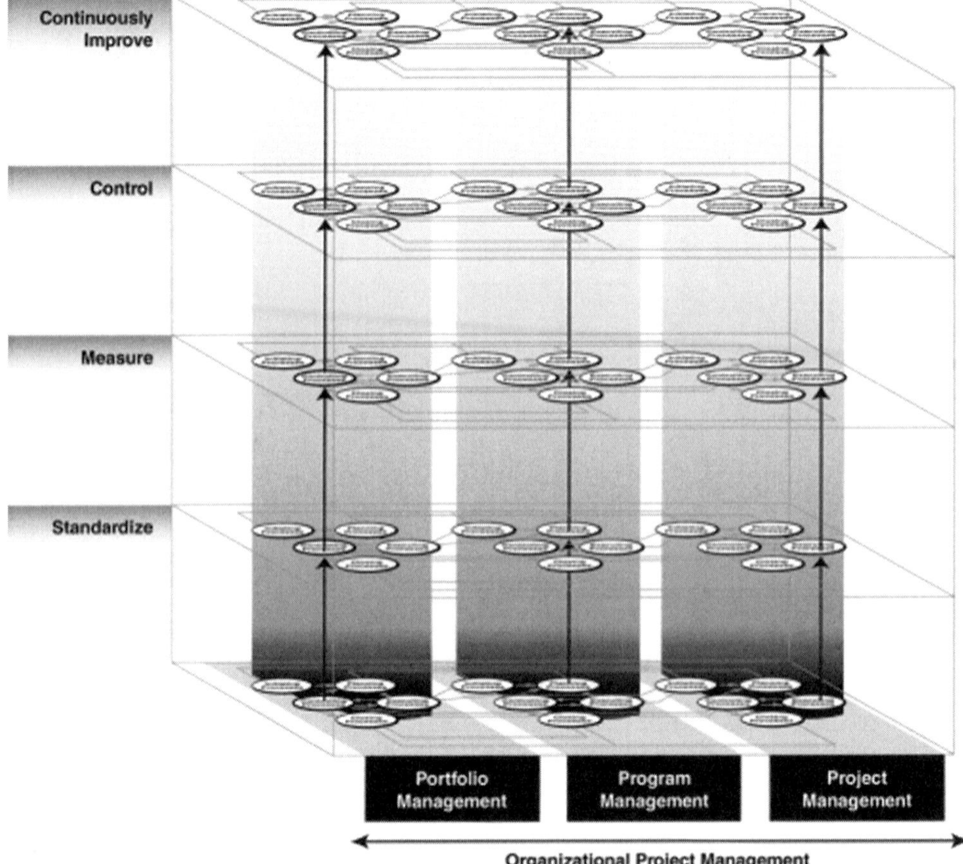

Source: © Project Management Institute, *Organizational Project Management Maturity Model* (*OPM3*), Project Management Institute, Inc., 2003. Copyright and all rights reserved. Material from this publication has been reproduced with the permission of PMI.

product is likely to take the form of both a book and a CD-ROM, and to contain both means of self-assessment and assessment by external consultants.

The model is designed to be used by an organization for any or all of four purposes:

- To understand what practices and processes have been consistently found to be useful by organizations seeking to undertake "organizational project management," which is defined as "the application of knowledge, skills, tools, and techniques to organisational and project activities to achieve the aims of an organisation through projects" (Fahrenkrog et al., 2003, p. 2).

- To assess its ability to implement its high-level strategic planning at the tactical level of managing individual projects and groups of projects.
- To drive business improvement.
- To integrate organizational practices and processes in the domains of portfolio management, program management, and project management.

Untangling the Vocabulary and Distinguishing Relevant Concepts

Earlier in this chapter, three areas of confusion and ambiguity were identified, each of which adds somewhat to the entangled nature of conversations about project management maturity models. This section attempts to reach toward answers to three fundamental questions: "What is the extent of practices and processes that are necessary to the effective and efficient management of projects?" "What is meant by the words 'practice' and 'process' as they apply to the management of projects?" and "What is meant by the words 'maturity' and 'capability' as they apply to the management of projects?"

What Topics Are Covered by the "Management of Projects"? Unless the scope of the topic can be agreed, it is unlikely that it will be possible to agree on what the "management of projects" might look like in its "perfected end-state," so this first conversation is fundamental to the topic of organizational project management maturity. (Similar discussions occur and points are made in the chapters by Crawford on project management standards and by Morris on the validity of project management knowledge.)

Perhaps the place to start in considering this question is with a review of the "bodies of knowledge" that are produced by several of the world's project management professional associations. Both the longest-established and the most widely distributed is undoubtedly the PMBOK® Guide produced by the Project Management Institute. First produced in 1976, and most recently updated in 2000, this document, which had over 270,000 copies in circulation in September 2001 (Crawford, 2002), seeks "to identify and describe the knowledge and practices that are applicable to most projects most of the time" (Duncan, 1996, p. 3). It recognizes that the management of projects also requires general management and specific application area (i.e., industry, market, or technology) knowledge and practice, but restricts itself to the knowledge and practices that are generally applicable to the management of individual projects.

The Association for Project Management in 1986 developed the framework for what was to become the *APM Body of Knowledge*, which is now in its fourth edition, having been updated in 2000 (Dixon, 2000) on the basis of research carried out by the Centre for Research in the Management of Projects (Morris, 2001). A much broader range of topics is covered by the *APM Body of Knowledge*, in line with the findings of research into project success, which suggests that a much broader range of factors is critical to project success than the knowledge and practices contained within PMBOK® Guide. (Baker, Murphy, and Fisher, 1974; Morris and Hough, 1987; Lechler, 1998; Pinto and Slevin, 1998; Crawford, 2000; Crawford, 2001; Cooke-Davies, 2001; Cooke-Davies, 2002) (See also the chapter by Cooke-Davies on project success.)

Following the development of "bodies of knowledge" by various European professional associations, the International Project Management Association in 1998 published in French, German, and English the "International Competency Baseline" (Caupin et al., 1999), offering a coordinated set of definitions to the terms used in the Swiss, German, French, and UK documents. Other professional associations (e.g., AIPM, PMISA) have their own "bodies of knowledge" and/or competency standards, usually resembling some combination of those that have already been discussed.

Most recently, a three-year joint academic/government/industry study in Japan has resulted in the production of an innovative standard for project management known as P2M (Project Management Professionals Certification Center, or PMCC, 2002). This is remarkable for the thoroughness with which it reexamines and redefines the practices, processes, and competencies that are necessary to deliver innovation through strategies, programs, and projects.

It has been argued forcefully and cogently (Morris, 2001; Morris, 2003; Crawford, 1998; Crawford, 2001; Crawford, 2002) that the absence of global standards works to the detriment of the practice of managing projects in multinational or global organizations. Precisely the same argument can be used with regard to maturity models—to the extent that enterprises are seeking to assess their organizational capability for managing projects, the absence of a generally accepted definition of what is involved inevitably inhibits the value of any maturity model to the whole of an organization.

What Do "Practices" and "Processes" Mean in Connection with Project Management? The second fundamental area that causes some confusion is precisely what is meant by the terms "process" and "practice." The 1980s and 1990s saw an emerging fashionable focus on adopting a "process" view of organizations—defining a process as "a specific ordering of work activities across time and place, with a beginning, an end, and clearly identified inputs and outputs: a structure for action" (Davenport, 1993, p. 6). The term is used in much the same way in the PMBOK® Guide, where the process groups and the "knowledge areas" are defined in terms of individual processes, described in terms of their inputs, outputs, and mechanisms.

The practice of identifying, describing, and then improving business processes lies at the heart of the quality movement, as has been seen earlier in this chapter. It accounts for the emphasis given to processes in models such as the EFQM "Business Excellence" model and the Baldridge award, to which reference has already been made. In parallel with the quality movement, the fashion for viewing organizations as collections of interlocking processes gave rise to disciplines such as business process reengineering (e.g., Hammer and Champy, 1993) and benchmarking (e.g., Camp, 1989). These movements didn't simply focus on making processes repeatable and predictable, but rather they sought to identify the specific work flows and working practices that lead to improved process performance. Camp's working definition of benchmarking as "the search for industry best practices that lead to superior performance" (1989, p. 12) makes this clear. And process performance is measured by physical characteristics such as throughput time, efficiency measured by unit of output produced for unit of input, or cost per unit of throughput.

Thus, in terms of a "process view" of work, a *process* describes how the organization's inputs are converted into outputs, and *practices* describe how the processes are carried out.

However, the process view is not the only way of describing work. It could be argued that projects, for example, are a specialized subset of processes—those that are carried out only once so as to produce a unique product, service, or beneficial change. Most of the definitions of a project make reference to the unique nature of each of them, whereas the essence of the process view is the repetitive nature of the work carried out.

Clearly the management of projects involves a large number of processes, as described in each of the bodies of knowledge, but equally clearly, different organizations employ their own practices to undertake these processes. And as research shows, different organizations in different industries recognize very different areas of practice as being appropriate to managing projects (Toney and Powers, 1997; Turner and Keegan, 1999; Turner and Keegan, 2000; Cooke-Davies and Arzymanow, 2002; Morris, 2003).

Outside of the process view of an organization, considerable fluidity characterizes the way literature on the management of projects uses the words "process," "practice," and even "discipline." For example, in a discussion of the "topic" of project management, the *APM Body of Knowledge* states that "Project Management is *the discipline* of managing projects successfully," and then on the following page includes a diagram with the caption "this diagram illustrates *the project management process*" (2000, pp. 14, 15; italics mine).

In contrast to the use of the term "practice" by proponents of benchmarking and other approaches to process management to denote how "processes" are carried out, PMBOK® Guide states that "Part II, the Project Management Knowledge Areas, describes project management *and practice* in terms of its component processes" (2000b, p. 7 "italics mine"). Thus, in some usage by project management practitioners, practice is a characteristic of processes, and elsewhere processes are components of practice.

Once more, as in the prior discussion about what is involved in the management of projects, the absence of generally accepted definitions for these two key terms creates confusion for the application of maturity models.

What Do "Maturity" and "Capability" Mean in Connection with Maturity Models? In Collins Dictionary, the adjective "mature," from which the noun "maturity" is derived, has a number of different meanings in common usage. It can, for example, mean "(1) fully-developed or grown up; (2) of plans or theories it can mean that they are fully considered, perfected; (3) of insurance policies or bills it can mean due or payable; and (4) of fruit, wine or cheese it can mean ripe or fully aged." The last two of these do not offer any obvious link to the world of project management (unless through projects that are described as "pear-shaped" when they are in serious trouble), so it is in either meaning (1) or (2) that the word is used in the term "maturity models."

According to SEI's CMM "a software process can be defined as a set of activities, methods, practices, and transformations that people use to develop and maintain software and the associated products (e.g., project plans, design documents, code, test cases, and user manuals). As an organization matures, the software process becomes better defined and more consistently implemented throughout the organization." (Paulk et al., 1996, p. 3) In other words, maturity is used in meaning (2) as a technical description of the state of definition and consistency of implementation of an end-to-end process.

There is a clear implication that this can be accomplished only through the willful application of process improvement effort over time, and this would seem to link the use of the term to its more general meaning: (1). However, a working definition of a mature organization or a mature process would seem to be one that has reached what is, to all intent and purposes, a perfected state—one that is capable of delivering the requisite outcomes consistently, efficiently, and effectively.

This working definition also builds a link to the second word being considered here: "capability." Once more, there are two rather different meanings in the dictionary: "(1) the quality of being capable, ability; and (2) potential aptitude".

Once more CMM states that "software process capability describes the range of expected results that can be achieved by following a software process. The software process capability of an organization provides one means of predicting the most likely outcomes to be expected from the next software project the organization undertakes" (Paulk et al., 1996, p. 3). This sounds more like meaning (2)—dealing with a potential rather than an actual quality possessed. The correlation is made more explicit when the explanation of important concepts goes on to define "software process performance" as being the actual results accomplished by following a software process.

All this looks logical and clear, but all is not as it seems. Having spoken about what happens to "the" software process as an organization matures, the CMM introduces as a fundamental concept "software process maturity"—the development through five stages from definition, through management, measurement, and control to effectiveness. The term "maturity" and the five stages of development are thus applied *both* to the software process and to the organization that is undertaking it.

Within CMM, where each term is tightly defined, this does not appear to cause difficulties. Indeed, the descriptions in the model and the assessment methods are sufficiently tight, and little room is left for ambiguity.

As the terms and concepts are translated into the world of project management, with all of its inherent uncertainties, ambiguities, and disagreements, the distinction between the maturity of a process and the "maturing" of an organization becomes more problematical. It seems reasonably intuitive and logical to describe the "maturity," the "capability," and the "performance" of a single process within the field of project management, such as "activity duration estimating," using precisely the same definitions as are used in CMM. It becomes less so when considering the "maturity," the "capability," or the "performance" of an organization that undertakes many projects and programs of different categories (such as engineering, marketing, or business development) in many different business units, for many different purposes, using many different criteria for success. And it is this consideration that gives rise to the fundamental tacit assumption behind project management maturity models: that there is an underlying development path or trajectory that must be followed by organizations as they seek to improve their ability to manage projects successfully.

But is there?

This question will be examined in more detail a little later in the chapter, when the potential contribution of maturity models is considered in relation to the field of project management. Before that, though, it is appropriate to review the "state of play" on the field of maturity models in the light of these three fundamental questions.

Maturity Models: The "State of Play" Reviewed

Starting with the most recent two models first, it would appear that both PMM and OPM3 avoid some of the inherent difficulties that have been sketched previously, but not all of them. Of course, by the time they are launched as potential standards, they will certainly undergo changes, which may alter the conclusions about them.

OPM3 is by far the largest and most complex of the project management maturity models, and it might well turn out to be the most comprehensive. It recognizes the heritage of maturity models in the quality movement and acknowledges that "practices" are components of processes or process groups. It is also clear in its recognition that a process can be described as "mature" when it has achieved its "perfected end-state," has a measured and defined "capability," and is subject to process improvement initiatives. The identification of both outcomes and KPIs for each "capability" represents a substantial achievement.

In itself, however, OPM3 is far from "mature" in its own terms. The choice of the terms "best practice" and "capability" for two of the five elemental components of the model is not the happiest, in view of the semantic confusion that already exists. In spite of extensive market research (mainly conducted within the community of practitioners that are familiar with PMBOK® Guide), neither the 600 or so "best practices" nor the 4,000 or more dependency paths that the model contains can be demonstrated empirically to describe the essential trajectory (or trajectories—the model incorporates great flexibility) to organizational maturity, as measured by the successful implementation of strategy through projects, which is the stated goal of the model.

PMMM is very different in terms of its detail (67 pages compared with what might turn out to be more than 800 for OPM3), its focus (on government departments undertaking projects with a high IT content, or other organizations embracing PRINCE2, rather than any organization undertaking projects), and its derivation directly from CMM. The 21 processes described in the five stages incorporate many of the broader areas of the field of project management that are included in the *APM Body of Knowledge*, and also the principles of quality management. The terminology of "maturity," "capability," and "process" is very close to that contained in CMM.

On the other hand, there are at present some surprising omissions. For example, neither program management nor portfolio management is included in the model. There is also, more than in OPM3, an implicit single "development path" or "trajectory" toward the state of "maturity" to which, by implication, most organizations should aspire.

To turn back to models that were considered in the literature review, it is clear that different categories of model have different strengths and weaknesses that can be reviewed briefly. The earliest maturity models that combine the concept of CMM's five stages of maturity with the PMBOK® Guide's project management processes (e.g., Couture and Russett, 1998; Ibbs and Kwak, 1997; Pennypacker, 2002) fail to distinguish between organizational maturity and process maturity (the organization is mature when every process is mature) and also omit from their consideration the extensive areas of practice that contribute to the successful management of projects that are not covered by PMBOK® Guide.

The CMM family of models itself and its derivatives, of course, are useful in terms of organizations to whom the software process is an important component of what APM's Body of Knowledge refers to as "technology management". Those organizations that are seeking to improve their overall excellence, using the Baldridge or EFQM models, certainly cover the whole field of practices necessary for the management of not only projects, but of everything else as well. They contain neither implicit nor explicit process or capability elements to assist organizations that are seeking explicitly to improve the maturity of their project management.

Each of the remaining models that has been considered contains its own assumptions about the processes that need to be added at each stage of maturity, and thus implies its own hypothesis about the appropriate "development path" that leads to maturity. Until empirical project management research is in a position to demonstrate the validity of one or more of these development paths to project management maturity, or the correlation of project management maturity to consistent project success, then the adoption of a particular model remains largely an act of faith. But is such an act of faith reasonable? Do maturity models, in their present form, add value to the field of project management, or do they simply add to the confusion? That is the question to which we can now hazard an answer.

Maturity Models: Silver Bullets or Unhelpful Distractions?

The purposes for which an organization might seek to use a maturity model have been variously described as the following:

1. To understand what practices and processes have been consistently found to be useful by organizations seeking to undertake organizational project management
2. To drive business improvement, for example, by understanding the key practices that need to be embedded within the organization to achieve the next level of maturity
3. To assess its ability to implement its high-level strategic planning at the tactical level of managing individual projects and groups of projects
4. To integrate organizational practices and processes in the domains of portfolio management, program management, and project management

Several limitations to the various different types of maturity model have already been described, but these do not necessarily prevent the models from providing value, or from helping organizations to accomplish any or all of the objectives in the list.

As demand has grown for the effective management of an increasing number projects, it has been helpful to the advancement of project management to identify those processes that are "applicable to most projects, most of the time" and to help a growing number of practitioners learn what those are. Maturity models, in a sense, seek to do for organizations seeking to implement strategy through projects what "bodies of knowledge" have done for individual practitioners seeking to improve their ability to manage projects.

But three factors make the practice of "organizational project management" considerably more problematical than the management of individual projects. First, it has long been recognized that the related product-oriented processes "are typically defined by the project life cycle, and vary by application area" (Project Management Institute, 2000a, p. 30). For example, the product-oriented processes involved in the development of new pharmaceutical products, with their inherent technical risk and consequent uncertainties of scope, will inevitably differ in many important respects from those involved in the construction of a new building or the development of software. Thus, different organizations will have a bias toward projects with a certain set of characteristics, and the strategies that lead to business success in different "application areas," industry, or markets may differ radically from each other. Morris' review (2003) of the differences between project management as practiced in four different industries (construction, information technology, defense/aerospace, and pharmaceuticals) describes these differences very clearly. These differences may be so great as to raise doubts about the concept of a single "development path" toward a "perfected end-state."

Second, regardless of the type of "application area," organizations themselves operate in many different industry and market environments. Banks, for example, operating in the financial services market, undertake construction projects, information technology projects, business development projects, process reengineering projects, and so on. The environmental differences can be characterized on a grid as shown in Figure 49.6. Project management

FIGURE 49.6. DIFFERENT "ENVIRONMENTS" FOR PROJECTS.

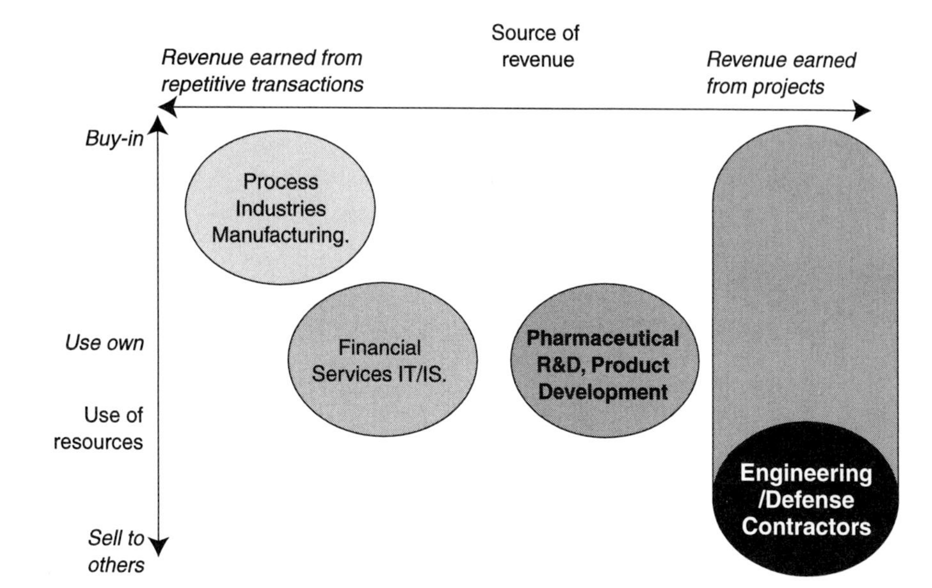

clearly has a different "voice" in the traditional project-based engineering environment, displayed on the right-hand side of the diagram, than it does in those nearer to the left. The vertical axis also plays a part—the more the revenue of an organization depends on the efficient and effective use of its own resources on projects (i.e., the nearer to the bottom of the diagram), the greater the commercial pressures on that organization to develop a world-class corporate capability in project management. This perhaps accounts for the evidence that engineering suppliers to the process industries have the highest project management "maturity" of any industry (Cooke-Davies and Arzymanow, 2002). It also suggests that there may be no common path to "project management maturity" in different industries and for different types of projects.

The third factor that complicates the question of "organizational project management maturity" is that in order to compete strategically in their markets, organizations have to pay attention to two very different aspects of business: "business as usual," in terms of extracting value from their current products and markets, and "business change," in terms of preparing the organization to compete in the future in new markets or with new products. Projects can play their part in either of these aspects, as illustrated in Figure 49.7. For example, one of the gurus of the quality movement, J. M. Juran, has maintained that quality of products or services can be improved only through projects (Anbari, 2003), and it is

FIGURE 49.7. THE RELATIONSHIP BETWEEN "BUSINESS AS USUAL" AND "BUSINESS CHANGE."

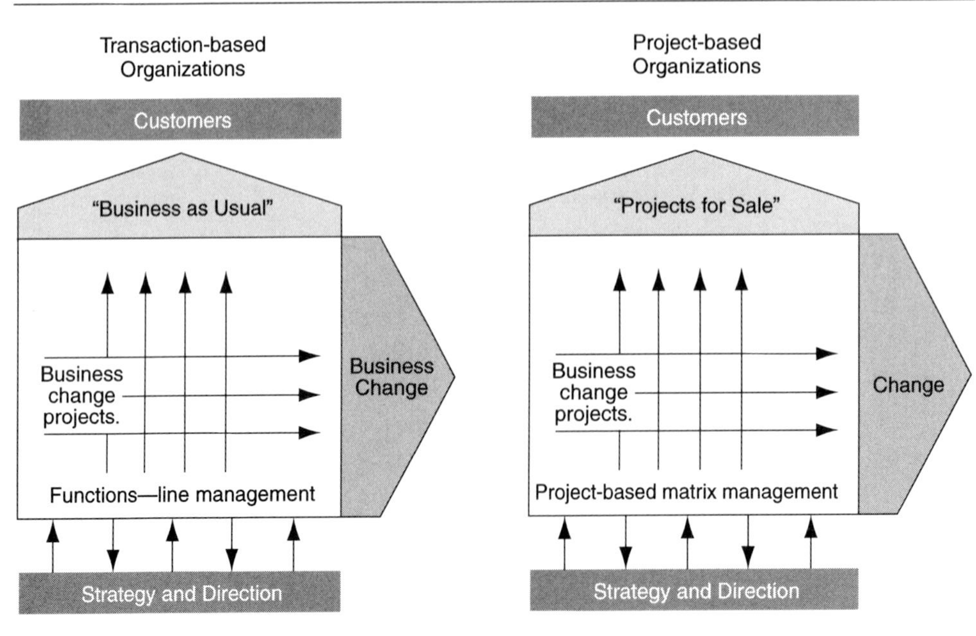

accepted wisdom among project management practitioners that project management is essentially the management of business change.

The reason that this complicates the question of project management maturity is that, as Mintzberg (1989) has argued, business strategy is "crafted" by managers in a way that is analogous to how a potter crafts a pot from the clay with which he or she is working. The strategy is a function of the present situation, with its inherent possibilities and the skills of the organization. But the present situation that organizations find themselves in varies widely depending on the nature of the organization's structure and culture, and the relative pressure felt to emanate from each of the two strategic aspects. It is these three factors that establish the unique environment of each organization as it seeks to accomplish any of the four purposes of a maturity model: to know what "good practice" looks like, to identify sensible improvement initiatives, to implement strategy more effectively, or to integrate the management of all the projects in the organization.

By definition, any model is a simplification of reality. If that were not the case, the model would serve no purpose, since it would be as complex and as opaque as reality itself. Where project management maturity models are concerned, however, there remains a question mark as to whether it is possible to strike a balance between a model that is an essentially accurate representation of the complex reality that covers the management of all projects by all organizations, and one that is simple enough to offer practical help to most organizations, most of the time.

Until that question is answered, each organization that is considering adopting a maturity model should ask itself a number of questions:

- Does the scope of the model cover all those arcas of the management of projects that we believe to be strategically important in our current competitive environment?
- Does the definition of maturity— thc perfected end-state described by the model—look like a state to which it is strategically important that our organization should aspire?
- Does the cost of assessing our current state of maturity and identifying our desirable path of development toward maturity look as if it is justified in the light of the benefits we could expect?
- Is the use of a maturity model the most cost-effective way that we can assess our current status and identify our future development needs?

The answers are likely to vary considerably from model to model, and from organization to organization.

Summary

There are three general conclusions that can be drawn from this discussion. First, Thomas and Jugdev's assertion (2002) that maturity models are not a holistic view of the field of project management appears to be justified. The field is broader than can be embraced simply within the process view that the models embody, and projects possess distinctive features that distinguish them from processes.

Second, maturity models make a positive contribution to the field because they

- at best broaden the discussion from a narrow definition of project management toward the the management of projects;
- recognise that developing a mature capability to manage projects requires an organizational commitment to the development of incremental capabilities along some kind of a development path, as well as the continual improvement of associated processes;
- bring into the field of managing projects core values that have proven themselves to be valuable in the field of quality management.

Third, maturity models are unlikely to be the silver bullet that some hope for because they

- lack an important precondition in terms of a well-researched and theoretically grounded understanding of just what is involved in the management of projects;
- are built, in many cases, on an unproven assumption that there is an ideal development path toward maturity that most organizations must follow most of the time, regardless of application area, project, and market environment or competitive strategy;
- must steer a perilous path between the "Scylla" of over-simplification and the "Charybdis" of excessive complexity.

Regardless of these conclusions, project management maturity models are a visible feature of the current landscape of project management. Their use as a means of comparing capability across organizations and industries seems likely to ensure their longevity.

The real question that project management practitioners, consultants, and academics should be asking is this: "Will they simply remain an interesting phenomenon of limited relevance and application, or will they provide the means of transforming the success rate of projects for which organizations are searching?" Any answer to this question remains largely a matter of conjecture at this time. In the long run, it will depend on whether or not the considerable investment of effort involved in assessing, adopting, adapting, and implementing a project management maturity model in an organization translates into sufficient value from improved results to justify its continued place in the armory of project management practices. The increase in the adoption of the software CMM suggests that IS/IT organizations are finding the investment worthwhile, but it does not follow that this will also hold true for the much more complex and diffuse world of organizational project management. Watch this space!

References

Anbari, F. T. 2003. Strategic implementation of six sigma and project management. *PMI Global Congress 2003—Europe.* Newtown Square, PA: Project Management Institute.

Baker, B. N., D. C. Murphy, and D. Fisher. 1974. *Determinants of project success*, NGR 22-03-028. National Aeronautics and Space Administration.

Camp, R. C. 1989 *Benchmarking*. Milwaukee: ASQC Quality Press.

Caupin, G., H. Knoepfel, P. W. G. Morris, E. Motzel, and O. Pannenbäcker. O. 1999. *ICB. IPMA Competence Baseline*. Monmouth IPMA.

Cooke-Davies, T. J. 2001. *Towards improved project management practice: Uncovering the evidence for effective practices through empirical research*. www.dissertation.com.

———. 2002. The "real" success factors on projects. *International Journal of Project Management* 20(3): 185–90.

Cooke-Davies, T. J., and A. Arzymanow. 2002. The maturity of project management in different industries. *Proceedings of IRNOP V. Fifth International Conference of the International Network of Organizing by Projects*. Rotterdam: Erasmus University.

Cooke-Davies, T. J., F. J. Schlichter, and C. Bredillet. 2001. Beyond the PMBOK Guide. *Proceedings of the 32nd Annual project Management Institute 2001 Seminars and Symposium*. Newtown Square, PA: Project Management Institute.

Couture, D., and R. Russett. 1998. Assessing project management maturity in a supplier environment. *Proceedings of the 29th Annual project Management Institute 1998 Seminars and Symposium*. Newtown Square, PA: Project Management Institute.

Crawford, L. 1998. Standards for a global profession—project management. *Proceedings of the 29th Annual project Management Institute 1998 Seminars and Symposium*. Sylva, NC: Project Management Institute.

———. 2000. Profiling the Competent Project Manager. *Proceedings of PMI Research Conference*. Newtown Square, PA: Project Management Institute.

———. 2001. Towards global project management standards. *Proceedings of the International Project Management Congress 2001, Project Management Development in the Asia-Pacific Region in the New Century*. Tokyo, November 18–21. Tokyo: ENAA and JPMF.

———. 2002. Developing project management competence for global enterprise. *Proceedings of ProMAC 2002 Conference*. July. Singapore: Nanyung Technical University.

Davenport, T. H. 1993. *Process innovation: Reengineering work through information technology*. Boston: Harvard Business School Press.

Deming, W. E. 1986. *Out of the crisis*. Cambridge, UK: Cambridge University Press.

Dixon, M. 2000. *Project Management Body of Knowledge*. 4th ed. High Wycombe, UK: Association for Project Management.

Duncan, W. R. 1996. *A guide to the Project Management Body of Knowledge*. Newtown Square, PA: Project Management Institute.

Fahrenkrog, S., C. M. Baca, L. M. Kruszewski, and P. R. Wesman. 2003. Project Management Institute's Organizational Project Management Maturity Model (OPM3). *PMI Global Congress 2003—Europe* Newtown Square, PA: Project Management Institute.

Friedrich, R., F. J. Schlichter, and W. Haeck. 2003. The history of OPM3. *PMI Global Congress 2003—Europe*. Newtown Square, PA: Project Management Institute.

Gareis, R. 2001. Assessment of competences of project-oriented companies: application of a process-based maturity model. *Proceedings of the 32nd Annual project Management Institute 2001 Seminars and Symposium*. Newtown Square, PA: Project Management Institute.

Hammer, M., and J. Champy, J. 1993. *Reengineering the corporation: A manifesto for business revolution*. London: Nicholas Brealey.

Hillson, D. 2001. Benchmarking organizational project management capability. *Proceedings of the 32nd Annual project Management Institute 2001 Seminars and Symposium*. Newtown Square, PA: Project Management Institute.

Ibbs, W. C., and Y. H. Kwak. 1997. *The benefits of project management. Financial and organizational rewards to corporations.* Newtown Square, PA: PMI Educational Foundation.

Ibbs, W. C., and J. Reginato. 2002. Can good project management actually cost less? *Proceedings of the 33rd Annual project Management Institute 2002 Seminars and Symposium.* Newtown Square, PA: Project Management Institute.

Kerzner, H. 2001. *Strategic planning for project management using a project management maturity model.* New York: Wiley.

Kwak, Y. H., and W. C. Ibbs. 2000. Calculating project management's return on investment. *Project Management Journal* 31(2):38–47.

Lechler, T. 1998. When it Comes to project management, it's the people that matter: An empirical analysis of project management in Germany. *IRNOP III. The Nature and Role of Projects in the Next 20 Years: Research Issues and Problems.* Calgary: University of Calgary.

Mintzberg, H. 1989. *Mintzberg on management: Inside our strange world of organisations.* New York: Free Press.

Morris, P. W. G. 2000. Researching the unanswered questions of project management. *Proceedings of PMI Research Conference 2000.* Newtown Square, PA: Project Management Institute.

———. 2001. Updating the project management bodies of knowledge. *Project Management Journal* 32(3): 21–30.

———. 2003. The irrelevance of project management as a professional discipline. *17th World Congress on Project Management.* Moscow: IPMA.

Morris, P. W. G., and G. H. Hough. 1987. *The anatomy of major projects: A study of the reality of project management.* London: Wiley.

Mullaly, M. 1998. 1997 Canadian project management baseline survey. *Proceedings of the 29th Annual project Management Institute 1998 Seminars and Symposium.* Newtown Square, PA: Project Management Institute.

Paulk, M. C., W. Curtis, M. Chrissis, and C. V. Weber. 1996. *Capability Maturity Model for Software, Version 1.1.* Technical Report: CMU/SEI-93-TR-024: ESC-TR-93-177: February 1993. Pittsburgh: Software Engineering Institute, Carnegie Mellon University.

Pennypacker, J. S. 2002. Benchmarking project management maturity: Moving to higher levels of performance. *Proceedings of the 33rd Annual Project Management Institute 2002 Seminars and Symposium.* Newtown Square, PA: Project Management Institute.

Pennypacker, J. S., and K. P. Grant. 2003. Project management maturity: An industry benchmark. *Project Management Journal* 34(1):4–11.

Peterson, A. S. 2000. The impact of PM maturity on integrated PM processes. *Proceedings of the 31st Annual project Management Institute 2000 Seminars and Symposium.* Newtown Square, PA: Project Management Institute.

Project Management Institute. 2000. *A guide to the Project Management Body of Knowledge.* 2000 ed. Newtown Square, PA: Project Management Institute.

Project Management Professionals Certification Center (PMCC). 2002. *P2M: A guidebook of project and program management for enterprise innovation. Summary translation.* Revision 1 August 2002 ed. Tokyo: PMCC.

Rosenstock, C., R. S. Johnston, and L. M. Anderson. 2000. Maturity model implementation and use: A case study. *Proceedings of the 31st Annual project Management Institute 2000 Seminars and Symposium.* Newtown Square, PA: Project Management Institute.

Schlichter, F. J. 2001. PMI's organizational project management maturity model: Emerging standards. *Proceedings of the 32nd Annual project Management Institute 2001 Seminars and Symposium.* Newtown Square, PA: Project management Institute.

Software Engineering Institute. 2003. Process Maturity Profile. www.sei.cmu.edu/sema/presentations.html (accessed May 14, 2003).

Standish Group. 1994. Chaos. www.standishgroup.com/sample_research/chaos_1994_1.php(accessed May 16, 2003).

Suares, I. 1998. A real world look at achieving project management maturity. *Proceedings of the 29th Annual project Management Institute 1998 Seminars and Symposium*. Newtown Square, PA: Project Management Institute.

Thomas, J., and K. Jugdev. 2002. Project management maturity models: The silver bullets of competitive advantage? *Project Management Journal* 33(4):4–14.

Toney, F., and R. Powers. 1997. *Best practices of project management groups in large functional organizations*. Newtown Square, PA: Project Management Institute.

Turner, J. R., and A. Keegan. 1999. The versatile project-based organization: Governance and operational control. *European Management Journal* 17(3):296–309.

———. 2000. Processes for operational control in the project-based organization. *Proceedings of PMI Research Conference*. Newtown Square, PA: Project Management Institute.

SECTION III

APPLICATIONS IN PRACTICE

INTRODUCTION

This book has made the case repeatedly that project management is contextual; successful project managers are quick to recognize that while some truths are universal, their effective application is often the result of tailoring these practices to best fit the special circumstances of the organization, geographical location, or business model of the firm engaged in the project. To suggest that context is critical, however, is not to suggest that each of us cannot gain valuable insight into the management of our own projects by looking at how these activities are undertaken in other locales or for other classes of project. Thus, construction and manufacturing have had a powerful impact on teaching us all the benefits of carefully managing project supply chains. Pharmaceutical project management is helping shape our knowledge of the linkage between portfolio, program, and project management. Aerospace and defense project management have taught us the importance of earned value and technology management. The list goes on. This section applies some very direct examples of project management within a variety of industry contexts. Readers will gain considerable inside knowledge of project management in these settings, as well as an understanding of why each has contributed to the body of knowledge in project management research and practice.

One of the first questions one comes up against in looking at the discipline across a broad range of contexts is how to categorize the application area. Aaron Shenhar and Dov Dvir have done as much work as anyone in this area, and they provide a stimulating discussion in Chapter 50 showing that there are in fact several different categorizations that are valid and that work well under different circumstances. Based on their research, Aaron and Dov believe that in order to select the appropriate management style, managers should first assess the environment, the product, and the task; second, classify the project by the levels of uncertainty, complexity, and pace; and third, select the right style to fit the specific project type.

The first grouping—environment, product, and task—largely drove us to selecting several industry sectors as examples of the practice of managing projects. The sectors we chose are new product development, pharmaceuticals, defense, autos, and construction. Probably the one major sector that is not included here is IT/organizational change projects. We'll say a bit about these in a moment.

Dragan Milosevic, in Chapter 51, looks at the broad range of projects that might be considered as new product development (NPD). Dragan compares NPD projects with software development and construction and contends that technological novelty, product visibility, speed, visibility, and the nature, and level, of risk especially distinguish NPD projects. Pipeline management is a key issue; Dragan discusses it at the strategic (portfolio), project, and functional levels. He then looks at how project management varies by project size, novelty, and type (which he further splits into routine, administrative, technical, and unique).

Janet Foulkes and Peter Morris, in Chapter 52, discuss an interesting class of new product development projects: drug development. Typically it may take seven to ten years or more to develop a drug and may cost over $600 million. Drug development projects are both fascinating in themselves and interesting as an example of new product development. Unlike just about all other projects, the product itself is not being designed. Essentially the chemical entity has already been found; the challenge is to find out how it reacts therapeutically in human beings. In addition, there is a very high likelihood that the product will not behave as expected and will fail (attrite). This possibility makes resource planning something of a nightmare, particularly in the large matrix environment that becomes typical in the later stages of drug development, and risk management particularly important. The "management of projects" challenge is to manage a stream of contending candidates simultaneously at the project, program, and portfolio levels.

John Roulston, in Chapter 53, discusses defense projects, another type of new product development, predominantly from the suppliers' perspective—defense contracting—and it is very much the technical/procurement viewpoint that tends to dominate the chapter. This background is described individually from U.S., UK, French, and other perspectives. Prime contracting and partnering trends are discussed together with the particular challenges of consortia management. Examples are given of project-planning challenges, along with life cycle and earned value management.

Perhaps what is missing in this and Dragan's NPD chapter from a rounded "management of projects" viewpoint, and that is also so important in IT/organizational change projects, is the people dimension. Whereas IT projects certainly share much of the characteristics of defense—strong technical management bias, strong process element (methodologies etc.)—and indeed incorporate a distinct systems orientation, a particular challenge is the difficulty in IT/change projects of (a) tying down the requirements (though many would say this applies in defense too), (b) managing software developers and their "internal" world of a largely invisible product, and (c) managing the client/user group in a way that will get the system/change successfully used.

Many people maintain that project management has its origins in construction. In the ancient history sense, this may be true, but not in the modern sense of the discipline, as

Peter Morris in Chapter 54 on construction demonstrates. In fact, he contends, while project management as a practice is widely recognized as important in that part of construction known as process engineering (oil and gas, paper, power, etc.), in building and civil engineering it is only weakly recognized even as a practice, let alone as an important discipline. Procurement has had a dominating effect on the management of construction, but as we have seen earlier in the book, practices here are changing significantly now (partnering, prime contracting, PFI/BOT). And, *qua* the above, people—individually and as teams—have, perhaps as a result, typically played a very big part in delivering projects successfully.

Christophe Midler and Christian Navarre take us back in Chapter 55 to the strategic level in their overview of developments in the auto industry and the role of not so much project management as projects. Christophe and Christian identify four phases of the industry: functional (till 1970), project coordination (1970 to 1985), project functions and the rise of concurrent engineering (1985 to 1995), and the latest (1995–2003+). They then review the challenges facing project management in the industry now: linking with the commercial business case, managing radical innovation, working with intercompany alliances and partnerships, managing programs (platforms), and managing colearning and codesign. In short, another example of our "management of projects" thesis.

Finally, in Chapter 56 Lynn Crawford briefly reviews the rise and role of project management associations around the world since the mid-1960s. She looks at the various international initiatives that are now working in parallel with these and concludes by looking at the future. She pulls no punches: There is still much debate as to whether or not it is a practice or a discipline, and how it fits with other subjects and disciplines. Nevertheless, many now recognize—indeed, demand—that standards exist for the discipline. Defining these at national and international levels is an important responsibility for us all.

About the Authors

Aaron J. Shenhar

Dr. Aaron J. Shenhar is the Institute Professor of Management and the founder of the Project Management Program at Stevens Institute of Technology. He is also a visiting professor at Tel-Aviv University and the Technion in Israel. He was named Engineering Manager of the Year by the Engineering Management Society of IEEE in 2000. Prior to his academic career, he has been involved in managing projects, innovation, R&D, and high-technology businesses for almost 20 years. Working for the Israel defense industry, he participated in all phases of engineering and management—from project manager up through the highest executive posts. As executive at Rafael, the Armament Development Authority of Israel, he was appointed Corporate Vice President, Human Resources, and later, President of the Electronic Systems Division. In his second career in academia, Dr. Shenhar's work focuses on research, teaching, and consulting in project management, strategic project leadership, technology and innovation management, product development, and the leadership of professionals in technology-based organizations. He is serving as a consultant to several major corporations. With more than 150 publications to his credit, his writings have influenced project management research and education throughout the world. Dr. Shenhar holds five academic degrees in engineering, statistics, and management, including a PhD in Electrical Engineering from Stanford University. In 2003 he became the first recipient of the PMI Research Achievement Award.

Dov Dvir

Dr. Dov Dvir is Senior Lecturer at the School of Management, Ben Gurion University, Israel. Formerly, he was the Head of the Management of Technology (MOT) department at the Holon Center for Technological Education. He holds a BSc in Electrical Engineering from the Technion–(Israel Institute of Technology), an MSc in Operations Research and an MBA from Tel Aviv University, and a PhD in management (specialization in MOT) from Tel Aviv University. His research interests include project management, technology transfer, technological entrepreneurship, and the management of technological organizations. Dr. Dvir has accumulated over 25 years of technical, management, and consulting experience in government and private organizations.

Dragan Z. Milosevic

A leading authority on program and project management, Dragan Milosevic earned his credentials as a project manager in the private sector managing large projects around the world. As Associate Professor of Engineering Management at Portland State University (Oregon), he has developed practical tools and innovative approaches to the traditional and current challenges of project management. And as a consultant with Rapidinnovation, LLC, an executive consulting company, he helps leading companies streamline their project and program management models to ensure profitability. Dragan has more than 20 years' experience in program and project management theory and practice. He has worked in this field at a wide range of blue-chip companies. He has managed projects worth more than $600 million with partners from over 50 countries. Dragan Milosevic has written extensively, and his work has been published in major academic and management publications around the world. His book, *Project Management Toolbox*, was published in 2003 by Wiley. Professor Milosevic holds a BS, an MBA, and a PhD in management, all from the University of Belgrade. He also conducted project management seminars for the Project Management Institute.

Janet Foulkes

Janet Foulkes was most recently the Vice President of the Worldwide Project Management and Medical Operations group of Pfizer Pharmaceuticals. The group includes in its mission "efficient Program & Lifecycle Management & improved resource utilization on all of Pfizer Pharmaceutical product teams" primarily post-launch and beyond. Having been with Pfizer for over 20 years (15 in product management), Janet has had broad therapeutic development experience in both the Research and Commercial divisions (on both sides of the Atlantic) and has enjoyed increasing responsibility through clinical research and now project management. Janet has been involved in over seven major FDA filings and built the Pfizer Global Pharmaceuticals PM group from a 9-person to a 42-person unit with much broader responsibility. Prior to joining Pfizer, Janet received a Biochemistry degree from Cardiff University and pursued postgraduate research in Biochemical Engineering at University College, London. After moving to the United States, Janet received her MBA in Pharmaceutical Marketing from St Joseph's University in Philadelphia.

Peter Morris

Peter Morris is Professor of Construction and Project Management at University College, London, Visiting Professor of Engineering Project Management at UMIST, and Director of the UCL/UMIST-based Centre for Research in the Management of Projects. He is also Executive Director of INDECO Ltd, an international projects-oriented management consultancy. He is a past Chairman and Vice President of the UK Association for Project Management and Deputy Chairman of the International Project Management Association. His research has focused significantly around knowledge management and organizational learning in projects, and in design management. Dr. Morris consults with many major companies on developing enterprise-wide project management competency. Prior to joining INDECO, he was a Main Board Director of Bovis Limited, the holding company of the Bovis Construction Group. Between 1984 and 1989 he was a Research Fellow at the University of Oxford and Executive Director of the Major Projects Association. Prior to his work at Oxford, he was with Arthur D. Little in Cambridge, Massachusetts, and previously with Booz Allen Hamilton in New York and with Sir Robert McAlpine in London. He has written approximately 100 papers on project management, as well as the books *The Anatomy of Major Projects* (Wiley, 1988), *The Management of Projects* (Thomas Telford, 1997) and *Translating Corporate Strategy into Project Strategy* (PMI, 2004). He is a Fellow of the Association for Project Management, Institution of Civil Engineers, and Chartered Institute of Building and has a PhD, MSc and BSc, all from UMIST.

John Roulston

John Roulston graduated from Queens University, Belfast in 1970 and embarked on an engineering career in the defense industry with Ferranti Edinburgh, reaching a board-level position as Technical Director before Ferranti joined the GEC group. He has continued with its successor companies and is currently Technical Director for the Avionics Group within BAE SYSTEMS. John has held responsibility in many major development projects, delivering radar and electro-optics equipment into the United Kingdom and export military fleets. He is a founding director of the GTDAR radar joint venture in France and nonexecutive director of two small and growing UK companies. In Europe John has contributed to the radar of the Gripen aircraft through subcontract to Ericsson and he is Chairman of the Euroradar and EuroDASS consortia, providing radar and countermeasure systems to the Eurofighter Typhoon aircraft. He is also Industrial Professor of Electronics at Edinburgh University. He is known for wide-ranging technical and management interests and contributes to national efforts to promote science and mathematics education as a social priority. He is a Fellow of a Royal Academy of Engineering, a Fellow of the Royal Society of Edinburgh, a Fellow of the Institution of Electrical Engineers and a Chartered Engineer.

Christophe Midler

Dr. Christophe Midler, is Research Director at the French National Research Council in the Polytechnique Management Research Center and teaches innovation and project management at Ecole des Mines de Paris, Ecole polytechnique and Marne la Vallée University. He is Doctor Honoris Causa at Umea University, Sweden. His research topics are product development, project, and innovation management, in relation with organizational learning theory. He has explored these topics in various industrial contexts. He developed intensive collaborations with European car manufacturers on various questions: project function, concurrent engineering and structure of automobile firm, codesign with suppliers, managing collaborative projects in global alliances, research, and predevelopment phase management. He also developed research with construction, electronics, chemistry, pharmaceutics sectors on questions as contracts, project portfolios, platform management, and human resources management in R&D. His favorite methodology is long-term interactive researches with firms. Some publications in the area are *L'auto qui n'existait pas* (Dunod, 1996), *Project as Arenas for Renewal and Learning Processes* (as a coeditor with R. A. Lundin; Kluwer Academic, 1998), *Innovation Based Competition & Design Systems Dynamics* (in collaboration with P. J. Benghozi and F. Charue Dubo (L'Harmattan, 2000).

Christian Navarre

Dr. Navarre has conducted several internationally recognized studies on project management methods and has international consulting experience to organizations throughout North America, Europe and Africa. A significant portion of this work has been focused in the automotive industry and includes new car product development for automobile manufacturers. Professor Navarre is currently Director of the Car Internet Research Program (CIRP), which conducts research and studies on the effects of new information and communication technologies within the automobile industry. CIRP research addresses all levels including corporate-level strategies, business models, organizations, and individuals. CIRP is funded by academia, government, and numerous industry participants. Since 1995, he has oriented his research on the impact of new information and communication technologies upon the automotive industry. Research has included numerous field studies and surveys (B2B, B2C, Telematics) in North America and Europe, including an extensive 18-country European study in 1999 on the impact of new technologies on car distribution channels.

Lynn Crawford

With a diverse background as architect, project manager, regional planner and policy adviser, and with qualifications in human resource management and business administration, Dr. Lynn Crawford has experience both as a project manager and an adviser to project-based organizations on human resources, strategic and business planning, and development issues. Ongoing research and practice includes working with leading organizations that are developing their organizational project management competence by sharing and developing knowledge and best practices as members of a global system of project management knowledge networks. A particular area of research and expertise is the assessment and development of individual and corporate project and program management capability. Lynn was until recently Director of the postgraduate Project Management Program at the University of Technology Sydney. She has been Project Director for several major research projects funded by the Australian Research Council in areas of project management competence and the management of multiple, interdependent, and soft projects and continues to conduct research in these areas, working with industry partners. Lynn was a member of the steering committee for the development of Australian National Competency Standards in Project Management and is currently leading initiatives aimed at development of global standards for project management. These involve all major project management professional associations, recognized leaders in project management, and representatives of global corporations.

CHAPTER FIFTY

HOW PROJECTS DIFFER, AND WHAT TO DO ABOUT IT

Aaron J. Shenhar, Dov Dvir

Much of the project management literature and training treats projects as universal; assuming one set of techniques and tools applies to all situations. In reality, however, projects differ in many ways, and "one size does not fit all." Many writers have suggested in recent years that the time has come to develop a standard framework in project management that would help managers, researchers, and teachers distinguish among projects, identify appropriate management styles for different types of projects, and direct the further development of project-specific techniques and tools. This chapter summarizes ten years of research on the differences among projects and project management styles. It suggests three frameworks for distinguishing among projects, and in each framework several dimensions for project classification. Each framework can be used by different managerial levels and for different purposes. And within each framework we provide guidelines for management on how to treat different projects in different ways.

Introduction

Ask any project manager about his or her project, and the project manager will tell you why it is unique, and how the project team must adapt their style to their specific challenge and problems. Indeed, no two projects are alike. As any experienced manager will tell you, you must adapt yourself to the situation, the circumstances, the environment, and the people—and you should not try adapting the environment to you.

As a formal managerial discipline, project management is still relatively young. It started only in the middle of the twentieth century, when the first PERT charts marked the beginning of a new discipline (Morris, 1997). Over the years, however, very little attention has been given to the differences between projects, and there is still no universal framework for distinguishing among them. According to most books, all projects are the same, and most project management training still uses an approach of "a project is a project is a project."

We have seen this problem in our ongoing study of project management around the world. Over the years we have studied more than 600 projects and conducted hundreds of interviews. We have come to realize that project success depends greatly on the proper project management style and on adapting the right style to the right project. We have seen projects fail because managers assumed that their current project would be the same as their previous one. And we realized there is a need for a framework that will help managers look at a project by first assessing the project type and then selecting an appropriate management style. After a decade of research, we propose here such a framework and show how to make it work for various project settings and environments.

As you will see later, different managerial levels will use different frameworks to distinguish among projects and for different needs. Therefore, we will present not one, but three frameworks. However, there is an overarching concept to all. It is described in the next theoretical section, which will be followed by the outline of the frameworks and what each framework means for project management.

The Theory of Contingency

Classical contingency theory asserts that different external conditions might require different organizational characteristics, and that the effectiveness of any organization is contingent upon the amount of congruence or goodness of fit between structural and environmental variables (Lawrence and Lorsch, 1967; Drazin and van de Ven, 1985; Pennings, 1992). Burns and Stalker (1961) were the first to introduce the concept of contingency to organizational theory. They presented what is now accepted as the traditional distinction between incremental and radical innovation, and between organic and mechanistic organizations. Organic organizations would better cope with uncertain and complex environments, while mechanistic organizations predominate in simple, stable, and more certain environments.

The project management literature has often ignored the importance of project contingencies, assuming that all projects share a universal set of managerial characteristics (Pinto and Covin, 1989; Shenhar, 1993; Yap and Souder, 1994). Yet, projects can be seen as "temporary organizations *within* organizations" and may exhibit variations in structure when compared to their mother organizations. Indeed, various authors have often expressed disappointment from the universal, one-size-fits-all idea and recommended a more contingent approach to the study of projects (Ahituv and Neumann, 1984; Pearson, 1990; Wheelwright and Clark, 1992; Yap and Souder, 1994; Balachandra and Friar, 1997; Brown and Eisenhardt, 1997; Eisenhardt and Tabrizi, 1995; Song, Souder, and Dyer, 1997; Souder and Song, 1997).

Although contingency studies have had only a limited impact on the literature of project management, some exceptions exist. Most have focused on the impact of uncertainty and change on the way organizations are conducting their project operations. For example,

Wheelwright and Clark (1992) have mapped in-house product development projects according to the degree of change in product portfolio. Some have adapted the radical versus incremental distinction (e.g., Yap and Souder, 1994; Eisenhardt and Tabrizi, 1995; Brown and Eisenhardt, 1997; Song et al., 1997; Souder and Song, 1997), while others suggested more refined frameworks (e.g., Steele, 1975; Ahituv and Neumann, 1984; Cash et al., 1988; Pearson, 1990).

The second dimension of focus is complexity. To deal with complexity, the hierarchical nature of systems and their subsystems has long been the cornerstone of general systems theory (Boulding, 1956; Van Gigch, 1978; Rechtin, 1991), and hierarchies in products are almost always addressed in practitioners' books and monographs, which deal with engineering design problems (e.g., Pahl and Beitz, 1984; Lewis and Samuel, 1989; Rechtin, 1991).

Finally, while classical theory was focused on sustaining organizations, projects as temporary organizations must be studied in the context of time and the constraints they put on project management. Given the high velocity with which decisions are made and the shortened life cycles of products and markets, time and urgency become central factors in any modern look at the organization (Eisenhardt, 1989; Eisenhardt and Tabrizi, 1995; Brown and Eisenhardt, 1997). Based on these observations, our conceptual model is discussed in the next section.

The Basic Frameworks: The UCP Model

Regardless of the industry or technology involved, our research identified three dimensions to distinguish among projects: uncertainty, complexity, and pace (Shenhar and Bonen, 1997; Dvir et al., 1998; Shenhar, 2001a). Together, we call them the UCP Model, and they form a context-free framework for selecting the proper management style (see Figure 50.1). As you will see later, almost all frameworks are based on at least one of these three dimensions.

FIGURE 50.1. THE UCP MODEL.

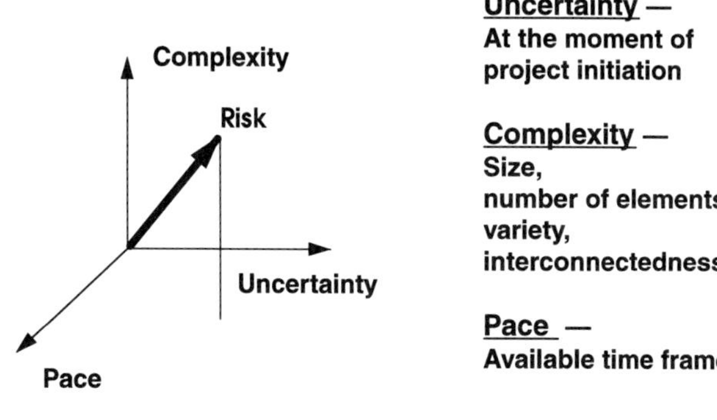

- *Uncertainty.* Different projects present, at the outset, different levels of uncertainty, and project execution can be seen as a process that is aimed at uncertainty reduction. Uncertainty determines, among other things, the length and timing of front-end activities, how well and how fast one can define and finalize product requirements and design, the degree of detail and extent of planning accuracy, and the level of contingency resources (time buffer and budget reserve). Uncertainties could be external or internal, depending on the environment and on the specific task.
- *Complexity.* Project complexity depends on product scope, number and variety of elements, and the interconnection among them. But it also depends on the complexity of the organization and the connections among its parties. Complexity will determine the organization and the process, as well as the formality with which the project will be managed.
- *Pace.* The third dimension for distinction among projects involves the urgency and criticality of time goals. The same goal with different time constraints may require different project structures and different management attention.

To select the appropriate management style, managers could follow a three-step process: First assess the environment, the product, and the task; second, classify a project by the levels of uncertainty, complexity, and pace; and third, select the right style to fit the specific project type (see Figure 50.2).

Here are some major factors that would impact each domain:

- *Environment.* The external environment includes the market, the industry, the customers, and the competitors. It may also involve the economical environment, as well as the political and the geographical environment where the project or task is being performed. But the environment is also the internal environment of the organization—comprising the organizational culture, the people, the procedures, the way projects are typically being managed, and available resources.
- *Product.* What is the exact product that this project is related to? What does the product do? How does it do it? What are the product operational requirements, and what are its specifications?

FIGURE 50.2. SELECTING PROJECT MANAGEMENT STYLE.

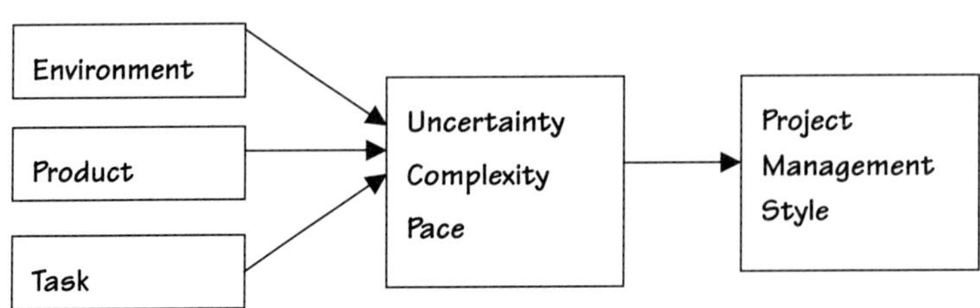

- *Task.* This refers to the exact work that needs to be done and the deliverables of the project. How difficult is this task? How well known is it? Have similar tasks been done before? How complex is it and how much time is available?

Mapping the Frameworks for Project Distinction

Since different managerial levels have different needs, we will use different frameworks for distinction among projects, and each framework will have several dimensions for classifying projects (see Table 50.1). The goal is to identify specific managerial activities, decisions, and styles that are best appropriate for each level and each project type (Shenhar, 2001b; Shenhar et al., 2002).

Although these frameworks proved the most useful, in no way are they unique. Some organizations may find it useful to develop their own frameworks that will fit their specific needs.

Strategic Portfolio Classification

This framework is based on the need to select projects in accordance with their strategic impact and to form a policy for project selection, since many projects compete for the same resources. To make the selection as rational as possible, we suggest (1) dividing projects into groups, based on their strategic impact, (2) allocating resources to each group based on the company or business strategic direction, and (3) selecting the individual projects in each group according to a set of criteria that was created in advance and carefully discussed.

We identified two dimensions to divide projects: the strategic goal dimension, which includes operational and strategic projects, and the customer dimension, which involves

TABLE 50.1. SUMMARY OF FRAMEWORKS TO DISTINGUISH AMONG PROJECT TYPES.

Framework	Major Users	Dimensions	Use
Strategic Portfolio Classification	Top management	• Strategic goal • User	Portfolio management
NCTP	Project managers	• Novelty • Complexity • Technology • Pace	Selecting project management style—leader, team, structure, processes
Work Package	Project teams Subcontractors	• Product type • Work Type	Assessing risk and time to completion of work packages

TABLE 50.2. STRATEGIC PORTFOLIO CLASSIFICATION.

	Operational	Strategic
External	• Product improvement	• New product development
Internal	• Maintenance • Improvement • Problem solving	• Utility and infrastructure • Research

external and internal customers (Shenhar et al., 2002). This results in four major groups of projects (see Table 50.2):

- *Operational projects* deal with existing businesses. They involve improvements in products, line extension, and cash cow projects, to gain more revenue from existing businesses.
- *Strategic projects* relate to new business. These are prime efforts that are made to create or sustain strategic positions in markets and businesses. Typically, strategic projects are initiated with a long-term perspective in mind. Examples are major automobile models introduction, new aircrafts such as Boeing 777, and so on.
- *External projects* are made for external customers, contracted or noncontracted.
- *Internal customers* are done within the organization, for internal departments or units.

FIGURE 50.3. THE NCTP FRAMEWORK.

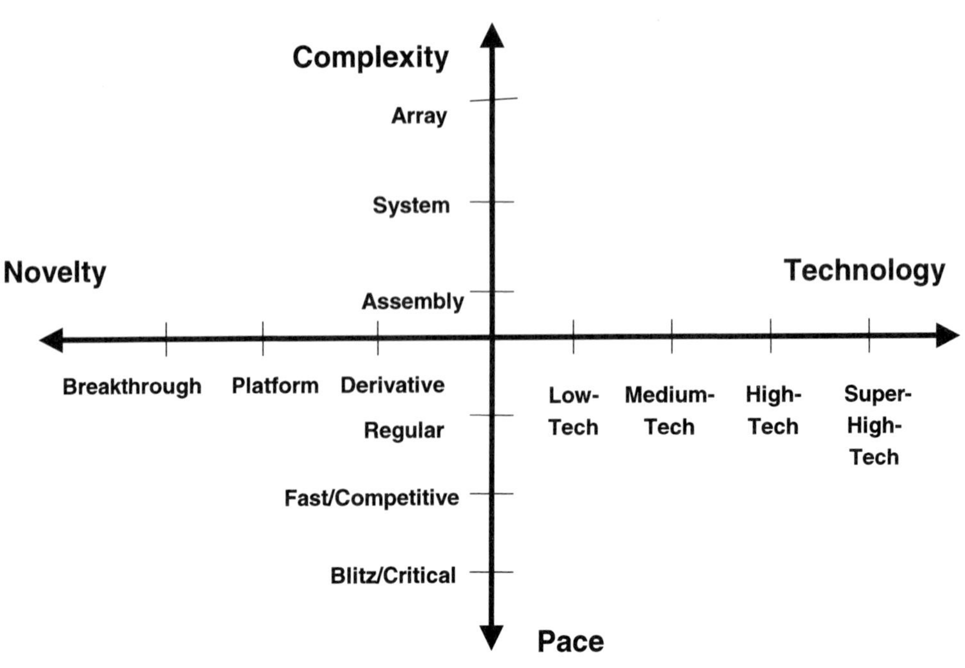

The policy for resource allocation among the groups is based on the expected impact from each group on the business, current strategy, infrastructure needs, and so on. The criteria for individual project selection are based among other things on risk and opportunity, difficulty, and available resources and skills.

The NCTP Model: Novelty, Complexity, Technology, Pace

The NCTP Model is the central framework that has evolved from our studies for distinction among projects (Shenhar and Dvir, 1996; Shenhar, 1998; Shenhar, 2001a; Shenhar, 2001b). It can guide project managers in selecting their project management style during project initiation, recruiting team members, determining structure and processes, and choosing the right tools. This framework involves four dimensions: novelty, complexity, technology, and pace. Each dimension includes at least three different project types, as shown in Figure 50.3.

Product Novelty

The first dimension is product novelty. It is defined by how new the product is to its potential users. It represents the extent to which customers are familiar with this kind of product, the way theywill use it, and its benefits. Product novelty will have an impact on project management, especially on product definition and market-related activities. It affects the accuracy and confidence with which one can define the end product—the "what" needs to be done. Different levels of product novelty will determine how accurate market research can be, how well the product can be defined upfront, and what kind of different marketing techniques will be required. When Sony considered introducing its first Walkman, the product presented a new concept to the market, and market research proved inconclusive. The decision to launch this most successful product was based on managerial intuition.

Wheelwright and Clark suggested three major new product categories to manage the company's product portfolio and to create an aggregate project plan: derivatives, platforms, and breakthroughs (Wheelwright and Clark, 1992). These categories will determine different management behavior regarding the marketing of new products and their impact on project management:

Derivative products are extensions and improvements of existing products. Projects that produce derivative products include cost reduction, product improvement, product modifications, and additions to existing lines. Previous products are well established in the marketplace, and market data is readily available. Predictions about product cost, as well as other product requirements, can be fairly accurate, and product requirements should be fixed as early as possible. Finally, marketing of derivatives is focused on product advantage in comparison to previous models, trying to serve existing customers, as well as potential new customers, with added product features and varieties.

Platform products are new generations in existing product families. Projects for platform products typically create new families of products to form the basis for numerous derivatives. Such products replace previous products in a well-established market sector. New combat aircraft designs or new automobile models of are typical platform examples. For these projects, companies should perform extensive market research, study data of previous generations, and make careful planning of product price. The final setting of product requirements will therefore take much longer, and it should be planned well into the project execution period. However, it should not be delayed too long to ensure timely product introduction and reasonable profitability. Marketing efforts for platforms should be focused on creating the product's image, emphasize product advantages, and differentiate it from its competitors.

Breakthrough products are new-to-the-world products. Breakthrough projects introduce a new concept or a new idea, or a new use of a product, that customers have never seen before. The first Post-It notes or the first Walkman are just two examples; so is the first personal computer. Breakthrough products may use new or mature technologies, but in each case the market does not know anything about the new product, nor do customers know how to use it. Market studies are therefore ineffective and product definition must be based on guessing, intuition, and market trial and error. Requirements must remain flexible until first market introduction is possible and until customer feedback is available. Fast prototyping is necessary and is much more critical than extensive market research. Changes in initial product specifications are inevitable, and while clear product definition is perceived to be critical to any project, management of breakthrough projects must realize that product definition will most likely change in time after initial market trials. Marketing of breakthrough products is completely different from marketing of previous types. It is focused on getting the attention of the "early adopter" customers. Its goal is to educate customers about the potential of the new product and often articulate hidden customer needs.

Table 50.3 summarizes the definitions of the three levels of product novelty and the major impact they will have on product definition and market-related activities.

Technological Uncertainty

The major source of task uncertainty is technological uncertainty. (Other types may involve team experience or tight budget constraints). Higher technological uncertainty at the time of project initiation requires longer development phases, more design cycles, more testing, and later design freeze. In general, we associate such uncertainty with the degree of using new (to the company) versus mature, or well-known, technology within the product or process produced. We found four distinct levels of technological uncertainty associated with different project categories: low-tech projects, medium-tech projects, high-tech projects, and super-high-tech projects (Shenhar, 1991; Shenhar and Dvir, 1996; Dvir et al., 1998; Shenhar, 1998; Shenhar, 2001a).

Type A: Low-Tech Projects. These projects rely on existing and well-established technologies. The most typical examples of such projects are certain types of construction or "build to print" efforts, namely, rebuilding an existing product. Projects in this category require no development work; their architecture, design, and resource planning are all completed prior to the project's implementation phase. In such projects, the product is entirely shaped, and the design frozen, prior to the project's formal inception. Management style in such projects must be firm, since profit margins are typically slim. Project managers should therefore stick to the initial plan, with a *no-nonsense, no-changes, get-the-job-done* attitude.

TABLE 50.3. PRODUCT NOVELTY LEVELS AND THEIR MAJOR IMPACT ON PROJECT MANAGEMENT.

Product Novelty	Derivative	Platform	Breakthrough
Definition	An extension or improvement of an existing product	A new generation in an existing product family.	A new-to-the-world product.
Data on Market	Accurate market data exists	Need extensive market research. Careful analysis of previous generations, cocmpetitors, and markets	Nonreliable market data. Market needs not clear. No experience with similar products
Product Definition	Clear understanding of required cost, functionality, features, etc. Early freeze of product requirements	Invest extensively in product definition. Involve potential customers in process. Freeze requirements later, usually at mid-project.	Product definition based on intuition and trial and error. Fast prototyping is necessary to obtain market feedback. Very late freeze of requirements
Marketing	Emphasize product advantage in comparison to previous model. Focus on existing as well as gaining new customers based on added product features and varieties.	Create product image. Emphasize product advantages. Differentiate from competitors.	Creating customer attention. Educating customers about potential of product. Articulate hidden customer needs. Extensive effort to create the standards.

Type B: Medium-Tech Projects. These use mainly existing or base technology, yet incorporate some new technology or a new feature that did not exist in previous products. Examples include improvements and modifications of existing products (derivatives), but also new generations of products in industries where technology is relatively stable, such as appliances, automobile, or heavy equipment. Although most of the technologies are not new, the project will involve some development and testing. Changes to the design, however, should be confined to a limited period, and the design should be frozen quite early, typically after two cycles. While still firm, management style could be more flexible at the initial phase of the project, since some changes and adjustment in design are needed. Later changes should only be allowed if absolutely necessary for success or for safety reasons. The policy and attitude could be "limit changes to minimum and freeze as early as possible."

Type C: High-Tech Projects. These projects represent situations in which most of the technologies employed are new but nevertheless exist when the project is initiated. Such technologies had been developed prior to project inception, and the project represents the first effort to integrate them into one product. Most defense development projects belong to this category, but also new generations of computers and many products in the high-tech industry. Such projects are characterized by long periods of design, development, testing, and redesign, and they require at least three design cycles. Design freeze must be scheduled, therefore, to a much later phase, normally during the second or even the third quarter of the project execution period. Management style must be more flexible in these projects, since numerous designs must be tested and will lead to changes and improvements during much longer periods.

Type D: Super High-Tech Projects. These projects are based on new technologies that do not exist at project initiation. While the mission is clear, the solution is not. This type of project is relatively rare and is usually carried out by only a few, probably large organizations or government agencies. One of the most famous examples of this type was the Apollo moon-landing program (Shenhar, 1992). At project inception, it had a well-defined mission and timetable, but no available technology was at hand, and nobody really knew how to get to the moon (see Case Study 1).

Case Study 1

Managing Uncertainty in Space Programs

Two of the most famous engineering efforts of this century belong to NASA. They are the moon-landing Apollo program in the 1960s and building the Space Shuttle in the 1970s and early 1980s. While these two programs had much in common in terms of uncertainty, complexity, and risk, they were managed in two different styles and with two different attitudes. Apollo was initiated in 1961 by President Kennedy, who set the goal of reaching the moon before the end of the decade. From inception, it was clear to NASA that the Apollo program was highly uncertain and much more risky than anything it has done before. Uncertainties ranged from radiation and meteoroid hazards to lunar surface extreme environment to unknown launch configurations. In the context of this chapter, the Apollo program was perceived and managed as a super-high-tech, Type D project. NASA spent enormous time and resources to develop the nonexistent technologies. It did so while embarking on a low-scale prototype, called the Gemini program. It had, in addition, installed numerous safety mechanisms to make sure everything was tested and guaranteed before freezing the system design and configuration, well into the program's execution period. The successful moon landing of Apollo 11 in July of 1969 symbolized a victory for humankind, not only on unknown space territories but also on new and far-reaching technologies.

It seems, however, that many of the lessons learned during the Apollo era were forgotten during the development of the Space Shuttle. This project was managed in a different way. Pressured by the Administration to limit the budget and making everything possible to win Congress approval, NASA preferred to treat the program as a safe bet, based on existing technologies and previous skills. The configuration and design were frozen as early as three month after project approval, and a final commitment was made to new and yet nontested technologies. In reality, project uncertainties were higher than in the Apollo program. Developing the Space Shuttle turned out to be one of the most difficult and exasperating engineering challenges of the space age, and the task essentially proved too difficult to accomplish using the "success-oriented" attitude employed by NASA. Eventually, the program suffered extreme overruns, amounting to almost three years in schedule delays and 60 percent in unexpected costs, all accumulated even before the tragic Challenger accident. The second loss of a shuttle in 2003 only added to the doubt whether the development project was managed correctly.

The Presidential Commission that investigated the Challenger accident concluded that the failure was the result of faulty joint design; that the decision to launch the shuttle on a cold January morning was wrong, given early warnings concerning the design problem and recommendations. Using the UCP framework, it is clear that NASA's management style toward the project was one that in our notation would be classified a Type C. In retrospect, it seems that the program had to be conducted using a different, Type D philosophy. In that case, the final configuration, the freezing point of the design, and the operational phase declaration would have been scheduled for a much later point, leaving open possibilities for using other, more thoroughly tested, and maybe improved technologies. In addition, a Type D style would have vested the program with a different attitude toward risks, possible development problems, and probability of failure and would have created the need for a much higher level of communication among various parties involved (Shenhar, 1992).

Unlike previous, lower-uncertainty-type projects, a great deal of the effort in this type is devoted to the development of completely new technologies and to testing and selecting among various alternatives. As our studies found, projects in this category use similar techniques to resolve the issue of unknown technologies. They institute an intermediate program to build a *small-scaled prototype* on which new technologies are built and tested. Type D projects thus require extensive development periods, very late design freeze, and intensive management of ideas and change until the final configuration is selected.

Management style must be extremely flexible, yet cautious. The attitude in these risky projects must be "look for trouble; it's there." And since things change so fast, extensive communication is essential among project teams and team members. The formal system of reporting could simply not accommodate the degree of changes and new information that is created.

A summary of characteristics and managerial styles of different levels of technological uncertainty is described in Table 50.4.

Project Complexity (System Scope)

In our study of project complexities, we observed three typical management styles. They are distinguished by the way the project is organized and its subelements are coordinated,

TABLE 50.4. PROJECT CHARACTERISTICS AND TECHNOLOGICAL UNCERTAINTY LEVELS.

Variable	Low-tech A	Medium-tech B	High-tech C	Super High-tech D
Technology	No new technology.	Some new technology.	New, but existing technologies.	Key technologies do not exist at project's initation.
Typical industries	Construction, production, utilities, public works.	Mechanical, electrical, chemical, some electronics.	High-tech and technology-based industries; computers, aerospace, electronics.	Advanced high-tech and leading industries; electronics, aerospace, computers, biotechnology.
Type of products	Buildings, bridges, telephone installation, build-to-print.	Non-revolutionary models, derivatives or improvement.	New, first of its kind family of products, new military systems (within state of the art).	New, nonproven concept beyond existing state of the art.
Development and testing	No development, no testing.	Limited development, some testing.	Considerable development and testing. Prototypes usually used during development.	Develop of key technologies needed. Small-scale prototype is used to test concepts and new technologies.
Design cycles and design freeze	Only one cycle. Design freeze before start of project execution.	One of two cycles. Early design freeze, in first quarter.	At least two to three cycles. Design freeze usually during second quarter.	Three to five cycles. Late design freeze, usually during third or even forth quarter.
Communication and interaction	Mostly formal communication during scheduled meetings.	More frequent communication, some informal interaction.	Frequent communication through multiple channels; informal interaction.	Many communication channels; informal interaction encouraged by management.
Project manager and project team	Adminisstrative skills. Mostly semiskilled workers, few academicians.	Some technical skills. Considerable proportion of academicians.	Manager with good technical skills. Many professionals and academicians on project team.	Project manager with exceptioinal technical skills. Highly skilled professionals and many academicians.
Management style and attitude	Firm style. Sticking to the initial plan.	Less firm style. Readiness to accept some changes.	More flexible style. Many changes are expected.	Highly flexible style. Living with continuous change, "looking for trouble."

and by the extent of formal versus informal interaction and documentation. As we found, a simple way to define and distinguish among different levels of complexity is to use a hierarchical framework of systems (Shenhar and Dvir, 1996; Dvir et al., 1998; Shenhar, 1998; Shenhar, 2001a). We call it *system scope,* and in most cases a lower scope level may be seen as a subsystem of the next-higher level. Project management practices, however, can be distinguished by three typical levels as follows.

Level 1: Assembly Projects. Assembly projects involve creating a collection of elements, components, and modules combined into a single unit or entity that is performing a single function. Assembly projects are relatively simple; they may produce a stand-alone product such as a CD player or a coffee machine, or create a subsystem of a larger system such as a computer hard drive or a radar receiver. They may also involve restructuring a functional organization or building a stand-alone service (see the discussion in Case Study 2 on reengineering projects).

Case Study 2

The Case of Reeingineering

Reengineering projects to restructure business and service process have become a common part of organizational life. Yet not all engineering projects are the same, and their execution should be adapted to the specific problem. Since typical reengineering projects build a new generation of business process, they can be classified as platform from the point of view of market uncertainty, their internal uncertainly is Type A (or at most B), and their pace is regular (time is generally not a critical factor, and it is not associated with a window of opportunity or meeting market demand). the distinctive measure in these projects, however, is complexity, and it should be determined by the level of the reengineered organization: When you are dealing with one department or a single process within a local unit, such as a bank's branch office, the effort is an *assembly* project. People know each other well, the process is simple to depict, and not much information technology is needed to construct and run the process. A reengineering effort of an entire business with various integrated functions, such as engineering, manufacturing, sales, and distribution, will be classified as a *system project.* In this case, coordination becomes much more critical, formal tools and extensive documentation must be employed, and information technology and software is unavoidable. The highest level of complexity involves large corporations, normally spread over the country or even the entire world. Reengineering, then, becomes an *array project.* It must be carried out in a very formal and bureaucratic way, with many subprojects devoted to different parts of the company.

Assembly projects are typically carried out by a single organizational function or a small cross-functional team, often within one organization and with a low level of formality or bureaucracy. The number of project activities or subtasks is normally in the range of tens, and as we learned, the planning of budget and schedule in this type of projects is simple, requires only basic tools of project management, and is often done manually. Interaction, communication, and much of the decision making is, to a large extent, informal; documentation and reporting is minimal, as is the need for administrative staff within the project.

Level 2: System Projects. System-type projects involve a complex collection of interactive elements and subsystems, jointly dedicated to a wide range of functions to meet a specific operational need. System projects may produce aircraft, cars, computers, or buildings, but they may also involve reengineering efforts of entire businesses (see Case Study 2). A main contractor or program office typically leads system projects, and the total effort is divided among numerous subcontractors—some in-house and some external. The program office is responsible for the final integrated result and for meeting overall performance, quality, time, and budget goals. Such projects are managed in a more formal way than assemblies, with extensive documentation and contracting between main and subcontractors.

Since most system projects extend beyond organizational borders, they usually need special administrative staff to handle the planning, budget management, contracting, and controlling issues. The number of project activities is in the range of hundreds to a few thousands. While there are many existing project management tools and applications, system projects and their corporations often find it necessary to develop their own tools and documents to meet their specific requirements.

Level 3: Array Projects. Array-type projects deal with large, widely dispersed collections of systems (sometimes called "supersystems") that function together to achieve a common purpose, such as city public transportation systems, national air defense systems, or interstate telecommunication infrastructures. Arrays are clearly large-scale projects, yet in most cases, they do not involve building the supersystem from scratch. Rather, such projects often involve gradual growth, addition, or modification to an existing infrastructure. Notable array projects were the upgrading of the New York Subway system in the early 1990s, the preparation of the city of Atlanta to the Olympic games of 1996, and building the Anglo-French Channel Tunnel.

Array projects require the administration of many separate programs, each one devoted to a different component or system. Projects are typically organized in the form of a central "umbrella organization" that is set up as a separate entity or company and that formally (and legally) coordinates the efforts of numerous subprojects in other organizations. The actual technical work is executed, however, within the suborganizations. The focus is on extensive documentation, contracting, and tight financial controls. The dispersed nature of the end product and the extent of subcontracting make it necessary to manage these programs in a highly formal way and to put a lot of effort into the legal aspects of the various contracts. Ordinary tools of planning and control are even less relevant here, and management must often develop its own system of contract coordination and program control.

A summary of characteristics and managerial styles of different levels of system scope is described in Table 50.5, and an example of different scope levels of reengineering projects is discussed in Case Study 2.

TABLE 50.5. PROJECT CHARACTERISTICS AND SYSTEM SCOPE LEVELS

	Assembly 1	System 2	Array 3
Definition	A collection of components and modules in one unit, performing a single function.	A complex collection of assemblies that is performing multiple functions.	A widespread collection of systems functioning together to achieve a common mission.
Examples	A system's power supply; a VCR, a single functional service.	A complete building; a radar; an aircraft; a business unit.	A city's highway system; an air fleet, a national communication network; a global corporation.
Customers	Consumers or a subcontractor of a larger project.	Consumers, industry, public, government or military agencies.	Public organizations, government or military agencies.
Form of purchase and delivery	Direct purchase or a simple contract; contract ends after delivery of product.	Complex contract; payments by milestones; Delivery accompanied by logistic support.	Multiple contracts; sequential and evolutionary delivery as various components are completed.
Project organization	Performed within one organization, usually under a single functional group; almost no administrative staff in project organization.	A main contractor, usually organized in a matrix or pure project form many internal and external subcontractors; technical and administrative staff.	An umbrella organization—usually a program office to coordinate subprojects; many staff experts: technical, administrative, finance, legal, etc.
Planning	Simple tools, often manual; rarely more than 100 activities in the network.	Complex planning; advanced computerized tools and software packages; hundreds or thousands activities.	A central master plan with separate plans for subprojects; advanced computerized tools; up to ten thousand activities.
Control and reporting	Simple, in-house control; reporting to management or main contractor.	Tight and formal control on technical, financial and schedule issues; reviews with customers and management.	Master or central control by program office; separate additional control for subprojects; many reports and meetings with contractors.
Documentation	Simple, mostly technical documents.	Many technical and managerial formal documents.	Mostly managerial documents at program office level; technical and managerial documents at lower level.
Management style, attitude, and concern	Mostly informal style; familylike atmosphere.	Formal and bureaucratic style; some informal relationship with subcontractors and customers; often political and interorganizational issues.	Formal, tight bureaucracy; high awareness to political, environmental, and social issues.

Project Pace: How Critical Is Your Time Frame?

On this scale projects will differ by urgency, or how much time is available, and what happens if the time goal is not met. We identified three different levels of urgency, or pace: regular, fast-competitive, and critical-blitz (Shenhar et al., 2002):

Regular projects are those efforts where time is not critical to immediate organizational success. They are typically initiated to achieve long-term or infrastructure goals. They may include some public projects, projects in a noncompetitive or nonprofit environment, organizational improvements, or technology build-up efforts. Unless specifically prioritized, such projects are managed in a relatively casual format and may often be delayed or pushed aside by more urgent assignments.

Fast-competitive projects are the most common projects carried out by industrial and profit-driven organizations. They are typically conceived to address market opportunities, create a strategic positioning, or form new business lines. Time-to-market is directly associated with competitiveness, and although missing the deadline may not be fatal, it could hurt profits and competitive positioning.

Fast-competitive projects must be managed strategically. Project managers should focus on meeting schedule, but also achieving profit goals and addressing market needs. Managing the time frame should be one of the main concerns.

Critical-blitz projects are the most urgent, most time-critical. Meeting schedule is critical to success, and project delay means failure. Such projects are often initiated during a crisis, or as a result of an unexpected event. Examples may be industrial crisis projects to respond to a surprising move by competition, military projects during wartime, or a natural disaster that needs immediate recovery. Famous cases include the Y2K problem or saving the lives of Apollo 13 crew.

To succeed, such projects must be managed completely differently from other forms. The work flow in these projects is very tight. It is performed almost around the clock, with nonstop interaction and continuous decision making. There is normally no time for detailed documentation or report writing. Project managers in critical-blitz projects must possess high autonomy; project organization must be "pure project," and all team members must report directly to the project leader. Top management must be continuously involved to support and monitor project progress.

Table 50.6 describes characteristics of different project pace levels.

Managing Highly Uncertain and Complex Programs

Perhaps the two most important dimensions of the NCTP model are system scope and technological uncertainty. According to our studies, they have the highest impact on differences that can be found among projects and project management styles (for more details, see Dvir et al., 1998). Of particular interest are those projects that involve high complexity together with high technological uncertainty. In fact, many defense and space programs fall in this category, as well as some commercial efforts such as the development of Boeing 777 or GM's electric vehicle. Project management in this case is even more complex, and additional concerns and techniques must be employed.

As we have seen earlier, higher uncertainty involves longer design periods, more design cycles, and later design freeze, as well as better technical skills and more frequent technical

TABLE 50.6. CHARACTERISTICS OF DIFFERENT PROJECT PACE LEVELS.

Pace Level	Regular	Fast-Competitive	Blitz-Critical
Definition	Time not critical tto organizational success.	Time-to-market is a competitive advantage and has an impact on business success.	Time is critical for project success. Delays mean project failure.
Examples	Public works, government initiative, internal projects.	Business related projects, new product introduction.	Crisis situations, war, fast response to natural disasters, fast response to business related surprises.
Organization	Matrix or functional.	Matrix, teams, subcontractors.	Pure project, special task force.
Personnel		Qualified to the job.	Specifically picked.
Focus	No particular focus.	Strategically focused on time to market.	Swift solution of the crisis.
Procedures	No specific attention.	Structured procedures.	Shortened, simple, nonbureaucratic.
Top management involvement	Management by exception.	Go ahead at stages.	Highly involved and constantly supportive.

decisions. Similarly, higher complexity (scope) involves increase in formality and documentation. When both dimensions are at their higher levels, an important addition to classical project management is the discipline of *systems engineering management.* Developed previously in the defense environment, systems engineering management is a multidisciplinary discipline dedicated to controlling requirements, specifications, design, development, building, testing, manufacturing, and operation so that all elements are integrated to provide an optimum, overall system. It requires a "holistic view"—the design of the whole, rather than the design of the parts—and it involve identifying and understanding the need; creating the system concept and architecture; and combining economic, societal, environmental, and political issues into the design of the system.

The problem of system integration mentioned earlier is typically critical in high-uncertainty, high-complexity projects. Subsystem qualification does not guarantee total system performance. Considerable time and effort must be allocated to solving the integration issue. Numerous problems of interface, fit, and mutual effects must be resolved before the entire systems is qualified. And, obviously, configuration management becomes extremely important, as any change in any part may impact numerous other components and subsystems.

Finally, although all projects involve a certain level of risk, higher-scope and higher-tech projects are more risky and are more prone to problems. Thus, they need more systematic risk management. The objective of risk management is not to eliminate risks, but to manage them across the project so as to avoid investing excessive resources in the solution of a given risk while neglecting others. Risk management is conducted in a rigorous and systematic way to identify, analyze, and mitigate all sources of program risk. Figure 50.4

FIGURE 50.4. TECHNOLOGICAL UNCERTAINTY AND SYSTEM SCOPE IMPACT ON PROJECT MANAGEMENT.

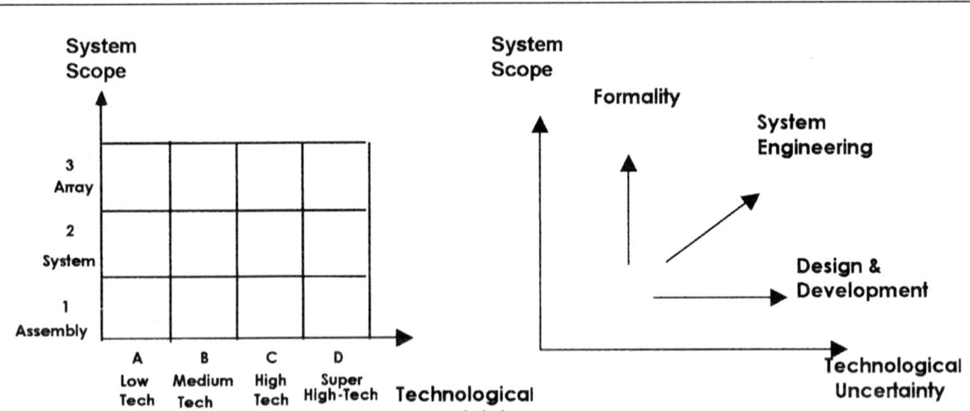

describes the two dimensions of technological uncertainty and system scope and the impact of a simultaneous increase of uncertainty and scope. Neglecting the impact of high uncertainty or high complexity may result in difficulties, delays, and even project failure (Dvir et al., 2003).

The Work Package Framework

This framework is particularly useful for team members and subcontractors when dealing with individual work packages. The work breakdown structure provides a tool to identify project tasks, allocate time and resources, create the product cost structure, develop a network-diagramming schedule (PERT and critical path), and assign responsibility for the execution and monitoring of task completion. The lowest level in the WBS is a work package, which may serve a basis for identifying project risks and deciding how to avoid or reduce them.

Our studies show that conventional treatment of all work packages as the same is unrealistic and must be replaced by a more adaptive approach. We use two dimensions to distinguish among packages: the type of task outcome and the type of activity or work that needs to be done to achieve it. Task outcomes may be tangible or intangible, and activity can be distinguished as requiring craft or intellectual work.

Task Outcome: Product of Work Package

Tangible outcomes produce a physical artifact (hardware). Any piece of equipment, such as a computer keyboard or a car's steering wheel, is a tangible product. Tangible products must be physically assembled and reproduced. Product and process design must be integrated to create a manufacturing process and is subject to quality control, assembly operations, and cost of production.

Intangible outcomes, in contrast, do not produce an artifact. Intangible products produce new information, which can be stored on different physical media such as a CD-ROMs. Software code, manuals, books, newspapers, blueprints, or movies are intangible products. Reproducing them in high quantities is easy, instant, and cheap and does not require dedicated production lines.

Type of Activity on Work Package

Craft activity involves repetitive efforts that have been done previously. Work outcomes are predictable, and duration is accurately anticipated and subjected to the classical learning curve. Craft examples include machine shop work, painting jobs, watch assembly, or car servicing.

Intellectual activity requires creative effort. Since such activities have not been done before, they are exploratory in nature and require new ideas and imagination. Artwork is obviously intellectual, but so are activities such as developing a new technology, new designs, or "building a better mousetrap."

As shown in Table 50.7, the four cells of our framework suggest different strategies for different work packages.

TABLE 50.7. ADJUSTING THE RISK OF WORK PACKAGES IN WORK BREAKDOWN STRUCTURES.

	Intangible	Tangible
Intellectual	New software code. Most effort in exploratory and creative work. Less risky than intellect tangible. Does not need process building.	New kind of hardware. Hardware never done before. Needs creativity, development, iterations, and testing. Must also plan and build new process. lowest accuracy in resources planning. Needs contingency resources and backup plans. Highest risk in product and process.
Craft	Writing routine plans or procedures. Minimal risk to produce routine text. No new ideas are required.	Building or producing well-known types of hardware. Repetitive tasks. Good estimation of work duration and other resources. Needs costly but predictable process building resources. Production quality becomes the main concern.

Using the Frameworks in Practice

How can managers and organizations benefit from adapting the conceptual framework presented here? An explicit, clear identification of the project type prior to execution should provide a basis for a proper adaptation of managerial attitudes and management style, for the selection of project managers and project team members, for establishing the proper project organization, and for a better choice of managerial tools. For example, identifying a project as high-tech will explicitly have to lead to a highly flexible style, to longer development periods, and to more design cycles. Dealing with a super-high-tech project will entail, in addition, the establishment of an intermediate program to be used as a testing bed for new technologies. Similarly, system projects will require the establishment of an efficient subcontracting procedure, as well as adequate time for system integration.

The framework suggested here might also be useful for identifying technical skill development needs, managerial development needs, and management training needs. Individuals would often gain compelling capabilities and expertise in managing specific kinds of projects. Moving into different, sometimes remote kinds of projects requires adopting a different style and the development of additional skills. Understanding the strategic, as well as the operational, differences between projects may help avoid potential errors and may considerably shorten the learning process. For example, moving into high- and especially into super-high-tech projects requires exceptional technical skills as well as the capability to assess potential value and risk involved in new or even not-yet-developed technology. Moving from assembly to system projects requires additionally a wealth of administrative skills. When managers are dealing with the system level for the first time, they need to develop the "system view"—being able to see the system as a whole and understand the effect of its separate components on the entire system. System project managers must be mature enough and able to utilize the accumulation of technical intuition from their original field to interdisciplinary problems of systems. When moving into the array level, managers must learn to back off from the technical matters and develop instead a broader view of the industry and its players. They must learn to deal with legal, environmental, and political issues, usually not addressed by managers of lower scope levels.

Summary

While the frameworks presented in this chapter may not be conclusive, they provide a good basis for organizations to distinguish among their project types. As mentioned, sometimes organizations will need to find their own way for classification. For example, the dimensions of uncertainty may identify other forms of uncertainty—political, economical, or environmental; and the dimension of complexity may be distinguished according to geography and spread of the project team—either in-house, across different organizations, or across international boundaries.

In conclusion, companies must realize that project management is not universal and adapting project management styles is critical to project success. Each company must identify its specific typical project types and select the right tools, procedures, and people for the specific project. The potential of improved project management is much too large to be neglected in this dynamic age of increased competition.

Acknowledgments

The authors are grateful for the help and support that the following institutions have provided to these studies over the years:

Israeli Institute for Business Research at Tel Aviv University; The Directorate for the Development of Armament and Production Infrastructure at the Israeli MOD; Center for the Development of Technological Leadership at the University of Minnesota; Center for Technology Management Research at Stevens Institute of Technology; and the National Science Foundation and its Management of Technological Innovation Program.

References

Ahituv, N., and S. Neumann. 1984. A flexible approach to information system development. *MIS Quarterly* (June): 69–78.

Balachandra, R., and J. H. Friar. 1997. Factors for success n R&D project and new product innovation: A contextual framework. *IEEE* Transactions on Engineering Management 44:276–287.

Boulding, K. 1956. General systems theory: The skeleton of science. *Management Science* (April): 197–208.

Brown, S. L., and K. M. Eisenhardt. 1997. The art of continuous change: Linking complexity theory and time-paced evolution in relentlessly shifting organizations. *Administrative Science Quarterly* 42: 1–34.

Burns, T., and G. Stalker. 1961. *The management of innovation.* London: Tavistock.

Cash, J. I. Jr., W. F. McFarlan, and J. L. McKenney. 1988. *Corporate information systems management.* Homewood, IL: Irwin.

Drazin, R., and A. H., van de Ven. 1985. Alternative forms of fit in contingency theory. *Administrative Science Quarterly* 30:514–539.

Dvir, D., S. Lipovetskey, A. J. Shenhar, and A. Tishler. 1998. In search of project classification: A non-uniform mapping of project success factors. *Research Policy* 27:915–935.

Dvir, D., A. J. Shenhar, and S. Alkaher. 2003. From a single discipline project to a multidisciplinary System: Adapting the right style to the right project. *System Engineering* 6(3):123–134.

Eisenhardt, K. M. 1989. Building theories from case study research. *Academy of Management Review* 14: 532–550.

Eisenhardt, K. M., and B. N. Tabrizi. 1995. Accelerating adaptive processes: Product innovation in the global computer industry. *Administrative Science Quarterly* 40:84–110.

Lawrence, P. R., and J. W. Lorch. 1967. *Organization and environment: Managing differentiation and integration,* Boston: Graduate School of Business Administration, Harvard University.

Lewis, W., and A. Samuel. 1989. *Fundamentals of engineering design.* Englewood Cliffs, NJ: Prentice Hall.

Morris, P. W. G. 1997. *The management of projects.* London: Thomas Telford.

Pahl, G., and W. Beitz. 1984. *Engineering design.* New York: Springer-Verlag.

Pearson, A. W. 1990. Innovation strategy. *Technovation* 10(3):185–192.

Pennings, J. M. 1992. Structural contingency theory: A reappraisal. *Research in Organizational Behavior* 14:267–309.

Pinto, J. K., and J. G. Covin. 1989. Critical factors in project implementation: A comparison of construction and R&D projects. *Technovation* 9:49–62.

Rechtin, E. 1991. *Systems architecting.* Englewood Cliffs, NJ: Prentice Hall.

Shenhar, A. J. 1991. Project management style and technological uncertainty: From low- to high-tech. *Project Management Journal* 22(4):11–14, 47.

———. 1992. Project management style and the space shuttle program: A retrospective look. *Project Management Journal* 23(1):32–37.

———. 1993. From low- to high-tech project management. *R&D Management* 23(3):199–214.

———. 1998. From theory to practice: Toward a typology of project management styles. *IEEE Transactions on Engineering Management* 41(1):33–48.

———. 2001a. One size does not fit all projects: Exploring classical contingency domains. *Management Science* 47(3):394–414.

———. 2001b. Contingent management in temporary organizations: The comparative analysis of projects. *Journal of High Technology Management Research* 12:230–271.

Shenhar, A. J., and Z. Bonen. 1997. A new taxonomy of systems: Toward an adaptive systems engineering framework. *IEEE Transactions on Systems, Man, and Cybernetics* 27(2):137–145.

Shenhar, A. J., and D. Dvir. 1996. Toward a typological theory of project management. *Research Policy* 25:607–632.

Shenhar, A. J., D. Dov, T. Lechler, and M. Poli. 2002. One size does not fit all: True for projects, true for frameworks. PMI Research Conference, Seattle, Washington.

Song, M. X., W. E. Souder, and B. Dyer. 1997. A casual model of the impact of skills, synergy, and design sensitivity on new product performance. *Journal of Innovation Management,* 14:88–101.

Souder, W. E., and M. X. Song. 1997. Contingent product design and marketing strategies influencing new product success and failure in U.S. and Japanese electronics firms. *Journal of Product Innovation Management* 14:21–34.

Steele, L. W. 1975. *Innovation in big business.* New York: Elsevier Publishing Company.

Van Gigch, J. P. 1978. *Applied general systems theory.* 2nd ed. New York: Harper & Row.

Wheelwright, S. C., and K. B. Clark. 1992. *Revolutionizing product development.* New York: Free Press.

Yap, C. M., and W. E. Souder. 1994. Factors influencing new product success and failure in small entrepreneurial high-technology electronics firms. *Journal of Product Innovation Management* 11:418–432.

CHAPTER FIFTY-ONE

MANAGING NEW PRODUCT DEVELOPMENT PROJECTS

Dragan Milosevic

The innovation process can be easily divided into three parts: The fuzzy front end, new product development (NPD), and commercialization (see Figure 51.1) (Smith and Reinertsen, 1991). The fuzzy front end, which precedes the well-structured NPD process, is a "messy" period that typically includes strategic planning, concept generating, and pre-technical evaluating (Belliveau et al., 2002). Then, during the commercialization phase, the new product is taken from development to market (Belliveau et al., 2002).

The Unique World of New Product Development Projects

Setting the first and third phases aside, this chapter focuses on the best practices of the middle phase, new product development, as it relates to the overall process of strategy, organization, product definition, product/marketing plan creation, execution, and evaluation (Rosenau et al., 1996). "New product" here refers both to manufactured goods and services new to the company marketing them.

New Product Development: The Key To Corporate Success

NPD is perhaps the most important activity in the modern corporation (Cooper, 2001). No wonder then that some experts view it as the engine of corporate success, survival, and renewal (Bowen et al., 1994; Brown and Eisenhardt, 1996). Supporting this view are numerous firms that owe their rapid rise and dominating position in the marketplace to NPD.

FIGURE 51.1. THREE PHASES MAKE UP THE ENTIRE INNOVATION PROCESS: THE FUZZY FRONT END, NEW PRODUCT DEVELOPMENT, AND COMMERCIALIZATION.

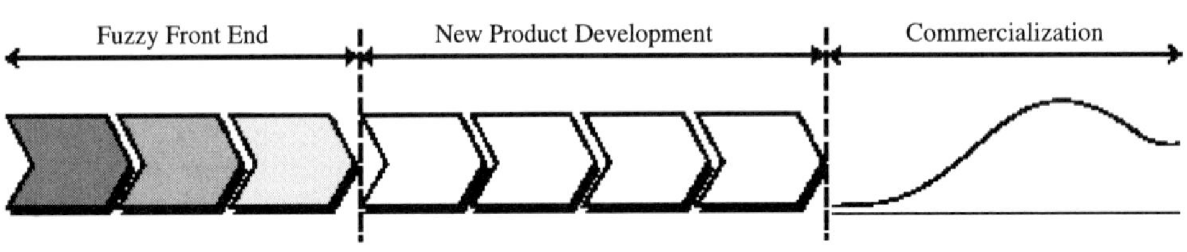

- Nike grew from a small start-up to an international juggernaut by launching a steady stream of imaginative new sneakers and other products into an array of market segments.
- Dell Computers, also at one time a small start-up, succeeded against giants such as IBM and Hewlett-Packard by creating a continuous supply of new products that offers customers the benefits of customization unmatched by its rivals.
- Rubbermaid pioneered an incredible number of new products in an industry often considered mature, prompting experts to select it as the most admired company in the United States a few years ago.

At the heart of NPD power lies its ability to create competitive advantages that lead to valuable rewards, some of which are reflected in corporate sales, profits, and long-term investment value. New products contributed to a stunning 33 percent of company sales, on average (Griffin, 1998). A second NPD reward is profits: New products accounted for 49.2 percent of the best companies' profits (Cooper, 2001). A third reward is the firm's standing in financial markets. Specifically, Cooper found that a firm's ability to create new products had the highest influence on the investment value of that firm (Cooper, 2001). In other words, NPD is important in determining the worth of the company as a long-term investment.

For companies whose business strategy hinges on the speed with which it brings new products to market, rapid NPD has additional benefits. For example, the earlier a product hits the market, the greater the chance the company has to command premium pricing, thus gaining market share and higher profits (Merrils, 1989; Nevens et al., 1990; Smith and Reinertsen, 1991; Adler et al., 1996; Calantone and Benedetto, 2000). When competing products appear, the price typically goes down, but the early company has been able to move down the manufacturing learning curve ahead of its competition. Clearly, there is a strong incentive to be first to market.

With such huge stakes in NPD, nations and corporations have no choice but to spend heavily on research and development projects. For example, the spending on industrial research and development in the United States grew in 1999 by 9 percent (Cooper, 2001). Similarly, some industries that have enjoyed substantial growth and profitability in the last several decades are noted for high investments in research and development. In particular, in 1998, the pharmaceutical industry spent 12.3 percent of sales revenues on research and development; communications equipment came in second with 12.1 percent; computer services third with 11.8 percent, and electronic components fourth with 10.3 percent of sales on research and development (Cooper, 2001). These impressive expenditures are additional testimony that NPD projects are central to corporate success.

What Makes NPD Projects Different

NPD projects are obviously the engine of corporate sales, market share, profits, and even share prices. But what is so special about them? In particular, which characteristics of NPD projects are fundamentally similar to other projects, and which differentiate them, ultimately leading to so much attention being devoted to the management of NPD? Five project characteristics help compare NPD projects with software development (SWD) and construction projects, two common types of projects in the business world (see Figure 51.2).

FIGURE 51.2. COMPARED CHARACTERISTICS OF NEW PRODUCT DEVELOPMENT, SOFTWARE DEVELOPMENT, AND CONSTRUCTION PROJECTS.

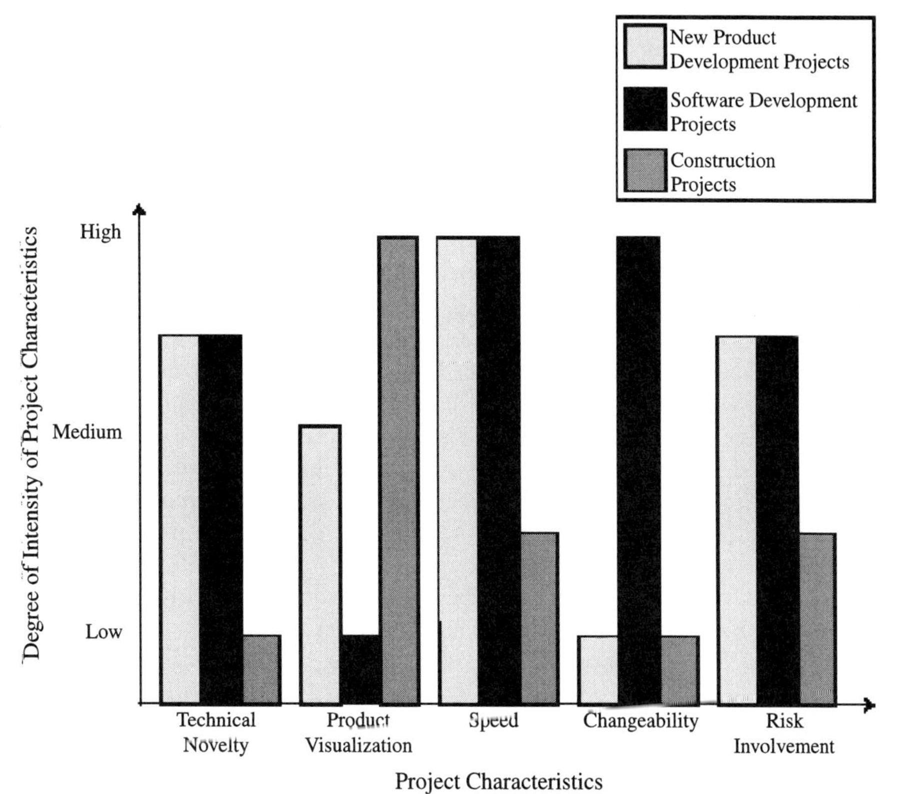

Technological Novelty. Projects involving more novel technologies are considered to have a higher uncertainty than those with more mature technologies (Shenhar, 1998). For example, NPD projects that create product platforms have a higher level of technological novelty than projects adapting the platform for a certain market niche (Wheelwright and Clark, 1992b). Since the essence of NPD and SWD projects is an innovation advantage, a large number of these projects deal with a medium-to-high level of technological uncertainty. In comparison, construction projects typically deal with mature technologies (low technical novelty) and less technological uncertainty. Generally, projects with higher uncertainty require a more flexible PM style (Shenhar, 2000). (See also the chapter by Shenhar.)

Product Visibility. To what extent can one physically see or touch a project product? This question addresses the characteristic of product visibility (Brooks, 1987; McDonald, 2001). For example, software products have low visibility. In contrast, NPD products and constructed facilities have concrete visibility, NPD projects medium, and construction projects high. This difference in product visibility produces differences in the PM process. Visible products are easier to visualize, which in the case of NPD and construction teams makes it easier to transform this visualization into a tangible project scope that can be quantified and estimated onto the PM process deliverables and milestones (Clark and Wheelwright, 1993; Cooper and Kleinschmidt, 1994; Kappel and Rubenstein, 1999). In contrast, it is much more difficult for a software development (SWD) team to visualize its product beyond a set of SWD requirements and specifications (Brooks, 1987). Such intangible visualization is more difficult to translate into a tangible project scope that can be similarly quantified and estimated onto the PM process deliverables and milestones (Findley, 1998).

Speed. NPD and SWD projects live off of high speed. In particular, the ability to accelerate these projects—ahead of competition and within the window of opportunity—is crucial to success (Eisenhardt and Brown, 1998). As already mentioned, speed enables premium pricing, higher market share, and higher profits. These are possible because speed creates a competitive advantage by responding to customers' needs first and changing markets faster than rivals can (Cooper, 2001). Short cycle time also means a lower probability that market conditions will radically change during project development, thus yielding fewer surprises. Note, however, that too much haste may be detrimental (Braun, 1990; Meyer and Utterback, 1995). For example, many mistakes may occur when an organization skips steps, a frequent approach to accelerating NPD projects (Crawford, 1992). Also, pressures from fast-project teams can chew up a firm's support resources, leaving other projects under-resourced and late. There are construction projects that face these high-speed pressures as well. In general, however, markets of many construction projects often tolerate low to medium speed. These differences in speed have their consequences. Faster projects need a more flexible PM process, in which the activities are overlapped to accelerate their execution and shorten duration (this approach is known as fast-tracking) (Zirger and Hartley, 1996).

Changeability. Changeability is the magnitude of the consequences of changing the project product design. Typically, the later in the project life cycle that the changes occur, the graver the consequences. (See the chapter by Cooper and Reichelt.) For NPD projects, the level of changeability is kept low because of the high costs of change, including changes in interfaces, tooling and fixtures, materials, the manufacturing process, and so on, all resulting in the product being late to market (Brooks, 1987). Similar is the case of construction projects. In contrast, if we characterize software as a pure thought-stuff, there are no tools, materials, and manufacturing process changes in SWD projects. Thus, changing the software is relatively easier. The differences in changeability in these classes of projects may result in different needs in managing the projects, for example, management of scope and customer involvement.

Risk Involvement. For NPD or SWD projects, new product and process technologies increase risks (Gupta and Wilemon, 1990; Raz, 1993). Also, the risk level increases if the project involves many personnel, has a high product and application complexity, and suffers from a lack of sufficient resources and team expertise (Little and Leverrick, 1995; Griffin, 1997; Handfield et al., 1999; Jiang et al., 2000; Tatikonda and Rosenthal, 2000). Consequently, many NPD and SWD projects face medium to high severity of risk. While some construction projects are exposed to the similar medium to high level of risk, often the by-product of new and untested architectural designs and construction techniques, NPD projects, as Pinto and Govin point out, ". . . often involve greater risks than construction projects" (Pinto and Govin, 1989). They also add that ". . . risk factors are generally thought to be more prevalent in NPD than in construction projects." Thus, in general, risk is often low to medium in construction projects. As a result, risk management has a more emphasized place in management of NPD and SWD projects than in construction projects.

This analysis of project characteristics in NPD, SWD, and construction projects points to a clear conclusion: While NPD projects have similarities with other types of projects, they also have differences. These differences and other features (to be discussed next) have prompted some experts to describe NPD projects as a microcosm of the whole organization (Bowen et al., 1994). In particular, because NPD projects typically are implemented under severe speed and financial pressures, they tend to expose the strengths and weaknesses of a company, including its culture, management systems, organizational structure, and people. Therefore, NPD projects are a comprehensive, real-time test of the whole corporation (Bowen et al., 1994).

Factors Affecting the Success of NPD Projects

Because of the basic differences between NPD projects and other projects, it is reasonable to expect differences in the critical success factors required for successful completion of these projects. As Figure 51.3 shows, these factors have an impact on NPD projects (Brown and Eisenhardt, 1995):

- The project team, leader, senior management, and suppliers all impact process performance (e.g., the speed and productivity of NPD projects).
- The project leader, customers, and senior management influence product effectiveness (e.g., the fit with market needs and competencies).
- The interplay of an efficient PM process with an effective product and a generous market molds financial success (i.e., revenue, profitability, and market share).

FIGURE 51.3. CRITICAL SUCCESS FACTORS IN NEW PRODUCT DEVELOPMENT PROJECTS.

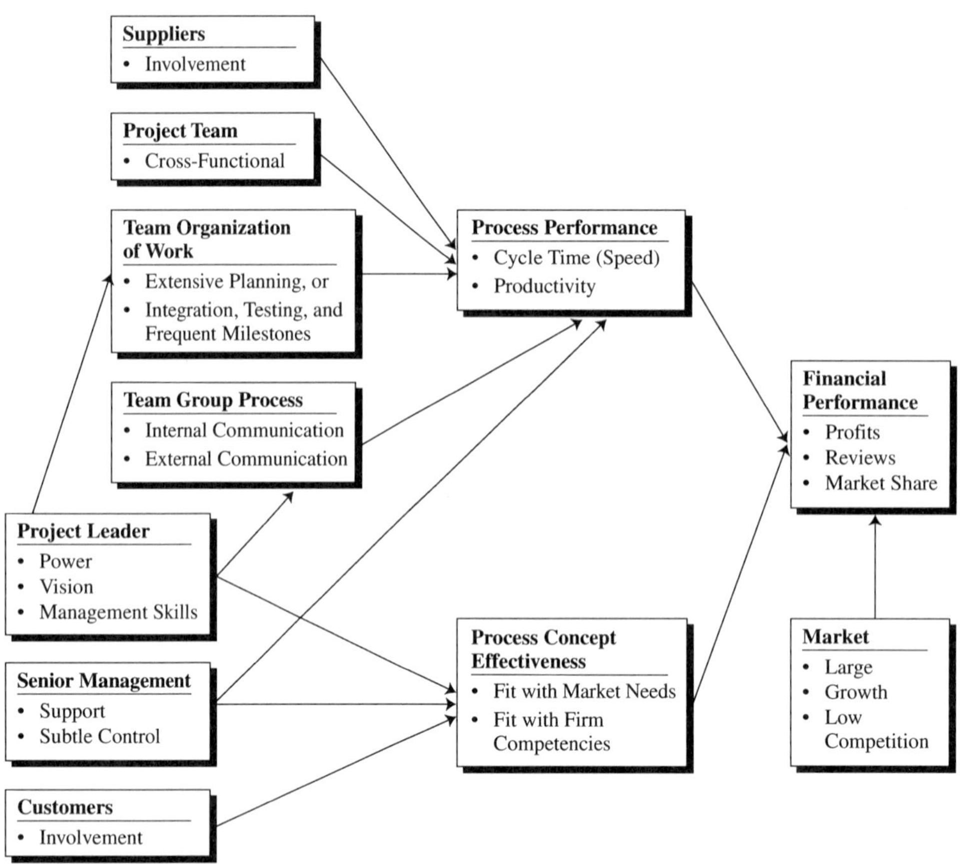

Source: Product Development: Past Research, Present Findings, and Future Directions, Vol. 20 by S. L. Brown and K. M. Eisenhardt. Copyright 1995 by Academy of Management Review. Reprinted by permission of Academy of Management Review via the Copyright Clearance Center.

The success factors are numerous.

Project Team. Project teams are absolutely critical to process performance in NPD projects (Zirger and Madique, 1990; Wheelwright and Clark, 1992b). They include the people who convert ambiguous ideas and concepts into workable product designs. For this, such teams need several features. First, teams need to be cross-functional, having members from Engineering and Manufacturing to Marketing, and so on. Such functional diversity increases the amount and variety of information that the members can use to enhance design process performance. Second, effective group processes—especially communication—are vital to the PM process. Internal communication within the team will then yield more information, build team cohesion, and reduce misunderstandings and hurdles, thus improving speed and productivity (Dougherty, 1992). A similar impact comes from frequent, task-oriented external communication with customers, suppliers, and other organizational personnel (Katz and Tushman, 1981). The third crucial team feature is work organization, which in the case of mature products (e.g., automobiles) tends to include extensive planning. This helps eliminate extra work, sequence PM activities, and avoid mistakes. In high-velocity industries, such as microcomputers, however, frequent iterations of product design and extensive testing augment process performance (Eisenhardt and Tabrizi, 1995).

Project Leader. As a pivotal person on the project team, the project leader critically impacts both the process performance and product effectiveness (see Figure 51.3). Several leadership characteristics matter here. First, powerful leaders are more effective in obtaining resources, attracting better team members, and keeping the team focused and motivated (Wheelwright and Clark, 1992b). Vision is another important characteristic. Leaders with the ability to subtly match corporate strategies and competencies with customer needs to shape an effective product concept are more effective (Jassawalla and Sashittal, 2000). Finally, project leaders with a complete skill set including experienced technical, administrative, interpersonal, and business skills make a stronger contribution to speed, productivity, and product effectiveness (Shenhar and Thamhain, 1994).

Senior Management. In addition to the team and project leader, senior management support is essential to fast and productive NPD projects (Cooper and Kleinschmidt, 1987; Zirger and Madique, 1990). This support is instrumental in securing resources that will attract good team members to the project, in receiving the project go-ahead, and in obtaining funding. Equally crucial to the process performance and effectiveness of the product is the ability of senior management to exercise subtle control. Such control includes working together with the project leader to develop and communicate a distinctive and coherent product concept. This control also means giving autonomy to the project team to be creative and motivated (Brown and Eisenhardt, 1995).

Suppliers and Customers. The last factor impacting the success of NPD projects involves suppliers and customers (see Figure 51.3). Early and extensive supplier involvement helps cut the design complexity, which then makes the PM process of NPD projects faster and more productive (Gupta and Wilemon, 1990; Handfield et al., 1999). Also, customer involvement can reduce the number of errors in the product design, helping improve the effectiveness of the product concept (Cooper and Kleinschmidt, 1987; Zirger and Madique, 1990).

Financial Success. So far I have emphasized that the project team, project leader, senior management, suppliers, and customers affect process performance and product effectiveness.

As Figure 51.3 shows, these two factors interact with characteristics of the market to predict the financial performance of the product (Brown and Eisenhardt, 1995). A productive process leads to lower costs, consequently to lower prices, thus to greater product success. Also, a rapid process provides strategic flexibility and shorter time to product launch, which may facilitate the financial success of products. When it comes to the impact of product effectiveness, it is the product characteristics, such as low cost and unique customer benefits, that lead to financially successful products (Cooper and Kleinschmidt, 1987; Zirger and Madique, 1990).

The financial performance of products is also driven by a healthy market, one that is large and growing, and with low competition (Brown and Eisenhardt, 1995). The logic here is that companies in such markets can enjoy large sales and opportunities to introduce new products. Overall, a strong PM process for NPD projects, an attractive product, and a healthy market are critical to a financially successful product.

Pipeline Management

What methods do firms use to mesh these success factors and deliver successful NPD projects? The next two sections outline two sets of such methods. The first focuses on the strategic level of managing NPD projects: pipeline management methods. Supporting them is the PM process, a set of methods on the operational level that offers a step-by-step roadmap for managing individual NPD projects.

Framework

For NPD pipeline management to succeed, there must exist a full alignment of the business strategy, the strategic pipeline, the project pipeline, the functional pipeline, and the PM process (see Figure 51.4). The alignment begins at the top, where four different business strategies are available (Miles and Snow, 1978):

- *The prospector strategy*. This strategy deals with leading in technology, product, and market development, even though an individual product may not be profitable. The strategic goal is to hit the market first with innovation.
- *The defender strategy*. Firms with this strategy concentrate on a product-market domain and protect it by all means, not necessarily through new product development. The strategic goal is placing high value on improvements in efficiency.
- *The analyzer strategy*. Analyzers pursue an imitative innovation strategy, rather than being first to market with new products. The strategic goal is to reach the market swiftly with an equivalent or slightly better product after another firm opens up the market (Belliveau et al., 2002).
- *The reactor strategy*. This strategy results from situations wherein one of the three other strategies is improperly pursued (Miles and Snow, 1978). Firms with this strategy develop new products only if forced to by competitive pressures.

FIGURE 51.4. STRATEGY-ALIGNMENT PYRAMID FOR NPD PROJECT MANAGEMENT.

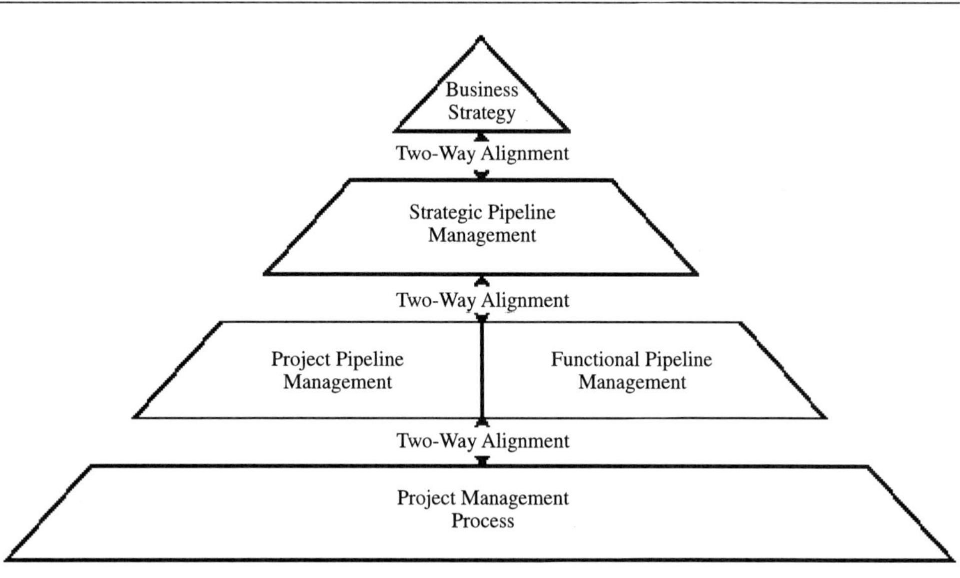

The choice of the business strategy drives the strategic pipeline approach, whose purpose is to select, prioritize, and balance NPD projects that are optimally aligned with the business strategy and then support them with rough-cut resource planning and skill sets. Such a pipeline drives the next-lower level in the pyramid (see Figure 51.4). On that level, project pipeline management works to fine-tune resource deployments (made by the strategic pipeline) in projects during ramp-up, ramp-down, and midcourse adjustments (Belliveau et al., 2002). Also on that level is functional pipeline management, which, in alignment with the project pipeline, optimizes the flow of projects through functional areas per the requirements of the strategic pipeline. Through these three pipelines, the business strategy dictates the PM process for individual NPD projects, the lowest level in the pyramid.

Strategic Pipeline Management

This part of pipeline management (also called project portfolio management—see the chapter by Archer and Ghasemzadeh) strives to achieve the strategic balance of NPD projects through five broad goals.

Maximize Value. The first goal is to select and prioritize NPD projects in such a way that the sum of their values gives the highest value to the portfolio. Typically, this value is expressed in terms of a business objective. If that objective is profitability, for example, the value can be stated as return on investment (ROI) (Baker, 1974: Bard et al., 1988). Other examples include likelihood of success and overall strategic value expressed in a scoring model.

Balance Projects. The point here is to develop a portfolio of projects that is balanced across multiple parameters. For example, one can aim at the right balance of those projects that are more difficult to execute versus those less difficult to execute, or high-risk projects versus low-risk ones, or across project types (e.g., platform projects, derivative projects, etc.) or market segments (market A, market B, etc.) (Cooper et al., 1998a; Archer and Ghasemzadeh, 1999).

Align with the Business Strategy. The key here is to have an "on-strategy" portfolio (Cooper et al., 1998b). This means that the final project portfolio is truly aligned with the business strategy—the breakdown of spending on projects across project types, markets, domains, product domains, and so on is in tune with strategic initiatives established by management and expressed in the business strategy.

Manage Rough-Cut Resource Capacity. Many firms tend to overcommit their resources; that is, they have more active projects than their resources can support. As a result, the pipeline is congested, projects take longer to get to the market, and corners are cut for lack of resources (Harris and McKay, 1996). The cure is in rough-cut resource capacity planning—for example, through creating an aggregate project plan that matches the resources required for active projects with the resources that are available (Wheelwright and Clark, 1992a). Part of this is done through adjusting the organization's skill sets in order to deliver products (Harris and McKay, 1996).

Set Boundaries. The final goal is to clearly articulate empowerment boundaries for project and functional management (Harris and McKay, 1996). Then, project and functional managers will know how to adjust resource deployments and schedules in line with the strategic pipeline priorities.

Project Pipeline Management

Local rationality can plague NPD pipeline flow (Fricke and Shenhar, 2000). In particular, overly autonomous project teams often make resource decisions based on what is good for their project rather than what is good for the business (Harris and McKay, 1996). This and other problems indicate that projects are managed with a "local-interests-first" attitude, ignoring the greater good of the NPD pipeline.

By eliminating this local rationality, project pipeline management aims at proactive and continual adjustments of resource deployments while ramping up—and down—project teams and responding to the always-present disruptions in the NPD process. In this effort, the guiding criteria are strategic priorities set by the strategic pipeline. It is of critical importance that all activities that consume resource hours of NPD personnel are included. Such activities may include work on NPD projects, technology development, support activities, and continuous improvement of the NPD system (Harris and McKay, 1996). For example, the support activities may involve sustaining engineering and marketing support. When all of this is done properly, project pipeline management should optimize the flow of projects in the pipeline.

The starting point in project pipeline management is rough-cut capacity allocations made in the strategic pipeline. These are adjusted based on allocations of nominal resource capacity, which includes the real level of available resources—that is, the sum of all FTEs (full-time employees) in each functional area of NPD. When the nominal capacity is reduced by the nonproject or overhead time (e.g., vacation, training, staff meetings, etc.) and pet time (when personnel work on their own new ideas), effective capacity is obtained. The effective capacity can be categorized as *critical* (project work with drop-dead deadlines that cannot be delayed) and *noncritical* commitments (project work that can be deferred). The relative size of each of these time "buckets" is the policy choice of each company. In one company, for example, overhead time is 15 percent, and pet time 10 percent, which means the effective capacity is 75 percent of an FTE's available time.

Figure 51.5 provides an example of balancing resources in the pipeline. For the first three months (balancing periods are usually two to three months long), the required resources exceeded the effective capacity. In particular, aggregate critical commitments were so high that all noncritical commitments were postponed and half of the pet commitment time was also consumed. In succeeding periods, the critical commitments descended (as drop-dead projects were completed), pet time was restored to normal, and noncritical commitments grew. Still, there were enough projects in the pipeline to keep the sum of critical and noncritical commitments at the effective capacity level. For this to function, however, functional managers and project managers should agree on the number of resource hours for each person involved in projects, so that everyone's resource commitments are within the aggregate nominal capacity (Nevison, 2002). To maximize everyone's effective resource capacity, it is important to establish the maximum number of projects a person is allowed to be involved in at any one time (e.g., three at a time). This helps control switchover time from project to project, often a significant resource drain (Fricke and Shenhar, 2000).

For project pipeline management to work effectively, companies need to make resource adjustments within the project teams. This requires continuous evaluation of the active projects in the context of the strategic priorities of the company. When the priorities change, so should the resource allocations. Popular stage-gate or end-of-phase reviews should also be designed and conducted so that their decisions reflect the strategic priorities of the pipeline and the aggregate nominal capacity of all functions. Most importantly, healthy stage-gate reviews should never add new projects to the pipeline unless the resources are available to support them sufficiently. In short, only the continual optimization of the pipeline will help achieve the maximum possible throughput.

Functional Pipeline Management

A significant role in achieving the maximum possible throughput also belongs to functional managers. In particular, they are expected to provide a smooth flow of projects through their functional areas. Too often, however, they exercise the local rationality approach,

FIGURE 51.5. BALANCING REQUIRED AND AVAILABLE RESOURCE CAPACITY.

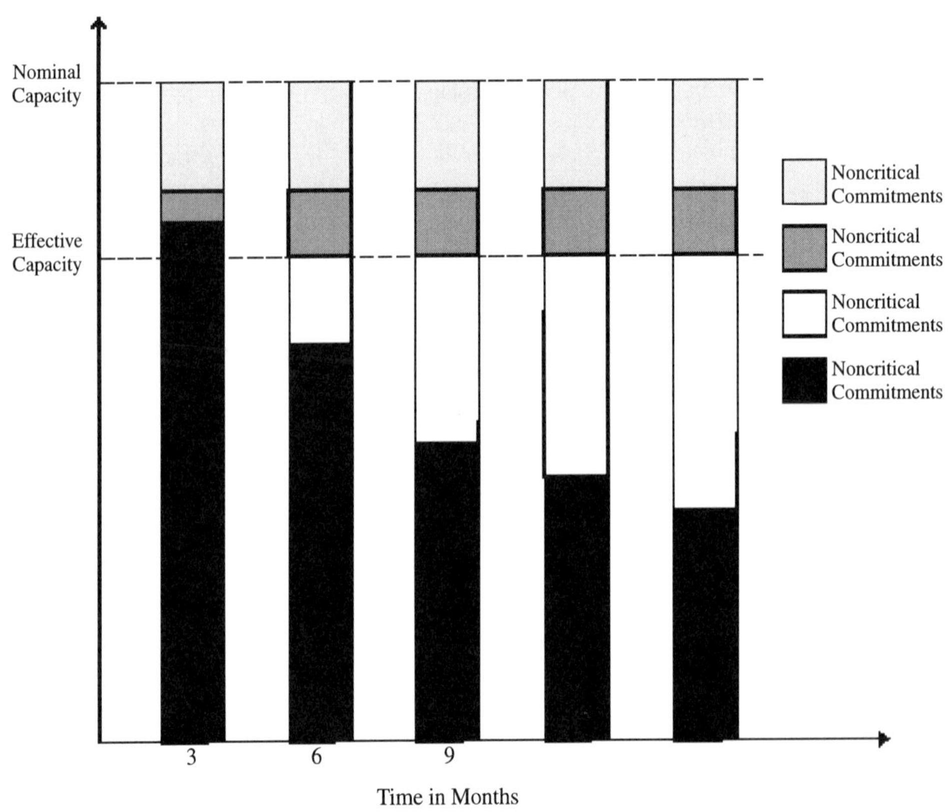

directing their efforts solely to their area and only rarely talking to other interfacing functions to solve specific problems (Harris and McKay, 1996). In doing so, functional managers optimize their area on the basis of that area's specific objectives and, at the same time, actually suboptimize their project work. To optimize truly, they need to handle project flow in full alignment with priorities set by the strategic pipeline. Only then will functional areas deliver what projects need from them: Functional excellence.

Functional excellence includes more than conformance to specifications of the functional tasks for the project. This "more" relates to functional realism and predictability. Since pipeline management is notorious for placing unrealistic project load on functional areas, it is crucial that functional managers understand what realistic demands are and build a predictable process to manage those demands. Such a process enables them to recognize functional bottlenecks—often the major killer of project throughput—in a timely way and to address them with various competencies.

One of these competencies is to cross-train functional experts to eliminate the bottlenecks. Another is to identify transferable expertise from other functional areas and deploy them when necessary to fight such bottlenecks. Timely hiring is also a vital competence of the process. In one microprocessor development project, the company required all functional areas to predict the number of each type of functional expert needed at least six months in advance. This gave enough lead time to hire the best experts and to maintain accelerated project speed. Sometimes, availing oneself of outsourcing is the only way to maintain the smooth flow of projects. This requires knowing exactly what pieces of the functional expertise can be outsourced and from which sources, and what process will be used for joint development with the supplier. Still another competence is continuously improving the overall predictability of the process.

When all of these competencies are developed and synchronized, functional pipeline management not only supplies functional excellence but also delivers the long-term capacity to handle the flow of projects through functional areas. Working closely with project pipeline management on strategic pipeline priorities, such functional pipeline management is able to share in achieving the maximum possible project throughput.

The NPD Pipeline Management Process

NPD projects are delivered in a dynamic environment fraught with market and technological changes. Consequently, pipeline management requires a dynamic process capable of constant adjustment and optimization. Logically, then, the process needs to be designed, deployed, and improved on a continuous basis as part of the larger system for management of NPD projects. The three components of this process, shown in Figure 51.6, include pipeline management teams, methodology, and tools.

FIGURE 51.6. THE PIPELINE MANAGEMENT PROCESS.

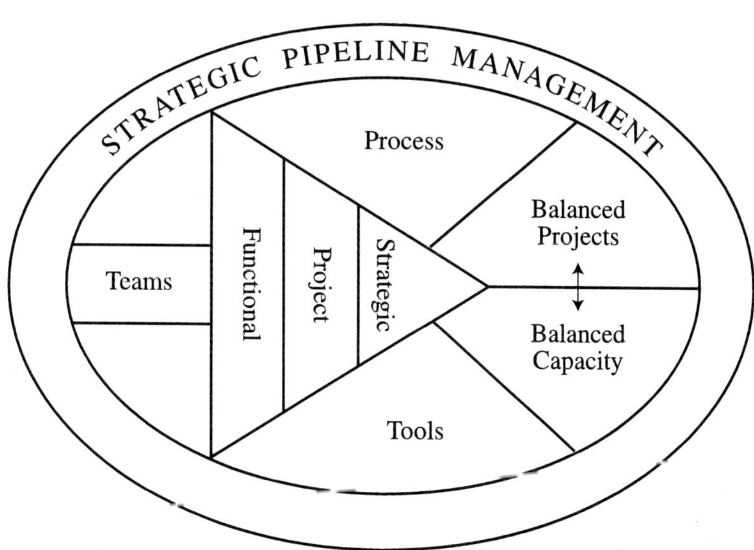

The crucial role in managing the pipeline belongs to teams, each with clearly defined decision-making boundaries. Functional managers who cannot solve a pipeline problem in their area are expected to bring it to the attention of the project pipeline management team, where they also take part in decision making. Problems still unresolved are then further escalated to the strategic pipeline team.

Because it is responsible for organizational strategic planning, the strategic pipeline management team includes executives who care about linking the competitive strategy to project resource allocations and functional budgeting (Harris and McKay, 1996). On the project level, the project pipeline management team controls resource deployment and modifications through end-of-phase reviews. Members of this team are functional managers and project leaders. It is a frequent practice in small organizations to merge the two teams into one and call it the portfolio management team (Platje et al., 1994).

Essentially, the task of these pipeline management teams is to balance projects and resource capacity. In particular, they must balance multiple NPD projects across multiple functions, primarily selecting the best mix of projects along many dimensions. Also, they must plan and adjust resource capacity to support the project mix (Cooper et al., 1998a). To perform this complex assignment, these teams need a standardized methodology that specifies a sequence of steps for pipeline management. Finally, the tools that support the methodology and teams include project selection tools, traditional charts for portfolio management, and portfolio mapping tools (Cooper et al., 1998a, 1998b).

Selecting and Customizing the Project Management Process for NPD Projects

What Is the Project Management Process for NPD Projects?

The purpose of a strategy for managing individual NPD projects is to enable the implementation of projects that will effectively support an organization in pursuing its pipeline management and, through this, its business strategy and goals. Since the strategy for managing individual NPD projects is often delivered by means of the PM process (the lowest level on the pyramid in Figure 51.4), this section will focus on that process. Its importance is obvious: Multiple research studies have confirmed that the process is the critical success factor in managing NPD projects (Cooper and Kleinschmidt, 1994; Lynn et al., 1999). Note that in Figure 51.3 I used the term *Team Organization of Work* in place of *PM process*. Also note that many companies choose the term *NPD project life cycle* as a synonym for *PM process*.

The PM process implements several elements (Kerzner, 2000):

- Project life cycle phases
- Managerial and technical activities
- Deliverables
- Milestones

The project life cycle is viewed as a collection of project phases determined by the control needs of the organizations involved in the project. Consequently, a variety of project life cycle models are in use in corporations today. Some of them are traditional models including the phases of concept, definition, execution, and finish. These models seem to be losing their appeal because of their rather generic nature, while new, more industry-specific ones have emerged and are gaining in popularity. One example is the concurrent engineering process, which for ease of understanding is illustrated in Figure 51.7 as sequential. (See the chapter by Thamhain.)

FIGURE 51.7. AN EXAMPLE OF THE PROJECT MANAGEMENT PROCESS FOR NEW PRODUCT DEVELOPMENT PROJECTS.

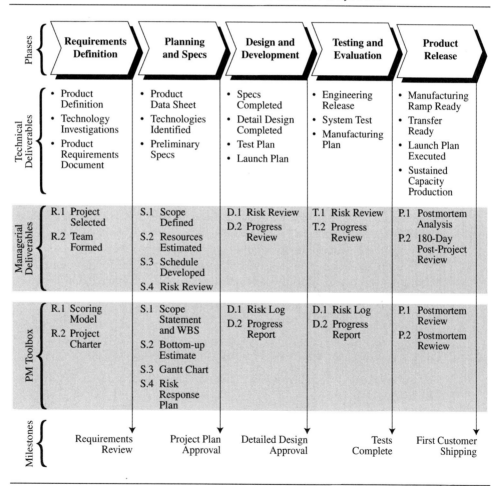

Phases	**Requirements Definition**	**Planning and Specs**	**Design and Development**	**Testing and Evaluation**	**Product Release**
Technical Deliverables	• Product Definition • Technology Investigations • Product Requirements Document	• Product Data Sheet • Technologies Identified • Preliminary Specs	• Specs Completed • Detail Design Completed • Test Plan • Launch Plan	• Engineering Release • System Test • Manufacturing Plan	• Manufacturing Ramp Ready • Transfer Ready • Launch Plan Executed • Sustained Capacity Production
Managerial Deliverables	R.1 Project Selected R.2 Team Formed	S.1 Scope Defined S.2 Resources Estimated S.3 Schedule Developed S.4 Risk Review	D.1 Risk Review D.2 Progress Review	T.1 Risk Review T.2 Progress Review	P.1 Postmortem Analysis P.2 180-Day Post-Project Review
PM Toolbox	R.1 Scoring Model R.2 Project Charter	S.1 Scope Statement and WBS S.2 Bottom-up Estimate S.3 Gantt Chart S.4 Risk Response Plan	D.1 Risk Log D.2 Progress Report	D.1 Risk Log D.2 Progress Report	P.1 Postmortem Review P.2 Postmortem Rewiew
Milestones	Requirements Review	Project Plan Approval	Detailed Design Approval	Tests Complete	First Customer Shipping

Notes:

1. Typical elements of the PM process (managerial and technical activities, and technical deliverables and their supporting tools) have been left out to simplify this figure.
2. The PM toolbox and managerial deliverables are shaded to indicate that the former support the latter.
3. The managerial deliverables and the PM tools that support the deliverables are similarly coded.

Project life cycle phases are composed of logically related project activities. These activities can be divided into two groups: managerial and technical activities. The former are activities by which projects are managed; *develop the project scope* and *construct the project schedule* are typical examples. Managerial activities are similar across project types, whether NPD, construction, software, marketing, or financial.

However, technical activities vary across project types. For example, technical activities characteristic of NPD projects include *technology investigations* and *ramping up for manufacturing* (see Figure 51.7). While they may be present in some construction projects, they are not as central to the construction projects as they are central to all the NPD projects. On the other hand, examples of typical technical activities in construction projects such as *pre-job meeting* and *punch list development* are much less present in NPD projects. In short, technical activities take care of managing the project product and are thus naturally project-type-specific, reflecting the nature of the project product. (*Note:* To simplify Figure 51.7, managerial and technical activities have been left out.)

Both PM and technical activities usually culminate in the completion of deliverables, that is, tangible products in the PM process. Managerial activities produce managerial deliverables (also called PM deliverables) such as *scope defined* or *risk review* (see Figure 51.7), while technical activities lead to technical deliverables, such as *product data sheet*, *system test*, and *manufacturing ramp-ready*. (Note that in Figure 51.7, only the *deliverables* and their milestones, which signify the end of a phase, are shown. Managerial and technical *activities* have been left out in order to simplify the figure.)

Supporting the managerial deliverables is the PM toolbox. Two principles are important in this support. First, each managerial deliverable is specifically supported by a specific tool or tools; in other words, each tool is selected because its systematic procedure helps produce the deliverable. For example, the S.1 tools *Scope Statement* and *WBS* (*Work Breakdown Structure*) support the accomplishment of the S.1 managerial deliverable *Scope Defined*. (In this figure, the tools and the managerial deliverables they support are numbered correspondingly.) Second, the PM toolbox is designed to contain all the tools needed to implement the whole PM process and accomplish its whole set of deliverables (emphasized in Figure 51.7 by the shaded areas). Technical deliverables and their tools are project-type-specific and are beyond the scope of this chapter.

Benefits of the Project Management Process for NPD Projects

The process creates value because it is able to deliver NPD projects that consistently have the following:

- Speed
- Repeatability
- Concurrency

Speed. This is the ability of an organization to deliver a project *fast*. While in actuality what is meant by *fast* may vary, it is always competitive. For example, *fast* may mean that the cycle time has to be reduced from 18 months to 9 months for the organization to keep up with the competition. For this to be possible, many components in the PM process have to be present; for instance, there must be within- and across-phase overlapping of project activities. Or non-speed-adding activities may have to be eliminated as well as any redundancies, such as unnecessary executive involvement in the day-to-day decision making activities. What this really means is having a PM process that is streamlined enough to deliver per customer demands.

Repeatability. Delivering a project with speed is not enough unless it is repeatable. This means the organization must be able to consistently deliver a stream of consecutive projects, one after another, per customer requirements every time. This is called *longitudinal repeatability*. If the customer requirement is speed, for example, then each delivered project must be consistently fast. Repeatable projects minimize variation in how they are executed, improving speed and quality. Improvements in quality lead to lower cost because they result in less rework, fewer mistakes, fewer delays and snags, and better use of time. With higher speed, better quality, and lower cost, the organization can better respond to customer demands and satisfy them.

Concurrency. In addition to speed and repeatable consecutive projects, responding to the customer's demands also requires an ability to deliver a host of simultaneous projects that are often interdependent. This is called *lateral repeatability*, a different challenge from the longitudinal one. Here, some projects are large, others small. Since they share the same pool of resources, the challenge is to execute all of them in parallel, as a concerted group. No variation is allowed in any one; each needs to maintain speed and quality. If they don't, they will cause other projects to slip, increase cost, and make the customer unhappy. Similar to longitudinal repeatability, minimizing the variation in each project will trigger speed and quality improvements that lead to lower cost, again contributing to meeting customer demands and increasing customer satisfaction.

Because companies pursue different business strategies, it is critical to customize the PM process per those strategic needs. A convenient framework can be used for selecting and customizing the NPD PM process. The following three of the many possible options for customizing a PM process are perhaps the most frequently used:

- Customization by project size
- Customization by project novelty
- Customization by project type

Customization by Project Size

Some NPD organizations use project size as the key variable when customizing the PM process. Their view is that larger projects are more complex than smaller ones. In other words, size is the measure of the PM process's complexity. The reasoning here is that as project size increases, so does the number of PM activities and resulting managerial deliverables, as well as the number of interactions among them. However, the number of interactions has compound growth, rather than linear growth (Smith and Reinertsen, 1991). Such increased complexity, then, has its price: Larger projects require more managerial work and deliverables to coordinate the increased number of interactions.

Because different project sizes require different PM processes, you first need a way to classify projects by size and then to customize their processes. For size classification, three real-life examples are presented in Table 51.1. All three of these companies created three classes of projects: small, medium, and large; and all are measured in dollars (person-hour budgets). Thus, on the basis of size, these companies determined the managerial complexity of their projects. Furthermore, this complexity dictated the PM process, a simplified example of which is illustrated in Table 51.2. (*Note:* For the sake of simplicity, only managerial deliverables are shown, leaving out technical deliverables, the toolbox, and milestones.)

As Table 51.2 indicates, some of the managerial deliverables for projects of different size are the same; others are different. For example, all use the *Progress reported* deliverable because all projects need to report on their performance. Since managerial complexity of the three project classes calls for different PM processes, some of the managerial deliverables in the processes differ. The *Risk tracked* deliverable, for example, is needed only in large projects.

The experience of these companies offers several guidelines for customizing the PM process by project size:

- Identify a few classes of projects and their PM processes.
- Define each class by the size parameter.
- Match the class complexity with the proper PM process, specifying those PM tools that support a specific managerial deliverable in the PM process.

Note that while customization by project size offers the advantages of simplicity, it also carries a risk of being generic, disregarding other situational variables. In some cases, these other variables may be of vital importance, as noted in the section on customization by project novelty that follows.

TABLE 51.1. EXAMPLES OF PROJECT CLASSIFICATION PER SIZE IN THREE COMPANIES.

	Project Size		
Company	**Small**	**Medium**	**Large**
New product development projects in a \$1B/year high-tech manufacturer	\$1–2 M	\$2–5 M	>\$5 M
New product development projects in a \$150M/year low-tech manufacturer	<\$50K	\$50–150K	>\$150K
Software development projects in a \$40M/year company	300–400 person-hours	1000–3000 person-hours	>3000 person-hours

TABLE 51.2. AN EXAMPLE OF CUSTOMIZATION OF PROJECT MANAGEMENT PROCESS BY PROJECT SIZE.

Project Size	*Project Phases*			
	Definition	**Planning**	**Execution**	**Closure**
Small	• Team formed	• Scope defined • Responsibilities assigned • Schedule developed	• Progress reported	• Progress reported
Medium	• Team formed • Team skill gaps identified	• Scope defined • Responsibilities assigned • Costs estimated • Schedule developed	• Progress reported • Schedule updated	• Progress reported
Large	• Team formed • Team skill gaps identified • Team development model chosen • Stakeholders analyzed	• Scope defined • Responsibilities assigned • Schedule developed • Costs estimated • Risks assessed • Team member commitments obtained	• Progress reported • Schedule updated • Changes tracked • Risks tracked • Team member commitments obtained	• Progress reported • Postmortem performed

Customization by Project Novelty

Companies can opt to customize the PM process by a project's technical novelty. In particular, they may recognize that their NPD projects have multiple project classes of different technical novelty, each calling for a customization of the PM process. Consider, for example, the case of a low-tech manufacturer (see Table 51.3). This company classifies all of its NPD projects into three groups: simple, medium, and complex. To distinguish them, the company uses 11 characteristics, mostly related to the project's technical novelty. Some of these include *customer*, *product features*, *end use application*, *manufacturing process*, and *assembly process*.

Generally, the higher the technical novelty, the more complex the projects (Tatikonda and Rosenthal, 2000). This is because the increasing technical novelty leads to more uncertainty, elevating the need for more flexibility in the PM process. In particular, as the technical novelty of projects increases, the PM process requires the following:

- More iterations and time to define the project scope
- Higher technical and managerial skills

TABLE 51.3. AN EXAMPLE OF CLASSIFYING PROJECTS BY TECHNICAL NOVELTY.

	Project Class		
Characteristic	**Simple**	**Medium**	**Complex**
Customer	Existing	Existing or new	Existing or new
Product features	0–5 feature changes to the existing product	>5 feature changes to the existing product	No similarity to an existing product
End use application	Same	Same	Same or new
Manufacturing process	Existing	Existing	New
Assembly process	Existing	Some changes to existing process	New

- Deeper communication
- More effective management of change

A simple example reflecting these trends in adapting the PM process for the three classes of NPD projects from Table 51.3 is illustrated in Table 51.4. As the table shows, the PM processes of the three classes of projects are similar in only some aspects. For example, all use *Schedule developed* and *Progress reported* deliverables. Others differ. Complex projects include deliverables such as *Team skill gaps identified* and *Postmortem performed* that are not required in projects of simple and medium complexity. Obviously, the variation in the technical novelty of the project is the source of the differences.

In customizing by project novelty, one has to be aware of certain advantages and risks. The advantages include an easy-to-understand PM process, because this customization relies on the technical aspects of projects, which lies at the center of the professional background of project managers. On the other hand, risks occur when this type of customization is used for NPD projects that are light on technology. In these projects, technical novelty is not a relevant issue, rendering the customization irrelevant.

In summary, there are several deployment guidelines an organization should follow when customizing the PM process by project novelty:

- Classify projects and their PM processes into a few classes.
- Define each class with a few technical parameters.
- Support each class with the proper PM process, defining PM tools backing a specific managerial deliverable in the PM process.

Customization by Project Type

While the previous two options for customization rely on one dimension, or variable, each customization by project type—project complexity (size) and project novelty, respectively—

TABLE 51.4. AN EXAMPLE OF CUSTOMIZATION OF PROJECT MANAGEMENT TOOLBOX IN A PROJECT FAMILY TYPE BY TECHNICAL NOVELTY.

Project Class	*Project Phases*			
	Definition	Planning	Execution	Closure
Simple	• Projects selected • Portfolio balanced • Team formed	• Schedule developed	• Progress reported	• Progress reported
Medium	• Projects selected • Portfolio balanced • Team formed	• Customers interviewed • Scope defined • Schedule developed	• Progress reported • Schedule updated	• Progress reported
Complex	• Projects selected • Portfolio balanced • Team formed • Team development model chosen • Team skill gaps identified	• Voice of the customer identified • Scope defined • Responsibilities assigned • Costs estimated • Schedule developed • Risk plan developed (qualitative) • Team member commitments obtained	• Progress reported • Costs monitored • Schedule updated • Changes tracked • Change log kept • Risks tracked • Team member commitments obtained	• Progress reported • Postmortem performed

uses both of the variables. This two-dimensional model is called by the name of its creator: the Shenhar model (Shenhar, 2001). (See also the chapter by Shenhar and Dvir.) To make it more pragmatic for this discussion, the model has been simplified.

Each of the two dimensions includes two levels:

- Technical novelty (low/high)
- Project complexity via system scope or assembly type (low/high)

This helps create a 2x2 matrix that makes it easy to discuss four major types of projects: *routine, administrative, technical,* and *unique*, as in Figure 51.8.

Routine projects—for example, upgrading an existing product or developing a new model of a traditional product, say a toaster—have a low level of technical novelty and assembly system scope. At the time of initiation, these projects mostly use existing or mature technologies or adapt familiar ones. Sometimes, some new technology or feature may be used, but it does not exceed more than 50 percent of the total number of technologies used.

FIGURE 51.8. FOUR TYPES OF NEW PRODUCT DEVELOPMENT PROJECTS.

Project Complexity ↑	Technical Novelty: Low	Technical Novelty: High
High	**Administrative Projects** Projects Featuring: • Mature Technologies, Some New (< 50%) • Scope Frozen Before or Early in Execution • Few Scope Changes • System That Requires Integration • Performed By Many Organizations	**Unique Projects** Projects Featuring: • > 50% New Technologies • Long Cycle Times • Scope Frozen in 2nd or 3rd Quarter • Lots of Scope Changes • System That Requires Integration • Performed By Many Organizations
Low	Projects Featuring: • Mature Technologies, Some New (< 50%) • Scope Frozen Before or Early in Execution • Few Scope Changes • Independent Single-Function Product/Subsystem • Performed Within One Organization/Function **Routine Projects**	Projects Featuring: • > 50% New Technologies • Long Cycle Times • Scope Frozen in 2nd or 3rd Quarter • Lots of Scope Changes • Independent Single-Function Product/Subsystem • Performed Within One Organization/Function **Technical Projects**

Because of these known technologies, the system scope is frozen before or early in the execution phase, and few scope changes occur. Essentially, these are low- to medium-tech projects. The system scope—the assembly type—means that the project can produce a product that is a subsystem of a larger system or a stand-alone product capable of performing a single function. Typically, the project is performed within a single organization or organizational function—for instance, Engineering or Marketing (Shenhar, 2001). Because they routinely deal with existing technologies and have a modest scope, these projects have been dubbed *routine.*

Administrative projects—for example, upgrading a new computer or developing a new automobile model—are similar to routine projects in terms of the technical novelty: They are low- to medium-tech. Therefore, these projects also use less than 50 percent new technologies, enjoy an early scope freeze, and have the benefit of few scope changes. However, they differ in the system scope domain. Unlike routine projects, they produce project products consisting of a collection of interactive subsystems (assemblies) that are capable of performing a wide range of functions (Shenhar, 1998). As a result, many organizations or organizational functions are involved, generating a strong need for the integration of both the subsystems and the organization. Such integration calls for more administrative work, hence the label *administrative* for these projects.

The major property of *technical* projects—for instance, developing a new model of computer or a new model of a computer game—is their technical content. Thus, more than 50 percent of the technologies that these projects use are new or have not yet been developed at project initiation. This naturally makes them high-tech and tends to create a lot of uncertainty, requiring long project cycle times. Because of the challenging nature of new technology deployment or development, the scope often changes and is typically frozen only in the second or third quarter of project implementation. Like routine projects, technical projects build single-function independent products or subsystems of larger systems. For this reason, they are of a low complexity, calling for a single organization to execute the project.

Like technical projects, *unique* projects, such as developing a new generation of microprocessors or developing a platform product in an internationally dispersed corporation, feature high-technology content. What makes them unique is that they push to the extreme on both system complexity and technological uncertainty. More than 50 percent of their technologies are new or nonexistent at the project start. This level of uncertainty, combined with high system complexity, is destined to prolong project cycle times and cause many scope changes. Adding to this challenge is the need for integration of the multiple uncertain technologies. In such an environment, normally the scope would freeze in the second or third quarter of the project cycle time. Further complications are instigated by the involvement of many performing organizations that also need to be managerially integrated.

How do the two dimensions—technical novelty and project complexity—impact the PM process of these four project types? Overall, the growing technical novelty in projects generates more uncertainty, which consequently requires more flexibility in the PM process. As a result, the PM process, according to Shenhar (1998), does the following:

- Takes more cycles and time to define and freeze the project scope
- Needs to make use of more technical skills
- Intensifies communication
- Requires more tolerance toward change

System scope as a measure of project complexity also has a unique impact on the PM process. In essence, an increasing system scope leads to an increased level of administration, requiring the PM process to feature the following (Shenhar, 1998):

- More planning and tighter control
- More subcontracting
- More bureaucracy
- More documentation

As a result of these influences, the PM process needs to be adapted. Figure 51.9 shows examples of such adaptation for several elements of the PM process.

Customization of the PM process by project type has its advantages and its risks. The advantages are that this customization approach is sufficiently comprehensive because it includes two dimensions that account for the major sources of the contingencies. Its risks include the possibility that the demanding framework with two dimensions may be difficult to apply in certain corporate cultures.

FIGURE 51.9. CUSTOMIZING PROJECT MANAGEMENT PROCESS BY PROJECT TYPE.

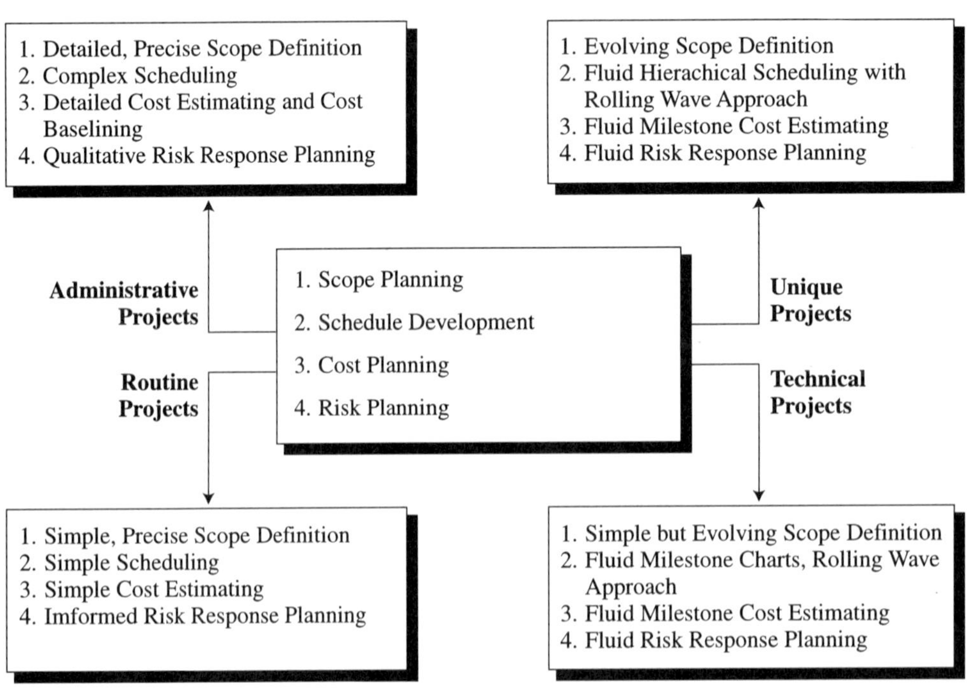

In summary, when customizing the PM process by project type, you may find the following guidelines helpful:

- Use the two dimensions as described or adapt them to your needs.
- Classify your projects and their PM processes into four types.
- Match each PM process with the proper PM tools, supporting each managerial deliverable in the process with specific tools.

Which Customization Option to Choose?

This discussion offers three options for the customization of the PM process. Each has its advantages and risks, and each fits some situations better than others. Table 51.5 offers a way to decide which one may be best. Customization by project size is the right option when an organization has projects of varying size and needs a simple start toward more mature forms of PM process customization. Projects of varying size characterized by mature technologies lend themselves well to this customization.

TABLE 51.5. PROJECT SITUATIONS THAT FAVOR EACH OF THE THREE OPTIONS FOR CUSTOMIZING THE PROJECT MANAGEMENT PROCESS.

Situation	Favoring Customization by Project Size	Favoring Customization by Project Novelty	Favoring Customization by Project Type
Simplest start in customizing the PM process.	√		
Projects of varying size with mature technologies.	√		
Projects with both mature and novel technologies; size is not an issue.		√	
Projects with strong industry or professional culture.		√	
Projects of varying size with both mature and novel technologies.			√
Firm needs a unifying framework for all organizational projects.			√

If an organization has a stream of projects that feature both mature and novel technologies but project size is not an issue, customization by project complexity may be the best option. This is also a good option when projects are dominated by a strong industry or professional culture.

Customization of the PM process by project type may well be best when the organization has lots of projects that vary significantly not only in size but also in technical novelty, spanning from low-tech to high-tech. Organizations searching for a unifying framework that can provide customization for all the types of their projects—from NPD to facilities to manufacturing to marketing to information systems—may find customization by project type the appropriate choice.

Summary

NPD is the engine of corporate success, survival, and renewal. Its characteristics, such as high technological novelty and risks, interplay to create a unique set of critical success factors—the project team and leader, senior management, customers, suppliers, process, and product effectiveness—that impact the process performance (speed and productivity) and financial performance (profits, revenues, and market share) of NPD projects.

Hoping to make full use of their competitive prowess, many companies are working hard to radically redesign the management of NPD projects. Their focus is on remaking the NPD from an unpredictable art form into a predictable, sophisticated, well-structured, and well-managed effort. Crucial roles in this effort belong to two processes. The first, the pipeline management process, is on the enterprise level, striving to integrate and continually optimize new product strategy, project management, and functional management. The second is the process for managing individual NPD projects. Although their implementation is difficult, together these two processes can help companies beat their competition.

References

Adler, P. S., A. Mandelbaum, V. Nguyen, and E. Schwerer. 1996. Getting the most out of your product development process. *Harvard Business Review* 74 (2, March–April): 134–152.

Archer, N. P., and F. Ghasemzadeh. 1999. An integrated framework for project portfolio selection. *International Journal of Project Management* 17(4):207–216.

Baker, N. R. 1974. R&D project selection models: An assessment. *IEEE Transactions on Engineering Management* 21(4):165–170.

Bard, J. F., R. Balachandra, and P. E. Kaufmann. 1988. An interactive approach to R&D project selection and termination. *IEEE Transactions on Engineering Management* 35(3):139–146.

Belliveau, P., A. Griffin, and S. Somermeyer, eds. 2002. *The PDMA toolbook for new product development.* New York: Wiley.

Bowen, H. K., K. B. Clark, C. A. Holloway, and S. C. Wheelwright. 1994. Development projects: The engine of renewal. *Harvard Business Review* 72(3):110–120.

Braun, C. F. V. 1990. The acceleration trap. *Sloan Management Review* 32(1):49–58.

Brooks, F. P. J. 1987. No silver bullet: Essence and accidents of software engineering. *IEEE Computer* 20(4):10–19.

Brown, S. L., and K. M. Eisenhardt. 1995. Product development: Past research, present findings, and future directions. *Academy of Management Review* 20(2):343–378.

———. 1996. The art of continuous change: Linking complexity theory and time-paced evolution in relentlessly shifting organization. *Administrative Science Quarterly* 42:1–34.

Calantone, R. J., and C. A. D. Benedetto. 2000. Performance and time to market: Accelerating cycle time with overlapping stages. *IEEE Transactions on Engineering Management* 47 (2, May): 232–244.

Clark, K. B., and S. C. Wheelwright. 1993. *Managing new product and process development text and cases.* New York: Free Press.

Cooper, R. G. 2001. *Winning at new products.* 3rd ed. Reading, MA: Perseus Books.

Cooper, R. G., and E. J. Kleinschmidt. 1987. Success factors in product innovation. *Industrial Marketing Management* 16(3):215–224.

———. 1994. Determinants of timeliness in product development. *Journal of Product Innovation Management* 11:381–396.

Cooper, R. G., S. J. Edgett, and E. J. Kleinschmidt, E. J. 1998a. Best practices for managing R&D portfolios: Lessons from the leaders—II. *Research Technology Management* 41(4):20–33.

———. 1998b. *Portfolio management for new products.* Reading, MA: Perseus Books.

Crawford, C. M. 1992. The hidden costs of accelerated product development. *Journal of Product Innovation Management* 9:188–199.

Dougherty, D. 1992. Interpretative barriers to successful product innovation in large firms. *Organization Science* 3:179–202.

Eisenhardt, K. M., and S. L. Brown. 1998. Time pacing: Competing in markets that won't stand still. *Harvard Business Review* 76(2):59–69.

Eisenhardt, K. M., and B. N. Tabrizi. 1995. Accelerating adaptive processes: Product innovation in the global computer industry. *Administrative Science Quarterly* 40:84–110.

Findley, D. A. 1998. *Controlling costs for a software development project.* AACE International, Morgantown, WV: 6–9.

Fricke, S. E., and A. J. Shenhar. 2000. Managing multiple engineering projects in a manufacturing support environment. *IEEE Transactions on Engineering Management* 47(2):258–268.

Griffin, A. 1997. The effect of project and process characteristics on product development cycle time. *Journal of Marketing Research* 34 (February): 24–35.

———. 1998. *Drivers of NPD success: The 1997 PDMA report.* Chicago: Product Development and Management Association.

Gupta, A. K., and D. L. Wilemon. 1990. Accelerating the development of technology-based new products. *California Management Review* 32 (2, Winter): 24–44.

Handfield, R. B., G. L. Ragatz., K. J. Petersen, and R. M. Monczka. Fall 1999. Involving suppliers in new product development. *California Management Review* 42(1):59–82.

Harris, J. R., and J. C. McKay. 1996. Optimizing product development through pipeline management. In *The PDMAA handbook of new product development*, ed. D. R. Rosenau, et al. 63–76. New York: Wiley.

Jassawalla, A. R., and H. C. Sashittal. 2000. Strategies of effective new product teams. *California Management Review* 42(2):34–51.

Jiang J. J., G. Klein, and T. L. Means. 2000. Project risk impact on software development team performance. *Project Management Journal* 31 (4, December): 19–26.

Kappel, T. A., and A. H. Rubenstein. 1999. Creativity in design: The contribution of information technology. *IEEE Transactions on Engineering Management* 46 (May): 132–143.

Katz, R., and M. L. Tushman. 1981. An investigation into the managerial roles and career paths of gatekeepers and project supervisors in a major R&D facility. *R&D Management* 11:103–110.

Kerzner, H. 2000. *Applied project management.* New York: Wiley.

Little, D., and F. Leverrick. 1995. Joint ventures for product development: Learning from experience. *Long Range Planning* 28:58–67.

Lynn, G., K. D. Abel, W. S. Valentine, and R. C. Wright. 1999. Key factors in increasing speed to market and improving new product success rate. *Industrial Marketing Management* 28:319–326.

McDonald, J. 2001. Why is software project management difficult and what that implies for teaching software project management. *Computer Science Education* 11 (1, January): 55–71.

Merrils, R. 1989. How Northern Telecom competes on time. *Harvard Business Review* 67 (4, July–August): 108–114.

Meyer, M. H., and J. M. Utterback. 1995. Product development cycle time and commercial success. *IEEE Transactions on Engineering Management* 42(4):297–304.

Miles, R. E., and C. C. Snow. 1978. *Organizational strategy, structure, and process.* New York: McGraw-Hill.

Nevens, T. M., G. L. Summe, and B. Uttal. 1990. Commercializing technology: What the best companies do. *Harvard Business Review* 68(3):154–163.

Nevison, J. M. 2001. Multiple project management: Responding to the challenge. In *Project management for business professionals: A comprehensive guide*, ed. J. Knutson. New York: Wiley.

Pinto, J. K., and J. G. Govin. 1989. Critical factors in project implementation: A comparison of construction and R&D projects. *Technovation* 9:49–62.

Platje, A., H. Seidel, and S. Wadman. 1994. Project and portfolio planning cycle. *International Journal of Project Management* 12(2):100–106.

Raz, T. 1993. Introduction of the project management discipline in a software development organization. *IBM Systems Journal* 32(2):265–277.

Rosenau, M. D., A. Griffin, G. A. Castellion, and N. F. Anschuetz. 1996. *The PDMA handbook of new product development.* p. 528. New York: Wiley.

Shenhar, A. J. 1998. From theory to practice: Toward a typology of project-management styles. *IEEE Transactions on Engineering Management* 45(1):33–48.

———. 2000. One size does not fit all projects: Exploring classical contingency domains. *Management Science* 47(3):394–414.

———. 2001. Contingent management in temporary, dynamic organizations: The comparative analysis of projects. *The Journal of High Technology Management Research* 12:239–271.

Shenhar, A., and H. J. Thamhain. 1994. A new mixture of project management skills. *Human Resource Management Journal* 13(1):27–40.

Smith, P., and D. Reinertsen. 1991. *Developing products in half the time.* New York: Van Nostrand Reinhold.

Tatikonda, M. V., and S. R. Rosenthal. 2000. Technology novelty, project complexity, and product development project execution success: A deeper look at task uncertainty in product innovation. *IEEE Transactions on Engineering Management* 47:74–87.

Wheelwright, S. C., and K. B. Clark. 1992a. Creating project plans to focus product development. *Harvard Business Review* 70(2):70–82.

———. 1992b. *Revolutionizing Product Development.* New York: Free Press.

Zirger, B. J., and J. L. Hartley. 1996. The effect of accelerating techniques on product development time. *IEEE Transactions on Engineering Management* 43: 143–152.

Zirger, B. J., and M. Madique. 1990. A model of new product development: An empirical test. *Management Science* 36: 867–883.

CHAPTER FIFTY-TWO

PHARMACEUTICAL DRUG DEVELOPMENT PROJECT MANAGEMENT

Janet Foulkes, Peter W. G. Morris

Project management in the pharmaceutical industry is a widely variable discipline both in maturity and deployment. Indeed, the pharmaceutical industry itself is composed of a diverse set of businesses. A virtual biotechnology company with a single product to develop, for example, has very different project management needs than a large, multinational pharmaceutical firm. In fact, in some virtual drug companies there may be only a handful of scientists and staff, all of whom are *de facto* project managers overseeing component parts of a single drug project that is fully outsourced for execution. Large pharmaceutical firms by contrast have many thousands of employees (especially with the ongoing consolidation and merger activity in the industry), a vast portfolio of products in various stages of development and commercialization, and manufacturing and business locations around the globe.

Within these different types of pharmaceutical firms there are at least four main approaches and uses of project management: drug development, manufacturing, information technology, and construction and facilities. As the considerations of managing projects in the other areas are covered elsewhere in this book (and as the virtual biotechnology firms only represent a fraction of the industry at this point in time), this chapter addresses the maturity and deployment of project management as it pertains to drug development primarily in the large pharmaceutical (pharma) firm setting.

Drug Development Projects

Large pharma drug development projects are of a duration and scale that probably compare best with major projects in aerospace. A drug typically takes 12 to 15 years from the time the candidate is identified in the lab to getting it fully tested and to market. The current estimate of the full cost of developing a drug is $800 million USD (DiMasi, 2003). The high-risk nature of the discovery and development phases of the projects, and the strict regulatory environment that drugs are managed under, leads to an extremely high rate of project attrition—especially in the early discovery phases. Of every 10,000 to 30,000 compounds entering discovery, only 250 can be expected to make it to preclinical evaluation. Of that number, only 5 to 10 will receive full clinical evaluation, and only one of that original 30,000 will get approved and reach the market to treat patients.

The other important consideration is that, in reality, the drug project is not the creation of a physical thing—a compound—so much as the building up of a vast package of data and knowledge around the chemical entity that has been discovered, and this data assembly happens over the many years of its development, via numerous established scientific processes and tests. That is, the project is not just "how to build" the compound itself, as in nearly all other types of projects (construction, systems, organizational change, etc.). Instead, it involves understanding how the chemical entity will behave in the body therapeutically—indeed, how to position the chemical entity best in its potential market (best both in therapeutic and commercial terms). In effect, it involves how to "build the information."

Building up information on the therapeutic efficacy of the entity is only part of the task. Manufacturing considerations, chemical stability, packaging, and administration (pill, solution, cream, etc.) of the drug compound are important factors in determining whether the project can be scaled up to commercial quantities and meet regulatory standards. Particularly important is the interaction between the marketing-commercialization and the scientific-technical dimensions of the emerging information package: For although the business of pharmaceutical drug development is curing medical conditions, it is in the end a business. The therapeutic has to be combined with the commercial.

Drug development is performed along a highly defined process comprising a set of well-established milestones and subproject deliverables. This process has been assessed and optimized by numerous consultants and every firm in the industry. The process (following a dramatic weeding-out of candidates in the discovery phase) is depicted in Figure 52.1. It moves from preclinical and chemical screening; to tests in healthy volunteers; to evaluation of the potential drugs' safety and efficacy in small numbers of patients with the targeted medical condition; and then to very large, multiyear studies with regular and very robust reports to the various government agencies around the world. Once all of the testing is complete, a final submission of the data package is compiled and filed with these global government agencies.

As the essence of the project is buildup of the scientific data and medical knowledge around the compound, the interpretation, reanalysis, and even additional testing required to reach governmental agreement of the conclusions of that data can be considerable. The negotiations to agree on a final product "labeling" (the summary of the drug data as typically given in the label insert that comes with the medicine) are complex enough that this stage is often viewed as another subproject in its own right. The project lead at this point often turns to a company specialist in the regulatory guidelines.

The competitive pressure to optimize, streamline, and make the process more efficient is paramount. Each day that a blockbuster is delayed from the market can equate to millions

FIGURE 52.1. THE DRUG DEVELOPMENT CYCLE.

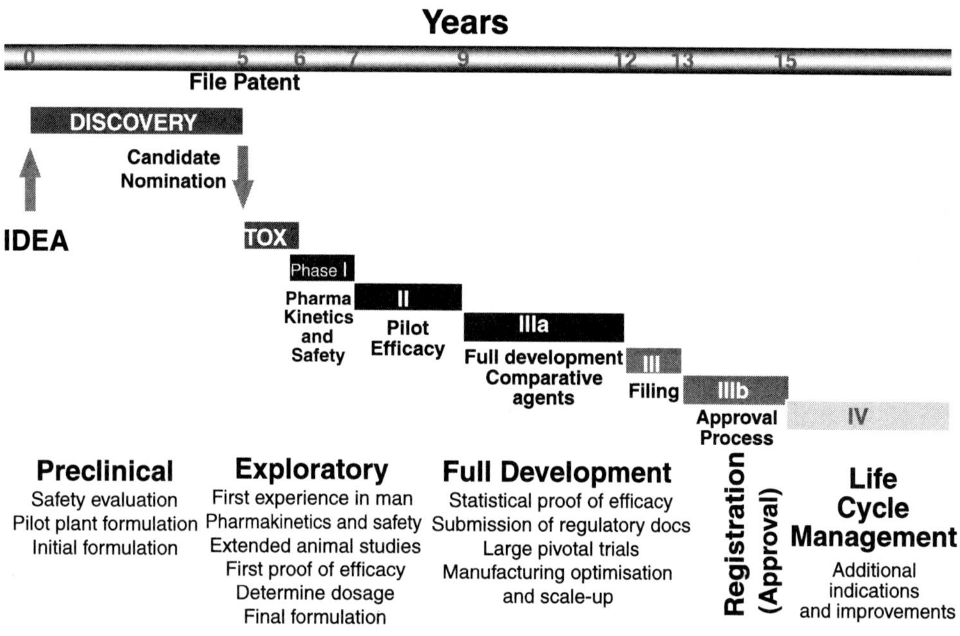

Years

0 5 6 7 9 12 13 15

File Patent

DISCOVERY

IDEA

Candidate Nomination

TOX

Phase I

Pharma Kinetics and Safety

II

Pilot Efficacy

IIIa

Full development Comparative agents

III

Filing

IIIb

Approval Process

IV

Preclinical
Safety evaluation
Pilot plant formulation
Initial formulation

Exploratory
First experience in man
Pharmakinetics and safety
Extended animal studies
First proof of efficacy
Determine dosage
Final formulation

Full Development
Statistical proof of efficacy
Submission of regulatory docs
Large pivotal trials
Manufacturing optimisation
and scale-up

Registration (Approval)

Life Cycle Management
Additional indications
and improvements

of dollars of lost revenue. The team that has to execute this process, and project optimization relies on a multitude of scientists, physicians, and operational staff (in addition to patients and external experts) working in a complex matrix throughout, and even beyond, the organization, with members across the globe—especially in the clinical phase of development. Many members of this project team will have only part-time status on any individual project and therefore have other goals; and in most large pharmaceutical firms, people frequently work on other, competing projects. Needless to say, this makes the organizational challenges much greater, as we shall see later in this chapter.

Companies have spent vast amounts of dollars and manpower trying to streamline and optimize the return on the investment needed to develop and launch a new drug. However, there are also established standards, guidelines, and minima that are set by governmental bodies (e.g., the Food and Drug Administration in the United States and the European Medical Evaluation Agency in Europe) and agreed by, or developed in conjunction with, medical experts and often the companies themselves. As process and technical experts, the companies often play a key role in providing the state-of-the-art data that drives these standards. So, however much the companies may seek to shorten timescales by processing activities in parallel and using other optimization tactics, at every stage and milestone of the drug development process, certain standards define and constrain all drug project timelines and scope. A mandated two-year toxicology study, for example, has obvious time constraints.

Like any long and complex—but routine (in the sense that it is consistently repeated)—process, drug development lends itself to segmentation by stage and functional specialization. Activities, metrics, templates, and resources can then be planned, tracked, reported, and managed by these segmentations. Much of the project management activity is about analyzing and improving plans within this basic development "template." (Project management in drug development is rarely about scheduling the principal development activities *de novo*.)

With this background of the multifaceted process of drug development, let's look at some of the components project management skills and roles in drug development in a little more detail.

The Traditional Project Management Discipline in Drug Development

Traditional project management skills of scope documentation and time, resource, and communication management are utilized broadly both within the specialized functional discipline areas (clinical, pharmaceutical sciences, manufacturing, etc.) as well as across the overall development project. The resulting matrix of project details and multilayered plans, while not unique to pharmaceuticals, is of both such an order of magnitude (in size and number of activities) and length of overall project duration that it is little wonder that project management has had to evolve over the last decade or two from its original tactical, task-oriented focus to a higher level of functioning, as we shall see in this chapter.

Even *within* drug development, the project management function is deployed in a number of distinct ways. Besides the formal project management department that most firms have, the skills, tools, and approaches of project management are typically used in several other functional departments of drug development—for example, in clinical research, pharmaceutical sciences (the physical development of a commercializable product), regulatory, manufacturing, and marketing. These differing areas lead not just to a variety of differing project management plans but also to a diffuse set of sub- or functional project practices—principally time plans but increasingly resource, risk, and other plans—that require considerable overall coordination and integration.

Traditionally, project managers were seen primarily in terms of these plans: as the schedulers, trackers, and minute/meeting managers of the actions associated with the emerging data set. These scheduling and administrative activities are still recognized as a valuable, indeed crucial, project role. However, over the last 20 years or so, the formal discipline has advanced to include the role of program directors and leaders who integrate and link the various departments and subteams across these large projects. Through its portfolio management contributions (discussed later in this chapter), project management is now widely seen as a key strategic influence on pharmaceutical companies' overall business performance.

The nature of project management is thus fairly distinctive in drug development, though it is evolving. Let's look in some more detail at both the impact and level of maturity of a number of traditional project management practices within drug development projects. Looking first at the subject of quality.

Quality

The nature of any highly regulated project is that, to some degree, the definition and robust nature of the regulation itself leads to inherent quality control. This is very true in the pharmaceutical industry. The various regulatory agencies dictate various quality practices, for example:

- Medical standards for the clinical testing (e.g., "good clinical practice," or GCP)
- Manufacturing standards ("good manufacturing practice," or GMP)
- "Good laboratory practice," or GLP
- Acceptable safety standards through federal guidelines
- Ethical approaches and minimums for interacting with patients and handling their personal medical information (e.g., the International Congress of Harmonization—a global agreement about clinical standards/practices in studies, independent review boards, and so on).

Additional quality gates are being layered on to the drug development process all the time. Recently, for example, 21CFR II (security and reliability of "electronic signatures" and patient data privacy) was enacted to ensure that the security and integrity of the clinical data submitted electronically by pharmaceutical companies for regulatory purposes is controlled. This standard is incorporated on top of GCP and ICH. *Center Watch*, an industry analyst journal, recently estimated that it would cost each company $100 million USD to comply with the standard (Center Watch, 2002). While this is predominantly a technical project for the IT function (and their project managers) within the company, it is a quality standard imposed that affects both the timing and potentially the cost (depending on size of the firm) of all ongoing drug projects.

With all of these guidelines, quality is not an area that is usually formally addressed within drug development plan documentation—it is assumed. Project quality management, in the sense of there being a project quality plan describing how it is intended that the project is to be managed, and quality control or quality assurance, as so typically found in systems and construction projects, is not a common practice in most drug development projects—yet. However, several companies are beginning to look at this as a core project management practice, and it would not be a surprise if it were to become more widely accepted soon.

Scope Management (and Value Management)

In traditional project management, a project is viewed as successful if it is delivered as originally specified or scoped. Significantly, this is sometimes not a strong—or even a preferred—characteristic of managing a pharmaceutical project, depending crucially on the nature of the change control and governance practices authorizing the scope change. These governance practices themselves reflect the broader business (or therapeutic) drivers emerging from the trials data—drivers that tend to emerge at certain phases of the project development process.

Many valuable drugs currently in use bear little resemblance to the original scope of their early development project plan. Sometimes this is a result of the changing nature of humankind's scientific knowledge. For example, a compound that could reasonably be expected to have a result on the heart (given structural insights and previous experience) enters into early clinical testing and is then found to cause an effect in a completely different physiological pathway. Rather than abandon the project, if it still meets a medical need, a new set of clinical practices may be born, and, if it makes business sense, it may be desirable to change the scope of the project to meet those practices. Again, it is important to remember that the project is the emerging knowledge and information around a given compound—not the development of the compound itself.

Sometimes an acceptable change to the scope of the project reflects more the nature of the competitive field. A project that looked promising in early testing and had acceptable side effects at the start of the project plan may suddenly become unacceptable as a physician discovers a different treatment regime (e.g., the impact of taking the test drug with another) that minimizes the said side effects. The originally scoped project suddenly has a whole new set of activities and milestones (e.g., potentially a new combination formulation or new clinical programs) to complete in order to be acceptable for approval. In addition to these scope changes, new external information, from a competitor, for example, may make the project no longer as medically or commercially attractive as originally planned (even though it could be delivered within scope). The project may have to be scaled back or dropped. Good scope management is thus an interesting debate for project management practitioners in the pharmaceutical industry.

Equally, if company resources allow and it would seem that performing additional tests would lead to a significantly broader label, and therefore more customers (patients) and financial returns, managing to the original project scope is clearly not a priority. (As already seen, the label is the insert that one finds with the medicine describing the drug, the conditions under which it should be taken, the dosages, etc.; the label is in effect the target product scope.) Value management-type reviews of such scope-enhancing possibilities are periodically carried out during the candidate's development cycle by most pharmaceutical companies, though generally without ever using the terms associated with the formal practice of value management. (Thiry's chapter on value management earlier in this book is quite apposite here.)

Risk Management

Given the delicate and emotive nature of pharma projects—in other words, drugs for human benefit—along with the projects' magnitude (in terms of dollars budgeted and exposed), it is surprising that formal risk management practices have been so slow to develop in this industry. Risk management however, is now a key area of focus for pharma project management.

Identifying, quantifying, and managing risk as early (and effectively) as possible, and thereafter on an ongoing basis, clearly has considerable advantages in improving that 1-in-

30,000 success rate, especially as project costs escalate dramatically in the later stages of development. Obviously, the purpose of the regulatory and scientific hurdles is to eliminate medical risk and ensure only the safest products possible reach patients. Project risk—that is, covering the various issues that can affect both the development of the candidate during its project cycle and the performance of the drug in use—is one of the most enthusiastically pursued practices in the industry today and is handled at a number of levels within both the individual project and across the company. It is a "key responsibility of the project manager leading any (drug) project" (Lewis, 1998). This is discussed later in this chapter.

Planning, Scheduling and Resource Management

Planning is, not surprisingly given the routine nature of the development process, a relatively well advanced practice in drug development, though accountability and the aggressive prosecution of schedules is sometimes wanting. Most projects have strategies formally developed and systematically reviewed by senior management at strategic gates in the process. Schedules generally comprise relatively standard templates. As we've already said, many project managers, in fact, do little original scheduling; critical path, critical chain, and the like may not figure high in the agenda.

However, one of the biggest problems facing planners is the unpredictability of resource availability. In big pharma, certainly, drug development projects generally operate as large and quite complicated matrix organizations competing for resources with other large, complex projects. Allocating resources from the functional lines to these projects is often a major challenge, not least, of course, when candidates attrite (fail) and priorities change. This can cause real difficulties in maintaining the plan. Too often, perhaps, there is insufficient commitment to and ownership of the schedule targets. It is more often the exception for project managers or teams in many big pharma companies to be rewarded for coming in early!

Again, this may be changing, prompted partly by the commercial pressures on companies to get products through the pipeline quickly (and to analysts' expectations) and the resultant pressure on optimizing resource deployment, but partly because of the recent growth in the practice on cross-company benchmarking. Most large pharma companies routinely compare their durations between, and efficiency in reaching, common milestones.

Overall Project Development or Life Cycle Management Plans

Let us now discuss the life cycle of the project and introduce a little more consideration of the management of programs and the portfolio as key drivers for successful drug development. Most large pharmaceutical firms can only succeed by having a diverse portfolio with a steady reliable stream of new—major—products reaching the market (via major projects). Pharma companies with larger market capitalization typically need to *average* three new chemical entities annually to sustain expected growth rates. There are two main internal tensions exerting pressure on this pipeline: high failure and attrition rates, especially in discovery and early development, and, usually, too many opportunities (both novel products and life cycle options of existing products) for available resources. It is clear, therefore, that managing the resources and business opportunities to make the right choices through portfolio decisions and life cycle management alternatives is critical for the firm's business success.

Focusing on the appropriate life cycle plan for a drug is a critical activity in both large pharma companies (where more flexibility of timing and resources may be feasible) and small firms and biotechnology companies, where the decisions may have a more profound impact on the company's overall success, not just the success of the project.

Projects are generally reckoned to have a defined beginning and end, and to be "unique." Though this holds true for drug projects, the concept of start and stop dates requires some definition and flexibility of understanding. The development process is, as we have seen, a well-characterized and staged set of activities that each candidate compound has to progress along. However, one of the big debates for a drug project team, fairly early on in this process, is the issue of defining scope and linking that with the overall business performance. For the original filing, that is, the first deliverable (which in effect defines the first major "project" for each drug), should the team study every possible use or indication? Should they produce many dosage strengths to enhance the drug's tolerability and efficacy? How many formulations should be developed and tested (e.g., injectable form for emergency use or tablets for long-term convenience—or both)? Or should the project team stagger these options over a much lengthier process and possibly submit multiple filings? Should this option then be treated and staffed as many projects or as one large program with a number of interim starts and stops. While this may seem like an academic difference, it is not. Both in terms of traditional project performance, and the success of the project in terms of the portfolio mix and overall business success, it has profound impacts. The interplay between portfolio, program, and project management is increasingly recognized as a key area of pharmaceutical industry performance (see Figure 52.2).

FIGURE 52.2. PROJECT, PROGRAM, AND PORTFOLIO MANAGEMENT IN DRUG DEVELOPMENT.

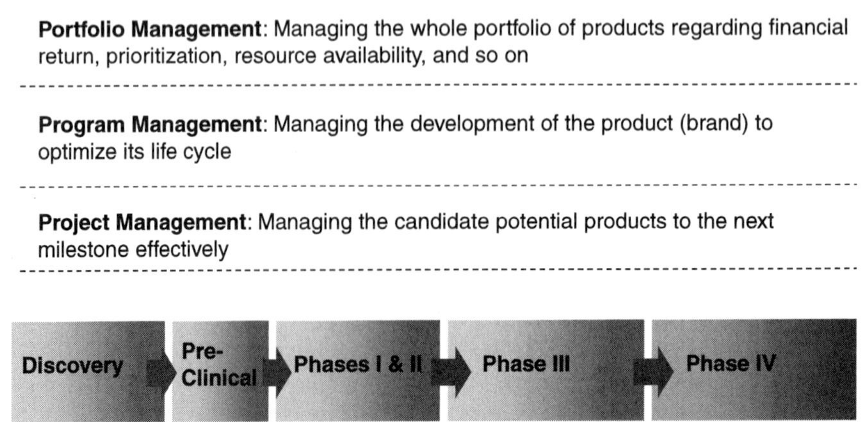

Portfolio Management: Managing the whole portfolio of products regarding financial return, prioritization, resource availability, and so on
Program Management: Managing the development of the product (brand) to optimize its life cycle
Project Management: Managing the candidate potential products to the next milestone effectively
Discovery
Pre-Clinical
Phases I & II
Phase III
Phase IV

There is no single best approach to managing the overall product life cycle for each drug or company. It is a blend of strategic risk management—balancing market and shareholder expectations (and reflecting the company's risk tolerance)—as well as commercial considerations (the portfolio and program management alternatives), along with operational factors (human and financial resources). Getting a drug to fulfil an unmet medical need as quickly as possible has to be weighed against getting that same new chemical entity project/program properly assessed and tested for other indications that, if pursued, could improve many more patients' lives and improve returns on the investment but that may take significantly longer to reach those patients via the market.

If the company is fortunate, and has the resources, a combination of many indications and a quick filing may be possible. However, even if these funds and human talent are available, the organization then has to consider the amount of resource drain and potential impact, or opportunity cost, on other critical drugs and projects in the overall portfolio. Understanding the risk from both a technical and commercial perspective is a critical skill for the successful management of not only drug projects but the overall company. Good, risk-adjusted, project, program, and portfolio management plans play a critical role in this process and thereby the company's overall success.

The effect of managing the intellectual property of the drug via patents is also a highly visible and critical element of managing the life cycle of pharma projects. If a new formulation or indication can extend or provide new patent protection and thereby extend the viable commercial life of the project, the timing of applying for that patent has to be carefully managed. The pressures on market exclusivity of proprietary drugs has been growing for many years. Drug patents that expired in 1989 to 90 lost an average of 47 percent of their market share to generics after 18 months; in the years 1991 to 1992, that figure had increased to 75 percent. Recent experiences and public debate over the access to cheaper versions of the products have only intensified this issue. While the patent per se is not a factor handled by most project managers, it is a critical milestone for a drug project that can be managed through careful life cycle planning. The considerations of factors influencing it (such as the timing of new indications or an additional formulation) are a primary responsibility of the project team. They do this in partnership with legal and patent experts and by assessing the various possibilities of the staggering of the development plan of the drug project and managing its risk.

Life cycle management is not a simple, one-time, "scope definition" exercise. It is repeated as new data is collected and understood on the drug compound and as the competitive and medical environments evolve. It is especially critical in balancing the investment of time and resources in Phase IV studies (post-approval) that can extend the life span of the commercially viable project well past the drug's initial introduction to the market.

Communicating within the complex, matrix world of drug development requires not only strategic management of the portfolio but also a constant flow of data and project analysis at the program and project level of the various options. This somewhat explains the multifaceted nature of project management skills and the divergent deployment of the role of project managers in the industry. It also provides a strong insight into why formal project management is embedded more in the day-to-day business life of the industry and is seen less as a dominating solo discipline. Equally—and potentially more—critical, the successful management of projects in this environment requires exquisite focus, deployment, and integration across all levels of portfolio, program, project, and even task management; a goal the industry continues to address and clearly can improve upon!

The Various Roles of the Project Manager

It comes as no surprise, therefore, that the title of project manager is widely used in the pharmaceutical industry and represents many varied duties, as well as a broad range of responsibilities and capabilities. While this may be obvious given the complex blend of IT, manufacturing, and development skills involved in delivering a drug project, it is true even within each of these areas, and not least in drug development itself. Again, let's delve into the development phase of drug projects and the various project management roles within this component.

It is fairly common to have multiple project managers interacting on one project. There is the project leader, whose main role is to ensure an integration of the ideas, issues, and plans on a particular drug project. Typically he or she will take the lead in formulating the strategy for the development of the project. In doing so, of course, he or she will work with many others—immediate members of their team such as specialists from clinical, pharmacology, manufacturing, regulatory affairs, statisticians, marketing, and so on, as well as other support groups such as portfolio analysis. From this key integrating position, these project leaders represent the integrated project view to senior management (for governance decisions and portfolio input) and to other key stakeholders such as line management (for time, resources, and quality input); external experts (such as medical advisory boards); other companies on joint venture teams; and even the internal, matrixed project team themselves, who may otherwise have a very segmented view of the project. Leadership and all the associated organizational skills are critical competencies for this person, involving communications, team building, negotiating and influencing, and so on. (And often these are the last skills this person was either selected for having or has been trained in.)

In drug development, this project leader is usually a senior scientist/clinician or a commercial leader, depending on the firm and the stage of development of the drug project. This individual may be part of a formal project management group, but frequently is not. This project leader, depending on the culture and practices of the firm, may even resent thinking of him- or herself as "project management," which they view as intellectually and professionally inferior to being a scientist. But, in effect, this is what the person is doing. And in performing this function, the person may, particularly in large pharma, work with a "number 2," often called the project manager, drawn from a formal project management group or function. It is this project manager who normally handles the technical project management aspects of the project integrations, including schedules, resources, risk management, and contingency planning; documentation; and action and issue management. This project manager is still frequently viewed as administrative support, however, and too frequently is not given any credibility as a technical resource. (In pharma, technical skills are usually considered the purely scientific or commercial contributions.)

Thus, one of the challenges several pharma organizations are now facing is how to integrate these two versions of the project's management so that the project leader appreciates the project management functions (for example, on risk and scheduling), and the project manager reaches more into the business, if not to say strategic, dimensions of the project. (The situation is reminiscent of the architect vis-à-vis the emerging construction project manager; or, slightly less accurately, the debate within IT between the program manager or sponsor and the project manager.)

In addition to these core team members, the team will consist of a whole raft of technical specialists, all headed up potentially by project leaders. Figure 52.3 shows many of the differing specialists typically found in a late Phase III project. There are many "leaders" from the various functional disciplines. These are typically drawn both from the R&D

FIGURE 52.3. A TYPICAL LATE PHASE III PROJECT TEAM.

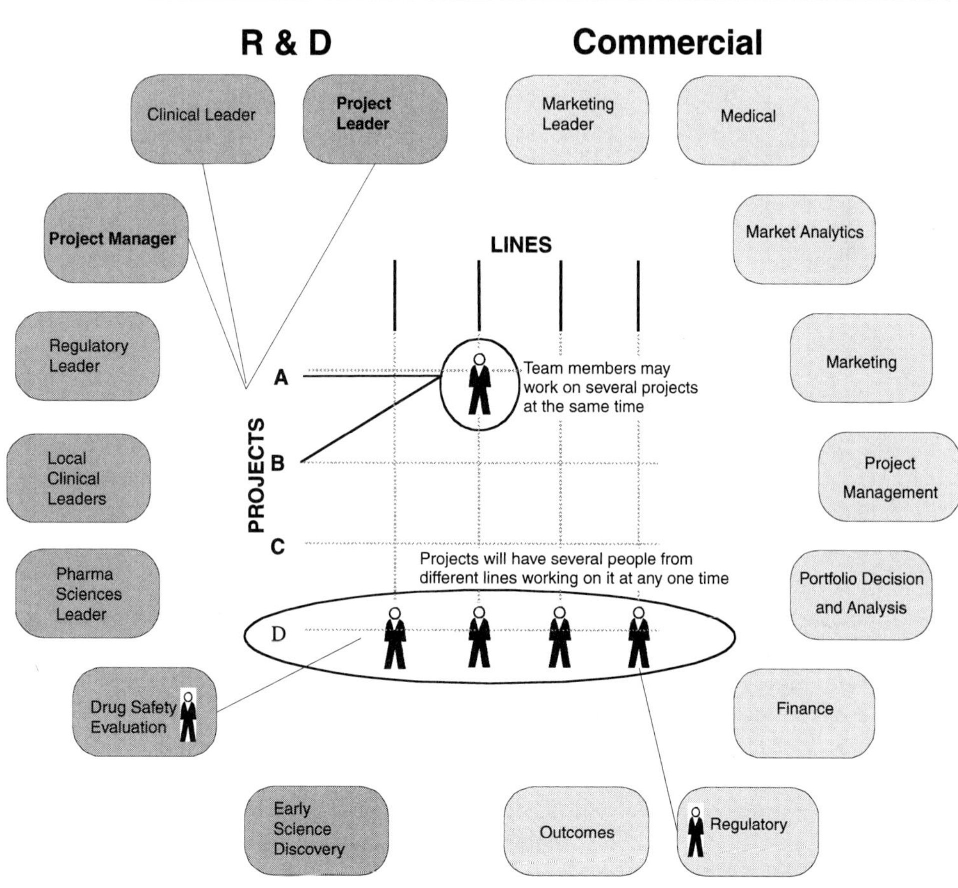

organization and the commercial organization (which will dominate in Phase IV). Both commercial and R&D may have their own project management support function. In Phase III the R&D project management essentially consists of the project leader and a project manager. From a technical viewpoint, the overall lead is often shared with a clinical leader. For the pharma development organization managing in an integrated way, this bigger ("management of projects") picture is central to its business success.

In addition, the clinical project leaders or managers (usually physicians) typically subcontract the management of all or parts of their patient studies to external vendors (contract research organizations, or CROs) to actually run and monitor parts or all of the clinical studies. These vendors in turn will assign a, or even many, project manager(s) to oversee their component of the project.

Many individuals in the pharmaceutical industry, therefore, feel that their role is that of project manager, whether full-time or not and whether part of a functional group called project management or not. This may explain the somewhat historical confusion and skepticism about the technical role of the project manager. At team meetings it is sometimes difficult to find anyone who does *not* consider him- or herself a project manager. (This said, one of the most frequent challenges facing these line representatives on the core project team is to fully and effectively argue for the project once they get back in their functional line.) In fact, as in all organizations, the description of the role (and the competencies required) will vary depending on the particular project management function in question.

In 1999 a survey was conducted of members of the Drug Information Association (DIA) (Foulkes, 2000). At the time of the survey, over 5,000 DIA members were registered as affiliated or actively interested in the project management field. (Indeed, the project management track is one of the largest and fastest-growing areas of the DIA.) The survey was sent to the 3,700 project management DIA members that had e-mail contacts. Only 35 percent of the respondents believed their role was full-time or purely project management, and only 16 percent had any formal project management qualifications!

Some interesting insights were gained when the areas of responsibility of these project managers were tested against traditional competencies of formal project managers from other industries (using the PMBOK principles as a guide) (Foulkes, 2000). Communication and time management were the most consistently aligned capabilities of the respondents in terms of their responsibilities. Most felt that quality, risk, and cost management were shared responsibilities across the project, or even the firm. Those who claimed to be full-time project managers reported that the management of risk was significantly more important to their roles than others who only had part-time project management responsibilities. As discussed, management of project risk is now a more recognized, and important, contribution of the project manager.

So, is there a subgroup of project managers in the pharmaceutical industry whose skills can be clearly identified and appropriately compared to those in other industries? In the subgroup in the survey that believed project management was their full-time role and responsibility, only 13 percent reported any procurement activities as part of their responsibility, and only 32 percent reported the management of project resources. Procurement and resource management are recognized across a broad range of other industries as key responsibilities of a project manager (100 percent alignment) (Morris, 2000, 2001).

It is clear that the discipline is either not yet very mature in the pharmaceutical industry or, just as plausibly, and probably more accurately, is developing to fit the nature of the organizations' operating models for dealing with the complex environments, long durations of drug projects, and heavy emphasis on the scientific—not project—effectiveness.

There is some indication that this evolution may be becoming more pronounced, however: The market consolidations and other external pressures on the industry are, for example, making resource management (both finance and human), and even contracting, more important success factors in managing projects within the pharmaceutical field.

Summary: The Evolving Role of Project Management in Drug Development Projects

During the 20 years or so that project management has been around as a recognized discipline in the pharmaceutical industry, there has been significant change in the nature of the industry's environment; this, together with the changes in the discipline itself, has led to a flexible and quite distinctive character to the discipline.

The most obvious indication of this change is the evolving role of the project manager. While originally deployed to handle project and team scheduling, arranging meetings, and maintaining progress project metrics, project management is now clearly recognized as a more strategic function, intimately involved with the portfolio and life cycle decisions of the various projects. As discussed, the two main roles of full-time or "pure" project management within drug development are the strategic team leader and the project manager. Both roles have gradually gained greater credibility and increased stature, providing as they do so greater consistency of deployment and increased measurable business value. Specifically:

- The "project leader" has evolved from being initially a purely scientific role (especially in early development) to one where it is now seen as managerially critical, with particular responsibilities for setting project strategy, team leadership, and portfolio management. The project leader must negotiate and present the optimal distribution of regulatory filings and options for the drug project as it evolves through its life cycle.
- The increased importance of the "project manager"—the project leader's project management number 2—is reflected in a growing responsibility for project cost and resource performance in addition to the traditional responsibilities for scheduling and administration. Project managers must capture good data on project current and forecast status and provide assessments of the risk of the various life cycle management alternatives of the drug. The project management function is now even providing impact analysis to improve the overall drug development productivity of the firm.

Both individuals play critical roles in the life cycle and success of the drug project. Indeed, in many ways it is the management of the life cycle plan that provides the perfect illustration of the interface between project leader and project manager in drug development. The optimization of this plan (i.e., a drug's possibilities across its life cycle) is still a more valued ability than simply ensuring accurate maintenance of the project's scope via start and stop dates, or ensuring that procurement, resources, or even time activity is as "originally" documented. In the pharmaceutical industry, it is still better to produce a blockbuster project than ensure that the project is delivered on budget and within scope.

As these roles mature, interestingly, from the discipline's viewpoint, certain of the more strategic dimensions of the management of projects are coming to be more formally articulated. Value management, for example, long practiced in pharma projects with varying degrees of (in)formality, is being recognized as critical in formally and systematically reviewing and enhancing the project development strategy (or the candidate's "configuration"—dosage, method of delivery, etc.). Risk management is being elevated into being looked at more formally as risk and opportunity management, where value-building possibilities are traded against risks. Quality management (plans and reviews) is beginning to be applied to project processes, particularly in crossing gate review points, to improve the rigor and quality of management. Program management (life cycle management)—as (multiproject) brand or platform (chemical entity) management—is coming more into prominence, particularly as candidates cross from discovery into exploratory development, and in Phase IV.

The pharmaceutical industry is beginning to emerge as one of the best exemplars of the benefits that accrue from managing portfolio, program, and project management in an integrated manner, where the linkage between strategy, technical/scientific, commercial, organization, and control issues is truly continuous and interactive. In light of this, project management—the management of projects—is beginning to be seen as a broader, more centrally important discipline to the business success of the industry than at any time in its history.

References

DiMasi, J. A., R. W. Hansen, and H. G. Grabowski. 2003. The price of innovation: New estimates of drug development costs. *Journal of Health Economics* 22:151–185.

Foulkes, J. 2000. Project management certification—Various views, various models: Survey results. *36th Annual Drug Information Association Meeting.* San Diego, June 11–15.

Foulkes, J. 2000. Art or science? How—and even does—the pharmaceutical industry apply the discipline of project management? *PMI Research Conference 2000.* Paris.

Gambrill, S. 2002. The unexpected outcomes of 21 CFR 11 compliance. *Center Watch* 9(6).

Lewis, N. 1999. Implementing risk management on pharmaceutical product development projects. *Project Management Annual Symposium Proceedings.*

Morris, P. W. G, M. B. Patel, and S. H. Wearne. 2000. Research into revising the APM Project Management Body of Knowledge. *International Journal of Project Management* 18(3):155–164.

Morris P. W. G. 2001. Updating the project management bodies of knowledge. *Project Management Journal* 32(3):21–30.

CHAPTER FIFTY-THREE

PROJECT MANAGEMENT IN THE DEFENSE INDUSTRY

John F. Roulston

"When all else fails, apply to the Department of Common Sense"

STEVE MARSH

There are factors encountered in defense contracting that may have a profound effect on project management yet are unlikely to be encountered elsewhere. When planning and executing projects, the project manager must understand the singular characteristics displayed in defense. In addition to possessing an awareness of the special defense environment, the project manager must be competent in a wide range of modern management techniques.

First, many contracts are obtained in competitive engagement, a concept that stems from free-market dynamics. The defense market does not behave as a free market, however. In a free market interaction, there should be plurality of customers and vendors. In defense, both customers and vendors are in short supply. This implies that the customer has to behave so as to induce competitive pressure, or to suppress it, if he or she judges that its result is undesirable for a compelling reason. It also implies that the follow-on consequences of success or failure for vendors are more severe than a plurality market would dictate. These factors may intrude into the management of contracts in regard to pricing, risk analysis, and schedule planning. There may also be a dangerous inducement for coupling similar but distinct undertakings to the same or different, though related, customers. Finally, many defense contracts are of long duration, exceeding the tenure of most of the population that contribute to them so that continuity of policy, behavior, and even strategy becomes an issue for management.

While the defense sector has its idiosyncrasies, which should be respected, these do not negate the principles of general contract management. Successful projects result from the application of sound logic in planning and measurement and control in execution, provided, of course, that the contract is practically founded in the first instance. It is unfortunately the case that a proportion of defense contracts will not be deliverable within the norms of their bidding, simply because of the pressures attendant at their acquisition. In this regard, conditions vary considerably, dependent on the national disposition of the procurement system involved.

It is necessary to distinguish variations in practice—that is, the broad classes of experience that may be encountered—in the United Kingdom, the United States, France, and the rest of Europe. It is also necessary to distinguish the special implication of contracting at various levels within the procurement chain, from prime contracts that deliver a weapon system to a cardinal point or capability specification, through subsystems governed by a specification that resides as a component of an integrated whole, to commodity parts that may or may not involve considerable technology ingredient.

It is not feasible within the constraints of a chapter to attempt to provide a comprehensive handbook of practice for defense contract management. Moreover, the procurement backdrop to defense contracting is in rapid change in the United Kingdom and Europe generally, so that, in terms of value to the reader, principles and understanding are transcendent over rules and regulations. Hence, the objective here is to alert the reader to the specific ambience of defense and to contribute to the all-important project-planning phase, wherein the logic of the project success must be found. If the logic of the project plan is robust, and the manpower and technological resources are affordable and available to match to the scope of the task, then the project will yield to conventional good management practice.

Contracts for system development tend to be among the most risky undertakings a defense company can face, so that any valuable contribution to defense contract management must address this matter. The national variations seen in procurement practice are, at least in part, an acknowledgment of this fact and its balance against the military risk inherent in a project outturn. Military risk can occur when industry satisfies a system specification but the resultant product fails to deliver the requisite military value. The military has a strong appetite for technology, and most advances are quickly applied to gaining improved performance or increasing functionality and complexity. Despite attempts to ameliorate this, there are few examples of conscious trade-off between technology and performance, and in fact, the processes for doing so are scarcely developed. The popular fiction of a competitive defense market dynamic tends to propagate this error, as contractors in the competitive phase are motivated to emphasize the technological advantages of their bid solutions, often in the process ignoring less obvious consequences that may be germane to ownership or indeed underplaying development risk.

The Contracting Environment

The United States houses the largest defense market in the world, larger than the aggregate of all Europe and typically 20 times the scale of individual European nations. Not surprisingly, the most sophisticated procurement processes apply there, and in general, new technology systems are fielded first in the United States. That is not to say, that U.S. procurement would be judged highly efficient on other nations' norms; it just means that the output of the United States cannot be sensibly judged that way. The United States carries the burden of being the first to field most new technology, and effectively demonstrates feasibility of new concepts for the rest of the world to follow. It can do so because of its scale, because of the importance it attaches nationally to defense technology, and because of the sophistication of the relationship between defense industry and government procurement agencies. Nations and industries that are prepared to follow closely on the heels of the United States benefit from its effort to establish feasibility. It is always safer and more efficient to embark on a technology-based development in the knowledge that it has been accomplished before. In general, the systems that come second are more refined, more optimal, and significantly better value for the money expended. This trait is evident in many European developments that follow in the wake of U.S. predecessors.

The United States

A designated defense contractor in the United States, whether at system or equipment level, is understood, in government and sociopolitical contexts, to contribute an important edge to the military might of the nation. Similarly, defense technology is legally controlled as a weapon in the national arsenal and treated as part of the international power projection of the United States. The industry is highly constrained in its freedom to supply raw technology outside of the U.S. home market, and developed systems are exported under tight government controls. In return, the terms of contracting that apply to national programs are designed to protect the contractor from unforeseen risk. This is possible because of the intrinsic technical sophistication available to the procurement agencies, which ensures that the technical risks underlying procurement of complex systems are understood and are allocated sensible costs within the procurement package.

The U.S. defense industry has a reputation for successfully completing complex development projects to the satisfaction of all stakeholders. This is certainly true within its national context, but less true when U.S. companies venture into contracts outside the protective shield of U.S. procurement. There are notable examples of export contracts getting into financial and performance difficulties at all levels, from prime contracting to equipment and component supply.

The large scale of U.S. defense contracting and the immense effort that has been applied there to develop standards for all aspects of the process has led many nations to model their procurement on tailored versions of U.S. practice. This has been less true in Britain and France than elsewhere.

On the engineering front, U.S. defense standards are rigorous and conservative and, when followed, lead to systems that can be relied upon to perform under the most arduous military conditions. Many U.S. standards have been adopted by the NATO alliance as common standards in the interests of promoting interoperability of forces. Since the 1950s, U.S. standards have been a stabilizing force in the international defense market and, unlike U.S. technology, are openly available to all countries inclined to exploit them. It is notable that major European programs, such as the Eurofighter project, have adopted them as a unifying convenience, while their product is in direct competition with U.S. counterparts.

Digital electronics is one important area where the stability of component standards is now highly suspect. In digital systems, where advances in the commercial market for electronic components are so rapid that component lifetimes are much shorter than system life cycles, it is already difficult to complete development and qualification of a substantial system without encountering component obsolescence and necessity for lifetime purchase of some components. Recognizing this, the United States has ceased to maintain its military component standards. As a consequence, many components cannot be purchased to regulated standard, with serious implications for the processes needed to apply reliability engineering in design and probably for equipment reliability itself. Moreover, the military market for electronic components has shrunk in relative terms from around 18 percent in 1980 to approximately 0.2 percent today. Within the United States, the Defense Logistics Command seeks to compensate by support of organizations such as the Defense MicroElectronics Activity (DMEA), which maintains the technical capability to reverse-engineer silicon microcircuits and to assay the quality of commercial parts. In extremis, DMEA could, and does, fabricate replacement parts to military quality, both by replication of original design and by substitution of design. There is no corresponding capability in Europe, a fact that should hold profound implication for future life cycle costs of European sourced equipment.

In recent times, following the procurement policy introduced in the United States in 1994, there has been a growing enthusiasm for customized off-the-shelf (COTS) solutions in parts of the electronics domain. Associated with this is a preference for commercial "open standards" as an alternative to bespoke or military standards. Undoubtedly COTS, meaning adaptation of commercial parts, has a role to play in economic satisfaction of military system requirements, and open standards will assist the obsolescence issue. However, the life cycle cost implication of procurements based on these principles is hard to predict. It seems necessary to accept that equipment replenishment will be necessary on a short cycle time relative to platform life cycle. Moreover, the system architecture needs to be conceived from the outset to permit local technology insertion without encountering the need for global requalification or recertification of performance; otherwise, the approach becomes unworkable. In fact, it is vital that military procurers accept that COTS and open standards are short-term expedients that may reduce acquisition cost against a higher uncertainty in life cycle cost overall. There are signs that in the United States this trade-off is becoming properly appreciated, and appropriate constraints have been applied to, for example, the F-35 fighter development to ensure that a manageable life cycle is obtained. Elsewhere in the world, COTS is invading platforms as an expedient basis for update programs with little forward analysis of consequence. It seems highly imprudent at this stage of its evolution for an industrial contractor to take life cycle responsibility for logistic support of a COTS-based system without very careful analysis of the risks involved.

There are fundamental threats to future application of COTS hardware in the military sector. The consumer electronics market drives inexorably toward finer transistor and interconnect geometries within silicon chips. This allows higher complexity per chip and has a positive effect on cost per function. Unfortunately, there is a physical limit to geometrical shrinkage for chips that must survive in a military environment, and this is currently within sight. The implication for program planning will be profound, since, in effect, military contractors will be denied the spin-off of the consumer market and military demand will not, of itself, be sufficient to sustain foundries to manufacture custom military chips at anything close to the prices experienced with COTS. The industry and the customer base has yet to face up to the consequence of this.

France and the United Kingdom

Next to the United States, France and the United Kingdom rank approximately equal in importance in defense market size and scope. France has retained a more nationalistic outlook, whereas the United Kingdom has placed more emphasis on partnering with European neighbors and indeed the U.S. in major procurements. The enthusiasm within the United Kingdom for partnering is driven by the overt benefit of reduction in acquisition costs and the covert benefit of stability following the unfortunate experience of a number of cancellations of national programs in the 1960s. International programs generally cost more in total, if less per nation, are longer in execution, probably hold compromises in specification, but are difficult, if not impossible, to cancel. Generally speaking, cooperating nations demand from each other strong commitments to continuity, backed by harsh financial penalties against dissenting parties.

France. The French contracting system is well developed and very prescriptive, requiring an in-depth technical analysis of project deliverables at the outset and demanding step-by-step expression of this through the "clauses techniques" documentation that is built into the contract. The accountability of industry and government officials to the *clauses techniques* is high, and variations, if required, involve at least as much rigorous analysis as supported by the original contract. While the technical sophistication available within government procurement circles in France is undoubtedly diminishing, it still remains at a relatively high level and contributes to close cooperation between the industry and the state. This may also be seen in the ability of France to conceive of platforms and systems that recover their development costs in export success. The Mirage family of aircraft are a good example, representing the only real competition to U.S. dominance in fighter exports of the last generation. Clearly the Rafale project is designed to follow in the same vein, but it remains to be seen if the historic pattern of French export successes will continue, Rafale so far having failed to attract export customers.

France engages in collaborative projects to defray costs but is generally unwilling to bend national imperatives to do so. Thus, we have witnessed the failure of France and the United Kingdom to find a formula to go forward on the Horizon frigate program despite the creation of the managing project office in London. France has reconstituted the project with Italy, presumably finding that partnership more accommodating. Similarly, some years earlier, France withdrew from the project that then became the four-nation Eurofighter (Germany, United Kingdom, Spain, and Italy), launching instead the national Rafale fighter program. In the technology sector, the United Kingdom and France have a number of long-running ventures that have yielded significant gains; prominent among these is GTDAR (GEC Thomson DASA Airborne Radar). This is an EEIG (European Economic Interest Grouping) that started bilaterally between GEC-Ferranti in the United Kingdom and Thomson of France. It later expanded to include Telefunken System Technik, then the electronics part of Daimler-Benz Aerospace of Germany, in a symmetrical arrangement. While all of the collaborating company names have changed, the GTDAR label remains as a reminder of the origins of a grouping that has created a European technology for active electronically scanned array radar for fighter applications. Without this cooperation, Europe would probably not have an indigenous supply option in this difficult and costly technology area. The original concept of the EEIG was to form the basis of a European radar business in supply of active arrays to fighters. While the grouping has been a technological success, its growth as an international business has been curtailed through evolutions of the competing Eurofighter and Rafale aircraft, both of which will need distinctive active array radars to attract export customers and will need to rely on the radar industries contributing to the respective aircraft programs. This evolved situation is just one example of how the equipment sector in Europe, by necessity, services the imperatives of the platform sector and can lose structurally in the process.

The United Kingdom. Defense science in the United Kingdom began out of almost nothing in 1935, in panicked reaction to the realization that Germany was in the process of rearming and that science and technology would have an impact in conflict. It was apparent to some that technological counters were needed to a growing bomber threat that had been largely ineffective in the previous war. Despite early antagonism between defense scientists and the military, wartime efforts were prodigious, giving birth to the aviation and electronics industries as we know them, and a remarkable capability existed by the end of the war. The years that followed saw the progressive transfer of technology and design competence from the wartime Defence Science Establishments into the hands of U.K. industry that previously had been concerned only with manufacturing.

By the 1960s, the British procurement practice had evolved toward fixed-price contracts with design authority vested in industry. In the mid-1960s a political row broke out over allegedly excessive profits earned by the Ferranti Company in development and supply of the Bloodhound surface-to-air missile system. Whether the claims by politicians were justified or not, the incident caused a change in procurement practice toward cost-reimbursement contracts with low industry risk but controlled profits. This situation maintained until a procurement overhaul under Margaret Thatcher's government in the early 1980s created a powerful civil service post with wide-ranging powers under the title of Chief of Defence Procurement (CDP). Thatcher's government had been embarrassed by the technical failure of the Nimrod Early Warning Radar development for the RAF. This prompted the procurement overhaul, though it is likely that Thatcher would have taken this path in any event. CDP returned to fixed-price contracting against competitive open tendering and moreover set out to avoid carrying project risk at government level. This involved a strategy of building prime contracting competence within a small number of major companies who would absorb project risk on behalf of the Ministry of Defence (MoD).

The processes introduced under the first CDP emphasized competition as the mechanism for determining value for money, and these evolved little until the major overhaul undertaken in the mid-1990s. This initiative, assisted by McKinsey consultants, operates under the somewhat obsequious title of "smart procurement." Its introduction was accompanied by relocation of the MoD Procurement Executive from premises scattered across central London to a custom-built complex at Abbeywood in Bristol. In the process of relocation and reorganization, the staff numbers involved in procurement were significantly reduced.

Smart procurement seeks to exercise better control of procurement decisions and a more innovative range of solutions through a staged process that starts by soliciting precompetitive contributions from suppliers who will later battle each other for the privilege of being the selected contractor. The MoD runs this process through an integrated project team (IPT) structure, combining the many stakeholder interests that a procurement process has to satisfy. McKinsey, in designing this arrangement, emphasized the necessity for high-quality IPT leadership, and indeed this has been shown to be vital to project success. In the early, precompetitive phases of a project, the IPT works with industry to define the defense requirements in a capability-focused document termed a user requirements document, or URD. Continuing this work, the URD is developed into a systems requirement document, or SRD. This has to take account of feasible implementation and is the basis on which competing bids are solicited. Generally, when a contract is placed, the defining document is the SRD and the URD loses authority. The project pivots on the quality of the work done at the URD/SRD stage, since after this, changes will be costly and may be constrained by strict budget control.

Almost at the same time as procurement practice headed toward smart procurement, the decision was taken to privatize the Defence Evaluation and Research Agency, which at this point contained the evolved totality of defense science establishments in the United Kingdom. The strength of MoD science reduced from some 9,000 staff to 3,000 retained within a newly formed Defence Science and Technology Laboratory (DSTL). The remainder formed a privatized entity, "QinetiQ," and passed into U.S. part-ownership with a mission to spread its interests into the civil sector. In consequence of this, technical support to procurement management and technological backup to the smart procurement process was reduced to levels well below that maintained by United States and France. There is bound to be an effect on the conduct of defense contracts, which industrial managers must take into account in planning and execution. In fact, it is necessary for industry to substitute for the diminished knowledge base in defense science and to emphasize partnership behaviors with MoD's IPTs. This requires MoD and industry to hold commonly to a clear vision of the dominating importance of product value in defense contracts and for both parties to respect the financial constraints and corresponding performance levels agreed at the placement of contracts.

Levels of Contracting

Generally speaking, a prime contract is fulfilled through a hierarchy of contracts, distinguished by the responsibilities and risks resident in the obligations undertaken at successive levels. Any complex undertaking will require success at all levels to register success overall and will require that the hierarchy of contracting does not mask the unity of objective.

The Prime Contract Position

The accountability underlying expenditure of public money drives toward devolution of risk from treasury to industry. A prime contractor accepts to manage a project to outturn, generally under fixed-price terms. Though there may be situations in which this is acknowledged to be unrealistic, it will always remain an objective of a government procurement agency to bound the risk to public funds.

It is very unlikely that a successful project will result if the complexity and resources needed are seriously underestimated at the outset. It serves no stakeholder's interests to allow this to happen, and in fact, a positive element of smart procurement is the willingness to spend up to 15 percent of the procurement budget in ensuring that risks are understood and conquered before a contract obligation is required.

Despite best efforts to analyze and define risks and master the costs of technological ingredients for inclusion in the bid price at the outset, all parties must acknowledge that after project launch, new risks will be encountered. In a good situation, the logic on which the project is launched will be well founded and survive, but plans may need revision to protect project goals with concomitant increments in forecast cost. There is also a possibility that a modification of the planning logic can become necessary as a result of matters learned in the course of the project. The prime contractor may incur additional costs within his or her own domain of action, or may have to acknowledge increases suffered further down the contracting chain. In any event, the reality of error in specification or project logic and misjudgment of technological difficulty requiring rework or extra work requires that the prime contractor and the contracting agency agree on some sensible level of contingency funding and mechanisms for its release. In some cases, unused contingency may be returned or shared as part of an incentive arrangement.

Whatever the rules for managing contingency, it is very important that all parties recognize that the needs of the specified product should dominate and there should be no constructs that create motivations in conflict with this. At each interface in the contracting hierarchy, there is potential for distortion to occur. For example, at the point at which a contractor applies for contingency, the procuring authority may have identified omissions in the specified product and wish to use contingency allocated against subcontract risk to mitigate what is essentially procurer's risk. In the interaction between a prime contractor and government procurement, this highlights the important distinction between industrial risk and military risk. Similarly, a prime contractor may be tempted to blur the boundaries between specification error on his or her part and technological difficulties in the subcontract domain. This kind of contention exists throughout the contracting chain from top to bottom, and if unmanaged, its impact on contracts can be catastrophic. The dictum of adherence to product value and the practical artifice of cultivating partnering behaviors across contracting boundaries are the useful management responses to stakeholder conflict.

In an ideal environment, each party holds true to his or her specified responsibility and absorbs the consequences of his or her own errors, whether regarding estimation of manpower or technology. Contingency flows across boundaries to finance the project consequence caused by effects external to the boundary. The responsibility of higher levels in the chain to promote success of lower levels is paramount. These simple principles need to be tempered by an overriding need to intervene in any circumstance that might cause project failure, even if the matter falls inside subcontracted responsibility. Clearly, in a practical situation, judgment will be needed, and integrity of the parties engaged is essential for efficient and lasting resolution of issues. The need for objective technical interpretation cannot be avoided. The technical skill and domain knowledge base underpinning the procurement process is fundamental to overcoming the practical difficulties that beset real projects.

Contracts Below the Prime Level

It frequently happens in fixed-price contracts that a prime contractor will attempt to pass down responsibilities that rightly belong at the prime level. This is, of course, a symptom of erroneous logic in the planning of the project and is motivated by a misunderstanding of risk. At each level in a contracting chain, a party can only be held responsible for matters that naturally fall inside his or her control. Contracts should be very clear in this respect. When a contracted objective needs accommodation outside the control of the contractor, the concept of "dependency" is engaged. The contract then contains a definition of the dependencies that must be rendered as deliverables into the contract for it to be valid. That is, the obligation on the contractor will be invalidated if dependencies are not honored as defined in the contract. This can turn a fixed-price contract obligation into a negotiation.

The use of dependencies in the contracting chain cannot be avoided, but they should be religiously honored by prime contractors and used by subcontractors, not for defense, but as planned constructs necessary to the composite logic of the project. If a prime contractor attempts to pass down illegitimate risk, it always results in a defensive dependency that is sure to be unfulfilled. Contract friction is inevitable in this circumstance, and the subcontractor may enjoy a strong position. This sort of friction endangers the fundamental project objective. It is a serious hazard of fixed-price contracts. In fact, the tensions of fixed prices demand extremely cooperative working at all levels from procurement downwards. This is the province of project partnering.

The Partnering Charter

Project partnering has its origins in the auto industry and then moved to the construction industry, where fixed-price practice and variation of costs are commonplace. (See the chapters by Langdon and Murray, and by Morris.) The construction industry has the advantage of a proper market dynamic but still has found that project goals have been lost in commercial friction, to the detriment of the industry overall. In consequence, the idea of a "partnering charter" has sprung up to advantage. It is particularly well developed in Australia, the United States, and the United Kingdom.

The essential principle is acceptance by all stakeholders that a successful project serves the interests of all and that, correspondingly, an unsuccessful project damages all. That is, the focus is the overall project goal, not the parts of the project that are separately contracted in a hierarchical manner. With this focus, all stakeholders combine efforts to achieve a successful project and boundaries are secondary to success. Of course, such sentiment cannot mask a flawed logic, erroneous plan, or misjudged price, but it can do a lot to inject efficiency that aids achievement and reduces cost. Moreover, the sentiment of partnering influences behaviors across the internal boundaries of the project, and this helps greatly to regulate the dependency issue. An enforced partnering charter is much more than sentiment; it becomes an embedded culture, assisting the people contributing to it to stretch their narrow company perspective to encompass the longer-term goal of project success, product reputation, and opportunity to repeat the business of the project.

A partnering charter is separate from a legal contract. It is not a legally enforceable undertaking and should never attempt to be such. It is a formalization of a commonsense undertaking. It should be signed by senior representatives of the engaged stakeholders and should reflect the dignity of the stakeholders. Some tailoring of text is to be expected in different applications, but the following representative extract illustrates the cardinal points:

- Project vision statement—The mission of the project.
- The affirmation of partnering behavior—"We will work together in the interest of achieving a successful project."
- Listing of values held in common:
  - Customer focus
  - Priority of product and program
  - Mutual trust
  - Candor
  - Openness
  - Integrity
  - Solution orientation and "can do" attitudes
  - Avoidance of surprises for all
- Implementation. Right to expect our stated partnering behaviors apply in all interactions.

Arguably, a fixed-price contracting chain supplemented by partnering principles will be more efficient than any form of cost reimbursement arrangement, provided that the founding logic of the enterprise is robust and that adequate attention has been paid to concept engineering to underpin the project cost estimates. That is, knowledge to support a robust logic is needed to make an efficient plan. This is fundamental; no project can succeed without it. After that, efficiency stems from behaviors that are regulated by partnering principles.

The Creation and Management of Consortia

In the previous section we discussed vertical relationships within a contracting hierarchy such as is commonly found in the execution of a large defense project. It is commonly and increasingly the case that at various and perhaps multiple levels within such a hierarchy we find not individual companies, but consortia of companies attempting to fulfill a contracted objective. These peer companies have to self-organize, splitting the work among them and bearing composite responsibility for their joint deliverable. It may be that the project has a group of international sponsors and that there are rules governing work-share, payment, intellectual property, security, and other matters that are predetermined by the international politics in the foundation of the project. In most internationally sponsored projects in Europe, work-share is allocated at each contracting level in proportion to product offtake by the participating nations. Moreover, money will not cross national boundaries, and generally speaking, companies will communicate only prices, not costs. This is true within the horizontal strata that compose consortia or in the vertical dimension through project hierarchy.

Consortium Principles

The horizontal dimension adds complexity. It must reduce efficiency, though, in mitigation, the combination of multiple forces and cultures can accomplish more. It brings a wider skill base to bear on problem solving, and a healthy competitive axis within a consortium which, with company and national pride, can add to quality of action. Collaborative projects must be organized to emphasize the advantages of collaboration and suppress the disadvantages. For the most part, collaborative projects are undertaken by consortia that form specifically for the purpose. Such consortia are, in effect, teaming arrangements that do impose legal responsibilities in respect of conduct of the project but stop short of constituting a legal identity. Normally one of the collaborating companies will be nominated the *primus inter pares* leader and will accept the contract, in turn placing subcontracts that reflect exactly the terms of the original. In this arrangement, subcontracting is a convenient way of distributing the legal obligation of the contract. The management is intended to be consensus-driven, and the normal hierarchy of prime-/subcontracting does not apply.

The guiding principle is that of the "single equivalent company." That is, a consortium should recognize that its deliverable product competes with corresponding deliverables emanating from single companies. Hence, its unity must emulate the unity of a single company. The major disparity is the distinct cost bases maintained by the collaborating companies. Outside of this, which should be accommodated in the initial bidding, there must be unity of purpose. In fact, there are situations where the disparity in cost bases cannot be tolerated and companies form joint venture entities that do have a legal personality. This is particularly important if the nature of the business requires product sales to multiple customers over a long time span, or if evolutions in the product require substantial work that falls outside of the initial planning.

It can be desirable to avoid individual liability assignments within a consortium, particularly if the product life cycle will be long and the joint business opportunity long-lived. Much inefficiency stems from effort to apportion blame when things go wrong. Frequently, the best intellects of contributing companies are engaged in canceling each other out, rather than adding value to the jointly owned product. A more robust and enlightened practice is to form a consortium around the principle of shared liability, where the sharing is in simple proportion to the work-share. With this principle, no member of the consortium can isolate him- or herself from problems; all are motivated to find solutions, and the total intellect of the consortium attacks the problem rather than being dissipated on defensive action. A good analogy is that of voyagers in a boat: If the boat leaks, everyone gets wet feet.

Since it is essential that consortia act in a unified manner and there is potential for duplication of role and effort, care must be given to constructing an organization that is matched to the contracted task rather than being a reflection of the internal mechanisms of the contributing companies. For major projects, the consortium should be constructed under a project board that lends authority and identity. The board should be populated by the business directors responsible for resource allocation in each company. Each of these may be assisted by an advisor if the duration of the project is such that continuity benefits from this arrangement. The board can be chaired by an independent director, provided that that person can be committed long-term to the project. If this cannot be achieved, rotating chairpersonship is an option, but that arrangement loses the identity value that a permanent chairperson confers in terms of external influence for the project.

Consortium Management Organization

There are three imperatives guiding the organization of a consortium. First, there must be openness and trust in commercial dealings and a mechanism for harmonization of commercial position. This implies the need for a commercial group led by the legal prime contractor and reporting directly to the joint board. Second, there must be an operational management group that effectively spans the resources of the contributing companies. This too needs leadership, and it should come from the prime company, but all assigned to this forum should have good insight into the resources their company contributes and the matching resources in their collaborators' domains. Specialist subgroups should form as needed and retire as needed and report to the operational management group, though some specialisms will be represented from beginning to end. Systems engineering tends to fall into this category. Finally, the operational structure must be composed to promote horizontal communication at working level. It is grossly inefficient if working-level issues have to propagate upward and outward before resolution. It is much better to sanction peer-to-peer communication across partner companies. This means that nominated contact points at the same level and in corresponding areas of interest are needed in each of the contributing companies.

It is generally good practice, but the more so in international collaborative programs, to ensure that each employee posts an open profile of his or her expected contribution to the project. This is quite distinct from job descriptions that an employee may have as part of his or her career progression in the company. A single-sheet pro forma can be drawn up for this and maintained by the operational project leader. The pro forma should contain the following type of information, tailored to the project need:

- Identity of incumbent and title of the role
- The position in the reporting chain
- The identity of the group reporting to the role
- The context and purpose of the role
- The major actions that the role performs
- The objectives the role delivers to the project
- The key outputs and decisions required of the role
- The interfaces the role services

- The inputs needed to perform the role
- The sources of inputs needed to perform the role

This is a mechanism of accountability to the project and can be used by management to reinforce project identity and project unity. It is invaluable when new members must be introduced into the management team, and it is a useful review tool when the project leader is considering the adaptation of management roles as projects evolve through phases.

U.S. Collaborative Programs

Collaborative programs under U.S. leadership tend to follow U.S. practice and are cost-reimbursed against audited effort with profit paid against milestone accomplishment. A measure of cooperative incentive is achieved by weighting profit milestones such that deliverables from a single company score low, while those demanding group effort score high. The U.S. government does not, in general, tie itself into collaboration as strongly as European governments do, and in most programs, it is the principal investor. Hence, these programs are dominated by U.S. practice and can rely on U.S. procurement resources. Internal U.S. industry collaboration within the U.S. procurement system seems to work well. To some degree this can be attributed to the sophistication of U.S. procurement management and to overt product alignment.

Project Operational Management

The operational management of defense projects differs little from management of projects in other spheres. The matters raised here as peculiar to defense tend to affect the strategy and policy of project management rather than its execution. The majority of defense projects deliver new products, be they weapon systems or black boxes, and in most cases dependent on the most recent technology it is practical to incorporate. Electronic component complexity[1] continues to escalate alarmingly and supports corresponding surges in the complexity of system software, where the dominating rationale for the software ingredient is its ability to mask complexity. While utility hardly reflects complexity, the embedded product risk most certainly does, and for this reason, it is necessary that project management practice is well developed.

The Basis of Planning

A project starts with recognition of an opportunity to win a contract. The first decision point within a company is reached at this stage of early interest. This is because the next

[1] The much quoted "Moore's Law" is really an observation that over 30 years, electronic microcircuits have increased in complexity by a factor of 100,000 times.

stage should involve significant cost that should not be undertaken lightly. There should be formal endorsement of the opportunity and allocation of resource against the task of forming a concept for realizing the desired product. The concept pulls in past experience, domain knowledge, demonstration of novel parts, analysis to predict performance, measurement to calibrate analysis, innovation to add competitive edge, investment to gain efficiency, and management to estimate and control the cost of realizing the concept. Estimation is usually a mixed process; parts can be estimated by similarity to past product, by parametric analysis, and by detailed task breakdown. It has elements of top-down and bottom-up. In addition to this, the process may not give the answer management wants, so part of the estimate may be a judgment on what can be done over the project duration to improve the estimate. At any rate, the quantification of concept can be a very dangerous stage of the project, because if improperly managed, it may lead to winning at an impossible price.

The more engineering work that can be afforded in the concept stage, the more certain will be the estimates on which bids are based and plans formed. Concept risk should be identified early and work programs initiated to limit this risk to acceptable low levels. There is no substitute for demonstration or precedent. Where demonstration of a critical feature is not practical, management must consider affording parallel paths to achieve the concept feature in question so that success probability is enhanced. When costs are understood and concept feasibility, admitting moderate and quantifiable risk, is established, the process of deciding whether to bid can be formally approached. It must be a formal approach. A contract is a legal obligation; it should never be considered an automatic consequence of the preparatory work undertaken. It is undoubtedly true that the most important contracts for any company are those bad contracts that the company has had the wisdom to avoid through the diligence of its bidding processes.

When a decision is made to pursue a bid, a full business analysis should be undertaken to establish acceptable terms, risks, and rewards. The keen interest of senior management does not substitute for objective business analysis. The terms on which a contract will be accepted should be established on business return and company strategy. If strategic management judgments are to be influential in this matter, they should be turned into monetary value and they must be managed to achieve the predicted value in parallel with the strategic contract. Traceability of decisions and supporting data is an essential requirement, since competitive procurement often involves several years from first investment by a company in an opportunity and the achievement of a contract.

There are differing views on the management of the bidding process. At one end of the scale sit those who believe the bid team should become the project team. At the other end lies the opinion that bidding is an art that demands a professional bid team that will later hand the job over to a project team and turn attention to winning another bid. Reality lies somewhere between but should be colored with the knowledge that continuity in the project is valuable, even from the earliest stage, while professional preparation of proposal documents can do much to enhance clarity of communication and allow the virtues of the chosen concept to be appreciated in the evaluation of bids.

Project Life Cycle

The life cycle of the project has really begun with the first expenditure of effort, and it serves well to have a management process that controls the quality of action at each of a series of waypoints or gates starting at this point. (For further discussion of life cycle development processes, see the chapters by Archer and Ghasemzadeh, by Cooper et al., and by Milosevic.) Hence, we may map a series of gates as follows:

- Recognition of opportunity—Authorize concept study
- Bid/no-bid on basis of evolved concept and business analysis, including estimate of available contract terms
- Bid submission—Reviews of winning strategy and bid quality
- Contract acceptance—Formal recognition of the responsibilities of contract and acceptance of its terms
- Contract mobilization—Review of resources and plans
- Preliminary design acceptance
- Design closeout acceptance
- Delivery (production) readiness—Aligned with design review cycle
- Support phase readiness
- Disposal

The implementation of a "gateway" or "life cycle" process requires senior-level commitment in a company. It is necessary to embed the process within the culture of the organization and ensure that the organization does not manipulate the process as a shield against the normal work pressures it experiences. For example, it is never constructive for an organization to cite that the timescales it can commit to an external customer are driven by its regulatory processes. If this occurs, it is symptomatic of the process being abused rather than being used as the enabler it should be.

The conduct of life cycle reviews requires a college of assessors with sufficient background in the subject areas to challenge project staff constructively. A healthy process leads to cross-fertilization among projects within a company and naturally spreads good practice. In dealing with the gates or phases of a project, the whole-life review cycle should be planned at the earliest stage, and if possible, the overall responsibility for conducting this should be entrusted to a senior manager independent of the project. Preferably this manager will stay with the project, chairing successive reviews and acting as a guiding influence. However, it is important to realize that the delivery of a successful review and a particular transit to the next phase of project activity is always the responsibility of project staff to deliver. The reviewers do not replace the project management; reviewers can make recommendations with the force of company policy behind them, and these recommendations must be considered rationally by project management. In extreme cases, a bad review may end in a recommendation to stop or suspend a project until remedial action has been taken, and if necessary, the authority of the CEO must be engaged to support this.

Alignment of Gateway Reviews and Design Reviews

There are three important review dimensions in managing contracts. They can be regarded as near to mutually orthogonal, guarding distinct though interacting features of the project (see Figure 53.1). Gateway reviews check the quality of management process and readiness to transit through project phases. Design reviews check the design solution against require-

FIGURE 53.1. THE ORTHOGONAL DIMENSIONS OF PROJECT REVIEWS.

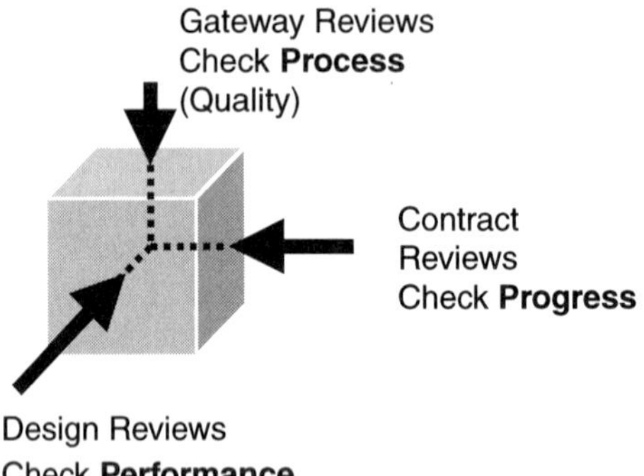

ments. Progress reviews check value earned against the financial plan and regularly update costs to complete the project. These three dimensions must be conducted so as to have a common focus in adding value to the product involved.

While contract reviews against the financial plan should be undertaken on a monthly cycle, design reviews follow the natural phases of the project and hence align with gateway reviews. It is necessary to plan these cycles to suit the constraints of specific projects, but some generic guidance can be given as shown in Figure 53.2.

In general, before gateway reviews can be held, supporting design reviews must be held and achieve a satisfactory result for the design. Moreover, design reviews should generate action lists, and these must be closed out or satisfactorily incorporated into the regular management process before the design review can be considered complete. The requirement specification is the baseline yardstick by which the project is measured in a design review. This is often augmented by additional requirements that a particular organization may promote—for instance, manufacturability, cost, growth potential, obsolescence, robustness, and so on. It is prudent to consider an interdisciplinary approach, engaging the stakeholders who will have to take the design forward as well as the originating design team. Some contracts require that design reviews be attended by the customer. Such reviews should never be accepted as qualifying reviews within a competent gateway process. The customer has every right to appraise the design, and customer inputs should be welcomed and weighed for design value against the contracted requirement. However, a competent design review must descend to a level of detail that would be impractical to present to a customer audience. The distinction is mainly on the practicality of appraising the detail without in-depth design knowledge, but there is also a danger of confusing responsibilities. Design decisions rest with the design team and should be supported by trade-off and optimization studies conditioned by specification constraints. They can be challenged in review on this objective basis.

FIGURE 53.2. ALIGNMENT OF DESIGN REVIEWS AND GATEWAY REVIEWS.

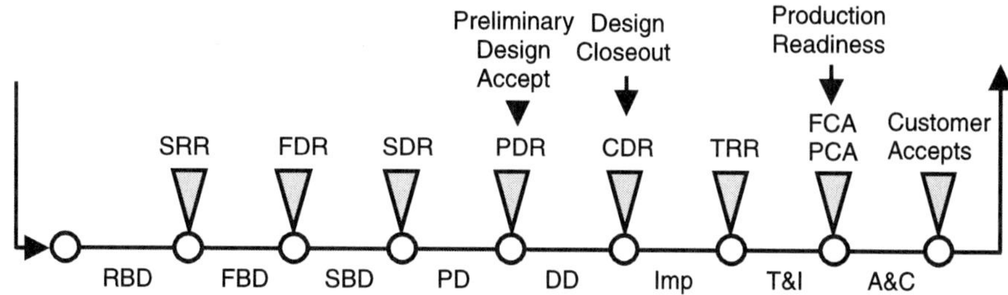

Activity	Review
RBD = Requirements Baseline Definition	SRR = System Requirements Review
FBD = Functional Baseline Definition	FDR = Functional Design Review
SBD = System Baseline Definition	SDR = System Design Review
PD = Preliminary Design	PDR = Preliminary Design Review
DD = Detailed Design	CDR = Critical Design Review
Imp = Implementation	TRR = Test Readiness Review
T&I = Test and Integration	PCA = Physical Configuration Audit
A&C = Acceptance and Certification	FCA = Functional Configuration Audit

Progress Reviews and Earned Value

The third review requirement is progress. This should be undertaken regularly, typically monthly, and with a focus on confirming or quantitatively adjusting the forecast of completion cost. It is always important to keep a regular emphasis on completion, even from the earliest stages of a project. Analyzing the requirement for progress reporting exposes certain essentials of project planning, staffing, and execution.

Work Breakdown Structure (WBS)

Progress can only be assessed within a defined structure. The accepted approach to providing this breaks the total project work down into packages of such atomicity as to allow progress data to be estimated with reasonable accuracy. At the appropriate level of atomicity, the law of large numbers works constructively and derived progress measures can be quite accurate. In defining work packages, each must have a clearly identifiable start condition and a measurable finish point. At the lowest level, the work package comprises a list of describable activities, each of unambiguous scope. An activity should exist in only one work package and ideally work package progress can be measured by the degree of completion of its constituent activity population. The work package is a useful building block from which planning packages can be assembled, and it is a fundamental requirement for assignment of staffing and allocation of budgets. The framework for control is complete when an organizational breakdown structure is mapped onto the work breakdown structure and budgets are assigned to the work package managers. Activities within a work package may be linked or disjoint and the same may be said for work packages within the plan. Linkage between activities or work packages means that a predecessor must finish before a linked successor can begin, so there is a concept of predecessor-successor within the planning reflecting the practical dependencies of the project. There are good software tools for capturing and associating the description, resource allocation, duration, and linkages of tasks within a project structure. These can also be loaded with charging rates for the resource types employed and hence provide forecast cost data as a function of project time using the task input data. Figure 53.3 shows how work packages may be represented using the popular Microsoft Project tool.

When a logically consistent and properly linked WBS has been achieved and allocation of resources made, a contract budget baseline can be drawn up, as a reference for progress accounting. Typically, the baseline will be composed as shown in Figure 53.4.

The aim of project management is to achieve the planned margin by managing earned value against the limit of authorized works cost. Monthly reporting against the WBS is the primary tool for doing this.

Earned Value Management (EVM)

Earned value is measured at the lowest practical level, preferably at task level. Task managers should report their judgment on percentage completion at the end of each reporting cycle, and this should be used to estimate work package earned value using the budget value as the scaling reference. The value obtained can then be compared with the actual costs

FIGURE 53.3. GANTT CHART ILLUSTRATING WBS FOR A SMALL PROJECT.

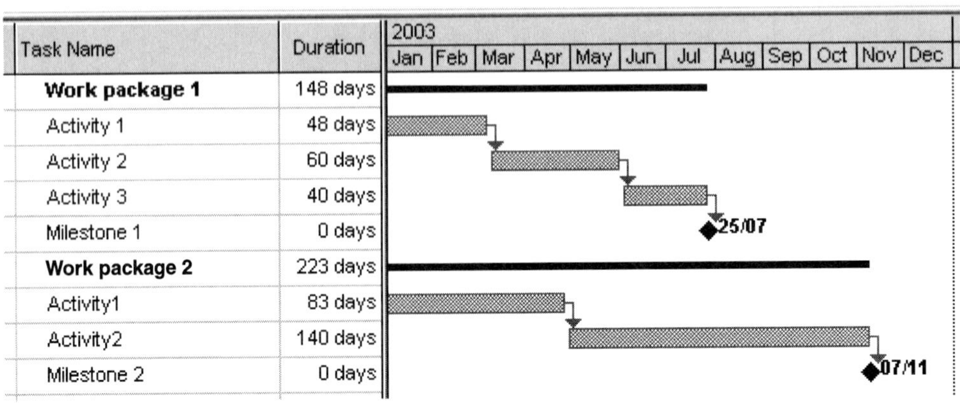

FIGURE 53.4. THE ELEMENTS COMPOSING A BASELINE BUDGET.

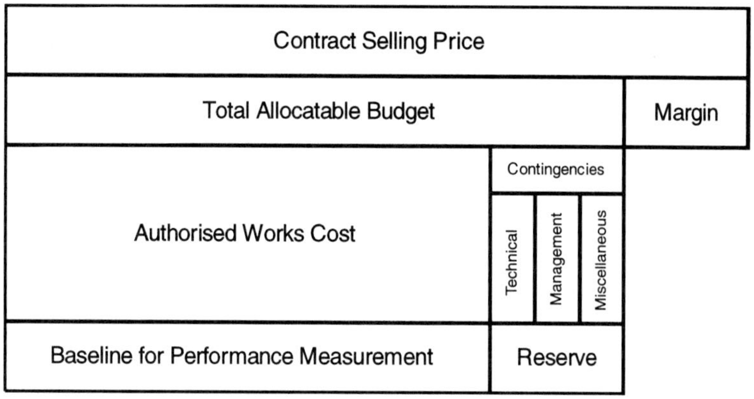

Contract Selling Price
Total Allocatable Budget
Margin
Contingencies
Authorised Works Cost
Technical
Management
Miscellaneous
Baseline for Performance Measurement
Reserve

consumed and a value variance computed. Project history is well reflected in the time graph of earned value. This can be compared to the actual cost incurred as a function of time and the budgeted cost forecast as a function of time. As an example of the power of the technique, consider the unfortunate project represented by Figure 53.5 and reflect how its woes are amply demonstrated by the curves presented.

The project planned a sharp buildup of resources, so budgeted costs rise steeply at the beginning. However, actual cost and earned value lag far behind. It is obvious that by month three, the plan cannot be realized and either there is a resource crisis that should

FIGURE 53.5. EVM CURVES FOR A HYPOTHETICAL PROJECT.

Cost
1.6
1.4
1.2
1
0.8
0.6
0.4
0.2
0
1 2 3 4 5 6 7 8 9 10 11 12
Project Time in Months
Actual Cost
Budgeted Cost
Earned Value

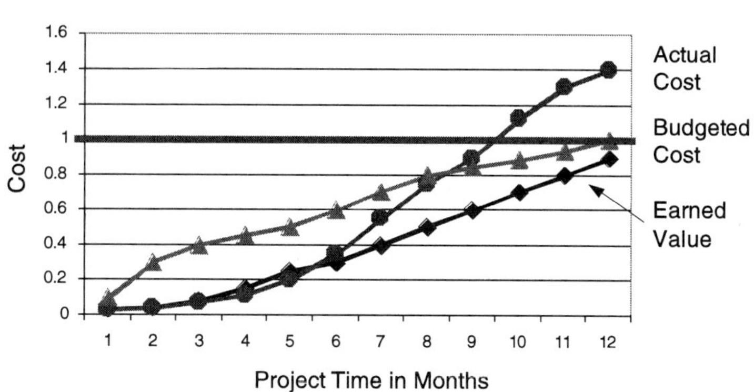

be properly addressed or the project should be replanned before more value is lost. However, myopia is a rampant disease in management, and fueled by the optimistic thought that earned value and actual cost are in line, the management ignore the warning that earned value is far below budgeted cost. Within the next quarter the inefficiencies of the first quarter strike, but resources become available and actual cost pulls away from earned value, exceeding budgeted cost in the eighth month. It is all too late at this stage; project slippage and overspend are now inevitable, and in an effort to contain them, the management will increase the rate of expenditure, sacrificing cost for time. By the forecast end of the project, earned value is still below budgeted cost, while actual cost to achieve this has soared away. Moreover, the fact that the slope of the actual cost curve at this time is greater than the slope of the earned value line indicates that overspend will increase further before the budgeted value will be earned and the project can be closed. (See the chapter by Brandon for a further treatment of earned value.)

A value variance between forecast and earned values immediately allows the forecast cost at completion to be updated. This is the primary indicator of project progress and should always prompt reaction.

While management judgment should be exercised continuously at the task level in order to generate the most reactive value variance estimate, it is prudent to defer contract payment claims until planned milestones have been fully achieved. In contracts where the customer demands visibility of earned value, an increasingly common situation, customer reporting can be based on milestone achievement while project management works to the lower granularity of task achievement.

The contingency element within the budget baseline may be used to fund risk mitigation activities aimed at avoiding overspend, or to fund unavoidable overspend. Careful, though proactive, management of contingency affords an opportunity for increased margin, and most projects will run an opportunity register identifying potential cost savings alongside a risk register with the aim of conserving contingency and promoting margin. The counterpoise of risks and opportunities forces innovative thinking into the management review and often contributes unexpected value.

Summary

Defense contracting demands a management approach cognizant of the political environment that inevitably accompanies an activity with strategic national and international connections. The basic management principles of definition and control are common to other industries. Because high complexity is frequently unavoidable, it is vitally important to achieve a robust logic to underpin project planning. This depends on excellent domain knowledge and alertness to the nuances of the contracting environment within the industry. When contracts are undertaken in partnership, consensual management processes are required and shared liability schemes are to be recommended. Role profiling is a useful tool in developing necessary and sufficient organization structures. The vertical contracting hierarchy between customers, prime contractors, and suppliers should recognize that focusing on product value is a uniting factor that resolves most contentious issues.

In addition, partnering charters may be useful mechanisms for conditioning behaviors. A strong and disciplined review culture enforced from the top down through the organization is highly beneficial to the management of complex projects. This should recognize the necessity for three independent but singly focused reviews covering management process quality, product design integrity, and project progress. In turn, these reviews are best serviced by a gateway approach for management phases, a hierarchical approach for design from system concept to production readiness, and an earned value approach to progress.

Acknowledgment

This chapter owes its existence to Steve Marsh, who retired from BAE SYSTEMS in poor health and sadly passed away before seeing it in print. He would approve of the emphases applied here, because many were generated in discussion with him. Steve belonged to the school of commonsense managers who could achieve untold cooperation, regardless of the tensions of the project situation.

References

Morris, R., and C. Ericsson, C. 2003. *Developing effective engineering leadership.* London: IEE Press.

Component Obsolescence Group (COG). 2003. Application of obsolescence strategies. Conference proceedings. May 19–22. Glasgow.

Defense MicroElectronic Activity. *www.dmea.osd.mil/.*

Guiliani, R. W., 2002. *Leadership*, London: TimeWarner Books.

McTaggart, J., P. Kontes, and M. Mankins, 1994. *The value imperative.* New York: Free Press.

Morris, P. W. G. 1994. *The management of projects.* London: Thomas Telford.

Pearson, B. 1990. *The profit driven manager* London: Hawksmere,

United States Department of Defense. Defense Standardization Program. *www.dsp.dla.mil/.*

Wagner, A. 1996. *The transactional manager.* London: The Industrial Society.

CHAPTER FIFTY-FOUR

PROJECT MANAGEMENT IN THE CONSTRUCTION INDUSTRY

Peter W. G. Morris

Construction is widely thought of as a major industry sector that spawned modern project management.

Construction is in fact a conglomeration of different sectors, ranging from house building, commercial property, institutional building (schools, hospitals, prisons, etc.), and other structures generally in the building sector; railways, roads, sewers, water containment, and other large civil engineering works (tunnels, bridges, etc.); and power, oil and gas, water treatment plants, food processing, pulp and paper, and other "process engineering" sectors.

Though all these construction sectors are project-based, project management as a significant formal *discipline* is hardly recognized in building and civil engineering, and while project management departments exist in most process engineering companies, there would be few who would see the discipline as having a coherent body of knowledge that is central to their business success. Nevertheless, the way construction projects are managed tells us a lot about the realities of "the management of projects."

Construction: Building, Civil Engineering, Process Engineering, and Other Sectors

Construction is more than a single industry sector. It comprises several different sectors. Though these do indeed share many characteristics, the differences between the process engineering (sometimes called engineering construction) industries and civil and building industries are significant, especially in the way project management is practiced.

First, the similarities. All construction projects essentially involve building facilities for people or things to occupy. There is a strong element of product design (some of it using 'off-the-shelf' products; some with a greater or lesser degree of bespoke design and manufacturing) prepared to some form of client brief or specification. The manufactured products (doors, pipes, bricks, concrete, etc.) are erected on-site: The logistics of getting these products to site on time and in an efficient manner are generally an important part of construction management's responsibility; ground conditions, weather, and spatial layout may be significant factors in that process.

Most construction projects are truly unique—that is, having a unique design, with a supply chain (owner, designers, manufacturers, main and subcontractors, suppliers, regulators, etc.) distinctly chosen for the project. (This is not necessarily true for all construction projects—for example, for some housing projects—but it generally is.) Crucially, the manner in which the members of the supply chain are 'contracted' to work together—when, under what terms and conditions, and how—exerts a major influence on the whole character of the construction project, as we have seen in several chapters already in this book, including Langford and Murray, and Lowe. Traditionally the commercial nature of these arrangements has been muscular, to say the least (and often darned right aggressive); this tendency is exacerbated when suppliers purposefully bid low in order to get work, in anticipation of subsequent change orders that they hope to price high. This mode of behavior has often led to poor integration between project participants, at least unless well managed. One good project management result is that cost control and contract administration, and more recently, value management, are probably as well practiced in construction as in any project management industry. Health and safety have always been a very high priority in construction (it is a genuinely risky industry); this is particularly true today, as we saw in the chapter by Gibb. The management of environmental issues also features highly.

None of these characteristics are necessarily unique to construction. What is distinctive overall is the combination of one-off bespoke designs and production processes and the importance of the supply chain configuration and the way suppliers are engaged (procured).

There is a widespread view that construction is something of a poorly managed industry. In fact, it receives a huge amount of management attention. In the United Kingdom, for example, there have been at least four major initiatives in the last 10 to 15 years. Two were in building and civil engineering: the Latham and Egan reports (Latham, 1994; Egan, 1998); and two were in process engineering: CRINE, an initiative in the late 1980s concerned with "Cost Reduction in the New Era" (www.logic-oil.com), and ACTIVE, essentially the onshore CRINE, which followed a few years later. (ACTIVE stands for Achieving Cost Reduction through Innovation and Value Engineering; see www.eci-online.org.) In the United States, there has been a sustained program of work carried out under the auspices of the Construction Industry Institute (CII), a body sponsored initially by the Business Roundtable (www.construction-institute.org/). *The Business Roundtable* is a grouping of over 40 U.S. companies (largely from the heavy-construction end of the industry) that together work with academia on research to improve construction performance. CII's work has spread to Europe, and there is now a European Construction Institute (www.eci-online.org/).

These initiatives have led directly to a raft of major improvements in the way construction is managed. To an extent there will always be a tendency for poor management to be present, given the ease of entry to the industry and the prevalence to bid low to secure work. Overall, however, the industry's project delivery performance is not at all bad (and has certainly improved in recent years), as various benchmarking data attempts to show. (See www.cii-benchmarking.org for data on the process engineering/heavy construction sector, and www.cbpp.org.uk/cbpp/for UK building and civil engineering benchmarking data.) Where it does fail, it probably does so more through poor application of good practice (for human or organizational reasons) rather than any industry-wide lack of skill, knowledge, or technique.

What then distinguishes the different construction sectors? In shorthand, buildings are structures with roofs on them; civil engineering structures tend to be more in the ground and relate more to infrastructure (roads, harbors, dams, etc.). Buildings are generally designed by architects, with the assistance of several types of engineers and other specialists. Civil engineering projects are rarely designed by architects. Building projects are typically more complex technically and organizationally than civil engineering projects, having more trades working in less space.

The process engineering industries cover, inter alia, the oil, gas, chemical, power, and pulp and paper sectors. (Interestingly, rail and water projects were historically within the purview of civil engineering; over the last few decades, however, they have migrated somewhat into process engineering, partly as technology has become more process-oriented—signaling systems in rail projects, water treatment systems in water, and so on—and partly as process engineering contract terms have been adopted as more appropriate.)

Building, as well as often being more complex organizationally than civil and process engineering, puts a higher premium on design originality (hence architects). It also typically has fewer experienced owners (clients) than the other sectors:

- In building the project leader has traditionally been the architect.
- In civil engineering this role has been assumed by the engineer.
- In process engineering, it is the project manager.

In building and civil engineering, there is not a tradition of project management—as a formal discipline, function, or even role—actively managing the integration of design and construction in the same way as there is in process engineering. In building, however, the architect's lead role has increasingly been challenged by others, claiming, with varying degrees of wishy-washiness, to be acting as project managers—from quantity surveyors (cost-control-type administrators) to construction and "program" managers (more technology-/production-oriented experts). This relationship between architect and project manager is comparable to the project leader/project manager debate currently going on in drug development (see the chapter by Foulkes and Morris). There is little evidence of a strong formal project management push in civil engineering. In process engineering, on the other hand, project management as a practice is strongly and expertly performed.

Project management has, in fact, been in process engineering for much longer than in building and civil engineering. Its application in construction in general, and building and civil engineering in particular, has been held up largely because of the challenges posed by (1) the separation of the designer from the overall management of the project, (2) the process of bringing in contractors and suppliers and playing out the terms and conditions of these contracts, and (3) the historic role of the designer in administering these contracts. The first and third of these do not apply in process engineering to anything like the same extent as in building and civil engineering.

The salient characteristics of the different sectors are summarized in Table 54.1.

Project Management in Construction

Project management is thought by many to have originated in the construction industry. This is not true, except in the more general of senses, largely for the historical reason associated with the split between design and construction.

Modern project management, in the sense that we use the language of the discipline today, originated primarily in the U.S. defense sector in the mid-1950 (Morris, 1997)—though, as noted later, Stone & Webster and Exxon both had project engineer functions in the 1920s (Stone & Webster) and 1930s (Exxon). It *is* true that the critical path method was developed by DuPont in 1957, almost at exactly the same time PERT was developed on Polaris, for planning plant construction. Though much of construction in the broadest sense—civil and building as well as process engineering—was by the mid-1960s using some of the new project management techniques, most notably network scheduling, the engineering project management practices developed in the DoD and NASA were far in advance of construction management practices at this time.

Building and civil engineering are among the oldest and most respected industries in the world. The architect of the first pyramid in Egypt, Imhotep, was later deified. Yet even the ancient project structures of the Greeks, Persians, Assyrians, Romans, Chinese, and others employed a construction procurement process that separated the designer from the builder. Though the "master mason" practice of the Middle Ages blurred this distinction somewhat with the mason acting both as craft designer and as builder, by the time of the formation of the engineering and architectural professions in Europe in the seventeenth to the nineteenth centuries, it had hardened, with results that still affect project management in these industries today (Straub, 1952; Watson, 1998). From the early nineteenth century onward, the designer was commissioned by the client to prepare plans for the structure; a contractor was then engaged (formally contracted to the commissioning employer) and overseen by the designer, on behalf of the employer. The architect and engineer were gentlemen; the builders were workmen. Changes were not supposed to happen, or if they did, were priced through a 'schedule of rates' or process of claims (and if these were approved by the client's designer, everyone benefited financially, except, of course, the client, who had had no say in the matter). There was no management integration of constructability and design other than through the designer's professional knowledge. Project management, in the sense of actively integrating all the work that needed to be done to ensure an optimal project, did not exist in building and civil engineering—arguably right up to somewhere between now and 10 to 15 years ago.

The process engineering industries, however, represent quite a different history, as well as structure. From the mid- to late nineteenth century, various types of large and complex

TABLE 54.1. CHARACTERISTICS OF BUILDING, CIVIL ENGINEERING AND PROCESS ENGINEERING.

House building; Commercial property; Institutional buildings	Roads, Bridges; Harbors, Rail; Dams, Sewers	Power, Paper and Pulp; Food processing, Chemicals; Oil and Gas
Similarities		
All are involved in creating facilities All involve unique design... .... And supply chains are generally unique too to the project Contracting is significant All involve significant logistics Weather and, sometimes, ground conditions tend to be an issue Health, Safety and Environmental issues are of major importance People (leadership & teams) play a significant part in project realization		
Differences		
Buildings typically have roofs on them May not be large always but are often organizationally complex Repetitious — Premium often on architectural design — Emphasis on value-for-money Environmental engineering important	Heavy emphasis on ground works and structures Projects often large, though not always organizationally complex	Standardization of equipment greater than in building, and civil Can be quite large projects. — "Fine" chemicals are more complex. — Oil and gas can be big, demanding projects.
Management Characteristics		
Historically, designer-led, and dominated Often "contracting" problems Project management only now emerging "Construction management" (focussing on site works) since the late 1960s	Project led by "The Engineer" Often "contracting" problems Project management hardly really significant	Strong "Engineer-Procure-Construct" management practice Project management seen as a core management practice

engineering projects—railway engineering in particular, but also mining and major plants such as steel, power generation, and other utilities—began to generate a more integrated approach to the management of projects. By the end of the century, a new sector had arisen: oil, gas, and chemicals. This sector took even further the management characteristics associated with (1) large, efficient owners having (2) integrated views of their projects with (3) substantial opportunities for standardization within the engineering design. These are quite different characteristics from, say, the prestige building market, where there is a premium on innovative, high-quality design. They lead much more readily to an approach to construction project management that is owner-oriented, integrating engineering, procurement, and construction (EPC) (Strassman and Wells, 1988; Linder, 1994).

Hence, the process engineering industries were able to develop a much more "through-project" integrated approach to engineering project management, with engineering (design), procurement, and construction typically being performed by the same management team. Thus, Stone & Webster, for example, had a project engineer function in the 1920s, as did Exxon in the 1930s—an engineer who could follow a project as it progressed through its various functional stages. At about the same time, the U.S. Navy Aircraft Factory and U.S. Air Corps' Materiel Division began developing the project office function to monitor the development and progress of aircraft (Pinney, 2000; Morris, 1997).

The building and civil engineering industries, on the other hand, with their effective split of management responsibility between the engineer (or architect) and the contractor, as formalized in contractual forms such as the ICE and JCT, and even FIDIC (see the chapter by Lowe) had no single person actively managing the shaping and delivery of the project from its earliest phases through design to construction and operation. It still hardly ever does.

(The obvious exception, apart from house building, which is addressed in the next paragraph, is design-build-type contracts. These are projects where the builder provides the design, almost always to either specifications or outline design concepts provided by someone else—see the chapters by Langford and by Lowe. The form is generally used for industrial-type buildings or property development projects where aesthetically original design is not highly valued. It most closely corresponds to the EPC model. Even here there is not quite the active project management of design from the earliest stages of the project, though in some cases it may come near.)

Within this broad schematization, there are, of course, important variations. Within building, for example, house building is a quite different sector from commercial office building: it offers much more opportunity for integrated design, procurement, and construction, with a much stronger role for the owner (sponsor); and in commercial building there is a difference in emphasis on quality of design between prestige projects and run-of-the-mill "to let" developments. That project management did not emerge in house building reflects the lack of a dominating, single-project orientation of the typical house-building undertaking: more of a program of production, in fact, than a project.

In recent years this has been changing somewhat, though there is still far to go. Building in particular (reflecting the challenge of managing complex projects better) has seen the emergence of first "construction management," or CM, and more recently both civil engineering and building have promoted formal "project management" and "program management" as a discipline, though rarely in the same hands-on sense as seen in process engineering. Construction management, however, is essentially an approach to managing the site end of the project (trades contractors), with bare recognition of the real challenge, and opportunities, of managing design and the "front end." Conversely, much of project management in building is a synonym for scheduling and cost management, and in civil engineering, for supplier management.

Program management is a newer and vaguer term tending to mean much more strategic and front-end-oriented work, managing the project from the very earliest stages in building (very akin to the "management of projects" model discussed in the Introduction and used in other chapters), while in civil engineering it tends to bring in more process engineering skills, as in the rail sector. Program management offers scope to focus on the client—"the program"—who has a series of projects. This helps improve project delivery through a closer understanding and articulation of client needs.

The truth is that project management in building and civil engineering has consistently been weakest at the front end largely because "management," as a discipline, has historically been a contractor's strength, not a designer's. The design professionals, though enjoying technical leadership, have rarely exercised active management leadership. This is strongly in contrast to practice in the process engineering industries and in the rest of project management. (Systems integrators, for example, are front-end, design-oriented managers who lead the technical-management integration activities within ICT projects: There is not—yet—the equivalent in building and civil engineering.)

The industry is still living with this legacy: Project management in building and civil engineering has rarely exhibited the kind of leadership qualities that we ought to expect from the discipline, such as insisting on proper project briefing, strategy development, and effective design management (including value management), or creating and implementing strikingly productive procurement arrangements. We see much more of these qualities in process engineering.

Construction Project Management Practices

We have therefore a few different situations in construction project management:

- *Building.* Generally a complex, fragmented industry but with a strong construction management orientation plus still largely embryonic project and program management services
- *Civil engineering.* Less developed CM and PM services, though with interesting inroads being made by process engineering EPCs, particularly in transportation
- *Process engineering.* A much more developed classical project management situation, with through engineer-procure-construct integration, though even here project management is seen as a general practice more than as a formal discipline.

And curiously, there has been little recognition at the policy end of the industry that project management, *as a discipline,* can really make a difference to performance and ought therefore to be promoted, as we shall now see.

Lack of Formal Recognition of Project Management as a Discipline

The late 1980s and 1990s saw major initiatives in all construction sectors in the United States and Europe aimed at improving the management of construction projects (the CII, ECI, CRINE, ACTIVE, Latham, and Egan reports already mentioned). Project management, notably, as a discipline is mentioned directly in hardly any of the publications coming from these studies, other than as a generic activity associated with managing projects.

The Egan report in the United Kingdom in the late 1990s, for example, emphasized five key drivers of change for setting the agenda for the construction industry at large: (1) committed leadership, (2) a focus on the customer, (3) integrated processes and teams, (4) a quality-driven agenda, and (5) commitment to people (Egan, 1998). The report reflected themes that have been rolled out for the last 50 years in a series of critical reports on the structure and performance of the UK building industry (Murray and Langford, 2003). Egan called for greater integration of project processes without mentioning project management per se. (Though the activity of "project implementation" is referred to, however, being positioned as project execution, following "project development."):

> The conventional construction process is generally sequential because it reflects the input of designers, constructors and key suppliers. This process may well minimize the risk to constructors by defining precisely, through specifications and contracts, what the next company in the process will do. Unfortunately, it is less clear that this strategy protects the clients and it often acts as an effective barrier to using the skills and knowledge of suppliers and constructors effectively in the design and planning of the projects. Moreover, the conventional processes assume that clients benefit from choosing a new team of designers, constructors and suppliers competitively for every project they do. . . . The repeated selection of new teams . . . inhibits learning, innovation and the development of skilled and experienced teams.
>
> *Rethinking Construction*, Chapter 3

What we have in Egan is the industry casting about for new management models with which to address the need to provide greater integration between design and production: precisely the job of project management. Yet there is no acknowledgment that there is a discipline in project management that, properly deployed, can proactively integrate business strategy, design, procurement, and construction, doing so in a way that meets the project's strategic objectives (KPIs). Management practices discussed in Egan such as value management, design management, procurement management, planning, cost management, teamwork, and so on—there is no mention of scope management—are not positioned as tools that form that part of a holistic discipline and that, as such, should be seen and used in an integrated manner.

The immediate reason for this lacuna was clear: The report's authors, like many in the construction industry, were, as many still are, profoundly unconvinced that project management is indeed an overall discipline at all—let alone one that is capable of shaping the design and delivery of buildings in a way that works for the project's many stakeholders. The reason for this skepticism lies in the historic split between design and construction and the failure so far of project management to appear as much more than either super construction management or dressed-up cost estimating and planning.

Winch makes an important, parallel critique. Winch traces the Egan perspective from the auto industry and "mass production" lean manufacturing paradigm (Winch, 2003). While appropriate to those types of construction which create standardized building types such as housing, schools, or retail units, it is less appropriate for more complex projects such as "those which push the envelope or create iconic architecture" (Winch, 2002). And tellingly, while many in construction were turning to manufacturing to provide a new organizational model, manufacturing, at least as far as product development is concerned, has been turning to the management of projects. Heavyweight project managers, or chief project managers, have emerged to coordinate the most important NPD projects, and they are required to have a broader perspective than just quality, cost, and time (see the chapters by Milosevic, and by Midler and Navarre, for example).

(A potentially interesting development that occurred during the period this chapter was being written, however, was a call by Egan's successor, Peter Rogers, for "project managers who have an understanding not only of the process but on design and the broader benefits of procuring a major project" (Building, 2003). Not only are project managers at last being named, the role they need to focus on is explicitly being identified with the broader technology, business, and commercial "management of projects" one.)

The process engineering studies, on the other hand, reflecting the more overt project management presence in this sector, have taken a more explicitly project management oriented approach. Thus, for example, in a series of major and influential joint industry-/ academia-based studies, CII in the United States and ECI in Europe identify the following practices as having the greatest impact on project performance (while still largely avoiding the term "project management"):

- *Pre-project planning.* Particularly technical but also commercial (e.g., contracting strategy) and logistical; budget matters and finance will also be important ingredients.
- *Scope definition.* Covering the phasing of the project (option development, definition, execution, verification), including scope development (including charter, vision and objectives, specs, regs, site plans, reliability) and definition (including "best value"), and overlapping with change management.
- *Design and technology.* Emphasizing reliability and quality, and ICT, which, via CAD-CAM and product and project modeling, is seen as central to future industry improvement.
- *Change management.* Reporting, authorization, and trending tools.
- *Constructability.* A key means of improving productivity (analogous to manufacturability) involving the activity of reviewing the design so that construction is more efficient but also bringing with it important process and organizational implications of how one gets that input (do you buy the advice for a fee, bring in the constructor early, do so on a fast track basis, etc.?).
- *Risk mitigation.*
- *Safety.* Interestingly, the projects with the best safety records (zero incidents) are the most effective.
- *Team building.*

- *Strategic alliancing.* There has been a huge move toward partnering and strategic alliances in the industry.
- *Construction start-up.* Effective start-up of the construction teams and processes.

At least the language here is recognizably "project management," though what is interesting is just how much of it relates explicitly to technology issues, and in fact how much of this discussion is clustered around the "front end."

As a further example, look at the management trends that CII and ECI see as central to the industry over the next 20 years (Construction Industry Institute, 1999):

- *More global digitized technology.* Design and procurement, leadership usage of IT/CAD-CAM, improved materials
- *Value-based procurement.* Value-based contracting, incentives aligned to owners' requirements, greater distribution of risk throughout the project team, more partnering, global outsourcing.
- *Broader, more intelligent planning, design and construction.* Greater emphasis on project selection, early emphasis on critical equipment, more "whole-life" design ("design to capacity/for plant life"), more standardization, greater single data entry, increased prefabrication, more intelligent tools, more benchmarking

What you see here is a view of managing projects that is clearly technology-led but that marries technical possibilities within a business context. This, in fact, reflects strongly the pattern of management in other industries where the effective management of projects requires a focus on technical, commercial, organizational, and control issues—as well as strategic ones, to which we turn next.

Strategy and the Front-End

It is an old project insight that there is merit in spending more time managing the early, front-end part of the project work better. (As long ago as 1958 the DoD concluded that more time needed to be spent at the front end of the project development process and created a Milestone 0 as a pre-Milestone 1 gate [Morris, 1997].) We see this strongly in manufacturing projects—aerospace, automobiles, pharmaceuticals, and so on—as described, for example, in the chapters by Roulston, Midler, and Navarre, and by Foulkes and Morris.

Over the last decade or so, all the construction sectors have seen a progressive shift in management focus from the site 'back-end' forward, through procurement, to the design, planning, legal, and financing front-end issues. In process engineering, this focus tends to be heavily oriented around design (engineering), legal, and financial issues; in building it also typically involves people issues more, along with disciplines associated with people such as marketing. In process engineering, this emphasis has become so widely accepted that the term "front-end loading" (FEL) is now widely accepted as connoting time usefully spent on front-end work. (Benchmarking data shows that there is a correlation between time spent in FEL and project outcomes. See www.cii-benchmarking.org.)

A key player in determining the way a project is set up, and how effectively it is going to be managed, is the project owner. The project owner's decisions on how he or she wants to engage the supply chain members and his or her views on how the strategy should be shaped will be among the most formative strategic influences on the project.

One of the major differences between building, civil engineering, and the process engineering industries is, as we have seen, the nature of the project owners in each:

- Many building projects have periodic, relatively, or even highly *inexperienced* owners. (Some *are* experienced—property developers and managers, and house builders, for example—but most commission a new building construction only infrequently and hence cannot be expected to retain the project management expertise in-house.)
- Civil engineering projects have typically been commissioned by state entities who historically have taken a very hands-off and diminished attitude to their role as an owner's project manager—though with the advent of privately funded infrastructure projects (see later in the chapter), this is beginning to change.
- Process engineering, on the other hand, is typically managed by large organizations who commission projects quite often. As a result, the level of owner sophistication can be much higher.

There has been growing recognition in recent years of the value in managing more systematically the linkage between business strategy and project strategy. (See the chapter by Jamieson and Morris.) Whereas some construction projects arise with little detailed work on their development strategy, others see a strong interaction between business and project strategy; where sponsors are investing for longer-term strategic reasons, the project will want to put more effort into shaping the strategic options available to it.

Process engineering is perhaps where this is done least, largely because the alignment between business strategy and project strategy is generally well rendered in terms of product positioning (demand, competitive position), and by financial (net present value (NPV)/internal rate of return (IRR) and cash flow) and technical measures (plant performance). There is less emphasis on how the project will interact at the business level with people (marketing, user reactions, etc.). Figure 54.1 is a generic version of the "Active" (onshore process engineering; www.eci-online.org/) view of front-end loading activities needed for effective project management. Note how much of the key practices are technical. A few are strategic (Project Objectives, Contracting Strategy), but virtually all the remainder refers to technical aspects of the proposed plant. In process engineering, project strategy is thus typically treated as Project Execution Strategy, reflecting the lesser importance of the interplay between the broader business elements of strategy that we see so clearly, say, in IT projects, and in building and even some civil engineering projects—rail, road, and other transportation projects, for example. (Interestingly, there is not such a formalized model of front-end work in building and civil engineering.)

FIGURE 54.1. PROCESS ENGINEERING FRONT-END LOADING ACTIVITIES.

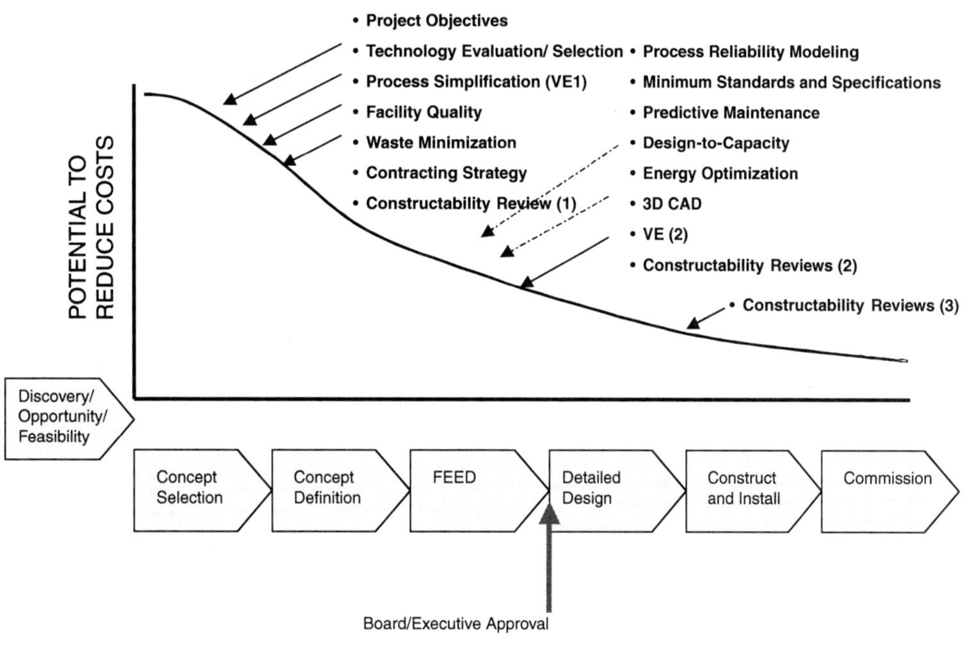

In building there is generally a greater need to look at the interaction between design and project delivery on the one hand, and business return and user satisfaction on the other. Commercial development projects, for example, will pay great attention to optimizing floor layouts and occupancy (lettable space) ratios (and possibly even attractiveness of design!) in order to improve the commercial return.

A number of project management practices are regularly and importantly applied in developing and optimizing the project's strategy in the various construction sectors. These run the whole gamut of project management, from risk management to stakeholder management, procurement strategy and logistics to finance and cash flow planning, resource planning to quality management. We shall briefly look at three before moving on to consideration of more procurement related issues: value engineering; briefing; and safety, health, and the environment (SHE)—each offers a special insight into the application of project management.

It would thus be wrong therefore to suggest that construction is not sensitive to strategic issues. In fact, the two biggest changes in the industry over the last decade have been in response to strategic developments, both of which have affected the practice of managing construction projects significantly—namely, the provision of private finance for public sector projects (power; water; transport, including railways, roads, and airports; hospitals; schools;

prisons; government offices; and so on.), and the greater use of partnering. (A third is arguably looming now: skills shortage, both at the craft and at the graduate intake levels.)

Value Engineering. With its emphasis on engineering and the management of technical issues, value engineering has emerged as an effective practice for systematically reviewing the effectiveness of the design in this context in all the construction sectors. (Cost savings of over 10 percent are commonly achieved through better use of equipment in process plant, and in better space utilization and choice of materials and equipment in building.) Value management, however, with its emphasis more on strategy, which as you have seen, often receives less formal management time and effort in process engineering, is generally done poorly if at all in process engineering; it is much more common in building.

Briefing. Briefing is a more complex activity in building than in process engineering. (Briefing received considerable attention in the 1994 Latham report.) The project's requirements are generally expressed as the client's requirements. In the more speculative type of projects, users' requirements are rarely specified in detail, as, say, in systems projects, for the simple reason that the user is only generally identified. Indeed, many speculative projects will be developed in two stages: first constructing the "shell and core"—that is, the structure—and then only later building to the occupier's requirements ("fit out") within the core structure, once an occupier is identified and his or her needs can be determined. Detailed requirements definition is an accepted practice more in projects having a strong facilities management orientation, such as hospitals, schools, and so on. User requirements may here come under the term "operational briefing," which would be distinguished from "strategic briefing" (reflecting the longer term view of the organizations and the facilities) and from "project briefing" (identifying the project's design requirements—analogous to system requirements).

There is rarely if ever the same emphasis on structuring requirements in a traceable and testable way (for verification and validation) as in systems projects (Barrett and Stanley, 1999; Blyth and Worthington, 2001; Kamara, Anumba, and Evbuomwan, 2002; Nutt, 1993) There is nothing like the same connect between business performance measurement and project management as there is in, say, ICT projects or product development products. (The language of the chapter by Davis, Hickey, and Zweig would be quite strange to most construction professionals.) New product development projects typically emphasize customer/user reaction and business performance, and integrate implications of these into project design much more than process engineering. Requirements management is hardly used at all in process engineering; instead, the engineering works to specifications derived from plant balances and piping and instrumentation diagrams.

Safety. Safety is a crucial dimension of project performance in all construction sectors. Legislation has tightened in recent years, making corporate responsibility for accidents much tighter. Safety planning is now integral to project management processes, whether for the project realization, as in the United Kingdom's CDM regulations (HSE, 1994) and the HAZCON procedure (Tubb and Gibb, 1995), or with regard to plant operations safety, as in HAZOP studies. (See the chapter by Gibb.) As a result, safety assurance systems have a high priority in construction project management. For most in the industry, the "iron triangle" of cost, time, and scope thus has a fourth topic in the middle: safety. Environmental protection is often extremely important, whether with regard to noise, dirt, and dust or to flora and fauna. (Safety and environment are typically and, perhaps unhelpfully, bundled in with health, to form HSE, or SHE, as a package of concerns.)

Strategic Change: Concession Contracting and Partnering

While the private sector has always funded projects in process engineering and commercial building, the state has historically been the primary source of funding for infrastructure building and civil engineering projects. Over the last decade or two, the state has begun to look for ways of involving private finance increasingly in the funding of infrastructure projects, generally using some form of project financing mechanism (i.e., where the repayment of the funding is guaranteed by the project's income stream rather than by recourse to non-project-related funds) (Morris, 1997). This trend comes under several names—concession contracting, build-operate-transfer, Private Finance Initiative (PFI), and so on—and has been promoted, with varying degrees of acceptance, in most countries in the world that have been able to offer reliable financing arrangements (which does in fact exclude a number of developing nations), with the United States curiously being one of the major exceptions. (See the chapters by Turner, and particularly by Ive.)

From a project management perspective, this development has had one major impact: It has forced the project participants to focus on the "whole-life" characteristics of the thing being built. (It has had many others of course; for example, it has changed dramatically the impact of infrastructure spending on countries' capital spending requirements and it has radically changed the businesses of those construction companies that have developed this type of work to any significant extent.)

Since the project's financial return is now a function of its long-term financial performance, design (including strategy and briefing) and construction have to become subservient to long-term operational performance. The "short-termism" that has so typically bedeviled building and civil engineering in particular is now increasingly challenged by the newer long-term perspectives of operational performance. Hence, we see a rising interest in construction in practices such as facilities management, costs-in-use, whole-life performance, and integrated logistics support (as Kirkpatrick et al. illustrate in their chapter), and even, though for different reasons, configuration management (though here probably under the guise of "asset management systems").

In parallel, owners have increasingly been realizing that operating with their suppliers on the basis of a series of short-term relationships—each one individually bid for resulting in often acrimonious relationships with uneven, inefficient patterns of work—was not the best way to maximize their suppliers' strategic contribution. Hence, as we have seen (Egan, CII, etc.) during the 1990s the industry generally moved towards patterns of more strategic relationships between key groupings in the supply chain. Not only should this facilitate the development of better alignment between the construction parties, it ought to provide more opportunities for strategic learning and improvements in productivity: CII metrics indicate that "partnering and strategic alliances contributed directly to the success of projects 80% of the time. On average, schedules were reduced by 15%; costs by 12%; and change control, safety, and quality were enhanced" (Jorberg, 1998)

Partnering, in fact, has proved both extremely successful for some and difficult for others. Partnering, which may be job-specific or reflecting a long-term commitment to work together, at heart requires trust—and cooperation. Various practices can be deployed to help generate this, ranging from forms of contractual agreement, through organizational practices such as chartering, to overt team building and other behavioral techniques. All these have been used, and successfully too. Significant cost savings and productivity improvements have resulted in many cases. But above all, trust requires appropriate attitudes, and in some cases the changes in traditional behavior required has simply not worked or proved sustainable, and for some, the promise has soured. For many others, however, partnering remains the absolute basis for business and continuing performance improvement.

Contract Management

Partnering has brought new challenges to construction procurement practices. Construction generally is, for obvious reasons, extremely formal in the way it procures equipment, materials, services, and other items. There are typically many different items to be bought on a construction project, and the chances of significantly impacting the project's profitability in doing so are great (up or down). Inevitably, some of this purchasing leverage has to come via hard-nosed negotiation, and this does not always sit comfortably with long-term arrangements built around trust and cooperation. Setting targets, agreeing responsibilities, and creating relationships within the partnering arrangement need to be carefully managed. (See the chapter by Steel.) Sometimes this does indeed end up being a problem area (in which case there needs to be even more effort put into team building after the contracts are let), but in general, opportunities for savings lie in better specification, reducing the number of separate interfaces, and better sourcing; partnering helps in achieving alignment of objectives and practices and in leveraging supply chain knowledge. (Similarly, there are many instances in all kinds of projects where poor selection of suppliers, or poor specifying, often introduced late in the procurement process, perhaps under severe negotiating pressure, lead to enormous difficulties later in the project, with sometimes disastrous results.) There is a real project management challenge here in managing the interplay between technical and commercial issues—and indeed behavioral ones. This is an example of where project management is much more than merely project control.

Also noteworthy during the 1990s has been the (very slow) move toward more overtly project management-based forms of contract. The building and civil engineering forms of contract have traditionally followed, not surprisingly, the historical role of the designer as the administrator of the construction contract. (This obviously varies between countries/legal systems but is generally true: Even the World Bank's forms of procurement basically follow this pattern.) The Latham report concentrated heavily on the need to improve procurement practices, recommending new documentation and revising legislation. One of its key planks was, interestingly, to recommend a new form of contract that more overtly followed project management principles: the New Engineering Contract (Institution of Civil Engineers, 1995). The process engineering industries, on the other hand, have used a much more consistently project-management-oriented set of contracts (for example, those in the United Kingdom of the Institute of Chemical Engineering.)

Teamwork and Leadership

Teamwork is generally very good in construction, despite the conflict that is inherent in the design-construction split, and its associated contract forms in building and civil engineering. The industry is largely an outdoors one where people are active and hardworking. Generally there is money to be made by getting on with the job. There was great interest in Japanese construction management practices a decade or so ago: One of the features this emphasized was the importance of 'toolbox' meetings on-site before the beginning of each day's activity when construction teams would go over key issues for the day. Safety invariably figured very highly.

Recently, more formal attention has been given to team-building techniques Project chartering, for example, is seen by many to be an important aspect of partnering (Hellard, 1995). High-performance teamwork—the generating, and subsequent follow through, to "stretch targets" largely through a combination of traditional facilitated team building coupled to value management as a tool for identifying process and engineering cost saving/value improvement opportunities—has been much touted in the process engineering industries as one of the keys to the performance achievements of the 1990s (Scott, 2000). There is some evidence that the excitement generated by the approach has diminished in recent years, however, primarily for commercial reasons where either operating costs were found to have suffered as a result of overemphasizing capex (i.e., capital expenditure) savings or a deterioration in team spirit when the mooted savings turned into contract losses.

If teamwork is generally good, leadership may perhaps be a different matter. There are certainly plenty of leaders doing plenty of leading; what is lacking is project leadership: integrating the whole supply chain for the benefit of the sponsor. This is less the case in process engineering projects than in traditional building and civil engineering, for the reasons you have seen:

- The greater recognition of project management as a practice, if not quite a formal discipline, in process engineering compared with building and civil engineering
- Increasingly, the recognition that project management really means, managing the project (nowhere is this now more clearly, and urgently, being argued for than in PFI-type projects)

There is perhaps an important opportunity for project management here. In the past, construction has been typified by lack of repeatable processes (because of the uniqueness of bespoke construction projects), negative procurement processes, and lack of overall management leadership. In the absence of such formal, "good management" aids, clients and their advisors have tended to rely on the quality of the people working on the job. Thus, personal and team characteristics are typically important in bid evaluation; interpersonal relationships make a huge difference in "working the contract."

Slowly all this is changing, however. Process is getting a little more formalized (as we saw, for example, in the chapter by Cooper et al.); procurement is becoming more integrated across the supply chain toward the owner's needs. People will always be important. But if we could truly get project managers to lead the project definition and delivery for business benefit, then we would have made a giant step forward.

Summary

Construction is not a single sector; and even within major sectors such as building, there is an enormous variation in types of project and contexts within which these will be managed. Project management is a more central practice in process engineering than in building or civil engineering, but in no case is the discipline as such formally or clearly recognized as central to its performance. The reality of managing construction projects is that there is a continuous interplay between technical, commercial, organizational, and control issues (and to a lesser extent, strategic ones); the way this interaction is managed has a major impact on the business performance of project stakeholders. The traditional model of project management as an execution-, control-oriented discipline is quite inadequate to the reality of managing construction projects. A fuller "management of projects" model is required.

References

Barrett, P., and C. Stanley. 1999. *Better construction briefing.* Oxford, UK: Blackwell Science.

Building. 2003. "Rogers slams "weak" project managers and contractors." July 4, p. 11.

Blyth, A., and J. Worthington. 2001. *Managing the brief for better design.* London: Spon.

Construction Industry Institute. 1999. *Vision 2000.*

Tubb, D., and A. G. F. Gibb., eds. *Total project management of construction safety, health and environment.* 2nd ed. European Construction Institute. London: Thomas Telford.

Egan, J. 1998. *Rethinking construction: The report of the construction task force.* London: Department of Trade and Industry.

Latham, M. 1994. *Constructing the team.* London: The Stationery Office.

Linder, M. 1994. *Projecting capitalism: A history of the internationalization of the construction industry.* Westport, CT: Greenwood Press.

Hardecker, J F 1926. The functions of a project engineer in a technical organization. *American Machinist* 64(16):641–642.

Hellard, R. B. 1995. *Project partnering.* London: Thomas Telford.

HSE. 1994. *CDM regulations: How the regulations affect you.* Health and Safety Executive. London: The Stationery Office.

Institution of Civil Engineers. 1995. *The Engineering and Construction Contract.* London: Thomas Telford.

Jorberg, R. F. 1998. *An assessment of the impact of CII.* www.construction-institute.org/.

Kamara, J. M., C. J. Anumba, and N. F. Evbuomwan. 2002. *Capturing client requirements in construction projects.* London: Thomas Telford.

Morris, P. W. G. 1997. *The management of projects.* London: Thomas Telford.

Murray, M., and D. Langford. 2003. *Construction reports.* Oxford, UK: Blackwell Science.

Nutt, B. 1993. The strategic brief. *Facilities* 11:28–32.

Pinney, B. W. 2000. Between craft and system: Explaining project-based enterprise in the United States, circa 1910. *Proceedings of the PMI Research Conference.* 401–406. Newtown Square, PA: Project Management Institute.

Scott, R. 2000. *Partnering in Europe.* London: Thomas Telford.

Strassman, W. P., and J. Wells. 1988. *The global construction industry.* London: Unwin Hyman.

Straub, H. 1952. *A history of civil engineering.* London: Hans E. Rockwell, Leonard Hill.

Watson, G. 1998. *The civils: The story of the Institution of Civil Engineers.* London: Thomas Telford.

Winch, G. M. 2003. Models of manufacturing and the construction process: The genesis of re-engineering construction. *Building Research and Information.*

Winch, G. M. 2002. *Managing construction projects.* p. 4. Oxford, UK: Blackwell Science.

CHAPTER FIFTY-FIVE

PROJECT MANAGEMENT IN THE AUTOMOTIVE INDUSTRY

Christophe Midler, and Christian Navarre

The automotive industry has always been a testing ground and a powerful specifier for managerial innovation—one need only think back to "Fordism," "Sloanism," and the "Japanese Model of Manufacturing", which was in fact quite simply the Toyota Model. The success of the book *The Machine That Changed the World* is typical of this point of view (Womack, Jones, and Ross, 1990). The assimilation of managerial techniques by the auto industry—Total Quality Management (TQM) and just-in-time (JIT), for example—certainly has transformed the way in which production is managed in car plants. However, much more than this, it completely goes beyond the dominant theories on the management of inventory and quality, and has radically changed perception of the relative importance of these disciplines throughout the various schools of thought on management.

This chapter explains how and why the concept of the project gradually became formalized and deployed in car firms, and how this development generated, directly or indirectly, profound changes in (1) corporate structures and the professional practice in their technical disciplines and (2) the relationships between carmakers and their subcontractors. In short, it details how the development of project management transformed the automotive industry.

Generally speaking, the strategic importance of project management methods is largely dependent on the importance both of product strategies and the competitive environment. The mass production of a small number of standardized, relatively undifferentiated products with a long life cycle does not require mastery of very sophisticated project management skills. Conversely, mass production of a large number of differentiated products has as a direct consequence the fact that the design and marketing of a large number of distinct products is hardly conceivable in the absence of the concomitant development of very sophisticated project management skills. The history of project management since World War II follows closely that of the markets (Morris, 1997).

We will trace the evolution of project management in the auto industry through four stages:

1. From the postwar period up to the 1970s, there was no differentiation between the "product strategies" of carmakers in North America and Europe. Disciplined management of projects was not a core component in competitive strategy.
2. During the 1970s and 1980s, the gradual saturation of markets changed the competitive environment radically. Japanese carmakers succeeded in breaking into the North American market using (novel) product proliferation strategies, and the direct consequence of this business model was an explosive increase in the number of projects to be managed. The management of projects for new vehicles now assumed strategic importance.
3. In the late 1980s and early 1990s, manufacturers radically reorganized their approach to the management of projects for new products in order to develop more quickly and at lower cost a greater number of products of increasingly high quality. The management of projects for new vehicles was now at the heart of corporate strategy.
4. By the late 1990s, the limits of the reorganization of the beginning of the decade were becoming blatantly obvious. In addition, new challenges emerged. Manufacturers initiated a second wave of reorganization. New vehicle project management became more complex in order to cope with the new challenges, namely: alliances, market globalization, and innovation.

In characterizing the specific position of the project function in firms at each stage, we will make use of the organizational diagrams of Clark and Wheelwright (1988). However, going beyond this framework, we shall show that the development of project-oriented logic was to bring about profound change in the permanent processes of companies, both internally and in their dealings with outside firms.

First Phase: From the Postwar Period to the 1960s

The "product strategies" of carmakers in North America and Europe were undifferentiated. Disciplined control of projects was not a core component in competitive strategy. The management of projects for new vehicles operated via functional structures, coordination was informal, and learning occurred within development projects.

From a strategic point of view, the 1950s and 1960s were typified, in Europe and even more so in North America, by a conventional approach to mass production. The development of car manufacturing firms exhibited a gradual formation of product ranges, namely: a small number of models, long life cycles for models (as long as 12 to 14 years), little product diversification, competitiveness focused on cost reduction through standardization, and longer series (over 250,000 a year for a model).

The design of these products was conducted using an organizational form of "project craft" in an essentially function-oriented corporate structure (see Figure 55.1). Firms were divided into powerful, compartmentalized, trade-focused entities: the product engineering office, the process engineering department, manufacturing, and so on. There was no direct linkage between functions. Projects passed in sequence from one function to the next, following a metaphorical relay race. Each project was handled on a case-by-case basis. The only player joining up functions and acting as arbiter between them was the senior management team and often the CEO himself or herself.

Technical learning took place within the projects themselves, each project being a genuine locus for the development of new skill sets relating not only to products but also to production processes. The buildings of new bodies of expertise occurred in and through successive project development programs. Given this process, the consequences of the risks associated with major technical learning often became more visible in projects.

The resulting level of performance in terms of duration, cost, and quality in the new products was mediocre: long development times (five to seven years), often with delays of one or two years. Product launches were frequently beset with unforeseen problems of industrial feasibility. It often took several years before nominal production rates could be achieved in plants. And, finally, even in a context in which competition was weak because of lack of market availability of products, it is manifest that a large number of unsuitable products reached the market, a sign that upstream project targeting and evaluation processes had been relatively ineffective.

Second Phase: From 1970 to 1985

During the 1970s and 1980s, the gradual saturation of markets brought about radical change in the competitive environment. The management of new vehicle projects assumed strategic importance. Firms learned how to steer their projects strategically and centralize their coordination.

The late 1960s saw, both in Europe and in the United States, the deployment of a new strategy that was to lead to the arrival of the modern, multiproduct vehicle model range, to the diversification of models (power trains, bodywork, and fittings), and to the international deployment of the companies.

In such circumstances, the "project craft" of the preceding period was incapable of coping with the new complexity of product strategies. It is at this point that we see the beginnings of a professionalization of project management:

- The first project functions were created in the early 1970s, along with periodic review systems involving corporate management.
- The careful guidance of projects to completion was gradually put in place, along with formalization of development timetables and the deployment of economic reporting tools integrating all the variables in the projects concerned.

FIGURE 55.1. THE FUNCTIONAL STRUCTURE.

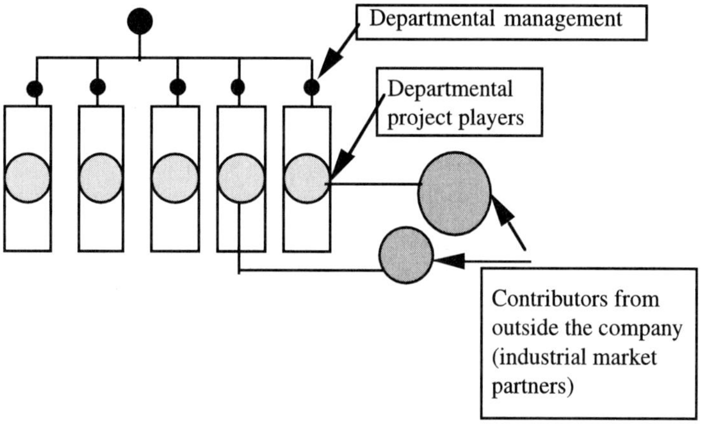

Source: Adapted from Clark, Hayes, and Wheelwright, 1988.

Other than this centralization of control, however, there was no change in the relationship between strategies for the building of technical skills occurring in engineering design offices and central process planning departments, on the one hand, and development policies, on the other (see Figure 55.2). Project teams had neither the political weight nor the expertise to defend their own logic against the strategies of technical departments. This period can be characterized as that of the "lightweight project manager," a notion given formal expression by Clark, Hayes, and Wheelwright (1988).

These new forms of project organization and instrumentation certainly brought with them improvements in new vehicle projects, but the limits of this form of coordination became clear as early as the beginning of the 1980s. The failures that could be seen in projects became increasingly prejudicial:

- Control of project profitability and lead times was often lost, signposting the limitations of the use of sequential input of trade-focused logic and an excessively hierarchical approach to the negotiation of compromises (Cabridain, 1988);
- Product quality at start-up was disappointing on occasion, reflecting an organizational balance in which there was no powerful internal actor capable of measuring and managing the risks generated by technical innovation strategies.

Last, despite a few hesitant attempts, there was no innovation in the area of project management on the part of American or European carmakers. The innovation, in fact, came from a small number of companies, Toyota and Honda in particular. Stalk and Hout (1990) have shown that by the end of the 1980s, certain Japanese firms were implementing highly

FIGURE 55.2. THE PROJECT COORDINATION STRUCTURE.

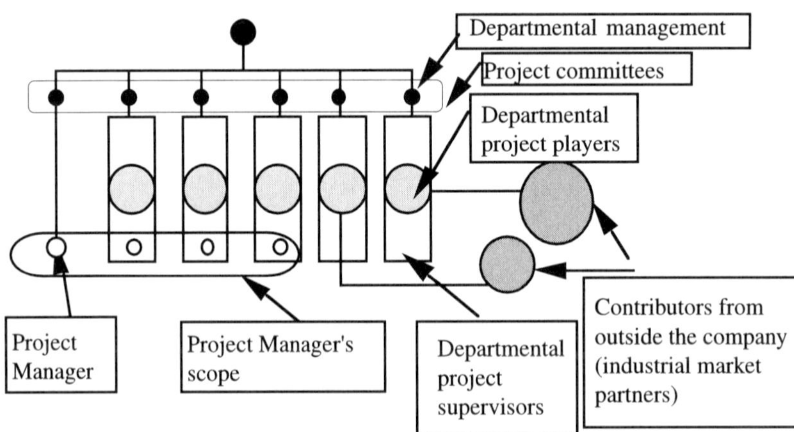

Source: Adapted from Clark, Hayes, and Wheelwright, 1988.

aggressive product proliferation strategies, the principle of which was to drown competitors in a flood of very rapidly replaced products. In such an environment, the products of slower competitors quickly go out of fashion. Stalk and Hout show how the use of such strategies by Honda and Yamaha won them dominance in the motorcycle market. A similar approach can be seen in the conquest of the North American market by Japanese carmakers. Table 55.1 (Womack, Jones, and Ross, 1990) shows that from 1955 to 1989:

- The number of vehicles offered to consumers was increased by a factor of five.
- Japanese producers, absent from the market in 1955, were, by the late 1980s, offering a vehicle range equivalent in variety to, or even slightly more diverse than, their North American competitors. It is worth noting that European manufacturers did not penetrate the North American market with additional automobile models during this period. Moreover, the same study shows that in 1989 the models brought to market by Japanese car firms were more recent and superseded more quickly by new models than those of their North American competitors.
- Average sales per vehicle declined substantially. As a consequence, it became increasingly difficult to provide input for assembly plants designed for volumes greater than 250,000 vehicles per year.

Such proliferation strategies were based on highly effective project management methods. Comparative studies (Clark and Fujimoto, 1987), updated in 1990 and published in 1995, highlight a significant differential in relation to the development performance achieved by Japanese firms according to the three metrics chosen by the researchers: lead time, project

TABLE 55.1. FRAGMENTATION OF AMERICAN AUTO MARKET.

	1955	1973	1986	1989
American Products:				
Number on sale	25	38	47	50
Sales/product (000s)	309	322	238	219
European Products:				
Number on sale	5	27	27	30
Sales/product (000s)	11	35	26	18
Japanese Products:				
Number on sale	0	19	41	58
Sales/product (000s)	0	55	94	73
Total:				
Products on sale	30	84	117	142
Sales/product (000s)	259	169	136	112
Market share captured by six largest-selling products	73	43	25	24

Source: Adapted from James P. Womack, Daniel T. Jones, Daniel Roos, 1990, Figure 5.6, p. 125.

team productivity as measured by the number of engineering hours required to develop the projects, and the quality of the vehicles placed on the market.

This work was widely disseminated and analyzed by industry professionals. These studies stimulated intense reflection among academics, researchers, and industry professionals. The conditions were in place for a radical change in the way projects were managed in North American and European auto firms in the late 1980s.

Third Phase: 1985–1995—The Rise of Project Functions and the Deployment of Concurrent Engineering

At the end of the 1980s, a new template emerged: concurrent engineering, characterized by the spectacular rise of project functions and the deployment of new development methodologies. (See the chapter by Thamhain.)

The most visible sign of this break with the past was the creation of project directors who were destined to become genuine entrepreneurs in automotive development. The time had come for the "heavyweight project manager" template described by Clark, Hayes, and Wheelwright (1988).

The heavyweight project manager structure, under the label "Susha," had existed for many years in the Japanese firm Toyota. (The first Susha was appointed at Toyota in 1953!) A Susha is an independent project director with wide-ranging powers, enjoying authority from the preliminary project stage right up to design and manufacture. As early as 1984, a

FIGURE 55.3. THE PROJECT DIRECTOR STRUCTURE.

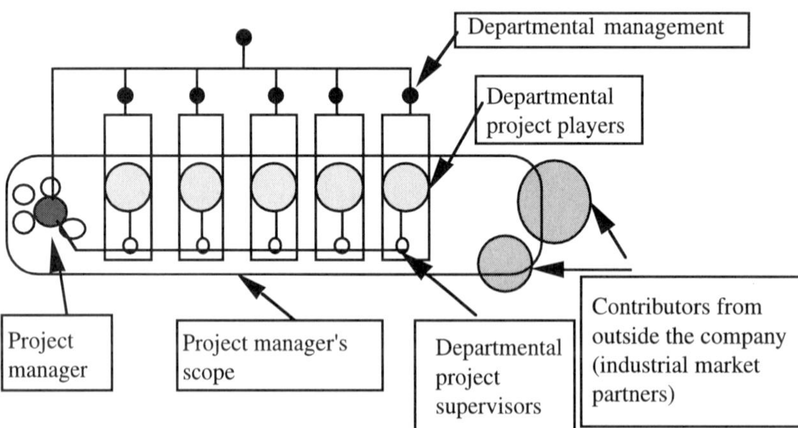

Source: Adapted from Clark, Hayes, and Wheelwright, 1988.

few perspicacious managers visiting Japan had discovered the Susha system and pointed, using the example of Toyota, to its strategic dimension for the management of new vehicle projects.

In the United States, the company that took the heavyweight project manager template furthest was Chrysler. Early in the 1980s, Chrysler was close to bankruptcy and its project performance was poor: Development was taking 60 months with little or no cost control, and vehicles devoid of strong identity were being launched, which was undermining the company's commercial credibility. The team around Lee Iacocca pointed to the inability of vertical functional groups (trades) to cooperate and to soft compromises on policy. Vertical functional groups were clearly incompatible with innovation and audacious "product policies."

The new management redefined how the development of new vehicles was to be organized on the basis of range segments functioning as autonomous enterprises in their own right. All engineering and project management resources, which previously had been conventionally organized by trade, were divided into five platforms: Top of the Range (Large Car), Entry Range (Small Car), Jeep, Truck, and Minivan. The entire 10,000-strong workforce employed in the development of products and engineering design was split into medium-sized units comprising between 2,000 and 3,000 people and provided with clear leadership.

In Europe, Renault was the first manufacturer to put in place powerful project directors with a genuine entrepreneurial dimension in the late 1980s (Midler, 1993).

These structural changes went on to drive profound modifications in project communication and decision processes. We summarize these modifications in five broad categories in the text that follows.

1. Overhaul of Project Control Processes

The previous stage was characterized by the development of sophisticated control of planning and project costs. The limitations of this bureaucratic template have been thoroughly analyzed in the literature, namely: inertia and cost of control, lack of accountability, and lack of solidarity on shared general goals. In contrast, the new template emphasized the importance of adherence to an overall vision and the meaning of the project—of the personal involvement of individuals in seeking collectively to achieve objectives of a more general nature. There was a shift from burdensome controls that removed any sense of personal accountability to an encouragement for individual responsibility and self-regulation by project participants, within a framework defined by "meta-rules" guaranteeing minimum coherence (Jolivet and Navarre, 1993).

At Chrysler, the new compact organization based on clear leadership created a new feeling of solidarity, favoring early resolution of problems, encouraging calculated risk taking, and reducing the inertia in decision making inherent in structures of the matrix type.

In Europe, at Renault, followed by PSA Peugeot Citroën, this logic of personal accountability for overall objectives, quality-cost-duration, was rolled out for all vehicle subassemblies: seats, engine, dashboard, and so on.

2. The Development of Concurrent Engineering

The new project actors laid down new game rules for the coordination of project contributors. These changes were at three levels. First, they related to the timing of the contributions by the various specialists: Traditional sequential processes gave way to a planning logic aimed at maximum anticipation of problems through the early involvement of the trades involved in production: Plants were involved in the manufacture of prototypes in order to validate process feasibility, future products were tested with the sales networks, and so on. Next, they concerned communications between functions: Previously, intertrade dialogue had occurred largely at the top of the management tree. Project management departments now began to promote such dialogue at the bottom in decentralized work groups responsible for all aspects of the development of a given part of the vehicle (seats, dashboard, etc.). And last, the changes related to the spatial organization of work, with systematic use of colocation of participants in project office suites, and development of tools to accelerate intertrade communications.

To point to the example of Chrysler once again, the construction of a single development center had the effect of locating transverse project communications within the workplace. "Chrysler's management realized that the company's conventional functional organization, which consisted of the usual design, engineering, manufacturing, marketing, and sales departments, was a prime culprit. Each functional group tended to operate as an independent fiefdom with its own goals . . . When there was a major decision to be made, it had to go up to the president, because each functional group had its own objectives, and it was difficult to get agreement among them. Chrysler's president often ended up having to arbitrate among groups that should have been cooperating instead of competing." The implementation of integrated multifunctional "platform design teams" allowed Chrysler to replace the conventional sequential engineering with simultaneous engineering. To enhance communications among engineers and to make concurrent product development possible, it was seen as extremely important to have all platform team members colocated. Despite limited financial resources at the time of the turnaround, the costly decision to relocate all the people developing new cars under one roof was made. Lee Iacocca, then the company chairman, announced the new technical center in 1984. Construction began two years later. When an economic recession hit in 1989, Wall Street and a lot of others called for Chrysler to abandon the project to save money, but Iacocca refused, believing it was too important to Chrysler's long-term health . . . The Chrysler Technical Center opened in 1991 and quickly became an important part of the enterprise.

The architecture of the design center placed each platform/project physically on the building's floors (horizontally), disposing trades vertically (along a vertical axis connected by escalators). The company implemented simultaneous engineering tools and, more generally, transverse communication systems (the center thus pioneered intranet and extranet development). Support programs were initiated to encourage cross-category culture and continuous learning processes: ongoing vocational training for staff and an aggressive policy of recruitment of young engineers, along with the overhaul of the system of personnel management and evaluation (notably by introducing coworker evaluation, etc.).

3. The Dynamic of the Technical Trades Involved in Projects

While project process efficiency is a necessary condition of the new product innovation strategies, the excellence of the technical disciplines is just as essential. However, the new types of organization implemented profoundly destabilized the processes of capitalization and development of expertise in the various trades involved in vehicle engineering. One of the crucial problems facing carmakers in the 1990s was therefore the issue of how to create a project management capability without this being detrimental to the key disciplines' expertise base.

The example of Chrysler illustrates an extreme case in which trade-focused logic has manifestly been subordinated to project platform logic. However, the company set up "expert clubs" to maintain cross-platform solidarity learning on key technical skills. Initially informal, this system was consolidated and officially instituted in 1994, under the sponsorship of a general manager.

In Europe, Renault preserved the trade-based structure of its vehicle engineering but thoroughly overhauled the boundaries between specialities to ensure their assimilation of product-process engineering logic by dividing up departments on the basis of vehicle subassemblies and functions.

In Japan, the issue of how to organize corporate technical departments was at the heart of Toyota's engineering reorganization in 1992.

4. A Change in Relations with the Outside World: From Subcontracting to Codevelopment

The development of the project concept did not undermine internal corporate mechanisms alone. Automotive projects usually unfold in a space much wider than that of a single company: Today, the proportion of production cost relating to parts bought in from suppliers is generally more than 70 percent. The new project actors have come to play a major role in the development of new types of co-development between carmakers and suppliers. Calls for bids from and subsequent selection of suppliers operate from the outset of a project, on the basis of agreement on core project objectives. The chosen supplier is then associated closely with the engineering study process: "co-location" in the project "office suite," participation in project progress meetings, and so on. Compared with the conventional template for competition between suppliers based around detailed project specifications, this new template for the relationship involves, for those taking part in it, the need to modify their organizational and contractual frameworks (Banville and Chanaron, 1991; Lamming, 1993; Garel, Kesseler, and Midler, 1997; Kesseler, 1998). The purchasing process in particular is completely revolutionized.

Once again, Chrysler can be seen to be, in North America, a precursor in its implementation of the "extended enterprise" concept, closely involving suppliers in projects. In Europe, Renault also consciously committed to codevelopment policies (Midler, 1993; Kesseler, 1998).

5. Developing Project Management Professionalism within Automobile Firms

The empowerment and generalization of project function called for institutionalization of project management professionalization processes. Beyond the variety of the programs and professional patterns that developed in the late 1980s and 1990s, we will pinpoint the following characteristic points (Boudes, Charue-Duboc, Midler, 1998):

- Project management was not generally developed into a specific professional pattern, as, for example, in construction. On the contrary, human relations rules organized transversal carrier trajectory, mixing skilled-based and project-based roles. Two ideas are behind that choice: (1) that a key difficult point is to maintain solidarity between project and functional populations and (2) that alternating the role will enhance the capitalization of inside project learning. Emphasizing the individual project management expertise appeared in that perspective not as important as developing a more collective project management competency.
- The maturity the project management approach of the firm can be correlated to the hard/soft orientation of project management learning programs: the deeper in organizational and strategic themes, the heavier the project management function.
- The more technical side of project management (planning tools, budgeting . . .) diffused through highly specialized staff people in the firm.
- Emphasizing on the entrepreneurial aspect of project function led to develop cross-practitioner learning and capitalizing programs. Such project managers exchanges not only occurred within each firm but also extensively in cross-sector clubs and associations.

The Results: The Spectacular Success of Western Carmakers

The implementation of these new project management modes led to significant progress in the performance levels achieved by Western manufacturers in terms of new vehicle launches.

Duplicating the study they had conducted in the 1980s, Ellison, Clark, Fujimoto, and Hun demonstrated that Western carmakers had very much caught up with the Japanese according to the various metrics they had defined as being most indicative of project management performance levels (see Table 52.2).

At Chrysler, the results were initially impressive. "The . . . Neon compact car model took 31 months to bring to market, while the current Dakota pickup truck made it in only 29 months." Over the first five years from the initial setting up of these new structures, Chrysler developed more new models than in the 20 previous years.

Unquestionably, from the mid-1980s to the mid-1990s, Chrysler has been an exemplar company, especially for the development of new products. During the 1990s and up to the time of the merger with Daimler, Chrysler was generating by far the highest average profit per vehicle of any North American auto manufacturer (Harbour and Associates, from 1995 to 1999). In addition, its assembly plants also ranked, according to the same studies, among the non-Japanese plants with the highest productivity.

TABLE 55.2. ADJUSTED LEAD TIME AND ENGINEERING HOURS.

	Japan	U.S.	Europe	Korea	Total	(w/o Korea) Total
Adjusted engineering hours (EHAD):						
1980s	1703	3366	2915		2507	2507
	(843)	(642)	(950)		(1084)	(1084)
1990s	2093	2297	2777	2127	2438	2477
	(500)	(947)	(723)	(926)	(739)	(755)
Total	1847	2880	2843	2127	2474	2493
	(745)	(936)	(822)	(926)	(926)	(941)
Adjusted lead time (LTAD):						
1980s	44.6	60.9	59.2		53.5	53.5
	(7.4)	(5.6)	(6.1)		(9.9)	(9.9)
1990s	54.5	51.6	56.1	54.5	54.7	54.7
	(12.6)	(3.8)	(12.2)	(3.6)	(10.3)	(10.9)
Total	48.6	56.7	57.6	54.5	54.1	54.1
	(10.7)	(6.7)	(9.7)	(3.6)	(10.1)	(10.3)
Total Product Quality (TPQ):						
1980s	53	35	60		52	52
	(29)	(29)	(21)		(27)	(27)
1990s	61	42	59	21	52	56
	(16)	(18)	(17)	(1)	(20)	(18)
Total	56	38	60	21	52	54
	(25)	(24)	(19)	(1)	(24)	(23)

Definitions:
Adjusted engineering hours-The number of hours required to develop a project of average project complexity.
Adjusted lead time-The number of months required to develop a project of average project complexity.
Total product quality (TPQ)-The TPQ index presented in Table 2 makes quality comparisons relative to other vehicles in the same class. As a consequence, the TPQ index has already been adjusted for project quality and is presented here to allow comparisons across the three measures of development performance.

Adapted from Ellison, David J., Kim B. Clark, Takahiro Fujimoto, and Young-suk Hyun. "Product Development Performance in the Auto Industry: 1990s Update.". #w-0060a, IMVP 1995, p.11.

In Europe during the first half of the 1990s, Renault reaped the benefits of its breakthrough in project management, bringing to market an innovative product range while also restoring its image in terms of product quality and improving its cost base. The success of vehicles such as the Twingo and the Megane Scenic—which were to create new market niches as the Espace minivan had done in years past—was the most spectacular result of the new project systems.

New Difficulties in the Late 1990s

The 1990s ended less happily than they had begun. Surprisingly, the effectiveness of the new systems put in place a few years before seemed to run out of steam. Chrysler got into trouble and was taken over by Daimler-Benz. Unfortunately, excellence in new product development is not sufficient by itself to sustain profitability and growth on a long-term basis. The inescapable turmoil induced by a merger, the disruptive effect of the 9-11 terrorist attack combined with a cutthroat competition, again made Chrysler vulnerable. Since the end of 2001, Chrysler is again in crisis.

In Europe, the most advanced carmakers evidenced slower product cycles, losses of control over lead times, and mismatches between vehicles and customer expectations. During this period, in Japan the landscape was very mixed. Toyota could do no wrong in any market around the world, while Nissan was mired in a profound crisis. This crisis ended only with its takeover by Renault, when the deep renewal was implemented triumphantly by the new managers of the company under the leadership of Carlos Ghosn.

Fourth Phase: 1995–2003+

Project management faces new challenges in the automotive industry.

Several factors serve to explain what might seem at first sight to be a "relapse," but that was in fact another stage in the intensification of innovation-based competition in the automotive industry; a stage described as one of "intensive innovation"—innovation that is both more radical in content and repeated at a faster rate (Hatchuel and Weil, 1999; Benghozi et al., 2000) and that has been increasingly deployed against the background of global alliances.

The comparative benefit of efficient project systems for the pioneers in the early 1990s tended to run out of steam as "best practices" spread rapidly to their competitors. To find new values for differentiation, firms went down the road of innovation policies that were far more radical in terms of both engineering and styling. On the one hand, in doing so they were adding a source of higher risk; on the other, they were facing problems for which "heavyweight" types of automotive project organization were unsuited, since the most relevant context for the deployment of technical innovation is not vehicle development but transverse learning covering both the preliminary project stages and whole product ranges. No cross-functional and cross-product project existed to address, coordinate, and control these learning tracks on radical innovative features and technologies (a good example is car telematics; Lenfle and Midler, 2003).

The increasing number of projects in firms clearly highlights problems associated with the heavyweight project management template, problems that were not burdensome when such forms of organization were the exception. As early as the beginning of the 1990s, Toyota had taken a critical look at its Susha system (Cusumano and Neoboka, 1998). By 1992 the number of platforms rose from 8 to 18, along with a decline in average production volume per platform.

Communication and coordination problems became critical. In 1991, a Susha was communicating and working with 48 departments in 12 engineering divisions, plus the R&D division. Trade divisions were conducting concurrent dialogues with 15 projects. The relationship between Sushas and senior management in the plan-product division was all the more strained because detailed supervision of 15 Sushas was impossible without setting up an enormous project-focused bureaucracy. Staff in the plan-product division were finding it difficult to monitor all 15 projects and became disconnected from reality. R&D was perceived as being distant—and its interfacing with projects became very problematic. The influence and power enjoyed by Sushas were considerable and difficult to control because of the failings of the system.

As crisis after crisis occurred, the powers of Sushas were strengthened, along with their responsibilities, which had the effect of "locking them in" to projects and encouraging the appearance of a blinkered silo mentality—albeit "transverse" silos. The rapidly expanding number of projects led to the appointment of young, inexperienced Sushas. Coordination became increasingly difficult, or even impossible. The increased number of departments and divisions led to narrow specialization in the engineers, making it more and more difficult for them to understand cross-category logic and interfacing, and this in turn led to less well-thought-out, less integrated products. Capitalization and transfer of expertise from one project to another became extremely problematic. The same solutions were continually being reinvented. The total workforce assigned to development was not far from 15,000.

The end of the 1990s was also a period of unprecedented strategic shifts in the auto industry. There were more and more alliances of all kinds. Projects needed to assimilate this new factor, which was a source of further difficulty and new constraints compared with the previous stage.

These strategic changes led eventually to new developments in automotive project management, which can be summarized in terms of five interdependent trends:

- Deployment of the project function downstream of initial product development
- Implementation of systems better suited to steering radical innovations, both upstream and between product development programs
- Increasing numbers of projects conducted by intermanufacturer partnerships
- Development of "platform" projects as a way of managing the plethora of multibrand product development programs
- Enlargement of supplier-manufacturer cooperation fields: modularization and colearning.

Deployment of the Project Function in the Commercial Phases

The intensification of innovation-based competition in the automotive industry had consequences. Downstream of development projects, it undermined the stability of the products' commercial life cycles. This was because, since the mid-1970s, the initial development of new products was a key stage that allowed the introduction of innovations, whether in products or processes. Between product launches, there was room for minor modifications. In the mid-1990s, the price war in the European and North American markets led to a new strategy involving more systematic deployment of development forces for products already in the market, in order to obtain immediately all the benefits of the innovations introduced between new product cycles. Transverse integration, formerly restricted to the development phase, now spread to the entire product life cycle. This led in European car firms to a rise in importance of new actors, the "series life project manager" and the "program director," with the task of coordinating the various trade components within a segment of the manufacturer's range throughout the life cycle.

The Management of Radical Innovation Projects: The Growing Importance of Predevelopment as a Base for Differentiation Strategies

At the same time as this increase in the rate of appearance of innovations, the differentiation strategy increasingly included research activities in order to find new competitive advantages. The "quality-cost-lead time" triangle no longer sufficed. It was necessary to introduce more strikingly radical innovations in the services offered to customers: hence, recent changes in upstream development disciplines, which had until this point remained relatively aloof from ongoing changes.

At Renault, by the end of the 1990s, the research division was totally overhauled, strengthened, and tightly interfaced with the preliminary project design departments. The logic of this reorganization was clearly expressed in the change of director: Previously led by a scientist with a past career in a French pure research body, the French National Research Council, the division was now led by a former vehicle project manager. The division's activities, previously based around scientific disciplines, were now guided by programs focused on areas for innovation allying services and technology and that have clear importance for the evolution of automotive transport. Within vehicle development programs, the post of "innovation project manager" was created to manage the convergence of complex technical innovations (the "keyless car," for example). At Peugeot-Citroën Group, an Innovation Division provides a focus for all upstream specialists.

By the early 2000s a new form of project organization was gradually being put in place to guide exploration upstream of vehicle projects (Lenfle and Midler, 2002). Although one can find in this type of project some of the features of vehicle projects (specifically the need to conduct such exploration by coordinating a plurality of forms of expertise—technical, marketing, design, etc.), these projects also have novel characteristics. First, the level of risk is much higher because of the extent of the innovation in terms of both technology and product features. This leads projects to be conducted within exploration portfolios in which efforts are made to maximize synergy and offset risks, rather than mobilizing teams on "one-shot" operations, which is the principle underlying conventional projects. (See the chapter by Jamieson and Morris.) Second, the direct "result" of such a project is not a product placed on the market but a concept that is validated and knowledge that is acquired, which it will be necessary to exploit later in actual products. (See the chapters by Artto, Thiry and others.) This virtual, intangible character of the result is undoubtedly one of the difficulties of this type of project: The tangible, practical nature of a new product launch is no longer present as a focus for the contributors to the project.

The Development of Projects by Intermanufacturer Partnerships and the Interfacing of Project Management and Strategic Alliances

The auto industry is one of the sectors that saw a spectacular wave of globalization and corporate restructuring in the 1990s (mergers, acquisitions, strategic alliances, industrial co-operation, as well as spin-offs and exit). Internationalization strategies and the importance of size are, of course, not new features of this sector, but the changes seen in the 1990s were unprecedented in their scale and generalized character, giving the impression of a fashion phenomenon that has swept all company strategies along in its wake. Such strategies for growth through alliances were reflected in two ways in projects. First, manufacturers increased the number of one-off cooperative programs in joint projects, in order to round out their model ranges in niche markets and gain access to new markets by pooling with others the costs and risks involved in such developments. Second, where alliances were more global in character, projects for new products provided useful leverage for the exploitation of synergy between merged companies through the sharing of platforms, systematic exchange of mechanical subassemblies, and so on.

Piron (2001) and Midler, Monnet, and Neffa (2002) have emphasized the importance of three problem sets that exist in cooperative projects, compared to the traditional automotive project culture:

Mutual understanding within joint project teams. In single-manufacturer project teams, coordination is based on numerous unarticulated bodies of expertise forged throughout the company's past history. Once projects begin to be conducted as cooperative endeavors by more than one manufacturer, the risks of misunderstanding will be high if the participants do not make substantial efforts to make themselves clear. Such effort is costly and difficult to gauge correctly, since it runs counter to the participants' need to protect their knowledge and expertise.

The management of fair and equal treatment of the partners. This becomes a prerequisite for cohesion and focus in joint project teams. Studying a joint project by General Motors and Renault, Midler, Neffa, and Monnet (2002), adapting Piron (2001), have demonstrated that the practical implementation of fair treatment in projects could take three forms: *Distributive justice* involves the search for a balanced proportionality between the partners, a "fair return," in Piron's words. The point is, for the firms, to find a fair distribution of goods and powers based on the goals sought and the resources committed by each. *Procedural justice* refers to the feeling that procedures have been fair. The point is for the participants to judge a decision-making process relative to a reference that is well known and considered legitimate. The factors that influence this include a feeling of participation in decision making, an explanation of decisions, and clarity concerning expectations and the rules of the game, all of which influence whether the participants feel they have been treated fairly and equitably.

Finally, *interactive justice* refers to individual interactions based on fairness in behavior, which makes it possible for a decision to be considered doable. Hence, respect and courtesy between allies prove to have an important contribution to make in the fostering of a positive atmosphere for interpersonal relations during the cooperation process.

Regulation of tensions between the project and the strategies of the parent companies. Cooperation projects are by their very nature unstable and vulnerable to exogenous events. In the case of the Renault-GM project, joint decisions have been upset by shocks coming from outside the project, such as the Renault-Nissan alliance in 1999 and the GM-Fiat alliance in 2000. However, the nomination of a project general manager representing the interests of both firms and the governance structures has strengthened the joint program and attenuated the impact of external events such as the Renault-Nissan and GME-Fiat Alliances or the high currency rate between the euro and sterling.

The Growing Importance of the Platform Concept in Vehicle Design Strategies

One of the core tensions in the automotive industry has always been the product-standardization/product-differentiation dilemma. In the 1980s, the myth of the "global car" was a subject much talked up by manufacturers, and a burden on the accounts of those American manufacturers who actually developed the concept. The development of highly entrepreneurial project managers allowed, in the early 1990s, the rapid launch of innovative products in market niches that conventional strategies were incapable of reaching. The successes achieved at that time by a firm like Renault can certainly be put down to this logic.

However, with the advent of strategies for growth by merger and acquisition in the 1990s, the advantages of design strategies based on platforms common to different products took on new relevance. Firms such as the VAG group, PSA, and, less successfully, Fiat used such strategies to drive their development in Europe.

Although the concept of the platform is an attractive one, putting it into practice involves, in project terms, a dual difficulty in interfacing the "platform-driven projects" with "product-driven projects." Specifically, the issues are as follows:

- The core issue is how to benefit from covert standardization while nevertheless preserving differentiation in the finished products. The boundary is, however, far from obvious here. In multibrand groups, it usually leads to the de facto domination of one brand identity over the others, which must fit in with the constraints imposed by the initial designer of the platform.
- The second issue is how to link up the replacement cycles for platform and products. If a requirement to use a platform is imposed on a product, it will often mean that recent innovations cannot be introduced, with the consequence that a risk is taken in a market characterized by swift obsolescence. But if the platform evolves at the same speed as the products derived from it, there is no longer much of a distinction between the concept and the conventional notion of "carry-over."

The New Boundaries of Manufacturer/Supplier Relationships and Vehicle Programs: Toward Colearning and Modular Codesign?

In the 1990s, the successes achieved in efforts at codevelopment led first-rank suppliers to gradually broaden their field of competence and learn trades that had in the past been the reserved domain of carmakers. Major value shifts were observed as subcontracting was generalized. How far should one go in this new allocation of the roles of the manufacturer of the vehicle as a whole and the suppliers of its components? The issue of extension of supplier responsibilities and involvement is apparent today in two main areas.

The first area relates to timing. In the 1980s, suppliers provided input in the later stages of projects, in the context of relationships that were precisely governed by detailed technical specifications laid down by the car manufacturer. In the 1990s, it became gradually possible to define effective arrangements for codevelopment (Garel and Midler, 2000; Kesseler, 1998)—that is to say, cooperation between the carmaker and its suppliers on the basis of overall functional objectives. However, with the increasing importance of innovation policies, carmaker/supplier cooperation sought to extend itself upstream within development projects, in "colearning" arrangements (Lenfle and Midler, 2001) for the design of innovative concepts for product features in which more effective coordination of the partners' respective "roadmaps" was sought. The expanding importance of concept competition phases compared with the more traditional competitive bid processes was the visible sign of this strong trend, which raises several questions: For example, how should one allocate the costs and risks involved in such upstream explorative programs, whose outcome is highly uncertain? What type of regulation of intellectual property issues might encourage the partners to provide the transparency imperative to the success of the partnership?

The second relates to the spatial and functional boundaries of supply. The last two decades have been ones in which the functional and spatial scope of the supplier's role has expanded: There has been a shift from the individual part to the component, followed, in the 1990s, by entire systems (complete functional assemblies) and modules (an assembly whose boundaries can be geographically isolated). The principle is to define interface standards to enable the producers of components (whether modules or systems) to develop items that are simultaneously less expensive (volume effects), more "versatile" (offering greater variety), and compatible with open-architecture interiors. Suppliers delivering similar or homogeneous items to more than one manufacturer are in a position to gradually define "commonalities" in the components they offer. They are also in a position to define the best ways to differentiate them at low cost and to contain, or even to eliminate, the extra cost because of variety. Volume also allows R&D costs to be spread over a number of client manufacturers.

This approach, summed up in the "black box sourcing" concept, suggested by Clark, forms part of a particularly active flow of work on modular design (Henderson and Clark, 1990; Ulrich, 1995; Baldwin and Clark, 2000). The area that has been revolutionized by the modular architecture concept is that of telecommunications and information technology. Is the auto industry ready to be rethought in terms of combinations of interchangeable components whose interaction can be predetermined through the adoption of common standards and integrated into platforms with a strong and sharply defined architecture? A highly futuristic concept car, Autonomy, presented by GM in 2002, is a step in this direction.

However, many authors rightly emphasize the "integrity" that is characteristic of the car as object: It is very difficult to uncouple components without penalizing overall performance, given that its functions are split between a number of component parts (vehicle behavior, weight, compactness, noise—all these are examples of highly distributed functions). On the other hand, industrial vehicles do allow much more advantageous use of modularity.

Finally, one of the original features of the car as product is that while this dual evolution toward platforms and modular black box design provides a rich mine of inspiration, it cannot be applied literally as it has been in information technology. (There is, in fact, a need to specify this area more precisely: Portable computers, like mobile telephones, have integrity characteristics associated with their compact dimensions, which place tight constraints on the deployment of modularity.) This observation is of major importance where project management is concerned because it forces partners to interact during the design phase in order to arrive at compromises on distributed functions. Given that fact, a template for contractualized coordination capable of handling black box logic in terms of functional performance is inadequate. It must be overlaid with procedures for interaction enabling the detection of problems and the negotiation of compromises aimed at resolving them.

Limits of Performance in New Product Development Are Constantly Surpassed

Today, in a nutshell, the performance envelope for new vehicle project management can probably be located in the following ranges in North America:

1. Cost of development of a new vehicle: between $1bn and $1.5bn (compared with $3bn to $5bn ten or so years ago).
2. Accelerating fragmentation of market supply and creation of innovative new product lines (Mini Rover, Beetle, Crossover, Hybrids, Nissan Cube, SUVs, etc.). In addition, some manufacturers, with Renault in the forefront, are taking the risk of introducing radical styling that breaks with the past, staking the visual identity of their products.
3. Development duration is of the order of 24 months or less (with a mean of around 36 months), compared to 60 months (and a year's delay) at the beginning of the decade. Lead times are shortening between concept cars and serial production.
4. Carmakers in North America are now generally close together in terms of measured quality, which, however, does not preclude wide variations in quality as perceived by the consumer. In addition, quality is no longer a competitive advantage but a prerequisite. However, a worrying rise in the cost of vehicle recalls can be seen.
5. A continuous decline is apparent in the costs of assembly and processes (productivity gains estimated in the region of 7 percent per year) as well as components (on the order of 20 percent to 30 percent for each new vehicle). In addition, manufacturers have included systematic annual volume-linked price reductions in their contracts with suppliers.
6. These savings are passed on to the consumer through constant enrichment of the features offered by vehicles at a constant price level. Moreover, for the last three years the beginnings of further reductions in prices targeted on certain model ranges have been observed. In short, there are now more features for a given price in constant, declining money terms.

7. Architecture is controlled in terms of platforms and modules in order to take advantage of scale effects on common portions while nevertheless preserving the diversity and identity of the vehicles. For example, Toyota sells, on the basis of the same platform, vehicles as different as a sedan, a van, a Lexus, and an SUV.

Summary

From the postwar period until the present time, the development of project management has radically changed structures and processes within car manufacturing companies. But on the reverse, we can say that project management has been changed by its implementation within the automotive context: from technique and tool orientation to more strategic and organizational approaches, from highly precise contractualized relation patterns to procedural open learning "meta-rules." The auto industry was a latecomer to project management, compared to military equipment or the construction industry. But these sectors are now trying to transform their project management tradition and adopt the project management practices that were developed in the late 1980s and 1990s in the auto sector. We can see various reasons behind such a dragging effect: the economic importance and symbolic notoriety of the auto sector, of course, but also the importance of management research in the field, that evaluated the performance of various project patterns and traced their transformations around the world.

This process is not yet complete. Performance limits are constantly increasing.

References

Ashley, S. (1997). Keys to Chrysler's comeback, *Mechanical Engineering Online*, The American Society of Mechanical Engineers.

Ayas, K. 1997. *Design for learning for innovation.* Delft, Netherlands: Eburon Publishers.

Banville E. de et Chanaron, J. J. (1991). *Vers un système automobile européen*, Economica, Paris.

Bayart, D., Y. Bonhomme, and C. Midler. 1999. Management tools for R&D project portfolios in complex organizations: The case of an international pharmaceutical firm. In *6th International Product Development Management Conference.* Cambridge, UK.

Benghozi, P. J., F. Charue-Duboc, and C. Midler. 2000. *Innovation based competition and design systems dynamics.* paris: L'Harmattan.

Ben Mammoud-Jouini, S. 1998. Stratégies d'offre innovantes et dynamiques des processus de conception: Le cas des grandes entreprises françaises de bâtiment. Université Paris IX Dauphine, Thèse de doctorat de gestion, Paris.

Boudes, T., F. Charue-Duboc, and C. Midler. 1998. Project management learning: A contingent approach. In *Projects as arenas for renewal and learning processes,* ed. R. A. Lundin and C. Midler. 61–71. Norwell, MA: Kluwer Academic Publishers.

Brown, S. L., and K. M. Eisenhardt. 1997. The art of continuous change: Linking complexity theory and time-paced evolution in relentlessly shifting organizations. *Administrative Science Quarterly* 42 (1, March).

Chapel, V. 1996. La croissance par l'innovation: de la dynamique d'apprentissage à la révélation d'un modèle industriel. Le cas Tefal. Thèse de doctorat spécialité Ingénierie et Gestion, École Nationale Supérieure des Mines de Paris, Paris.

Charue-Duboc, F. 1997. Maîtrise d'oeuvre, maîtrise d'ouvrage et direction de projet: des catégories pour comprendre l'évolution des fonctionnements en projet dans le secteur chimie de Rhône Poulenc. *Gérer et Comprendre* (September): 54–64.

———. 1998. The role of research departments in focusing innovative projects and understanding customer usage and needs. In *International Research Network on Project Management and Temporary Organization (IRNOP Conference)*. Calgary, Canada.

———. 2001. What are the organizational artefacts implemented to foster learning processes in a context of intensive innovation? In *17th EGOS Colloquium: The Odyssey of Organizing*. Lyon.

Charue-Duboc, F., and C. Midler. 1999. Impact of the development of integrated project structures on the management of research and development. In *6th International Product Development Management Conference*. Cambridge, UK.

Ciavaldini, B. 1996. Des projets à l'avant projet: l'incessante quète de la réactivité. Thèse de doctorat spécialité Ingénierie et Gestion, Ecole des Mines de Paris, Paris.

Clark, K. B., and T. Fujimoto. 1991. *Product development performance: Strategy, organization and management in the world auto industry*. Cambridge, MA: Harvard Business School Press.

Clark, K. B., and S. C. Wheelwright. 1992. *Revolutionizing product development*. New York: Free Press.

Cooper, R. G., S. J. Edgett, and E. J. Kleinschmidt. 1998. Best practices for managing R&D portfolios. *Research Technology Management* (July–August).

Cusumano, M., and K. Nebeoka. 1998. *Thinking beyond lean*. New York: Free Press.

Ellison, D. J., K. B. Clark, T. Fujimoto, and Y. S. Hyun. 1995. Product development performance in the auto industry: 1990s Update. #w-0060a, IMVP.

Garel, G. and C. Midler. 2001. Front-loading problem-solving in co-development: Managing the contractual, organizational and cognitive dimensions. *Journal of Automotive Technology and Management*. no. 3.

Hatchuel, A., V. Chapel, X. Deroy, and P. Le Masson. 1998. *Innovation répétée et croissance de la firme*. Paris: Rapport de recherche CNRS.

Hatchuel, A., and B. Weil. 1999. Design oriented organisations. Towards a unified theory of design activities. In *6th New Product Development Conference*. Cambridge, UK.

Jolivet, F., and C. Navarre. 1996. Large-scale projects, self-organizing and meta-rules: towards new forms of management. *International Journal of Project Management* 14(5):265–271

Harbour and Associates. *The Harbour Report North America 1995, 1996, 1997, 1998, 1999*. Harbour and Associates Publication.

Kesseler, A. 1998. *The creative supplier: A new model for strategy, innovation, and customer relationships in concurrent design and engineering processes: The case of the automotive industry*. Thèse de Doctorat, spéc. Gestion, Ecole Polytechnique, Paris.

Kogut, B., and Kulatilaka, N. 1994. Options thinking and platform investments: Investing in opportunity. *California Management Review* (Winter): 52–71.

Lamming, R. (1993). *Beyond Partnership: Strategies for Innovation and Lean Supply*. Englewood Cliffs, NJ: Prentice Hall.

Lemasson, P., and B. Weil. 1999. Nature de l'innovation et pilotage de la recherche industrielle. *Cahiers de recherche du CGS, Ecole des Mines de Paris*.

Lenfle, S., and C. Midler. 2002. Innovation-based competition and the dynamics of design in upstream suppliers. *International Journal of Automotive Technology and Management*. vol 2, n°5.

———. 2003. Innovation in automotive telematics services: Characteristics of the field and management principles. *International Journal of Automotive Technology & Management* 3(1/2):144,159.

Lundin, R. A., and A. Söderholm. 1995. A theory of the temporary organization. *Scandinavian Journal of Management* 11(4):437–455.

Lundin, R. A., and C. Midler. 1998. *Projects as arenas for renewal and learning processes.* Norwell, MA: Kluwer Academic Publishers.

Midler, C. 1993. *L'auto qui n'existait pas; management des projets et transformation de l'entreprise.* Paris: Dunod.

———. 1995. "Projectification" of the firm: The Renault case. *Scandinavian Journal of Management* 11(4): 363–375.

Midler, C., J. C. Monnet, P. Neffa. 2002. Globalizing the firm through projects: The Case of Renault. *International Journal of Automotive Technology & Management* Vol 2, n°1, pp 24, 46.

Morris, P. W. G. *The management of projects.* 1997 London: Thomas Telford.

Navarre C. 1992. De la bataille pour mieux produire à la bataille pour mieux concevoir. *Gestion 2000* 6:13–30.

Piron, P. (2001). *L'alliance en convergence. Développer conjointement dans l'industrie europééne des missiles tactiques.* Ph.D Thesis, Ecole Polytechnique, Paris.

Ponssard, J. P., and H. Tanguy. 1992. Planning in firms: An interactive approach. *Theory and Decision* 34:139–159.

Schön, D. A. 1983. *The reflective practitioner: How professionals think in action.* New York: Basic Books.

Sharpe, P., and T. Keelin. 1998,. How SmithKline Beecham makes better resource-allocation decisions. *Harvard Business Review* (March–April).

Simon, H. A. 1969. *The sciences of the artificial.* Cambridge, MA: MIT Press.

Söderlund, J., and N. Andersson. 1998. A framework for analyzing project dyads: The case of discontinuity, uncertainty and trust. In *Projects as arenas for renewal and learning processes,* ed. R.A. Lundin and C. Midler. 181–189. Norwell, MA: Kluwer Academic Publishers.

Stalk, G., T. M. Hout. 1990. *Competing against time: How time-based competition is reshaping global markets* New York: Free Press.

Weil, B. 1999. Conception collective, coordination et savoirs. Les rationalisations de la conception automobile. Thèse de doctorat spécialité ingénierie et gestion, École Nationale Supérieure des Mines de Paris, Paris.

Womack, P. J. P., D. T. Jones, and D. Ross. 1990, *The machine that changed the world.* New York: Rawson Associates.

CHAPTER FIFTY-SIX

PROFESSIONAL ASSOCIATIONS AND GLOBAL INITIATIVES

Lynn Crawford

Communities of practice (Wenger, 1998) are formed when people doing similar things realize they have shared interests. They recognize that there are opportunities to improve both their practices and their performance by sharing knowledge and experience. The project management professional associations as we know them today began in this way: as informal gatherings and forums for networking and exchanging ideas and information.

INTERNET, now known as the International Project Management Association, IPMA, was initiated in 1965 (IPMA 2003; Stretton, 1994) as a forum for European network planning practitioners to exchange knowledge and experience. The Project Management Institute originated in North America in 1969, as "an opportunity for professionals to meet and exchange ideas, problems and concerns with regard to project management, regardless of the particular area of society in which managers function" (Cook, 1981). The UK national project management association began in 1972 as the Association of Project Managers and was subsequently renamed the Association for Project Management. The Australian Institute of Project Management was initially formed as the Project Managers Forum in 1976. The early focus of these project management professional associations on exchange of knowledge and experience between practitioners clearly illustrates their origins as communities of practice.

Recognition of shared interests results in fairly informal gatherings, often referred to as a forum for meeting, networking, and exchange of ideas. At some point members of this community of practice express a need or desire to define their areas of common interest or practice. They begin to think of themselves as a community, and then sometimes, as in the case of project management, as a profession, and to attempt to define and delineate that profession in order to make it visible and acceptable to those outside the community. This is the point when the community begins to put in place the building blocks of a profession, which include the following (Dean, 1997):

- A store or body of knowledge that is "more than ordinarily complex"
- A theoretical understanding of the basis of the area of practice
- An ability to apply theoretical and complex knowledge to the practice in solving human and social problems
- A desire to add to and improve the stock or body of knowledge (research)
- A formal process for handing on to others the stock or body of knowledge and associated practices (education and training)
- Established criteria for admission, legitimate practice, and proper conduct (standards, certification and codes of ethics/practice)
- An altruistic spirit.

From this list it is easy to identify a number of the issues that have preoccupied the various national project management professional associations as they have emerged from the more informal forums of practitioners with shared interests. Primary preoccupations have been as follows:

- Definition of a distinct body of knowledge
- Development of standards
- Development of certification programs

In focusing on these areas, project management professional associations have tended to develop proprietorial or vested interests in the products they have produced. This has resulted in a proliferation of competing project management standards and certification programs, largely local in their origin, if not in their application. By the second half of the 1990s, members of the project management community began to realize that the nature of their community was changing and that a more unified approach would be needed to promote project management as a practice and profession and to meet the needs of increasing numbers of corporate adopters of project management with increasingly global scope to their operations.

From Local to Global

Modern project management may be considered to have had its genesis in the international arena when, in the 1950s (Stretton, 1994; Morris, 1994), companies such as Bechtel began to use the term "project manager" in their international work, primarily on remote sites. Communities of project management practice, however, with their focus on interactions between people, developed locally, becoming formalized in national project management professional associations. Even international projects were considered as endeavors conducted by national organizations, offshore. During the period in which project management professional associations were emerging, the focus was on techniques for planning, scheduling, and control. Practitioners who were forming these local and essentially national communities of practice were primarily involved in major projects in the engineering, construction, defense, and aerospace industries, and the interests of these practitioners strongly influenced the activities and focus of project management professional associations from their emergence in the 1960s through to the early to mid-1990s.

A change of focus became evident in the early 1990s. For some time the application of project management had been spreading beyond its traditional origins, to a wider range of application areas, particularly information technology. In a rapidly changing and responsive environment, where more and more of the endeavors of organizations are unique and could benefit from being identified and managed as projects, interest in project management grew progressively stronger, extending to project management as an approach to enterprise management (Dinsmore, 1996). However, these "projects" are often internal, without physical end products or clearly identified "clients," although they will often have many interested stakeholders both internal and external to the parent organization. In an era of networking, alliances, and partnerships, there may not even be a single or clearly identified parent organization, and resources are often shared across multiple projects and, in some cases, multiple organizations. Further, the application of project management extended beyond international projects, managed offshore by nationally based companies to use by global corporations through globally distributed operations and projects. Hence, the communities of project management practice, formalized in primarily local or national professional associations, faced, and continue to face, a dual challenge.

One aspect of this challenge is that the expertise that underpins the development and definition of project management practice is founded primarily in the management of clearly recognized and defined stand-alone projects such as those in engineering and construction. These practices were first transferred and minimally transformed in application to information systems and technology projects, but they retain a strong and identifiable legacy from their origins in major engineering projects. They do not recognize or offer a response to the systemic and complex nature of business projects, including implementation of corporate strategy, management of organizations by projects, and enterprise innovation.

The other aspect of the challenge facing project management as a community of practice, with aspirations to being a recognized profession, has been fragmentation through internal competition between the professional associations as the formal manifestations of the community, which have tended to remain locally focused and proprietorial about the knowledge created by their communities. In contrast, aided by the development of information technology; easier, faster, and less expensive travel; and an increasingly global economy, informal communities of project management practice have developed through online communities of practice, attendance at an increasing number of project management conferences and other initiatives for exchange occurring outside the official channels. This development has a distinctly global dimension, encouraged by global corporations and facilitated by global communication technologies.

The movement toward globalization, like modern project management, is considered to have begun in the 1950s, and by 1996, international business had become "the fastest-growing field in the business world, just behind technology" (Lenn, 1997, p. 1), leading to the suggestion that "to be successful in the twenty-first century, global professionals will require an education that guarantees their competence in their individual country and competitiveness in an international marketplace" (Lenn and Campos, 1997, p. 9). Indeed, the North American Free Trade Agreement (NAFTA, 1993) and the World Trade Organization's General Agreement on Trade in Services (GATS, 1994) removed many barriers to professional mobility and required the "development of policies that evaluate professional competence based on fair, objective criteria and transparent (publicly known) procedures" (Lenn, 1997, p. 2). These agreements have put pressure on established professions and their professional associations to consider mutually acceptable standards in cooperation with other countries and to actively plan for reciprocal recognition as a minimum.

By the mid-1990s, these wider external economic and social pressures, as well as internal pressures from the increasing number and range of users of project management, lead to the emergence of a number of initiatives aimed at enhancing the global communication and cooperation between the professional associations as formal manifestations of the project management communities of practice and providing a more globally relevant and rational framework for the definitions of practice and recognition of professional competence through standards and certification.

This chapter briefly identifies and describes key project management professional associations and provides a review of initiatives relating to enhancement of global communications, research, education, standards, and certification in project management.

The Professional Associations

There was relatively little growth in membership and in number of project management professional associations from their emergence in the mid-1960s and 1970s through to the start of 1990s, which heralded a decade of unprecedented growth. The most significant membership growth was that of the Project Management Institute, which experienced a growth from 8,500 members in 1990, located primarily in the United States, to nearly 110,000 members in 2003, of which 69 percent were located in the United States, 11 percent in Canada, and 20 percent in other parts of the world (Project Management Institute, 2003a).

At the start of the twenty-first century, the majority of people who want to participate in project management professional associations can do so through national associations, many of which are members of the International Project Management Association, which describes itself as an "international network of national project management societies" (International Project Management Association, 2003); through one of the 207 Chartered and 52 Potential PMI Chapters located in 125 countries (Project Management Institute, 2003a); or through one of a number of online project management communities.

Only a few of the project management professional associations will be briefly described here: the two organizations that purport to be global or international in their reach (PMI and IPMA), a few of the more influential or active of the associations that are neither PMI chapters nor members of IPMA (AIPM, PMSA), and a small number of other national associations and/or PMI chapters that have characteristics of particular interest or have made specific contributions to the promotion and development of project management practice.

International Project Management Association (IPMA)

The International Project Management Association began as a discussion group comprising managers of international projects and has evolved into a network or federation comprising 30 national project management associations representing approximately 20,000 members, primarily in Europe but also in Africa and Asia (International Project Management Association, 2003). The International Project Management Association has developed its own standards and certification program (see my chapter earlier in the book), which maintains a central framework and quality control process but encourages development of conforming national programs by national association members. The International Project Management Association and member national associations promote their standards and certification program in competition with those of others, primarily the Project Management Institute. The IPMA is hampered by its structure as a federation, by vested interests and priorities of its national association membership, and by lack of funds available for international and global development, which is a particular issue regarding the large number of member associations representing transitional economies who require subsidization of their membership and services.

As membership of the IPMA is subject to change, anyone interested in current membership should refer to the IPMA Web Site—*www.ipma.ch.*

Project Management Institute

The Project Management Institute began as the national project management association for the United States. By the late 1990s the Institute realized that with over 15 percent of its members and a number of chapters located worldwide, it was rapidly becoming an international organization. In September 1997 the PMI Board established a Globalization Subcommittee and then a Globalization Project Action Team (PMI Globalization Project Action Team, 1998) to assist the board in establishing its position on globalization. It has subsequently refocused its activities as a global organization rather than a national association with international or offshore chapters. PMI's headquarters continues to be located in the United States (in Philadelphia, PA), and the organization remains subject to the law of that state. However, in May 2003 the Institute held its first Global Congress in Europe (The Hague) and in June 2003 opened a PMI Regional Service Centre for Europe, Middle East, and Africa (EMEA Regional Service Centre) in Brussels, Belgium (Project Management Institute, 2003b).

The PMI approach, positioning itself as a global organization, plus its significant membership—109,117 individual members as of July 2003 (Project Management Institute, 2003a)—suggests that it is the organization that provides the primary representation and voice for the project management community, globally. The Institute itself claims to be "the leading nonprofit professional association in the area of Project Management" (www.pmi.org, June 2003). However, while headquarters and nearly 70 percent of membership remain located in the United States, its products and services, despite recent efforts to increase globally representative involvement are, or are perceived to have been, primarily developed in the United States to suit the needs of that market, and there is considerable reluctance on the part of project management professionals in some countries outside the United States to relinquish their independence and genuinely national representation. A further issue is economic. Practitioners in many countries cannot afford professional membership fees that are acceptable in the United States. In many cases it has been necessary to establish fully national associations in order to meet the needs of local jurisdictions and/or to provide a more affordable alternative. A notable example is South Africa, where project management practitioners were for many years represented by a PMI Chapter. The PMI South Africa Chapter continues to exist, but a separate national association (Project Management South Africa) was established in 1997 to satisfy local economic and regulatory requirements.

American Society for the Advancement of Project Management (asapm)

While project management practitioners in some countries outside the United States prefer to retain local representation through national associations rather than having their professional interests addressed through a local PMI chapter, a number of those in the United States took the view that if the Project Management Institute is a global organization, there is no longer any national association representation in the United States. As a result, the American Society for the Advancement of Project Management (asapm) was formed in July 2001.

PMINZ: PMI New Zealand Chapter

The Project Management Institute, New Zealand (PMINZ) is an example of a PMI chapter that has been accepted, to date unopposed, as the national project management association. It was established in 1994, as a chapter of the Project Management Institute and by 2003 had 850 members (www.pmi.org.nz, June 2003).

Association for Project Management, UK (APM)

The Association for Project Management (UK), although a member of the IPMA, deserves mention in its own right, as it has more members (14,000) than any of the other member organizations of IPMA and has done considerable influential work in definition of the project management body of knowledge and development of certification programs. The *APM Body of Knowledge* was one of the key documents referenced in writing of the *ICB: IPMA Competence Baseline*, as well as in the development of certification programs.

A PMI chapter was established in the United Kingdom in 1995 and reported membership of 2,000 by 2003, claiming that many of their members "are also members of local UK project management associations and groups" (UK PMI Chapter, 2003)

Australian Institute of Project Management (AIPM)

The Australian Institute of Project Management, begun in 1976 as the Project Managers Forum, is the Australian national project management association and by 2003 had 4,000 members distributed over eight state and territory chapters. As AIPM is not a member of the International Project Management Association, it has been well placed to offer an independent voice and in some cases act as an intermediary between the Project Management Institute and the International Project Management Association in the interests of global cooperation. The AIPM is also notable for having secured government support for development of performance-based competency standards for project management, recognized within the Australian Qualifications Framework (see my earlier chapter). Another role of AIPM has been to encourage development of national project management associations in the Asia Pacific Region and cooperation among them. AIPM has cooperative agreements with a number of Asian professional associations and has participated in a fairly loose Asia Pacific Forum for some years. In 2002 the AIPM took a leading role in the formation of the Asia Pacific Federation for Project Management (APFPM), an umbrella organization of national project management institutes.

The Australian Institute of Project Management remained unopposed as the national project management association until 1996, when the first of a number of PMI chapters was chartered in Australia. By 2003 there were PMI chapters established in Sydney, Melbourne, Canberra, Adelaide, Queensland, and Western Australia (Project Management Institute, 2003c), with a total membership of 1,500, with 700 of these being members of the Sydney Chapter (Project Management Institute, 2003d). Relationships between the Australian Institute of Project Management and the Australian PMI Chapters varies from time to time, and from state to state, between friendly cooperation and active competition.

Project Management South Africa (PMSA)

Project Management South Africa (PMSA) has already been mentioned in discussion of the Project Management Institute. It is worthy of separate mention because, like the Australian Institute of Project Management, it is not a member of the IPMA and therefore also has the opportunity to offer an independent voice in the global arena. However, because the PMI South African Chapter was formed first (1982) and was very active for a long time before formation of PMSA and because PMSA was essentially formed by members of the PMI South Africa Chapter (which continues to exist), there is a far closer and more consistently cooperative relationship between the two organizations. Membership of PMSA increased from 400 at formation in 1997 to over 1,200 in 2003.

The drive to create PMSA came from a need for a cross-sector forum for practitioners to meet and work together and for a national body to work with local organizations and the South African government in developing effective project management within South Africa. Another argument in favor of formation of the national association was that PMI membership fees had become prohibitive for many in South Africa with the decline in the value of the South African rand relative to U.S. currency.

PMSA has taken an active role in working with the South African government in development of performance-based competency standards for project management, similar in format to those developed in Australia and recognized within the South African National Qualifications Framework.

Japan Project Management Forum (JPMF)

Japan is another country that embraces both a national association, the Japan Project Management Forum (JPMF) and a PMI chapter (Tokyo), both established in 1998. The JPMF is a division of the Engineering Advancement Association (ENAA), which was founded in 1978 as a nonprofit organization based on corporate rather than individual membership, dedicated capability enhancement, and promotion of the Japanese engineering services industry. ENAA enjoys government support through the Ministry of Economy, Trade and Industry (METI), and membership encompasses 250 engineering and project-based companies. Since inception, ENAA has had a Project Management Committee, and 46 member companies are involved in this committee. While ENAA engages and addresses the needs of industry and corporations, JPMF acts as the professional association for individual practitioners.

To advance the use of project management approaches in Japan, the ENAA was commissioned by the government (METI) over a period of three years from 1999 to develop a "Japanese style project management knowledge system," which has been developed and published as a standard guidebook under the *title P2M: Project and Program Management for Enterprise Innovation* (ENAA, 2002). Subsequently, in April 2002, the Project Management Professionals Certification Center (PMCC) was established, to widely disseminate the P2M in Japan and beyond and to establish a related certification process (see my earlier chapter).

Project Management Research Committee, China (PMRC)

Established in 1991, the PMRC was the first, and claims to be the only national, cross-industry project management professional association in China, supported by over 100 universities and companies and 3,500 active individual members form universities, industries, and government.

In 1994 the PMRC initiated, with support from China Natural Science Fund, the development of a *Chinese Project Management Body of Knowledge* (C-PMBOK), which was published together with the *China-National Competence Baseline* (C-NCB) in May 2001. Over 160,000 copies of this document had been issued by the start of 2003.

PMRC supports both the IPMA, of which it has been a member since 1996, and the Project Management Institute. It helped introduce the PMP certification into China and started an IPMA certification program in July 2001 (Yan, 2003).

In 2003, there were three PMI Potential Chapters in China (Project Management Institute, 2003c). Also active in China is the Project Management Committee (PMC), established in November 2001 as a branch of the China Association of International Engineering Consultants (CAIEC), an organization guided by the Ministry of Foreign Trade and Economic Cooperation, China (MOFTEC). The PMC promotes project management training and certification among the 255 member companies of CAIEC (Yan, 2003).

Society of Project Managers, Singapore

The Society of Project Managers, Singapore is worthy of separate mention because of its active role in the Asian region, strong links with other parts of the region, notably Japan, and as it purports to be "an amalgam of learned society and a professional body." The Society was formed in 1994 by a group of professionals as a vehicle for advancing the development of project management. It has held a number of well-regarded research-based conferences in the region (The Society of Project Managers, 2002).

Global Initiatives

As outlined in the previous section, there is strong evidence of considerable and increasing interest in project management from individuals, corporations, and governments. Responses have originated at the local level, giving rise to many competing membership opportunities, conferences, standards, and certifications, not only between countries but often within a single country. Two key organizations have attempted to achieve a more unified and global approach: the International Project Management Association, as a federation of national associations, and the Project Management Institute in taking an active stance as a global organization. To a large extent this has merely increased the dilemma of choice for individuals and organizations, especially as national governments such as those in Australia, South Africa, the United Kingdom, and Japan have recently begun to take an active role in support and recognition of project management standards and certifications. The International Project Management Association, on behalf of member associations, promotes the *ICB: IPMA Competence Baseline* and aligned National Competence Baselines, along with its 4-Level Certification Program. At the same time, the Project Management Institute promotes its PMBOK Guide, PMP Certification, and a range of other standards and certification products, globally. Individuals and organizations must therefore not only decide between one national and one "global" or international membership and set of standards and association certifications, but between several products available at both the national and global levels.

A frustration for the "users" of the products of professional associations, primarily in the form of standards and certification, has been the apparent unwillingness or inability of the project management associations, despite the signing of "cooperative agreements," to really work together to enhance global cooperation and communication and to bring a sense of global unity to the profession, resolving the dilemmas of decision in terms of investment in project management association membership, standards, and certification. As a result, efforts toward global communication, cooperation, and alignment of standards and certification have primarily been informal and unfunded, occurring outside the project management professional associations. Some of the primary initiatives are reviewed here.

Global Project Management Forum (GPMF)

A response to the independent and largely local development of standards and certification, as well as research and education, first came to a head in 1994. At the PMI Symposium in

Vancouver, Canada, there was a meeting of representatives of PMI, IPMA, APM, and the AIPM, at which "formal cooperation on several global issues, including standards, certification and formation of a global project management organization or 'confederation'" were discussed (Pells, 1996, p. ix). Another, informal meeting, of individuals from about a dozen countries, also gathered, intensely discussing cooperation and communication among project management professionals around the world and formulating a declaration of intent. The first of a series of Global Project Management Forums, held in association with the PMI Symposium in New Orleans in 1995, was the result.

There were nearly 200 attendees, representing over 30 countries, at the first Global Project Management Forum. There were high hopes that the energy and enthusiasm evidenced at this meeting would be "an opportunity for the world's leading project management associations to take another major step towards achieving agreements on international standards, recognition of project management certifications, and development of a global core Project Management Body of Knowledge" (Pells, 1996, p. x). Breakout sessions were held to discuss five topics:

1. International project management standards
2. Globally-recognized project management certification
3. Global communication among project management professional organizations
4. Global cooperation and organization of the project management profession
5. Development of a global core project management body of knowledge

This first meeting of the Global Project Management Forum was a high point for this initiative and will be a lasting and important memory for those privileged to be there. However, by the time the thirteenth Global Project Management Forum was held in Moscow in June 2003, little real progress had been achieved. The GPMF had evolved into an informal association with a slogan "Toward the globalization of project management" and a stated mission "To advance globalization of the project management profession by promoting communications and cooperation between and among project management organizations and professionals around the world" (www.pmforum.org). The forums have been held as a one-day event associated with the annual or biannual conferences of project management professional associations. Until 1999 the key venues were the annual Symposia of the Project Management Institute and the biannual IPMA World Congress, but from 1999 onward, the Project Management Institute declined to provide for the forums in association with their symposia, preferring instead to offer a PMI Global Assembly aimed primarily at representatives of their globally distributed chapters.

The agenda for each forum has become relatively predictable, with presentations and breakout sessions on Research, Standards, Education, and Certification. The initial ideal that the initiative would bring together people from all over the world in an open forum to keep touch with developments in the field of project management has been fulfilled, but it became clear within a couple of years of the first GPMF meeting that meaningful cooperation between the project management professional associations was far from being realized. Informal cooperation and lip service were possible. Formal cooperation and real progress in the interests of a strong and unified project management profession were hampered by political issues and vested proprietorial interests.

Recognizing that real achievements were required to maintain the momentum begun by the GPMF, the IPMA established and convened a series of Global Working Groups that first met in East Horsley, UK, in February 1999 (IPMA, 1999).

Global Working Groups

The Working Groups were established in six areas, namely (IPMA, 1999):

- Standards
- Education
- Certification
- Accreditation/Credentialing
- Research
- The Global Forum

Certification and Accreditation/Credentialing were recognized as related and were merged. The subsequent five Working Groups were tasked with delivering a progress report at the next Global Project Forum, the last held in association with a Project Management Institute Symposium, in Philadelphia in September 1999. These five Working Groups have continued to present progress reports at each of the following Global Forums. The Working Group on the Global Forum confirmed that it should remain informal, independent of established professional associations. Because of their nature, Education and Research have generated considerable interest, and breakout sessions have provided excellent opportunities for sharing ideas. A research project concerning the benchmarking of the degree of project-orientation of societies was originated in 1999 and has been furthered under the auspices of the IPMA through Projekt Management Austria (PMA) (IPMA, 2003). The Certification Working Groups and GPMF breakout session have facilitated ongoing discussion of global issues surrounding certification and credentialing, but little progress is possible without formal involvement of the professional associations that "own" the existing certification products and processes. A global approach to certification also needs, as a prerequisite, a global approach to standards.

Progress arising from the Working Group: Standards is reviewed in detail in my previous chapter. The most active and promising of the initiatives generated by the Global Working Group: Standards is the development of a global framework of performance-based standards for project management personnel (see my earlier chapter and http://www.globalPMstandards.org).

Toward a Global Body of PM Knowledge (OLCI)

This initiative, which began in 1998 by bringing together those working at the time on various representations of the body of project management knowledge (*ICB: IPMA Competence*

Baseline, reviews of the *APM Body of Knowledge* and the PMBOK Guide), has progressed through a series of annual workshops hosted by organizations, including NASA, Telenor, ESC Lille and the Project Management Professionals Certification Center, and JPMF. This initiative is discussed in detail in my earlier chapter.

International Research Network on Organizing by Projects (IRNOP)

IRNOP, the International Research Network for Organizing by Projects, is somewhat different from the initiatives outlined previously, as it is not directly associated with project management professional associations but is an important expression of the informal interactions of the global communities of project management practice and has been an important initiative in achieving global communication and cooperation in project management research. It has no formal organization but comprises a loosely coupled group of researchers. It was initiated in 1993 to support and enhance efforts aiming at the development of a theory on temporary organizations and project management. Its main activity is the facilitation of research conferences, and these have been held in Sweden (1994), Paris (1996), Calgary (1998), Sydney (2000), and the Netherlands (2002). The next conference will be held in Turku, Finland, in 2004.

Prior to staging the first Project Management Research Conference by PMI in Paris in 2000, IRNOP provided the only real impetus and opportunity for students, researchers, and academics in project management to come together to share ideas and information about their research.

Summary

By referring to the formal manifestations of project management communities of practice as "professional associations," we assume that project management is in fact a profession. Although there is a strong sense of aspiration among project management practitioners and their representative associations to professional status, this remains a matter of debate and has been powerfully questioned by Zwerman and Thomas (2001), who have highlighted the "barriers on the road to professionalization." They maintain that although project management has been moving toward satisfying various criteria indicative of professional status, it is still some distance away and achievement will require significant effort on the part of the professional associations and members.

Even recognition of project management as an occupation or field of practice is vulnerable, as many see it merely as an aspect of general management, and there is a growing view that project management should form part of a wide range of skill sets. Much of the knowledge base of project management is shared with or has been annexed from bodies of knowledge of other professions, and Turner (1999) suggests that in order to be a mature profession, project management must develop a sound theoretical basis indicating that considerable further research is required to establish a sound foundation for professional status.

A challenge that professional associations are likely to face is the increasing involvement of government in defining practice standards for project management as evidenced by the project management standards forming part of National Qualifications Frameworks in Australia, South Africa, and the United Kingdom; government support of project management standards development in Japan and China; and active leadership by of the Office of Government Commerce in the United Kingdom in developing standards and certification programs for individuals and organizations involved in projects and programs. Influenced by trade agreements such as the World Trade Organization's General Agreement on Trade in Services (GATS, 1994), governments are motivated to seek mutual recognition in areas of standards, accreditation, and certification. Further, unless project management professional associations take the lead, global corporations will do so in order to satisfy their own needs.

Clearly, there is considerable work to be done to establish project management as a profession, and this suggests the need for project management's formal representation to adopt a globally unified stance rather than pursue the fragmentation and internal competition that has characterized development to date. It is interesting to note that the majority of activity promoting global communication and cooperation, the free sharing and exchange of ideas among members of the global community of project management practice, is occurring outside the formal structures, through online discussion groups and through informally structured initiatives.

References

Cook, D. L. 1981. Certification of project managers: Fantasy or reality? In *A decade of project management: Selected readings from* Project Management Quarterly—*1970 through 1980,* ed. J. R. Adams and N. S. Kirchoff. Newtown Square, PA: Project Management Institute.

Dean, P. J. 1997. Examining the profession and the practice of business ethics. *Journal of business ethics* 16(15):1637–1649.

Dinsmore, P. 1996. On the leading edge of management: Managing organizations by projects. *PM Network* (March): 9–11.

ENAA 2002. *P2M: A guidebook of project and program management for enterprise innovation: Summary translation.* Revision 1. Tokyo: Project Management Professionals Certification Center (PMCC).

International Project Management Association. 2003. About IPMA. Available at www.ipma.ch/ (accessed June 26, 2003).

IPMA 1999. Documentation of Meeting: Global Working Groups. East Horsley, UK, February 27, 1999.

IPMA. 2003.Research and development. www.ipma.ch/ (accessed June 26, 2003).

Lenn, M.P. 1997. Introduction. In *Globalization of the professions and the quality imperative: professional accreditation, certification and licensure,* ed. M. P. Lenn and L. Campos. Madison WI: Magna Publications, Inc.

Lenn, M. P., and L. Campos. 1997. International organizations. In *Globalization of the professions and the quality imperative: professional accreditation, certification and licensure,* ed. M. P. Lenn and L. Campos, 9–10. Madison, WI: Magna Publications, Inc.

Morris, P. W. G. 1994. *The management of projects.* London: Thomas Telford.

Pells, D. L. 1996. Introduction. In *The global status of the project management profession,* ed. J. S. Pennypacker, ix–xii. Sylva, NC: PMI Communications.

PMI Globalization Project Action Team. 1998. Project definition. www.pmforum.org/featindex.htm/gpatp1.htm (accessed April 13, 2003).

Project Management Institute. 2003a. The PMI Member FACT sheet—July 2003. www.pmi.org/prod/groups/public/documents/info/gmc_memberfactsheet.asp. (accessed August 28, 2003).

Project Management Institute. 2003b. PMI opens regional service centre in Brussels, Belgium. www.pmi.org/prod/groups/public/documents/info/ap_news-emeaopen.asp (accessed June 28, 2003).

Project Management Institute. 2003c. PMI Chapters Outside the United States. www.pmi.org/info/GMC_ChapterListingOutsideUS.asp#P128_1923. (accessed June 26, 2003).

Project Management Institute, Sydney Chapter. 2003d. PMI Sydney Chapter. Available at http://sydney.pmichapters-australia.org.au/ (accessed June 26, 2003d).

Stretton, A. 1994. A short history of project management. Part one: The 1950s and 60s. *The Australian Project Manager* 14(1):36–37.

The Society of Project Managers. 2002. The society. www.sprojm.org.sg/web/socpm/ (accessed June 29, 2003).

Turner, J. R. 1999. Project management: A profession based on knowledge or faith? Editorial, *International Journal of Project Management* 17(6):329–330.

UK PMI Chapter. 2003. Welcome to the UK PMI Web page. www.pmi.org.uk/ (Accessed June 26, 2003).

Wenger, E. 1998. *Communities of practice: Learning, meaning and identity*. Cambridge, UK:Cambridge University Press.

Yan, Xue. 2003. PMRC and other organizations in China. E-mail from Xue Yan (xue_yan@cvicse.com.cn) to Lynn Crawford (Lynn.Crawford@uts.edu.au). January 2, 2003.

Zwerman, B., and J. Thomas. 2001. Barriers on the road to professionalization. *PM Network* 15 (4, April): 50–62.

INDEX